BASIC FINANCIAL

Frequently Use

α_t	Certainty equivalent coefficient in period *t*
AROR	Accounting rate of return
ß	Beta of an asset, the slope of the regression or characteristic line
$\mathbf{CAF_t}$	Annual after-tax expected cash flow in time period *t*
DCL	Degree of combined leverage
DFL	Degree of financial leverage
DOL	Degree of operating leverage
EAA	Equivalent annual annuity
EBIT	Earnings before interest and taxes
EOQ	Economic order quantity
EPS	Earnings per share
FV	Future value
FVIF	Future value interest factor
FVIFA	Future value interest factor for an annuity
g	Annual growth rate
IO	the initial cash outlay
IRR	Internal rate of return
$\mathbf{K_c}$	Cost of internal common equity (also *Kc*)
$\mathbf{K_d}$	After-tax cost of debt
$\mathbf{K_o}$	Weighted cost of capital
$\mathbf{K_p}$	Cost of preferred stock
M/B	Market-to-book ratio
MCC	Marginal cost of capital
MIRR	Modified internal rate of return
NPV	Net present value
PMT	Periodic level payment of an annuity
P/E	Price/earnings ratio
PV	Present value
PVIF	Present value interest factor
PVIFA	Present value interest factor for an annuity
R	Investor's required and/or expected rate of return
$\mathbf{R_f}$	Risk free rate of return
ROA	Return on assets
ROE	Return on common equity
RP	Risk premium
SML	Security market line
σ	Standard deviation (lowercase sigma)
σ^2	Variance (standard deviation squared)
TIE	Times interest earned
T	Tax rate
WCC	Weighted cost of capital
$\mathbf{W_d}$, $\mathbf{W_c}$	Percentage (weights) of funds provided by debt and common equity respectively
YTM	Yield to maturity

Basic Financial Management

SECOND CANADIAN EDITION

Arthur J. Keown
Virginia Polytechnic and
State University

David F. Scott, Jr.
University of Central Florida

John D. Martin
University of Texas at Austin

J. William Petty
Baylor University

David W. McPeak
University of Ottawa

PRENTICE HALL CANADA INC., SCARBOROUGH, ONTARIO

Canadian Cataloguing in Publication Data

Main entry under title:

Basic Financial management

2nd Canadian ed.
ISBN 0-13-570441-3

1. Business enterprises – Finance. 2. Corporations – Finance. I. Keown, Arthur J.

HG4026.B38 1997 658.15 C96-930615-6

Prentice-Hall, Inc., Englewood Cliffs, New Jersey
Prentice-Hall International (UK) Limited, London
Prentice-Hall of Australia, Pty. Limited, Sydney
Prentice-Hall Hispanoamericana, S.A., Mexico City
Prentice-Hall of India Private Limited, New Delhi
Prentice-Hall of Japan, Inc., Tokyo
Simon & Schuster Asia Private Limited, Singapore
Editora Prentice-Hall do Brasil, Ltda., Rio de Janeiro

ISBN 0-13-570441-3

Acquisitions Editor: Patrick Ferrier
Developmental Editor: Lesley Mann
Copy Editor: Shirley Corriveau
Production Editor: Valerie Adams
Production Coordinator: Deborah Starks
Art Direction: Kyle Gell
Cover Design: Monica Kompter
Cover Image: Scott Barrow, Inc., Photography
Page Layout: Michael Kelley

2 3 4 5 CC 01 00 99 98 97

Printed and bound in the U.S.A.

Every reasonable effort has been made to obtain permissions for all articles and data used in this edition. If errors or omissions have occurred, they will be corrected in future editions provided written notification has been received by the publisher.

We welcome readers' comments, which can be sent by e-mail to
phcinfo_pubcanada@prenhall.com

BRIEF CONTENTS

PART FIVE WORKING-CAPITAL MANAGEMENT

PART SIX FINANCIAL ANALYSIS, PLANNING, AND CONTROL

PART SEVEN SPECIAL TOPICS IN FINANCIAL MANAGEMENT

CONTENTS

FINANCIAL ANALYSIS, PLANNING, AND CONTROL

PART SIX

SPECIAL TOPICS IN FINANCIAL MANAGEMENT

PART SEVEN

PREFACE

Finance continues to evolve because of changes in the business environment and developments in the academic world. New financing and risk management techniques seem to appear on almost a daily basis. How do you prepare for a field as dynamic as finance? The answer is to go beyond the answers and understand the logic that drives those answers. Our objective is to provide the student with a relevant understanding of financial decision making that is based on theory. The first-time student of finance will find that corporate finance builds upon both the disciplines of economics and accounting. Economics provides much of the theory that underlies our techniques, while accounting provides the input or data on which decision making is based.

Unfortunately, it is all too easy for students to lose sight of the logic that motivates finance and focus on memorizing formulas and procedures. As a result, students have trouble understanding the interrelationships among the topics covered. Moreover, later in life when problems encountered do not match the textbook presentation, students may find themselves unprepared to abstract from what they have learned. To overcome this problem, the opening Chapter 1 presents ten principles or axioms of finance, which are presented in an intuitively appealing manner and thereafter are tied to all that follows. In essence, the student is presented with a cohesive, interrelated perspective from which future problems can be approached.

With a focus on the big picture, we provide an introduction to financial decision making rooted in current financial theory and in the current state of world economic conditions. This focus can be seen in a number of ways, perhaps most obvious being the attention paid to both valuation and to the capital markets and their influence on corporate financial decisions. What results is an introductory treatment of a discipline rather than the treatment of a series of isolated finance problems. The goal of this text is to go beyond teaching the tools of a discipline or of a trade, and to help the student gain a complete understanding of the subject. This will provide the student with the ability to apply what is learned to new and yet unforeseen problems—in short, to educate the student in finance.

DISTINCTIVE FEATURES

Basic Financial Management is not simply another introductory finance text. Its structure reflects the vitality and expanding nature of the discipline. Finance has grown too complex not to teach with an eye on the big picture, focusing on the interrelationships among the topics that are covered. Listed below are some of the distinctive pedagogical features presented in this book.

Ten Axioms of Finance: The principles that motivate the practices of corporate finance, are presented in the ten axioms in Chapter 1. These axioms appear throughout the text as inserts entitled "Back to the Basics" and are placed to remind students of the underlying principles and to keep students from becoming so immersed in specific calculations that the interrelationships among topics are lost.

Chapter Reorganization and Consolidation: In response to both the continued development of financial thought and reviewers' comments, the text has been dramatically reorganized, including the consolidation and elimination of old chapters and the creation of new ones. In addition to the introduction of the ten axioms, the main features of this reorganization and consolidation include two new chapters on capital budgeting, two chapters on valuation of bonds and shares, as well as expanded coverage of financial markets and interest rates in Chapter 2.

Perspectives in Finance: The inserts entitled "Perspectives in Finance" appear throughout the text to redirect students' attention to the "big picture." Although tools, techniques, and calculations are treated extensively, the use of these Perspectives, as well as the ten axioms, keep the student from losing sight of the interrelationships and motivating factors behind the tools.

Basic Financial Management in Practice: The inserts entitled "Basic Financial Management in Practice" have been incorporated into the text to reflect Canadian business practices and to add life to chapter presentations.

Ethics in Financial Management: Within the inserts entitled "Ethics in Financial Management," we have identified several important ethical issues and examined their relation to financial decision making.

Introductory Examples: Each chapter opens with an introductory example, setting the stage for what is to follow with the related experiences of an actual company. In this way the relevance, use and importance of the material to be presented can easily be understood by the student.

Chapter Learning Objectives: Each chapter begins by stating the learning objectives for the chapter, and highlighting what that chapter will enable the student to achieve.

World Wide Web Site

To help students make use of the exciting Internet resources now available, Prentice Hall has developed a special "Web links" icon. It appears in the margin, together with an Internet address, opposite passages of the text that discuss companies or organizations with interesting and informative home pages.

Video Case

The CBC icon identifies points in the text at which a video supplement can be used. Prentice Hall Canada and the CBC have worked together to bring you dynamic and challenging segments from the *Venture* and *Market Place* series. The tapes in this video supplement have substantial content and extremely high production quality. A detailed synopsis, teaching outline, and discussion questions for each segment are included in the Instructor's Manual.

International Financial Management: In view of globalization of world markets, we have integrated international finance into the text through the inserts entitled "International Financial Management." In addition, we acknowledge that many instructors approach the teaching of international finance in different ways; hence, we include a chapter specifically on international financial management.

Financial Calculators: The use of calculators has been integrated into this text, especially with respect to the presentation of the time value of money. Where appropriate, calculator solutions appear in the margin for the student. We have developed a more in-depth application for bond solutions and home loan amortizations by applying the worksheets from the Texas Instruments BA II Plus calculator. We have also designed an insert card to the text that shows how to find the compound value interest factors and the present value factors without the use of the table values. In the *Study Guide* for *Basic Financial Management,* we provide a complete article on the use of all the popular financial calculators as well as a separate appendix that demonstrates the use of the Texas Instruments BA II Plus calculator.

In addition to the broader changes just mentioned, modifications have also been made in each chapter. The following list includes the major additions that are new to *Basic Financial Management* in the 2nd edition:

- New relatively simple, problems have been added to almost every chapter to give the instructor more choice in homework assignments.
- Chapter 2 has been substantially revised and streamlined with regard to both organization and to the financial markets. The chapter examines several key topics including (a) why financial markets exist, (b) historical rates of return in financial markets, (c) an overview of the interest rate determination, and (d) the term structure of interest rates.
- The subject material on the tax environment focuses on how taxes affect financial decision making with consideration of small businesses.
- Chapter 3, Risk and Rates of Return, has additional practical and real-world insights into our understanding of the capital markets. We have updated the material on the capital asset pricing model so as to provide a better understanding of its limitations. We have added material which provides insight as to how modern portfolio theory can assist in choosing among portfolios.
- Coverage of the effective annual rate and annuities due is now provided in Chapter 4.
- Chapter 5 begins with a description of the types and characteristics of bonds and then follows with an explanation of how to value bonds.
- Chapter 6 begins with an introduction to the types and characteristics of preferred shares and common shares. This material is then followed with an explanation of the various methods used to value shares.
- Chapter 9, Planning the Firm's Financing Mix, contains new material on (a) agency costs, (b) the static-tradeoff and pecking order theories of capital structure determination, (c) free cash flow theory and (d) actual financing decisions.
- Sections of Chapter 12, Cost of Capital, have been based on time spent with practitioners so as to understand how firms use this concept in practice. As a result we have examined how PepsiCo, Inc., managers work with the firm's cost of capital to increase shareholder value. We will show the student just how the management of PepsiCo computes the company's cost of capital.
- Capital budgeting is now covered in four chapters. Chapter 14 includes a section on the use of the modified internal rate of return (MIRR). Whereas, Chapter 16 provides several examples of how the capital-budgeting process can be applied to financial management decisions.
- Chapter 17, Introduction to Working-Capital Management, has been streamlined to focus on the concept of hedging, with the application of hedging techniques to interest rates.
- Coverage of the concept of total quality management (TQM), which is a company-wide systems approach to quality, is now included in Chapter 19. This new order in customer-supplier relationships, as well as the implications for inventory control, are examined.
- Chapter 21, Financial Forecasting, Planning and Budgeting, examines how consideration for the need of discretionary financing and sustainable growth are integrated into the financial forecasting process.
- Chapter 22 presents additional material that provides an introduction to the valuation of options and the application of currency swaps.
- Chapter 23, Corporate Restructuring, has been revised to include material on divestiture methods such as reorganization and liquidation.
- Chapter 25, Financial Management: Small and Mid-Sized Business Perspective, has been updated to reflect the importance of these types of businesses to the Canadian economy.

A NOTE TO THE INSTRUCTOR

The emphasis of this text is to stress the decision-making process. As a result, we have given readability a high priority by providing a concise writing style, especially in the treatment of concepts requiring the use of mathematics. In addition, we have provided complete, step-by-step examples that are frequently used to clarify issues in the student's mind. In summary, the pedagogical approach taken, particularly for the more difficult topics, progresses from an intuitive presentation of the problem to the introduction and illustration of the appropriate decision-making process.

SUPPLEMENTS

Instructor's Manual with Solutions: Written by the authors, this supplement includes a chapter orientation and outline, as well as fully worked out solutions to the end-of-chapter materials, including both Problem Sets and the Integrative Problems. Also included are a detailed synopsis, teaching outline, and discussion questions for the CBC video cases, and a special section on how to use the Texas Instruments Financial Calculator. (ISBN 0-13-577404-7)

Test Item File: Revised and updated for this edition, the test item file includes approximately 2,000 questions. (ISBN 0-13-577446-2)

PH Custom Test (Windows): Prentice Hall's computerized test file uses a state-of-the-art software program that provides fast, simple, and error-free test generation. Entire tests can be previewed on-screen before printing. PH Custom Test can print multiple variations of the same test, scrambling the order of questions and multiple-choice answers. Tests can be saved to ASCII format and revised in your word-processing system. (ISBN 0-13-577479-9)

Electronic Transparencies: Tabular material from the book, lecture notes for use in class, extra graphs, and end-of-chapter solutions are available on PowerPoint disks. (ISBN 0-13-577438-1)

CBC/Prentice Hall Video Library: Prentice Hall and CBC help you take full advantage of the video medium by offering documentary videos that directly relate to the concepts and applications in *Basic Financial Management*. The package offers seven video clips from CBC's *Venture* and *Market Place* series. Points in the text at which the videos may be used most effectively are identified by the CBC logo, and a synopsis with teaching suggestions is included in the Instructor's Manual. (ISBN 0-13-577412-8)

Study Guide: To assist students in preparing class lectures, the *Study Guide* for *Basic Financial Management* provides a concise review of important topics with both questions and problem examples to increase understanding of the material. The *Study Guide* has been prepared to enable students to take additional notes in the margins, but made complete enough to enable students to concentrate on the material which is presented in class. The margins include calculator solutions to problem material where appropriate. (ISBN 0-13-577420-9)

TESTIMONIALS

For the Second Canadian Edition of *Basic Financial Management,* I acted as problem reviewer. In this role, I checked every integrative problem for terminology and procedure and then solved each of these integrative problems. A substantial number of new end-of-chapter problems have been added to this edition and these have all been thoroughly checked. Any new illustrative examples in the text have been checked for content and accuracy. In all cases, suggestions for changes and corrections have been forwarded to the publisher for further review by the author.

Norman J. Bell

As requested I have read the proof copy of *Basic Financial Management,* Second Canadian Edition and have checked the arithmetic in all examples and exhibits as well as ensuring that references to those examples and exhibits within the text were accurate.

Christine Watt

ACKNOWLEDGEMENTS

We gratefully acknowledge the support, encouragement and assistance of those who have contributed to the successful completion of *Basic Financial Management.* In addition, we would like to express our appreciation to the CGA Research Centre at the University of Ottawa for their support in providing annual reports.

Particular thanks are due to the following people for reviewing the manuscript and offering many useful suggestions: Norman Bell; Melanie Cao, University of Toronto; John D'Amato, Sheridan College; Peggy Hedges, University of Calgary; Melanie Russell; and Hao Zhang, University of Victoria.

We are also grateful to the scores of instructors across the country who took the time to respond to a survey conducted while this edition was being planned. The thoughts and opinions of these respondents were a valuable guide as we mapped out a strategy for improving the text and its ancillaries.

We also appreciate the technical checking done by Norman Bell and Christine Watt, and the careful proofreading by Michael Rowan, Gordon Sova, Robyn Packard, and Gail Copeland.

Readers wishing further information on excerpts provided through the cooperation of Statistics Canada may obtain copies of related publications by mail from: Publications Sales, Statistics Canada, Ottawa, Ontario K1A 0T6.

Finally, we would like to thank the numerous members of the Prentice Hall staff with whom we had an opportunity to work, including Patrick Ferrier, Lisa Penttila, Lesley Mann, and Valerie Adams. Their extraordinary patience and fine support throughout writing, production and marketing of the text were extremely valuable. Their guidance and help has had a significant and positive impact on the quality of the text.

D.W.M.
A.J.K.
D.F.S.
J.W.P.
J.D.M.

CHAPTER 1

AN INTRODUCTION TO FINANCIAL MANAGEMENT

LEARNING OBJECTIVES

After reading this chapter you should be able to

1. Define financial management and explain the role of the financial executive.
2. Explain the goal of the firm.
3. Explain the ten axioms that form the basics of financial management.
4. State the general topics to be covered in the remainder of this text.

INTRODUCTION

In 1994, Frank Stronach, the chairman of Magna Corporation, received total compensation of over $40 million. That works out to over $100,000 per day, or over $13,000 per hour with no time off. What is interesting about this is the fact that few shareholders complained. Magna International Inc. showed a one-year return on common equity for the 1994 fiscal year of 22 percent, but a five-year return on equity of 1.2 percent. Stronach's total compensation was 17.4 percent of 1994 profits. The majority of his total compensation was earned on conversion of stock options.

In this chapter, we are not so concerned with how much a firm's CEO earns, but whether the compensation package helps align the interests of the shareholders with those of the managers. We will see that this alignment between managers and shareholders is necessary in order to achieve the goal of the firm. We will also see that aligning the interests of both shareholders and managers serves as one of the basic axioms of finance which we develop in this chapter. Moreover, the tax laws encourage corporations to incorporate pay for performance into management contracts, thereby aligning the interests of shareholders and managers.

CHAPTER PREVIEW

In this chapter we briefly define financial management in order to introduce the goal of the firm—the maximization of shareholder wealth. The goal of the firm enables financial executives to develop rules for decision making. We then examine the ten axioms that form the basics of financial management. While these axioms may seem quite simple, they provide the driving force behind all that follows. These axioms provide the threads that tie together the concepts and techniques introduced in the chapters, all directed to create shareholder wealth.

OBJECTIVE 1

FINANCIAL MANAGEMENT AND THE ROLE OF THE FINANCIAL EXECUTIVE

Economic value or wealth

Economic value relates to the concept of consumption over the planning horizon of the decision maker. In financial management, we will come to know value as the discounted amount of future expected cash flows.

Financial management is concerned with the creation and maintenance of **economic value or wealth** of a firm's shareholders. Consequently, this text focuses on decision making with an eye toward creating wealth.

To illustrate, consider two firms, Merck and IBM. At the end of 1993, the total market value of Merck, a large pharmaceutical company, was $46 billion. Over the life of the business, Merck's investors had invested about $17 billion in the business. In other words, management created $29 billion in additional wealth for the shareholders. IBM, on the other hand, was valued at $54 billion in December 1993. But over the years, IBM's investors had actually invested $70 billion—a loss in value of $16 billion. Thus we see that Merck created wealth for its shareholders, while IBM lost it.

Financial executives actively participate in both the creation and maintenance of the economic value of the firm. As managers of the firm, financial executives assume decisional, interpersonal, and informational roles.[1] For example, a company's treasurer makes decisions concerning the acquisition of financing for investment projects; whereas, the controller monitors the financial transactions of the firm and disseminates information to the public through the company's reporting function. Other financial executives make decisions on the acquisition or replacement of the company's fixed assets. Needless to say, the financial executive must develop decision-making techniques to perform each of these roles.

In introducing decision-making techniques, we will emphasize the logic behind those techniques, thereby insuring that we do not lose sight of the concepts when dealing with the calculations. To the first-time student of finance, this may sound a bit overwhelming. But as we will see, the techniques and tools introduced in this text are all motivated by ten underlying principles or axioms that will guide us through the decision-making process.

OBJECTIVE 2

GOAL OF THE FIRM

Maximization of shareholder wealth

The maximization of the total market value of shareholders' common shares.

In financial management, the goal of the firm is the **maximization of shareholder wealth**, which implies the maximization of the total market value of shareholders' common shares. Not only is this goal in the best interest of the shareholders, but it also provides the most benefits to society. For example, society will benefit if scarce resources are allocated to firms that provide the most productive use for these

[1]Hellreigel, Slocum and Woodman have indicated that there are three categories of roles that managers assume: interpersonal, informational and decisional. For more information, see: Hellreigel, D., J. W. Slocum, and R. W. Woodman. *Organizational Behaviour* (New York: West Publishing Company, 1986).

resources. Moreover, a firm's productive use of resources creates wealth for its shareholders. To better understand this goal, we will first examine the deficiencies of profit maximization as a goal for the firm. Then we will compare profit maximization to the maximization of shareholder wealth to see why, in financial management, the latter is the more appropriate goal for the firm.

The Deficiencies of Profit Maximization

In microeconomic courses, profit maximization is frequently given as the goal of the firm. Profit maximization stresses the efficient uses of capital resources, but it is not specific with respect to the time frame over which profits are to be measured. Do we maximize profits over the current year, or do we maximize profits over some longer period? A financial executive could easily increase current profits by eliminating research and development expenditures and cutting down on routine maintenance. In the short run, this might result in increased profits, but this clearly is not in the best long-run interests of the firm. If we are to base financial decisions on a goal, that goal must be precise, not allow for misinterpretation, and deal with all the complexities of the real world.

In microeconomics, profit maximization functions largely as a theoretical goal, with economists using it to prove how firms behave rationally to increase profit. Unfortunately, it ignores many real-world complexities that financial executives must address in their decisions. In the more applied discipline of financial management, firms must deal every day with three major factors not considered by the goal of profit maximization: uncertainty, timing and the cost of equity.[2]

Uncertainty of Returns

Projects and investment alternatives are compared by examining their **expected values or weighted average profits**. Whether or not one project is riskier than another does not enter these calculations; economists do discuss risk, but only tangentially.[3] In reality, projects do differ with respect to risk characteristics, and failing to recognize these differences can result in incorrect decisions in the practice of financial management. To better understand the implications of ignoring risk, let us look at two **mutually exclusive** investment alternatives (that is, only one can be accepted). The first project involves the use of existing plant to produce plastic combs, a product with an extremely stable demand. The second project uses existing plant to produce electric vibrating combs. This latter product may catch on and do well, but it could also fail. The possible outcomes (optimistic prediction, pessimistic prediction, and expected outcome) are given in Table 1-1.

No variability is associated with the possible outcomes for the plastic comb project. If things go well, poorly, or as expected, the outcome will still be $10,000. With

Expected value or weighted average

The average of all possible outcomes, where those outcomes are weighted by the probability that each outcome will occur.

Mutually exclusive

The acceptance of one alternative which necessarily means rejection of the other alternatives.

Table 1-1 Possible Project Outcomes

	Profit	
	Plastic Comb	Electric Comb
Optimistic prediction	$10,000	$20,000
Expected outcome	10,000	10,000
Pessimistic prediction	10,000	0

[2]Profit maximization focuses only on the amount and not on the risk assumed in obtaining the expected profits.

[3]For example, see Robert S. Pindyck and Daniel L. Rubenfeld, *Microeconomics*, 2nd ed. (New York: Macmillan, 1992), pp. 244–246.

the electric comb, however, the range of possible outcomes goes from $20,000 if things go well, to $10,000 if they go as expected, to zero if they go poorly. If we look just at the expected outcomes, the two projects appear equivalent. They are not. The returns associated with the electric comb involve a much greater degree of uncertainty or risk.

An evaluation of these two projects in terms of maximizing profits ignores uncertainty and considers these projects equivalent in terms of desirability. In reality, projects differ a great deal with respect to risk characteristics, and to disregard these differences in the practice of financial management can result in incorrect decisions. As we will discover later in this chapter, there is a very definite relationship between risk and expected return—that is, investors demand a higher expected return for taking on added risk—and to ignore this relationship would lead to improper decisions.

Timing of Returns

Another problem with using the goal of profit maximization is that it ignores the timing of the project's returns. To illustrate, let us re-examine our plastic comb versus electric comb investment decision. This time let us ignore risk and say that each of these projects is going to return a profit of $10,000 for one year; however, it will be one year before the electric comb can go into production, while the plastic comb can begin production immediately. The timing of the profits from these projects is illustrated in Table 1-2.

In this case the total profits from each project are the same, but the timing of the returns differs. As we will see later, money has a definite time value. Thus, the plastic comb project is the better of the two. After one year the $10,000 profit from the plastic combs could be invested in a savings account earning 5 percent interest. At the end of the second year it would have grown to $10,500. Since investment opportunities are available for money in hand, we are not indifferent to the timing of the returns. Given equivalent cash flows from profits, we want those cash flows sooner rather than later.[4] Therefore, the financial manager must always consider the possible timing of cash flows in financial decision making.

Cost of Equity

Shareholders provide funds to a firm at a cost. This cost is not considered in defining profits, nor is it considered in using profit maximization as a goal of the firm. Although accounting profits consider interest expense as a cost of borrowing money, the calculation ignores the cost of funds provided by the firm's shareholders (owners). Are a company's managers acting in the best interest of the firm's shareholders if a firm earns a 4 percent return on shareholders' equity, when short-term government securities (a risk-free investment) are paying a 6 percent return? Although management is certainly increasing the firm's profits, they are destroying shareholder wealth by investing in projects that do not even earn a rate of return

Table 1-2 Timing of Profits

	Profit	
	Plastic Comb	Electric Comb
Year 1	$10,000	$ 0
Year 2	0	10,000

[4]The goal of profit maximization also ignores the dividend decision. In fact, if the goal of the firm was solely profit maximization, it would not pay any dividends at all and thereby maximize profits.

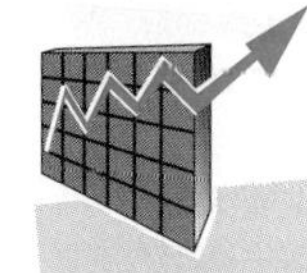

What Should Corporations Maximize?

The following excerpt from the article "What Should Corporations Maximize?" by Chamberlain and Gordon examines the traditional assumption that a firm's investment and financing decisions should maximize the current market value of its shares.

When Are Managements Servants of Shareholders?

When the ownership of shares in public corporations is widely distributed among portfolio investors, their interest is served by investment and financing policies that maximize current share price. But in this case, immediate control of the corporation is in the hands of management and its interest may be served by different policies. What are these policies? To get at them, we must first establish the conditions under which management will follow policies which maximize current share price.

It requires a view of the corporation that it is engaged solely in production. Opportunities to earn an abnormal return on capital are random accidents of nature. They are independent of management's performance, and the positive or negative windfalls that results accrue to the residual shareholders' equity in the firm. Furthermore, shareholders are completely informed on a manager's performance exactly and managers who are dismissed due to bankruptcy or contraction of the firm move to another job without loss of human capital.

A More Realistic View

A more realistic view of the firm is that it engages in two classes of activities. One is production and the other is a range of activities intended to make the production profitable. The development of new products and methods of production, marketing, advertising, employee relations and organizational innovations are among the non-production activities undertaken to maintain and increase the spread between price and production cost and to expand sales at these favourable margins. Success in these activities generates incremental cashflow, and shareholders benefit through increases in the market value of the firm's shares.

Management participates too through higher base compensation, bonuses and stock options plans. However, total compensation for a given managerial position varies little among firms. Even bonuses and stock options plans, though sometimes large in comparison with a manager's salary, capture relatively little of the differential performance among individuals and firms. The other way in which management can share in the rents generated by successful performance is through promotions or more attractive jobs elsewhere which are made possible by the firm's growth. In contrast, dismissal due to bankruptcy or contraction of the firm seriously impairs the manager's human capital due to some period of unemployment followed by a possible inferior new position.

Conclusion: Manager and Shareholder Goals May Clash

The managements of corporations owned by portfolio investors thus have a particular interest in the firm's long-run growth and survival. A high rate of growth in compensation and indirectly through a reduction in the probability of bankruptcy. This objective is ordinarily served by a lower debt-equity ratio than that which maximizes current share price. A higher ratio creates a higher probability of bankruptcy in the near future. Indifference to short term share price maximization would also result in a policy of paying no dividends and using earnings to maximize the firm's growth rate.

However, the larger the spread between actual dividend and debt policies and those that maximize share price, the greater the threat of a hostile takeover. Hence, management will set the dividend rate and the debt-equity ratio to achieve a balance between the desire for growth and survival in the long run and the avoidance of a hostile takeover in the short run. The likely result still is a lower dividend rate and a higher rate of growth in capital than under a policy of maximizing share price. Also, expenditures to facilitate growth, such as marketing, research and development, are also likely at a higher rate.

Source: Adapted from: Trevor W. Chamberlain and Myron J. Gordon, "What Should Corporations Maximize?" *Canadian Investment Review* (Fall 1991), pp. 41–45.

equal to that paid on government securities. As a result, we now turn to the examination of a more robust goal for the firm: the maximization of shareholder wealth.

Maximization of Shareholder Wealth

In formulating our goal of maximization of shareholder wealth we are doing nothing more than modifying the goal of profit maximization to deal with the complexities of the operating environment. We have chosen maximization of shareholder wealth—that is, maximization of the market value of the firm's common shares—because the effects of all financial decisions are thereby included. The shareholders react to poor investment or dividend decisions by causing the total value of the firm's common shares to fall and they react to good decisions by pushing the price of the shares up. In effect, under this goal, good decisions are those that create wealth for the shareholder.[5]

Obviously, there are some serious practical problems in implementing this goal and in using the firm's shares to evaluate financial decisions. Many factors affect share prices; to attempt to identify a reaction to a particular financial decision would simply be impossible. Fortunately, this is not necessary. In order to employ this goal, we need not consider every share price change to be a market interpretation of the worth of our decisions. Other factors, such as economic expectations, also affect share price movements. What we do focus on is the effect that our decision should have on share price if everything were held constant. The market price of the firm's shares reflects the value of the firm as seen by its owners. It takes into account uncertainty or risk, time, and any other factors that are important to the owners. Thus, the framework of maximization of shareholder wealth allows for a decision environment that includes the complexities and complications of the real world. As we follow this goal throughout our discussions, we have to keep in mind one more question: who exactly are the shareholders? Shareholders are the legal owners of the firm.

The Agency Problem

Although our goal for the firm will be the maximization of shareholder wealth, in reality the **agency problem** may interfere with the implementation of this goal. Several forms of the agency problem exist: management versus shareholders, shareholders versus bondholders, and management versus government.

Agency problem

Problem resulting from conflicts of interest between the manager (the shareholders' agent) and the shareholders.

Management versus Shareholders

The first form of the agency problem occurs because of the separation of management and the ownership (shareholders) of the firm. The firm's managers can be properly thought of as agents for the firm's shareholders. To ensure that agent-managers act in the shareholders' best interests requires that they have (1) proper incentives to do so and (2) that their decisions are monitored. The incentives usually take the form of executive compensation plans and managerial "perks." For example, perks may take the form of a bloated support staff, country club memberships, luxurious corporate planes, or other amenities of a similar nature. Monitoring requires that certain costs be borne by the shareholders, such as (1) bonding the managers, (2) auditing financial statements, (3) structuring the organization in unique ways that limit useful managerial decisions, and (4) reviewing the costs and benefits of managerial perks. The main point is that these costs are ultimately cov-

[5]The principle of maximization of shareholder wealth is guided by the Fisher separation principle and economic utility theory. The Fisher separation principle indicates that there is separation of investment/operating decisions of firms from shareholder preferences or tastes. As a result, the current shareholder's expected utility is maximized if a firm maximizes the price of current shares. For a more detailed discussion, see: Copeland, Thomas E. and J. Fred Weston, *Financial Theory and Corporate Policy*, 2nd ed. (Don Mills: Addison-Wesley Publishing Company, 1983), pp. 123–127.

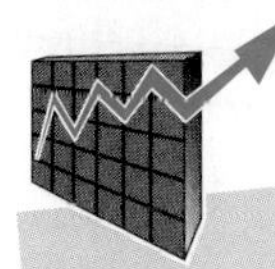

Bradford Cornell and Alan C. Shapiro on Stockholder-Manager Agency Conflicts

Managers, like all other economic agents, are ultimately concerned with maximizing their own utility, subject to the constraints they face. Although management is legally mandated to act as the agent of shareholders, the laws are sufficiently vague that management has a good deal of latitude to act in its own behalf. This problem, together with the separation of ownership and control in the modern corporation, results in potential conflicts between the two parties. The agency conflict between managers and outside shareholders derives from three principal sources.[6]

The first conflict arises from management's tendency to consume some of the firm's resources in the form of various perquisites. But the problem of overconsumption of "perks" is not limited to corporate jets, fancy offices, and chauffeur-driven limousines. It also extends, with far greater consequences for shareholders, into corporate strategic decision-making. As Michael Jensen points out, managers have an incentive to expand the size of their firms beyond the point at which shareholder wealth is maximized.[7] Growth increases managers' power and perquisites by increasing the resources at their command. Because changes in compensation are positively related to sales growth, growth also tends to increase managerial compensation.[8]

As Jensen has argued persuasively, the problem of overexpansion is particularly severe in companies that generate substantial amounts of "free cash flow"—that is, cash flow in excess of that required to undertake all economically sound investments (those with positive net present values).[9] Maximizing shareholder wealth dictates that free cash flow be paid out to shareholders. The problem is how to get managers to return excess cash to the shareholders instead of investing it in projects with negative net present values or wasting it through organizational inefficiencies.

A second conflict arises from the fact that managers have a greater incentive to shirk their responsibilities as their equity interest falls. They will trade off the costs of putting in additional effort against the marginal benefits. With a fixed salary and a small equity claim, professional managers are unlikely to devote energy to the company equivalent to that put forth by an entrepreneur.

Finally, their own risk aversion can cause managers to forgo profitable investment opportunities. Although the risk of potential loss from an investment may be diversified in the capital markets, it is more difficult for managers to diversify the risks associated with losing one's salary and reputation. Forgoing profitable, but risky, projects amounts to the purchase by management of career insurance at shareholder expense.

Source: Bradford Cornell and Alan C. Shapiro, "Financing Corporate Growth," *Journal of Applied Corporate Finance* 1 (Summer 1988), pp. 6–22.

---

[6]The agency conflict is discussed in Michael C. Jensen and William H. Meckling, "Theory of the Firm: Managerial Behavior, Agency Costs and Ownership Structure," *Journal of Financial Economics* (October 1976), pp. 305–360.

[7]Michael C. Jensen, "Agency Costs of Free Cash Flow, Corporate Finance, and Takeovers," *American Economic Review* (May 1986), pp. 323–329.

[8]Evidence on this point is supplied by Kevin J. Murphy, "Corporate Performance and Managerial Remuneration: An Empirical Analysis," *Journal of Accounting and Economics* (April 1985), pp. 11–42.

[9]We define the net present value concept in Chapter 13. Very briefly, an investment's present value represents its expected contribution to the value of the firm's common stock and, consequently, shareholder wealth.

ered by the owners of the company—its common shareholders. It is difficult to measure exactly how significant a problem this is. However, while this problem may interfere with the implementation of the goal of maximization of shareholder wealth in some firms, it does not affect the goal's validity.

Shareholders versus Bondholders

Another form of the agency problem stems from the conflict of interest between shareholders and bondholders. Acting in the shareholders' best interests might cause management to invest in extremely risky projects. Existing investors in the firm's bonds could logically take a dim view of such an investment policy. In fact, this type of investment policy could lead to a downward revision of the bond rating the firm currently enjoys. A lowered bond rating in turn would lower the current market value of the firm's bonds. To reduce this conflict of interest, the creditors (bond investors) will require that several protective covenants be included in the bond contract. These bond covenants are discussed in more detail in Chapter 5, but essentially they may be thought of as restrictions on managerial decision making. Typical covenants restrict payment of cash dividends on common shares, limit the acquisition or sale of assets, or limit further debt financing. To make sure that the protective covenants are complied with by management means that monitoring costs are incurred. Like all monitoring costs, they are borne by common shareholders. In addition, shareholders will incur indirect costs because management may forgo higher-return projects because of bond covenants.

The costs associated with the agency problem are also difficult to measure, but occasionally we can see the effect of this problem in the marketplace. For example, if the market feels that the management of a firm is damaging shareholder wealth, we might see a positive reaction in the share price to the removal of that management. On the day following the death of John Dorrance, Jr., chairman of Campbell Soup, in 1989, Campbell's share price rose about 15 percent. Some investors felt that Campbell's relatively small growth in earnings might be improved with the departure of Dorrance. There was also speculation that Dorrance was the major obstacle to a possible positive reorganization.

Management versus Government

Management's actions may be constrained because of government regulations. For example, environmental regulations on emissions of pollutants may be imposed on a firm's manufacturing plants by governments. As a result, the firm will incur monitoring costs in the form of fees for environmental audits or compliance audits, so as to ensure that regulations are met. Tax regulations may also affect the firm's results since management may postpone the recognition of income in order to defer taxes payable to future years. Each of these types of regulations cause shareholders either to forgo benefits or incur direct costs. In addition, shareholders incur indirect costs because management's actions may be constrained by government regulations.

OBJECTIVE 3

TEN AXIOMS THAT FORM THE BASICS OF FINANCIAL MANAGEMENT

To the first-time student of finance, the subject matter may seem like a collection of unrelated decision rules. This could not be further from the truth. In fact, our decision rules, and the logic that underlies them, spring from ten simple axioms that do not require knowledge of finance to understand. *However, while it is not necessary to understand finance in order to understand these axioms, it is necessary to understand these axioms in order to understand finance.* Keep in mind that although these axioms may at first appear simple or even trivial, they will provide the driving force behind all that follows. These axioms will weave together concepts and techniques presented in this text, thereby allowing us to focus on the logic underlying the practice of financial management.

Axiom 1: The Risk-Return Tradeoff—We Won't Take on Additional Risk Unless We Expect to Be Compensated with Additional Return

At some point we have all saved some money. Why have we done this? The answer is simple: to expand our future consumption opportunities. We are able to invest those savings and earn a return on our dollars because some people would rather forgo future consumption opportunities to consume more now. Assuming there are a lot of different people that would like to use our savings, how do we decide where to put our money?

First, investors demand a minimum return for delaying consumption that must be greater than the anticipated rate of inflation. If they didn't receive enough to compensate for anticipated inflation investors would purchase whatever goods they desired ahead of time or invest in assets that were subject to inflation and earn the rate of inflation on those assets. There isn't much incentive to postpone consumption if your savings are going to decline in terms of purchasing power.

Investment alternatives have different amounts of risk and expected returns. Investors sometimes choose to put their money in risky investments because these investments offer higher expected returns. The more risk an investment has, the higher will be its expected return. This relationship between risk and expected return is shown in Figure 1-1.

Notice that we keep referring to *expected* return rather than *actual* return. We may have expectations of what the returns from investing will be, but we can't peer into the future and see what those returns are actually going to be. If investors could see into the future no one would invest money in the shares of a company that would eventually file for bankruptcy. Until after the fact, you are never sure what the return on an investment will be. That is why Sears Canada bonds pay more interest than Government of Canada bonds of the same maturity. The additional interest convinces some investors to take on the added risk of purchasing a Sears Canada bond.

This risk-return relationship will be a key concept as we value stocks, bonds, and proposed new projects throughout this text. We will also spend some time determining how to measure risk. Interestingly, much of the work for which the 1990 Nobel Prize for Economics was awarded centred on the graph in Figure 1-1

Figure 1-1 The Risk-Return Relationship

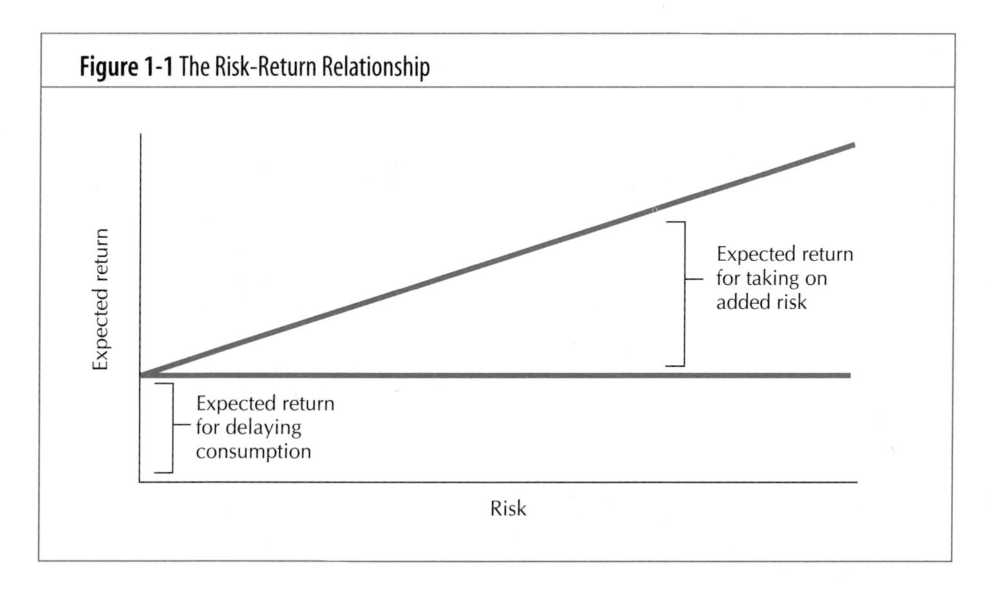

and how to measure risk. Both the graph and the risk-return relationship it depicts will reappear often in this text.

Axiom 2: The Time Value of Money—A Dollar Received Today Is Worth More Than a Dollar Received in the Future

A fundamental concept in finance is that money has a time value associated with it: A dollar received today is worth more than a dollar received a year from now. Because we can earn interest on money received today, it is better to receive money earlier rather than later. In your economics courses, this concept of the time value of money is referred to as the opportunity cost of passing up the earning potential of a dollar today.

In this text we focus on the creation and measurement of wealth. To measure wealth or value we will use the concept of the time value of money to bring the future benefits and costs of a project back to the present. Then, if the benefits outweigh the costs the project creates wealth and should be accepted; if the costs outweigh the benefits the project does not create wealth and should be rejected. Without recognizing the existence of the time value of money, it is impossible to evaluate projects with future benefits and costs in a meaningful way.

To bring future benefits and costs of a project back to the present, we must assume a specific opportunity cost of money, or interest rate. Exactly what interest rate to use is determined by ***Axiom 1, The Risk-Return Tradeoff,*** which states investors demand higher returns for taking on more risky projects. Thus, when we determine the present value of future benefits and costs, we take into account that investors demand a higher return for taking on added risk.

Axiom 3: Cash—Not Profits—Is King: Measuring the Timing of Costs and Benefits

In measuring wealth or value we will use cash flows, not accounting profits, as our measurement tool. Cash flows are received by the firm and can be reinvested. Accounting profits, on the other hand, are shown when they are earned rather than when the money is actually in hand. A firm's cash flows and accounting profits may not occur together. For example, capital expenses, such as the purchase of new equipment or a building, is depreciated over several years, with the annual depreciation subtracted from profits. However, the cash flow associated with this expense generally occurs immediately. Therefore cash outflows involving paying money out and cash inflows that can be reinvested correctly reflect the timing of the benefits and costs.

At this point, the first three axioms we have presented can be used to determine the value of any asset, be it a business, a new project, or a financial asset like a share of stock or a bond. In future chapters we will provide the techniques for determining the value of an asset based on these axioms.

Axiom 4: Incremental Cash Flows—It's Only What Changes That Counts

Incremental cash flow

The difference between the cash flows if the project is taken on versus what they will be if the project is not taken on.

In making business decisions, we are concerned with the results of those decisions: What happens if we say yes versus what happens if we say no? ***Axiom 3*** states that we should use cash flows to measure the benefits that accrue from taking on a new project. We are now fine tuning our evaluation process so that we only consider **incremental cash flows**. The incremental cash flow is the difference between the cash flows if the project is taken on versus what they will be if the project is not taken on.

Not all cash flows are incremental. For example, when Leaf Inc., a manufacturer of sports cards, introduced Donruss Triple Play Baseball Cards in 1992, the product competed directly with the company's Leaf and Donruss baseball cards. There is no doubt that some of the sales dollars that ended up with Donruss Triple Play

Cards would have been spent on Donruss or Leaf Cards if Triple Play cards had not been available. Although the Leaf corporation meant to target the low-cost end of the baseball cards market held by Topps, there was no question that Triple Play sales bit into—actually cannibalized—sales from the company's existing product lines. The *difference* between revenues generated by introducing the new cards versus maintaining the original series are the incremental cash flows. This difference reflects the true impact of the decision.

What is important is that we *think* incrementally. Our guiding rule in deciding whether a cash flow is incremental is to look at the company with and without the new product. In fact, we will take this incremental concept beyond cash flows and look at all consequences from all decisions on an incremental basis.

Axiom 5: The Curse of Competitive Markets—Why It's Hard to Find Exceptionally Profitable Projects

Our job as financial managers is to create wealth. Therefore, we will look closely at the mechanics of valuation and decision making. We will focus on estimating cash flows, determining what the investment earns, and valuing assets and new projects. But it will be easy to get caught up in the mechanics of valuation and lose sight of the process of creating wealth. Why is it so hard to find projects and investments that are exceptionally profitable? Where do profitable projects come from? The answers to these questions tell us a lot about how competitive markets operate and where to look for profitable projects.

In reality, it is much easier evaluating profitable projects than finding them. If an industry is generating large profits, new entrants are usually attracted. The additional competition and added capacity can result in profits being driven down to the required rate of return. Conversely, if an industry is returning profits below the required rate of return, then some participants in the market drop out, reducing capacity and competition. In turn, prices are driven back up. This is precisely what happened in the VCR video rental market in the mid-1980s. This market developed suddenly with the opportunity for extremely large profits. Because there were no barriers to entry, the market quickly was flooded with new entries. By 1987 the competition and price cutting produced losses for many firms in the industry, forcing them to flee the market. As the competition lessened with firms moving out of the video rental industry, profits again rose to the point where the required rate of return could be earned on invested capital.

In competitive markets, extremely large profits simply cannot exist for very long. Given that somewhat bleak scenario, how can we find good projects—that is, projects that return more than the required rate of return? Although competition makes them difficult to find, we have to invest in markets that are not perfectly competitive. The two most common ways of making markets less competitive are to differentiate the product in some key way or to achieve a cost advantage over competitors.

Product differentiation insulates a product from competition, thereby allowing prices to stay sufficiently high to support large profits. If products are differentiated, consumer choice is no longer made by price alone. For example, in the pharmaceutical industry, patents create competitive barriers. SmithKline Beecham's Tagamet, used in the treatment of ulcers, and Hoffman-La Roche's Valium, a tranquilizer, are protected from direct competition by patents.

Service and quality are also used to differentiate products. For example, Caterpillar Tractor has long prided itself on the quality of its construction and earth-moving machinery. As a result, it has been able to maintain its market share. Similarly, much of Toyota and Honda's brand loyalty is based on quality. Service can also create product differentiation, as shown by McDonald's fast service, cleanliness, and consistency of product that brings customers back.

Whether product differentiation occurs because of advertising, patents, service, or quality, the more the product is differentiated from competing products, the less competition it will face and the greater the possibility of large profits.

Economies of scale and the ability to produce at a cost below competition can effectively deter new entrants to the market and thereby reduce competition. The retail hardware industry is one such case. In the hardware industry there are fixed costs that are independent of the store's size. For example, inventory costs, advertising expenses, and managerial salaries are essentially the same regardless of annual sales. Therefore, the more sales that can be built up, the lower the per-sale dollar cost of inventory, advertising, and management. Restocking from warehouses also becomes more efficient as delivery trucks can be used to full potential.

Regardless of how the cost advantage is created—by economies of scale, proprietary technology, or monopolistic control of raw materials—the cost advantage deters new market entrants while allowing production at below industry cost. This cost advantage has the potential of creating large profits.

The key to locating profitable investment projects is to first understand how and where they exist in competitive markets. Then the corporate philosophy must be aimed at creating or taking advantage of some imperfection in these markets, either through product differentiation or creation of a cost advantage, rather than looking to new markets or industries that appear to provide large profits. Any perfectly competitive industry that looks too good to be true won't be for long.

Axiom 6: Efficient Capital Markets—The Markets Are Quick and the Prices Are Right

Our goal as financial executives is the maximization of shareholder wealth. Decisions that maximize shareholder wealth lead to an increase in the market price of the existing common stock. To understand this relationship, as well as how securities such as bonds and stocks are valued or priced in the financial markets, it is necessary to have an understanding of the concept of **efficient markets**.

Efficient market

A market in which the values of all assets and securities at any instant in time fully reflect all available public information

Whether a market is efficient has to do with the speed with which information is impounded into security prices. An efficient market is characterized by a large number of profit-driven individuals who act independently. In addition, new information regarding securities arrives in the market in a random manner. Given this setting, investors adjust to new information immediately and buy and sell the security until they feel the market price correctly reflects the new information. Under the efficient market hypothesis, information is reflected in security prices with such speed that there are no opportunities for investors to profit from publicly available information.[10] Investors competing for profits ensure that security prices appropriately reflect the expected earnings and risks involved and thus the true value of the firm.

What are the implications of efficient markets for us? First, the price is right. Stock prices reflect all publicly available information regarding the value of the company. This means we can implement our goal of maximization of shareholder wealth by focusing on the effect each decision *should* have on the stock price if everything else were held constant. Second, earnings manipulations through accounting changes will not result in price changes. Stock splits and other changes in accounting methods that do not affect cash flows are not reflected in prices. Market prices reflect expected cash flows available to shareholders. Thus, our preoccupation with cash flows to measure the timing of the benefits is justified.

As we will see, it is indeed reassuring that prices reflect value. It allows us to look at prices and see value reflected in them. While it may make investing a bit less exciting, it makes corporate finance much less uncertain.

---

[10]The efficient market hypothesis has been divided into three forms according to information type: weak, semi-strong, and strong. The weak form evaluates whether all historical pricing information has been incorporated into current share prices. The semi-strong form evaluates whether both historical pricing information and public information have been incorporated into current share prices. Finally, the strong form evaluates whether all information, public and nonpublic, is incorporated into current share prices and whether any type of investor can make an excess profit. For a more detailed discussion, see Edwin J. Elton and Martin J. Gruber, *Modern Portfolio Theory And Investment Analysis,* 4th ed. (New York: John Wiley & Sons, Inc., 1991).

Axiom 7: The Agency Problem—Managers Won't Work for the Owners Unless It's in Their Best Interest

Although the goal of the firm is the maximization of shareholder wealth, in reality the agency problem may interfere with the implementation of this goal. The *agency problem* results from the separation of management and the ownership of the firm. For example, a large firm may be run by professional managers who have little or no ownership in the firm. Because of this separation of the decision makers and owners, managers may make decisions that are not in line with the goal of maximization of shareholder wealth. They may approach work less energetically and attempt to benefit themselves in terms of salary and perquisites at the expense of shareholders.

If the management of the firm works for the owners, who are actually the shareholders, why doesn't the management get fired if they don't act in the shareholder's best interest? *In theory*, the shareholders pick the corporate board of directors and the board of directors in turn picks the management. Unfortunately, *in reality* the system frequently works the other way around. Management selects the board of director nominees and then distributes the ballots. In effect, shareholders are offered a slate of nominees selected by the management. The end result is management effectively selects the directors, who then may have more allegiance to managers than to shareholders. This in turn sets up the potential for agency problems with the board of directors not monitoring managers on behalf of the shareholders as they should.

A considerable amount of effort is spent in monitoring managers and trying to align their interests with shareholders. Managers can be monitored by auditing financial statements and managers' compensation packages. The interests of managers and shareholders can be aligned by establishing management stock options, bonuses, and perquisites that are directly tied to how closely their decisions coincide with the interest of shareholders. The agency problem will persist unless an incentive structure is set up that aligns the interests of managers and shareholders. In other words, what's good for shareholders must also be good for managers. If that is not the case, managers will make decisions in their best interest rather than maximizing shareholder wealth.

Axiom 8: Taxes Bias Business Decisions

Hardly any decision is made by the financial manager without considering the impact of taxes. When we introduced ***Axiom 4***, we said that only incremental cash flows should be considered in the evaluation process. More specifically, the cash flows we will consider will be *after-tax incremental cash flows to the firm as a whole.*

When we evaluate new projects, we will see income taxes playing a significant role. When the company is analyzing the possible acquisition of a plant or equipment, the returns from the investment should be measured on an after-tax basis. Otherwise the company will not truly be evaluating the true incremental cash flows generated by the project.

The government also realizes taxes can bias business decisions and uses taxes to encourage spending in certain ways. If the government wanted to encourage spending on research and development projects it might offer an *investment tax credit* for such investments. This would have the effect of reducing taxes on research and development projects, which would in turn increase the after-tax cash flows from those projects. The increased cash flow would turn some otherwise unprofitable research and development projects into profitable projects. In effect, the government can use taxes as a tool to direct business investment to research and development projects, to the inner cities, and to projects that create jobs.

Taxes also play a role in determining a firm's financial structure, or mix of debt and stock. Although this subject has been the focus of intense controversy for over three decades, one aspect remains constant: The tax laws give debt financing a definite cost advantage over stock. As we noted when we examined how taxes are computed, *interest payments are a tax-deductible expense, whereas dividend payments to shareholders may not be used as deductions in computing a corporation's taxable profits.* Interest payments lower profits, which are not a cash flow item, and this in turn lowers taxes due, which are a cash flow item. In effect, paying interest as opposed to dividends reduces taxes.

Axiom 9: All Risk Is Not Equal—Some Risk Can Be Diversified Away, and Some Cannot

Much of finance centres around ***Axiom 1, The Risk-Return Tradeoff.*** But before we can fully use ***Axiom 1*** we must decide how to measure risk. As we will see, risk is difficult to measure. ***Axiom 9*** introduces you to the process of diversification and demonstrates how it can reduce risk. We will also provide you with an understanding of how diversification makes it difficult to measure a project or an asset's risk.

You are probably already familiar with the concept of diversification. There is an old saying, "don't put all your eggs in one basket." Diversification allows good and bad events or observations to cancel each other out, thereby reducing total variability without affecting expected return.

To see how diversification complicates the measurement of risk, let us look at the difficulty Ocelot Energy Inc., has in determining the level of risk associated with a new natural gas well drilling project. Each year Ocelot Energy Inc. might drill several hundred wells, with each well having only a 1 in 10 chance of success. If the well produces, the profits are quite large, but if it comes up dry, the investment is lost. Thus, with a 90 percent chance of losing everything, we would view the project as being extremely risky. However, if Ocelot Energy Inc. each year drills 2,000 wells all with a 10 percent, independent chance of success, then they would typically have 200 successful wells. Moreover, a bad year may result in only 190 successful wells, while a good year may result in 210 successful wells. If we look at all the wells together the extreme good and the bad results tend to cancel each other out and the well drilling projects taken together do not appear to have much risk or variability of possible outcome.

The amount of risk in a gas well project depends upon our perspective. Looking at the well standing alone, it looks like a lot; however, if we consider the risk that each well contributes to the overall firm risk, it is quite small. This occurs because much of the risk associated with each individual well is diversified away within the firm. From the point of view of a diversified shareholder, much of the risk that a project contributes to the firm is further diversified away as the shareholder adds the Ocelot Energy Inc. stock to other stocks in his or her portfolio. The risk of an investment varies depending upon the perspective of the individual considering the risk.

Perhaps the easiest way to understand the concept of diversification is to look at it graphically. Consider what happens when we combine two projects, as depicted in Figure 1-2. In this case, the cash flows from these projects move in opposite directions, and when they are combined, the variability of their combination is totally eliminated. Notice that the return has not changed—both the individual project's and their combination's return averages 10 percent. In this case the extreme good and bad observations cancel each other out. The degree to which the total risk is reduced is a function of how the two sets of cash flows or returns move together.

As we will see for most projects and assets, some risk can be eliminated through diversification, while some risk cannot. This will become an important distinction later in our studies. *For now, we should realize that the process of diversification can reduce risk, and as a result, measuring a project or an asset's risk is very difficult.* A project's risk changes depending on whether you measure: (1) the project's risk when it is standing alone, (2) the amount of risk a project contributes to a firm, or (3) the amount of risk that this project contributes to the shareholder's portfolio.

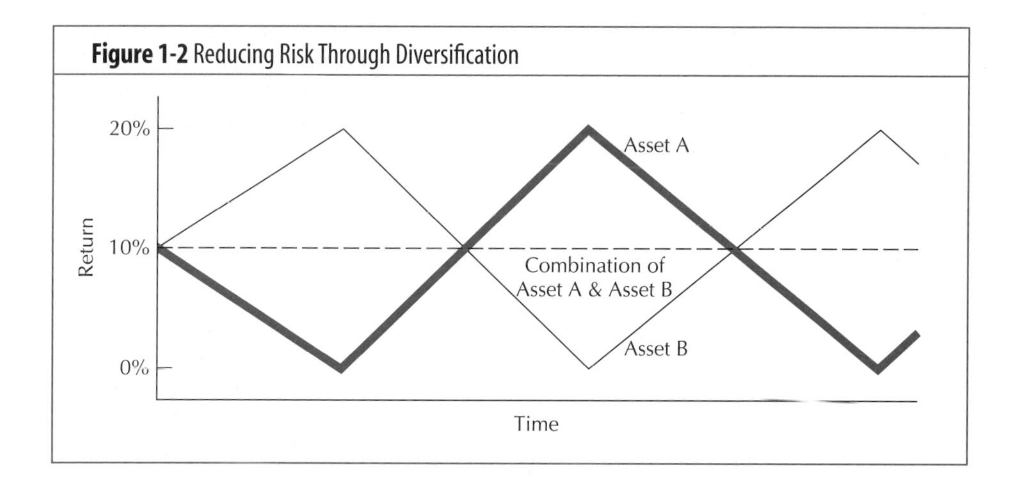

Figure 1-2 Reducing Risk Through Diversification

Axiom 10: Ethical Behaviour Is Doing the Right Thing, and Ethical Dilemmas Are Everywhere in Finance

Ethics, or rather a lack of ethics in finance, is a recurring theme in the news. During the late 1980s and early 1990s the fall of Ivan Boesky and Drexel, Burnham, Lambert, and the near collapse of Salomon Brothers seemed to make continuous headlines. Meanwhile, the movie *Wall Street* was a hit at the box office and the book *Liar's Poker*, by Michael Lewis, chronicling unethical behaviour in the bond markets, became a best seller. As the lessons of Salomon Brothers and Drexel, Burnham, Lambert illustrate, ethical errors are not forgiven in the business world. Not only is acting in an ethical manner morally correct, it is congruent with our goal of maximization of shareholder wealth.

Ethical behaviour means "doing the right thing." A difficulty arises, however, in attempting to define "doing the right thing." The problem is that each of us has his or her own set of values, which forms the basis for our personal judgments about what is the right thing to do. However, every society adopts a set of rules or laws that prescribe what it believes to be "doing the right thing." In a sense, we can think of laws as a set of rules that reflect the values of the society as a whole, as they have evolved. For purposes of this text, we recognize that individuals have a right to disagree about what constitutes "doing the right thing," and we will seldom venture beyond the basic notion that ethical conduct involves abiding by society's rules. However, we will point out some of the ethical dilemmas that have arisen in recent years with regard to the practice of financial management. These dilemmas generally arise when some individual behaviour is found to be at odds with the wishes of a large portion of the population, even though that behaviour is not prohibited by law. Ethical dilemmas can therefore provide a catalyst for debate and discussion, which may eventually lead to a revision in the body of the law. So as we embark on our study of finance and encounter ethical dilemmas, we encourage you to consider the issues and form your own opinion.

Many students ask, "Is ethics really relevant?" This is a good question and deserves an answer. First, although business errors can be forgiven, ethical errors tend to end careers and terminate future opportunities. Why? Because *unethical behaviour eliminates trust, and without trust businesses cannot interact.* Second, *the most damaging event a business can experience is a loss of the public's confidence in its ethical standards.* In finance we have seen several recent examples of such events. It was the ethical scandals involving insider trading at Drexel, Burnham, Lambert that brought down that firm. In 1991 the ethical scandals involving attempts by Salomon Brothers to corner the Treasury bill market led to the removal of its top executives and nearly put the company out of business.

Beyond the question of ethics is the question of social responsibility. In general, corporate social responsibility means that a corporation has responsibilities to society beyond the maximization of shareholder wealth. It asserts that a corporation answers to a broader constituency than shareholders alone. As with most debates that centre on ethical and moral questions, there is no definitive answer. One opinion is that because financial managers are employees of the corporation, and the corporation is owned by the shareholders, the financial managers should run the corporation in such a way that shareholder wealth is maximized and then allow the shareholders to decide if they would like to fulfill a sense of social responsibility by passing on any of the profits to deserving causes. Very few corporations consistently act in this way. For example, in 1992 Bristol-Myers Squibb Co. announced it would start an ambitious program to give away heart medication to those who cannot pay for it. This announcement came in the wake of an American Heart Association report showing that many of the nation's working poor face severe health risks because they cannot afford heart drugs. Clearly, Bristol-Myers Squibb felt it had a social responsibility to provide this medicine to the poor at no cost.

How do you feel about this decision?

A Final Note on the Axioms

Hopefully, these axioms are as much statements of common sense as they are theoretical statements. These axioms provide the logic behind what is to follow. We will build on them and attempt to draw out their implications for decision making. As we continue, try to keep in mind that while the topics being treated may change from chapter to chapter, the logic driving our treatment of them is constant and is rooted in these ten axioms.

OBJECTIVE 4

OVERVIEW OF THE TEXT

This text is divided into seven parts, each dealing with one major area of financial concern:

- The Scope and Environment of Financial Management
- The Basics of Valuation
- Financial Structure and the Cost of Capital
- The Valuation of Capital Investments
- Working-Capital Management
- Financial Analysis, Planning and Control
- Special Topics in Financial Management

We now describe these parts in greater detail.

Part I: The Scope and Environment of Financial Management

Chapter 1 introduces the goal of the firm, which provides a guide for decision-making by financial executives. This chapter also describes the ten axioms that form the basics of financial management. Chapter 2 examines the financial markets and interest rates, looking at both the determinants of interest rates and their effect on business decisions. This chapter also presents the legal and tax environment in which financial decisions are made.

Part II: The Basics of Valuation

Chapters 3 through 6 examine how financial assets are valued. In Chapter 3, we define and measure the risk and expected return of financial assets. Chapter 4 presents the mathematics of finance and examines the concept of the time value of money. An understanding of this topic allows us to compare benefits and costs that occur in different time periods. We apply these valuation techniques to both fixed-income securities in Chapter 5 and common shares in Chapter 6. Chapter 6 also examines various valuation models that explain how financial decisions affect the firm's share price.

Part III: Financial Structure and the Cost of Capital

Chapters 7 through 12 examine the implications of short- and long-term financing on the firm. In Chapter 7, we begin by examining the role of financial markets in the firm's decision to raise capital. Chapter 8 describes the impact of leverage on returns to the firm and its implications on a firm's capital structure. Once these relationships between leverage and valuation are developed, we move to the process of planning the firm's financing mix in Chapter 9. Chapter 10 examines the implications of the dividend-retained earnings decision on the firm's financing strategy. In Chapter 11, we explore the various sources of short-term financing. Chapter 12 examines what type of costs are associated with raising new funds.

Part IV: The Valuation of Capital Investments

Chapters 13 through 16 make use of the valuation principles in earlier chapters to examine the capital budgeting decision of the firm. Chapter 13 discusses the capital budgeting decision, which involves the financial evaluation of investment proposals in fixed assets. We then examine the measurement of cash flows in Chapter 14 and introduce methods to incorporate risk in the analysis in Chapter 15. Finally, in Chapter 16, we examine how valuation principles are applied to several examples of financial decision making.

Part V: Working-Capital Management

Chapters 17 through 19 deal with working-capital management, the management of current assets and liabilities. We discuss methods for determining the appropriate investment in cash, marketable securities, inventory, and accounts receivable, as well as the risks associated with these investments and the control of these risks.

Part VI: Financial Analysis, Planning and Control

Chapters 20 and 21 introduce the basic financial tools that financial executives use to maintain control over the firm and its operations. These tools enable the financial executive to locate potential problem areas and plan for the future.

Part VII: Special Topics in Financial Management

Chapter 22 presents a discussion on the use of convertibles, warrants, options, futures and currency swaps. In Chapter 23, we discuss the following methods for corporate restructuring: mergers, spinoffs and leveraged buyouts. Chapter 24 deals with international financial management and focuses on how financial decisions are affected by the international environment. The final chapter in the text, Chapter 25, examines how the principles of financial management are applied to small and mid-sized businesses.

SUMMARY

This chapter outlines a framework for the maintenance and creation of wealth. In introducing decision-making techniques aimed at creating wealth, we have emphasized the logic behind these techniques. This chapter examines the goal of the firm. The commonly accepted goal of profit maximization is contrasted with the more complete goal of maximization of shareholder wealth. Because it deals well with uncertainty and time in a real-world environment, the goal of maximization of shareholder wealth is found to be the proper goal for the firm.

Ten Axioms That Form the Basics of Financial Management

This chapter also examines the ten axioms on which finance is built and which motivate the techniques and tools introduced in this text:

Axiom 1: *The Risk-Return Tradeoff—We Won't Take on Additional Risk Unless We Expect to Be Compensated with Additional Return*

Axiom 2: *The Time Value of Money—A Dollar Received Today Is Worth More Than a Dollar Received in the Future*

Axiom 3: *Cash—Not Profits—Is King: Measuring the Timing of Costs and Benefits*

Axiom 4: *Incremental Cash Flows—It's Only What Changes That Counts*

Axiom 5: *The Curse of Competitive Markets—Why It's Hard to Find Exceptionally Profitable Projects*

Axiom 6: *Efficient Capital Markets—The Markets Are Quick and the Prices Are Right*

Axiom 7: *The Agency Problem—Managers Won't Work for the Owners Unless It's in Their Best Interest*

Axiom 8: *Taxes Bias Business Decisions*

Axiom 9: *All Risk Is Not Equal—Some Risk Can Be Diversified Away, and Some Cannot*

Axiom 10: *Ethical Behaviour Is Doing the Right Thing, and Ethical Dilemmas Are Everywhere in Finance*

Finally, the seven parts of the book are briefly previewed.

STUDY QUESTIONS

1-1. Define financial management. What roles do financial executives play within financial management?

1-2. What are some of the problems involved in the use of profit maximization as the goal of the firm? How does the goal of shareholder wealth maximization deal with those problems?

1-3. Compare and contrast the goals of profit maximization and maximization of shareholder wealth.

1-4. Firms often involve themselves in projects that do not result directly in profits; for example, Canadian corporations frequently support public television broadcasts. Do these projects contradict the goal of maximization of shareholder wealth? Why or why not?

1-5. What is the minimum return required by a consumer for delaying consumption? Explain why a company's bonds provide a greater return than Government of Canada bonds.

1-6. What are the implications of efficient markets for investors?

1-7. What must investors account for when projecting their savings to future years?

1-8. Explain why decision makers must consider incremental cash flows when evaluating a project.

1-9. Explain why successful companies seek imperfections in the markets for investment opportunities.

1-10. Explain how governments can bias business decisions.

1-11. Explain what doing "the right thing" means for the purposes of this text.

THE FINANCIAL ENVIRONMENT

CHAPTER 2

LEARNING OBJECTIVES

After reading the chapter you should be able to

1. Explain the five stages required in developing a financial market system.
2. Explain the fundamentals of interest rate determination.
3. Explain the popular theories of the term structure of interest rates.
4. Compare the various legal forms of business and explain why the corporate form of business is the most logical choice for a firm that is large or growing.
5. Identify the corporate tax features that affect business decisions.

INTRODUCTION

In the summer of 1995, The Financial Post *and Arthur Andersen and Co. reported the results of a survey that examined the concerns of entrepreneurs when conducting business in Canada.[1] The following list summarizes by order of importance which issues concerned entrepreneurs the most: taxes, government regulations, securing financing, recession, political uncertainty, interest rates, availability of skilled labour, fluctuations in the Canadian dollar, global competition and availability of skilled management. Many entrepreneurs indicate that too many regulations mean too many bureaucrats. Such a combination keeps government costs up and taxes high. Several of these issues can be traced to the financial environment and have an impact on the decision-making process of both entrepreneurs and financial executives alike. In this chapter, we will examine the effects of taxes, government regulations, and interest rates on decision making by different business organizations.*

[1]Adapted from *The Financial Post,* "Entrepreneurs Speak Out," July 22, 1995, p. 10.

CHAPTER PREVIEW

This chapter focuses on the various factors that affect decision making in a market environment. We begin by examining the logic behind the determination of interest rates and required rates of return in the capital markets. Several theories are presented to explain the shape of the term structure of interest rates. We then compare the various legal forms of business so as to highlight the advantages and disadvantages of each form. Finally, we examine how business income is affected by income taxes.

PERSPECTIVE IN FINANCE

When you borrow money from a commercial bank to finance a purchase of stereo equipment or take out a mortgage to pay for a new home, you have used this country's financial market system. The average individual (and you are not that because of your study of financial management) tends to take the reality of the financial markets for granted. But they were not always there, and they have evolved amidst much controversy over a long period of time.

Recent years have produced heated debates over alterations in our financial system—this just illustrates the dynamic nature of these markets. Much of that debate followed the collapse of the equity markets on Monday, October 19, 1987, when both the Toronto Stock Exchange 300 Index and the Dow Jones Industrial Average fell by an unprecedented 407 and 508 points, respectively. Like banks, the equity markets are another piece of our complex financial market system. The financial market system must have the confidence of those who must use it. If the financial market system did not exist, real assets (like your home) would not get produced at an adequate rate and the populace would suffer. We now examine the structure of the financial market.

OBJECTIVE 1

THE STRUCTURE OF THE FINANCIAL MARKET

Financial markets are institutions and procedures that facilitate transactions in all types of financial claims. The purchase of your home, the common shares you may own, and your life insurance policy all took place in some type of financial market. Why do financial markets exist? What would the economy lose if our complex system of financial markets were not developed? We will address these questions here.

Financial markets

Those institutions and procedures that facilitate transactions in all types of financial claims (securities).

World Wide Web Site

http://www.worldbank.org
Comment: Information on financial markets compiled by the World Bank.

Some economic units, such as governments and business enterprises, spend more during a given period of time than they earn. Other economic units, such as persons and unincorporated businesses, spend less on current consumption than they earn. For example, business firms in the aggregate usually spend more during a specific time period than they earn. Households in the aggregate spend less on current consumption than they earn. As a result, some mechanism is needed to facilitate the transfer of savings from those economic units with a surplus to those with a deficit. That is precisely the function of financial markets. Financial markets exist in order to allocate the supply of savings in the economy to the demanders of those savings. The central characteristic of a financial market is that it acts as the vehicle through which the forces of demand and supply for a specific type of financial claim (such as a corporate bond) are brought together.

Now, why would the economy suffer without a developed financial market system? The answer is simple. The wealth of the economy would be less without the financial markets. The rate of capital formation would not be as high if financial markets did not exist.[2] This means that the net additions during a specific period to

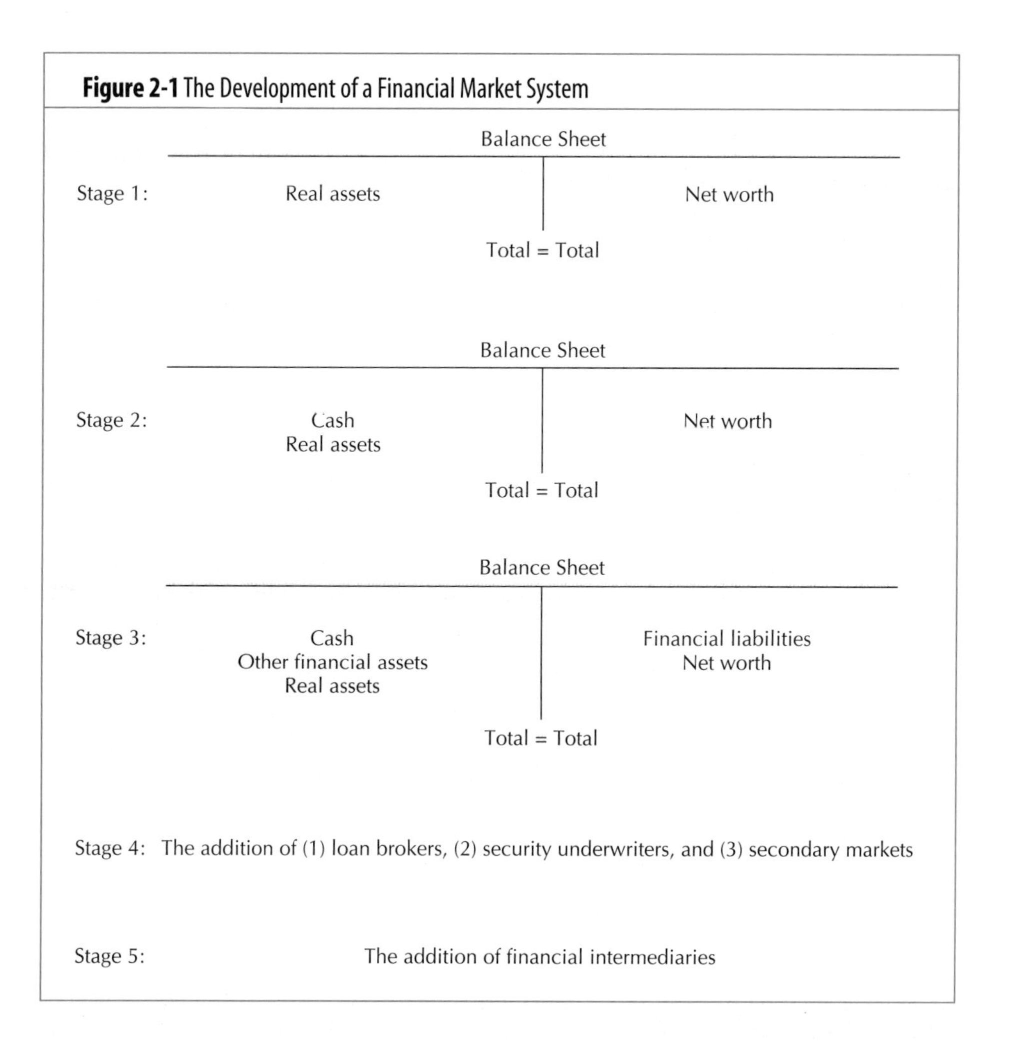

Figure 2-1 The Development of a Financial Market System

the stocks of (1) dwellings, (2) productive plant and equipment, (3) inventory, and (4) consumer durables would occur at lower rates.[3] Figure 2-1 clarifies the rationale behind this assertion. The abbreviated balance sheets in the figure refer to firms or any other type of economic units that operate in the private as opposed to governmental sectors of the economy. This means that such units cannot issue money to finance their own activities.

At stage 1 in Figure 2-1 only real assets exist in the hypothetical economy. **Real assets** are tangible assets like houses, equipment, and inventories. They are distinguished from **financial assets**, which represent claims for future payment on other economic units. Common and preferred shares, bonds, bills, and notes all are types of financial assets. If only real assets exist, then savings for a given economic unit, such as a firm, must be accumulated in the form of real assets. If the firm has a great idea for a new product, that new product can be developed, produced, and distributed only out of company savings (retained earnings). Furthermore, all investment in the new product must occur simultaneously as the savings are gener-

Real assets

Tangible assets such as houses, equipment, and inventories.

Financial assets

Claims for future payments by one economic unit on another.

[2]The rate of capital formation in Canada is dependent on the efficiency of its financial markets to provide a mechanism that facilitates the process of borrowing and lending.

[3]Canada is considered a capital-scarce country, having to offer a premium on interest rates to attract investors. Consequently, the capital investments of firms which desire to modernize, become more expensive as compared to firms located in capital-rich countries.

ated. If you have the idea, and we have the savings, there is no mechanism to transfer our savings to you. This is not a good situation.

At stage 2 paper money comes into existence in the economy. Here, at least, you can store your own savings in the form of money. Thus, you can finance your great idea by drawing down your cash balances. This is an improvement over stage 1, but there is still no effective mechanism to transfer our savings to you. You see, we will not just hand you our dollar bills. We will want a receipt.

The concept of a receipt which represents the transfer of savings from one economic unit to another is a monumental advancement. The economic unit with excess savings can lend the savings to an economic unit that needs them. To the lending unit these receipts are identified as "other financial assets" in stage 3 of Figure 2-1. To the borrowing unit, the issuance of financial claims (receipts) shows up as "financial liabilities" on the stage 3 balance sheet. The economic unit with surplus savings will earn a rate of return on those funds. The borrowing unit will pay that rate of return, but it has been able to finance its great idea.

In stage 4 the financial market system becomes more sophisticated because loan brokers help locate pockets of excess savings and channel such savings to economic units needing the funds. Some economic units will actually purchase the financial claims of borrowing units and sell them at a higher price to other investors; this process is called **underwriting.**[4] Further evidence for the sophistication of the financial markets is the development of a secondary market which enables trading in already-existing financial claims. For example, the sale of Northern Telecom common shares between a friend and yourself represents a secondary market transaction. **Secondary markets** reduce the risk of investing in financial claims. Should you need cash, you can liquidate your claims in the secondary market. This induces savers to invest in securities.

Underwriting

The purchase and subsequent resale of a new security issue. The risk of selling the new issue at a profitable price is assumed (underwritten) by an investment banker.

Secondary market

Transactions in currently outstanding securities.

Stage 5 represents the progression of the financial markets toward specialization as seen by the different functions of financial intermediaries. You can think of financial intermediaries as the major financial institutions with which you are used to dealing. These include commercial banks, trusts and mortgage and loan companies, credit unions, life insurance companies, and mutual funds. Financial intermediaries share a common characteristic: they offer their own financial claims, called **indirect securities**, to economic units with excess savings. The proceeds from selling their indirect securities are then used to purchase the financial claims of other economic units. These latter claims can be called **direct securities**. Thus, a mutual fund might sell mutual fund shares (their indirect security) and purchase the common shares (direct securities) of some major corporations. A life insurance company sells life insurance policies and purchases huge quantities of corporate bonds. Financial intermediaries thereby involve many small savers in the process of capital formation. This means there are more "good things" for everybody to buy.

Indirect securities

The unique financial claims issued by financial intermediaries. Mutual fund shares are an example.

Direct securities

The pure financial claims issued by economic units to savers. These claims can later be transformed into indirect securities.

A developed financial market system provides for a greater level of wealth in the economy. In the absence of financial markets, savings are not transferred to the economic units most in need of those funds. Financial markets bring together net savers and net users of funds. For example, banks accumulate the savings of individual investors and in return pay interest to these investors for use of these funds. On the other hand, growing companies are frequently net users of funds because their expenditures often exceed their income. A growth company might borrow funds (i.e., sell debt securities) from the insurance company. This act is a typical financial market transaction.

World Wide Web Sites

http://www.bmo.com
Comment: Bank of Montreal Home Page

http://www.RoyalBank.com
Comment: Royal Bank Home Page

http://www.tdbank.com
Comment: Toronto Dominion Bank Home Page

In the following article, we profile one of the more important financial institutions in Canada, the banking industry.[5]

Video Case 1

[4]Underwriting will be discussed in more detail in Chapter 7.

[5]Other profiles of Canadian financial institutions are presented in Chapters 7 and 25.

Banking

Banks are the main sources of financial services in Canada and provide over 100 retail financial services. These services normally include savings and chequing accounts, loans with or without security, lines of credit, credit cards, financial leasing and bank-ing-related data processing services. Banks also handle foreign trade-related transactions, maintain safe-deposit facilities and accept deposits.

Canadian banks are chartered under the Bank Act, which is subject to regular parliamentary review. Banking is one of the most heavily regulated industries in Canada: the goals of such legislation are to ensure bank solvency, to help banks remain effective, efficient and competitive and to maintain public confidence in the banking system. While the federal government has sole jurisdiction over banks and the banking system, provincial governments can influence banking activities through their securities laws, contract law and consumer protection legislation.

The Bank Act lists banks under its Schedule I or Schedule II, according to the diversity of the public distribution of their shares. Schedule I banks are widely held; no individual investor can own more than 10 percent. Moreover, no group of non-residents, except U.S. investors, can own more than 25 percent of any class of shares. Therefore, the Schedule I banks are Canadian-owned. Most Schedule II banks are closely held subsidiaries of foreign banks.

Retail banking traditionally requires a local presence. Schedule I banks operate nation-wide branch systems in a diverse range of markets, from very small villages to large metropolitan areas. These branches represent over 60 percent of the retail outlets provided by all financial institutions.

Chartered banks are the major suppliers of consumer and commercial credit in Canada. Business loans constitute the largest element of their asset holdings, followed by mortgage loans and consumer loans. The banks' domestic commercial credit operations encompass a wide range of borrowers, including small businesses, large firms operating across Canada and different levels of government. Banks are the primary source of financing to small businesses and provide a range of support services. Larger Canadian companies rely on banks for short-term working capital and a number of other services such as treasury, cash management, trade finance services and the underwriting and distribution of corporate securities. Banks also underwrite and distribute the debt issues of Canadian Governments and Crown corporations and invest their surplus funds in government and corporate securities.

Banks face intense competition in commercial markets, as trust companies, life insurance companies and securities dealers also seek to expand their share of commercial financing. With the slower growth in commercial lending, Schedule I banks have focused their efforts further on the small business market. While the intense competition is beneficial for corporations seeking funds, it has put pressure on the earnings of several Schedule II banks. Schedule I banks also have an advantage in that their retail deposit bases often provide funds at lower costs than the wholesale money markets upon which the Schedule II banks rely.

While Schedule I banks are the main suppliers of commercial credit, foreign banks have had a presence in this market in Canada for many years. The 1980 Bank Act acknowledged their influence and the increased influence and the increased competition that their operations have provided in the Commercial credit markets. In allowing them to operate in Canada, the Bank Act requires the foreign banks to establish subsidiaries (Schedule II banks) or representative offices. Foreign bank subsidiaries can be closely held indefinitely, but their combined Canadian dollar assets, excluding those held by U.S.-controlled Schedule II banks, cannot exceed 12 percent of the domestic assets of the Canadian banking system. The size of individual banks is controlled through regulatory approval of their capital bases. Schedule II banks are small relative to domestic banks because of the competition from Schedule I banks and other sources of commercial financing.

To compete with Schedule I banks, Schedule II banks tend to seek niche markets. Many focus on wholesale banking operations and try to attract Canadian customers by providing services that are not generally available at local banks. Schedule II banks are also oriented towards commercial banking services because of the practical difficulties in establishing large retail branch networks. Some Schedule II banks are in the Canadian market to better serve their international clients rather than to seek a share of the Canadian market.

Canadian-owned banks are very active abroad. They operate through foreign branches and many subsidiary and agency offices. They also have correspondent relationships with financial institutions worldwide. They accept deposits, extend loans, participate in underwriting of corporate securities and loan syndications, and provide new products such as interest rate and foreign exchange swaps.

Source: Adapted from: "Industry Profile, 1990–1991, Banking" (Communications Branch, Industry, Science and Technology Canada, 1991).

BACK TO THE BASICS

Our first axiom, ***Axiom 1: The Risk-Return Tradeoff—We Won't Take on Additional Risk Unless We Expect to Be Compensated with Additional Return,*** *describes the tradeoff between risk and return which governs the financial markets. We now examine the kinds of risks that are rewarded in the risk-return tradeoff presented in Axiom 1.*

OBJECTIVE 2

INTEREST RATE DETERMINANTS IN A NUTSHELL

Net users of funds must compete with one another for the funds supplied by net savers. Consequently, to obtain financing a firm must offer the supplier a rate of return which is competitive with the next best investment alternative.[6] We refer to the rate of return on this next best investment alternative as the supplier's opportunity cost of capital. The opportunity cost concept is crucial in financial management and we will refer to it often.

The price of using someone else's savings is expressed in terms of interest rates. The financial market system helps to move funds to the most productive end use. Those economic units with the most-promising projects should be willing to bid the highest (in terms of rates) to obtain the savings. When a rate is quoted, it is generally the nominal, or the observed rate.

Several different types of risks have an impact on the general level of interest rates. Equation (2-1) describes the relationship between the different types of risk and the output variable, the **nominal interest rate**.

Nominal rate of interest

The observed rate of interest on a specific fixed-income security.

$$k = k^* + IRP + DRP + MP + LP \tag{2-1}$$

where k = the nominal or observed rate of interest on a specific fixed-income security

k^* = the real rate of interest on a fixed-income security that has no risk and is in an economic environment of zero inflation—this can be reasonably thought of as the rate of interest demanded by investors on a three-month Government of Canada Treasury bill during periods of no inflation

[6]Financial managers should make investment and financing decisions on current or marginal information and not on historical or average costs.

IRP = the inflation-risk premium: the additional return required by investors to compensate for the inflation rate

DRP = the default-risk premium: the additional return required to compensate investors for the uncertainty of expected returns from a security attributable to possible changes in financial capacity of the issuer to make future payments to the security owner

LP = the liquidity premium: the additional return required by investors in securities that cannot be quickly converted into cash at a reasonably predictable price

MP = the maturity premium: the additional return needed to compensate investors of longer-term securities for the greater risk of price fluctuations on those securities caused by interest rate changes

The nominal interest rate is the rate that you would read about in *The Financial Post* for a specific fixed-income security. Sometimes in analyzing interest rate relationships over time it is of use to focus on what is called the "nominal risk-free rate of interest." So let us designate the nominal risk-free interest rate as k_f. Drawing, then, on our discussions and notation from above we can write this for k_f:

$$k_f = k^* + IRP \tag{2-2}$$

This equation just says that the nominal risk-free rate of interest is equal to the real risk-free interest rate plus the inflation-risk premium. It also provides a quick and *approximate* way of estimating the risk-free rate of interest, k^*, by solving directly for this rate. This basic relationship in Equation (2-2) contains important information for the financial decision maker. It has also for years been the subject of fascinating and lengthy discussions among financial economists. We will look more at the substance of the real rate of interest in the next section. In this following section we will improve on equation (2-2) by making it more *precise.*

The Effects of Inflation on Rates of Return and the Fisher Effect

Real rate of interest

The nominal rate of interest less the expected rate of inflation over the maturity of the fixed-income security. This adjustment represents the expected increase in actual purchasing power to the investor.

The **real rate of interest**, represents the rate of increase in actual purchasing power after adjusting for the inflation rate.

For example, if you have \$100 today and loan it to someone for a year at a nominal rate of interest of 11.3 percent, you will get back \$111.30 in one year. But, if during the year prices of goods and services increase by 5 percent, it will take \$105 at year end to purchase the same goods and services that \$100 purchased at the beginning of the year. What was your increase in purchasing power over the year? The quick and dirty answer is found by subtracting the inflation rate from the nominal rate, 11.3%−5%=6.3%, but this is not exactly correct. To be more precise, let the nominal rate of interest be represented by k, the anticipated rate of inflation by i, and the real rate of interest by k^*. Using these notations, we can express the relationship between the nominal interest rate, the rate of inflation and the real rate of interest as follows:

$$1 + k = (1 + k^*)(1 + i) \tag{2-3}$$
$$\text{or} \quad k = k^* + i + ik^*$$

Consequently, the nominal rate of interest (k) is equal to the sum of the real rate of interest (k^*), the inflation rate (i), and the product of the real rate and the inflation premium.[7] This relationship between nominal rates, real rates, and the rate of inflation has come to be called the *Fisher effect.*[8] Equation (2-3) describes how the

[7]The inflation premium is not tied to the actual inflation rate at a given time, but rather to the expected inflation for the duration of the security's life.

[8]This relationship was analyzed many years ago by Irving Fisher.

observed nominal rate of interest includes both the real rate of interest and the inflation rate.

Substituting into equation (2-3) using a nominal rate of 11.3 percent and an inflation rate of five percent, we calculate the real rate of interest, k^*, as follows:

$$k = k^* + i + ik^*$$
$$.113 = k^* + .05 + (.05)(k^*)$$
$$k^* = 6\%$$

Thus, at new higher prices, your purchasing power will have increased by only 6 percent although you have $11.30 more than you had at the start of the year. To see why, let's assume that at the outset of the year one unit of the market basket of goods and services costs $1, so you could purchase 100 units with your $100. Now, at the end of the year you have $11.30 more, but each unit now costs $1.05 (remember the 5 percent rate of inflation). How many units can you buy at the end of the year? The answer is 106 units ($111.30/$1.05), which represents a 6 percent increase in real purchasing power.[9]

OBJECTIVE 3

THE TERM STRUCTURE OF INTEREST RATES

Term structure of interest rates

The relationship between interest rates and the term to maturity, where the risk of default is held constant.

World Wide Web Sites

http://www.altamira.com
Comment: Altamira's web site contains daily market reports, research reports and a glossary of financial terms.

http://www.RoyalBank.com/english/fin/fin_daily/index.html
Comment: Royal Bank Treasury Economics department provides a daily market summary.

The relationship between a debt security's rate of return and the length of time until the debt matures is known as the **term structure of interest rates**. For the relationship to be meaningful to us, all the factors other than maturity, such as the chance of the bond defaulting, must be held constant. Thus, the term structure reflects observed rates or yields on similar securities, except for the length of time until maturity, at a particular moment in time. Figure 2-2 shows a Government of Canada yield curve that captures the term structure of interest rates on Friday, October 27, 1995. Four days before the Quebec Referendum, the curve is upward sloping, indicating that longer terms to maturity command higher returns, or yields. In this period of political uncertainty, the rate of interest on a five-year bond was 7.5 percent, whereas the comparable rate on a twenty-year bond was 8.2 percent.

Figure 2-2 The Term Structure of Interest Rates

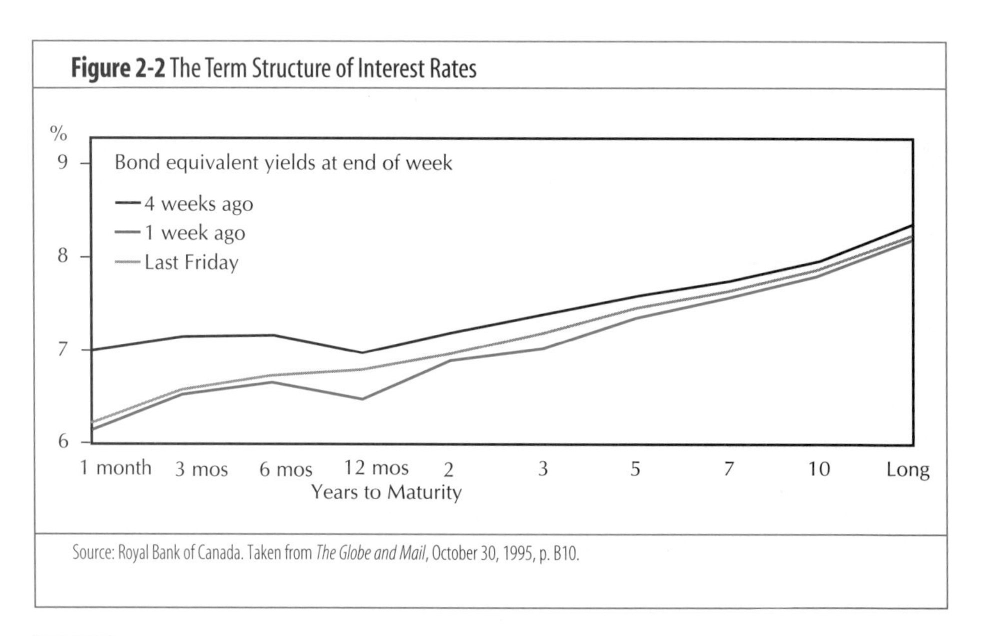

Source: Royal Bank of Canada. Taken from *The Globe and Mail*, October 30, 1995, p. B10.

[9]The time value of money will be discussed in greater detail in Chapter 4.

Observing Historical Term Structures of Interest Rates

The term structure of interest rates is dependent on the financial environment and as a result will change over time. The particular term structure observed today may be quite different from either the term structure of a month ago or the term structure of a month from now.[10] A perfect example of the changing term structure, or the yield curve, was witnessed during the early days of the Persian Gulf crisis in August 1990. Figure 2-3 shows the yield curves one day prior to the Iraqi invasion of Kuwait and then again just three weeks later. The change is noticeable, particularly for long-term interest rates. Investors quickly developed new fears about the prospect of increased inflation to be caused by the crisis and consequently increased their required rates of return.

Although the upward-sloping curve in Figure 2-2 is the one most commonly observed, yield curves can assume several shapes. Sometimes the term structure is downward-sloping; at other times it rises and then falls (hump-backed); and at still other times it may be relatively flat. Figure 2-4 shows some yield curves at different points in time.

Figure 2-3 Changes in the Term Structure of Interest Rates at the Outbreak of the Persian Gulf Crisis

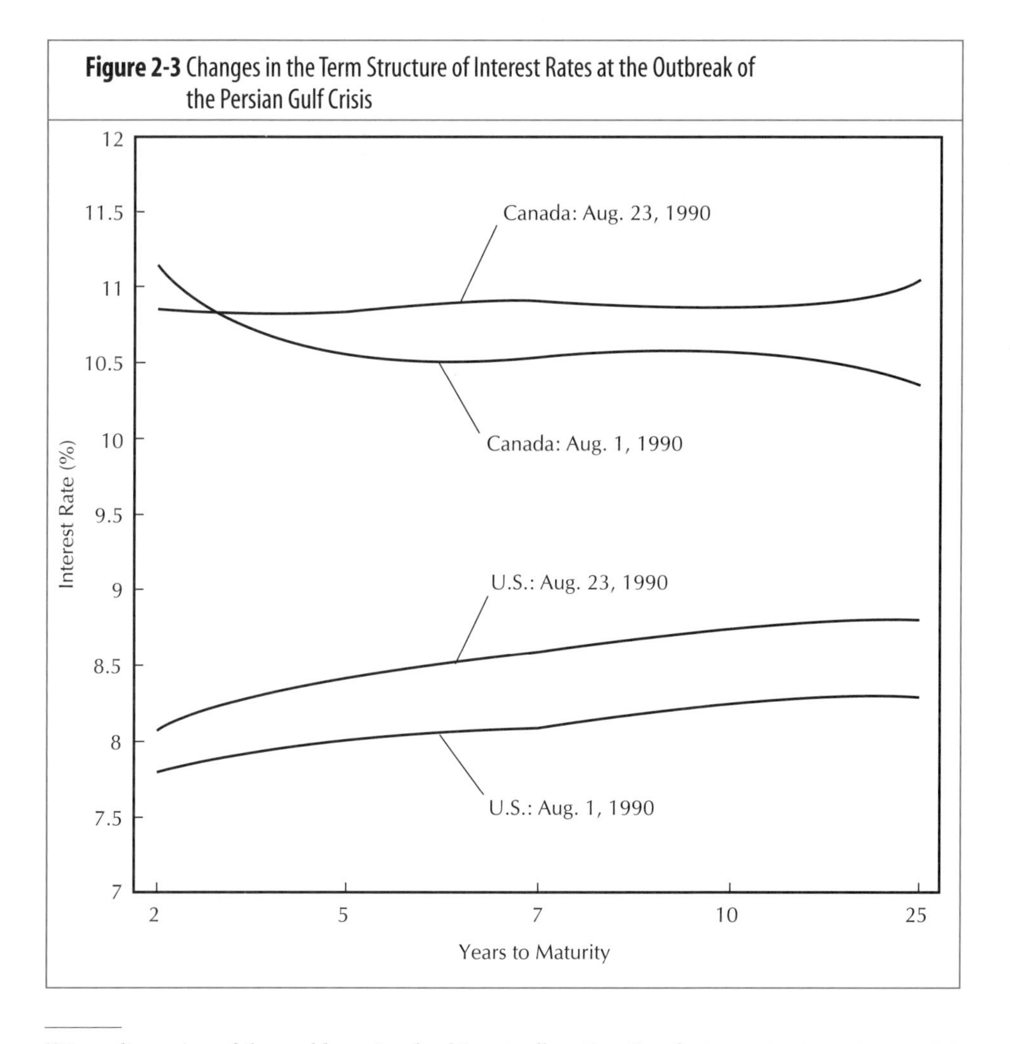

[10]For a discussion of the problems involved in actually estimating the term structure at any point in time, see Willard T. Carleton and Ian A. Cooper, "Estimation and Uses of the Term Structure of Interest Rates," *Journal of Finance* 31 (September 1976), 1067–1084.

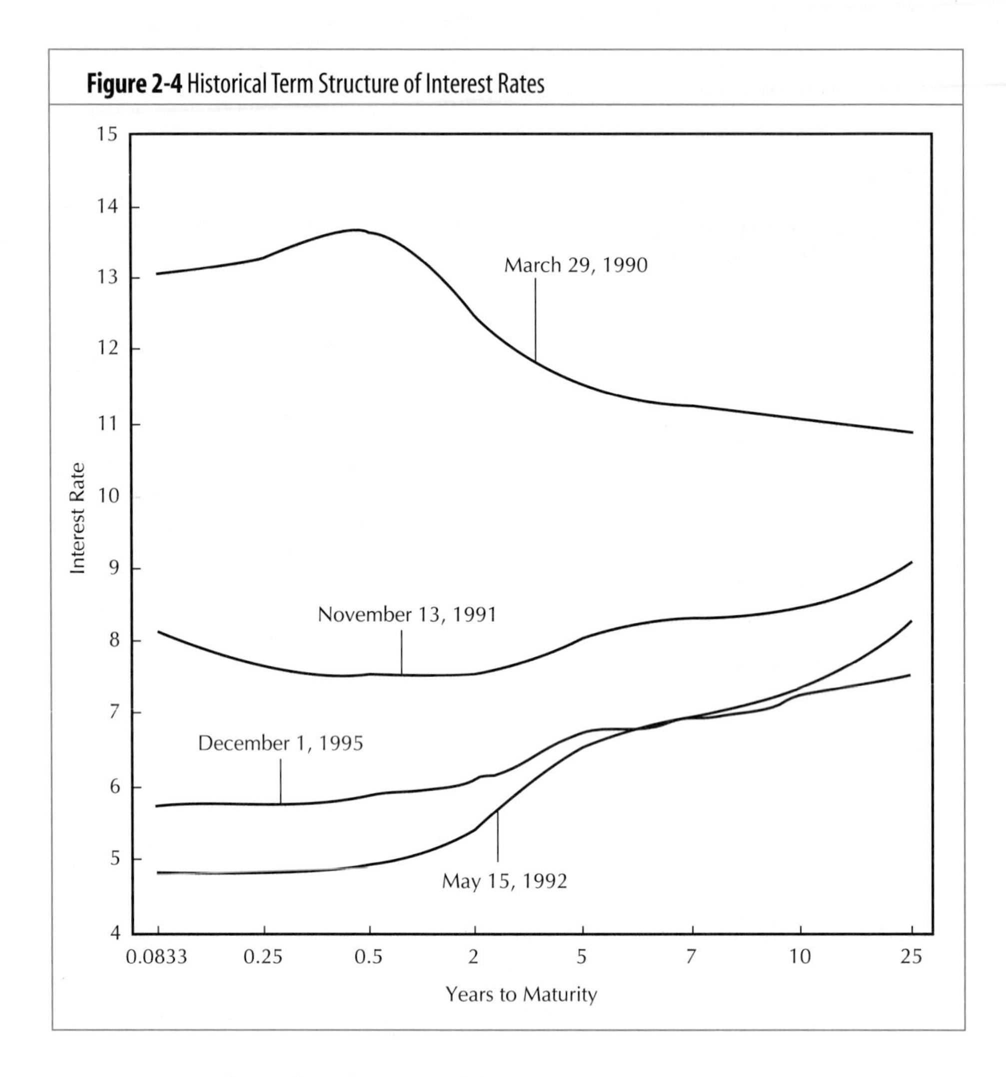

Trying to Explain the Shape of the Term Structure

A number of theories may explain the shape of the term structure of interest rates at any point in time. Three possible explanations are presented: (1) the unbiased expectations theory, (2) the liquidity preference theory, and (3) the market segmentation theory.[11] We now examine each of these theories in greater detail.

The Unbiased Expectations Theory

Unbiased expectations theory

The shape of the term structure is determined by investor's expectations about future interest rates.

The **unbiased expectations theory** says that the term structure is determined by an investor's expectations about future interest rates.[12] To see how this works, consider the following investment problem faced by Mary Maxell. Mary has $10,000 that she wants to invest for two years, at which time she plans to use her savings to make a down payment on a new home. Wanting not to take any risk of losing her savings, she decides to invest in Government of Canada securities. She has two choices. First, she can purchase a government security that matures in two years,

[11] See Richard Roll, *The Behavior of Interest Rates: An Application of the Efficient Market Model to U.S. Treasury Bills* (New York: Basic Books, Inc., 1970).

[12] Irving Fisher thought of this idea in 1896. The theory was later refined by J. R. Hicks in: *Value and Capital* (London: Oxford University Press, 1946) and F. A. Lutz and V. C. Lutz, *The Theory of Investment in the Firm* (Princeton, NJ: Princeton University Press, 1951).

which offers her an interest rate of 9 percent per year.[13] If she does this, she will have $11,881 in two years, calculated as follows:[14]

Principal amount	$10,000
Plus: year 1 interest (.09 × $10,000)	900
Principal plus interest at the end of year 1	$10,900
Plus: year 2 interest (.09 × $10,900)	981
Principal plus interest at the end of year 2	$11,881

Alternatively, Mary could buy a government security maturing in one year that pays an 8 percent rate of interest. She would then need to purchase another one-year security at the end of the first year. Which alternative Mary will prefer obviously depends in part upon the rate of interest she expects to receive on the government security she will purchase a year from now. We cannot tell Mary what the interest rate will be in a year; however, we can at least calculate the rate that will give her the same two-year total savings she would get from her first choice, or $11,881. The interest rate can be calculated as follows:

Savings needed in two years:	$11,881
Savings at the end of the year 1: $10,000 × (1 + .08)	$10,800
Interest needed in year 2:	$ 1,081

For Mary to receive $1,081 in the second year, she would have to earn about 10 percent on her second-year investment, computed as follows:

$$\frac{\text{Interest received for year 2}}{\text{Investment made at beginning of year 2}} = \frac{\$1{,}081}{\$10{,}800} = 10.01\%$$

So the term structure of interest rates for our example consists of the one-year interest rate of 8 percent and the two-year rate of 9 percent which is shown in Figure 2-5. This Figure also provides information about the expected one-year rate for investments made one year hence. In a sense, the term structure contains implications about investor expectations of future interest rates and is called the unbiased expectations theory of the term structure of interest rates.

Although we can see a relationship between current interest rates with different maturities and the investor's expectations about future interest rates, is this the whole story? Are there influences other than the investor's expectations about future interest rates? Probably, so let's continue to think about Mary's dilemma.

Liquidity Preference Theory

In presenting Mary's choices, we have suggested that she would be indifferent between buying the two-year government security offering a 9 percent return and two consecutive one-year investments, offering 8 and 10 percent, respectively. However, that would be so only if she is unconcerned about the risk associated with not knowing the rate of interest on the second security at the time she chooses it. If

---

[13]When the Government of Canada and other governments issue securities to help finance government expenses, they issue these securities with different times to maturity, depending on the government's funding needs.

[14]We could also calculate the principal plus interest for Mary's investment using the following compound interest equation: $\$10{,}000(1 + .09)^2 = \$11{,}881$. The mathematics will be examined in greater detail in Chapter 4.

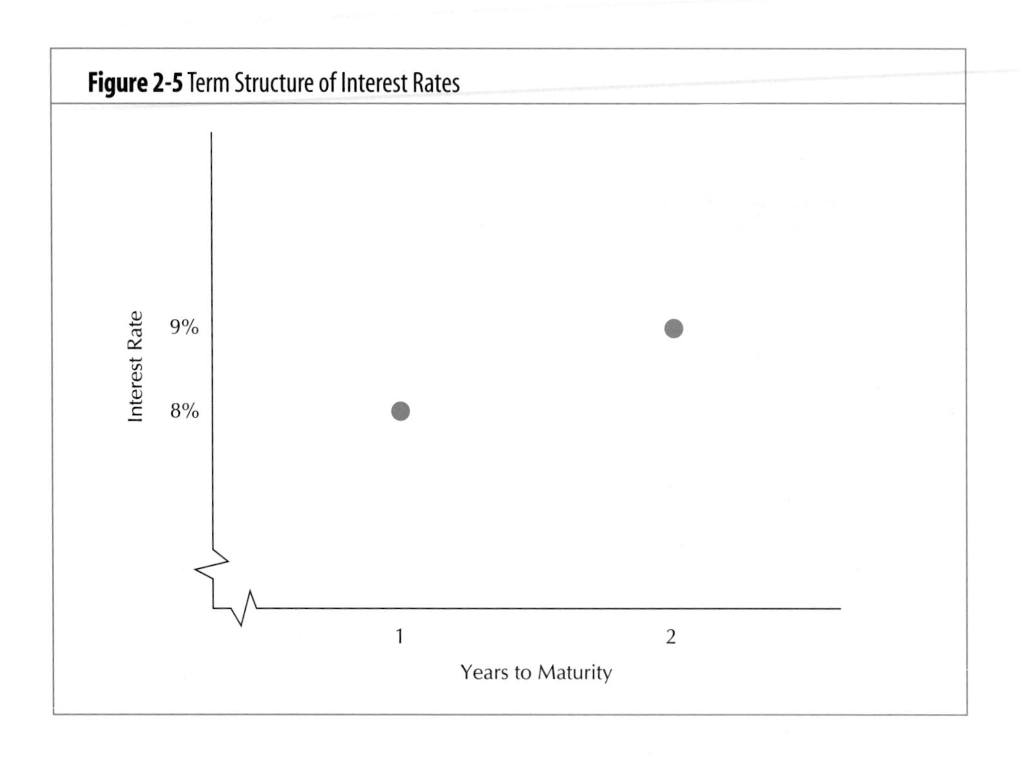

Figure 2-5 Term Structure of Interest Rates

Mary is risk-averse (that is, she dislikes risk), she might not be satisfied with a 10 percent return on the second one-year government security. She might require some additional expected return to be truly indifferent. Mary may in fact decide that she will expose herself to the uncertainty of future interest rates only if she can reasonably expect to earn an added one-half percent in interest (0.5%), or 10.5 percent on the second one-year investment. This risk premium (additional required interest rate) to compensate Mary for the risk of changing future interest rates is nothing more than the *maturity premium* (*MP*) introduced earlier, and it is this idea that underlies the **liquidity preference theory** of the term structure.[15] In the liquidity preference theory, investors require liquidity premiums to compensate them for buying securities that expose them to the risks of fluctuating interest rates.

Liquidity preference theory

The shape of the term structure is determined by investors requiring a maturity premium in compensation for the added risks of fluctuating interest rates.

Market Segmentation Theory

The **market segmentation theory** explains the term structure of interest rates by building on the idea that legal restrictions and personal preferences limit choices for investors to certain ranges of maturities.[16] For example, commercial banks prefer short- to medium-term maturities because of the short-term nature of their deposit liabilities. Investors prefer not to invest in long-term securities. Life insurance companies, on the other hand, have longer term liabilities, so this type of investor prefers longer maturities in investments. At the extreme, the market segmentation theory implies that the rate of interest for a particular security is determined solely by demand and supply for a given maturity, and that it is independent

Market segmentation theory

The shape of the term structure is determined by the demand for and the supply of securities having a specific maturity.

[15]This theory was first presented by John R. Hicks in *Value and Capital* (London: Oxford University Press, 1946), pp. 141–45, with the risk premium referred to as the liquidity premium. For our purposes, we will use the term *maturity premium (MP)* to describe this risk premium, thereby keeping our terminology consistent within this chapter.

[16]An early advocate of this theory was J. M. Culbertson, "The Term Structure of Interest Rates," *Quarterly Journal of Economics* 71 (November 1957), pp. 489–505.

of the demand and supply for securities having different maturities. A more moderate version of the theory allows investors strong maturity preferences, but it also allows them to modify their feelings and preferences if significant yield inducements occur.[17]

In Chapter 7 we will examine more closely the process of issuing shares and bonds in the financial markets. For now we have introduced the importance of the financial markets and the role of interest rates in the marketplace.

PERSPECTIVE IN FINANCE

The yield curve provides an investor with the following financial information.[18] *(1) The shape of the yield curve provides a prediction of future interest rates—for example, an upward-sloping yield curve implies a forecast of higher interest rate levels. (2) The yield curve provides lenders and borrowers with a method to estimate the levels of risk and return when choosing the terms of maturity for debt transactions—for example, a lender can evaluate the additional risk incurred by estimating the incremental yield that occurs with a longer term of maturity for the loan. (3) The yield curve can be used as an instrument to select individual securities—for example, securities that are priced too low (i.e., yield-to-maturity too high) lie above the yield curve. (4) The yield curve provides assistance in the pricing of new securities—for example, the issuer of new debt will set the coupon rate slightly above the yield of a comparable instrument that is located on the yield curve so as to make it attractive to the market. (5) The yield curve provides information to participants which "ride the yield curve" in the money market—for example, an upward-sloping yield curve would provide an investor with the opportunity to invest in a security whose maturity is longer than the investor's holding period, and then to sell the security prior to maturity in order to obtain a greater return.*

LEGAL FORMS OF BUSINESS ORGANIZATION

OBJECTIVE 4

The legal forms of business organization are quite diverse and numerous. However, there are three basic categories: the sole proprietorship, the partnership, and the corporation. We will first examine the nature of these forms and how to judge the best form for a particular company.

World Wide Web Site

http://www.dfait_maeci.gc.ca/english/invest/menu.htm

Comment: Frequently asked questions about doing business and investing in Canada.

The Nature of the Organizational Forms

In coming to understand the basic differences between each legal organizational form, we ought to know the definition of each form, the procedures necessary for originating the form, and the advantages and disadvantages of each. The article on the next page examines each of the organizational forms and describes their differences.

Comparison of Organizational Forms

Owners of new businesses have some important decisions to make in choosing an organizational form. Not only must they consider a number of different factors, new owners might find that the variables they examine to make their decision may

[17]See for example Franco Modigliani and Richard Sutch, "Innovations in Interest Rate Policy," *American Economic Review* (May 1966), pp. 178–197.

[18]S. Kerry Cooper and Donald R. Fraser, "The Term Structure of Interest Rates"; In: *The Financial Marketplace*, 3rd ed. (Don Mills, ON: Addison-Wesley Publishing Company, 1990), p. 172.

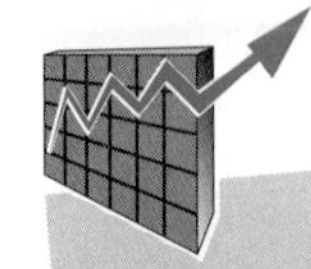

Carefully Evaluate Your Firm's Ownership Structure

One of the questions most frequently asked by those starting a business is: "Which type of ownership structure should I use?" This important choice should be carefully evaluated, not only by those wishing to go into business, but also by those who are already in business. What follows is a general overview of the advantages and disadvantages of the three basic types of business ownership: The sole proprietorship, the partnership and the corporation.

The Sole Proprietorship

The simplest and the most common form of business ownership is the sole proprietorship. Sole proprietorship means that the business is owned and operated by one individual. This individual has complete freedom of operation: when an action must be taken or when a business decision is required, it is not necessary to obtain the consent of another person. Also, the proprietor receives all the profits generated by the business. Initial start-up costs are minimal, and the payment of fees for registration and verification of the company name is usually all that is required to create the business. The proprietor can sell the business or cease operating whenever he wishes to do so.

However, there are certain disadvantages to this form of business ownership, the greatest of which is the owner's unlimited liability. This means that the owner is responsible for all of his company's debts. If the business fails, therefore, creditors may seize his personal assets in order to satisfy these debts. Another disadvantage is limited capital-raising ability. The personal assets of the average individual do not predispose lenders to take risks, and may not by themselves provide the capital required to finance the business. Also, one person may not have the skills adequate to cope with all the potential situations the company may encounter. Finally, the owner is personally responsible for all losses. Gains and losses are included on the individual's personal income tax return and are subject to personal tax rates.

The Partnership

A partnership is formed when two or more individuals share in the ownership of a business. When the same individuals share in the management or control of the business, the organization is legally referred to as a general partnership. Shared control allows partners' abilities to complement each other (two heads are better than one). Start-up costs for a general partnership are the same as those of a sole proprietorship, however it is advisable that a partnership agreement be drawn up, even among relatives and friends. This agreement should clearly spell out all of the responsibilities, rights and duties of the parties involved and should include the following: duration of the partnership, salaries and withdrawals of each partner; amount of capital invested by each partner, settlement in the event of disability or death; distribution of assets in the event of dissolution; division of profits and losses; provision for settlement of disputes among partners; and, provision for the addition and withdrawal of partners.

This agreement should be drawn up by a legal representative and reviewed by each partner's own counsel in order to ensure that the partners' best interests have been taken into consideration. (This will, of course, add to the cost of forming a partnership.) In a partnership, the assets of all the partners are combined, thus making it easier to acquire capital to finance operations. Financial institutions will be more willing to lend money to partnerships because the partners share the debt. Any partner can bind the entire partnership in a business arrangement even though the remaining partners are not aware of, or in agreement with, the decision. Partners are liable for all business debts. Creditors can pursue all of the general partners for these debts, regardless of stipulations in a partnership agreement. It is up to the partners to settle any differences among themselves.

The disadvantages of the unlimited personal liability imposed by a general partnership arrangement can be avoided by forming a limited partnership.[19] A limited partnership has

Partnership

An association of two or more individuals joining together as co-owners to operate a business for profit.

Sole proprietorship

A business owned by a single individual.

[19]The limited partnership has been a major vehicle for financing ventures with limited lives, such as real estate developments or oil and gas exploration in Canada.

one or more partners whose losses are limited to their initial investment. Creditors cannot pursue a limited partner, and such a partner can have no say in the management of the business. However, a limited partnership must have at least one general partner with unlimited liability. Transfer of ownership in the case of a partnership is more complicated than a sole proprietorship. A limited partner can sell his or her interest at any time, however for a general partner this may be more difficult. Depending on the provisions of the partnership agreement, transfer of a general partner's interest to another party may be subject to the approval of the other general partners. Profits and losses are distributed among the partners in the fashion agreed upon and are included in their personal income tax returns, and are thus subject to personal tax rates.

The Corporation

A corporation is an association of shareholders created under law, and is regarded as a legal entity, separate and distinct from its owners and directors. This separation creates limited liability for the owners. Any legal liability as a result of business activities is limited to the amount of their investment in the company. A corporation is the most attractive form of business for raising capital because of the limited liability of the owners and because it provides access to alternative methods of raising funds. These methods include borrowing money, selling bonds, and issuing preferred or common stock. It is, in fact, the shareholders that own the company. Ownership of common stock generally carries with it the right to vote at the annual shareholders' meeting. It is the common shareholders who elect the board of directors, who in turn appoint the officers of the corporation. Furthermore, publicly traded shares are easily transferred whenever the shareholder wishes to do so. The corporation is therefore long-lived and able to continue in existence indefinitely, unless restricted by its charter. Thus the business would not cease to exist abruptly, as may be the case with the death of a partner or proprietor. Finally, corporations generally pay a lower tax rate than individuals.

Incorporation is not without its share of disadvantages, however. Federal or provincial law governs the provisions under which the company's charter (raison d'être) is granted. Provincial registration restricts the company to within a particular province. Another disadvantage is that corporate activities are limited to those specifically granted by its charter. The costs associated with incorporation are generally higher than for other types of business ownership due to the amount of paperwork required by governments and legal implications associated with share offerings. Finally, the major disadvantage of incorporation is the non-deductibility of dividends in determining the company's net income. Shareholders must pay taxes on the dividends they receive which are paid from after-tax earnings. Dividends are thus subject to double taxation.

Keep in mind that there is no one type of ownership structure that is best. Each situation must be looked at individually to determine the most advantageous choice, and once this choice has been made, it is not irrevocable. The type of structure can be changed as circumstances dictate. It is often a good idea to obtain professional advice before making such an important decision.

Source: Concordia University Business Consultants, "Carefully Evaluate Your Firm's Ownership Structure," in *Profits*, by Federal Business Development Bank, 11 (Autumn 1991), p. 4.

Corporation

An entity that *legally* functions separate and apart from its owners.

be in conflict. One consideration might suggest that a partnership is the best route, while another might indicate that a corporation is best. Table 2-1 provides an overview of the most important criteria. The bottom line in the table suggests a favoured form of business organization given the particular list of factors above.

PERSPECTIVE IN FINANCE

The decision as to which organizational form is most suitable should include an evaluation of the relative costs involved in carrying out the firm's business affairs. The decision to minimize organizational costs complies with the firm's goal of maximizing shareholder wealth.

Table 2-1 Selection of Legal Form of Organization

Form of Organization	Organizational Requirements and Costs	Liability of the Owners	Continuity of the Business
Sole proprietorship	Minimum requirements: Generally no registration or filing fee.	Unlimited liability.	Dissolved upon proprietor's death.
General partnership	Minimum requirements: Generally no registration or filing fee. Partnership agreement not legally required but is strongly suggested.	Unlimited liability.	Unless partnership agreement specifies differently, dissolved upon withdrawal or death of partner.
Limited partnership	Moderate requirements: Written agreement often required, including identification of general and limited partners.	General partners: Unlimited liability.	General partners: Same as general partnership.
		Limited partners: Liability limited to investment in company.	Limited partners: Withdrawal or death does not affect continuity of business.
Corporation	Most expensive and greatest requirements: Filing fees; compliance with federal and provincial regulations for corporations.	Liability limited to investment in company.	Continuity of business unaffected by shareholder withdrawal or death.
Form of organization normally favoured	Proprietorship or general partnership	Limited partnership or corporation	Corporation

Management Control and Regulations

The sole proprietor has absolute control of the firm and is not restrained by government regulation. With the few exceptions relating to assumed or fictitious names and special licensing, the sole proprietorship may operate in any province without complying with registration and qualification requirements. The general partnership is likewise not impeded by any significant government regulations. However, since control within this legal form of business is normally based upon the majority vote, an increase in the number of partners reduces each partner's voice in management. The limited partnership is characterized by a restricted separation of ownership from control: there is no such separation in the sole proprietorship and the general partnership. As to government regulation, the limited partnership requires detailed registration to inform the public of the authority of the individual partners. Within the corporation, the control factor has two dimensions: (1) the formal control vested in the shareholders having the majority of the voting common shares and (2) the functional control exercised by the corporate officers in conducting the daily operations. For the small corporation, these two controls usually rest in the same individuals. However, as the size of the corporation increases, the two facets become distinctly separate.[20] Finally, the corporation is encumbered with substantial government regulation in terms of registrations as well as compliance with statutory requirements. For businesses wanting to operate in different provinces of Canada, the provisions governing incorporation are found in the Canada Business Corporation Act (CBCA). Under this Act, the corporation is treated as a natural person, which has rights, powers and privileges of a natural person. Furthermore, the CBCA vests the corporate director with managing the business

[20]Considerable attention has been given to the potential conflict between management and shareholders. Approaches which can be used to obtain more homogeneous expectations between investors and managers include encouraging managers to adopt more of a private owner's perspective by relating their compensation more closely to share-price performance and increasing the use of debt to buy back the firm's common shares.

Table 2-1 (Continued)

Organization	Form of of Ownership	Transferability and Regulations	Management Control Raising Capital	Attractiveness for Income Taxes
Sole proprietorship	May transfer ownership in company name and assets.	Absolute management freedom, negligible formal requirements.	Limited to proprietor's personal capital.	Income from the business is taxed as personal income to the proprietor.
General partnership	Requires the consent of all partners.	Majority vote of partners required for control; negligible formal requirements.	Limited to partners' ability and desire to contribute capital.	Income from the business is taxed as personal income to the partners.
Limited Partnership	General partners: Same as general partnership.	General partners: Same as general partnership.	General partners: Same as general partnership.	General partners: Same as general partnership.
	Limited partners: May sell interest in the company.	Limited partners: Not permitted any involvement in management.	Limited partners: Limited liability provides a stronger inducement in raising capital	Limited partners: Same as general partnership.
Corporation	Easily transferred by transferring shares.	Shareholders have final control, but usually board of directors controls company policies.	Usually the most attractive form for raising capital.	The corporation is taxed on its income and the shareholder is taxed when dividends are paid.
Form of organization normally favoured	Depends upon the circumstances.	Control: Depends upon the circumstances. Regulation: Proprietorship and general partnership.	Corporation	Depends upon the circumstances.

and affairs of the corporation. However, the Act recognizes that the power of the director's management is subject to the corporation's bylaws or unanimous shareholder agreement.[21] In 1994, the Toronto Stock Exchange committee on corporate governance proposed 15 guidelines which address the governance practices of corporate Canada. These guidelines examine both the responsibilities and the functions of a board of directors. Although these guidelines are beyond the scope of this text, they provide a baseline against which governance practices can be evaluated in Canada.

Organization Requirements and Costs

In every instance, the sole proprietorship is the "cheapest" organization to organize. Generally, no legal requirement must be satisfied; the owner simply begins operating. (Exceptions do exist, depending upon the nature of the product or service.) The general partnership may possibly be as inexpensive to create as the proprietorship, in that no legal criterion must be met. However, if a partnership is to be functional in the long term, a written agreement is usually advisable. The importance of this contract cannot be overemphasized. This document, if properly prepared, may serve to avoid personal misunderstandings and may even minimize several disadvantages usually associated with the partnership form of organization. The limited partnership is more expensive, owing to statutory requirements. The partners must provide a certification of the general partners and the limited partners and indicate the rights and responsibilities of each. Also, a written agreement is compulsory. The corporation is typically the most expensive form of business. As stated earlier, compliance with numerous statutory provisions is required. The legal costs and the time involved in creating a corporation exceed those for the other legal types of organization. In short, the organizational requirements increase as the formality of the organization increases. However, this consideration is of mini-

[21]CCH Editorial Staff. *Canada Business Corporations Act with Regulations 1991*, 8th ed. (Don Mills, ON: CCH Canadian Limited, 1991).

mum importance, and to forgo a choice because of its initial cost may prove to be expensive in the long run.

Liability of the Owners

The sole proprietorship and the general partnership have an inherent disadvantage: the feature of unlimited liability. For these organizations, there is no distinction between business assets and personal assets. The creditors lending money to the business can require the owners to sell personal assets if the firm is financially unable to repay its loans. The limited partnership alleviates this problem for the limited partner. However, a limited partner must be careful to maintain this protection. Failure to give due notice or to refrain from actively participating in management may result in the loss of this privilege. The corporation has a definite advantage in terms of limited liability, since creditors are able only to look at corporate assets in resolving claims. However, this advantage for the corporation may not always be realized. If a firm is small, its president may be required to guarantee a loan personally. Also, if the corporate form is being used to defraud creditors, the courts may "pierce the corporate veil" and hold the owners personally liable. In fact, the CBCA has extended the principle of statutory liability to cover a greater range of actions concerning the payment of corporate funds by corporate directors.[22] Nevertheless, the limitation of liability is usually an important concern in the selection of the legal organization.

Continuity of Business

The continuity of the business is largely a function of the legal form of organization. The sole proprietorship is immediately dissolved upon the owner's death. Likewise, the general partnership is terminated upon the death or withdrawal of a partner. This weakness can be minimized in the partnership through a written agreement which specifies what is to occur if a partner dies or desires to withdraw. Failure to incorporate such a provision into the agreement may result in a forced liquidation of the firm, possibly at an inopportune time. Finally, the corporation offers the greatest degree of continuity. The status of the investor simply does not affect the corporation's existence. The corporate business organization has a distinct advantage in its perpetual nature.

Transferability of Ownership

Transferability of ownership is intrinsically neither good nor bad; its desirability depends largely upon the owners' preferences. In certain businesses the owners may want the option to evaluate any prospective new investors. In other circumstances, unrestricted transferability may be preferred. The sole proprietor has complete freedom to sell any interest in the business. At the other extreme, members of a general partnership may not sell or assign their interest without the prior consent of the remaining partners. However, this limitation may be removed by providing otherwise in the partnership agreement. The limited partnership has a twofold nature: the assignment of interest by general partners requires the prior consent of the other partners, while the limited partners have unrestricted transferability. The corporation affords the investors complete flexibility in transferring their interest.

---

[22]The CBCA specifies situations in which a director is severally liable because of a vote on a corporate resolution that has led to a corporate loss. These situations include: the consideration received for the issue of shares is less than fair market value; payments of commissions which are not reasonable to any person who is involved with the repurchase of shares; payments of indemnity to corporate directors or officers, which are inappropriate to those specified in the act; and payments (e.g., financial assistance) to shareholders, which are inappropriate to those specified in the act.

Attractiveness for Raising Capital

As a result of the limited liability, the ease of transferring ownership through the sale of common shares, and the flexibility in dividing the shares, the corporation is the most effective business entity in terms of attracting new capital. In contrast, the unlimited liability of the sole proprietorship and general partnership are deterrents to raising equity capital. Between these extremes, the limited partnership does pro vide limited liability for the limited partners, which has a tendency to attract wealthy investors. Limited partners who have invested in oil, gas, mining or real estate investments benefit from tax savings created by large losses which occur in the early years of such investments. The losses flow through to the individual tax returns of the limited partners. Such an arrangement is not possible if these investments are organized as corporations. The limited partnership is restricted as a financing arrangement to individuals who are subject to maximum rates of personal taxation and who can absorb the significant risk associated with investments of this type. The limited appeal of this form of business or organization prevents it from competing effectively with the corporation.

Income Taxes

Income taxes frequently have a major impact upon the owner's selection of the legal form of business. The sole proprietorship and the partnership organization are not taxed as separate entities; the owners report business profits on their personal tax returns. The earnings from the company are taxable to the owner, regardless of whether these profits have been distributed to the investors. This feature may place the owner in a cash squeeze if taxes are due but income has been retained within the company. On the other hand, the corporation is taxed as a separate and distinct entity. This same income is taxed again when distributed to the shareholder in the form of dividends. Determination of the best form of legal entity with respect to taxes should be based upon the objective of maximizing the after-tax profits to the owners.[23] This decision depends partly upon the tax rates for individuals relative to rates for the corporation. Also, whether or not the profits are to be retained in the business or paid as dividends to the common shareholders also is important.

Taxes and Capital Investment Decisions

As will be explained in more depth later, income taxes are a significant element in evaluating the firm's investment decisions. When the company is analyzing the possible acquisition of a plant or equipment, the returns from the investment should be measured on an after-tax basis. Otherwise the company will be omitting an important variable. For example, suppose management is considering the purchase of production equipment costing $1,000. If the $1,000 is spent, the financing of the expenditure must come from after-tax dollars. Stated differently, the firm may keep this $1,000 without having to be concerned about any tax consequences. However, if the capital is expended on a capital project (plant or equipment), a portion of the cash inflows to be received from the investment will be taxed. Ignoring the time value of money, assume the project, if accepted, is expected to generate $1,200 in cash inflows before taxes, which at first might appear to be satisfactory. However, this $1,200 is before-tax dollars, which simply means that the firm has not paid the taxes that will be owed as a result of receiving these funds. If the company eventually has to pay $300 in taxes, only $900 will be received in cash flows after-tax, which is the amount directly comparable with the $1,000 investment cost. Clearly, the project is undesirable, but the taxes had to be included in the analysis before this fact could be determined.

[23]The firm's objective is to maximize the shareholder wealth. However, since no risk is involved in selecting the form of business, maximizing profits will also maximize wealth.

In computing the after-tax cash flows which result from an investment decision, we should note that the method of depreciation affects both the amount of capital cost allowance to be declared and the timing of the capital cost allowance to be declared. Assuming that a firm has positive taxable income, a capital cost allowance reduces taxable income which in turn reduces taxes payable and increases the amount of cash flow after taxes. A tax depreciation method, such as the declining balance method, will have an impact upon the timing of cash flows since there is a greater amount of capital cost allowances available to reduce taxable income during the earlier years of the project's life as compared to the smaller amounts of capital cost allowances available in later years. In this manner, lower taxes are paid in the initial years with counterbalancing higher taxes in later years. If the time value of money is recognized, this shift in taxes to later time periods is beneficial.

Taxes and the Firm's Capital Structure

The second major policy variable for the financial manager is to determine the appropriate mix between debt and equity financing. Extensive controversy on this issue has continued for well over three decades. However, regardless of the different views maintained, the tax laws do give debt financing a definite cost advantage over preferred shares and common shares.[24] As already noted, interest payments are a tax-deductible expense, while dividend payments to preferred shareholders and to common shareholders may not be used as deductions in computing a corporation's taxable profits.

Taxes and Corporate Dividend Policies

The importance of taxes with respect to the firm's dividend policy is recognized primarily at the common shareholder level rather than at the corporate level. However, since the financial manager's objective is to maximize the common shareholder's wealth, the impact of taxes upon the shareholder is important. Remember that corporate earnings are taxed, whether or not the earnings are paid out in dividends or retained to be reinvested. Yet if the dividends are paid, the investor will be required to report this income. On the other hand, if the profits are retained and reinvested, the price of the company's share should increase. However, until the share is actually sold at a gain, the shareholder is not required to recognize the income. Hence, the opportunity for the firm's common investors to delay the tax payment might influence their preference between capital gains and dividend income. In turn, this preference may affect the corporation's dividend policy.

OBJECTIVE 5

THE CANADIAN INCOME TAX SYSTEM

Objectives of the Income Tax System

The major objective of the tax reform measures of 1987 was to generate revenues in order to finance the federal government's expenditures. Tax reform was also implemented to achieve social and economic objectives such as (1) to stimulate the economy, (2) to encourage investment in certain regions or economic sectors, and (3) to encourage particular economically disadvantaged groups within our society. For example, the federal government can use tax legislation to stimulate the econo-

[24]The tax advantage of debt for the corporation is a tax disadvantage to the investor holding the firm's debt. This issue is addressed more fully in Chapter 9.

my by reducing the taxes payable of an individual. The underlying assumption of this type of economic stimulation is that an increase in discretionary income would increase the demand for products and thereby generate new jobs. Tax reform provided an opportunity for Revenue Canada to replace many of the existing deductions with a tax credit system for individuals. For example, tax credits have been created for individuals who are dependents or individuals who attend university and pay university tuition. In addition to the changes resulting from the tax reform measures of 1987, the federal budgets that have been tabled over recent years have made changes to the corporate tax structure in order to make the system fairer, protect the revenue base and better target the tax assistance delivered to certain businesses. In summary, the objectives of both the tax reform bill of 1987 and recent federal budgets have been to create a taxation system which fulfills the following goals: stability and dependability of revenues, fairness, progressiveness, simplicity and ease of compliance, social and economic goals, balance between sectors, and international competitiveness.[25]

World Wide Web Site

http://www.fin.gc.ca

Comment: Department of Finance site includes news releases, speeches, budget documents, and policy papers.

Types of Taxpayers

In order to understand the tax system, we must first ask what constitutes a "taxpayer." The Canadian Income Tax Act includes within the definition of a taxpayer, any person whether or not liable to pay tax.[26] Before the introduction of amendments

---

[25]The Honourable Michael H. Wilson, "Income Tax Reform. Tax Reform 1987" (Department of Finance, Canada, 1987).

[26]CCH Canadian Limited, "Interpretations," *Canadian Income Tax Act With Regulations 1995.* Part XVII Section 248(1): 65th edition (North York, ON, 1995).

BASIC FINANCIAL MANAGEMENT IN PRACTICE

Corporate Taxation

Some of the consequences of levying taxes on taxpayers have been summarized in the 1995 *Annual Report* of Cambridge Shopping Centres Limited.

The vocal opposition that Cambridge has expressed to high levels of taxation is based on common business sense—taxation expenditures reduce cash flow that is otherwise available for general business growth. Moreover, it diminishes the profitability of tenants, which in turn reduces their ability to expand and provide employment opportunities. Excessive taxation also decreases the disposable income available for retail spending.

Cambridge will continue to review and comment on tax legislation at all levels of government. In doing so the Company will make every effort to assist government and industry to develop a fair and equitable tax system that is appropriate to the economic realities in Canada.

Source: *Annual Report* 1995 of Cambridge Shopping Centres Limited.

in 1993, the concept of a person included an individual, a corporate or political body, heirs, executors, administrators, or other legal representatives of such a person. However in 1993, Revenue Canada broadened the definition of a person to include entities such as municipal authorities, municipal or provincial corporations, registered charities, associations of universities and non-profit organizations.[27] The definition of an individual applies not only to the employees of a company and self-employed persons owning their own business, but also to the members of a partnership.[28]

These individuals report their income to Revenue Canada by filing personal tax returns. In a similar manner, a corporation reports its income by filing a corporate tax return as a separate legal entity with Revenue Canada. The owners (shareholders) of a corporation need not report corporate earnings in their personal tax returns, except when all or part of the profits are distributed in the form of dividends. Finally, estates and trusts must also file a tax return with Revenue Canada but these entities pay taxes on any income generated using the same rules as an individual.[29]

Although the taxes levied on income earned by an estate or a trust represent an important source of income to the federal government, neither is especially relevant to the financial executive. Since most firms of any size are corporations, we will restrict our interest to the corporation. In doing so, a caveat is necessary. Tax legislation can be quite complex, with numerous exceptions to most general rules. The laws can also change quickly, and certain details discussed here may no longer apply. It could be said appropriately that "a little knowledge is a dangerous thing."

World Wide Web Site

http://www.kpmg.ca
Comment: KPMG offers tax information, tips, surveys and publications.

Tax Considerations

There are several aspects of existing tax legislation that have relevance for the financial manager, these being (1) the effects of depreciation on the taxpayer's taxes, (2) the tax treatment of non-capital losses, (3) the recognition of net capital gains and losses, and (4) the deductions and credits for dividends. Let us look at each of these tax provisions in turn.[30]

Depreciation

The means by which an asset's value is expensed over its useful life for accounting purposes.

Capital cost allowance (CCA)

The amount of depreciation (write-off) claimed by a taxpayer for tax purposes on an asset.

Capital cost (C_0)

The sum of the purchase cost of the asset and any incremental cost associated with the installation , engineering, training or administration of the purchase.

Undepreciated capital cost (UCC)

The capital cost of the asset net of any capital cost allowance claimed by the firm. The UCC of an asset class is the sum of the UCCs for each depreciable asset in the class.

Depreciation

The Canadian government encourages taxpayers to invest in fixed assets for business use by permitting them to write off (depreciate) the investment value of an asset through the use of the capital cost allowance system. The amount of **depreciation** (write-off) claimed by a taxpayer for tax purposes on an asset is called a **capital cost allowance (CCA)**.

Revenue Canada uses the term, **capital cost (C_0)**, to describe the concept of investment value for an asset at time zero. Hence, the capital cost of an asset is the sum of the purchase cost of the asset and any incremental cost associated with the installation, engineering, training or administration of the purchase. In addition, Revenue Canada defines **undepreciated capital cost (UCC)** of an asset as the capital cost of the asset net of any capital cost allowance claimed by the firm.

The capital cost allowance system classifies all depreciable assets into groups, called assets classes, which specify both the method of depreciation and the capital

---

[27]Ibid.

[28]Each partner reports their individual share of the partnership's income and pays taxes on this share of the income.

[29]CCH Canadian Limited, "Trusts And Their Beneficiaries," *Canadian Income Tax Act, With Regulations 1995.* Part I Section 104(2): 65th ed. (North York, ON, 1995).

[30]The effect of each of these considerations on taxes payable is examined in Appendix 2A.

cost allowance rate.[31] Revenue Canada administers the capital cost allowance system on the basis of asset classes and not on the basis of an individual depreciable asset. As a result, we define the UCC of an asset class as the sum of UCCs for each depreciable asset in the class.

When a taxpayer reports a capital cost allowance for an asset class, the amount represents the depreciation claimed for all assets that belong to the asset class. In other words, a taxpayer does not report separately the depreciation of individual assets. Each taxpayer is entitled to claim any amount of capital cost allowance up to a maximum which is limited by the CCA rate of the asset class. The capital cost allowance is determined as follows:

$$\text{CCA} = (\text{UCC}_n)(d) \qquad (2\text{-}4)$$

where CCA = the capital cost allowance in year n

UCC_n = the undepreciated capital cost of the asset class after adjustments for dispositions or purchases made during year n

d = the capital cost allowance rate

The majority of depreciable assets belong to asset classes which require the declining balance method to write off the asset's value. As long as a depreciable asset is being employed for business use, a taxpayer is entitled to claim the capital cost allowance from the depreciation of the asset. By reporting the capital cost allowance from each asset class, the taxpayer reduces the taxable income of the business, and in turn, the taxpayer pays less in taxes.

For example, assume your firm has two asset classes, 10 and 8, where the maximum allowable capital cost allowance rates are 30 and 20 percent respectively. If the beginning UCC balances for classes 10 and 8 are $20,000 and $50,000, how much CCA can be claimed for each of the next two years?

Table 2-2 UCC and CCA

Class	10	8	Total
CCA rate (d)	30%	20%	CCA
UCC	$20,000	$ 50,000	
CCA (year 1)	−6,000	−10,000	$16,000
UCC	14,000	40,000	
CCA (year 2)	−4,200	−8,000	12,200
UCC	$9,800	$ 32,000	

As shown in Table 2-2, the declining balance method is used to calculate CCA. Notice that although CCA decreases in subsequent years, it could be calculated indefinitely as long as the UCC balance is positive and there are assets in the class. In other words, CCA can be calculated irrespective of the asset's expected useful life.

Half-Net-Amount Rule

Since 1981, Revenue Canada has allowed a firm to claim only half of the total CCA eligible in the year the asset is put into use. We will assume that an asset is put into use in the year of acquisition. The other half is carried over to the UCC for the following period's calculation. This tax rule is known as the half-net-amount rule or the half-year convention and was adopted to discourage corporations from purchasing assets at or close to their fiscal year end and claiming CCA for the entire year. To illustrate the half-year convention, assume a beginning UCC balance of $10,000 for an asset class where an additional asset worth $80,000 is purchased. The CCA rate is 20 percent. How much CCA can be claimed in year 1 and 2?

[31]For a more detailed discussion of asset classes, see Chapter 14.

Table 2-3 Half-Year Convention

Acquisition	$80,000
CCA rate(d)	20%
UCC beginning	$10,000
Half of the acquisition	40,000
Balance	$50,000
CCA (year 1)	(10,000)
UCC ending	$40,000
Half of the acquisition	40,000
Balance	$80,000
CCA (year 2)	(16,000)
UCC ending	$64,000

As shown in Table 2-3 the CCA claims for year 1 and 2 are $10,000 and $16,000. CCA for subsequent years is computed on 100% of the declining balance.

The sale of a depreciable asset causes the taxpayer to lose its claim to any future capital cost allowances from this asset. In addition, Revenue Canada requires the taxpayer to adjust the undepreciated capital cost of the asset class by subtracting the lesser of the proceeds of the sale or the capital cost of the asset.

For example, referring to Table 2-3, if the $80,000 acquisition was sold in year 3 for $54,000, the CCA in year 3 would be based on a UCC balance of $10,000 ($64,000 – $54,000) and $2,000 CCA could be claimed. Also, note that if acquisitions are accompanied by sales within an asset class over the same period, the half-year convention applies only if net acquisitions are positive. Net acquisitions are the difference between assets added and assets disposed of within the same class.

Table 2-4 Data for Net Acquisition Problem

Sale proceeds of old asset	$ 7,000
Capital cost of sold asset	10,000
New acquisition	31,000
UCC beginning	28,000
CCA rate(d)	20%

Consider the information provided in Table 2-4. What is the maximum CCA that can be claimed in years 1 and 2? What is the incremental CCA?

Net Acquisition ($31,000–$7,000)	$24,000
CCA rate(d)	20%
UCC beginning	$28,000
Half of the net acquisition	12,000
Balance	$40,000
CCA (year 1)	(8,000)
UCC ending	$32,000
Half of the net acquisition	12,000
Balance	$44,000
CCA (year 2)	(8,800)
UCC ending	$35,200

The depreciable balance would include the opening UCC $28,000, plus half of net acquisitions ($31,000 – $7,000)(.5) = $12,000. The maximum CCA for year 1 is $8,000 ($40,000 × .2), and for year 2, it is $8,800 [($40,000 – $8,000 + $12,000)(.2)].

The incremental CCA is calculated on the net acquisition figure ignoring the opening UCC balance. In other words, for year 1, incremental CCA is $2,400 ($12,000 × .2) and for year 2, it is $4,320 [($12,000 − $2,400 + $12,000)(.2)].

For tax purposes, the amount of recaptured depreciation to be reported is determined from the difference between the lesser of either the salvage value (SV_n) or the original capital cost (C_o) of the asset to be salvaged and the undepreciated capital cost (UCC_n) of the asset class. Since both the salvage value and the original capital cost are greater than the UCC_n of the asset class in this situation, the recaptured depreciation is added to the taxpayer's taxable income.

For example, a firm terminates an asset class for tax purposes by selling an asset in the asset class for a salvage value of $50,000. The original capital cost of the asset was $150,000 and the balance for the UCC_n account at this time was $25,000. The recaptured depreciation from the termination of the asset class is $25,000 ($50,000–$25,000), the difference between the lesser of either the salvage value ($50,000) or the original capital cost ($150,000) and the UCC_n ($25,000) at the time of the sale. In this situation, the recaptured depreciation is added to taxable income to determine the firm's taxes payable.

If the salvage value is less than the undepreciated capital cost of the asset class, the taxpayer has written off the asset too slowly. In this situation, the taxpayer may report a terminal loss because the asset class is being terminated. For tax purposes, the amount of terminal loss to be reported is determined from the difference between the undepreciated capital cost (UCC_n) of the asset class and the lesser of the salvage value (SV_n) of the asset or the capital cost. The terminal loss is subtracted from the taxpayer's taxable income since either the salvage value (SV_n) or the original capital cost (C_o) is less than the UCC_n of the asset class.

For example, a firm terminates an asset class for tax purposes by selling the only asset in the asset class for $25,000. The original capital cost of the asset was $150,000 and the balance for the UCC_n account at this time was $50,000. The terminal loss from the termination of the asset class is $25,000 ($50,000 – $25,000), the difference between the UCC_n at the time of the sale ($50,000) and the lesser of either the salvage value ($50,000) or the original capital cost ($150,000). In this situation, the terminal loss is subtracted from taxable income to determine the firm's taxes payable.[32]

Non-Capital Loss Deduction

If a corporation incurs an operating loss in a given fiscal year, this corporation is permitted to apply this type of non-capital loss as a deduction against income earned in other years.[33] In other words, Revenue Canada permits a **carryback** or a

[32]It may be helpful for the purposes of capital budgeting to view the amount of recaptured depreciation as follows:

$$\text{recaptured depreciation} = UCC_n - \text{lesser of} \begin{cases} SV_n \\ \text{or} \\ C_0 \end{cases}$$

This type of equation would intuitively capture the appropriate sign of the flow and determine its contribution to the value of the project. For example, recaptured depreciation will decrease the value of a project because the salvage value of the asset to be sold is greater than the undepreciated capital cost of the asset class and taxes will be incurred on the recapture. In a similar manner, a terminal loss can be viewed for the purposes of capital budgeting as follows:

$$\text{terminal loss} = UCC_n - SV_n$$

Again this type of equation would intuitively capture the appropriate sign of the flow and determine its contribution to the value of the project. For example, a terminal loss would increase the value of the project because the asset is sold at salvage value that is less than the undepreciated capital cost of the asset class and taxes will be saved on the terminal loss.

[33]A non-capital loss is defined in section 118(8)(b) of the Income Tax Act. Eligible losses include losses from business, losses from property, and allowable business investment losses.

Non-capital loss carryback and forward

A tax provision that permits the taxpayer first to apply the loss against the profits in the three prior years (carryback). If the loss has not been completely absorbed by the profits in these three years, it may be applied to taxable profits in each of the seven following years (carryforward).

carryforward of a non-capital loss. The taxpayer should first apply the loss against income in three prior years (carryback). If this loss has not been completely absorbed by the income in these three years, then the loss may be carried forward to each of the seven following years (carryforward). At that time, any loss still remaining may no longer be used as a tax deduction. To illustrate, a 1996 operating loss may be used as a deduction against income earned in 1993, 1994 and 1995. As a result, this deduction lowers taxable income which in turn can be used to reclaim taxes paid, in whole or in part, during 1993, 1994, and 1995. If any part of the loss still remains, this amount may be used to reduce taxable income, if any, during the seven-year period of 1997 through 2003. It should be noted that the income or losses from business operations for both the partnership and the proprietorship flow through to the individual and are taxed at this level. In other words, non-capital losses of the partnership or the proprietorship are combined with the individual's other non-capital losses and are eligible for the carryback or the carryforward provisions.[34]

Net Capital Gains and Losses

Capital gain or loss

For tax purposes, a gain or loss resulting from the sale or exchange of a capital asset.

A corporation incurs a **capital gain** if the selling price is greater than the purchase cost for both depreciable and capital assets; conversely, a **capital loss** occurs if the selling price is less than the purchase cost of a nondepreciable asset. As a result of 1987 tax reform measures, both capital gains and capital losses continue to be given preferential tax treatment. Revenue Canada levies a tax on the following portions of both capital gains and capital losses: one-half for years prior to 1987, two-thirds for 1988 and 1989, and three-quarters for years after 1989. A corporation that has capital losses which exceed capital gains in any year may not deduct these net capital losses from ordinary income. Revenue Canada requires that net capital losses be first carried back and applied against net capital gains in each of the three years before the current year.[35] If the loss is not completely used in the three prior years, any remaining loss may be carried forward and applied against any net capital gains indefinitely. For example, if a corporation has an $80,000 net capital loss in 1996, it may apply this loss against any net capital gains in 1993, 1994, and 1995. If any net capital loss remains, it may be carried forward indefinitely.

We should also note that any net capital gains or losses which are incurred by either a partnership or a proprietorship flow through to the individual and are taxed at this level. As a result, the net capital gains or losses which are incurred by either the partnership or the proprietorship are combined with the individual's other net capital gains or losses.

Deductions and Credits for Dividends

The earnings generated by a corporation are potentially subject to "double taxation," first at the corporate level and then at the shareholder level as the firm's profits are distributed in the form of dividends. However, Revenue Canada permits corporations to deduct from their taxable income all dividends received from a taxable Canadian corporation or a subsidiary corporation that is a Canadian resident. For instance, if corporation A receives dividends of $1,000 in its fiscal year from ownership of common shares in a taxable Canadian corporation B, the $1,000 in dividends received will be deducted from taxable income of corporation A and result in no corporate taxes payable on intercorporate dividends received. It should be not-

[34]During the early years of a small business, it is highly likely that the business may incur non-capital losses. This flow-through provision in the Income Tax Act becomes an important consideration when choosing whether to incorporate or to remain either as a proprietorship or partnership. As a proprietorship or partnership, the losses would be immediately deductible by the shareholder, but may not be utilized by the corporation for a number of years.

[35]A net capital loss is defined in section 111(8)(a) of the Income Tax Act. Capital losses exclude losses on listed property and allowable business investment losses.

ed that dividends received by either the partnership or the proprietorship from other Canadian corporations are taxed at the individual's level and not at the level of the business. However, individuals are able to deduct a dividend tax credit from their taxes payable. The dividend tax credit is determined as follows:

$$\text{Dividend tax credit} = 13.33\% \text{ of grossed-up amount of dividends} \quad (2\text{-}5)$$

$$\text{where, Grossed-up amount} = (\text{dividend amount}) + (25\% \text{ of dividend amount}) \quad (2\text{-}6)$$

As a result, Revenue Canada enables individuals to diminish the effect of "double taxation" on their dividend income by deducting the dividend tax credit from their taxable income.

Other Tax Considerations

The Income Tax Act identifies the following corporations when determining how income is to be taxed: public corporation, private corporation and the Canadian-controlled private corporation. The public corporation is differentiated from the private corporation on the basis of whether any of the corporation's shares are listed on a Canadian stock exchange. Whereas the difference between the Canadian-controlled private corporation and the private corporation is based on the control of the firm.[36]

Canadian-controlled private corporations (CCPCs) enjoy a lower basic corporate tax rate on active business income as compared to public corporations.[37] Revenue Canada enables CCPCs to benefit from a lower basic corporate tax rate through the use of a "Small Business Deduction." This deduction is calculated as 16 percent of the least of three following amounts:

1. The corporation's active business income for a taxation year.
2. The corporation's taxable income earned in Canada for the year achieved by subtracting from the total taxable income any foreign-source income.[38]
3. The corporation's "business limit" for the year. In most cases the "business limit" for a CCPC is $200,000 less any portion allocated to associated corporations.

The net effect of the small business deduction is to reduce the net federal tax rate to 13.12 percent on active business income for a CCPC.[39] Another advantage of being classified as a CCPC is if the shares of this type of corporation meet the requirements of a qualified small business corporation share. The owners of such shares would benefit from a capital gains exemption of $500,000 or a taxable gains exemption of $375,000.

Given the globalization of the economy, income may originate in a foreign country. As a financial executive, you would obviously consider the firm's place of residence for tax purposes when deciding where to locate your business, since a firm can minimize its taxes by reporting as much income in the low-tax-rate countries and as little as possible in the high-tax-rate countries. Of course, other factors, such as political risk, may discourage your efforts to minimize taxes across national

---

[36]For tax purposes, control usually means majority ownership. The Income Tax Act provides additional rules on control [ITA 256(5.1)] and deemed control [ITA 251 (5)]. For further discussion, see: C. Byrd, I. Chen, and M. Jacobs, *Canadian Tax Principles* 1994-95 ed., Vol. 2 (Scarborough, ON: Prentice-Hall Canada, 1994), p.8.

[37]The federal budget of 1994 has gradually phased out the "Small Business Deduction" for CCPCs having taxable capital employed in Canada of between $10 and $15 million. The net effect of this legislation is to reduce the $200,000 business limit over time using a straight line approach.

[38]The Income Tax Act provides a mechanism to estimate the foreign source income. For further explanation, see R. Beam and S. Laiken, *Introduction To Federal Income Taxation In Canada*, 11th ed. (Don Mills, ON: CCH Canadian Limited, 1991), p. 425.

[39]The term "active business" is defined in section 125(7)(a) of the Income Tax Act. The term includes any business of a corporation other than that of a specified investment business or a personal services business. The definition excludes income derived from investments in property located in Canada.

Table 2-5 Comparison of Foreign Corporate Taxes

Country	Income Tax Rate[1]	Value Added Tax
France	34–42%	5.5–18.6%
Japan	28–37.5%	3% Consumption Tax
Korea	20–34%	10%
United Kingdom	25–33%	17.5%
United States	15–34%	none; state and local taxes vary up to 8%

[1]Range includes preferential tax treatment for small business

Source: *International Tax Summaries*, Coopers & Lybrand International Tax Network (New York: John Wiley, 1994).

borders. Most Canadian tax treaties with foreign countries require that income be taxed first in the country where it was earned. If income originates in a foreign country, the tax rates and the method of taxing the firm frequently vary. Table 2-5 sheds some light as to basic differences in tax rates in several industrialized countries. In some cases, Canada levies a higher tax rate than that of the country where it was earned. If under these circumstances the firm is a resident of Canada, it will pay taxes to Revenue Canada at the higher tax rate. However, Revenue Canada provides tax relief on the portion of taxes owed to the country where the income is earned through the use of the foreign tax credit.

In addition to this type of tax credit, the federal government provides investment tax credits to corporations in order to stimulate new investment in activities such as manufacturing and processing, operating oil or gas wells, extracting petroleum or natural gas, extracting minerals, processing ore, logging, farming, fishing, storing grain and producing industrial minerals. In addition, investment tax credits are provided to all Canadian taxpayers for qualified scientific research and experimental development expenditures. Many of these investment tax credits have been restricted to certain regions of the country; for example, Gaspé and the Atlantic provinces. The importance of investment tax credits to financial decision making is twofold: (1) there is a direct reduction in the amount of taxes that are payable by a specified percentage of the capital expenditure and (2) there is a one-year delay in the reduction of the cost of capital by the amount of the investment tax credit when determining the appropriate capital cost allowance. In other words, a corporation will reduce its taxes payable by the investment tax credit in the year when the capital expenditure was made. Moreover, investment tax credits reduce all future CCA claims because the capital cost net of the investment tax credit is placed in the CCA asset class.[40]

PERSPECTIVE IN FINANCE

Few financial managers know and understand complex tax law. The fact is you need not be a tax wizard to be a good financial manager. However, you do need to know how taxes relate to investment decisions, financing decisions, and dividend policies of the firm.

---

[40]Generally, the Canadian government enacts legislation for investment tax credits when capital investment needs to be accelerated. The investment tax credit on net qualifying expenditures can be as high as 35 percent depending on the type of corporation. CCPCs enjoy an investment tax credit rate of 35 percent on the first $2 million of qualified expenditures on scientific research and experimental development provided taxable income does not exceed $400,000. This is a direct tax credit; that is, it can be used to offset taxes that are due. A firm can carry back these unused tax credits up to three years and carry forward these unused tax credits up to ten years. It should be noted that the federal budget of 1994 imposed restrictions on the firm when claiming credits from the previous years and either eliminated or reduced the investment tax credit rate for certain types of expenditures. In addition, the tax credit has disappeared for CCPCs whose taxable capital has reached $10 million.

SUMMARY

Financial managers should be aware of the external influences affecting the company. This chapter has examined three key elements: the financial markets, the legal forms of business organization and the tax structure.

The Financial Markets

The financial markets provide the avenue for bringing together savers and borrowers. Equally important, the markets provide management an indication of the investor's opportunity costs, which suggest the rates of return being required by the investors.

The interest rates in the financial markets are based on the demand and supply of money in the economy, as reflected in the risk-free rate of return. Also, the investor is rewarded for the potential loss of purchasing power resulting from inflation. The rates are further influenced by the risk level of the investment and the length of time until the security matures.

Legal Forms of Organization

The sole proprietorship is a business operation owned and managed by a single individual. The initiation of this form of business is extremely simple and generally does not involve any substantial organizational costs. The proprietor has complete control of the firm but must be willing to assume full responsibility for the outcome.

The general partnership, which is simply a coming together of two or more individuals, is quite similar to the sole proprietorship. The limited partnership has been created to permit all but one of the partners to have limited liability if this is agreeable to all partners.

The corporation has served to increase the flow of capital from the investment public to the business community. Although larger organizational costs and regulations are imposed upon this legal entity, the corporation is more conducive to raising large amounts of capital. Limited liability, continuity of life, and ease of transfer in ownership, which increases the marketability of the investment, have contributed greatly in attracting large numbers of investors into the business. The formal control of the corporation is vested in the parties having the greatest number of shares. However, day-to-day operations are determined by the corporate officers, who theoretically serve on behalf of the common shareholders.

Taxes

Several forms of taxation exist; however, the primary tax concern in a business relates to income taxation. The Canadian income tax system has the following objectives: (1) to provide a revenue source for the federal government, (2) to stimulate the economy, (3) to encourage particular economically disadvantaged groups within our society and (4) to encourage investment in certain regions or economic sectors. The Canadian Income Tax Act includes within the definition of a taxpayer the following entities, among others: individuals, corporate and political bodies, heirs, executors, administrators, or other legal representatives of such a person. Only information on the corporate entity has been given here.

A corporation's net income for accounting purposes must be adjusted to obtain a corporation's taxable income. The adjustments to net income account for the preferential tax treatment of capital gains and capital losses, the difference between depreciation for accounting purposes and depreciation for tax purposes, and the treatment of dividends received from other Canadian corporations. The taxation system influences the investment decision through both the method and rate of depreciation. The decision of choosing the appropriate mix of financing for a firm is influenced by the cost advantage of debt over the costs of preferred and common share financing. Finally, the decision of choosing whether the firm should distribute dividends or retain earnings is influenced by the tax status of the firm's shareholders.

STUDY QUESTIONS

2-1. What are financial markets? What function do they perform? How would an economy be worse off without them?

2-2. Explain the term *opportunity cost* with respect to the cost of funds to the firm.

2-3. Explain the Fisher effect in terms of the impact of inflation on nominal rates of returns.

2-4. Define the *term structure of interest rates.*

2-5. Explain the popular theories for the rationale of the *term structure of interest rates.*

2-6. Define (a) sole proprietorship, (b) partnership, and (c) corporation.

2-7. Identify the primary characteristics of each form of legal organization.

2-8. Using the following criteria, specify the legal form of business that is favoured: (a) organizational requirements and costs, (b) liability of the owners, (c) continuity of business, (d) transferability of ownership, (e) management control and regulations, (f) capability to raise capital, and (g) income taxes.

2-9. Does a partnership pay taxes on its income? Explain.

2-10. When a corporation receives a dividend from another corporation, how is it taxed?

2-11. What is the difference between a non-capital loss and a net capital loss?

STUDY PROBLEMS (SET A)

2-1A. (*Inflation and Interest Rates*) What would you expect the nominal rate of interest to be if the real rate is 4 percent and the expected inflation rate is 7 percent? What would the "inflation premium" be?

2-2A. (*Inflation and Interest Rates*) Assume the expected inflation rate to be 4 percent. If the current real rate of interest is 6 percent, what ought the nominal rate of interest be? What about the "inflation premium"?

2-3A. (*Inflation and Interest Rates*) What would you expect the nominal rate of interest to be if the real rate is 5 percent and the expected inflation rate is 3 percent? What would the "inflation premium" be?

2-4A. (*Inflation and Interest Rates*) Assume the expected inflation rate to be 2 percent. If the current real rate of interest is 6 percent, what ought the nominal rate of interest be? What about the "inflation premium"?

2-5A. (*Term Structure of Interest Rates*) You want to invest your savings of $20,000 in government securities for the next two years. Currently, you can invest either in a security that pays interest of 8 percent per year for the next two years or in a security that matures in one year but pays only 6 percent interest. If you make the latter choice, you would then reinvest your savings at the end of the first year for another year.

- **a.** Why might you choose to make the investment in the one-year security that pays an interest rate of only 6 percent, as opposed to investing in the two-year security paying 8 percent? Provide numerical support for your answer. Which theory of term structure have you supported in your answer?
- **b.** Assume your required rate of return on the second-year investment is 11 percent; otherwise, you will choose to go with the two-year security. What rationale could you offer for your preference?

2-6A. (*Capital Cost Allowance*) Target Inc., owns equipment that belongs to asset class 8. The undepreciated capital cost of this asset class at the beginning of the tax year was $10,000. Class 8 requires that assets be depreciated using a declining balance method at a CCA rate of 20 percent. Determine the amount of capital cost allowance and the undepreciated capital cost balance for asset class 8 at the end of the tax year if no acquisitions or dispositions were made during the year.

2-7A. (*Capital Cost Allowance*) Remains Inc., owns and uses equipment in its operations that belongs to asset class 8. The undepreciated capital cost of this equipment at the beginning of the tax year was $250,000. During the tax year, Remains Inc., bought and put into use new equipment in asset class 8 that had a capital cost of $50,000. Class 8 requires that assets be depreciated using a declining-balance method at a CCA rate of 20 percent. Determine the amount of capital cost allowance and the undepreciated capital cost balance for asset class 8 at the end of the tax year.

2-8A (*Capital Cost Allowance*) Monet Inc., owns and uses computer equipment (hardware) which belongs to asset class 10. The undepreciated capital cost of this equipment at the beginning of the tax year was $400,000. During the tax year, Monet Inc., bought and put into use new equipment in asset class 10 that had a capital cost of $150,000. In addition, Monet Inc., disposed of equipment for $30,000 during the same tax year. Class 10 requires that assets be depreciated using a declining balance method at a CCA rate of 30 percent.

a. Determine the amount of capital cost allowance and the undepreciated capital cost balance for asset class 10 at the end of the tax year.

b. Determine the amount of capital cost allowance and the undepreciated capital cost balance for asset class 10 at the end of the next tax year if no acquisitions or dispositions occurred in the asset class.

2-9A (*Capital Cost Allowance System*) Tapis Inc., owns and uses computer equipment that belongs to asset class 10. The undepreciated capital cost of the equipment at the beginning of the tax year is $250,000. Class 10 requires that assets be depreciated using a declining-balance method at a CCA rate of 30 percent. The original capital cost of all the equipment in asset class 10 was $375,000. During the tax year, management at Tapis Inc. have decided to dispose all of the equipment in asset class 10 and then lease new equipment from Lease-A-Computer Inc., on a yearly basis.

a. Determine the amount of capital cost allowance for asset class 10 at the end of the tax year for Tapis Inc.

b. Assume that Tapis Inc. disposed all of the equipment in asset class 10 for $100,000 during the tax year. Explain the tax consequences.

c. Assume that Tapis Inc. disposed all of the equipment in asset class 10 for $300,000 during the tax year. Explain the tax consequences.

2-10A. (*Non-capital Losses; Carryback–Carryforward*) The non-capital losses and taxable income for Maness, Inc., for the past seven years are given below. As a consultant, indicate the optimal use of the non-capital losses to management given the information provided.

Year	Non-capital loss	Taxable income
1989	0	$4,500,000
1990	0	5,500,000
1991	0	3,500,000
1992	$8,500,000	−1,200,000
1993	1,500,000	− 800,000
1994	0	3,000,000
1995	0	4,000,000

2-11A. (*Capital Gains versus Dividends*) In 1995, Mr. Smith, a proprietor of a small business, earns $2,500 in taxable income during the tax year. Assume that the federal tax rate for Mr. Smith is 17 percent.

a. Determine Mr. Smith's taxable income if he receives an additional $20,000 in capital gains in 1995. Assume that Mr. Smith did not receive any other capital gains or capital losses in 1995.

b. Determine Mr. Smith's taxable income if he receives $20,000 in dividends from a taxable Canadian corporation.

c. Determine the amount of capital gains that Mr. Smith must receive to pay the same amount of federal taxes payable as in Question b. of this problem.

2-12A. (*Investment tax credits and capital cost allowance*) CB Investo, Inc., has made an acquisition of a new machine which will be used in one of the company's research and development projects. CB Investo, Inc., has purchased and put into use this machine for a capital cost of $100,000. The undepreciated capital cost of this asset class at the beginning of the tax year was $150,000. The equipment belongs to asset class 8 which requires that assets be depreciated using a declining-balance method at a CCA rate of 20 percent. The firm is entitled to a 20 percent investment tax credit for this capital expenditure for its research and development project. Determine the amount of capital cost allowance and the undepreciated capital cost balance for asset class 8 for the next two tax years if no acquisitions or dispositions were made during these years.

APPENDIX 2A

THE CALCULATION FOR TAXES PAYABLE

In finance, the concept of "free cash flows" is important because of its use in establishing the financial position of the firm. One of the components required in determining free cash flows is the amount of taxes payable by the firm. The amount of taxes payable by the firm is dependent on the firm's taxable income. A financial executive must realize that accounting income cannot be used as a substitute for taxable income because of the following differences: the depreciation methods used to determine the capital cost allowances; the tax treatment of capital gains or losses; and the treatment of incorporate dividends. As a result, in the next section we outline the major adjustments used to convert accounting income into taxable income.

COMPUTING TAXABLE INCOME

The accounting income of a corporation, as derived from the Income Statement, must be adjusted to obtain taxable income for tax purposes. A corporation's net income, for tax purposes, is based upon the gross income from all sources, except for allowable exclusions, less any tax-deductible expenses. Gross income equals the firm's dollar sales from its product less the cost of producing or acquiring the product. Tax-deductible expenses include any operating expenses, such as marketing expenses and administrative expenses. Also, the interest expense paid on the firm's outstanding debt is a tax-deductible expense. However, dividends paid to the firm's shareholders, either preferred or common shareholders, are not deductible expenses. Other taxable income includes interest income and dividend income.

To demonstrate how to compute a corporation's net income for accounting purposes, consider the J and S Corporation, a manufacturer of home accessories. The firm, originally established by Kelly Stites, has sales of $5,000,000 for the year. The cost of producing the accessories totalled $2,300,000. Operating expenses were $1,000,000. The firm sold marketable securities at the beginning of the year for $1,550,000 that originally cost $750,000, but incurred a capital loss of $500,000 on the sale of shares. In addition, this firm received $200,000 in dividends from shares owned in another taxable Canadian corporation. This firm has $1,250,000 in debt outstanding, with a 16 percent interest rate, which resulted in an interest expense of ($1,250,000)(.16) or $200,000. Management paid $100,000 in dividends to the firm's common shareholders. The firm's combined tax rate is 45 percent. The J and S Corporation reported net income of $1,100,000 for accounting purposes as shown in Table 2A-1.

Revenue Canada requires the corporation to report taxable income in order to calculate taxes payable. Table 2A-2 shows the procedure that enables the corporation to adjust the net income for accounting purposes to obtain taxable income.

In extending our example of J and S Corporation, this corporation will claim depreciation for tax purposes as a capital cost allowance of $250,000. The corporation also incurred a recaptured capital cost allowance of $250,000 on the sale of its only fixed asset. The taxable income for J and S would be $1,875,000 as shown in Table 2A-3.

Once we know J and S Corporation's taxable income, we can next determine the amounts of taxes payable by the firm.

Table 2A-1 Net Income Statement of J and S Corporation

Sales		$5,000,000
Cost of Goods Sold		−2,300,000
Gross Profit		2,700,000
Operating Expenses:		
Administrative Expenses	−$400,000	
Depreciation Expenses	−150,000	
Marketing Expenses	−450,000	−1,000,000
Operating Income		$1,700,000
Other Income:		
Dividends from Other Corporations		200,000
Gain on Investment		800,000
Loss on Shares		−500,000
Earnings before Interest and Taxes		$2,200,000
Interest Expense		−200,000
Earnings before Taxes		$2,000,000
Tax Provision (@45%)		−900,000
Net Income		$1,100,000

Note: Dividends paid to common shareholders ($100,000) are not tax-deductible expenses.

Table 2A-2 Corporation's Taxable Income

	Net Income (Accounting purposes)
Add:	Tax Provision
	Taxable Capital Gain
	Accounting Depreciation Expense
	Recaptured Depreciation
	Loss on Investment
Deduct:	Gain from Investment
	Capital Cost Allowance
	Terminal Loss
	Taxable Capital Losses
	Net Income for Tax Purposes
Deduct:	Dividends from Other taxable Canadian Corporations
	Taxable Income

Table 2A-3 Taxable Income of J and S Corporation

	Net Income (Accounting purposes)	$1,100,000
Add:	Tax Provision	900,000
	Taxable Capital Gain (.75)($800,000)	600,000
	Accounting Depreciation Expense	150,000
	Loss on Sale of Shares	500,000
	Recaptured Capital Cost Allowance	250,000
Deduct:	Gain from Sale of Marketable Shares	−800,000
	Capital Cost Allowance	−250,000
	Taxable Capital Losses (.75)($500,000)	−375,000
	Net Income for Tax Purposes	$2,075,000
Deduct:	Dividends from other Canadian Corporations	−200,000
	Taxable Income	$1,875,000

Financial Perspective

Returns from a capital investment must be examined on an after-tax basis in order to improve the estimation of the investment's cash flows. As we examine various issues in financial management, the appropriate tax rate to be used will be a tax rate that combines both federal and provincial corporate tax rates. To understand the use of the combined corporate tax rate, we must examine the computation required to determine the total taxes payable by a corporation. Furthermore, we should note the importance of how both revenue size and type of corporation affects the taxes payable by a corporation.

COMPUTING THE TAXES OWED

Marginal tax rate

The tax rate that would be applied to the next dollar of income.

Revenue Canada recognizes net business income from either a proprietorship or a partnership by taxing taxable income at the level of the individual. The federal **marginal tax rates** which are applied to net business income for the 1994 and the 1995 tax years, vary according to the level of an individual's taxable income; for example, the federal tax rate for levels of taxable income of $29,590 or less is 17 percent; for levels of taxable income between $29,591 and $59,180 the tax rate is 26 percent; and for levels of taxable income more than $59,180, the tax rate is 29 percent.

The calculation which determines federal tax payable for a corporation requires the use of the basic corporate tax rate of 38 percent. However, Revenue Canada allows the basic corporate tax rate to be reduced through various deductions. An example of this type of deduction is the provincial abatement which enables corporations residing in one of the Canadian provinces or territories to reduce the basic corporate tax rate of 38 percent by 10 percent to 28 percent.

Another important deduction that is used to reduce the basic corporate tax rate is the small business deduction (SBD) which can be claimed by Canadian-controlled private corporations (CCPCs). In 1995, the net effect of the SBD is to reduce the net federal tax rate to 13.12 percent on active business income for most CCPCs. For example, the net federal tax rate of a CCPC is determined as follows:

	Basic corporate tax rate	38.00%
Deduct:	Provincial abatement	−10.00%
	Federal tax rate	28.00%
Deduct:	Small business deduction	−16.00%
Add:	Federal surtax (4% of 28%)	1.12%
	Net federal tax rate	13.12%

In an attempt to improve on productivity and to facilitate the restructuring of operations, Revenue Canada requires manufacturing and processing profits to be taxed the net federal tax rate of 23.84 percent in 1992, 22.84 percent in 1993, 21.84 percent in 1994, 22.08 percent in 1995. The corporate tax rate is levied on manufacturing and processing profits other than profits eligible for the small business deduction.

The Government of Canada also levies a corporate surtax on the basic federal tax after the provincial abatement. Effective February 27, 1995, this rate has increased from 3 percent to 4 percent. This corporate surtax is determined as follows:

$$
\begin{aligned}
\text{Corporate surtax} &= .04 \times (\text{basic corporate tax rate} - \\
&\qquad \text{provincial abatement}) \times \text{taxable income} \\
&= .04 \times (.38 - .10) \times \text{taxable income} \\
&= .0112 \times \text{taxable income}
\end{aligned}
$$

A summary of the calculation used to determine the total taxes payable by a corporation is shown in Table 2A-4.

The provincial abatement enables the provinces to impose their own tax rate as shown in Table 2A-5.

The federal government allows all taxpayers, including corporations, certain tax credits such as the foreign business income tax credit and the investment tax credit. The purpose of each of these tax credits is to reduce federal taxes payable.

Most provinces depend on Revenue Canada to collect their corporate taxes. However three provinces, Quebec, Ontario, and Alberta, collect their own corpo-

Table 2A-4 Total Tax Payable for Corporations

	Taxable Income (A)
	Tax at 38 percent on the amount A
Deduct:	Provincial Abatement (10 percent of A)
	Small Business Deduction (16 percent of A)
	Manufacturing & Processing Profit Deduction (5 percent of A)
Add:	Corporate Surtax [(0.04)(.38A − .10A)]
Deduct:	Foreign Business Income Tax Credit
	Investment Tax Credit
	Federal Tax Payable (Part I Tax)
Add:	Provincial Tax Payable (17 percent of A)
	Total Tax Payable

Note: The 17 percent tax rate represents an example of a provincial tax rate.

Table 2A-5 Provincial Corporate Tax Rates for 1995

Province	Corporate Tax Rate
Newfoundland	5–16 percent
Nova Scotia	5–16 percent
Prince Edward Island	7.5–15 percent
New Brunswick	9–17 percent
Quebec	5.75–16.25 percent
Ontario	9.5–15.5 percent
Manitoba	9.5–17 percent
Saskatchewan	8.5–17 percent
Alberta	6.0–15.5 percent
British Columbia	10–16.5 percent
Yukon Territory	2.5–15.5 percent
Northwest Territories	5–14 percent

Table 2A-6 Total Taxes for J and S Corporation

	Taxable Income	$1,875,000	(A)
	Tax at 38 percent on the amount A	712,500	
Deduct:	Provincial Abatement (10% of A)	−187,500	
Add:	Corporate Surtax [(0.04)(.38A − .10A)]	21,000	
	Federal Tax Payable (Part I Tax)	$ 546,000	
Add:	Provincial Tax Payable (16% of A)	300,000	
	Total Tax Payable	$ 846,000	

rate taxes. Provinces which permit the Canadian government to collect their corporate taxes will apply their own corporate tax rate to taxable income to obtain provincial taxes payable as shown in Table 2A-5. Both federal tax payable and provincial tax payable are then added together to obtain the total tax payable by the firm.

For example, suppose that the headquarters for J and S Corporation were situated in Nova Scotia where the provincial corporate tax rate is 16 percent, the total tax payable for this corporation is shown in Table 2A-6.

Corporate Taxes: An Example

To illustrate certain portions of the tax laws for a corporation, assume that the Griggs Corporation is a public corporation with a net income of $1,200,000 for accounting purposes. The firm reported in its Income Statement a tax provision of $800,000 and included in other income $100,000 in dividends received from another Canadian corporation. In turn, it paid $40,000 in interest and $75,000 in dividends. Also included in other income is an accounting gain of $100,000 for the sale of an old machine. For tax purposes, this old machine had originally cost $200,000 and is now sold for $300,000. The undepreciated capital cost of the asset class in which the old machine belongs is $50,000. Since this asset class is being closed upon the sale of this old machine, the firm is required to declare a recaptured depreciation of

Table 2A-7 Griggs Corporation Tax Computations

	Net Income (Accounting purposes)	$1,200,000	
Add:	Tax Provision	800,000	
	Taxable Capital Gain (.75)($100,000)	75,000	
	Accounting Depreciation Expense	100,000	
	Loss on Sale of Land	50,000	
	Recaptured Depreciation	150,000	
Deduct:	Gain from Sale of Old Machine	−100,000	
	Capital Cost Allowance	−110,000	
	Taxable Capital Losses (.75)($50,000)	− 37,500	
	Net Income for Tax Purposes	$2,127,500	
Less:	Dividends from other Canadian Corporations	− 100,000	
	Taxable Income (A)	$2,027,500	
	Taxable Income	$2,027,500	(A)
	Tax at 38 percent on the amount A	770,450	
Deduct:	Provincial Abatement (10 percent of A)	−202,750	
Add:	Corporate Surtax [(0.04)(.38A − .10A)]	22,708	
	Federal Tax Payable (Part I Tax)	$ 590,408	
Add:	Provincial Tax Payable (15 percent of A)	304,125	
	Total Tax Payable	$ 894,533	

Note: The $75,000 Griggs paid in dividends is not tax deductible.

$150,000 in the fiscal year. The firm reported $100,000 in depreciation expenses for accounting purposes but incurred capital cost allowances of $110,000. Finally, the company sold a piece of land for $50,000 that had cost $100,000 six years ago.

Based upon a federal corporate tax rate of 38 percent and a provincial tax rate of 15 percent, Griggs's tax liability is $894,533, as shown in Table 2A-7.

As mentioned earlier, the appropriate rate to be used in financial analyses is the combined corporate tax rate, the tax rate that combines federal and provincial corporate tax rates. Thus, when making financial decisions involving taxes, always use the combined corporate tax rate in your calculations.

STUDY PROBLEMS (SET A)

2A-1A. (*Taxable Income*) The William B. Waugh Corporation is a large clothing retailer. During the most recent year the company generated sales of $3 million. The combined cost of goods sold and the operating expenses were $2.1 million. The firm reported neither depreciation expense nor capital cost allowance for the fiscal year. Also, $400,000 in interest expense was paid during the year. The firm received $6,000 during the year in dividend income from 1,000 common shares of a Canadian corporation that had been purchased three years previously. However, these common shares were sold toward the end of the year for $100 per share; its initial cost was $80 per share. The company also sold land that had been recently purchased and had been held for only four months. The selling price was $50,000; the cost was $45,000.

a. Given that William B. Waugh's combined corporate tax rate is 45 percent, determine the corporation's net income by illustrating the Income Statement.

b. Determine the corporation's taxable income for tax purposes.

2A-2A. (*Taxable Income*) Sales for L. B. Menielle, Inc., during the past year amounted to $5 million. Gross profits for the year were $3 million. Operating expenses totalled $1 million. The firm reported neither depreciation expense nor capital cost allowance for the fiscal year. The interest and dividend income from securities owned of a Canadian corporation were $20,000 and $25,000, respectively. The firm's interest expense was $100,000. The firm sold securities on two occasions during the year, realizing a gain of $40,000 on the first sale but losing $50,000 on the second.

a. Given that L. B. Menielle's combined corporate tax rate is 45 percent, determine the corporation's net income by illustrating the Income Statement.

b. Determine the corporation's taxable income for tax purposes.

2A-3A. (*Corporate Income Tax*) Sandersen, Inc., sells minicomputers. During the past year the company's sales were $3 million. The cost of its merchandise sold came to $2 million, and cash operating expenses were $400,000. The firm reported a depreciation expense of $100,000, but had capital cost allowance of $100,000 for the fiscal year. Sandersen, Inc., paid $150,000 in interest on bank loans. Also, the corporation received $50,000 in dividend income from another Canadian corporation but paid $25,000 in the form of dividends to its own common shareholders.

a. Given that Sandersen's combined corporate tax rate is 45 percent, determine the corporation's net income by illustrating the Income Statement.

b. Determine the corporation's taxable income for tax purposes.

c. Determine the corporation's total taxes payable assuming the federal tax rate is 38 percent, the corporate surtax is 4 percent, the provincial abatement is 10 percent and the provincial tax rate is 16 percent.

2A-4A. (*Corporate Income Tax*) A. Don Drennan, Inc., had sales of $6 million during the past year. The company's cost of goods sold was 70 percent of sales, cash operating expenses were $750,000, and depreciation expense was $50,000. The firm claimed a capital cost allowance of $75,000 for tax purposes. In addition, the firm sold a capital asset (stock) for $75,000, which had been purchased five months earlier at a cost of $80,000.

a. Given that A. Don Drennan's combined corporate tax rate is 45 percent, determine the corporation's net income by illustrating the Income Statement.

b. Determine the corporation's taxable income for tax purposes.

c. Determine the corporation's total taxes payable assuming the federal tax rate is 38 percent, the corporate surtax is 4 percent, the provincial abatement is 10 percent and the provincial tax rate is 15 percent.

2A-5A. *(Taxable Income)* The Robbins Corporation is an oil wholesaler. The company's sales last year were $1 million, with the cost of goods sold equal to $600,000. The firm paid interest of $200,000, and its cash operating expenses were $100,000. Depreciation expense was reported at $150,000, but the firm claimed $125,000 in capital cost allowance during the fiscal year. Furthermore, the firm received $40,000 in dividend income while paying only $10,000 in dividends to its preferred shareholders.

a. Given that Robbins' combined corporate tax rate is 45 percent, determine the corporation's net income by illustrating the Income Statement.

b. Determine the corporation's taxable income for tax purposes.

2A-6A. *(Corporate Income Tax)* The Fair Corporation had sales of $5 million this past year. The cost of goods sold was $4.3 million and cash operating expenses were $100,000. Depreciation expense was reported at $100,000, but the firm claimed $125,000 in capital cost allowance. Dividend income totalled $50,000. The firm sold an old machine to terminate the asset class and incurred a recaptured capital cost allowance of $120,000. This sale of the old machine also created a capital gain of $50,000 and an accounting gain of $70,000. The firm received $150 per share from the sale of 1,000 shares of stock. The stock was purchased for $100 per share three years ago.

a. Given that Fair's combined corporate tax rate is 45 percent, determine the corporation's net income by illustrating the Net Income Statement.

b. Determine the corporation's taxable income for tax purposes.

c. Determine the corporation's total taxes payable assuming the federal tax rate is 38 percent, the corporate surtax is 4 percent, the provincial abatement is 10 percent, and the provincial tax rate is 15 percent.

2A-7A. *(Taxable Income)* Sales for J. P. Hulett, Inc., during the past year amounted to $4 million. The firm supplies statistical information to engineering companies. Gross profits totalled $1 million while cash operating and depreciation expense were $500,000 and $350,000, respectively. The firm reported $340,000 in capital cost allowance for tax purposes. Dividend income for the year was $12,000.

a. Given that J. P. Hulett's combined corporate tax rate is 45 percent, determine the corporation's net income by illustrating the Income Statement.

b. Determine the corporation's taxable income for tax purposes.

2A-8A. *(Corporate Income Tax)* Anderson, Inc., sells computer software. The company's past year's sales were $5 million. The cost of its merchandise sold came to $3 million. Cash operating expenses were $175,000, plus depreciation expense totaling $125,000. The firm reported $135,000 in capital cost allowance for tax purposes. An old machine was sold which created a terminal loss of $25,000 and an accounting loss of $30,000. Furthermore, the firm paid $200,000 interest on loans. The firm sold stock during the year, receiving a $40,000 gain on a stock owned for six months but losing $60,000 on stock held four months.

a. Given that Anderson's combined corporate tax rate is 45 percent, determine the corporation's net income by illustrating the Income Statement.

b. Determine the corporation's taxable income for tax purposes.

c. Determine the corporation's total taxes payable assuming the federal tax rate is 38 percent, the corporate surtax is 4 percent, the provincial abatement is 10 percent, and the provincial tax rate is 15 percent.

2A-9A. *(Corporate Income Tax)* G. R. Edwin, Inc., reported both a net income of $203,500 and a tax provision of $166,500 for accounting purposes. The firm reported neither depreciation expense nor capital cost allowance.

a. Determine the corporation's total taxes payable assuming the federal tax rate is 38 percent, the corporate surtax is 4 percent, the provincial abatement is 10 percent and the provincial tax rate is 15 percent.

b. Given the information in Part (a) of this problem, determine the firm's combined corporate tax rate.

STUDY PROBLEMS (SET B)

2A-1B. *(Taxable Income)* The M. M. Roscoe Corporation is a large clothing retailer. During the most recent year the company generated sales of $4 million. The combined cost of goods sold and the operating expenses were $3.2 million. The firm reported neither depreciation expense nor capital cost allowance for the fiscal year. Also, $300,000 in interest expense was paid during the year. The firm received $5,000 during the year in dividend income from 1,000 common shares of a Canadian corporation that had been purchased three years previously. However, these common shares were sold toward the end of the year for $100 per share; its initial cost was $80 per share. The company also sold land that had been recently purchased and had been held for only four months. The selling price was $55,000; the cost was $45,000.

a. Given that M. M. Roscoe's combined corporate tax rate is 45 percent, determine the corporation's net income by illustrating the Income Statement.

b. Determine the corporation's taxable income for tax purposes.

2A-2B. *(Taxable Income)* Sales for J. P. Enterprises during the past year amounted to $5 million. Gross profits for the year were $2.5 million. Operating expenses totalled $900,000. The firm reported neither depreciation expense nor capital cost allowance for the fiscal year. The interest and dividend income from securities owned of a Canadian corporation were $15,000 and $20,000, respectively. The firm's interest expense was $100,000. The firm sold securities on two occasions during the year, realizing a gain of $45,000 on the first sale but losing $60,000 on the second.

a. Given that J. P. Enterprises' combined corporate tax rate is 45 percent, determine the corporation's net income by illustrating the Income Statement.

b. Determine the corporation's taxable income for tax purposes.

2A-3B. *(Corporate Income Tax)* Carter B. Daltan, Inc., sells minicomputers. During the past year the company's sales were $3.5 million. The cost of its merchandise sold came to $2 million, and cash operating expenses were $500,000. The firm reported depreciation expense was $150,000, but had a capital cost allowance of $150,000 for the fiscal year. Carter B. Daltan, Inc., paid $165,000 in interest on bank loans. Also, the corporation received $55,000 in dividend income from another Canadian corporation but paid $25,000 in the form of dividends to its own common shareholders.

a. Given that Carter B. Daltan's combined corporate tax rate is 45 percent, determine the corporation's net income by illustrating the Income Statement.

b. Determine the corporation's taxable income for tax purposes.

c. Determine the corporation's total taxes payable assuming the federal tax rate is 38 percent, the corporate surtax is 4 percent, the provincial abatement is 10 percent, and the provincial tax rate is 16 percent.

2A-4B. *(Corporate Income Tax)* Kate Menielle, Inc., had sales of $8 million during the past year. The company's cost of goods sold was 60 percent of sales; cash operating expenses of $750,000; and depreciation expense of $150,000. The firm claimed a capital cost allowance of $275,000 for tax purposes. In addition, the firm sold a capital asset (stock) for $75,000, which had been purchased five months earlier at a cost of $80,000.

a. Given that Kate Menielle's combined corporate tax rate is 45 percent, determine the corporation's net income by illustrating the Income Statement.

b. Determine the corporation's taxable income for tax purposes.

c. Determine the corporation's total taxes payable assuming the federal tax rate is 38 percent, the corporate surtax is 4 percent, the provincial abatement is 10 percent, and the provincial tax rate is 15 percent.

2A-5B. *(Taxable Income)* The Burgess Corporation is an oil wholesaler. The company's sales last year were $2.5 million, with the cost of goods sold equal to $700,000. The firm paid interest of $200,000, and its cash operating expenses were $150,000. Depreciation expense was reported at $50,000, but the firm claimed $25,000 in capital cost allowance during the fiscal year. Furthermore, the firm received $50,000 in dividend income while paying only $15,000 in dividends to its preferred shareholders.

a. Given that Burgess's combined corporate tax rate is 45 percent, determine the corporation's net income by illustrating the Income Statement.

b. Determine the corporation's taxable income for tax purposes.

2A-6B. *(Corporate Income Tax)* The Aka Corporation had sales of $5.5 million this past year. The cost of goods sold was $4.6 million and cash operating expenses were $125,000. Depreciation expense was reported at $75,000, but the firm claimed $100,000 in capital cost allowance. Dividend income totalled $50,000. The firm sold an old machine to terminate the asset class and incur a recaptured capital cost allowance of $140,000. This sale of the old machine also created a capital gain of $50,000 and an accounting gain of $70,000. The firm received $125 per share from the sale of 1,000 shares of stock. The stock was purchased for $100 per share three years ago.

a. Given that Aka's combined corporate tax rate is 45 percent, determine the corporation's net income by illustrating the Income Statement.

b. Determine the corporation's taxable income for tax purposes.

c. Determine the corporation's total taxes payable assuming the federal tax rate is 38 percent, the corporate surtax is 4 percent, the provincial abatement is 10 percent, and the provincial tax rate is 15 percent.

2A-7B. *(Taxable Income)* Sales for Diana Schubert, Inc., during the past year amounted to $5 million. The firm supplies statistical information to engineering companies. Gross profits totalled $1.2 million while cash operating and depreciation expense was $500,000 and $400,000, respectively. The firm reported $300,000 in capital cost allowance for tax purposes. Dividend income for the year was $15,000.

a. Given that Diana Schubert's combined corporate tax rate is 45 percent, determine the corporation's net income by illustrating the Income Statement.

b. Determine the corporation's taxable income for tax purposes.

2A-8B. *(Corporate Income Tax)* Williams, Inc., sells computer software. The company's past year's sales were $4.5 million. The cost of its merchandise sold came to $2.2 million. Cash operating expenses were $175,000, plus depreciation expense totaling $130,000. The firm reported $135,000 in capital cost allowance for tax purposes. An old machine was sold which created a terminal loss of $25,000 and an accounting loss of $30,000. Furthermore, the firm paid $200,000 interest on loans. The firm sold stock during the year, receiving a $40,000 gain on a stock owned for six months but losing $60,000 on stock held four months.

a. Given that Williams' combined corporate tax rate is 45 percent, determine the corporation's net income by illustrating the Income Statement.

b. Determine the corporation's taxable income for tax purposes.

c. Determine the corporation's total taxes payable assuming the federal tax rate is 38 percent, the corporate surtax is 4 percent, the provincial abatement is 10 percent, and the provincial tax rate is 15 percent.

2A-9B. *(Corporate Income Tax)* J. Johnson, Inc., reported both a net income of $198,000 and a tax provision of $162,000 for accounting purposes. The firm reported neither depreciation expense nor capital cost allowance.

a. Determine the corporation's total taxes payable assuming the federal tax rate is 38 percent, the corporate surtax is 4 percent, the provincial abatement is 10 percent, and the provincial tax rate is 15 percent.

b. Given the information in Part (a) of this problem, determine the firm's combined corporate tax rate.

Risk and Rates of Return

CHAPTER 3

LEARNING OBJECTIVES

After reading this chapter you should be able to

1. Examine the three elements that define the value of an asset.
2. Define and measure the expected rate of return of an individual investment.
3. Define and measure the riskiness of an individual investment.
4. Explain how diversifying investments affects the riskiness and expected rate of return of a portfolio of assets.
5. Measure the *market risk* of an individual asset.
6. Calculate the *market risk* of a portfolio of investments.
7. Explain the relationship between an investor's *required* rate of return on an investment and the riskiness of the investment.
8. Explain recent criticisms of the capital asset pricing model.
9. Examine the historical relationship between risk and rates of return for various types of securities.

INTRODUCTION

In 1987 Arnab Banerji started recommending investments in the stock markets of developing countries to his clients. He might as well have been trying to sell a cure for baldness.

"Some people thought I was a smooth salesman," says the chief investment officer of Foreign & Colonial Emerging Markets, Ltd., in London. "Others thought I had a good idea, but intrinsically they couldn't believe it, because it didn't fit with their experiences. There was a lot of curiosity, but nobody was putting money on the table."

He recalls one potential investor who, after arguing that countries like Malaysia, Indonesia, Mexico, and Argentina had been falling behind the industrialized West for three centuries, told him: "I don't think they are going to reverse that in the next thirty years."

Another potential client called Mr. Banerji into his office and showed him a wall covered with defaulted bonds issued in the nineteenth and early twentieth centuries by Russia, China, and several South American countries. Paraphrasing Karl Marx, the man warned: "History repeats itself—the first time as tragedy, the second as farce." Attitudes have changed a lot since 1987. Mr. Banerji now heads a twelve-person team that manages $1.5 billion in twenty-seven emerging markets, including Portugal, Hungary, China, Thailand, Chile and Uruguay. But the fundamental issue is still the same: How do investors assess risk and expected returns from their investments—in this case, investments in companies in developing countries where the uncertainty is truly significant.[1]

[1]Adapted from Michael R. Sesit, "Flocking to the Frontier," *The Wall Street Journal*, September 24, 1993, R4.

CHAPTER PREVIEW

In the next four chapters, we examine the principles of valuation and then apply these principles to the valuation of financial instruments. In Chapter 3, we begin by defining the concepts of cash flow, expected return and risk. We show how each of these concepts is measured quantitatively. We then explain how diversifying our investment, and how the length of time we hold an investment, can affect the expected return and riskiness of our investments. We also examine some of the recent criticisms of our approach to measuring risk. We consider how the riskiness of an investment should affect the required rate of return on an investment. Finally, we show what historical rates of return could have earned on different types of investments if we had invested over a long period of time.

BACK TO THE BASICS

This chapter has one primary objective, that of helping us understand ***Axiom 1: The Risk-Return Tradeoff—We Won't Take on Additional Risk Unless We Expect to Be Compensated with Additional Return.***

OBJECTIVE 1

THE BASICS OF VALUATION

Intrinsic value

The present value of the asset's expected future cash flows. This value is the amount the investor considers to be fair value, given the amount, timing, and the riskiness of future cash flows.

In the following chapters, we will refer to the value of an asset as its **intrinsic value**, which is the present value of its expected future cash flows, where these cash flows are discounted back to the present using the investor's required rate of return. In other words, value is affected by three elements:

1. The amount and timing of the asset's expected cash flows.
2. The riskiness of these cash flows.
3. The investor's required rate of return for undertaking the investment.

The first two factors are characteristics of an asset, while the required rate of return is the minimum rate necessary to attract an investor to purchase or hold a security. This rate must be high enough to compensate the investor for the risk perceived in the asset's future cash flows.

These factors are depicted in Figure 3-1. As the figure shows, the procedure for finding the value of an asset involves the following:

1. Assessing the asset's characteristics, which include the amount and timing of the expected cash flows and the riskiness of these cash flows.

2. Determining the investor's required rate of return, which embodies the investor's attitude for assuming risk and perception of the riskiness of the asset.

3. Discounting the expected cash flows back to the present, using the investor's required rate of return. In this chapter, we examine both the riskiness of cash flows and the investor's required rate of return. In the next chapter, we examine the mechanics of how cash flows are brought back to the present.

OBJECTIVE 2

EXPECTED RETURN

The expected benefits or returns that an investment generates come in the form of cash flows. Cash flows, not accounting profits, are the relevant variable that the financial manager uses to measure returns. Although accounting information is vital in our study of a company's finances, a large body of evidence suggests that

BACK TO THE BASICS

Our discussion should remind us of three of our axioms that help us understand finance:

Axiom 1: The Risk-Return Tradeoff—We Won't Take on Additional Risk Unless We Expect to Be Compensated with Additional Return.

Axiom 2: The Time Value of Money—A Dollar Received Today Is Worth More Than a Dollar Received in the Future.

Axiom 3: Cash—Not Profits—Is King: Measuring the Timing of Costs and Benefits

Determining the economic worth or value of an asset always relies on these three axioms. Without them, we would have no basis for explaining value. With them, we can know that the amount and timing of cash, not earnings, drive value. Also, we must be rewarded for taking risk; otherwise, we will not invest.

investors or shareholders look behind accounting numbers to discover the underlying economic reality of an investment's or a firm's performance. For example, the economic reality of a firm's operations is ultimately reflected in the cash flow it provides to its investors, not its reported earnings. Specifically, we will identify the firm's free cash flow to be the amount of cash that is available for distributing to its investors not only after operating expenses have been paid, but after any investments have been made that will generate future cash flows. Free cash flow is defined as follows:

$$\text{Free cash flow} = \text{revenues} - \text{expenses} - \text{additional investment in assets} \quad (3\text{-}1)$$

In the case of an investment into a firm, the free cash flow available for distribution to investors is:

$$\text{Free cash flow for a firm} = \text{operating income} + \text{depreciation and amortization expenses} - \text{income taxes} - \text{additional investment in assets} \quad (3\text{-}1a)$$

We note here that there is a difference between the ideas of free cash flow and incremental cash flow. An incremental cash flow measures the difference between

Figure 3-1 Basic Factors Determining an Asset's Value

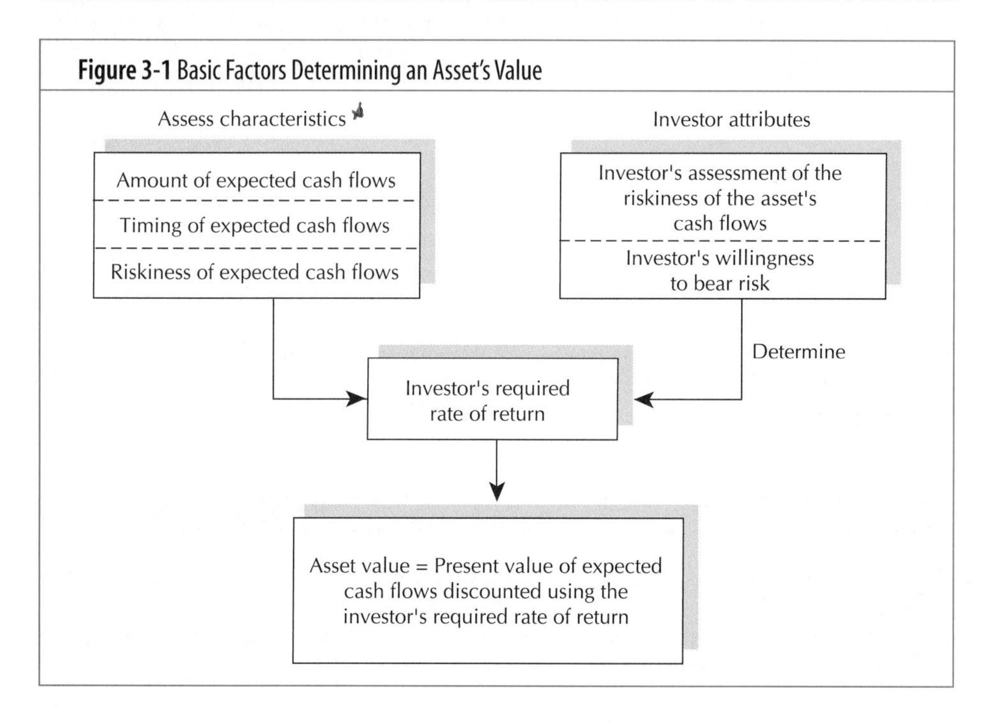

the cash flows if the project is taken on versus what they will be if the project is not taken on. Remember **Axiom 4: Incremental Cash Flows—It's Only What Changes That Counts.** In contrast, free cash flow measures the difference that occurs when subtracting expenses and any additional investment in assets from revenues. By measuring the benefits and costs of an investment in terms of cash, we are able to determine an investment's return. This principle of using cash flows to measure returns holds true regardless of the type of investment, whether it is an investment into the firm's operations or an investment into a debt instrument, preferred share, or common share.

BACK TO THE BASICS

*Think cash flows, not reported earnings, when calculating rates of return. Remember **Axiom 3: Cash—Not Profits—Is King: Measuring the Timing of Costs and Benefits.***

In an uncertain world, an accurate measurement of expected future cash flows is frequently not easy to ascertain for an investor. To illustrate: Assume you are considering an investment costing $10,000, where the future cash flows from owning the security depend upon the state of the economy, and these cash flows have been estimated in Table 3-1.

Thus, in any given year, the investment could produce any one of three possible cash flows depending upon the particular state of the economy. With this information, how should we select the cash flow estimate that means the most in measuring the investment's expected rate of return? Our approach to answering this question relies on descriptive statistics. (A brief review of probability distributions and descriptive statistics is presented in Appendix 3A.) This approach provides an objective method of calculating the expected cash flow.[2] The expected cash flow is simply the weighted average of the possible cash flow outcomes where the weights are the probabilities of the occurrence of the various states of the economy. Let X_i designate the ith possible cash flow; N reflects the number of possible states of the economy, and $P(X_i)$ indicates the probability that the ith cash flow or state of economy will occur. The expected cash flow, $\overline{X}$, may then be calculated as follows:

$$\overline{X} = X_1P(X_1) + X_2P(X_2) + \ldots + X_NP(X_N) \tag{3-2}$$

$$\text{or } \overline{X} = \sum_{i=1}^{N} X_iP(X_i)$$

For the present illustration

$$\overline{X} = (.5)(\$1{,}400) + (.3)(\$1{,}200) + (.2)(\$1{,}000) = \$1{,}260$$

Expected rate of return

The weighted average of all possible returns where the returns are weighted by the probability that each will occur.

In addition to computing an expected dollar return from an investment, we can also calculate an **expected rate of return** earned on the $10,000 investment on the security. The last column in Table 3-1 shows that the $1,400 cash inflow, assuming strong economic growth, represents a 14 percent return ($1,400 ÷ $10,000). Similarly, the $1,200 and $1,000 cash flows result in 12 percent and 10 percent returns, respectively. Using these percentage returns in place of the dollar amounts, the expected rate of return, $\overline{k}$, is expressed as follows:

$$\overline{k} = k_1P(k_1) + k_2P(k_2) + \ldots + k_nP(k_n) \tag{3-3}$$

$$\text{or } \overline{k} = \sum_{i=1}^{N} k_iP(k_i)$$

In our example:

$$\overline{k} = (.5)(14\%) + (.3)(12\%) + (.2)(10\%) = 12.6\%$$

[2]The expected value of cash flows is dependent on the different possible cash flows that occur because of an event and the likelihood or probability of the event occurring.

Table 3-1 Measuring the Expected Return

State of the Economy	Probability of the States[a]	Cash Flows from the Investment	Percentage Returns (Cash Flow/ Investment Cost)
Economic recession	20%	$1,000	10% ($1,000/$10,000)
Moderate economic growth	30%	1,200	12% ($1,200/$10,000)
Strong economic growth	50%	1,400	14% ($1,400/$10,000)

[a]The probabilities assigned to the three possible economic conditions have to be determined subjectively, which requires management to have a thorough understanding of both the investment cash flows and the general economy.

With our concept and measurement of expected returns, let's consider the other side of the investment coin: risk.

RISK

OBJECTIVE 3

To gain a basic understanding of investment risk, there are at least three fundamental questions that we must ask:

1. What is risk?
2. How do we know the amount of risk associated with a given investment. That is, how do we measure risk?
3. If we choose to diversify our investments by owning more than one asset, as most of us do, will such diversification impact the riskiness of our combined portfolio of investments?[3]

PERSPECTIVE IN FINANCE

Risk is the potential variability in future cash flows. That variability reflects risk can easily be shown by examining a coin toss game. Consider the possibility of flipping a coin—heads you win, tails you lose—for 25 cents with your finance professor. Most likely you would be willing to take on this game, indicating that the utility gained from winning 25 cents is about equal to the utility lost if you lose 25 cents. On the other hand, if the flip is for $1,000 you may be willing to play only if you are offered more than $1,000 if you win, say $1,500 if it turns out heads and lose $1,000 if it turns out tails. In each case the probability of winning and losing is the same; that is, there is an equal chance that the coin will land heads or tails; but the width of the dispersion changes, and that is why the second coin toss is more risky and why it will not be taken unless the payoffs are altered. The key here is the fact that only the dispersion changes; the probability of winning or losing is the same in each case. The wider the range of possible events that can occur, the greater the risk.

To help us grasp the fundamental meaning of risk, consider two possible investments:

1. The first investment is a Government of Canada Treasury bill, a government security that matures in three months and promises to pay an annualized return of 8 percent. If we purchase and hold this security for three months, we are virtu-

[3]The logic in this section is the same as that used in Chapter 15 for capital budgeting under uncertainty.

ally assured of receiving no more and no less than an annualized return of 8 percent. For all practical purposes, the risk of loss is nonexistent.[4]

2. The second investment involves the purchase of common shares of a local publishing company. Looking at the past returns of this firm's common shares, we have made the following estimate of the annual returns from the investment:

Chance of Occurrence	Rate of Return on Investment
1 chance in 10 (10%)	0%
2 chances in 10 (20%)	5%
4 chances in 10 (40%)	15%
2 chances in 10 (20%)	25%
1 chance in 10 (10%)	30%

Investing in the publishing company could conceivably provide a return as high as 30 percent if all goes well or no return (zero percent) if everything goes against the firm. However, in future years, both good and bad, we could expect a 15 percent return on average.[5]

$$\bar{k} = (.10)(0\%) + (.20)(5\%) + (.40)(15\%) + (.20)(25\%) + (.10)(30\%) = 15\%$$

Comparing the Treasury bill investment with the publishing company investment, we see that the Treasury bills offer an expected 8 percent rate of return, while the publishing company has an expected rate of return of 15 percent. However, our investment in the publishing firm is clearly more "risky"—that is, it has greater uncertainty as to the final outcome. Stated somewhat differently, there is a greater variation or dispersion of possible returns, which in turn implies greater **risk**.[6] Figure 3-2 shows these differences in the form of discrete probability distributions.

Risk

The prospect of an unfavourable outcome. This concept has been measured operationally as the standard deviation or beta, which will be explained later.

Standard deviation

A measure of the spread or dispersion about the mean of a probability distribution. We calculate it by squaring the difference between each outcome and its expected value, weighting each squared difference by its associated probability, summing over all possible outcomes, and taking the square root of this sum.

Coefficient of variation

A measure of relative dispersion of a probability distribution. This measure is defined as the standard deviation divided by the expected value.

Although the return from investing in the publishing firm is clearly less certain than for Treasury bills, quantitative measures of risk are useful when the difference between two investments is not so evident. Examples of such measures are the **standard deviation** (σ) and the **coefficient of variation** (γ) (see Appendix 3A for a review of these descriptive statistics). The standard deviation is simply the square root of the "average squared deviation of each possible return from the expected return."

That is

$$\sigma = \sqrt{\sum_{i=1}^{N} (k_i - \bar{k})^2 P(k_i)} \qquad (3\text{-}4)$$

where N = the number of possible outcomes or different rates of return on the investment

k_i = the value of the ith possible rate of return

$\bar{k}$ = the expected value of the rates of return

$P(k_i)$ = the chance or probability that the ith outcome or return will occur.

[4]A three-month Treasury bill is essentially riskless in terms of any change of default by the Government of Canada, but the presence of inflation places the investor at some risk in knowing with certainty the eventual real return, that is, the return after adjusting for loss of purchasing power.

[5]We assume that the particular outcome or return earned in one year does not affect the return earned in the subsequent year. Technically speaking, the distribution of returns in any year is assumed to be independent of the outcome in any prior year.

[6]How can we possibly view variations above the expected return as risk? Should we not be concerned only with the negative deviations? Some would agree and view risk as only the negative variability in returns from a predetermined minimum acceptable rate of return. However, as long as the distribution of returns is symmetrical, the same conclusions will be reached.

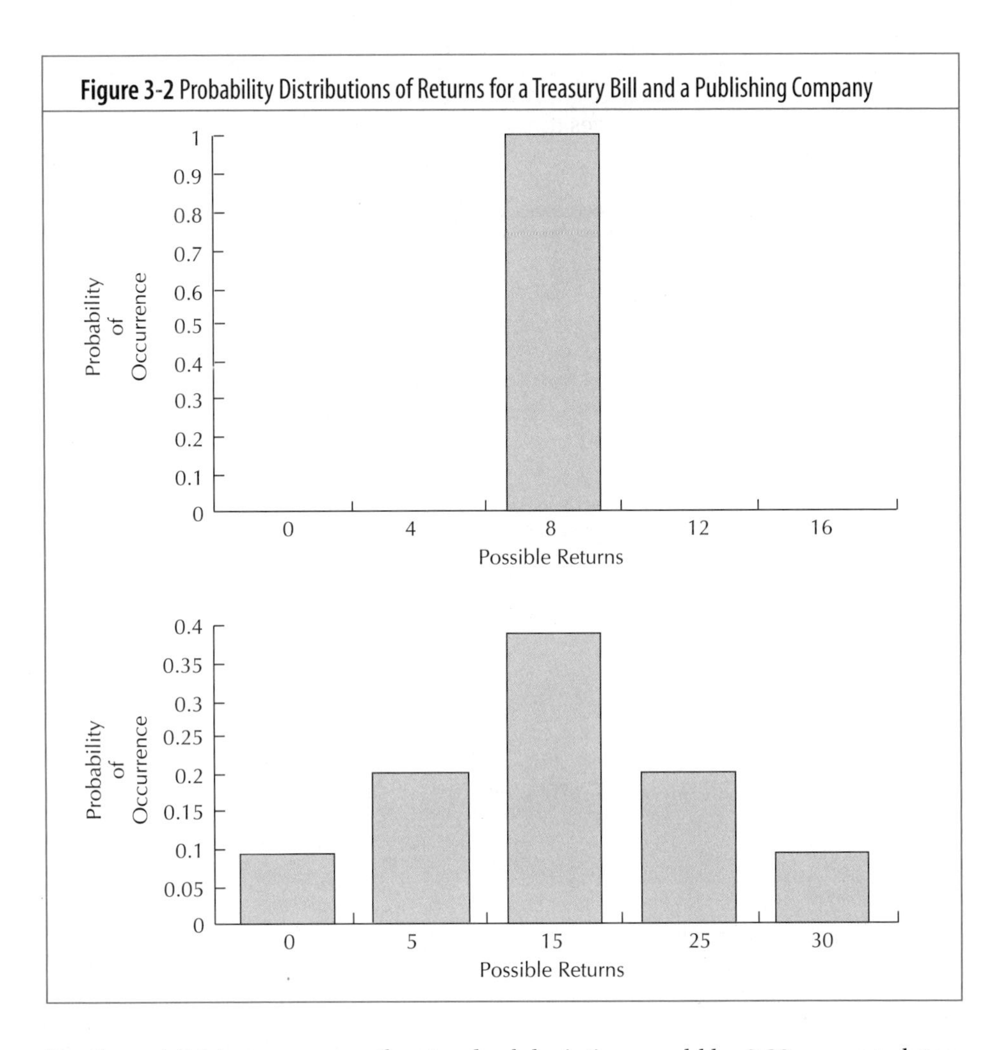

For the publishing company, the standard deviation would be 9.22 percent, determined as follows:

$$\sigma = [(0\% - 15\%)^2(.10) + (5\% - 15\%)^2(.20) + (15\% - 15\%)^2(.40) + (25\% - 15\%)^2(.20) + (30\% - 15\%)^2(.10)]^{(1/2)}$$
$$= \sqrt{85\%} = 9.22\%$$

Although the standard deviation of returns provides us with a quantitative measure of an asset's riskiness, how should we interpret the results? What does it mean? Is the 9.22 percent standard deviation for the publishing company investment good or bad? We can answer this question only by comparing the investment in the publishing firm against other investments. The attractiveness of a security with respect to its return and risk cannot be determined in isolation. Only by examining other available alternatives can we reach a conclusion about a particular investment's risk. For example, if another investment, say an investment in a firm that owns a local radio station, has the same expected return as the publishing company (15 percent) but with a standard deviation of 7 percent, we would consider the risk associated with the publishing firm (9.22) to be excessive. This type of behaviour is characterized as risk aversion in finance such that risk-averse investors seek to maximize their returns given investment choices of equal risk and the corollary, they seek to minimize risk given investment choices of equal return. We could verify this decision quantitatively by determining the coefficient of variation which provides us with a measure of the relative dispersion of a probability distribution—that is, the risk per unit of return. Mathematically, it is defined as the standard deviation divided by the expected value, or

$$\gamma = \frac{\sigma}{\bar{k}} \quad (3\text{-}5)$$

In our example for the publishing firm,

$$\gamma = \frac{9.22}{15} = .61$$

In our example for the firm which owns the local radio station,

$$\gamma = \frac{7}{15} = .47$$

As a result, the investment into the publishing firm has a greater risk per unit return than that of the investment into the firm which owns the local radio station. In common sense terms, this condition is due to the fact that the radio company investment has the same expected return as the publishing company investment but is less risky.

The attractiveness of the radio company investment over the publishing company investment, which has the same expected return but greater risk, is readily apparent. The real difficulty in selecting the "better" investment comes when one investment has a higher expected return but also exhibits greater risk. The coefficient of variation is especially attractive if the projects have unequal returns and standard deviations. This quantitative measure uses the ratio of standard deviation to expected return to compare the relative riskiness of two projects. Thus, the use of the coefficient of variation along with the standard deviation is especially important when two projects of unequal returns are being compared. To ignore the coefficient of variation in this case could lead to misconceptions as to the relative level of uncertainty contained in each project.

PERSPECTIVE IN FINANCE

The first Chinese symbol represents danger, the second stands for opportunity. The Chinese define risk as the combination of danger and opportunity. Greater risk, according to the Chinese, means we have greater opportunity to do well, but also greater danger to do badly.

OBJECTIVE 4

RISK AND EXPECTED RETURN OF A PORTFOLIO

Our analysis has focused on the risk and return associated with each individual investment or project under consideration. However, firms do not just invest in a single investment but are generally involved in a number of investments at any one time. As a result, we examine how the expected return and the risk of a portfolio of investments can be quantified.

Expected Return of a Portfolio

Suppose we are considering combining two assets, putting two-thirds of our money in investment A which has an expected return of 12 percent and the remaining third in investment B which has an expected return of 18 percent. The expected return of this two-asset portfolio is the weighted average of the expected returns from each investment. In this case of portfolios the weighting is defined in terms of the percentages of money invested in each investment: two-thirds of our money in investment A and one-third in investment B. The equation for the expected return of a portfolio is defined as follows:

$$\bar{k}_p = (W_A)(\bar{k}_A) + (W_B)(\bar{k}_B) \tag{3-6}$$

where $\bar{k}_p$ = the expected return of portfolio

W_A = the weight or fraction of total funds invested in asset A

$\bar{k}_A$ = the expected return of investment A

W_B = the weight or fraction of total funds invested in asset B

$\bar{k}_B$ = the expected return of investment B

For our example,

$$\bar{k}_p = (.67)(12\%) + (.33)(18\%)$$

$$\bar{k}_p = 14\%$$

Thus, our portfolio of two assets has an expected return of 14 percent.

Risk of a Portfolio

Risk has been defined as the variability of future returns and can be measured by the standard deviation. Let's consider for the moment how risk is affected if we diversify our investment by holding a variety of investments. The degree to which the diversification affects a portfolio is a function of the relationship between the pattern of cash flows from the new investment and the existing investment. For example, let us look at a company that doubles its size by adding a new division. First, assume that the cash flows from both the original company and the new division move directly with the business cycle. Time Chart A of Figure 3-3 illustrates that the cash flows of the firm and the division move together and there is no diversification effect. In other words, the effect of cash flows moving together causes their percentage variability to remain the same. In this case, diversification has no effect on the portfolio. If, on the other hand, the new division produced cash flows that were countercyclical, the two series of cash flows would move in opposite

Figure 3-3 Diversification Effects on Cash Flow

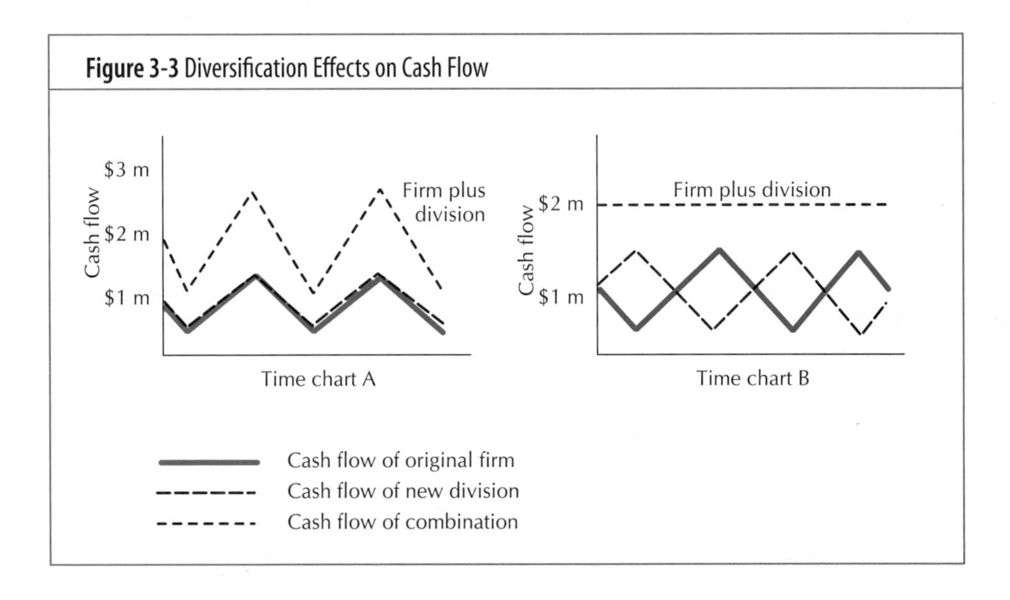

directions, and diversification would affect the portfolio. Time Chart B of Figure 3-3 illustrates the diversification effects on the cash flows of the firm and the division. In this case, because the variability of the cash flows of the original firm and the new division move in opposite directions, the variability of their combination is totally eliminated.

Obviously, the degree to which the total risk is reduced becomes a function of how the two sets of cash flows or returns move together. In order to measure this relationship, we use the concept of correlation. Correlation measures the linear relationship between any two sets of values. The correlation coefficient, a statistical measure, tells us two things about this relationship. First, it can take on either a positive or a negative sign. A positive sign indicates that the variables tend to move in the same direction; a negative sign, that they move in opposite directions. Also, the value that the correlation coefficient takes on, ranging from +1.0 down to –1.0, provides us with information about the degree or strength of this relationship. If its value is near zero, the linear relationship is weak; however, if the relationship is close to either +1.0 or –1.0, there is a strong linear relationship between the two variables. Letting ρ denote the correlation coefficient, k_p the return on the new investment, and k_{Firm} the return on the firm, Figure 3-4 graphically illustrates several possible situations.

The degree of correlation between investments takes on importance because it determines the extent to which the diversification portfolio effect will reduce the risk level of the combination. Negatively correlated combinations will provide the greatest level of risk reduction, whereas perfectly positively correlated investments, those with correlation coefficients of +1.0, will not provide any diversification effect. If one thinks about it, it makes sense. Any impact from diversification results from combining investments with divergent return or cash flow patterns. If the correlation coefficient is 1.0, these patterns are identical; hence, there is no diversification value present. If, on the other hand, the correlation coefficient is –1.0, then the patterns move in opposite directions; hence, the diversification value is maximum.

Choosing Among Portfolios

We now illustrate how modern portfolio theory provides an investor with valuable insight for choosing among portfolios. The selection process requires that we account for return, as measured by expected return, and risk, as measured by the standard deviation. For purposes of simplicity, our portfolio will be formed from a combination of two assets.

The mathematical formula for the standard deviation of this two-asset combination is given in Equations (3-7a) and (3-7b).

$$\sigma_{A+B} = \sqrt{Variance\,(A+B)} \tag{3-7a}$$

$$\sigma_{A+B} = \sqrt{Variance\,(A) + Variance\,(B) + Covariance\,(A+B)} \tag{3-7b}$$

It should be noted that the mathematical formula for covariance is given in Equation (3-8).

$$\text{Covariance (A + B)} = \sqrt{\sigma_{\text{A}}\sigma_{\text{B}}\rho_{\text{AB}}} \tag{3-8}$$

As a result of Equation (3-8), the mathematical formula from Equation (3-7b) can be rewritten to give the mathematical formula for the standard deviation of a two-asset combination as shown in Equation (3-9).

$$\sigma_{A+B} = \sqrt{W_A^2\sigma_A^2 + W_B^2\sigma_B^2 + 2W_A W_B\,\rho_{AB}\sigma_A\sigma_B} \tag{3-9}$$

where

σ_{A+B} = the standard deviation of the portfolio that combines asset A and asset B

W_A = the proportion or fraction of total funds invested in asset A

W_B = the proportion or fraction of total funds invested in asset A

Figure 3-4 Illustrations of Various Degrees of Correlation

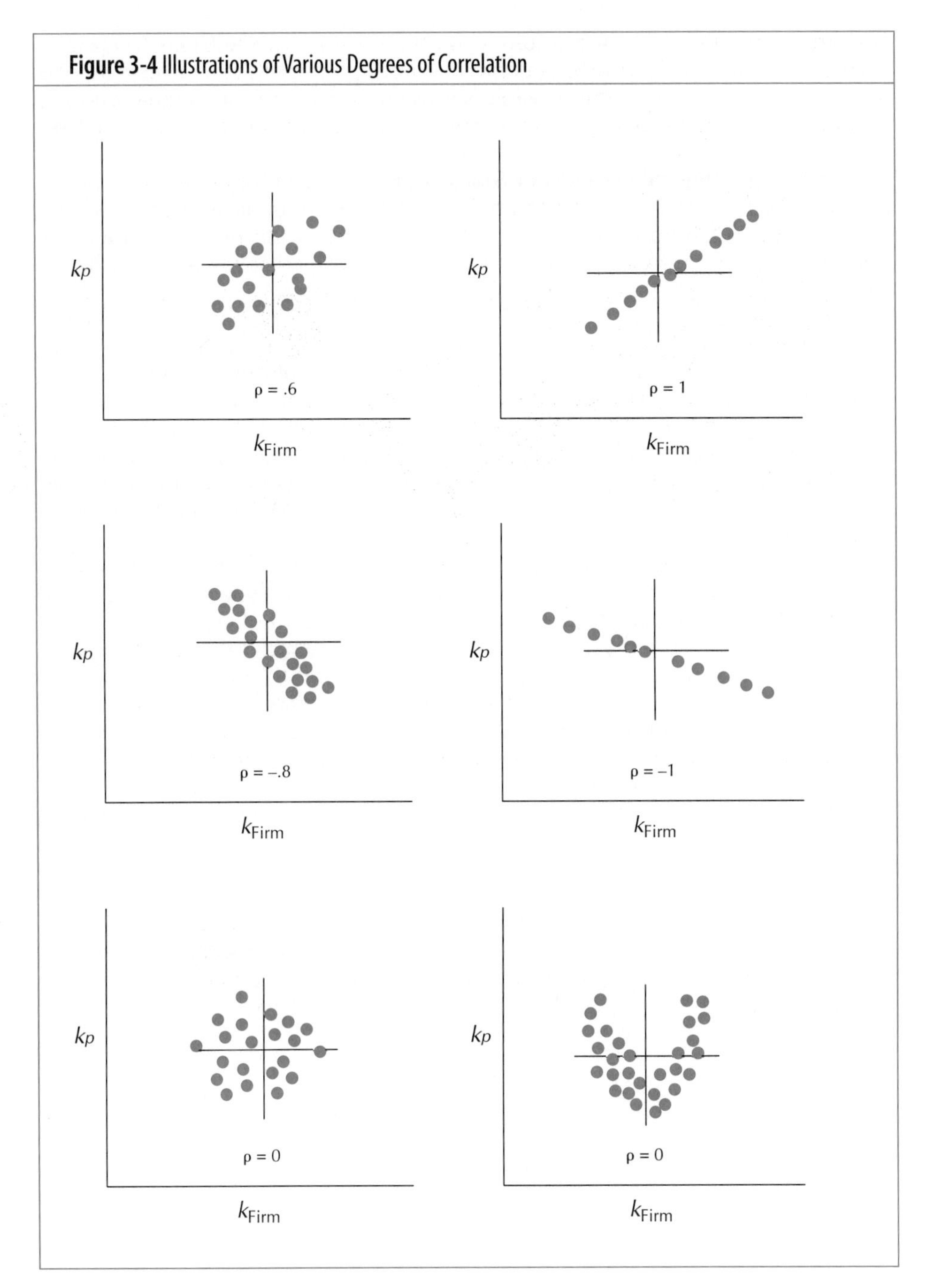

σ_A = the standard deviation of asset A

σ_B = the standard deviation of asset B

ρ_B = the correlation coefficient between assets A and B

In forming a two-asset portfolio, an investor can create an infinite number of portfolios by varying the proportion of funds to be invested between the first asset and the second asset. For example, an investor may invest 60 percent of the funds to be invested in the first asset, while investing the remaining portion, 40 percent, of the amount in the second asset. With a two-asset portfolio, the number of possible portfolios that can be created by an investor forms a feasible set and is described by either a line or a curve. However, a feasible set or opportunity set is extended to a larger space when we consider all the portfolios that can be formed from a group of

N securities. In Figure 3-5, the feasible set is characterized by all the possible portfolios which lie on or within the boundaries denoted by points K, M, U, and V. A point outside the boundaries of the feasible set is not considered attainable because investors cannot replicate the proportions of funds invested in N assets that are required to generate the appropriate levels of expected return and risk.

Although the feasible set of portfolios represents a considerable number of investment opportunities, investors are only interested in those portfolios that are efficient.[7] For a set of portfolios to be efficient it must meet *both* of the following criteria:

1. Offer maximum expected return for varying levels of risk.
2. Offer minimum risk for varying levels of expected return.[8]

In Figure 3-6, we see that point K offers the least amount of risk of all the points in the feasible set. Point K offers minimum risk as compared to points along the line horizontal to point K, which have the same levels of expected return in the feasible set. As a result, the portfolio represented by point K is said to be efficient. In a similar manner, point V offers the maximum expected return of all the points in the feasible set. Point V offers the maximum expected return as compared to points along a line vertical to point V, which have the same levels of risk in the feasible set. As a result, the portfolio represented by point V is also said to be efficient. In fact, all points which rest on the boundary between points K and V represent portfolios that are efficient. In contrast, point W lies within the feasible set but represents an inefficient portfolio because portfolio O shows a greater expected return for the same amount of risk; whereas portfolio P shows a lesser amount of risk for the same amount of expected return. In modern portfolio theory, both portfolios O and P are said to dominate over portfolio W. In this example, portfolios W, M, and U are inefficient. Whereas, all portfolios on the boundary between points K and V form the

Figure 3-5 The Feasible Set

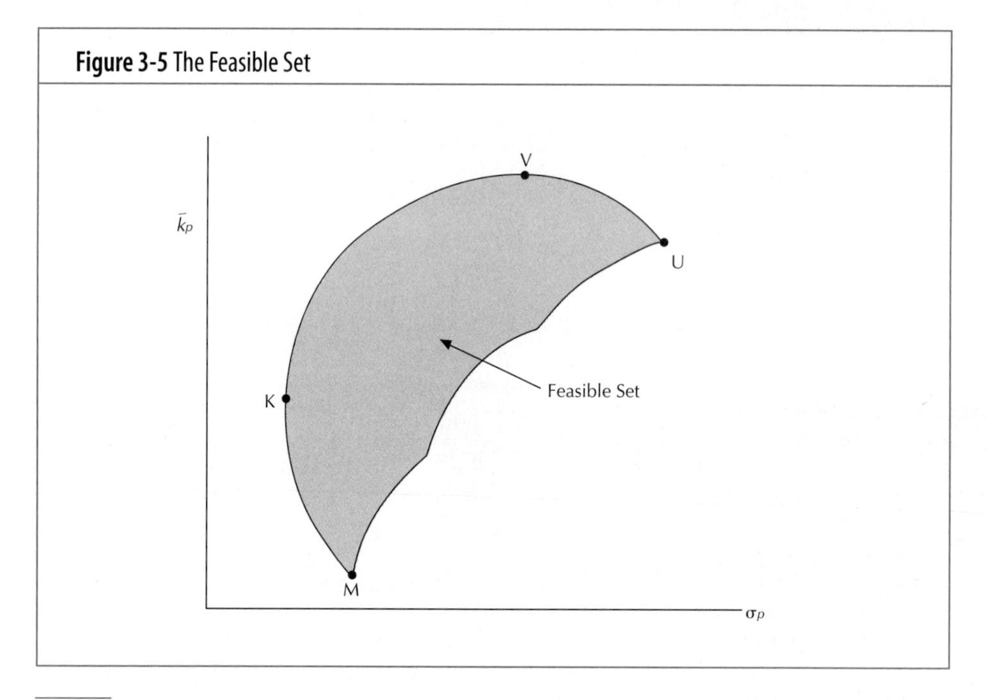

[7]The efficient set theorem makes two important assumptions about an investor: non-satiation and risk aversion. The non-satiation assumption indicates that investors always prefer higher levels of terminal wealth to lower levels of terminal wealth. The risk aversion assumption indicates that an investor will always choose, between two portfolios having the same level of expected return, that portfolio which has the smaller standard deviation.

[8]For further discussion on portfolio analysis, see: W. Sharpe, G. Alexander and D. Fowler, *Investments*, first Canadian ed. (Scarborough, ON: Prentice-Hall Canada, 1993).

Figure 3-6 The Efficient Frontier and the Feasible Set

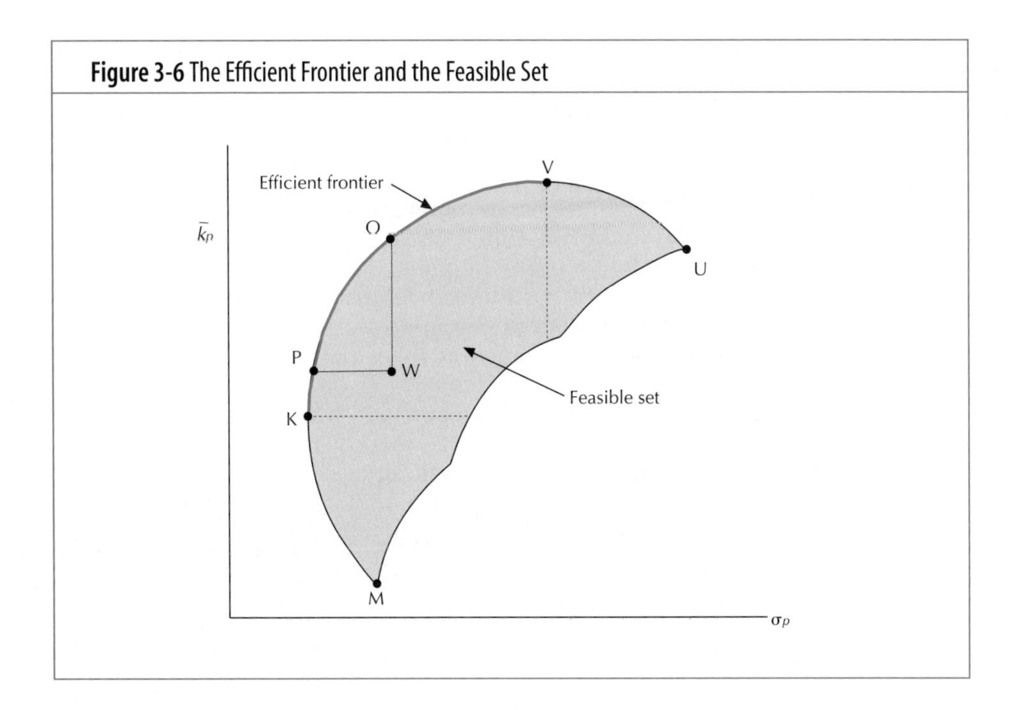

set of efficient portfolios or the efficient frontier.

Risk and Diversification

You may recall the disaster at the Union Carbide plant in Bhopal, India, that occurred in the summer of 1984. Thousands of individuals were killed or injured. While the financial considerations cannot compare to the loss of life, if you had invested your life savings of $400,000 in Union Carbide shortly before the Bhopal tragedy, the value of your savings would have declined by 40 percent to $240,000 almost overnight.[9] Or what if you had been fortunate enough to buy some Pennzoil shares in November 1985 immediately before Pennzoil was awarded $10.5 billion by the courts against Texaco for alleged misconduct in "tortuously" outbidding Pennzoil in Texaco's acquisition of Getty Oil Company? Based upon this single event, Pennzoil shares almost doubled from $46 per share to $90 per share in a matter of days. Texaco stock, on the other hand, declined from $38 to $25.[10]

The events affecting Union Carbide, Texaco, and Pennzoil were unique to those firms; they had little, if any, impact on other companies. Other examples of events that are unique to a single company include labour strikes, the discovery of a new product or the sudden obsolescence of an old one, errors in judgment by a firm's management regarding a large capital investment, and the resignation or death of a key executive.

How do you feel about these unique events? Obviously, if you had been a shareholder in Pennzoil you would have felt great, but you might have decided to invest in Texaco (not a bad company), or even worse, Union Carbide. Most of us would like to avoid such fluctuations; that is, we are risk-averse. Wouldn't it be great if we could reduce the risk associated with our portfolio, without having to accept a lower expected return? Good news: it is possible!

---

[9]See "Union Carbide Shares Acquired by Bass Group," *The Wall Street Journal,* January 21, 1985, pp. 2–3. We must add that the price of the stock recovered in about six months.

[10]See "Pennzoil Rejects Texaco Offer," *The Wall Street Journal,* January 8, 1986, p. 3. The final settlement after Texaco filed for bankruptcy was $3 billion.

Partitioning the Risk

If we diversify our investments across different securities, rather than investing in only one security, the variability in the returns of our portfolio should decline. The reduction in risk will occur if the security returns within our portfolio do not move precisely together over time—that is, if they are not perfectly correlated. Figure 3-7 shows graphically what we might expect to happen to the variability of returns as we add additional securities to the portfolio.[11] The reduction occurs because some of the volatility in returns of a security are unique to that security. The unique variability of a single security tends to be countered by the uniqueness of another security.[12] However, we should not expect to eliminate all risk from our portfolio. In practice, it would be rather difficult to cancel all the variations in returns of a portfolio because share prices have some tendency to move together. Thus, we can divide the total risk (total variability) of our portfolio into two types of risk: (1) **firm-specific** or **company-unique risk** and (2) **market-related risk**. Company-unique risk might also be called **diversifiable risk**, since it can be diversified away. Market risk is **nondiversifiable risk**; it cannot be eliminated, no matter how much we diversify. Figure 3-7 illustrates these two types of risk. Total risk decreases until we have approximately 20 securities, and then the decline becomes very slight.[13] The remaining risk is the portfolio's market risk. At this point, our portfolio is highly correlated with all securities in the marketplace. Events that affect our portfolio now are not so much unique events, but changes in the general economy, major political events, and sociological changes. Examples would include changes in general interest rates, or changes in tax legislation that affect companies, or increasing concerns by the public about business practices on the environment.

Firm-specific risk or company-unique risk (diversifiable risk or unsystematic risk)

The portion of the variation of the investment returns that can be eliminated through investor diversification. The diversifiable risk is the result of factors that are unique to the particular firm.

Market-related risk (systematic risk or nondiversifiable risk)

The portion of variations in investment returns that cannot be eliminated through investor diversification. This variation results from factors that affect all stocks.

Since we can remove the unique, or unsystematic risk, there is no reason to believe that the market will reward us with additional returns for assuming risk that

Figure 3-7 Variability of Returns Compared with Size of Portfolio

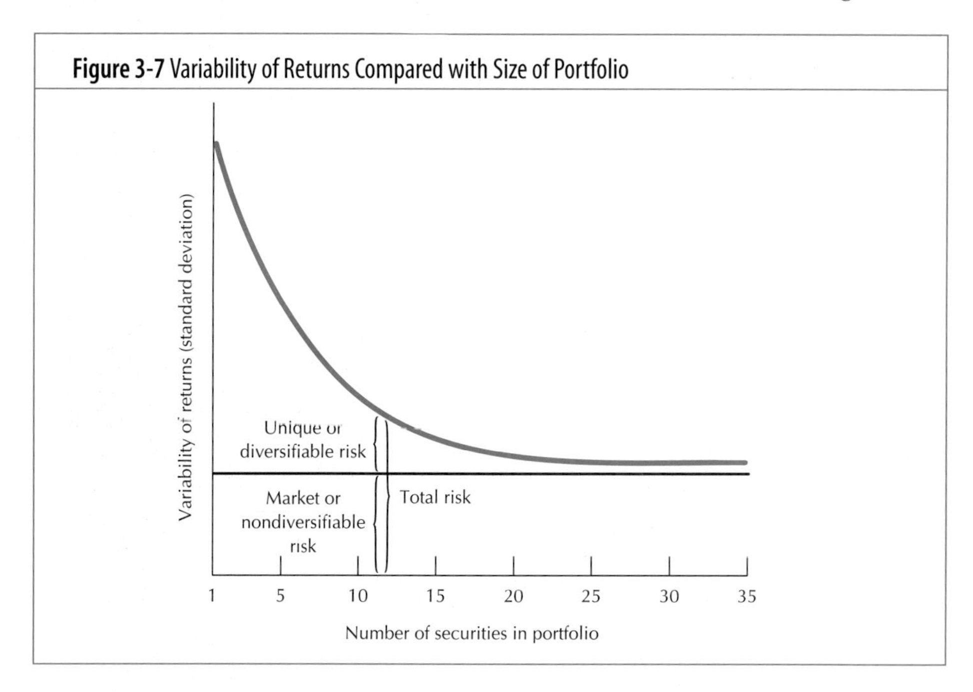

[11]The desire to reduce risk through diversification is the underlying motivation for the creation of mutual funds.

[12]This concept will also be presented in Chapter 15 with respect to the risk of capital investments.

[13]A number of studies have noted that portfolios consisting of approximately 20 randomly selected common securities have virtually no unique or diversifiable risk. See Robert C. Klemkosky and John D. Martin, "The Effect of Market Risk on Portfolio Diversification," *The Journal of Finance*, 14 (March 1975), pp. 147–154.

could be avoided by simply diversifying. Our new measure of risk should therefore measure how responsive a security or portfolio is to changes in a "market portfolio," such as the TSE 300 Index.[14] This relationship could be determined by plotting past returns, say on a monthly basis, of a particular security or a portfolio of shares against the returns of the "market portfolio" for the same period.

Measuring Market Risk: An Example

OBJECTIVE 5

To help clarify the idea of systematic risk, let us examine the relationship between the common stock returns of Spar Aerospace and the returns of the TSE 300 Index. Table 3B-1 of Appendix B presents the monthly returns for Spar Aerospace and for the TSE 300 Index for 60 months ending December 1995. These monthly returns, or **"holding-period returns,"** as we oftentimes call them, are calculated as follows:[15]

Holding-period return

The return an investor would receive from holding a security for a designated period of time. For example, a monthly holding-period return would be the return for holding a security for a month.

$$k_t = \left(\frac{P_t}{P_{t-1}}\right) - 1 \qquad (3\text{-}10)$$

where, k_t = the holding-period return in month t for Spar Aerospace (or for the TSE 300 Index) and
P_t = the price of Spar Aerospace's share (or the TSE 300 Index) at the end of month t.

For instance, the holding-period return for Spar Aerospace and the TSE 300 Index for February 1994 is computed as follows:

$$\text{Spar Aerospace return for February 1994} = \frac{\text{Share price end of February 94}}{\text{Share price end of January 94}} - 1$$

$$= \left(\frac{18.50}{20.25}\right) - 1 = -8.64\%$$

$$\text{TSE 300 Index return for February 1994} = \frac{\text{Index value end of February 94}}{\text{Index value end of January 94}} - 1$$

$$= \left(\frac{4423.8}{4554.9}\right) - 1 = -2.88\%$$

Table 3-2 shows the computed averages of the returns and the standard deviation of these returns for both Spar Aerospace and the TSE 300 Index. Since we are using historical data for returns, we assume that each observation has an equal probability of occurrence. Thus, the average return, $\bar{k}$, is found by summing the returns and dividing by the number of months,

$$\text{Average return} = \frac{\sum_{t=1}^{n} \text{returns in month } t}{\text{number of months}}$$

and the standard deviation is computed as:

$$\text{Standard deviation} = \sqrt{\frac{\sum_{t=1}^{n} (\text{return in month } t - \text{average return})^2}{\text{number of months} - 1}}$$

[14]The Toronto Stock Exchange (TSE 300) Composite Index is a stock index that measures the combined performance of 300 companies in Canada. The Toronto Stock Exchange sets the criteria for inclusion in the TSE 300 Composite Index. In contrast, the New York Stock Exchange Index is an index that reflects the performance of all stocks listed on the New York Stock Exchange.

[15]For simplicity's sake, we are ignoring the dividend that the investor receives from the share as part of the total return. That is, letting D_t equal the dividend received by the investor in month t, the holding-period return would more accurately be measured as:

$$k_t = \frac{P_t + D_t}{P_{t-1}} - 1 \qquad (3\text{-}10A)$$

Table 3-2 Statistics for Monthly Holding-Returns for TSE 300 Index and Spar Aerospace, January 1991 to December 1995

	TSE 300 Index	Spar Aerospace
Average Monthly Holding-Return	0.66%	1.72%
Standard Deviation	3.02%	11.98%
Correlation Coefficient	0.18	

The average monthly return for Spar Aerospace and the TSE 300 Index are found to be 1.72 percent and 0.66 percent, respectively. We also see that Spar Aerospace has experienced greater volatility of returns over the five years, or a standard deviation of 11.98 percent compared to 3.02 percent for the TSE 300 Index. As explained earlier, we could avoid some of this risk (reduce the standard deviation) by diversifying our portfolio and owning common shares other than Spar Aerospace.

Characteristic line

The line of "best fit" through a series of returns for a firm's stock relative to the market returns. The slope of the line, frequently called beta, represents the average movement of the firm's stock returns in response to a movement in the market's returns.

Beta

A measure of the relationship between an investment's returns, and the market returns. This is a measure of the investment's nondiversifiable risk.'

Figure 3-8 illustrates the relation between Spar Aerospace's returns and the TSE 300 Index returns. If we draw a line of "best fit" through these returns, the slope of the line is 0.72. The slope of this line of best fit, which we call the **"characteristic line,"** indicates the average movement in the share price of Spar Aerospace in response to a movement in the general market portfolio (TSE 300 Index). The slope of the characteristic line, the ratio of the rise of the line relative to the run of the line, is called **"beta"** in finance.[16]

PERSPECTIVE IN FINANCE

The slope of the characteristic line is called "beta" and it is a measure of a systematic or market risk. The slope of the characteristic line indicates the average response of a share's returns to the change in the market as a whole.

OBJECTIVE 6

Measuring a Portfolio's Beta

Figure 3-8 shows that on average a 1.0 percent change in the return of the TSE 300 Index causes a change of 0.72 percent in the return of Spar Aerospace. Yet, we also see a lot of fluctuation around the characteristic line.[17] However by diversifying our holdings to own shares of 20 companies with betas of about 0.72, we could essentially eliminate the variation of the line. We would remove almost all the volatility in returns, except for what is caused by the general market, which is represented by the slope of the line. If we plotted the returns of our 20-stock portfolio against the TSE 300 Index, the points of our new graph would fit nicely along a straight line with a slope of 0.72. The new graph would look something like the one shown in Figure 3-9.

One remaining question needs to be addressed. Assume we want to diversify our portfolio, as we have just suggested, but instead of acquiring securities with the same beta as Spar Aerospace (0.72), we buy shares of 8 companies with betas of 0.72 and shares of 12 companies with betas of 1.5. What would the beta of our portfolio become? As it works out, the **"beta portfolio"** is simply the average of the individual betas of the companies' shares. Actually, the portfolio beta is a weighted average of the individual security's betas, whereby the weights are equal to the pro-

[16]For our purposes, we are primarily interested in understanding the concept of beta. Appendix 3B demonstrates how to calculate the beta of Spar Aerospace. Betas are also reported by investment services, for example: the *Polymetric Report* lists betas for Canadian companies and *Merril Lynch, Pierce, Fenner & Smith Investment Survey* lists betas for companies in the United States.

[17]A security with a zero beta has no systematic risk, while a security with a beta of one has systematic or market risk equivalent with the "typical" security in the marketplace. Most securities have betas between 0.60 and 1.60.

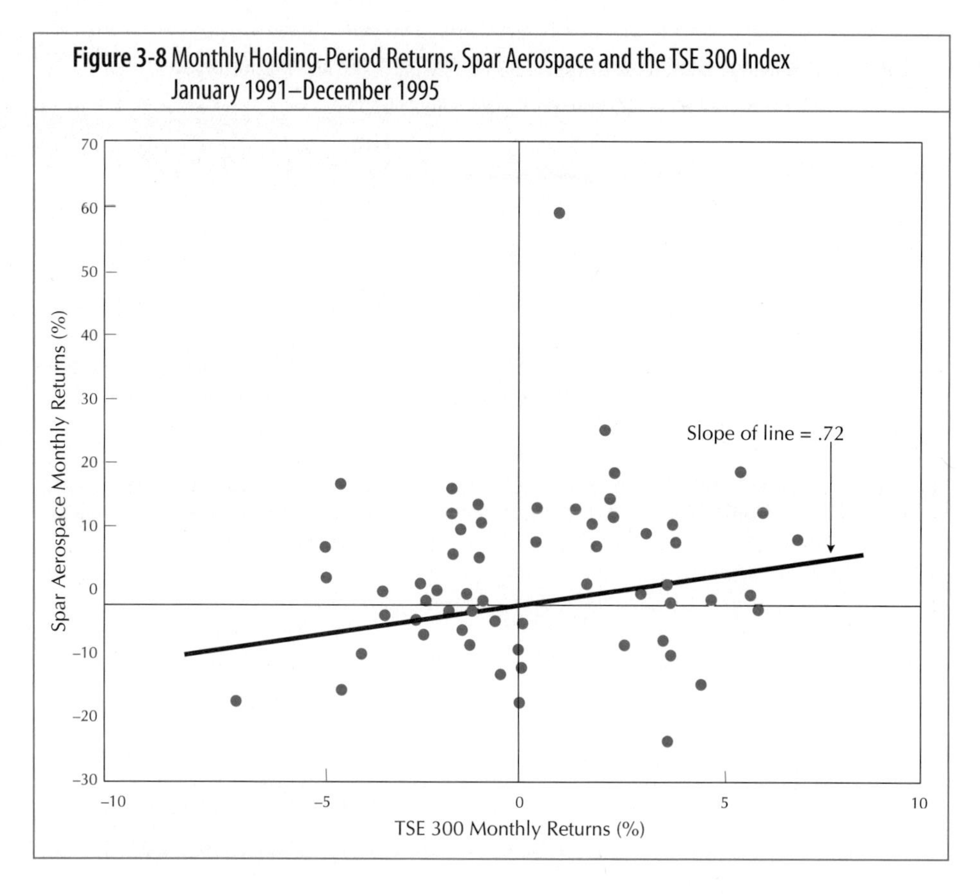

Figure 3-8 Monthly Holding-Period Returns, Spar Aerospace and the TSE 300 Index January 1991–December 1995

portion of the portfolio invested in each security. Thus, the beta (β) of a portfolio consisting of *N* securities is equal to:

$$\beta_{\text{Portfolio}} = \sum_{j=1}^{N} (\text{percentage invested security } j) \times (\text{security } j\text{'s } \beta) \quad (3\text{-}11)$$

Portfolio beta

The relationship between a portfolio's returns and the market returns. Beta is a measure of the portfolio's nondiversifiable risk.

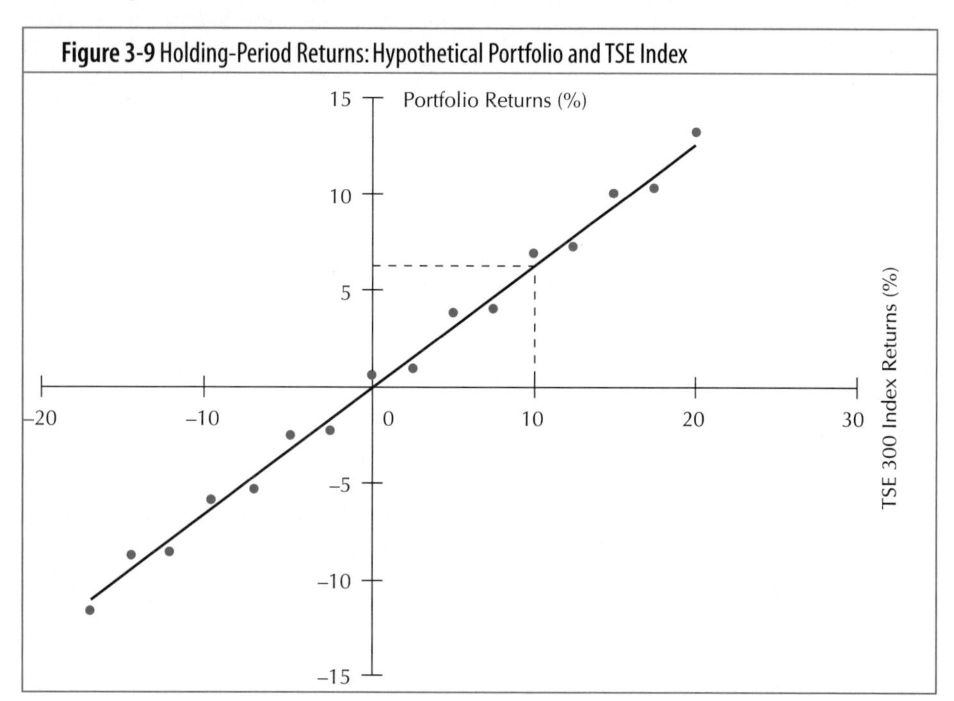

Figure 3-9 Holding-Period Returns: Hypothetical Portfolio and TSE Index

Figure 3-10 Holding Period Returns: High and Low Beta Portfolios and TSE 300 Index

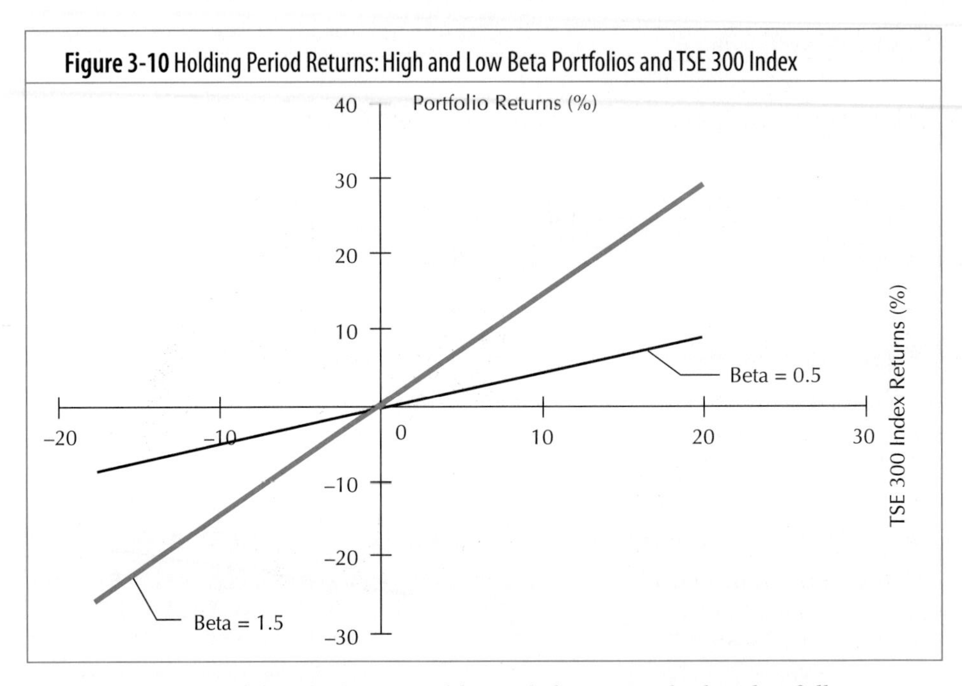

So for our new portfolio, the beta would simply be 1.19, calculated as follows:

$$\beta_{\text{Portfolio}} = \left(\frac{8}{20} \times .72\right) + \left(\frac{12}{20} \times 1.50\right)$$
$$= 1.19$$

An increase or decrease of one percent in the returns of the general market portfolio (TSE 300 Index) would on average change 1.19 percent, meaning that our new portfolio has more systematic or market risk than has the market portfolio.

So we can conclude that the beta of a portfolio is determined by the betas of the individual security betas. If we have a portfolio consisting of securities with low betas, then our portfolio will have a low beta. The reverse of this observation also holds true. Figure 3-10 illustrates these situations.

Although portfolio betas tend to be stable, individual betas are not necessarily stable and not always particularly meaningful. For example, in the 1960s Meade Johnson had a negative beta following its takeover by Bristol-Myers. Apparently, Meade Johnson introduced a product called "Metrecal," a dietary supplement that Meade Johnson sold to customers, who drank this instead of eating their lunches. In any case, the public loved it, and Meade Johnson's shares shot up in price just as the market sank into a deep slump. As the market rebounded in 1963 and 1964, the "Metrecal" fad died and Meade Johnson dropped in price, again moving in an opposite direction from the market. Later in the 1960s, just as the market began to drop, Meade Johnson reintroduced the exact same product, this time called "Nutrament," telling consumers to buy it and drink it in addition to their normal lunch to put on weight. The result of all this was that Meade Johnson had a negative beta. Needless to say, it would be unfortunate if capital budgeting decisions were made using Meade Johnson's beta as the yardstick by which they were measured. The point here is that betas for individual stocks are not always reliable. In fact, many individual securities have coefficients of determination of 5 percent or less, indicating that the characteristic line with slope of beta explains only 5 percent or less of the variability of the security's returns. The average coefficient of determination is 30 percent.

The concept of beta is an underlying concept used in measuring a security's risk. It also proves useful when we attempt to specify what the relationship should be between an investor's required rate of return and the security's market risk.

BACK TO THE BASICS

We can reduce risk through diversifying our portfolio, but only to a point. What we remove is company-unique or specific risk (also known as diversifiable or unsystematic risk). Systematic risk or market risk (also termed nondiversifiable risk) cannot be eliminated. Thus, we are reminded of ***Axiom 9: All Risk Is Not Equal—Some Risk Can Be Diversified Away, and Some Cannot.***

THE INVESTOR'S REQUIRED RATE OF RETURN

OBJECTIVE 7

In this section we examine the concept of the investor's required rate of return, especially as it relates to the riskiness of the asset, and then we see how the required rate of return is measured.

The Required Rate of Return Concept

The **investor's required rate of return** may be defined as the minimum rate of return necessary to attract an investor to purchase or hold a security. This definition considers the investor's opportunity cost of making an investment; that is, if an investment is made, the investor must forgo the return available from the next best investment. This "forgone return" is an **opportunity cost** of undertaking the investment and consequently is the investor's required rate of return. In other words, we invest with the intention of achieving a rate of return sufficient to warrant making the investment. The investment will be made only if the purchase price is low enough relative to the expected future cash flows to provide a rate of return greater than or equal to our required rate of return.

Investor's required rate of return

The minimum rate of return necessary to attract an investor to purchase or hold a security. This rate of return is also the discount rate that equates the present value of the cash flows with the value of the security.

Opportunity cost of funds

The next best rate of return available to the investor for a given level of risk.

To help us better understand the nature of an investor's required rate of return, we may separate the return into its basic components: the risk-free rate of return plus a risk premium. The investor's required rate of return is expressed as follows:

$$k = k_f + k_{rp} \tag{3-12}$$

where, k = the investor's required rate of return
k_f = the risk-free return
k_{rp} = the risk premium

The **risk-free or riskless rate of return** rewards us for deferring consumption, and not for assuming risk; that is, the risk-free return reflects the basic fact that we invest today so that we can consume more later. By itself, the risk-free rate should be used only as the required rate of return, or discount rate, for riskless investments. Typically, our measure for the risk-free rate of return is the Government of Canada Treasury bill rate.

Risk-free or riskless rate of return

The rate of return on risk-free investments. The interest rate on three-month Government of Canada Treasury bills is commonly used to measure this rate.

The **risk premium**, k_{rp}, is the additional return we must expect to receive for assuming risk. As the level of risk increases, we will demand additional expected returns. Even though we may or may not actually receive this incremental return, we must have reason to expect it; otherwise, why should we expose ourselves to the chance of losing all or part of our money?

Risk premium

The additional rate of return we expect to earn above the risk-free rate for assuming risk.

EXAMPLE

To demonstrate the concept of required rate of return, let us take Bell Canada, which has bonds that mature on May 1, 2000. Based upon the market price of these bonds in May of 1995, we can determine that investors are expecting an 8 percent return. The three-month Treasury bill rate at that time was about 7

percent, which means that Bell Canada bondholders were requiring a risk premium of 1 percent.[18] Stated as an equation, we have

$$\text{required rate} = \text{risk-free rate } (k_f) + \text{risk premium } (k_{rp})$$
$$= 7\% + 1\%$$
$$= 8\%$$

Measuring the Required Rate of Return

Capital asset pricing model (CAPM)

An equation stating that the expected rate of return on a project is a function of (1) the risk-free rate, (2) the investment's systematic risk, and (3) the expected risk premium for the market portfolio of all risky securities.

The investor's required rate of return has been described in Equation (3-12) by the following terms: (1) systematic risk is the relevant source of risk and (2) the required rate of return, k, equals the risk-free rate, k_f, plus a risk premium, k_{rp}. However, we have difficulty in applying this equation to real-world examples because of the problems in estimating the risk premium. In an attempt to overcome these difficulties, the **capital asset pricing model (CAPM)** offers an intuitive approach for thinking about the return that an investor requires on an investment, given the asset's systematic or market risk.

Equation (3-12) provides the natural starting point for measuring the investors' required rate of return. Rearranging this equation to solve for the risk premium (k_{rp}), we have

$$k_{rp} = \bar{k} - k_f \tag{3-13}$$

which simply says that the risk premium for a security, k_{rp}, equals the security's expected return, $\bar{k}$, less the risk-free rate existing in the market, k_f. For example, if the expected return for a security is 15 percent and the risk-free rate is 7 percent, the risk premium is 8 percent. Also, if the expected return for the market, k_m, is 12 percent, and the risk-free rate, k_f, is 7 percent, the risk premium, k_{rp}, for the general market would be 5 percent. This 5 percent risk premium would apply to any security having systematic (nondiversifiable) risk equivalent to the general market, or a beta of 1.

In this same market, a security with a beta of 2 should provide a risk premium of 10 percent, or twice the 5 percent risk premium existing for the market as a whole.[19] Hence, in general, the appropriate required rate of return for the jth security, k_j, should be determined by

$$k_j = k_f + \beta_j(k_m - k_f) \tag{3-14}$$

Security market line

The return line reflects the attitudes of investors regarding the minimal acceptable return for a given level of systematic risk.

Equation (3-14) is the capital asset pricing model (CAPM). This equation designates the risk-return tradeoff existing in the market, where risk is defined in terms of beta. Figure 3-11 illustrates the CAPM as the **security market line.**[20] As presented in this figure, securities with betas equal to 0, 1, and 2 should have required rates of return as follows:

If $\beta_j = 0.0$: $k_j = 7\% + 0.0(12\% - 7\%) = 7$ percent
If $\beta_j = 1.0$: $k_j = 7\% + 1.0(12\% - 7\%) = 12$ percent
If $\beta_j = 2.0$: $k_j = 7\% + 2.0(12\% - 7\%) = 17$ percent

---

[18]The risk premium here can be thought of as a composite of a "default risk premium" (reflected in the difference in the bond's rate of return and the rate on a similar maturity government bond) and "term structure" premium (reflected in the difference in the three-month Treasury bill rate and the long-term government bond rate).

[19]Empirical evidence suggests that betas for individual securities are highly unstable over time. Thus, when you calculate a beta for a firm based on historical data, you have no assurance that the beta will remain the same in the next period.

[20]Two key assumptions are made in using the security market line. First, we assume that the marketplace where securities are bought and sold is highly efficient. Market efficiency indicates that the price of an asset responds quickly to new information, thereby suggesting that the price of a

where the risk-free rate, k_f, is 7 percent and the expected market return, k_m, is 12 percent.[21]

BACK TO THE BASICS

The conclusion of the matter is that ***Axiom 1*** *is alive and well. It tells us,* ***We Won't Take an Additional Risk Unless We Expect to Be Compensated with Additional Return.*** *That is, there is a risk-return tradeoff in the market.*

THE FAMA AND FRENCH ATTACK ON THE CAPM

OBJECTIVE 8

The primary implication of the capital asset pricing model (CAPM) is that higher returns accrue to securities which have higher levels of nondiversifiable or systematic risk (measured by the security's beta coefficient). However, recent evidence by Fama

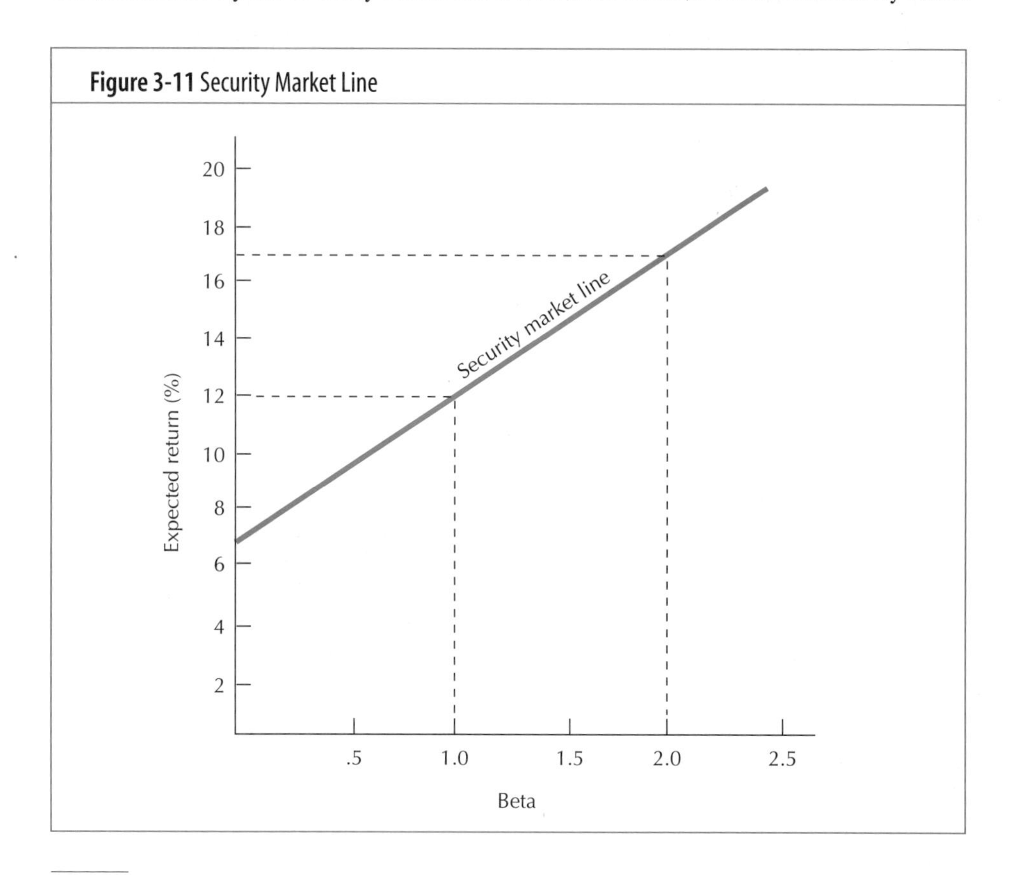

Figure 3-11 Security Market Line

security reflects all available information. As a result, the current price of a security is considered to represent the best estimate of its future price. Second, the model assumes that a perfect market exists. A perfect market is one in which information is readily available to all investors at a nominal cost. Also, securities are assumed to be infinitely divisible, with any transaction costs incurred in purchasing or selling a security being negligible. Furthermore, investors are assumed to be single-period wealth maximizers who agree as to the meaning and the significance of the available information. Finally, within the perfect market, all investors are "price takers," which simply means that a single investor's actions cannot affect the price of a security. These assumptions are obviously not descriptive of reality. However, from the perspective of positive economics, the mark of a good theory is the accuracy of its predictions and not the validity of the simplifying assumptions that underlie its development.

[21]For more detailed treatment of this model, see B. Rosenberg, "The Capital Asset Pricing Model and the Market Model," *Journal of Portfolio Management* (Winter 1981) pp. 5–16.

and French indicates that over the period 1963–90, differences in beta do not explain differences in the performance (rates of returns) of stocks.[22] Further, they found that the total market value of the firm's equity and the ratio of the firm's equity book value to its equity market value explain a large portion of the variation in stock returns, much larger than beta! From this evidence they conclude that "beta is dead."

The Defence of the CAPM

Two counter-arguments have been offered in response to the Fama and French criticisms of CAPM and beta. First, some argue that Fama and French's research methodology was flawed; others contend that Fama and French's contentions are theoretically incomplete. The bottom line is that CAPM is not dead but our faith in its ability to explain the world of risk and return has been shaken.

Two research studies have addressed the "empirical refinement" line of defence of the CAPM. Chan and Lakonishok look at a much longer series of returns than Fama and French.[23] They evaluate the entire period 1926–91 and found that for the period ended in 1982, higher betas were indeed associated with higher returns. However, for the period after 1982 they found again that beta and stock returns were unrelated. Furthermore, when they looked at the returns of the ten worst and ten best months in terms of the stock market's performance, they found that stocks with higher betas did worse in the worst months and better in the best months than their lower-beta counterparts. Thus, we have some limited evidence supporting the basic CAPM prediction that higher betas are associated with higher stock returns.

The second empirical study is by Kothari, Shankin and Sloan. These authors reexamined the issue as to whether beta explains variation in average returns over the post-1940 period as well as the longer post-1926 period.[24] They, like Fama and French before them, tested the notion that book-to-market value of the firm's equity captures the variation in average returns over a longer period, 1947–87, using a somewhat different data set than used by Fama and French. Two of their conclusions are as follows:

- They found a 6 percent risk premium in the sample of stocks used, compared to Fama and French's results indicating there was no risk premium relative to increasing systematic risk.
- Using a different sample of firms in their study, they found that book-to-market value was only weakly related to average stock returns, which sharply disagrees with Fama and French. Based upon this observation they argue that the Fama-French results are limited to their sample of companies, and do not necessarily apply to a broader sample of companies.

The theoretical argument against the Fama and French allegations was offered by Roll and Ross.[25] They argue that using an index, such as the Standard & Poor 500 Index or New York Stock Exchange Index, to calculate the market returns may not accurately reflect the returns of the true market portfolio. Since the market portfolio proxy used by Fama and French may not be the correct one, there is no reason to believe that the betas and rates of return should be positively related.[26] Thus the Fama and French results may either be the result of using an incorrect market port-

---

[22] Eugene Fama and Kenneth French, "The Cross-Section of Expected Stock Returns," *Journal Of Finance* 48, no. 2 (June 1992), pp. 427–465.

[23] Louis Chan and Joseph Lakonishok, "Are the Reports of Beta's Death Premature?" University of Illinois Working Paper (December 1992).

[24] S. P. Kothari, J. Shankin and R. Sloan, "Another Look at the Cross-Section of Expected Stock Returns," *Journal of Finance* 50, no. 1 (March 1995).

[25] Richard Roll and Stephen A. Ross, "On the Cross-Sectional Relation Between Expected Returns and Betas," UCLA Working Paper (July 8, 1992).

[26] Technically, Roll and Ross noted that where the market portfolio is mean-variance efficient, beta and expected returns are exactly linear and positively related. Where the market portfolio is even slightly inefficient this positive relation no longer holds.

folio proxy or due to a failure of the model to explain any relationship between risk and return. We cannot tell which.

Weighing the Evidence

So what are we to conclude? Is beta dead? At the very least this latest salvo of criticism has forced the academic community to again come to grips with the fundamental shortcomings of the CAPM. The model, like all models which attempt to explain complex real-world phenomena using simplifying assumptions, is an abstraction and does not completely and perfectly "fit the facts" as to the way the world works. Does this mean that the model lacks usefulness? We think not. The model points toward the need to diversify and identifies the source of the market risk premium as being tied to the risk of the security which cannot be diversified away. Is the model a complete guide to the underlying determinants of risk premiums? Probably not—we know that insolvency risk is ignored by the CAPM and this risk is a significant fact of life in the way investors evaluate and value securities. Then, just how useful is the model? We are reminded by the Fama and French results that the CAPM is, at best, only a rough approximation to the relationship between risk and return. Thus, we are reminded to treat beta estimates and corresponding market risk premium estimates with great care.

The attacks and counterattacks on the effectiveness of the CAPM will, we believe, continue until a more appealing theory comes along that better explains the relationship between risk and returns. In fact, to this point, only one alternative theory has been offered as a substitute, or as a possible complement, for the CAPM. This newer theory, the **arbitrage pricing model (APM)** considers multiple economic factors when explaining required rates of return, rather than looking at systematic risk or general market returns as a single determinant of an investor's required rate of return.[27]

Despite having some desirable features and potential, the arbitrage pricing model has yet to be put to widespread use. We have nevertheless, provided a brief treatment of the model in Appendix 3C.

Whatever method is used to assess an appropriate required rate of return for an investment, one key point remains. To formulate a complete concept or understanding of security valuation, we must understand the nature of the investor's required rate of return. The required rate of return, which serves as the discount rate in the valuation process, is the investor's minimum acceptable return that would induce him or her to purchase or hold a security. Despite the problem with establishing a link between risk and return, the financial manager is given no choice but to make a good faith effort to recognize and adapt to the risk-return relationship.

Arbitrage pricing model

An alternative theory to the capital asset pricing model for relating stock returns and risk. The theory maintains that security returns vary from their expected amounts when there are unanticipated changes in basic economic forces. Such forces would include unexpected changes in industrial production, inflation rates, term structure of interest rates, and the difference between interest rates of high- and low-risk bonds (see Appendix 3C).

RATES OF RETURN: THE INVESTOR'S EXPERIENCE

OBJECTIVE 9

We now examine the historical relationship that exists between risk and expected after-tax return for various types of securities held by investors. Figure 3-12 summarizes the risk-return relationship for various types of securities. The risk-return relationship simply indicates that investors experience increasing levels of risk when investing in securities such as the 3-month Government of Canada Treasury bill, mortgage bonds, subordinated debentures, preferred shares, common shares and derivative securities such as options and futures.

The historical relationship between risk and return supports ***Axiom 1: The Risk-Return Tradeoff—We Won't Take on Additional Risk Unless We Expect to Be Compensated with Additional Return.*** In other words, securities such as preferred shares and common shares reward investors, on average, with an additional return that

[27] A description of the APM is found in Dorothy H. Bower, Richard S. Bower, and Denis E. Logue, "A Primer on Arbitrage Pricing Theory," in Joel M. Stern and Donald H. Chew, Jr., eds., *The Revolution in Corporate Finance* (New York: Basil Blackwell, 1986), pp. 69–77.

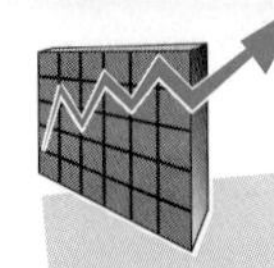

Risk and Diversification

What can diversification across different portfolios accomplish? To demonstrate the effect of diversification on risk and rate of return, we examine data from Ibbotson Associates which uses 1970–93 returns to compare three portfolios (A, B, C) which were constructed as shown below:

The results show that an investor can use diversification to improve the risk-return characteristics of a portfolio. Specifically, we see that:

1. Portfolio A, which consists entirely of long-term government bonds, had an average annual return of 5.5 percent with a standard deviation of 11.3 percent.
2. In Portfolio B, we have diversified across all three types of securities, with the majority of the funds (63 percent) now invested in Treasury bills and a lesser amount (25 percent) in stocks. The effects are readily apparent. The average returns of the two portfolios, A and B, are identical, but the risk associated with Portfolio B is almost half that of portfolio A—standard deviation of 6.1 percent for Portfolio B compared with 11.3 percent for Portfolio A. In other words, the coefficient of variation for Portfolio B, 1.1, is lower than the coefficient of variation for Portfolio A, 2.1. Notice that risk has been reduced in portfolio B even though stocks, a far more risky security, have been included in the portfolio. How can this be? Simple: Stocks behave differently than both government bonds and Treasury bills, with the effect being a less risky (lower standard deviation) portfolio.
3. While Portfolio B demonstrated how an investor can reduce risk while keeping returns constant, Portfolio C, with its increased investment in stocks (52 percent), shows how an investor can increase average returns while keeping risk constant. This portfolio has a risk level identical to that of long-term government bonds alone (Portfolio A), but achieves a higher average return of 8 percent, compared with 5.5 percent for the government bond portfolio.

Clearly, investors are rewarded for diversifying their investments. Note that the diversification in the above example is across different asset types—Treasury bills versus long-term government bonds versus common shares. Diversifying among different kinds of assets is called **asset allocation**, as compared to diversification within the different asset classes. The benefit we receive from diversifying is far greater through effective asset allocation than through merely selecting individual securities to include within an asset category.

The market is also seen to reward investors for assuming risk. Then why would we not always invest in common shares, especially of smaller companies, rather than in bonds? An investor in common stocks must often wait longer to earn the higher returns than those provided by bonds—maybe as long as 20 years.

The Vanguard Group, a mutual fund company, compared

Asset allocation

Identifying and selecting the asset classes appropriate for a specific investment portfolio and determining the proportions of these assets within the given portfolio.

	Investment Mix in Portfolio		
Types of securities	A	B	C
Short-term Treasury bills (1926–93)	0%	63%	34%
Long-term government bonds (1970–93)	100%	12%	14%
Large-company stocks (1926–93)	0%	25%	52%

The average returns and standard deviations of the three portfolios were reported as follows:

Portfolio	Average Annual Return	Risk (Standard Deviation)	Coefficient of Variation
A	5.5%	11.3%	2.1
B	5.5%	6.1%	1.1
C	8.0%	11.3%	1.4

the rates of return you would have earned on the Standard and Poor 500 stocks for different holding periods, such as one year or five years. The time frame of their study included the years from 1950 through 1980. The results of this study are as shown below:

	1 Year	5 Years	10 Years	15 Years	20 Years	25 Years
High	52.3%	20.1%	16.4%	13.9%	11.6%	10.2%
Average	11.4%	9.3%	8.5%	8.4%	8.4%	8.9%
Low	−26.3%	−2.4%	−1.2%	4.3%	6.5%	7.9%

The results indicate that if we want to reduce the variability, we should invest for 15, 20 or 25 years—a long time, but not much of a problem for people in their twenties or thirties.

The conclusions to be drawn from these examples are clear: (1) the market rewards diversification and (2) the market rewards investors for being patient. By diversifying our investments, we can indeed lower risk without sacrificing expected return, and/or we can increase expected return without having to assume more risk. Furthermore, returns tend to converge toward the average as we lengthen our holding period, a principle our statistics professor told us about, calling it the central tendency theorem. Now we see it in real life in the capital markets.

compensates investors for any additional risk. For example, Government of Canada long-term bonds provide a greater risk premium than Government of Canada Treasury bills because of the difference incurred in maturity risk by the investor.[28] In a similar manner, common shares provide a greater risk premium than preferred shares which in turn provide a greater risk premium than debentures or mortgage bonds because of differences in default risk. We note that the 3-month Government of Canada Treasury bill is considered the least risky of the securities because maturi-

Figure 3-12 The Relationship Between Risk and Expected After-tax Return for Various Security Types

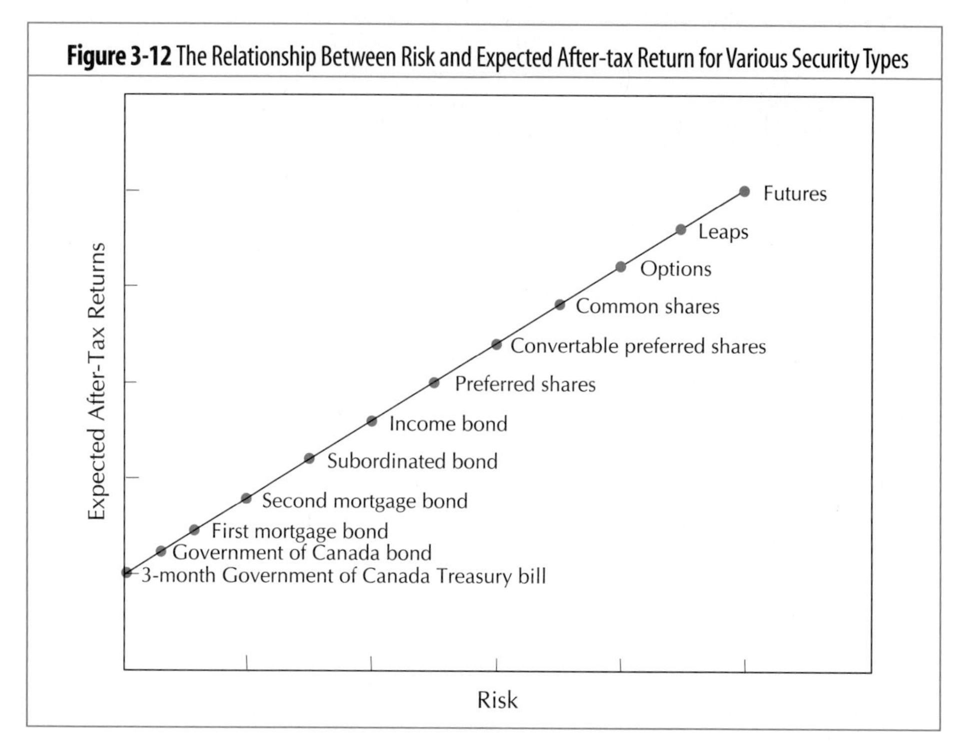

[28]The greater volatility for long-term bonds relative to short-term bonds will be discussed in detail in Chapter 5.

ty risk and default risk are essentially nonexistent for this type of security. Whereas, derivative securities such as options and futures provide the greatest risk premium of all the securities listed because of the chance that an investor may lose all of his or her investment if the underlying security does not meet the investment's conditions.

It should also be noted that the historical relationship between risk and return is sometimes distorted because of economic conditions. During the early eighties, increases in the inflation rate caused greater interest rate risk. As a result, Government of Canada bonds showed greater price volatility and were observed to be more risky than corporate bonds. However, when longer time intervals are examined, our historical relationship between risk and return has proven to be accurate.[29]

International Financial Management

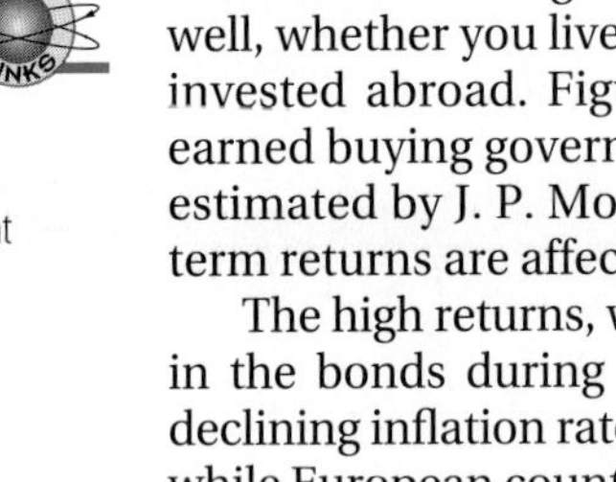

World Wide Web Site

http://www.canadianfinance.com
Comment: Canadian Financial Network provides access to Canadian investment information and resources.

An investment in government bonds in the early nineties would have done quite well, whether you lived in Canada and invested in Government of Canada bonds or invested abroad. Figure 3-13 shows the rates of return an investor would have earned buying government bonds in different countries and the average return, as estimated by J. P. Morgan's Global Index of government bonds. (Note: U.S.-dollar term returns are affected by currency rates at the time of conversion.)

The high returns, which consisted of the interest received and the price increases in the bonds during the year, were attributable to decreasing interest rates and declining inflation rates. Australia, Japan and Canada clearly had the highest returns, while European countries, with the exception of Spain, produced lower returns.

Figure 3-13 Returns on Government Bonds in Different Countries

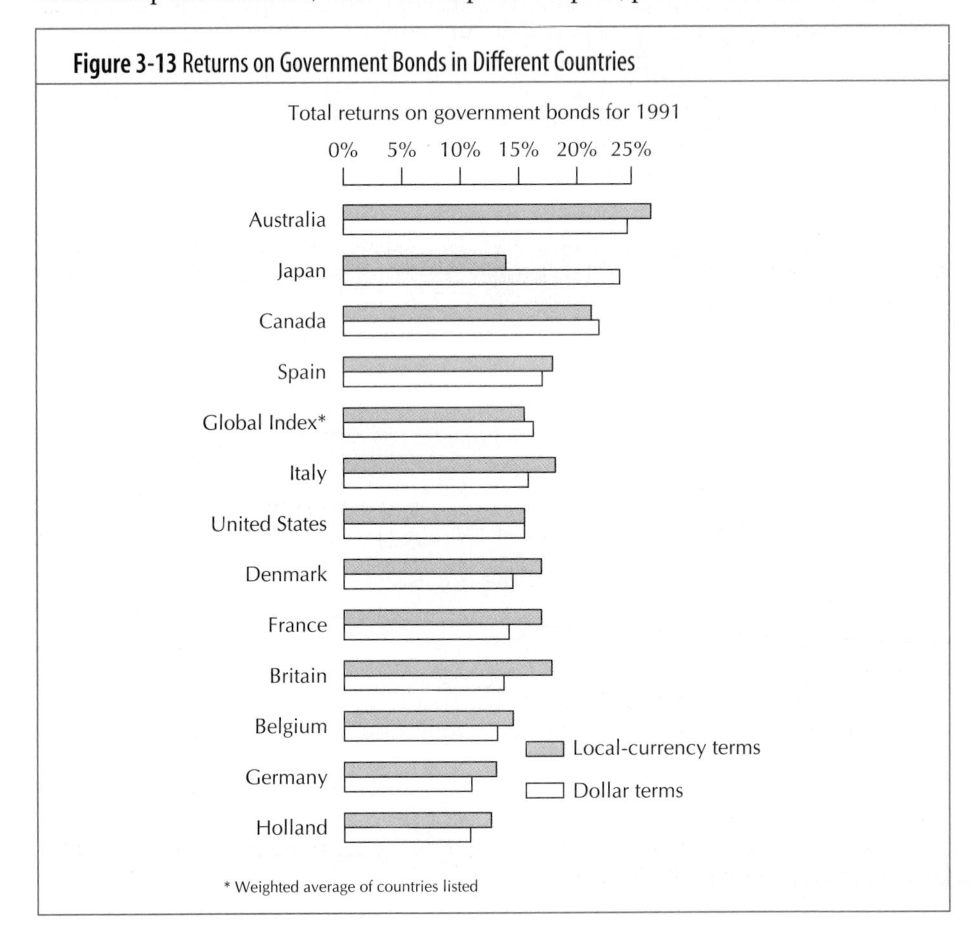

[29]James E. Hatch and Robert W. White, *Canadian Stocks, Bonds, Bills, and Inflation: 1950–1987* (Charlottesville, Virginia: The Research Foundation of the Institute of Chartered Financial Analysts, 1988), pp. 161–66.

SUMMARY

In chapter 2, we referred to the discount rate as the interest rate of the opportunity cost of funds. At that point, we considered a number of important factors that influenced interest rates, including the price of time, expected or anticipated inflation, the risk premium related to maturity, liquidity and variability of future returns.

In this chapter, we returned to our study of the rates of return, and looked carefully at the relationship between risk and rates of returns.

Expected Returns

In a world of uncertainty, we cannot make forecasts with certainty. Thus we must speak in terms of expected events. The expected return on an investment may, therefore, be stated as the average of all possible outcomes where those outcomes are weighted by the probability that each will occur.

Risk

Risk, for our purposes, is the prospect of an unfavourable outcome and may be measured by the standard deviation.

Diversifying Away the Risk

We have made an important distinction between nondiversifiable risk and diversifiable risk. We concluded that the only relevant risk given the opportunity to diversify our portfolio is a security's nondiversifiable risk, which we called by two other names: systematic risk and market risk.

Measuring a Security's Market Risk

A security's market risk is represented by beta, the slope of the characteristic line. Beta measures the average responsiveness of a security's returns to the movement of the general market, such as the TSE Composite 300 Index. If beta is one, the security's returns move 1-to-1 with the market returns; if beta is 1.5, the security's returns move up and down 1.5 percent for every 1 percent change in the market's returns.

Measuring a Portfolio's Beta

A portfolio's beta is simply a weighted average of the individual stocks' betas, where the weights are the percentage of funds invested in each stock. The portfolio beta measures the average responsiveness of the portfolio's returns to the movement of the general market, such as the TSE 300 Index.

Fama and French

For several years, the capital asset pricing model (CAPM) was touted as the "new investment technology" and received the blessings of the vast majority of professional investors and finance professors. The model, like any abstract theory, creates some unresolved issues. For example, we might question whether the risk of an asset can be totally captured in a single dimension of sensitivity to the market, as the CAPM proposes. Fama and French, two financial economists, argue that they have developed evidence that the CAPM does not explain why stock returns differ. Thus beta, at least according to Fama and French, does not adequately explain the risk-return relationship in the markets.

The Investor's Required Rate of Return

The capital asset pricing model, even with its weaknesses, provides an intuitive framework for understanding the risk-return relationship. The CAPM suggests that investors determine an appropriate required rate of return, depending upon the amount of systematic risk inherent in a security. This minimum acceptable rate of return is equal to the risk-free rate plus a return premium for assuming risk.

The Investor's Experience in the Capital Markets

Annual rates of return for the following types of portfolios show a positive relationship between risk and return:

1. Government of Canada Treasury bills.
2. Long-term Government of Canada bonds.
3. Long-term industrial bonds.
4. Equities.

Government of Canada Treasury bills were the least risky; whereas, the common shares of large companies were the most risky. The market rewards investors who diversify and who are patient.

STUDY QUESTIONS

3-1. Explain what the value of a security is.

3-2. Explain the three elements that determine the value of an asset.

3-3. **a.** What is meant by "the investor's required rate of return"?
b. How have we measured the riskiness of an asset?
c. How should the proposed measurement of risk be interpreted?

3-4. What is (a) unique risk and (b) systematic risk (market or nondiversifiable risk)?

3-5. What is the meaning of beta? How is it used to calculate k, the investor's required rate of return?

3-6. Define the term *security market line.* What does it represent?

3-7. How do we measure the beta for a portfolio?

3-8. If we were to graph the returns of a share against the returns of the TSE 300 Index, and the points did not follow a very neat pattern, what could we say about that share? If the share's returns tracked the TSE 300 returns very closely, then what could we say?

3-9. Over the past six decades, we have had the opportunity to observe the rates of return and variability of these returns for different types of securities. Summarize these observations.

SELF-TEST PROBLEMS

ST-1. (*Expected Return and Risk*) Universal Corporation is planning to invest in a security that has several possible rates of return. Given the following probability distribution of returns, what is the expected rate of return on the investment? Also compute the standard deviation of the returns. What do the resulting numbers represent?

Probability	Return
.10	−10%
.20	5%
.30	10%
.40	25%

ST-2. (*Capital Asset Pricing Model*) Using the CAPM, estimate the appropriate required rate of return for the three securities listed in the following table, given that the risk-free rate is 5 percent, and the expected return for the market is 17 percent.

Security	Beta
A	.75
B	.90
C	1.40

ST-3. (*Expected Return and Risk*) Given the holding period returns shown below, calculate the average returns and the standard deviations for the Kaifu Corporation and for the market.

Month	Market	Kaifu Corp.
1	2%	4%
2	3%	6%
3	1%	0%
4	−1%	2%

ST-4. (*Holding-Period Returns*) From the price data below, compute the holding period returns.

Time	Share Price
1	$10
2	13
3	11
4	15

ST-5. (*Security Market Line*) Determine the expected return and the beta for the following portfolio.

Security	Percentage of Portfolio	Beta	Expected Return
1	40%	1.00	12%
2	25%	0.75	11%
3	35%	1.30	15%

The risk-free rate is 8 percent. Also, the expected return on the market portfolio is 12 percent.
Given the information above, draw the security market line and show where your securities fit on the graph. How would you interpret your findings?

STUDY PROBLEMS (SET A)

3-1A. (*Expected Rate of Return and Risk*) Pritchard Press, Inc., is evaluating a security. One-year Treasury bills are currently paying 9.1 percent. Calculate the investment's expected return and its standard deviation. Should Pritchard invest in this security?

Probability	Return
.15	5%
.30	7%
.40	10%
.15	15%

3-2A. (*Expected Rate of Return and Risk*) Syntex, Inc., is considering an investment in one of two common shares. Given the information below, which investment is better, based upon risk (as measured by the standard deviation) and return?

Common Share A		Common Share B	
Probability	Return	Probability	Return
		.20	−5%
.30	11%	.30	6%
.40	15%	.30	14%
.30	19%	.20	22%

3-3A. (*Expected Rate of Return and Risk*) Friedman Manufacturing, Inc., has prepared the following information regarding two investments under consideration. Which investment should be accepted?

Common Share A		Common Share B	
Probability	Return	Probability	Return
.20	−2%	.10	4%
.50	18%	.30	6%
.30	27%	.40	10%
		.20	15%

3-4A. **a.** (*Required Rate of Return Using CAPM*) Compute a "fair" rate of return for the common shares of IBM given a Beta of 1.2. The risk-free rate is 6 percent and the market portfolio (Toronto Stock Exchange 300 Index) has an expected return of 16 percent.

b. Why is the rate you have computed a "fair" rate?

3-5A. (*Estimating Beta*) The figure below illustrates the relationship between the holding-period returns for Aram, Inc., and the TSE 300 Index. Estimate the firm's beta.

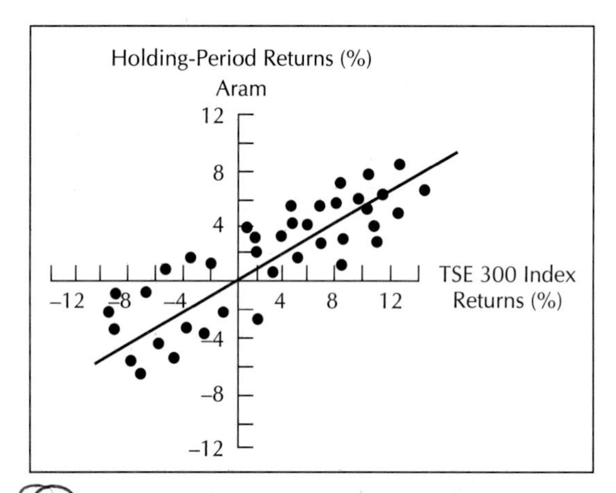

risk free rate.

3-6A. (*Capital Asset Pricing Model*) Johnson Manufacturing, Inc., is considering several investments. The rate on Treasury bills is currently 6.75 percent, and the expected return for the market is 12 percent. What should be the required rates of return for each investment (using the CAPM)?

rm

Security	Beta
A	1.50
B	0.82
C	0.60
D	1.15

3-7A. (*Capital Asset Pricing Model*) CSB, Inc., has a beta of 0.765. If the expected market return is 11.5 percent and the risk-free rate is 7.5 percent, what is the appropriate required return of CSB (using the CAPM)?

3-8A. (*Capital Asset Pricing Model*) The expected return for the general market is 12.8 percent, and the risk premium in the market is 4.3 percent. Tasaco, LBM, and Exxos have betas of 0.864, 0.693, and 0.575, respectively. What are the corresponding required rates of return for the three securities?

3-9A. (*Computing Holding-Period Returns*) From the price data below, compute the holding-period returns for Asman and Salinas.

Time	Asman	Salinas
1	$10	$30
2	12	28
3	11	32
4	13	35

How would you interpret the meaning of a holding-period return?

3-10A. (*Measuring Risk and Rates of Return*) Given the holding-period returns shown below, calculate the average returns and the standard deviations for the Zemin Corporation and for the market.

Month	Market	Zemin Corp.
1	4%	6%
2	2%	3%
3	1%	−1%
4	−2%	−3%
5	2%	5%
6	2%	0%

If Zemin's beta is 1.54, and the risk-free rate is 8 percent, what would be an appropriate required return for an investor owning Zemin? (Note: Since the above returns are based on monthly data, you will need to annualize the returns to make them compatible with the risk-free rate. For simplicity, you may convert from monthly to yearly returns by multiplying the average monthly returns by 12.)

How does Zemin's historical average return compare with the return you believe to be a "fair" return, given the firm's systematic risk?

3-11A. (*Portfolio Beta and Security Market Line*) You own a portfolio consisting of the following shares.

Security	Percentage of Portfolio	Beta	Expected Return
1	20%	1.00	16%
2	30%	0.85	14%
3	15%	1.20	20%
4	25%	0.60	12%
5	10%	1.60	24%

The risk-free rate is 7 percent. Also, the expected return on the market portfolio is 15.5 percent.

a. Calculate the expected return and the beta for your portfolio.

b. Given the information above, plot the security market line on paper. Plot the shares from your portfolio on your graph.

c. From your plot in part (b), which shares appear to be your "winners" and which ones appear to be "losers"?

d. Why should we consider our conclusions in part (c) to be less than certain?

3-12A. (*Expected Return, Standard Deviation, and Capital Asset Pricing Model*) Below you will find the end-of-month prices, both for the TSE 300 Index and for Exxon common shares.

a. Using the data below, calculate the holding-period returns for each month in 1993.

Month and Year	Prices TSE 300 Index	Exxon	Month and Year	Prices TSE 300 Index	Exxon
1992			May	3,882.7	66.50
December	3,350.0	$61.00	June	3,966.4	66.13
			July	3,967.2	65.63
1993			August	4,137.6	65.50
January	3,305.5	$46.38	September	3,990.6	65.50
February	3,451.7	63.63	October	4,255.5	65.38
March	3,602.4	66.25	November	4,180.2	62.75
April	3,789.4	65.13	December	4,321.4	63.00

b. Calculate the average monthly return and the standard deviation of these returns both for the TSE 300 Index and Exxon.

c. Develop a graph that shows the relationship between the Exxon stock returns and the TSE 300 Index (show the Exxon returns on the vertical axis and the TSE 300 Index returns on the horizontal).

d. From your graph, describe the nature of the relationship between Exxon stock returns and the returns for the TSE 300 Index.

INTEGRATIVE PROBLEM

Note: Although not absolutely necessary, you are advised to use a computer spreadsheet to work the following problem.

a. Use the following price data for the Toronto Stock Exchange 300 Composite Index, Ford Motor Company, and General Electric to calculate the holding-period returns for the 24 months during 1992 and 1993.

Month and Year	TSE 300	Prices Exxon	GE	Month and Year	TSE 300	Prices Exxon	GE
1991				December	3,350.0	$42.75	$85.50
December	3,512.4	$28.13	$76.50				
				1993			
1992				January	3,305.5	46.38	86.13
January	3,596.1	30.50	75.00	February	3,451.7	45.88	84.13
February	3,582.0	37.00	78.63	March	3,602.4	52.00	89.13
March	3,412.1	38.38	75.75	April	3,789.4	54.75	90.63
April	3,355.6	45.35	76.63	May	3,882.7	52.13	92.75
May	3,387.9	44.50	76.38	June	3,966.4	52.38	95.63
June	3,387.7	45.75	77.75	July	3,967.2	52.88	98.50
July	3,443.4	45.75	76.50	August	4,137.6	51.00	98.25
August	3,402.9	40.75	74.00	September	3,990.6	55.25	95.75
September	3,297.9	39.63	78.25	October	4,255.5	61.88	97.00
October	3,336.1	36.50	76.63	November	4,180.2	60.75	98.25
November	3,283.0	42.00	83.25	December	4,321.4	64.50	104.88

b. Calculate the average monthly holding-period return and the standard deviation of these returns for the TSE 300 Index, Ford and General Electric.

c. Plot (1) the holding-period returns for Ford against the TSE 300 Index, and (2) plot the General Electric holding-period returns against the TSE 300 Index. (Use Figure 3-8 as the format for your graph.)

d. From your graphs in part (c), describe the nature of the relationship between the Ford share returns and the returns from the TSE 300 Index. Make the same comparison for General Electric.

e. Assume that you have decided to invest one-half of your money in Ford and the rest in General Electric. Calculate the monthly holding-period returns for your two-stock portfolio. (Hint: In this case, the monthly return for the portfolio is the average of the two stock's monthly returns.)

f. Plot the returns of your two-stock portfolio against the TSE 300 Index as you did for the individual stocks in part (c). How does the graph compare to the graphs for the individual stocks? Explain the difference.

g. Below you are provided the returns on an annualized basis that were realized from holding long-term government bonds during 1992 and 1993. Calculate the average monthly holding-period return and the standard deviations of these returns. (Hint: You will need to convert the annual returns to monthly returns by dividing each return by 12 months.)

Month and Year	Annualized Rate of Return	Month and Year	Annualized Rate of Return
1992		*1993*	
January	10.02	January	9.84
February	10.08	February	9.37
March	10.37	March	9.41
April	10.56	April	9.42
May	10.22	May	9.17
June	9.96	June	8.88
July	9.34	July	8.65
August	9.31	August	8.37
September	9.71	September	8.48
October	9.57	October	8.25
November	9.90	November	8.32
December	9.70	December	8.02

h. Now assume that you have decided to invest equal amounts of money in Ford, General Electric, and the long-term government securities. Calculate the monthly returns for your three-asset portfolio. What are the average return and the standard deviation?

i. Make a comparison of the average returns and the standard deviations for all the individual assets and the two portfolios that we designed. What conclusions can be reached by your comparisons?

j. The betas for Ford and General Electric are 1.11 and 1.23, respectively. Compare the meaning of these betas relative to the standard deviations calculated above.

k. The Government of Canada Treasury bill rate at the end of 1993 was approximately 4 percent. Given the betas for Ford and General Electric and using the above data for the TSE 300 Index as a measure for the market portfolio expected return, estimate an appropriate required rate of return given the level of systematic risk for each stock.

STUDY PROBLEMS (SET B)

3-1B. (*Expected Rate of Return and Risk*) B. J. Gautney Enterprises is evaluating a security. One-year Treasury bills are currently paying 8.9 percent. Calculate the investment's expected return and its standard deviation. Should Gautney invest in this security?

Probability	Return
.15	6%
.3	5%
.4	11%
.15	14%

3-2B. *(Expected Rate of Return and Risk)* Kelly B. Stites, Inc., is considering an investment in one of two common shares. Given the information below, which investment is better, based upon risk (as measured by the standard deviation) and return?

Common Share A		Common Share B	
Probability	Return	Probability	Return
		.15	6%
.20	10%	.30	8%
.60	13%	.40	15%
.20	20%	.15	19%

3-3B. *(Expected Rate of Return and Risk)* Clevenger Manufacturing, Inc., has prepared the following information regarding two investments under consideration. Which investment should be accepted?

Common Share A		Common Share B	
Probability	Return	Probability	Return
.20	−2%	.10	4%
.50	19%	.30	7%
.30	25%	.40	12%
		.20	14%

3-4B. **a.** *(Required Rate of Return Using CAPM)* Compute a "fair" rate of return for the common shares of IBM given a beta of 1.5. The risk-free rate is 8 percent and the market portfolio (Toronto Stock Exchange 300 Index) has an expected return of 16 percent.

b. Why is the rate you have computed a "fair" rate?

3-5B. *(Estimating Beta)* The following figure illustrates the relationship between the holding-period returns for Bram, Inc., and the TSE 300 Index. Estimate the firm's beta.

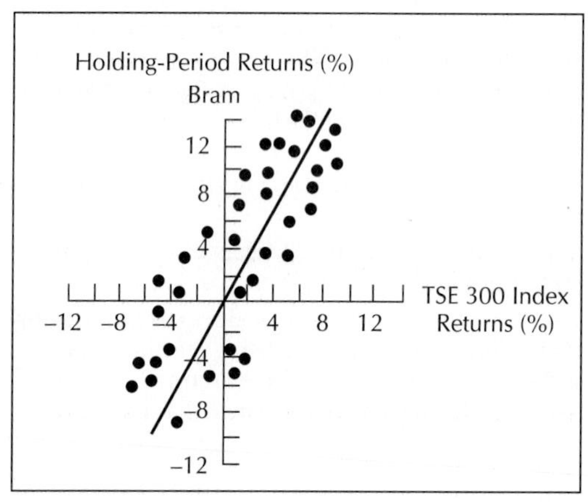

3-6B. *(Capital Asset Pricing Model)* Bobbi Manufacturing, Inc., is considering several investments. The rate on Treasury bills is currently 6.75 percent, and the expected return for the market is 12 percent. What should be the required rates of return for each investment (using the CAPM)?

Security	Beta
A	1.40
B	0.75
C	0.80
D	1.20

3-7B. *(Capital Asset Pricing Model)* Breckenridge, Inc., has a beta of 0.85. If the expected market return is 10.5 percent and the risk-free rate is 7.5 percent, what is the appropriate required return of Breckenridge (using the CAPM)?

3-8B. *(Capital Asset Pricing Model)* The expected return for the general market is 12.8 percent, and the risk premium in the market is 4.3 percent. NCNB, Toyota, and Macintosh have betas of 0.82, 0.57, and 0.68, respectively. What are the corresponding required rates of return for the three securities?

3-9B. *(Computing Holding-Period Returns)* From the price data below, compute the holding-period returns for O'Toole and Shamrock.

Time	O'Toole	Shamrock
1	$22	$45
2	24	50
3	20	48
4	25	52

How would you interpret the meaning of a holding-period return?

3-10B. *(Measuring Risk and Rates of Return)* Given the holding-period returns shown below, calculate the average returns and the standard deviations for the Sugita Corporation and for the market.

Month	Market	Sugita Corp.
1	1.50%	1.80%
2	1.00%	−0.50%
3	0.00%	2.00%
4	−2.00%	−2.00%
5	4.00%	5.00%
6	3.00%	5.00%

If Sugita's beta is 1.18, and the risk-free rate is 8 percent, what would be an appropriate required return for an investor owning Sugita? (Note: Since the above returns are based on monthly data, you will need to annualize the returns to make them compatible with the risk-free rate. For simplicity, you may convert from monthly to yearly returns by multiplying the average monthly returns by 12.)

How does Sugita's historical average return compare with the return you believe to be a "fair" return, given the firm's systematic risk?

3-11B. *(Portfolio Beta and Security Market Line)* You own a portfolio consisting of the following shares.

Security	Percentage of Portfolio	Beta	Expected Return
1	10%	1.00	12%
2	25%	0.75	11%
3	15%	1.30	15%
4	30%	0.60	9%
5	20%	1.20	14%

The risk-free rate is 8 percent. Also, the expected return on the market portfolio is 11.6 percent.

a. Calculate the expected return and the beta for your portfolio.

b. Given the information above, plot the security market line on paper. Plot the shares from your portfolio on your graph.

c. From your plot in part (b), which shares appear to be your "winners" and which ones appear to be "losers"?

d. Why should we consider our conclusions in part (c) to be less than certain?

3-12B. *(Expected Return, Standard Deviation, and Capital Asset Pricing Model)* Below you will find the end-of-month prices, both for the TSE 300 Index and for Ford common shares.

a. Using the data below, calculate the holding-period returns for each month in 1993.

	Prices			Prices	
Month and Year	TSE 300 Index	Ford	Month and Year	TSE 300 Index	Ford
1992			May	3,882.7	52.13.
December	3,350.0	$42.75	June	3,966.4	52.38
			July	3,967.2	52.88
1993			August	4,137.6	51.00
January	3,305.5	$46.38	September	3,990.6	55.25
February	3,451.7	45.88	October	4,255.5	61.88
March	3,602.4	52.00	November	4,180.2	60.75
April	3,789.4	54.75	December	4,321.4	64.50

b. Calculate the average monthly return and the standard deviation of these returns both for the TSE 300 Index and Ford.

c. Develop a graph that shows the relationship between the Ford stock returns and the TSE 300 Index (show the Ford returns on the vertical axis and the TSE 300 returns on the horizontal).

d. From your graph, describe the nature of the relationship between Ford stock returns and the returns for the TSE 300 Index.

SELF-TEST SOLUTIONS

SS-1.

(A) Probability $P(k_i)$	(B) Return (k_i)	Expected Return $(\bar{k})$ (A) × (B)	Weighted Deviation $(k_i - \bar{k})^2 p(k_i)$
.10	−10%	−1%	52.9%
.20	5%	1%	12.8%
.30	10%	3%	2.7%
.40	25%	10%	57.6%
		$\bar{k}$ = 13%	σ^2 = 126.0%
			σ = 11.22%

From our studies in statistics, we know that if the distribution of returns was normal, then Universal could expect a return of 13 percent with a 67 percent possibility that this return would vary up or down by 11.22 percent between 1.78 percent (13% − 11.22%) and 24.22 percent (13% + 11.22%). However, the distribution is not normal, as is apparent from the probabilities.

SS-2. Security A $5\% + .75(17\% - 5\%) = 14\%$

Security B $5\% + .90(17\% - 5\%) = 15.8\%$

Security C $5\% + 1.40(17\% - 5\%) = 21.8\%$

SS-3. Market:

$$\text{Average return} = \frac{2\% + 3\% + 1\% - 1\%}{4}$$

$$= 1.25\%$$

Standard deviation =

$$\sqrt{\frac{(2\% - 1.25\%)^2 + (3\% - 1.25\%)^2 + (1\% - 1.25\%)^2 + (-1\% - 1.25\%)^2}{4 - 1}}$$

$$= 1.71\%$$

Kaifu:

$$\text{Average return} = \frac{4\% + 6\% + 0\% + 2\%}{4} = 3.00\%$$

$$\text{Standard deviation} = \sqrt{\frac{(4\% - 3\%)^2 + (6\% - 3\%)^2 + (0\% - 3\%)^2 + (2\% - 3\%)^2}{4 - 1}}$$

$$= 2.58\%$$

SS-4.

Time	Share Price		Holding-Period Return
1	\$10		
2	13	(\$13/\$10) – 1	30.0%
3	11	(\$11/\$13) – 1	–15.4%
4	15	(\$15/\$11) – 1	36.4%

SS-5. Portfolio beta:

$(.4 \times 1) + (.25 \times .75) + (.35 \times 1.3) = 1.04$

Portfolio expected return:

$(.4 \times 12\%) + (.25 \times 11\%) + (.35 \times 15\%) = 12.8\%$

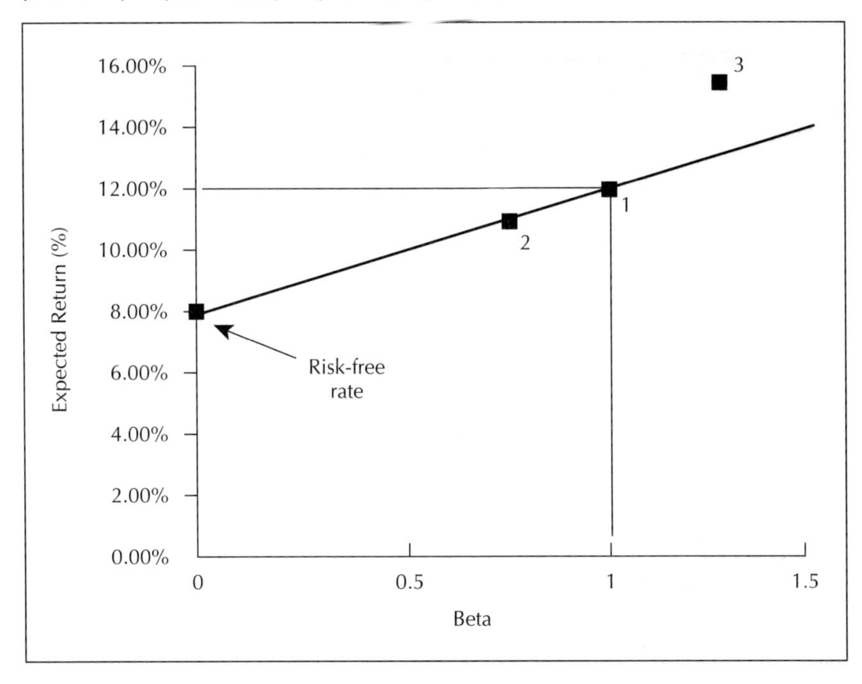

Securities 1 and 2 seem to be on the security market line, which suggests that they earn a fair return, given their systematic risk. Security 3, on the other hand, is earning more than a fair return (above the security market line). However, we may be seeing an illusion; it is possible to misspecify the security market line by using bad estimates in our data.

APPENDIX 3A

REVIEW OF PROBABILITY DISTRIBUTIONS AND DESCRIPTIVE STATISTICS

Probability distributions illustrate the complete set of probabilities over all possible outcomes for that particular event. There are two general types of probability distributions—discrete and continuous. A discrete probability distribution is one in which a probability is assigned to each possible outcome in the set of all possible outcomes. For example, if we are considering a coin toss, there would be two possible outcomes, each with a .50 probability of occurring; if we were considering a horse race with ten horses, there would be ten possible outcomes, with the probability of each outcome depending on the speed of each horse. An investment opportunity with five possible outcomes, as shown in Table 3A-1, is another example of a discrete distribution.

This distribution is illustrated graphically in Figure 3A-1, giving us a picture of all the possible outcomes and the probabilities associated with them.

Note that the sum of all probabilities attached to all possible outcomes must add up to 1.0. This is because one of these outcomes must occur; otherwise all possible outcomes have not been identified.

Whereas in a discrete probability distribution probabilities are assigned to specific outcomes, in a continuous probability distribution there are an infinite number of possible outcomes, in which the probability of an event is related to a range of possible outcomes. The easiest way to illustrate a continuous distribution is graphically, as in Figure 3A-2. Probability in this case is measured by the area under the curve.

Table 3A-1 A Discrete Probability Distribution

Possible Outcome (X_1)	Probability of Occurrence P(X_1)
$ 5,000	0.10
$ 7,000	0.25
$ 8,000	0.30
$ 9,000	0.25
$11,000	0.10

Figure 3A-1 A Discrete Probability Distribution

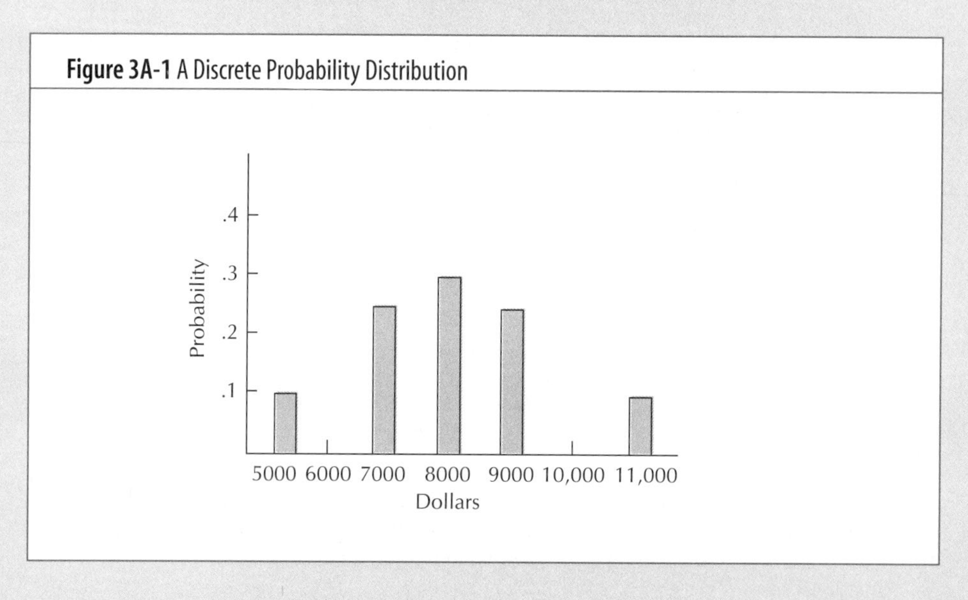

Figure 3A-2 A Continuous Probability Distribution

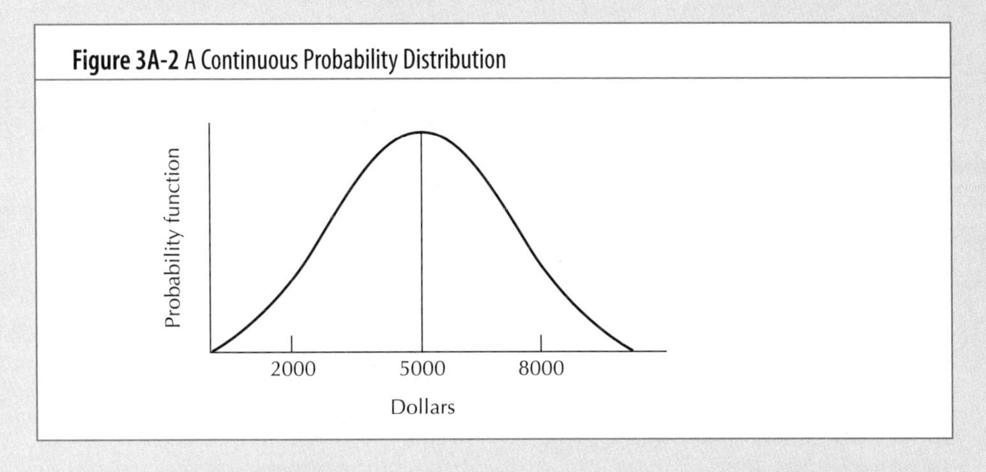

Thus, when provided with a continuous distribution, we can answer the question, What is the probability that an outcome will fall within a certain range of values? A discrete distribution, on the other hand, lets us answer the question, What is the probability that a specific outcome will occur? The key to understanding the difference between these questions—and the underlying differences between these classifications of distributions—is that for discrete distributions the number of possible outcomes is finite, while for continuous distributions the number is infinite.

Continuous distributions are quite common and valuable, but in general they require a degree of calculus that might be confusing. Thus, in the remainder of the book we will be concerned primarily with discrete probability distributions.

Expected Value

The expected value of a distribution is the arithmetic mean or average of all possible outcomes, where those outcomes are weighted by the probability that each outcome will occur. Thus, the expected value, $\overline{X}$, can be calculated as follows:

$$\overline{X} = \sum_{i=1}^{N} X_i P(X_i) \qquad \text{(3A-1)}$$

where N = the number of possible outcomes
X_i = the value of the ith possible outcome
$P(X_i)$ = the probability that the ith outcome will occur

To illustrate the computation of the expected value, consider the investment opportunity given in Table 3A-1. To determine the expected value of this distribution, we need only multiply each possible outcome by its probability of occurrence and add the products together as follows:

$$\begin{aligned}\overline{X} &= \$5{,}000(.10) + \$7{,}000(.25) + \$8{,}000(.30) \\ &\quad + \$9{,}000(.25) + \$11{,}000(.10) \\ &= \$8{,}000\end{aligned}$$

When we are estimating cash flows that are to occur in the future, we cannot estimate them with certainty. The expected value provides us with only the mean of the probability distribution from which that future cash flow will come. While the expected value provides us with a measure of central tendency, it tells us nothing about the amount of dispersion contained in the probability distribution. To measure dispersion, we will examine the standard deviation and coefficient of variation.

Standard Deviation

The standard deviation provides a measure of the spread of the probability distribution. The larger the standard deviation, the greater the dispersion of the distribution. It is calculated by squaring the difference between each outcome and its expected value, weighting each value by its probability, adding all possible outcomes, and taking the square root of this sum. Thus, the standard deviation becomes

$$\sigma = \sqrt{\sum_{i=1}^{N} (X_i - \overline{X})^2 P(X_i)} \tag{3A-2}$$

where N = the number of possible outcomes
X_i = the value of the ith possible outcome
$\overline{X}$ = the expected value
$P(X_i)$ = the probability that the ith outcome will occur

The calculations for the standard deviation for the distribution given in Table 3A-1 are provided in Table 3A-2.

In the case of a normal distribution, the standard deviation gives us even more than a measure of the absolute dispersion of that distribution. In this case we can, by consulting a table of areas under the normal curve, determine the probability that the outcome will be above or below a specific value. For example, in a normal distribution there is a 68.3 percent probability that the outcome will be within one standard deviation (plus or minus) of the expected value and a 95.0 percent probability that it will be within two standard deviations. Thus, if we are looking at two normal distributions, one with an expected value of $1,000 and a standard deviation of $200 and the other with an expected value of $2,000 and a standard deviation of $300, we can conclude that the second distribution has more dispersion, and hence more absolute risk associated with it. This does not mean that the second distribution is less desirable, just that it has more absolute risk. While the standard deviation gives us a good measure of a distribution's absolute dispersion, it is also desirable to measure dispersion or risk relative to return to provide the manager with a relative measure of risk. This is the purpose of the coefficient of variation.

Coefficient of Variation

The coefficient of variation provides us with a measure of the relative dispersion of a probability distribution—that is, the risk per unit of return. Mathematically, it is defined as the standard deviation divided by the expected value, or

$$\gamma = \frac{\sigma}{\overline{X}} \tag{3A-3}$$

This measure derives its value from the fact that using the standard deviation to compare the riskiness of two investments can be misleading when the investments

Table 3A-2 Calculation of the Standard Deviation for the Distribution in Table 3A-1

Step 1: Calculate Differences $(X_i - \overline{X})$	Step 2: Square Differences $(X_i - \overline{X})^2$	Step 3: Squared Differences Times Probabilities $(X_i - \overline{X})^2P(X_i)$
$5,000 − $8,000 = −$3,000	$9,000,000	$9,000,000(.10) = $900,000
7,000 − 8,000 = − 1,000	1,000,000	1,000,000(.25) = 250,000
8,000 − 8,000 = 0	0	0(.30) = 0
9,000 − 8,000 = 1,000	1,000,000	1,000,000(.25) = 250,000
11,000 − 8,000 = 3,000	9,000,000	9,000,000(.10) = 900,000

Step 4: Sum	Step 5: Take the Square Root
$\sum_{i=1}^{N} (X_i - \overline{X})^2 P(X_i) = \$2,300,000$	$\sigma = \sqrt{\sum_{i=1}^{N} + (X_i - \overline{X})^2 P(X_i)} = \underline{\underline{\$1,517}}$

Table 3A-3 Coefficient of Variation

		Investment A	Investment B
Expected value	$\bar{X}$	\$1,000	\$2,000
Standard deviation	σ	\$ 200	\$ 300
Coefficient of variation	$\sigma/\bar{X}$	0.20	0.15

are of unequal size.

For example, let us look at two normal distributions given in Table 3A-3. In this case distribution B has more absolute risk, as indicated by its standard deviation; however, distribution A has more risk associated with it relative to its expected value. The value of examining the relative risk can easily be seen when examining distributions containing the same level of absolute risk or standard deviation but dramatically different levels of expected value. For example, consider two investments, both with standard deviations of \$1,000, one having an expected value of \$1,000, and the other an expected value of \$1,000,000. In each case the absolute risk is the same, as measured by the standard deviation, but the relative risk is not. Thus, the use of the coefficient of variation along with the standard deviation is especially important when two investments of unequal size are being compared. To ignore the coefficient of variation in this case could lead to misconceptions as to the relative level of uncertainty contained in each investment.

omit

Appendix 3B

MEASURING A SECURITY'S RETURN AND RISK

The capital asset pricing model (CAPM) draws heavily on our ability to measure a security's risk statistics, such as beta. We must determine an accurate beta for the CAPM to yield good results. An estimate of the appropriate required rate of return is only as accurate as the data used to obtain it.

As noted in this chapter, there are some difficulties in using the CAPM, most of which relate to empirical problems. Specifically, from where do we draw our data in making our estimates? Simply put, for us to use the model correctly, we need to know the investors' expectations about future returns. However, only historical rates of returns are available to us. Nevertheless, our best hope at seeing the future is by looking at the past. Essentially, we use the historical return data and hope that the past will closely reflect the investor's expectations about the future. Thus, we rely on historical returns when we compute the beta for a security or a portfolio.

To compute risk statistics, professional analysts typically use five years of monthly return data, or 60 months, and make certain adjustments to their computations. For instance, we know that betas over time tend to move toward a value of one. If the beta is, at the present, substantially greater than one, it will gradually decline over time, and betas that are substantially below one will tend to increase in future periods. So forecasters adjust for this observed tendency. However, for our purposes, we simply will use the historical return data of the past 12 months.

We now compute the risk statistics for Spar Aerospace's common share using historical returns for the 60 months ending December 1995. These returns, along with the corresponding returns for the TSE 300 Index, are shown in Table 3B-1.

We determine the beta for asset *j* which can take the form of a security or a portfolio by regressing its returns on the returns of the market portfolio. This relationship is described by the following equation:

$$k_{jt} = \alpha_j + \beta_j(k_{Mt}) + e_{jt}$$

where k_{jt} = the monthly holding-period return for asset j in month t
α_j = alpha, the point where the regression line intercepts the vertical axis
β_j = beta for asset j; the slope of the regression line
k_{Mt} = the return on the market portfolio in month t
e_{jt} = the error term; the difference between the actual return in month t and the expected return given the market's return (that is, the distance the actual return lies away from the regression line).

Our objective is to find the alpha and beta values that minimize the sum of the squares of the error terms. This type of calculation enables us to find the line that best fits the data; that is, it best describes the average relationship between the returns of asset j and the returns of the market portfolio. There are three ways for determining alpha and beta.

1. We could plot the return data on graph paper and then "eyeball" the regression line, trying to get what appears to be the best fit to the data. The point where the line crosses the vertical axis would indicate our estimate for the alpha value. We would then estimate the beta by measuring how steeply the line increases vertically (return of asset j) relative to a change on the horizontal axis (return of market portfolio). That is, we would find:

$$\beta_j = \frac{\text{Rise}}{\text{Run}} = \frac{\Delta k_{jt}}{\Delta k_{Mt}}$$

The obvious limitation with this approach is its potential for error since accuracy of the estimation depends in part on the sharpness of the eye. Your estimate would most likely be somewhat different from another person's, especially if the data did not have a tight fit.

2. The beta of asset j can be calculated from the following equation:

$$\beta_j = \frac{(\rho)(\sigma_j)(\sigma_M)}{(\sigma_M)^2} = \frac{(\rho)(\sigma_j)}{(\sigma_M)} \tag{3B-1}$$

where ρ = the correlation coefficient of the regression of returns of asset j on returns of the market portfolio
σ_j = the standard deviation of the returns of asset j
σ_M = the standard deviation of the returns of the market portfolio

Table 3-2 shows that the standard deviations of returns of Spar Aerospace and the TSE 300 Index are 11.98 and 3.02, respectively. The correlation coefficient (ρ) of this regression equation is 0.18. Thus, the beta of Spar Aerospace can be computed by substituting these values in Equation (3B-1), as follows:

$$\beta = \frac{(\rho)(\sigma_j)}{(\sigma_M)} = \frac{(0.18)(11.98)}{(3.02)} = 0.72$$

Thus, the beta for Spar Aerospace is 0.72.

3. We can also use a statistical package from a calculator (see student's *Study Guide*) or a computer spreadsheet to determine alpha and beta. This option is preferred because of the ability to view the original return data in order to check for accuracy in data entry and the ease of working with the calculations that are repetitious.

Table 3B-1 Monthly Holding-Period Returns, Spar Aerospace, and the TSE 300 Index December 1990–December 1995

	Spar Aerospace		TSE 300 Index	
	Price	Return	Value	Return
Month and Year	$	%		%
December 1990	10.88	6.10	3,256.8	3.36
January 1991	12.13	11.49	3,272.9	0.49
February	12.5	3.09	3,462.4	5.79
March	14.38	15.00	3,495.7	0.96
April	12.5	−13.04	3,468.8	−0.77
May	12.63	1.00	3,546.1	2.23
June	12.88	1.98	3,465.8	−2.26
July	14.25	10.68	3,539.6	2.13
August	15.75	10.53	3,517.9	−0.61
September	14.38	−8.73	3,387.9	−3.70
October	15.38	6.96	3,515.8	3.78
November	16.00	4.07	3,448.5	−1.91
December	16.00	0.00	3,512.4	1.85
January 1992	18.38	14.84	3,596.1	2.38
February	18.25	−0.68	3,582.0	−0.39
March	19.13	4.79	3,412.1	−4.74
April	19.00	−0.65	3,355.6	−1.66
May	18.25	−3.95	3,387.9	0.96
June	15.50	−15.07	3,387.7	−0.01
July	15.75	1.61	3,443.4	1.64
August	16.75	6.35	3,402.9	−1.18
September	16.38	−2.24	3,297.9	−3.09
October	15.25	−6.87	3,336.1	1.16
November	15.00	−1.64	3,283.0	−1.59
December	15.75	5.00	3,350.0	2.04
January 1993	17.25	9.52	3,305.5	−1.33
February	17.25	0.00	3,451.7	4.42
March	17.38	0.75	3,602.4	4.37
April	15.13	−12.95	3,789.4	5.19
May	14.25	−5.82	3,882.7	2.46
June	16.63	16.70	3,966.4	2.16
July	15.0	−9.80	3,967.2	0.02
August	13.88	−7.47	4,137.6	4.30
September	16.25	17.07	3,990.6	−3.55
October	17.63	8.49	4,255.5	6.64
November	16.75	−4.99	4,180.2	−1.77
December	18.25	8.96	4,321.4	3.38
January 1994	20.25	10.96	4,554.9	5.40
February	18.5	−8.64	4,423.8	−2.88
March	17.875	−3.38	4,329.6	−2.13
April	18.125	1.40	4,267.1	−1.44
May	16.875	−6.90	4,326.8	1.40
June	14.375	−14.81	4,025.3	−6.97
July	15.0	4.35	4,179.0	3.82
August	11.625	−22.50	4,349.5	4.08
September	11.125	−4.30	4,354.2	0.11
October	10.625	−4.49	4,291.7	−1.44
November	9.125	−14.12	4,093.4	−4.62
December	9.75	6.85	4,213.6	2.94
January 1995	10.0	2.56	4,017.5	−4.65
February	12.375	23.75	4,124.8	2.67
March	14.5	17.17	4,313.6	4.58

Table 3B-1 (continued)

Month and Year	Spar Aerospace Price $	Spar Aerospace Return %	TSE 300 Index Value	TSE 300 Index Return %
April	13.75	−5.17	4,279.5	−0.79
May	13.375	−2.73	4,448.6	3.95
June	13.125	−1.87	4,527.2	1.77
July	15.25	16.19	4,615.1	1.94
August	15.0	−1.64	4,516.7	−2.13
September	14.0	−6.67	4,529.8	0.29
October	14.25	1.79	4,459.2	−1.56
November	13.125	−7.89	4,661.2	4.53
December	20.75	58.10	4,713.5	1.12

STUDY PROBLEMS (SET A)

3B-1A. *(Risk and Return Statistics for a Security)* Compute the following statistics for Arka's common share: (1) average return, (2) standard deviation of the returns, (3) alpha, and (4) beta.

Yearly Holding-Period Returns

Year	Arka	TSE 300
1990	10%	12%
1991	6%	7%
1992	18%	24%
1993	15%	18%

3B-2A. *(Risk and Return Statistics for a Security)* Compute the following statistics for the common shares of Son Sen, Inc.: (1) average return, (2) standard deviation of the returns, (3) alpha, and (4) beta.

Yearly Holding-Period Returns

Month	Son Sen	TSE 300
January	0.6%	0.9%
February	0.3	1.0
March	−1.5	−1.0
April	−0.7	0.0
May	1.9	3.0
June	0.5	−1.5
July	2.5	3.0
August	0.8	1.2
September	−2.5	−4.0
October	3.0	4.5
November	0.4	1.4
December	0.1	0.2

Appendix 3C OMIT

MEASURING THE REQUIRED RATE: THE ARBITRAGE PRICING MODEL[30]

The basic theme of the arbitrage pricing model (APM) can be stated as follows:

1. Actual security returns vary from their expected amounts because of unanticipated changes in a number of basic economic forces, such as unexpected changes in industrial production, inflation rates, the term structure of interest rates, and the difference in interest rates between high- and low-risk bonds.[31]
2. Just as the CAPM defined a portfolio's systematic risk to be its sensitivity to the general-market returns (i.e., its beta coefficient), APM suggests that the risk of a security is reflected in its sensitivity to the unexpected changes in important economic forces.
3. Any two stocks or portfolios that have the same sensitivity to the meaningful economic forces (that is, the same relevant or systematic risk) must have the same expected return. Otherwise, we could replace some of the stocks in our portfolio with other stocks having the same sensitivities, but higher expected returns, and earn riskless profits.
4. We would expect portfolios that are highly sensitive to unexpected changes in macroeconomic forces to offer investors high expected returns. This relationship may be represented quantitatively as follows:

$$E(k_i) = k_f + (S_{i1})(RP_1) + (S_{i2})(RP_2) + \ldots + (S_{ij})(RP_j) + \ldots + (S_{im})(RP_m) \qquad \text{(3C-1)}$$

where $E(k_i)$ = the expected return for stock or portfolio i
k_f = the risk-free rate
S_{ij} = the sensitivity of stock i returns to unexpected changes in economic force j
RP_j = the market risk premium associated with an unexpected change in the jth economic force
m = the number of relevant economic forces

To help understand the APM model, we will draw from the actual research of Bower, Bower, and Logue (BBL).[32] After computing monthly returns for 815 stocks from 1970 through 1979, BBL used a technique called factor analysis to study the general movement of monthly returns for the 815 stocks. The technique identified four factors that helped to explain the movement of the returns, and also used Equation (3C-1) to represent the risk-return relationship in an APM format. The actual model appears as follows:

$$\text{Expected return for Stock } i = 6.2\% - 185.5\%(S_{i1}) + 144.5\%(S_{i2}) + 12.4\%\,(S_{i3}) - 274.4\%(S_{i4})$$

The value of 6.2% in the equation is an estimate of the risk-free rate, as determined by the model; the remaining values, –185.5, ... –274.4, signify the market risk premiums for each of the four factors; and the $S_{i1}, \ldots, S_{i4}$ represent the sensitivities of stock i to the four factors.

---

[30]The following description of the APM is taken in part from Dorothy H. Bower, Richard S. Bower, and Dennis E. Logue, "A Primer on Arbitrage Pricing Theory," *The Revolution in Corporate Finance*, eds. Joel M. Stern and Donald H. Chew, Jr. (New York: Basil Blackwell, 1986), pp. 69–77.

[31]See R. Roll and S. Ross, "The Arbitrage Pricing Theory Approach to Strategic Portfolio Planning," *Financial Analysts Journal* (May–June 1984), pp. 14–26.

[32]Dorothy A. Bower, Richard S. Bower, and Dennis E. Logue, "Equity Screening Rates Using Arbitrage Pricing Theory," in C. F. Lee, ed., *Advances in Financial Planning* (Greenwich, CT: JAI Press), 1984.

The BBL factors are determined statistically from past return data and were not intuitively or economically identified. That is, the technique provides "factors" that tell us more about the movement in the returns than would any other factors.[33] Each factor could conceivably relate to a single economic variable; however, it is more likely that each factor represents the influence of several economic variables.[34]

Having specified the APM risk-return relationship, BBL then used the model to estimate the expected (required) rates of return for 17 stocks. This estimation required BBL to use regression analysis to study the relationships of the returns of the 17 stocks to the four factors. From this regression analysis, they were able to measure the sensitivity of each security's return to a particular factor. These "sensitivity coefficients" for a particular stock may then be combined with the APM model in Equation (3C-1) to estimate the investors' required rate of return.

Using 3 of the 17 stocks to demonstrate the calculation, the regression coefficients (sensitivity coefficients) for American Hospital Supply, CBS, and Western Union are shown below:

Stock Sensitivity Coefficients (S)

	Factor 1	Factor 2	Factor 3	Factor 4
American Hospital Supply	−0.050	0.010	0.040	0.020
CBS	−0.050	0.002	0.005	0.010
Western Union	−0.050	−0.020	−0.010	0.009

Using these stock sensitivity coefficients and the APM model, as developed by BBL, we can estimate the investors' expected (required) rate of return for each stock as follows:

$$\begin{aligned}
\text{American Hospital Supply} &= 6.2\% - 185.5\%(-0.050) + 144.5\%(0.010) \\
&\quad + 12.4\%(0.040) - 274.4\%(.020) \\
&= 11.93\% \\
\text{CBS} &= 6.2\% - 185.5\%(-0.050) + 144.5\%(0.002) \\
&\quad + 12.4\%(0.005) - 274.4\%(0.010) \\
&= 13.08\% \\
\text{Western Union} &= 6.2\% - 185.5\%(-0.050) + 144.5\%(-0.020) \\
&\quad + 12.4\%(-0.010) - 274.4\%(0.009) \\
&= 9.99\%
\end{aligned}$$

We have now used two approaches for measuring the required rate of return (or expected rate of return) of investors: the CAPM and the APM. The capital asset pricing model (CAPM) is a single-index model that considers the general market to be the only important factor in setting the values of individual stocks. The arbitrage pricing model (APM), on the other hand, recognizes several macroeconomic factors in estimating returns (and values).

---

[33]A more detailed analysis of this material is provided by: N. Chen, "Some Empirical Tests of the Theory of Arbitrage Pricing," *Journal of Finance*, (December 1983), 1393–1414; and Richard Roll, "A Critique of the Asset Pricing Theory's Tests," *Journal of Financial Economics* (March 1977), pp. 129–176.

[34]As noted earlier in the chapter, work is under way to determine the economic factors that most influence security returns. See R. Roll and S. Ross, "The Arbitrage Pricing Theory Approach to Strategic Portfolio Planning," *Financial Analysts Journal* (May–June 1984), pp. 14–26.

STUDY PROBLEMS (SET A)

3C-1A. *(Arbitrage Pricing Model)* Using the APM, along with the results of the Bower, Bower, and Logue research given in the chapter, estimate the appropriate required rates of return for the following three securities:

Sensitivity Factor				
Security	Factor 1	Factor 2	Factor 3	Factor 4
A	−0.070	−0.020	0.010	0.003
B	−0.070	0.030	0.005	0.010
C	−0.070	−0.010	0.006	0.009

3C-2A.a. *(Required Rate of Return Using APM)* If we use the arbitrage pricing model to measure investors' required rate of return, what is our concept of risk?

b. Using the results of the Bower, Bower, and Logue study of return variability, as captured in Equation (3C-1), calculate the investors' required rate of return for the following securities:

Sensitivity Factor				
Security	Factor 1	Factor 2	Factor 3	Factor 4
A	−0.070	0.030	0.005	0.010
B	−0.070	−0.010	0.006	0.009
C	−0.050	0.004	−0.010	−0.007
D	−0.060	−0.007	−0.006	0.006

c. Assuming (i) a risk-free rate of 6.1 percent and (ii) an expected market return of 17.25 percent, estimate the investors' required rate of return for the four securities in part (b), using the CAPM.

Security	Beta
A	1.40
B	1.70
C	1.20
D	1.10

d. What might explain the differences in your answers to parts (b) and (c)?

THE MATHEMATICS OF FINANCE

CHAPTER 4

LEARNING OBJECTIVES

After reading this chapter you should be able to

1. Explain the mechanics of compounding; that is, how money grows over time when it is invested.
2. Determine the future or present value of a sum when there are nonannual compounding periods.
3. Explain the relationship between compounding and bringing back to the present.
4. Define an ordinary annuity and calculate its compound or future value.
5. Calculate the annual percentage yield or effective annual rate of interest and explain how it differs from the nominal or stated interest rate.
6. Differentiate between an ordinary annuity and an annuity due, and determine the future and present value of an annuity due.

INTRODUCTION

In business, there is probably no other single concept with more power or applications than that of the time value of money. Sidney Homer, in his landmark book, A History of Interest Rates, *noted that if $1,000 was invested for 400 years at 8 percent interest it would grow to $23 quadrillion—that would work out to approximately $5 million per person on earth. He was not giving a plan to make the world rich, but effectively pointing out the power of the time value of money.*

The power of the time value of money can also be illustrated through a story Andrew Tobias tells in his book Money Angels. *There he tells of a peasant who wins a chess tournament put on by the king. The king then asks the peasant what he would like as the prize. The peasant answers that he would like for his village one piece of grain to be placed on the first square of his chessboard, two pieces of grain on the second square, four pieces on the third, eight on the fourth and so forth. The king, thinking he was getting off easy, pledged on his word of honour that it would be done. Unfortunately for the king, by the time all 64 squares on the chess board were filled, there were 18.5 million trillion grains of wheat on the board—the kernels were compounding at a rate of 100 percent over the 64 squares of the chessboard. Needless to say, no one in the village ever went hungry. In fact, that is so much wheat that if the kernels were one-quarter inch long, if laid end to end they could stretch to the sun and back 391,320 times.*

Understanding the techniques of compounding and moving money through time are critical to almost every business decision. It will help you to understand such varied things as how shares and bonds are valued, how to determine the value of a new project, how much you should save for the children's education, and how much your mortgage payments will be.

CHAPTER PREVIEW

In this chapter, we examine the procedures used to make cash flows comparable. A key concept that underlies these procedures is the time value of money; that is, a dollar today is worth more than a dollar received a year from now. Intuitively this idea is easy to understand. We are all familiar with the concept of interest. This concept illustrates what economists call an opportunity cost of passing up the earning potential of a dollar today. This opportunity cost is the time value of money.

In evaluating and comparing financial assets, we need to examine how dollar values might accrue from investing in these financial assets. To do this, all dollar values must first be comparable; since a dollar received today is worth more than a dollar received in the future, we must move all dollar flows back to the present or out to a common future date. An understanding of the time value of money is essential, therefore, to an understanding of financial management, whether basic or advanced.

BACK TO THE BASICS

In this chapter we develop the tools to incorporate ***Axiom 2: The Time Value of Money—A Dollar Received Today Is Worth More Than a Dollar Received in the Future*** *into our calculations. In coming chapters, we will use this concept to measure value by bringing the benefits and costs from a project back to the present.*

OBJECTIVE 1

THE VALUATION OF SINGLE PERIOD PAYMENTS

Future Value

Compound interest

Interest that is earned on the sum of the principal and any interest reinvested during the prior periods.

Most of us encounter the concept of future value or the compounding of interest at an early age. Anyone who has ever had a savings account at a commercial bank or trust company or purchased a Government of Canada savings bond has received compound interest. **Compound interest** occurs when interest paid on the investment during the first period is added to the principal and, during the second period, interest is earned on this new sum.

For example, suppose we place \$100 in a savings account that pays 6 percent interest, compounded annually. How will our savings grow? At the end of the first year we have earned 6 percent, or \$6 on our initial deposit of \$100, giving us a total of \$106 in our savings account. The future value of the investment is determined by the following equation:

$$FV_1 = PV_0(1 + i) \tag{4-1}$$

where FV_1 = the future value of the investment at the end of one year
i = the annual compound interest rate

PV_0 = the present value or original amount invested at the beginning of the first year

In our example

$$FV_1 = PV_0(1 + i) \tag{4-1}$$
$$= \$100(1 + .06)$$
$$= \$100(1.06)$$
$$= \$106$$

Carrying these calculations one period further, we find that we now earn the 6 percent interest on a principal of \$106, which means we earn \$6.36 in interest during the second year. Why do we earn more interest during the second year than we did during the first? Simply because we now earn interest on the sum of the original principal (or present value) and the interest we earned in the first year. In effect we are now earning interest on interest; this is the concept of compound interest. The future value of the investment in the second year is determined by the following equation:

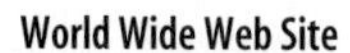

http://www.mathpro.com/math
Comment: Mathpro improves your mathematical problem-solving ability.

$$FV_2 = FV_1(1 + i) \tag{4-2}$$

which, for our example, gives

$$FV_2 = \$106(1.06)$$
$$= \$112.36$$

Looking back at Equation (4-1), we can see that FV_1, or \$106, is actually equal to $PV_0(1 + i)$, or \$100 (1 + .06). If we substitute these values into Equation (4-2), we get

$$FV_2 = PV_0(1 + i)(1 + i) \tag{4-3}$$
$$= PV_0(1 + i)^2$$

Carrying this forward into the third year, we find that we enter the year with \$112.36 and we earn 6 percent, or \$6.74 in interest, giving us a total of \$119.10 in our savings account. Expressing this mathematically:

$$FV_3 = FV_2(1 + i) \tag{4-4}$$
$$= \$112.36(1.06)$$
$$= \$119.10$$

If we substitute the value in Equation (4-3) for FV_2 into Equation (4-4), we find

$$FV_3 = PV_0(1 + i)(1 + i)(1 + i) \tag{4-5}$$
$$FV_3 = FV_2(1 + i)$$
$$FV_3 = PV_0(1 + i)^3$$

By now a pattern is beginning to be evident. We can generalize this formula to illustrate the value of our investment if it is compounded annually at a rate of i for n years to be[1]

$$FV_n = PV_0(1 + i)^n \tag{4-6}$$

where FV_n = the future value of the investment at the end of n years
n = the number of years during which the compounding occurs
i = the annual interest rate
PV_0 = the present value or original amount invested at the beginning of the first year

Table 4-1 illustrates how this investment of \$100 would continue to grow for the first ten years at a compound interest rate of 6 percent. It is easy to see that the amount of interest earned annually increases each year. Again, the reason is that each year interest is received on the sum of the original investment plus any interest earned in the past.

[1]The use of years as a unit of time is arbitrary. Any time period may be used. As a result, n is the number of periods and i is the rate per period.

Table 4-1 Illustration of Compound Interest Calculations

Year	Beginning Value	Interest Earned	Ending Value
1	$100.00	$6.00	$106.00
2	106.00	6.36	112.36
3	112.36	6.74	119.10
4	119.10	7.15	126.25
5	126.25	7.57	133.82
6	133.82	8.03	141.85
7	141.85	8.51	150.36
8	150.36	9.02	159.38
9	159.38	9.57	168.95
10	168.95	10.13	179.08

When we examine the relationship between the number of years an initial investment is compounded for and its future value graphically, as shown in Figure 4-1, we see that we can increase the future value of an investment by increasing the number of years we let it compound or by compounding it at a higher interest rate. We can also see this by examining Equation (4-6), since an increase in either i or n while PV_0 is held constant will result in an increase in FV_n.

PERSPECTIVE IN FINANCE

Keep in mind that future cash flows are assumed to occur at the end of the time period during which they accrue. For example, if a cash flow of $100 occurs in time period 5, it is assumed to occur at the end of time period 5 which is also the beginning of time period 6. In addition, cash flows that occur in time t = 0 *occur right now. These cash flows are already in present dollars.*

EXAMPLE

If we place $1,000 in a savings account paying 5 percent interest compounded annually, how much will our account accrue to in ten years? Substituting PV_0 = $1000, i = 5 percent, and n = 10 years into Equation (4-6), we get

$$\begin{aligned} FV_n &= PV_0(1+i)^n \qquad (4\text{-}6)\\ &= \$1{,}000(1+.05)^{10}\\ &= \$1{,}000(1.6289)\\ &= \$1{,}628.89 \end{aligned}$$

Thus at the end of ten years we will have $1,628.89 in our savings account.

Future-value interest factor ($FVIF_{i,n}$)

The value $(1+i)^n$ used as a multiplier to calculate an amount's future value.

As the determination of future value can be quite time-consuming when an investment is held for a number of years, tables have been compiled for the **future value interest factor** for i and n, ($FVIF_{i,n}$), which is defined as $(1+i)^n$. Table 4-2 is an abbreviated compound interest or future value interest factor table. A more comprehensive version of this table appears in Appendix A in the back of this book. The $FVIF_{i,n}$ values could also be easily determined using a calculator. Note that the compounding factors given in these tables represent the value of $1 compounded at rate i at the end of the nth year. Thus, to calculate the future value of an initial investment we need only to determine the $FVIF_{i,n}$ using a calculator or the tables at the end of the text and multiply this by the initial investment. In fact, we can rewrite Equation (4-6) as follows:

$$FV_n = PV_0(FVIF_{i,n}) \qquad (4\text{-}6a)$$

Figure 4-1 Future Value of $100 Initially Deposited and Compounded at 0, 5, and 10 Percent

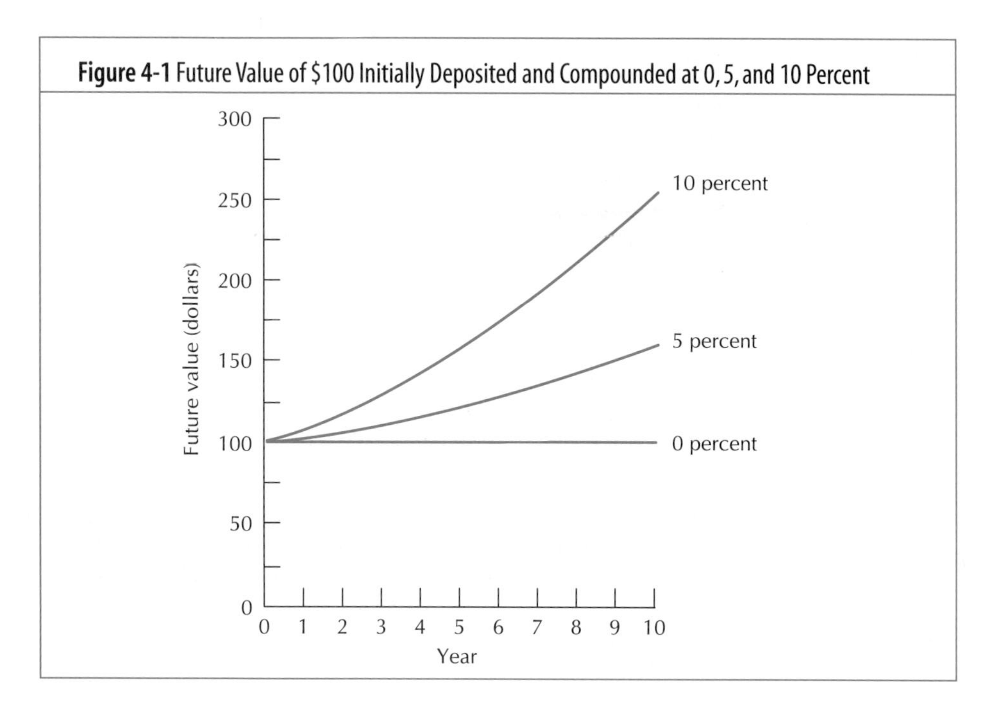

EXAMPLE

If we invest $500 in the bank where it will earn 8 percent interest compounded annually, how much will it be worth at the end of seven years? Looking in Table 4-2 in the row $n = 7$ and column $i = 8\%$, we find that the $FVIF_{8\%,7yr}$ has a value of 1.7138. Substituting this in Equation (4-6), we find

$$\begin{aligned} FV_n &= PV_0(1 + i)^n \\ &= \$500(1.7138) \\ &= \$856.90 \end{aligned} \tag{4-6}$$

Thus, we will have $856.90 at the end of seven years.

Table 4-2 $FVIF_{i,n}$ or the Compound Sum of $1

n	1%	2%	3%	4%	5%	6%	7%	8%	9%
1	1.0100	1.0200	1.0300	1.0400	1.0500	1.0600	1.0700	1.0800	1.0900
2	1.0201	1.0404	1.0609	1.0816	1.1025	1.1236	1.1449	1.1664	1.1881
3	1.0303	1.0612	1.0927	1.1249	1.1576	1.1910	1.2250	1.2597	1.2950
4	1.0416	1.0824	1.1255	1.1699	1.2155	1.2625	1.3108	1.3605	1.4116
5	1.0510	1.1041	1.1593	1.2167	1.2763	1.3382	1.4026	1.4693	1.5386
6	1.0615	1.1262	1.1941	1.2653	1.3401	1.4185	1.5007	1.5869	1.6771
7	1.0721	1.1487	1.2299	1.3159	1.4071	1.5036	1.6058	1.7138	1.8280
8	1.0829	1.1717	1.2668	1.3686	1.4775	1.5938	1.7182	1.8509	1.9926
9	1.0937	1.1951	1.3048	1.4233	1.5513	1.6895	1.8385	1.9990	2.1719
10	1.1046	1.2190	1.3439	1.4802	1.6289	1.7908	1.9672	2.1589	2.3674
11	1.1157	1.2434	1.3842	1.5395	1.7103	1.8983	2.1049	2.3316	2.5804
12	1.1268	1.2682	1.4258	1.6010	1.7959	2.0122	2.2522	2.5182	2.8127
13	1.1381	1.2936	1.4685	1.6651	1.8856	2.1329	2.4098	2.7196	3.0658
14	1.1495	1.3195	1.5126	1.7317	1.9799	2.2609	2.5785	2.9372	3.3417
15	1.1610	1.3459	1.5580	1.8009	2.0789	2.3966	2.7590	3.1722	3.6425

In the future we will find several uses for Equation (4-6); not only will we find the future value of an investment, but we can also solve for PV_0, i, or n. In any case, we will be given three of the four variables and will have to solve for the fourth.

PERSPECTIVE IN FINANCE

As you read through the chapter it is a good idea to solve the problems as they are presented. If you just read the problems, the principles behind them often do not sink in. The material presented in this chapter forms the basis for the rest of the course. Therefore a good command of the concepts underlying the time value of money is extremely important.[2]

EXAMPLE

How many years will it take for an initial investment of \$300 to grow to \$774 if it is invested at 9 percent compounded annually? In this problem we know the initial investment (PV_0 = \$300), the future value (FV_n = \$774), the compound growth rate (i = 9%), and we are solving for the number of years it must compound for, *n*. Substituting the known values in Equation (4-6), we find

$$FV_n = PV_0(1 + i)^n \tag{4-6}$$
$$\$774 = \$300(1 + .09)^n$$
$$2.58 = (1 + .09)^n$$

Thus we are looking for a value of 2.58 in the $FVIF_{i,n}$ table, and we know it must be in the 9% column. Looking down the 9% column for the value closest to 2.58, we find that it occurs in the $n = 11$ row. Thus, it will take 11 years for an initial investment of \$300 to grow to \$774 if it is invested at 9 percent compounded annually.

EXAMPLE

At what rate must \$100 be compounded annually for it to grow to \$179.10 in ten years? In this case we know the initial investment (PV_0 = \$100), the future value of this investment at the end of *n* years (FV_n = \$179.10), and the number of years that the initial investment will compound for (n = 10 years). Substituting into Equation (4-6), we get

$$FV_n = PV_0(1 + i)^n \tag{4-6}$$
$$\$179.10 = \$100(1 + i)^{10}$$
$$1.7910 = (1 + i)^{10}$$

We know we are looking in the $n = 10$ row of the $FVIF_{i,n}$ table for the value closest to 1.7910, and we find this in the $i = 6\%$ column. Thus, if we want our initial investment of \$100 to accrue to \$179.10 in ten years, we must invest it at 6 percent.

Moving Money Through Time with the Aid of a Financial Calculator

Time value of money calculations can be made simple with the aid of a financial calculator. In solving time value of money problems with a financial calculator you will be given three or four variables and will have to solve for the unknown variable. Before presenting any solution using a financial calculator we will introduce the calculator's five most common keys. (In most time value of money problems, only four of these keys are relevant.) These keys are:

[2]In 1624, Peter Minuit purchased Manhattan from the Indians for \$24 in "knick-knacks" and jewellery. If banks had existed then and the sum was invested at 8% compounded annually at the end of 1624, it would be worth \$4.433 trillion at the end of 1993.

Menu Key	Description
N	Stores (or calculates) the total number of payments or compounding periods
I/Y	Stores (or calculates) the interest or discount rate.
PV	Stores (or calculates) the present value of a cash flow or series of cash flows.
FV	Stores (or calculates) the future value, that is, the dollar amount of a final cash flow or the compound value of a single flow or series of cash flows.
PMT	Stores (or calculates) the dollar amount of each annuity payment deposited or received at the end of each year.

One thing you must keep in mind when using a financial calculator is that outflows generally have to be entered as negative numbers. In general, each problem will have two cash flows, one an outflow with a negative value and one an inflow with a positive value. The idea is that you deposit money in the bank at some point in time (an outflow), and at some other point in time you take money out of the bank (an inflow). Also, every calculator operates a bit differently with respect to entering variables. Needless to say, it is a good idea to familiarize yourself with exactly how your calculator functions.

As stated above, in any problem you will be given three or four variables. These variables will always include *N* and *I/Y*; in addition, two out of the final three variables, *PV*, *FV*, and *PMT*, will also be included. To solve a time value of money problem using a financial calculator, all you need to do is enter the appropriate numbers for three of the four variables and then press the key of the final variable to calculate its value. It is also a good idea to enter zero for any of the five variables not included in the problem in order to clear that variable.

Now let's solve the previous example using a financial calculator. We were trying to find at what rate must $100 be compounded annually for it to grow to $179.10 in ten years. The solution using a financial calculator would be as follows:

Step 1 Input Values of Known Variables

Data Input	Function Key	Description
10	N	Stores $N = 10$ years
100	+/- PV	Stores $PV = -\$100$
179.10	FV	Stores $FV = \$179.10$
0	PMT	Clears *PMT* to 0

Step 2 Calculate the value of the unknown variable

Function Key	Answer	Description
CPT I/Y	6.00%	Calculates $I/Y = 6.00\%$

Note: If you are using a TI BA II calculator, you must ensure that payments per year = 1.0 (2nd P/Y 1 ENTER).

Any of the problems in this chapter can easily be solved using a financial calculator; and the solutions to many examples using a financial calculator are provided in the margins. One final point: you will notice that solutions using the present-value tables versus solutions using a calculator may vary slightly—a result of rounding errors in the tables.[3]

[3]For further explanation, see the student's *Study Guide* which provides a tutorial on the use of financial calculators.

PERSPECTIVE IN FINANCE

The concepts of compound and present value will follow us through the remainder of this book. Not only will they allow us to determine the future value of any investment, but they will allow us to bring the benefits and costs from new investment proposals back to the present and thereby determine the value of the investment in today's dollars.

OBJECTIVE 2

Future Value with Nonannual Periods

Until now we have assumed the compounding period was always annual; however, it need not be, as we can see by examining chartered banks and trust companies that compound on a quarterly, daily, and in some cases continuous basis. Fortunately, this adjustment of the compounding period follows the same format. If we invest our money for five years at 8 percent interest compounded semiannually, we are really investing our money for 10 six-month periods during which we receive 4 percent interest each period. If it is compounded quarterly, we receive 2 percent interest per period for 20 three-month periods. This process can easily be generalized, giving us the following equation to find the future value of an investment where interest is compounded in nonannual periods:

$$FV_n = PV_0\left(1 + \frac{i}{m}\right)^{mn} \tag{4-7}$$

where FV_n = future value of the investment at the end of n years
n = number of years during which the compounding occurs
i = annual interest rate
PV_0 = present value or original amount invested at the beginning of the first year
m = the number of times compounding occurs during the year

In the case of continuous compounding, the value of m in Equation (4-7) is allowed to approach infinity. In fact, the effect of continuous compounding causes interest to earn interest immediately. As this happens, the value of $[1 + (i/m)]^{mn}$ approaches e^{in}, with e being defined as follows and having a value of approximately 2.7183:

$$e = \lim_{m \to \infty}\left(1 + \frac{1}{m}\right)^{m} \tag{4-8}$$

where ∞ indicates infinity. Thus the future value of an investment compounded continuously for n years can be determined from the following equation:

$$FV_n = PV_0 e^{(i)(n)} \tag{4-9}$$

where FV_n = the future value of the investment at the end of n years
e = 2.7183
n = the number of years during which the compounding occurs
i = the annual interest rate
PV_0 = the present value or original amount invested at the beginning of the first year

While continuous compounding appears quite complicated, it is used frequently and is a valuable theoretical concept. Continuous compounding takes on this importance because it allows interest to be earned on interest more frequently than any other compounding period. We can easily see the value of intra-year compounding by examining Table 4-3. Since interest is earned on interest more frequently as the length of the compounding period declines, there is an inverse relationship between the length of the compounding period and the effective annual interest rate.

Table 4-3 The Value of $100 Compounded at Various Intervals

For One Year at *i* Percent

i =	2%	5%	10%	15%
Compounded annually	$102.00	$105.00	$110.00	$115.00
Compounded semiannually	102.01	105.06	110.25	115.56
Compounded quarterly	102.02	105.09	110.38	115.87
Compounded monthly	102.02	105.12	110.47	116.08
Compounded weekly (52)	102.02	105.12	110.51	116.16
Compounded daily (365)	102.02	105.13	110.52	116.18
Compounded continuously	102.02	105.13	110.52	116.18

For Ten Years at *i* Percent

i =	2%	5%	10%	15%
Compounded annually	$121.90	$162.89	$259.37	$404.56
Compounded semiannually	122.02	163.86	265.33	424.79
Compounded quarterly	122.08	164.36	268.51	436.04
Compounded monthly	122.12	164.70	270.70	444.02
Compounded weekly (52)	122.14	164.83	271.57	447.20
Compounded daily (365)	122.14	164.87	271.79	448.03
Compounded continuously	122.14	164.87	271.83	448.17

EXAMPLE

If we place $100 in a savings account that yields 12 percent compounded quarterly, what will our investment grow to at the end of five years? Substituting $n = 5$, $m = 4$, $i = 12$ percent, and $PV_0 = \$100$ into Equation (4-7), we find

$$FV_5 = \$100\left(1 + \frac{.12}{4}\right)^{(5)(4)}$$
$$= \$100(1 + .03)^{20}$$
$$= \$100(1.8061)$$
$$= \$180.60$$

Thus, we will have $180.60 at the end of five years.

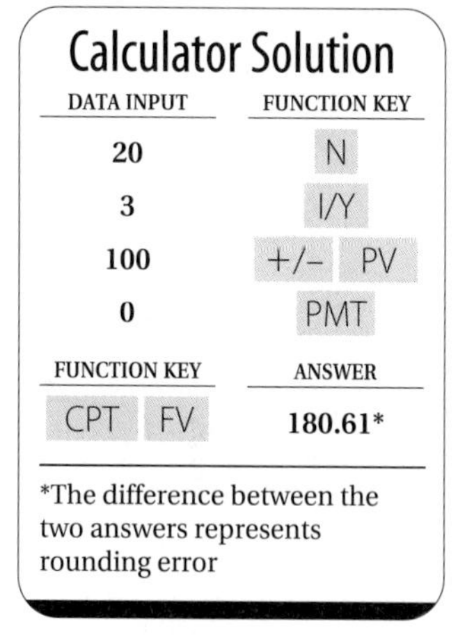

EXAMPLE

How much money will we have at the end of 20 years if we deposit $1,000 in a savings account yielding 10 percent interest continuously compounded? Substituting $n = 20$, $i = 10$ percent, and $PV_0 = \$1,000$ into Equation (4-9) yields

$$FV_{10} = \$1,000(2.7183)^{(.10)(20)}$$
$$= \$1,000(2.7183)^{2}$$
$$= \$1,000(7.3892)$$
$$= \$7,389.20$$

Thus, we will have $7,389.20 at the end of 20 years.

Present Value

OBJECTIVE 3

Up until this point we have been moving money forward in time. We know how much we have at the moment and are trying to determine how much that amount will grow to in a certain number of years when compounded at a specific rate. We are now going to look at the reverse question: What is the value in today's dollars of a sum of money to be received in the future? The answer to this question will help us determine the desirability of investment projects in Chapters 13, 14 and 15. In this case we are moving money to be received in the future, back to the present. We will be determining the **present value** of a lump sum, which in simple terms is the current value of a future payment. What we will be doing is, in fact, nothing other than inverse compounding. The differences in these techniques come about merely from the investor's point of view. In compounding we used the compound interest rate and the initial investment; in determining the present value we use the discount rate and present value. Determination of the discount rate was defined in Chapter 3 as the rate of return available on an investment of equal risk to the investment that is being discounted. Other than that, the technique and the terminology remain the same, and the mathematics is simply reversed. In Equation (4-6) we were attempting to determine the future value of an initial investment. We now want to determine the initial investment or present value. By dividing both sides of Equation (4-6) by $(1 + i)^n$, we get

Present value

The current value of a future sum.

$$PV_0 = FV_n\left[\frac{1}{(1+i)^n}\right] \tag{4-10}$$

where FV_n = the future value of the investment at the end of n years
n = the number of years until the payment will be received[4]
i = the opportunity rate or discount rate
PV_0 = the present value of the future sum of money

Since the mathematical procedure for determining the present value is exactly the inverse of determining the future value, we also find that the relationships among n, i, and PV_0 are just opposite of those we observed in the future value. The present value of a future sum of money is inversely related to both the number of years until the payment will be received and the discount rate. Figure 4-2 illustrates this relationship.

PERSPECTIVE IN FINANCE

While the present value equation [Equation (4-10)] will be used extensively in evaluating new investment proposals, it should be stressed that the present value equation is actually the same as the future value or compounding equation [Equation (4-6)], where it is solved for PV_0.

Calculator Solution

DATA INPUT	FUNCTION KEY
10	N
6	I/Y
500	+/- PV
0	PMT
FUNCTION KEY	**ANSWER**
CPT FV	279.20

EXAMPLE

What is the present value of \$500 to be received ten years from today if our discount rate is 6 percent? Substituting FV_{10} = \$500, n = 10, and i = 6 percent into Equation (4-10), we find

$$PV_0 = \$500\left[\frac{1}{(1+.06)^{10}}\right]$$

$$= \$500\left(\frac{1}{1.7908}\right)$$

$$= \$500\,(.5584)$$

$$= \$279.20$$

Thus, the present value of the \$500 to be received in ten years is \$279.20.

Present value interest factor ($PVIF_{i,n}$)

The value $1/(1+i)^n$ used as a multiplier to calculate an amount's present value.

To aid in the computation of present values, the **present value interest factor** for i and n, **($PVIF_{i,n}$)**, which is defined as $[1/(1+i)^n]$ has been compiled for various com-

[4]Note that a period does not need to be one year. Any period is allowed, with i being the rate per period and n the number of periods.

Figure 4-2 Present Value of $100 to Be Received at a Future Date and Discounted Back to the Present at 0, 5, and 10 percent

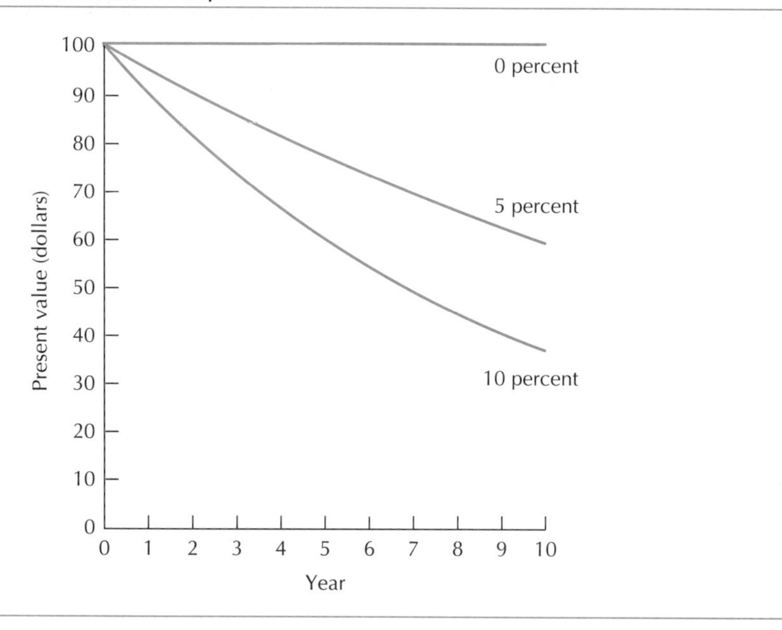

binations of i and n. The present value interest factors appear in Appendix B in the back of this book. An abbreviated version of Appendix B appears in Table 4-4. A close examination shows that the values in Table 4-4 are merely the inverses of those found in Appendix A. This, of course, is as it should be, as the values in Appendix A are $(1 + i)^n$ and those in Appendix B are $[1/(1 + i)^n]$. Now, to determine the present value of a sum of money to be received at some future date, we need only determine the value of the appropriate $PVIF_{i,n}$, by either using a calculator or consulting the tables and multiplying it by the future value. In fact, we can use our new notation and rewrite Equation (4-10) as follows:

$$PV_0 = FV_n(PVIF_{i,n}) \tag{4-10a}$$

EXAMPLE

What is the present value of $1,500 to be received at the end of 10 years if our discount rate is 8 percent? By looking in the $n = 10$ row and $i = 8\%$ column of Table 4-4, we find the value of $[1/(1 + .08)^{10}]$ is .4632. Substituting this value into Equation (4-10), we find

$$PV_0 = \$1{,}500(.4632)$$
$$= \$694.80$$

Thus, the present value of this $1,500 payment is $694.80.

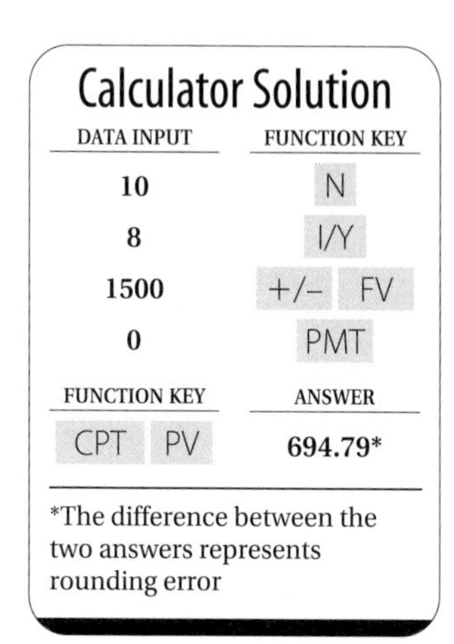

Again, we only have one present value–future value equation; that is, Equations (4-6) and (4-10) are identical. We have introduced them as separate equations to simplify our calculations; in one case we are determining the value in future dollars and in the other case the value in today's dollars. In either case the reason is the same: In order to compare values on alternative investments and to recognize that the value of a dollar received today is not the same as that of a dollar received at some future date, we must measure the dollar values in dollars of the same time period. Since all present values are comparable, we can add and subtract the present value of inflows and outflows to determine the present value of an investment.

Table 4-4 $PVIF_{i,n}$ **or the Present Value of $1**

n	1%	2%	3%	4%	5%	6%	7%	8%	9%
1	.9901	.9804	.9709	.9615	.9524	.9434	.9346	.9259	.9174
2	.9803	.9612	.9426	.9246	.9070	.8900	.8734	.8573	.8417
3	.9706	.9423	.9151	.8890	.8638	.8396	.8163	.7938	.7722
4	.9610	.9238	.8885	.8548	.8227	.7921	.7629	.7350	.7084
5	.9515	.9057	.8626	.8219	.7835	.7473	.7130	.6806	.6499
6	.9420	.8880	.8375	.7903	.7462	.7050	.6663	.6302	.5963
7	.9327	.8706	.8131	.7599	.7107	.6551	.6227	.5835	.5470
8	.9235	.8535	.7894	.7307	.6768	.6274	.5820	.5403	.5019
9	.9143	.8368	.7664	.7026	.6446	.5919	.5439	.5002	.4604
10	.9053	.8203	.7441	.6756	.6139	.5584	.5083	.4632	.4224
11	.8963	.8043	.7224	.6496	.5847	.5268	.4751	.4289	.3875
12	.8874	.7885	.7014	.6246	.5568	.4970	.4440	.3971	.3555
13	.8787	.7730	.6810	.6006	.5303	.4688	.4150	.3677	.3262
14	.8700	.7579	.6611	.5775	.5051	.4423	.3878	.3405	.2992
15	.8613	.7430	.6419	.5553	.4810	.4173	.3624	.3152	.2745

EXAMPLE

What is the present value of an investment that yields $500 to be received in five years and $1,000 to be received in ten years if our discount rate is 4 percent? Substituting the values of $n = 5$, $i = 4$ percent, and $FV_5 = \$500$ and $n = 10$, $i = 4$ percent, and $FV_{10} = \$1{,}000$ into Equation (4-10) and adding these values together, we find

$$\begin{aligned} PV_0 &= \$500\left[\frac{1}{(1+.04)^5}\right] + \$1{,}000\left[\frac{1}{(1+1.04)^{10}}\right] \\ &= \$500(.8219) + \$1000(.6756) \\ &= \$410.95 + \$675.60 \\ &= \$1{,}086.55 \end{aligned}$$

Again, present values are comparable because they are measured in the same time period's dollars.

OBJECTIVE 4

THE VALUATION OF ANNUITIES

Annuity

A series of equal dollar payments for a specified number of years.

Ordinary annuity

An annuity in which the payments occur at the end of each period.

Annuity due

An annuity in which the payments occur at the beginning of each period.

An **annuity** is a series of equal dollar payments for a specified number of years. Because annuities occur frequently in finance—for example, as bond interest payments—we will treat them specially.[5] Although the future and the present value of an annuity can be dealt with using the methods we have just described, these processes can be time-consuming, especially for larger annuities. Thus, we have modified the equations and will deal directly with annuities.

Although all annuities involve a series of equal dollar payments for a specified number of years, there are two basic types of annuities: an **ordinary annuity** and an **annuity due**. With an ordinary annuity, we assume that the payments occur at the end of each period; with an annuity due, the payments occur at the beginning of each period. Because an annuity due provides the payments earlier (at the beginning of each period instead of the end of the period as with an ordinary annuity), it has a greater present value. After we master ordinary annuities, we will examine annuities due. However, in finance, ordinary annuities are used much more fre-

[5]Other examples of annuities include retirement payments, lottery disbursements, mortgage payments, and auto-loan payments.

quently than are annuities due. Thus, in this text, whenever the term, "annuity" is used, you should assume that we are referring to an ordinary annuity unless otherwise specified.

The Future Value of Annuities

Future value of annuities

Depositing an equal sum of money at the end of each year for a certain number of years and allowing it to grow.

In determining the **future value of an annuity**, we are determining the value of an investment in which we deposited an equal sum of money at the end of each year for a certain number of years and allowed it to grow. Perhaps we are saving money for education, a new car, or a vacation home. In any case we want to know how much our savings will have grown by some point in the future.

Actually, we can find the answer by using Equation (4-6), our future value equation, and compounding each of the individual deposits to its future value. For example, if to provide for a college education we are going to deposit \$500 at the end of each year for the next five years in a bank where it will earn 6 percent interest, how much will we have at the end of five years? Compounding each of these values using Equation (4-6), we find that we will have \$2,818.50 at the end of five years.

$$
\begin{aligned}
FV_5 &= \$500(1.06)^4 + \$500(1.06)^3 + \$500(1.06)^2 + \$500(1.06)^1 + \$500 \\
&= \$500(1.2625) + \$500(1.1910) + \$500(1.1236) + \$500(1.0600) + \$500 \\
&= \$631.25 + \$595.50 + \$561.80 + \$530.00 + \$500.00 \\
&= \$2,818.55
\end{aligned}
$$

From examining the mathematics involved and the graph of the movement of money through time in Table 4-5, we can see that this procedure can be generalized to

$$FV_n = PMT\left[\sum_{t=0}^{n-1}(1+i)^t\right] \tag{4-11}$$

where FV_n = the future value of the annuity at the end of the nth year[6]
PMT = the annuity payment deposited or received at the end of each year
i = the annual interest rate
n = the number of years for which the annuity will last

To simplify the process of determining the future value of an annuity, the **Future Value Interest Factor of an Annuity** for i and n, ($FVIFA_{i,n}$) is defined as

$$\left[\sum_{t=0}^{n-1}(1+i)^t\right]$$

Future-value interest factor of an annuity ($FVIFA_{i,n}$)

The value

$$\left[\sum_{t=0}^{n-1}(1+i)^t\right]$$

used as a multiplier to calculate the future value of an annuity.

and has compiled for various combinations of n and i in Appendix C. An abbreviated version is given in Table 4-6. Using this new notation, we can rewrite Equation (4-11) as follows:

$$FV_n = PMT(FVIFA_{i,n}) \tag{4-11a}$$

Table 4-5 Illustration of a Five-Year \$500 Annuity Compounded at 6 Percent

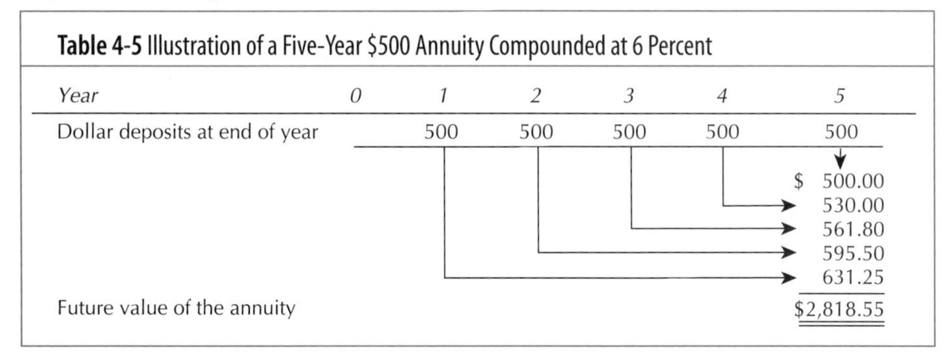

Year	0	1	2	3	4	5
Dollar deposits at end of year		500	500	500	500	500
						\$ 500.00
						530.00
						561.80
						595.50
						631.25
Future value of the annuity						\$2,818.55

[6]The future value of the annuity at the end of the nth year can also be expressed as follows:

$$FV_n = PMT\left[\frac{(1+i)^n - 1}{i}\right]$$

Table 4-6 $FVIFA_{i,n}$ or the Sum of an Annuity of $1 for *n* Years

n	1%	2%	3%	4%	5%	6%	7%	8%	9%
1	1.0000	1.0000	1.0000	1.0000	1.0000	1.0000	1.0000	1.0000	1.0000
2	2.0100	2.0200	2.0300	2.0400	2.0500	2.0600	2.0700	2.0800	2.0900
3	3.0301	3.0604	3.0909	3.1216	3.1525	3.1836	3.2149	3.2464	3.2781
4	4.0604	4.1216	4.1836	4.2465	4.3101	4.3746	4.4399	4.5061	4.5731
5	5.1010	5.2040	5.3091	5.4163	5.5256	5.6371	5.7507	5.8666	5.9847
6	6.1520	6.3081	6.4684	6.6330	6.8019	6.9753	7.1533	7.3359	7.5233
7	7.2135	7.4343	7.6625	7.8983	8.1420	8.3938	8.6540	8.9228	9.2004
8	8.2857	8.5830	8.8923	9.2142	9.5491	9.8975	10.2598	10.6366	11.0285
9	9.3685	9.7546	10.1591	10.5828	11.0266	11.4913	11.9780	12.4876	13.0210
10	10.4622	10.9497	11.4639	12.0061	12.5779	13.1808	13.8164	14.4866	15.1929
11	11.5668	12.1687	12.8078	13.4864	14.2068	14.9716	15.7836	16.6455	17.5603
12	12.6825	13.4121	14.1920	15.0258	15.9171	16.8699	17.8885	18.9771	20.1407
13	13.8093	14.6803	15.6178	16.6268	17.7130	18.8821	20.1406	21.4953	22.9534
14	14.9474	15.9739	17.0863	18.2919	19.5986	21.0151	22.5505	24.2149	26.0192
15	16.0969	17.2934	18.5989	20.0236	21.5786	23.2760	25.1290	27.1521	29.3609

Calculator Solution

DATA INPUT	FUNCTION KEY
5	N
6	I/Y
500	+/– PMT
0	PV
FUNCTION KEY	**ANSWER**
CPT FV	2,818.55

Solving the previous example to determine the value of $500 deposited at the end of each of the next five years in the bank at 6 percent interest after five years, we would look in the $n = 5$ year row and $i = 6\%$ column and find the value of $FVIFA_{6\%,5yr}$ to be 5.6371. Substituting this value into Equation (4-11a), we get

$$FV_5 = \$500(5.6371) = \$2,818.55$$

This is the same answer we obtained earlier.

Rather than asking how much we will accumulate if we deposit an equal sum in a savings account each year, a more common question is how much we must deposit each year in order to accumulate a certain amount of savings. This problem frequently occurs with respect to saving for large expenditures and pension funding obligations.

For example, we may know that we need $10,000 for education in eight years; how much must we deposit at the end of each year in the bank at 6 percent interest in order to have the college money ready? In this case we know the values of *n, i,* and FV_n in Equation (4-11); what we do not know is the value of *PMT.* Substituting these example values in Equation (4-11), we find

$$\$10,000 = PMT\left[\sum_{t=0}^{8-1}(1 + .06)^t\right]$$

$$\$10,000 = PMT(9.8975)$$

$$\frac{10,000}{9.8975} = PMT$$

$$PMT = \$1,010.36$$

Calculator Solution

DATA INPUT	FUNCTION KEY
8	N
6	I/Y
10,000	+/– FV
0	PV
FUNCTION KEY	**ANSWER**
CPT PMT	1,010.36

Thus, we must deposit $1,010.36 in the bank at the end of each year for eight years at 6 percent interest in order to accumulate $10,000 at the end of eight years.

EXAMPLE

How much must we deposit in an 8 percent savings account at the end of each year in order to accumulate $5,000 at the end of ten years? Substituting the values $FV_{10} = \$5,000$, $n = 10$, and $i = 8$ percent into Equation (4-11), we find

$$\$5,000 = PMT\left[\sum_{t=0}^{10-1}(1 + .08)^t\right]$$

$$\$5{,}000 = PMT(14.4866)$$

$$\frac{\$5{,}000}{14.4866} = PMT$$

$$PMT = \$345.15$$

Thus, we must deposit $345.15 per year for ten years at 8 percent in order to accumulate $5,000.

Calculator Solution

DATA INPUT	FUNCTION KEY
10	N
8	I/Y
5,000	+/- FV
0	PV

FUNCTION KEY	ANSWER
CPT PMT	345.15

PERSPECTIVE IN FINANCE

A time line often makes it easier to understand the time value of money problems. By visually plotting the flow of money you can better determine which formula to use. Arrows placed above the line are inflows, and arrows placed below the line represent outflows. One thing is certain, time lines reduce errors.

Present Value of an Annuity

Pension funds, insurance obligations, and interest received from bonds all involve annuities. To compare them, we need to know the present value of each. While we can find this by using the present value table in Appendix B, this can be time-consuming, particularly when the annuity lasts for several years. For example, if we wish to know what $500 received at the end of the next five years is worth to us today given the appropriate discount rate of 6 percent, we can simply substitute the appropriate values into Equation (4-10), such that

$$\begin{aligned} PV_0 &= \$500\left[\frac{1}{(1+.06)^1}\right] + \$500\left[\frac{1}{(1+.06)^2}\right] + \$500\left[\frac{1}{(1+.06)^3}\right] \\ &\quad + \$500\left[\frac{1}{(1+.06)^4}\right] + \$500\left[\frac{1}{(1+.06)^5}\right] \\ &= \$500(.9434) + \$500(.8900) + \$500(.8396) + \$500(.7921) + \$500(.7473) \\ &= \$2{,}106.20 \end{aligned}$$

Thus, the present value of this annuity is $2,106.20. From examining the mathematics involved and the graph of the movement of these funds through time in Table 4-7, we see that this procedure can be generalized to

$$PV_0 = PMT\left[\sum_{t=1}^{n} \frac{1}{(1+i)^t}\right] \tag{4-12}$$

where, PMT = the annuity payments deposited or received at the end of each year
i = the annual interest or discount rate
PV_0 = the present value of the future annuity[7]
n = the number of years for which the annuity will last

To simplify the process of determining the present value of an annuity, the **Present Value Interest Factor of an Annuity** for i and n, **($PVIFA_{i,n}$)** is defined as

$$\left[\sum_{t=1}^{n} \frac{1}{(1+i)^t}\right]$$

and has been compiled for various combinations of n and i in Appendix D with an abbreviated version given in Table 4-8. Using this new notation, we can rewrite equation 4-12 as follows:

Present value interest factor of an annuity ($PVIFA_{i,n}$)

The value

$$\left[\sum_{t=1}^{n} \frac{1}{(1+i)^t}\right]$$

used as a multiplier to calculate the future value of an annuity.

[7]The present value of this future annuity can also be expressed as follows:

$$PV_0 = PMT\,\frac{1 - \left[\dfrac{1}{(1+i)^n}\right]}{i}$$

Table 4-7 Illustration of a Five-Year $500 Annuity Discounted Back to the Present at 6 Percent

Year	*0*	*1*	*2*	*3*	*4*	*5*
Dollars received at the end of year		500	500	500	500	500
	$ 471.70 ← (year 1)					
	445.00 ← (year 2)					
	419.80 ← (year 3)					
	396.05 ← (year 4)					
	373.65 ← (year 5)					
Present value of the annuity	$2,106.20					

$$PV_0 = PMT(PVIFA_{i,n}) \tag{4-12a}$$

Solving the previous example to find the present value of $500 received at the end of each of the next five years discounted back to the present at 6 percent, we look in the $n = 5$ year row and $i = 6\%$ column and find the $PVIFA_{6\%,5yr}$ to be 4.2124. Substituting the appropriate values into Equation (4-12a), we find

$$\begin{aligned} PV_0 &= \$500(4.2124) \\ &= \$2106.20 \end{aligned}$$

This, of course, is the same answer we calculated when we individually discounted each cash flow to the present. The reason is that we really only have one table; the Table 4-8 value for an n-year annuity for any discount rate i is merely the sum of the first n values in Table 4-4. We can see this by comparing the value in the present-value-of-an-annuity table (Table 4-8) for $i = 8\%$ and $n = 6$ years, which is 4.6229, with the sum of the values in the $i = 8\%$ column and $n = 1, \ldots, 6$ rows of the present value table (Table 4-4), which is equal to 4.6229, as shown in Table 4-9.

Calculator Solution

DATA INPUT	FUNCTION KEY
10	N
5	I/Y
1,000	+/- PMT
0	FV

FUNCTION KEY	ANSWER
CPT PV	7,721.73*

*The difference between the two answers represents rounding error

EXAMPLE

What is the present value of a ten-year $1,000 annuity discounted back to the present at 5 percent? Substituting $n = 10$ years, $i = 5$ percent, and $PMT = \$1,000$ into Equation (4-12), we find

$$PV_0 = \$1,000\left[\sum_{t=1}^{10} \frac{1}{(1 + .05)^t}\right] = \$1,000(PVIFA_{5\%,10yr})$$

Table 4-8 $PVIFA_{i,n}$ or the Present Value of an Annuity of $1

n	1%	2%	3%	4%	5%	6%	7%	8%	9%
1	0.9901	0.9804	0.9709	0.9615	0.9524	0.9434	0.9346	0.9259	0.9174
2	1.9704	1.9416	1.9135	1.8861	1.8594	1.8334	1.8080	1.7833	1.7591
3	2.9410	2.8839	2.8286	2.7751	2.7232	2.6730	2.6243	2.5771	2.5313
4	3.9020	3.8077	3.7171	3.6299	3.5460	3.4651	3.3872	3.3121	3.2397
5	4.8534	4.7135	4.5797	4.4518	4.3295	4.2124	4.1002	3.9927	3.8897
6	5.7955	5.6014	5.4172	5.2421	5.0757	4.9173	4.7665	4.6229	4.4859
7	6.7282	6.4720	6.2303	6.0021	5.7864	5.5824	5.3893	5.2064	5.0330
8	7.6517	7.3255	7.0197	6.7327	6.4632	6.2098	5.9713	5.7466	5.5348
9	8.5660	8.1622	7.7861	7.4353	7.1078	6.8017	6.5152	6.2469	5.9952
10	9.4713	8.9826	8.5302	8.1109	7.7217	7.3601	7.0236	6.7101	6.4177
11	10.3676	9.7868	9.2526	8.7605	8.3064	7.8869	7.4987	7.1390	6.8052
12	11.2551	10.5753	9.9540	9.3851	8.8633	8.3838	7.9427	7.5361	7.1607
13	12.1337	11.3484	10.6350	9.9856	9.3936	8.8527	8.3577	7.9038	7.4869
14	13.0037	12.1062	11.2961	10.5631	9.8986	9.2950	8.7455	8.2442	7.7862
15	13.8651	12.8493	11.9379	11.1184	10.3797	9.7122	9.1079	8.5595	8.0607

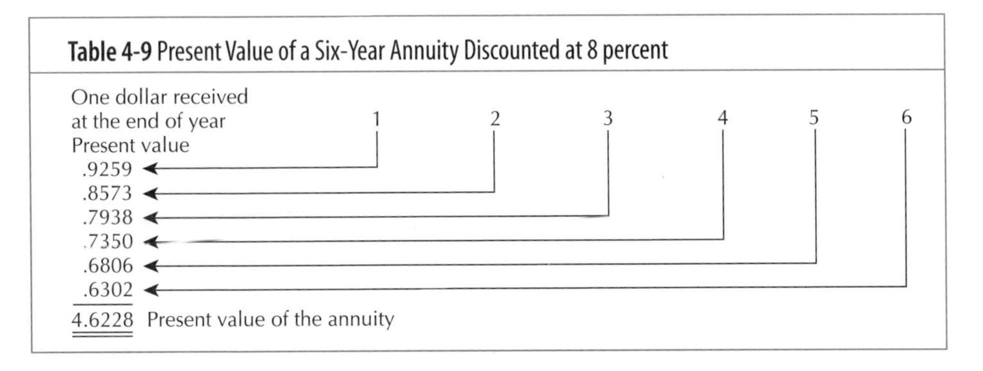

Table 4-9 Present Value of a Six-Year Annuity Discounted at 8 percent

One dollar received at the end of year	Present value
1	.9259
2	.8573
3	.7938
4	.7350
5	.6806
6	.6302
Present value of the annuity	4.6228

Determining the value for ($PVIFA_{5\%,10yr}$) from Table 4-8, row $n = 10$, column $i = 5\%$, and substituting it into the equation, we get

$$PV_0 = \$1{,}000(7.7217)$$
$$= \$7{,}721.70$$

Thus, the present value of this annuity is $7,721.70.

As with our other compounding and present value tables, given any three of the four unknowns in Equation (4-12), we can solve for the fourth. In the case of the present-value-of-an-annuity table we may be interested in solving for *PMT*, if we know *i*, *n*, and PV_0. The financial interpretation of this action would be: How much can be withdrawn, perhaps as a pension or to make loan payments, from an account that earns *i* percent compounded annually for each of the next *n* years if we wish to have nothing left at the end of *n* years? For an example, if we have $5,000 in an account earning 8 percent interest, how large an annuity can we draw out each year if we want nothing left at the end of five years? In this case the present value, PV_0, of the annuity is $5,000, $n = 5$ years, $i = 8$ percent, and *PMT* is unknown. Substituting this into Equation (4-12), we find

$$\$5{,}000 = PMT(3.9927)$$
$$\$1{,}252.29 = PMT$$

Thus, this account will fall to zero at the end of five years if we withdraw $1,252.29 at the end of each year.

Calculator Solution

DATA INPUT	FUNCTION KEY
5	N
8	I/Y
5,000	+/– PV
0	FV
FUNCTION KEY	**ANSWER**
CPT PMT	1,252.28*

*The difference between the two answers represents rounding error

MAKING INTEREST RATES COMPARABLE

OBJECTIVE 5

In order to make intelligent decisions on where to invest or borrow money, it is important that we make the quoted (nominal) interest rate comparable. Unfortunately, some rates are quoted as compounded annually, while others are quoted as compounded quarterly or compounded daily. But we already know that it is not fair to compare interest rates with different compounding periods to each other. Thus, the only way interest rates can logically be compared is to convert them to some common compounding period and then compare them. The **effective annual rate** (EAR) enables an investor to compare interest rates with different compounding periods. In order to understand the process of making different interest rates comparable, it is necessary to define the nominal or quoted interest rate.

Effective annual rate or annual percentage yield

The effective annual rate expresses the nominal rate in terms of an annual compounded rate.

Nominal or quoted interest rate

The stated rate of interest on the debt contract.

The **nominal or quoted interest rate** is the rate of interest stated on the contract. For example, if you shop around for loans and are quoted interest rates of 8 percent compounded annually and 7.85 percent compounded quarterly, then 8 percent and 7.85 percent would be nominal rates. Unfortunately, because on one the interest rate is compounded annually while on the other interest is compounded quarterly, they are not comparable. In fact, it is never appropriate to compare nominal rates unless they include the same number of compounding periods per year. To make these interest rates comparable, we must calculate an equivalent rate at some common compounding period. We do this by calculating the effective annual rate or annual percentage yield. The effective annual rate expresses the nominal rate in terms of an annual compounded rate.

For example, the loan that was quoted by the bank at 7.85 percent compounded quarterly can be converted to an effective annual rate as follows:

$$(1 + EAR)^f = \left(1 + \frac{i}{m}\right)^m \tag{4-13}$$

where, EAR = effective annual rate

f = frequency at which payments are made during the year; the effective annual rate is determined as if payments are made on an annual basis, that is $f = 1$

i = nominal (quoted) interest rate

m = number of times compounding occurs during the year

$$(1 + EAR)^1 = \left(1 + \frac{0.0785}{4}\right)^4$$

$$EAR = (1 + 0.01963)^{(4/1)} - 1$$

$$EAR = 1.08084 - 1 = 0.08084 \text{ or } 8.08\%$$

The EAR for a quoted interest rate compounded annually would be calculated in a similar manner as above and is 8 percent.

Generalizing on this process, we can calculate the EAR using the following equation:

$$EAR = \left(1 + \frac{i}{m}\right)^m - 1 \tag{4-14}$$

where EAR is the effective annual rate, i is the nominal rate of interest per year, and m is the number of compounding periods within a year. Given the wide variety of compounding periods used by business and banks, it is important to know how to make these rates comparable so that logical decisions can be made.

Calculator Solution

DATA INPUT	FUNCTION KEY
4	N
15	I/Y
6000	+/- PV
0	FV
FUNCTION KEY	**ANSWER**
CPT PMT	2,101.59*

*The difference between the two answers represents rounding error

TERM LOANS

Term loans generally share three common characteristics: they (1) have maturities of one to ten years, (2) are repaid in periodic instalments (such as quarterly, semi-annual, or annual payments) over the life of the loan, and (3) are usually secured by a chattel mortgage on equipment or a mortgage on real property. The principal suppliers of term credit are commercial banks, insurance companies, and to a lesser extent pension funds. For example, commercial banks restrict their term lending to one- to five-year maturities. These shorter-maturity term loans are often secured by chattel mortgages, or a mortgage on machinery and equipment, or with securities such as shares and bonds. Insurance companies and pension funds with their longer-term liabilities generally make loans with five- to fifteen-year maturities. The longer-maturity term loans are often secured by mortgages on real estate.

Loan amortization

A procedure by which a loan is paid off in equal instalments.

Loan Amortization

Term loans are generally repaid with periodic instalments, which include both an interest and a principal component. Thus, the loan is repaid over its life with equal annual, semiannual, or quarterly payments. The procedure by which loans are paid off in equal periodic payments is called **loan amortization**. This procedure requires that we solve for PMT, the annuity payment value when i, n, and PV_0 are known.

EXAMPLE

Suppose a firm wants to purchase a piece of machinery. To do this, it borrows \$6,000 to be repaid in four equal payments at the end of each of the next four years, and the interest rate that is paid to the lender is 15 percent on the outstanding portion of the loan. To determine what the annual payments associated with the repayment of this debt will be, we simply use Equation (4-12) and solve for the value of *PMT*, the annual annuity. Again we know three of the four values in the equation, PV_o, *i*, and *n*. The present value (PV_o) of the future annuity is \$6,000; *i*, the annual interest rate, is 15 percent; and *n*, the number of years for which the annuity will last, is four years. *PMT*, the annuity payment deposited or received at the end of the year, is unknown. Substituting these values in Equation (4-12) we find

$$\$6{,}000 = PMT\left[\sum_{t=1}^{4} \frac{1}{(1+0.15)^t}\right]$$

$$\$6{,}000 = PMT(PVIFA_{15\%,4yr})$$

$$\$6{,}000 = PMT(2.8550)$$

$$PMT = \$2{,}101.58$$

In order to repay the principal and interest on the outstanding loan in four years the annual payments would be \$2,101.58. It should be noted that in some cases term loans require a larger final payment which is called a balloon payment. One approach that can be used to determine either the amount of balance left to be paid off or the total interest paid after *n* payments in a term loan, is to separate the interest and principal payments for each period. Such a breakdown of interest and principal payments is given in the loan amortization schedule. Table 4-10 shows the loan amortization schedule which provides a breakdown of interest and principal payments for our example of a \$6,000 term loan. As you can see, the interest payments decline each year as the loan outstanding declines.

Balloon payment

The final debt payment that repays the outstanding balance of the loan, typically larger than the loan's other payments.

Calculator Solution

Amortization Worksheet

KEYSTROKES	DISPLAY
2nd AMORT	P1=(old contents)
2nd CLR Work	P1=1.00*
↓	P2=1.00
↓	BAL=–4,798.41
↓	PRN=1,201.59
↓	INT=900.00
↓	P1=1.00
2 ENTER	P1=2.00
↓	P2=1.00
2 ENTER	P2=2.00
↓	BAL=–3,416.58
↓	PRN=1,3818.83
↓	INT=719.76
↓	P1=2.00
3 ENTER	P1=3.00
↓	P2=2.00
3 ENTER	P2=3.00
↓	BAL=–1,827.47
↓	PRN=1,589.11
↓	INT=512.49
↓	P1=3.00
4 ENTER	P1=4.00
↓	P2=3.00
4 ENTER	P2=4.00
↓	BAL=0.00
↓	PRN=1827.47
↓	INT=274.12

*Note that the balance left to be paid off, as well as the total interest paid can be determined after a range of *n* payments by entering the coordinates of the range in P1 and P2.

Amortization of a Home Mortgage Loan

The payments of a home mortgage loan are affected by Government of Canada regulations which require the interest rate of a home mortgage loan to be compounded on a semiannual basis. The amount of loan payments is also influenced by the

Table 4-10 Loan Amortization Schedule Involving a \$6,000 Loan at 15 Percent to Be Repaid in Four Years

Year	Annuity	Interest Portion of the Annuity[a]	Repayment of the Principal Portion of the Annuity[b]	Outstanding Loan Balance after the Annuity Payment
1	\$2,101.58	\$900.00	\$1,201.58	\$4,798.42
2	2,101.58	719.76	1,381.82	3,416.60
3	2,101.58	512.49	1,589.09	1,827.51
4	2,101.58	274.07	1,827.51	

[a]The interest portion of the annuity is calculated by multiplying the outstanding loan balance at the beginning of the year by the interest rate of 15 percent. Thus, for year 1 it was \$6,000.00 × 0.15 = \$900.00, for year 2 it was \$4,798.42 × 0.15 = \$719.76, and so on.
[b]Repayment of the principal portion of the annuity was calculated by subtracting the interest portion of the annuity (column 2) from the annuity (column 1).

Loan Rates Not What They Appear: Compounding Can Significantly Hike the Interest You Pay

The rate you pay depends on how often the rate compounds. The more often a rate compounds, the higher the cost.

As shown in the accompanying table, 7% compounded semi-annually equates to a true(effective) annual rate of 7.123%—nearly, one-eighth of a percentage point more than the posted, or nominal, rate. With monthly compounding, it's 7.229%—nearly a quarter point more.

A quarter of a percentage point may not sound like much but it costs you 3.27% more in cold hard cash.

This discrepancy was the crux of a hearing last Friday in which the Supreme Court of Canada put an end to a long running case.

Dunphy Leasing Enterprises Ltd., a Calgary company that went under in 1982, claimed its bank was overcharging on a line of credit because the agreement called for a rate of 24% while the true(effective) rate, after monthly compounding, was 26.8%.

In a rare move, the high court ruled against Dunphy as soon as the arguments were heard. It's been a well-established practice for Canadian lenders to just quote the nominal, or base rate with the compounding frequency disclosed only in the agreement itself.

There are, however, compounding conventions and some loans are regulated by legislation.

For example, consumer contracts such as car loans and lines of credit compound monthly. The typical fixed-rate home mortgage with blended payments compounds semi-annually. But variable rate mortgages compound monthly, although the contract will probably include the semi-annual equivalents. If you have a second mortgage, it might also compound monthly.

Surprisingly, Visa and Mastercard credit cards issued by financial institutions don't compound interest. If your rate is 15.5%, that's what you pay. But credit cards issued by retailers normally compound monthly. For years, their standard rate has been 28.8%, equating to a true (effective) annual rate of 32.923%.[8]

The Edmonton-based Borrowers Action Society (BAS) has been pressing class suits against several retail card issuers, claiming it's illegal to quote 28.8% while really collecting 32.923%. Their cause has been financed with $100 donations from people who were promised a share of any judgment. But it's highly doubtful BAS will get anywhere after last week's Supreme Court decision.

BAS's efforts would be better directed at Parliament and the provincial legislatures. For example, the U.S. and Britain have legislation requiring lenders to quote the true(effective) annual rate.

Yes, just as borrowers pay different true(effective) rates on different loans, savers earn different rates on different investments. For example, a 5.5% GIC yields less than a 5.5% strip bond. That's because the GIC compounds annually while the strip bond's yield is based on semi-annual compounding.

Source: Excerpts from: Bruce Cohen, "Loan Rates Not What They Appear: Compounding Can Significantly Hike the Interest You Pay," *The Financial Post*, March 24, 1994, p. 21.

How Rates Are

Nominal rate	Semiannual	Monthly
5.00%	5.062%	5.116%
5.50	5.576	5.641
6.00	6.090	6.168
6.50	6.606	6.697
7.00	7.123	7.229
7.50	7.641	7.763
8.00	8.160	8.300
8.50	8.681	8.839
9.00	9.203	9.381
9.50	9.726	9.925
10.00	10.250	10.471
11.00	11.302	11.572
12.00	12.360	12.683
13.00	13.423	13.803
14.00	14.490	14.934
15.00	15.562	16.075
16.00	16.640	17.227
28.80	30.874	32.923

[8]In 1994, a group representing 9,000 people lost a class-action lawsuit for $720 million in interest refunds from the credit granting arm of Canadian Tire Ltd. This company was charging 2.4 percent monthly on its credit accounts and told its customers that the rate was 28.8 percent annually. As its defence, Canadian Tire Ltd. argued successfully that 28.8% percent is considered an equivalent. *The Ottawa Citizen*, "Canadian Tire Wins Court Case," August 24, 1994, p. D12.

frequency of loan payments that the home-buyer chooses to make during the year. Canadian lending institutions have provided home-buyers with greater flexibility in developing their loan repayment schedules by offering packages that permit home-buyers to vary the frequency of the loan payments during the year—for example, weekly, bimonthly, monthly, quarterly and even semi-annually. Fortunately, the calculation to determine the amount of loan payment is similar to that of the annuity payment. However, we must adjust for the frequency of loan payments by using the effective interest rate which is calculated as follows:

$$k_{eff} = \left(1+\frac{i}{m}\right)^{(m/f)} - 1 \tag{4-15}$$

where k_{eff} = the effective interest rate for the frequency of payments during the year
f = the frequency of annuity payments during the year
i = the annual interest rate
m = the number of times compounding occurs during the year

Now the payment can be determined by substitution into the equation for the present value of an annuity as follows:

$$PV_0 = PMT\left[\sum_{t=1}^{(n)(f)} \frac{1}{(1 + k_{eff})^t}\right] \tag{4-16}$$

where PMT = the annuity payment required at the end of each period
f = the frequency of annuity payments during the year
n = the number of years for which the annuity will last
k_{eff} = the effective interest rate for the frequency of payments during the year
PV_0 = the present value of the annuity.

For example, suppose a home-buyer negotiates a home mortgage loan with his local bank over a 25-year period in order to purchase a $150,000 home. The local bank requires a 25 percent down payment on the purchase price and charges an annual interest rate of 9.80% for the loan. The home-buyer has decided to pay his loan payments on a monthly basis. The loan's terms are renegotiable after three years. Determine the amount of the monthly payment owed by the home-buyer.

The first step of this calculation is to determine the effective interest rate on a monthly basis since the frequency of the loan payment is being repaid on a monthly basis ($f = 12$).

$$k_{eff} = \left(1 + \frac{0.098}{2}\right)^{(2/12)} - 1$$

$$k_{eff} = (1 + .049)^{0.1667} - 1$$

$$k_{eff} = 0.80\% \text{ per month}$$

The second step is to determine the amount of the loan after the down payment is deducted. The down payment represents (.25)($150,000) or $37,500; the home-buyer will require a loan of ($150,000 − $37,500) or $112,500.

The third step is to find the amount of the annuity payment that is required over the 300 months of the loan using the present value annuity formula:

$$112{,}500 = PMT\left[\sum_{t=1}^{(12)(25)} \frac{1}{(1 + 0.008)^t}\right]$$

$$PMT = \frac{112{,}500}{113.5515}$$

$$PMT = \$991.19 \text{ per month}$$

As shown in Table 4-11, the home-buyer will be required to pay a monthly payment of $991.19 over the next three years, at which time the home-buyer can renegotiate the loan terms.

Table 4-11 Home Mortgage Loan Amortization Schedule for the First Four Periods

Period	Beginning Balance	Amount of Payment	Interest	Amount of Repayment	Ending Balance
1	$112,500.00	$991.19	$900.54	$90.65	$112,409.35
2	112,409.35	991.19	899.81	91.38	112,317.97
3	112,317.97	991.19	899.08	92.11	112,225.86
4	112,225.86	991.19	898.34	92.85	112,133.01
.	.	.	.	.	.
.	.	.	.	.	.
.	.	.	.	.	.
.	.	.	.	.	.
297	3,886.66	991.19	31.11	960.08	2,926.59
298	2,926.59	991.19	23.43	967.76	1,958.82
299	1,958.82	991.19	15.68	975.51	983.32
300	983.32	983.32	7.87	983.32	0.00

OBJECTIVE 6

Annuities Due

An annuity due is a series of equal payments deposited at the beginning of each period. For example, leasing is a long-term form of financing, which requires a series of payments to be deposited at the beginning of each period. In a similar manner, we can determine the present value or the future value of an annuity due. In determining the future value of an annuity due, we determine the value of an investment in which we deposit an equal sum of money at the beginning of each year for a certain number of years and allow it to grow. We now redo the earlier example in which we provide for a college education by depositing $500 at the beginning of each year for the next five years in a bank where it will earn 6 percent interest. How much will we have at the end of five years?

$$
\begin{aligned}
FV_5 &= \$500(1+.06)^5 + \$500(1+.06)^4 + \$500(1+.06)^3 + \$500(1+.06)^2 \\
&\quad + \$500(1+.06)^1 \\
&= \$500(1.3382) + \$500(1.2625) + \$500(1.1910) + \$500(1.1236) + \$500(1.0600) \\
&= \$669.11 + \$631.24 + \$595.51 + \$561.80 + \$530.00 \\
&= \$2{,}987.66
\end{aligned}
$$

Compounding each of these values using Equation (4-6), we find that we will have $2,987.66 at the end of five years.

From examining the mathematics involved and the graph of the movement of money through time in Table 4-12, we can see that this procedure can be generalized to

Table 4-12 Illustration of a Five-Year $500 Annuity Due Compounded at 6 Percent

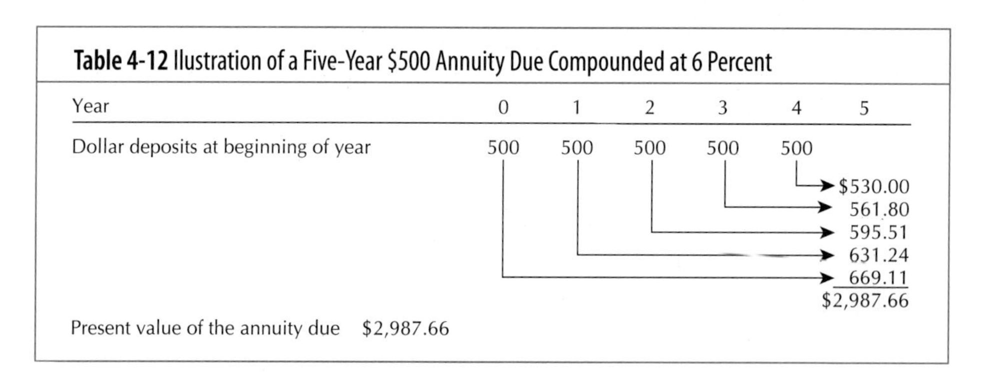

$$FV_n = PMT\left[\sum_{t=0}^{n-1}(1+i)^t\right][(1+i)] \tag{4-17}$$

or

$$FV_n = PMT(FVIFA_{i,n})(1+i) \tag{4-18}$$

To solve for our example, we use Equation (4-18) to find the future value of this annuity due

$$FV_n = \$500(FVIFA_{6\%,5})(1+.06)$$
$$FV_n = \$500(5.6371)(1.06)$$
$$FV_n = \$2{,}987.66$$

Thus, the future value of this annuity due is $2,987.66.

In determining the present value of an annuity due, we are determining today's value of a sum of equal payments which have been deposited at the beginning of each period. For example, if we wish to know what $500 received at the beginning of the next five years is worth to us today given the appropriate discount rate of 6 percent, we can simply substitute the appropriate values into Equation (4-10), such that

$$PV_0 = \$500 + \$500\left[\frac{1}{(1+.06)^1}\right] + \$500\left[\frac{1}{(1+.06)^2}\right] + \$500\left[\frac{1}{(1+.06)^3}\right] + \$500\left[\frac{1}{(1+.06)^4}\right]$$
$$= \$500 + \$500(.9434) + \$500(.8900) + \$500(.8396) + \$500(.7921)$$
$$= \$2{,}232.55$$

Thus, the present value of this annuity due is $2,232.55. From examining the mathematics involved and the graph of the movement of these funds through time in Table 4-13, we see that this procedure can be generalized to

$$PV_0 = PMT\left[\sum_{t=1}^{n}\frac{1}{(1+i)^t}\right][(1+i)] \tag{4-19}$$

or

$$PV_0 = PMT(PVIFA_{i,n})(1+i) \tag{4-20}$$

To solve for our example, we use Equation (4-20) to find the present value of this annuity due

$$PV_0 = \$500(PVIFA_{6\%,5})(1+.06)$$
$$= \$500(4.2124)(1.06)$$
$$= 2{,}232.57$$

Thus, the present value of this annuity due is $2,232.57.

Table 4-13 Illustration of a Five-Year $500 Annuity Due Discounted at 6 Percent

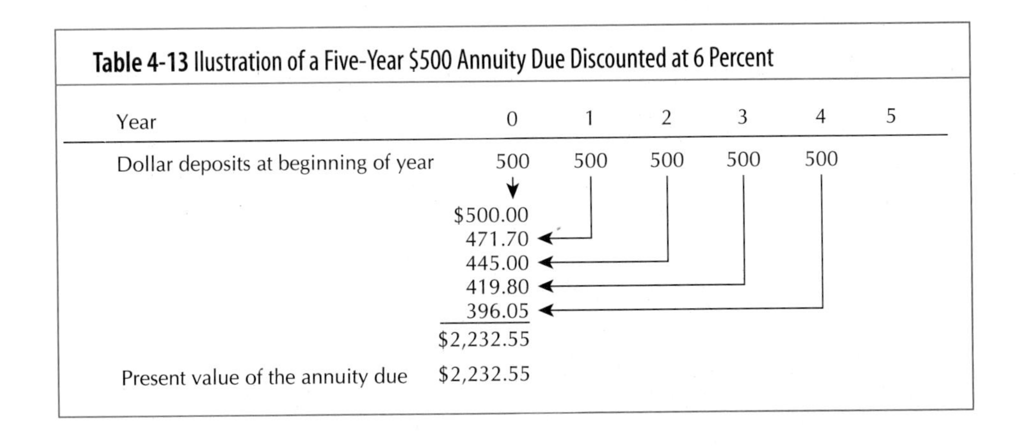

PERPETUITIES

A perpetuity is an annuity that continues forever; that is, every year from its establishment, this investment pays the same dollar amount. An example of a perpetuity is a preferred share that yields a constant dollar dividend infinitely. Determining the present value of a perpetuity is delightfully simple; we merely need to divide the constant flow by the discount rate.[9] For example, the present value of a $100 perpetuity discounted back to the present at 5 percent is $100/.05 = $2,000. Thus, the equation representing the present value of a perpetuity is

$$PV_0 = \frac{PP}{i} \tag{4-21}$$

where PV_0 = the present value of perpetuity
PP = the constant dollar amount provided by the perpetuity
i = the annual interest or discount rate

EXAMPLE

What is the present value of a $500 perpetuity discounted back to the present at 8 percent? Substituting *PP* = $500 and *i* = .08 into Equation (4-21), we find

$$PV_0 = \frac{\$500}{.08} = \$6,250$$

Thus, the present value of this perpetuity is $6,250.

PRESENT VALUE OF AN UNEVEN STREAM

Although some projects will involve a single cash outflow and a series of cash inflows of equal amounts, many projects will involve uneven cash flows over several years. For the analysis of projects we will be comparing not only the present value of cash flows between projects, but also the cash inflows and outflows within a particular project. Again we will be required to determine that project's present value. However, this will not be difficult because the present value of any cash flow is measured in today's dollars and thus can be compared, through addition for inflows and subtraction for outflows, to the present value of any other cash flow also measured in today's dollars. For example, suppose we wished to find the present value of the following cash flows:

Year	Cash Flow
1	$500
2	200
3	−400
4	500
5	500
6	500
7	500
8	500
9	500
10	500

[9]The concept of valuing a perpetuity is used to determine the intrinsic value of a preferred share.

Table 4-14 Illustration of an Example of Present Value of an Uneven Stream Involving One Annuity Discounted to Present at 6 percent

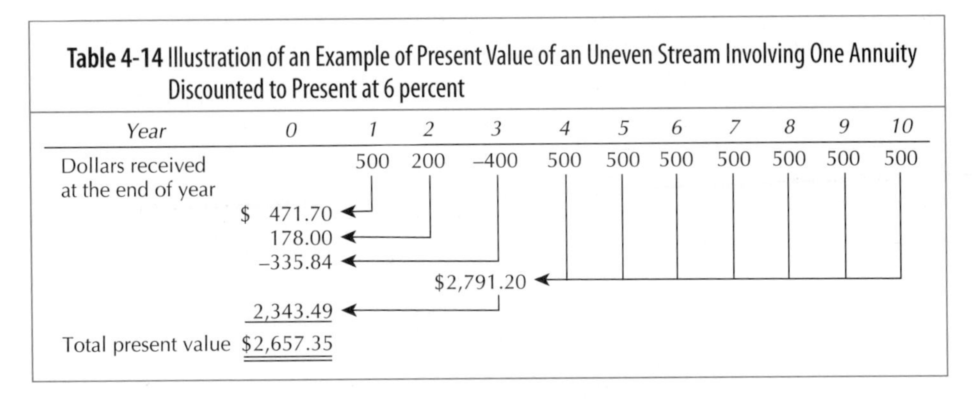

If given a 6 percent discount rate, we would merely discount the flows back to the present and total them by adding in the positive flows and subtracting the negative ones. However, this problem is complicated by the annuity of $500 that runs from years 4 through 10. To accommodate this factor, we can first discount the annuity back to the beginning of period 4 (or end of period 3) by using the present-value-of-an-annuity table and get its present value at that point in time. We then use the present-value-of-a-single-flow table (Table 4-4) and discount the present value of this annuity at the end of period 3 back to the present. In effect we discount twice, first back to the end of period 3 then back to the present. This is shown graphically in Table 4-14 and numerically in Table 4-15. Thus, the present value of this uneven stream of cash flows is $2,657.35.

EXAMPLE

What is the present value of an investment involving $200 received at the end of years 1 through 5, a $300 cash outflow at the end of year 6, and $500 received at the end of years 7 through 10 given a 5 percent discount rate? Here we have two annuities, one that can be discounted directly back to the present by multiplying it by the value of the $PVIFA_{5\%,5yr}$ and one that must be discounted twice to bring it back to the present. This second annuity must first be discounted back to the beginning of period 7 or end of period 6, by multiplying it by the value of the $PVIFA_{5\%,4yr}$; then, the present value of the annuity at the end of period 6 must be discounted back to the present, by multiplying it by the value of the $PVIFA_{5\%,6yr.}$ To arrive at the total present value of this investment, we subtract the present value of the $300 cash outflow at the end of year 6 from the sum of the present value of the two annuities. Table 4-16 shows this graphically; Table 4-17 gives the calculations. Thus the present value of this series of cash flows is $1,965.05.

Table 4-15 Determination of Present Value of an Example with Uneven Stream Involving One Annuity Discounted to Present at 6 Percent

1. Present value of $500 received at the end of one year	=	$500(.9434)	= $ 471.70
2. Present value of $200 received at the end of two years	=	$200(.8900)	= 178.00
3. Present value of a $400 outflow at the end of three years	=	–$400(.8396)	= –335.84
4. (a) Value at the end of year 3 of a $500 annuity, years 4 through 10:		$500(5.5824) = $2,791.20	
(b) Present value of $2,791.20 received at the end of year 3		= $2,791.20(.8396)	= 2,343.49
5. Total present value			= $2,657.35

Table 4-16 Illustration of an Example of Present Value of an Uneven Stream Involving Two Annuities Discounted at 5 percent

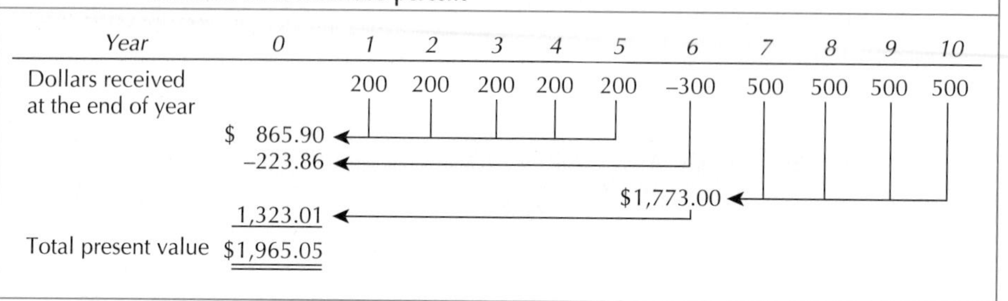

Year	*0*	*1*	*2*	*3*	*4*	*5*	*6*	*7*	*8*	*9*	*10*
Dollars received at the end of year		200	200	200	200	200	−300	500	500	500	500
	$ 865.90 ← (years 1–5)										
	−223.86 ← (year 6)										
							$1,773.00 ← (years 7–10)				
	1,323.01 ← ($1,773.00)										
Total present value	$1,965.05										

Table 4-17 Determination of Present Value of an Example with Uneven Stream Involving Two Annuities Discounted at 5 Percent

1. Present value of first annuity, years 1 through 5	= $200(4.3295)	=	$ 865.90
2. Present value of $300 cash outflow	= −300(.7462)	=	−223.86
3. (a) Value at end of year 6 of second annuity, years 7 through 10	= $500(3.5460) = $1,773.00		
(b) Present value of $1,773.00 received at the end of year 6	= $1,773.00(.7462)	=	1,323.01
4. Total present value		=	$1,965.05

Table 4-18 Summary of Time Value of Money Equations

Calculation	Equation	Related Table
Future value of a single payment	$FV_n = PV(1 + i)^n = PV(FVIF_{i,n})$	Appendix A
Present value of a single payment	$PV = FV_n\left[\frac{1}{(1+i)^n}\right] = FV_n(PVIF_{i,n})$	Appendix B
Future value of an annuity	$FV_n = PMT\left[\sum_{t=0}^{n-1}(1 + i)^t\right] = PMT(FVIFA_{i,n})$	Appendix C
Present value of an annuity	$PV = PMT\left[\sum_{t=1}^{n}\frac{i}{(1 + i)^t}\right] = PMT(PVIFA_{i,n})$	Appendix D
Present value of a perpetuity	$PV = \frac{PP}{i}$	
Future value of a single payment with nonannual compounding	$FV_n = PV_0\left(1 + \frac{i}{m}\right)^{(n)(m)}$	

where FV_n = the future value of the investment at the end of the n years
n = the number of years for which the annuity will last
i = the annual interest or discount rate
PV_0 = the present value of the future sum of money
PMT = the annuity payments deposited or received at the end of each year
PP = the constant dollar amount provided by the perpetuity
m = the number of times compounding occurs during the year.

SUMMARY

To make decisions, financial managers must compare the costs and benefits of alternatives that do not occur during the same time period. Whether for profitable investments or favourable interest rates, financial decision making requires the understanding of the time value of money. Managers who use the time value of money in all their financial calculations assure themselves of more logical decisions. The use of the time value of money ensures that all dollar values are comparable; since money has a time value, it moves all dollar flows either back to the present or out to a common future date. All time-value equations presented in this chapter originate from the single compounding equation: $FV_n = PV_0(1 + i)^n$. The formulae are used to deal simply with common financial situations—for example discounting single cash flows, compounding annuities, and discounting annuities. Table 4-18 provides a summary of these calculations.

STUDY QUESTIONS

4-1. What is the time value of money? Why is it so important?

4-2. The processes of discounting and compounding are related. Explain this relationship.

4-3. How would an increase in the interest rate (i) or a decrease in the holding period (n) affect the future value (FV_n) of a sum of money? Explain why.

4-4. Suppose you were considering depositing your savings in one of three banks, all of which paid 5 percent interest; bank A compounded annually, bank B compounded semiannually, and bank C compounded continuously. Which bank would you choose? Why?

4-5. What is the relationship between the $PVIF_{i,n}$ and the $PVIFA_{i,n}$? What is the $PVIFA_{10\%,10}$? Add up the values of the $PVIF_{10\%,n}$ for $n = 1,...,10$. What is this value? Why do these values have the relationship they do?

4-6. What is an annuity? Give some examples of annuities. Distinguish between an annuity and a perpetuity.

4-7. What does continuous compounding mean?

SELF-TEST QUESTIONS

ST-1. (*Future Value*) You place $25,000 in a savings account paying annual interest of 8 percent for three years and then move it into a savings account that pays 10 percent interest compounded annually. How much will your money have grown at the end of six years?

ST-2. (*Future Value*) Answer the following parts independently of one another.

a. $5,000 is invested at 10 percent compounded annually. After 10 years, what is the value of this investment?

b. How many years will it take for $100.00 to grow to $298.60, if it is invested at 20 percent compounded annually?

c. At what rate does $300 have to be invested in order to grow to $422.10 in seven years?

ST-3. (*Present Value*) Suppose you have just celebrated your 20th birthday. Your benefactor will pay you $50,000 when you are 30 years old or $25,000 today. If the relevant interest rate is 8 percent, what is your choice?

ST-4. (*Future Value, Multiple Cash Flows*) Suppose you deposit $6,000 today, another $2,000 in two years, withdraw $4,000 in four years and deposit $3,000 in six years. How much will you have in ten years if the interest rate is 6 percent?

ST-5. (*Present Value, Multiple Cash Flows*) What is the present value of the series of deposits made in ST-4 above?

ST-6. (*Loan Amortization*) You purchase a boat for $35,000 and pay $5,000 down and agree to pay the rest over the next 10 years in 10 equal annual payments that include principal payment plus 13 percent compound interest on the unpaid balance. What will be the amount of each payment? How much of the first payment is interest and how much represents the repayment of principal?

ST-7. (*Rate of Return*) You have the opportunity to buy a bond for $1,000 that will pay no interest during its ten-year life and have a value of $3,106 at maturity. What rate of return or yield does this bond pay?

ST-8. (*Future Value Annuity vs. Annuity Due*) You are considering an investment that would pay you $2,400 per year for each of the next six years. If you require a 12 percent return, how much will you have accumulated in six years? How much will you have accumulated in six years if the investment cash flows were made at the beginning of each year?

ST-9. (*Present Value vs. Annuity Due*) Consider the information in the preceding problem, ST-8. What is the most you can pay (today) for the expected end-of-period cash flows? How much more would you pay if the cash flows were made at the beginning of each period?

STUDY PROBLEMS (SET A)

4-1A. (*Future Value*) What will the following investments accumulate to?

a. $5,000 invested for 10 years at 10 percent compounded annually
b. $8,000 invested for 7 years at 8 percent compounded annually
c. $775 invested for 12 years at 12 percent compounded annually
d. $21,000 invested for 5 years at 5 percent compounded annually

4-2A. (*Future Value Solving for* n) How many years will it take for the following?

a. $500 to grow to $1,039.50 if invested at 5 percent compounded annually
b. $35 to grow to $53.87 if invested at 9 percent compounded annually
c. $100 to grow to $298.60 if invested at 20 percent compounded annually
d. $53 to grow to $78.76 if invested at 2 percent compounded annually

4-3A. (*Future Value Solving for* i) At what annual rate would the following have to be invested?

a. $500 in order to grow to $1,948.00 in 12 years
b. $300 in order to grow to $422.10 in 7 years
c. $50 in order to grow to $280.20 in 20 years
d. $200 in order to grow to $497.60 in 5 years

4-4A. (*Present Value*) What is the present value of the following future amounts?

a. $800 to be received 10 years from now discounted back to present at 10 percent
b. $300 to be received 5 years from now discounted back to present at 5 percent
c. $1,000 to be received 8 years from now discounted back to present at 3 percent
d. $1,000 to be received 8 years from now discounted back to present at 20 percent

4-5A. (*Future Value Annuity*) What is the accumulated sum of each of the following streams of payments?

a. $500 a year for 10 years compounded annually at 5 percent
b. $100 a year for 5 years compounded annually at 10 percent
c. $35 a year for 7 years compounded annually at 7 percent
d. $25 a year for 3 years compounded annually at 2 percent

4-6A. (*Present Value of an Annuity*) What is the present value of the following annuities?

a. $2,500 a year for 10 years discounted back to the present at 7 percent
b. $70 a year for 3 years discounted back to the present at 3 percent
c. $280 a year for 7 years discounted back to the present at 6 percent
d. $500 a year for 10 years discounted back to the present at 10 percent

4-7A. (*Future Value*) Mark Cox, who recently sold his Porsche, placed $10,000 in a savings account paying annual interest of 6 percent.

a. Calculate the amount of money that will have accrued if he leaves the money in the bank for 1, 5, and 15 years.
b. If he moves his money into an account that pays 8 percent or one that pays 10 percent, rework part (a) using these new interest rates.
c. What conclusions can you draw from the relationship between interest rates, time, and future sums from the calculations you have done above?

4-8A. (*Future Value with Nonannual Periods*) Calculate the amount of money that will be in each of the following accounts at the end of the given deposit period:

Account	Amount Deposited	Annual Interest Rate	Compounding Period (Compounded Every *N* Months)	Deposit Period (Years)
Rod Wheeler	$ 1,000	10%	12	10
Bimbo Coles	95,000	12	1	1
Frankie Allen	8,000	12	2	2
Wayne Robinson	120,000	8	3	2
Jimmy Carruth	30,000	10	6	4
Del Curry	15,000	12	4	3

4-9A. (*Future Value with Nonannual Periods*)

a. Calculate the future sum of $5,000, given that it will be held in the bank five years at an annual interest rate of 6 percent.

b. Recalculate part (a) using a compounding period that is (1) semiannual and (2) bimonthly.

c. Recalculate parts (a) and (b) for a 12 percent annual interest rate.

d. Recalculate part (a) using a time horizon of 12 years (annual interest rate is still 6 percent).

e. With respect to the effect of changes in the stated interest rate and holding periods on future sums in parts (c) and (d), what conclusions do you draw when you compare these figures with answers found in parts (a) and (b)?

4-10A. (*Solving for* i *in Annuities*) Nicki Johnson, a sophomore mechanical engineering student, receives a call from an insurance agent, who believes that Nicki is an older woman ready to retire from teaching. He talks to her about several annuities that she could buy that would guarantee her an annual fixed income. The annuities are as follows:

Annuity	Initial Payment into Annuity (at $t = 0$)	Amount of Money Received per Year	Duration of Annuity (Years)
A	$50,000	$8,500	12
B	$60,000	$7,000	25
C	$70,000	$8,000	20

If Nicki could earn 11 percent on her money by placing it in a savings account, should she place it instead in any of the annuities? Which ones, if any? Why?

4-11A. (*Future Value*) Sales of a new finance book were 15,000 copies this year and were expected to increase by 20 percent per year. What are expected sales during each of the next three years? Graph this sales trend and explain.

4-12A. (*Future Value*) Reggie Jackson, formerly of the New York Yankees, hit 41 home runs in 1980. If his home run output grew at a rate of 10 percent per year, what would it have been over the following five years?

4-13A. (*Amortization of Home Mortgage Loan*) Todd Dodge purchased a new house for $80,000. He paid $20,000 down and agreed to pay the rest on a bimonthly basis over the next 25 years. The bank charges a nominal interest rate of 9.75 percent for this type of loan and the terms of the loan can be renegotiated after five years. Determine what the bimonthly payments will be equal to. (Note: Remember that the interest rate must be compounded semi-annually.)

4-14A. (*Solving for PMT in an Annuity*) To pay for your child's education you wish to have accumulated $15,000 at the end of 15 years. To do this you plan on depositing an equal amount into the bank at the end of each year. If the bank is willing to pay 6 percent compounded annually, how much must you deposit each year in order to obtain your goal?

4-15A. (*Solving for i in Compound Interest*) If you were offered $1,079.50 ten years from now in return for an investment of $500 currently, what annual rate of interest would you earn if you took the offer?

4-16A. (*Present Value and Future Value of an Annuity*) In ten years you are planning on retiring and buying a house in Ovida, Florida. The house you are looking at currently costs $100,000 and is expected to increase in value each year at a rate of 5 percent. Assuming you can earn 10 percent annually on your investments, how much must you invest at the end of each of the next ten years in order to be able to buy your dream home when you retire?

4-17A. (*Future Value*) The Aggarwal Corporation needs to raise $10 million to retire a $10 million mortgage that matures on December 31, 2002. To retire this mortgage, the company plans to put a fixed amount into an account at the end of each year for ten years, with the first payment occurring on December 31, 1993. The Aggarwal Company expects to earn 9 percent annually on the money in this account. What equal annual contribution must it make to this account in order to accumulate the $10 million by December 31, 2002?

4-18A. (*Future Value with Nonannual Periods*) After examining the various personal loan rates available to you, you find that you can borrow funds from a finance company at 12 percent compounded monthly or from a bank at 13 percent compounded annually. Which alternative is the most attractive?

4-19A. (*Present Value of an Uneven Stream of Payments*) You are given three investment alternatives to analyze. The cash flows from these three investments are as follows:

	Investment		
End of Year	A	B	C
1	$10,000		$10,000
2	10,000		
3	10,000		
4	10,000		
5	10,000	$10,000	
6		10,000	50,000
7		10,000	
8		10,000	
9		10,000	
10		10,000	10,000

Assuming a 20 percent discount rate, find the present value of each investment.

4-20A. (*Present Value*) The Kumar Corporation is planning on issuing bonds that pay no interest but can be converted into $1,000 at maturity, seven years from their purchase. In order to price these bonds competitively with other bonds of equal risk, it is determined that they should yield 10 percent, compounded annually. At what price should the Kumar Corporation sell these bonds?

4-21A. (*Perpetuities*) What is the present value of the following?

- **a.** A $300 perpetuity discounted back to the present at 8 percent
- **b.** A $1,000 perpetuity discounted back to the present at 12 percent
- **c.** A $100 perpetuity discounted back to the present at 9 percent
- **d.** A $95 perpetuity discounted back to the present at 5 percent

4-22A. (*Present Value of an Annuity Due*) What is the present value of a ten-year annuity due of $1,000 annually given a 10 percent discount rate?

4-23A. (*Solving for* n *with Nonannual Periods*) About how many years would it take for your investment to grow fourfold if it were invested at 16 percent compounded semiannually?

4-24A. (*Amortization of Home Mortgage Loan*) Eric Doe purchased a new house for $225,000. He paid $25,000 down and agreed to pay the rest on a monthly basis over the next 25 years. The bank charges a nominal interest rate of 10.5 percent for this type of loan and the terms of the loan can be renegotiated after five years. Determine what the monthly payments will be equal to. (Note: Remember that the interest rate must be compounded semiannually.)

4-25A. (*Complex Present Value*) How much do you have to deposit today so that beginning 11 years from now you can withdraw $10,000 a year for the next five years (periods 11 through 15) plus an additional amount of $20,000 in that last year (period 15)? Assume an interest rate of 6 percent.

4-26A. (*Loan Amortization*) On December 31, Beth Klemkosky bought a yacht for $50,000, paying $10,000 down and agreeing to pay the balance in 10 equal annual installments that include both the principal and 10 percent interest on the declining balance. How big would the annual payments be?

4-27A. (*Solving for* i *in an Annuity*) You lend a friend $30,000, which your friend will repay in five equal annual payments of $10,000, with the first payment to be received one year from now. What rate of return does your loan receive?

4-28A. (*Solving for* i *in Compound Interest*) You lend a friend $10,000, for which your friend will repay you $27,027 at the end of five years. What interest rate are you charging your "friend"?

4-29A. (*Loan Amortization*) A firm borrows $25,000 from the bank at 12 percent compounded annually to purchase some new machinery. This loan is to be repaid in equal annual installments at the end of each year over the next five years. How much will each annual payment be?

4-30A. (*Present Value Comparison*) You are offered $1,000 today, $10,000 in 12 years, or $25,000 in 25 years. Assuming that you can earn 11 percent on your money, which should you choose?

4-31A. (*Future Annuity*) You plan on buying some property in Florida five years from today. To do this you estimate that you will need $20,000 at that time for the purchase. You would like to accumulate these funds by making equal annual deposits in your savings account, which pays 12 percent annually. If you make your first deposit at the end of this year and you would like your account to reach $20,000 when the final deposit is made, what will be the amount of your deposits?

4-32A. (*Loan Amortization*) On December 31, Son-Nan Chen borrowed $100,000 agreeing to repay this sum in 20 equal annual instalments that include both the principal and 10 percent interest on the declining balance. How large will the annual payments be?

4-33A. (*Amortization of Home Mortgage Loan*) To buy a new house you must borrow $150,000 from a bank. To do this you borrow $150,000 on a 30-year home mortgage loan at a nominal interest rate of 10 percent. Your mortgage payments are made on a semiannual basis and the terms of the loan can be renegotiated every five years. How large will your semiannual payments be? (Remember that the interest rate must be compounded semiannually.)

4-34A. (*Present Value*) The provincial lottery's 1 million dollar payout provides for $1 million to be paid over 19 years in $50,000 payments. The first $50,000 payment is made immediately and the 19 remaining $50,000 payments occur at the end of each of the next 19 years. If 10 percent is the appropriate discount rate, what is the present value of this stream of cash flows? If 20 percent is the appropriate discount rate, what is the present value of this stream of cash flows?

4-35A. (*Future Value of Annuity Due*) Find the future value at the end of year10 of an annuity due of $1,000 per year for ten years compounded annually at ten percent. What would be the future value of this annuity if it were compounded annually at 15 percent?

4-36A. (*Present Value of an Annuity Due*) Determine the present value of an annuity due of $1,000 per year for ten years discounted back to the present at an annual rate of 10 percent. What would be the present value of this annuity due if it were discounted at an annual rate of 15 percent?

4-37A. (*Present Value of a Future Annuity*) Determine the present value of an ordinary annuity of $1,000 per year for ten years with the first cash flow from the annuity coming at the end of year 8 (that is, no payments at the end of year 1 through 7 and annual payments at the end of year 8 through year 17) given a 10 percent discount rate.

4-38A. (*Solving for* i *in Compound Interest—Financial Calculator Needed*) In September 1963 the first comic book *X-MEN* was issued. The original price for the issue was 12 cents. By September 1995, 32 years later, the value of this comic book had risen to $3,500. What annual rate of interest would you have earned if you had bought the comic in 1963 and sold it in 1995?

4-39A. (*Complex Present Value*) You would like to have $50,000 in 15 years. To accumulate this amount you plan to deposit each year an equal sum in the bank, which will earn 7 percent interest compounded annually. Your first payment will be made at the end of the year.

a. How much must you deposit annually to accumulate this amount?

b. If you decide to make a large lump-sum deposit today instead of the annual deposits, how large should this lump-sum deposit be? (Assume you can earn 7 percent on this deposit.)

c. At the end of five years you will receive $10,000 and deposit this in the bank toward your goal of $50,000 at the end of 15 years. In addition to this deposit, how much must you deposit in equal annual deposits in order to reach your goal? (Again assume you can earn 7 percent on this deposit.)

4-40A. (*Comprehensive Present Value*) You are trying to plan for retirement in ten years, and currently you have $100,000 in a savings account and $300,000 in shares. In addition you plan on adding to your savings by depositing $10,000 per year in your savings account at the end of each of the next five years and then $20,000 per year at the end of each year for the final five years until retirement.

a. Assuming your savings account returns 7 percent compounded annually while your investment in stocks will return 12 percent compounded annually, how much will you have at the end of 10 years? (Ignore taxes.)

b. If you expect to live for 20 years after you retire, and at retirement you deposit all of your savings in a bank account paying 10 percent, how much can you withdraw each year after retirement (20 equal withdrawals beginning one year after you retire) in order to end up with a zero balance at death?

4-41A. (*Comprehensive Present Value*) You have just inherited a large sum of money and you are trying to determine how much you should save for retirement and how much you can spend now. For retirement you will deposit today (January 1, 1997) a lump sum in a bank account paying 10 percent compounded annually. You don't plan on touching this deposit until you retire in five years (January 1, 2002), and you plan on living for 20 additional years and then dropping dead on December 31, 2021. During your retirement you would like to receive income of $50,000 per year to be received the first day of each year, with the first payment on January 1, 2002, and the last payment on January 1, 2021. Complicating this objective is your desire to have one final three-year fling during which time you'd like to play golf in Europe. To finance this you want to receive $250,000 on January 1, 2018 and nothing on January 1, 2019, and January 1, 2020, as you will be on the road. In addition, after you pass on (January 1, 2022), you would like to have a total of $100,000 to leave to your children.

a. How much must you deposit in the bank at 10 percent on January 1, 1997, in order to achieve your goal? (Use a time line in order to answer this question.)

b. What kinds of problems are associated with this analysis and its assumptions?

INTEGRATIVE PROBLEM

For your job as the business reporter for a local newspaper, you are given the task of writing a series of articles that explain the power of the time value of money to your readers. Your editor would like you to address several specific questions in addition to demonstrating for the readership the use of the time-value-of-money techniques by applying them to several problems. What would be your response to the following memorandum from your editor?

TO: Business Reporter

FROM: Perry White, Editor, *Daily Planet*

RE: Upcoming Series on the Importance and Power of the Time Value of Money

In your upcoming series on the time value of money, I would like to make sure you cover several specific points. In addition, before you begin this assignment, I want to make sure we are all reading from the same script, as accuracy has always been the cornerstone of the *Daily Planet.* In this regard, I'd like a response to the following questions before we proceed:

a. What is the relationship between discounting and compounding?

b. What is the relationship between the $PVIF_{i,n}$ and $PVIFA_{i,n}$?

c. 1. What will $5,000 invested for ten years at 8 percent compounded annually grow to?
   2. How many years will it take $400 to grow to $1,671, if it is invested at 10 percent compounded annually?
   3. At what rate would $1,000 have to be invested to grow to $4,046 in ten years?

d. Calculate the future sum of $1,000, given that it will be held in the bank for five years and earn 10 percent compounded semiannually.

e. What is an annuity due? How does this differ from an ordinary annuity?

f. What is the present value of an ordinary annuity of $1,000 per year for seven years discounted back to the present at 10 percent? What would be the present value if it were an annuity due?

g. What is the future value of an ordinary annuity of $1,000 per year for seven years compounded at 10 percent? What would be the future value if it were an annuity due?

h. You have just borrowed $100,000, and you agree to pay it back over the next 25 years in 25 equal end-of-year annual payments that include the principal payments plus 10 percent compound interest on the unpaid balance. What will be the size of these payments?

i. What is the present value of a $1,000 perpetuity discounted back to the present at 8 percent?

j. What is the present value of a $1,000 annuity for ten years with the first payment occurring at the end of year 10 (that is, ten $1,000 payments occurring at the end of year 10 through year 19) given an appropriate discount rate of 10 percent?

k. Given a 10 percent discount rate, what is the present value of a perpetuity of $1,000 per year if the first payment does not begin until the end of year 10?

l. What is the effective annual rate (*EAR*) on an 8 percent loan compounded quarterly?

STUDY PROBLEMS (SET B)

4-1B. (*Future Value*) What will the following investments accumulate to?

a. $4,000 invested for 11 years at 9 percent compounded annually
b. $8,000 invested for 10 years at 8 percent compounded annually
c. $800 invested for 12 years at 12 percent compounded annually
d. $21,000 invested for 6 years at 5 percent compounded annually

4-2B. (*Future Value Solving for* n) How many years will it take for the following?

a. $550 to grow to $1,043.90 if invested at 6 percent compounded annually
b. $40 to grow to $88.44 if invested at 12 percent compounded annually
c. $110 to grow to $614.79 if invested at 24 percent compounded annually
d. $60 to grow to $78.30 if invested at 3 percent compounded annually

4-3B. (*Future Value Solving for* i) At what annual rate would the following have to be invested?

a. $550 in order to grow to $1,898.60 in 13 years
b. $275 in order to grow to $406.18 in 8 years
c. $60 in order to grow to $279.66 in 20 years
d. $180 in order to grow to $486.00 in 6 years

4-4B. (*Present Value*) What is the present value of the following future amounts?

a. $800 to be received 10 years from now discounted back to present at 10 percent
b. $400 to be received 6 years from now discounted back to present at 6 percent
c. $1,000 to be received 8 years from now discounted back to present at 5 percent
d. $900 to be received 9 years from now discounted back to present at 20 percent

4-5B. (*Future Value Annuity*) What is the accumulated sum of each of the following streams of payments?

a. $500 a year for 10 years compounded annually at 6 percent
b. $150 a year for 5 years compounded annually at 11 percent
c. $35 a year for 8 years compounded annually at 7 percent
d. $25 a year for 3 years compounded annually at 2 percent

4-6B. (*Present Value of an Annuity*) What is the present value of the following annuities?

a. $3,000 a year for 10 years discounted back to the present at 8 percent
b. $50 a year for 3 years discounted back to the present at 3 percent
c. $280 a year for 8 years discounted back to the present at 7 percent
d. $600 a year for 10 years discounted back to the present at 10 percent

4-7B. (*Future Value*) Mark Cox, who recently sold his Porsche, placed $20,000 in a savings account paying annual interest of 7 percent.

a. Calculate the amount of money that will have accrued if he leaves the money in the bank for 1, 5, and 15 years.
b. If he moves his money into an account that pays 9 percent or one that pays 11 percent, rework part (a) using these new interest rates.
c. What conclusions can you draw from the relationship between interest rates, time, and future sums from the calculations you have done above?

4-8B. (*Future Value with Nonannual Periods*) Calculate the amount of money that will be in each of the following accounts at the end of the given deposit period:

Account	Amount Deposited	Annual Interest Rate	Compounding Period (Compounded Every *N* Months)	Deposit Period (Years)
Chuck Jones	$ 2,000	12%	2	2
Alonzo Spellman	50,000	12	1	1
Mark Bean	7,000	18	2	2
Steve Tovar	130,000	12	3	2
Ty Morrison	20,000	14	6	4
Roger Harper	15,000	15	4	3

4-9B. (*Future Value with Nonannual Periods*)

a. Calculate the future sum of $6,000, given that it will be held in the bank five years at an annual interest rate of 6 percent.
b. Recalculate part (a) using a compounding period that is (1) semiannual and (2) bimonthly.
c. Recalculate parts (a) and (b) for a 12 percent annual interest rate.
d. Recalculate part (a) using a time horizon of 12 years (annual interest rate is still 6 percent).
e. With respect to the effect of changes in the stated interest rate and holding periods on future sums in parts (c) and (d), what conclusions do you draw when you compare these figures with answers found in parts (a) and (b)?

4-10B. (*Solving for* i *in Annuities*) Matt Denis, a sophomore mechanical engineering student, receives a call from an insurance agent who believes that Matt is an older man ready to retire from teaching. He talks to him about several annuities that he could buy that would guarantee him an annual fixed income. The annuities are as follows:

Annuity	Initial Payment into Annuity (at $t = 0$)	Amount of Money Received per Year	Duration of Annuity (Years)
A	$50,000	$8,500	12
B	$60,000	$7,000	25
C	$70,000	$8,000	20

If Matt could earn 12 percent on his money by placing it in a savings account, should he place it instead in any of the annuities? Which ones, if any? Why?

4-11B. (*Future Value*) Sales of a new marketing book were 10,000 copies this year and were expected to increase by 15 percent per year. What are expected sales during each of the next three years? Graph this sales trend and explain.

4-12B. (*Future Value*) Reggie Jackson, formerly of the New York Yankees, hit 41 home runs in 1980. If his home run output grew at a rate of 12 percent per year, what would it have been over the following five years?

4-13B. (*Amortization of Home Mortgage Loan*) Doug Musgrave purchased a new house for $150,000. He paid $30,000 down and agreed to pay the rest on a bimonthly basis over the next 15 years. The bank charges a nominal interest rate of 10 percent for this type of loan and the terms of the loan can be renegotiated after five years. Determine what the bimonthly payments will be equal to. (Note: Remember that the interest rate must be compounded semiannually.)

4-14B. (*Solving for* PMT *in an Annuity*) To pay for your child's education you wish to have accumulated $25,000 at the end of 15 years. To do this you plan on depositing an equal amount into the bank at the end of each year. If the bank is willing to pay 7 percent compounded annually, how much must you deposit each year in order to obtain your goal?

4-15B. (*Solving for* i *in Compound Interest*) If you were offered $2,376.50 ten years from now in return for an investment of $700 currently, what annual rate of interest would you earn if you took the offer?

4-16B. (*Present Value and Future Value of an Annuity*) In 10 years you are planning on retiring and buying a house in Marco Island, Florida. The house you are looking at currently costs $125,000 and is expected to increase in value each year at a rate of 5 percent. Assuming you can earn ten percent annually on your investments, how much must you invest at the end of each of the next 10 years in order to be able to buy your dream home when you retire?

4-17B. (*Future Value*) The Knutson Corporation needs to raise $15 million to retire a $15 million mortgage that matures on December 31, 2005. To retire this mortgage, the company plans to put a fixed amount into an account at the end of each year for ten years, with the first payment occurring on December 31, 1996. The Knutson Company expects to earn 10 percent annually on the money in this account. What equal annual contribution must it make to this account in order to accumulate the $15 million by December 31, 2005?

4-18B. (*Future Value with Nonannual Periods*) After examining the various personal loan rates available to you, you find that you can borrow funds from a finance company at 24 percent compounded monthly or from a bank at 26 percent compounded annually. Which alternative is the most attractive?

4-19B. (*Present Value of an Uneven Stream of Payments*) You are given three investment alternatives to analyze. The cash flows from these three investments are as follows:

	Investment		
End of Year	A	B	C
1	$15,000		$20,000
2	15,000		
3	15,000		
4	15,000		
5	15,000	$15,000	
6		15,000	60,000
7		15,000	
8		15,000	
9		15,000	
10		15,000	20,000

Assuming a 20 percent discount rate, find the present value of each investment.

4-20B. (*Present Value*) The Shin Corporation is planning on issuing bonds that pay no interest but can be converted into $1,000 at maturity, eight years from their purchase. In order to price these bonds competitively with other bonds of equal risk, it is determined that they should yield 9 percent, compounded annually. At what price should the Shin Corporation sell these bonds?

4-21B. (*Perpetuities*) What is the present value of the following?

a. A $400 perpetuity discounted back to the present at 9 percent

b. A $1,500 perpetuity discounted back to the present at 13 percent

c. A $150 perpetuity discounted back to the present at 10 percent

d. A $100 perpetuity discounted back to the present at 6 percent

4-22B. (*Present Value of an Annuity Due*) What is the present value of a five-year annuity due of $1,000 annually given a 10 percent discount rate?

4-23B. (*Solving for* n *with Nonannual Periods*) About how many years would it take for your investment to grow sevenfold if it were invested at 10 percent compounded semiannually?

4-24B. (*Amortization of Home Mortgage Loan*) Jane Doe purchased a new house for $250,000. She paid $55,000 down and agreed to pay the rest on a monthly basis over the next 25 years. The bank charges a nominal interest rate of 9 percent for this type of loan and the terms of the loan can be renegotiated after five years. Determine what the monthly payments will be. (Note: Remember that the interest rate must be compounded semiannually).

4-25B. (*Complex Present Value*) How much do you have to deposit today so that beginning 11 years from now you can withdraw $10,000 a year for the next five years (periods 11 through 15) plus an additional amount of $15,000 in that last year (period 15)? Assume an interest rate of 7 percent.

4-26B. (*Loan Amortization*) On December 31, Matt Denis bought a yacht for $60,000, paying $15,000 down and agreeing to pay the balance in ten equal annual installments that include both the principal and 9 percent interest on the declining balance. How big would the annual payments be?

4-27B. (*Solving for* i *in an Annuity*) You lend a friend $45,000, which your friend will repay in five equal annual payments of $9,000 with the first payment to be received one year from now. What rate of return does your loan receive?

4-28B. (*Solving for* i *in Compound Interest*) You lend a friend $15,000, for which your friend will repay you $37,313 at the end of five years. What interest rate are you charging your "friend"?

4-29B. (*Loan Amortization*) A firm borrows $30,000 from the bank at 13 percent compounded annually to purchase some new machinery. This loan is to be repaid in equal annual installments at the end of each year over the next four years. How much will each annual payment be?

4-30B. (*Present Value Comparison*) You are offered $1,000 today, $10,000 in 12 years, or $25,000 in 25 years. Assuming that you can earn 11 percent on your money, which should you choose?

4-31B. (*Compound Annuity*) You plan on buying some property in Florida five years from today. To do this you estimate that you will need $30,000 at that time for the purchase. You would like to accumulate these funds by making equal annual deposits in your savings account, which pays 10 percent annually. If you make your first deposit at the end of this year and you would like your account to reach $30,000 when the final deposit is made, what will be the amount of your deposits?

4-32B. (*Loan Amortization*) On December 31, Eugene Chung borrowed $200,000 agreeing to repay this sum in twenty equal annual instalments that include both the principal and 10 percent interest on the declining balance. How large will his annual payments be?

4-33B. (*Amortization of Home Mortgage Loan*) To buy a new house you must borrow $250,000 from a bank. To do this you borrow $250,000 on a 30-year home mortgage loan at a nominal interest rate of 9 percent. Your mortgage payments are made on a semiannual basis and the terms of the loan can be renegotiated every five years. How large will your semiannual payments be? (Remember that the interest rate must be compounded semiannually)

4-34B. (*Present Value*) The provincial lottery's million-dollar payout provides for $1 million to be paid over 24 years in $40,000 payments. The first $40,000 payment is made immediately and the 24 remaining $40,000 payments occur at the end of each of the next 24 years. If 10 percent is the appropriate discount rate, what is the present value of this stream of cash flows? If 20 percent is the appropriate discount rate, what is the present value of this stream of cash flows?

4-35B. (*Future Value of Annuity Due*) Find the future value at the end of year 5 of an annuity due of $1,000 per year for five years compounded annually at 5 percent. What would be the future value of this annuity if it were compounded annually at 8 percent?

4-36B. (*Present Value of an Annuity Due*) Determine the present value of an annuity due of $1,000 per year for 15 years discounted back to the present at an annual rate of 12 percent. What would be the present value of this annuity due if it were discounted at an annual rate of 15 percent?

4-37B. (*Present Value of a Future Annuity*) Determine the present value of an ordinary annuity of $1,000 per year for ten years with the first cash flow from the annuity coming at the end of year 8 (that is, no payments at the end of year 1 through year 7 and annual payments at the end of year 8 through year 17) given a 15 percent discount rate.

4-38B. (*Solving for i in Compound Interest—Financial Calculator Needed*) In March 1963, issue number 39 of *Tales of Suspense* was issued. The original price for the issue was 12 cents. By March 1995, 32 years later, the value of this comic book had risen to $2,500. What annual rate of interest would you have earned if you had bought the comic in 1963 and sold it in 1995?

4-39B. (*Complex Present Value*) You would like to have $75,000 in 15 years. To accumulate this amount you plan to deposit each year an equal sum in the bank, which will earn 8 percent interest compounded annually. Your first payment will be made at the end of the year.

a. How much must you deposit annually to accumulate this amount?

b. If you decide to make a large lump-sum deposit today instead of the annual deposits, how large should this lump-sum deposit be? (Assume you can earn 8 percent on this deposit.)

c. At the end of five years you will receive $20,000 and deposit this in the bank toward your goal of $75,000 at the end of 15 years. In addition to this deposit, how much must you deposit in equal annual deposits in order to reach your goal? (Again assume you can earn 8 percent on this deposit.)

4-40B. (*Comprehensive Present Value*) You are trying to plan for retirement in ten years, and currently you have $150,000 in a savings account and $250,000 in stocks. In addition you plan on adding to your savings by depositing $8,000 per year in your savings account at the end of each of the next five years and then $10,000 per year at the end of each year for the final five years until retirement.

a. Assuming your savings account returns 8 percent compounded annually while your investment in stocks will return 12 percent compounded annually, how much will you have at the end of ten years? (Ignore taxes.)

b. If you expect to live for 20 years after you retire, and at retirement you deposit all of your savings in a bank account paying 11 percent, how much can you withdraw each year after retirement (20 equal withdrawals beginning one year after you retire) in order to end up with a zero balance at death?

4-41B. (*Comprehensive Present Value*) You have just inherited a large sum of money and you are trying to determine how much you should save for retirement and how much you can spend now. For retirement you will deposit today (January 1, 1997) a lump sum in a bank account paying 10 percent compounded annually. You don't plan on touching this deposit until you retire in 5 years (January 1, 2002), and you plan on living for 20 additional years and then dropping dead on December 31, 2021. During your retirement you would like to receive income of $60,000 per year to be received the first day of each year, with the first payment on January 1, 2002, and the last payment on January 1, 2021. Complicating this objective is your desire to have one final three-year fling during which time you'd like to play tennis in Europe. To finance this you want to receive $300,000 on January 1, 2018, and nothing on January 1, 2019, and January 1, 2020, as you will be on the road. In addition, after you pass on (January 1, 2022), you would like to have a total of $100,000 to leave to your children.

a. How much must you deposit in the bank at 10 percent on January 1, 1997, in order to achieve your goal? (Use a time line in order to answer this question.)

b. What kinds of problems are associated with this analysis and its assumptions?

SELF-TEST SOLUTIONS

SS-1. This is a compound interest problem in which you must first find the future value of \$25,000 growing at 8 percent compounded annually for three years and then allow that future value to grow for an additional three years at 10 percent. First, the value of the \$25,000 after three years growing at 8 percent is

$$FV_3 = PV_0(1 + i)^n$$
$$FV_3 = \$25{,}000(1 + .08)^3$$
$$FV_3 = \$25{,}000(1.2597)$$
$$FV_3 = \$31{,}492.50$$

Thus, after three years you have \$31,492.50 and now this amount is allowed to grow for three years at 10 percent. Plugging this into Equation (4-6), $PV_0 = \$31{,}492.50$, $i = 10\%$, $n = 3$ years, and we solve for FV_3:

$$FV_3 = \$31{,}492.50(1 + .10)^3$$
$$FV_3 = \$31{,}492.50(1.3310)$$
$$FV_3 = \$41{,}916.52$$

Thus, after six years the \$25,000 will have grown to \$41,926.52.

SS-2. The future value equation is: $FV_n = PV_0(1+i)^n$

a.

$$FV_{10} = \$5{,}000(1+0.10)^{10}$$
$$FV_{10} = \$5{,}000(FVIF_{10\%,10\text{ yr}})$$
$$FV_{10} = \$5{,}000(2.5937)$$
$$FV_{10} = \$12{,}968.50$$

b.

$$\$298.60 = \$100(1.20)^n$$
$$\$298.60 = \$100(FVIF_{20\%,n\text{ yr}})$$
$$(FVIF_{20\%,n\text{ yr}}) = 2.9860$$

From the *FVIF* table under 20%, $n = 6$ years

c.

$$\$422.10 = \$300(1+i)^7$$
$$1.407 = FVIF_{(i\%,7\text{ yr})}$$

From the *FVIF* table under 7 periods, $i = 5\%$

SS-3. The present value of receiving \$50,000 ten years from now at 8 percent is $PV = \$50{,}000 \times [1/(1.08)]^{10} = \$23{,}159.67$. Therefore, choose \$25,000 today. Note: \$25,000 today is worth \$53,973.12 ten years from now using the future value equation $FV = \$25{,}000 \times (1.08)^{10} = \$53{,}973.12$

SS-4. Calculate the future value of each deposit and withdrawal (negative amount) at year 10.

$\$6{,}000 \times (1.06)^{10} =$	\$10,745.09
$+\ \$2{,}000 \times (1.06)^8 =$	+ 3,187.69
$-\ \$4{,}000 \times (1.06)^6 =$	− 5,674.08
$+\ \$3{,}000 \times (1.06)^4 =$	+ 3,787.43
Total (year 10)	\$12,046.13

SS-5. We could take the present value of each amount individually and sum the result. A more efficient method is to take the present value of the total amount in year 10.

$\$12{,}046.13 \times [1/(1.06)^{10}] = \$6{,}726.49$

SS-6. This loan amortization problem requires that you determine the present value of annuity in which we know the values of i, n, and PV_0 and are solving for *PMT*. In this case the value of i is 13 percent, n is 10 years, and PV_0 is \$30,000. Substituting these values into Equation (4-12) we find:

$$\$30{,}000 = PMT\left[\sum_{t=1}^{10} \frac{1}{(1 + .13)^t}\right]$$

$$\$30{,}000 = PMT(5.4262)$$
$$PMT = 5{,}528.73$$

The interest portion of this payment is \$3,900 (\$30,000 × 0.13) and the repayment of principal is \$1,628.73 (\$5,528.73 − \$3,900).

SS-7.

$$FV_n = PV_0(1+i)^n = PV_0(FVIF_{i\%,n\,yr})$$
$$\$3{,}106 = \$1{,}000(FVIF_{i\%,10\,yr})$$
$$(FVIF_{i\%,10\,yr}) = 3.106$$

Looking in the $n = 10$ row of Appendix A, we find a value of 3.106 in the 12% column. Therefore, $i = 12\%$.

SS-8. The equation for the future value of an ordinary annuity is:

$$FV_n = PMT\left[\frac{(1+i)^n - 1}{i}\right]$$

substituting,

$$FV_6 = \$2{,}400\left[\frac{(1.12)^6 - 1}{0.12}\right]$$

$$FV_6 = \$19{,}476.45$$

If the series is an annuity due, the ordinary annuity factor is multiplied by $(1 + i)$.

$$FV_6 = \$19{,}476.45\,(1.12) = \$21{,}813.63$$

The additional interest earned is \$2,337.18 (\$21,813.63 − \$19,476.45).

SS-9. The equation for the present value of an ordinary annuity is:

$$PV_0 = PMT\left[\frac{1 - \frac{1}{(1+i)^n}}{i}\right]$$

substituting,

$$PV_0 = \$2{,}400\left[\frac{1 - \frac{1}{(1.12)^6}}{.12}\right]$$

$$PV_0 = \$9{,}867.38$$

If the series is an annuity due, the ordinary annuity factor is multiplied by $(1 + i)$.

$$PV_0 = \$9{,}867.38\,(1.12) = \$11{,}051.46$$

The additional interest earned is \$1,184.08 (\$11,051.46 − \$9,867.38).

BOND VALUATION

CHAPTER 5

LEARNING OBJECTIVES

After reading this chapter you should be able to

1. Explain the more popular characteristics of bonds.
2. Distinguish among the various types of bonds.
3. Distinguish among the various ways to measure value.
4. Describe the basic process for valuing financial assets.
5. Estimate the value of a bond.
6. Compute expected rate of return for a bondholder.
7. Explain the five important relationships that exist in bond valuation.

INTRODUCTION

In late January 1994, a $500,000 investment in a Canadian bond fund would have been considered a comfortable nest egg for most Canadians to plan for their retirement. Think again, five months later this same investment was worth $401,300, a decrease of $98,700. This scenario was true for a lot of bond investors across North America during this period. The reason: interest rates increased. The Bank of Canada rate increased from 3.88 percent to 7.09 percent; whereas the interest rates on ten-year corporate bonds increased from 6.32 percent to 9.43 percent during this same five month period. As we shall see in this chapter, when interest rates increase, bond values decline—sometimes by a lot.

CHAPTER PREVIEW

The market price of a company's bonds can fluctuate dramatically within a few months. Managers must know what drives the value of financial assets so as to maximize the value of the firm. As a result, managers will act in the best interest of the firm's investors by understanding how bonds and shares are valued in the marketplace.

The various types of bonds are one of several forms of long-term debt instruments that a company uses to raise capital. In this chapter, we identify the various types of bonds. We then examine the characteristics of most bonds. We also present the concepts and the procedures for valuing assets; these procedures are then applied to the valuation of bonds. Finally, we examine some of the relationships that exist in the valuation process.

OBJECTIVE 1

THE CHARACTERISTICS OF BONDS

Bond

A type of debt or a long-term promissory note, issued by the borrower, promising to pay its holder a predetermined and fixed amount of interest each year.

A **bond** is a long-term promissory note, issued by the borrower, promising to pay its holder a predetermined and fixed amount of interest each year. We now examine some of the terminology and characteristics of bonds.

Par Value

Par value of a bond

The bond's face value that is returned to the bondholder at maturity, usually $1,000.

The **par value** or the face value of a bond is the amount that is returned to the bondholder at maturity. In general, most corporate bonds are issued in denominations of $1,000, although there are some exceptions to this rule. When bond prices are quoted by either financial managers or in the financial press, prices are generally expressed as a percentage of the bond's par value. For example, a Noranda Mines bond that pays $105 per year interest, and matures in 1999, was quoted in *The Financial Post* as selling for 99.75. That does not mean you can buy the bond for $99.75. It does mean that this bond is selling for 99.75 percent of its par value, which is always $1,000. Hence, the market price of this bond is actually $997.50. At maturity in 1999, the bondholder will receive the $1,000.

Coupon Interest Rate

Coupon interest rate

A bond's coupon interest rate indicates what percentage of the par value of the bond will be paid out annually in the form of interest.

The **coupon interest rate** on a bond indicates what percentage of the par value of the bond will be paid out annually in the form of interest. Thus, regardless of what happens to the price of a bond with an 8 percent coupon interest rate and a $1,000 par value, it will pay out $80 annually in interest until maturity.

Maturity

Maturity

The length of time until the bond issuer returns the par value to the bondholder and terminates the bond.

The **maturity** of a bond indicates the length of time until the bond issuer returns the par value to the bondholder and terminates the bond.

Indenture

Indenture

The legal agreement or contract between the firm issuing the bonds and the bond trustee who represents the bondholders.

An **indenture** is the legal agreement between the firm issuing the bonds and the bond trustee who represents the bondholders.[1] The indenture provides the specific terms of the loan agreement, including a description of the bonds, the rights of the bondholders, the rights of the issuing firm, and the responsibilities of the trustees.

[1]The indenture agreement is also referred to as a *trust deed.*

This legal document may run 100 pages or more in length, with the majority of it devoted to defining protective provisions for the bondholder. The bond trustee, usually a banking institution or trust company, is then assigned the task of overseeing the relationship between the bondholder and the issuing firm, protecting the bondholder, and seeing that the terms of the indenture are carried out.

Typically, the restrictive provisions included in the indenture attempt to protect the bondholder's financial position relative to that of other outstanding securities. Common provisions involve (1) restrictions on the sale of accounts receivable, (2) constraints on the issuance of common shares dividends, (3) restrictions on the purchase or sale of fixed assets, and (4) constraints on additional borrowing. Restrictions on the sale of accounts receivable are specified because such sales would benefit the firm's short-run liquidity position at the expense of its future liquidity position. Constraints on common share dividends generally means limiting their distribution when working capital falls below a specified level, or simply limiting the maximum dividend payout to 50 or 60 percent of earnings under any circumstance. Fixed-asset restrictions generally require lender permission before the liquidation of any fixed asset or prohibit the use of any existing fixed asset as collateral on new loans. Constraints on additional borrowing are usually in the form of restrictions or limitations on the amount and type of additional long-term debt that can be issued. All these restrictions have one thing in common: they attempt to prohibit action that would improve the status of other securities at the expense of bonds and to protect the status of bonds from being weakened by any managerial action.

Current Yield

The **current yield** on a bond refers to the ratio of the annual interest payment to the bond's market price. If, for example, we are examining a bond with an 8 percent coupon interest rate, a par value of $1,000, and a market price of $700, it would have a current yield of

Current yield

The ratio of the annual interest payment to the bond's market price.

$$\text{current yield} = \frac{\text{annual interest payments}}{\text{market price of the bond}} \tag{5-1}$$

$$= \frac{.08 \times \$1{,}000}{\$700} = \frac{\$80}{\$700} = 11.4 \text{ percent}$$

Yield to Maturity

The **yield to maturity** refers to the bond's internal rate of return. It incorporates into the analysis both the annual interest payments and capital gains or losses. Mathematically, the yield to maturity is the discount rate that equates the present value of the interest and principal payments with the current market price of the bond.

Yield to maturity

The rate or return the investor will earn if the bond is held to maturity.

Bond Ratings

Both the Canadian Bond Rating Service and the Dominion Bond Rating Service provide ratings on bonds issued in Canada, whereas Moody's, and Standard and Poor's, provide ratings on bonds issued in North America. These ratings involve a judgment about the future risk potential of the bond.[2] Although they deal with expectations, several historical factors seem to play a significant role in their determination.[3] Bond ratings are favourably affected by (1) a low utilization of financial leverage, (2) profitable operations, (3) a low variability in past earnings, (4) large

[2]Bond ratings are extremely important in that a firm's bond rating tells much about (a) the cost of funds and (b) the firm's access to the debt markets.

[3]See Thomas F. Pogue and Robert M. Soldofsky, "What's in a Bond Rating?" *Journal of Financial and Quantitative Analysis* 4 (June 1969), pp. 201–228; and George E. Pinches and Kent A. Mingo, "A Multivariate Analysis of Industrial Bond Ratings," *Journal of Finance* 28 (March 1973), pp. 1–18.

firm size, and (5) little use of subordinated debt. In turn, the rating a bond receives affects the rate of return demanded on the bond by the investors. The poorer the bond rating, the higher the rate of return demanded in the capital markets. An example and description of these ratings is given in Table 5-1. Thus, for the financial manager, bond ratings are extremely important. They provide an indicator of default risk that in turn affects the rate of return that must be paid on borrowed funds. Bennett Stewart of Stern Stewart Management Services provides some perspective on just how bond ratings are actually determined in the next Basic Financial Management in Practice feature, "Bennett Stewart on Bond Ratings."

BACK TO THE BASICS

When we say that a lower bond rating means a higher interest rate charged by the investors (bondholders), we are observing an application of ***Axiom 1: The Risk-Return Tradeoff—We Won't Take on Additional Risk Unless We Expect to Be Compensated with Additional Return.***

Determinants of the Cost of Long-Term Debt

In Chapter 3 we defined the investor's required rate of return. We now focus on the factors that determine the required rate of return that investors demand from debt financing. The total cost of debt that the firm will pay depends primarily upon five factors: (1) the size of the issue, (2) the issue's maturity, (3) the issue's riskiness or rating, (4) the restrictive requirements of the issue, and (5) the current riskless interest rate. To a large extent the administrative costs of issuing debt are fixed, and they will decrease in percentage terms as the size of the issue increases. In effect, economies of scale are associated with the issuance of debt. The issue's maturity also affects the cost of debt. Borrowers prefer to borrow for long periods in order to lock in interest rates and avoid the problem of frequent refinancing, while lenders would rather not tie up their money for long periods. In order to bring about equilibrium between supply and demand for funds, long-term debt generally carries a higher interest rate than short-term debt. This tends to encourage some borrowers to borrow for shorter periods and some investors to lend for longer periods.

Investors also desire additional return for taking on added risk. Thus, the less risky the bond or the higher the bond rating, the lower will be the interest rate. We can see this by looking at the movement of bond yields for different ratings from 1986 to 1994, as shown in Figure 5-1.[4]

The restrictive requirements and rights of both issuer and holder also affect the interest rate paid on the bond. The more the bondholder requires in the way of protection and rights, the lower the rate of return earned will be.[5] On the other hand, the more rights the issuer demands—for example, the right to repurchase the debt at a predetermined price (called a *call provision* and discussed later in detail)—the higher will be the rate that the issuer will have to offer in order to convince investors to purchase the bond.

Finally, the current riskless interest rate plays a major role in determining the cost of debt. In fact, corporate bond rates move in the same direction as does the

[4]Bond prices seem to be particularly vulnerable to price downswings during takeovers and management buyouts. These price downswings occur in cases in which a company takes on large amounts of debt in order to accomplish leveraged buyouts. This was the case for RJR Nabisco bonds which fell 10 percent when the leveraged buyout was proposed.

[5]Bondholders are extremely concerned about the possibility of the firm taking on additional debt, particularly associated with management buyouts or restructuring. As a result, many bondholders are pushing for companies to issue new bonds with covenants that limit losses in the event of future restructuring. An example of such a feature is a poison put which allows the bondholder to sell the bonds back to the company at face value if a restructuring lowers their rating to less than investment grade.

Table 5-1 Bond Ratings Used by the Canadian Bond Rating Service

A++

This category encompasses bonds of outstanding quality. They possess the highest degree of protection of principal and interest. Companies with debt rated A++ are generally large national and/or multi-national corporations whose products or services are essential to the Canadian economy.

These companies are the acknowledged leaders in their respective industries and have clearly demonstrated their ability to best withstand adverse economic or trade conditions either national or international in scope. Characteristically these companies have had a long and creditable history of superior debt protection, in which the quality of their assets and earnings has been constantly maintained or improved, with strong evidence that this will continue.

A+

Bonds rated A+ are very similar in characteristics to those rated A++ and can also be considered superior in quality. These companies have demonstrated a long and satisfactory history of growth with above average protection of principal and interest on their debt securities.

These bonds are generally rated lower in quality because the margin of asset or earning protection may not be as large or as stable as those rated A++. In both these categories the nature and quality of the asset and earnings coverages are more important than the numerical values of the ratios.

A

Bonds rated A are considered to be good quality securities and to have favourable long-term investment characteristics. The main feature which distinguishes them from their higher rated securities is that these companies are more susceptible to adverse trade or economic conditions. The protection is consequently lower than for the categories of A++ and A+.

In all cases the A rated companies have maintained a history of adequate asset and earning protection. There may be certain elements that may impair this protection sometime in the future. Our confidence that the current overall financial position will be maintained or improved is slightly lower than for the above rated securities.

B++

Issues rated B++ are classified as medium or average grade credits and are considered to be investment grade. These companies are generally more susceptible than any of the higher rated companies to swings in economic or trade conditions which would cause a deterioration in protection should the company enter a period of poor operating conditions.

There may be factors present either from within or without the company which may adversely affect the long-term level of protection of the debt. These companies bear closer scrutiny but in all cases both interest and principal are adequately protected at the present time.

B+

Bonds which are rated B+ are considered to be lower medium grade securities and have limited long-term protective investment characteristics. Asset and earning coverage may be modest or unstable.

A significant deterioration in interest and principal protection may occur during the periods of adverse economic or trade conditions. During periods of normal or improving economic conditions, asset and earning protection are adequate; however, the company's ability to continually improve its financial position and level of debt protection is at present limited.

B

Securities rated B lack most qualities necessary for long-term fixed income investment. Companies in this category have a general history of volatile operating conditions during which time the assurance that principal and interest protection will be maintained at an adequate level has been in doubt. Current coverages may be below industry standards and there is little assurance that debt protection will significantly improve.

C

Securities in this category are clearly speculative. The companies are generally junior in many respects and there is little assurance that the adequate coverage of principal and interest can be maintained uninterruptedly over a period of time.

D

Bonds in this category are in default of some provisions in their trust deed and the companies may or may not be in the process of liquidation.

Source: Canadian Bond Rating Service, "An Introduction to The CBRS Method of Rating Securities," Montreal, 1985, pp. 14–15.

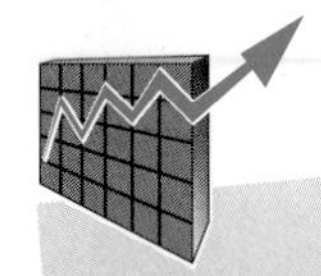

Bennett Stewart on Bond Ratings

It is useful to classify financial risk in terms of bond ratings since these provide a framework for analyzing the risks important to all creditors. Furthermore, relating the financial posture adopted to a particular bond rating facilitates a determination of the availability and cost of funds in both the private and public debt markets.

When selecting a particular bond rating and financial posture to pursue, management is really choosing a desired level of financing flexibility. We define financing flexibility to be the degree to which a company's financing options are kept open for raising funds while keeping the equity holders of the company within an acceptable risk level in any economic climate. A company with too low a level of financing flexibility could be forced to raise common equity at an inopportune time to meet a need for funds.

Financial planners must decide what bond rating is right for their company. There is really no one correct bond rating for any particular company. The choice will depend on the levels of financing flexibility and financial risk the company feels comfortable with. We only say that a company should not seek a financial risk level that would endanger management's financing flexibility and give rise to an undue risk of insolvency. In our opinion, companies with ratings of less than Baa/BBB [Moody's or Standard and Poor's Bond Rating] on their long-term senior debt run this risk.

While countless factors affect companies' debt capacity, we have found five which together are most important in explaining bond ratings. Stern Stewart has developed a bond rating simulation framework which focuses on these five factors. Such a framework can aid the long-range financial planning process by showing the projected effect of a company's operating performance and alternative financing strategies on its credit standing.

Briefly, the technique, which is based on a statistical concept known as "discriminant analysis," involves identifying the key quantifiable characteristics that play a major role in determining the financial strength of a company. Each of these characteristics is assigned a "weight" based on its relative importance, and then a weighted "score" of financial strength is calculated. The score is then put on the Bond Rating Score Range to determine the estimated rating (AAA, AA, etc.).

A description of the factors and an interpretation of the significance of each in determining financial strength is presented below.

1. **Size:** Company size, as measured by total assets, is the single most important factor. It acts as a proxy for the company's ease of access to capital markets, a proven track record, continuity and depth of management, control over markets, and diversification of activities by product line and geography. In general, the larger the company, the higher the bond score.
2. **Risk-Adjusted Return**: The risk-adjusted rate of return measures both the level and stability of the economic earnings of a company. The level of earnings is important because it provides a good indication of the profitability of a company. And, for any given leverage implied, the higher a company's return on total capital, the more likely it will be able to pay interest expense out of earnings. The stability of earnings is also important to creditors because they do not participate in the capital gain rewards accorded equity investors. Consequently, creditors wish to avoid the earnings fluctuations caused by economic cycles.

   The risk-adjusted return, which is computed by subtracting one standard deviation from the five year average rate of return on total capital (assets), provides an indication of the minimum rate of return a company may be expected to earn approximately 83 percent of the time. To calculate the return on total capital each year, we divide NOPAT (Net Operating Profits After Tax) by total capital employed. Higher risk- adjusted returns strengthen a company's bond score.
3. **Long-term Debt-to-Capital (Assets) Ratio:** The three-year average of a company's long-term debt-to-capital ratio

is a measure of the financial risk of a company. This measure averaged over a three-year period is a better measure of permanent leverage policy than the debt-to-capital ratio in any one year; hence, it is a better measure of the expected future financial risk of a company. The higher the long-term debt-to-capital ratio, the weaker the bond score will tend to be.

4. **Pension-Adjusted Liabilities--to-Net-Worth Ratio**: Pension-adjusted liabilities to net worth is calculated as total liabilities plus the lesser of 50 percent of unfunded pension liabilities or 30 percent of net worth, divided by net worth. This is a good measure of asset protection in the event of bankruptcy, since pension-adjusted liabilities include the off-balance sheet liability of unfunded pension liabilities. This liability is senior to almost all other liabilities and can be substantial. It therefore seems reasonable that the higher this ratio is, the lower the bond score will be.
5. **Investments and Advances to Unconsolidated Subsidiaries-to-Capital Ratio:** Calculated as a percentage of total capital, this measure indicates the proportion of capital of a company that is already supporting debt. The equity in unconsolidated subsidiaries supports debt on its own books and cannot be relied upon to support parent debt as well. Higher ratios of investments and advances to unconsolidated subsidiaries-to-capital will pull down the bond score, all else constant.

Source: Bennett Stewart, "Commentary: A Framework for Setting Required Rates of Return by Line of Business," in *Six Roundtable Discussions of Corporate Finance with Joel Stern*, ed., Donald H. Chew, Jr. (New York: Quorum Books, 1986).

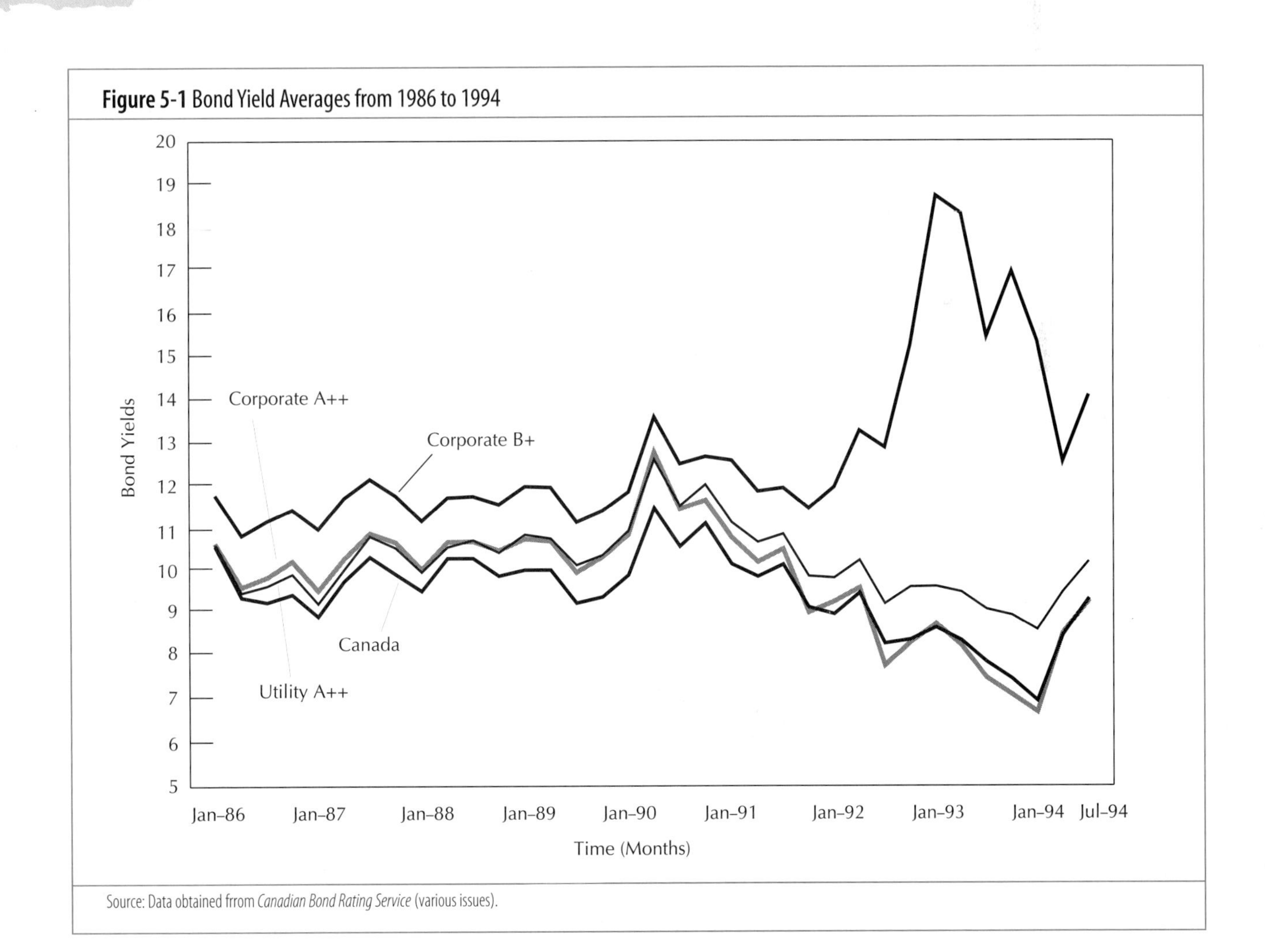

Figure 5-1 Bond Yield Averages from 1986 to 1994

Source: Data obtained frrom *Canadian Bond Rating Service* (various issues).

riskless rate of interest. The determination of the riskless interest rate is generally deferred to courses on money and banking, and financial institutions. We note here that one of the major factors affecting the riskless rate of interest is the anticipated rate of inflation.

Claims on Assets and Income

In the case of insolvency, claims of debt are honoured prior to those of both common shares and preferred shares. However, different types of debt may also have a hierarchy among themselves as to the order of their claim on assets.

Bonds also have a claim on income that is prior to those of both common shares and preferred shares. If interest on bonds (other than income bonds, to be discussed later) is not paid, the bond trustees can classify the firm insolvent and force it into bankruptcy. Thus, the bondholder's claim on income is more likely to be honoured than that of common and preferred shareholders, whose dividends are paid at the discretion of the firm's management.

OBJECTIVE 2

THE VARIOUS TYPES OF BONDS

Debentures

Debenture

Any unsecured long-term debt.

The term **debenture** applies to any unsecured long-term debt. Because these bonds are unsecured, the earning ability of the issuing corporation is of great concern to the bondholder. Debentures are also viewed as being riskier than secured bonds and as a result must provide investors with a higher yield than secured bonds provide. Often the issuing firm attempts to provide some protection to the holder through the prohibition of any additional encumbrance of assets. This prohibits the future issuance of secured long-term debt that would further tie up the firm's assets and leave the bondholders less protected. A major advantage of issuing debentures is that a firm is not required to provide property as security. This allows the firm to issue debt and still preserve some future borrowing power.

Subordinated Debentures

Subordinated debenture

A debenture that is subordinated to other debentures in being paid in the case of insolvency.

Many firms have more than one issue of debentures outstanding. In this case a hierarchy may be specified, in which some debentures are given subordinated standing in the case of insolvency. The claims of the **subordinated debentures** are honoured only after the claims of secured debt and unsubordinated debentures have been satisfied.

Income Bonds

Income bond

A bond that requires interest payments only if earned. Failure to meet these interest payments will not result in bankruptcy.

An **income bond** requires interest payments only if earned, and failure to meet these interest payments cannot lead to bankruptcy. In this sense income bonds are more like preferred shares (which will be discussed in Chapter 6) than bonds. They are generally issued during the reorganization of a firm facing financial difficulties. The maturity of income bonds is usually much longer than that of other bonds in order to relieve pressure associated with the repayment of principal. While interest payments may be passed, unpaid interest is generally allowed to accumulate for some period of time and must be paid prior to the payment of any common share dividends. This cumulative interest feature provides the bondholder with some security.

Mortgage Bonds

A **mortgage bond** is a bond secured by a lien on real property. Typically, the value of the real property is greater than that of the mortgage bonds issued. This provides the mortgage bondholders with a margin of safety in the event that the market value of the secured property declines. In the case of foreclosure, the trustees have the power to sell the secured property and use the proceeds to pay the bondholders. In the event that the proceeds from this sale do not cover the bonds, the bondholders become general creditors, similar to debenture bondholders, for the unpaid portion of the debt.

Mortgage bond

A bond secured by a lien on real property.

Equipment Trust Certificates

An **equipment trust certificate** is an example of a financial instrument which uses equipment as security.[6] This type of financing is a cross between lease financing and debt financing. Equipment trust financing is used primarily by railroads and some airlines and oil companies to finance the purchase of rolling stock or airplanes. This type of debt has a fairly short maturity, ranging from 1 to 17 years. The short maturity occurs because of the nature of the collateral, which is subject to substantial wear and tends to deteriorate rapidly. Under this financing method, the railroad or issuing firm first orders the railroad cars or equipment. When the equipment is received, the title is transferred to a trustee. The manufacturer is then paid in full by the trustee, which obtains funds primarily through the issuance of equipment trust certificates. These trust certificates have minimal risk because they are secured by the equipment being purchased, and this is generally quite standardized.

Equipment trust certificate

A financial instrument which uses equipment as collateral.

Floating Rate or Variable Rate Bonds

Floating rate or **variable rate bonds** are appealing to issuers and investors during periods of unstable interest rates. Banks and trust companies issue floating rate bonds to eliminate some of the risk and variability in earnings that accompany interest rate swings. To the investor, it eliminates major swings in the market value of the debt that would otherwise have occurred if interest rates had changed.

Floating rate or variable rate bonds

Bonds in which interest payments are tied to the prevailing interest rate (London Interbank Offered Rate) level at the time of distribution.

For example, the Royal Bank of Canada issued U.S. $350 million (or CDN $408,345,000) of floating rate debentures due in July 2005. The interest rate on these debentures is subject to semiannual adjustments of 6.25 basis points above the one month London Interbank Offered Rate. However, the Royal Bank guarantees a minimum rate of 5 percent during the first five years. The London Interbank Offered Rate, or LIBOR as it is often called, is the rate that most creditworthy international banks charge each other for large loans and is often used as a benchmark to which variable rate loans are pegged. Thus, if the LIBOR was 6.00 percent, the Royal Bank debentures would pay 6.0625 percent and if the LIBOR rose to 6.50% percent at the time of the quarterly adjustment, the Royal Bank debentures would be adjusted to pay 6.5625 percent.

While every floating rate bond is a bit different, these bonds generally possess common characteristics:

1. After an initial period of 3 to 60 months during which a minimum interest rate is guaranteed, the coupon is allowed to float. Then weekly, monthly, quarterly, or semiannually, the coupon rate changes to a new level, usually 0.5 to 3.0 percent above the preceding week's average treasury bill rate or the LIBOR.

---

[6]In general, real property includes fixed assets such as land, buildings, and equipment. However, collateral trust bonds are secured by common shares and/or bonds, which are held by a trustee who has the power to liquidate them in the event of default. The indenture will normally contain a clause that the market value must be at least 20 to 50 percent greater than the value of the bonds so as to ensure ample coverage.

2. The bondholder generally has the option of redeeming the bond at par value every 6 months.
3. The issuer is generally, although not always, a bank or a trust company whose revenues are subject to swings with interest rate fluctuations.

There are, of course, exceptions and variations in the concept of floating rate bonds. For example, the Toronto Dominion Bank raised U.S. $150 million in financing through a variation on the floating-rate debenture called the *collared, subordinated floating rate note.* The coupon rate equals the six-month LIBOR less .125 percent but, the bank is required to pay holders a minimum rate of 5 percent and a maximum rate of 10 percent over the next ten years. In effect, the Toronto Dominion Bank has capped its interest payments between 5 and 10 percent over ten years. In the North American market, there are other innovative examples of variations of this type of feature. For example, bonds issued by Petro-Lewis, a Texas oil firm, and Sunshine Mining Company, a silver-mining firm. In 1980 Petro-Lewis issued bonds in which the interest rate was tied to the price of crude oil from West Texas. If the oil price increased more than 10 percent, the bond's interest rate would also increase. The Sunshine Mining Company debt was issued in February 1983 and carried a coupon interest rate of 8 percent, with the principal being tied to the price of silver. The common feature of all the variable rate bonds is that an attempt is made to counter uncertainty by allowing the interest rate (or in the case of the Sunshine Mining Bonds, the principal) to float. In this way a decline in cash inflows to the firm should be offset by a decline in interest payments. Table 5-2 gives several examples of variations of floating rate bonds.

BACK TO THE BASICS

Some have thought junk bonds were fundamentally different from other securities, but they are not. They are bonds with a great amount of risk, and therefore promise high expected returns. Thus, ***Axiom 1: The Risk-Return Tradeoff—We Won't Take on Additional Risk Unless We Expect to Be Compensated with Additional Return.***

Table 5-2 Floating Rate (or Floating Principal) Bonds

Issuer	Maturity (Year)	Initial Interest Rate	Terms of Adjustment
National Bank of Canada	2087	6-month LIBOR plus 0.125%	Adjusted quarterly to 20 basis points (0.20%) above the 6-month LIBOR
National Bank of Canada	2083	30-day Bankers' Acceptances Rate plus 0.5%	Adjusted quarterly to 50 basis points (0.5%) above the 30-day Bankers' Acceptance Rate.
Bank of Montreal	1998	London Eurodollar Deposit Rate plus 0.05%. Minimum Rate 6.05%	Adjusted quarterly to 5 basis points (0.05%) above the London Eurodollar Rate.
Bank of Montreal	2089	Canadian 90-day Bankers' Acceptance Rate	Adjusted quarterly to 40 basis points (0.4%) above the Canadian 90-day Bankers' Acceptance Rate.
Sunshine Mining Co.	1995 or whenever redemption value exceeds $2,000 for a period of 30 consecutive days.	8.00%	Interest rate is fixed; however, the redemption value at maturity is equal to the greater of $1,000 or the market price of 50 ounces of silver.

Junk Bonds

Junk or **low-rated bonds** are bonds rated by Standard and Poor's Bond Rating Service as BB or below. Originally, the term was used to describe bonds issued by "fallen angels"; that is, firms with sound financial histories that were facing severe financial problems and suffering from poor credit ratings. The most likely participants in the junk bond market are new firms that do not have an established record of performance. Prior to the mid-seventies these new firms simply did not have access to the capital markets because of the reluctance of investors to accept speculative grade bonds. However, this trend changed during the eighties because of the shortage of capital which was needed to finance corporate takeovers, such as the financing of Campeau's acquisition of Federated Department Stores Inc.

The size of the junk bond market in Canada reached about $3 billion in 1990 but was about 100 times smaller than that of the United States.[7] The Canadian junk bond market was smaller because of restrictive covenants which protected the bondholder's investment and the greater **due diligence** required by institutional investors. Furthermore, the Canadian junk bond market was driven by the needs of the Canadian institutional investors and not by those of the issuers. For example, Canadian institutional investors required the coupons of the junk bonds issued to Sceptre Resources Ltd. to carry an interest rate that was 6.5 percent more than the equivalent Government of Canada bond (A++ grade) rate. Nevertheless, many new firms without established performance records used this financing alternative to secure debt financing through a public offering, rather than being forced to rely on commercial bank loans.

Junk bonds

Any bond rated BB or below.

Due diligence

The obligation of an underwriter to ensure that the prospectus is factually accurate and has no material omissions.

PERSPECTIVE IN FINANCE

The cost advantages that North American firms have been able to achieve in the Eurobond market are very difficult to explain given the fact that the financial markets are relatively well integrated. In integrated markets the flow of capital between countries should keep interest rates approximately equal in different countries.

Eurobonds

Eurobonds are bonds, issued in a country different from the one in whose currency the bond is denominated. For example, a Eurobond may be issued in France by a Canadian company and will pay interest and principal to the lender in Canadian dollars. The Eurobond market actually has its roots in the 1950s and 1960s as the U.S. dollar became increasingly popular because of its role as the primary international reserve. The characteristics of Eurobonds differ from those issued in the Canadian bond market. For example, Eurobonds have coupon interest payments which are issued on an annual basis, Eurobonds are issued on a bearer basis. Eurobonds have a term-to-maturity of approximately 10–12 years and Eurobonds do not have to be registered with any Security Commission in Canada. The primary attraction to borrowers, aside from favourable rates,[8] in the Eurobond market is the relative lack of regulation, less rigorous disclosure requirements, and the speed with which they can be issued.

Eurobonds

Bonds issued in a country different from the one in whose currency the bond is denominated—for instance, a bond issued in Europe or Asia by a Canadian company that pays interest and principal to the lender in Canadian dollars.

---

[7]Susan Gittins, "Junk Bonds: Handle With Care," *The Financial Post,* April 11, 1990, p. 25.

[8]Two empirical studies have documented the savings. They are D.S. Kidwell, M.W. Marr, and G.R. Thompson, "Eurodollar Bonds: Alternative Financing for U.S. Companies," *Financial Management* (Winter 1985), 18–27. A correction to the above study appears in *Financial Management* (Spring 1986), 78–79; and M.W. Marr and J. Trimble, "Domestic versus Euromarket Bond Sale: A Persistent Borrowing Cost Advantage," University of Tennessee, Department of Finance Working Paper, 1988.

The Modern Role of Bond Covenants

"Because corporate debt is a major source of investment funds, knowledge about types of bond covenants, their intended protection, and their use in practice is extremely important for any investor contemplating a bond portfolio." (Ileen Malitz)

Of course, it is impossible to have covenants that protect against every risk associated with debt. Bond contracting is a negotiation process involving the give-and-take between the debtor and the creditor.

Covenants should not be overly restrictive, thus preventing management from properly doing its job. On the other hand, the indenture should not be so loose as to provide little or no control over management actions that may weaken the position of the creditor. There should be a happy medium satisfactory to both parties. Bond indentures and their covenants are like insurance. Many people would rather do without and most do not take the time to read them nor do they care to pay for them. But when something untoward occurs, it may be comforting to know that covenants are there.

Some feel that the decline in covenant protection is because sophisticated investors require less protection than previously. they have greater and better facilities for the analysis and monitoring of the issues in their portfolios. While that may be true, history has shown that these knowledgeable investors and other market participants can and do make mistakes in judgment. They have been "taken to the cleaners" just as the smaller and less sophisticated investors have.

It is naive to believe that just because an issue is rated at the top of the investment grade ladder and has minimal default risk that it will always have those characteristics. If one were certain that there would be no ratings drift, then one could invest in unprotected issues.

"The lesson for creditors is to require protection or be willing to accept the consequences if debt value declines. Creditors cannot have it both ways. They must make a choice between high yield and decreased risk. Each potential creditor must decide which is more important and live with the decision." (Ileen Malitz)

Source: Adapted from Richard S. Wilson, "Review of the Modern Role of Bond Covenants," *Financial Management Collection* (Spring/Summer 1994), pp. 20–21.

The use of Eurobonds by Canadian firms to raise funds has fluctuated dramatically. Their use is dictated by both interest rates and the amount of available capital in the European markets.

Zero Coupon or Stripped Bonds

Zero coupon or stripped bonds
Bonds issued at a substantial discount from their $1,000 face value that pay no interest

Zero coupon or **stripped bonds** offer an investor no coupon interest payments over the maturity of the bond. As a result, an investor buys the zero coupon or stripped bond at a discount and receives one large payment at the bond's term to maturity. For example in 1987, the Government of Canada issued $150 million of zero coupon bonds, called Canada Zeros, which provided a yield to maturity of 8.99% when the bonds expired in the year 2001. These bonds were sold at a 70.8% discount from their par value. In other words, investors who purchased these bonds for $292 and who hold them until they mature in the year 2001 will receive an 8.99% yield to maturity, with all of this yield coming from appreciation of the bond. To solve for the yield to maturity (*YTM*) on a Canada Zero, one assumes semiannual compounding:

$$B_0 = \frac{1{,}000}{\left[1 + \left(\frac{YTM}{2}\right)\right]^{2n}} \qquad (5\text{-}2)$$

$$1 + \left(\frac{YTM}{2}\right) = \left(\frac{1{,}000}{292}\right)^{1/28}$$

$$YTM = 8.99\%$$

On the other hand, the Government of Canada has no cash outflows until the bonds mature. However, at that time the Government of Canada will have to pay back $150 million even though it only received $43.8 million when the bonds were first issued.

Zero coupon or stripped bonds first appeared on the Canadian bond market in 1982. Most of the bonds occurred as a result of investment dealers stripping the interest coupons from high-quality government bond issues, such as Government of Canada bonds, and selling both the interest component and the principal component separately. However, in 1991 Bell Canada became the first company in Canada to issue stripped corporate bonds. These bonds were the product of the investment dealer stripping the interest coupons from a corporate bond issue. Both components were then sold separately using receipts: a receipt was sold for the stripped bond or residual and a receipt was sold for the interest coupons. In 1992, Laidlaw Inc. issued $170 million of stripped corporate bonds using a private placement. This issue was sold at a premium of 3% to its par value, offered a coupon rate of 12.08% and matures in the year 2032. The debt offering was issued in the form of receipts: receipts were issued for the principal payment or residual and 2 classes of receipts were issued for the interest coupons. In other words, the first class of receipts was issued for the interest coupons over the first five years and the second class of receipts was issued for interest coupons over the term of the bond.

As with any form of financing, there are both advantages and disadvantages of issuing zero coupon or stripped bonds. The disadvantages are: first, when zero coupon bonds mature the issuing institution will face an extremely large nondeductible cash outflow, much greater than the cash inflow it experienced when the bonds were first issued. Second, zero coupon bonds are not callable and can only be retired at maturity. Thus, if interest rates fell, the issuing institution could not benefit. The advantages of zero coupon or stripped bonds are: first, that annual cash outflows associated with interest payments do not occur with zero coupon bonds; and second, zero coupon bonds have strong investor demand with the price of the bond being bid up causing the yield to fall. As a result, the issuing institution can issue zero coupon or stripped bonds at a lower yield than the traditional coupon bonds. Finally, the issuing institution is able to deduct the annual amortization of the discount on the principal as an interest expense. Zero coupon or stripped bonds offer an advantage to investors who place these investments in registered retirement savings plans (RRSP), since interest income is sheltered from both federal and provincial taxes during high-income earning years.

DEFINITIONS OF VALUE

OBJECTIVE 3

Book value

The value of an asset as shown on a firm's balance sheet. The book value represents the historical cost of the asset rather than its current market value or replacement cost.

The term *value* is often used in different contexts, depending on the application being made. Examples of different uses of this term include book value, liquidation value, market value, and intrinsic value.

Book value is the value of an asset as shown on a firm's balance sheet. It represents a historical value rather than a current worth.[9] For example, the book value for a company's common shares is the sum of the capital shares value, the contributed

[9]The book value of a firm will vary with the accounting practices used by the firm, and consequently may bear only a weak relationship to the firm's economic value.

surplus, adjustments for translation, and the retained earnings of the firm. The capital shares value and the contributed surplus are equal to the amount the company received when the securities were issued. Retained earnings represents the amount of net income retained within the firm rather than being paid out as dividends to common shareholders. Thus, the retention of earnings is an indirect way for the shareholders to invest more money in the firm.[10] If we want to know the book value for each common share, we simply divide the total book value by the number of shares outstanding. For example, the book value for the Canadian Marconi Company common shares on March 31, 1995, was $14.41, computed as follows:

Common shares ($000)	$ 11,924
Retained earnings ($000)	324,047
Cumulative translation adjustment	8,864
Total book value of common ($000)	$344,835
Number of shares outstanding	23,938,468
Book value per share	
($344,835,000/23,938,468 shares)	$14.41

Liquidation value

The amount that could be realized if an asset were sold individually and not as a part of a going concern.

Liquidation value is the dollar sum that could be realized if an asset were sold individually and not as a part of a going concern. For example, if a product line is discontinued, the machinery used in its production might be sold. The sale price would be its liquidation value and would be determined independently of the firm's value. Similarly, if the firm's operations were discontinued and its assets were sold as a separate collection, the sales price would represent the firm's liquidation value.

Market value

The observed value for the asset in the marketplace.

The **market value** of an asset is the observed value for the asset in the marketplace. This value is determined by demand and supply forces working together in the marketplace, where buyers and sellers negotiate a mutually acceptable price for the asset. For instance, the range of market prices for which Canadian Marconi Company traded during the fiscal year 1994–95 was from $12.75 to $17.50. On March 31, 1995 the market price of Canadian Marconi's common shares closed at $13.13. The price range was determined by a large number of buyers and sellers working through the Toronto Stock Exchange. In theory, a market price exists for all assets. However, many assets have no readily observable market price because trading seldom occurs. For instance, the market price for the common shares of McCain Ltd., a New Brunswick-based family-owned firm, would be more difficult to establish than the market value of the Seagram company's common shares.

Intrinsic or economic value

The present value of the asset's expected future cash flows. The intrinsic value is the amount the investor considers to be a fair value, given the amount, timing and riskiness of future cash flows.

The **intrinsic or economic value** of an asset can be defined as the present value of the asset's expected future cash flows. This value is also called the *fair value*, as perceived by the investor, given the amount, timing, and riskiness of future cash flows. In essence, intrinsic value is like the "value in the eyes of the investor." Given the risk, the investor determines the appropriate discount rate, sometimes called the *capitalization rate*, to use in computing the present or intrinsic value of the asset. Once the investor has estimated the intrinsic value of a security, this value can be compared with its market value. If the intrinsic value is greater than the market value, then the security is undervalued in the eyes of the investor. Should the market value exceed the investor's intrinsic value, then the security is overvalued.

Efficient market

A market in which the values of securities at any instant in time fully reflects all available information, which results in the market value and the intrinsic value being the same.

We hasten to add that if the securities market is working efficiently, the market value and the intrinsic value of a security will be equal. That is, whenever a security's intrinsic value differs from its current market price, the competition among investors seeking opportunities to make a profit will quickly drive the market price back into equilibrium with intrinsic value. Thus, we may define an **efficient market** as one in which the values of all securities at any instant in time fully reflect all

[10]Note that 96.5 percent of Canadian Marconi's book value is attributable to the reinvestment of the firm's profits and 3.5 percent represents new investments in the firm.

available information, which results in the market value and the intrinsic value being the same.[11] If the markets are truly efficient, it is extremely difficult for an investor to make extra profits from an ability to predict prices.

BACK TO THE BASICS

The fact that investors have difficulty identifying stocks that are undervalued relates to ***Axiom 6: Efficient Capital Markets—The Markets Are Quick and the Prices Are Right.*** *In an efficient market, prices reflect all available public information about the security, and therefore it is priced fairly.*

The idea of market efficiency has been the backdrop for an intense battle between professional investors and university professors. The academic community has contended that a blindfolded monkey throwing darts at the list of securities in *The Financial Post* could do as well as a professional money manager. Market professionals, on the other hand, retort that academicians are so immersed in research they could not recognize potential profits even if they were delivered to their offices. The war has been intense, but one that the student of finance should find intriguing.

PERSPECTIVE IN FINANCE

Intrinsic value is the present value of expected future cash flows. This fact is true regardless of what type of asset we are valuing. If you remember only one thing from this chapter, remember that intrinsic value is the present value of expected future cash flows.

THE PROCESS OF VALUING FINANCIAL SECURITIES

OBJECTIVE 4

The valuation process involves calculating the present value of an asset's expected future cash flows using the investor's required rate of return. The investor's required rate of return, k, is determined by the level of the risk-free rate of interest, k_f, and the risk premium, k_{rp}, that the investor feels is necessary to compensate for the risks assumed in owning the asset. Therefore, the basic security valuation model can be defined mathematically as follows:

$$V = \frac{C_1}{(1+k)^1} + \frac{C_2}{(1+k)^2} + \dots + \frac{C_n}{(1+k)^n} \quad (5\text{-}3)$$

or

$$V = \sum_{t=1}^{n} \frac{C_t}{(1+k)^t}$$

where V = the intrinsic value or present value of an asset producing expected future cash flows, C_t, in years 1 through n

k = the investor's required rate of return

Using Equation (5-3), there are three basic steps in the valuation process:

1. Estimate the C_t in Equation (5-3), which is the amount and timing of the future cash flows the security is expected to provide.
2. Determine k, the investor's required rate of return, using $k = k_f + k_{rp}$. We do this by evaluating the riskiness of the security's future cash flows and determining an appropriate risk premium, k_{rp}. We then observe the risk-free rate (such as the rate of interest on three-month Treasury bills) and add the two together for the required rate of return.

[11]Market efficiency is an equilibrium condition. This condition implies that intrinsic and market values are not equal at every instant, but are equal on average.

3. Calculate the intrinsic value, V, as the present value of expected future cash flows discounted at the investor's required rate of return.

PERSPECTIVE IN FINANCE

There are three easy rules about cash flows and valuation.
1. We prefer more cash over less cash.
2. We prefer cash sooner rather than later.
3. We prefer a less risky cash flow over a more risky cash flow.
These three rules capture the essence of what we do in determining an asset's value.

With these general principles of valuation as a foundation, we can investigate the procedures for valuing particular types of securities. Specifically, we will learn how to value bonds, preferred shares, and common shares.

OBJECTIVE 5

BOND VALUATION

Video Case 2 

Valuation Procedure

The valuation process for a bond, as depicted in Figure 5-2, requires knowledge of three essential elements: (1) the amount of the cash flows to be received by the investor, (2) the maturity date of the loan, and (3) the investor's required rate of return. The amount of cash flows is dictated by the periodic interest to be received and by the par value to be paid at maturity. Given these elements, we can compute the value of the bond, or the present value.

PERSPECTIVE IN FINANCE

The value of a bond is the present value of the future interest payments to be received and the par or maturity value of the bond. Simply list these cash flows, use your discount rate of return, and find the value.

World Wide Web Site 

http://info.ic.gc.ca/opengov/cbsc/english/
Comment: This site provides infomration about federal and provincial government programs, services, and regulations. This site also provides the contact addresses for the Canadian Business Centres in each province.

Figure 5-2 Data Requirements for Bond Valuation

(A) Cash Flow Information

Periodic interest payments
For example, $105 per year

Principal amount or par value
For example, $1,000

(B) Term to Maturity

For example, 7 years

(C) Investor's Required Rate of Return

For example, 7 percent

EXAMPLE

Assume that the Government of Canada issues an annual coupon bond,[12] having a maturity date of January 1st, 2003 and a stated nominal interest rate of 10.5% percent. The bonds were originally issued at a price of $1,000. By 1996, however, since the investors' required rate of return had changed, the bond value was more than $1,000. In fact, in early 1996, investors were requiring a return of 7.00 percent for this type of security. At this time the value of the security was approximately $1,189, which can be calculated using the following three-step valuation procedure.

1. **Estimate the amount and timing of the expected future cash flows. Two types of cash flows are received by the bondholder:**
   a) **Annual interest payments equal to the nominal interest rate times the face value of the bond. In this example the interest payments equal $105 or (.105)($1,000). Assuming that 1996 interest payment has already been made, these cash flows will be received by the bondholder in each of the 7 years before the bond matures (1997 through 2003 = 7 years).**
   b) **The face value of the bond of $1,000 to be received in 2003.**

   To summarize, the cash flows received by the bondholder are as follows:

1	2	3	4	5	6	7
$105	$105	$105	$105	$105	$105	$ 105
						+$1,000
						$1,105

2. **Determine the investor's required rate of return by evaluating the riskiness of the bond's future cash flows. A 7.00 percent required rate of return for the Government of Canada bondholders is given.**
3. **Calculate the intrinsic value of the bond as the present value of the expected future interest and principal payments discounted at the investor's required rate of return.**

The present value of Government of Canada bonds is found as follows:

$$\text{Bond value} = V_b = \frac{\$\text{ interest in year 1}}{(1 + \text{required rate of return})^1} + \frac{\$\text{ interest in year 2}}{(1 + \text{required rate of return})^2} + \ldots + \frac{\$\text{ interest in year 7}}{(1 + \text{required rate of return})^7} + \frac{\$\text{ par value of bond}}{(1 + \text{required rate of return})^7}$$

or, summing over the interest payments,

$$V_b = \underbrace{\sum_{t=1}^{7} \frac{\$\text{ interest in year } t}{(1 + \text{required rate of return})^t}}_{\text{present value of interest}} + \underbrace{\frac{\$\text{ par value of bond}}{(1 + \text{required rate of return})^7}}_{\text{present value of par value}} \quad (5\text{-}4a)$$

[12]In actuality, the Government of Canada remits the interest to its bondholders on a semiannual basis such as January 15 and July 15. However, for the moment assume the interest is to be received annually. The effect of semiannual payments will be examined later.

PERSPECTIVE IN FINANCE

This equation is a restatement in a slightly different format of Equation (5-3). Recall that Equation (5-3) states that the value of an asset is the present value of future cash flows to be received by the investor.

Using I_t *to represent the interest payment in year* t, M *to represent the bond's maturity value, and* k_b *to equal the bondholder's required rate of return, we may express the value of a bond maturing in year* n *as follows:*

$$V_b = \sum_{t=1}^{n} \frac{\$I_t}{(1 + k_b)^t} + \frac{\$M}{(1 + k_b)^n} \qquad (5\text{-}4b)$$

Calculator Solution

DATA INPUT	FUNCTION KEY
7	N
7.00	I/Y
105	PMT
1000	FV
FUNCTION KEY	**ANSWER**
CPT PV	−1,188.63*

*The difference between the two answers represents rounding error

Since the interest payments I_t are an annuity for n years, and the maturity value is a one-time amount in year n, we can use the present value factors in Appendixes B and D to solve for the present value of the bond as follows:

$$V_b = \$I_t\,[PVIFA_{k_b,n}] + \$M\,[PVIF_{k_b,n}] \qquad (5\text{-}4c)$$

For our Government of Canada bond,

$$\begin{aligned} V_b &= \$105(5.3893) + \$1{,}000(.6228) \\ &= \$565.88 + \$622.80 \\ &= \$1{,}188.68 \end{aligned}$$

Thus, if investors consider 7.00 percent to be an appropriate required rate of return in view of the risk level associated with these Government of Canada bonds, paying a price of $1,188.68 would satisfy their return requirement.[13] We emphasize that the bondholder's required rate of return is the minimum rate of return that an investor must earn in order to invest in this type of financial instrument. In fact, the required rate of return is an opportunity cost which is determined by the rate of return from a comparable financial instrument in the market. In other words, the required rate of return is determined from the interest rates of bonds selling in the market, which have the same risk characteristics and the same term-to-maturity. We describe an investor who is buying a bond at the required rate as buying the bond "at margin," since the investor is just willing to buy the bond at the market price and is not willing to pay a higher price for it.

Semiannual Interest Payments

In the preceding illustration, the coupons were assumed to be paid annually. However, companies typically forward an interest cheque to bondholders semiannually. For example, rather than disbursing $105 in interest at the conclusion of each year, Government of Canada pays $52.50 (half of $105) on January 1 and July 1.

Several steps are involved in adapting Equation (5-4b) to recognize semiannual interest payments.[14] First, thinking in terms of "periods" instead of years, a bond with a life of n years paying coupons semiannually should be thought to have a life of $2n$ periods. In other words, a seven-year bond ($n = 7$) that remits its coupons on a semiannual basis actually makes 14 payments. Although the number of periods has doubled, the dollar amount of interest being sent to the investors for each period and the bondholders' required rate of return are half of the equivalent annual figures. I_t becomes $I_t/2$ and k_b is changed to $k_b/2$; thus, for semiannual compounding, Equation (5-4b) becomes

[13]Instead of using the present value tables in the appendix, it is easier to use a financial calculator. Also, the calculator is not as restrictive in that you are not limited to integer percentages. See student's *Study Guide* for a tutorial on using the Texas Instruments BAII Plus calculator.

[14]The logic for calculating the value of a semiannual coupon bond is similar to the material presented in Chapter 4, in which compound interest with nonannual periods was discussed.

$$V_b = \sum_{t=1}^{2n} \frac{\frac{\$I_t}{2}}{\left(1+\frac{k_b}{2}\right)^t} + \frac{\$M}{\left(1+\frac{k_b}{2}\right)^{2n}} \quad (5\text{-}5a)$$

However, if the present value tables are used, V_b is calculated as

$$V_b = \frac{\$I_t}{2}\left[PVIFA_{\frac{k_b}{2},2n}\right] + \$M\left[PVIF_{\frac{k_b}{2},2n}\right] \quad (5\text{-}5b)$$

EXAMPLE

To see the effect of semiannual coupon payments upon the value of a bond, assume the Government of Canada has issued a 9.75 percent, $1,000 bond. The debt is to mature in nine years, and the investor's required rate of return is currently 8.16 percent. If interest is paid semiannually, the number of periods is 18 (9 years $\times$ 2); the dollar interest to be received by the investor at the end of each six-month period is $48.75 ($97.50/2); and the required rate of return for six months is 4.08 percent (8.16 percent/2). Therefore, the valuation equation is

$$V_b = \sum_{t=1}^{18} \frac{\$48.75}{(1+.0408)^t} + \frac{\$1{,}000}{(1+.0408)^{18}}$$

Equivalently,

$$\begin{aligned} V_b &= \$48.75\,[PVIFA_{4.08\%,18}] + \$1{,}000\,[PVIF_{4.08\%,18}] \\ &= \$48.75(12.5774) + \$1{,}000(.4868) \\ &= \$613.15 + \$486.80 \\ &= \$1{,}099.95 \end{aligned}$$

Thus, the value (present value) of a bond paying $48.75 semiannually, and $1,000 at maturity (six years), is $1,099.95, provided the investor's required rate of return is 8.16 percent.[15]

Calculator Solution

DATA INPUT	FUNCTION KEY
18	N
4.08	I/Y
48.75	PMT
1,000	FV

FUNCTION KEY	ANSWER
CPT PV	–1,099.99*

*The difference between the two answers represents rounding error

BONDHOLDER'S EXPECTED RATE OF RETURN AND YIELD TO MATURITY

OBJECTIVE 6

Mathematically, the internal rate of return (IRR) is the discount rate that equates the present value of the inflows with the present value of the outflows. The idea of comparing the present value of inflows to that of outflows is especially valuable when evaluating an investment project such as a security investment. For this type of security investment, the investor expectations of the actual rate of return from future cash flows are defined as the expected rate of return. For example, the expected rate of return implicit in the current bond price is the discount rate that equates the present value of the future cash flows with the current market price of the bond.[16] The Yield-To-Maturity (YTM) represents an estimate of the expected rate of return for a bond if the bond is held to maturity. Thus, when referring to bonds, the terms **expected rate of return** and *yield to maturity* are often used interchangeably when discussing the rate of return which equates the present value of the future interest payments received until maturity to the current bond price.

Expected rate of return

The discount rate that equates the present value of the future cash flows (interest and maturity value) with the current market price of the bond. The expected rate of return is the rate of return that an investor will earn if a bond is held to maturity.

[15]A slightly different answer may result from differences in rounding.

[16]When we speak of computing an expected rate of return, we are not describing the situation very accurately. Expected rates of return are *ex ante* (before the fact) and are based upon "expected and unobservable future cash flows" and therefore can only be "estimated."

Calculator Solution

DATA INPUT*	FUNCTION KEY
10	N
40	+/– PMT
1,120	PV
1,000	+/– FV

FUNCTION KEY	ANSWER
CPT I/Y	5.24**

*Payments must be set for semiannual coupons as follows:

CE/C CE/C

**The difference between the two answers represents rounding error

Calculator Solution

Bond Worksheet

DATA INPUT*	FUNCTION KEY
01.0198	STD =
ENTER	01-01-1998
↓ 8 ENTER	CPN = 8.00
↓ 01.0103	RDT =
ENTER	01-01-2003
↓ 100 ENTER	RV = 100.00
↓	360
↓	2/Y
↓	YLD = 0.00
↓ 112 ENTER	PRI = 112.00
↑ CPT	YLD = 5.24**

*To access the Bond Worksheet and clear its old contents, enter

2nd Bond 2nd CLR Work

and payments must be set for semiannual coupons as follows:

2nd P/Y 2 ENTER

CE/C CE/C

**The difference between the two answers represent rounding error

Table 5-3 Computing the YTM for a Bond Without a Financial Calculator

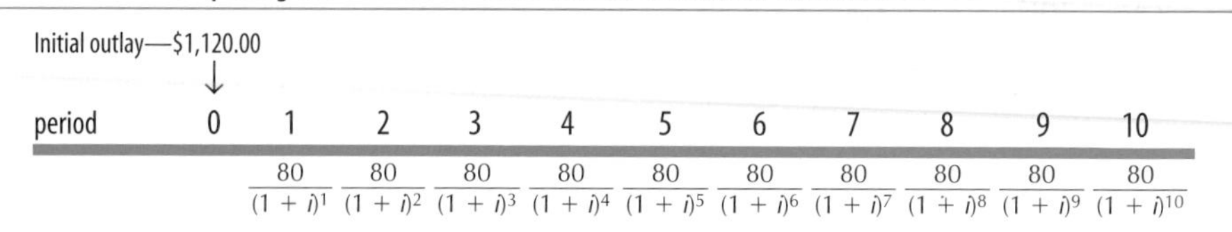

EXAMPLE

You are considering an investment into the bonds of Able Corporation which are currently selling for $1,120. These bonds provide semiannual coupons with an annual interest rate of 8.0 percent and mature in five years.

In determining the expected rate of return implicit in the current market price, we need to find the rate that discounts the cash flows back to the present value of $1,120, the existing market price for the bond. Table 5-3 shows the method that should be used to estimate the YTM of the bond.

***Solution*: Pick an arbitrary discount rate and use it to determine the present value of the inflows. Compare the present value of the inflows with the initial outlay; if they are equal, you have determined the YTM. Otherwise, raise the discount rate if the present value of the inflows is larger than the initial outlay or lower the discount rate if the present value of the inflows is less than the initial outlay. Repeat this process until the YTM is found.**

1. **Try semiannual interest rate = 2 percent**

$$\begin{aligned} V_b &= \$40(PVIFA_{2.0\%,10}) + \$1{,}000\,(PVIF_{2.0\%,10}) \\ &= \$40(8.9826) + \$1{,}000(0.8203) \\ &= \$359.30 + \$820.30 \\ &= \$1{,}179.60 \end{aligned}$$

2. **Try semiannual interest rate = 3 percent**

$$\begin{aligned} V_b &= \$40(PVIFA_{3\%,10}) + \$1{,}000(PVIF_{3\%,10}) \\ &= \$40(8.5302) + \$1{,}000(0.7441) \\ &= \$341.21 + \$744.10 \\ &= \$1{,}085.31 \end{aligned}$$

The actual rate can be more precisely approximated through interpolation as follows:

Discount rate	Present value
2%	$1,179.60
YTM	$1,120.00
3%	$1,085.31

$$\frac{(YTM-3\%)}{(3\%-2\%)} = \frac{1{,}120.00-1{,}085.31}{1{,}085.31-1{,}179.60}$$

$$\frac{(YTM-3\%)}{(1\%)} = \frac{(34.69)}{(-94.29)}$$

$$\begin{aligned} YTM-3\% &= (1\%)\,(-0.368) \\ YTM &= 2.632\% \text{ on a semiannual basis} \end{aligned}$$

Thus, the expected rate of return on Able Corporation's bonds for an investor who purchases the bonds for $1,120 is approximately 5.3 percent or 6.9 percent compounded annually. A more precise answer can be obtained by using a calculator. (In the margin, we have provided two alternative methods to solve for the problem.)

Suppose that these bonds are callable at $1,040.00 (par value + one semiannual coupon) by Able Corporation after three years. Determine the yield-to-call if you hold the bond until Able Corporation exercises its call.

Solution:

Again, pick an arbitrary discount rate and use it to determine the present value of the inflows. Compare the present value of the inflows with the initial outlay; if they are equal, you have determined the yield-to-call.

1. Try semiannual interest rate = 2 percent

$$\begin{aligned} V_b &= \$40(PVIFA_{2.0\%,6}) + \$1{,}040\,(PVIF_{2.0\%,6}) \\ &= \$40(5.6014) + \$1{,}040(0.8880) \\ &= \$224.06 + \$923.52 \\ &= \$1{,}147.58 \end{aligned}$$

2. Try semiannual interest rate = 3 percent

$$\begin{aligned} V_b &= \$40(PVIFA_{3.0\%,6}) + \$1{,}040\,(PVIF_{3.0\%,6}) \\ &= \$40(5.4172) + \$1{,}040(0.8375) \\ &= \$216.69 + \$871.00 \\ &= \$1{,}087.69 \end{aligned}$$

The actual rate can be more precisely approximated through interpolation as follows:

Discount rate	Present value
2%	$1,147.58
YTM	$1,120.00
3%	$1,087.69

$$\frac{(YTM - 3\%)}{(3\% - 2\%)} = \frac{1{,}120.00 - 1{,}087.69}{1{,}087.69 - 1{,}147.58}$$

$$\frac{(YTM - 3\%)}{1\%} = \frac{32.31}{-59.89}$$

$$\begin{aligned} YTM - 3\% &= (1\%)\,(-0.539) \\ YTM &= 2.461\% \quad \text{on a semiannual basis} \end{aligned}$$

Thus, the expected rate of return on Able Corporation's bonds for an investor who purchases the bonds for $1,120 is approximately 4.9 percent or 6.1 percent compounded annually. A more precise answer can be obtained by using the bond worksheet that is built into the Texas Instruments BAII Plus Financial Calculator.

Calculator Solution

Bond Worksheet

DATA INPUT*	FUNCTION KEY
01.0198	STD =
ENTER	01-01-1998
↓ 8 ENTER	CPN = 8.00
↓ 01.0101	RDT =
ENTER	01-01-2001
↓ 104 ENTER	RV = 104.00
↓	360
↓	2/Y
↓	YLD = 0.00
↓ 112 ENTER	PRI = 112.00
↑ CPT	YLD = 4.9**

*To access the Bond Worksheet and clear its old contents, enter the following keystrokes:

2nd Bond 2nd CLR Work

**The difference between the two answers represents rounding error

BOND VALUATION: FIVE IMPORTANT RELATIONSHIPS

OBJECTIVE 7

We have now examined how to find the value of a bond (V_b), given (1) the amount of interest payments (I_t), (2) the maturity value (M), (3) the length of time to maturity (n years), and (4) the investor's required rate of return, k_b. We also know how to compute the expected rate of return, which also happens to be the current interest rate on the bond, given (1) the current market value (V_b), (2) the amount of interest payments (I_t), (3) the maturity value (M), and (4) the length of time to maturity (n years). These computations represent the basics of bond valuation; however, a more complete understanding of bond valuation requires that we examine five additional key relationships:

First Relationship

The value of a bond is inversely related to changes in the investor's present required rate of return (the current interest rate). That is, as interest rates increase (decrease), the value of the bond decreases (increases).

To illustrate, assume that an investor's required rate of return for a given bond is 12 percent. The bond has a par value of $1,000 and annual interest payments of $120, indicating a 12 percent coupon interest rate ($120 ÷ $1,000 = 12%). Assuming a five-year maturity date, the bond would be worth $1,000, computed as follows:

$$V_b = \frac{I_1}{(1+k_b)^1} + \ldots + \frac{I_n}{(1+k_b)^n} + \frac{M}{(1+k_b)^n}$$

$$V_b = \sum_{t=1}^{n} \frac{I_t}{(1+k_b)^t} + \frac{M}{(1+k_b)^n} \qquad (5\text{-}4b)$$

$$V_b = \sum_{t=1}^{n} \frac{\$120}{(1+.12)^t} + \frac{\$1{,}000}{(1+.12)^5}$$

$$V_b = \$120(PVIFA_{12\%,5}) + \$1{,}000(PVIF_{12\%,5})$$

$$V_b = \$120(3.605) + \$1{,}000(.567)$$
$$= \$432.60 + \$567.00$$
$$= \$999.60 \cong \$1{,}000.00$$

If, however, the investor's required rate of return (going interest rate) increases from 12 to 15 percent, the value of the bond would decrease to $899.24:

$$V_b = \$120(PVIFA_{15\%,5}) + \$1{,}000(PVIF_{15\%,5})$$
$$V_b = \$120(3.352) + \$1{,}000(.497)$$
$$= \$402.24 + \$497.00$$
$$= \$899.24$$

On the other hand, if the investor's required rate of return decreases to 9 percent, the bond would increase in value to $1,116.80:

$$V_b = \$120(PVIFA_{9\%,5}) + \$1{,}000(PVIF_{9\%,5})$$
$$V_b = \$120(3.890) + \$1{,}000(.650)$$
$$= \$466.80 + \$650.00$$
$$= \$1{,}116.80$$

This inverse relationship between the investor's required rate of return and the value of a bond is presented in Figure 5-3. Clearly, as an investor demands a higher rate of return, the value of the bond decreases. The higher rate of return which the investor desires, can be achieved only by paying less for the bond. Conversely, a lower required rate of return yields a higher market value for the bond.

Changes in bond prices represent an element of uncertainty for the bond investor. If the current interest rate (required rate of return) changes, the price of the

Figure 5-3 Value and Required Rates for a Five-Year Bond at 12 Percent Coupon Rate

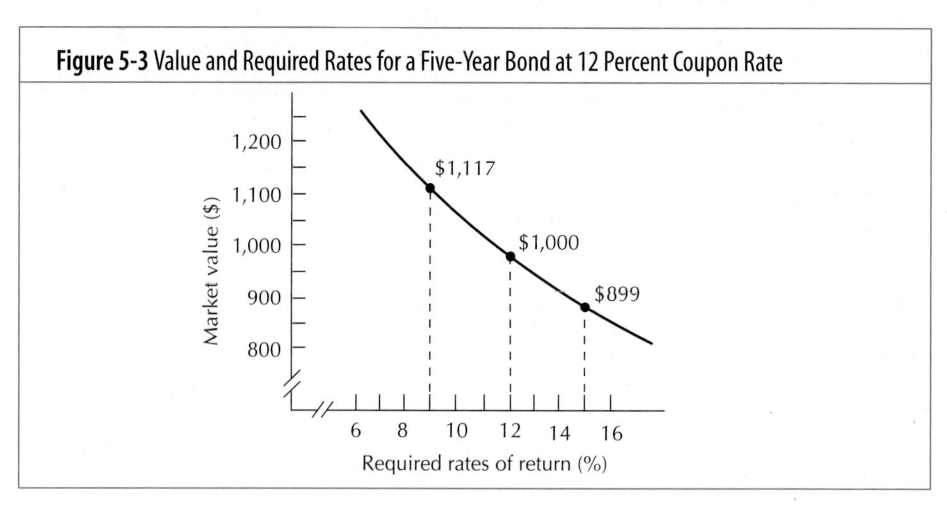

bond also fluctuates. An increase in interest rates causes the bondholder to incur a loss in market value. Since future interest rates and the resulting bond value cannot be predicted with certainty, a bond investor is exposed to the risk of changing values as interest rates vary. This risk has come to be known as **interest rate risk.**

Interest rate risk

The variability in a bond's value caused by changing interest rates.

Second Relationship

The market value of a bond will be less than the par value if the investor's required rate is above the coupon interest rate; but it will be valued above par value if the investor's required rate of return is below the coupon interest rate.

Using the previous example, we observed that:

1. The bond has a market value of $1,000, equal to the par or maturity value, when the investor's required rate of return equals the 12 percent coupon interest rate. In other words, if

required rate	=	*coupon rate,*	then	*market value*	=	*par value*
12%	=	12%	then	$1,000	=	$1,000

2. When the required rate is 15 percent, which exceeds the 12 percent coupon rate, the market value falls below par value to $899.24. That is if

required rate	>	*coupon rate,*	then	*market value*	<	*par value*
15%	>	12%	then	$899.24	<	$1,000

In this case the bond sells at a discount below par value; thus it is called a **discount bond.**

Discount bond

A bond that is selling below its par value.

3. When the required rate is 9 percent, or less than the 12 percent coupon rate, the market value, $1,116.80, exceeds the bond's par value. In this instance, if

required rate	<	*coupon rate,*	then	*market value*	>	*par value*
9%	<	12%	then	$1,116.80	>	$1,000

The bond is now selling at a premium above par value; thus, it is a **premium bond.**

Premium bond

A bond that is selling above its par value.

Third Relationship

As the maturity date approaches, the market value of a bond approaches its par value.

Continuing to draw from our example, the bond has five years remaining until the maturity date. At that time, the bond sells at a discount below par value ($899.24) when the required rate is 15 percent, and it sells at a premium above par value ($1,116.80) when the required rate is only 9 percent.

In addition to knowing value today, an investor would also be interested in knowing how these values would change over time, assuming no change in the current interest rates. For example, how will these values change when only two years remain until maturity rather than five years? Table 5-4 shows the values with five years remaining to maturity, the values as recomputed with only two years left until the bonds mature, along with the changes in values between the five-year bonds and the two-year bonds. The following conclusions can be drawn from these results:

1. The premium bond sells for less as maturity approaches. The price decreases from $1,116.80 to $1,053.08 over the three years.

Table 5-4 Values Relative to Maturity Dates

Required Rate	Market Value if Maturity Is 5 Years	2 Years	Change in Value
9 %	$1,116.80	$1,053.08	–$63.72
12	1,000.00	1,000.00	.00
15	899.24	951.12	51.88

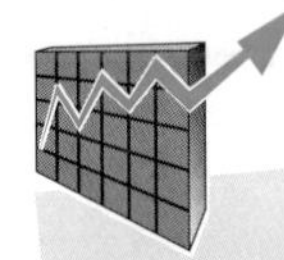

The Bond Market in the 1990's: The Revolution Will Roll On

With the advent of the 1990's, the bond revolution has now entered its second decade, with five major trends shaping the bond market and fixed income management in the 1990's. These trends will impact institutional investors, investment dealers and borrowers alike.

Trend #1: Globalization of the Securities Market

"Globalization" has ceased to be a buzzword. It is a stark reality that affects all capital market participants. To date, it has had its greatest impact on financing patterns. Governments and corporations are now able to raise funds in their choice markets and currencies, through a choice of domestic or international dealers. These alternatives have allowed borrowers to raise funds more cheaply, while customizing debt issues to meet their requirements. At the end of 1989, over 38% of the outstanding issues of Canadian provinces, municipalities and corporations were denominated in other currencies. The Euro-Canadian market had grown to 15.6% of the domestic market, with outstandings increasing ninefold between 1986 and 1990.

Globalization has benefited borrowers, but has had little direct impact on pension fund lenders. The 10% limit on foreign holdings has effectively prevented Canadian bond managers from utilizing the global market, especially while the bulk of foreign content is directed to equities. Canadian bond managers have watched with envy as foreign investors switch back and forth between liquid U.S. Treasuries and our Canadian domestic market. In North America and offshore, they purchase issues that offer the best expected returns, while our domestic pension fund managers search for Deutschmarks or U.S. dollar bonds that qualify as Canadian property.

Thus the proposed changes in the foreign content legislation for pension plans has been greeted with as much enthusiasm by bond managers as by equity managers and fund sponsors. The outcome will be increased flexibility for fixed income managers and greater potential returns for their clients.

Trend #2: More Foreign Participation in Canadian Bond Market

Globalization of markets has occurred at the same time as huge demands were placed on the Canadian market by deficit financing. The natural outcome has been greater participation in our domestic market by non-Canadian investors. Non-residents held $48 billion, or 36%, of publicly owned Government of Canada marketable bonds in 1989. Only ten years earlier, non-resident holdings amounted to only $5.4 billion, or 15% of the total.

The capital flows controlled by foreign investors can now overwhelm Canadian economic considerations for significant periods of time. This forces Canadian bond managers to monitor international conditions as well as domestic fundamentals. In fact, most 8:30 a.m. conversations with dealers typically start by discussing "what happened overnight in Tokyo and London."

Trend #3: Financing Methods Are Changing

In 1980, Government of Canada issues were underwritten by the Bank of Canada and distributed by Canadian dealers. Provincial issuers had lead managers and large syndicates. Corporate bonds were issued by prospectus, using banking groups to underwrite the risk. These methods have all but disappeared under the combination of increased competition among dealers, globalization and the speed of execution now required in markets.

Approximately 65% of Government of Canada bonds are now distributed through auctions. The "bought deal" accounts for 80% of provincial and corporate issues. Again, the winner in this change has been the borrower, generally at the expense of the investment dealers.

For the bond manager, this era of dealer competition and

fast-paced new issues has meant staying ready and alert. Properly priced new issues are sold out in a matter of minutes. Unfortunately, there are as many overpriced issues as properly priced ones to tempt the innocent.

Trend #4: The Growth of the "Quant"

In 1980, Martin Liebowitz of Salomon Brothers had just begun to popularize concepts such as immunization and dedication. Stripped coupons were about to appear. A few dealers offered bond analysis systems that subtracted two yields to find a spread, and bond traders couldn't spell duration, much less define it. In short, quantitative techniques were dirty words and not used in portfolio management. Players with quantitative skills and an interest in bond markets resided in the back room where their efforts focused on writing articles to convince a largely reluctant audience.

This situation changed quickly and dramatically. On the dealer side, the driving forces were a combination of market volatility and increased competition. Market volatility needed carefully orchestrated hedging strategies. Increased competition, including the loss of "old style" business, needed new products to fill the void. Futures, options, warrants, mortgage backed securities and swaps are the result of these forces.

Investment managers were also responding to competitive factors. Quantitative approaches could add extra value to boost relative performance. Moreover, clients were becoming increasingly aware of these new techniques and products. Business retention often depended on the ability to offer a quantitative approach or, at least, discuss its relative merits with key pension fund clients.

Trend #5: The Growth of the Consultant

Years ago, actuaries were concerned with only the liability side of the pension equation. Performance measurement services did only what the name implied. These roles were extended to the lucrative field of manager searches some time ago. When the confusion caused by rapid changes in investment alternatives and practices is added to the requirement of recent legislation, consultants saw an opportunity to create a whole new array of billable services. These typically include an analysis of both the asset side of the pension equation, determining an asset mix, and selecting investment managers to execute the overall strategy.

This could be a risky business for the consultant if investment managers fail to deliver good performance. To sidestep this business risk, the existence of efficient capital markets is recognized, and managers are compared as much by risk and approach, as they are by historical or projected returns. In other words, "philosophy" and "style" are now buzzwords of investment manager marketing. This trend has provided growth and profitability for the consulting community.

Source: Adapted from: Don Webster and Maureen Stapleton, "The Bond Market In The 1990's: The Revolution Will Roll On," *Canadian Investment Review* (Fall 1990), pp. 49–53.

2. The discount bond sells for more as maturity approaches. The price increases from $899.24 to $951.12 over the three years.

The change in prices over the entire life of the bond is shown in Figure 5-4. The graph clearly demonstrates that the value of a bond, either a premium or a discount bond, approaches par value as the maturity date becomes closer in time.

Fourth Relationship

Long-term bonds have greater interest rate risk than do short-term bonds. As already noted, a change in current interest rates (required rate of return) causes a change in the market value of a bond. However, the impact on value is greater for long-term bonds than it is for short-term bonds.

In Figure 5-3 we observed the effect of interest rate changes on a five-year bond paying a 12 percent coupon interest rate. What if the bond did not mature until ten years from today instead of five years? Would the changes in market value be the same? Absolutely not. The changes in value would be more significant for the

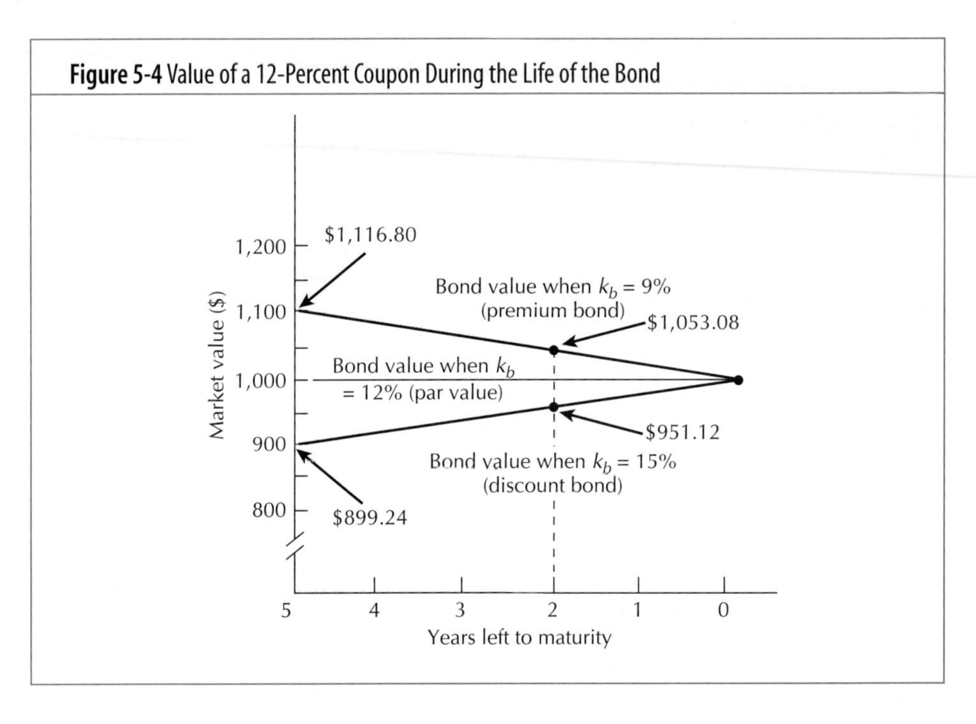

Figure 5-4 Value of a 12-Percent Coupon During the Life of the Bond

ten-year bond. For example, if we vary the current interest rates (the bondholder's required rate of return) from 9 percent to 12 percent and then to 15 percent, as we did earlier with the five-year bond, the values for both the five-year and the ten-year bonds would be as follows:

	Market Value for a 12% Coupon Rate Bond Maturing in	
Required Rate	5 Years	10 Years
9%	$1,116.80	$1,192.16
12	1,000.00	1,000.00
15	899.24	849.28

Using these values and the required rates, we can graph the changes in values for the two bonds relative to different interest rates. These comparisons are provided in Figure 5-5. The figure clearly illustrates that the price of a long-term bond (say ten years) is more responsive or sensitive to interest rate changes than the price of a short-term bond (say five years).

The reason long-term bond prices fluctuate more than short-term bond prices in response to interest rate changes is simple. Assume an investor bought a ten-year bond yielding a 12 percent interest rate. If the current interest rate for bonds of similar risk increased to 15 percent, the investor would be locked into the lower rate for ten years. If, on the other hand, a shorter-term bond had been purchased, say one maturing in two years, the investor would have to accept the lower return for only two years and not the full ten years. At the end of year 2, the investor would receive the maturity value of $1,000 and could buy a bond offering the higher 15 percent rate for the remaining eight years. Thus, interest rate risk is determined, at least in part, by the length of time an investor is required to commit to an investment. However, the holder of a long-term bond may take some comfort from the fact that long-term interest rates are usually not as volatile as short-term rates. If the short-term rate changed one percentage point, for example, it would not be unusual for the long-term rate to change only .3 percentage points.

Figure 5-5 Market Value of a 5-Year and a 10-Year Bond and Different Required Rates

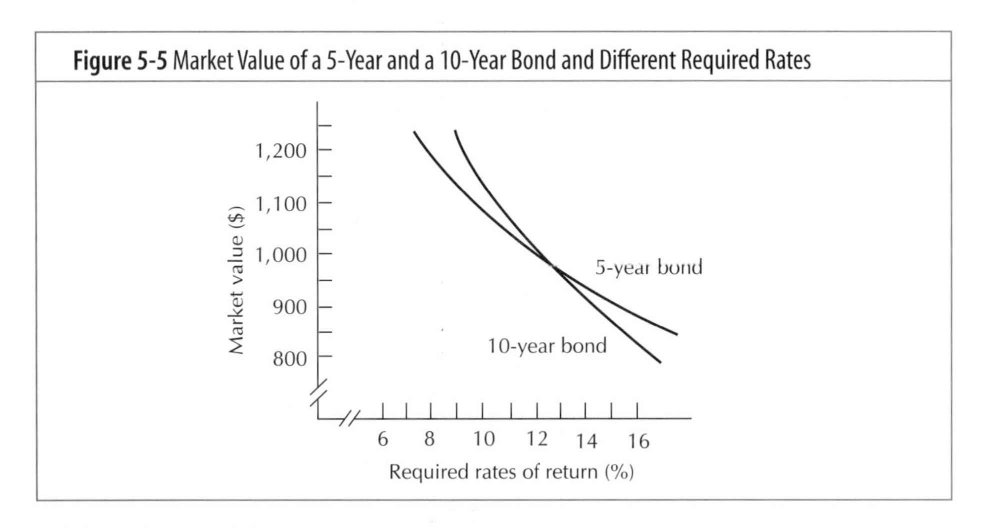

Fifth Relationship

The sensitivity of a bond's value to changing interest rates depends not only on the length of time to maturity, but also on the pattern of cash flows provided by the bond.

It is not at all unusual for two bonds with the same maturity to react differently to a change in interest rates. Consider two bonds, A and B, both with ten-year maturities and the same 10 percent interest rate. While the bonds are similar in terms of maturity date and the contractual interest rate, the structure of the interest payments is different for each bond. Bond A pays \$100 interest annually, with the \$1,000 principal being repaid at the end of the tenth year. Bond B is called a zero coupon bond; it pays no interest until the bond matures. At that time the bondholder receives \$1,593.70 in interest plus \$1,000 in principal. The value of both bonds, assuming a market interest rate (required rate of return) of 10 percent, is \$1,000. However, if interest rates fell to 6 percent, bond A's market value would be \$1,294, compared with \$1,447 for bond B. Why the difference? Both bonds have the same maturity, and each promises the same 10 percent rate of return. The answer lies in the differences in their cash flow patterns. Bond B's cash flows are received in the more distant future on average than are the cash flows for bond A. Since a change in interest rates always has a greater impact on the present value of later cash flows than on earlier cash flows (due to the effects of compounding), bonds with cash flows coming later on average will be more sensitive to interest rate changes than will bonds with earlier cash flows. This phenomenon was recognized in 1938 by Macaulay, who devised the concept of duration.

Duration

A measure of how responsive a bond's price is to changing interest rates. Duration is a weighted average time to maturity in which the weight attached to each year is the present value of cash flow for that year.

The **duration** of a bond is simply a measure of the responsiveness of its price to a change in interest rates. The greater the relative percentage change in a bond price in response to a given percentage change in the interest rate, the longer the duration. In computing duration, we consider not only the maturity or term over which cash flows are received but also the time pattern of interim cash flows. Specifically, duration is a weighted average time to maturity in which the weight attached to each period is the present value of the cash flow for that period. A measurement of duration may be represented as follows:

$$\text{duration} = \frac{\sum_{t=1}^{n} \frac{tC_t}{(1+k_b)^t}}{V_b} \qquad (5\text{-}6)$$

where t = the period the cash flow is to be received
n = the number of periods to maturity
C_t = the cash flow to be received in period t
k_b = the bondholder's required rate of return
V_b = the bond's present value

For our two bonds, A and B, duration would be calculated as follows:

$$\text{duration bond A} = \frac{\left[(1)\frac{\$100}{(1.1)^1} + (2)\frac{\$100}{(1.1)^2} + (3)\frac{\$100}{(1.1)^3} + \ldots + (9)\frac{\$100}{(1.1)^9} + (10)\frac{\$1{,}100}{(1.1)^{10}}\right]}{\$1{,}000}$$

$$= 6.759$$

$$\text{duration bond B} = \frac{\left[(1)\frac{0}{(1.1)^1} + (2)\frac{0}{(1.1)^2} + (3)\frac{0}{(1.1)^3} + \ldots + (9)\frac{0}{(1.1)^9} + (10)\frac{\$2{,}593.70}{(1.1)^{10}}\right]}{\$1{,}000}$$

$$= 10$$

Thus, while both bonds have the same maturity, ten years, the zero coupon bond (bond B) is more sensitive to interest rate changes, as suggested by its higher duration, which in this instance equals its maturity.[17] The lesson learned: In assessing a bond's sensitivity to changing interest rates, the bond's duration is the appropriate measure, not the term to maturity.

SUMMARY

In this chapter, the valuation procedure seen in Chapter 4 has been applied to value bonds. We have first examined the characteristics of bonds and then defined the various types of bonds.

The Characteristics of Bonds

Some of the more important terms and characteristics that are associated with bonds include the following:

- par value
- coupon interest rate
- maturity
- indenture
- current yield
- yield to maturity
- bond ratings
- claims on assets and income

The Various Types of Bonds

There are a variety of types of bonds, for example:

- debentures
- subordinated debentures
- income bonds
- mortgage bonds
- equipment trust certificates
- floating rate or variable rate bonds
- junk bonds
- Eurobonds
- zero coupon or stripped bonds

Definitions of Value

Valuation is an important issue if we are to manage a company effectively. An understanding of the concepts and how to compute the value of a security underlie much that we do in finance and in making correct decisions for the firm as a whole. Only if we know what matters to our investors can we maximize the firm's value.

[17]Maturity and duration are the same only for zero coupon bonds or securities that pay only one cash flow, and it occurs at maturity.

Value is defined differently depending on the context. But for us, value is the present value of future cash flows expected to be received from an investment discounted at the investor's required rate of return.

The Process of Valuing Financial Securities

The valuation process involves assigning a value to an asset by calculating the present value of its expected future cash flows using the investor's required rate of return as the discount rate. The investor's required rate of return, k, equals the risk-free rate of interest plus a risk premium to compensate the investor for assuming risk.

Bond Valuation

The value of a bond is the present value both of the future interest payments to be received and the par or maturity value of the bond.

Bondholder's Expected Rate of Return

To measure the bondholder's expected rate of return, we find the discount rate that equates the present value of the future cash flows (interest and maturity value) with the current market price of the bond. The expected rate of return for a bond is also the rate of return that the investor will earn if the bond is held to maturity, or the *yield to maturity.*

Bond Valuation: Five Important Relationships

Five key relationships exist in bond valuation:

1. A decrease in interest rate (required rate of return) will cause the value of a bond to increase; an interest rate increase will cause a decrease in value. The change in value caused by changing interest rate is called interest rate risk.
2. If the bondholder's required rate of return (current interest rate):
   a. equals the coupon interest rate, the bond will sell at par or maturity value.
   b. exceeds the bond's coupon rate, the bond will sell below par value, or at a *discount.*
   c. is less than the bond's coupon rate, the bond will sell above par value, or at a *premium.*
3. As a bond approaches maturity, the market price of the bond approaches the par value.
4. A bondholder owning a long-term bond is exposed to greater interest rate risk than one owning a short-term bond.
5. The sensitivity of a bond's value to interest rate changes is not only affected by the time to maturity, but also by the time pattern of interim cash flows, or its *duration.*

STUDY QUESTIONS

5-1. **a.** How does a bond's par value differ from its market value?
b. Explain the difference between the coupon interest rate and a bondholder's required rate.

5-2. Describe the bondholder's claim on the firm's assets and income.

5-3. What factors determine a bond's rating? Why is the rating important to the firm's manager?

5-4. Distinguish between debentures and mortgage bonds.

5-5. Define (a) Eurobonds, (b) zero coupon bonds, and (c) junk bonds.

5-6. What is a general definition of the term *intrinsic value* or *economic value* of an asset?

5-7. What are the basic differences between book value, liquidation value, market value, and intrinsic value?

5-8. Explain the relationship between an investor's required rate of return and the value of a security.

5-9. Define the bondholder's expected rate of return.

5-10. How does the market value of a bond differ from its par value when the coupon interest rate does not equal the bondholder's required rate of return?

5-11. Differentiate between a premium bond and a discount bond. What happens to the premium or discount for a bond over time?

5-12. Why is the value of a long-term bond more sensitive to a change in interest rates than a short-term bond?

5-13. Explain duration.

SELF-TEST QUESTIONS

ST-1. (*Bond Valuation*) Trico has issued annual coupon bonds which pay a nominal interest rate of 8 percent, have a par value of $1,000, and will mature in 20 years. If you require a return of 7 percent, what price would you be willing to pay for the bond? What happens if you pay more for the bond? What happens if you pay less for the bond?

ST-2. (*Bond Valuation*) Sunn Co. have issued semiannual coupon bonds which pay a nominal interest rate of 8 percent on a $1,000 face value and mature in seven years. If your required rate of return is 10 percent, what is the value of the bond? How would your answer change if these were annual coupon bonds?

ST-3. (*Bondholder's Expected Rate of Return*) Sharp Co. has issued annual coupon bonds which are selling in the market for $1,045. These 15-year bonds pay a nominal interest rate of 7 percent on a $1,000 par value. If they are purchased at the market price, what is the expected rate of return?

ST-4. (*Duration*) Calculate the value and the duration for the following bonds:

Bond	Years to Maturity	Annual Interest	Maturity Value
Argile	10	$80	$1,000
Terathon	15	65	1,000

Your required rate of return is 8 percent.

STUDY PROBLEMS (SET A)

5-1A. (*Bond Valuation*) Calculate the value of an annual coupon bond that expects to mature in 12 years with a $1,000 face value. The coupon pays a nominal interest rate of 8 percent and the investors' required rate of return is 12 percent.

5-2A. (*Bond Valuation*) Enterprise, Inc., has issued semiannual coupon bonds which pay a nominal interest rate of 9 percent. The bonds mature in eight years and their par value is $1,000. If your required rate of return is 8 percent, what is the value of the bond? What is the value of this bond if coupons are distributed on an annual basis?

5-3A. (*Bondholder's Expected Rate of Return*) The market price is $900 for a ten-year bond ($1,000 par value) that pays 8 percent interest (4 percent semiannually). What is the bond's expected rate of return?

5-4A. (*Bond Valuation*) Exxon 20-year, annual coupon bonds pay a nominal interest rate of 9 percent on a $1,000 par value. If you buy the bonds at $945, what is your expected rate of return?

5-5A. (*Bond Valuation*) Zenith Co. has issued annual coupon bonds which mature in 12 years and pay a nominal interest rate of 7 percent on a par value of $1,000. If you purchase the bonds for $1,150, what is your expected rate of return?

5-6A. (*Bond Valuation*) National Steel 15-year, annual coupon bonds pay a nominal interest rate of 8 percent on a par value of $1,000. The market price of the bonds is $1,085, and your required rate of return is 10 percent.

a. Compute the bond's expected rate of return.

b. Determine the value of the bond to you, given your required rate of return.

c. Should you purchase the bond?

5-7A. (*Bond Valuation*) You own a bond that pays $100 in annual interest, with a $1,000 par value. It matures in 15 years. Your required rate of return is 12 percent.

a. Calculate the value of the bond.

b. How does the value change if your required rate of return (1) increases to 15 percent or (2) decreases to 8 percent?

c. Explain the implications of your answers in part (b) as they relate to interest rate risk, premium bonds and discount bonds.

d. Assume that the bond matures in 5 years instead of 15 years. Recompute your answers in part (b).

e. Explain the implications of your answers in part (d) as they relate to interest rate risk, premium bonds and discount bonds.

5-8A. (*Bondholder's Expected Rate of Return*) Abner Corporation's bonds mature in 15 years and pay 9 percent interest annually. If you purchase the bonds for $1,250, what is your expected rate of return?

5-9A. (*Bond Valuation*) Telink Corporation bonds pay $110 in annual interest, with a $1,000 par value. The bonds mature in 20 years. Your required rate of return is 9 percent.

a. Calculate the value of the bond.

b. How does the value change if your required rate of return (1) increases to 12 percent or (2) decreases to 6 percent?

c. Interpret your findings in parts (a) and (b).

5-10A. (*Duration*) Calculate the value and the duration for the following bonds:

Bond	Years to Maturity	Annual Interest	Maturity Value
P	5	$100	$1,000
Q	5	70	1,000
R	10	120	1,000
S	10	80	1,000
T	15	65	1,000

Your required rate of return is 8 percent.

INTEGRATIVE PROBLEM

Below you will find data on $1,000 par value bonds issued by Thomson Corporation, Molson and Canadian Tire at the end of 1995. Assume you are thinking about buying these bonds as of January 1996. Answer the following questions for each of these bonds:

a. Calculate the value of the bonds if your required rates of return are as follows: Thomson Corporation 7 percent, Molson 8 percent, and Canadian Tire 7 percent where

	Thomson	Molson	Canadian Tire
Coupon interest rates	7.70%	8.40%	10.40%
Years to maturity	8	22	5

b. At the end of 1995, the bonds were selling for the following amounts:

Thompson Corporation	$1,015.50
Molson	$1,011.30
Canadian Tire	$1,117.50

What were the expected rates of return for each bond?

c. How would the values of the bonds change if (1) your required rate of return (k) increases 3 percentage points or (2) decreases 3 percentage points?

d. Explain the implications of your answers in part (b) as they relate to interest rate risk, premium bonds, and discount bonds.

e. Compute the duration for each of the bonds. Interpret your results.

f. What conclusions can you draw from the above computations?

g. Should you buy these bonds? Explain.

STUDY PROBLEMS (SET B)

5-1B. (*Bond Valuation*) Calculate the value of an annual coupon bond that expects to mature in ten years with a $1,000 face value. The coupon pays a nominal interest rate of 9 percent and the investors' required rate of return is 15 percent.

5-2B. (*Bond Valuation*) Enterprise, Inc., has issued semiannual coupon bonds which pay a nominal interest rate of 10 percent. The bonds mature in 11 years and their par value is $1,000. If your required rate of return is 9 percent, what is the value of the bond? What is the value of this bond if coupons are distributed on an annual basis?

5-3B. (*Bondholder's Expected Rate of Return*) The market price is $950 for an eight-year bond ($1,000 par value) that pays 9 percent interest (4.5 percent semiannually). What is the bond's expected rate of return?

5-4B. (*Bond Valuation*) Doisneau 20-year, annual coupon bonds pay a nominal interest rate of 10 percent on a $1,000 par value. If you buy the bonds at $975, what is your expected rate of return?

5-5B. (*Bond Valuation*) Hoyden Co. has issued annual coupon bonds which mature in 15 years and pay a nominal interest rate of 8 percent on a par value of $1,000. If you purchase the bonds for $1,175, what is your expected rate of return?

5-6B. (*Bond Valuation*) Fingen 14-year, annual coupon bonds pay a nominal interest rate of 9 percent on a par value of $1,000. The market price of the bonds is $1,100, and your required rate of return is 10 percent.

a. Compute the bond's expected rate of return.

b. Determine the value of the bond to you, given your required rate of return.

c. Should you purchase the bond?

5-7B. (*Bond Valuation*) Gas Utilities Inc., issued a bond that pays $80 in annual interest, with a $1,000 par value. It matures in twenty years. Your required rate of return is 7 percent.

a. Calculate the value of the bond.

b. How does the value change if your required rate of return (1) increases to 10 percent or (2) decreases to 6 percent?

c. Explain the implications of your answers in part (b) as they relate to interest rate risk, premium bonds and discount bonds.

d. Assume that the bond matures in 10 years instead of 20 years. Recompute your answers in part (b).

e. Explain the implications of your answers in part (d) as they relate to interest rate risk, premium bonds and discount bonds.

5-8B. (*Bondholder's Expected Rate of Return*) Zebner Corporation's bonds mature in 14 years and pay 7 percent interest annually. If you purchase the bonds for $1,110, what is your expected rate of return?

5-9B. (*Bond Valuation*) Visador Corporation bonds pay $70 in annual interest, with a $1,000 par value. The bonds mature in 17 years. Your required rate of return is 8.5 percent.

a. Calculate the value of the bond.

b. How does the value change if your required rate of return (1) increases to 11 percent or (2) decreases to 6 percent?

c. Interpret your findings in parts (a) and (b).

5-10B. *(Duration)* Calculate the value and the duration for the following bonds:

Bond	Years to Maturity	Annual Interest	Maturity Value
A	5	$ 90	$1,000
B	5	60	1,000
C	10	120	1,000
D	15	90	1,000
E	15	75	1,000

Your required rate of return is 7 percent.

SELF-TEST SOLUTIONS

SS-1.

$$\text{Price } (P_0) = \sum_{t=1}^{20} \frac{\$80}{(1.07)^t} + \frac{\$1,000}{(1.07)^{20}}$$

Thus,

$$\begin{aligned} \$80(10.5940) &= \$\ \ 846.72 \\ \$1,000(0.2584) &= \underline{\$\ \ 258.40} \\ \text{price } (P_0) &= \$1,105.12 \end{aligned}$$

If you pay more for the bond, your required rate of return will not be satisfied. In other words, by paying an amount for the bond that exceeds $1,105.12, the expected rate of return for the bond is less than the required rate of return. If you have the opportunity to pay less for the bond, the expected rate of return exceeds the 7 percent required rate of return.

SS-2. If interest is paid semiannually:

$$\text{price } (P_0) = \sum_{t=1}^{14} \frac{\$40}{(1+0.05)^t} + \frac{\$1,000}{(1+0.05)^{14}}$$

Thus,

$$\begin{aligned} \$40(9.8986) &= \$395.94 \\ \$1,000(0.5051) &= \underline{\$505.10} \\ \text{price } (P_0) &= \$901.04 \end{aligned}$$

If interest is paid annually:

$$\text{price } (P_0) = \sum_{t=1}^{7} \frac{\$80}{(1.10)^t} + \frac{\$1{,}000}{(1.10)^7}$$

$$P_0 = \$80(4.8684) + \$1{,}000(0.5132)$$
$$P_0 = \$902.67$$

Calculator Solution

DATA INPUT	FUNCTION KEY
15	N
70	+/− PMT
1,000	+/− FV
1,045	PV

FUNCTION KEY	ANSWER
CPT I/Y	6.52*

*The difference between the two answers represents rounding error

SS-3.

$$\$1{,}045 = \sum_{t=1}^{15} \frac{\$70}{(1 + k_b)^t} + \frac{\$1{,}000}{(1 + k_b)^{15}}$$

At 6%: $70(9.7122) + $1,000(0.4173) = $1,097.15
At 7%: $70(9.1079) + $1,000(0.3624) = $1,000.00

Interpolation:

Expected rate of return: $k_b = 6\% + \frac{\$52.15}{\$97.15}(1\%) = 6.54\%$

SS-4.

Bond	Argile $1,000		Terathon $872	
Years	$C_t \times$	$tPV(C_t)$	$C_t \times$	$tPV(C_t)$
1	$ 80	$ 74	$ 65	$ 60
2	80	137	65	111
3	80	191	65	155
4	80	235	65	191
5	80	272	65	221
6	80	302	65	246
7	80	327	65	265
8	80	346	65	281
9	80	360	65	293
10	1,080	5,002	65	301
11			65	307
12			65	310
13			65	311
14			65	310
15			1,065	5,036
Sum of $tPV(C_t)$		$ 7,246		$ 8,398
Duration		7.25		9.63

SHARE VALUATION

CHAPTER 6

LEARNING OBJECTIVES

After reading this chapter you should be able to:

1. Identify the basic features of preferred shares.
2. Value preferred shares.
3. Calculate a preferred shareholder's expected rate of return.
4. Explain the features used to retire preferred shares.
5. Identify the advantages and disadvantages of using preferred shares.
6. Identify the basic features of common shares.
7. Value common shares.
8. Calculate a common shareholder's expected rate of return.
9. Identify the advantages and disadvantages of using common shares.
10. Value a right.
11. Explain the nature of the relationship between a firm's earnings and the value of its common shares.

INTRODUCTION

In November 1990, Robert Monks of Lens Inc. launched a proxy battle for a board seat at Sears. He wanted the giant to sell its financial services unit, focus on retailing, and reform board practices. Although he lost, he and several institutions kept pressing for change. Then, in 1993, Monks bought more shares and again started talking with the company's management. This time Sears management responded by spinning off the firm's financial units, remodelling its stores, and cutting costs. The result: from 1991 through 1993, the firm shares jumped almost 140 percent, versus 33 percent for the TSE 300 Index.

A similar story occurred in 1992 when Monks wrote Eastman Kodak CEO, Kay Whitmore, to say that he, along with several large institutional investors in the firm, wanted Kodak to restructure and reduce debt. In response, Kodak began cutting costs, trimming staff, and expanding pay for performance. The stock market reacted favourably. In slightly over a year's time, Eastman Kodak's shares rose 31 percent.

This type of investing is called "relationship investing," and it is changing the way management views the shareholder. Simply put, whenever there is an established, committed link between a company and one or more shareholders, that's relationship investing. "It begins with shareholders taking a real interest in a company—asking questions of the board and not trading the shares like pork bellies—and goes all the way to taking a big stake, a board seat, and maybe even some debt," says Ira M. Millstein, a corporate-governance expert at Weil, Gotshal & Manges, a New York law firm.[1]

[1]Adapted from Richard A. Melcher and Patrick Oster, "Relationship Investing," *Business Week* (March 15, 1993), 68–75.

Will "relationship investing" be with us as we approach the next century? Only time will tell, but one thing is for certain—managers are paying more attention to their firm's shareholders and the value of their shares. Consequently, share valuation becomes a matter of increasing importance to all managers.

CHAPTER PREVIEW

In the last chapter, we examined the process of valuing financial assets. We applied the valuation process by determining the value of bonds. In this chapter, we apply the valuation process to both preferred and common shares. As already noted at the outset of our study of financial management, the financial executive's objective is to maximize the value of the firm's common shares. As a result, we need to understand what determines share value. An understanding of share valuation will also enable us to compute the firm's cost of capital which is essential to making effective capital investment decisions.

OBJECTIVE 1

FEATURES OF PREFERRED SHARES

Preferred share

A hybrid security with characteristics of common shares and bonds. Preferred shares are similar to common shares because they have no fixed maturity date, the nonpayment of dividends does not bring on bankruptcy, and dividends are deductible for tax purposes. Preferred shares are similar to bonds in that the payment of dividends is limited in amount.

Preferred shares are often referred to as hybrid securities because they have many characteristics of both common shares and bonds. Preferred shares are similar to common shares in that they have no fixed maturity date, the nonpayment of dividends does not bring on bankruptcy, and dividends are not deductible for tax purposes. On the other hand, preferred shares are similar to bonds in that dividends are limited in amount. Although the nonpayment of dividends does not bring on bankruptcy, the increased risk of bankruptcy creates an implicit cost which affects the firm's cost of raising funds in the marketplace.

The size of the dividend is generally fixed either as a dollar amount or as a percentage of the par value. For example, the Royal Bank of Canada has issued $1.88 preferred shares, while the Maritime Telegraph and Telephone (MTT) company has some 7.00 percent preferred shares outstanding. The par value on the MTT preferred shares is $10, hence each share pays 7.00% × $10, or $0.70 in dividends annually. Since these dividends are fixed, preferred shareholders do not share in the residual earnings of the firm but are limited to their stated annual dividend.

Although each issue of preferred shares is unique, several characteristics are common to almost all issues. These traits include the ability to issue multiple classes of preferred shares, the claim on assets and income, and the cumulative and protective features.

Multiple Classes

If a company desires, it can issue more than one series or class of preferred shares, and each class can have different characteristics. In fact, it is quite common for firms that issue preferred shares to issue more than one series. For example, the Royal Bank of Canada has seven different issues of preferred shares outstanding, while MTT has six different issues of preferred shares outstanding. These issues can be further differentiated by the fact that some are convertible into common shares while others are not, and they have varying priority status with respect to assets in the event of bankruptcy.

Floating Rate

In the early 1980s, another new financing alternative was developed in order to provide investors with some protection against wide swings in principal that occur when interest rates fluctuate up or down. This financing vehicle is called **floating rate preferred shares.** With floating rate preferred, quarterly dividends fluctuate with interest rates under a formula that ties the dividend payment to (1) a chartered bank's prime rate, (2) the Bank of Canada rate, or (3) the rate of bankers' acceptances. For example, Canadian Imperial Bank of Commerce class A preferred shares, series 5, have floating and adjustable cumulative dividends which are determined by applying one-quarter of 69 percent of the average of the prime rate of the bank in effect on each day during the period. While floating rate preferred shares allow dividends to be tied to the interest rates, they also provide a maximum and a minimum level to which they can climb or fall called the *dividend rate band.* The purpose of allowing the dividend rate to fluctuate on these preferred shares, is to minimize the fluctuation in the principal of the preferred shares. In times of high and fluctuating interest rates, this is a very appealing feature indeed.

Floating rate preferred share

A floating rate preferred share is intended to provide investors with some protection against fluctuations in the share value that occur when interest rates move up and down. The dividend rate changes along with prevailing interest rates.

In the North American market, there have been other examples of setting the dividend rate. For example, auction rate preferred shares began to appear in the United States in the late 1980s and early 1990s. Auction rate preferred shares are actually variable rate preferred shares in which the dividend rate is set by an auction process. In the case of auction rate preferred, the dividend rate is set every 49 days. At each auction, buyers and sellers place bids for shares, specifying the yield they are willing to accept for the next seven-week period. The yield is then set at the lowest level necessary to match buyers and sellers. As a result, the yield offered on auction rate preferred shares accurately reflects current interest rates, while keeping the market price of these securities at par.

Convertibility

Many of the preferred shares that are issued today are **convertible** at the discretion of the holder into some other class of shares, usually common, at a predetermined price over some stated period of time. As an example of a convertible preferred share issue, the preferred share series G of the Royal Bank of Canada is convertible to the bank's common shares only after October 31, 1999. The conversion price of convertible preferred shares is set at a premium of 10 to 15 percent of the market price of the common share when the preferred shares are issued. Investors benefit from conversion because there is no commission charge and any capital gain is deferred until the sale of the common shares. However, the decision to exercise the convertible feature of preferred shares is irreversible. The convertibility feature is, of course, desirable to the investor and enables the issuer to reduce the amount of dividend distribution on the preferred shares. It should also be noted that most preferred shares are redeemable. Holders of **retractable** preferred shares can force the company to redeem their shares on a predetermined date at a predetermined price. However, if the election is not exercised by the preferred shareholder, the retraction feature will expire and become a straight preferred share.

Convertible preferred share

A convertible preferred share allows the preferred shareholder to convert a preferred share into a predetermined number of common shares, if the person so chooses.

Retractable preferred share

A retractable preferred share enables preferred shareholders to redeem their shares at a predetermined price on a predetermined date.

Claim on Assets and Income

Preferred shareholders have priority over common shareholders with respect to claims on assets in the case of bankruptcy. The preferred shareholder's claim is honoured after that of bonds and before that of common shareholders. Multiple issues of preferred shares differ according to the priority of their claim. Preferred shareholders also have priority over common shareholders with respect to their claim on income. As a result, a firm must pay dividends to its preferred shareholders before it can distribute other dividends. In terms of risk, preferred shares are safer than common shares because preferred shareholders have priority with respect to their claim on assets and income. However, preferred shares are riskier than long-term debt because preferred shareholders' claim on assets or income comes after those of bonds.

Cumulative Feature

Cumulative preferred share

Requires all past unpaid preferred share dividends be paid before any common share dividends are declared.

Most preferred shares carry a **cumulative feature** that requires all unpaid dividends in the past to be paid before any new dividends are declared. The purpose of this feature is to provide some degree of protection for the preferred shareholder. Without a cumulative feature a firm could treat both preferred shareholders and common shareholders equally with respect to their priority of receiving a dividend. However, the cumulative feature ensures that unpaid dividends to preferred shareholders be paid before any dividends are distributed to common shareholders.

Protective Provisions

Protective provisions

Protective provisions for preferred shares are included in the terms of issue to protect the investor's interest. For instance, provisions generally allow for voting rights in the event of nonpayment of dividends, or they restrict the payment of common share dividends if sinking fund payments are not met or if the firm is in financial difficulty.

In addition to cumulative features, **protective provisions** are common to preferred shares. The following list provides examples of protective provisions:

- allow for voting rights in the event of nonpayment of dividends or dividends in arrears
- restrict the payment of dividends if sinking fund payments are not met
- restrict the payment of dividends if the firm is in financial difficulty
- restrict asset sales
- restrict any changes in the terms of the preferred shares
- restrict the issuing of other preferred shares

In effect, the protective features included with preferred shares are similar to the restrictive provisions included with long-term debt.

The preferred shares of the Royal Bank of Canada and the Reynolds Metals company illustrate some typical protective provisions. For example, various issues of preferred shares from the Royal Bank have a protective provision that provides preferred shareholders with voting rights whenever eight quarterly dividends are in arrears. At that point the preferred shareholders are given the power to elect a majority of the board of directiors. The Reynolds Metals company is also another example of a company which precludes the payment of dividends during any period in which the preferred shares sinking fund is in default. These provisions are desirable because they provide protection beyond that provided for by the cumulative provision and thereby reduce the shareholders' risk. In addition, these types of features reduce the cost of preferred shares to the issuing firm.

Participation

Participating preferred shares

A participating preferred share allows the preferred shareholder to participate in earnings beyond the payment of the stated dividend.

Although participating features are infrequent in preferred shares, their inclusion can greatly affect their desirability and cost. The **participation feature** allows the preferred shareholder to participate in earnings beyond the payment of the stated dividend. This is usually done in accordance with some set formula. For example, MacMillan Bloedel Ltd. has class B participating preferred shares which include the right of dividend only if revenues are received from oil and gas interests. These privately held shares provide the feature of participating in dividends distributed to the common shareholders. Although a participating feature is certainly desirable from the point of view of the investor, it is infrequently included in preferred shares.

PIK Preferred

One by-product of the levered buyout boom of the late 1980s was the creation of pay-in-kind or **PIK preferred shares.** With PIK preferred, investors receive no dividends initially; they receive more preferred shares, which in turn pay dividends in even more preferred shares. After five or six years the issuing company can replace the payment-in-kind with cash dividends. Needless to say, the issuing firm has to offer hefty dividends, generally ranging from 12 percent to 18 percent, to entice investors to purchase PIK preferred.

PIK preferred share

Investors receive no dividends initially, they simply get more preferred shares, which in turn pay dividends in even more preferred shares.

BACK TO THE BASICS

Valuing preferred shares relies on three of our axioms presented in Chapter 1, namely:

Axiom 1: The Risk-Return Tradeoff—We Won't Take on Additional Risk Unless We Expect to Be Compensated with Additional Return.
Axiom 2: The Time Value of Money—A Dollar Received Today Is Worth More Than a Dollar Received in the Future.
Axiom 3: Cash—Not Profits—Is King: Measuring the Timing of Costs and Benefits.

Determining the economic worth or value of a preferred share always relies on these three axioms. Without them, we would have no basis for explaining value. With them, we know that the amount and timing of cash dividends drives our determination of value. Also, we must be rewarded for taking risk in investing in shares which have uncertain returns; otherwise, we will not invest.

PREFERRED SHARE VALUATION

OBJECTIVE 2

Like a bondholder, the owner of a preferred share should receive a constant income from the investment in each period. However, the return from a preferred share comes in the form of dividends rather than interest. In addition, while bonds generally have a specific maturity date, most preferred shares are perpetuities (nonmaturing). In this instance, finding the value (present value) of a preferred share, V_p, with a level cash flow stream continuing indefinitely, may best be explained by an example.

EXAMPLE

To illustrate the valuation of a preferred share, consider the Royal Bank of Canada's preferred share issue. Once again we use the three-step valuation procedure.

1. **Estimate the amount and timing of the receipt of the future cash flows the preferred share is expected to provide.**
   Royal Bank of Canada's preferred share pays an annual dividend of $1.88. The shares do not have a maturity date; that is, they go to perpetuity.
2. **Evaluate the riskiness of the preferred share's future dividends and determine the investor's required rate of return.**
   The investor's required rate of return is assumed to equal 8.35 percent.
3. **Calculate the intrinsic value of a preferred share, which is the present value of the expected dividends discounted at the investor's required rate of return.**

The valuation model for a preferred share, V_p, is defined as follows:

$$V_p = \frac{\text{dividend in year 1}}{(1 + \text{required rate of return})^1} + \frac{\text{dividend in year 2}}{(1 + \text{required rate of return})^2}$$

$$+\ldots+\frac{\text{dividend in infinity}}{(1+\text{required rate of return})^{\infty}}$$

$$=\frac{D_1}{(1+k_p)^1}+\frac{D_2}{(1+k_p)^2}+\ldots+\frac{D_\infty}{(1+k_p)^\infty}$$

$$V_p=\sum_{t=1}^{\infty}\frac{D_t}{(1+k_p)^t} \quad (6\text{-}1)$$

Since the dividends in each period are equal for preferred share, Equation (6-1) can be reduced to the following relationship.[2]

$$V_p=\frac{\text{annual dividend}}{\text{required rate of return}}=\frac{D}{k_p} \quad (6\text{-}2)$$

Equation (6-2) represents the present value of an infinite stream of cash flows, where the cash flows are the same each year. We can determine the value of the Royal Bank's preferred shares using Equation (6-2), as follows:

$$V_p=\frac{D}{kp}=\frac{\$1.88}{.0835}=\$22.51$$

PERSPECTIVE IN FINANCE

The value of a preferred share is the present value of all future dividends. But since most preferred shares are nonmaturing—the dividends continue to infinity—we have to come up with a shortcut for finding value.

OBJECTIVE 3

THE PREFERRED SHAREHOLDER'S EXPECTED RATE OF RETURN

In computing the preferred shareholder's expected rate we use the valuation equation for preferred shares. Earlier, Equation (6-2) specified the value of a preferred share, V_p, as

$$V_p=\frac{\text{annual dividend}}{\text{required rate of return}}=\frac{D}{k_p} \quad (6\text{-}2)$$

Solving (6-2) for k_p,

$$k_p=\frac{\text{annual dividend}}{\text{value}}=\frac{D}{V_p} \quad (6\text{-}3)$$

---

[2]To verify this result, consider the following equation

(i) $$V_p=\frac{D_1}{(1+k_p)}+\frac{D_2}{(1+k_p)^2}+\ldots+\frac{D_n}{(1+k_p)^n}$$

If we multiply both sides of this equation by $(1+k_p)$, we have

(ii) $$V_p(1+k_p)=D_1+\frac{D_2}{(1+k_p)}+\ldots+\frac{D_n}{(1+k_p)^{n-1}}$$

Subtracting (i) from (ii) yields

$$V_p(1+k_p-1)=D_1-\frac{D_n}{(1+k_p)^n}$$

As n approaches infinity, $D_n/(1+k_p)^n$ approaches zero. Consequently,

$$V_pk_p=D_1 \text{ and } V_p=\frac{D_1}{k_p}$$

Since $D_1=D_2=\ldots=D_n$, we need not designate the year. Therefore,

(iii) $$V_p=\frac{D}{k_p}$$

which simply indicates that the expected rate of return of a preferred security equals the dividend yield (dividend/price). For example, if the present market price of a preferred share is $50 and it pays a $3.64 dividend, the expected rate of return implicit in the present market price is

$$k_p = \frac{D}{V_p} = \frac{\$3.64}{\$50} = 7.28 \text{ percent}$$

Therefore, investors at the margin (who pay $50 per share for a preferred share that is paying $3.64 in annual dividends) are expecting a 7.28 percent rate of return.

RETIRING PREFERRED SHARES

OBJECTIVE 4

Although preferred shares do not have a set maturity associated with them, issuing firms generally provide for some method of retirement. The retirement of preferred shares enables a firm to take advantage of falling interest rates.

The Use of Call Provisions

Most preferred shares have some type of **call provision** associated with them. The call feature on preferred shares usually involves an initial premium above the par value or issuing price of the preferred of approximately 10 percent. Then, over time, the call premium generally falls. For example, Quaker Oats in 1976 issued $9.56 cumulative preferred shares with no par value for $100 per share. This issue was not callable until 1980 and then was callable at $109.56. After that the call price gradually drops to $100 in the year 2000, as shown in Table 6-1.

Call provision

A call provision lets the company buy its preferred shares back from the investor, usually at a premium price above the share's par value.

By setting the initial call price above the initial issue price and allowing it to decline slowly over time, the firm protects the investor from an early call that carries no premium. A call provision also allows the issuing firm to plan the retirement of its preferred shares at predetermined prices.

The Use of a Sinking Fund Provision

A **sinking fund** provision requires the firm periodically to set aside an amount of money for the retirement of its preferred shares. This money is then used to purchase the preferred shares in the open market or through the use of the call provision, whichever method is cheaper. Although preferred shares do not have a maturity date associated with them, the use of a call provision in addition to a sinking fund can effectively create a maturity date. For example, the Quaker Oats issue we just examined has associated with it an annual sinking fund, operating between the years 1981 and 2005, which requires the annual elimination of a minimum of 20,000 shares and a maximum of 40,000. The minimum payments are designed so that the entire issue will be retired by the year 2005. If any sinking-fund payments are made above the minimum amount, the issue will be retired prior to 2005. The

Sinking fund

A sinking fund provision requires the firm periodically to set aside an amount of money for the retirement of its preferred shares. This money is then used to purchase the preferred shares in the open market or through the use of a call provision, whichever method is cheaper.

Table 6-1 Call Provision of Quaker Oats $9.56 Cumulative Preferred Shares

Date	Call Price
Date of issue until 7/19/80	Not callable
7/20/80 until 7/19/85	$109.56
7/20/85 until 7/19/90	107.17
7/20/90 until 7/19/95	104.78
7/20/95 until 7/19/00	102.39
After 7/19/00	100.00

issue of Quaker Oats preferred shares has a maximum life of 30 years, and the size of the issue outstanding decreases each year after 1981.

OBJECTIVE 5

ADVANTAGES AND DISADVANTAGES OF PREFERRED SHARES

Preferred shares are a hybrid of bonds and common shares. As a result, preferred shares offer a firm several advantages and disadvantages when compared with bonds and common shares.

Advantages to the Firm

1. Preferred shares do not have any default risk to the issuer. The nonpayment of dividends does not force the firm into bankruptcy, as does the nonpayment of interest on debt.
2. The dividend payments are generally limited to a stated amount. Thus, preferred shareholders do not participate in excess earnings as do common shareholders.
3. Preferred shareholders generally do not have voting rights except in the case of financial distress. Therefore, the issuance of preferred shares does not create a challenge to the owners of the firm.
4. Although preferred shares do not carry a specified maturity, the inclusion of call features and sinking funds provides the ability to replace the issue if interest rates decline.

Disadvantages to the Firm

1. The cost of financing a project with preferred shares is higher than that of bonds because preferred shares are riskier than bonds and their dividends are not tax deductible.
2. Although preferred dividends can be omitted, their cumulative nature makes dividend payment almost mandatory.

OBJECTIVE 6

FEATURES OF COMMON SHARES

Common Share

A common share is a claim of ownership in a corporation.

A **common share** is a claim of ownership in a corporation. In effect, bondholders and preferred shareholders can be viewed as creditors, while the common shareholders are the true owners of the firm. Common shares do not have a maturity date, but exist as long as the firm does. In addition, common shares have no upper limit on their dividend payments. Dividend payments must be declared by the firm's board of directors before they are distributed to common shareholders. In the event of bankruptcy the common shareholders, as owners of the corporation, cannot exercise their claims on assets until the claims of the bondholders and preferred shareholders have been satisfied.

Common shares are distinguished from other financial instruments by their claim on income and assets, their voting rights, and the limited-liability feature.

Claim on Income

As the owners of the corporation, the common shareholders have the right to the residual income after bondholders and preferred shareholders have been paid. This income may be paid directly to the shareholders in the form of dividends or retained and reinvested by the firm. While it is obvious the shareholder benefits immediately from the distribution of income in the form of dividends, the reinvestment of earnings also benefits the shareholder. Plowing back earnings into the firm results in an increase in the value of the firm, in its earning power, and in its future dividends. This action results in an increase in the value of the shares. In effect, residual income is distributed directly to shareholders in the form of dividends or indirectly in the form of capital gains on their common shares.

The right to residual income is advantageous to the common shareholder because the potential return is limitless. Once the claims of the most senior securities—that is, bonds and preferred shares—have been satisfied, the remaining income flows to the common shareholders in the form of dividends or capital gains. The disadvantage of having a right to residual income occurs if the claims of both bondholders and preferred shareholders on income absorb all of the earnings and common shareholders receive nothing. In years when earnings fall, it is the common shareholder who suffers first.

Claims on Assets

Common shareholders also have a residual claim on assets in the case of liquidation. Only after the claims of debt holders and preferred shareholders have been satisfied do the claims of common shareholders receive attention. Unfortunately, when bankruptcy does occur, the claims of the common shareholders generally go unsatisfied. In effect, this residual claim on assets adds to the risk of common shares.

Voting Rights

The common shareholders are entitled to elect the board of directors and are in general the only security holders given a vote. In corporate Canada, many firms have two or more classes of common shares. One or more of the classes may contain restricted shares.[3] Canadian Tire, Molson and Bombardier are examples of large corporations which have two classes of common shares. This practice is common in Canada because many of these corporations are either family-owned or closely held. In contrast, the practice of issuing two classes of common shares was virtually eliminated in the United States because of the regulations implemented by the Securities and Exchange Commission and the New York Stock Exchange's refusal to list common shares without voting privileges. However, with the merger boom of the eighties dual classes of common shares with different voting rights again emerged, this time as a defensive tactic used to prevent takeovers.

Common shareholders not only have the right to elect the board of directors, they also must approve any change in the corporate charter. A typical charter change might involve the authorization to issue new shares or perhaps a merger proposal.

Voting for directors and charter changes occurs at the corporation's annual meeting. While shareholders may vote in person, the majority generally vote by proxy. A **proxy** gives a designated party the temporary power of attorney to vote for the signee at the corporation's annual meeting. The firm's management generally solicits proxy votes and, if the shareholders are satisfied with its performance, has little problem securing them. However, in times of financial distress or when management takeovers are threatened, **proxy fights**—battles between rival groups for proxy votes—occur.

Although each share carries the same number of votes, the voting procedure is not always the same from company to company. The two procedures commonly used are majority and cumulative voting. Under **majority voting**, each share allows the shareholder one vote, and each position on the board of directors is voted on

Proxy

A proxy gives a designated party the temporary power of attorney to vote for the signee at the corporation's annual meeting.

Proxy fight

When rival groups compete for proxy votes in order to control the decisions made in a shareholders' meeting.

Majority voting

Each share allows the shareholder one vote, and each position on the board of directors is voted on separately. As a result, a majority of shares has the power to elect the entire board of directors.

[3]The Ontario Securities Act recognizes three types of restricted shares: restricted voting shares, subordinated voting shares, and nonvoting shares. For a greater discussion on the types of restricted shares, see: William F. Sharpe, Gordon J. Alexander, and David J. Fowler, *Investments*, 1st ed. (Scarborough, ON: Prentice-Hall Canada, 1993), pp. 391–392.

separately. Since each member of the board of directors is elected by a simple majority, a majority of shares has the power to elect the entire board of directors.

With **cumulative voting**, each share allows the shareholder a number of votes equal to the number of directors being elected. The shareholder can then cast all of his or her votes for a single candidate or split them among the various candidates. The advantage of a cumulative voting procedure is that it gives minority shareholders the power to elect a director.

Cumulative voting

Each share allows the shareholder a number of votes equal to the number of directors being elected. The shareholder can then cast all of his or her votes for a single candidate or split them among the various candidates.

BACK TO THE BASICS

In theory, the shareholders pick the corporate board of directors, generally through proxy voting, and the board of directors in turn hires management. Unfortunately, in reality the system frequently works the other way around. Shareholders are offered a slate of nominees selected by management from which to choose. The end result is that management effectively selects the directors, who then may have more allegiance to the managers than to shareholders. This in turn sets up the potential for agency problems in which a divergence of interests between managers and shareholders is allowed to exist, with the board of directors not monitoring the managers on behalf of the shareholders as they should. The result: ***Axiom 7: The Agency Problem—Managers Won't Work for the Owners Unless It's in Their Best Interest.***

Limited Liability

Although the common shareholders are the actual owners of the corporation, their liability in the case of bankruptcy is limited to the amount of their investment. The advantage is that investors who might not otherwise invest their funds in the firm become willing to do so. This limited-liability feature aids the firm in raising funds.

Preemptive Rights

The **preemptive right** entitles the common shareholder to maintain a proportionate share of ownership in the firm. When new shares are issued, common shareholders have the first right of refusal. If a shareholder owns 25 percent of the corporation's shares, then he or she is entitled to purchase 25 percent of the new shares. Certificates issued to the shareholders giving them an option to purchase a stated number of new shares at a specified price during a two- to ten-week period are called **rights**. These rights can be exercised (generally at a price below the common share's current market price), can be allowed to expire, or can be sold in the open market. Appendix 6A will examine the process of raising funds and the valuation of a rights offering.

Preemptive right

The right of a common shareholder to maintain a proportionate share of ownership in the firm. When new shares are issued, common shareholders have the first right of refusal.

Rights

Certificates issued to shareholders giving them an option to purchase a stated number of new shares at a specified price during a two-to-ten week period.

BACK TO THE BASICS

Valuing common shares is no different from valuing preferred shares; only the pattern of the cash flows changes, but nothing else. Thus, the valuation of common shares relies on the same three axioms developed in Chapter 1 that were used in valuing preferred shares:

Axiom 1: The Risk-Return Tradeoff—We Won't Take on Additional Risk Unless We Expect to Be Compensated with Additional Return.
Axiom 2: The Time Value of Money—A Dollar Received Today Is Worth More Than a Dollar Received in the Future.
Axiom 3: Cash—Not Profits—Is King: Measuring the Timing of Costs and Benefits.

Determining the economic worth or value of a common share always relies on these three axioms. Without them, we would have no basis for explaining value. With them, we can know that the amount and timing of cash dividends drives the valuation of the common share. Also, we must be rewarded for taking greater risk in investing in common shares as compared to the risk taken in investing in preferred shares; otherwise, we will not invest.

COMMON SHARE VALUATION

OBJECTIVE 7

The third and last security we will learn to value is the common share. Like both a bond and a preferred share, a common share's value is equal to the present value of all future cash flows expected to be received by the shareholder. However, in contrast to a bond, a common share does not guarantee interest income or a maturity payment at some specified time in the future. Nor does a common share entitle the holder to a predetermined constant dividend, as does a preferred share. For a common share, the dividend is based upon the profitability of the firm and upon management's decision to pay dividends or to retain the profits for reinvestment purposes. As a consequence, the dividend stream tends to increase with the growth in corporate earnings. Thus, the growth of future dividends is a prime distinguishing feature of a common share.

The Growth Factor in Valuing a Common Share

What is meant by the term "growth" when used in the context of valuing a common share? A company can grow in a variety of ways. It can become larger by borrowing money to invest in new projects. Likewise, it can issue new shares for expansion. Management could also acquire another company to merge with the existing firm; the firm's assets would clearly be increased. In all these cases, the firm is growing through the use of new financing, by issuing debt or common shares. While management could accurately say that the firm has grown, the original shareholders may or may not participate in this growth. Growth is being realized through the infusion of new capital.[4] The firm size has clearly increased, but unless the original investors increase their investment in the firm, they will own a smaller portion of the expanded business.

Another means of growing is internal growth, which requires that management retain some or all of the firm's profits for reinvestment in the firm, resulting in the growth of future earnings and hopefully the value of the common shares. This process underlies the essence of potential growth for the firm's current shareholders, and what we may call the only "relevant growth," for our purposes in valuing a firm's common shares.[5]

EXAMPLE

To illustrate the nature of internal growth, assume that the return on equity for PepsiCo is 16 percent.[6] If PepsiCo's management decides to pay all the profits out in dividends to its shareholders, the firm will experience no growth internally. It might become larger by borrowing more money or issuing new shares, but internal growth will come only through the retention of profits. If, on the other hand, PepsiCo retained all the profits, the shareholders' investment in the firm will grow by the amount of profits retained, or by 16 percent. If, however, management kept only 50 percent of the profits for reinvestment, the common shareholders' investment would increase only by half of the 16 percent return on equity, or by 8 percent. Generalizing this relationship, we have

$$g = ROE \times r \tag{6-4}$$

[4]Growth in and of itself does not mean that we are creating value for shareholders. Only if we reinvest at a rate of return that is greater than the investor's required rate of return will growth cause an increase in the value of the firm. In fact, any investment at a rate less than the required rate of return for our investors causes the value of the firm to actually decline.

[5]We are not arguing that the existing common shareholders never benefit from the use of external financing; however, such benefit is more evasive when dealing with efficient capital markets.

[6]The return on equity is the percentage return on the common shareholder's investment in the company, and is computed as follows:

$$\text{Return on equity} = \frac{(\text{net income} - \text{preferred dividends})}{(\text{common shares} + \text{contributed surplus} + \text{retained earnings})}$$

where g = **the growth rate of future earnings and the growth in the common shareholders' investment in the firm**

ROE = **the return on equity (net income/common book value)**

r = **the company's percentage of profits retained, which is called the profit-retention rate**[7]

Thus, if only 25 percent of the profits were retained by PepsiCo, we would expect the common shareholders' investment in the firm and the value of the share price to increase or grow by 4 percent; that is,

$$g = 16\% \times .25 = 4\%$$

In summary, common shareholders frequently rely upon share price as a source of return. If the company is retaining a portion of its earnings for reinvestment, future profits and dividends should grow. This growth should be reflected in an increased market price of the common share in future periods. Therefore, both types of return (dividends and price appreciation) are necessary in the development of a valuation model for common shares.

To explain this process, let us begin by examining how an investor might value a common share that is to be held for only one year.

Common Share Valuation—Single Holding Period

For an investor holding a common share for only one year, the value of the share should equal the present value of both the expected dividend to be received in one year, D_1, and the anticipated market price of the share at year-end, P_1. If k_c represents a common shareholder's required rate of return, the value of the security, V_c, would be

$$\begin{aligned} V_c &= \text{present value of dividend } (D_1) \\ &\quad + \text{present value of market price } (P_1) \\ &= \frac{D_1}{(1 + k_c)} + \frac{P_1}{(1 + k_c)} \end{aligned} \tag{6-5}$$

Suppose an investor is contemplating the purchase of RMI common shares at the beginning of this year. The dividend at year-end is expected to be \$1.64 and the market price by the end of the year is projected to be \$22. If the investor's required rate of return is 18 percent, the value of the security would be

$$\begin{aligned} V_c &= \frac{\$1.64}{1 + .18} + \frac{\$22}{1 + .18} \\ &= \$1.39 + \$18.64 \\ &= \$20.03 \end{aligned}$$

Once again we see that valuation is the same three-step process. First we estimate the expected future cash flows from common share ownership (a \$1.64 dividend and a \$22 end-of-year expected share price). Second, we estimate the investor's required rate of return after assessing the riskiness of the expected cash flows (assumed to be 18 percent). Finally, we discount the expected dividend and end-of-year share price back to the present at the investor's required rate of return.

PERSPECTIVE IN FINANCE

The intrinsic value of a common share is similar to that of a preferred share. Both these financial instruments have an intrinsic value which represents the present value of all future dividends. Furthermore, we have the same problem as we had with the valuation of preferred shares; that is, the difficulty of valuing cash flows that continue in perpetuity. So we must make some assumptions about the expected growth of future dividends. If, for example, we assume that dividends grow at a constant rate forever, we can then calculate the present value of the share.

[7]The retention rate is also equal to (1 – the percentage of profits paid out in dividends). The percentage of profits paid out in dividends is often called the dividend payout ratio.

Common Share Valuation—Multiple Holding Periods

Since common shares have no maturity date and are frequently held for many years, a multiple-holding-period valuation model is needed. The general common share valuation model can be defined as follows:

$$V_c = \frac{D_1}{(1 + k_c)^1} + \frac{D_2}{(1 + k_c)^2} + \ldots + \frac{D_n}{(1 + k_c)^n} + \ldots + \frac{D_\infty}{(1 + k_c)^\infty} \tag{6-6}$$

PERSPECTIVE IN FINANCE

Turn back to chapter 5, and compare Equation (5-3). Equation (6-6) is a restatement in a slightly different format of Equation (5-3). Recall that Equation (5-3) states that the value of an asset is the present value of future cash flows to be received by the investor. Equation (6-6) is simply applying Equation (5-3) to value a common share.

This equation simply indicates that we are discounting the dividend at the end of the first year, D_1, back one year, the dividend in the second year, D_2, back two years, ..., the dividend in the nth year back n years, ..., and the dividend in infinity back an infinite number of years. The required rate of return is k_c. In using Equation (6-6), note that the value of the share is being established at the beginning of the year, say January 1, 1996. The most recent past dividend D_0 would have been paid the previous day on December 31, 1995. Thus, if we purchased the share on January 1, the first dividend would be received in 12 months, on December 31, 1996, which is represented by D_1.

Fortunately, Equation (6-6) can be reduced to a much more manageable form if dividends grow each year at a constant rate, g. The "constant growth" common share valuation model is defined as follows:[8]

$$\text{common share value} = \frac{\text{dividend in year 1}}{(\text{required rate of return}) - (\text{growth rate})}$$

$$V_c = \frac{D_1}{k_c - g} \tag{6-9}$$

Consequently, the intrinsic value (present value) of a common share whose dividends grow at a constant annual rate can be calculated using Equation (6-9).

---

[8]Where common share dividends grow at a constant rate of g every year, we can express the dividend in any year in terms of the dividend paid at the end of the previous year, D_0. For example, the expected dividend one year hence is simply $D_0(1 + g)$. Likewise, the dividend at the end of t years is $D_0(1 + g)^t$. Using this notation, the common share valuation Equation (6-6) can be rewritten as follows:

$$Vc = \frac{D_0(1 + g)^1}{(1 + k_c)^1} + \frac{D_0(1 + g)^2}{(1 + k_c)^2} + \ldots + \frac{D_0(1 + g)^n}{(1 + k_c)^n} + \ldots + \frac{D_0(1 + g)^\infty}{(1 + k_c)^\infty} \tag{6-7}$$

If both sides of Equation (6-7) are multiplied by $(1 + k_c)/(1 + g)$ and then Equation (6-7) is subtracted from the product, the result is

$$\frac{V_c(1 + k_c)}{1 + g} - V_c = D_0 - \frac{D_0(1 + g)^\infty}{(1 + k_c)^\infty} \tag{6-8}$$

If $k_c > g$, which normally should hold, $[D_0(1+ g)^\infty/(1 + k_c)^\infty]$ approaches zero. As a result,

$$\frac{V_c(1 + k_c)}{1 + g} - V_c = D_0$$

$$V_c\left(\frac{1 + k_c}{1 + g}\right) - V_c\left(\frac{1 + g}{1 + g}\right) = D_0$$

$$V_c\left[\frac{(1 + k_c) - (1 + g)}{1 + g}\right] = D_0$$

$$V_c(k_c - g) = D_0(1 + g)$$

$$V_c = \frac{D_1}{k_c - g}$$

Although the interpretation of this equation may not be intuitively obvious, we should remember that it solves for the present value of the future dividend stream growing at a rate, g, to infinity, assuming that k_c is greater than g.

EXAMPLE

Consider the valuation of a common share that paid a \$2 dividend at the end of the last year and is expected to pay a cash dividend every year from now to infinity. Each year the dividends are expected to grow at a rate of 10 percent. Based upon an assessment of the riskiness of the common share, the investor's required rate of return is 15 percent. Using this information, we would compute the value of the common share as follows:

1. **Since the \$2 dividend was paid last year (actually yesterday), we must compute the next dividend to be received, that is, D_1, where**

$$\begin{aligned} D_1 &= D_0(1 + g) \\ &= \$2(1 + .10) \\ &= \$2.20 \end{aligned}$$

2. **Now using Equation (6-9),**

$$\begin{aligned} V_c &= \frac{D_1}{k_c - g} \\ &= \frac{\$2.20}{.15 - .10} \\ &= \$44 \end{aligned}$$

Finally, we examine the model for common share valuation in which dividends grow at non-constant rates. Non-constant growth rates are typical for the majority of real-world applications. To determine the value of common shares with this type of growth rate, we assume that at some future period, the expected dividends will begin to grow at a constant growth rate or at a zero growth rate. As a result, we will use either Equation (6-9) of the constant growth rate model or Equation (6-2) of the zero growth rate model to determine the value of the common share when the expected dividends begin to grow at a constant growth rate or at a zero growth rate. The value of the common share at this time will then be discounted back to the present. The valuation procedure involves three steps:

1. Forecast the dividends until they revert to zero growth or constant growth;
2. Forecast the price at that point in time;
3. Determine the present value the relevant cash flows at the required rate of return.

We summarize the steps of this valuation procedure through the following equation:

$$V_c = \frac{D_0(1 + g_1)}{(1 + k_c)^1} + \frac{D_1(1 + g_2)}{(1 + k_c)^2} + \frac{D_2(1 + g_3)}{(1 + k_c)^3} + \ldots + \frac{\frac{D_{(n-1)}(1 + g_{cte})}{(k_c - g_{cte})}}{(1 + k_c)^n} \qquad (6\text{-}10)$$

$$V_c = \frac{D_0(1 + g_1)}{(1 + k_c)^1} + \frac{D_1(1 + g_2)}{(1 + k_c)^2} + \frac{D_2(1 + g_3)}{(1 + k_c)^3} + \ldots + \frac{P_n}{(1 + k_c)^n} \qquad (6\text{-}11)$$

In order to solve for the value of this common share, we have assumed that after some future period ($t = n$), the dividends will begin to grow at a constant rate. As a result, the value of the common share at period $t = n$, can be determined using Equation (6-9) from the "constant growth" model for common share valuation. The value of the common share at $t = n$ can then be discounted back to the present.

EXAMPLE

Prospect Inc., which announced a \$2.00 dividend last week, is expected to grow at 10 percent compounded annually next year, 9 percent compounded annual-

ly in year 2, and 8 percent compounded annually in year 3; after which a normal 3 percent growth is forecasted indefinitely. If your required return is 16 percent given the risk level of this company, should you buy its shares at the prevailing market price of $20.00?

Step 1.

$$D_0 = \$2.00$$
$$D_1 = \$2.00\,(1.10) = \$2.20$$
$$D_2 = \$2.20\,(1.09) = \$2.40$$
$$D_3 = \$2.40\,(1.08) = \$2.59$$

Step 2.

$$P_3 = D_4/(k_c - g) = \$2.59\,(1.03)/(.16 - .03) = \$20.52$$

Step 3.

$$P_0 = \$2.20/(1.16) + \$2.40/(1.16)^2 + \$2.59/(1.16)^3 + \$20.52/(1.16)^3 = \$18.49$$

Since the share of prospect is worth less than market value ($18.49 < $20.00), do not purchase.

We have argued that the value of a common share is equal to the present value of all future dividends, which is without question a fundamental premise of finance. However, in practice, managers, along with many security analysts, often talk about the relationship between share value and earnings, rather than dividends. Appendix 6B provides some insights into the rationale and propriety of using earnings to value a firm's share. We encourage you to be very cautious in using earnings to value a share. Even though it may be a popular practice, the fact remains that investors look to the cash flows generated by the firm, not its earnings, for value. A firm's value truly is the present value of the cash flows it produces.

ETHICS IN FINANCIAL MANAGEMENT

Ethics: Keeping Perspective

Ethical and moral lapses in business and financial community, academia, politics and religion fill the daily press. But the rash of insider-trading cases on Wall Street against recent graduates of top business and law schools seems particularly disturbing because the cream of the crop, with six-figure incomes and brilliant careers ahead, is being convicted.

Most appear to have been bright, highly motivated overachievers, driven by peer rivalries to win a game in which the score had a dollar sign in front of it. While there have been a few big fish, most sold their futures for $20,000 to $50,000 of illicit profits. They missed the point—that life is a marathon, not a sprint.

In fact, most business school graduates become competent executives, managing people and resources for the benefit of society. The rewards—the titles and money—are merely byproducts of doing a good job.

Source: John S. R. Shad, "Business's Bottom Line: Ethics," *Ethics in American Business: A Special Report*, Touche Ross & Co., 1988, p 56.

OBJECTIVE 8

THE COMMON SHAREHOLDER'S EXPECTED RATE OF RETURN

The valuation equation for a common share was defined earlier as

$$\text{value} = \frac{\text{dividend in year 1}}{(1 + \text{required rate of return})^1} + \frac{\text{dividend in year 2}}{(1 + \text{required rate of return})^2} + \ldots + \frac{\text{dividend in year infinity}}{(1 + \text{required rate of return})^\infty}$$

$$V_c = \frac{D_1}{(1 + k_c)^1} + \frac{D_2}{(1 + k_c)^2} + \ldots + \frac{D_\infty}{(1 + k_c)^\infty} \quad (6\text{-}6)$$

$$V_c = \sum_{t=1}^{\infty} \frac{D_t}{(1 + k_c)^t}$$

Owing to the difficulty of discounting to infinity, a key assumption was made that the dividends, D_t, increase at a constant annual compound growth rate of g. If this assumption is valid, Equation (6-6) was shown to be equivalent to

$$\text{value} = \frac{\text{dividend in year 1}}{(\text{required rate of return} - \text{growth rate})}$$

$$V_c = \frac{D_1}{(k_c - g)} \quad (6\text{-}9)$$

Thus, V_c represents the maximum value that an investor having a required rate of return of k_c would pay for a security having an anticipated dividend in year 1 of D_1 which is expected to grow in future years at rate g.[9] Solving Equation (6-6) for k_c, we can compute the expected rate of return for a common share implicit in its current market price as follows:

$$k_c = \underbrace{\left(\frac{D_1}{V_c}\right)}_{\text{dividend yield}} + \underbrace{g}_{\text{annual growth rate}} \quad (6\text{-}13)$$

From this equation, the common shareholder's expected rate of return is equal to the dividend yield plus a growth factor. Although the growth rate, g, applies to the growth in the company's dividends, both the firm's earnings per share and its share price may also be expected to increase at the same rate. For this reason, g represents the annual percentage growth in the share price. In other words, the investors' expected rate of return is satisfied by receiving dividends, expressed as a percentage of the price of the share (dividend yield), and capital gains, as reflected by the percentage growth rate.[10]

We now compute the expected rate of return for a common share. Assume a firm's common share has a current market price of $44 and its dividends grow at a

---

[9]At times the expected dividend at year-end (D_1) is not given. Instead we might only know the most recent dividend (paid yesterday), that is, D_0. If so, Equation (6-6) must be restated as follows:

$$V_0 = \frac{D_1}{k_c - g} = \frac{D_0(1 + g)}{k_c - g} \quad (6\text{-}12)$$

[10]If a firm's dividends, earnings per share, and share price increase at the same rate, the growth rate is equal to the firm's capital gains yield, which is expressed as follows: $g = (V_c - V_0)/V_c$. Hence, the investor's expected rate of return can now be determined as follows:

investor's expected rate of return = capital gains yield + dividend yield.

$$= \frac{(V_c - V_0)}{V_c} + \frac{D}{V_c} \quad (6\text{-}14)$$

constant rate to infinity. If the expected dividend at the conclusion of this year is \$2.20 and dividends and earnings are growing at a 10 percent annual rate (last year's dividend was \$2), the expected rate of return implicit in the \$44 share price is as follows:

$$k_c = \frac{\$2.20}{\$44} + 10\% = 15\%$$

We can also compute the expected rate of return for a common share using the Capital Asset Pricing Model. The required rate of return on the common share would be determined as follows:

$$k_c = k_f + \beta(k_m - k_f) \qquad (6\text{-}15)$$

Equation (6-15) indicates that the required rate of return is the sum of the risk-free rate and a risk premium for the common share. This risk premium is determined from the product of the beta for the common share and the difference between the required rate of return on the market portfolio and the risk-free rate. As a result, we now have two approaches to determine the expected rate of return for a common share, the capital asset pricing model and the model combining dividend yield and the growth rate.

BACK TO THE BASICS

We have just learned that on average, the expected return will be equal to the investor's required rate of return. This equilibrium condition is achieved by investors paying for an asset only the amount that will exactly satisfy their required rate of return. Thus, finding the expected rate of return based on the current market price for the security relies on two of the axioms given in Chapter 1.

Axiom 1: The Risk-Return Tradeoff—We Won't Take on Additional Risk Unless We Expect to Be Compensated with Additional Return.

Axiom 2: The Time Value of Money—A Dollar Received Today Is Worth More Than a Dollar Received in the Future.

EXAMPLE

Presently, the common shares for Shriver's Inc. and Ito's Inc. are selling in the market for \$25 and \$38, respectively. Uncertainty in the market has caused stock analysts to revise their beta estimates for these companies as follows:

	Shriver Inc.	Ito Inc
Beta	\$1.15	1.40
Recent dividend distributed	\$2.00	\$0.90
Projected constant growth rate	5.5%	4.5%
Three-month T-bill rate	7.5%	
Return on the TSE 300 Index	9.0%	

Determine whether these stocks are overvalued or undervalued given the revised estimates for their betas. Explain the mechanism by which the share price will return to equilibrium.

Shriver's Common Shares

Given the revised estimates for beta, we calculate the investor's required rate of return using the security market line. Here, we assume that under equilibrium conditions the expected rate of return equals the investor's required rate of return.

$$k_{Shriver} = k_f + \beta(k_m - k_f)$$
$$k_{Shriver} = 7.5 + 1.15\,(9.0 - 7.5)$$
$$k_{Shriver} = 9.23\%$$

We now determine the equilibrium price using the model to value common shares with a constant growth rate.

$$P_0 = \frac{(1+g)(D_0)}{k_{Shriver} - g}$$

$$P_0 = \frac{(1+0.055)(\$2.00)}{0.0923 - 0.055} = \$56.57$$

Given an equilibrium price of \$56.57, Shriver's common share is currently undervalued. In this case, investors will realize that the asset is undervalued and buy so as to create a demand in the market for this asset. This increase in demand will cause a shortage of shares in the market. As a result, investors will bid up the price of the shares until their required rate of return equals the expected rate of return or equilibrium.

Ito's Common Shares

For Ito's shares we again calculate the investor's required rate of return using the security market line. Here, we assume that under equilibrium conditions the expected rate of return equals the investor's required rate of return.

$$k_{Ito} = k_f + \beta(k_m - k_f)$$
$$k_{Ito} = 7.5 + 1.40\,(9.0 - 7.5)$$
$$k_{Ito} = 9.60\%$$

We now determine the equilibrium price using the model to value common shares with a constant growth rate.

$$P_0 = \frac{(1+g)(D_0)}{k_{Ito} - g}$$

$$P_0 = \frac{(1+0.045)(\$0.90)}{0.0960 - 0.045} = \$18.44$$

Given an equilibrium price of \$18.44, Ito's common share is currently overvalued. In this case, investors will realize that the asset is overvalued and sell so as to create an excess supply for this asset in the market. This increase in supply will cause excess of shares in the market. As a result, investors will bid down the price of the shares until their required rate of return equals the expected return or equilibrium.

OBJECTIVE 9

ADVANTAGES AND DISADVANTAGES OF COMMON SHARES

The raising of new funds with common shares offers the firm several advantages and disadvantages when compared to bonds and preferred shares.

Advantages

1. The firm is not legally obligated to pay dividends. Thus, in times of financial distress there need not be a cash outflow associated with common shares, while there must be with bonds.
2. Because common shares have no maturity date, the firm does not have a cash outflow associated with its redemption. If the firm desires, it can repurchase its shares in the open market, but it is under no obligation to do so.
3. By issuing common shares the firm increases its financial base and thus its future borrowing capacity. On the other hand, issuing debt increases the financial base of the firm, but also cuts into the firm's borrowing capacity. If the firm's capital structure is already overburdened with debt, a new debt offering may

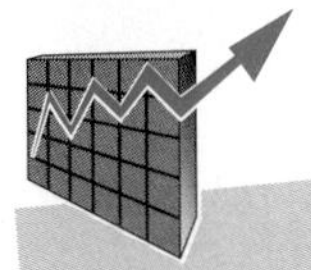

Stock Prices Adjust Slowly to Earnings Jolt

Sometimes share prices do just the opposite of what you might expect from reading a company's earnings announcement. Each financial-reporting period sees hundreds of such "earnings surprises" —cases in which companies' reported earnings are stunningly different from what Wall Street's financial analysts predicted.

Although a company's share price quickly reacts to such earnings surprises, studies show that the full impact doesn't filter through for months. This is especially true for smaller, lesser known issues that aren't widely held by institutions.

Moreover, the more extreme the expectations that analysts have for a company, the bigger the impact of any unexpected turn in earnings. When the market's expectations are either very high or very low, an earnings surprise is likely to force a broad reassessment of the share, says Al Gould, senior vice president at Lynch, Jones & Ryan, a New York brokerage firm.

Helpful Factor

Shearson Lehman Brothers Inc. found that earnings surprises were a helpful factor in identifying shares that would outperform the market. In a Shearson study of 22 quantitative models used in evaluating share price movements since August 1985, a model based on earnings surprises was the third best in determining which shares would outshine the market average. Also, Shearson found that the companies in the top 20% in terms of having the most-positive earnings surprises saw their share prices gain an average of 80% by mid-May 1987 —about twice the market's gain.

Research by Richard Rendleman and the late Henry Latane of the University of North Carolina and Charles Jones of North Carolina State University also indicates that share-price changes attributed to earnings surprises tend to be substantial. In the study, they sorted companies into 10 groups, ranging from those that had the most-positive earnings surprises to those that had the biggest earnings disappointments, between 1971 and 1980.

The researchers found that the share prices of the companies with the biggest positive surprises outperformed the market by an average of 5.6% in the three months after earnings were posted. For the group that had the biggest disappointments, share prices lagged the market by an average of 5.4%.

Several studies have shown that in most cases share prices only gradually adjust to earnings surprises. Profs. Jones, Rendleman and Latane's research, for instance, found that even if investors had waited a full month to react to a particular earnings surprise, they still could have captured more than 35% of the share-price change.

Gradual Correction

What's more, one earnings surprise tends to lead to another in the next quarter as analysts gradually raise estimates to account for the unexpected change in a company's fortune. A study by Thomas Kerrigan, president of Horizon Investment Research Inc., Boston, found that when analysts revise estimates upward in a given quarter, there is a 65% chance the same trend will appear in the next two quarters. For downward adjustments, the odds are 70% that the pattern will be repeated.

Mr. Gould says earnings surprises tend to repeat themselves because analysts are slow to acknowledge that their forecasts were incorrect. Most analysts "tend to rely heavily on management forecasts, which tend to be notoriously wrong," he says. When it comes time to adjust those estimates to actual earnings, the changes tend to be made in small increments.

No matter what an investor's basic strategy, studies indicate that the earnings-surprise approach can be included to improve returns. Drexel Burnham Lambert Inc. found that if this approach had been used from 1984 to 1986 with an institutional client's low price-earnings strategy—which already was outperforming the market—annual returns would have been improved by an average of six percentage points. Similar gains would have occurred with high-growth and small-capitalization strategies, Drexel found.

Source: An excerpt from: Barbara Donnelly, "Stock Prices Adjust Slowly to Earnings Jolt," *The Wall Street Journal*, June 19, 1987, p. 19.

preclude any future debt offering until the existing equity base is enlarged. Thus, financing with common shares increases the firm's financing flexibility.

Disadvantages

1. Because dividends are not tax deductible, as are interest payments, and because flotation costs on equity are larger than those on debt, common shares have a higher cost than does debt.
2. The issuance of new common shares may result in a change in the ownership and control of the firm. While the owners have a preemptive right to retain their proportionate control, they may not have the funds to do so. If this is the case, the original owners may see their control diluted by the issuance of new shares.

SUMMARY

Valuation is an important process in financial management. An understanding of valuation, both the concepts and procedures, supports the financial executive's objective of maximizing the value of the firm.

Features of Preferred Shares

Preferred shares have no fixed maturity date and the dividends are fixed in amount. Some of the more common features of preferred shares include:

- multiple classes
- a priority of claim on assets and income over common shares
- cumulative dividends
- convertible to common shares

For some types of preferred shares:

- dividend rate is adjustable as interest rates change
- participation in the firm's earnings in certain situations
- payment in kind

In addition, there are provisions frequently used to retire an issue of preferred shares, such as the ability for the firm to call its preferred shares or to use a sinking fund provision.

Valuing Preferred Shares

Value is the present value of future cash flows discounted at the investor's required rate of return. For securities with cash flows that are constant in each year but where there is no specified maturity, such as preferred shares, the present value equals the dollar amount of the annual dividend divided by the investor's required rate of return; that is

$$\text{preferred share value} = \frac{\text{dividend}}{\text{requred rate of return}}$$

The expected rate of return for preferred shares is computed as follows:

$$\text{expected rate of return for preferred shares} = \frac{\text{annual dividend}}{\text{share market price}}$$

Features of Common Shares

Common shareholders have ownership in the firm. Whereas, bondholders and preferred shareholders can be viewed as creditors. Common shares do not have a maturity date, and exist as long as the firm does. Dividend payments must be declared by the firm's board of directors before they are issued. In the event of bankruptcy, the common shareholders, as owners of the corporation, cannot exercise claims on assets until the firm's creditors, which includes both bondholders and preferred shareholders, have been satisfied. However, the liability of common shareholders is limited to the amount of their investment.

The common shareholders are entitled to elect the board of directors and are in general the only security holders given a vote. Common shareholders also have the right to approve any change in the corporate charter. Although each share carries the same number of votes, the voting procedure differs from company to company.

The preemptive right entitles the common shareholder to maintain a proportionate share of ownership in the firm.

For common shares which are expected to pay dividends at a constant growth rate, their value is determined by the following equation:

$$\text{common share value} = \frac{\text{dividend in year 1}}{\text{required rate of return} - \text{growth rate}}$$

Growth here relates to *internal* growth only, where management retains the firm's profits to be reinvested and thereby grow the firm—as opposed to growth through issuing new stock or acquiring another firm.

Growth in and of itself does not mean that we are creating value for the shareholders. Only if we are reinvesting at a rate of return that is greater than the investor's required rate of return will growth result in increased value to the firm. In fact, if we are investing at rates less than the required rates of return for our investors, the value of the firm will actually decline.

The expected rate of return on a security is the required rate of return of investors who are willing to pay the present market price for the security, but no more. This rate of return is important to the financial executive because it equals the required rate of return of the firm's investors.

The expected rate of return for common shares is calculated as follows:

$$\begin{array}{c}\text{expected rate of return} \\ \text{for common shares}\end{array} = \frac{\text{dividend in year 1}}{\text{share market price}} + \text{dividend growth rate}$$

STUDY QUESTIONS

6-1. Why are preferred shares referred to as a hybrid security? This type of share is often said to combine the worst features of common shares and bonds. What is meant by this statement?

6-2. Inasmuch as preferred share dividends in arrears must be paid before common share dividends, should they be considered a liability and appear on the right-hand side of the balance sheet?

6-3. Why would a preferred shareholder want their shares to have a cumulative dividend feature and other protective provisions?

6-4. Distinguish between fixed rate preferred shares and floating rate preferred shares. What is the rationale for a firm issuing floating rate preferred shares?

6-5. Why are preferred shares frequently convertible? Why would they be callable?

6-6. Compare the valuation processes of preferred shares and common shares.

6-7. The common shareholders receive two types of return from their investment. What are they?

6-8. State how the investor's required rate of return is computed.

6-9. Define the investor's expected rate of return.

SELF-TEST QUESTIONS

ST-1. (*Preferred Share Valuation*) What is the value of a preferred share if the dividend rate is 16 percent of a $100 par value? The appropriate discount rate for a share of this risk level is 12 percent.

ST-2. (*Preferred Shareholder Expected Rate of Return*) You own 250 shares of Dalton Resources' preferred shares, which currently sell for $38.50 per share and pay annual dividends of $3.25 per share.

- **a.** What is your expected return?
- **b.** If you require an 8 percent return, given the current price, should you sell or buy more shares?

ST-3. (*Preferred Share Valuation*) The preferred share of Armlo pays a $2.75 dividend. What is the value of the share if your required return is 9 percent?

ST-4. (*Common Share Valuation*) Crosby Corporation's common share paid $1.32 in dividends last year and is expected to grow indefinitely at an annual 7 percent rate. What is the value of the share if you require an 11 percent return?

ST-5. (*Common Shareholder Expected Rate of Return*) Blackburn & Smith's common shares currently sell for $23 per share. The company's executives anticipate a constant growth rate of 10.5 percent and an end-of-year dividend of $2.50.

- **a.** What is your expected rate of return?
- **b.** If you require a 17 percent return, should you purchase the shares?

STUDY PROBLEMS (SET A)

6-1A. (*Preferred Share Valuation*) Calculate the value of the preferred share that pays a dividend of $6 per share if your required rate of return is 12 percent.

6-2A. (*Measuring Growth*) Pepperdine Inc.'s return on equity is 16 percent and management plans to retain 60 percent of earnings for investment purposes. What will be the firm's growth rate?

6-3A. (*Preferred Share Valuation*) What is the value of a preferred share where the dividend rate is 14 percent on a $100 par value? The appropriate discount rate for a share of this risk level is 12 percent.

6-4A. (*Preferred Share Valuation*) Solitron's preferred share is selling for $42.16 and pays $1.95 in dividends. What is your expected rate of return if you purchase the security at the market price?

6-5A. (*Preferred Share Valuation*) You own 200 preferred shares of Somner Resources. Currently, a preferred share of Somner Resources sells for $40 and pays annual dividends of $3.40.

- **a.** What is your expected return?
- **b.** If you require an 8 percent return given the current price, should you sell or buy more shares?

6-6A. (*Common Share Valuation*) You intend to purchase a common share of Marigo at $50 per share, hold it one year, and sell after a dividend of $6 is paid. How much will the share price have to appreciate if your required rate of return is 15 percent?

6-7A. (*Common Share Valuation*) Made-It's common share currently sells for $22.50. The company's executives anticipate a constant growth rate of 10 percent and an end-of-year dividend of $2.

- **a.** What is your expected rate of return?
- **b.** If you require a 17 percent return, should you purchase the shares?

6-8A. (*Common Share Valuation*) Header Motor, Inc., paid a $3.50 dividend last year. At a growth rate of 5 percent, what is the value of a common share if the investors require a 20 percent rate of return?

6-9A. (*Measuring Growth*) Given that a firm's return on equity is 18 percent and management plans to retain 40 percent of earnings for investment purposes, what will be the firm's growth rate?

6-10A. (*Common Shareholder's Expected Rate of Return*) The common share of Zaldi Co. is selling for $32.84. This share recently paid dividends of $2.94 per share and has a projected growth rate of 9.5 percent. If you purchase the share at the market price, what is your expected rate of return?

6-11A. (*Common Share Valuation*) Honeywag's common share is expected to pay $1.85 in dividends next year, and the market price is projected to be $42.50 by year end. If the investor's required rate of return is 11 percent, what is the current value of the share?

6-12A. (*Common Share Valuation*) The market price for a Hobart common share is $43. The price at the end of one year is expected to be $48, and dividends for next year should be $2.84. What is the expected rate of return?

6-13A. (*Preferred Share Valuation*) Pioneer's preferred share is selling for $33 in the market and pays a $3.60 annual dividend.

- **a.** What is the expected rate of return on the share?
- **b.** If an investor's required rate of return is 10 percent, what is the value of the share for that investor?
- **c.** Should the investor acquire the preferred share?

6-14A. (*Common Share Valuation*) The common share of NCP paid $1.32 in dividends last year. Dividends are expected to grow at an 8 percent annual rate for an indefinite number of years.

- **a.** If NCP's current market price is $23.50, what is the share's expected rate of return?
- **b.** If your required rate of return is 10.5 percent, what is the value of the share for you?
- **c.** Should you make the investment?

INTEGRATIVE PROBLEM

You are considering three investments. The first is a bond that is selling in the market at $1,200. The bond has a $1,000 par value, pays interest at 14 percent, and is scheduled to mature in 12 years. For bonds of this risk class, you believe that a 12 percent rate of return should be required. The second investment that you are analyzing is a preferred share with par value of $100 that sells for $90 and pays an annual dividend of $12. Your required rate of return for this share is 14 percent. The last investment is a common share that recently paid a $3 dividend. The firm's earnings per share has increased from $4 to $8 in 10 years, which also reflects the expected growth in dividends per share for the indefinite future. The share is selling for $25, and you think a reasonable required rate of return for this share is 20 percent.

- **a.** Calculate the value of each security based on your required rate of return.
- **b.** Which investment(s) should you accept? Why?
- **c.** If your required rates of return changed to 14 percent for the bond, 16 percent for the preferred share, and 18 percent for the common share, how would your answers change to parts (a) and (b)?
- **d.** Assuming again that your required rate of return for the common share is 20 percent, but the anticipated constant growth rate changed to 12 percent, how would your answers to parts (a) and (b) change?

STUDY PROBLEMS (SET B)

6-1B. (*Preferred Share Valuation*) Calculate the value of the preferred share that pays a dividend of $7 per share if your required rate of return is 10 percent.

6-2B. (*Measuring Growth*) The Stanford Corporation's return on equity is 24 percent and management plans to retain 70 percent of earnings for investment purposes. What will be the firm's growth rate?

6-3B. (*Preferred Share Valuation*) What is the value of a preferred share where the dividend rate is 16 percent on a $100 par value? The appropriate discount rate for a share of this risk level is 12 percent.

6-4B. (*Preferred Share Valuation*) Shewmaker's preferred share is selling for $55.16 and pays $2.35 in dividends. What is your expected rate of return if you purchase the security at the market price?

6-5B. (*Preferred Share Valuation*) You own 250 preferred shares of McCormick Resources. Currently, a preferred share of McCormick Resources sells for $38.50 and pays annual dividends of $3.24.

- **a.** What is your expected return?
- **b.** If you require an 8 percent return and given the current price, should you sell or buy more shares?

6-6B. (*Common Share Valuation*) You intend to purchase a common share of Bama, Inc., at $52.75, hold it one year, and sell after a dividend of $6.50 is paid. How much will the share price have to appreciate if your required rate of return is 16 percent?

6-7B. (*Common Share Valuation*) A common share of Blackburn & Smith's currently sells for $23. The company's executives anticipate a constant growth rate of 10.5 percent and an end-of-year dividend of $2.50.

- **a.** What is your expected rate of return?
- **b.** If you require a 17 percent return, should you purchase the share?

6-8B. (*Common Share Valuation*) Gilliland Motor, Inc., paid a $3.75 dividend last year. At a growth rate of 6 percent, what is the value of a common share if the investors require a 20 percent rate of return?

6-9B. (*Measuring Growth*) Given that a firm's return on equity is 24 percent and management plans to retain 60 percent of earnings for investment purposes, what will be the firm's growth rate?

6-10B. (*Common Share Valuation*) A common share of Bouncy-Bob Moore Co. is selling for $33.84. This share recently paid dividends of $3 and has a projected growth rate of 8.5 percent. If you purchase the share at the market price, what is your expected rate of return?

6-11B. (*Common Share Valuation*) Honeywag's common share is expected to pay $1.85 in dividends next year, and the market price is projected to be $40 by year end. If the investors' required rate of return is 12 percent, what is the current value of the share?

6-12B. (*Common Share Valuation*) The market price for M. Simpson & Co.'s common share is $44. The price at the end of one year is expected to be $47, and dividends for next year should be $2. What is the expected rate of return?

6-13B. (*Preferred Share Valuation*) Gree's preferred share is selling for $35 in the market and pays a $4 annual dividend.

- **a.** What is the expected rate of return on the share?
- **b.** If an investor's required rate of return is 10 percent, what is the value of the share for that investor?
- **c.** Should the investor acquire the preferred share?

6-14B. (*Common Share Valuation*) The common share of KPD paid $1 in dividends last year. Dividends are expected to grow at an 8 percent annual rate for an indefinite number of years.

- **a.** If KPD's current market price is $25, what is the share's expected rate of return?
- **b.** If your required rate of return is 11 percent, what is the value of the share for you?
- **c.** Should you make the investment?

SELF-TEST SOLUTIONS

SS-1.

$$\text{Value } (V_p) = \frac{0.16 \times \$100}{0.12}$$

$$V_p = \frac{\$16}{0.12}$$

$$V_p = \$133.33$$

SS-2.

a.

$$\text{Expected return} = \frac{\text{dividend}}{\text{market price}}$$

$$\text{Expected return} = \frac{\$3.25}{\$38.50}$$

$$\text{Expected return} = 8.44\%$$

b. Given your 8 percent required rate of return, the shares are worth $40.62 to you:

$$\text{Value} = \frac{\text{dividend}}{\text{required rate of return}}$$

$$\text{Value} = \frac{\$3.25}{0.08}$$

$$\text{Value} = \$40.62$$

Because the expected rate of return (8.44 percent) is greater than your required rate of return (8 percent) or because the current market price ($38.50) is less than $40.62, the shares are undervalued and you should buy.

SS-3.

$$\text{Value} = \frac{\text{dividend}}{\text{required rate of return}}$$

$$\text{Value} = \frac{\$2.75}{0.09}$$

$$\text{Value} = \$30.56$$

SS-4.

$$\text{Value} = \left[\frac{\text{(last year's dividend) (1 + growth rate)}}{\text{required rate of return} - \text{growth rate}}\right]$$

$$\text{Value} = \frac{\$1.32(1.07)}{0.11 - 0.07}$$

$$\text{Value} = \$35.31$$

SS-5.

a.

$$\text{Expected rate of return} = \frac{\text{dividend in year 1}}{\text{market price}} + \text{growth rate}$$

$$\text{Expected rate of return} = \frac{\$2.50}{\$23.00} + 0.105$$

$$\text{Expected rate of return} = 21.37\%$$

b. The value of the share would be $38.46. Thus, the expected rate of return exceeds your required rate of return, which means that the value of the security to you is greater than the current market price. Thus you should buy the shares.

$$\text{Value} = \frac{\$2.50}{0.17 - 0.105}$$

$$\text{Value} = \$38.46$$

APPENDIX 6A

THE RIGHTS OFFERING

We will look first at the dates surrounding a rights offering and then examine the process of raising funds and the value of a right.

OBJECTIVE 10

DATES SURROUNDING A RIGHTS OFFERING

Announcement date

The date a firm announces when "holders of record" will be issued rights, when they will be mailed, and when they will expire.

Holder-of-record date

The date when rights are officially issued to shareholders.

Expiration date

The date when the rights offer officially expires.

Ex rights date

The date on or after which the shares sell without rights.

Rights-on price

The price of the shares prior to the ex rights date.

Ex rights price

The price of the shares on or after the ex-rights date.

Let us examine the announcement of a rights offering by a hypothetical corporation. On March 1 the firm announces that all "holders of record" as of April 6 will be issued rights, which will expire on May 30 and will be mailed to them on April 25. In this example March 1 is the **announcement date,** April 6 the **holder-of-record date**, and May 30 the **expiration date**. While this seems rather straightforward, it is complicated by the fact that if the shares are sold a day or two before the holder-of-record date, the corporation may not have time to record the transaction and replace the old owner's name with that of the new owner; the rights might then be sent to the wrong person. To deal with this problem an additional date has been created, the ex rights date. The **ex rights date** occurs four trading days prior to the holder-of-record date.

On or after the ex rights date, the shares sell without the rights. Whoever owns the shares on the day prior to the ex rights date receives the rights. Thus, if the holder-of-record date is April 6 and four trading days earlier is April 2, anyone purchasing the shares on or before April 1 will receive the rights, while anyone purchasing the shares on or after April 2 will not. The price of the shares prior to the ex rights date is referred to as the **rights-on price**, while the price on or after the ex rights date is the **ex rights price**. The timing of this process and the terminology are shown in Table 6A-1.

RAISING FUNDS THROUGH RIGHTS OFFERINGS

Subscription price

The price for which the security may be purchased in a rights offering.

Three questions and theoretical relationships must be dealt with if we are to understand rights offerings. First, how many rights are required to purchase a new share? Second, what is the theoretical value of a right? Finally, what effect do rights offerings have on the value of the common shares outstanding?

Let us continue with the example of our hypothetical corporation and assume it has 600,000 shares outstanding, currently selling for \$100 per share. In order to finance new projects, this firm needs to raise an additional \$10,500,000 and wishes to do so with a rights offering. Moreover, the subscription price on the new shares is \$70 per share. The **subscription price** is set below the current market price of the shares in order to ensure a complete sale of the new shares. To determine how many shares must be sold to raise the desired funds, we merely divide the desired funds by the subscription price:

Table 6A-1 Illustration of the Timing of a Rights Offering

	Date		Event	
Shares sell rights-on	March 1	—	Announcement date	
	April 1	—	The owner of the shares as of this date receives the rights	
Shares sell ex rights	April 2	—	Ex rights date	four trading days
	April 6	—	Holder-of-record date	
	May 30	—	Expiration date	

$$\text{new shares to be sold} = \frac{\text{desired funds to be raised}}{\text{subscription price}} \tag{6A-1}$$

$$= \frac{\$10{,}500{,}000}{\$70}$$

$$= 150{,}000 \text{ shares}$$

We know that each common share receives one right, and 150,000 new common shares must be sold. Therefore, to determine the number of rights necessary to purchase one share, we merely divide the original number of shares outstanding by the new shares to be sold:

$$\begin{array}{c}\text{number of rights}\\ \text{necessary to purchase one share}\end{array} = \frac{\text{original number of shares outstanding}}{\text{new shares to be sold}} \tag{6A-2}$$

$$= \frac{600{,}000 \text{ shares}}{150{,}000 \text{ shares}}$$

$$= 4 \text{ rights}$$

This indicates that if a current shareholder wishes to purchase a new share, he or she needs four rights plus $70:

$$\begin{array}{c}\text{price of a}\\ \text{new share}\end{array} = \begin{array}{c}\text{subscription}\\ \text{price}\end{array} + \begin{array}{c}\text{number of rights necessary}\\ \text{to purchase one share}\end{array} \tag{6A-3}$$

$$= \$70 + 4 \text{ rights}$$

Since the subscription price is well below the current market value, there is clearly some value to a right.

The theoretical value of the right obviously depends upon (1) the relationship between the market price of the shares and the subscription price and (2) the number of rights necessary to purchase one share. To determine the value of a right in the preceding example, first let us determine the market value of the corporation. Originally, the firm had 600,000 shares outstanding, selling at $100 each, for a total value of $60,000,000. Let us now assume that the market value of the firm went up by exactly the amount raised by the rights offering, $10,500,000, making the new market value of the firm $70,500,000. In reality the market value of the firm will go up by more than this amount if investors feel the firm will earn more than its required rate of return on these funds.

Taking the new market value for the firm, $70,500,000, and dividing by the total number of shares outstanding, 750,000, we find that the new market value for the shares will be $94 per share. That is to say, after all the new shares have been subscribed to, the market value of the shares will fall to $94 per share. Since it takes four rights and $70 to purchase one share, the value of the share is worth $94 and the value of a right is equal to the savings made ($24—that is, you can buy a $94 share for $70), divided by the number of rights necessary to purchase one share:

$$\begin{array}{c}\text{theoretical value}\\ \text{of one right}\end{array} = \frac{(\text{market price of shares ex rights}) - (\text{subscription price})}{\text{number of rights necessary to purchase one share}} \tag{6A-4}$$

$$R = \frac{P_{ex} - S}{N} \tag{6A-5}$$

$$= \frac{\$94 - \$70}{4}$$

$$= \$6$$

where R = value of one right
P_{ex} = ex rights price of the shares
S = subscription price
N = number of rights necessary to purchase one share

If the shares were selling rights-on—that is, prior to the ex rights date—the theoretical value of a right could be determined from the following equation:[11]

$$R = \frac{P_{on} - S}{N + 1} \qquad \text{(6A-11)}$$

where P_{on} is the rights-on price of the shares. Substituting in the values from our example, we find:

$$R = \frac{\$100 - \$70}{4 + 1} = \$6$$

We found the same theoretical value for the right from both equations because the second equation is derived directly from the first.

Although a rights offering causes the value of an investor's common shares to decrease, an investor rerceives the exact amount of value lost in the form of the rights received. For example, an investor who own 200 shares prior to the rights offering has an investment value of $20,000. Prior to the expiration date of the rights offering, this investor owns 200 shares worth $18,800 (200 shares at $94 per share) and 200 rights worth $1,200 (200 shares × 1 right per share × $6 per right) for a total investment value of $20,000. In fact, an investor will not lose investment value from a rights offering unless the investor fails to exercise or sell the rights by the expiration date. For the case in which the investor sells the 200 rights at a theoretical value of $6 per right, the total value of this investment position ($20,000) is the sum of the value of common shares, $18,800 (200 shares at $94 per share) and the proceeds from the rights, $1,200 (200 rights at $6 per right). For the case in which the investor exercises the 200 rights to buy an additional 50 shares, the total investment value of this position is $20,000, the difference between the value of the 250 common shares, $23,500 (250 shares at $94 per share) and the cost of purchasing the new shares, $3,500 (50 shares at $70). In other words, investors will only incur a loss if they neglect to exercise or sell the rights.

At this point it should be clear that the subscription price is irrelevant in a rights offering. The main concern in selecting a subscription price is setting it low enough that the price of the shares will not fall below it. As long as the subscription price is set far enough below the price of the shares to ensure that the rights maintain a positive value, the offering should be successful and the shareholders should not benefit or gain from the offering.

You have probably noticed that in our discussion of rights valuation we have said we are looking at the theoretical value of a right. The actual value may differ from the theoretical value for several reasons. One common reason is that transactions costs can limit investor arbitrage that would otherwise push the market price of the right to its theoretical value. A second reason is that speculation and irregular sale of rights over the subscription period may cause shifts in supply that push the market price of the right above or below its theoretical value. Flotation costs associated with rights offerings are largely fixed costs; hence, the larger the rights offering, the lower the flotation costs as a percent of the amount of funds raised. Flotation costs also vary with the diffusion of share ownership.

---

[11]This equation is derived from the preceding equation as follows. We know that

$$P_{ex} = P_{on} - R \qquad \text{(6A-6)}$$

Substituting $(P_{on} - R)$ for P_{ex} in Equation (6A-5) yields

$$R = \frac{P_{on} - R - S}{N} \qquad \text{(6A-7)}$$

Simplifying,

$$RN = P_{on} - R - S \qquad \text{(6A-8)}$$

$$RN + R = P_{on} - S \qquad \text{(6A-9)}$$

$$R(N + 1) = P_{on} - S \qquad \text{(6A-10)}$$

$$R = \frac{P_{on} - S}{N + 1} \qquad \text{(6A-11)}$$

The primary benefit for the firm is that issuing new shares via a rights offering has a high probability of success. The rights must be exercised or sold to someone who will exercise them, or the shareholder will lose money. However, one of the questions that has haunted finance over the past decade is this: Why do firms avoid rights offerings? It was thought that the cheapest way to issue shares was to make a rights issue and avoid paying an underwriter for a standby agreement to guarantee the issue's success. This notion that rights offerings are less expensive than underwritten offerings has now been dismissed. Recent findings show that rights offerings are advantageous only for closely held corporations. Moreover, firms with diffused share ownership incur greater costs from rights offerings than from underwritten offerings.[12] Therefore, while rights offerings are not of direct value to the shareholder, in some cases they offer advantages to the financial manager.

STUDY QUESTIONS

6A-1. Since a rights offering allows common shareholders to purchase common shares at a price below the current market price, why is it not of value to the common shareholder?

6A-2. Define the following terms:

- **a.** *Subscription price*
- **b.** *Holder-of-record date*
- **c.** *Preemptive right*

6A-3. Why may the market value of a right differ from its theoretical value?

SELF-TEST PROBLEM

ST-1. (*Rights Offering*) A firm is considering a rights offering to raise $30 million. Currently this firm has three million shares outstanding selling for $60 per share. The subscription price on the new shares would be $40 per share.

- **a.** How many shares must be sold to raise the desired funds?
- **b.** How many rights are necessary to purchase one share?
- **c.** What is the value of one right?

STUDY PROBLEMS (SET A)

6A-1A. (*Rights Offering*) The L. Turner Corporation is considering raising $12 million through a rights offering. It has one million shares outstanding, currently selling for $84 per share. The subscription price on the new shares will be $60 per share.

- **a.** How many shares must be sold to raise the desired funds?
- **b.** How many rights are necessary to purchase one share?
- **c.** What is the value of one right?

6A-2A. (*Rights Offering*) The E. Muransky Corporation is considering a rights offering to raise $35 million to finance new projects. It currently has two million common shares outstanding, selling for $50 per share. The subscription price on the new shares would be $35 per share.

- **a.** How many shares must be sold to raise the desired funds?
- **b.** How many rights are necessary to purchase one share?
- **c.** What is the value of one right?

[12]See Robert S. Hansen and John M. Pinkerton, "Direct Equity Financing: A Resolution of a Paradox," *Journal of Finance* 37 (June 1982), pp. 651–665.

6A-3A. (*Rights Offerings*) The B. Fuller Corporation is in the process of selling common shares through a rights offering. Prior to the rights offering, the firm had 500,000 common shares outstanding. Through the rights offering, it plans on issuing an additional 100,000 shares at a subscription price of $30. After the shares went ex rights, the market price was $40. What was the price of the B. Fuller common shares just prior to the rights offering? [*Hint:* Set Equation (6A-5) equal to Equation (6A-11) and solve for P_{on}.]

STUDY PROBLEMS (SET B)

6A-1B. (*Rights Offering*) The Good Gravy Corporation is considering raising $15 million through a rights offering. It has one million common shares outstanding, currently selling for $86 per share. The subscription price on the new shares will be $60 per share.

a. How many shares must be sold to raise the desired funds?

b. How many rights are necessary to purchase one share?

c. What is the value of one right?

6A-2B. (*Rights Offering*) The Utlonghorns Corporation, a film company responsible for the hit movie "Revenge of the Nightmare Police Academy the Thirteenth Part VI," is considering a rights offering to raise $75 million to finance new projects. It currently has 1.5 million common shares outstanding, selling for $35 per share. The subscription price on the new shares would be $25 per share.

a. How many shares must be sold to raise the desired funds?

b. How many rights are necessary to purchase one share?

c. What is the value of one right?

6A-3B. (*Rights Offerings*) The Gold Castle Corporation is in the process of selling common shares through a rights offering. Prior to the rights offering, the firm had 600,000 common shares outstanding. Through the rights offering, it plans on issuing an additional 100,000 shares at a subscription price of $37. After the shares went ex rights, the market price was $42. What was the price of the Gold Castle common shares just prior to the rights offering? [*Hint:* Set Equation (6A-5) equal to Equation (6A-11) and solve for P_{on}.]

SELF-TEST SOLUTION

SS-1. a.

$$\text{new shares to be sold} = \frac{\text{desired funds to be raised}}{\text{subscription price}} = \frac{\$30{,}000{,}000}{\$40} = 750{,}000 \text{ shares}$$

b.

$$\begin{array}{c}\text{number of rights}\\ \text{necessary to purchase}\\ \text{one share}\end{array} = \frac{\text{original number of shares outstanding}}{\text{new shares to be sold}} = \frac{3{,}000{,}000}{750{,}000} = 4 \text{ rights}$$

c.

$$\text{value of one right} = \frac{P_{on} - S}{N+1} = \frac{\$60 - \$40}{4+1} = \frac{\$20}{5} = \$4$$

APPENDIX 6B

THE RELATIONSHIP BETWEEN VALUE AND EARNINGS

In understanding the relationship between a firm's earnings and the market price of its share, it is helpful to look first at the relationship for a nongrowth firm and then expand our view to include the growth firm.

THE RELATIONSHIP BETWEEN EARNINGS AND VALUE FOR THE NONGROWTH FIRM

OBJECTIVE 11

When we speak of a nongrowth firm, we mean one that retains no profits for the purpose of reinvestment. The only investments made are for the purpose of maintaining status quo—that is, investing the amount of the depreciation taken on fixed assets so that the firm does do not lose its current earnings capacity. The result is both constant earnings and a constant dividends stream to the common shareholder, since we are paying all earnings out in the form of dividends (i.e., dividend in year *t* equals earnings in year *t*). This type of common share is essentially no different from a preferred share. Recalling from our earlier discussion about valuing a preferred share, we may value the nongrowth common share similarly, expressing our valuation in one of two ways:

$$\text{value of a nongrowth firm} = \frac{\text{earnings per share}_1}{\text{required rate of return}} \tag{6B-1}$$

$$= \frac{\text{dividend per share}_1}{\text{required rate of return}} \tag{6B-2}$$

or

$$V_{cng} = \frac{EPS_1}{k_c} = \frac{D_1}{k_c}$$

EXAMPLE

The Reeves Corporation expects its earnings per share this year to be $12, which is to be paid out in total to the investors in the form of dividends. If the investors have a required rate of return of 14 percent, the value of the share would be $85.71:

$$V_{cng} = \frac{\$12}{.14} = \$85.71$$

In this instance, the relationship between value and earnings per share is direct and unmistakable. If earnings per share increase (decrease) 10 percent, then the value of the share should increase (decrease) 10 percent. That is, the ratio of price to earnings will be a constant, as will the ratio of earnings to price. A departure from the constant relationship would occur only if the investors change their required rate of return, owing to a change in their perception about such things as risk or anticipated inflation. Thus, there is good reason to perceive a relationship between next year's earnings and share price for the nongrowth company.

THE RELATIONSHIP BETWEEN EARNINGS AND VALUE FOR THE GROWTH FIRM

Turning our attention now to the growth firm, one that does reinvest its profits back into the business, we will recall that our valuation model depended on dividends and earnings, for that matter, increasing at a constant growth rate. Returning to Equation (6-9), we valued a common share where dividends were expected to increase at a constant growth rate as follows:

$$\text{value} = \frac{\text{dividend}_1}{\text{required rate of return} - \text{growth rate}} \tag{6-9}$$

or

$$V_c = \frac{D_1}{k_c - g}$$

Although Equation (6-9) is certainly the conventional way of expressing value of the growth share, it is not the only means. We could also describe the value of a share as the present value of the dividend stream provided from the firm's existing assets plus the present value of any future growth resulting from the reinvestment of future earnings. We could represent this concept notationally as follows:

$$V_c = \frac{EPS_1}{k_c} + PVDG, \tag{6B-3}$$

where EPS_1 / k_c is the present value of the cash flow stream provided by the existing assets and *PVDG* is the net present value of any dividend growth resulting from the reinvestment of future earnings. The first term, EPS_1 / k_c, is immediately understandable given our earlier rationale about nongrowth shares. The second term, the present value of future dividend growth (*PVDG*), needs some clarification.

To begin our explanation of *PVDG*, let r equal the fraction of a firm's earnings that are retained in the business, which implies that the dividend in year 1 (D_1) would equal $(1 - r) \times EPS_1$. Next, assume that any earnings that are reinvested yield a rate of ROE (return on equity). Thus, from the earnings generated in year 1, we would be investing the percentage of earnings retained, r, times the firm's earnings per share, EPS_1, or $r \times EPS_1$. In return, we should expect to receive a cash flow in all future years equal to the expected return on our investment, ROE, times the amount of our investment, or $r \times EPS_1 \times \text{ROE}$. Since the cash inflows represent an annuity continuing in perpetuity, the net present value from reinvesting a part of the firm's earnings in year 1 (NPV_1) would be equal to the present value of the new cash flows less the cost of the investment:

$$NPV_1 = \underbrace{\left(\frac{rEPS_1ROE}{k_c}\right)}_{\substack{\text{present value}\\ \text{of increased}\\ \text{cash flows}}} - \underbrace{rEPS_1}_{\substack{\text{amount}\\ \text{of cash}\\ \text{retained and}\\ \text{reinvested}}} \tag{6B-4}$$

If we continued to reinvest a fixed percentage of earnings each year and earned ROE on these investments, there would also be a net present value in all the following years. That is, we would have NPV_2, NPV_3, NPV_4, $\ldots NPV_\infty$. Also, since r and ROE are both constant, the series of *NPV*s will increase at a constant growth rate of $r \times \text{ROE}$. We may therefore use the constant-growth valuation model to value *PVDG* as follows:

$$PVDG = \frac{NPV_1}{k_c} - g \tag{6B-5}$$

Thus, we may now establish the value of a common share as the sum of (1) a present value of a constant stream of earnings generated from the firm's assets already in place, and (2) the present value of an increasing dividend stream coming from the retention of profits. That is,

$$V_c = \frac{EPS_1}{k_c} + \frac{NPV_1}{k_c} - g \qquad \text{(6B-6)}$$

EXAMPLE

The Upp Corporation should earn $8 per share this year, of which 40 percent will be retained with the firm for reinvestment and 60 percent paid in the form of dividends to the shareholders. Management expects to earn an 18 percent return on any funds retained. Let us use both the constant-growth dividend model and the *PVDG* model to compute Upp's share value.

Constant-Growth Dividend Model

Since we are assuming that the firm's ROE will be constant and that management faithfully intends to retain 40 percent of earnings each year to be used for new investments, the dividend stream flowing to the investor should increase by 7.2 percent each year, which we know by solving for (r)(ROE), or (.4)(18%). The dividend for this year will be $4.80, which is the dividend-payout ratio of (1 − r) times the expected earnings per share of $8 (.60 × $8 = $4.80). Given a 12 percent required rate of return for the investors, the value of the security may be shown to be $100.

$$V_c = \frac{D_1}{k_c - g} \qquad \text{(6-9)}$$

$$= \frac{\$4.80}{.12 - .072}$$

$$= \$100$$

PVDG Model

Restructuring the problem to compute separately the present value of the no-growth stream and the present value of future growth opportunities, we may again determine the value of the share to be $100. Solving first for value assuming a no-growth scenario,

$$V_{cng} = \frac{EPS_1}{k_c} \qquad \text{(6B-1)}$$

$$= \frac{\$8}{12}$$

$$= \$66.67$$

We next estimate the value of the future growth opportunities coming from reinvesting corporate profits each year, which is

$$PVDG = \frac{NPV_1}{k_c - g} \qquad \text{(6B-5)}$$

Knowing k_c to be 12 percent and the growth rate to be 7.2 percent, we lack knowing only NPV_1, which can easily be determined using Equation (6B-4):

$$NPV_1 = \left(\frac{rEPS_1ROE}{k_c}\right) - rEPS_1 \qquad \text{(6B-4)}$$

$$= \left(\frac{(.4)(\$8)(.18)}{.12}\right) - (.4)(\$8)$$

$$= \$4.80 - \$3.20$$

$$= \$1.60$$

The *PVDG* may now be computed:

$$PVDG = \frac{\$1.60}{.12 - .072}$$

$$= \$33.33$$

Thus, the value of the combined streams is $100:

$$V_c = \$66.67 + \$33.33 = \$100$$

From the preceding example, we see that the value of the growth opportunities represents a significant portion of the total value, 33 percent to be exact. Further, in looking at the *PVDG* model, we observe that value is influenced by the following: (1) the size of the firm's beginning earnings per share, (2) the percentage of profits retained, and (3) the spread between the return generated on new investments and the investor's required rate of return. The first factor relates to firm size; the second to management's decision about the investor's required rate of return. While the first two factors are not unimportant, the last one is the key to wealth creation by management. Simply because management retains profits does not mean that wealth is created for the shareholders. Wealth comes only if the return on equity from the investments, ROE, is greater than the investor's required rate of return, k_c. Thus, we should expect the market to assign value not only to the reported earnings per share for the current year, but also to the anticipated growth opportunities that have a marginal rate of return that exceeds the required rate of return of the firm's investors.

RAISING CAPITAL IN THE FINANCIAL MARKET

CHAPTER 7

LEARNING OBJECTIVES

After reading this chapter you should be able to

1. Understand the historical relationship between internally generated corporate sources of funds and externally generated sources of funds.
2. Understand the financing mix that tends to be used by firms raising long-term financial capital.
3. Explain the financing process by which savings are supplied and raised by major sectors in the economy.
4. Describe the key components of the Canadian financial market system.
5. Understand the role of the investment banking business in the context of raising corporate capital.
6. Distinguish between privately placed securities and publicly offered securities.
7. Explain the concepts of securities flotation costs and securities markets regulations.

INTRODUCTION

In 1995, several large, well-established Canadian companies, such as Trizec, Royal Bank, Rogers Communications and Alberta Energy, were listed on the New York Stock Exchange.[1] *Between 1992 and 1995, the number of inter-listed Canadian companies on the NYSE increased from 26 to 37. This increase in the number of companies was reflected in an increase in trading volume of Canadian-based interlisted securities from 820 million in 1992 to 2.1 billion in 1995 on the NYSE. For inter-listed companies like Trizec, a listing on both the Toronto Stock Exchange (TSE) and the New York Stock Exchange (NYSE) means that the company can broaden its financing base and establish its common shares as a basis for making acquisitions. In the case of Trizec, a real estate developer, 70 percent of its operations and cash flow are based in the United States. Despite this type of presence, American financial institutions would not hold Trizec's shares because they were not listed on the NYSE.*

Companies must attract investment capital to finance new projects and acquisitions. Moreover, these companies must compete against their peer groups to attract capital investment. The financial market provides companies like Trizec a mechanism to compete against other competitors to attract investment capital. In this chapter you will learn about (a) the importance of financial markets to a developed economy, (b) how capital is raised in the financial markets, and (c) some of the factors that a company must consider in raising capital in the financial markets. This type of information will help you to understand the role of the financial markets in a company's business operations.

---

[1]Adapted from: A. Willis, "Why Trizec Went on the NYSE," *The Globe and Mail*, January 19, 1996, p. B12.

CHAPTER PREVIEW

In the next six chapters, we examine the various approaches that a firm can use to finance its expenditures. In this chapter, we focus on the market environment in which long-term capital is raised. We begin by examining both the levels and the types of capital that are used by nonfinancial private corporations to finance their expenditures. We trace the movement of funds through the economy to understand how businesses are being financed. We then examine the various methods used to raise capital and the financial intermediaries which facilitate these financing methods. Finally, we discuss two important factors that must be considered in raising capital in the market environment: flotation costs and market regulations.

BACK TO THE BASICS

In this chapter, we cover material that introduces the financial executive to the process involved in raising funds in the nation's capital markets. We will see that Canada has a highly developed, complex, and competitive system of financial markets that allow for the quick transfer of savings from those economic units with a surplus of savings to those economic units with a savings deficit. Such a system of highly developed financial markets allows great ideas (like the personal computer) to be financed and increases the overall wealth of the economy. Consider your wealth, for example, compared to that of the average family in Russia. Russia lacks a complex system of financial markets to facilitate transactions in financial claims (securities). As a result, real capital formation has suffered.

Thus we return now to ***Axiom 6: Efficient Capital Markets—The Markets Are Quick and the Prices Are Right.*** *Financial executives like our system of capital markets because they trust them. This trust stems from the fact that the markets are efficient. Financial executives trust prices in the securities markets because those prices quickly and accurately reflect all available information about the value of underlying market participants, more so than do simpler things like accounting changes and the sequence of past price changes in a specific security. With security prices and returns (like interest rates) competitively determined, more financial executives (rather that fewer) participate in the markets and help ensure the basic concept of efficiency.*

OBJECTIVE 1

THE FINANCIAL MARKET SYSTEM

Financial markets

All institutions and procedures that facilitate transactions in all types of financial claims and securities.

Capital market

All institutions and procedures that facilitate transactions in long-term financial instruments.

A firm may not be able to meet its proposed expenditures with internally generated funds. In these situations, the corporation may find it necessary to attract large amounts of financial capital externally.[2] In this chapter we define **financial markets** as the institutions and procedures that facilitate transactions in all types of financial claims or securities. However, all institutions and procedures that facilitate transactions in long-term financial instruments (i.e., common shares and bonds) will be defined as **capital markets**. Thus, long-term funds are raised in the capital markets.

A financial market system has to be organized and resilient. Periods of economic recession, for instance, test the financial markets and those firms that continually use the markets. Economic conditions are especially challenging to financial decision-makers because all recessions are unique. These types of economic forces also cause corporate financing policies to become unique.

[2]By externally, we mean that the funds are "obtained" other than through retentions or depreciation. These latter two "sources" are commonly called internally generated funds.

Table 7-1 Nonfinancial Private Corporation Sector Sources of Funds: Internal and External

Year	Total Sources ($ millions)	Percent Internal Cash flows	Percent External Funds
1995	$152,443	26.3	73.7
1994	173,945	20.1	79.9
1993	139,938	19.2	80.8
1992	73,155	26.4	73.6

Source: Adapted from Statistics Canada, *Financial Flow Accounts*, Catalogue 13-001, various issues. Reproduced with the permission of the Minister of Industry, 1996.

During the 1981–82 recession interest rates remained high in North America during the worst phases of the economic downturn. High interest rates occurred because policy-makers at both the Bank of Canada and the Federal Reserve in the United States decided to wring a high rate of inflation out of the economy by means of a tight monetary policy. Simultaneously, share prices were depressed. These business conditions induced firms to forgo raising funds by external means. Table 7-1 indicates that during 1992 the nonfinancial private corporation sector raised only $73,155 million in funds as compared to a high of $173,945 million in funds raised during 1994.[3]

In the recession of the early nineties, realized inflation and inflationary expectations were not the main culprits of the economic contraction. In fact, interest rates moved to low levels because of monetary policy and low inflation rates. For example, Treasury bills of three- and six-month maturities sold at prices which produced yields that were lower than 6.10 percent during the first quarter of 1994. These yields contrast to Treasury bill yields of 16.3 percent during 1982 and 20.8 percent during 1981.[4] Accordingly, corporate financial managers made more trips to the financial markets with new security issues. The cost of corporate capital was perceived as being low. In the midst of such business conditions firms turn to the financial markets and raise a greater proportion of their funds externally. As economic policy shapes the environment of the financial markets, managers must both understand the meaning of the economic changes and remain flexible in their decision-making processes.

The amount of capital raised in the capital markets can be vast. For example in 1995, the Government of Canada sold 83.8 million common shares of Canadian National Railroad to the public at $27.00 per share. This common share offering was sold on an instalment basis with the first instalment priced at $16.25 and the second instalment priced at $10.75. As an initial public offering, the common share offering was expected to raise approximately $2.26 billion. The sale of Canadian National Railroad shares continued a government-launched privatization program of government-owned corporations. The underwriting process was led by Nesbitt Burns Inc. and ScotiaMcLeod Inc., from Toronto, as well as Goldman, Sachs & Co. and Morgan Stanley from New York. Other examples of companies using the capital markets to raise capital through different types of securities include the following: Conwest Exploration issuing 3 million common shares at a price of $23.25 per share for a value of $69.8 million; the Loewen Group issuing 8.8 million series C, preferred shares at a par value of $25 to raise $220 million; Magna International Inc., issuing US$ 345 million of 5 percent convertible, subordinated debentures;

[3]Statistics Canada, *Financial Flow Accounts*, Catalogue 13-001, various issues. Reproduced with the permission of the Minister of Industry, 1996.

[4]Bank of Canada Review, "Selected Canadian and International Interest Rates, including Bond Yields and Interest Arbitrage," Table F1, various issues.

and Southam Inc., issuing $75 million of 8.47% unsecured notes. Clearly, these are important amounts of money to the companies selling the related securities.

To be able to distribute and absorb security offerings of this size, an economy must have a well-developed financial market system. To use that system effectively, the financial officer must have a basic understanding of its structure. Accordingly, this chapter explores the rudiments of raising capital in the financial markets.

OBJECTIVE 2

THE MIX OF CORPORATE SECURITIES SOLD IN THE CAPITAL MARKET

When corporations decide to raise cash in the capital market, what type of financing vehicle is most favoured? Many individual investors think that issuing common shares is the answer to this question. This is understandable, given the coverage of the level of common share prices by the popular news media. All the major television networks, for instance, quote the closing prices for the Toronto Stock Exchange 300 Index, the Montreal Stock Exchange Index and the Dow Jones Industrial Average on their nightly news broadcasts. Common shares, though, are not the financing method relied on most heavily by corporations. Corporate bonds are usually chosen. The corporate debt markets clearly dominate the corporate equity markets when new funds are being raised. This is a long-term relationship—it occurs year after year. Table 7-2 supports this relationship.

In Table 7-2 we see the total volume (in millions of dollars) of corporate securities sold for cash over the 1974–1995 period. The percentage breakdown among common shares, preferred shares, and bonds is also displayed. We know from our previous discussion in Chapter 2 that the Canadian tax system inherently favours debt as a means of raising capital. Quite simply, interest expense is deductible from other income when computing the firm's tax liability, whereas the dividends paid on both preferred and common shares are not.

Financial managers responsible for raising corporate cash know this. When they have a choice between marketing new bonds and marketing new preferred shares, the outcome is usually in favour of bonds. The after-tax cost of capital on the debt is less than that incurred on the preferred shares. Likewise, if the firm has unused debt capacity and the general level of equity prices is depressed, financial managers favour the issuance of debt securities over the issuance of new common shares. It is always good to keep some benchmark figures in your head. The average (unweighted) mix of corporate securities sold for cash over the 1974–1995 period is as follows:

Common shares	27%
Preferred shares	17%
Bonds and notes	56%
Total	100.0%

PERSPECTIVE IN FINANCE

The movement of financial capital (funds) throughout the economy just means the movement of savings to the ultimate user of those savings. Some sectors of the economy save more than other sectors. As a result, these savings are moved to a productive use—say to manufacture that Corvette you want to buy.

The price of using someone else's savings is expressed in terms of interest rates. The financial market system helps to move funds to the most productive end use. Those economic units with the most promising projects should be willing to bid the highest (in terms of rates) to obtain the savings. The concepts of financing and moving savings from one economic unit to another are now explored.

Table 7-2 Corporate Securities Offered for Cash

Year	Total Volume ($ millions)	Percent Common Shares	Percent Preferred Shares	Percent Bonds and Notes
1995	$33,183	21.2%	6.8%	72.0%
1994	38,713	46.8	3.9	49.4
1993	43,229	39.9	6.1	54.0
1992	32,192	36.4	8.1	55.5
1991	30,310	32.1	11.0	56.9
1990	22,226	24.7	9.6	65.7
1989	35,368	27.6	17.7	54.7
1988	26,557	24.3	4.8	70.9
1987	33,822	42.8	13.0	44.2
1986	35,043	33.3	16.1	50.6
1985	24,637	26.8	27.4	45.8
1984	16,626	29.3	33.9	36.9
1983	16,606	51.1	14.3	34.6
1982	15,194	26.2	22.3	51.5
1981	18,695	27.3	25.8	46.9
1980	12,718	29.3	24.0	46.7
1979	9,316	31.7	19.0	49.2
1978	13,050	8.6	45.0	46.4
1977	9,805	7.6	25.8	66.6
1976	6,568	9.2	11.3	79.4
1975	5,438	10.2	13.9	75.9
1974	3,647	8.7	14.0	77.3

Source: Canadian Economic Observer, "Security Issues and Retirements," Statistics Canada, Catalogue 11-210, 1991/1992, and the Bank of Canada Review, "Gross New Issues and Retirements: Corporations, Other Institutions and Foreign Debtors," Winter 1996 Reproduced with the permission of the Minister of Industry, 1996.

FINANCING BUSINESS: THE MOVEMENT OF FUNDS THROUGH THE ECONOMY

OBJECTIVE 3

The Financing Process

We now understand the crucial role that financial markets play in a capitalist economy. At this point we will take a brief look at how funds flow across some selected sectors of the Canadian economy. In addition, we will focus a little more closely on the process of financial intermediation that was introduced in Chapter 2. Some actual data are used to sharpen our knowledge of the financing process. We will see that financial institutions play a major role in bridging the gap between savers and borrowers in the economy. Nonfinancial corporations, we already know, are significant borrowers of financial capital.

Table 7-3 shows how funds were supplied and raised by the major sectors of our economy in 1995. The sector of persons and unincorporated business was the largest net supplier of funds to the financial markets. This is the case, by the way, year in and year out. In 1995, the sector of persons and unincorporated business made available $23,852 million in funds to other sectors. That was the excess of their funds supplied over their funds raised in the markets. In the jargon of economics, the sector of persons and unincorporated business is a savings-surplus sector.

In 1995, the corporate and government business enterprise sector supplied to the financial markets $142 million. This type of surplus breaks with a historical trend

Table 7-3 A Sector View of the Flow of Funds in the Canadian Financial Markets for 1995 ($ millions)

Sector	Funds Supplied [1]	Funds Used [2]	Net Funds Supplied [1] − [2]
Persons and unincorporated business	63,899	40,047	23,852
Corporate and government business enterprise	85,983	85,841	142
Government	−14,959	17,686	−32,645
Nonresidents	9,097	878	8,219

Source: Statistics Canada, *Financial Flow Accounts,* Catalogue 13-001 (1996).

in which this sector was consistently reported as a savings-deficit sector. Over the last four years, the net funds supplied by the corporate and government business sector has decreased from –$14,617 million in 1992 to $142 million in 1995. It would appear that the high level of corporate bankruptcies and the trend towards corporate restructuring of the early nineties has caused business enterprises to be more cautious when considering their use of funds as compared with amount of funds available or supplied from the financial markets.

The government sector was observed to be a savings-deficit sector in 1995. The government sector used $32,645 million in excess of the funds it supplied to the financial markets. This highlights a serious problem for the entire economy and for the financial executive. Government deficits have increased the role of the government sector in the market for borrowed funds, making the government a "quasi-permanent" savings-deficit sector. Most financial economists agree that this tendency puts upward pressure on interest rates in the financial marketplace and thereby raises the general (overall) cost of capital to corporations. This phenomenon has become known as "crowding-out": The private borrower is pushed out of the financial markets in favour of the government borrower.

Table 7-3 further highlights how important foreign financial investment is to the activity of the Canadian economy. As the government sector has become more of a "confirmed" savings-deficit sector, the need for funds has been increasingly supplied by foreign interests. Thus, in 1995, the non-resident sector supplied a net $8,219 million to the domestic capital markets.

Table 7-3 demonstrates that the financial market system must exist to facilitate the orderly and efficient flow of savings from the surplus sectors to the deficit sectors of the economy. The result over long periods of time is that the corporate and government business enterprise sector is typically dependent on the sector of persons and unincorporated business to finance its investment needs; whereas the government sector is quite reliant on non-resident financing.

As we noted in the preceding section, the financial market system includes a complex network of intermediaries that assist in the transfer of savings among economic units. Two intermediaries will be highlighted here: life insurance companies and mutual funds. They are especially important participants in the capital market of North America.

Because of the nature of their business, life insurance companies can invest heavily in long-term financial instruments. This investment tendency arises for two key reasons: (1) life insurance policies usually include a savings element in them, and (2) their liabilities liquidate at a very predictable rate. Thus, life insurance companies invest in the "long end" of the securities markets. This means that they favour (1) mortgages and (2) corporate bonds as investment vehicles rather than shorter term-to-maturity financial instruments like Treasury bills. To a lesser extent, they acquire corporate shares for their portfolios. In the article on page 221, we profile the life and health insurance industry.

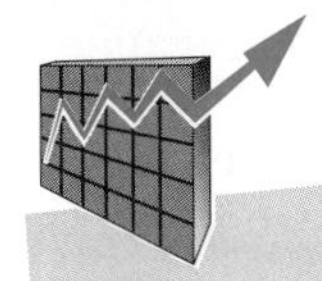

Life and Health Insurance

The life and health insurance industry consists of both companies incorporated in Canada, either federally or provincially, and non-resident companies incorporated outside Canada that operate branches in Canada. The companies may be either stock companies owned by their shareholders (they are not necessarily publicly traded) or mutual companies whose policy holders have ownership rights in the companies.

The industry is highly regulated. Federal and provincial regulations require companies to have the financial resources to meet obligations to their policyholders. Other consumer protection measures are also included in the legislation. The federal Office of the Superintendent of Financial Institutions (OSFI) has been given primary responsibility for supervising the financial condition of federally incorporated companies and Canadian branches of non-resident companies. The Superintendent of Insurance of each province has primary responsibility for supervising the financial condition of companies incorporated in its jurisdiction. Provincial superintendents and regulatory bodies, in addition to their supervisory functions, have jurisdiction over the licensing of agents and brokers. All companies, whether federally or provincially incorporated, must be licensed in each province in which they conduct business.

The basic products the life and health insurance industry supplies are related to the risks associated with premature death and loss of income due to disease or accident and the accumulation of assets for savings and retirement purposes. The savings component of some insurance policies, as well as the wide variety of products offered, put life insurance companies in direct competition with a broad range of financial institutions for personal savings.

Until the 1970s, life insurance companies sold primarily whole life insurance (permanent life insurance payable on the death of the life insured, whenever that occurs). They also sold annuities, generally group annuities for pension plans. This situation changed as a result of the high and volatile interest rates and inflationary pressures of the 1970s, an aging Canadian population and the growing sophistication of consumers, factors that reduced the marketability of whole life policies as long-term investments.

In response to these pressures, Canadian life insurance companies have had to shift towards shorter-term and more interest sensitive products, principally annuities. Canadian life insurance companies have also developed a number of new variable-return life insurance products, such as new money policies and universal life policies. Life and health insurance companies have also shifted their emphasis from a traditional insurance function towards a savings function. The gradual erosion of demarcations among the four pillars of the financial services market (banks, insurance, trusts and securities) has increased competition among them for a share of personal savings. Life insurers offer a broad range of asset accumulation vehicles, including new money policies, RRSPs, Registered Retirement Income Funds and deferred annuities. Many of the annuities sold to individuals by life insurance companies are effectively substitutes for term deposits at a deposit-taking institution. By continuing to develop new and competitive products such as these, life insurers may be able to retain their share of the personal savings market in Canada.

Source: Adapted from: "Industry Profile, 1990–1991, Life and Health Insurance" (Ottawa: Communications Branch, Industry, Science and Technology Canada, 1991).

World Wide Web Site 

http://www.insurancecanada.ca/insurcan
Comment: Links to other insurance resources on the Internet.

Table 7-4 shows the financial asset mix of life insurance companies for the years 1991–1994. The box drawn around their holdings of corporate shares, claims, and other bonds demonstrates that these firms are major suppliers of financial capital to the corporate and government enterprise sector. The sum of corporate shares, claims, and other bonds held by life insurance companies expressed as a percent of their total investment in financial assets for each of the four selected years in Table 7-4 is listed as follows:

Year	Percent
1991	47.0%
1992	26.2
1993	31.7
1994	−2.3

Over the four-year period, about 25.7 percent of the financial assets of life insurance firms are represented by corporate shares, claims, and other bonds. We see that life insurance companies are an important financial intermediary. By issuing life insurance policies (indirect securities), they can acquire direct securities (corporate shares, claims and other bonds) for their investment portfolios. Their preference, by far, is for bonds over shares.

Let us now direct our attention to another financial intermediary, mutual funds in Canada. Table 7-5 shows the financial asset mix of mutual funds in Canada. It is constructed in a manner similar to that of Table 7-4. In comparison with life insurance companies, two factors stand out. First, mutual funds have significantly greater amounts of financial assets than do life insurance companies. Second, mutual funds invest more heavily in corporate shares than they do in corporate bonds. The sum of corporate shares, claims, and other bonds held by mutual funds expressed as a percent of their total investment in financial assets for each of the selected years in Table 7-5 is listed below:

Year	Percent
1991	21.4%
1992	31.6
1993	25.2
1994	39.2

Over the four-year period, then, about 29.4 percent of the financial assets of mutual funds have been tied up in corporate shares, claims, and bonds. Financial institutions are also a significant source of financing for corporate and government enterprise in this country. Mutual funds play the same intermediary role as the life insurance subsector of the economy.

The Movement of Savings

Figure 7-1 provides a useful way to summarize our discussion of (1) why financial markets exist and (2) the movement of funds through the economy. It also serves as an introduction to the role of the investment dealer—a subject discussed in detail later in this chapter.

We see that savings are ultimately transferred to the business firm in need of cash in three ways:

1. **The direct transfer of funds.** Here the firm seeking cash sells its securities directly to savers (investors) who are willing to purchase them in hopes of earning a reasonable rate of return. New business formation is a good example of this process at work. The new business may go directly to a saver or group of savers called "venture capitalists." The venture capitalists will lend funds to the firm or

Table 7-4 The Financial Asset Mix of Life Insurance Companies (in millions)

	1991		1992		1993		1994	
Assets	$	%	$	%	$	%	$	%
Currency & bank deposits	−104	−1.0	304	3.3	131	1.7	112	1.7
Other deposits	12	0.1	−53	−0.6	32	0.4	83	1.2
Foreign currency deposits	−1	−0.1	5	0.1	19	0.2	−4	−0.1
Consumer credit	182	1.8	45	0.5	159	2.1	108	1.6
Other loans	89	0.9	60	0.7	69	0.9	156	2.3
Canada short-term paper	212	2.1	353	3.9	659	8.5	−303	−4.5
Other short-term paper	−549	−5.3	−632	−6.9	−91	−1.2	281	4.1
Mortgages	3,294	32.0	2,501	27.3	575	7.4	1,136	16.7
Canada bonds	334	3.2	1,188	13.0	1,475	19.0	3,077	45.5
Provincial bonds	1,363	13.2	2,130	23.4	1,529	19.7	1,200	17.7
Municipal bonds	376	3.6	169	1.9	264	3.4	115	1.7
Other bonds[a]	3,205	31.1	2,227	24.4	2,196	28.4	−487	−7.2
Corporate claims	974	9.4	183	2.0	549	7.1	243	3.6
Shares	667	6.5	−20	−0.2	−297	−3.8	87	1.3
Foreign Investments	20	0.2	234	2.6	37	0.5	199	2.9
Other financial assets	233	2.3	418	4.6	442	5.7	782	11.5
Total assets	10,307	100.0	9,112	100.0	7,748	100.0	6,785	100.0

[a] Other bonds includes corporate bonds and debentures.

Source: Statistics Canada, *Financial Flow Accounts, Life Insurance Business*, Cat. No. 13-014 (1995), Table 15.

Table 7-5 The Financial Asset Mix of Trusteed Pension Plans (in millions)

	1991		1992		1993		1994	
Assets	$	%	$	%	$	%	$	%
Currency & bank deposits	71	0.6	−57	−0.4	787	2.2	−556	−2.5
Other deposits	16	0.1	−83	−0.5	8	0.1	223	1.0
Foreign currency deposits	2	0.1	31	0.2	94	0.3	699	3.1
Consumer credit	—	—	—	—	—	—	—	—
Other loans	—	—	—	—	—	—	—	—
Canada short-term paper	4,125	34.4	1,883	12.4	3,005	85	−1,851	−8.2
Other short-term paper	873	7.3	754	5.0	3,296	9.3	−110	−0.5
Mortgages	1,524	12.7	2,564	16.9	4,630	13.0	779	3.5
Canada bonds	444	3.7	1,796	11.8	1,946	5.5	5,927	26.3
Provincial bonds	199	1.7	1,111	7.3	2,364	6.7	679	3.0
Municipal bonds	332	2.8	74	0.5	27	0.1	18	0.1
Other bonds[a]	658	5.5	963	6.3	1,360	3.8	2,412	10.7
Corporate claims	—	—	—	—	—	—	—	—
Shares	1,896	15.9	3,838	25.3	7,622	21.4	6,426	28.5
Foreign Investments	1,475	12.3	2,328	15.3	8,642	24.1	6,704	29.7
Other financial assets	347	2.9	−20	−0.1	1,760	5.0	1,188	5.3
Total assets	11,962	100.0	15,182	100.0	35,541	100.0	22,538	100.0

[a] Other bonds includes corporate bonds and debentures.

Source: Statistics Canada, *Financial Flow Accounts, Mutual Funds*, Cat. No. 13-014 (1995), Table 19.

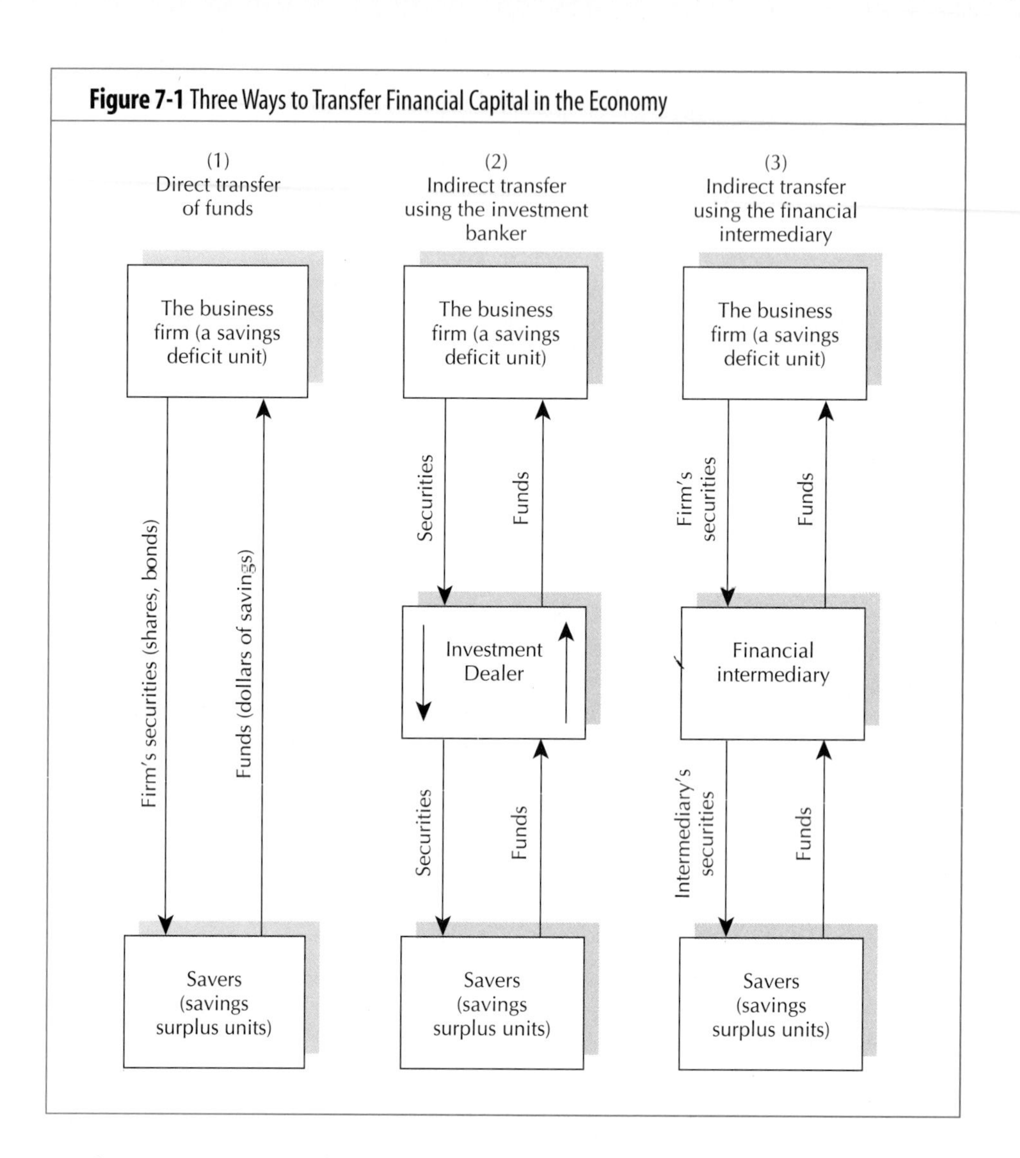

Figure 7-1 Three Ways to Transfer Financial Capital in the Economy

take an equity position in the firm if they feel the product or service the new firm hopes to market will be successful.

2. **Indirect transfer using the investment dealer.** This system emphasizes a common arrangement whereby the managing investment dealer will form a syndicate of several investment dealers. The syndicate will buy the entire issue of securities from the firm that is in need of financial capital. The syndicate will then sell the securities at a higher price than it paid for them to the investing public (the savers). RBC Dominion Securities Inc. and ScotiaMcLeod Inc. are examples of investment dealers. Notice that under this second method of transferring savings, the securities being issued just pass through the investment dealer. They are not transformed into a different type of security.
3. **Indirect transfer using the financial intermediary.** This is the type of system that life insurance companies and trusteed pension funds operate within. The financial intermediary collects the savings of individuals and issues its own (indirect) securities in exchange for these savings. The intermediary then uses the funds collected from the individual savers to acquire the business firm's (direct) securities, such as shares and bonds.

We all benefit from the three transfer mechanisms displayed in Figure 7-1. Capital formation and economic wealth are greater than they would be in the absence of this financial market system.

PERSPECTIVE IN FINANCE

Since Canada enjoys such a developed system of financial markets, the terms used to discuss operations in those markets are numerous—some would say limitless. The financial executive, and those who work close to the financial executive need to master a basic understanding of the commonly used terms, and financing situations. In the next section we describe the components of the Canadian financial market system and define the terminology used in financing situations.

COMPONENTS OF THE CANADIAN FINANCIAL MARKET SYSTEM

OBJECTIVE 4

Numerous approaches exist for classifying the securities markets. At times, the array can be confusing. An examination of four sets of dichotomous terms can help provide a basic understanding of the structure of the Canadian financial market.

Public Offerings and Private Placements

When a corporation decides to raise external capital, those funds can be obtained by making a public offering or a private placement. In a **public offering** both individual and institutional investors have the opportunity to purchase the securities. The securities are usually made available to the public at large by a managing investment dealer and its underwriting risk-taking syndicate. The firm does not meet the ultimate purchasers of the securities in a public offering; the public market is an impersonal market.

Public offering

A security offering where all investors have the opportunity to acquire a portion of the financial claims being sold.

In a **private placement**, also called a **direct placement**, the securities are offered and sold to a limited number of investors. The firm will usually hammer out, on a face-to-face basis with the prospective buyers, the details of the offering. In this setting the investment dealer may act as a finder by bringing together potential lenders and borrowers. The private placement market is a more personal market than its public counterpart. Both public offerings and private placements are explored in more detail later in this chapter.

Private placement

A security offering limited to a small number of potential investors.

Primary Markets and Secondary Markets

Primary markets are financial markets in which securities are offered for the first time to potential investors. A new issue of common shares by Northern Telecom is a primary market transaction. This type of transaction increases the total stock of financial assets outstanding in the economy.

Primary market

Transactions in securities offered for the first time to potential investors.

As mentioned in our discussion of the development of the financial market system, **secondary markets** represent transactions in currently outstanding securities.[5] If the first buyer of the Northern Telecom shares subsequently sells it, he or she does so in the secondary market. All transactions after the initial purchase take place in the secondary market. The sales do not affect the total stock of financial assets that exist in the economy. Both the money market and the capital market, described next, have primary and secondary sides.

Money Market and Capital Market

Money Market

The key distinguishing feature between the money and capital markets is the maturity period of the securities traded in them. The **money market** refers to all institu-

Money market

All institutions and procedures that facilitate transactions in short-term credit instruments.

[5]Note that the "secondary market" includes trading in outstanding securities on both the organized exchanges and over-the-counter markets.

tions and procedures that provide for transactions in short-term debt instruments generally issued by borrowers with very high credit ratings. By financial convention, "short-term" means maturity periods of one year or less. Notice that equity instruments, either common or preferred, are not traded in the money market. The major instruments issued and traded are Government of Canada Treasury bills, various federal and provincial short-term bonds and notes, bankers' acceptances, negotiable certificates of deposit, and commercial paper. Keep in mind that the money market is an intangible market. You do not walk into a building on Bay Street that has the words "Money Market" etched in stone over its arches. Rather, the money market is primarily a telephone market where trading does not occur at any specific location.

Capital Market

The **capital market** refers to all institutions and procedures that provide for transactions in long-term financial instruments. Long-term here means having maturity periods that extend beyond one year. In the broad sense this encompasses term loans and financial leases, corporate equities, and bonds. The funds that comprise the firm's capital structure are raised in the capital market. Important elements of the capital market are the organized security exchanges and the over-the-counter markets.

Organized security exchanges

Formal organizations involved in the trading of securities. These security exchanges are tangible entities that conduct auction markets in listed securities.

Over-the-counter markets

All security markets except the organized exchanges.

Organized Security Exchanges and Over-the-Counter Markets

Organized security exchanges are tangible entities; they physically occupy space (such as a building or part of a building), and financial instruments are traded on their premises. The **over-the-counter markets** include all security markets except the organized exchanges. The money market, then, is an over-the-counter market. Because both markets are important to financial managers concerned with raising long-term capital, some additional discussion is warranted.

Organized Security Exchanges

For practical purposes there are four security exchanges in Canada. These are the (1) Toronto Stock Exchange, (2) Montreal Stock Exchange, (3) Vancouver Stock Exchange, and (4) Alberta Stock Exchange.[6] However, it should be noted that the Winnipeg Stock Exchange does offer trading facilities, but on a much smaller scale as compared to other stock exchanges in Canada. Both the Alberta Stock Exchange and the Winnipeg Stock Exchange are considered to be regional exchanges within Canada. Generally, the Vancouver Stock Exchange is known for trading penny stocks, although the other exchanges also trade penny stocks. In the United States the major exchanges are: (1) New York Stock Exchange, (2) American Stock Exchange, (3) Chicago Stock Exchange, (4) Pacific Stock Exchange, (5) Philadelphia Stock Exchange, (6) Boston Stock Exchange, and (7) Cincinnati Stock Exchange. Of all the North American stock exchanges, the New York Stock Exchange (NYSE) and the Toronto Stock Exchange (TSE) ranked one and two respectively for total volume of trading during the period of 1991–1995.[7]

The Toronto Stock Exchange offers investors several market indices to track stock performance in Canada; for example, the TSE 300 Composite Index, the Toronto 35 Index, the TSE 100 Index and the TSE 200 Index. The TSE 300 Composite Index is the best-known Canadian benchmark which provides investors with a broad market indicator. The TSE 300 Index is a value-weighted index since weights are based on the total number of shares outstanding minus the number of shares in

[6]Many Canadian firms list their shares on more than one stock exchange. For example, the Seagram's Company lists on the Toronto Stock Exchange, Montreal Stock Exchange, and the New York Stock Exchange.

[7]"North American Trading," *Toronto Stock Exchange Review* (July 1995), p.6.

any control blocks. The Toronto 35 Index comprises a cross-section of major publicly listed Canadian stocks. The Toronto 35 Index serves to track the performance of the TSE 300 Index, but is a modified value-weighted index which is based on the number of shares outstanding with no one stock permitted to represent more than 10 percent of the index. In October 1993, the exchange launched the TSE 100 Index made up of 100 of the largest and most liquid TSE stocks. The TSE 100 Index serves as a performance benchmark for institutional investors. The TSE 200 Index is made up of the 200 stocks in the TSE 300 Composite Index that are not included in the TSE 100 Index and serves as a small to mid-cap index. Both the Toronto 35 Index and the TSE 100 Index are the basis for derivatives trading.

The Toronto Stock Exchange continues to play a major role in the Canadian financial market as demonstrated by this exchange's 81 percent share, or 207,664 million dollars, of the total value traded in 1995. Both the Montreal Exchange and the Vancouver Stock Exchange recorded 15.2 and 2.5 percent share of the total value traded in the Canadian financial markets. The Montreal Exchange's share of trading value represented $38,833 million, whereas the Vancouver Stock Exchange's share of trading value was recorded at $6,422 million.[8] The Alberta Stock Exchange plays a smaller role in the Canadian financial market in that it captured only 1.3 percent, or 3,370 million dollars, of the total trading value.

The business of an exchange, including securities transactions, is conducted by its members. Members are said to occupy "seats." Major brokerage firms own seats on the exchanges. An officer of the firm is designated to be the member of the exchange, and this membership permits the brokerage firm to use the facilities of the exchange to effect trades. The price of seats on the exchange can range from $50,000 in 1995 to as high as $370,000 in October of 1987.[9] This price range for an exchange seat reflects the unsettled nature of the markets during economic slowdowns.

Stock Exchange Benefits Both corporations and investors enjoy several benefits provided by the existence of organized security exchanges. These include the following:

1. **Providing a continuous market.** This may be the most important function of an organized security exchange. A continuous market provides a series of continuous security prices. Price changes from trade to trade tend to be smaller than they would be in the absence of organized markets. The reasons are that there is a relatively large sales volume in each security, trading orders are executed quickly, and the range between the price asked for a security and the offered price tends to be narrow. The result is that price volatility is reduced. This enhances the liquidity of security investments and makes them more attractive to potential investors. While it is difficult to prove, it seems logical that this market feature probably reduces the cost of capital to corporations.
2. **Establishing and publicizing fair security prices.** An organized exchange permits security prices to be set by competitive forces. They are not set by negotiations off the floor of the exchange, where one party might have a bargaining advantage. The bidding process flows from the supply and demand underlying each security. This means the specific price of a security is determined in the manner of an auction. In addition, the security prices determined at each exchange are widely publicized. Just read the pages of most newspapers, and the information is available to you. By contrast, the prices and resulting yields on the private placements of securities are more difficult to obtain.
3. **Helping business raise new capital.** Because a continuous secondary market exists where prices are competitively determined, it is easier for firms to float new security offerings successfully. This continuous pricing mechanism also facilitates the determination of the offering price of a new issue. This means that comparative values are easily observed.

World Wide Web Sites

http://www.vse.ca
Comment: Vancouver Stock Exchange Official Home Page supplies daily market reports.

http://www.canadianfinance.com
Comment: Canadian Financial Network is a free Internet-based service which provides access to Canadian investment information and resources.

http://www.quote.com
Comment: Quote.com Financial Market Data is a commercial service that provides quotes on financial instruments.

[8]Allan Swift, "1995 record year at ME," *The Globe and Mail* (January 9, 1996) p. B12.

[9]Barry Critchley and Susan Gittins, "TSE Seat Prices Reflect Hard Times," *The Financial Post,* June 7, 1991, p. 44.

Listing Requirements To receive the benefits provided by an organized exchange, the firm must seek to have its securities listed on the exchange. An application for listing must be filed and a fee paid. The requirements for listing vary for industrial, mining, oil and gas, and foreign companies from exchange to exchange. The minimum listing requirements for the TSE touch on financial criteria such as profitability, size, and working capital. The TSE also imposes other requirements on a company's public distribution, quality of management and affiliation or sponsorship with another company. To give you a flavour of the actual set of listing requirments, those set forth by the TSE are displayed in Table 7-6.

Over-the-Counter Markets

Many publicly held firms do not meet the listing requirements of major stock exchanges. Others may want to avoid the (1) reporting requirements and (2) fees required to maintain listing. As an alternative, their securities may trade in the over-the-counter markets. On the basis of sheer numbers (not dollar volume), more securities are traded over-the-counter than on organized exchanges. As far as secondary trading in corporate bonds is concerned, the over-the-counter markets are where the action is. In a typical year, more than 90 percent of corporate bond business takes place over-the-counter.

Most over-the-counter transactions are done through a loose network of security traders who are known as broker-dealers and brokers. Brokers do not purchase securities for their own account, whereas dealers do. Broker-dealers stand ready to buy and sell specific securities at selected prices. They are said to "make a market" in those securities. Their profit is the spread or difference between the price they will pay for a security (bid price) and the price at which they will sell the security (ask price).

Table 7-6 TSE Listing Requirements for Industrial Companies

Minimum Requirements:

1. Financial Requirements

i) **Profitability**

Pre-tax cash flow of $400,000 in the fiscal year immediately preceding the filing of the listing application.

ii) **Size**

Net tangible assets must be at least $1 million. Earnings of at least $100,000 before taxes and extraordinary items, in the fiscal year immediately preceding the filing of the listing application; Industrial companies may also qualify if they have net tangible assets of $5,000,000 and provide evidence satisfactory to the exchange indicating reasonable likelihood of future profitability.

iii) **Working capital**

Industrial companies must show adequate working capital and capitalization to carry on business.

Other Requirements:

2. Public Distribution

At least 1,000,000 freely tradeable shares having an aggregate market value of $2,000,000 must be held by at least 300 public holders, each holding one hundred shares or more.

3. Quality of management

The exchange will consider the background and expertise of management in the context of the business of the company. This requirement implies that management (1) submit full, true and plain disclosure; and (2) ensure that the business be conducted with integrity and in the best interests of its security holders and the investing public, as well as comply with the rules and regulations of the exchange and all other regulatory bodies having jurisdiction.

4. Sponsorship and Affiliation

Although not mandatory, sponsorship of an applicant company by a member firm of the exchange, or an affiliation with an established enterprise, can be a significant factor in the determination of the suitability of the company for listing, particularly where the company only narrowly meets the prescribed minimum listing requirements.

Source: The Toronto Stock Exchange, *Listing Requirements*, p. 2, 3.

Price Quotes In 1986 the Ontario Securities Commission established an over-the-counter trading system, called the Canadian Over-the-counter Automated Trading System (COATS). However in 1991, the Toronto Stock exchange assumed administrative responsibility and changed the name to the Canadian Dealing Network Inc. (CDN). The trading system is a telecommunications system that provides a national information link among the brokers and dealers operating in the over-the-counter markets. Subscribing traders use computer terminals which enable them to obtain representative bids and ask prices for securities traded over-the-counter. Canadian Dealing Network Inc. provides a quotation system, not a transactions system. The final trade is still consummated by direct negotiation between traders.

CDN price quotes for Canadian unlisted stocks are published daily in *The Globe and Mail* and *The Financial Post.* In the United States, the National Quotation Bureau publishes daily "pink sheets," which contain prices on over-the-counter securities; these sheets are available in the offices of most security dealers.

PERSPECTIVE IN FINANCE

The investment dealer is to be distinguished from the commercial banker in that the former's organization is not a permanent deposit-taking organization for funds. However, deregulation of the investment industry in this country enables commercial banks to perform more functions and services that have belonged almost exclusively to the investment dealers industry. For the moment, it is important for you to learn about the role of the investment dealer in the funding of commercial activity.

THE INVESTMENT DEALER

OBJECTIVE 5

Most corporations do not raise long-term capital frequently. The activities of working-capital management go on daily, but attracting long-term capital is, by comparison, episodic. The sums involved can be huge, so these situations are considered of great importance to financial managers. And since most managers are unfamiliar with the subtleties of raising long-term funds, they will enlist the help of an expert. That expert is an investment dealer.

Definition

The **investment dealer** is a financial specialist involved as an intermediary in the merchandising of securities. This type of financial intermediary acts by facilitating the flow of savings from those economic units that want to invest to those units that want to raise funds. We use the term "investment dealer" to refer both to a given individual and to the organization for which such a person works. Although these firms are called "investment dealers," they perform no deposit-taking or lending functions. Just what is the role of the financial intermediary? That is most easily understood in terms of the basic functions of an investment dealer.

Investment dealer

A financial specialist who underwrites and distributes new securities and advises corporate clients about raising external financial capital.

Functions

The investment dealer performs three basic functions: (1) underwriting, (2) advising, and (3) distributing.

Underwriting

The term *underwriting* is borrowed from the field of insurance. It means "assuming a risk." The investment dealer assumes the risk of selling a security issue at a satisfactory price.[10] A satisfactory price is one that will generate a profit for the investment dealer.

The procedure goes like this. The managing investment dealer and its syndicate will buy the security issue from the corporation in need of funds. The **syndicate** is a group of other investment dealers who are invited to help buy and resell the issue. The managing investment dealer is the firm that originated the business because its corporate client decided to raise external funds. On a specific day, the firm that is raising capital is presented with a cheque in exchange for the securities being issued. At this point the underwriting syndicate owns the securities. The corporation has its cash and can proceed to use it. The firm is now immune from the possibility that the security markets might turn sour. If the price of the newly issued security falls below that paid to the firm by the syndicate, the syndicate will suffer a loss. The syndicate, of course, hopes that the opposite situation will result. Its objective is to sell the new issue to the investing public at a price per security greater than its cost.

Syndicate

A group of investment dealers who contractually assist in the buying of a new security issue.

Advising

The investment dealer is an expert in the issuance and marketing of securities. The expertise of an investment dealer includes an understanding of prevailing market conditions and an ability to relate those conditions to the particular type of security that should be sold at a given time. Business conditions may be pointing to a future increase in interest rates. The investment dealer might advise the firm to issue its bonds in a timely fashion to avoid the higher yields that are forthcoming. The dealer can analyze the firm's capital structure and make recommendations as to what general source of capital should be issued. In many instances the firm will invite its investment dealer to sit on the board of directors. This permits the dealer to observe corporate activity and make recommendations on a regular basis.[11]

Distributing

Once the syndicate owns the new securities, it must get them into the hands of the ultimate investors. This is the distribution or selling function of investment banking. The investment dealer may have branch offices across both Canada and the United States, or it may have an informal arrangement with several security dealers who regularly buy a portion of each new offering for final sale. It is not unusual to have 100 to 600 dealers involved in the selling effort across North America. The syndicate can properly be viewed as the security wholesaler, while the dealer organization can be viewed as the security retailer.

Distribution Methods[12]

Several methods are available to the corporation for placing new security offerings in the hands of final investors. The investment dealer's role is different in each of these. Sometimes, in fact, it is possible to bypass the investment dealer. These methods are described in this section. Private placements, because of their importance, are treated separately later in the chapter.

---

[10]This "assumed" risk can be of varied degrees. The investment dealer may assume total risk (a firm commitment for the entire security issue), some risk (a standby agreement), or no risk (a best efforts agreement).

[11]Investment dealers have access to more information since they are frequently in the market and have staff to accumulate and analyze relevant data. On this basis, investment dealers are able to price and sell advice.

[12]The discussion on distribution methods draws heavily on material from James E. Hatch and Michael J. Robinson, *Investment Management*, 2nd ed. (Scarborough, ON: Prentice-Hall Canada, 1989).

Public Offering

The most prevalent method of distributing securities in both Canada and the United States is the public offering. The public offering of securities is either negotiated with a specific investment dealer or the offering is auctioned off among several investment dealers. In each case, the issuing institution reduces the risk of selling new securities to the public by negotiating the underwriting of the new issue with an investment dealer. The objective of the underwriting agreement is to determine the type and the price of the securities to be sold to the public. Once the terms of the underwriting agreement are negotiated between the two parties, the investment dealer buys the firm's securities and then resells them to the public. In this type of underwriting arrangement, the investment provides the firm with its expertise in the issuing, marketing, and selling of the securities.

The issuing institution may use the auction process to choose the investment dealer which underwrites and distributes the new securities issue. This type of auction process is commonly used by the Government of Canada and provincial governments to issue government bonds. The investment dealers willing to pay the greatest dollar amount per new security will win the competitive bid. The argument in favour of competitive bids is that any undue influence of the investment dealer over the firm is mitigated and the price received by the firm for each security should be higher. Thus, we would intuitively suspect that the cost of capital in a competitive bid situation would be less than in a negotiated purchase situation. Evidence on this question, however, is mixed.[13] One problem with the competitive bid purchase as far as the fundraising firm is concerned is that the benefits gained from the advisory function of the investment dealer are lost.

Commission or Best-Efforts Basis

In some cases, the investment dealer acts as an agent rather than as a principal in the distribution process. The securities are not underwritten. The investment dealer attempts to sell the issue in return for a fixed commission on each security actually sold. Unsold securities are returned to the corporation. This arrangement is typically used for more speculative issues. The issuing firm may be smaller or less established than the investment dealer would like. Because the underwriting risk is not passed on to the investment dealer, this distribution method is less costly to the issuer than a public offering. On the other hand the investment dealer only has to give it his or her "best effort." A successful sale is not guaranteed.

Privileged Subscription or Rights Offerings

Occasionally, the firm may feel that a distinct market already exists for its new securities. When a new issue is marketed to a definite and select group of investors, it is called a **privileged subscription**. Three target markets are typically involved: (1) current shareholders, (2) employees, or (3) customers. Of these, distributions directed at current shareholders are the most prevalent. In a privileged subscription the investment dealer may act only as a selling agent. It is also possible that the issuing firm and the investment dealer might sign a standby agreement, which would obligate the investment dealer to underwrite the securities that are not accepted by the privileged investors.

Privileged subscription

The process of marketing a new security issue to a select group of investors.

Bought Deal

In certain cases, an underwriter or a group of underwriters can negotiate the sale of an entire issue with large institutional or retail investors before the issue comes to market. Since the underwriter has assured itself of reselling the securities when they are bought from the issuing firm, the distribution method is

[13]Gary D. Tallman, David F. Rush, and Ronald W. Melicher, "Competitive versus Negotiated Underwriting Cost of Regulated Industries," *Financial Management* 3 (Summer 1974), pp. 49–55.

Bought deal

A distribution method in which securities are bought from the issuing firm.

called a **bought deal**. The success of this type of distribution method is dependent on the terms of the securities issue and the reputations of both the issuing firm and the underwriter.

The Public Offering Process

The public offering is the distribution method most likely to be used by the private corporation. It makes sense, then, to cover in some detail the sequence of events that comprise the negotiated underwriting.

1. **Selection of an investment dealer.** The firm initiates the fundraising process by choosing an investment dealer. On some occasions a third party, called a finder, may direct the firm to a specific investment dealer. Typically, the finder receives a fee from the fundraising firm if an underwriting agreement is eventually signed. Use of a finder is one way of gaining access to a quality investment dealer.
2. **Preunderwriting conferences.** A series of preunderwriting conferences are held between the firm and the investment dealer. Key items discussed are (1) the amount of capital to be raised, (2) whether the capital markets seem to be especially receptive at this time to one type of financing instrument over another, and (3) whether the proposed use of the new funds appears reasonable. At these conferences, confirmation is obtained that a flotation will occur and the particular investment dealer will manage the underwriting. The investment dealer then undertakes a complete financial analysis of the corporation and assesses its future prospects. The final outcome of the preunderwriting conferences is the *tentative underwriting agreement.* This agreement details (1) the approximate price the investment dealer will pay for the securities and (2) the upset price. The upset price is a form of escape mechanism for the benefit of the issuing firm. If the market price of the firm's securities should drop in a significant fashion just before the new securities are to be sold, the price the investment dealer is to pay the firm might drop below the upset price. In this situation the new offering would be aborted.
3. **Formation of the underwriting syndicate.** An underwriting syndicate is a temporary association of investment dealers formed to purchase a security issue from a corporation for subsequent resale, hopefully at a profit to the underwriters. The syndicate is formed for very sound reasons. First, the managing investment dealer probably could not finance the entire underwriting itself. Second, the use of a syndicate reduces the risk of loss to any single underwriter. Third, the use of a syndicate widens the eventual distribution effort. Each underwriter has its own network of security dealers who purchase a portion of the participation in the offering. Most syndicates contain 5 to 20 investment dealers. Since each dealer has its own distribution network, the selling group can often consist of 100 to 600 dealers across North America. All members of the syndicate sign the agreement which binds them to certain performance standards, provides detail as to their participation (amount they can purchase) in the offering and determines their liability for any portion of the issue unsold by their fellow underwriters.
4. **Registering the securities.** Most provinces in Canada have enacted legislation through their Securities Acts to protect investors against fraud, to ensure fair conduct within the marketplace, and to encourage full and true disclosure by corporations. Each of these provinces has also empowered its own Securities Commission to oversee the regulations of the Securities Act. For example in Ontario, a new public issue cannot be offered to prospective investors before the new issue is registered with the Ontario Securities Commission (OSC). However, for well-established firms, the registration process has been streamlined to make the process more expedient. We will now examine both the long-form registration and the short-form registration processes.
   a) **Long-form registration process**
   The distribution of a new issue of securities across Canada requires that the firm comply with the regulations of the Securities Acts in each of the

provinces where the security is to be distributed. To facilitate the process of meeting all the regulations of each province, the firm would generally register the issue where distribution will be the greatest, for example Ontario. The Ontario Securities Act dictates that a registration statement must be submitted to the OSC; the objective of this statement is disclosure of all facts relevant to the new issue that will permit an investor to make an informed decision. The OSC itself does not judge the investment quality of securities. The registration statement is a lengthy document containing (1) historical, (2) financial, and (3) administrative facts about the firm. The details of the new offering are presented, as well as the proposed use of the funds that are being raised. During a three- to six-week waiting period, the OSC examines the statement for errors or omissions. While this examination process is being carried out, the underwriting syndicate cannot offer the security for sale. Part of the package presented to the OSC for scrutiny is a document called a **prospectus**. Once approved, it is the official advertising vehicle for the offering. During the waiting period a **preliminary prospectus**, outlining the important features of the new issue, may be distributed to potential investors. The preliminary prospectus contains no selling price information or offering date. In addition, a stamped red-ink statement on the first page tells the reader that the document is not an official offer to sell the securities. In the jargon of finance, this preliminary prospectus is called a **red herring**. Once the registration statement is approved by the OSC, a series of provincial agreements enable the security issue to be registered with the provincial Security Commissions and to be offered for sale in all provinces provided the prospectus is made available to all concerned parties.

Prospectus

A condensed version of the full registration statement filed with the Securities Commission that describes a new security issue.

Preliminary prospectus

A document outlining the important features of a new issue, which may be distributed to potential investors. The preliminary prospectus contains no selling price information or offering date.

Red herring

Finance jargon for a preliminary prospectus, which outlines the important features of a new issue without specifying the selling price and offering date.

b) **Prompt Offering Qualification System (POP System)**
In 1982, security regulators agreed to implement the POP system which streamlines the process of public distribution of securities to Canadian capital markets. The POP system streamlines the procedures for a public offering through the use of a short-form prospectus. However, issuing firms must satisfy the following conditions to be eligible for distribution of securities by the POP system:
- must satisfy the reporting period test which requires an issuing firm to have reported for at least 12 months;
- must satisfy the public float test which requires an issuing firm to have an aggregate market value of Canadian listed equity shares held by non-insiders of at least $75 million or more outstanding during the last calendar month of the issuer's most recently completed financial year;
- must satisfy the requirements of securities legislation for provinces or territories where securities will be distributed;
- must satisfy either of the top three categories of recognized ratings organizations for issues of investment grade nonconvertible debt securities and preferred shares.

The compliance of seasoned issuers with these conditions enables a public offering to be cleared by security regulators through an expedited review which is designed to shorten the period for clearance from 5 to 3 working days.[14] This type of delay contrasts with the 21 to 30 days required for the traditional long-form prospectus. Investor's risk is reduced by the strict disclosure rules which require firms using the short-form prospectus method to file financial information with security regulators on annual basis. Such eligibility conditions ensure that the transactions of an issuing firm are reported by the financial media and analyzed by financial specialists for their financial information content.

The POP system can be used for the public offering of common shares, as well as for the public offerings of preferred shares and investment grade nonconvertible debt securities. The latter two securities may be offered under more relaxed

[14]Ontario Securities Commission, *Annual Report* (Toronto: Queen's Printer for Ontario, 1995), pp. 11–12.

conditions; for example, the issuing firms of preferred shares may be required to satisfy the reporting period test but not have to satisfy the public float test.

5. **Formation of the selling group.** In order to distribute the securities to final investors, a **selling group** is formed. The dealers comprising the selling group purchase portions of the new issue from the syndicate members with whom they regularly do business. A dealer of the selling group pays a higher price than that paid by the syndicate member, but less than the offering price to the investing public. The responsibilities and rewards of the selling group are outlined in the selling group agreement. Formation of the selling group completes the distribution network for the new offering. The structure of such a network is illustrated in Figure 7-2.

Selling group

A collection of securities dealers that participate in the distribution of new issues to final investors. A selling group agreement links these dealers to the underwriting syndicate.

6. **The due diligence meeting.** This is a "last-chance" gathering to get everything in order before taking the offering to the public. Usually, all members of the syndicate are present, along with key officers of the issuing firm. Any omissions from the prospectus should be corrected at this meeting. Most important, the final price to be paid by the syndicate to the firm is settled at this meeting. As we said earlier, if the issue represents additional common shares, the underwriting price may be a fixed amount below the closing price of the firm's outstanding shares on the day prior to the new offering. Capital market conditions are discussed, and the security's offering price to the public is set in light of those conditions. The "go ahead" is given to print the final prospectus, which now contains all relevant price information. The offering is usually made the day after the meeting.
7. **Price-pegging.** Once the issue has been offered for sale, the managing underwriter attempts to mitigate downward price movements in the secondary market for the security by placing orders to buy at the agreed-upon public offering price. The objective is to stabilize the market price of the issue so it can be sold at the initial offering price. The syndicate manager's intention to perform this price maintenance operation must be disclosed in the registration statement.
8. **Syndicate termination.** A contractual agreement among the underwriters identifies the duration of the syndicate. Pragmatically, the syndicate is dissolved when the issue has been fully subscribed. If the demand for the issue is great, it may be sold out in a few days. If the issue lingers on the market, without much buyer

Figure 7-2 Distributing New Securities

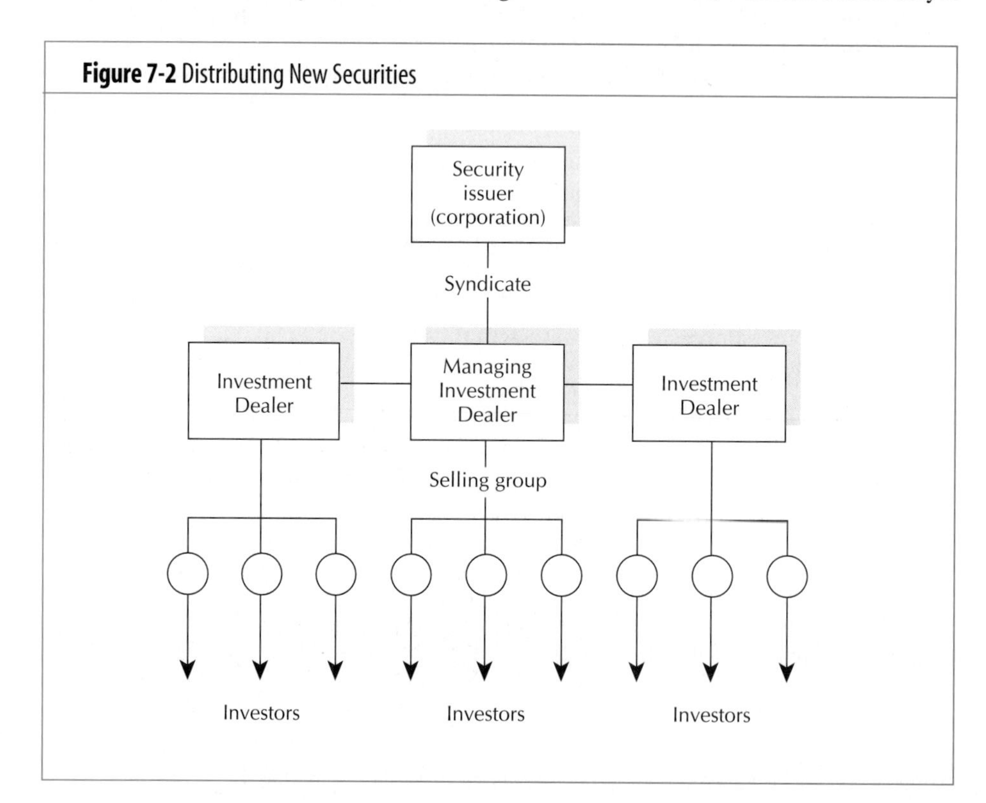

interest, the remaining inventory may be sold at the existing secondary market price. The name of the game in underwriting is not margin, but turnover. Should the issue go sour, the underwriters will quickly absorb the loss and get on to the next underwriting project.

Industry Leaders

Investment dealers in Canada provide underwriting services for both domestic and international issues of debt and equity financing. Domestic issues include both public offerings and private placements. In addition, federal crown corporations, municipalities, provincial governments, and corporations use Canadian investment dealers for underwriting debt issues. As an example of the magnitude of financings in the early nineties, RBC Dominion Securities Inc. leads the industry as a manager of $9.4 billion worth of financings: $3.6 billion of corporate issues and $5.8 billion of provincial issues.[15]

PRIVATE PLACEMENTS

OBJECTIVE 6

Private placements offer an alternative to the sale of securities to the public or to a restricted group of investors through a privileged subscription. Any type of security can be privately placed (directly placed). However, the private placement market is clearly dominated by debt issues.

In arranging a private placement the firm may either avoid the use of an investment dealer and work directly with the investing institutions or engage the services of an investment dealer. If the firm does not use an investment dealer, of course, it does not have to pay a fee. On the other hand, investment dealers can provide valuable advice in the private placement process. They are usually in contact with several major institutional investors; thus, the investment dealer will know if a firm is in a position to invest in a proposed offering, and can help with evaluating the terms of the new issue.

Private placements have both advantages and disadvantages as compared with those of public offerings. The financial manager must carefully evaluate both sides of the question. The advantages associated with private placements are these:

1. **Speed.** The firm usually obtains funds more quickly through a private placement than a public offering.
2. **Reduced flotation costs.** Underwriting and distribution costs are lower than those of public offerings because it is not necessary to file a long-form prospectus.
3. **Financing flexibility.** In a private placement, the firm deals on a face-to-face basis with a small number of investors. This means that the terms of the issue can be tailored to meet the specific needs of the company. For example, all of the funds need not be taken by the firm at once. In exchange for a commitment fee the firm can "draw down" against the established amount of credit with the investors. This provides some insurance against capital market uncertainties, and the firm does not have to borrow the funds if the need does not arise. There is also the possibility of renegotiation. The terms of the debt issue can be altered. The term to maturity, the interest rate, or any restrictive covenants can be discussed among the affected parties.

The following disadvantages of private placements must be evaluated:

1. **Interest costs.** It is generally conceded that interest costs on private placements exceed those of public issues. Whether this disadvantage is enough to offset the reduced flotation costs associated with a private placement is a determination that the financial manager must make. There is some evidence that on smaller

[15]"1991 A Big Year For The Underwriters," *The Financial Post,* June 29, 1992, p. 13.

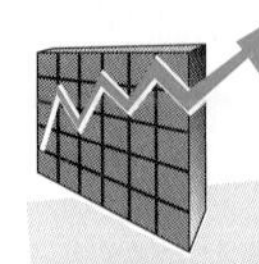

Investment Banking and Commercial Banking

One of the more important events that has affected our financial markets is the deregulation of the investment industry. The following article describes some of the realities that chartered banks have faced since legislation was enacted in 1987.

Banks Face More Uncertainty on Brokerage Front

Back in 1987, after finally getting permission to own securities dealers, four of the country's five largest chartered banks lost little time buying into some of the country's largest investment houses. Shortly after four of the five major Canadian banks entered the securities business in the late 1980s, the brokerage industry took a nosedive. In 1990, possibly the worst year in its history, the securities industry lost $226 million. And although the industry turned things around with a profit of $338 million in 1991, uncertain times are still ahead.

But Mike Andrews, associate director for financial service research at the Conference Board of Canada in Ottawa, says as long as bankers remain bullish on their investments, they'll wait it out.

"[Restructuring] is inevitable in any business," says Peter Godsoe, Toronto-based president and chief operating officer of the Bank of Nova Scotia. "The industry is constantly renewing itself as new firms evolve and others rationalize."

Many changes have been made since 1987, and more are on the way. But experts agree that the banks' infusion of equity into the industry did much to strengthen it.

"The banks injected a fair bit of capital into the industry," Andrews says. "If the banks weren't there, its hard to say what would have happened."

New Relationship

His report, *The Canadian Securities Industry: A Decade of Transition*, says regulatory capital grew $1 billion in 1987 to $2.5 billion, thanks to the banks' purchase of ownership in the largest security firms.

The new relationship between banks and their securities subsidiaries is hardly one way. Andrews notes that it gives banks greater access to capital markets, while their subsidiaries gain a foothold in the credit market. Together, it gives them more clout in the competitive market.

Godsoe adds that as financial markets move toward greater integration of product and services, "we had no choice but to develop better skills in [shares] distribution and underwriting". And having ScotiaMcLeod Inc., a wholly owned subsidiary since 1988, on board helps. "We continue to build [our businesses] in the U.S., Europe, and the Far East," he says. When clients require more than banking services, they can be referred to the brokerage firm, which operates independently of its parent. In Canada, "we have transferred some operations from the bank—mostly in short-term trading—to ScotiaMcLeod because they have greater expertise in that area." A discount brokerage service, offered through another unit, Scotia Securities Inc., uses ScotiaMcLeod's "expertise in processing and running the back office, and that makes us more competitive."

Paul Taylor, executive vice-president of treasury and investment banking, at Royal Bank of Canada, which bought 75% of RBC Dominion Securities Inc. (DS) in 1988, is similarly pleased with the bank's investment in DS. "It's improved our knowledge of the investment industry and market. [The DS team] were more advanced in operations, such as bond trading, providing corporate acquisition advice, right down the line. So we chose to leave that business in their group."

Big Players Now Bigger

Industry analysts note that with the new ownership structure, big players have become bigger, middle size firms have all but disappeared, replaced by boutique firms that address niche markets.

However, Charles Winograd, chairman and chief executive officer of Richardson Greenshields of Canada Ltd., in Toronto, notes that competition from bankers can hurt small players. "Banks are suppliers of capital to brokerage business," he adds. While there hasn't been much reason for concern, "its never a comfortable situation when you're dependent on your competitor for capital." The smaller the player, the greater the discomfort. While bank-owned securities firms can now use their parent's extensive retail outlets to push their own sales, it isn't a major concern to independents, Winograd says.

The issue is whether they can get the right people to offer competent investment advice at those outlets. Taylor notes that while some DS investment advisors have moved into various bank premises across Canada, "we don't think it's necessary or logical to have stockbrokers in all the banks' outlets." For the most part, the way the industry has operated hasn't changed much, mainly because the new players have retained the same managers to run their companies.

"All you've done is altered ownership," says Charles Caty, president of the Toronto-based Investment Dealers Association. In most cases, "management of these subsidiaries are still run by the securities people, not bankers. The banks were buying management [skills], and so long as they perform, they'll keep the management."

Source: Adapted from: Miriam Cu-Uy-Gam, "Banks Face More Uncertainty On Brokerage Front," *The Financial Post*, September 14, 1992, p. s21.

issues, say $500,000 as opposed to $30 million, the private placement alternative would be preferable.[16]

2. **Restrictive covenants.** Dividend policy, working-capital levels, and the raising of additional debt capital may all be affected by provisions in the private placement debt contract. That is not to say that such restrictions are always absent in public debt contracts. Rather, the financial officer must be alert to the tendency for these covenants to be especially burdensome in private contracts.

Until 1990 Canadian access to the U.S. private placement market was considered expensive because U.S. security laws restricted the resale of privately sold securities. In April of 1990 the U.S. Securities and Exchange Commission passed a security law enabling a large institution to resell debt securities. As a result, large Canadian corporations are using the private placement market to place corporate debt in the U.S. capital markets. For example, both CP Forest Products and Hees International have used the private placement market to issue U.S. $102 million and U.S. $200 million in corporate debt in the early nineties. The most likely Canadian companies to use the U.S. private placement market are companies with U.S. revenues that can use the U.S. debt as a hedge against currency fluctuations. The major investors in private placements are large financial institutions. Based on the volume of securities purchased, the three most important investor groups are (1) life insurance companies, (2) registered retirement funds, and (3) private pension funds.

FLOTATION COSTS

OBJECTIVE 7

The firm raising long-term capital incurs two types of **flotation costs**: (1) the underwriter's spread and (2) issuing costs. Of these two costs, the underwriter's spread is the larger. The underwriter's spread is simply the difference between the gross and net proceeds from a given security issue expressed as a percent of the gross proceeds. The issue costs include (1) printing and engraving, (2) legal fees, (3) accounting fees, (4) trustee fees, and (5) several other miscellaneous components. The two

Flotation costs

The underwriter's spread and issuing costs associated with the issuance and marketing of new securities

[16]John D. Rea and Peggy Brockschmidt, "The Relationship Between Publicly Offered and Privately Placed Corporate Bonds," *Monthly Review*, Federal Reserve Bank of Kansas City (November 1973), p. 15.

most significant issue costs are printing and engraving and legal fees.

Flotation costs vary according to the type of security issued and the size of the issue. For example, the costs associated with issuing common shares are notably greater than the costs associated with preferred share offerings. In turn, costs of preferred shares exceed those of bonds. Furthermore, flotation costs (expressed as a percent of gross proceeds) decrease as the size of the security issue increases.

This variation in flotation costs indicates that these costs are sensitive to the risks involved in successfully distributing a security issue. Common shares are riskier to own than corporate bonds. Underwriting risk is, therefore, greater with common shares than with bonds. Thus, flotation costs just reflect these risk relationships. Furthermore, a portion of the flotation costs is fixed. Legal fees and accounting costs are good examples. So, as the size of the security issue rises, the fixed component is spread over a larger gross proceeds base. As a consequence, average flotation costs vary inversely with the size of the issue.

PERSPECTIVE IN FINANCE

Since late 1986, there has been a renewal of public interest in the rules which regulate the North American financial markets. The concern was raised because of a massive insider trading scandal that made the name "Ivan F. Boesky" one of almost universal recognition—but, unfortunately, in a negative sense. This event was followed by the October 19, 1987, crash of the equity markets. More recently, in early 1990, the investing community (both institutional and individual) became increasingly concerned over a weakening in the so-called "junk bond market" and the collapse of the real estate market. The upshot of all of this enhanced awareness is a new appreciation of the crucial role that security regulation plays in the financial system. The basics are presented below.

REGULATION

The regulation of securities is legislated both federally and provincially through the Canada Business Corporation Act and provincial Securities Acts which administer security, commodity, futures, and deposits regulations. Although the Canadian Securities Administrators work with provincial securities commissions and regulators to achieve compatibility among provincial securities regulations, the provincial securities regulators are responsible for administering the Securities Act in their provinces. For example in Ontario, the Ontario Securities Commission (OSC) has both administrative and enforcive responsibility over the Securities Act, the Commodity Futures Act and the Deposits Regulation Act. As of January 1, 1995, the OSC has been given statutory authority to make rules that have the effect of securities regulations. This instrument is in addition to the variety of instruments such as policy statements, blanket orders and rulings, and notices that the OSC has already at its disposal to participate in Ontario's regulatory scheme.[17] However, the actual regulation is administered through self-regulatory organizations (SRO) such as the Investment Dealers Association of Canada (IDA) and the various Canadian Exchanges. These self-regulatory organizations impose financial and trading rules on their members which are enforced through independent audits and compliance checks. In addition, the OSC will oversee self-regulatory organizations such as the Toronto Stock Exchange, the Toronto Futures Exchange and the Investment Dealers Association of Canada to ensure the following activities:

1. Licensing and registering persons and firms that trade in securities and commodity futures contracts.

[17]Ontario Securities Commission, *Annual Report* (Toronto: Queen's Printer for Ontario, 1995), p. 8.

2. Reviewing and receipting prospectuses and other offering documents.
3. Monitoring continuous disclosure documents.
4. Conducting compliance reviews of registrants, issuers and other market participants.
5. Exercising discretion respecting securities regulatory requirements.
6. Investigating possible violations of securities regulatory requirements.
7. Commencing proceedings before the Commission or prosecutions before the Ontario Court of Justice.[18]

The introduction of new issues on the Toronto Stock Exchange is governed by the Securities Act of Ontario. The intent of this act is important. It aims to provide potential investors with accurate, truthful disclosure about the firm and the new securities being offered to the public. This does not prevent firms from issuing highly speculative securities. The OSC says nothing whatsoever about the possible investment worth of a given offering. It is up to the investor to separate the junk from the jewels. The OSC does have the legal power and responsibility to enforce the Securities Act.

Full public disclosure is achieved by the requirement that the issuing firm file a registration statement with the OSC containing requisite information. The statement details particulars about the firm and the new security being issued. During a minimum 21-day waiting period, the OSC examines the submitted document. In numerous instances the 21-day wait has been extended by several weeks. The OSC can ask for additional information that was omitted in order to clarify the original document. The OSC can also order that the offering be stopped.

During the registration process a preliminary prospectus (the red herring) may be distributed to potential investors. When the registration is approved, the final prospectus must be made available to the prospective investors. The prospectus is actually a condensed version of the full registration statement. If, at a later date, the information in the registration statement and the prospectus is found to be lacking, purchasers of the new issue who incurred a loss can sue for damages. Officers of the issuing firm and others who took part in the registration and marketing of the issue may suffer both civil and criminal penalties.

Recent changes in the Canadian equity market have caused regulatory developments to occur to ensure that Canadian firms have access to foreign sources of capital. The article on the following page discusses one of these important regulatory developments.

Secondary Market Regulation

The trading of securities on the secondary market is regulated by both the securities exchange, for example the Toronto Stock Exchange, and the provincial Securities Act. The policy of the Toronto Stock Exchange concerning persons investing in securities listed on this exchange is to ensure equal access to information that may affect their investment decisions. Thus, both the Toronto Stock Exchange and the OSC require companies listed on the exchange to provide timely, accurate and efficient disclosure of material information concerning their business and affairs. In this respect, the Toronto Stock Exchange in its goal to achieve public protection and to promote integrity and honesty in the capital markets may review the conduct of an officer, director or major shareholder that holds sufficient ownership to materially affect control. The purpose of this type of review is to ensure that the business of the company is being conducted with integrity and in the best interests of its security holders and the investing public; as well as to satisfy itself that the rules and regulations of the exchange and all other regulatory bodies having jurisdiction are being complied with.[19] In addition, the OSC has the following powers to fulfill its mandate for effective regulation:

[18]Ibid., p. 5.

[19]The Toronto Stock Exchange, *Listing Requirements*, p. 9.

Canadian Equity Markets: Taking Advantage of Change

The emerging global economy is challenging many industries, and the securities business is no exception. In recent years, international equity trading has been encouraged by a number of factors, including the evolution of technology and the deregulation/regulation of security markets. One measure of this trend is the surge in gross cross-border equity flows: the value of overall trading in domestic shares conducted in the domestic market by or on behalf of foreign non-residents rose from U.S. $73 billion in 1979 to $1.44 trillion in 1990.

For Canadian corporations, the implications of this development are varied. As barriers fall in the global equity market, large corporations are benefiting from improved access to foreign capital. Many mid-sized and small firms, however, still find it difficult to tap international markets. Thus the Canadian securities industry faces a twofold challenge: to ensure that Canadian companies benefit from new opportunities for access to foreign capital, and to maintain a viable domestic securities market for Canadian corporations in an increasingly competitive global environment.

Regulatory Changes

One important step toward increasing access to foreign capital is the Multi Jurisdictional Disclosure System (MJDS). Five years in the making, MJDS was negotiated by the U.S. Securities and Exchange Commission (SEC) and the Canadian provincial securities regulators. Essentially, the system allows large Canadian and U.S. issuers to do public offerings of their securities in another country using the disclosure documents required at home.

Since its implementation in July 1991, a number of Canadian corporations have used MJDS. An example is Magna International Inc., which offered an equity issue in the United States and Canada using MJDS. Because of the agreement, Magna was able to offer the issue in both countries without a U.S. SEC review, by using documents vested by the Ontario Securities Commission and other provincial regulators. Magna officials say the new system allowed the company to access the U.S. market faster and less expensively than before. In addition, Magna notes that although MJDS has only recently come into effect, U.S. underwriters had no difficulty marketing the issue using Canadian-style documents.

Experiences like these show that MJDS can help Canadian companies to obtain a foreign source of capital at a competitive price, and to quickly take advantage of favourable market conditions in the United States. Although at present the system is being used mainly by companies like Magna, which have previously sought capital abroad, the coming years likely will see a growing use of MJDS by Canadian corporations that might not otherwise have made offerings in the United States.

Despite the success of this arrangement, no initiatives are yet in place to extend the benefits of regulatory harmonization to smaller corporations seeking financing in countries other than the United States. Although the International Organization of Securities Commissions (IOSCO) contemplates the possible development of a system analogous to MJDS for a large group of countries including Canada, such a project is unlikely to bear fruit for many years.

Source: Adapted from: K. Michael Edwards, "Canadian Equity Markets: Taking Advantage of Change," *Canadian Business Review* (Spring 1992), pp. 26–28.

- to investigate
- to suspend, to cancel or impose terms on registration, or reprimand registrants
- to grant or deny exemptions from provisions concerning prospectuses, registration, take-over bids or issuer bids
- to cease trading of any security
- to freeze funds or apply to courts for the appointment of receivers
- to order audits of registrants, reporting issuers, mutual fund custodians and clearing houses of commodity futures and stock exchanges
- to grant relief from financial reporting requirements
- to grant relief from proxy solicitation and insider reporting requirements[20]

SUMMARY

This chapter centred on the market environment in which corporations raise long-term funds, including the structure of the Canadian financial markets, the institution of investment banking, and the various methods for distributing securities.

The Mix of Corporate Securities Sold

When corporations go to the capital market for cash, the most favoured financing method is debt. The corporate debt markets clearly dominate the equity markets when new funds are raised. The Canadian tax system inherently favours debt capital as a fund-raising method. In an average year over the 1974–1995 period, bonds and notes made up 56 percent of external cash that was raised.

Why Financial Markets Exist

The function of financial markets is to allocate savings efficiently in the economy to the ultimate demander (user) of the savings. In a financial market the forces of supply and demand for a specific financial instrument are brought together. The wealth of an economy would not be as great as it is without a fully developed financial market system.

Financing Business

Every year households are a net supplier of funds to the financial markets. The nonfinancial business sector is always a net borrower of funds. Both life insurance companies and private pension funds are important buyers of corporate securities. In the economy, savings are ultimately transferred to the business firm seeking cash by means of (1) the direct transfer, (2) the indirect transfer using the investment dealer, or (3) the indirect transfer using the financial intermediary.

Components of the Canadian Financial Market System

Corporations can raise funds through public offerings or private placements. The public market is impersonal in that the security issuer does not meet the ultimate investors in the financial instruments. In a private placement, the securities are sold directly to a limited number of institutional investors.

The primary market is the market for new issues. The secondary market represents transactions in currently outstanding securities. Both the money and capital markets have primary and secondary sides. The money market refers to transactions in short-term debt instruments. The capital market, on the other hand, refers

[20]Ontario Securities Commission, *Annual Report* (Toronto: Queen's Printer for Ontario, 1995), p. 5.

to transactions in long-term financial instruments. Trading in the money and capital markets can take place in either the organized security exchanges or the over-the-counter market. The money market is exclusively an over-the-counter market.

The Investment Dealer

The investment dealer is a financial specialist involved as an intermediary in the merchandising of securities. He or she performs the functions of (1) underwriting, (2) advising, and (3) distributing. Major methods for the public distribution of securities include (1) the public offering, (2) the commission or best-efforts basis, (3) privileged subscriptions, and (4) bought deal. The bought deal may bypass the use of an investment dealer. The public offering provides the greatest amount of investment underwriting services to the corporate client.

Private Placements

Privately placed debt provides an important market outlet for corporate bonds. Major investors in this market are (1) life insurance firms, (2) retirement funds, and (3) private pension funds. Several advantages and disadvantages are associated with private placements. The financial officer must weigh these attributes and decide if a private placement is preferable over a public offering.

Flotation Costs

Flotation costs consist of the underwriter's spread and issuing costs. The flotation costs of common shares exceed those of preferred shares, which, in turn, exceed those of debt. Moreover, flotation costs as a percent of gross proceeds are inversely related to the size of the security issue.

Regulation

The primary market for new issues is regulated at the provincial level by their Securities Acts. A national distribution of securities must meet the regulations of every province before it can be sold to the public. A firm that is distributing a new issue of securities for the first time must use the long-form registration process to register the issue with the provincial Security Commissions. However, large and well-established firms may use the Prompt Offering Qualification System (POP System) to register a security issue instead of the long-form registration process which is more costly and time-consuming. Secondary market trading is also regulated by the provincial Securities Acts. In Canada, members of the secondary market are self-regulated. The Security Acts empower their Security Commissions to oversee the self-regulatory organizations to ensure the licensing and registration of persons and firms trading in securities and commodity futures contracts, to review prospectuses and other disclosure documents, and to enforce the regulations of the Securities Acts and the Commodity Futures Act.

STUDY QUESTIONS

7-1. What are financial markets? What function do they perform? How would an economy be worse off without them?

7-2. Define in a technical sense what we mean by the term *financial intermediary.* Give an example of your definition.

7-3. Distinguish between the money and capital markets.

7-4. What major benefits do corporations and investors enjoy because of the existence of organized security exchanges?

7-5. What are the general categories examined by an organized exchange in determining whether an applicant firm's securities can be listed on it? (Specific numbers are not needed here, but rather "areas" of investigation.)

7-6. Why do you think most secondary market trading in bonds takes place over-the-counter?

7-7. What is an investment dealer, and what major functions does he or she perform?

7-8. Explain the Prompt Offering Qualification System (POP System).

7-9. Why is an investment dealer syndicate formed?

7-10. Why might a large corporation want to raise long-term capital through a private placement rather than a public offering?

7-11. As a recent business school graduate, you work directly for the corporate treasurer. Your corporation is going to issue a new security and is concerned with the probable flotation costs. What tendencies about flotation costs can you relate to the treasurer?

7-12. You own a group of five clothing stores, all located in southern Ontario. You are about to market $200,000 worth of new common shares. Your shares trade over-the-counter. The shares will be sold only to Ontario residents. Your financial advisor informs you that the issue must be registered with the OSC. Is the advisor correct?

7-13. When corporations raise funds, what type of financing vehicle (instrument or instruments) is most favoured?

7-14. What is the major (most significant) savings-surplus sector in the Canadian economy?

7-15. Identify three distinct ways that savings are ultimately transferred to business firms in need of cash.

CHAPTER 8

ANALYSIS AND IMPACT OF LEVERAGE

LEARNING OBJECTIVES

After reading this chapter you should be able to

1. Understand the difference between business risk and financial risk.
2. Use the technique of breakeven analysis in a variety of analytical settings.
3. Calculate the firm's breakeven point in terms of units produced and sold, and in sales dollars.
4. Distinguish among the financial concepts of operating leverage, financial leverage, and combined leverage.
5. Calculate the degrees of operating leverage, financial leverage, and combined leverage.
6. Explain why a firm with a high business risk exposure might logically choose to employ a low degree of financial leverage in its financial structure.

INTRODUCTION

In 1995, Bombardier Inc., reported an increase in sales revenue of 24.8 percent over the level of reported sales for 1994. The firm's change in net income, however, rose 40.3 percent over the same one-year period.[1] Such a disparity in the relationship between sales fluctuations and net income fluctuations is not peculiar to companies like Bombardier Inc. Furthermore, these types of fluctuations do not always occur in the same positive direction. For example, Electrohome Inc., reported an increase of 23.4 percent in sales revenue, but reported a decrease of 20.4 percent in net income for the year 1995.[2]

What is it about the nature of businesses that causes changes in sales revenues to translate into larger variations in net income and finally the earnings available to the common shareholders? It would actually be a good planning tool for managements to be able to decompose such fluctuations into those policies that are associated with the operating side of the business, as distinct from those policies associated with the financing side of the business. Such knowledge could be put to effective use when the firm builds its strategic plan. This chapter will show you how to do just that.

---

[1]Bombardier Inc., *Annual Report* (1995), p. 1.

[2]Electrohome Ltd., *Annual Report* (1995), p. 1.

CHAPTER PREVIEW

A firm's decision to employ a certain amount of financial leverage is influenced by the asset structure of the firm. In turn, the decision to employ financial leverage affects the determination of the firm's financial structure. Figure 8.1 illustrates that the costs of the various sources of financing used for capital budgeting, or the cost of capital, provides a direct link between the formulation of the firm's asset structure and its financial structure. In other words, the cost of capital represents the hurdle rate at which the rates of return for the firm's capital investments are compared. However, we should also note that the cost of capital is affected by the composition of the right-hand side of the firm's balance sheet—that is, its financial structure.

This chapter examines which tools a financial executive can use to determine the firm's proper financial structure. First, we distinguish between two types of risk: operating risk and financial risk. We then review the technique of breakeven analysis. This technique provides the foundation for the relationships to be highlighted in the remainder of the chapter. Finally, we examine the concept of operating leverage, some consequences of the firm's use of financial leverage, and the impact on the firm's earnings stream when operating and financial leverage are combined in various patterns.

PERSPECTIVE IN FINANCE

In this chapter we become more precise in assessing the causes of variability in the firm's expected revenue streams. Business risk occurs as a result of the firm's investment decisions. For example, the composition of the firm's assets determines its exposure to business risk. This type of risk is described by what appears on the left-hand side of the company's balance sheet. Financial risk occurs as a result of the choices used to arrange the right-hand side of the company's balance sheet. The choice to use more financial leverage means that the firm will experience greater exposure to financial risk. The tools developed here will help you quantify the firm's business and financial risk. A solid understanding of these tools will make you a better financial manager.

OBJECTIVE 1

BUSINESS RISK AND FINANCIAL RISK

Risk

The likely variability associated with expected revenue or income streams.

Business risk

The relative dispersion in the firm's expected earnings before interest and taxes.

In studying capital-budgeting techniques we referred to **risk** as the likely variability associated with expected cash flows. As our attention is now focused on the firm's financing decision rather than its investment decision, it is useful to separate the income stream variations attributable to (1) the company's exposure to business risk[3] and (2) its decision to incur financial risk.

Business risk refers to the relative dispersion (variability) in the firm's expected earnings before interest and taxes (EBIT).[4] Figure 8-2 shows a subjectively estimated probability distribution of next year's EBIT for the Pierce Grain Company and the same type of projection for Pierce's larger competitor, the Blackburn Seed Company. The expected value of EBIT for Pierce is $100,000, with an associated standard deviation of $20,000. If next year's EBIT for Pierce fell one standard deviation short of the expected $100,000, the actual EBIT would equal $80,000. Blackburn's expect-

[3]Examples of firms that have a low degree of business risk include electric power corporations, whereas examples of firms that have a high degree of business risk include computer software companies.

[4]If what the accountants call "other income" and "other expenses" are equal to zero, then EBIT is equal to net operating income. These terms will be used interchangeably.

Figure 8-1 Cost of Capital as a Link Between the Firm's Asset Structure and Financial Structure

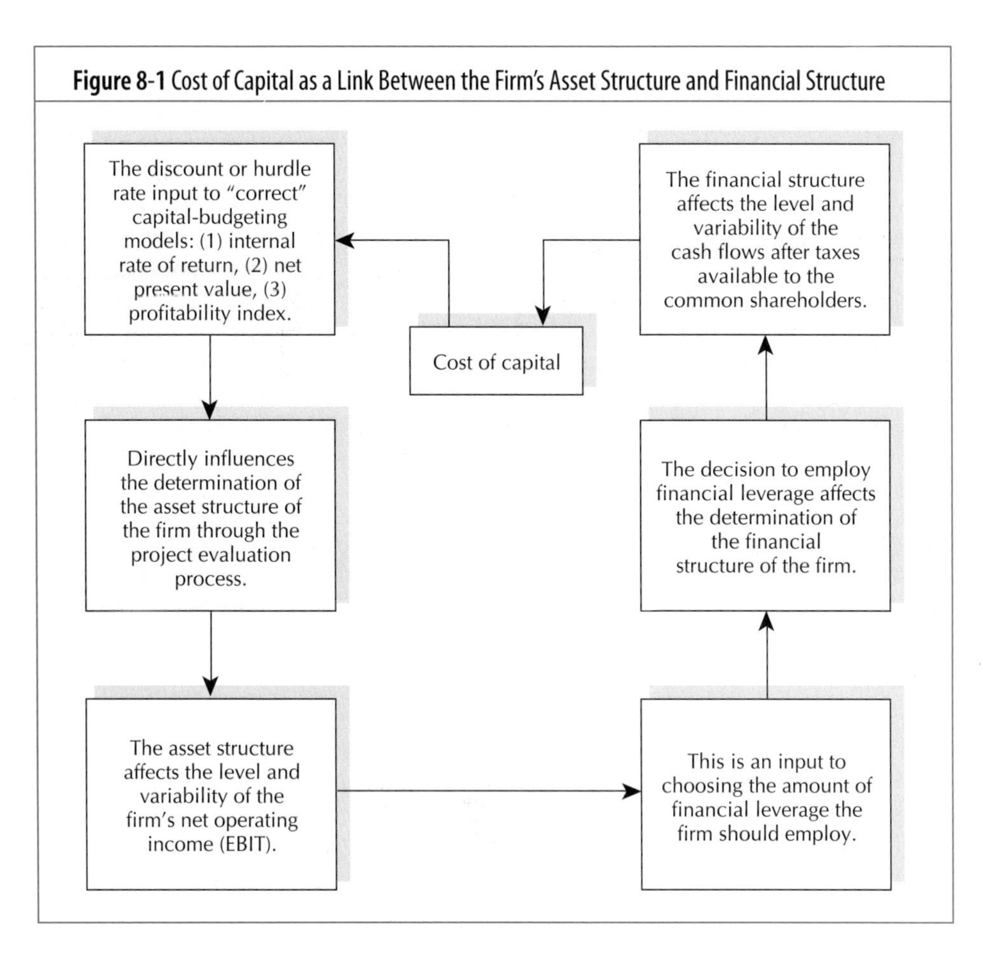

ed EBIT is $200,000, and the size of the associated standard deviation is $20,000. The standard deviation for the expected level of EBIT is the same for both firms. We would say that Pierce's degree of business risk exceeds Blackburn's owing to its larger coefficient of variation of expected EBIT, as follows:

Figure 8-2 Subjective Probability Distribution of Next Year's EBIT

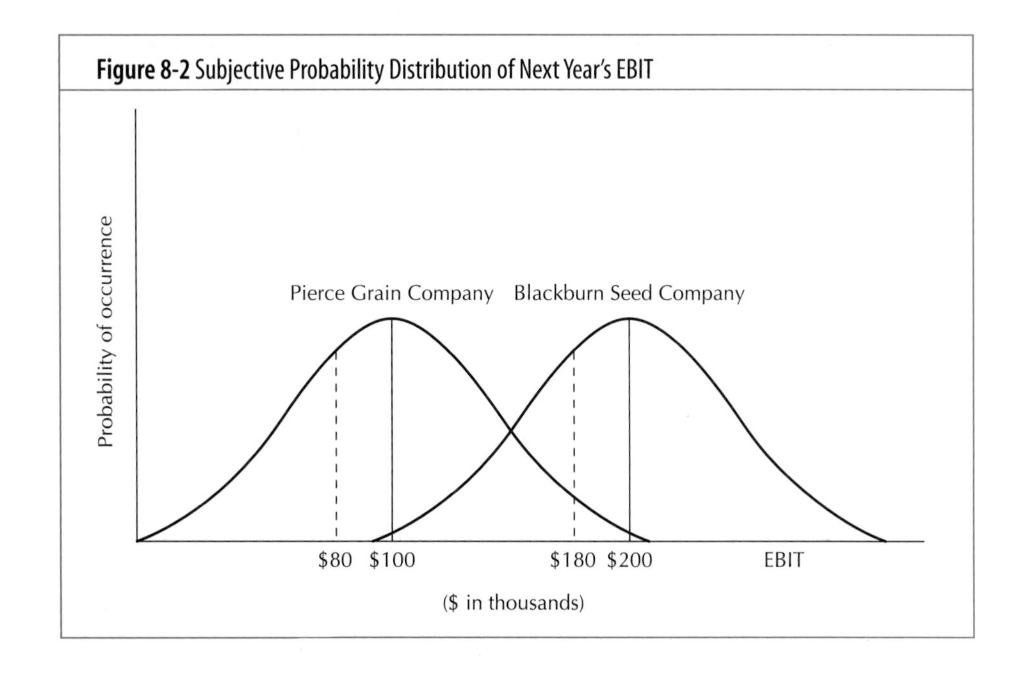

$$\text{Pierce's coefficient of variation of expected EBIT} = \frac{\$20,000}{\$100,000} = .20$$

$$\text{Blackburn's coefficient of variation of expected EBIT} = \frac{\$20,000}{\$200,000} = .10$$

The relative dispersion in the firm's EBIT stream, measured here by its expected coefficient of variation, is the residual effect of several causal influences. Dispersion in operating income does not cause business risk; rather, this dispersion which we call business risk is the result of several influences. Some of these are listed in Table 8-1, along with an example of each particular attribute. Notice that the company's cost structure, product demand characteristics, and intra-industry competitive position all affect its business risk exposure. Such business risk is a direct result of the firm's investment decision. It is the firm's asset structure, after all, that gives rise to both the level and variability of its operating profits.[5]

Financial risk

The additional variability in earnings available to the firm's common shareholder, and the additional chance of insolvency borne by the common shareholder caused by the use of financial leverage.

Financial leverage

The use of securities bearing a fixed (limited) rate of return to finance a portion of the firm's fixed assets. Financial leverage can arise from the use of either debt or preferred share financing. The use of financial leverage exposes the firm to financial risk.

Financial risk, on the other hand, is a direct result of the firm's financing decision. In the context of selecting a proper financing mix, this risk applies to (1) the additional variability in earnings available to the firm's common shareholders and (2) the additional chance of insolvency borne by the common shareholder caused by the use of financial leverage.[6] **Financial leverage** means financing a portion of the firm's assets with securities bearing a fixed (limited) rate of return in hopes of increasing the ultimate return to the common shareholders. The decision to use debt or preferred shares in the financial structure of the corporation means that those who own the common shares of the firm are exposed to financial risk. Any given level of variability in EBIT will be magnified by the firm's use of financial leverage, and such additional variability will be embodied in the variability of earnings available to the common shareholder and earnings per share.[7] If these magnifications are negative, the com-

Table 8-1 The Concept of Business Risk

Business Risk Attribute	Example[a]
1. Sensitivity of the firm's product demand to general economic conditions	If GNP declines, does the firm's sales level decline by a greater percentage?
2. Degree of competition	Is the firm's market share small in comparison with other firms that produce and distribute the same product(s)?
3. Product diversification	Is a large proportion of the firm's sales revenue derived from a single major product or product line?
4. Operating leverage	Does the firm utilize a high level of operating leverage resulting in a high level of fixed costs?
5. Growth prospects	Are the firm's product markets expanding and (or) changing, making income estimates and prospects highly volatile?
6. Size	Does the firm suffer a competitive disadvantage due to lack of size in assets, sales, or profits that translates into (among other things) difficulty in tapping the capital market for funds?

[a]Affirmative responses indicate greater business risk exposure.

[5]Portfolio theory suggests that shareholders can diversify away much of the business risk through the choice of the assets in their portfolios.

[6]Note that the concept of financial risk used here differs from that used in our examination of cash and marketable securities management in Chapter 18.

[7]Sometimes a firm's management is burdened with financial problems because of its policy towards high financial risk, even though management may have a good track record. For example, the Campeau Corporation used huge amounts of debt to acquire Allied Stores and Federated Department Stores. This type of financial decision made life very uncomfortable for the operating side of this company.

mon shareholder has a higher chance of insolvency than would have existed had the use of fixed charge securities (debt and preferred shares) been avoided.

The closely related concepts of business and financial risk are crucial to the problem of financial structure design. This follows from the impact of these types of risk on the variability of the earnings stream flowing to the company's common shareholders. In the rest of this chapter we will study techniques that permit a precise assessment of the earnings stream variability caused by (1) operating leverage and (2) financial leverage. We have already defined financial leverage. Table 8-1 shows that the business risk of the enterprise is influenced by the use of what is called operating leverage. **Operating leverage** refers to the occurrence of fixed operating costs in the firm's income stream. To understand the nature and importance of operating leverage, we need to draw upon the basics of cost-volume-profit analysis or breakeven analysis.

Operating leverage

The responsiveness to sales changes of the firm's earnings before interest and taxes. This responsiveness arises from the firm's use of fixed operating costs.

PERSPECTIVE IN FINANCE

The breakeven analysis concepts presented in the next section are often covered in many of your other classes such as basic accounting principles and managerial economics. This just shows you how important and accepted this tool is within the realm of business decision making. The "Objective and Uses" section below identifies five typical uses of the breakeven model. You can probably add an application or two of your own. Hotels and motels, for instance, know exactly what their breakeven occupancy rate is. This breakeven occupancy rate gives them an operating target. This operating target, in turn, often becomes a crucial input to the hotel's advertising strategy. You may not want to become a financial manager—but you do want to understand how to compute breakeven points.

BREAKEVEN ANALYSIS

OBJECTIVE 2

The technique of breakeven analysis is familiar to legions of people in business. It is usefully applied in a wide array of business settings, including both small and large organizations. This tool is widely accepted by the business community for two reasons: it is based on straightforward assumptions and companies have found the information gained from the breakeven model beneficial in decision-making situations.

Objective and Uses

The objective of breakeven analysis is to determine the breakeven quantity of output by studying the relationships among the firm's cost structure, volume of output, and profit. Alternatively, the firm ascertains the breakeven level of sales dollars that corresponds to the breakeven quantity of output. We will develop the fundamental relationships by concentrating on units of output. Then, we will extend the procedure to calculate the breakeven sales level.

What is meant by the breakeven quantity of output? It is that quantity of output, denominated in units, that results in an EBIT level equal to zero. Use of the breakeven model, therefore, enables the financial officer (1) to determine the quantity of output that must be sold to cover all operating costs, as distinct from financial costs, and (2) to calculate the EBIT that will be achieved at various output levels.

There are many actual and potential applications of the breakeven approach. Some of these include:

1. **Capital expenditure analysis**. As a complementary technique to discounted cash flow evaluation models, the breakeven model locates in a rough way the sales volume needed to make a project economically beneficial to the firm. It should not be used to replace the time-adjusted evaluation techniques.
2. **Pricing policy**. The sales price of a new product can be set to achieve a target EBIT level. Furthermore, should market penetration be a prime objective, a price

could be set that would cover slightly more than the variable costs of production and provide only a partial contribution to the recovery of fixed costs. The negative EBIT at several possible sales prices can then be studied.

3. **Labour contract negotiations.** The effect of increased variable costs resulting from higher wages on the breakeven quantity of output can be analyzed.
4. **Cost structure**. The choice of reducing variable costs at the expense of incurring higher fixed costs can be evaluated. Management might decide to become more capital-intensive by performing tasks in the production process through use of equipment rather than labour. Application of the breakeven model can indicate what the effects of this tradeoff will be on the breakeven point for the given product.
5. **Financing decisions**. Analysis of the firm's cost structure will reveal the proportion that fixed operating costs bear to sales. If this proportion is high, the firm might reasonably decide not to add any fixed financing costs on top of the high fixed operating costs.

Essential Elements of the Breakeven Model

To implement the breakeven model, we must separate the production costs of the company into two mutually exclusive categories: fixed costs and variable costs. You will recall from your study of basic economics that in the long run all costs are variable. Breakeven analysis, therefore, is a short-run concept.

Assumed Behaviour of Costs

Fixed costs (indirect costs)

Costs that do not vary in total dollar amount as sales volume or quantity of output changes.

Fixed Costs **Fixed costs**, also referred to as **indirect costs**, do not vary in total amount as sales volume or the quantity of output changes over some relevant range of output.[8] Total fixed costs are independent of the quantity of product produced and equal to some constant dollar amount. As the production volume increases, the fixed cost per unit of product decreases because the fixed costs are being spread over larger and larger quantities of output. Figure 8-3 illustrates the behaviour of total fixed costs with respect to the company's relevant range of output. This total is shown to be unaffected by the quantity of product that is manufactured and sold. Over some other relevant output range, the amount of total fixed costs might be higher or lower for the same company.

Variable costs (direct costs)

Costs that are fixed per unit of output but vary in total as output changes.

In a manufacturing setting, some specific examples of fixed costs are:

1. Administrative salaries
2. Depreciation

Figure 8-3 Fixed Cost Behaviour over the Relevant Range of Output

Costs ($)
Fixed costs
0
Units produced and sold

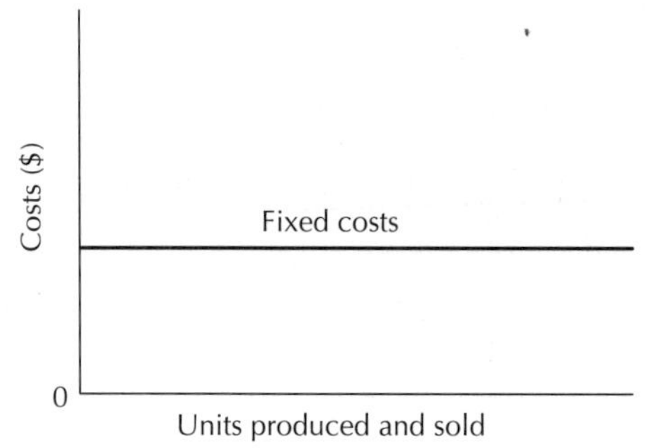

[8]A firm would normally produce several products in the same plant. As a result, some fixed costs will be allocated across a number of different end products. The method of assigning the fixed costs to the products will vary with company practice; typical approaches are to allocate on the basis of labour inputs, square footage used in the production process, or even by selling price.

Figure 8-4 Variable Cost Behaviour over the Relevant Range of Output

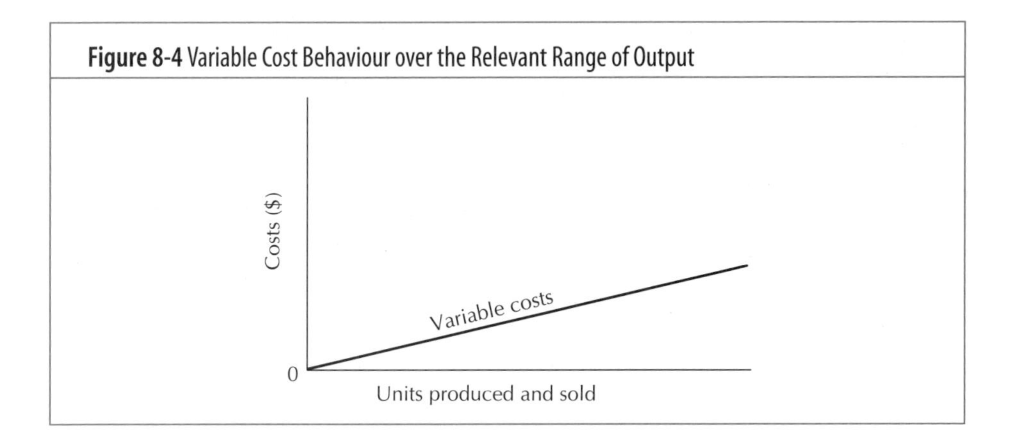

3. Insurance
4. Lump sums spent on intermittent advertising programs
5. Property taxes
6. Rent

Variable Costs Variable costs are sometimes referred to as **direct costs. Variable costs** are fixed per unit of output but vary in total as output changes. Total variable costs are computed by taking the variable cost per unit and multiplying it by the quantity produced and sold. The breakeven model assumes proportionality between total variable costs and sales. Thus, if sales rise by 10 percent, it is assumed that variable costs will rise by 10 percent. Figure 8-4 illustrates the behaviour of total variable costs with respect to the company's relevant range of output. Total variable costs are seen to depend on the quantity of product that is manufactured and sold. Notice that if zero units of the product are manufactured, then variable costs are zero but fixed costs are greater than zero. This implies that some contribution to the coverage of fixed costs occurs as long as the selling price per unit exceeds the variable cost per unit. This helps explain why some firms will operate a plant even when sales are temporarily depressed—that is, to provide some increment of revenue toward the coverage of fixed costs.

For a manufacturing operation, some examples of variable costs include:

1. Direct labour
2. Direct materials
3. Energy costs (fuel, electricity, natural gas) associated with the production area
4. Freight costs for products leaving the plant
5. Packaging
6. Sales commissions

More on Behaviour of Costs No one really believes that all costs behave as neatly as we have illustrated the fixed and variable costs in Figures 8-3 and 8-4. Nor does any law or accounting principle dictate that a certain element of the firm's total costs always be classified as fixed or variable. This will depend on each firm's specific circumstances. One firm's energy costs may be predominantly fixed, whereas another firm's energy costs may vary with output.[9]

Semivariable costs (semifixed costs)

Costs that exhibit the joint characteristics of both fixed and variable costs over different ranges of output.

Furthermore, some costs may be fixed for a while, then rise sharply to a higher level as a higher output is reached, remain fixed, and then rise again with further increases in production. Such costs may be termed either (1) **semivariable** or (2) **semifixed**. The label is your choice, since both are used in industrial practice. An

[9]In a greenhouse operation, where plants are grown (manufactured) under strictly controlled temperatures, heat costs will tend to be fixed whether the building is full or only half-full of seedlings. In a metal stamping operation, where levers are being produced, there is no need to heat the plant to as high a temperature when the machines are stopped and the workers are not there. In this latter case, the heat costs will tend to be variable.

Figure 8-5 Semivariable Cost Behaviour over the Relevant Range of Output

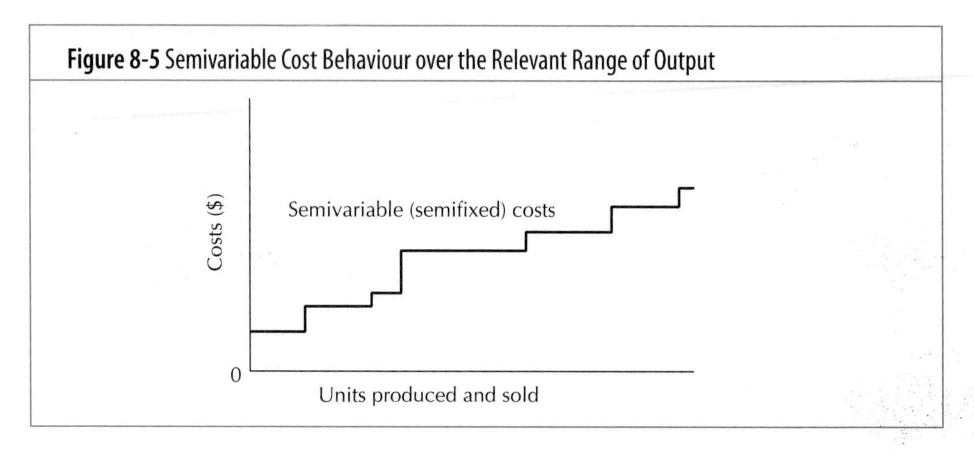

example might be the salaries paid production supervisors. Should output be cut back by 15 percent for a short period, the management of the organization is not likely to lay off 15 percent of the supervisors. Similarly, commissions paid to salespeople often follow a stepwise pattern over wide ranges of success. This sort of cost behaviour is shown in Figure 8-5.

To implement the breakeven model and deal with such a complex cost structure, the financial manager must (1) identify the most relevant output range for planning purposes and then (2) approximate the cost effect of semivariable items over this range by segregating a portion of them to fixed costs and a portion to variable costs. In an actual business application the analyst who deals with semivariable items will spend considerably more time allocating costs to fixed and variable categories than in carrying out the actual breakeven calculations.

Total Revenue and Volume of Output

Besides fixed and variable costs, the essential elements of the breakeven model include total revenue from sales and volume of output. **Total revenue** means sales dollars and is equal to the selling price per unit multiplied by the quantity sold. The **volume of output** refers to the firm's level of operations and may be indicated either as a unit quantity or as sales dollars.

Total revenue
Total sales dollars

Volume of output
The firm's level of operations expressed either in sales dollars or as units of output.

Finding the Breakeven Point

Finding the breakeven point in terms of units of production can be accomplished in several ways. All approaches require the essential elements of the breakeven model just described. The breakeven model is a simple adaptation of the firm's income statement expressed in the following analytical format:[10]

$$\text{sales} - (\text{total variable cost} + \text{total fixed cost}) = \text{profit} \qquad (8\text{-}1)$$

On a units of production basis, it is necessary to introduce (1) the price at which each unit is sold and (2) the variable cost per unit of output. Because the profit item studied in breakeven analysis is EBIT, we will use that acronym instead of the word "profit." In terms of units, the income statement shown in Equation (8-1) becomes the breakeven model by setting EBIT equal to zero:

$$\begin{pmatrix}\text{sales price}\\ \text{per unit}\end{pmatrix}\begin{pmatrix}\text{units}\\ \text{sold}\end{pmatrix} - \left[\begin{pmatrix}\text{variable cost}\\ \text{per unit}\end{pmatrix}\begin{pmatrix}\text{units}\\ \text{sold}\end{pmatrix} + \begin{pmatrix}\text{total fixed}\\ \text{cost}\end{pmatrix}\right] = \text{EBIT} = \$0 \qquad (8\text{-}2)$$

We must now determine the number of units that must be produced and sold in order to satisfy Equation (8-2)—that is, to arrive at an EBIT = \$0. This can be done by (1) trial-and-error analysis, (2) contribution margin analysis, or (3) algebraic analysis. Each approach will be illustrated using the same set of circumstances.

[10]Note: Sales − Variable costs = Contribution to overhead and profits.

The Problem Situation

OBJECTIVE 3

Even though the Pierce Grain Company manufactures several different products, the product mix for this company is rather constant if examined over a lengthy period. This allows management to conduct its financial planning by use of a "normal" sales price per unit and "normal" variable cost per unit. The "normal" sales price and variable cost per unit are calculated from the constant product mix. The assumption for this calculation is that the product mix is considered as one product. The selling price is $10 and the variable cost is $6. Total fixed costs for the firm are $100,000 per year. What is the breakeven point in units produced and sold for the company during the coming year?

Trial-and-Error Analysis

The most cumbersome approach to determining the firm's breakeven point is to employ the trial-and-error technique illustrated in Table 8-2. The process simply involves the arbitrary selection of an output level and the calculation of a corresponding EBIT amount. When the level of output is found that results in an EBIT = $0, the breakeven point has been located. Notice that Table 8-2 is just Equation (8-2) in worksheet form. For the Pierce Grain Company, total operating costs will be covered when 25,000 units are manufactured and sold. This tells us that if sales equal $250,000, the firm's EBIT will equal $0.

Contribution Margin Analysis

Unlike trial and error, use of the contribution margin technique permits direct computation of the breakeven quantity of output. The **contribution margin** is the difference between the unit selling price and unit variable costs, as follows:

Contribution margin

Unit sales price minus unit variable cost.

Unit sales price
− Unit variable cost
= Unit contribution margin

The use of the word "contribution" in the present context means contribution to the coverage of fixed operating costs. For the Pierce Grain Company, the unit contribution margin is

Unit sales price	$10
Unit variable cost	−6
Unit contribution margin	$ 4

Table 8-2 The Pierce Grain Company Sales, Cost, and Profit Schedule

	(1) Units Sold	(2) Unit Sales Price	(3)=(1)×(2) Sales	(4) Unit Variable Cost	(5)=(1)×(4) Total Variable Cost	(6) Total Fixed Cost	(7)=(5)+(6) Total Cost	(8)=(3)−(7) EBIT
1.	10,000	$10	$100,000	$6	$ 60,000	$100,000	$160,000	−$60,000
2.	15,000	10	150,000	6	90,000	100,000	190,000	−40,000
3.	20,000	10	200,000	6	120,000	100,000	220,000	−20,000
4.	25,000	10	250,000	6	150,000	100,000	250,000	0
5.	30,000	10	300,000	6	180,000	100,000	280,000	20,000
6.	35,000	10	350,000	6	210,000	100,000	310,000	40,000

Input Data	Output Data
Unit sales price = $10	Breakeven point in units = 25,000 units produced and sold
Unit variable cost = $6	Breakeven point in sales = $250,000
Total fixed cost = $100,000	

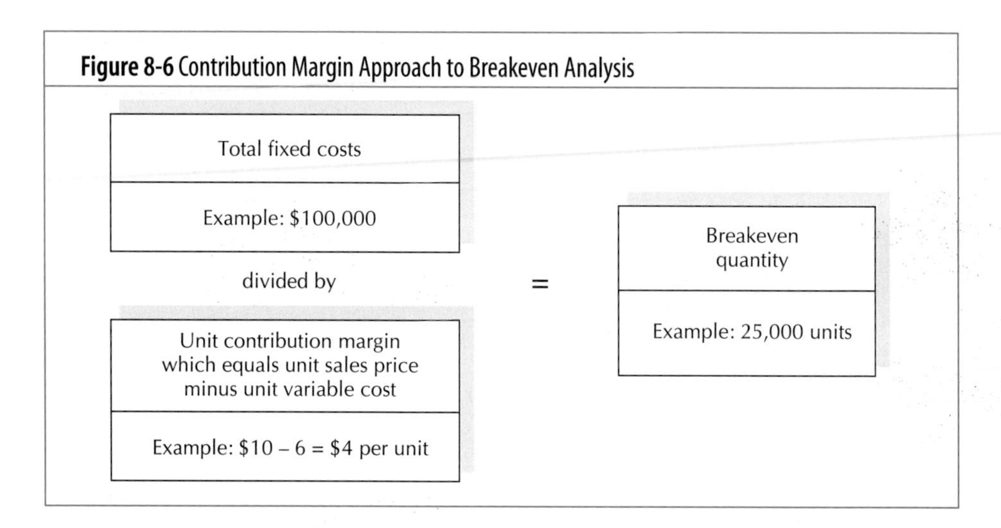

Figure 8-6 Contribution Margin Approach to Breakeven Analysis

If the annual fixed costs of $100,000 are divided by the unit contribution margin of $4, we find the breakeven quantity of output for Pierce Grain is 25,000 units. With much less effort, we have arrived at the identical result found by trial and error. Figure 8-6 portrays the contribution margin technique for finding the breakeven point.

Algebraic Analysis

To explain the algebraic method for finding the breakeven output level we need to adopt the following notation. Let

Q = the number of units sold
Q_B = the breakeven level of Q
P = the unit sales price
F = total fixed costs anticipated over the planning period
V = the unit variable cost

Equation (8-2), the breakeven model, is repeated below as Equation (8-2a) with the model symbols used in place of words. The breakeven model is then solved for Q, the number of units that must be sold in order that EBIT will equal $0. We label the breakeven point quantity Q_B.

$$(P \cdot Q) - [(V \cdot Q) + F] = EBIT = \$0 \quad \text{(8-2a)}$$
$$(P \cdot Q) - (V \cdot Q) - F = \$0$$
$$Q(P - V) = F$$
$$Q_B = \frac{F}{P-V} \quad \text{(8-3)}$$

Observe that Equation (8-3) says: divide total fixed operating costs, F, by the unit contribution margin, $P - V$, and the breakeven level of output, Q_B, will be obtained. The contribution margin analysis is nothing more than Equation (8-3) in different garb.

Application of Equation (8-3) permits direct calculation of Pierce Grain's breakeven point, as follows:

$$Q_B = \frac{F}{P - V} = \frac{\$100{,}000}{\$10 - \$6} = 25{,}000 \text{ units}$$

The Breakeven Point in Sales Dollars

In dealing with the multi-product firm, it is convenient to compute the breakeven point in terms of sales dollars rather than units of output.[11] Sales, in effect, become

[11]This approach is also useful if we relax one of the basic assumptions of the breakeven analysis and allow price to vary over some output range; in this form, price is a function of the volume produced and assumed to be sold.

a common denominator associated with a particular product mix. Furthermore, an outside analyst may not have access to internal unit cost data. He or she may, however, be able to obtain annual reports for the firm. If the analyst can separate the firm's total costs as identified from its annual reports into their fixed and variable components, he or she can calculate a general breakeven point in sales dollars.

We will illustrate the procedure using the Pierce Grain Company's cost structure, contained in Table 8-2. Suppose that the information on line 5 of Table 8-2 is arranged in the format shown in Table 8-3. We will refer to this type of financial statement as an **analytical income statement**. This distinguishes it from audited income statements published, for example, in the annual reports of public corporations. If we are aware of the simple mathematical relationships on which cost-volume-profit analysis is based, we can use Table 8-3 to find the breakeven point in sales dollars for the Pierce Grain Company.

Analytical income statement

A financial statement used by internal analysts that differs in composition from audited or published financial statements.

First, let us explore the logic of the process. Recall from Equation (8-1) that

$$\text{sales} - (\text{total variable cost} + \text{total fixed cost}) = \text{EBIT}$$

If we let total sales = S, total variable cost = VC, and total fixed cost = F, the above relationship becomes

$$S - (VC + F) = \text{EBIT}$$

Since variable cost per unit of output and selling price per unit are assumed constant over the relevant output range in breakeven analysis, the ratio of total sales to total variable cost, VC/S, is a constant for any level of sales. This permits us to rewrite the previous expression as

$$S - \left[\left(\frac{VC}{S}\right)S\right] - F = \text{EBIT}$$

and

$$S\left(1 - \frac{VC}{S}\right) - F = \text{EBIT}$$

At the breakeven point, however, EBIT = 0, and the corresponding break-even level of sales can be represented as S^*. At the breakeven level of sales, we have

$$S^*\left(1 - \frac{VC}{S}\right) - F = 0$$

or

$$S^*(1 - \frac{VC}{S}) = F$$

Therefore,

$$S^* = \frac{F}{1 - \dfrac{VC}{S}} \qquad (8\text{-}4)$$

The application of Equation (8-4) to Pierce Grain's analytical income statement[12] in Table 8-3 permits the breakeven sales level for the firm to be directly computed, as follows:

Table 8-3 The Pierce Grain Company Analytical Income Statement

Sales	$300,000
Less: Total variable costs	180,000
Revenue before fixed costs	$120,000
Less: Total fixed costs	100,000
EBIT	$ 20,000

[12]The derivation of Equation (8-4) requires that we assume VC/S is constant over the relevant output range.

$$S^* = \frac{\$100{,}000}{1 - \frac{\$180{,}000}{\$300{,}000}}$$

$$= \frac{\$100{,}000}{1 - .60} = \$250{,}000$$

Notice that this is indeed the same breakeven sales level for Pierce Grain that is indicated on line 4 of Table 8-2.

Graphic Representation, Analysis of Input Changes, and Cash Breakeven Point

In making a presentation to management, it is often effective to display the firm's cost-volume-profit relationships in the form of a chart. Even those individuals who truly enjoy analyzing financial problems find figures and equations dry material at times. Furthermore, by quickly scanning the basic breakeven chart, the manager can approximate the EBIT amount that will prevail at different sales levels.

Such a chart has been prepared for the Pierce Grain Company. Figure 8-7 has been constructed for this firm using the input data contained in Table 8-2. Total fixed costs of $100,000 are added to the total variable costs associated with each production level to form the total costs line. When 25,000 units of product are manufactured and sold, the sales line and total costs line intersect. This means, of course, the EBIT that would exist at that volume of output is zero. Beyond 25,000 units of output, notice that sales revenues exceed the total costs line. This causes a positive EBIT. This positive EBIT or profits is labelled "original EBIT" in Figure 8-7.

The unencumbered nature of the breakeven model makes it possible to quickly incorporate changes in the requisite input data and generate the revised output. Suppose a favourable combination of events causes Pierce Grain's fixed costs to decrease by $25,000. This would put total fixed costs for the planning period at a level of $75,000 rather than the $100,000 originally forecast. Total costs, being the sum of fixed and variable costs, would be lower by $25,000 at all output levels. The revised

Figure 8-7 The Pierce Grain Company's Breakeven Chart

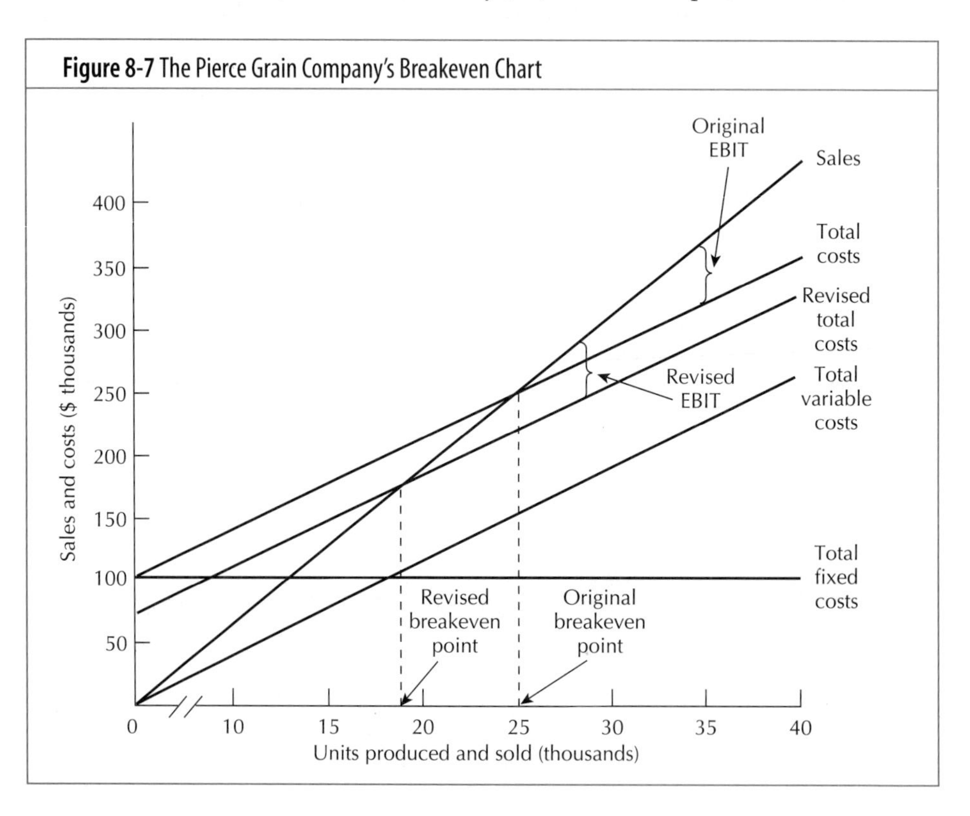

total costs line in Figure 8-7 reflects Pierce Grain's reduction in fixed costs. Under these revised conditions, the new breakeven point in units would be as follows:

$$Q_B = \frac{\$75{,}000}{\$10 - \$6} = 18{,}750 \text{ units}$$

The revised breakeven point of 18,750 units is identified in Figure 8-7, along with the revised EBIT amounts that would prevail at differing output and sales levels. The chart clearly indicates that at any specific production and sales level, the revised EBIT would exceed the original EBIT. This must be the case, as the revised total costs line lies below the original total costs line over the entire relevant output range. The effect on the breakeven point caused by other changes in (1) the cost structure or (2) the pricing policy can be analyzed in a similar fashion.

The data in Figure 8-7 can be used to demonstrate another version of basic cost-volume-profit analysis. This can be called **cash breakeven analysis**. If the company's fixed or variable cost estimates contain allowance for any noncash expenses, then the resultant breakeven point is higher on an accounting profit basis than on a cash basis. This means the firm's production and sales levels do not have to be as great to cover the cash costs of manufacturing the product.

Cash breakeven analysis

A variation from traditional breakeven analysis that removes (deducts) noncash from the cost items.

What are these noncash expenses? The largest and most significant is depreciation expense. Another category is prepaid expenses. Insurance policies are at times paid to cover a three-year cycle. Thus, the time period for which the breakeven analysis is being performed might not involve an actual cash outlay for insurance coverage.

For purposes of illustration, assume that noncash expenses for Pierce Grain amount to $25,000 over the planning period and that all these costs are fixed. We can compare the revised total costs line in Figure 8-7, which implicitly assumes a lower fixed cash cost line, with the sales revenue line to find the cash breakeven point. Provided Pierce Grain can produce and sell 18,750 units over the planning horizon, revenues from sales will be equal to cash operating costs.

PERSPECTIVE IN FINANCE

The "limitations" of the breakeven models that we have examined are really the underlying assumptions of the models. All models rest upon some set of assumptions and knowing those assumptions is requisite to effective application of the technique. This does not mean the tool is of no value in decision making; it only means that you need to be aware of exactly what you are doing when you use the tool to make a business decision. If you can explain the assumptions to someone else, you can effectively use the technique.

Limitations of Breakeven Analysis

Earlier we identified some of the applications of breakeven analysis. This technique is a useful tool in many settings. It must be emphasized, however, that breakeven analysis provides a beneficial guide to managerial action, not the final answer. The use of cost-volume-profit analysis has limitations, which should be kept in mind.[13] These include the following:

1. The cost-volume-profit relationship is assumed to be linear. This is realistic only over narrow ranges of output.
2. The total revenue curve (sales curve) is presumed to increase linearly with the volume of output. This implies any quantity can be sold over the relevant output range at that single price. To be more realistic, it is necessary in many situations to compute several sales curves and corresponding breakeven points at differing prices.

[13]The utilities industry and the hotel and lodging industry are examples of industries in which the breakeven analysis is likely to be very useful because of the stable cost and pricing structure. In contrast, the computer industry and related software industry are examples of industries which do not show long-term stability in their cost and pricing structure.

3. A constant production and sales mix is assumed. Should the company decide to produce more of one product and less of another, a new breakeven point would have to be found. Only if the variable cost-to-sales ratios were identical for products involved would the new calculation be unnecessary.
4. The breakeven chart and the breakeven computation are static forms of analysis. Any alteration in the firm's cost or price structure dictates that a new breakeven point be calculated. Breakeven analysis is more helpful, therefore, in stable industries than in dynamic ones.
5. From the perspective of finance, the analysis of accounting costs does not consider the timing and the riskiness of cash outflows made by the firm. For example, how much should the business report costs for depreciation, given the timing difference between the accounting methods of depreciation and the methods used for the capital cost allowance system.
6. The breakeven point occurs when the level of EBIT equals zero. The implication of the breakeven point is that there are no funds available to service interest charges. In reality the firm would be considered insolvent and creditors could paralyse the firm's operations by placing the firm in receivership. Thus the notions of breaking-even or "just making it" are not really reassuring when operating income is zero.

OBJECTIVE 4

OPERATING LEVERAGE

Operating leverage occurs if fixed operating costs are present in the firm's cost structure. Fixed operating costs do not include interest charges incurred from the firm's use of debt financing. Those costs will be incorporated into the analysis when financial leverage is discussed.

So operating leverage arises from the firm's use of fixed operating costs. But what is operating leverage? **Operating leverage** is the responsiveness of the firm's EBIT to fluctuations in sales. By continuing to draw upon our data for the Pierce Grain Company, we can illustrate the concept of operating leverage. Table 8-4 contains data for a case in which a possible fluctuation occurs in the firm's sales level. It is assumed here that Pierce Grain is currently operating at an annual sales level of $300,000. This is referred to in the tabulation as the base sales level at t (time period zero). The question is: How will Pierce Grain's EBIT level respond to a positive 20 percent change in sales? A sales volume of $360,000, referred to as the forecast sales level at $t + 1$, reflects the 20 percent sales rise anticipated over the planning period. Assume that the planning period is one year.

Operating leverage

The responsiveness to sales changes of the firm's earnings before interest and taxes. This responsiveness arises from the firm's use of fixed operating costs.

Operating leverage relationships are derived within the mathematical assumptions of cost-volume-profit analysis. In the present example, this means that Pierce Grain's variable cost-to-sales ratio of .6 will continue to hold during time period $t +$ 1, and the fixed costs will hold steady at $100,000.

Given the forecasted sales level for Pierce Grain and its cost structure, we can measure the responsiveness of EBIT to the upswing in volume. Notice in Table 8-4 that EBIT is expected to be $44,000 at the end of the planning period. The percentage change in EBIT from t to $t + 1$ can be measured as follows:

Table 8-4 The Concept of Operating Leverage: An Increase in Pierce Grain Company Sales

Item	Base Sales Level, t	Forecast Sales Level, $t + 1$
Sales	$300,000	$360,000
Less: Total variable costs	180,000	216,000
Revenue before fixed costs	$120,000	$144,000
Less: Total fixed costs	100,000	100,000
EBIT	$ 20,000	$ 44,000

$$\text{percentage change in EBIT} = \frac{\$44{,}000_{t+1} - \$20{,}000_t}{\$20{,}000_t}$$

$$= \frac{\$24{,}000}{\$20{,}000}$$

$$= 120\%$$

We know that the projected fluctuation in sales amounts to 20 percent of the base period, t, sales level. This is verified below:

$$\text{percentage change in sales} = \frac{\$360{,}000_{t+1} - \$300{,}000_t}{\$300{,}000_t}$$

$$= \frac{\$60{,}000}{\$300{,}000}$$

$$= 20\%$$

OBJECTIVE 5

By relating the percentage fluctuation in EBIT to the percentage fluctuation in sales, we can calculate a specific measure of operating leverage.[14] Thus, we have

$$\begin{array}{c}\text{degree of operating leverage}\\ \text{from the base sales level(s)}\end{array} = DOL_s = \frac{\text{percentage change in EBIT}}{\text{percentage change in sales}} \quad (8\text{-}5)$$

Applying Equation (8-5) to our Pierce Grain data gives

$$DOL_{\$300{,}000} = \frac{120\%}{20\%} = 6 \text{ times}$$

Unless we understand what the specific measure of operating leverage tells us, the fact that we may know it is equal to six times is nothing more than sterile information. For Pierce Grain, the inference is that for any percentage fluctuation in sales from the base level, the percentage fluctuation in EBIT will be six times as great. If Pierce Grain expected only a 5 percent rise in sales over the coming period, a 30 percent rise in EBIT would be anticipated as follows:

$$(\text{percentage change in sales}) \times (DOL_s) = \text{percentage change in EBIT}$$
$$(5\%) \times (6) = 30\%$$

We will now return to the postulated 20 percent change in sales. What if the direction of the fluctuation is expected to be negative rather than positive? What is in store for Pierce Grain? Unfortunately for Pierce Grain, but fortunately for the analytical process, we will see that the operating leverage measure holds in the negative direction as well. This situation is displayed in Table 8-5.

At the $240,000 sales level, which represents the 20 percent decrease from the base period, Pierce Grain's EBIT is expected to be –$4,000. How sensitive is EBIT to this sales change? The magnitude of the EBIT fluctuation is calculated as[15]

Table 8-5 The Concept of Operating Leverage: A Decrease in Pierce Grain Company Sales

Item	Base Sales Level, t	Forecast Sales Level, $t + 1$
Sales	$300,000	$240,000
Less: Total variable costs	180,000	144,000
Revenue before fixed costs	$120,000	$ 96,000
Less: Total fixed costs	100,000	100,000
EBIT	$ 20,000	$ –4,000

[14]The degree of operating leverage (DOL) is an elasticity equation which has the general form:

$$\text{Elasticity coefficient} = \frac{\%\text{ change in some dependent variable}}{\%\text{ change in some independent variable}}$$

[15]Some students have conceptual difficulty in computing these percentage changes when negative amounts are involved. Notice by inspection in Table 8-5 that the difference between an EBIT amount of +$20,000 at t and –$4,000 at $t + 1$ is –$24,000.

$$\text{percentage change in EBIT} = \frac{-\$4{,}000_{t+1} - \$20{,}000_t}{\$20{,}000_t}$$

$$= \frac{-\$24{,}000}{\$20{,}000}$$

$$= -120\%$$

Making use of our knowledge that the sales change was equal to –20 percent permits us to compute the specific measure of operating leverage as

$$DOL_{\$300,000} = \frac{-120\%}{-20\%} = 6 \text{ times}$$

What we have seen, then, is that the degree of operating leverage measure works in the positive or negative direction. A negative change in production volume and sales can be magnified severalfold when the effect on EBIT is calculated.

To this point our calculations of the degree of operating leverage have required two analytical income statements: one for the base period and a second for the subsequent period that incorporates the possible sales alteration. This cumbersome process can be simplified. If unit cost data are available to the financial manager, the relationship can be expressed directly in the following manner:

$$DOL_s = \frac{Q(P - V)}{Q(P - V) - F} \tag{8-6}$$

Observe in Equation (8-6) that the variables were all previously defined in our algebraic analysis of the breakeven model. Recall that Pierce sells its product at $10 per unit, the unit variable cost is $6, and total fixed costs over the planning horizon are $100,000. Still assuming that Pierce is operating at a $300,000 sales volume, which means output (Q) is 30,000 units, we can find the degree of operating leverage by application of Equation (8-6):

$$DOL_{\$300,000} = \frac{30{,}000(\$10 - \$6)}{30{,}000(\$10 - \$6) - \$100{,}000} = \frac{\$120{,}000}{\$20{,}000} = 6 \text{ times}$$

Whereas Equation (8-6) requires us to know unit cost data to carry out the computations, the next formulation we examine does not. If we have an analytical income statement for the base period, then Equation (8-7) can be employed to find the firm's degree of operating leverage:

$$DOL_s = \frac{\text{revenue before fixed costs}}{EBIT} = \frac{S - VC}{S - VC - F} \tag{8-7}$$

Use of Equation (8-7) in conjunction with the base period data for Pierce Grain shown in either Table 8-4 or 8-5 gives

$$DOL_{\$300,000} = \frac{\$120{,}000}{\$20{,}000} = 6 \text{ times}$$

The three versions of the operating leverage measure all produce the same result. Data availability will sometimes dictate which formulation can be applied. The crucial consideration, though, is that you grasp what the measurement tells you. For Pierce Grain, a 1 percent change in sales will produce a 6 percent change in EBIT.[16]

PERSPECTIVE IN FINANCE

Before we complete our discussion of operating leverage and move on to the subject of financial leverage, ask yourself "which type of leverage is more under the control of management?" You will probably (and correctly) come to the conclusion that the firm's managers have less control over its operating cost structure and almost complete control over its financial structure. What the firm actually produces, for example, will determine to a significant degree the division between fixed and variable costs. There is more room for substitution among the various sources of financial capital than there is among the labour and real capital inputs that enable the firm to meet its production requirements. Thus, you can anticipate more arguments over the choice to use a given degree of financial leverage than the corresponding choice over operating leverage use.

[16]Note that each of the underlying parameters can be assessed in terms of its effect on DOL.

Table 8-6 The Pierce Grain Company Degree of Operating Leverage Relative to Different Sales Bases

Units Produced and Sold	Sales Dollars	DOL_S
25,000	$ 250,000	Undefined
30,000	300,000	6.00
35,000	350,000	3.50
40,000	400,000	2.67
45,000	450,000	2.25
50,000	500,000	2.00
75,000	750,000	1.50
100,000	1,000,000	1.33

Implications

As the firm's scale of operations moves in a favourable manner above the break-even point, the degree of operating leverage at each subsequent (higher) sales base will decline. In short, the greater the sales level, the lower the degree of operating leverage. This is demonstrated in Table 8-6 for the Pierce Grain Company. At the breakeven sales level for Pierce Grain, the degree of operating leverage is undefined, since the denominator in any of the computational formulas is zero. Notice that beyond the breakeven point of 25,000 units, the degree of operating leverage declines. It will decline at a decreasing rate and asymptotically approach a value of 1.00. As long as some fixed operating costs are present in the firm's cost structure, however, operating leverage exists, and the degree of operating leverage (DOL_S) will exceed 1.00. Operating leverage is present, then, whenever the firm faces the following situation:

$$\frac{\text{percentage change in EBIT}}{\text{percentage change in sales}} > 1.00$$

The data in Table 8-6 are presented in graphic form in Figure 8-8.

Figure 8-8 The Pierce Grain Company Degree of Operating Leverage Relative to Different Sales Bases

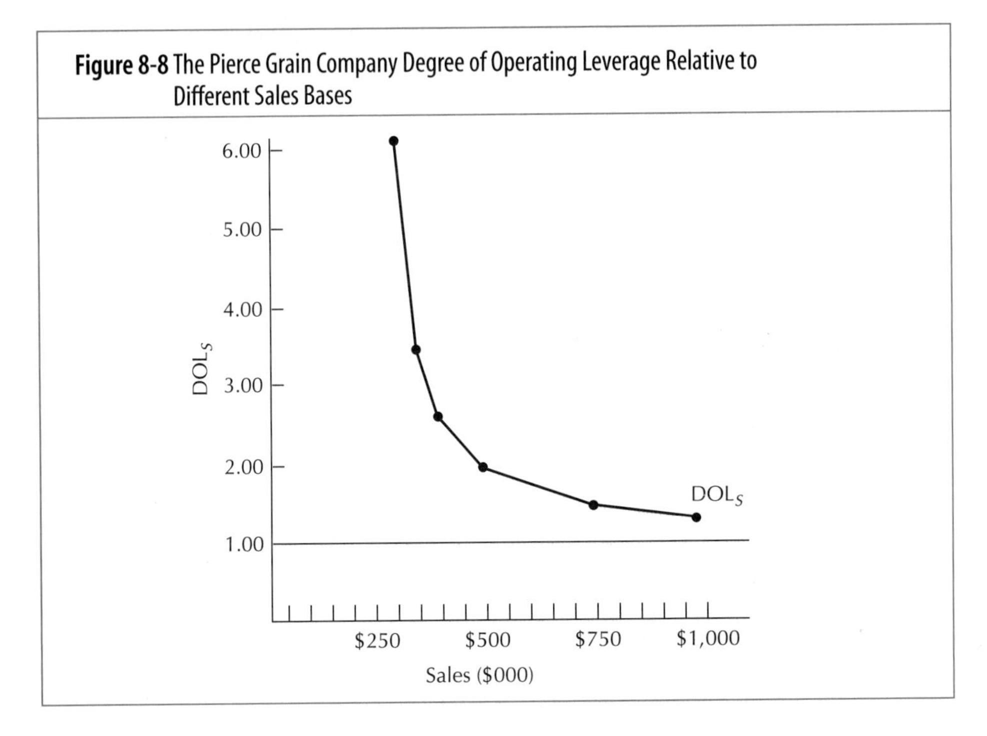

Using Operating Leverage to Increase Small Business Profits

In a manufacturing company, two types of leverage generally exist: operating leverage and financial leverage.

Operating leverage or "first-level leverage" measures the effect of a sales variation on the net operating income. With operating leverage, the percentage change in the operating income exceeds that of sales volume.

Financial leverage or "second-level leverage" measures the effect of a variation of the net operating income on the earnings per share.

In the following discussion, we are dealing mainly with operating leverage. If we ignore financial leverage, it is simply because it has more limited use in smaller companies than in larger companies. With regard to the debt-equity ratio, it is generally the lenders rather than the owners who decide on the financial ratio to observe and the latter's room to manoeuvre is therefore fairly limited.

How to Visualize Your Operating Leverage

The manager of a small or medium-sized business can realize a number of advantages from the proper interpretation of his or her company's break-even point (B/P) graph.

For example, let us see what information can be obtained from analyzing and comparing the graphs of Companies A, B, C, and D (see Table 1). First of all, if we use their sales as a yardstick, we have to take into consideration that we are comparing companies of identical size.

We can easily see that the B/P and sales of Companies A, B, C are identical ($750,000 and $1 million) but the profits are 5%, 10%, and 15% due solely to the operating leverage.

A second, just as important observation is the increase in operating risk for companies whose fixed costs are higher. For example, if the forecast sales of $1 million are reduced to $600,000, Company A will lose $30,000, Company B, $60,000, and Company C, $90,000.

Taking a close look at Company D's graph, we can see that the company will record operating losses if its revenues fall below $874,300. And here we have to carry our line of reasoning further, as this company has three types of revenue which produce different profit ratios. The $874,300 break-even point is determined from an average profit of 45.75% when Product 1 accounts for 80% of revenues, Product 2 for 15%, and Product 3 for 5%. Since the profit ratios are quite different, it becomes obvious that any change in the composition or mix of revenues will have an effect on the break-even point.

On the basis of these preliminary observations, we can already conclude that a principal advantage of using a break-even point graph for a small or medium-sized business is that it enables it to set objectives that will guarantee the best profitability possible, given the company's business plan (its resources, size, industry segment, etc.). The B/P graph also allows us to forecast the important controls on which the business will have to focus. For example:

- Company A will first have to control its variable costs well.
- Company B will also have to monitor its fixed costs.
- Companies C and D will have to control their fixed costs regularly and also ensure that their sales objectives are met, in view of the much greater leverage and risk for these two types of companies.

By pursuing this line of reasoning, we can deduce that a work stoppage in Companies C and D will be more costly than in Companies A and B.

And what about incentives the small and medium-sized business manager can use to have a positive impact on the attainment of a company's objectives? Bonuses, premiums, incentives, various types of participation, etc. obtainable from the break-even point, will certainly be welcome in Companies C and D if emphasis is placed on profitability.

Industry Segment Image

The size of the small or medium-sized business apart, we can also identify our business with its industrial segment and establish a certain relationship between

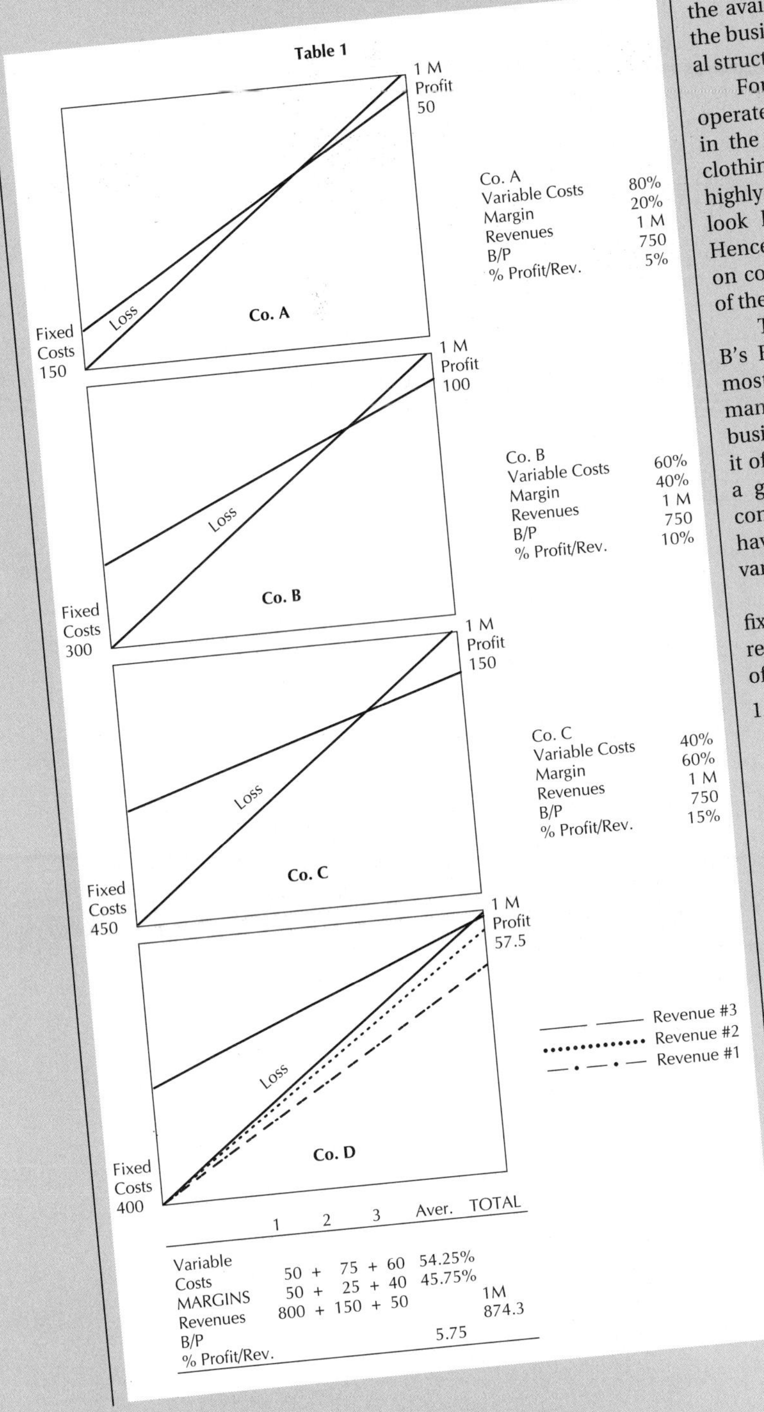

the available external ratios and the business's own organizational structure.

For example, if my company operates as a sub-contractor in the automobile parts, wood, clothing, or other industries, it is highly probable that my B/P will look like Company A's graph. Hence, I will want to concentrate on controlling my costs, in view of the marginal profitability.

The structure of Company B's B/P is the same as that of most small and medium-sized manufacturing businesses. All businesses that have a gross profit of between 20% and 30% have a graph that is similar to this company's B/P. They therefore have to monitor their fixed and variable costs constantly.

Companies C and D have a fixed cost graph line that can represent at least four different types of businesses:

1. Businesses that are high in capital investments (fixed assets) representing high maintenance, insurance, financing, and depreciation costs.
2. Businesses that are heavy on human resources with fixed compensation (salesmen on salary, engineers, appraisers, specialists, stylists, etc.).
3. Companies that specialize in research and development.
4. Small and medium-sized businesses that take care of their own distribution to individuals, which increases their distribution and administration costs.

These businesses will have to keep close track of changes in their fixed costs and place the emphasis on achieving sales and profitability objectives. Of course, these categories of businesses also need to stress "quality". The "marketing" approach should help them to absorb high fixed costs more quickly.

How to Ensure Follow-up

The best method to employ to ensure follow-up between the B/P graph and reality is undoubtedly the variable costing (direct costing) method, simply because this method draws its inspiration from the same line of reasoning that we apply when we wish to draw a B/P graph.

From the standpoint of affordability, however, the small manufacturing business will use a very different approach from that followed by a medium-sized business. The accountant in a small business can hardly go into the amount of detail and will not have the analyses and related reports available as that of a more developed organization.

Small Business

To determine the profit ratio obtained and the burden of fixed costs from period to period, the small manufacturing business will have to have an accounting plan that is comparable to direct costing. This plan must make it possible to draw up interim statements as required that will unambiguously show the profits achieved by revenue category (product or customer) and the corresponding fixed cost burden.

It will obviously be easier for a small business that mass produces to use standards for determining its costs and evaluate regularly. The small business that does custom work needs to make controlling its labour for each order manufactured a priority (e.g., printing).

Given the personnel constraints in small business, it is not realistic to require too elaborate reports from the person in charge of the accounting information. He or she generally has to support the employer in other areas of daily management and obviously has less time for accounting duties. The production of interim statements, accompanied by a profit analysis (either per order, per product, or per customer), should allow the business to achieve acceptable follow-up with the forecast objectives. Data processing is now increasingly accessible to this type of business.

Medium-Sized Business

We are referring here to a manufacturing business that is generally better structured. This type of business often has experienced managers, each with a clearly defined role. They generally use data processing in their management.

In addition to regular interim financial statements, the business comptroller or management accountant will be able to provide additional information related to the statements. The directors will be able to know the hourly contributions obtained either by product, customer or machine, depending on the type of operation of the production equipment.

This information will obviously have to be interpreted correctly in terms of effective and efficient management.

When the product is for current consumption, the cycle of the products will have to be taken into account before drawing conclusions about the profit ratio obtained (principally if the ratio is low). The decision will depend on the answer to the question: "Are we at the beginning or the end of the cycle?"

The use of flexible budgets in mass production plants will take on its full significance. The explanation of the variances will become a variable aid for decision-making. The medium-sized business that produces by order and by bid will profit from knowing the variations between the actual and the estimated.

Omnipresent Tools

In addition to the close relationship between the B/P graph and the variable costs (direct costing) graph, they have another dimension in common. Both are highly recommended for short-term management.

With or without tariff barriers, with or without free trade, these tools will always be with us because they are based on behavior of the costs (fixed or variable) from which a B/P chart can be drawn and our business's leverage can be visualized.

According to a National Association of Accountants study, the use of the direct costing method (variable costs) has contributed to the making of better decisions. In reality, it is only an extension of the B/P chart into day-to-day management.

Source: Adapted from C. Lortie, "Using operating leverage to increase small business profits," *CMA Management Accounting Magazine* 63(9) (November 1989), pp. 32–34. Reprinted from *CMA* magazine by permission of the Society of Management Accountants of Canada.

The greater the firm's degree of operating leverage, the more its profits will vary with a given percentage change in sales. Thus, operating leverage is definitely an attribute of the business risk that confronts the company. From Table 8-6 and Figure 8-8 we have seen that the degree of operating leverage falls as sales increase past the firm's breakeven point. The sheer size and operating profitability of the firm, therefore, affect and can lessen its business risk exposure.

A manager who is considering an alteration in the firm's cost structure will benefit from an understanding of the operating leverage concept. It might be possible to replace part of the labour force with capital equipment (machinery). A possible result is an increase in fixed costs associated with the new machinery and a reduction in variable costs attributable to a lower labour bill. This conceivably could raise the firm's degree of operating leverage at a specific sales base. If the prospects for future sales increases are high, then increasing the degree of operating leverage might be a prudent decision. The opposite conclusion will be reached if sales prospects are unattractive.

PERSPECTIVE IN FINANCE

As you are introduced to the topic of "financial leverage" keep in mind that this is one of the most crucial policy areas on which a financial executive spends his or her time. We describe and measure here what happens to the firm's earnings per share when financial risk is assumed. Try to understand this effect. We demonstrate how to actually measure this effect in the next section. By now you should be realizing that variability of all types—be it in an earnings stream or in stock returns—is a central element of financial thought and the practice of financial management.

FINANCIAL LEVERAGE

OBJECTIVE 4

We have defined **financial leverage** as the practice of financing a portion of the firm's assets with securities bearing a fixed rate of return in hope of increasing the ultimate return to the common shareholders. In the present discussion we focus on the responsiveness of the company's earnings per share to changes in its EBIT. For the time being, then, the return to the common shareholder being concentrated

Financial leverage

The use of securities bearing a fixed (limited) rate of return to finance a portion of the firm's fixed assets. Financial leverage can arise from the use of either debt or preferred share financing. The use of financial leverage exposes the firm to financial risk.

Table 8-7 The Pierce Grain Company Possible Financial Structures

Plan A: 0% debt			
		Total debt	$ 0
		Common equity	200,000 [a]
Total assets	$200,000	Total liabilities and equity	$200,000
Plan B: 25% debt at 8% interest rate			
		Total debt	$ 50,000
		Common equity	150,000 [b]
Total assets	$200,000	Total liabilities and equity	$200,000
Plan C: 40% debt at 8% interest rate			
		Total debt	$ 80,000
		Common equity	120,000 [c]
Total assets	$200,000	Total liabilities and equity	$200,000

[a] 2,000 common shares outstanding
[b] 1,500 common shares outstanding
[c] 1,200 common shares outstanding

upon is earnings per share. We are not saying that earnings per share is the appropriate criterion for all financing decisions. In fact, the weakness of such a contention will be examined in the next chapter. The effect of financial leverage on the firm earnings can be examined by concentrating on an earnings per share criterion.

Let us assume that the Pierce Grain Company is in the process of getting started. The firm's potential owners have calculated that $200,000 is needed to purchase the necessary assets to conduct the business. Three possible financing plans have been identified for raising the $200,000; they are presented in Table 8-7. In plan A no financial risk is assumed: the entire $200,000 is raised by selling 2,000 common shares. In plan B a moderate amount of financial risk is assumed: 25 percent of the assets are financed with a debt issue that carries an 8 percent annual interest rate. Plan C would use the most financial leverage: 40 percent of the assets would be financed with a debt issue costing 8 percent.[17]

Table 8-8 presents the impact of financial leverage on earnings per share associated with each fund-raising alternative. If EBIT should increase from $20,000 to $40,000, then earnings per share would rise by 100 percent under plan A. The same positive fluctuation in EBIT would occasion an earnings per share rise of 125 percent under plan B, and 147 percent under plan C. In plans B and C the 100 percent increase in EBIT (from $20,000 to $40,000) is magnified to a greater than 100 percent increase in earnings per share. The firm is employing financial leverage, and exposing its owners to financial risk, when the following situation exists:

$$\frac{\text{percentage change in earnings per share}}{\text{percentage change in EBIT}} > 1.00$$

Table 8-8 The Pierce Grain Company Analysis of Financial Leverage at Different EBIT Levels

(1) EBIT	(2) Interest	(3) = (1) – (2) EBT	(4) = (3) × .5 Taxes	(5) = (3) – (4) Net Income to Common	(6) Earnings per Share	
Plan A: 0% debt; $200,000 common equity; 2000 shares						
$ 0	$ 0	$ 0	$ 0	$ 0	$ 0	
20,000	0	20,000	10,000	10,000	5.00	}100%
40,000	0	40,000	20,000	20,000	10.00	
60,000	0	60,000	30,000	30,000	15.00	
80,000	0	80,000	40,000	40,000	20.00	
Plan B: 25% debt; 8% Interest rate; $150,000 common equity; 1500 shares						
$ 0	$4,000	$ (4,000)	$ (2,000)[a]	$ (2,000)	$ (1.33)	
20,000	4,000	16,000	8,000	8,000	5.33	}125%
40,000	4,000	36,000	18,000	18,000	12.00	
60,000	4,000	56,000	28,000	28,000	18.67	
80,000	4,000	76,000	38,000	38,000	25.33	
Plan C: 40% debt; 8% Interest rate; $120,000 common equity; 1200 shares						
$ 0	$6,400	$ (6,400)	$ (3,200)[a]	$ (3,200)	$ (2.67)	
20,000	6,400	13,600	6,800	6,800	5.67	}147%
40,000	6,400	33,600	16,800	16,800	14.00	
60,000	6,400	53,600	26,800	26,800	22.33	
80,000	6,400	73,600	36,800	36,800	30.67	

[a]The negative tax bill recognizes the deduction arising from the loss carryback and carryforward provision of the Income Tax Act. See Chapter 2.

[17]In actual practice, moving from a 25 to a 40 percent debt ratio would probably result in a higher interest rate on the additional bonds. That effect is ignored here to let us concentrate on the ramifications of using different proportions of debt in the financial structure.

By following the same general procedures that allowed us to analyze the firm's use of operating leverage, we can lay out a precise measure of financial leverage. Such a measure deals with the sensitivity of earnings per share to EBIT fluctuations. The relationship can be expressed as

OBJECTIVE 5

$$\text{degree of financial leverage from base EBIT level} = DFL_{EBIT} = \frac{\text{percentage change in earnings per share}}{\text{percentage change in EBIT}} \tag{8-8}$$

Use of Equation (8-8) with each of the financing choices outlined for Pierce Grain is shown below. The base EBIT level is $20,000 in each case.

$$\text{Plan A: } DFL_{\$20,000} = \frac{100\%}{100\%} = 1.00 \text{ times}$$

$$\text{Plan B: } DFL_{\$20,000} = \frac{125\%}{100\%} = 1.25 \text{ times}$$

$$\text{Plan C: } DFL_{\$20,000} = \frac{147\%}{100\%} = 1.47 \text{ times}$$

Like operating leverage, the degree of financial leverage concept performs in the negative direction as well as the positive. Should EBIT fall by 10 percent, the Pierce Grain Company would suffer a 12.5 percent decline in earnings per share under plan B. If plan C were chosen to raise the necessary financial capital, the decline in earnings would be 14.7 percent. Observe that the greater the degree of financial leverage, the greater the fluctuations (positive or negative) in earnings per share. The common shareholder is required to endure greater variations in returns when the firm's management chooses to use more financial leverage rather than less. The degree of financial leverage measure allows the variation to be quantified.

Rather than taking the time to compute percentage changes in EBIT and earnings per share, the degree of financial leverage can be found directly, as follows:

$$DFL_{EBIT} = \frac{\text{EBIT}}{\text{EBIT} - \text{I}} \tag{8-9}$$

In Equation (8-9) the variable, I, represents the total interest expense incurred on all the firm's contractual debt obligations. If six bonds are outstanding, I is the sum of the interest expense on all six bonds. If the firm has preferred shares in its financial structure, the dividend on such issues must be inflated to a before-tax basis and included in the computation of I.[18] In this latter instance, I is in reality the sum of all fixed financing costs.

Equation (8-9) has been applied to each of Pierce Grain's financing plans (Table 8-8) at a base EBIT level of $20,000. The results are as follows:

$$\text{Plan A: } DFL_{\$20,000} = \frac{\$20,000}{\$20,000 - 0} = 1.00 \text{ times}$$

$$\text{Plan B: } DFL_{\$20,000} = \frac{\$20,000}{\$20,000 - \$4000} = 1.25 \text{ times}$$

$$\text{Plan C: } DFL_{\$20,000} = \frac{\$20,000}{\$20,000 - \$6400} = 1.47 \text{ times}$$

As you have probably suspected, the measures of financial leverage shown above are identical to those obtained by use of Equation (8-8).[19] This will always be the case.

[18]Suppose (1) preferred dividends of $4,000 are paid annually by the firm and (2) it faces a combined corporate tax rate of 45 percent. How much must the firm earn before taxes to make the $4,000 payment out of after-tax earnings? Since preferred dividends are not tax-deductible to the paying company, we have $4,000/(1 – .45) = $7,272.73. Note that from a financial policy viewpoint, corporate debt financing is less expensive than preferred share and common share financing, because the interest cost of debt financing is generally tax deductible.

[19]The degree of financial leverage (DFL) increases as the interest costs (I) to the firm increase. Further, as EBIT increases, DFL decreases (holding interest costs constant).

The Coca-Cola Company Financial Policies

The fact that financial leverage effects can be measured provides management the opportunity to shape corporate policy formally around the decision to use or avoid the use of leverage-inducing financial instruments. The Coca-Cola Company has very specific policies on the use of financial leverage. The learning objectives of this chapter, then, comprise more than mere academic, intellectual exercises. The material is, in fact, the stuff of boardroom-level discussion.

Management's primary objective is to maximize shareowner value over time. To accomplish this objective, the Coca-Cola Company and subsidiaries (the Company) have developed a comprehensive business strategy that emphasizes maximizing long-term cash flows. This strategy focuses on continuing aggressive investment in the high-return soft drink business, increasing returns on existing investments and optimizing the cost of capital through appropriate financial policies. The success of this strategy is evidenced by the growth in the Company's cash flows and earnings, its increased returns on total capital and equity and the total return to its share owners over time.

Management seeks investments that strategically enhance existing operations and offer cash returns that exceed the Company's long-term after-tax weighted average cost of capital, estimated by management to be approximately 11 percent as of January 1, 1994. The Company's soft drink business generates inherent high returns on capital, providing an attractive area for continued investment.

Maximizing share-owner value necessitates optimizing the Company's cost of capital through appropriate financial policies.

The Company maintains debt levels considered prudent based on the Company's cash flows, interest coverage, and the percentage of debt to the Company's total capital. The Company's overall cost of capital is lowered by the use of debt financing, resulting in increased return to share owners.

The Company's capital structure and financial policies have resulted in long-term credit ratings of "AA" from Standard & Poor's and "Aa3" from Moody's, as well as the highest credit ratings available for its commercial paper programs. The Company's strong financial position and cash flows allow for opportunistic access to financing in financial markets around the world.

Source: The Coca-Cola Company, *Annual Report* (1993), 44–46.

PERSPECTIVE IN FINANCE

The effect on the earnings stream available to the firm's common shareholders from combining operating and financial leverage in large degrees is dramatic. When the use of both leverage types is indeed heavy, a large sales increase will result in a very large rise in earnings per share. Be aware, though, that the very same thing happens in the opposite direction should the sales change be negative! Piling heavy financial leverage use upon a high degree of operating leverage, then, is a very risky way to do business. This is why you will find leveraged buyouts are not concentrated in heavy, durable goods industries. The firms in such industries have major (real) capital spending requirements. Rather, the firms favoured in leveraged buyouts will be those with comparatively low levels of operating leverage; retail operations are good examples. As a result, many national retail chains and department stores have been involved in leveraged buyouts during the past few years. The leveraged buyout, by its very nature, translates into a high degree of financial leverage use.

Figure 8-9 Leverage and Earnings Fluctuations

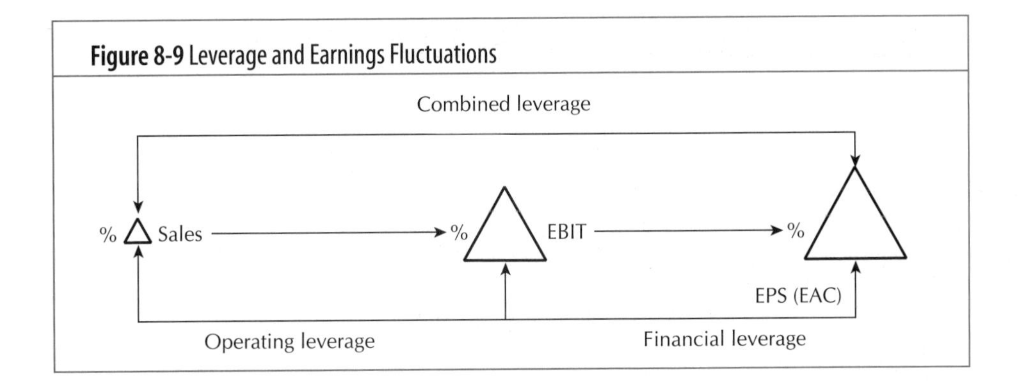

COMBINING OPERATING AND FINANCIAL LEVERAGE

OBJECTIVE 4

Changes in sales revenues cause greater changes in EBIT. Additionally, changes in EBIT translate into larger variations in both earnings per share (EPS) and total earnings available to the common shareholders (EAC), if the firm chooses to use financial leverage. It should be no surprise, then, to find out that combining operating and financial leverage causes rather large variations in earnings per share. This entire process is visually displayed in Figure 8-9.

Since the risk associated with possible earnings per share is affected by the use of combined or total leverage, it is useful to quantify the effect. For an illustration, we refer once more to the Pierce Grain Company. The cost structure identified for Pierce Grain in our discussion of breakeven analysis still holds. Furthermore, assume that plan B, which carried a 25 percent debt ratio, was chosen to finance the company's assets. Table 8-9 shows the combined leverage analysis for the Pierce Grain Company.

In Table 8-9 an increase in output for Pierce Grain from 30,000 to 36,000 units is analyzed. This increase represents a 20 percent rise in sales revenues. From our earlier discussion of operating leverage and the data in Table 8-9, we can see that this 20 percent increase in sales is magnified into a 120 percent rise in EBIT. From this base sales level of $300,000 the degree of operating leverage is 6 times.

The 120 percent rise in EBIT induces a change in earnings per share and earnings available to the common shareholders of 150 percent. The degree of financial leverage is therefore 1.25 times.

The upshot of the analysis is that the 20 percent rise in sales has been magnified to 150 percent, as reflected by the percentage change in earnings per share. The formal measure of combined leverage can be expressed as follows:

$$\left(\begin{array}{c}\text{degree of}\\ \text{combined leverage from}\\ \text{the base sales level}\end{array}\right) = DCL_s = \left(\frac{\text{percentage change in earnings per share}}{\text{percentage change in sales}}\right) \tag{8-10}$$

This equation was used in the bottom portion of Table 8-9 to determine that the degree of combined leverage from the base sales level of $300,000 is 7.50 times. Pierce Grain's use of both operating and financial leverage will cause any percentage change in sales (from the specific base level) to be magnified by a factor of 7.50 when the effect on earnings per share is computed. A 1 percent change in sales, for example, will result in a 7.50 percent change in earnings per share.

Notice that the degree of combined leverage is actually the product (not the simple sum) of the two independent leverage measures. Thus, we have

$$(DOL_s) \times (DFL_{EBIT}) = DCL_s \tag{8-11}$$

or $$(6) \times (1.25) = 7.50 \text{ times}$$

Table 8-9 The Pierce Grain Company Combined Leverage Analysis

Item	Base Sales Level, t	Forecast Sales Level, $t+1$	Selected Percentage Changes
Sales	\$300,000	\$360,000	+20
Less: Total variable costs	180,000	216,000	
Revenue before fixed costs	\$120,000	\$144,000	
Less: Total fixed costs	100,000	100,000	
EBIT	\$ 20,000	\$ 44,000	+120
Less: Interest expense	4,000	4,000	
Earnings before taxes (EBT)	\$ 16,000	\$ 40,000	
Less: Taxes at 50%	8,000	20,000	
Net income	\$ 8,000	\$ 20,000	+150
Less: Preferred dividends	0	0	
Earnings available to common (EAC)	\$ 8,000	\$ 20,000	+150
Number of common shares	1,500	1,500	
Earnings per share (EPS)	\$ 5.33	\$ 13.33	+150

$$\text{Degree of operating leverage} = DOL_{\$300{,}000} = \frac{120\%}{20\%} = 6 \text{ times}$$

$$\text{Degree of financial leverage} = DFL_{\$20{,}000} = \frac{150\%}{120\%} = 1.25 \text{ times}$$

$$\text{Degree of combined leverage} = DCL_{\$300{,}000} = \frac{150\%}{20\%} = 7.50 \text{ times}$$

OBJECTIVE 5

It is possible to ascertain the degree of combined leverage in a direct fashion, without determining any percentage fluctuations or the separate leverage values. We need only substitute the appropriate values into Equation (8-12):[20]

$$DCL_s = \frac{Q(P - V)}{Q(P - V) - F - I} \tag{8-12}$$

The variable definitions in Equation (8-12) are the same ones that have been employed throughout this chapter. Use of Equation (8-12) with the information in Table 8-9 gives

$$\begin{aligned} DCL_{\$300{,}000} &= \frac{30{,}000(\$10 - \$6)}{30{,}000(\$10 - \$6) - \$100{,}000 - \$4000} \\ &= \frac{\$120{,}000}{\$16{,}000} \\ &= 7.5 \text{ times} \end{aligned}$$

OBJECTIVE 6

Implications

The total risk exposure the firm assumes can be managed by combining operating and financial leverage in different degrees. Knowledge of the various leverage measures assists the financial officer in determining the proper level of overall risk that should be accepted. If a high degree of business risk is inherent to the specific line of commercial activity, then a low posture with regard to financial risk would minimize additional earnings fluctuations stemming from sales changes. On the other hand, the firm that by its very nature incurs a low level of fixed operating costs might choose to use a high degree of financial leverage in the hope of increasing earnings per share and the rate of return on the common equity investment.

[20]As was the case with the degree of financial leverage metric, the variable, I, in the combined leverage measure must include the before-tax equivalent of any preferred dividend payments when preferred shares are in the financial structure.

BACK TO THE BASICS

Our analysis of business risk, financial risk, and the three measurements of leverage use all relate directly to ***Axiom 1: The Risk-Return Tradeoff—We Won't Take on Added Risk Unless We Expect to Be Compensated With Additional Return.*** *Should the firm decide to "pile on" heavy financial leverage use on top of a high degree of business risk exposure, then we would expect the firm's overall cost of capital to rise and its share price to fall. These effects emphasize the critical nature of designing the firm's financing mix to both the financial executive and the shareholders. This central area of financial decision making is explored further in the next chapter, "Planning the Firm's Financing Mix."*

SUMMARY

In this chapter we began to study the process of arriving at an appropriate financial structure for the firm. Tools that can assist the financial manager in this task were examined. We were mainly concerned with assessing the variability in the firm's residual earnings stream (either earnings per share or earnings available to the common shareholders) induced by the use of operating and financial leverage. This assessment built upon the tenets of breakeven analysis.

Breakeven Analysis

Breakeven analysis permits the financial manager to determine the quantity of output or the level of sales that will result in an EBIT level of zero. This means that the firm has neither a loss nor a profit before any tax consideration. The effect of price changes, cost structure changes, or volume changes upon profits (EBIT) can be studied. To make the technique operational, it is necessary that the firm's costs be classified as fixed or variable. Not all costs fit neatly into one of these two categories. Over short planning horizons, though, the preponderance of costs can be assigned to either the fixed or variable classification. Once the cost structure has been identified, the breakeven point can be found by use of (1) trial-and-error analysis, (2) contribution margin analysis, or (3) algebraic analysis.

Operating Leverage

Operating leverage is the responsiveness of the firm's EBIT to changes in sales revenues. It arises from the firm's use of fixed operating costs. When fixed operating costs are present in the company's cost structure, changes in sales are magnified into even greater changes in EBIT. The firm's degree of operating leverage from a base sales level is the percentage change in EBIT divided by the percentage change in sales. All types of leverage are two-edged swords. When sales decrease by some percentage, the negative impact upon EBIT will be even larger.

Financial Leverage

A firm employs financial leverage when it finances a portion of its assets with securities bearing a fixed rate of return. The presence of debt and/or preferred shares in the company's financial structure means that it is using financial leverage. When financial leverage is used, changes in EBIT translate into larger changes in earnings per share. The concept of the degree of financial leverage dwells on the sensitivity of earnings per share to changes in EBIT. The degree of financial leverage from a base EBIT level is defined as the percentage change in earnings per share divided by the percentage change in EBIT. All other things equal, the more fixed-charge securities the firm employs in its financial structure, the greater its degree of financial leverage. Clearly, EBIT can rise or fall. If it falls, and financial leverage is used, the firm's shareholders endure negative changes in earnings per share that are larger than the relative decline in EBIT. Again, leverage is a two-edged sword.

Combining Operating and Financial Leverage

Firms use operating and financial leverage in various degrees. The joint use of operating and financial leverage can be measured by computing the degree of combined leverage, defined as the percentage change in earnings per share divided by the percentage change in sales. This measure allows the financial manager to ascertain the effect on total leverage caused by adding financial leverage on top of operating leverage. Effects can be dramatic, because the degree of combined leverage is the product of the degrees of operating and financial leverage. Table 8-10 summarizes the salient concepts and calculation formats discussed in this chapter.

Table 8-10 Summary of Leverage Concepts and Calculations

Technique	Description or Concept	Calculation	Text Reference
Breakeven Analysis			
1. Breakeven point quantity	Total fixed costs divided by the unit contribution margin.	$Q_B = \frac{F}{P - V}$	(8-3)
2. Breakeven sales level	Total fixed costs divided by 1 minus the ratio of total variable costs to the associated level of sales.	$S^* = \frac{F}{1 - \frac{VC}{S}}$	(8-4)
Operating Leverage			
3. Degree of operating leverage	Percentage change in EBIT divided by the percentage change in sales; or revenue before fixed costs divided by revenue after fixed costs.	$DOL_s = \frac{Q(P - V)}{Q(P - V) - F}$	(8-6)
Financial Leverage			
4. Degree of financial leverage	Percentage change in earnings per share divided by the percentage change in EBIT; or EBIT divided by EBT.[a]	$DFL_{EBIT} = \frac{EBIT}{EBIT - I}$	(8-9)
Combined Leverage			
5. Degree of combined leverage	Percentage change in earnings per share divided by the percentage change in sales; or revenue before fixed costs divided by EBT.[a]	$DCL_s = \frac{Q(P - V)}{Q(P - V) - F - I}$	(8-12)

[a]The use of EBT here presumes no preferred dividend payments. In the presence of preferred dividend payments replace EBT with earnings available to common shares (EAC).

STUDY QUESTIONS

8-1. Distinguish between business risk and financial risk. What gives rise to, or causes, each type of risk?

8-2. Define the term *financial leverage*. Does the firm use financial leverage if preferred shares is present in the capital structure?

8-3. Define the term *operating leverage*. What type of effect occurs when the firm uses operating leverage?

8-4. What is the difference between the (ordinary) breakeven point and the cash breakeven point? Which will be the greater?

8-5. A manager in your firm decides to employ breakeven analysis. Of what shortcomings should this manager be aware?

8-6. What is meant by total risk exposure? How may a firm move to reduce its total risk exposure?

8-7. If a firm has a degree of combined leverage of 3.0 times, what does a negative sales fluctuation of 15 percent portend for the earnings available to the firm's common share investors?

8-8. Breakeven analysis assumes linear revenue and cost functions. In reality these linear functions over large output and sales levels are highly improbable. Why?

SELF-TEST QUESTIONS

ST-1. (*Breakeven Point*) You are a hard-working analyst in the office of financial operations for a manufacturing firm that produces a single product. You have developed the following cost structure information for this company. All of it pertains to an output level of 10 million units. Using this information, find the breakeven point in units of output for the firm.

Return on operating assets	= 30%
Operating asset turnover	= 6 times
Operating assets	= $20 million
Degree of operating leverage	= 4.5 times

ST-2. (*Leverage Analysis*) You have developed the following analytical income statement for your corporation. It represents the most recent year's operations, which ended yesterday.

Sales	$20,000,000
Variable costs	12,000,000
Revenue before fixed costs	$ 8,000,000
Fixed costs	5,000,000
EBIT	$ 3,000,000
Interest expense	1,000,000
Earnings before taxes	$ 2,000,000
Taxes (0.50)	1,000,000
Net income	$ 1,000,000

Your supervisor in the financial studies office has just handed you a memorandum that asked for written responses to the following questions:

a. At this level of output, what is the degree of operating leverage?

b. What is the degree of financial leverage?

c. What is the degree of combined leverage?

d. What is the firm's breakeven point in sales dollars?

e. If sales should increase by 30 percent, by what percent would earnings before taxes (and net income) increase?

f. Prepare an analytical income statement that verifies the calculations from part (e) above.

ST-3. (*Fixed Costs and the Breakeven Point*) Bonaventure Manufacturing expects to earn $210,000 next year after taxes. Sales will be $4 million. The firm's single plant is located on the outskirts of Lockport, NS. The firm manufactures a combined bookshelf and desk unit used extensively in college dormitories. These units sell for $200 each and have a variable cost per unit of $150. Bonaventure experiences a combined corporate tax rate of 30 percent.

a. What are the firm's fixed costs expected to be next year?

b. Calculate the firm's breakeven point in both units and dollars.

STUDY PROBLEMS (SET A)

8-1A. (*Breakeven Point*) Napa Valley Winery (NVW) is a boutique winery that produces from organically grown cabernet sauvignon grapes a high-quality nonalcoholic red wine. Each bottle sells for $30. NVW's chief financial officer, Jackie Cheng, has estimated variable costs to be 70 percent of sales. If NVW's fixed costs are $360,000, how many bottles of its wine must NVW sell to break even?

8-2A. (*Operating Leverage*) In light of a sales agreement that Napa Valley Winery just signed with a national chain of health food restaurants, NVW's CFO Jackie Cheng is estimating that NVW's sales in the next year will be 50,000 bottles at $30 per bottle. Given the information in 8-1A and that variable costs are expected to be 70 percent of sales, what is NVW's expected degree of operating leverage?

8-3A. (*Breakeven Point and Operating Leverage*) Some financial data for each of three firms are given below:

	Jake's Lawn Chairs	Sarasota Sky Lights	Jefferson Wholesale
Average selling price per unit	$32.00	$875.00	$97.77
Average variable cost per unit	$17.38	$400.00	$87.00
Units sold	18,770	2,800	11,000
Fixed costs	$120,350	$850,000	$89,500

a. What is the profit for each company at the indicated sales volume?
b. What is the breakeven point in units for each company?
c. What is the degree of operating leverage for each company at the indicated sales volume?
d. If sales were to decline, which firm would suffer the largest relative decline in profitability?

8-4A. (*Leverage Analysis*) You have developed the following analytical income statement for your corporation. It represents the most recent year's operations, which ended yesterday.

Sales	$45,750,000
Variable costs	22,800,000
Revenue before fixed costs	$22,950,000
Fixed costs	9,200,000
EBIT	$13,750,000
Interest expense	1,350,000
Earnings before taxes	$ 12,400,000
Taxes (0.50)	6,200,000
Net income	$ 6,200,000

Your supervisor in the controller's office has just handed you a memorandum asking for written responses to the following questions:

a. At this level of output, what is the degree of operating leverage?
b. What is the degree of financial leverage?
c. What is the degree of combined leverage?
d. What is the firm's breakeven point in sales dollars?
e. If sales should increase by 25 percent, by what percent would earnings before taxes (and net income) increase?

8-5A. (*Breakeven Point and Operating Leverage*) Footwear Inc. manufactures a complete line of men's and women's dress shoes for independent merchants. The average selling price of its finished product is $180 per pair. The variable cost for the same pair of shoes is $126. Footwear Inc. incurs fixed costs of $540,000 per year.

a. What is the breakeven point in pairs of shoes for the company?
b. What is the dollar sales volume the firm must achieve in order to reach the breakeven point?

c. What would be the firm's profit or loss at the following pairs of production sold: 12,000 pairs of shoes? 15,000 pairs of shoes? 20,000 pairs of shoes?

d. Find the degree of operating leverage for the production and sales levels given in part (c).

8-6A. (*Breakeven Point and Operating Leverage*) Zeylog Corporation manufactures a line of computer memory-expansion boards used in microcomputers. The average selling price of its finished product is $180 per unit. The variable cost for these same units is $110. Zeylog incurs fixed costs of $630,000 per year.

a. What is the breakeven point in units for the company?

b. What is the dollar sales volume the firm must achieve in order to reach the breakeven point?

c. What would be the firm's profit or loss at the following units of production sold: 12,000 units? 15,000 units? 20,000 units?

d. Find the degree of operating leverage for the production and sales levels given in part (c).

8-7A. (*Breakeven Point and Operating Leverage*) Some financial data for each of three firms are shown below:

	Blacksburg Furniture	Lexington Cabinets	Williamsburg Colonials
Average selling price per unit	$15.00	$400.00	$40.00
Average variable cost per unit	$12.35	$220.00	$14.50
Units sold	75,000	4,000	13,000
Fixed costs	$35,000	$100,000	$70,000

a. What is the profit for each company at the indicated sales volume?

b. What is the breakeven point in units for each company?

c. What is the degree of operating leverage for each company at the indicated sales volume?

d. If sales were to decline, which firm would suffer the largest relative decline in profitability?

8-8A. (*Fixed Costs and the Breakeven Point*) A & B Beverages expects to earn $50,000 next year after taxes. Sales will be $375,000. The company's average product sells for $27 a unit. The variable cost per unit is $14.85. The company experiences a corporate tax rate of 45 percent.

a. What are the company's fixed costs expected to be next year?

b. Calculate the company's breakeven point in both units and dollars.

8-9A. (*Breakeven Point and Profit Margin*) Ms Thompson is planning to open a new wholesaling operation. Her target operating profit margin is 26 percent. Her unit contribution margin will be 50 percent of sales. Average annual sales are forecast to be $3,250,000.

a. How large can fixed costs be for the wholesaling operation and still allow the 26 percent operating profit margin to be achieved?

b. What is the breakeven point in dollars for the firm?

8-10A. (*Leverage Analysis*) You have developed the following analytical income statement for your corporation. It represents the most recent year's operations, which ended yesterday.

Sales	$30,000,000
Variable costs	13,500,000
Revenue before fixed costs	$16,500,000
Fixed costs	8,000,000
EBIT	$ 8,500,000
Interest expense	1,000,000
Earnings before taxes	$ 7,500,000
Taxes (0.50)	3,750,000
Net income	$ 3,750,000

Your supervisor in the controller's office has just handed you a memorandum that asked for written responses to the following questions:

a. At this level of output, what is the degree of operating leverage?

b. What is the degree of financial leverage?

c. What is the degree of combined leverage?

d. What is the firm's breakeven point in sales dollars?

e. If sales should increase by 25 percent, by what percent would earnings before taxes (and net income) increase?

8-11A. (*Breakeven Point*) You are a hard-working analyst in the office of financial operations for a manufacturing firm that produces a single product. You have developed the following cost structure information for this company. All of it pertains to an output level of 10 million units. Using this information, find the breakeven point in units of output for the firm.

Return on operating assets	= 25%
Operating asset turnover	= 5 times
Operating assets	= $20 million
Degree of operating leverage	= 4 times

8-12A. (*Breakeven Point and Operating Leverage*) Allison Radios manufactures a complete line of radio and communication equipment for law enforcement agencies. The average selling price of its finished product is $200 per unit. The variable cost for these same units is $125. Allison Radios incurs fixed costs of $620,000 per year.

a. What is the breakeven point in units for the company?

b. What is the dollar sales volume the firm must achieve in order to reach the breakeven point?

c. What would be the firm's profit or loss at the following units of production sold: 12,000 units? 15,000 units? 20,000 units?

d. Find the degree of operating leverage for the production and sales levels given in part (c).

8-13A. (*Breakeven Point and Operating Leverage*) Some financial data for each of three firms are shown below:

	Oviedo Seeds	Gainesville Sod	Athens Peaches
Average selling price per unit	$14.00	$200.00	$25.00
Average variable cost per unit	$11.20	$130.00	$17.50
Units sold	100,000	10,000	48,000
Fixed costs	$25,000	$100,000	$35,000

a. What is the profit for each company at the indicated sales volume?

b. What is the breakeven point in units for each company?

c. What is the degree of operating leverage for each company at the indicated sales volume?

d. If sales were to decline, which firm would suffer the largest relative decline in profitability?

8-14A. (*Fixed Costs and the Breakeven Point*) D. Q. Kirk Inc., expects to earn $40,000 next year after taxes. Sales will be $400,000. The micro-brewery sells only kegs of beer for $20 a keg. The variable cost per keg is $8. This manufacturing company experiences a combined corporate tax rate of 17 percent.

a. What are the company's fixed costs expected to be next year?

b. Calculate the firm's breakeven point in both units and dollars.

8-15A. (*Fixed Costs and the Breakeven Point*) Albert's Cooling Equipment hopes to earn $80,000 next year after taxes. Sales will be $2 million. The firm's single plant is located on the edge of Slippery Rock, NB, and manufactures only small refrigerators. These are used in many of the dormitories found on college campuses. The refrigerators sell for $80 per unit and have a variable cost of $56. Albert's has a combined corporate tax rate of 45 percent.

a. What are the firm's fixed costs expected to be next year?

b. Calculate the firm's breakeven point both in units and dollars.

8-16A. (*Breakeven Point and Selling Price*) Gerry's Tool and Die Company will produce 200,000 units next year. All of this production will be sold as finished goods. Fixed costs will total $300,000. Variable costs for this firm are relatively predictable at 75 percent of sales.

a. If Gerry's Tool and Die wants to achieve a level of earnings before interest and taxes of $240,000 next year, at what price per unit must it sell its product?

b. Based upon your answer to part (a), set up an analytical income statement that will verify your solution.

8-17A. (*Breakeven Point and Selling Price*) Parks Castings, Inc., will manufacture and sell 200,000 units next year. Fixed costs will total $300,000, and variable costs will be 60 percent of sales.

a. The firm wants to achieve a level of earnings before interest and taxes of $250,000. What selling price per unit is necessary to achieve this result?

b. Set up an analytical income statement to verify your solution to part (a).

8-18A. (*Breakeven Point and Profit Margin*) A recent business graduate is planning to open a new wholesaling operation. His target operating profit margin is 28 percent. His unit contribution margin will be 50 percent of sales. Average annual sales are forecast to be $3,750,000.

a. How large can fixed costs be for the wholesaling operation and still allow the 28 percent operating profit margin to be achieved?

b. What is the breakeven point in dollars for the firm?

8-19A. (*Operating Leverage*) Rocky Mount Metals Company manufactures an assortment of wood-burning stoves. The average selling price for the various units is $500. The associated variable cost is $350 per unit. Fixed costs for the firm average $180,000 annually.

a. What is the breakeven point in units for the company?

b. What is the dollar sales volume the firm must achieve in order to reach the breakeven point?

c. What is the degree of operating leverage for a production and sales level of 5,000 units for the firm? (Calculate to three decimal places.)

d. What will be the projected effect upon earnings before interest and taxes if the firm's sales level should increase by 20 percent from the volume noted in part (c)?

8-20A. (*Breakeven Point and Operating Leverage*) The Rockland Recreation Company manufactures a full line of lawn furniture. The average selling price of a finished unit is $25. The associated variable cost is $15 per unit. Fixed costs for Rockland average $50,000 per year.

a. What is the breakeven point in units for the company?

b. What is the dollar sales volume the firm must achieve in order to reach the breakeven point?

c. What would be the company's profit or loss at the following units of production sold: 4,000 units? 6,000 units? 8,000 units?

d. Find the degree of operating leverage for the production and sales levels given in part (c).

e. What is the effect on the degree of operating leverage as sales rise above the breakeven point?

8-21A. (*Fixed Costs*) Harting Heat Treating forecasts that next year its fixed costs will total $120,000. Its only product sells for $12 per unit, of which $7 is a variable cost. The management of Harting Heat Treating is considering the purchase of a new machine that will lower the variable cost per unit to $5. The new machine, however, will add to fixed costs through an increase in depreciation expense. How large can the addition to fixed costs be in order to keep the firm's breakeven point in units produced and sold unchanged?

8-22A. (*Operating Leverage*) The management of Harting Heat Treating did not purchase the new piece of equipment (see Problem 8-21A). Using the existing cost structure, calculate the degree of operating leverage at 30,000 units of output. Comment on the meaning of your answer.

8-23A. (*Leverage Analysis*) An analytical income statement for Harting Heat Treating is shown below. It is based upon an output (sales) level of 40,000 units. You may refer to the original cost structure data in Problem (8-21A).

Sales	$480,000
Variable costs	280,000
Revenue before fixed costs	$200,000
Fixed costs	120,000
EBIT	$ 80,000
Interest expense	30,000
Earnings before taxes	$ 50,000
Taxes	25,000
Net income	$ 25,000

a. Calculate the degree of operating leverage at this output level.
b. Calculate the degree of financial leverage at this level of EBIT.
c. Determine the combined leverage effect at this output level.

8-24A. (*Breakeven Point*) You are employed as a financial analyst for a single-product manufacturing firm. Your supervisor has made the following cost structure information available to you, all of which pertains to an output level of 1,600,000 units.

Return on operating assets	= 15%
Operating asset turnover	= 5 times
Operating assets	= $3 million
Degree of operating leverage	= 8 times

Your task is to find the breakeven point in units of output for the firm.

8-25A. (*Fixed Costs*) Percy's Printing Services is forecasting fixed costs next year of $300,000. The firm's single product sells for $20 per unit and incurs a variable cost per unit of $14. The firm may acquire some new binding equipment that would lower variable cost per unit to $12. The new equipment, however, would add to fixed costs through the price of an annual maintenance agreement on the new equipment. How large can this increase in fixed costs be and still keep the firm's present breakeven point in units produced and sold unchanged?

8-26A. (*Leverage Analysis*) Your firm's cost analysis supervisor supplies you with the following analytical income statement and requests answers to the four questions listed below the statement.

Sales	$12,000,000
Variable costs	9,000,000
Revenue before fixed costs	3,000,000
Fixed costs	2,000,000
EBIT	$ 1,000,000
Interest expense	200,000
Earnings before taxes	$ 800,000
Taxes	400,000
Net income	$ 400,000

a. At this level of output, what is the degree of operating leverage?
b. What is the degree of financial leverage?
c. What is the degree of combined leverage?
d. What is the firm's breakeven point in sales dollars?

8-27A. (*Leverage Analysis*) You are supplied with the following analytical income statement for your firm. It reflects last year's operations.

Sales	$16,000,000
Variable costs	8,000,000
Revenue before fixed costs	$8,000,000
Fixed costs	4,000,000
EBIT	$ 4,000,000
Interest expense	1,500,000
Earnings before taxes	$ 2,500,000
Taxes	1,250,000
Net income	$ 1,250,000

a. At this level of output, what is the degree of operating leverage?

b. What is the degree of financial leverage?

c. What is the degree of combined leverage?

d. If sales should increase by 20 percent, by what percent would earnings before taxes (and net income) increase?

e. What is your firm's breakeven point in sales dollars?

8-28A. (*Sales Mix and the Breakeven Point*) Crash Components produces four lines of auto accessories for the major Windsor automobile manufacturers. The lines are known by the code letters A, B, C, and D. The current sales mix for Crash and the contribution margin ratio (unit contribution margin divided by unit sales price) for these product lines are as follows:

Product Line	Percent of Total Sales	Contribution Margin Ratio
A	33.33%	40%
B	41.67	32
C	16.67	20
D	8.33	60

Total sales for next year are forecast to be $120,000. Total fixed costs will be $29,400.

a. Prepare a table showing (1) sales, (2) total variable costs, and (3) the total contribution margin associated with each product line.

b. What is the aggregate contribution margin ratio indicative of this sales mix?

c. At this sales mix, what is the breakeven point in dollars?

8-29A. (*Sales Mix and the Breakeven Point*) Because of production constraints, Crash Components (see Problem 8-28A) may have to adhere to a different sales mix for next year. The alternative plan is outlined below:

Product Line	Percent of Total Sales
A	25%
B	36.67
C	33.33
D	5

a. Assuming all other facts in Problem 8-28A remain the same, what effect will this different sales mix have on Crash's breakeven point in dollars?

b. Which sales mix will Crash's management prefer?

INTEGRATIVE PROBLEM

Imagine that you were hired recently as a financial analyst for a relatively new, highly leveraged ski manufacturer located in Jasper's Rocky mountains. Your firm manufactures only one product, a state-of-the-art snow ski. The company has been operating up to this point without much quantitative knowledge of the business and financial risks it faces.

Ski season just ended, however, so the president of the company has started to focus more on the financial aspects of managing the business. He has set up a meeting for next week with the CFO, Maria Sanchez, to discuss matters such as the business and financial risks faced by the company. Accordingly, Maria has asked you to prepare an analysis to assist her in discussions with the president.

As a first step in your work, you compiled the following information regarding the cost structure of the company.

Output level	50,000 units
Operating assets	$2,000,000
Operating asset turnover	7 times
Return on operating assets	35 %
Degree of operating leverage	5 times
Interest expense	$ 400,000
Tax rate	35%

Your next step is to determine the breakeven point in units of output for the company. One of your strong points has been that you always prepare supporting working papers, which show how you arrive at your conclusions. You know Maria would like to see such working papers for this analysis to facilitate her review of your work.

Thereafter you will have the information you require to prepare an analytical income statement for the company. You are sure that Maria would like to see this statement; in addition, you know you need it to be able to answer the following questions. You also know Maria expects you to prepare, in a format that is presentable to the president, answers to questions to serve as a basis for her discussions with the president.

a. What is the degree of financial leverage?

b. What is the degree of combined leverage?

c. What is the firm's breakeven point in sales dollars?

d. If sales should increase by 30 percent (as the president expects), by what percent would EBT (earnings before taxes) and net income increase?

e. Prepare another analytical income statement, this time to verify the calculations from part (d) above.

STUDY PROBLEMS (SET B)

8-1B. (*Breakeven Point*) Roberto Martinez is the chief financial analyst at New Wave Pharmaceutical (NWP), a company that produces a vitamin claimed to prevent the common cold. Roberto has been asked to determine the breakeven point in units. He obtained the following information from the company's financial statements for the year just ended. In addition, he found out from NWP's production manager that the company produced 40 million units in that year. What will Roberto determine the breakeven point to be?

Sales	$20,000,000
Variable costs	16,000,000
Revenue before fixed costs	4,000,000
Fixed costs	2,400,000
EBIT	$ 1,600,000

8-2B. (*Leverage Analysis*) New Wave Pharmaceutical is concerned that recent unfavourable publicity about the questionable medicinal benefits of other vitamins will temporarily hurt NWP's sales, even though such assertions do not apply to NWP's vitamin. Accordingly,

Roberto has been asked to determine the company's level of risk based on the financial information for the year ended. Given the information in 8-1B, Roberto learned from the company's financial statements that the company incurred $800,000 of interest expense in the year just ended. What will Roberto determine the (a) degree of operating leverage, (b) degree of financial leverage, and (c) degree of combined leverage to be?

8-3B. (*Breakeven Point and Operating Leverage*) Avitar Corporation manufactures a line of computer memory-expansion boards used in microcomputers. The average selling price of its finished product is $175 per unit. The variable cost for these same units is $115. Avitar incurs fixed costs of $650,000 per year.

- **a.** What is the breakeven point in units for the company?
- **b.** What is the dollar sales volume the firm must achieve in order to reach the breakeven point?
- **c.** What would be the firm's profit or loss at the following units of production sold: 10,000 units? 16,000 units? 20,000 units?
- **d.** Find the degree of operating leverage for the production and sales levels given in part (c).

8-4B. (*Breakeven Point and Operating Leverage*) Some financial data for each of three firms are given below:

	Durham Furniture	Raleigh Cabinets	Charlotte Colonials
Average selling price per unit	$20.00	$435.00	$35.00
Average variable cost per unit	$13.75	$240.00	$15.75
Units sold	80,000	4,500	15,000
Fixed costs	$40,000	$150,000	$60,000

- **a.** What is the profit for each company at the indicated sales volume?
- **b.** What is the breakeven point in units for each company?
- **c.** What is the degree of operating leverage for each company at the indicated sales volume?
- **d.** If sales were to decline, which firm would suffer the largest relative decline in profitability?

8-5B. (*Fixed Costs and the Breakeven Point*) Cypress Books expects to earn $55,000 next year after taxes. Sales will be $400,008. The company's average product sells for $28 a unit. The variable cost per unit is $18. The company experiences a combined tax rate of 45 percent.

- **a.** What are the company's fixed costs expected to be next year?
- **b.** Calculate the company's breakeven point in both units and dollars.

8-6B. (*Breakeven Point and Profit Margin*) Ms Neeley is planning to open a new wholesaling operation. Her target operating profit margin is 28 percent. Her unit contribution margin will be 45 percent of sales. Average annual sales are forecast to be $3,750,000.

- **a.** How large can fixed costs be for the wholesaling operation and still allow the 28 percent operating profit margin to be achieved?
- **b.** What is the breakeven point in dollars for the firm?

8-7B. (*Leverage Analysis*) You have developed the following analytical income statement for your corporation. It represents the most recent year's operations, which ended yesterday.

Sales	$40,000,000
Variable costs	16,000,000
Revenue before fixed costs	$24,000,000
Fixed costs	10,000,000
EBIT	$14,000,000
Interest expense	1,150,000
Earnings before taxes	$12,850,000
Taxes	3,750,000
Net income	$ 9,100,000

Your supervisor in the controller's office has just handed you a memorandum asking for written responses to the following questions:

a. At this level of output, what is the degree of operating leverage?
b. What is the degree of financial leverage?
c. What is the degree of combined leverage?
d. What is the firm's breakeven point in sales dollars?
e. If sales should increase by 20 percent, by what percent would earnings before taxes (and net income) increase?

8-8B. (*Breakeven Point*) You are a hard-working analyst in the office of financial operations for a manufacturing firm that produces a single product. You have developed the following cost structure information for this company. All of it pertains to an output level of 7 million units. Using this information, find the breakeven point in units of output for the firm.

Return on operating assets	= 25%
Operating asset turnover	= 5 times
Operating assets	= $20 million
Degree of operating leverage	= 4 times

8-9B. (*Breakeven Point and Operating Leverage*) Matthew Electronics manufactures a complete line of radio and communication equipment for law enforcement agencies. The average selling price of its finished product is $175 per unit. The variable cost for these same units is $140. Matthew incurs fixed costs of $550,000 per year.

a. What is the breakeven point in units for the company?
b. What is the dollar sales volume the firm must achieve in order to reach the breakeven point?
c. What would be the firm's profit or loss at the following units of production sold: 12,000 units? 15,000 units? 20,000 units?
d. Find the degree of operating leverage for the production and sales levels given in part (c).

8-10B. (*Breakeven Point and Operating Leverage*) Some financial data for each of three firms are given below:

	Farm City Seeds	Empire Sod	Golden Peaches
Average selling price per unit	$15.00	$190.00	$28.00
Average variable cost per unit	$11.75	$145.00	$19.00
Units sold	120,000	9,000	50,000
Fixed costs	$30,000	$110,000	$33,000

a. What is the profit for each company at the indicated sales volume?
b. What is the breakeven point in units for each company?
c. What is the degree of operating leverage for each company at the indicated sales volume?
d. If sales were to decline, which firm would suffer the largest relative decline in profitability?

8-11B. (*Fixed Costs and the Breakeven Point*) Keller Inc., expects to earn $38,000 next year after taxes. Sales will be $420,002. The company's average product sells for $17 a unit. The variable cost per unit is $9. The company experiences a combined corporate tax rate of 45 percent.

a. What are Keller's fixed costs expected to be next year?
b. Calculate the firm's breakeven point in both units and dollars.

8-12B. (*Fixed Costs and the Breakeven Point*) Mini-Kool hopes to earn $70,000 next year after taxes. Sales will be $2,500,000. The firm's single plant manufactures only small refrigerators. These are used in many recreational campers. The refrigerators sell for $75 per unit and have a variable cost of $58. Mini-Kool experiences a combined corporate tax rate of 45 percent.

a. What are the firm's fixed costs expected to be next year?

b. Calculate the firm's breakeven point both in units and dollars.

8-13B. (*Breakeven Point and Selling Price*) Heritage Chain Company will produce 175,000 units next year. All of this production will be sold as finished goods. Fixed costs will total $335,000. Variable costs for this firm are relatively predictable at 80 percent of sales.

a. If Heritage Chain wants to achieve a level of earnings before interest and taxes of $270,000 next year, at what price per unit must it sell its product?

b. Based upon your answer to part (a), set up an analytical income statement that will verify your solution.

8-14B. (*Breakeven Point and Selling Price*) Thomas Appliances will manufacture and sell 190,000 units next year. Fixed costs will total $300,000, and variable costs will be 75 percent of sales.

a. The firm wants to achieve a level of earnings before interest and taxes of $250,000. What selling price per unit is necessary to achieve this result?

b. Set up an analytical income statement to verify your solution to part (a).

8-15B. (*Breakeven Point and Profit Margin*) A recent business graduate is planning to open a new wholesaling operation. His target operating profit margin is 25 percent. His unit contribution margin will be 60 percent of sales. Average annual sales are forecast to be $4,250,000.

a. How large can fixed costs be for the wholesaling operation and still allow the 25 percent operating profit margin to be achieved?

b. What is the breakeven point in dollars for the firm?

8-16B. (*Operating Leverage*) The B. H. Williams Company manufactures an assortment of wood-burning stoves. The average selling price for the various units is $475. The associated variable cost is $350 per unit. Fixed costs for the firm average $200,000 annually.

a. What is the breakeven point in units for the company?

b. What is the dollar sales volume the firm must achieve in order to reach the breakeven point?

c. What is the degree of operating leverage for a production and sales level of 6,000 units for the firm? (Calculate to three decimal places.)

d. What will be the projected effect upon earnings before interest and taxes if the firm's sales level should increase by 13 percent from the volume noted in part (c)?

8-17B. (*Breakeven Point and Operating Leverage*) The Palm Patio Company manufactures a full line of lawn furniture. The average selling price of a finished unit is $28. The associated variable cost is $17 per unit. Fixed costs for Palm Patio average $55,000 per year.

a. What is the breakeven point in units for the company?

b. What is the dollar sales volume the firm must achieve in order to reach the breakeven point?

c. What would be the company's profit or loss at the following units of production sold: 4,000 units? 6,000 units? 8,000 units?

d. Find the degree of operating leverage for the production and sales levels given in part (c).

e. What is the effect on the degree of operating leverage as sales rise above the breakeven point?

8-18B. (*Fixed Costs*) Tropical Sun projects that next year its fixed costs will total $135,000. Its only product sells for $13 per unit, of which $6 is a variable cost. The management of Tropical is considering the purchase of a new machine that will lower the variable cost per unit to $5. The new machine, however, will add to fixed costs through an increase in depreciation expense. How large can the addition to fixed costs be in order to keep the firm's breakeven point in units produced and sold unchanged?

8-19B. (*Operating Leverage*) The management of Tropical Sun did not purchase the new piece of equipment (see Problem 8-18B). Using the existing cost structure, calculate the degree of operating leverage at 40,000 units of output. Comment on the meaning of your answer.

8-20B. (*Leverage Analysis*) An analytical income statement for Tropical Sun is shown below. It is based upon an output (sales) level of 50,000 units. You may refer to the original cost structure data in Problem (8-18B).

Sales	$650,000
Variable costs	300,000
Revenue before fixed costs	$350,000
Fixed costs	135,000
EBIT	$215,000
Interest expense	60,000
Earnings before taxes	$155,000
Taxes	70,000
Net income	$ 85,000

a. Calculate the degree of operating leverage at this output level.
b. Calculate the degree of financial leverage at this level of EBIT.
c. Determine the combined leverage effect at this output level.

8-21B. (*Breakeven Point*) You are employed as a financial analyst for a single-product manufacturing firm. Your supervisor has made the following cost structure information available to you, all of which pertains to an output level of 1,700,000 units.

Return on operating assets	= 16%
Operating asset turnover	= 6 times
Operating assets	= $3.25 million
Degree of operating leverage	= 9 times

Your task is to find the breakeven point in units of output for the firm.

8-22B. (*Fixed Costs*) Saskatoon Silkscreen is forecasting fixed costs next year of $375,000. The firm's single product sells for $25 per unit and incurs a variable cost per unit of $13. The firm may acquire some new binding equipment that would lower variable cost per unit to $11. The new equipment, however, would add to fixed costs through the price of an annual maintenance agreement on the new equipment. How large can this increase in fixed costs be and still keep the firm's present breakeven point in units produced and sold unchanged?

8-23B. (*Leverage Analysis*) Your firm's cost analysis supervisor supplies you with the following analytical income statement and requests answers to the four questions listed below the statement.

Sales	$13,750,000
Variable costs	9,500,000
Revenue before fixed costs	$ 4,250,000
Fixed costs	3,000,000
EBIT	$ 1,250,000
Interest expense	250,000
Earnings before taxes	$ 1,000,000
Taxes	430,000
Net income	$ 570,000

a. At this level of output, what is the degree of operating leverage?
b. What is the degree of financial leverage?
c. What is the degree of combined leverage?
d. What is the firm's breakeven point in sales dollars?

8-24B. (*Leverage Analysis*) You are supplied with the following analytical income statement for your firm. It reflects last year's operations.

Sales	$18,000,000
Variable costs	7,000,000
Revenue before fixed costs	$11,000,000
Fixed costs	6,000,000
EBIT	$ 5,000,000
Interest expense	1,750,000
Earnings before taxes	$ 3,250,000
Taxes	1,250,000
Net income	$ 2,000,000

a. At this level of output, what is the degree of operating leverage?

b. What is the degree of financial leverage?

c. What is the degree of combined leverage?

d. If sales should increase by 15 percent, by what percent would earnings before taxes (and net income) increase?

e. What is your firm's breakeven point in sales dollars?

8-25B. (*Sales Mix and the Breakeven Point*) Wayne Automotive produces four lines of auto accessories for the major Ontario automobile manufacturers. The lines are known by the code letters A, B, C, and D. The current sales mix for Wayne and the contribution margin ratio (unit contribution margin divided by unit sales price) for these product lines are as follows:

Product Line	Percent of Total Sales	Contribution Margin Ratio
A	25.67%	40 %
B	41.33	32
C	19.67	20
D	13.33	60

Total sales for next year are forecast to be $150,000. Total fixed costs will be $35,000.

a. Prepare a table showing (1) sales, (2) total variable costs, and (3) the total contribution margin associated with each product line.

b. What is the aggregate contribution margin ratio indicative of this sales mix?

c. At this sales mix, what is the breakeven point in dollars?

8-26B. (*Sales Mix and the Breakeven Point*) Because of production constraints, Wayne Automotive (see Problem 8-25B) may have to adhere to a different sales mix for next year. The alternative plan is outlined below:

Product Line	Percent of Total Sales
A	33.33 %
B	41.67
C	16.67
D	8.33

a. Assuming all other facts in Problem 8-25B remain the same, what effect will this different sales mix have on Wayne's breakeven point in dollars?

b. Which sales mix will Wayne's management prefer?

CASE PROBLEM

HAMILTON GENERAL PRODUCERS

Breakeven Analysis, Operating Leverage, Financial Leverage

Hamilton General Producers (HGP) is a medium-sized public corporation that until recently consisted of two divisions. The Retail Furniture Group (RFG) has eight locations in the Ontario area, mostly concentrated around Toronto. These retail outlets generate sales of contemporary, traditional, and early Canadian furniture. In addition, casual and leisure furniture lines are carried by the stores. The other (old) division of HGP is its Concrete Group (CG). The CG operates three plants in the Hull area of western Quebec. These plants produce precast concrete wall panels and concrete stave farm silos. The company headquarters of HGP is located in Hamilton, Ontario. This community touches Lake Ontario and is about 40 miles west of Toronto. Since Hamilton is almost equidistant from Toronto and Niagara Falls and is a connecting hub for several major highways, it makes a sensible spot for the firm's home offices.

HGP was started ten years ago as Hamilton Producers by its current president and board chairman, Anthony Toscano. During the firm's existence it has enjoyed periods of both moderate and strong growth in sales, assets, and earnings. Key managerial decisions have always been dominated by Toscano, who openly boasts of the fact that his company has never suffered through a year of negative earnings despite the often cyclical nature of both the retail furniture (RFG) and concrete (CG) divisions.

Recently Mr. Toscano decided to acquire a third division for his firm. The division manufactures special machinery for the seafood-processing industry and is appropriately called the Seafood Industry Group (SIG). The financial settlement for the acquisition took place yesterday. Currently, the SIG consists of one manufacturing plant in Hamilton. A single product is to be manufactured, assembled, and shipped from the facility. That product, however, represents a design breakthrough and carries with it a projected contribution margin ratio of .4000. This is greater than that enjoyed by either of HGP's other two divisions.

HGP's manufacturing operation in Hamilton will produce a new machine called "The Picker." The Picker was invented and successfully tested by Ben Pinkerton, the major shareholder and manager of a small seafood-processing firm in Port Hope, Ontario. Pinkerton plans and supervises all operations at Eastern Shore Processors. Eastern Shore specializes in freezing and pasteurizing New Zealand mussel meat. Freezing and pasteurizing procedures have been a boon to the seafood industry, for they permit the processor to retain a product without spoilage in hopes of higher prices at a later date. In comparing his industry to agriculture, Pinkerton aptly states: "The freezer is our grain elevator."

The seafood-processing industry is characterized by a notable lack of capital equipment and a corresponding heavy use of human labour. Pinkerton will tell you that at Eastern Shore Processors a skilled mussel meat picker will produce about 30 pounds of meat per day. No matter how skilled the human picker, however, he will leave about 5 pounds of meat per day in the top piece of the mussel shell. This past year, after five years of trying, Pinkerton perfected a machine that would recover about 30 percent of this otherwise lost meat. In exchange for cash, Eastern Shore Processors sold all rights to The Picker to HGP. Thus, Toscano established the SIG and immediately made plans to manufacture The Picker.

Recent income statements for the older divisions of HGP are contained in Exhibits 1 and 2. Toscano has now decided to assess more fully the probable impact of the decision to establish the SIG on the financial condition of HGP. Toscano knew that such figures should have been generated prior to the decision to enter this special field, but his seasoned judgment led him to a quick choice. He has requested several pieces of information, detailed below, from his chief financial officer.

Exhibit 1 Hamilton General Producers Retail Furniture Group Income Statement, December 31, Last Year

Sales	$12,000,000
Variable costs	7,920,000
Revenue before fixed costs	$4,080,000
Fixed costs	2,544,000
EBIT	$ 1,536,000
Interest expense	192,000
Earnings before taxes	$ 1,344,000
Taxes	672,000
Net income	$ 672,000

Exhibit 2 Hamilton General Producers Concrete Group Income Statement, December 31, Last Year

Sales	$8,000,000
Variable costs	5,920,000
Revenue before fixed costs	$2,080,000
Fixed costs	640,000
EBIT	$1,440,000
Interest expense	80,000
Earnings before taxes	$1,360,000
Taxes	680,000
Net income	$ 680,000

Questions and Problems

1. Using last year's results, determine the breakeven point in dollars for HGP (i.e., before investing in the new SIG). The breakeven point is defined here in the traditional manner where EBIT = \$0. (Hint: Use an aggregate contribution margin ratio in your analysis.)
2. Using last year's results, determine what volume of sales must be reached in order to cover all before-tax costs.
3. Next year's sales for the SIG are projected to be \$4,000,000. Total fixed costs will be \$640,000. This division will have no outstanding debt on its balance sheet. HGP uses a 50 percent tax rate in all its financial projections. Using the format of Exhibits 1 and 2, construct a pro forma income statement for the SIG.
4. After the SIG begins operations, what will be HGP's breakeven point in dollars (a) as traditionally defined and (b) reflecting the coverage of all before-tax costs? Base the new sales mix on last year's sales performance for the older divisions plus that anticipated next year for the SIG. Using your answer to part (b) of this question, construct an analytical income statement demonstrating that earnings before taxes = \$0.
5. Using next year's anticipated sales volume for HGP (including the SIG), compute (a) the degree of operating leverage, (b) the degree of financial leverage, and (c) the degree of combined leverage. Comment on the meaning of each of these statistics.
6. Using projected figures, determine whether acquisition of the SIG will increase or decrease the vulnerability of HGP's earnings before interest and taxes (EBIT) to cyclical swings in sales. Show your work.
7. Review the key assumptions of cost-volume-profit analysis.

SELF-TEST SOLUTIONS

SS-1: *STEP 1.* Compute the operating profit margin:

$$(\text{margin}) \times (\text{turnover}) = \text{return on operating assets}$$

$$(\text{M}) \times (6) = 0.30$$

$$\text{M} = \frac{0.30}{6} = 0.05$$

STEP 2. Compute the sales level associated with the given output level:

$$\frac{\text{sales}}{\$20{,}000{,}000} = 6$$

$$\text{sales} = \$120{,}000{,}000$$

STEP 3. Compute EBIT:

$$(.05)(\$120{,}000{,}000) = \$6{,}000{,}000 = \text{EBIT}$$

STEP 4. Compute revenue before fixed costs. Since the degree of operating leverage is 4.5 times, revenue before fixed costs (RBF) is 4.5 times EBIT, as follows:

$$\text{RBF} = (4.5)\ (\$6{,}000{,}000) = \$27{,}000{,}000$$

STEP 5. Compute total variable costs:

$$(\text{sales}) - (\text{total variable costs}) = \$27{,}000{,}000$$

$$\$120{,}000{,}000 - (\text{total variable costs}) = \$27{,}000{,}000$$

$$\text{total variable costs} = \$93{,}000{,}000$$

STEP 6. Compute total fixed costs:

$$\text{RBF} - \text{fixed costs} = \text{EBIT}$$

$$\$27{,}000{,}000 - \text{fixed costs} = \$6{,}000{,}000$$

$$\text{Fixed costs} = \$21{,}000{,}000$$

STEP 7. Find the selling price per unit (P), and the variable cost per unit (V):

$$P = \frac{\text{sales}}{\text{output in units}} = \frac{\$120{,}000{,}000}{10{,}000{,}000} = \$12.00$$

$$V = \frac{\text{total variable costs}}{\text{output in units}} = \frac{\$93{,}000{,}000}{10{,}000{,}000} = \$9.30$$

STEP 8. Compute the breakeven point:

$$Q_B = \frac{F}{P - V} = \frac{\$21{,}000{,}000}{\$12.00 - \$9.30}$$

$$= \frac{\$21{,}000{,}000}{\$2.70} = 7{,}777{,}778 \text{ units}$$

The firm will break even when it produces and sells 7,777,778 units.

SS-2: a. $\frac{\text{Revenue before fixed costs}}{\text{EBIT}} = \frac{\$8{,}000{,}000}{\$3{,}000{,}000} = 2.67 \text{ times}$

b. $\frac{\text{EBIT}}{\text{EBIT} - I} = \frac{\$3{,}000{,}000}{\$2{,}000{,}000} = 1.50 \text{ times}$

c. $\text{DCL}_{\$20,000,000} = (2.67)(1.50) = 4.00 \text{ times}$

d. $S^* = \frac{F}{1 - \frac{VC}{S}} = \frac{\$5{,}000{,}000}{1 - \frac{\$12M}{\$20M}} =$

$\frac{\$5{,}000{,}000}{1 - 0.60} = \frac{\$5{,}000{,}000}{0.40} = \$12{,}500{,}000$

e. $(30\%)(4.00) = 120\%$

f.

Sales	\$26,000,000
Variable costs	15,600,000
Revenue before fixed costs	\$10,400,000
Fixed costs	5,000,000
EBIT	\$ 5,400,000
Interest expense	1,000,000
Earnings before taxes	\$ 4,400,000
Taxes (0.50)	2,200,000
Net income	\$ 2,200,000

We know that sales have increased by 30 percent to \$26 million from the base sales level of \$20 million.

Let us focus now on the change in earnings before taxes. We can compute that change as follows:

$$\frac{\$4{,}400{,}000 - \$2{,}000{,}000}{\$2{,}000{,}000} = \frac{\$2{,}400{,}000}{\$2{,}000{,}000} = 120\%$$

Since the tax rate was held constant, the percentage change in net income will also equal 120 percent. The fluctuations implied by the degree of combined leverage measure are therefore accurately reflected in this analytical income statement.

SS-3. a.

$$\{(P \cdot Q) - [V \cdot Q + (F)]\}\,(1 - T) = \$210{,}000$$
$$[(\$4{,}000{,}000) - (\$3{,}000{,}000) - F]\,(.7) = \$210{,}000$$
$$(\$1{,}000{,}000 - F)\,(.7) = \$210{,}000$$
$$\$700{,}000 - .7F = \$210{,}000$$
$$F = \$700{,}000$$

Fixed costs next year, then, are expected to be \$700,000.

b. $Q_B = \frac{F}{P - V} = \frac{\$700{,}000}{\$50} = 14{,}000 \text{ units}$

$$S^* = \frac{F}{1 - \frac{VC}{S}} = \frac{\$700{,}000}{1 - .75} = \frac{\$700{,}000}{.25} = \$2{,}800{,}000$$

The firm will break even (EBIT = 0) when it sells 14,000 units. With a selling price of \$200 per unit, the breakeven sales level is \$2,800,000.

CHAPTER 9

Planning the Firm's Financing Mix

LEARNING OBJECTIVES

After reading this chapter you should be able to

1. Understand the concept of an optimal capital structure.
2. Explain the main underpinnings of capital structure theory.
3. Distinguish between the independence hypothesis and dependence hypothesis as these concepts relate to capital structure.
4. Explain the moderate position on capital structure.
5. Discuss the concepts of agency costs and free cash flow in the context of capital structure management.
6. Apply the basic tools of capital structure management.
7. Explain how corporate financing policies work in practice.

INTRODUCTION

North America was recession-free from 1983 to 1990, when a general business contraction did occur. During the early nineties, many business enterprises underwent a once in a lifetime global restructuring. This period was a challenging one for many North American business firms that had loaded their balance sheets with debt over the "good times."

Financial executives had to delicately manage cash flows to service existing debt or face bankruptcy. These same executives had to give considerable thought as to how to finance the next (incremental) capital project.

One company that had to pay attention to managing its financing mix was Dominion Textile Inc. This Canadian-based textile manufacturer has both domestic and international facilities. The company's operations extend to places such as the United States, Europe, South America and the Far East. During the late eighties, Dominion Textile enjoyed strong positive earnings like most North American companies. This strong earnings performance contrasted to the company's record losses reported during the early nineties.

The company's response to both the financial environment and its difficult financial position was summarized in Dominion Textile's 1991 Annual Report. *The chairperson stated how the company would respond to its debt exposure as follows: "The objective of the restructuring is a substantial reduction of total debt which will significantly reduce future annual interest costs and will enable the corporation to concentrate its efforts and resources on fewer core businesses." The company's strategic thrust was continued into 1992 with the objective of restoring strength to the company's balance sheet and emerging with leaner operations so as to achieve leadership positions in selected fields on a global basis.*[1]

The material of this chapter will enable you to make positive contributions to strategies such as those that deal with a company's financing mix. You can also help a company avoid making serious financial errors, the consequences of which can last for several years.

CHAPTER PREVIEW

In this chapter, we direct our attention to the determination of an appropriate financing mix for the firm. First, we distinguish between financial and capital structure. We also discuss the theory of capital structure to understand the mix of long-term sources of funds used by the firm. Then, we examine the basic tools of capital structure management. We conclude with a real-world look at actual capital structure management.

OBJECTIVE 1

THE IMPORTANCE OF CAPITAL STRUCTURE

Financial structure

The mix of all funds sources that appear on the right-hand side of the balance sheet.

Capital structure

The mix of long-term sources of funds used by the firm. Basically, this concept omits short-term liabilities.

Financial structure design

The management activity of seeking the proper mix of all financing components in order to minimize the cost of raising a given amount of funds.

Hedging concept

A working-capital management policy which states that the cash flow generating characteristics of a firm's investments should be matched with the cash flow requirements of the firm's sources of financing. Very simply, short-lived assets should be financed with short-term sources of financing while long-lived assets should be financed with long-term financing.

The fundamental difference between financial structure and capital structure is that **financial structure** is the mix of all items that appear on the right-hand side of the company's balance sheet and **capital structure** is the mix of the long-term sources of funds used by the firm.[2] The relationship between financial and capital structure can be expressed through the following equation:

$$\text{(financial structure)} - \text{(current liabilities)} = \text{capital structure} \quad (9\text{-}1)$$

Prudent **financial structure design** requires answers to the following two questions:

1. What should be the maturity composition of the firm's sources of funds; in other words, how should a firm divide its total sources of funds between short- and long-term components?
2. In what proportions relative to the total should the various forms of permanent financing be utilized?

The major influence on the maturity structure of the financing plan is the nature of the assets owned by the firm. A company heavily committed to real capital investment, represented primarily by fixed assets on its balance sheet, should finance those assets with permanent (long-term) types of financial capital. Furthermore, the permanent portion of the firm's investment in current assets should likewise be financed with permanent capital. Alternatively, assets held on a temporary basis are to be financed with temporary sources.[3] The present discussion assumes that the bulk of the company's current liabilities are comprised of temporary capital.

This **hedging concept** is discussed in Chapter 17. Accordingly, our focus in this chapter is to examine the process that is usually called *capital structure management.*

The objective of capital structure management is to mix the permanent sources of funds used by the firm in a manner that will maximize the company's common

[1]Adapted from Dominion Textile Inc., *Annual Report* (1991, 1992), pp. 3, 5.

[2]For the purposes of finance, capital structure should be expressed in terms of market value and not book value.

[3]The matching of asset and liability maturities is known as the matching principle and its implications will be discussed in greater detail in Chapter 17.

Table 9-1 Balance Sheet

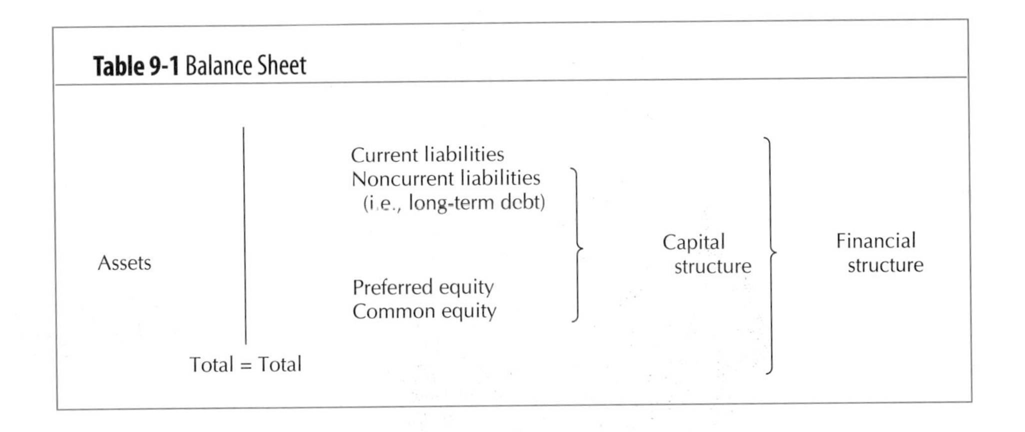

share price. Alternatively, this objective may be viewed as a search for the funds mix that will minimize the firm's composite (weighted) cost of capital. We can call this proper mix of funds sources the **optimal capital structure**.

Optimal capital structure

The unique capital structure that minimizes the firm's composite cost of long-term capital.

It makes economic sense for the firm to strive to minimize the cost of using financial capital. Both capital costs and other costs, such as manufacturing costs, share a common characteristic in that they potentially reduce the size of the cash dividend that could be paid to common shareholders.

We saw in Chapter 6 that the ultimate value of a common share depends in part on the returns investors expect to receive from holding these shares. Cash dividends comprise all (in the case of an infinite holding period) or part (in the case of a holding period less than infinity) of these expected returns. Now, hold constant all factors that could affect share price except capital costs. If these capital costs could be kept at a minimum, the dividend stream flowing to the common shareholders would be maximized. This, in turn, would maximize the firm's common share price.

If the firm's cost of capital can be affected by its capital structure, then capital structure management is clearly an important subset of business financial management.

Table 9-1 looks at Equation (9-1) in terms of a simplified balance sheet format. It helps us visualize the overriding problem of capital structure management. The sources of funds that give rise to financing fixed costs (long-term debt and preferred equity) must be combined with common equity in the proportions most suitable to the investing marketplace. If that mix can be found, then holding all other factors constant, the firm's common share price will be maximized.

While Equation (9-1) quite accurately indicates that the corporate capital structure may be viewed as an absolute dollar amount, the real capital structure problem is one of balancing the array of funds sources in a proper manner. Our use of the term *capital structure* emphasizes this latter problem of relative magnitude, or proportions.

The rest of this chapter will cover three main areas. First, we discuss the theory of capital structure to provide a perspective. Second, we examine the basic tools of capital structure management. We conclude with a real-world look at actual capital structure management.

PERSPECTIVE IN FINANCE

It pays to understand the essential components of capital structure theory. The assumption of excessive financial risk can put the firm into bankruptcy proceedings. Some argue that the decision to use little financial leverage results in an undervaluation of the firm's shares in the marketplace. Financial managers must know how to find the area of optimum financial leverage use which will enhance share value, all other considerations held constant. Thus, grasping the theory will make you better able to formulate a sound financial structure policy.

OBJECTIVE 2

A GLANCE AT CAPITAL STRUCTURE THEORY

An enduring controversy within financial theory concerns the effect of financial leverage on the overall cost of capital to the enterprise. The heart of the argument may be stated in the form of a question:[4]

> Can the firm affect its overall cost of funds, either favourably or unfavourably, by varying the mixture of financing sources used?

This controversy has taken many elegant forms in the finance literature. Most of these presentations appeal more to academics than financial management practitioners. To emphasize the ingredients of capital structure theory that have practical applications for business financial management, we will pursue an intuitive, or nonmathematical, approach to reach a better understanding of the underpinnings of this *cost of capital–capital structure argument.*

The Analytical Setting

The essentials of the capital structure controversy are best highlighted within a framework that economists would call a "partial equilibrium analysis." In a partial equilibrium analysis changes that do occur in a number of factors and have an impact upon a certain key item are ignored in order to study the effect of changes in a main factor upon that same item of interest. Here, two items are simultaneously of interest: (1) K_0, the firm's weighted cost of capital, and (2) P_0, the market price of the firm's common share. The firm's use of financial leverage is the main factor that is allowed to vary in the analysis. This means that important financial decisions, such as investment policy and dividend policy, are held constant throughout the discussion. We are concerned with the effect of changes in the financing mix on share price and capital costs.

Our analysis will be facilitated if we adopt a simplified version of the basic dividend valuation model presented in Chapter 6 in our study of valuation principles. That model is shown below as Equation (9-2):

$$P_0 = \sum_{t=1}^{\infty} \frac{D_t}{(1 + k_c)^t} \tag{9-2}$$

where P_0 = the current price of the firm's common share
D_t = the cash dividend per share expected by investors during period t
k_c = the cost of common equity capital

We can strip away some complications by making the following assumptions concerning the valuation process implicit in Equation (9-2):

1. Cash dividends paid will not change over the infinite holding period. Thus, $D_1 = D_2 = D_3 = \ldots = D_\infty$. There is no expected growth by investors in the dividend stream.
2. The firm retains none of its current earnings. This means that all of each period's per-share earnings are paid to shareholders in the form of cash dividends. The firm's dividend payout ratio is 100 percent. Cash dividends per share in Equation (9-2), then, also equal earnings per share for the time period.

Under these assumptions, the cash dividend flowing to investors can be viewed as a level payment over an infinite holding period. The payment stream is perpetual, and according to the mathematics of perpetuities, Equation (9-2) reduces to Equation (9-3), where E_t represents earnings per share during time period t.

$$P_0 = \frac{D_t}{k_c} = \frac{E_t}{k_c} \tag{9-3}$$

[4]The question can also be stated as follows: Can firm value be affected by the design of the company's capital structure? This type of question recognizes that all cost-of-capital estimation techniques have a valuation underpinning.

In addition to the suppositions noted above, the analytical setting for the discussion of capital structure theory includes the following assumptions:

1. Corporate income is not subject to any taxation. The major implication of removing this assumption is discussed later.
2. Capital structures consist of only shares and bonds. Furthermore, the degree of financial leverage used by the firm is altered by the issuance of common shares with the proceeds used to retire existing debt, or the issuance of debt with the proceeds used to repurchase shares. This permits leverage use to vary but maintains constancy of the total book value of the firm's capital structure.
3. The expected values of all investors' forecasts of the future levels of net operating income (EBIT) for each firm are identical. Say that you forecast the average level of EBIT to be achieved by General Motors of Canada over a very long period ($n \rightarrow \infty$). Your forecast will be the same as our forecast, and both will be equal to the forecasts of all other investors interested in General Motors common shares. In addition, we do not expect General Motors' EBIT to grow over time. Each year's forecast is the same as any other year's. This is consistent with our assumption underlying Equation (9-3), where the firm's dividend stream is not expected to grow.
4. Securities are traded in perfect or efficient financial markets. This means that transaction costs and legal restrictions do not impede any investors' incentives to execute portfolio changes that they expect will increase their wealth. Information is freely available. Moreover, corporations and individuals that are equal credit risks can borrow funds at the same rates of interest.

This completes our description of the analytical setting. We now discuss three differing views on the relationship between use of financial leverage and common share value.

PERSPECTIVE IN FINANCE

The discussion and illustrations of the two "extreme" positions on the importance of capital structure that follow are meant to highlight the critical differences between differing viewpoints. This is not to say that the markets really behave in strict accordance with either position—they don't. The point is to identify polar positions on how things might work. Then by relaxing various restrictive assumptions, a more useful theory of how financing decisions are actually made becomes possible. That results in the third, or "moderate," view.

Extreme Position 1: The Independence Hypothesis (NOI Theory)[5]

OBJECTIVE 3

The crux of this position is that the firm's weighted cost of capital, K_0, and common share price, P_0, are both independent of the degree to which the company chooses to use financial leverage. In other words, no matter how modest or excessive the firm's use of debt financing, its common share price will not be affected. Let us illustrate the mechanics of this point of view.

Suppose that Rix Camper Manufacturing Company has the following financial characteristics:

[5]The net operating income and net income capitalization methods, which are referred to here as "extreme positions 1 and 2," were first presented in comprehensible form by Durand. See David Durand, "Costs of Debt and Equity Funds for Business: Trends and Problems of Measurement," *Conference on Research in Business Finance* (New York: National Bureau of Economic Research, 1952, reprinted in *The Management of Corporate Capital*, Ezra Solomon, ed. New York: Free Press, 1959, pp. 91–116.) The leading proponents of the independence hypothesis in its various forms are Professors Modigliani and Miller. See Franco Modigliani and Merton H. Miller, "The Cost of Capital, Corporation Finance and the Theory of Investment," *American Economic Review* 48 (June 1958, pp. 261–297; Franco Modigliani and Merton H. Miller, "Corporate Income Taxes and the Cost of Capital: A Correction," *American Economic Review* 53 (June 1963), pp. 433–443; and Merton H. Miller, "Debt and Taxes," *Journal of Finance* 32 (May 1977), pp. 261–275.

Expected level of net operating income	(EBIT) =	\$2,000,000
Common shares outstanding	=	2,000,000 shares
Common share price, P_0	=	\$10 per share
Expected level of net operating income	(EBIT) =	\$2,000,000
Dividend payout ratio	=	100 percent

Currently the firm uses no financial leverage; its capital structure consists entirely of common equity. Earnings per share and dividends per share equal \$1 each. When the capital structure is all common equity, the cost of common equity, k_c, and the weighted cost of capital, K_0, are equal. If Equation (9-3) is restated in terms of the cost of common equity, we have for Rix Camper

$$k_c = \frac{D_t}{P_0} = \frac{\$1}{\$10} = 10\%$$

Now, the management of Rix Camper decides to use some debt capital in its financing mix. The firm sells \$8 million worth of long-term debt at an interest rate of 6 percent. With no taxation of corporate income, this 6 percent interest rate is the cost of debt capital, k_b. The firm uses the proceeds from the sale of the bonds to repurchase 40 percent of its outstanding common shares. After the capital structure change has been accomplished, Rix Camper Manufacturing Company has the financial characteristics displayed in Table 9-2.

On the basis of the predicted data described above, we notice that the recapitalization (capital structure change) of Rix Camper will result in a dividend paid to owners that is 26.7 percent higher than it was when the firm used no debt in its capital structure. Will this higher dividend result in a lower weighted cost of capital to Rix and a higher common share price? According to the principles of the independence hypothesis, the answer is no.

Modigliani and Miller Hypothesis–Value Invariance

Proposition I

Although the independence hypothesis suggests that the total market value of the firm's outstanding securities is unaffected by the manner in which the right-hand side of the balance sheet is arranged, none of the proponents of this hypothesis offered a formal proof. In 1958, Modigliani and Miller used intuition from macroeconomics to provide a formal and rigorous proof that showed the value of the firm to be independent of its capital structure. The intuition of this proof was founded on concepts which are used to construct a nation's balance sheet. This type of bal-

Table 9-2 Rix Camper Company Financial Data Reflecting the Capital Structure Adjustment

Capital Structure Information	
Common shares outstanding = 1,200,000	
Bonds at 6 percent = \$8,000,000	
Earnings Information	
Expected level of net operating income (EBIT) =	\$2,000,000
Less: Interest expense	480,000
Earnings available to common shareholders	\$1,520,000
Earnings per share (E_t)	\$1.267
Dividends per share (D_t)	\$1.267
Percentage change in both earnings per share and dividends per share relative to the unlevered capital structure	26.7 percent

ance sheet summarizes the nation's income and wealth and is constructed from the balance sheets of both business and household sectors. The balance sheets of both business firms and households are summarized as follows:

Balance Sheet of Business Firms

Assets	Liabilities
Productive Capital	Debts owed to households
	Equity in the firms owned by households

Balance Sheet of Households

Assets	Liabilities
Debt of firms	Households' net worth
Equity in firms	

From our knowledge of macroeconomics, the consolidation of the balance sheets of these two sectors will result in the following balance sheet:

Nation's Balance Sheet

Assets	Liabilities
Productive Capital	Households' net worth

The net effect of this type of consolidation is to eliminate the debt and equity securities from the balance sheets of both business firms and households. In other words, the nation's productive capital is financed by the net worth of households. However, the elimination of both debt and equity from the nation's balance sheet provides the following intuition: (1) the value of the nation's assets is independent of the value of the securities that business firms issue; and (2) the transactions of selling and buying securities between business firms and households are only intermediary transactions that partition risk and return of the nation's productive capital (i.e., national assets) among the different financing claims which make up the net worth of households (i.e., debt and equity). This latter viewpoint enabled Modigliani and Miller to formulate a proof for the value invariance hypothesis by using an individual's capability to use arbitrage transactions to wash out the debt-to-equity ratios (i.e., the firm's capital structure) found on the balance sheets of both business firms and households. In other words, an undervalued firm which used debt in its capital structure could be returned to its proper value if an arbitrageur buys an appropriate portion of this firm's debt and equity. For example, the balance sheets of the business firm and arbitrageur would be shown as follows:

Business Firm's Balance Sheet

Assets	Liabilities
Total Assets	Debt of firm
	Equity

Arbitrageur's Balance Sheet

Assets	Liabilities
Debt	Net worth
Equity	

The consolidation of both of these balance sheets leaves the arbitrageur with the following balance sheet.

Arbitrageur's Balance Sheet

Assets	Liabilities
Original debt	Net worth
Original equity	
Equity investment in firm	
Assets	

The net effect of consolidating both balance sheets is to make the arbitrageur a shareholder who owns a pure equity stream of the firm. Note that the interest stream is paid by the firm which is owned by both shareholders and the arbitrageur. This interest stream is the same interest stream that is received by both the debtholders and the arbitrageur. The net effect is that both interest streams cancel each other out and we wash away the firm's debt from the arbitrageur's balance sheet.

In a similar fashion, an unlevered firm which is undervalued as compared to a levered firm, that is a firm with corporate debt in its capital structure, can be returned to its proper value by arbitrageurs borrowing on their individual accounts (i.e., use of homemade leverage).

These two arbitrage proofs suggest that the total market value of the firm's outstanding securities is unaffected by the manner in which the right-hand side of the balance sheet is arranged. In other words, the total market value of the outstanding equity will always be the same regardless of how much or how little debt is actually used by the company. Figure 9-1 illustrates the crux of this position on financing choice. If the capital structure has no impact on the total market value of the company, the firm's value is determined by the marketplace capitalizing (discounting) the firm's expected net operating income stream, as follows:

$$V_U = \frac{NOI}{k_c^U} = \frac{NOI}{K_0} \tag{9-4}$$

where, V_U = the value of the unlevered firm
NOI = the expected level of net operating income
k_c^U = the cost of equity of the unlevered firm
K_0 = the weighted cost of capital

NOI approach to valuation

The concept from financial theory that suggests the firm's capital structure has no impact on its market valuation.

For Rix Camper Manufacturing Company, the firm's value is determined as follows:

$$V_U = \frac{\$2{,}000{,}000}{0.10} = \$20{,}000{,}000$$

Therefore, the first proposition of the value invariance model is sometimes called the independence hypothesis, as firm value is independent of capital structure design. Therefore the independence hypothesis rests on what is called the **net operating income (NOI) approach to valuation**.

Figure 9-1 Firm Value and Capital Structure Design

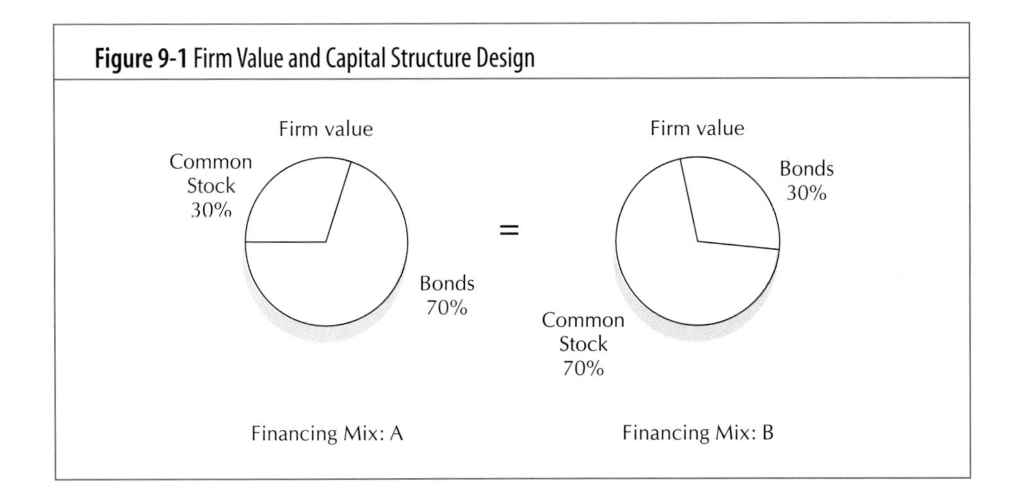

Proposition II

Modigliani and Miller's model without corporate taxes also asserted that the shareholder's required rate of return for the levered firm is the sum of the shareholder's required rate of return for the unlevered firm and a risk premium. The size of the risk premium is dependent upon the amount of financial leverage, the shareholder's required rate of return for the unlevered firm and the cost of debt to the levered firm. We summarize the shareholder's required rate of return for the levered firm through the following equation:

$$k_c^L = k_c^U + \left(\frac{M}{S}\right)(k_c^U - k_b) \tag{9-5}$$

where, k_c^L = shareholder's required rate of return for the levered firm
k_c^U = shareholder's required rate of return for the unlevered firm
k_b = cost of debt
M = level of debt
S = level of equity

As a result of these two propositions, we assert that financial leverage does not affect either the value of the levered firm or the weighted cost of capital to the levered firm. Although increases in financial leverage which have been created by greater amounts of cheaper debt, provide a benefit to the firm. This benefit is offset by the shareholder requiring a higher rate of return on equity due to the increase in risk caused by the increase in financial leverage.

The valuation process derived from the invariance model is a simple one, whereby the market value of the firm's common shares turns out to be a residual of the valuation process.

Recall that before Rix Camper's recapitalization, the total market value of the firm was \$20 million (2,000,000 common shares times \$10 per share). The firm's cost of common equity, k_c, and its weighted cost of capital, K_0, were each equal to 10 percent. The composite discount rate, K_0, is used to arrive at the market value of the firm's securities. After the recapitalization, we have for Rix Camper

Expected level of net operating income		\$ 2,000,000
Capitalized at $K_0 = 10$ percent		
=	Market value of debt and equity	\$20,000,000
−	Market value of the new debt	8,000,000
=	Market value of the common shares	\$12,000,000

With this valuation format, what is the market price of each common share? Since we know that 1,200,000 common shares are outstanding after the capital structure change, the market price per share is $10 ($12,000,000/ 1,200,000). This is exactly the market value per share, P_0, that existed before the change.

Now, if the firm is using some debt that has an explicit cost of 6 percent, k_b, and the weighted (composite) cost of capital, K_0, is still 10 percent, it stands to reason that the cost of common equity for the levered firm has risen above its previous level of 10 percent. What will the cost of common equity be in this situation? As we did previously, we can take Equation (9-3) and restate it in terms of k_c^L, the cost of common equity for the levered firm. After the recapitalization, the cost of common equity for Rix Camper is shown to rise to 12.67 percent:

$$k_c^L = \frac{D_t}{P_0} = \frac{\$1.267}{10} = 12.67\%$$

The cost of common equity for Rix Camper is 26.7 percent higher than it was before the capital structure shift. Notice in Table 9-2 that this is exactly equal to the percentage increase in earnings and dividends per share that accompanies the same capital structure adjustment. This highlights a fundamental relationship that is an integral part of the independence hypothesis. It concerns the perceived behaviour in the firm's cost of common equity as expected dividends (earnings) increase relative to a financing mix change:

$$\text{percentage change in } k_c^L = \text{percentage change in } D_t$$

In this framework the use of a greater degree of financial leverage may result in greater earnings and dividends, but the firm's cost of common equity will rise at precisely the same rate as the earnings and dividends do. Thus, the inevitable tradeoff between the higher expected return in dividends and earnings (D_t and E_t) and increased risk that accompanies the use of debt financing manifests itself in a linear relationship between the cost of common equity, k_c^L, and financial leverage use. This view of the relationship between the firm's cost of funds and its financing mix is shown graphically in Figure 9-2. Figure 9-3 relates the firm's share price to its financing mix under the same set of assumptions.

In Figure 9-2 the firm's overall cost of capital, K_0, is shown to be unaffected by an increased use of financial leverage.[6] If more debt is used in the capital structure, the cost of common equity will rise at the same rate that additional earnings are generated. This will keep the weighted cost of capital to the corporation unchanged. Figure 9-3 shows that since the cost of capital will not change with the leverage use, neither will the firm's share price.

Debt financing, then, has two costs—its **explicit cost of capital**, k_b, and an implicit cost. The **implicit cost of debt** is the change in the cost of common equity brought on by using financial leverage (additional debt). The real cost of debt is the sum of these explicit and implicit costs. In general, the real cost of any source of capital is its explicit cost, plus the change that it induces in the cost of any other source of funds.

Explicit cost of capital

The cost of capital for any funds source considered in isolation from other fund sources.

Implicit cost of debt

The change in the cost of common equity caused by the choice to use additional debt.

Followers of the independence hypothesis argue that the use of financial leverage brings a change in the cost of common equity large enough to offset the benefits of higher dividends to investors. Debt financing is not as cheap as it first appears to be. This will keep the weighted cost of funds constant. The implication for management is that one capital structure is as good as any other; financial officers should not waste time searching for an optimal capital structure. One capital structure, after all, is as beneficial as any other, because all result in the same weighted cost of capital.

[6]The firm's overall cost of capital is unaffected by the increase in financial leverage: $K_0 = (.4)(6.00\%) + (.6)(12.67\%) = 10\%$.

Figure 9-2 Capital Costs and Financial Leverage: No Taxes—The Independence Hypothesis (NOI Theory)

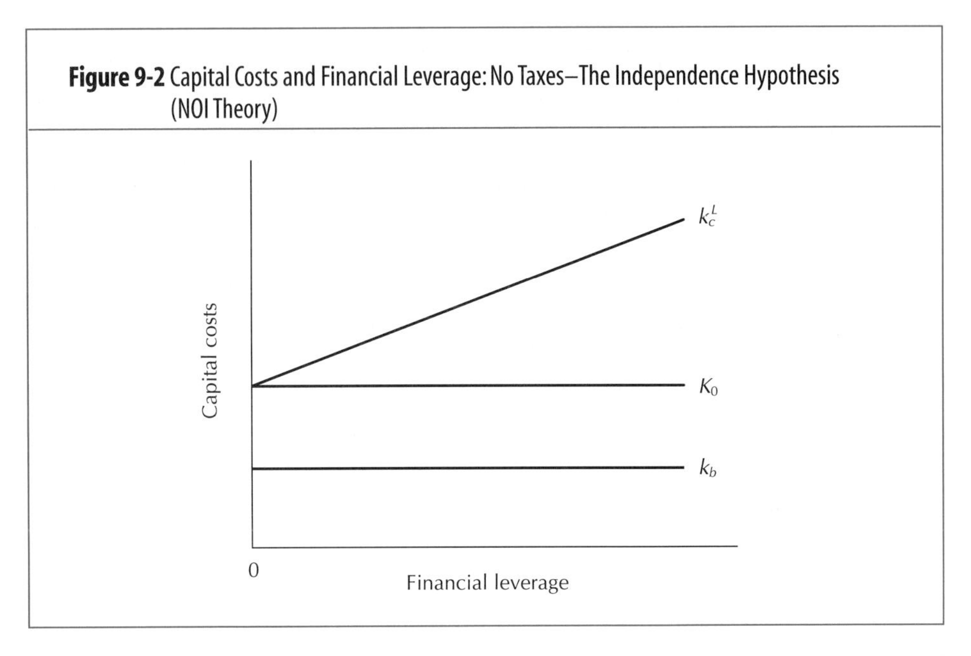

Figure 9-3 Share Price and Financial Leverage: No Taxes—The Independence Hypothesis (NOI Theory)

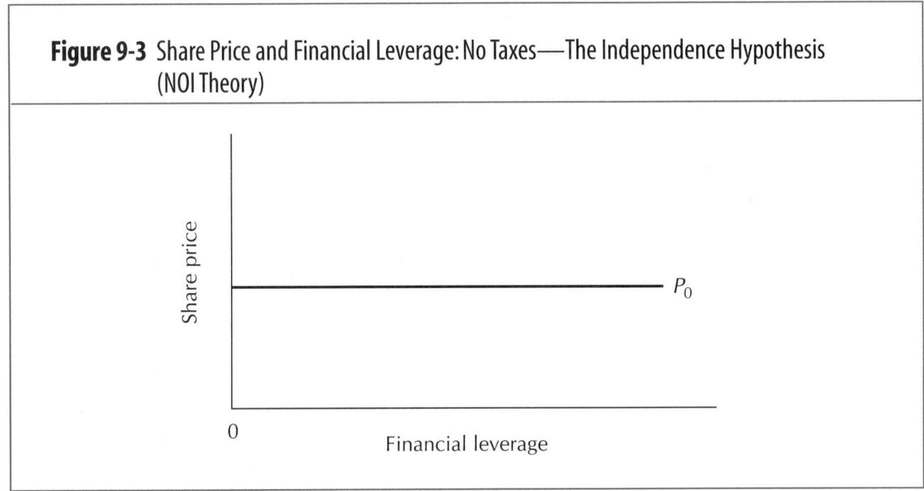

BACK TO THE BASICS

The suggestions from capital structure theory that one capital structure is just as good as any other within a perfect ("pure") market framework relies directly on ***Axiom 1: The Risk-Return Tradeoff—We Won't Take on Additional Risk Unless We Expect to Be Compensated with Additional Return.*** *This means that using more debt in the capital structure will not be ignored by investors in the financial markets. These rational investors will require a higher return on common stock investments in the firm that uses more leverage (rather than less), to compensate for the increased uncertainty stemming from the addition of the debt securities in the capital structure.*

Extreme Position 2: The Dependence Hypothesis (NI Theory)

OBJECTIVE 3

The dependence hypothesis is at the opposite pole from the independence hypothesis. It suggests that both the weighted cost of capital, K_0, and common share price, P_0, are affected by the firm's use of financial leverage. No matter how modest or excessive the firm's use of debt financing, both its cost of debt capital, k_b, and cost of equity capital, k_c^L, will not be affected by capital structure management. Since the cost of debt is less than the cost of equity, greater financial leverage will lower the firm's weighted cost of capital indefinitely. Greater use of debt financing will thereby have a favourable effect on the company's common share price. By returning to the Rix Camper situation, we can illustrate this point of view.

The same capital structure shift is being evaluated. That is, management will market \$8 million of new debt at a 6 percent interest rate and use the proceeds to purchase its own common shares. Under this approach the market is assumed to capitalize (discount) the expected earnings available to the common shareholders in order to arrive at the aggregate market value of the common shares. The market value of the firm's common equity is not a residual of the valuation process. After the recapitalization, the firm's cost of common equity, k_c^L, will still be equal to 10 percent. Thus, a 10 percent cost of common equity is applied in the following format:

	Expected level of net operating income	\$ 2,000,000
−	Interest expense	480,000
=	Earnings available to common shareholders capitalized at k_c^L = 10 percent	\$ 1,520,000
=	Market value of the common shares	\$15,200,000
+	Market value of the new debt	8,000,000
=	Market value of debt and equity	\$23,200,000

NI approach to valuation

The concept from financial theory that suggests the firm's capital structure has a direct impact upon and can increase its market valuation.

When we assume that the firm's capital structure consists only of debt and common equity, earnings available to the common shareholders is synonymous with net income. In the valuation process outlined above, it is net income that is actually capitalized to arrive at the market value of the common equity. Because of this, the dependence hypothesis is also called the **net income approach to valuation.**

Notice that the total market value of the firm's securities has risen to \$23,200,000 from the \$20 million level that existed before the firm moved from the unlevered to the levered capital structure.[7] The per-share value of the common shares is also shown to rise under this valuation format. With 1,200,000 shares outstanding, the market price per share is \$12.67 (\$15,200,000/1,200,000).

This increase in the share price to \$12.67 represents a 26.7 percent rise over the previous level of \$10 per share. This is exactly equal to the percentage change in earnings per share and dividends per share calculated in Table 9-1. This permits us to characterize the dependence hypothesis in a very succinct fashion:

percentage change in k_c^L = 0 percent < percentage change in D_t
(over all degrees of leverage)
percentage change in P_0 = percentage change in D_t

The dependence hypothesis suggests that the explicit and implicit costs of debt are one and the same. The use of more debt does not change the firm's cost of common equity.[8] Using more debt, which is explicitly cheaper than common equity, will lower the firm's weighted cost of capital, K_0. If you take the market value of Rix Camper's common share according to the net income theory of \$15,200,000 and express it as a percent of the total market value of the firm's securities, you get a market value weight of .655 (\$15,200,000/\$23,200,000). In a similar fashion, the

[7]Although the aggregate equity value decreases by only \$4.8 million (\$20.0 – \$15.2); at the same time, the value of the debt claim has increased from zero to \$8 million. As a result, the value of the firm increases from \$20 million to \$23.2 million.

[8]The NI model assumes that the choice to use debt does not affect the cost of common equity. In practice this may be true over certain ranges of leverage use. Note that some firms do increase their debt usage and maintain a given bond rating.

market value weight of Rix Camper's debt is found to be .345 ($8,000,000/ $23,200,000). After the capital structure adjustment, the firm's weighted cost of capital becomes

$$K_0 = (.345)(6.00\%) + (.655)(10.00\%) = 8.62\%$$

So, changing the financing mix from all equity to a structure including both debt and equity lowered the weighted cost of capital from 10 percent to 8.62 percent. The ingredients of the dependence hypothesis are illustrated in Figures 9-4 and 9-5.

The implication for management from Figures 9-4 and 9-5 is that the firm's weighted cost of capital, K_0, will decline as the debt-to-equity ratio increases. This also implies that the company's common-share price will rise with increased leverage use. Since the cost of capital decreases continuously with leverage, the firm should use as much leverage as is possible. Next, we will move toward reality in the analytical setting of our capital structure discussion. This is accomplished by relaxing some of the major assumptions that surrounded the independence and dependence hypotheses.

Figure 9-4 Capital Costs and Financial Leverage: No Taxes—The Dependence Hypothesis (NI Theory)

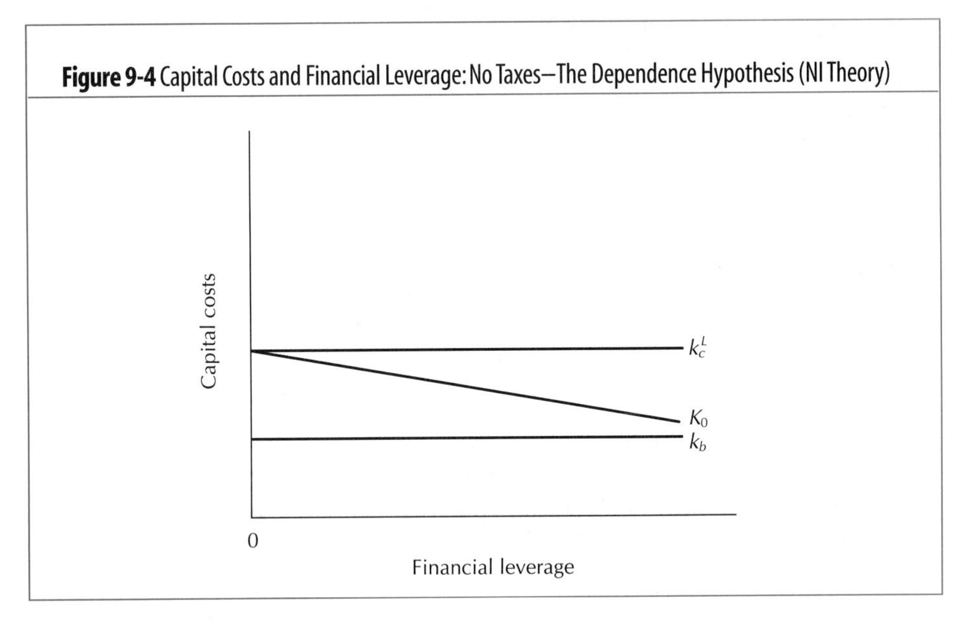

Figure 9-5 Share Price and Financial Leverage: No Taxes—The Dependence Hypothesis (NI Theory)

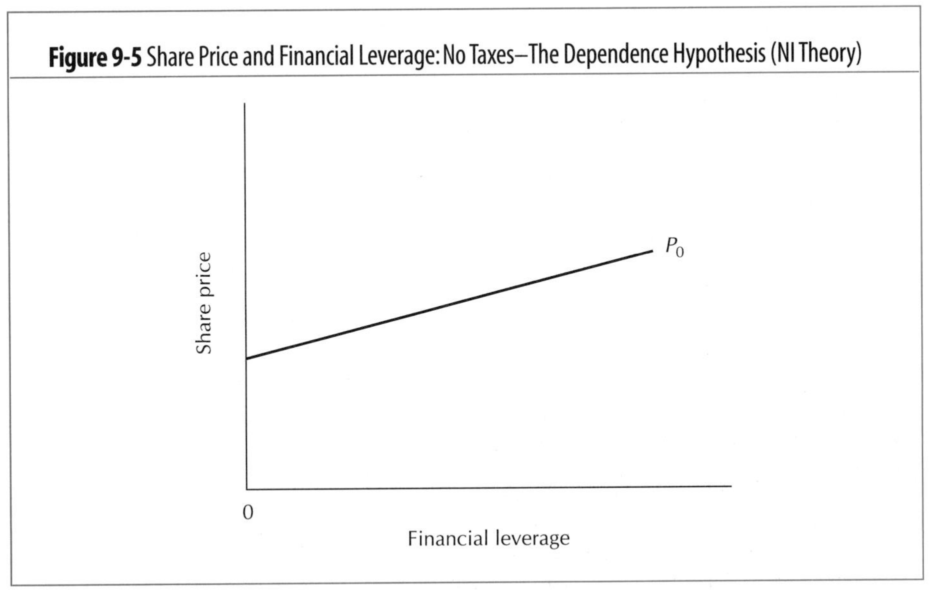

A Moderate Position: Corporate Income Is Taxed and Firms May Fail

In general, an analysis of extreme positions may be useful in that you are forced to sharpen your thinking not only about the poles, but also the situations that span the poles. In microeconomics the study of perfect competition and monopoly provides a better understanding of the business activity that occurs in the wide area between these two model markets. In a similar fashion, the study of the independence and dependence hypotheses of the importance of capital structure helps us formulate a more informed view of the possible situations between those polar positions.

We turn now to a description of the cost of capital–capital structure relationship that has rather wide appeal to both business practitioners and academics. This moderate view (1) admits to the fact that interest expense is tax deductible and (2) acknowledges that the probability of the firm's suffering bankruptcy costs are directly related to the company's use of financial leverage.

Modigliani and Miller Hypothesis with Corporate Taxes

This portion of the analysis recognizes that corporate income is subject to taxation. Furthermore, we assume that interest expense is tax deductible for purposes of computing the firm's tax bill. In this environment the use of debt financing should result in a higher total market value for the firm's outstanding securities. We will see why subsequently.

We continue with our Rix Camper Manufacturing Company example. In the case of no taxes and all equity financing, the value of the firm to the shareholders was determined as follows:

$$V_U = \frac{NOI}{k_c^U} \tag{9-4}$$

$$= \frac{\$2,000,000}{0.10}$$

$$= \$20 \text{ million}$$

Furthermore, the total cash distribution to the shareholders is $2 million in cash dividends. In the second case of no taxes with debt and equity financing, the value of the levered firm is determined by the following equation:

$$V_L = S_L + B_0 \tag{9-6}$$

$$= \frac{(NOI - I)}{k_c^L} + B_0$$

$$= \frac{(\$2,000,000 - \$480,000)}{0.1267} + \$8,000,000$$

$$= \$20,000,000$$

Again, the sum of cash dividends paid to common shareholders and the interest expense paid to bondholders is $2 million. Thus, both the value of the firm and the sum of the cash flows paid to contributors of debt and equity capital are not affected by the financing mix of the firm.

Tax shield

The element from the federal tax code that permits interest costs to be deductible when computing the firm's tax bill. The dollar difference (the shield) flows to the firm's security holders.

When corporate income is taxed by Revenue Canada, however, the sum of the cash flows made to all contributors of financial capital is affected by the firm's financing mix. Table 9-3 illustrates this point.

If Rix Camper makes the capital-structure adjustment identified in the preceding sections of this chapter, the total payments to equity and debt holders will be $240,000 greater than under the all-common-equity capitalization. Where does this $240,000 come from? The government's take, through taxes collected, is lower by that amount. This difference, which flows to the Rix Camper security holders, is called the **tax shield** on interest. In general, it may be calculated as follows:

Table 9-3 Rix Camper Company Cash Flows to All Investors—The Case of Taxes

	Unlevered Capital Structure	Levered Capital Structure
Expected level of net operating income	\$2,000,000	\$2,000,000
Less: Interest expense	0	480,000
Earnings before taxes	\$2,000,000	\$1,520,000
Less: Taxes at 50%	1,000,000	760,000
Earnings available to common shareholders	\$1,000,000	\$ 760,000
Expected payments to *all* security holders	\$1,000,000	\$1,240,000

$$PV(\text{interest tax shields}) = \frac{(R_b)(M)(T)}{k_b} \tag{9-7}$$

where R_b = the interest rate paid on the debt
M = the principal amount of the debt
T = the firm's combined tax rate
k_b = the required yield on the firm's debt

If we assume that both the interest rate paid on the firm's debt (R_b) and the required yield on the firm's debt (k_b) are equal, Equation (9-7) is simplified as follows:

$$PV(\text{interest tax shields}) = (M)(T) \tag{9-8}$$

Proposition I Modigliani and Miller asserted that the value of the levered firm was dependent on the value of leverage resulting from the firm's use of debt. In other words, the moderate position on the importance of capital structure presumes the tax shield must have value in the marketplace. The value of a levered firm in an environment of corporate taxes is determined as follows:

$$V_L = V_U + PV(\textit{Interest Tax Shields}) \tag{9-9}$$

To gain further insight into this equation, we examine both components of the right-hand side of the equation. Modigliani and Miller asserted that the value of the unlevered firm in an environment of corporate taxes is proportional to the firm's expected return. As a result, the value of the unlevered firm is determined as follows:

$$V_U = \frac{(NOI)(1 - T)}{k_c^U} \tag{9-10}$$

where, V_U = the value of the unlevered firm
NOI = the expected level of net operating income
k_c^U = the cost of equity of the unlevered firm
T = the corporate tax rate

Equation (9-7) defined the present value of interest tax shields; this tax benefit will increase the total market value of the firm, as shown through the following equation:

$$V_L = \frac{(NOI)(1 - T)}{k_c^U} + \frac{(R_b)(M)(T)}{k_b} \tag{9-11}$$

If the cost of debt to the levered firm, k_b, and the interest rate paid on debt are assumed equal, Equation (9-11) is simplified, such that the value of the levered firm is described as follows:

$$V_L = V_U + MT \tag{9-12}$$

In other words, the gain from leverage is proportional to the level of debt, *M*. As a result, we can now assert that financial leverage does affect the firm's value.

Proposition II Modigliani and Miller's model with corporate taxes also described the effect of financial leverage on the shareholder's required rate of return for a levered firm using the following equation:

$$k_c^L = k_c^U + \left(\frac{M}{S}\right)\left(k_c^U - k_b\right)\left(1 - T\right) \tag{9-13}$$

where, k_c^L = shareholder's required rate of return for the levered firm
k_c^U = shareholder's required rate of return for the unlevered firm
k_b = cost of debt
M = level of debt
S = level of equity
T = corporate tax rate

The shareholder's required rate of return for the levered firm is the sum of the shareholder's required rate of return for the unlevered firm and a risk premium. The size of the risk premium is dependent upon the amount of financial leverage, the shareholder's required rate of return for the unlevered firm, the cost of debt to the levered firm and the firm's tax rate.

As a result of these two propositions, we assert that the financial leverage affects the value of the levered firm, the shareholder's required rate of return and the weighted cost of capital to the levered firm. However, does this imply that a firm can increase its value indefinitely and lower its cost of capital continuously by using more and more financial leverage? Common sense would tell us no! So would most financial managers and academicians. The acknowledgement of bankruptcy costs provides one possible rationale. In addition, Modigliani and Miller have also indicated that the gain from leverage may not increase in direct proportion to the level of debt. In other words, the value of the levered firm may increase with debt but at a decreased rate, at least beyond some level of debt used. Furthermore, the treatment of taxes has neglected the role of different personal taxation rates for different sources of return which can be generated by a corporation; for example, the difference in tax treatment of interest and dividends at the personal level.

BACK TO THE BASICS

The section above on the "Modigliani and Miller Hypothesis with Corporate Taxes" is a compelling example of ***Axiom 8: Taxes Bias Business Decisions.*** *We have just seen that corporations have an important incentive provided by the Income Tax Act to finance projects with debt securities rather than new issues of common stock. The interest expense on the debt issue will be tax deductible. The common stock dividends will not be tax deductible. So firms can indeed increase their total after-tax cash flows available to all investors in their securities by using financial leverage. This element of the Canadian Income Tax Act should also remind you of* ***Axiom 3: Cash—Not Profits—Is King: Measuring the Timing of Costs and Benefits.***

Likelihood of Firm Failure

The probability that the firm will be unable to meet the financial obligations identified in its debt contracts increases as more debt is employed. The highest costs would be incurred if the firm actually went into bankruptcy proceedings. Here, assets would be liquidated. If we admit that these assets might sell for something less than their perceived market values, equity investors and debt holders could both suffer losses. Other problems accompany bankruptcy proceedings. Lawyers and accountants have to be hired and paid. Managers must spend time preparing lengthy reports for those involved in the legal action.

Milder forms of financial distress also have their costs. As their firm's financial condition weakens, creditors may take action to restrict normal business activity. Suppliers may not deliver materials on credit. Profitable capital investments may have to be forgone, and dividend payments may even be interrupted. At some

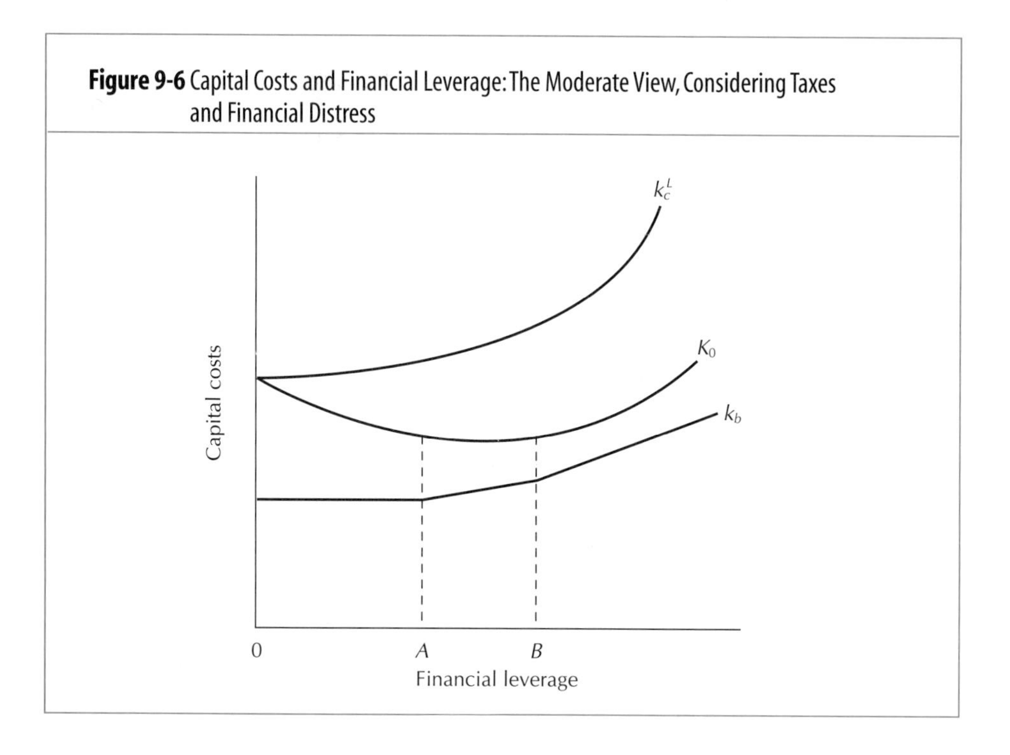

Figure 9-6 Capital Costs and Financial Leverage: The Moderate View, Considering Taxes and Financial Distress

point the expected cost of default will be large enough to outweigh the tax shield advantage of debt financing.[9] The firm will turn to other sources of financing, mainly common equity. At this point the real cost of debt is thought to be higher than the real cost of common equity.

Moderate View: Saucer-Shaped Cost of Capital Curve

OBJECTIVE 4

This moderate view of the relationship between financing mix and the firm's cost of capital is depicted in Figure 9-6. The result is a saucer-shaped (or U-shaped) weighted cost of capital curve, K_0.[10] The firm's average cost of equity, k_c^L, is seen to rise over all positive degrees of financial leverage use. For a while the firm can borrow funds at a relatively low cost of debt, k_b. Even though the cost of equity is rising, it does not rise at a fast enough rate to offset the use of the less expensive debt financing. Thus, between points 0 and A on the financial leverage axis, the weighted cost of capital declines and the share price increases.

Eventually, the threat of financial distress causes the cost of debt to increase. In Figure 9-6 this increase in the cost of debt shows up in the average cost of debt curve, k_b, at point *A*. Between points *A* and *B*, mixing debt and equity funds produces an average cost of capital that is (relatively) flat. The firm's **optimal range of financial leverage** lies between points *A* and *B*. All capital structures between these two points are optimal because they produce the lowest weighted cost of capital. As we said in the introduction to this chapter, finding this optimal range of financing mixes is the objective of capital structure management.

Point B signifies the firm's debt capacity. **Debt capacity** is the maximum proportion of debt the firm can include in its capital structure and still maintain its

Optimal range of financial leverage

The range of various financial structure combinations that generate the lowest composite cost of capital for the firm.

Debt capacity

The maximum proportion of debt that the firm can include in its capital structure and still maintain its lowest composite cost of capital.

[9]Even this argument that the tradeoff between bankruptcy costs and the tax shield benefit of debt financing can lead to an optimal structure has its detractors. See Robert A. Haugen and Lemma W. Senbet, "The Insignificance of Bankruptcy Costs to the Theory of Optimal Capital Structure," *Journal of Finance* 33 (May 1978), pp. 383–393.

[10]In reality these cost curves, just as valuations, are unique to each firm. This makes "the" optimal capital structure difficult to estimate. As a result, we settle for identifying an optimal range of leverage use.

lowest cost of capital. Beyond point B, additional fixed-charge capital can be attracted only at very costly interest rates. At the same time, this excessive use of financial leverage would cause the firm's cost of equity to rise at a faster rate than previously. The weighted cost of capital would then rise quite rapidly, and the firm's share price would decline.

This version of the moderate view as it relates to the firm's share price is characterized below. The notation is the same as that found in our discussion of the independence and dependence hypotheses.

1. Between points 0 and A:
   $0 <$ percentage change in $P_0 <$ percentage change in D_t
2. Between points A and B:
   percentage change in $P_0 = 0$
3. Beyond point B:
   percentage change in $P_0 < 0$

PERSPECTIVE IN FINANCE

Given the same task or assignment, it is quite likely that you will do it better for yourself than for someone else. If you are paid well enough, you might do the job about as effectively for that other person. Once you receive compensation, your work will be evaluated by someone. This process of evaluation is called "monitoring" within most discussions on agency costs.

This describes the heart of what is called the "agency problem." The large size of North American firms has caused the owners and managers of these firms to become (for the most part) separate groups of individuals. An inherent conflict exists, therefore, between managers and shareholders for whom managers act as agents in carrying out their objectives (for example, corporate goals). The following discussion relates the agency problem to the financial decision-making process of the firm.

Firm Value and Agency Costs

Although we have stated that the maximization of shareholder wealth is the goal of the firm, in reality the agency problem may interfere with the implementation of this goal. One form of the agency problem occurs because of the separation of management and the ownership of the firm. The firm's managers can be thought of as agents for the firm's shareholders.[11] To ensure that agent-managers act in the shareholders' best interests requires that they have (1) proper incentives to do so and (2) that their decisions are monitored. The incentives usually take the form of executive compensation plans and managerial "perks." For example, perks may take the form of a bloated support staff, country club memberships, luxurious corporate planes, or other amenities of a similar nature. Monitoring requires that certain costs be borne by the shareholders, such as (1) bonding the managers, (2) auditing financial statements, (3) structuring the organization in unique ways that

[11]Economists have studied the problems associated with control of the corporation for decades. An early, classic work on this topic was A. A. Berle, Jr., and G. C. Means, *The Modern Corporation and Private Property* (New York: Macmillan, 1932). The recent emphasis in corporate finance and financial economics stems from the important contribution of Michael C. Jensen and William H. Meckling, "Theory of the Firm: Managerial Behavior, Agency Costs and Ownership Structure," *Journal of Financial Economics* 3 (October 1976), pp. 305–360. Professors Jensen and Smith have analyzed the bondholder-shareholder conflict in a very clear style. See Michael C. Jensen and Clifford W. Smith, Jr., "Stockholder, Manager, and Creditor Interests: Applications of Agency Theory," in *Recent Advances in Corporate Finance*, eds. Edward I. Altman and Marti G. Subrahmanyam (Homewood, IL: Richard D. Irwin, 1985) pp. 93–131. An entire volume dealing with agency problems, including those of capital structure management, is Amir Barnea, Robert A. Haugen, and Lemma W. Senbet, *Agency Problems and Financial Contracting* (Englewood Cliffs, NJ: Prentice-Hall, 1985).

Figure 9-7 Agency Costs of Debt: The Tradeoffs

No Protective Bond Covenants	Many Protective Bond Covenants
High interest rates	Low interest rates
Low monitoring costs	High monitoring costs
No lost operating efficiencies	Many lost operating efficiencies

limit useful managerial decisions, and (4) reviewing the costs and benefits of managerial perks. This list is indicative, not exhaustive. The main point is that monitoring costs are ultimately covered by the owners of the company—its common shareholders.

Capital structure management also gives rise to agency costs. Agency problems stem from conflicts of interest, and capital structure management encompasses a natural conflict between shareholders and bondholders. Acting in the shareholders' best interests might cause management to invest in extremely risky projects. Existing investors in the firm's bonds could logically take a dim view of such an investment policy. A change in the risk structure of the firm's assets would change the business risk exposure of the firm. This could lead to a downward revision of the bond rating the firm currently enjoys. A lowered bond rating in turn would lower the current market value of the firm's bonds. Clearly, bondholders would be unhappy with this result.

To reduce this conflict of interest, the creditors (bond investors) and shareholders may agree to include several protective covenants in the bond contract. These bond covenants were discussed in detail in Chapter 5, but essentially they may be thought of as restrictions on managerial decision making. Typical covenants restrict payment of cash dividends on common shares, limit the acquisition or sale of assets, or limit further debt financing. To make sure that the protective covenants are complied with by management means that monitoring costs are incurred. Like all monitoring costs, they are borne by common shareholders. Further, like many costs, they involve the analysis of an important tradeoff.

Figure 9-7 displays some of the tradeoffs involved with the use of protective bond covenants. Note (in the left panel of Figure 9-7) that the firm might be able to sell bonds that carry no protective covenants only by incurring very high interest rates. With no protective covenants, there are no associated monitoring costs. Also, there are no lost operating efficiencies, such as being able to move quickly to acquire a particular company in the acquisitions market. Conversely, the willingness to submit to several covenants could reduce the explicit cost of the debt contract, but would involve incurring significant monitoring costs and losing some operating efficiencies (which also translates into higher costs). When the debt issue is first sold, then, a tradeoff will be arrived at between incurring monitoring costs, losing operating efficiencies, and enjoying a lower explicit interest cost.

Next, we have to consider the presence of monitoring costs at low levels of leverage and at higher levels of leverage. When the firm operates at a low debt-to-equity ratio, there is little need for creditors to insist on a long list of bond covenants. The financial risk is just not there to require that type of activity. The firm will likewise benefit from low explicit interest rates when leverage is low. When

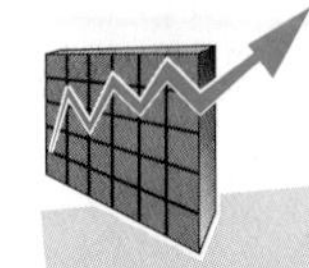

Financial Managers and Theory

Earlier in this chapter we discussed a moderate view of capital structure theory. The saucer-shaped cost of capital curve implied by this theory (Figure 9-6) predicts that managers will add debt to the firm's current capital structure if leverage is below the firm's optimal level of debt. Conversely, managers will add equity if leverage is above the firm's optimal level of debt. In either case, both of these financing activities lower the cost of capital of the firm and increase shareholder's wealth.

A recent survey of chief financial officers of the top (largest) nonfinancial, nonregulated U.S. firms addressed these predicted activities. Of the 800 firms surveyed, 117 responded for a response rate of 14.6 percent. These decision makers were asked how their firms would respond if confronted with certain specific financing activities.

It should be noted that the moderate view does not distinguish between internal equity (retained earnings and depreciation) and external equity (the sale of common shares). The questions posed to the financial managers, however, do make the distinction. Based on our financial asset valuation models, common equity is generally considered the most expensive source of funds, exceeding the costs of both debt and preferred shares. The cost of external equity exceeds internal equity by the addition of flotation costs .

Addressing the downward sloping portion of the cost of capital curve, managers were asked what their financing choice would be if (1) the firm has internal funds sufficient for investment requirements (capital budgeting needs), but (2) the debt ratio is below the level preferred by the firm. The moderate theory predicts that managers will add debt to the firm's capital structure in this situation. However, Table 9-4 indicates that 81 percent of the respondents said they would use internal equity to finance their investments. Only 17 percent suggested that they would use long-term debt, and 11 percent selected short-term debt. In this situation, most managers indicated they would choose to use the most expensive internal equity rather than the (seemingly) less expensive debt for investments.

Similarly, when internal funds are sufficient for new investment and the debt level exceeds the firm's optimum range of debt, managers again chose to use internal equity in 81 percent of the responses. Seventeen percent would issue new equity if market conditions were favourable. Since the debt level is currently excessive, the moderate theory would predict that external equity would be added to minimize the cost of capital. Instead, most firms preferred to fund investments with their internal funds. This is a prevalent tendency in corporate North America.

Responses to the preceding questions imply that, if managers follow the financing activity prescribed by the moderate view, it is only after the internal funds have been exhausted.[12] Since these internal funds are relatively expensive (compared with debt), their use as the initial financing option indicates that either: (1) managers do not view their financing goal to be the minimization of the firm's cost of capital, or (2) the explicit and implicit costs of new security issues are understated. If this is actually the case, internal funds may be the least expensive sources of funds from the perspective of financial managers.[13]

The chief financial officers were also asked what their financing decision would be under these conditions: (1) the firm requires external funds, (2)

[12]These observations are consistent with the findings of Donaldson, which indicated that a pecking order existed for financial policy: (1) use internal funds first, until these have been exhausted; (2) to the extent that external funds must be relied upon, issue debt first, the less risky the better; (3) issue common shares only as a last resort, after all debt capacity has been exhausted. These rules for financial policy form the basis of the pecking order hypothesis which offers an alternative view of capital structure theory.

[13]The suggestion that internal funds may be the least expensive source of funds to financial managers would also be an example of the agency problem that exists between shareholders/owners and managers of the firm.

financial debt exceeds the desired level, and (3) equity markets are underpricing the firm's shares. Table 9-4 summarizes the responses to this question.

In this difficult situation, Table 9-4 shows that 68 percent of the managers indicated that they would reduce their investment plans. In effect, managers would reduce their capital budgets. The primary reasons for this type of investment restriction were: (1) maximize shareholder's wealth, (2) control the firm's exposure to risk.

The next favoured choice (28 percent) was the use of short-term debt with a possible reduction of investment. This suggests that managers attempt to wait out difficult market conditions by adopting short-term solutions. The timing of security issues with favourable market conditions is a major objective of financial managers which is entirely consistent with the optimal capital structure range defined under the moderate theory (Figure 9-6).

Table 9-4 indicates that only four percent of the respondents indicated that they would issue equity despite the underpricing of the firm's shares. Thirteen percent of the firms indicated that they would attempt to correct the adverse pricing by providing the market with adequate information before attempting to issue equity. Another 13 percent of the respondents indicated that they would add long-term debt which moved the debt level beyond the optimal range.

A small percentage (4 percent) of the executives stated that they would obtain the needed funds by reducing the cash dividend payout. To managers, the shareholders' cash dividend payment is quite important. Firms prefer to forgo profitable projects rather than to reallocate the shareholders' expected cash dividends to investment. Dividend policies are extensively discussed in Chapter 10.

You can see that our ability as analysts to predict financing choices, as opposed to prescribing them, is far from perfect. In many instances managers appear to react as suggested by popular capital structure theories. In other instances, though, managers either are rejecting some aspects of the theories or the theories need more work. Understanding these aberrations is useful to the analyst.

A thorough grounding in both the theory of financing decisions and in the tools of capital structure management will assist you in making sound choices and ask some very perceptive questions when faced with a decision-making situation.

Source: Derived from David F. Scott, Jr., and Nancy Jay, "Financial Managers and Capital Structure Theories," Working Paper 9203, Orlando, FL: Dr. Phillips Institute for the Study of American Business Activity, University of Central Florida, March 1992.

Table 9-4 Managers and Theory: Some Responses

Question

Your firm requires external funds to finance the next period's capital investments. The use of financial leverage exceeds the level preferred by the firm. Equities markets are underpricing your securities. Would your firm:

Choices	Responses
Reduce your investment plans?	68%
Obtain short-term debt?	28%
Attempt to provide the market with adequate information to correctly price your securities before issuing equity?	13%
Issue long-term debt?	13%
Issue equity anyway	4%
Reduce your dividend payout?	4%

the debt-to-equity ratio is high, however, it is logical for creditors to demand a great deal of monitoring. This increase in agency costs will raise the implicit cost (the true total cost) of debt financing. It seems logical, then, to suggest that monitoring costs will rise as the firm's use of financial leverage increases. Just as the likelihood of firm failure (financial distress) raises a company's overall cost of capital (K_0), so do agency costs. On the other side of the coin, this means that total firm value (the total market value of the firm's securities) will be lower owing to the presence of agency costs. Taken together, the presence of agency costs and the costs associated with financial distress argue in favour of the concept of an optimal capital structure for the individual firm.

This discussion can be summarized by introducing Equation (9-14) for the market value of the levered firm.

$$\begin{aligned} \text{market value of levered firm} &= \text{market value of unlevered firm} + \text{present value of interest tax shields} \\ &\quad - \left(\text{present value of financial distress costs} + \text{present value of agency costs} \right) \end{aligned} \tag{9-14}$$

The relationship expressed in Equation (9-14) is presented graphically in Figure 9-8. There we see that the tax shield effect is dominant until point A is reached. After point A, the rising costs of the likelihood of firm failure (financial distress) and agency costs cause the market value of the levered firm to decline.[14] The objective for the financial manager here is to find point B by using all of his or her analytical skill; this must also include a good dose of seasoned judgment. At point B the actual market value of the levered firm is maximized, and its weighted cost of capital (K_0) is at a minimum. The implementation problem is that the precise costs of financial distress and monitoring can only be estimated by subjective means; a definite mathematical solution is not available. Thus, planning the firm's financing mix always requires good decision-making by management.

An alternative view aimed at predicting how managers will finance the firm's capital budgets is now known in the financial economics literature as the "pecking order theory." This capital structure is summarized by the following points:

1. Firms adapt dividend policy to investment opportunities. Note that this assumption is close to the concept of the *residual dividend theory* which is discussed in Chapter 10.
2. Firms prefer to finance investment opportunities with internally generated funds first, then external financial capital will be sought.
3. When external financing is needed, the firm will first choose to issue debt securities; issuing equity-type securities will be done last.
4. As more external financing is required to fund projects with positive net present values, the financing pecking order will be followed. This means a preference for more risky debt, then convertibles, preferred equity, and common equity as the last preference.

The crux of the pecking order theory is that no precisely defined target leverage ratio really exists. In other words, the observed leverage ratio (that is, total debt to total assets) merely reflects the cumulative external financing needs of the firm over time.

[14]Note that point A in Figure 9-8 could represent the point where a B++ bond rate firm may lie. Since the firm has used its perceived optimal debt capacity, any further leverage use will be severely penalized.

Figure 9-8 Firm Value Considering Taxes, Agency Costs, and Financial Distress Costs

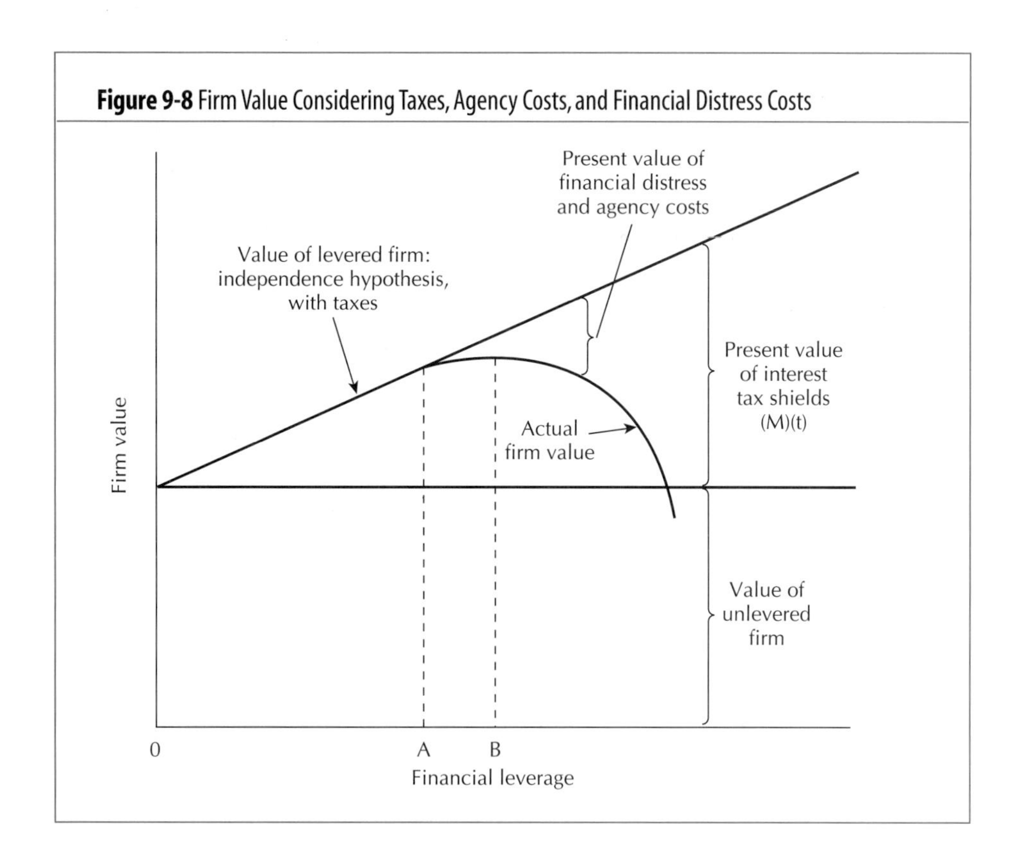

Agency Costs, Free Cash Flow, and Capital Structure

OBJECTIVE 5

In 1986, Professor Michael C. Jensen further extended the concept of agency costs into the area of capital structure management. The contribution revolves around a concept that Jensen labels "free cash flow."

Professor Jensen defines free cash flow as "cash flow in excess of that required to fund all projects that have positive net present values when discounted at the relevant cost of capital."[15]

Jensen indicated that substantial free cash flow can lead to misbehaviour by managers and poor decisions that are not in the best interests of the firm's common shareholders. In other words, managers have an incentive to hold on to the free cash flow and have "fun" with it, rather than "disgorging" it, say, in the form of higher cash dividend payments.

But, all is not lost. This leads to what Jensen calls his "control hypothesis" for debt creation. This means that by levering up, the firm's shareholders will enjoy increased control over their management team. For example, if the firm issues new debt and uses the proceeds to retire outstanding common shares, then management is obligated to pay out cash to service the debt—this simultaneously reduces the amount of free cash flow available to management with which to have fun.

We can also refer to this motive for financial leverage use as the "threat hypothesis." Management works under the threat of financial failure—therefore, according to the "free cash flow theory of capital structure," it works more efficiently. This is supposed to reduce the agency costs of free cash flow—which will in turn be recognized by the marketplace in the form of greater returns on the common shares.

---

[15]Michael C. Jensen, "Agency Costs of Free Cash Flow, Corporate Finance, and Takeovers," *American Economic Review* 76 (May 1986), pp. 323–329.

BACK TO THE BASICS

The discussions on agency costs, free cash flow and the control hypothesis for debt creation return us to ***Axiom 7: The Agency Problem—Managers Won't Work for the Owners Unless It's in Their Best Interest.*** *The control hypothesis put forth by Jensen suggests that managers will work harder for shareholder interests when they have to "sweat it out" to meet contractual interest payments on debt securities. But we also learned that managers and bond investors can have a conflict that leads to agency costs associated with using debt capital. Thus, the theoretical benefits that flow from minimizing the agency costs of free cash flow by using more debt will cease when the rising agency costs of debt exactly offset those benefits. You can see how very difficult it is, then, for financial managers to identify precisely their true optimal capital structure.*

Note that the free cash flow theory of capital structure does not give a theoretical solution to the question of just how much financial leverage is enough. Nor does it suggest how much leverage is too much. It is a way of thinking about why shareholders and their boards of directors might use more debt to control management behaviour and decisions. The basic decision tools of capital structure management still have to be utilized. They will be presented later in this chapter.

Professor Ben Bernanke indicates that not all analysts totally buy into the control hypothesis for debt creation (See "Basic Financial Management in Practice"). Bernanke reviews Jensen's free cash flow theory and also comments on the buildup in corporate leverage that has occurred in recent years.

Managerial Implications

Where does our examination of capital structure theory leave us? The upshot is that the determination of the firm's financing mix is centrally important to the financial manager. The firm's shareholders are affected by capital structure decisions.

At the very least, and before bankruptcy costs and agency costs become detrimental, the tax shield effect will cause the shares of a levered firm to sell at a higher price than they would if the company had avoided debt financing. Owing to both the risk of failure and agency costs that accompany the excessive use of leverage, the financial manager must exercise caution in the use of fixed-charge capital. This problem of searching for the optimal range of use of financial leverage is our next task.[16]

PERSPECTIVE IN FINANCE

You have now developed a workable knowledge of capital structure theory. This makes you better equipped to search for your firm's optimal capital structure. Several tools are available to help you in this search process and simultaneously help you make prudent financing choices. These tools are decision-oriented. They assist us in answering this question, The next time we need $20 million, should we issue common shares or sell long-term bonds?

[16]The relationship between capital structure and enterprise valuation by the marketplace continues to stimulate considerable research output. The complexity of the topic is reviewed in Stewart C. Myers, "The Capital Structure Puzzle," *Journal of Finance* 39 (July 1984), pp. 575–592. Ten useful papers are contained in *Corporate Capital Structures in the United States*, ed. Benjamin M. Friedman (Chicago: National Bureau of Economic Research and The University of Chicago Press, 1985).

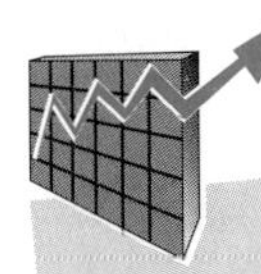

Ben Bernanke on the Free Cash FlowTheory of Capital Structure and the Buildup in Corporate Debt

Introduction

The idea is that the financial structure of firms influences the incentives of "insiders" (managers, directors, and large shareholders with some operational interest in the business) and that, in particular, high levels of debt may increase the willingness of insiders to work hard and make profit-maximizing decisions. This incentive-based approach makes a valuable contribution to our understanding of a firm's capital structure. But while this theory might explain why firms like to use debt in general, does it explain why the use of debt has increased so much in recent years?

Michael Jensen, a founder and leading proponent of the incentive-based approach to capital structure, argues that it can. Jensen focuses on a recent worsening of what he calls the "free cash flow" problem. Free cash flow is defined as the portion of a corporation's cash flow that it is unable to invest profitably within the firm. Companies in industries that are profitable but no longer have much potential for expansion—the U.S. oil industry, for example—have a lot of free cash flow.

Why is free cash flow a problem? Jensen argues that managers are often tempted to use free cash flow to expand the size of the company, even if the expansion is not profitable. This is because managers feel that their power and job satisfaction are enhanced by a growing company; so given that most managers' compensation is at best weakly tied to the firm's profitability, Jensen argues that managers will find it personally worthwhile to expand even into money-losing operations. In principle, the board of directors and shareholders should be able to block these unprofitable investments; however, in practice, the fact that the management typically has far more information about potential investments than do outside directors and shareholders makes it difficult to second-guess the managers' recommendations.

How More Leverage Can Help

The company manager with lots of free cash flow may attempt to use that cash to increase his power and perquisites, at the expense of the shareholders. Jensen argues that the solution to the free-cash-flow problem is more leverage. For example, suppose that management uses the free cash flow of the company, plus the proceeds of new debt issues, to repurchase stock from the outside shareholders—that is, to do a management buyout. This helps solve the free-cash-flow problem in several ways. The personal returns of the managers are now much more closely tied to the profits of the firm, which gives them incentives to be more efficient. Second, the re-leveraging process removes the existing free cash from the firm, so that any future investment projects will have to be financed externally; thus, future projects will have to meet the market test of being acceptable to outside bankers or bond purchasers. Finally, the high interest payments implied by re-leveraging impose a permanent discipline on the managers; in order to meet these payments, they will have to ruthlessly cut money-losing operations, avoid questionable investments, and take other efficiency-promoting actions.

According to Jensen, a substantial increase in free-cash-flow problems—resulting from deregulation, the maturing of some large industries, and other factors—is a major source of the recent debt expansion. Jensen also points to a number of institutional factors that have promoted increased leverage. These include relaxed restrictions on mergers, which have lowered the barriers to corporate takeovers created by the antitrust laws, and increased financial sophistication, such as the greatly expanded operations of takeover specialists like Drexel Burnham Lambert Inc. and the development of the market for "junk bonds." Jensen's diagnosis is not controversial: it's quite plausible that these factors, plus changing norms about what constitutes an "acceptable" level of debt, explain at least part of the trend toward increased corporate debt. One important piece of evidence in favour of this explanation is that net equity issues have been substantially negative since 1983. This suggests that much of the proceeds of the new debt issues is being used to repurchase outstanding shares. This is what we would expect if corporations are attempting to re-leverage their existing assets, rather than using debt to expand their asset holdings. However, the implied conclusion—that the debt buildup is beneficial overall to the economy—is considerably more controversial.

Criticisms of the Incentive-Based Rationale for Increased Debt

Jensen and other advocates of the incentive-based approach to capital structure have made a cogent theoretical case for the beneficial effects of debt finance, and many architects of large-scale restructurings have given improved incentives and the promise of greater efficiency as a large part of the rationale for increased leverage. The idea that leverage is beneficial has certainly been embraced by the stock market: even unsubstantiated rumours of a potential leveraged buy-out (LBO) have been sufficient to send the stock price of the targeted company soaring, often by 40 percent or more. At a minimum, this indicates that stock market participants believe that higher leverage increases profitability. Proponents of restructuring interpret this as evidence that debt is good for the economy.

There are, however, criticisms of this conclusion. First, the fact that the stock market's expectations of company profitability rise when there is a buy-out is not proof that profits will rise in actuality. It is still too soon to judge whether the increased leverage of the 1980s will lead to a sustained increase in profitability. One might think of looking to historical data for an answer to this question. But buy-outs in the 1960s and 1970s were somewhat different in character from more recent restructurings, and, in any case, the profitability evidence on the earlier episodes is mixed.

Even if the higher profits expected by the stock market do materialize, there is contention over where they are likely to come from. The incentive-based theory of capital structure says they will come from improved efficiency. But some opponents have argued that the higher profits will primarily reflect transfers to the shareholders from other claimants on the corporation—its employees, customers, suppliers, bondholders, and the government. Customers may be hurt if takeovers are associated with increased monopolization of markets. Bondholders have been big losers in some buyouts, as higher leverage has increased bankruptcy risk and thus reduced the value of outstanding bonds. The government may have lost tax revenue, as companies, by increasing leverage, have increased their interest deductions (although there are offsetting effects here, such as the taxes paid by bought-out shareholders on their capital gains). The perception that much of the profits associated with re-leveraging and buyouts comes from "squeezing" existing beneficiaries of the corporation explains much of the recent political agitation to limit these activities.

The debt buildup can also be criticized from the perspective of incentive-based theories themselves. Two points are worth noting: first, the principal problem that higher leverage is supposed to address is the relatively weak connection between firms' profits and managers' personal returns, which reduces managers' incentives to take profit-maximizing actions. But if this is truly the problem, it could be addressed more directly—without subjecting the company to serious bankruptcy risk—simply by changing managerial compensation schemes to include more profit-based incentives.

The Downside of Debt Financing

Increased debt is not the optimal solution to all incentive problems. For example, it has been shown, as a theoretical proposition, that managers of debt-financed firms have an incentive to choose riskier projects over safe ones; this is because firms with fixed-debt obligations enjoy all of the upside potential of high-risk projects but share the downside losses with the debt holders, who are not fully repaid if bad investment outcomes cause the firm to fail.

That high leverage does not always promote efficiency can be seen when highly leveraged firms suffer losses and find themselves in financial distress. When financial problems hit, the need to meet interest payments may force management to take a very short-run perspective, leading them to cut back production and employment, cancel even potentially profitable expansion projects, and sell assets at fire-sale prices. Because the risk of bankruptcy is so great, firms in financial distress cannot make long-term agreements; they lose customers and suppliers who are afraid they cannot count on an ongoing relationship, and they must pay wage premiums to hire workers.

These efficiency losses, plus the direct costs of bankruptcy (such as legal fees), are the potential downside of high leverage.

Source: Ben Bernanke, "Is There Too Much Corporate Debt?" *Business Review*, Federal Reserve Bank of Philadelphia (September–October 1989), pp. 5–8.

BASIC TOOLS OF CAPITAL STRUCTURE MANAGEMENT

OBJECTIVE 6

Recall from Chapter 8 that the use of financial leverage has two effects upon the earnings stream flowing to the firm's common shareholders. For clarity of exposition Tables 8-7 and 8-8 are repeated here as Tables 9-5 and 9-6. Three possible financing mixes for the Pierce Grain Company are contained in Table 9-5, and an analysis of the corresponding financial leverage effects is displayed in Table 9-6.

The first financial leverage effect is the added variability in the earnings-per-share stream that accompanies the use of fixed-charge securities in the company's capital structure. By means of the degree-of-financial-leverage measure (DFL_{EBIT}) we explained how this variability can be quantified. The firm that uses more financial leverage (rather than less) will experience larger relative changes in its earnings per share (rather than smaller) following EBIT fluctuations. Assume that Pierce Grain elected financing plan C rather than plan A. Plan C is highly levered and plan A is unlevered. A 100 percent increase in EBIT from $20,000 to $40,000 would cause earnings per share to rise by 147 percent under plan C, but only 100 percent under plan A. Unfortunately, the effect would operate in the negative direction as well. A given change in EBIT is magnified by the use of financial leverage. This magnification is reflected in the variability of the firm's earnings per share.

The second financial leverage effect concerns the level of earnings per share at a given EBIT under a given capital structure. Refer to Table 9-6. At the EBIT level of $20,000, earnings per share would be $5, $5.33, and $5.67 under financing arrangements A, B, and C, respectively. Above a critical level of EBIT, the firm's earnings per share will be higher if greater degrees of financial leverage are employed. Conversely, below some critical level of EBIT, earnings per share will suffer at greater degrees of financial leverage. Whereas the first financial leverage effect is quantified by the degree-of-financial-leverage measure (DFL_{EBIT}), the second is quantified by what is generally referred to as EBIT-EPS analysis. EPS refers, of course, to earnings per share. The rationale underlying this sort of analysis is simple. Earnings is one of the key variables that influences the market value of the firm's common shares. The effect of a financing decision on EPS, then, should be understood because the decision will probably affect the value of the shareholders' investment.

Table 9-5 The Pierce Grain Company Possible Capital Structures

Plan A: 0% debt			
		Total debt	$ 0
		Common equity	200,000[a]
Total assets	$200,000	Total liabilities and equity	$200,000
Plan B: 25% debt at 8% interest rate			
		Total debt	$ 50,000
		Common equity	150,000[b]
Total assets	$200,000	Total liabilities and equity	$200,000
Plan C: 40% debt at 8% interest rate			
		Total debt	$ 80,000
		Common equity	120,000[c]
Total assets	$200,000	Total liabilities and equity	$200,000

[a]2,000 common shares outstanding
[b]1,500 common shares outstanding
[c]1,200 common shares outstanding

Table 9-6 The Pierce Grain Company Analysis of Financial Leverage at Different EBIT Levels

(1) EBIT	(2) Interest	(3) = (1) − (2) EBT	(4) = (3) × .5 Taxes	(5) = (3) − (4) Net Income to Common	(6) Earnings per Share	
Plan A: 0% debt; $200,000 common equity; 2,000 shares						
$ 0	$ 0	$ 0	$ 0	$ 0	$ 0	
20,000	0	20,000	10,000	10,000	5.00	100%
40,000	0	40,000	20,000	20,000	10.00	
60,000	0	60,000	30,000	30,000	15.00	
80,000	0	80,000	40,000	40,000	20.00	
Plan B: 25% debt; 8% Interest rate; $150,000 common equity; 1,500 shares						
$ 0	$4,000	$(4,000)	$ (2,000)[a]	$ (2,000)	$ (1.33)	
20,000	4,000	16,000	8,000	8,000	5.33	125%
40,000	4,000	36,000	18,000	18,000	12.00	
60,000	4,000	56,000	28,000	28,000	18.67	
80,000	4,000	76,000	38,000	38,000	25.33	
Plan C: 40% debt; 8% Interest rate; $120,000 common equity; 1,200 shares						
$ 0	$6,400	$ (6,400)	$ (3,200)[a]	$ (3,200)	$ (2.67)	
20,000	6,400	13,600	6,800	6,800	5.67	147%
40,000	6,400	33,600	16,800	16,800	14.00	
60,000	6,400	53,600	26,800	26,800	22.33	
80,000	6,400	73,600	36,800	36,800	30.67	

[a]The negative tax bill recognizes the deduction arising from the loss carryback and carryforward provision of the Income Tax Act. See Chapter 2.

EBIT-EPS Analysis

EXAMPLE

Assume that plan B in Table 9-6 is the existing capital structure for the Pierce Grain Company. Furthermore, the asset structure of the firm is such that EBIT is expected to be $20,000 per year for a very long time. A capital investment is available to Pierce Grain that will cost $50,000. Acquisition of this asset is expected to raise the projected EBIT level to $30,000, permanently. The firm can raise the needed cash by (1) selling 500 common shares at $100 each or (2) selling new bonds that will net the firm $50,000 and carry an interest rate of 8.5 percent. These capital structures and corresponding EPS amounts are summarized in Table 9-7.

At the projected EBIT level of $30,000, the EPS for the common shares and debt alternatives are $6.50 and $7.25, respectively. Both are considerably above the $5.33 that would occur if the new project were rejected and the additional financial capital were not raised. Based on a criterion of selecting the financing plan that will provide the highest EPS, the bond alternative is favoured. But what if the basic business risk to which the firm is exposed causes the EBIT level to vary over a considerable range? Can we be sure that the bond alternative will always have the higher EPS associated with it? The answer, of course, is no. When the EBIT level is subject to uncertainty, a graphic analysis of the proposed financing plans can provide useful information to the financial manager.

Graphic Analysis

The EBIT-EPS analysis chart allows the decision maker to visualize the impact of different financing plans on EPS over a range of EBIT levels. The relationship

Table 9-7 The Pierce Grain Company Analysis of Financing Choices

Part A: Capital Structures

Existing Capital Structure		With New Common Share Financing		With New Debt Financing	
Long-term debt at 8%	$ 50,000	Long-term debt at 8%	$ 50,000	Long-term debt at 8%	$ 50,000
Common shares	150,000	Common shares	200,000	Long-term debt at 8.5%	50,000
				Common shares	150,000
Total liabilities and equity	$200,000	Total liabilities and equity	$250,000	Total liabilities and equity	$250,000
Common shares outstanding	1,500	Common shares outstanding	2,000	Common shares outstanding	1,500

Part B: Projected EPS Levels

	Existing Capital Structure	With New Common Share Financing	With New Debt Financing
EBIT	$20,000	$30,000	$30,000
Less: Interest expense	4,000	4,000	8,250
Earnings before taxes (EBIT)	$16,000	$26,000	$21,750
Less: Taxes at 50%	8,000	13,000	10,875
Net Income	$ 8,000	$13,000	$10,875
Less: Preferred dividends	0	0	0
Earnings available to common	$ 8,000	$13,000	$10,875
EPS	$ 5.33	$ 6.50	$ 7.25

between EPS and EBIT is linear. All we need, therefore, to construct the chart is two points for each alternative. Part B of Table 9-7 already provides us with one of these points. The answer to the following question for each choice gives us the second point: At what EBIT level will the EPS for the plan be exactly zero? If the EBIT level just covers the plan's financing costs (on a before-tax basis), then EPS will be zero. For the share plan, an EPS of zero is associated with an EBIT of $4,000. The $4,000 is the interest expense incurred under the existing capital structure. If the bond plan is elected, the interest costs will be the present $4,000 plus $4,250 per year arising from the new debt issue. An EBIT level of $8,250, then, is necessary to provide a zero EPS with the bond plan.

The EBIT-EPS analysis chart representing the financing choices available to the Pierce Grain Company is shown as Figure 9-9. EBIT is charted on the horizontal axis and EPS on the vertical axis. The intercepts on the horizontal axis represent the before-tax equivalent financing charges related to each plan. The straight lines for each plan tell us the EPS amounts that will occur at different EBIT amounts.

Notice that the bond plan line has a steeper slope than the share plan line. This ensures that the lines for each financing choice will intersect. Above the intersection point, EPS for the plan with greater leverage will exceed that for the plan with lesser leverage. The intersection point, encircled in Figure 9-9, occurs at an EBIT level of $21,000 and produces EPS of $4.25 for each plan. When EBIT is $30,000, notice that the bond plan produces EPS of $7.25 and the share plan, $6.50. Below the intersection point, EPS with the share plan will exceed that with the more highly levered bond plan. The steeper slope of the bond plan line indicates that with greater leverage, EPS is more sensitive to EBIT changes. This same concept was discussed in Chapter 8 when we derived the degree of financial leverage measure.

Computing Indifference Points

The point of intersection in Figure 9-9 is called the **EBIT-EPS indifference point**. It identifies the EBIT level at which the EPS will be the same regardless of the financing plan chosen by the financial manager. This indifference point, sometimes

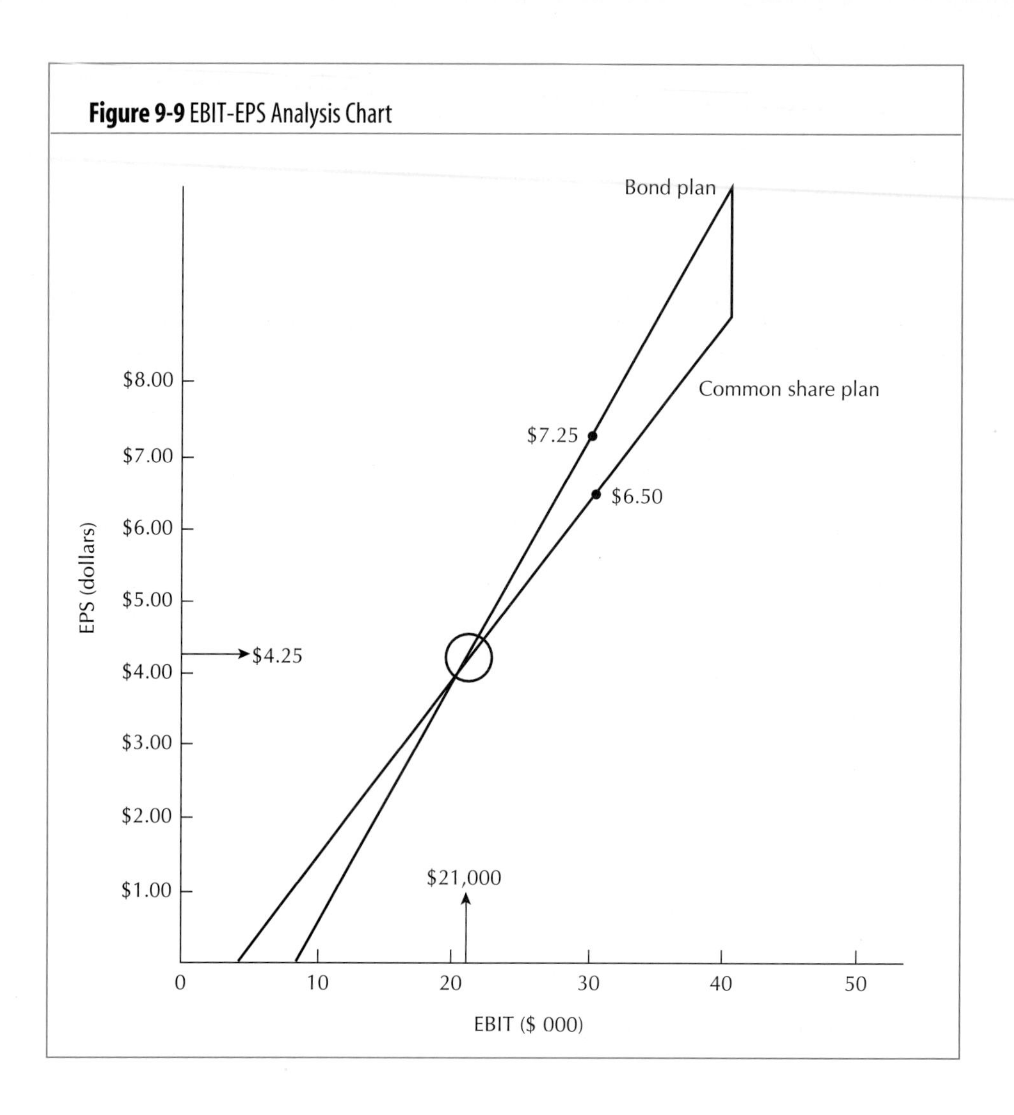

Figure 9-9 EBIT-EPS Analysis Chart

EBIT-EPS indifference point

The level of earnings before interest and taxes (EBIT) that will equate earnings per share (EPS) between two different financing plans.

called the breakeven point, has major implications for financial planning. At EBIT amounts in excess of the EBIT indifference level, the more heavily levered financing plan will generate a higher EPS. At EBIT amounts below the EBIT indifference level, the financing plan involving less leverage will generate a higher EPS. It is important, then, to know the EBIT indifference level.

We can find it graphically, as in Figure 9-9. At times it may be more efficient, though, to calculate the indifference point directly. This can be done by using the following equation:

$$\underset{\text{EPS: Share Plan}}{\frac{(EBIT - I)(1 - T) - D_p}{S_s}} = \underset{\text{EPS: Bond Plan}}{\frac{(EBIT - I)(1 - T) - D_p}{S_b}} \tag{9-15}$$

where S_s and S_b are the number of common shares outstanding under the share and bond plans, respectively, I is interest expense, T is the firm's income tax rate, and D_p is preferred dividends paid. In the present case D_p is zero, because there are no preferred shares outstanding. If preferred shares are associated with one of the financing alternatives, keep in mind that the preferred dividends, D_p, are not tax deductible. Equation (9-15) does take this fact into consideration.

For the present example, we calculate the indifference level of EBIT as

$$\frac{(\text{EBIT} - \$4{,}000)(1 - 0.5) - 0}{2{,}000} = \frac{(\text{EBIT} - \$8{,}250)(1 - 0.5) - 0}{1{,}500}$$

When the expression above is solved for EBIT, we obtain $21,000. If EBIT turns out to be $21,000, then EPS will be $4.25 under both plans.

Uncommitted Earnings per Share and Indifference Points

The calculations that permitted us to solve for Pierce Grain's EBIT-EPS indifference point made no explicit allowance for the repayment of the bond principal. This procedure is not that unrealistic. It only presumes the debt will be perpetually outstanding. This means that when the current bond issue matures, a new bond issue will be floated. The proceeds from the newer issue would be used to pay off the maturity value of the older issue.

Many bond contracts, however, require that sinking fund payments be made to a bond trustee. A **sinking fund** is a real cash reserve that is used to provide for the orderly and early retirement of the principal amount of the bond issue. Most often the sinking fund payment is a mandatory fixed amount and is required by a clause in the bond indenture. Sinking fund payments can represent a sizable cash drain on the firm's liquid resources.[17] Moreover, sinking fund payments are a return of borrowed principal, so they are not tax deductible to the firm.

Sinking fund

A cash reserve used for orderly and early retirement of the principal amount of a bond issue. Payments into the fund are known as sinking fund payments.

Because of the potentially serious nature of the cash drain caused by sinking fund requirements, the financial manager might be concerned with the uncommitted earnings per share (UEPS) related to each financing plan. The calculation of UEPS recognizes that sinking fund commitments have been honoured. UEPS can be used, then, for discretionary spending—such as the payment of cash dividends to common shareholders or investment in capital facilities.

If we let *SF* be the sinking fund payment required in a given year, the EBIT-UEPS indifference point can be calculated as

$$\underset{\textit{UEPS}\text{: Share Plan}}{\frac{(\text{EBIT} - I)(1 - T) - D_p - SF}{S_s}} = \underset{\textit{UEPS}\text{: Bond Plan}}{\frac{(\text{EBIT} - I)(1 - T) - D_p - SF}{S_b}} \tag{9-16}$$

If several bond issues are already outstanding, then *I* in Equations (9-15) and (9-16) for the share plan consists of the sum of their related interest payments. For the bond plan, *I* would be the sum of existing plus new interest charges. In Equation (9-16) the same logic applies to the sinking fund variable, *SF*. The indifference level of EBIT based on UEPS will always exceed that based on EPS.

A Word of Caution

Above the EBIT-EPS indifference point, a more heavily levered financial plan promises to deliver a larger EPS. Strict application of the criterion of selecting the financing plan that produces the highest EPS might have the firm issuing debt most of the time it raised external capital. Our discussion of capital structure theory taught us the dangers of that sort of action.

The primary weakness of EBIT-EPS analysis is that it disregards the implicit costs of debt financing. The effect of the specific financing decision on the firm's cost of common equity capital is totally ignored. Investors should be concerned with both the level and variability of the firm's expected earnings stream. EBIT-EPS analysis considers only the level of the earnings stream and ignores the variability (riskiness) inherent in it. Thus, this type of analysis must be used in conjunction with other basic tools in reaching the objective of capital structure management.

[17]As the sinking fund retires the bond issue faster, the issue is typically viewed as less risky than a non-sinking fund issue.

BACK TO THE BASICS

The comparison techniques of EBIT-EPS analysis and Uncommitted Earnings per Share analysis are well known within the corporate financial planning groups of corporations. It is useful to emphasize that these tools of capital structure management are best utilized if we relate them to both ***Axiom 3: Cash—Not Profits—Is King: Measuring the Timing of Costs and Benefits*** *and* ***Axiom 6: Efficient Capital Markets—The Markets Are Quick and the Prices Are Right.***

Thus the cash flows, as opposed to accounting profits, that are available to the firm after a financing choice is made, will drive market prices. Recall from Chapter 1 that we said efficient markets will not be fooled by accounting changes that merely manipulate reported earnings. In the context of using these tools, then, the proper way to think of earnings per share and uncommitted earnings per share is on a cash basis rather than on an accounting accrual basis. The firm services its debt contracts not out of accounting earnings, but out of cash flows.

Comparative Leverage Ratios

In Chapter 20 we will explore the overall usefulness of financial ratio analysis. Leverage ratios are one of the categories of financial ratios identified in that chapter. We emphasize here that the computation of leverage ratios is one of the basic tools of capital structure management.

Two types of leverage ratios must be computed when a financing decision faces the firm. We call these **balance sheet leverage ratios** and **coverage ratios**. The firm's balance sheet provides the required accounts for computing the balance sheet leverage ratios. In various forms these balance sheet ratios compare the firm's use of funds supplied by creditors with those supplied by owners.

Balance sheet leverage ratios

Financial ratios used to measure the extent of a firm's use of borrowed funds and calculated using information found in the firm's balance sheet.

Coverage ratios

A group of ratios that measure a firm's ability to meet its recurring fixed charge obligations, such as interest on long-term debt, lease payments, and/or preferred share dividends.

The accounts required to calculate the coverage ratios generally come from the firm's income statement. At times the external analyst may have to consult balance sheet information to construct some of these needed estimates. On a privately placed debt issue, for example, some fraction of the current portion of the firm's long-term debt might have to be used as an estimate of that issue's sinking fund. Coverage ratios provide estimates of the firm's ability to service its financing contracts. High coverage ratios, compared with a standard, imply unused debt capacity.

A Worksheet

Table 9-8 is a sample worksheet used to analyze financing choices. The objective of the analysis is to determine the effect each financing plan will have on key financial ratios. The financial officer can compare the existing level of each ratio with its projected level, taking into consideration the contractual commitments of each alternative.

In reality we know that EBIT might be expected to vary over a considerable range of outcomes. For this reason the coverage ratios should be calculated several times, each at a different level of EBIT. If this is accomplished over all possible values of EBIT, a probability distribution for each coverage ratio can be constructed. This provides the financial manager with much more information than simply calculating the coverage ratios based on the expected value of EBIT.

Industry Norms

The comparative leverage ratios calculated according to the format laid out in Table 9-8, or in a similar format, have additional utility to the decision maker if they can be compared with some standard. Generally, corporate financial analysts, investment dealers, commercial bank loan officers, and bond-rating agencies rely upon industry classes from which to compute "normal" ratios. Although industry groupings may actually contain firms whose basic business risk exposure differs

Table 9-8 Comparative Leverage Ratios: Worksheet for Analyzing Financing Plans

Ratios	Computation Method	Existing Ratio	Ratio with New Common Share Financing	Ratio with New Debt Financing
Balance sheet leverage ratios				
1. Debt ratio	$\frac{\text{total liabilities}}{\text{total assets}}$	______ %	______ %	______ %
2. Long-term debt to total capitalization	$\frac{\text{long-term debt}}{\text{long-term debt} + \text{net worth}}$	______ %	______ %	______ %
3. Total liabilities to net worth	$\frac{\text{total liabilities}}{\text{net worth}}$	______ %	______ %	______ %
4. Common equity ratio	$\frac{\text{common equity}}{\text{total assets}}$	______ %	______ %	______ %
Coverage ratios				
1. Times interest earned	$\frac{\text{EBIT}}{\text{annual interest expense}}$	______ times	______ times	______ times
2. Times burden covered	$\frac{\text{EBIT interest} + \text{sinking fund}}{\text{interest} + \frac{\text{sinking fund}}{1 - T}}$	______ times	______ times	______ times
3. Cash flow overall coverage ratio	$\frac{(\text{EBIT} + \text{lease expense} + \text{depreciation})}{\left(\text{interest} + \text{lease expense} + \frac{\text{preferred dividends}}{1 - T} + \frac{\text{principal payments}}{1 - T}\right)}$	______ times	______ times	______ times

widely, the practice is entrenched in North American business behaviour.[18] At the very least, then, the financial officer must be interested in industry standards because almost everybody else is.

Several published studies indicate that capital structure ratios vary in a significant manner among industry classes.[19] For example, random samplings of the common equity ratios of large retail firms differ statistically from those of major steel producers. The major steel producers use financial leverage to a lesser degree than do the large retail organizations. On the whole, firms operating in the same industry tend to exhibit capital structure ratios that cluster around a central value, which we call a *norm*. Business risk will vary from industry to industry. As a consequence, the capital structure norms will vary from industry to industry.

This is not to say that all companies in the industry will maintain leverage ratios "close" to the norm. For instance, firms that are very profitable may display high coverage ratios and high balance sheet leverage ratios. The moderately profitable

---

[18]An approach to grouping firms based on several component measures of business risk, as opposed to ordinary industry classes, is reported in John D. Martin, David F. Scott, Jr., and Robert F. Vandell, "Equivalent Risk Classes: A Multidimensional Examination," *Journal of Financial and Quantitative Analysis* 14 (March 1979), pp. 101–118.

[19]See, for example, Eli Schwartz and J. Richard Aronson, "Some Surrogate Evidence in Support of the Concept of Optimal Financial Structure," *Journal of Finance* 22 (March 1967), pp. 10–18; David F. Scott, Jr., "Evidence on the Importance of Financial Structure," *Financial Management* 1 (Summer 1972), pp. 45–50; and David F. Scott, Jr., and John D. Martin, "Industry Influence on Financial Structure," *Financial Management* 4 (Spring 1975), pp. 67–73.

firm, though, might find such a posture unduly risky. Here the usefulness of industry normal leverage ratios is clear. If the firm chooses to deviate in a material manner from the accepted values for the key ratios, it must have a sound reason.

Companywide Cash Flows: What Is the Worst That Could Happen?

In Chapter 20 we will note that liquidity ratios are designed to measure the ability of the firm to pay its bills on time. Financing charges are just another type of bill that eventually comes due for payment. Interest charges, preferred dividends, lease charges, and principal payments all must be paid on time, or the company risks being caught in bankruptcy proceedings. To a lesser extent, dispensing with financing charges on an other than timely basis can result in severely restricted business operations. We have just seen that coverage ratios provide a measure of the safety of one general class of payment—financing charges. Coverage ratios, then, and liquidity ratios are very close in concept.

A more comprehensive method is available for studying the impact of capital structure decisions on corporate cash flows. The method is simple but nonetheless very valuable. It involves the preparation of a series of cash budgets under (1) different economic conditions and (2) different capital structures.[20] The net cash flows under these different situations can be examined to determine if the financing requirements expose the firm to a degree of default risk too high to bear.

In work that has been highly acclaimed, Donaldson has suggested that the firm's debt-carrying capacity (defined in the broad sense here to include preferred dividend payments and lease payments) ought to depend upon the net cash flows the firm could expect to receive during a recessionary period.[21] In other words, target capital structure proportions could be set by planning for the "worst that could happen." An example will be of help.

Suppose that a recession is expected to last for one year.[22] Moreover, the end of the year represents the bottoming-out, or worst portion of the recession. Equation (9-17) defines the cash balance, CB_r, the firm could expect to have at the end of the recession period.[23]

$$CB_r = C_0 + (C_s + OR) - (P_a + RM + \ldots + E_n) - FC \tag{9-17}$$

where C_0 = the cash balance at the beginning of the recession
C_s = collection from sales
OR = other cash receipts
P_a = payroll expenditures
RM = raw material payments
E_n = the last of a long series of expenditures over which management has little control (nondiscretionary expenditures)
FC = fixed financial charges associated with a specific capital structure

If we let the net of total cash receipts and nondiscretionary expenditures be represented by NCF_r, then Equation (9-17) can be simplified to

$$CB_r = C_0 + NCF_r - FC \tag{9-18}$$

The inputs to Equation (9-18) come from a detailed cash budget. The variable representing financing costs, FC, can be changed in accordance with several alterna-

[20]Cash budget preparation is discussed in Chapter 21.

[21]Refer to Gordon Donaldson, "New Framework for Corporate Debt Policy," *Harvard Business Review* 40 (March–April 1962), pp. 117–131; Gordon Donaldson, *Corporate Debt Capacity* (Boston: Division of Research, Graduate School of Business Administration, Harvard University, 1961), Chap. 7; and Gordon Donaldson, "Strategy for Financial Emergencies," *Harvard Business Review* 47 (November–December 1969), pp. 67–79.

[22]The analysis can readily be extended to cover a recessionary period of several years. All that is necessary is to calculate the cash budgets over a similar period.

[23]For the most part, the present notation follows that of Donaldson.

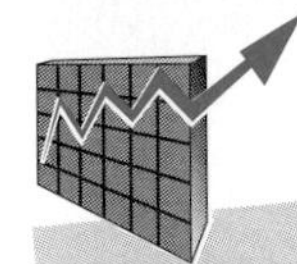

Corporate Policies on Using Financial Leverage

Managements continually face the challenge of determining how much financial leverage is enough. Cambridge Shopping Centres Ltd. deals with this difficult financial policy question by outlining the company's financial guidelines to their stakeholders in their 1995 *Annual Report*.

Financial Guidelines

Cambridge adheres to conservative financial and operating practices that enable the Company to not only operate effectively but also position itself for growth. These practices are founded on adherence to the guidelines below, which management believes to be appropriate disciplines and measures of performance for commercial real estate companies:

- The ratio of debt to equity on a current value basis will be less than 2:1;
- Floating-rate debt will not exceed 25% of total debt;
- Land held for future development will not exceed 10% of total property assets;
- Cash flow before debt service will be at least 1.1 times debt-service requirements;
- Gross overhead expenses will not exceed 5% of total managed revenues;
- The ratio of cash dividends to net cash flow will not exceed 30%.

Source: Cambridge Shopping Centres Ltd., *Annual Report* (1995), p. 26.

tive financing plans to ascertain if the net cash balance during the recession, CB_r, might fall below zero.

Suppose that some firm typically maintains $500,000 in cash and marketable securities. This amount would be on hand at the start of the recession period. During the economic decline, the firm estimates that its net cash flows from operations, NCF_r, will be $2 million. If the firm currently finances its assets with an unlevered capital structure, its cash balance at the worst point of the recession would be

$$CB_r = \$500{,}000 + \$2{,}000{,}000 - \$0 = \$2{,}500{,}000$$

This procedure allows us to study many different situations.[24] Assume that the same firm is considering a shift in its capitalization such that annual interest and sinking fund payments will be $2,300,000. If a recession occurred, the firm's cash balance at the end of the adverse economic period would be

$$CB_r = \$500{,}000 + \$2{,}000{,}000 - \$2{,}300{,}000 = \$200{,}000$$

The firm ordinarily maintains a liquid asset balance of $500,000. Thus, the effect of the proposed capital structure on the firm's cash balance during adverse circumstances might seem too risky for management to accept. When the chance of being

[24]It is not difficult to improve the usefulness of this sort of analysis by applying the technique of simulation to the generation of the various cash budgets. This facilitates the construction of probability distributions of net cash flows under differing circumstances. Simulation will be discussed in Chapter 15.

out of cash is too high for management to bear, the use of financial leverage has been pushed beyond a reasonable level. According to this tool, the appropriate level of financial leverage is reached when the chance of being out of cash is exactly equal to that which management will assume.

OBJECTIVE 7

A GLANCE AT ACTUAL CAPITAL STRUCTURE MANAGEMENT

In this chapter we have discussed (1) the concept of an optimal capital structure, (2) the search for an appropriate range of financial leverage, and (3) the fundamental tools of capital structure management. We now examine the opinions and practices of financial executives who work in the area of capital structure management.

Cheung, Roy and Gordon have surveyed Canadian financial managers about their firms' capital structure practices.[25] Each of these financial managers worked for companies listed on the Financial Post Survey of Industrials. The results of this survey indicated that 72 percent of these financial managers thought that their firm's common equity would be maximized at a particular debt ratio. This type of response is consistent with the results of other surveys (Scott and Johnson, 1982 and Walsh, 1972) which indicated that managers believe there is an optimum capital structure for the corporation. One respondent summarized his opinion on optimum capital structure of a firm as follows:

> In my opinion, there is an optimum capital structure for companies. However, this optimum capital structure will vary by individual companies, industries, and then is subject to changing economies, by money markets, earnings trends, and prospects... the circumstances and the lenders will determine an optimum at different points in time.[26]

This survey and others consistently point out that (1) financial officers set target debt ratios for their companies, and (2) the values for those ratios are influenced by a conscious evaluation of the basic business risk to which the firm is exposed.

Target Debt Ratios

Selected comments from financial executives of large international companies point to the widespread use of target debt ratios. A vice-president and treasurer of the American Telephone and Telegraph Company (AT&T) described his firm's debt ratio policy in terms of a range:

> All of the foregoing considerations led us to conclude, and reaffirm for a period of many years, that the proper range of our debt was 30% to 40% of total capital. Reasonable success in meeting financial needs under the diverse market and economic conditions that we have faced attests to the appropriateness of this conclusion.[27]

Cheung, Roy and Gordon have reported that only 11% of the Canadian firms surveyed have target ratios greater than 50%; the most popular target range among firms is 40–50%, with 21–50% being the most predominant.[28] These authors also reported that the predominant factor in establishing target ratios among Canadian

[25]Joseph K. Cheung, S. Paul Roy and Irene Gordon, "Financing Policies of Large Canadian Corporations," *CMA Management Accounting Magazine* 63(4) (May 1989), pp. 26–31.

[26]Francis J. Walsh, Jr., *Planning Corporate Capital Structures* (New York: The Conference Board, 1972), p. 14.

[27]John J. Scanlon, "Bell System Financial Policies," *Financial Management* 1 (Summer 1972), pp. 16–26.

[28]Joseph K. Cheung, S. Paul Roy and Irene Gordon, "Financing Policies of Large Canadian Corporations," *CMA Management Accounting Magazine* 63(4) (May 1989), pp. 26–31.

industrial companies was the desire to maintain financial flexibility.[29] Other factors which were mentioned in this study were: the need to maintain a minimum credit rating on the firm's long-term debt, the concern for covenant violation and share price maximization. A similar survey of American companies mentioned the following influences on the level of the target debt ratio: the ability of the firm to meet financing charges, maintaining a desired bond rating, providing an adequate borrowing reserve, and exploiting the advantage of financial leverage.[30]

Who Sets Target Debt Ratios?

From the discussion above, we know that firms do use target debt ratios in arriving at financing decisions. But what is the relative influence of external parties on the firm setting these target ratios? Table 9-9 summarizes the results of Cheung, Roy, and Gordon which indicate that managers ranked commercial bankers first as an outside party which influenced the setting of target ratios. Other parties which were ranked by managers were investment bankers and comparative ratios; trade creditors and security analysts were thought to have little or no influence in determining the target financial structure. The role of the investment dealers in this country's capital market system was explored in detail in Chapter 7.

Debt Capacity

Previously in this chapter we noted that the firm's debt capacity is the maximum proportion of debt that it can include in its capital structure and still maintain its lowest weighted cost of capital. It is of interest to have some understanding of how financial executives make the concept of debt capacity operational. Table 9-10 is adapted from a survey of the 1,000 largest industrial firms in the United States (as ranked by total sales dollars) and involves the responses from 212 financial executives. These executives defined debt capacity in a wide variety of ways. The most popular approach was as a target percentage of total capitalization.

Twenty-seven percent of the respondents thought of debt capacity in this manner. Forty-three percent of the participating executives remarked that debt capacity is defined in terms of some balance-sheet-based financial ratio (see the first three items in Table 9-10). Maintaining a specific bond rating was also indicated to be a popular approach to implementing the debt capacity concept.

Financial Leverage

Cheung, Roy and Gordon have examined how managers rank measures of financial leverage in terms of their importance.[31] These financial measures were derived from both the balance sheet and the income statement. These measures can also be used to determine the desired debt-equity mix of the firm. The results of this survey indicated that firms prefer to use the long-term debt to total capitalization ratio which is derived from the balance sheet to measure financial leverage. Other measures of financial leverage which were preferred by these firms include: long-term debt to net worth ratio, times interest earned and cash flow coverage ratio.

---

[29]We emphasize the following ideas on capital structure design: First, there exists empirical support for the assertion that firms identify and attempt to move toward target debt (leverage) ratios. Managers tell us this is so, and survey studies have confirmed the tendency. Second, there is also empirical support for the assertion that firms prefer internal finance before using external finance and debt is preferred to equity financing. The second assertion is now often referred to as the "pecking order theory" of capital structure.

[30]Francis J. Walsh, *Planning Corporate Capital Structures* (New York: The Conference Board, 1972), p. 17.

[31]Joseph K. Cheung, S. Paul Roy and Irene Gordon. "Financing Policies of Large Canadian Corporations," *CMA Management Accounting Magazine* 63(4) (May 1989), p. 26–31.

Table 9-9 External Influences on the Determination of Target Financial Structure

Type of Influence	Rank 1	Rank 2	Rank 3
Investment bankers	15	25	24
Commercial bankers	30	21	18
Trade creditors supplies	0	6	9
Outside security analysts	8	20	22
Comparative ratios	18	24	24
Other	29	4	3
Total	100%	100%	100%

Source: Joseph K. Cheung, S. Paul Roy and Irene Gordon, "Financing Policies of Large Canadian Corporations," *CMA Management Accounting Magazine* 63(4) (May 1989), p. 27.

Table 9-10 Definitions of Debt Capacity in Practice

Standard or Method	1,000 Largest Corporations (Percent Using)
Target percent of total capitalization (long-term debt to total capitalization)	27%
Long-term debt to net worth ratio (or its inverse)	14
Long-term debt to total assets	2
Interest (or fixed charge) coverage ratio	6
Maintain bond ratings	14
Restrictive debt covenants	4
Most adverse cash flow	4
Industry standard	3
Other	10
No response	16
Total	100%

Source: Derived from David F. Scott, Jr., and Dana J. Johnson, "Financing Policies and Practices in Large Corporations," *Financial Management* 11 (Summer 1982), pp. 51–59.

Table 9-11 presents some results from a study of financial structure tendencies over the 1970–86 period. The 100 largest nonfinancial corporations were studied in selected years with regard to several financial leverage measures.

Size was based on the book value of total assets. Two of those measures are shown in Table 9-11. Part A of the table shows the coverage and long-term debt to total capitalization ratios for all 100 firms for the selected years. Part B of the table shows the results for only those firms that existed across all years of the study. This means 47 firms from the 1970 list also existed in identifiable form in 1986. The second ratio in each panel was measured at book value.

Two main points can be made. First, there is a definite, observed weakening in interest coverage ratios since 1970. We can offer, then, that the likelihood of firm failure has increased over the past two decades for the sample firms. Second, the long-term debt to total capitalization ratios are rather stable over the period. Other researchers have noted this same set of relationships. They are not contradictory—it merely emphasizes that more than one approach has to be taken to assess leverage use. The debt and stock metrics have grown at about the same rates over time—but earnings available to service the interest payments on the debt have lagged.

Most analysts agree that the next recession will be a real (tough) test of recent capital structure policies. We can see how crucial the analysis of financial risk is both to the firm and to the aggregate economy.[32]

Business Cycles

Effective financial managers—those who assist in creating value for the firm's shareholders—are quite perceptive about and in constant communication with the financial marketplace. When market conditions change abruptly, company financial policies and decisions must adapt to the new conditions. However, there are some firms that do a better job of adjusting than others. Companies that are slow to adapt to changes in the aggregate business environment face a lower level of cash flow generation and increased risk of financial distress.

Changes in the aggregate business environment are often referred to as **business cycles**. There are many useful definitions of such cycles. For example, Dr. Fischer Black, a former finance professor and later an executive with the well-known investment banking firm of Goldman, Sachs and Co., defined business cycles as follows:[33]

Business cycles

A series of commercial adjustments to unanticipated new information accentuated by both public policy decisions and private-sector decisions.

> Business cycles are fluctuations in economic activity. Business cycles show up in virtually all measures of economic activity—output, income, employment, unemployment, retail sales, new orders by manufacturers, even housing starts. When times are good, they tend to be good all over; and when times are bad, they tend to be bad all over.

Video Case 3 

A slightly different but compatible definition of business cycles that we will use is the following:

> Business cycles are a series of commercial adjustments to unanticipated, new information. These adjustments are accentuated by both public and private sector decisions. When these decisions, on balance, are correct the economy expands. When these decisions, on balance, are incorrect, the economy contracts.

Table 9-11 Indicators of Financial Leverage Use

Part A: 100 Largest (by total assets) Nonfinancial Corporations				
Year/Ratio	*1986*	*1980*	*1975*	*1970*
Coverage ratio (times)	4.98	8.64	7.85	9.09
Long-term debt to total capitalization (%)	31.9	26.3	30.9	31.2
Part B: 47 "Survivors" of the 100 Largest Nonfinancial Corporations				
Year/Ratio	*1986*	*1980*	*1975*	*1970*
Coverage ratio (times)	3.91	8.91	9.97	13.22
Long-term debt to total capitalization (%)	30.4	26.0	27.8	26.1

Source: Derived from David F. Scott, Jr., and Nancy Jay, "Leverage Use in Corporate America," Working Paper 9011, Orlando, FL: Dr. Phillips Institute for the Study of American Business Activity, University of Central Florida, January 1990.

[32]Ben Bernanke, "Is There Too Much Corporate Debt?" *Business Review*, Federal Reserve Bank of Philadelphia (September–October 1989), 10. Similar findings on increased leverage use are found in Margaret Mendenhall Blair, "A Surprising Culprit Behind the Rush to Leverage," *The Brookings Review* (Winter 1989/90), 19–26.

[33]Fischer Black, "The ABCs of Business Cycles," *Financial Analysts Journal* 37 (November/December 1981), pp. 75–80. Dr. Black touched upon similar and other far-reaching points in his compelling presidential address to the American Finance Association; see Fischer Black, "Noise," *Journal of Finance* 41 (July 1986), pp. 529–543.

Since 1950, Canada has endured nine recessions. Recessions are the contractionary or negative phase of the entire business cycle.

Typically, different stages of the business cycle induce a different set of relationships in the financial markets. These different relationships are reflected in the different capital structure decisions managers make. For example, relationships between interest rates and equity prices may differ sharply over different phases of the cycle. Some phases of the cycle favour the issuance of debt securities over equity instruments, and vice versa. Complicating the decision-making process for the manager is the fact that financial relationships will be dissimilar over each cycle.[34]

The recession that began in 1990 produced its own set of unique financial characteristics. Accordingly, financial managers altered their firm's capital structures in response to new information that included capital cost relationships in the financial markets. For example, managers began to reverse the financial leverage buildup that occurred in the 1980s. Whereas in the early 1990s, corporations began to take advantage of an improved market for common equities and brought to the marketplace substantial amounts of new common share issues.

Business Risk

The single most important factor that should affect the firm's financing mix is the underlying nature of the business in which it operates. In Chapter 8 we defined business risk as the relative dispersion in the firm's expected stream of EBIT. If the nature of the firm's business is such that the variability inherent in its EBIT stream is high, then it would be unwise to impose a high degree of financial risk on top of this already uncertain earnings stream.

Corporate executives are likely to point this out in discussions of capital structure management. A financial officer in a large steel firm related:

> The nature of the industry, the marketplace, and the firm tend to establish debt limits that any prudent management would prefer not to exceed. Our industry is capital intensive and our markets tend to be cyclical.... The capability to service debt while operating in the environment described dictates a conservative financial structure.[35]

Notice how that executive was concerned with both his firm's business risk exposure and its cash flow capability for meeting any financing costs. The AT&T financial officer referred to earlier also has commented on the relationship between business and financial risk:

> In determining how much debt a firm can safely carry, it is necessary to consider the basic risks inherent in that business. This varies considerably among industries and is related essentially to the nature and demand for an industry's product, the operating characteristics of the industry, and its ability to earn an adequate return in an unknown future.[36]

It appears clear that the firm's capital structure cannot be properly designed without a thorough understanding of its commercial strategy.

---

[34]Available discussions and studies on the relationship between business cycles and corporate financing patterns are not plentiful. One such study is Robert A. Taggart, Jr., "Corporate Financing: Too Much Debt," *Financial Analysts Journal* 42 (May/June 1986), pp. 35–42.

[35]Francis J. Walsh, Jr., *Planning Corporate Capital Structures* (New York: The Conference Board, 1972), p. 18.

[36]John J. Scanlon, "Bell System Financial Policies," *Financial Management* 1 (Summer 1972), pp. 16–26.

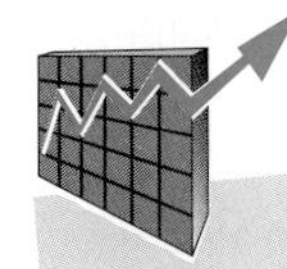

Yes, Managers, There is Still a Business Cycle

There are a number of ways in which managers may be able to profitably counter the business cycle.

First, inventory. The optimal time to begin reducing inventory levels is not when the recession hits; if inventories are reduced in anticipation of a slowdown, then carrying and associated costs are minimized and the organization does not experience the trauma of a knee-jerk inventory adjustment. Interestingly, one of the bright spots of the current recession is the relatively low inventory levels that most manufacturers, retailers and wholesalers were carrying when the economy turned down; this is in sharp contrast to 1981–82 when a huge inventory overhang played a major role in making that recession so bad.

Managers should reduce inventory in anticipation of contraction; they should accordingly expand in anticipation of expansion. Profits are maximized if product and capacity are available when the market wants them, but this requires foresight and planning. Organizations that count on building inventory after the expansion has commenced will miss those major opportunities that generally occur in the early months of expansion.

Second, accounts receivable. Bad debt losses track the business cycle. Debtors are more likely to encounter difficulties servicing their obligations in contractions than expansions. Managers can capitalize on this tendency by tightening credit conditions ahead of contraction and easing ahead of expansion. Optimal accounts receivable management requires careful consideration of the business cycle.

Third, cash. John Keynes mused at length on the countercyclical management of the liquid part of the balance sheet. When the economy is strong and security values are high, managers should increase liquidity so they might capitalize on the lower asset values that usually accompany the business cycle's inevitable downturn; the reverse similarly applies. Most people do not appreciate that John Keynes was far more than a theoretical economist; he was an extraordinary successful investor who relied heavily on the contrarian notion of buying and selling against the crowd. To some, John Keynes is the father of contrarian investing.

Fourth, investment. Managers who do not invest ahead of the business cycle in the state-of-the-art plant, equipment, machinery, research, development and training needed to assure internationally competitive workplaces, work processes and employees, will be surpassed. The time to put the facilities in place for the next expansion is many months or even years before it begins; similarly, the time to ease up to position for the next contraction is well before it begins. Few decisions are more important to an organization's long term prosperity than those involving investment; the business cycle is an important consideration in investment decision making.

Fifth, mergers and acquisitions. There is a popular one-liner to the effect that the stock market has correctly forecast 12 of the last four recessions. The stock market is not perfect but it does tend to peak and trough several months before the economy does. Correctly anticipating the business cycle is crucial to merger and acquisition success. The time to act on mergers and acquisitions is before the crowd moves equity prices to levels that make deals marginal.

Sixth, payables. The dull business of payables management is given short shrift at many organizations, but few organizations would not benefit from a review of payable policies, procedures and practices. When the cycle is turning down, payables management can be very important, especially if an organization is in jeopardy. Artfully using funds provided suppliers and even employees can be the difference in making it through rough spots. One thing is certain: the cash represented by the bills that an organization pays is not available for other purposes; when the economy is turning down, the precarious firm should assure it is making the right payments.

Seventh, debt. The common thread through most corporate crises is the inability to meet required payments associated with borrowed capital. Debt policy that misreads the business cycle can be lethal. In the face of contraction, managers should be reducing fixed charges by paying down debt, particularly short term debt. Heading into recession is no time to be commencing major debt-financed expansion programs; it is also no time to be seriously mismatching the balance sheet by financing longer term assets with short term liabilities. One of the most important management functions is synchronizing the liability side of the balance sheet with the business cycle.

The current downturn graphically illustrates the consequences of debt policies that misread the business cycle. Bankruptcies are skyrocketing; debt beleaguered lenders have turned credit tap off to all but the most creditworthy; the old notion that bridge financing through downturns would always be available to at least average balance sheets has died; the willingness of lenders to make funds available has become a major factor in this recession; for those who did not see the credit crunch coming and position their balance sheets accordingly, the road back is going to be rough.

Eighth, cost control. In today's fiercely competitive business world, few enterprises will prosper without relentless control of costs. At no time is cost control more important than heading into a downturn. Managers who wait for recession to eliminate waste, duplication and excess may find it too late. Nothing exposes ineffective management like the unexpected arrival of recession.

Ninth, promotion. Countercyclical promotion policy can pay huge dividends to the firm that calls the business cycle correctly. The time to have the marketplace enthusiastically aware of your product is when the economy is growing and confidence is high. Just as it takes nerve to spend on promotion when business conditions are soft, promotion spending is easiest to justify when the economy is peaking, and probably represents a highly inefficient use of the scarce promotion dollar. Correctly projecting the business cycle is key to maximizing promotion effectiveness.

Finally, product development. The time to have new products in the marketplace is when consumers are willing to buy. As with promotion spending, managers often find it difficult to convince the board of directors of the merits of product development spending in a downturn, but if it is not done in the downturn, success in the upturn is unlikely. Similarly, the optimal time to spend on product development is not when contraction is near.

It Is Worth Trying to Assess the Business Cycle?

Although economists have identified numerous business cycles, managers do not find it easy to identify where their given industry is in what cycle; a bit of luck probably beats all the analytical ammunition that can be mustered, but it is nonetheless useful for managers to attempt business cycle forecasting.

The exercise of trying, on a systematic basis, to assess the business cycle has organizational benefits regardless of how the predictions themselves prove out. It is hard to think of an important managerial decision that is not seriously impacted by how an enterprise fits through time in the phases of the economy. If nothing else, managers will be ahead if they come away from the current recession with newfound respect for the business cycle; the extraordinary 1982 to 1990 boom to the contrary, the business cycle clearly matters, and clearly should not be ignored and assumed away.

Source: John S. McCallum, "Yes, Managers, There is Still a Business Cycle," *Business Quarterly* 56(1) (Summer 1991), pp. 38–40.

SUMMARY

This chapter examines the firm's financing mix and emphasizes the importance of managing the firm's permanent sources of funds—that is, its capital structure. The objective of capital structure management is to arrange the company's sources of funds so that the price of its common share will be maximized, all other factors held constant.

Capital Structure Theory

Can the firm affect its weighted cost of capital by altering its financing mix? Attempts to answer this question have comprised a significant portion of capital structure theory for over three decades. Extreme positions show that the firm's share price is either unaffected or continually affected as the firm increases its reliance upon leverage-inducing funds. In the real world, an operating environment where interest expense is tax deductible and market imperfections operate to restrict the amount of fixed-income obligations a firm can issue, most financial officers and financial academics subscribe to the concept of an optimal capital structure. The optimal capital structure minimizes the firm's weighted cost of capital. Searching for a proper range of financial leverage, then, is an important financial management activity.

Complicating the manager's search for an optimal capital structure are conflicts that lead to agency costs. A natural conflict exists between shareholders and bondholders (the agency costs of debt). To reduce excessive risk taking by management on behalf of shareholders, it may be necessary to include several protective covenants in bond contracts that serve to restrict managerial decision making.

Another type of agency cost is related to free cash flow. For example, managers have the incentive to hold on to free cash flow and enjoy it, rather than paying it out in the form of higher cash dividend payments. This conflict between managers and shareholders leads to the concept of the free cash flow theory of capital structure. This same theory is also known as the control hypothesis and the threat hypothesis. The ultimate resolution of these agency costs affects the specific form of the firm's capital structure.

The Tools of Capital Structure Management

The decision to use senior securities in the firm's capitalization causes two types of financial leverage effects. The first is the added variability in the earnings per share stream that accompanies the use of fixed-charge securities. We explained in Chapter 8 how this could be quantified by use of the degree of financial leverage metric. The second financial leverage effect relates to the level of earnings per share (EPS) at a given EBIT under a specific capital structure. We rely upon EBIT-EPS analysis to measure this second effect. Through EBIT-EPS analysis the decision maker can inspect the impact of alternative financing plans on EPS over a full range of EBIT levels.

A second tool of capital structure management is the calculation of comparative leverage ratios. Balance sheet leverage ratios and coverage ratios can be computed according to the contractual stipulations of the proposed financing plans. Comparison of these ratios with industry standards enables the financial officer to determine if the firm's key ratios are materially out of line with accepted practice.

A third tool is the analysis of corporate cash flows. This process involves the preparation of a series of cash budgets that consider different economic conditions and different capital structures. Useful insight into the identification of proper target capital structure ratios can be obtained by analyzing projected cash flow statements that assume adverse operating circumstances.

Capital Structure Practices

Surveys indicate that the majority of financial officers in large firms believe in the concept of an optimal capital structure. The optimal capital structure is approximated by the identification of target debt ratios. The targets reflect the firm's ability to service fixed financing costs and also consider the business risk to which the firm is exposed.

Several studies have provided information on who sets or influences the firm's target leverage ratios. The firm's own management group and staff of analysts are the major influence, followed in importance by investment dealers. Studies also show that executives operationalize the concept of debt capacity in many ways. The most popular approach was to maintain financial flexibility. Other factors which were mentioned in this study were: the need to maintain a minimum credit rating on the firm's long-term debt, the concern for covenant violation, and share price maximization.

Financing policies change in significant ways over time. For example during the 1980s, most studies confirm that North American companies "levered up" when compared with past decades. Specifically, interest coverage ratios deteriorated during the 1980s when compared with the 1970s.

The early 1990s are displaying a reversal of this trend. Effective financial managers have a sound understanding of the business cycle. The recession that started in 1990 produced a unique set of financial characteristics that led financial managers to reverse the buildup of leverage in the 1980s and bring substantial amounts of new common shares to the marketplace.

Other studies of managers' financing tendencies suggest that (1) a tremendous preference for the use of internally generated equity to finance investments exists, (2) firms prefer to forgo seemingly profitable projects rather than reduce shareholders' expected cash dividends to finance a greater part of the capital budget, and (3) security issues are timed with favourable market conditions.

STUDY QUESTIONS

9-1. Define the following terms:
- **a.** *Financial structure*
- **b.** *Capital structure*
- **c.** *Optimal capital structure*
- **d.** *Debt capacity*

9-2. What is the primary weakness of EBIT-EPS analysis as a financing decision tool?

9-3. What is the objective of capital structure management?

9-4. Distinguish between (a) balance sheet leverage ratios and (b) coverage ratios. Give two examples of each and indicate how they would be computed.

9-5. Why might firms whose sales levels change drastically over time choose to use debt only sparingly in their capital structures?

9-6. What condition would cause capital structure management to be a meaningless activity?

9-7. What does the term *independence hypothesis* mean as it applies to capital structure theory?

9-8. Who have been the foremost advocates of the independence hypothesis?

9-9. A financial manager might say that the firm's weighted cost of capital is saucer-shaped or U-shaped. What does this mean?

9-10. Define the EBIT-EPS indifference point.

9-11. What is *uncommitted earnings per share* (UEPS)?

9-12. Explain how industry norms might be used by the financial manager in the design of the company's financing mix.

9-13. Define the term *free cash flow*.

9-14. What is meant by the free cash flow theory of capital structure?

9-15. Why should the financial manager be familiar with the business cycle?

SELF-TEST PROBLEMS

ST-1. *(Analysis of Recessionary Cash Flows)* The management of Story Enterprises is considering an increase in its use of financial leverage. The proposal on the table is to sell $6 million of bonds that would mature in 20 years. The interest rate on these bonds would be 12 percent. The bond issue would have a sinking fund attached to it requiring that one-twentieth of the principal be retired each year. Most business economists are forecasting a recession that will affect the entire company in the coming year. Story's management has been saying, "If we can make it through this, we can make it through anything." The firm prefers to carry an operating cash balance of $750,000. Cash collections from sales next year will total $3 million. Miscellaneous cash receipts will be $400,000. Raw material payments will be $700,000. Wage and salary costs will be $1,200,000 on a cash basis. On top of this, Story will experience nondiscretionary cash outlays of $1.2 million, including all tax payments. The firm faces a combined tax rate of 45 percent.

- **a.** At present, Story is unlevered. What will be the total fixed financial charges the firm must pay next year?
- **b.** If the bonds are issued, what is your forecast for the firm's expected cash balance at the end of the recessionary year (next year)?
- **c.** As Story's financial consultant, do you recommend that it issue the bonds?

ST-2. *(Assessing Leverage Use)* Some financial data and the appropriate industry norm for three companies are shown in the following table:

Measure	Firm X	Firm Y	Firm Z	Industry Norm
Total debt to total assets	20%	30%	10%	30%
Times interest and preferred dividend coverage	8 times	16 times	19 times	8 times
Price/earnings ratio	9 times	11 times	9 times	9 times

- **a.** Which firm appears to be employing financial leverage to the most appropriate degree?
- **b.** In this situation, which "financial leverage effect" appears to dominate the market valuation process?

ST-3. *(EBIT-EPS Analysis)* Four engineers from Martin-Bowing Company are leaving that firm in order to form their own corporation. The new firm will produce and distribute computer software on a national basis. The software will be aimed at scientific markets and at businesses desiring to install comprehensive information systems. Private investors have been lined up to finance the new company. Two financing proposals are being studied. Both of these plans involve the use of some financial leverage; however, one is much more highly levered than the other. Plan A requires the firm to sell bonds with an effective interest rate of 14 percent. One million dollars would be raised in this manner. In addition, under plan A, $5 million would be raised by selling common shares at $50 each. Plan B also involves raising $6 million. This would be accomplished by selling $3 million of bonds at an interest rate of 16 percent. The other $3 million would come from selling common shares at $50 each. In both cases the use of financial leverage is considered to be a permanent part of the firm's capital structure, so no fixed maturity date is used in the analysis. The firm considers a 50 percent tax rate appropriate for planning purposes.

a. Find the EBIT indifference level associated with the two financing plans, and prepare an EBIT-EPS analysis chart.

b. Prepare an analytical income statement that demonstrates that EPS will be the same regardless of the plan selected. Use the EBIT level found in part (a).

c. A detailed financial analysis of the firm's prospects suggests that long-term EBIT will be above $1,188,000 annually. Taking this into consideration, which plan will generate the higher EPS?

d. Suppose that long-term EBIT is forecast to be $1,188,000 per year. Under plan A, a price/earnings ratio of 13 would apply. Under plan B, a price/earnings ratio of 11 would apply. If this set of financial relationships does hold, which financing plan would you recommend be implemented?

e. Again, assume an EBIT level of $1,188,000. What price/earnings ratio applied to the EPS of plan B would provide the same share price as that projected for plan A? Refer to your data from part (d).

STUDY PROBLEMS (SET A)

9-1A. *(Analysis of Recessionary Cash Flows)* Ontherise, Inc., is considering the an expansion of its bagel bakery with the acquisition of new equipment to be financed entirely with debt. The company does not have any other debt or preferred shares outstanding. The company currently has a cash balance of $200,000 which is the minimum Baruch Chavez, the CFO of Ontherise, believes to be desirable, Baruch has determined that the following relationships exist among the company's various items of cash flow (except as noted, all are expressed as a percentage of cash collections on sales):

Other cash receipts	5%
Cash Disbursements for:	
Payroll	30%
Raw Materials	25%
Nondiscretionary expenditures (essentially fixed, thus not percent of sales)	$500,000

The new debt would carry fixed financial charges of $140,000 the first year (interest, $90,000, plus principal—sinking fund, $50,000). To evaluate the sensitivity of the proposed debt plan to economic fluctuations, Baruch would like to determine how low cash collections from sales could be in the next year while ensuring that the cash balance at the end of the year is the minimum he considers necessary.

9-2A. *(EBIT-EPS Analysis with Sinking Fund)* Due to his concern over the effect of the "worst that could happen" if he finances the equipment acquisition only with debt (he believes cash collection on sales could be as low as $1,100,000 in the coming year), Baruch Chavez, the CFO of Ontherise, Inc., also has decided to consider a part debt/part equity alternative to the proposed all-debt plan described above. The combination would include 60 percent equity and 40 percent debt. The equity part of the plan would provide $20 per share to the company for 30,000 new shares. The debt portion of this plan would include $400,000 of new debt with fixed financial charges of $52,000 for the first year (interest, $32,000, plus principal—sinking fund $20,000). The company is in the 22 percent tax bracket. The company currently has 100,000 shares outstanding. Baruch has asked you to determine the EBIT indifference level associated with the two financing alternatives.

9-3A. *(EBIT-EPS Analysis)* A group of college professors have decided to form a small manufacturing corporation. The company will produce a full line of contemporary furniture. Two financing plans have been proposed by the investors. Plan A is an all common equity alternative. Under this arrangement 1,000,000 common shares will be sold to net the firm $20 per share. Plan B involves the use of financial leverage. A debt issue with a 20-year maturity period will be privately placed. The debt issue will carry an interest rate of 10 percent and the principal borrowed will amount to $6 million. In addition, $14 million will be issued in common shares at a price of $20 per share. The corporate tax rate is 50 percent.

a. Find the EBIT indifference level associated with the two financing proposals.

b. Prepare an analytical income statement that proves EPS will be the same regardless of the plan chosen at the EBIT level found in part (a).

c. Prepare an EBIT-EPS analysis chart for this situation.

d. If a detailed financial analysis projects that long-term EBIT will always be close to $2,400,000 annually, which plan will provide for the higher EPS?

9-4A. *(Capital-Structure Theory)* Deep End Pools & Supplies has an all-common-equity capital structure. Some financial data for the company are shown below:

Common shares outstanding	= 900,000
Common share price, P_0	= $30 per share
Expected level of EBIT	= $5,400,000
Dividend payout ratio	= 100 percent

In answering the following questions, assume that corporate income is not taxed.

a. Under the present capital structure, what is the total value of the firm?

b. What is the cost of common equity capital, k_c? What is the weighted cost of capital, K_0?

c. Now, suppose Deep End sells $1.5 million of long-term debt with an interest rate of 8 percent. The proceeds are used to retire outstanding common shares. According to the net operating income theory (the independence hypothesis), what will be the firm's cost of common equity after the capital structure change?

1. What will be the dividend per share flowing to the firm's common shareholders?
2. By what percentage has the dividend per share changed owing to the capital structure change?
3. By what percentage has the cost of common equity changed owing to the capital structure change?
4. What will be the weighted cost of capital after the capital structure change?

9-5A. *(EBIT-EPS Analysis)* Four recent liberal arts graduates have interested a group of venture capitalists in backing a small business enterprise. The proposed operation would consist of a series of retail outlets that would distribute and service a full line of vacuum cleaners and accessories. These stores would be located in Alberta. Two financing plans have been proposed by the graduates. Plan A is an all-common-equity capital structure. Two million dollars would be raised by selling 80,000 common shares. Plan B would involve the use of long-term debt financing. One million dollars would be raised by marketing bonds with an interest rate of 12 percent. Under this alternative another $1 million would be raised by selling 40,000 common shares. With both plans, then, $2 million is needed to launch the new firm's operations. The debt funds raised under plan B are considered to have no fixed maturity date, in that this proportion of financial leverage is thought to be a permanent part of the company's capital structure. The fledgling executives have decided to use a 22 percent tax rate in their analysis.

a. Find the EBIT indifference level associated with the two financing proposals.

b. Prepare an analytical income statement that proves EPS will be the same regardless of the plan chosen at the EBIT level found in part (a).

9-6A. *(EBIT-EPS Analysis)* Three recent graduates of the computer science program are forming a company to write and distribute software for various personal computers. Initially, the corporation will operate in southern Ontario. Twelve serious prospects for retail outlets have already been identified and committed to the firm. The firm's software products have been tested and displayed at several trade shows and computer fairs in the perceived operating region. All that is lacking is adequate financing to continue with the project. A small group of private investors in the Toronto area is interested in financing the new company. Two financing proposals are being evaluated. The first (plan A) is an all-common-equity capital structure. Two million dollars would be raised by selling common shares at $20 per share. Plan B would involve the use of financial leverage. One million dollars would be raised selling bonds with an effective interest rate of 11 percent (per annum). Under this second plan, the remaining $1 million would be raised by selling common shares at the $20 price per share. The use of financial leverage is considered to be a permanent part of the firm's capitalization, so no fixed maturity date is needed for the analysis. A 22 percent tax rate is appropriate for the analysis.

a. Find the EBIT indifference level associated with the two financing plans.

b. A detailed financial analysis of the firm's prospects suggests that the long-term EBIT will be above $300,000 annually. Taking this into consideration, which plan will generate the higher EPS?

c. Suppose the long-term EBIT is forecast to be $300,000 per year. Under plan A, a price/earnings ratio of 19 would apply. Under plan B, a price/earnings ratio of 15 would apply. If this set of financial relationships does hold, which financing plan would you recommend?

9-7A. *(EBIT-EPS Analysis)* Three recent liberal arts graduates have interested a group of venture capitalists in backing a new business enterprise. The proposed operation would consist of a series of retail outlets to distribute and service a full line of personal computer equipment. These stores would be located in New Brunswick. Two financing plans have been proposed by the graduates. Plan A is an all-common-equity structure. Three million dollars would be raised by selling 75,000 common shares. Plan B would involve the use of long-term debt financing. One million dollars would be raised by marketing bonds with an effective interest rate of 15 percent. Under this alternative, another $2 million would be raised by selling 50,000 common shares. With both plans, then, $3 million is needed to launch the new firm's operations. The debt funds raised under plan B are considered to have no fixed maturity date, in that this proportion of financial leverage is thought to be a permanent part of the company's capital structure. The fledgling executives have decided to use a 22 percent tax rate in their analysis, and they have hired you on a consulting basis to do the following:

a. Find the EBIT indifference level associated with the two financing proposals.

b. Prepare an analytical income statement that proves EPS will be the same regardless of the plan chosen at the EBIT level found in part (a).

9-8A. *(EBIT-EPS Analysis)* Two recent graduates of the computer science program are forming a company to write, market, and distribute software for various personal computers. Initially, the corporation will operate in Western Canada. Twelve serious prospects for retail outlets in these different provinces have already been identified and committed to the firm. The firm's software products have been tested and displayed at several trade shows and computer fairs in the perceived operating region. All that is lacking is adequate financing to continue the project. A small group of private investors in Vancouver are interested in financing the new company. Two financing proposals are being evaluated. The first (plan A) is an all-common-equity capital structure. Four million dollars would be raised by selling common shares at $40 per share. Plan B would involve the use of financial leverage. Two million dollars would be raised by selling bonds with an effective interest rate of 16 percent (per annum). Under this second plan, the remaining $2 million would be raised by selling common shares at the $40 price per share. This use of financial leverage is considered to be a permanent part of the firm's capitalization, so no fixed maturity date is needed for the analysis. A 50 percent tax rate is appropriate for the analysis.

a. Find the EBIT indifference level associated with the two financing plans.

b. Prepare an analytical income statement that proves EPS will be the same regardless of the plan chosen at the EBIT level found in part (a).

c. A detailed financial analysis of the firm's prospects suggests that long-term EBIT will be above $800,000 annually. Taking this into consideration, which plan will generate the higher EPS?

d. Suppose that long-term EBIT is forecast to be $800,000 per year. Under plan A, a price/earnings ratio of 12 would apply. Under plan B, a price/ earnings ratio of 10 would apply. If this set of financial relationships does hold, which financing plan would you recommend be implemented?

9-9A. *(Analysis of Recessionary Cash Flows)* The management of PEI Produce is considering an increase in its use of financial leverage. The proposal on the table is to sell $10 million of bonds that would mature in 20 years. The interest rate on these bonds would be 15 percent. The bond issue would have a sinking fund attached to it requiring that one-twentieth of the principal be retired each year. Most business economists are forecasting a recession that will affect the entire economy in the coming year. PEI's management has been saying, "If we can make it through this, we can make it through anything." The firm prefers to carry an operating cash balance of $1 million. Cash collections from sales next year will total $4 million. Miscellaneous cash receipts will be $300,000. Raw material payments will be $800,000. Wage and salary costs will total $1.4 million on a cash basis. On top of this, PEI will experience nondiscretionary cash outflows of $1.2 million including all tax payments. The firm faces a 50 percent tax rate.

- **a.** At present, PEI is unlevered. What will be the total fixed financial charges the firm must pay next year?
- **b.** If the bonds are issued, what is your forecast for the firm's expected cash balance at the end of the recessionary year (next year)?
- **c.** As PEI's financial consultant, do you recommend that it issue the bonds?

9-10A. *(EBIT-EPS Analysis)* Four recent business school graduates have interested a group of venture capitalists in backing a small business enterprise. The proposed operation would consist of a series of retail outlets that would distribute and service a full line of energy-conservation equipment. These stores would be located in Alberta. Two financing plans have been proposed by the graduates. Plan A is an all-common-equity capital structure. Three million dollars would be raised by selling 60,000 common shares. Plan B would involve the use of long-term debt financing. One million dollars would be raised by marketing bonds with an interest rate of 10 percent. Under this alternative another $2 million would be raised by selling 40,000 common shares. With both plans, then, $3 million is needed to launch the new firm's operations. The debt funds raised under plan B are considered to have no fixed maturity date, in that this proportion of financial leverage is thought to be a permanent part of the company's capital structure. The fledgling executives have decided to use a 22 percent tax rate in their analysis.

- **a.** Find the EBIT indifference level associated with the two financing proposals.
- **b.** Prepare an analytical income statement that proves EPS will be the same regardless of the plan chosen at the EBIT level found in part (a).

9-11A. *(EBIT-EPS Analysis)* A group of college professors have decided to form a small manufacturing corporation. The company will produce a full line of contemporary furniture. Two financing plans have been proposed by the investors. Plan A is an all-common-equity alternative. Under this arrangement 1,400,000 common shares will be sold to net the firm $10 per share. Plan B involves the use of financial leverage. A debt issue with a 20-year maturity period will be privately placed. The debt issue will carry an interest rate of 8 percent and the principal borrowed will amount to $4 million. The corporate tax rate is 50 percent.

- **a.** Find the EBIT indifference level associated with the two financing proposals.
- **b.** Prepare an analytical income statement that proves EPS will be the same regardless of the plan chosen at the EBIT level found in part (a).
- **c.** Prepare an EBIT-EPS analysis chart for this situation.
- **d.** If a detailed financial analysis projects that long-term EBIT will always be close to $1,800,000 annually, which plan will provide for the higher EPS?

9-12A. *(EBIT-EPS Analysis)* The professors in problem 9-11A contacted a financial consultant to provide them with some additional information. They felt that in a few years the shares of the firm would be publicly traded over the counter, so they were interested in the consultant's opinion as to what the share price would be under the financing plan outlined in problem 9-11A. The consultant agreed that the projected long-term EBIT level of $1,800,000 was reasonable. He also felt that if plan A were selected, the marketplace would apply a price/earnings ratio of 12 times to the company's share; for plan B he estimated a price/earnings ratio of 10 times.

- **a.** According to this information, which financing alternative would offer a higher share price?
- **b.** What price/earnings ratio applied to the EPS related to plan B would provide the same share price as that projected for plan A?
- **c.** Comment upon the results of your analysis of problems 9-11A and 9-12A.

9-13A. *(Analysis of Recessionary Cash Flows)* Cavalier Agriculture Supplies is undertaking a thorough cash flow analysis. It has been proposed by management that the firm expand by raising $5 million in the long-term debt markets. All of this would be immediately invested in new fixed assets. The proposed bond issue would carry an 8 percent interest rate and have a maturity period of 20 years. The bond issue would have a sinking fund provision that one-twentieth of the principal would be retired annually. Next year is expected to be a poor one for Cavalier. The firm's management feels, therefore, that the upcoming year would serve well as a model for the worst possible operating conditions that the firm can be expected to encounter. Cavalier ordinarily carries a $500,000 cash balance. Next year sales collections are forecast to be $3 million. Miscellaneous cash receipts will total $200,000. Wages and salaries will amount to $1 million. Payments for raw materials used in the production process will be $1,400,000. In addition, the firm will pay $500,000 in nondiscretionary expenditures including taxes. The firm faces a 50 percent tax rate.

- **a.** Cavalier currently has no debt or preferred shares outstanding. What will be the total fixed financial charges that the firm must meet next year?
- **b.** What is the expected cash balance at the end of the recessionary period (next year), assuming the debt is issued?
- **c.** Based on this information, should Cavalier issue the proposed bonds?

9-14A. *(Assessing Leverage Use)* Some financial data for three corporations are displayed below.

Measure	Firm A	Firm B	Firm C	Industry Norm
Debt ratio	20%	25%	40%	20%
Times burden covered	8 times	10 times	7 times	9 times
Price/earnings ratio	9 times	11 times	6 times	10 times

- **a.** Which firm appears to be excessively levered?
- **b.** Which firm appears to be employing financial leverage to the most appropriate degree?
- **c.** What explanation can you provide for the higher price/earnings ratio enjoyed by firm B as compared with firm A?

9-15A. *(Assessing Leverage Use)* Some financial data and the appropriate industry norm are shown in the following table:

Measure	Firm X	Firm Y	Firm Z	Industry Norm
Total debt to total assets	35%	30%	10%	35%
Times interest and preferred dividend coverage	7 times	14 times	16 times	7 times
Price/earnings ratio	8 times	10 times	8 times	8 times

- **a.** Which firm appears to be using financial leverage to the most appropriate degree?
- **b.** In this situation which "financial leverage effect" appears to dominate the market's valuation process?

9-16A. (*Capital Structure Theory)* Titian Textiles has an all-common-equity capital structure. Pertinent financial characteristics for the company are shown below:

Common shares outstanding	= 1,000,000
Common share price, P_0	= $20 per share
Expected level of EBIT	= $5,000,000
Dividend payout ratio	= 100 percent

In answering the following questions, assume that corporate income is not taxed.

- **a.** Under the present capital structure, what is the total value of the firm?
- **b.** What is the cost of common equity capital, k_c? What is the weighted cost of capital, K_0?

c. Now, suppose that Titian Textiles sells $1 million of long-term debt with an interest rate of 8 percent. The proceeds are used to retire outstanding common shares. According to net-operating-income theory (the independence hypothesis), what will be the firm's cost of common equity after the capital structure change?
   1. What will be the dividend per share flowing to the firm's common shareholders?
   2. By what percent has the dividend per share changed owing to the capital structure change?
   3. By what percent has the cost of common equity changed owing to the capital structure change?
   4. What will be the weighted cost of capital after the capital structure change?

9-17A. *(Capital Structure Theory)* Easy Blend Auto Parts has an all-common-equity capital structure. Some financial data for the company are shown below:

Common shares outstanding	= 600,000
Common share price, P_0	= $40 per share
Expected level of EBIT	= $4,200,000
Dividend payout ratio	= 100 percent

In answering the following questions, assume that corporate income is not taxed.

a. Under the present capital structure, what is the total value of the firm?

b. What is the cost of common equity capital, k_c? What is the weighted cost of capital, K_0?

c. Now, suppose Easy Blend sells $1 million of long-term debt with an interest rate of 10 percent. The proceeds are used to retire outstanding common shares. According to the net operating income theory (the independence hypothesis), what will be the firm's cost of common equity after the capital structure change?
   1. What will be the dividend per share flowing to the firm's common shareholders?
   2. By what percentage has the dividend per share changed owing to the capital structure change?
   3. By what percentage has the cost of common equity changed owing to the capital structure change?
   4. What will be the weighted cost of capital after the capital structure change?

9-18A. *(EBIT-EPS Analysis)* Victoria Golf Equipment is analyzing three different financing plans for a newly formed subsidiary. The plans are described below:

Plan A	Plan B	Plan C
Common shares:	Bonds at 9%: $20,000	Preferred shares at 9%: $20,000
$100,000	Common shares: $80,000	Common shares: $80,000

In all cases the common shares will be sold to net Victoria $10 per share. The subsidiary is expected to generate an average EBIT per year of $22,000. The management of Victoria places great emphasis on EPS performance. Income is taxed at a 50 percent rate.

a. Where feasible, find the EBIT indifference levels between the alternatives.

b. Which financing plan do you recommend that Victoria pursue?

INTEGRATIVE PROBLEM

Several biking enthusiasts recently left their federal public service jobs and grouped together to form a corporation, Freedom Cycle., (FCI), which will produce a new type of bicycle. These new bicycles are to be constructed using space age technologies and materials so they will never need repairs or maintenance. The FCI founders believe there is a need for such a bicycle due to their perception that many people today, especially middle-aged working couples like themselves, really would like to ride bicycles for transportation as well as for pleasure, but are put off by the perceived high maintenance requirements of most bicycles today.

The founders believe such people would be quite willing to buy a maintenance-free bicycle for themselves as well as for their children, particularly after observing the repair and maintenance needs (for example, keeping spoked wheels trued and derailleurs and brakes adjusted) of the bikes they already have purchased for their children. Accordingly, the FCI group feels certain that their new age bicycles will meet the needs of this market and will be a tremendous hit.

To assist them with the financial management of the company, the FCI founders hired Mabra Jordan to be CFO. Mabra has considerable experience with start-up companies such as FCI, and is well respected in the venture capital community. Indeed, based on the strength of her business plan, Mabra has convinced a local venture capital partnership to provide funding for FCI. Two alternatives have been proposed by the venture capitalists: a high-leverage plan primarily using "junk" bonds (HLP), and a low-leverage plan (LLP) primarily using equity.

HLP consists of $6 million of bonds carrying a 14 percent interest rate and $4 million of $20-per-share common stock. LLP, on the other hand, consists of $2 million of bonds with an interest rate of 11 percent and $8 million of common shares at $20 per share. Under either alternative, FCI is required to use a sinking fund to retire 10 percent of the bonds each year. FCI's tax rate is expected to be 50 percent.

a. Find the EBIT indifference level associated with the two financing alternatives, and prepare an EBIT-EPS analysis graph.

b. Prepare an analytical income statement that demonstrates that EPS will be the same regardless of the alternative selected. Use the EBIT level computed in part (a) above.

c. If an analysis of FCI's long-term prospects indicates that long-term EBIT will be $1,300,000 annually, which financing alternative will generate the higher EPS?

d. If the analysis of FCI's long-term prospects also shows that at a long-term EBIT of $1,300,000 a price/earnings ratio of 18 likely would apply under LLP, and a ratio of 14 would apply under HLP, which of the two financing plans would you recommend and why?

e. At an EBIT level of $1,300,000, what is the price/earnings ratio that would have to be obtained under HLP for the EPS of HLP to provide the same stock price as that projected for LLP in part (d) above?

A concern of the venture capitalists, of course, is whether FCI would be able to survive its first year in business if for some reason—such as an economic recession or just an overly optimistic sales projection—the cash flow targets in FCI's business plan were not met. To allay such fears, Mabra included in the FCI business plan a worst-case scenario based on the following pessimistic projections.

Mabra believes FCI should maintain a $500,000 cash balance. Starting initially with zero cash, the company would obtain cash of $10,000,000 from either of the two financing alternatives described above. Financing of $9,500,000 would be used for capital acquisitions; the balance is intended to be available to provide initial working capital. The pessimistic sales forecasts indicates cash receipts would be $4 million. Miscellaneous cash receipts (for example, from the sale of scrap titanium and other materials) would be worth $200,000. Cash payments on the purchases of raw materials would be $1 million; wage and salary cash outlays would be $1,500,000; nondiscretionary cash costs (not including tax payments) would be $700,000; and estimated tax payments would be $265,000 under LLP and $54,000 under HLP (note that the difference in estimated tax payments is attributable to the variation in taxable income, which reflects the difference in deductible interest expense).

f. What would be the total fixed financial charges under each of the two alternative financing plans being considered by FCI?

g. A significant issue is whether FCI will have a sufficient cash balance at the end of the possible recessionary year. What is your estimate of FCI's cash balance under each of the two financing plans at the end of such a year?

h. In light of the above and your knowledge of FCI's desired cash level, which financing plan, LLP or HLP, would you recommend?

STUDY PROBLEMS (SET B)

9-1B. *(Analysis of Recessionary Cash Flows)* Cappuccino Express, Inc., is considering the expansion of its cafe business by adding a number of new stores. Strong consideration is being given to financing the expansion entirely with debt. The company does not have any other debt or preferred shares outstanding. The company currently has a cash balance of $400,000 which is the minimum Vanessa Jefferson, the CFO of Cappuccino Express, Inc., believes to be desirable, Vanessa has determined that the following relationships exist among the company's various items of cash flow (except as noted, all are expressed as a percentage of cash collections on sales):

Other cash receipts	5%
Cash Disbursements for:	
Payroll	40%
Raw Materials	20%
Nondiscretionary expenditures (essentially fixed, thus not percent of sales)	$500,000

The new debt would carry fixed financial charges of $300,000 the first year (interest, $200,000, plus principal—sinking fund, $100,000). To evaluate the sensitivity of the proposed debt plan to economic fluctuations, Vanessa would like to determine how low cash collections from sales could be in the next year while ensuring that the cash balance at the end of the year is the minimum she considers necessary.

9-2B. *(EBIT-EPS Analysis with Sinking Fund)* Due to her concern over the effect of the "worst that could happen" if she finances the equipment acquisition only with debt (she believes cash collection on sales could be as low as $1,300,000 in the coming year), Vanessa Jefferson, the CFO of Cappuccino Express, Inc., also has decided to consider a part debt/part equity alternative to the proposed all-debt plan described above. The combination would include 70 percent equity and 30 percent debt. The equity part of the plan would provide $20 per share to the company for 70,000 new shares. The debt portion of this plan would include $600,000 of new debt with fixed financial charges of $78,000 for the first year (interest, $48,000, plus principal—sinking fund $30,000). The company is in the 22 percent tax bracket. The company currently has 100,000 shares outstanding. Vanessa has asked you to determine the EBIT indifference level associated with the two financing alternatives.

9-3B. *(EBIT-EPS Analysis)* Three recent graduates of the computer science program at the University of Toronto are forming a company to write and distribute software for various personal computers. Initially, the corporation will operate in southern Ontario. Twelve serious prospects for retail outlets have already been identified and committed to the firm. The firm's software products have been tested and displayed at several trade shows and computer fairs in the perceived operating region. All that is lacking is adequate financing to continue with the project. A small group of private investors in the Toronto area are interested in financing the new company. Two financing proposals are being evaluated. The first (plan A) is an all common equity capital structure. Three million dollars would be raised by selling common shares at $20 per share. Plan B would involve the use of financial leverage. Two million dollars would be raised selling bonds with an effective interest rate of 11 percent (per annum). Under this second plan, the remaining $1 million would be raised by selling common shares at the $20 price per share. The use of financial leverage is considered to be a permanent part of the firm's capitalization, so no fixed maturity date is needed for the analysis. A 22 percent tax rate is appropriate for the analysis.

- **a.** Find the EBIT indifference level associated with the two financing plans.
- **b.** A detailed financial analysis of the firm's prospects suggests that the long-term EBIT will be above $450,000 annually. Taking this into consideration, which plan will generate the higher EPS?
- **c.** Suppose the long-term EBIT is forecast to be $450,000 per year. Under plan A, a price/earnings ratio of 19 would apply. Under plan B, a price/earnings ratio of 12.39 would apply. If this set of financial relationships does hold, which financing plan would you recommend?

9-4B. *(EBIT-EPS Analysis)* Three recent liberal arts graduates have interested a group of venture capitalists in backing a new business enterprise. The proposed operation would consist of a series of retail outlets to distribute and service a full line of personal computer equipment. These stores would be located in New Brunswick. Two financing plans have been proposed by the graduates. Plan A is an all-common-equity structure. Four million dollars would be raised by selling 80,000 common shares. Plan B would involve the use of long-term debt financing. Two million dollars would be raised by marketing bonds with an effective interest rate of 16 percent. Under this alternative, another $2 million would be raised by selling 50,000 common shares. With both plans, then, $4 million is needed to launch the new firm's operations. The debt funds raised under plan B are considered to have no fixed maturity date, in that this proportion of financial leverage is thought to be a permanent part of the company's capital structure. The fledgling executives have decided to use a 22 percent tax rate in their analysis, and they have hired you on a consulting basis to do the following:

- **a.** Find the EBIT indifference level associated with the two financing proposals.
- **b.** Prepare an analytical income statement that proves EPS will be the same regardless of the plan chosen at the EBIT level found in part (a).

9-5B. *(EBIT-EPS Analysis)* Two recent graduates of the computer science program are forming a company to write, market, and distribute software for various personal computers. Initially, the corporation will operate in Western Canada. Eight prospects for retail outlets in these different provinces have already been identified and committed to the firm. The firm's software products have been tested. All that is lacking is adequate financing to continue the project. A small group of private investors are interested in financing the new company. Two financing proposals are being evaluated. The first (plan A) is an all-common-equity capital structure. Three million dollars would be raised by selling common shares at $40 per share. Plan B would involve the use of financial leverage. One million dollars would be raised by selling bonds with an effective interest rate of 14 percent (per annum). Under this second plan, the remaining $2 million would be raised by selling common shares at the $40 price per share. This use of financial leverage is considered to be a permanent part of the firm's capitalization, so no fixed maturity date is needed for the analysis. A 50 percent tax rate is appropriate for the analysis.

- **a.** Find the EBIT indifference level associated with the two financing plans.
- **b.** Prepare an analytical income statement that proves EPS will be the same regardless of the plan chosen at the EBIT level found in part (a).
- **c.** A detailed financial analysis of the firm's prospects suggests that the long-term EBIT will be above $750,000 annually. Taking this into consideration, which plan will generate the higher EPS?
- **d.** Suppose that long-term EBIT is forecast to be $750,000 per year. Under plan A, a price/earnings ratio of 12 would apply. Under plan B, a price/earnings ratio of 9.836 would apply. If this set of financial relationships does hold, which financing plan would you recommend be implemented?

9-6B. *(Analysis of Recessionary Cash Flows)* The management of Calgary Collectibles (CC) is considering an increase in its use of financial leverage. The proposal on the table is to sell $11 million of bonds that would mature in 20 years. The interest rate on these bonds would be 16 percent. The bond issue would have a sinking fund attached to it requiring that one-twentieth of the principal be retired each year. Most business economists are forecasting a recession that will affect the entire economy in the coming year. CC's management has been saying, "If we can make it through this, we can make it through anything." The firm prefers to carry an operating cash balance of $500,000. Cash collections from sales next year will total $3.5 million. Miscellaneous cash receipts will be $300,000. Raw material payments will be $800,000. Wage and salary costs will total $1.5 million on a cash basis. On top of this, CC will experience nondiscretionary cash outflows of $1.3 million including all tax payments. The firm faces a 50 percent tax rate.

- **a.** At present, CC is unlevered. What will be the total fixed financial charges the firm must pay next year?
- **b.** If the bonds are issued, what is your forecast for the firm's expected cash balance at the end of the recessionary year (next year)?
- **c.** As CC's financial consultant, do you recommend that it issue the bonds?

9-7B. *(EBIT-EPS Analysis)* Four recent business school graduates have interested a group of venture capitalists in backing a small business enterprise. The proposed operation would consist of a series of retail outlets that would distribute and service a full line of energy-conservation equipment. These stores would be located in Alberta. Two financing plans have been proposed by the graduates. Plan A is an all-common-equity capital structure. Five million dollars would be raised by selling 75,000 common shares. Plan B would involve the use of long-term debt financing. Two million dollars would be raised by marketing bonds with an interest rate of 12 percent. Under this alternative, another $3 million would be raised by selling 55,000 common shares. With both plans, then, $5 million is needed to launch the new firm's operations. The debt funds raised under plan B are considered to have no fixed maturity date, in that this proportion of financial leverage is thought to be a permanent part of the company's capital structure. The fledgling executives have decided to use a 22 percent tax rate in their analysis.

a. Find the EBIT indifference level associated with the two financing proposals.

b. Prepare an analytical income statement that proves EPS will be the same regardless of the plan chosen at the EBIT level found in part (a).

9-8B. *(EBIT-EPS Analysis)* A group of college professors have decided to form a small manufacturing corporation. The company will produce a full line of contemporary furniture. Two financing plans have been proposed by the investors. Plan A is an all common equity alternative. Under this arrangement 1,200,000 common shares will be sold to net the firm $10 per share. Plan B involves the use of financial leverage. A debt issue with a 20-year maturity period will be privately placed. The debt issue will carry an interest rate of 9 percent and the principal borrowed will amount to $3.5 million. The corporate tax rate is 50 percent.

a. Find the EBIT indifference level associated with the two financing proposals.

b. Prepare an analytical income statement that proves EPS will be the same regardless of the plan chosen at the EBIT level found in part (a).

c. Prepare an EBIT-EPS analysis chart for this situation.

d. If a detailed financial analysis projects that long-term EBIT will always be close to $1,500,000 annually, which plan will provide for the higher EPS?

9-9B. *(EBIT-EPS Analysis)* The professors in Problem 9-8B contacted a financial consultant to provide them with some additional information. They felt that in a few years the shares of the firm would be publicly traded over the counter, so they were interested in the consultant's opinion as to what the share price would be under the financing plan outlined in Problem 9-8B. The consultant agreed that the projected long-term EBIT level of $1,500,000 was reasonable. He also felt that if plan A were selected, the marketplace would apply a price/earnings ratio of 13 times to the company's share; for plan B he estimated a price earnings ratio of 11 times.

a. According to this information, which financing alternative would offer a higher share price?

b. What price/earnings ratio applied to the EPS related to plan B would provide the same share price as that projected for plan A?

c. Comment upon the results of your analysis of Problems 9-8B and 9-9B.

9-10B. *(Analysis of Recessionary Cash Flows)* Seville Cranes, Inc., is undertaking a thorough cash flow analysis. It has been proposed by management that the firm expand by raising $6 million in the long-term debt markets. All of this would be immediately invested in new fixed assets. The proposed bond issue would carry a 10 percent interest rate and have a maturity period of 20 years. The bond issue would have a sinking fund provision that one-twentieth of the principal would be retired annually. Next year is expected to be a poor one for Seville. The firm's management feels, therefore, that the upcoming year would serve well as a model for the worst possible operating conditions that the firm can be expected to encounter. Seville ordinarily carries a $750,000 cash balance. Next year sales collections are forecast to be $3.5 million. Miscellaneous cash receipts will total $200,000. Wages and salaries will amount to $1.2 million. Payments for raw materials used in the production process will be $1,500,000. In addition, the firm will pay $500,000 in nondiscretionary expenditures including taxes. The firm faces a 50 percent tax rate.

a. Seville currently has no debt or preferred shares outstanding. What will be the total fixed financial charges that the firm must meet next year?

b. What is the expected cash balance at the end of the recessionary period (next year), assuming the debt is issued?

c. On the basis of this information, should Seville issue the proposed bonds?

9-11B. *(Assessing Leverage Use)* Some financial data for three corporations are displayed below:

Measure	Firm A	Firm B	Firm C	Industry Norm
Debt ratio	15%	20%	35%	25%
Times burden covered	9 times	11 times	6 times	9 times
Price/earnings ratio	10 times	12 times	5 times	10 times

a. Which firm appears to be excessively levered?

b. Which firm appears to be employing financial leverage to the most appropriate degree?

c. What explanation can you provide for the higher price/earnings ratio enjoyed by firm B as compared with firm A?

9-12B. *(Assessing Leverage Use)* Some financial data and the appropriate industry norm are shown in the following table:

Measure	Firm X	Firm Y	Firm Z	Industry Norm
Total debt to total assets	40%	35%	10%	35%
Times interest and preferred dividend coverage	8 times	13 times	16 times	7 times
Price/earnings ratio	8 times	11 times	8 times	8 times

a. Which firm appears to be using financial leverage to the most appropriate degree?

b. In this situation which "financial leverage effect" appears to dominate the market's valuation process?

9-13B. *(Capital Structure Theory)* Whittier Optical Labs has an all-common-equity capital structure. Pertinent financial characteristics for the company are shown below:

Common shares outstanding	= 1,000,000
Common share price, P_0	= \$22 per share
Expected level of EBIT	= \$4,750,000
Dividend payout ratio	= 100 percent

In answering the following questions, assume that corporate income is not taxed.

a. Under the present capital structure, what is the total value of the firm?

b. What is the cost of common equity capital, k_c? What is the weighted cost of capital, K_0?

c. Now, suppose that Whittier sells \$1 million of long-term debt with an interest rate of 9 percent. The proceeds are used to retire outstanding common shares. According to net-operating-income theory (the independence hypothesis), what will be the firm's cost of common equity after the capital structure change?

1. What will be the dividend per share flowing to the firm's common shareholders?
2. By what percentage has the dividend per share changed owing to the capital structure change?
3. By what percentage has the cost of common equity changed owing to the capital structure change?
4. What will be the weighted cost of capital after the capital structure change?

9-14B. *(Capital Structure Theory)* Fernando Hotels has an all-common-equity capital structure. Some financial data for the company are shown below:

Common shares outstanding	= 575,000
Common share price, P_0	= \$38 per share
Expected level of EBIT	= \$4,500,000
Dividend payout ratio	= 100 percent

In answering the following questions, assume that corporate income is not taxed.

a. Under the present capital structure, what is the total value of the firm?

b. What is the cost of common equity capital, k_c? What is the weighted cost of capital, K_0?

c. Now, suppose Fernando sells \$1.5 million of long-term debt with an interest rate of 11 percent. The proceeds are used to retire outstanding common shares. According to the net operating income theory (the independence hypothesis), what will be the firm's cost of common equity after the capital-structure change?

1. What will be the dividend per share flowing to the firm's common shareholders?
2. By what percent has the dividend per share changed owing to the capital structure change?
3. By what percent has the cost of common equity changed owing to the capital structure change?
4. What will be the weighted cost of capital after the capital structure change?

9-15B. *(EBIT-EPS Analysis)* Mount Rosemead Health Services, Inc., is analyzing three different financing plans for a newly formed subsidiary. The plans are described below:

Plan A	Plan B	Plan C
Common shares: \$150,000	Bonds at 10%: \$ 50,000 Common shares: \$100,000	Preferred shares at 10%: \$ 50,000 Common shares: \$100,000

In all cases the common shares will be sold to net Mount Rosemead \$10 per share. The subsidiary is expected to generate an average EBIT per year of \$36,000. The management of Mount Rosemead places great emphasis on EPS performance. Income is taxed at a 50 percent rate.

a. Where feasible, find the EBIT indifference levels between the alternatives.

b. Which financing plan do you recommend that Mount Rosemead pursue?

SELF-TEST SOLUTIONS

SS-1.

a. $FC = \text{interest} + \text{sinking fund}$

$$FC = (\$6{,}000{,}000)\,(.12) + \left(\frac{\$6{,}000{,}000}{20}\right)$$

$$FC = \$720{,}000 + \$300{,}000 = \$1{,}020{,}000$$

b.

$$CB_r = C_0 + NCF_r - FC$$

where $C_0 = \$750{,}000$
$FC = \$1{,}020{,}000$

and

$$NCF_r = \$3{,}400{,}000 - \$3{,}100{,}000 = \$300{,}000$$

so

$$CB_r = \$750{,}000 + \$300{,}000 - \$1{,}020{,}000$$
$$CB_r = \$30{,}000$$

c. We know that the firm has a preference for maintaining a cash balance of $750,000. The joint impact of the recessionary economic environment and the proposed issue of bonds would put the firm's recessionary cash balance (CB_r) at $30,000. Since the firm desires a minimum cash balance of $750,000 ($C_0$), the data suggest that the proposed bond issue should be postponed.

SS-2. **a.** Firm Y seems to be using financial leverage to the most appropriate degree. Notice that its price/earnings ratio of 16 times exceeds that of firm X (at 9 times) and firm Z (also at 9 times).

b. The first financial leverage effect refers to the added variability in the earnings per share stream caused by the use of leverage-inducing financial instruments. The second financial leverage effect concerns the level of earnings per share at a specific EBIT associated with a specific capital structure.

Beyond some critical EBIT level, earnings per share will be higher if more leverage is used. Based on the company data provided, the marketplace for financial instruments is weighing the second leverage effect more heavily. Firm Z, therefore, seems to be under-levered (is operating below its theoretical leverage capacity).

SS-3. **a.**

Plan A		Plan B
EPS: Less-Levered Plan		EPS: More-Levered Plan
$\frac{(\text{EBIT} - I)(1 - T) - D_p}{S_A}$	$=$	$\frac{(\text{EBIT} - I)(1 - T) - D_p}{S_B}$
$\frac{(\text{EBIT} - \$140{,}000)(1 - 0.5)}{100{,}000 \text{ (shares)}}$	$=$	$\frac{(\text{EBIT} - \$480{,}000)(1 - 0.5)}{60{,}000 \text{ (shares)}}$
$\frac{0.5\ \text{EBIT} - \$70{,}000}{10}$	$=$	$\frac{0.5\ \text{EBIT} - \$240{,}000}{6}$
EBIT = $990,000		

b. The EBIT-EPS analysis chart for Martin-Bowing is presented in Figure 9-10.

	Plan A	Plan B
EBIT	$990,000	$990,000
I	140,000	480,000
EBT	$850,000	$510,000
T (.5)	425,000	255,000
NI	$425,000	$255,000
D_p	0	0
EAC	$425,000	$255,000
÷ No. of common shares	100,000	60,000
EPS	$ 4.25	$ 4.25

c. Since $1,188,000 exceeds the calculated indifference level of $990,000, the more highly levered plan (plan B) will produce the higher EPS.

d. At this stage of the problem it is necessary to compute earnings per share (EPS) under each financing alternative. Then the relevant price/earnings ratio for each plan can be applied to project the common share price for the plan at a specific EBIT level.

Figure 9-10 EBIT-EPS Analysis Chart for Martin-Bowing

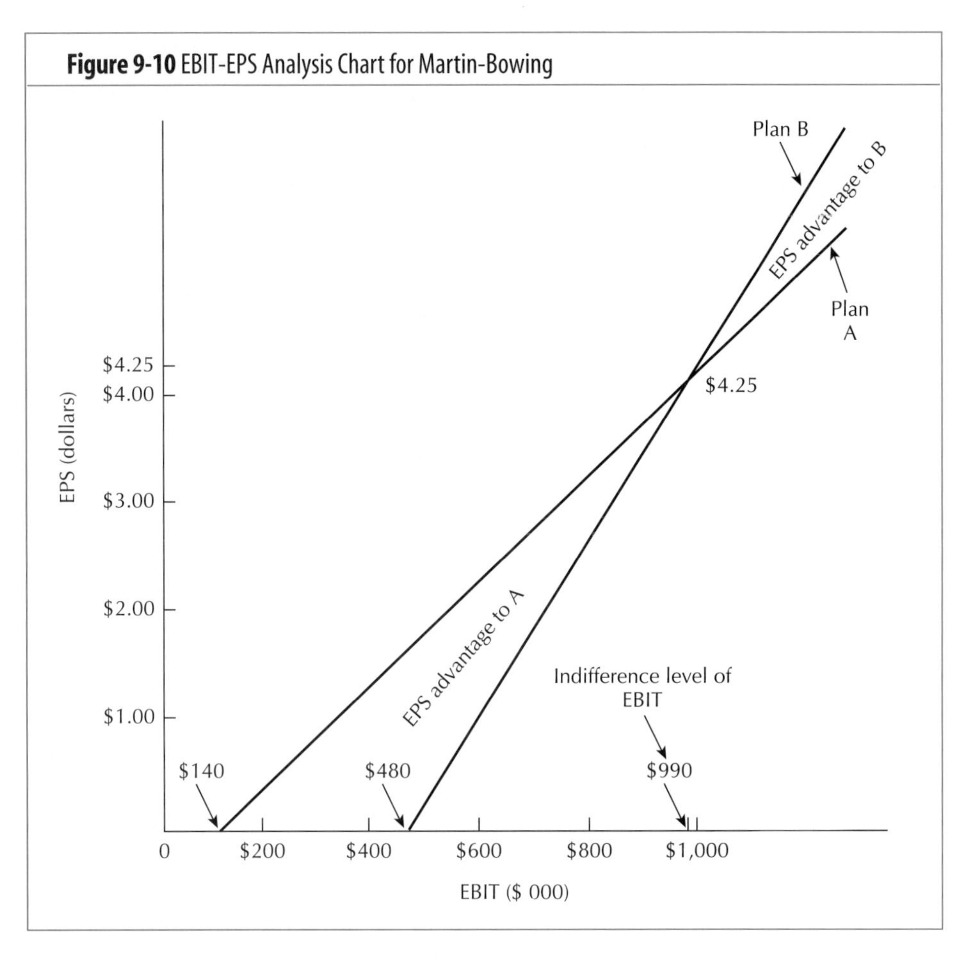

		Plan A	Plan B
	EBIT	$1,188,000	$1,188,000
	I	140,000	480,000
	EBT	$1,048,000	$708,000
	T(.5)	524,000	354,000
	NI	$ 524,000	$ 354,000
	D_p	0	0
	EAC	$ 524,000	$ 354,000
÷	No. of common shares	÷ 100,000	÷ 60,000
	EPS	$ 5.24	$ 5.90
×	P/E ratio	13	11
=	Projected share price	$ 68.12	$ 64.90

Notice that the greater riskiness of plan B results in the market applying a lower price/earnings multiple to the expected EPS. Therefore, the investors would actually enjoy a higher share price under plan A ($68.12) than they would under plan B ($64.90).

e. Here, we want to find the price/earnings ratio that would equate the common share prices for both plans at an EBIT level of $1,188,000. All we have to do is take plan B's EPS and relate it to plan A's share price. Thus:

$$\$5.90\left(\frac{P}{E}\right) = \$68.12$$

$$\left(\frac{P}{E}\right) = \frac{\$68.12}{\$5.90} = 11.546.$$

A price/earnings ratio of 11.546 when applied to plan B's EPS would give the same share price as that of plan A ($68.12).

CHAPTER 10

DIVIDEND POLICY AND INTERNAL FINANCING

LEARNING OBJECTIVES

After reading this chapter you should be able to

1. Describe the tradeoff between paying dividends and retaining profits within the company.
2. Explain the relationship between a corporation's dividend policy and the market price of its common shares.
3. Describe practical considerations that may be important to the firm's dividend policy.
4. Distinguish between the types of dividend policy corporations frequently use.
5. Specify the procedures a company follows in administering the dividend payment.
6. Describe why and how a firm might choose to pay noncash dividends (stock dividends and stock splits) instead of cash dividends.
7. Explain the purpose and procedures related to share repurchases.

INTRODUCTION

In 1994, Corporate Canada recorded increases in earnings of 140 percent as compared to the earnings of 1993.[1] *Yet, it was only in the third and fourth quarters of 1995 that investors began to see any increases in their rates of dividend distribution. A comparison of dividend yields from shares of the Toronto Stock Exchange 300 Composite Index shows that dividend yields have not changed significantly between 1993 and 1995. In early December, 1993, the dividend yield of the shares on the Toronto Stock Exchange 300 Composite Index was 2.32 percent, compared to a dividend yield of 2.27 percent at the end of December, 1995. Why is there a lag between corporate profits increases and increases in rates of dividend distribution? First, companies tend not to increase their dividends unless they are sure that higher earnings will remain. Second, cash dividends have become a sign of a mature business, since most fast-growing, highly profitable companies retain their profits and use cash for expansion. Finally, many companies are deciding to use their cash reserves to buy back their shares, especially if the shares are considered to be undervalued by management.*[2]

In this chapter, we examine the different types of dividend policy that a company can use to reward its investors. We also examine whether a firm's dividend policy affects the market price of the company's shares or shareholders value.

[1] Susan Bourette and Andrew Bell, "Canadian firms boosting dividend payouts," *The Globe and Mail*, March 4, 1995, pp. B1, B4.

[2] Dunnery Best, "Pay homage to the dividend," *The Globe and Mail*, June 10, 1995, p. B18.

CHAPTER PREVIEW

This chapter evaluates management's decisions on the firm's dividend and internal financing policies in light of the primary goal of the firm, the maximization of shareholder value or the price of the firm's common shares. We rephrase the same basic issue that we examined in Chapter 9 on the firm's financing decisions: Can management influence the price of the firm's shares, in this case, through its dividend policies? After addressing this important question, we then look at the practical side of the question: What are the practices commonly followed by managers in making decisions about paying or not paying a dividend to the firm's shareholders?

PERSPECTIVE IN FINANCE

Given what we have studied in earlier chapters, we would certainly expect that dividends, and therefore the firm's dividend policy, would be important to shareholders. However, when we consider the whole scheme of things, whether a firm pays a dividend may not matter much to investors, in terms of affecting the firm's stock price. In fact, we have difficulty explaining a firm's actions in regard to paying dividends. Yet chief financial officers, from time immemorial, have acted as if dividend policy is important. In this chapter, we will try to resolve this question, but will not do so as completely as we would honestly like.

OBJECTIVE 1

DIVIDEND PAYMENTS VERSUS PROFIT RETENTION

Before taking up the particular issues relating to dividend policy, we must understand several key terms and interrelationships.

Dividend payout ratio

The amount of dividends relative to the company's net income or earnings per share.

A firm's dividend policy can be described in terms of two basic characteristics. First, the **dividend payout ratio** indicates the amount of dividends paid relative to the company's earnings. For instance, if the dividend per share is $2 and the earnings per share is $4, the payout ratio is 50 percent ($2/$4). The second characteristic of a firm's dividend policy relates to the *stability* of the dividends over time. As will be observed later in the chapter, dividend stability may be almost as important to the investor as the amount of dividends received.

The financial executive faces an important tradeoff in formulating a firm's dividend policy. The decision to pay a large dividend means that the firm will retain little, if any, profits, which results in heavy reliance on external equity financing. Conversely, the decision to pay a small dividend payment is equivalent to high profit retention with less need for externally generated equity funds. Both these cases assume that management has already decided how much to invest and the debt-equity mix to be used in financing these investments. These tradeoffs, which are fundamental to our discussion, are summarized in Figure 10-1.

OBJECTIVE 2

DOES DIVIDEND POLICY AFFECT SHARE PRICE?[3]

The fundamental question to be resolved in our study of the firm's dividend policy may be stated simply: What is a sound rationale or motivation for dividend pay-

[3]The concepts of this section draw heavily from Donald H. Chew, Jr. (ed.), *Six Roundtable Discussions of Corporate Finance with Joel Stern,* "Do Dividends Matter? A Discussion of Corporate Dividend Policy" (New York: Quorum Books, 1986), pp. 67–101; and a book of readings edited by Joel M. Stern and Donald H. Chew, Jr., *The Revolution in Corporate Finance* (New York: Basil Blackwell,

Figure 10-1 Dividend-Retention-Financing Tradeoffs

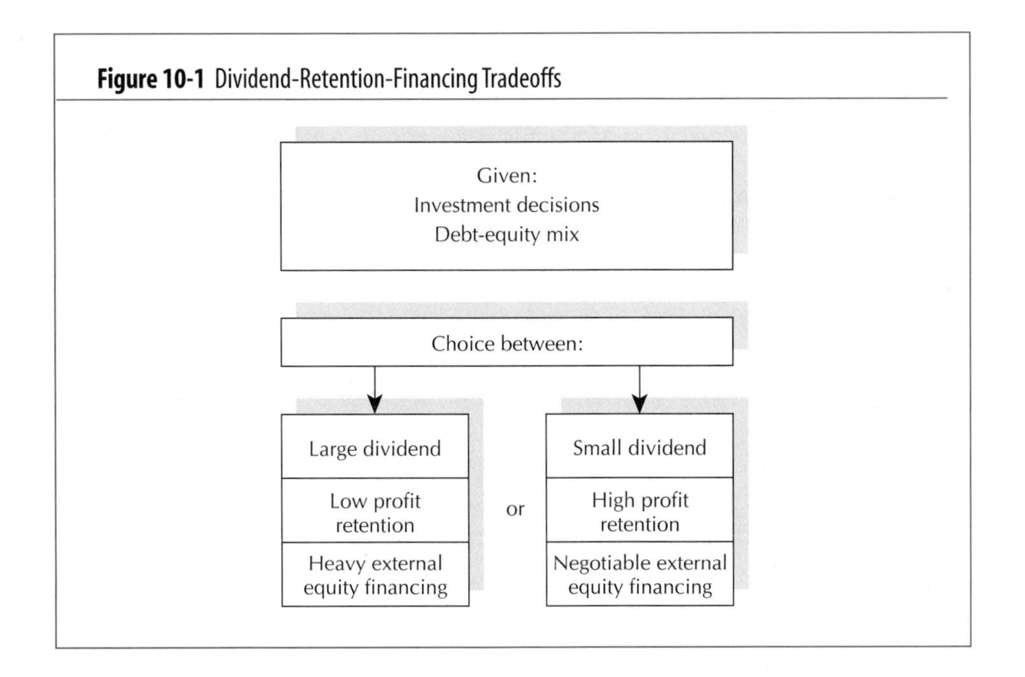

ments? If we believe our objective should be to maximize the market value of the common shares, we may restate the question as follows: Given the firm's capital-budgeting and borrowing decisions, what is the effect of the firm's dividend policies on the share price? Does a high dividend payment decrease share value, increase it, or make no real difference?

At first glance, we might reasonably conclude that a firm's dividend policy is important. We have already (in Chapter 6) defined the value of a share to be equal to the present value of future dividends. How can we now suggest that dividends are not important? Why do so many companies pay dividends, and why is space in prominent Canadian finance newspapers devoted to dividend announcements? Based upon intuition, we could quickly conclude that dividend policy is important. However, we might be surprised to learn that the dividend question has been a controversial issue for well over three decades. It has even been called the "dividend puzzle."[4]

Three Basic Views

Some would argue that the amount of the dividend is irrelevant, and any time spent on the decision is a waste of energy. Others contend that a high dividend will result in a high share price. Still others take the view that dividends actually hurt the share value. Let us look at these three views in turn.

View 1: Dividend Policy Is Irrelevant

Much of the controversy about the dividend issue is based on the time-honoured arguments between the academic and professional communities. Experienced practitioners perceive share price changes as resulting from dividend announcements, whereas professors argue that most dividend policies have not been careful-

1986). Specific readings include Merton Miller, "Can Management Use Dividends to Influence the Value of the Firm?" pp. 299–303; Richard Brealey, "Does Dividend Policy Matter?" pp. 304–309; and Michael Rozeff, "How Companies Set Their Dividend Payout Ratios," pp. 320–326.

[4]See Fischer Black, "The Dividend Puzzle," *Journal of Portfolio Management* 2 (Winter 1976), pp. 5–8.

ly defined, and as a result the perceived relationship between dividend announcements and share prices is an illusion.[5]

The position that dividends are not important rests on two preconditions. First, we assume that investment and borrowing decisions have already been made, and that these decisions will not be altered by the amount of any dividend payments. Second, **"perfect" capital markets** are assumed to exist, which means that (1) investors can buy and sell shares without incurring any transaction costs, such as brokerage commissions; (2) companies can issue shares without any cost of doing so; (3) there are no corporate or personal taxes; (4) complete information about the firm is readily available; and (5) there are no conflicts of interest between managements and shareholders.

Perfect capital markets

Capital markets in which (1) investors can buy and sell shares without incurring any transaction costs, such as brokerage commissions; (2) companies can issue shares without any cost of doing so; (3) there are no corporate or personal taxes; (4) complete information about the firm is readily available; (5) there are no conflicts of interest between management and shareholders; and (6) financial distress and bankruptcy costs are nonexistent.

The first assumption—that we have already made the investment and financing decisions—simply keeps us from confusing the issues. We want to know the effect of dividend decisions on a stand-alone basis, without mixing in other decisions. The second assumption of perfect capital markets allows us to study the effect of dividend decisions in isolation, much like a physicist studies motion in a vacuum to avoid the influence of friction.

Given these assumptions, the effect of a dividend decision on share price may be stated unequivocally: There is no relationship between dividend policy and share value. One dividend policy is as good as another one. Under this hypothesis, the investor is concerned only with total returns which come from the firm's investment decisions; they are indifferent whether these returns come from capital gains or dividend income. Given the firm's investment policy, they recognize that the firm's dividend decision is really a choice of financing strategy. The principles which enable a firm to grow through its investments have been discussed in Chapter 6. To finance growth, the firm (1) may choose to issue shares, allowing internally generated funds (profits) to be used to pay dividends; or (2) it may use internally generated funds (profits) to finance its growth, while paying less in dividends, but not having to issue new shares. In the first case, shareholders receive dividend income; in the second case, the value of their shares should increase, providing capital gains. The nature of the return is the only difference: total returns should be about the same. Thus, to argue that paying dividends can make shareholders better off is to argue that paying out cash with one hand and taking it back with the other hand is a worthwhile activity for management.

The firm's dividend payout could affect share price if the shareholder has no other way to receive income from the investment. In this case shareholders will sell their shares since capital markets have been assumed to be efficient. In the case that the firm pays a dividend, the investor could eliminate any dividend received, in whole or in part, by using the dividend to purchase shares. The investor can thus personally create any desired dividend stream, no matter what dividend policy is chosen by management. The dividend policy is therefore an inconsequential activity for management.

An Example of Dividend Irrelevance To demonstrate the argument that dividends may not matter, come to the Land of Ez (pronounced "ease"), where the environment is quite simple. First, the king, being a kind soul, has imposed no income taxes on his subjects. Second, investors can buy and sell securities without paying any sales commissions. In addition, when a company issues new securities (shares or bonds), there are no flotation costs. Furthermore, the Land of Ez is completely computerized, so that all information about firms is instantaneously available to the public at no cost. Next, all investors realize that the value of a company is a function of its investment opportunities and its financing decisions. Therefore, the dividend policy offers no new information about either the firm's ability to

[5]For an excellent presentation of this issue, see Merton Miller, "Can Management Use Dividends to Influence the Value of the Firm?" in *The Revolution in Corporate Finance*, edited by Joel M. Stern and Donald H. Chew, Jr. (New York: Basil Blackwell, 1986), pp. 299–305.

generate earnings or the riskiness of its earnings. Finally, all firms are owned and managed by the same parties; thus, we have no potential conflict between owners and managers.

Within this financial utopia, would a change in a corporation's dividend stream have any effect upon the price of the firm's shares? The answer is no. To illustrate, consider Dowell Venture, Inc., a corporation that received a charter at the end of 1995 to conduct business in the Land of Ez. The firm is to be financed by common shares only. Its life is to extend for only two years (1996 and 1997), at which time it will be liquidated.

Table 10-1 presents Dowell Venture's balance sheet at the time of its incorporation, as well as the projected cash flows from the short-term venture. The anticipated cash flows are based upon an expected return on investment of 20 percent, which is exactly what the common shareholders require as a rate of return on their investment in the firm's shares.

At the end of 1996 an additional investment of $300,000 will be required, which may be financed by (1) retaining $300,000 of the 1996 profits, or (2) issuing new common shares, or (3) some combination of both of these. In fact, two dividend plans for 1996 are under consideration. The investors would receive either $100,000 or $250,000 in dividends. If $250,000 is paid out of 1996's $400,000 in earnings, the company would be required to issue $150,000 in new shares to make up the difference in the total amount of funds needed of $300,000 less the $150,000 that is retained. Table 10-2 depicts these two dividend plans and the corresponding new share issue. Our objective in analyzing the data is to answer this question: Which dividend plan is preferable to the investors? In answering this question, we must take three steps: (1) Calculate the amount and timing of the dividend stream for the original investors. (2) Determine the present value of the dividend stream for each dividend plan. (3) Select the dividend alternative providing the higher value to the investors.

Step 1: Computing the Dividend Streams. The first step in this process is presented in Table 10-3. The dividends in 1996 (line 1, Table 10-3) are readily apparent from the data in Table 10-2. However, the amount of the dividend to be paid to the

Table 10-1 Dowell Venture Inc. Financial Data

	December 31, 1995
Total assets	$2,000,000
Common shares (100,000 shares)	$2,000,000

	1996	1997
Projected cash available from operations for paying DIVIDENDS or for REINVESTING	$400,000	$460,000

Table 10-2 Dowell Venture Inc. 1996 Proposed Dividend Plans

	Plan 1	Plan 2
Internally generated cash flow	$400,000	$400,000
Dividend for 1996	100,000	250,000
Cash available for reinvestment	$300,000	$150,000
Amount of investment in 1996	300,000	300,000
Additional external financing required	$ 0	$150,000

Table 10-3 Dowell Venture Inc. Step 1—Measurement of the Proposed Dividend Streams

	Plan 1 Total Amount	Amount Per Share[a]	Plan 2 Total Amount	Amount Per Share[a]
Year 1 (1996)				
(1) Dividend Year 2 (1997)	$ 100,000	$1.00	$ 250,000	$2.50
Total dividend consisting of:				
(2) Original investment:				
(a) Old investors	$2,000,000		$2,000,000	
(b) New investors	0		150,000	
(3) Retained earnings	300,000		150,000	
(4) Profits for 1997	460,000		460,000	
(5) Total dividend to all investors in 1996	$2,760,000		$2,760,000	
(6) Less dividends to new investors:				
(a) Original investment	0		(150,000)	
(b) Profits for new investors (20% of $150,000 investment)	0		(30,000)	
(7) Liquidating dividends available to original investors in 1997	$2,760,000	$27.60	$2,580,000	$25.80

[a]Number of original shares outstanding equals 100,000.

present shareholders in 1997 has to be calculated. To do so, we assume that investors receive (1) their original investments (line 2, Table 10-3), (2) any funds retained within the business in 1996 (line 3, Table 10-3), and (3) the profits for 1997 (line 4, Table 10-3). However, if additional shareholders invest in the company, as with plan 2, the dividends to be paid to these investors must be subtracted from the total available dividends (line 6, Table 10-3). The remaining dividends (line 7, Table 10-3) represent the amount that current shareholders will receive in 1997. Therefore, the amounts of the dividend may be summarized as follows:

Dividend Plan	Year 1	Year 2
1	$1.00	$27.60
2	2.50	25.80

Step 2: Determining the Present Value of the Cash Flow Streams. For each of the dividend payment streams the resulting common share value is

$$\text{share price (plan 1)} = \frac{\$1.00}{(1 + .20)} + \frac{\$27.60}{(1 + .20)^2} = \$20$$

$$\text{share price (plan 2)} = \frac{\$2.50}{(1 + .20)} + \frac{\$25.80}{(1 + .20)^2} = \$20$$

Therefore, the two approaches provide the same end product; that is, the market price of Dowell Venture's common share is $20 regardless of the dividend policy chosen.

Step 3: Select the Best Dividend Plan. If the objective is to maximize the shareholders' wealth, either plan is equally acceptable. Alternatively, shifting the dividend payments between years by changing the dividend policy does not affect the value of the security. Thus, only if investments are made with expected returns exceeding 20 percent will the value of the share increase. In other words, the only wealth-creating activity in the Land of Ez, where companies are financed entirely by equity, is management's investment decisions.

View 2: High Dividends Increase Share Value

The belief that a firm's dividend policy is unimportant implicitly assumes that an investor should use the same required rate of return, whether income comes through capital gains or through dividends. However, dividends are more predictable than capital gains; management can control dividends, but it cannot dictate the price of the share. Investors are less certain of receiving income from capital gains than from dividends. The incremental risk associated with capital gains relative to dividend income implies a higher required rate for discounting a dollar of capital gains than for discounting a dollar of dividends. In other words, we would value a dollar of expected dividends more highly than a dollar of expected capital gains. We might, for example, require a 14 percent rate of return for a share that pays its entire return from dividends, but a 20 percent return for a high-growth share that pays no dividend. In so doing, we would give a higher value to the dividend income than we would to the capital gains. This view, which says dividends are more certain than capital gains, has been called the **bird-in-the-hand dividend theory**.

Bird-in-the-hand dividend theory

The belief that dividend income has a higher value to the investor than do capital gains, since dividends are more certain than capital gains.

The position that dividends are less risky than capital gains, and should therefore be valued differently, is not without its critics. If we hold to our basic decision not to let the firm's dividend policy influence its investment and capital-mix decisions, the company's operating cash flows, both in expected amount and variability, are unaffected by its dividend policy. Since the dividend policy has no impact on the volatility of the company's overall cash flows, it has no impact on the riskiness of the firm.

Increasing a firm's dividend does not reduce the basic riskiness of the share; rather, if a dividend payment requires management to issue new shares, it only transfers risk and ownership from the current owners to new owners. We would have to acknowledge that the current investors who receive the dividend trade an uncertain capital gain for a "safe" asset (the cash dividend). However, if risk reduction is the only goal, the investor could have kept the money in the bank and not bought the shares in the first place.

We might find fault with this "bird-in-the-hand" theory, but there is still a strong perception among many investors and professional investment advisors that dividends are important. They frequently argue their case based upon their own personal experience. For example, one investment advisor who is an advocate of dividend policy stated the following:

> In advising companies on dividend policy, we're absolutely sure on one side that the investors in companies like the utilities and the banks want dividends. We're absolutely sure on the other side that...the high-technology companies should have no dividends. For the high earners—the ones that have a high rate of return like 20 percent, or more than their cost of capital—we think they should have a low payout ratio. We think a typical industrial company which earns its cost of capital—just earns its cost of capital—probably should be in the average [dividend payout] range of 40 to 50 percent.[6]

View 3: Low Dividends Increase Share Value

The third view of how dividends affect share price proposes that dividends actually hurt the investor. This argument has largely been based on the difference in tax treatment for dividend income and capital gains. Unlike the investors in the great

[6]Donald H. Chew, Jr. (ed.), "Do Dividends Matter? A Discussion of Corporate Dividend Policy," in *Six Roundtable Discussions of Corporate Finance with Joel Stern* (New York: Quorum Books, 1986), pp. 83–84.

Land of Ez, most other investors do pay income taxes. For these taxpayers, the objective is to maximize the after-tax return on investment relative to the risk assumed. This objective is realized by minimizing the effective tax rate on the income and, whenever possible, by deferring the payment of taxes.

Both the federal budget of 1994 and tax reform legislation of 1987 have changed the tax treatment of capital gains and dividends for individuals. Revenue Canada requires an individual to be taxed on only 75 percent of capital gains realized from the sale of capital property for 1990 and subsequent years. With respect to dividends, Revenue Canada requires an individual's dividend income to be grossed up by 25 percent and taxed at the individual's marginal tax rate. However, an individual is eligible to claim a dividend tax credit of 13.33 percent on the grossed up amount. This dividend tax credit is used to reduce the taxes payable of an individual. As a result, the after-tax return from dividend income will be greater than the after-tax return from capital gains income.[7] However, we should note that taxes on dividend income are paid in the year of receipt; whereas, taxes on price appreciation (capital gains) are deferred until the share is actually sold. Thus, when it comes to tax considerations, investors that earn large amounts of investment income will prefer the retention of a firm's earnings as opposed to the payment of cash dividends despite the higher after-tax return from dividend income. If earnings are retained within the firm, the share price increases, but the increase is not taxed until the shares are sold.

To summarize, when it comes to taxes, we want to maximize our after-tax return, as opposed to the before-tax return. As a result, investors try to defer taxes whenever possible. Shares that allow tax deferral (low dividends–high capital gains) will possibly sell at a premium relative to shares that require us to pay taxes immediately (high dividends–low capital gains). In this way, the two shares may provide comparable after-tax returns. This suggests that a policy of paying low dividends will result in a higher share price. That is, high dividends hurt investors, while low dividends and high retention help investors. This is the logic of advocates of the low-dividend policy.

Improving Our Thinking

We have now looked at three views on dividend policy. Which is right? The argument that dividends are irrelevant is difficult to refute, given the perfect market assumptions. However, in the real world, it is not always easy to feel comfortable with such an argument. On the other hand, the high-dividend philosophy, which measures risk by how we split the firm's cash flows between dividends and retention, is not particularly appealing when studied carefully. The third view, which is essentially a tax argument against high dividends, is persuasive, given that the "deferral advantage" of capital gains is still alive and well. However, if low divi-

[7]Examples of the effective tax rates for different amounts of investment income for 1995 are shown below:

	Effective tax rate[a]	
Income	Capital gains	Dividends
$ 23,650	8.4%[b]	0.0%[c]
47,300	17.7	13.9
125,000	22.0	19.7

[a] The tax calculation assumed no other type of income, federal tax and surtax rates for 1995, and a personal tax credit of $1,098.
[b] The tax calculation assumed that 75% of the capital gains is taxable.
[c] The tax calculation assumed a gross-up of 25% and a dividend tax credit of 13.33% on grossed-up taxable income.

dends are so advantageous and generous dividends are so hurtful, why do companies continue to pay dividends?[8] It is difficult to believe that managers would forgo such an easy opportunity to benefit their shareholders. What are we missing?

The need to find the missing elements in our "dividend puzzle" has not been ignored. In an effort to gain a better understanding of this puzzle, we need to improve our thinking and/or to gather more evidence about the topic. While no single definitive answer has yet been found that is acceptable to all, a number of plausible extensions have been developed. Several of the more popular additions include (1) the residual dividend theory, (2) the clientele effect, (3) information effects, (4) agency costs, and (5) expectations theory.

The Residual Dividend Theory

Within the Land of Ez, companies were blessed with professional consultants who were essentially charitable in nature; they did not seek any compensation when they helped a firm through the process of issuing shares. (Even in the Land of Ez, managers needed help from investment dealers, accountants, and attorneys to sell a new issue.) In reality, the process is quite expensive, and may cost as much as 20 percent of the dollar issue size.[9]

If a company incurs flotation costs, they may have a direct bearing upon the dividend decision. As a result of the high flotation costs, a firm must issue a larger amount of securities in order to receive the amount required for investment. For example, if $300,000 is needed to finance proposed investments, an amount exceeding the $300,000 will have to be issued to offset flotation costs incurred in the sale of the new share issue. This means, very simply, that new equity capital raised through the sale of common shares will be more expensive than capital raised through the retention of earnings.

Flotation costs eliminate our indifference between financing by internal capital and by new common shares. Earlier, the company could pay dividends and issue common shares or retain profits. However, internal financing is preferred to the high flotation costs. Dividends are paid only if profits are not completely used for investment purposes; that is, only when there are "residual earnings" after the financing of new investments. This policy is called the **residual dividend theory**.[10]

Residual dividend theory

A theory asserting that the dividends to be paid should equal capital left over after the financing of profitable investments.

With the assumption of no flotation costs removed, the firm's dividend policy would be as follows:

1. Maintain the optimum debt ratio in financing future investments.
2. Accept an investment if the net present value is positive. That is, the expected rate of return exceeds the cost of capital.
3. Finance the equity portion of new investments first by internally generated funds. Only after this capital is fully utilized should the firm issue new common shares.
4. If any internally generated funds still remain after making all investments, pay dividends to the investors. However, if all internal capital is needed for financing the equity portion of proposed investments, pay no dividend.

[8]In fact, most successful companies pay dividends. Furthermore, most financial managers think dividend policy is important. See H. Kent Baker, Gail E. Farrelly, and Richard B. Edelman, "A Survey of Management Views on Dividend Policy," *Financial Management* (Autumn 1985), pp. 78–84

[9]We discussed the costs of issuing securities in Chapter 7.

[10]The residual dividend theory is consistent with the "pecking order" theory of finance as described by Stewart Myers, "The Capital Structure Puzzle," *The Journal of Finance* (July 1984), pp. 575–592.

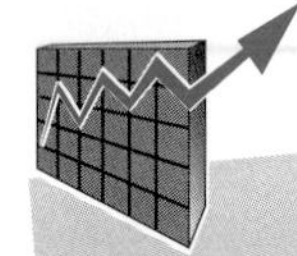

Excerpts from "Do Dividends Matter? A Discussion of Corporate Dividend Policy"

January 6, 1982

Joseph T. Willett, Moderator: I would like to welcome the participants and guests to this discussion, the subject of which is Corporate Dividend Policy. The general questions we want to address are these: Does dividend policy matter? And if so, why and how does it matter? Certain people argue that the theory of finance, combined with the treatment of dividends under U.S. tax law, would suggest that low dividends benefit investors. Others argue that because of the demand by some investors for current income, high dividends benefit investors. In the presence of these widely held views, I think it is fair to say that most carefully executed research has revealed no consistent relationship between dividends and share prices. From these studies, the market collectively appears to be "dividend neutral." That is, while individual investors may have preferences between dividends and capital gains, the results suggest neither a preference for nor an aversion to dividends. Which, of course, doesn't satisfy either the pro-dividend or anti-dividend group. Amid all this confusion, one observation stands out: nearly all successful firms pay dividends. And, furthermore, dividend policy is an important concern of most chief financial officers and financial managers generally. These facts of corporate practice, in light of all the evidence on the subject, present us with a puzzle—one which has continued to baffle the academic finance profession. In a paper written in 1976, entitled "The Dividend Puzzle," Fischer Black of MIT—one of the most widely respected researchers in the field—posed the question: "What should the individual investor do about dividends in his portfolio? What should the corporation do about dividend policy?"

Joel Stern: I'd like to point out that the major reason why people like Fischer Black believe they don't know the answer to the question of the appropriate dividend policy is this: the evidence that has been accumulated in the academic community by serious researchers—by people that we have a lot of respect for, who are on the faculties of the premiere business schools—almost without exception, these academics find that there is no evidence to suggest that investors at the margin, where prices are set, have any preference for dividends over capital gains. This supports the point of view that the price-setting, marginal investor is "dividend-neutral," which means that a dollar of dividends gained is equal to a dollar of capital gains returned, while being indifferent how that return was divided between dividends and price appreciation. There is a second point of view, that has been expressed recently in research, which shows that investors who receive dividends cannot undo the harmful tax consequences of receiving that dividend. And, as a result, the market is actually "dividend averse," marking down prices of shares that pay cash dividends, so that the pretax returns that investors earn are high enough such that, post-tax, the returns are what they would have been had the company not paid cash dividends in the first place. But there is no creditable evidence that I am aware of—none that has been accepted by the academic finance community—that shows that investors prefer dividends over capital gains.

If the evidence that has been published to date says that investors are dividend neutral or dividend averse, then how is it that somebody with the esteem of Fischer Black can come along and say: "We don't know what the right dividend policy is." The problem is that he is what we call a "positive economist." That doesn't mean that he is an economist who is positive about things. It means that he says the job of the economist is to account for what we see around us. He believes that markets behave in a sensible fashion at the margin; that under the guidance of the dominant price-setting investors, the market behaves in a rational manner, making the right choices for itself. Therefore, he is saying that there must be a reason why almost all companies for all time have been paying cash dividends. If a few companies paid dividends for all time, or almost all companies paid dividends only occasionally, then one could make the case that it is possible dividends are really not important. But, if we find that almost all companies pay dividends for almost all time, there must be a good reason why they are paying the dividends. Therefore, who are we, as financial advisors, to say to a company, "No, don't pay cash dividends. After all, it won't harm you very much despite the fact that almost all companies are paying cash dividends"? That wouldn't make very much sense.

Source: Donald H. Chew, Jr. (ed.), "Do Dividends Matter? A Discussion of Corporate Dividend Policy," in *Six Roundtable Discussions of Corporate Finance with Joel Stern* (New York: Quorum Books, 1986), pp. 67–101.

EXAMPLE

Assume that the Krista Corporation finances 40 percent of its investments with debt and the remaining 60 percent with common equity. Two million dollars have been generated from operations and may be used to finance the common equity portion of new investments or to pay common dividends. The firm's management is considering five investment opportunities. Figure 10-2 illustrates the expected rate of return for these investments, along with the firm's weighted marginal cost of capital curve. From the information contained in the figure, we would accept projects A, B, and C, requiring $2.5 million in total financing. Therefore, $1 million in new debt (40% of $2.5 million) would be needed, with common equity providing $1.5 million (60% of $2.5 million). In this instance the dividend payment decision would be to pay $500,000 in dividends, which is the residual or remainder of the $2 million internally generated capital.

To illustrate further, consider the dividend decision if project D had also been acceptable. If this investment were added to the firm's portfolio of proposed capital expenditures, then $4 million in new financing would be needed. Debt financing would constitute $1.6 million (40% × $4 million) and common equity would provide the additional $2.4 million (60% × $4 million). Since only $2 million is available internally, $400,000 in new common shares would be issued. The residual available for dividends would be zero, and no dividend would be paid.

In summary, dividend policy is influenced by (1) the company's investment opportunities, (2) the capital structure mix, and (3) the availability of internally generated capital. In the Krista Corporation example, dividends were paid only after all acceptable investments had been financed. This logic, called the residual dividend theory, implies that the dividends to be paid should equal the equity capital remaining after financing investments. According to this theory, dividend policy is a passive influence, having by itself no direct influence on the market price of the common shares.

Figure 10-2 Krista Corporation Investment Schedule

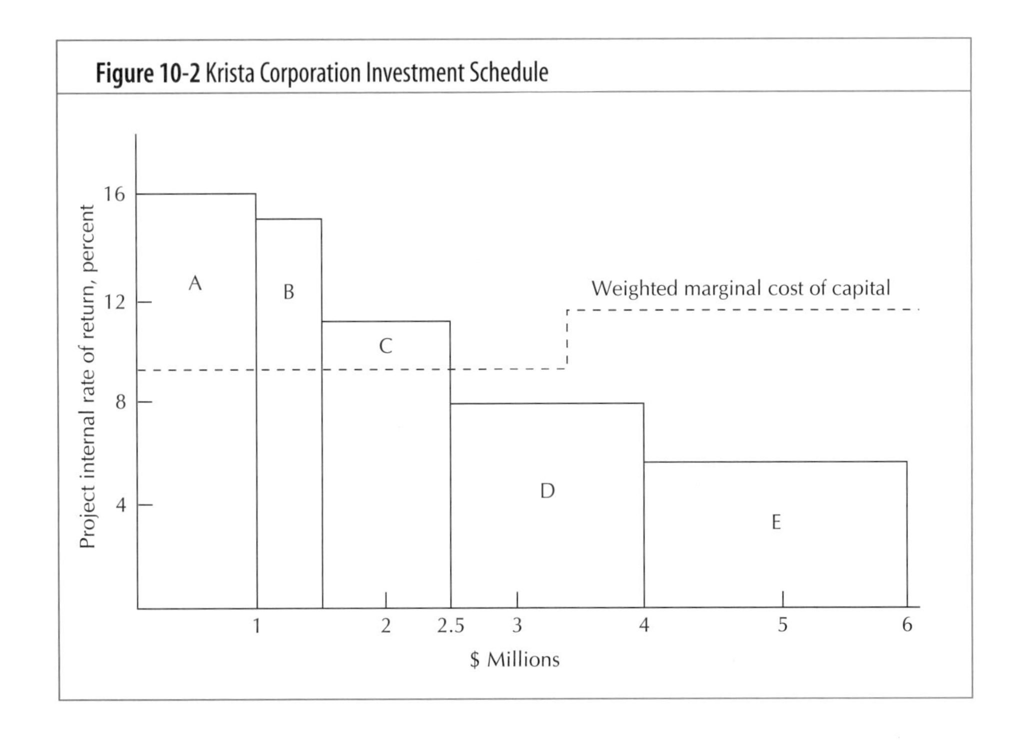

The Clientele Effect

What if the investors living in the Land of Ez did not like the dividend policy chosen by Dowell's management? No problem. They could simply satisfy their personal income preferences by purchasing or selling securities when the dividends received did not satisfy their current needs for income. If an investor did not view the dividends received in any given year to be sufficient, he or she could simply sell a portion of shares, thereby "creating a dividend." In addition, if the dividend were larger than the investor desired, he or she could purchase shares with the "excess cash" created by the dividend. However, once we leave the Land of Ez, we find that such adjustments in share ownership are not cost-free. For example, an investor who buys or sells shares incurs the following costs: brokerage fees, taxes on the cash dividends, and the cost of acquiring information needed to make the decision to buy or sell the shares. Finally, some institutional investors, such as university endowment funds, are precluded from selling shares and "spending" the proceeds.

As a result of these considerations, investors may not be too inclined to buy shares that require them to "create" a dividend stream more suitable to their purposes. Rather, if investors do in fact have a preference between dividends and capital gains, we could expect them to seek firms that have a dividend policy consistent with these preferences. They would, in essence, "sort themselves out" by buying shares that satisfy their preferences for dividends and/or capital gains. Individuals and institutions that need current income would be drawn to companies that have high dividend payouts. Other investors, such as wealthy individuals, would much prefer holding securities that offer no or small dividend income but large capital gains because of the "deferral advantage" of capital gains. In other words, there would be a **clientele effect**: Firms draw a given clientele, given their stated dividend policy.

Clientele effect

The belief that individuals and institutions that need current income will invest in companies that have high dividend payouts. Other investors prefer to avoid taxes by holding securities that offer only small dividend income, but large capital gains. Thus, we have a "clientele" of investors.

The possibility that clienteles of investors exist might lead us to believe that the firm's dividend policy matters. However, unless there is a greater aggregate demand for a particular policy than the market can satisfy, dividend policy is still unimportant: one policy is as good as the other. As a result of the clientele effect, a firm should avoid making capricious changes in their dividend policy. Given that the firm's investment decisions are already made, the level of the dividend is still unimportant. The change in the policy matters only when it requires clientele to shift to another company.

The Information Effect

The investor in the Land of Ez would argue with considerable persuasion that a firm's value is determined strictly by its investment and financing decisions, and that the dividend policy has no impact on value. Yet we know from experience that a large, unexpected change in dividends can have a significant impact on the share price. For instance, in 1988 TransCanada Pipelines Ltd. cut its dividend from $1.12 to $0.67, or a 39 percent decline. Despite management's insistence that the reduction was necessary because of the company's acquisition of Encor Energy Corporation, the response of the market to the dividend announcement caused the share price to decrease from $15.50 to $13.75, a decline of 11 percent. Other more dramatic reactions in North American markets to dividend announcements include the following: in November of 1990, Occidental Petroleum cut its dividend from $2 to $1, and the response of the market to the dividend announcement caused the share price to decrease from about $32 to $12; in July of 1984, ITT cut its dividend from $.69 to $.25, and the response of the market to the dividend announcement caused the share price to decrease from $31 to $21.13, a 32 percent decline. Thus, how can we suggest that dividend policy matters little, when we can cite numerous such examples of a change in dividend affecting the share price, especially when the change is negative?[11]

Market response to dividend announcements must be examined for its cause, that is, for its information content. In effect investors may understand the announcement of a change in dividend policy as a signal about the firm's financial condition, especially its earning power.[12] Thus, a dividend increase that is larger than expected "signals" to investors that management expects significantly higher earnings in the future. Conversely, a dividend decrease, or even a less than expected increase, "signals" that management is forecasting less favourable future earnings.

Some would argue that management frequently has inside information about the firm that it cannot make available to investors. This difference in access to information between management and investors, called **information asymmetry**, may result in a lower share price than would occur under conditions of certainty. However, a regularly increasing dividend may be interpreted by an investor that management is making a commitment to continue these cash flows to the shareholders for the foreseeable future. If we recognize the fact that increasing dividends impinge on management's flexibility, we can conclude that management is not concealing some negative information about the company. So in a risky marketplace, dividends become a means to minimize any "drag" on the share price that might come from differences in the level of information available to managers and investors.

Information asymmetry

The difference in access to information between management and investors may result in a lower share price than would occur under conditions of certainty.

Dividends may therefore be important only as a communication tool; management may have no other credible way to inform investors about future earnings, or at least no convincing way that is less costly.

Agency Costs

Let us return again to the Land of Ez. We had avoided any potential conflict between the firm's investors and managers by assuming them to be one and the same. With only a cursory look at the real marketplace, we can see that managers and investors are typically not the same people, and as noted in the preceding section, these two groups do not have the same access to information about the firm. If the two groups are not the same, we must then assume that management is dedicated to maximizing shareholder wealth.[13] In other words, we are making a presupposition that the market values of companies with separate owners and managers will not differ from those of owner-managed firms.

BACK TO THE BASICS

Axiom 7, The Agency Problem, *warned us there may be conflict between management and owners, especially in large firms where managers and owners have different incentives. That is,* ***Managers Won't Work for the Owners Unless It's in Their Best Interest.*** *As we shall see in this section, the dividend policy may be one way to reduce this problem.*

Two possibilities should help managers see things as the equity investors see them: (1) Low market values may attract takeover bids; and (2) a competitive labour

[11]Unanticipated changes in dividends do have an impact on share price. However, the price change is most likely the consequence of the investor using the dividend change as "new information" about the company.

[12]It is possible that dividends are used by investors as a proxy for a measure of risk. We know that high-dividend-paying companies are often less risky and lower-dividend-paying firms frequently have greater risk. However, it is difficult to measure the causal relationship between dividend decisions and shareholder value, because of other seemingly unrelated factors.

[13]This issue was addressed briefly in Chapter 1.

market may allow investors to replace uncooperative managers. That is, if management is not sensitive to the need to maximize shareholder wealth, new investors may buy the shares, take "control" of the firm, and remove management.[14] If current management is being less than supportive of the owners, these owners can always seek other managers who will work in the investors' best interest. If these two market mechanisms worked perfectly without any cost, the potential conflict would be nonexistent. In reality, however, conflicts may still exist, and the share price of a company owned by investors who are separate from management may be less than the share value of a closely held firm. The difference in price is the cost of the conflict to the owners, which has come to be called **agency costs**.[15]

Agency costs

The costs, such as a reduced share price, associated with potential conflict between managers and investors when these two groups are not the same.

Recognizing the possible problem, management, acting independently or at the insistence of the board of directors, frequently takes action to minimize the cost associated with the separation of ownership and management control. Such action, which in itself is costly, includes auditing by independent accountants, assigning supervisory functions to the company's board of directors, creating covenants in lending agreements that restrict management's powers, and providing inventive compensation plans for management that help "bond" the management with the owners.

A firm's dividend policy may be perceived by owners as a tool to minimize agency costs. Assuming that the payment of a dividend requires management to issue shares to finance new investments, new investors will be attracted to the company only if management provides convincing information that the capital will be used profitably. Thus, the payment of dividends indirectly results in a closer monitoring of management's investment activities. In this case, dividends may make a meaningful contribution to the value of the firm.

Expectations Theory[16]

A common thread throughout much of our discussion of dividend policy, particularly as it relates to information effects, is the word "expected." We should not overlook the significance of this word when we are making any financial decision within the firm. No matter what the decision area, how the market price responds to management's actions is not determined entirely by the action itself; it is also affected by investors' expectations about the ultimate decision to be made by management. This concept is called the **expectations theory**.

Expectations theory

The effect of new information about a company on the firm's share price depends more on how the new information compares to expectations than on the actual announcement itself.

As the time approaches for management to announce the amount of the next dividend, investors form expectations as to how much that dividend will be. These expectations are based on a number of factors internal to the firm, such as past dividend decisions, current and expected earnings, investment strategies, and financing decisions. They also consider such things as the condition of the general economy, the strength or weakness of the industry at the time, and possible changes in government policies.

When the actual dividend decision is announced, the investor compares the actual decision with the expected decision. If the amount of the dividend is as expected, even if it represents an increase from prior years, the market price of the share will remain unchanged. However, if the dividend is higher or lower than expected, investors will reassess their perceptions about the firm. They will ques-

[14]The "corporate control hypothesis," especially as it relates to companies merging or being acquired, has generated a great amount of interest in recent years. For example, see the April 1983 issue of *Journal of Financial Economics*.

[15]See M. C. Jenson and W. H. Meckling, "Theory of the Firm: Managerial Behavior, Agency Costs, and Ownership Structure," *Journal of Financial Economics* (October 1976), pp. 305–360.

[16]Much of the thoughts in this section came from Merton Miller, "Can Management Use Dividends to Influence the Value of the Firm?" in *The Revolution in Corporate Finance*, edited by Joel M. Stern and Donald H. Chew, Jr. (New York: Basil Blackwell, 1986), pp. 299–303.

tion the meaning of the unexpected change in the dividend. They may use the unexpected dividend decision as a clue about unexpected changes in earnings; that is, the unexpected dividend change has information content about the firm's earnings and other important factors. In short, management's actual decision about the firm's dividend policy may not be terribly significant, unless it departs from investors' expectations. If there is a difference between actual and expected dividends, we will more than likely see a movement in the share price.

The Empirical Evidence

Our search for an answer to the question of dividend relevance has been less than successful. We have given it our best thinking, but still no single definitive position has emerged. Maybe we could gather evidence to show the relationship between dividend practices and security prices. We might also inquire into the perceptions of financial managers who make decisions about dividend policies, with the idea that their beliefs affect their decision-making process. Then we could truly know that dividend policy is important or that it does not matter.

To test the relationship between dividend payments and security prices, we could compare a firm's dividend yield (dividend/share price) and the share's total return. The following question can then be asked: Do shares that pay high dividends provide higher or lower returns to investors? Such tests have been conducted with the use of highly sophisticated statistical techniques.[17] Despite the use of these extremely powerful analytical tools, which involve intricate and complicated procedures, the results have been mixed.[18] However, over long periods of time, the results have given a slight advantage to the low-dividend shares; that is, shares that pay lower dividends appear to have higher prices. The findings are far from conclusive, however, owing to the relatively large standard errors of the estimates. The apparent differences may be the result of random sampling error and not real differences. We simply have been unable to disentangle the effect of dividend policy from other influences.

Several reasons may be given for our inability to get conclusive results. First, to be accurate, we would need to know the amount of dividends investors expected to receive. Since these expectations cannot be observed, we can only use historical data, which may or may not relate to current expectations. Second, most empirical studies have assumed a linear relationship between dividend payments and share prices. The actual relationship may be nonlinear, possibly even discontinuous. Whatever the reasons, the evidence to date is inconclusive and the vote is still out.

What Are We to Conclude?

We have now looked carefully at the importance of a firm's dividend policy as management seeks to increase the shareholders' wealth. We have gone to great lengths

[17]Some of the difficulties which occur in this type of research include: (1) the inability to disentangle the effect of the dividend decision from other influences and (2) the need to know the investor's expected returns for holding the shares in order to test the relationship between dividend policy and the share price. However, the investor's expectations are unknown, forcing us to rely on historical returns in any testing of relationships.

[18]See F. Black and M. Scholes, "The Effects of Dividend Yield and Dividend Policy on Common Stock Prices and Returns," *Journal of Financial Economics* 1 (May 1974), pp. 1–22; P. Hess, "The Ex-Dividend Behavior of Stock Returns: Further Evidence on Tax Effects," *Journal of Finance* 37 (May 1982), pp. 445–456; R. H. Litzenberger and K. Ramaswamy, "The Effect of Personal Taxes and Dividends on Capital Asset Prices: Theory and Empirical Evidence," *Journal of Financial Economics* 7 (June 1979), pp. 163–195; and M. H. Miller and M. Scholes, "Dividends and Taxes: Some Empirical Evidence," *Journal of Political Economy* 90 (1982), pp. 1118–1141.

to gain insight and understanding from our best thinking. We have even drawn from the empirical evidence on hand to see what the findings suggest.

From our examination of the empirical evidence, the following conclusions would appear reasonable:

1. As a firm's investment opportunities increase, the dividend payout ratio should decrease. In other words, an inverse relationship should exist between the amount of investments with an expected rate of return that exceeds the cost of capital and the dividends remitted to investors. Since flotation costs are incurred with raising external capital, the retention of internally generated equity financing is preferable to selling shares in terms of the wealth of the current common shareholders.
2. The firm's dividend policy appears to be important; however, appearances may be deceptive. The real issue may be the firm's expected earning power and the riskiness of these earnings. Investors may be using the dividend payment as a source of information about the company's expected earnings. Management's actions regarding dividends may carry greater weight than a statement by management that earnings will be increasing.
3. Any influence that dividends have on share price may be attributed to the investor's desire to minimize and/or defer taxes and from the role of dividends in minimizing agency costs.
4. The expectations theory indicates to management that it must maintain a consistent dividend policy. The firm's dividend policy might effectively be treated as a long-term residual. Rather than projecting investment requirements for a single year, management could anticipate financing needs for several years. Based upon the expected investment opportunities during the planning horizon, the firm's debt-equity mix, and the funds generated from operations, a target dividend payout ratio could be established. If internal funds remained after projection of the necessary equity financing, dividends would be paid. However, the planned dividend stream should distribute residual capital evenly to investors. Conversely, if over the long term the entire amount of internally generated capital is needed for reinvestment in the company, then no dividend should be paid.

OBJECTIVE 3

THE DIVIDEND DECISION IN PRACTICE

Financial managers will take into consideration the concepts set forth in this chapter to select the firm's dividend policy. While these concepts do not provide an equation that explains the key relationships, they certainly give us a more complete view of the finance world, which can only help us make better decisions. Other considerations of a more practical nature affect the decision-making process for a firm's dividend policy.

Other Practical Considerations

A large number of considerations may influence a firm's decision about its dividends, some of them unique to that company. Some of the more general considerations are given below.

Legal Restrictions

Certain legal restrictions may limit the amount of dividends a firm may pay. These legal constraints fall into two categories. First, statutory restrictions may prevent a company from paying dividends. The distribution of dividends is permitted by the Canada Business Corporation Act if a company meets the conditions of a solvency test which requires that either before or after the dividend payment (1) the firm is

Dividends Must Be Adequately Funded

Companies make money in order to distribute it, eventually, to their shareholders. There is no benefit from making an investment in a company if the profits are neither distributed nor reinvested wisely.

If the business is successful, some of the money must be kept by the company to maintain its market position, its productivity, and in some cases, to comply with legal and social obligations. Still more money may be retained if the company has opportunities to reinvest it more profitably than the shareholders can manage on their own. Beyond that, the profits should be distributed. The choice is a simple one of allocating capital where it can be best used: by the company or by its owners.

Normally, companies tell their shareholders they intend to pay a fixed portion of their after-tax income as dividends. This is typically about 40%.

To be sure, net income is important, but a better starting point would be cash flow from operations—earnings before deducting non-cash charges.

These charges include depreciation (an accounting approximation for the wear and tear on machinery, equipment and buildings), amortization of intangible assets (which usually arise from a business acquisition), and increases in deferred taxes.

Obligatory Expenditures

Operating cash flow is sometimes treated as if were freely disposable income; it is not. To arrive at the cash flow figure, we must also deduct those obligatory expenditures that a company must make to enable it to at least maintain its existing productive capability and market position. After these expenditures are subtracted from operating cash flow, what remains is available for expansion, for new acquisitions and to pay dividends. This is called available cash earnings.

Operating cash flow will invariably be greater than earnings, but often available cash earnings are not. Indeed, in capital intensive businesses, available cash earnings are frequently zero or even negative. Throughout most of the 1980s, for example, Stelco Inc. enjoyed a large and positive operating cash flow. But its available cash earnings, after providing for its substantial cash investments, was either negative or at least not large enough to cover its dividends. In 1990, the company endured a substantial loss. If we look at 1989, after-tax income was $94 million and operating cash flow was $250 million; but capital needs totalled $285 million, so available cash earnings were really a negative $35 million. In addition, even though the company had negative cash earnings, it still saw fit to pay cash dividends of $35 million to its common shareholders. Although earnings and operating cash flow appeared to cover the dividend, the available cash earnings clearly did not. This had been true during the previous four years as well.

Debt Increased

It is this point that highlights how important it is to look at available cash earnings. Stelco paid out more than $165 million in dividends in the years 1986–1990. It had to borrow to pay these dividends. It did not eliminate the dividend until late in 1990, by which time its debt had increased substantially. The borrowing required just to pay the dividend was an important ingredient in the deterioration of the financial position.

One of the few aspects of dividend policy which is universally agreed on is that a dividend rate, once established, should not be cut. Unless the regular dividend is set at a level within the reasonably expected available cash flow, the company will be borrowing to pay the dividend at the best of times; when earnings drop, the dividend may have to be eliminated to avoid placing a financial hardship on the company. In this scenario, no one benefits.

On the other hand, companies that focus on cash earnings and formulate dividend policy consistent with the availability of cash earnings are allocating the capital in the most economically efficient manner. They are thereby demonstrating their loyalty to and respect for their shareholders. It is these companies which create real economic value.

Source: R. Fogler, "Dividends must be adequately funded," *Financial Post Daily*, February 21, 1992, p. 11.

able to pay its liabilities and (2) the firm's assets have a realizable value not less than the aggregate of its liabilities and equity capital.

The second type of legal restriction is unique to each firm and results from restrictions in debt and preferred-share contracts. To minimize their risk, investors frequently impose restrictive provisions upon management as a condition to their investment in the company. These constraints may include the provision that dividends may not be declared prior to the debt being repaid. Also, the corporation may be required to maintain a given amount of working capital. Preferred shareholders may stipulate that common dividends may not be paid when any preferred dividends are delinquent.

Liquidity Position

Contrary to common opinion, the mere fact that a company shows a large amount of retained earnings in the balance sheet does not indicate that cash is available for the payment of dividends. The firm's current position in liquid assets, including cash, is basically independent of the retained earnings account. Historically, a company with sizable retained earnings has been successful in generating cash from operations. Yet these funds are typically either reinvested in the company within a short time period or used to pay maturing debt. Thus, a firm may be extremely profitable and still be cash poor. Since dividends are paid with cash, and not with retained earnings, the firm must have cash available for dividends to be paid. Hence, the firm's liquidity position has a direct bearing on its ability to pay dividends.

Absence or Lack of Other Sources of Financing

As already noted, a firm may (1) retain profits for investment purposes and/or (2) pay dividends and issue new debt or equity securities to finance investments. For many small or new companies, this second option is not realistic. These firms do not have access to the capital markets, so they must rely more heavily upon internally generated funds. As a consequence, the dividend payout ratio is generally much lower for a small or newly established firm than for a large, publicly owned corporation.

Earnings Predictability

A company's dividend payout ratio depends to some extent upon the predictability of the firm's profits over time. If earnings fluctuate significantly, management cannot rely on internally generated funds to meet future needs. When profits are realized, the firm may retain larger amounts to ensure that money is available when needed. Conversely, a firm with a stable earnings trend will typically pay a larger portion of its earnings out in dividends. This company has less concern about the availability of profits to meet future capital requirements.

Ownership Control

For many large corporations, control through the ownership of common shares is not an issue. However, for many small and medium-sized companies, maintaining voting control takes a high priority. If the present common shareholders are unable to participate in a new offering, issuing new shares is unattractive, in that the control of the current shareholders is diluted. The owners might prefer that management finance new investments with debt and through profits rather than by issuing new common shares. This firm's growth is then constrained by the amount of debt capital available and by the company's ability to generate profits.

Inflation

Before the late 1970s, inflationary pressures had not been a significant problem for either consumers or businesses. However, over the last two decades, the deterioration of the dollar's purchasing power has had a direct impact upon the replacement of fixed assets. For most firms, the funds generated from depreciation are used to finance the replacements of fixed assets that have become worn and obsolete. During a period of inflation, the cost of replacing equivalent equipment increases and the depreciation funds are insufficient. These cost increases require a greater retention of profits which implies that dividends will be adversely affected.

Alternate Dividend Policies

OBJECTIVE 4

Regardless of a firm's long-term dividend policy, most firms choose one of the following dividend payment patterns:

1. **Constant dividend payout ratio.** In this policy, the percentage of earnings paid out in dividends is held constant. Although the dividend-to-earnings ratio is stable, the dollar amount of the dividend naturally fluctuates from year to year as profits vary.
2. **Stable dollar dividend per share.** This policy maintains a relatively stable dollar dividend over time. An increase in the dollar dividend usually does not occur until management is convinced that the higher dividend level can be maintained in the future. Management also will not reduce the dollar dividend until the evidence clearly indicates that a continuation of the present dividend cannot be supported.
3. **Small, regular dividend plus a year-end extra.** A corporation following this policy pays a small regular dollar dividend plus a year-end extra dividend in prosperous years. The extra dividend is declared toward the end of the fiscal year, when the company's profits for the period can be estimated. Management's objective is to avoid the connotation of a permanent dividend. However, this purpose may be defeated if recurring "extra" dividends come to be expected by investors. Ford Motor of Canada Ltd. and Budd Canada Inc. are examples of firms that in recent years have employed the low stable dividend plus an extra dividend at the end of the year.

Constant dividend payout ratio

A dividend payment policy in which the percentage of earnings paid out in dividends is held constant. The dollar amount fluctuates from year to year as profits vary.

Stable dollar dividend payout

A dividend policy that maintains a relatively stable dollar dividend per share over time.

Small regular plus year-end extra dividend payout

A dividend payment policy where the firm pays a small regular dividend plus an extra dividend only if the firm has experienced a good year.

Of the three dividend policies, the stable dollar dividend is by far the most common. Figure 10-3 illustrates the general tendency of companies to pay stable, but increasing, dividends, even though the profits fluctuate significantly. In a study by Lintner, corporate managers were found to be reluctant to change the dollar amount of the dividend in response to "temporary" fluctuations in earnings from year to year. This aversion was particularly evident when it came to decreasing the amount of the dividend from the previous level.[19] Smith explained stable dividends in terms of his **"increasing-stream hypothesis of dividend policy."**[20] He proposed that dividend stability is essentially a smoothing of the dividend stream in order to minimize the effect of other types of company reversals. Thus, corporate managers make every effort to avoid a dividend cut, attempting instead to develop a gradually increasing dividend series over the long-term future.[21] However, if a dividend

Increasing-stream hypothesis of dividend policy

A smoothing of the dividend stream in order to minimize the effect of company reversals. Corporate managers make every effort to avoid a dividend cut, attempting instead to develop a gradually increasing dividend series over the long-term future.

[19]John Lintner, "Distribution of Income of Corporations Among Dividends, Retained Earnings, and Taxes," *American Economic Review* 46 (May 1956), pp. 97–113.

[20]Keith V. Smith, "Increasing-Stream Hypothesis of Corporate Dividend Policy," *California Management Review* 15 (Fall 1971), pp. 56–64.

[21]Boards of directors, as a general rule, view the dividend as a fixed cost. Once they raise it, they will not cut it, unless there is no other alternative.

Figure 10-3 Corporate Earnings and Dividends

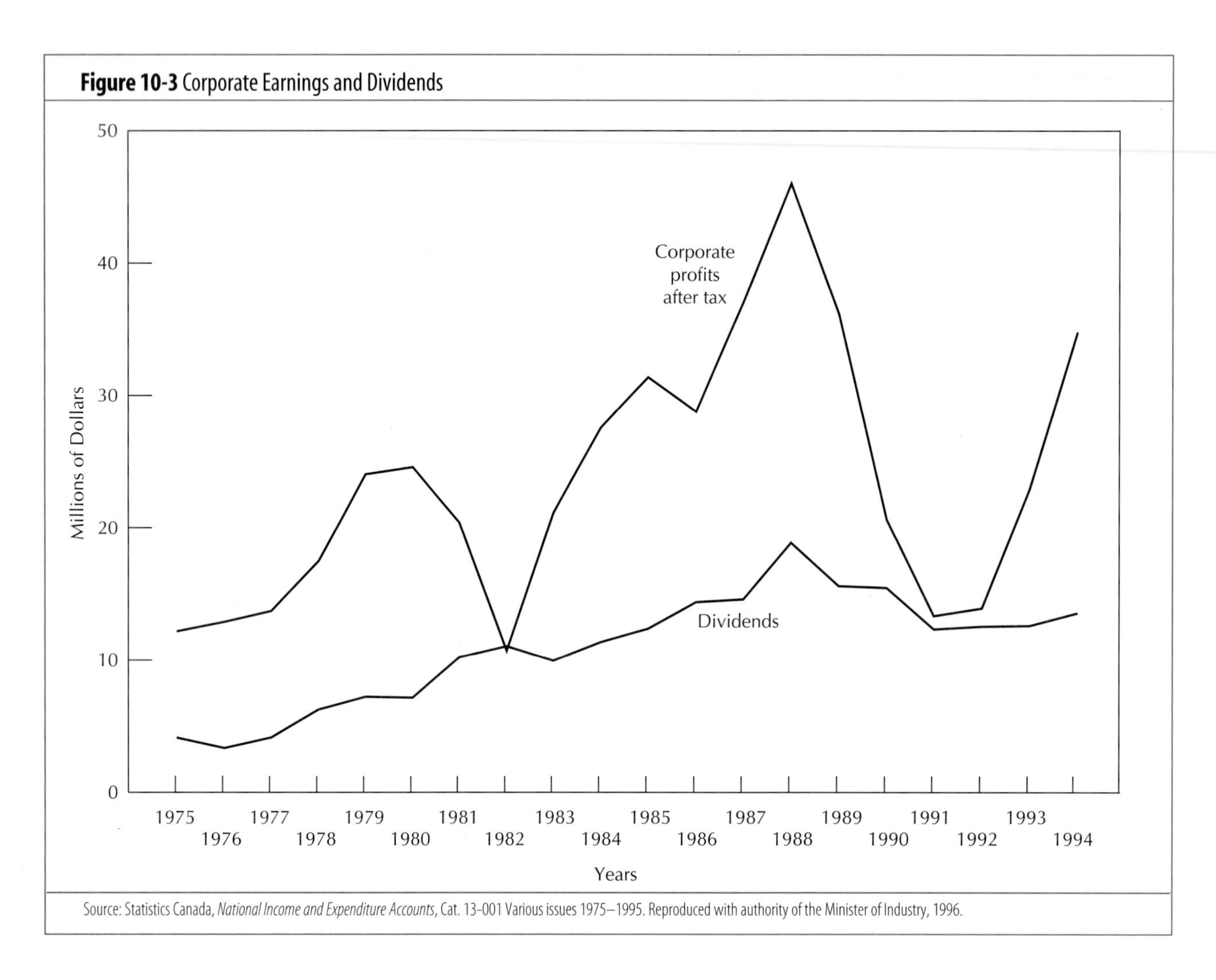

Source: Statistics Canada, *National Income and Expenditure Accounts*, Cat. 13-001 Various issues 1975–1995. Reproduced with authority of the Minister of Industry, 1996.

reduction is absolutely necessary, the cut should be large enough to reduce the probability of future cuts.[22]

As an example of a stable dividend policy, Figure 10-4 compares the Bank of Montreal's earnings per share and dividends per share for 1974 through 1995. Ignoring 1989, when earnings were exceptionally low, the firm has, on average, paid approximately 42 percent of its earnings out in dividends. The historical dividends and earnings patterns for the firm clearly demonstrate management's hesitancy to change dividends in response to short-term fluctuations in earnings.[23]

---

[22]In examining this issue, the financial manager should answer the following questions: (1) Should management dare to decrease the dividend to provide equity capital for financing future investments? and (2) If a decrease in dividends were to be made, how should this decision be communicated to the market?

[23]The empirical evidence suggests that the following variables have a significant impact on management's dividend policy. In other words, the next dividend to be paid will largely be the result of (1) management's target dividend payout ratio; (2) the current earnings per share; (3) the amount of the prior dividends paid, with the more recent dividend being weighted more heavily than earlier dividends; and (4) management's expectations about future earnings per share.

Figure 10-4 Earnings per Share and Dividends per Share for the Bank of Montreal

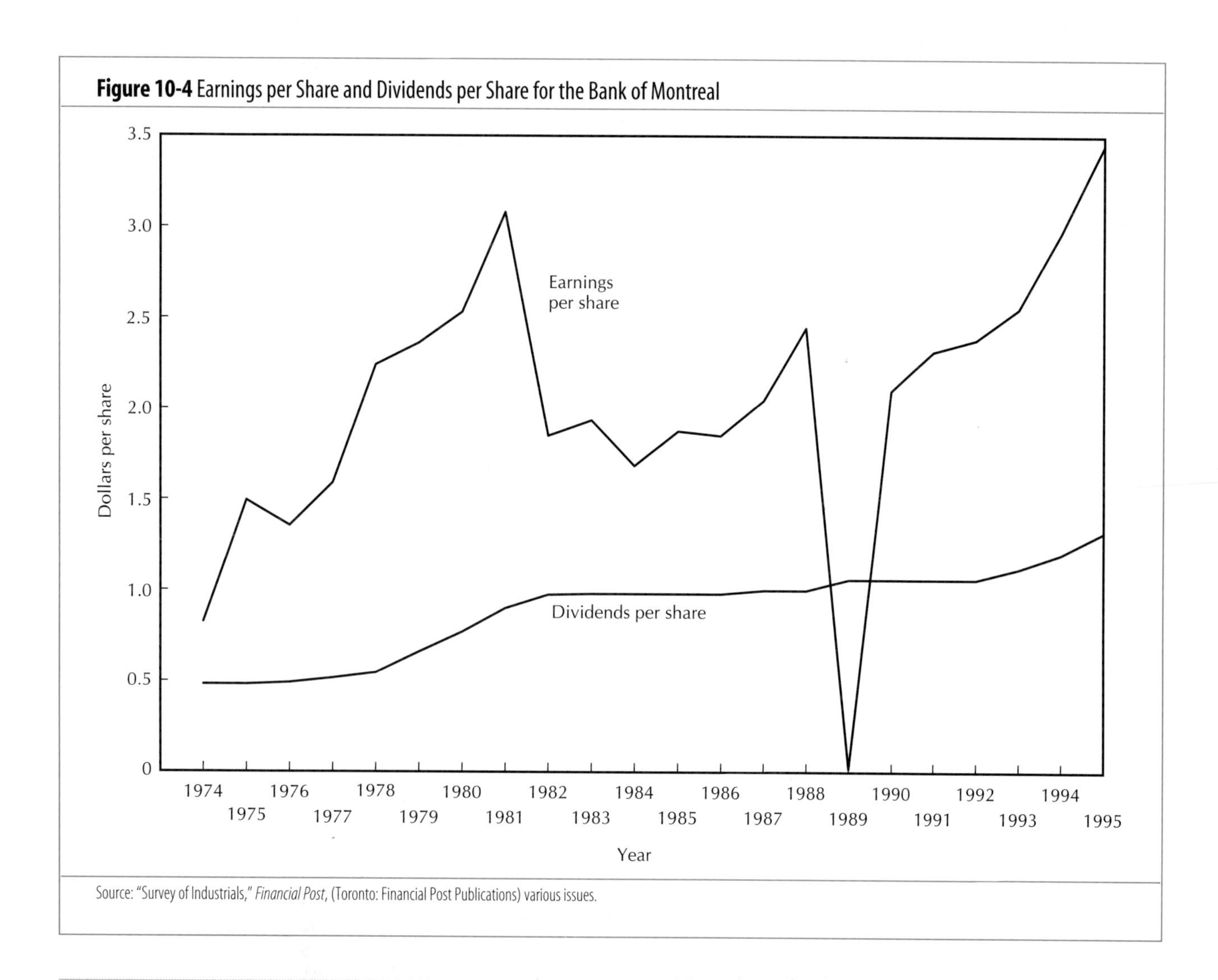

Source: "Survey of Industrials," *Financial Post*, (Toronto: Financial Post Publications) various issues.

PERSPECTIVE IN FINANCE

We don't know much with certitude about dividend policy and its impact on the firm's share price, but we do know that managers hate, even despise, the thought of cutting the dividend. It will be done only as a last resort. Count on it.

DIVIDEND PAYMENT PROCEDURES

OBJECTIVE 5

After the firm's dividend policy has been structured, several procedural details must be arranged.[24] For instance, how frequently are dividend payments to be made? If a shareholder sells the shares during the year, who is entitled to the dividend? To answer these questions, we need to understand the dividend payment procedures.

[24]Given that the dividend policy may be important because of its informational content, management ought to be as concerned about how the announcement is communicated as it is about the actual announcement.

Declaration date

The date upon which a dividend is formally declared by the board of directors.

Date of record

Date at which the share transfer books are to be closed for determining which investor is to receive the next dividend payment.

Ex-dividend date

The date upon which stock brokerage companies have uniformly decided to terminate the right of ownership to the dividend, which is four working days prior to the record date

Payment date

The date on which the company mails a dividend cheque to each investor.

Generally, companies pay dividends on a quarterly basis. To illustrate, Scott Paper pays $0.10 per share in annual dividends. However, the firm actually issues a $0.025 quarterly dividend for a total yearly dividend of $0.10 ($0.025 × 4 quarters).

The final approval of a dividend payment comes from the board of directors. As an example, B.C. Sugar Refinery Ltd., on January 9, 1996, announced that holders of record as of February 2 would receive a $0.10 dividend. The dividend payment was to be made on February 26. January 9 is the **declaration date**—the date when the dividend is formally declared by the board of directors. The **date of record**, February 2, designates when the share transfer books are to be closed. Investors shown to own the share on this date receive the dividend. If a notification of a transfer is recorded subsequent to February 2, the new owner is not entitled to the dividend. However, a problem could develop if the share were sold on February 1, one day prior to the record date. Time would not permit the sale to be reflected on the shareholder list by the February 2 date of record. To avoid this problem, stock brokerage companies have uniformly decided to terminate the right of ownership to the dividend four working days prior to the record date. This prior date is the **ex-dividend date**. Therefore, any acquirer of the B.C. Sugar Refinery Class A shares on January 29 or thereafter does not receive the dividend. Finally, the company mails the dividend cheque to each investor on February 26, the **payment date**.

OBJECTIVE 6

STOCK DIVIDENDS AND STOCK SPLITS

Stock dividend

A distribution of shares issued on a pro rata basis to the current shareholders.

Stock split

A transaction that increases the number of shares outstanding in proportion to the number of shares currently owned by shareholders.

An integral part of dividend policy is the use of **stock dividends** and **stock splits**. Both involve issuing new shares of stock on a pro rata basis to the current shareholders, while the firm's assets, its earnings, and the risk assumed and the investor's percentage of ownership in the company remain unchanged. The only definite result from either a stock dividend or stock split is the increase in the number of shares outstanding.

To illustrate the effect of a stock dividend, assume that the Katie Corporation has 100,000 shares outstanding.[25] The firm's after-tax profits are $500,000, or $5 in earnings per share. At present, the company's shares are selling at a price/earnings multiple of 10, or $50 per share. Management is planning to issue a 20 percent stock dividend, so that a shareholder owning 10 shares would receive two additional shares. We might immediately conclude that this investor is being given an asset (two shares) worth $100; consequently, his or her personal worth should increase by $100. This conclusion is erroneous. The firm will be issuing 20,000 new shares (100,000 shares × 20 percent). Since the $500,000 in after-tax profits does not change, the new earnings per share will be $4.167 ($500,000/120,000 shares). If the price/earnings multiple remains at 10, the market price of the share after the dividend should fall to $41.67 ($4.167 earnings per share × 10). The investor now owns 12 shares worth $41.67, which provides a $500 total value; thus he or she is neither better nor worse off than before the stock dividend.

This example may make us wonder why a corporation would even bother with a stock dividend or stock split if no one benefits. However, before we study the rationale for such distributions, we should understand the differences between a stock split and a stock dividend.

Stock Dividend versus Split

The only difference between a stock dividend and a stock split relates to their respective accounting treatment. Stated differently, there is absolutely no difference on an economic basis between a stock dividend and a stock split. Both represent a proportionate distribution of additional shares to the present shareholders.

[25]The logic of this illustration is equally applicable to a stock split.

Dividend Pay-Outs: Different Practices in Different Countries

How much of the earnings do most firms pay out in dividends? It depends on the country. As shown in the graph here, British firms pay out a lot more than German and Japanese companies.

Sources: Datastream; BZW; IDD Information Services; Securities Data Company; Paribas, Referenced in *The Economist*, January 25, 1992, p. 73.

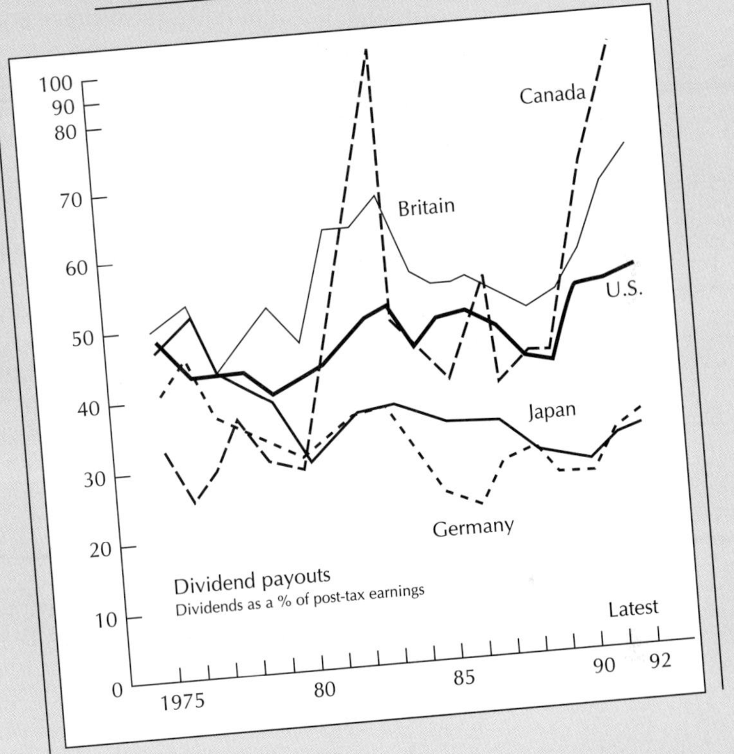

The accounting treatment for a dividend in common shares requires the issuing firm to capitalize the "market value" of the dividend.[26] In other words, the dollar amount of the dividend is transferred from retained earnings to the common shares account. This procedure may best be explained by an example. Assume that the L. Bernard Corporation is preparing to issue a 15 percent dividend in common shares. Table 10-4 presents the equity portion of the firm's balance sheet prior to the distribution. The market price for the share has been $14. Thus, the 15 percent stock dividend increases the number of shares by 150,000 (1,000,000 shares × 15

[26]Canadian Institute of Chartered Accountants, *CICA Handbook*, 3240.01–04.

percent). The "market value" of this increase is $2,100,000 (150,000 shares × $14 market price). To record this transaction, $2,100,000 would be transferred from retained earnings, resulting in a $2,100,000 increase in the value of common shares. Table 10-5 shows the revised balance sheet.

What if the management of L. Bernard Corporation changed the plan and decided to split the stock two-for-one? In other words, a 100 percent increase in the number of shares would result. In accounting for the split, the changes to be recorded are (1) the increase in the number of shares and (2) the decrease in the per-share value from $2 to $1. The dollar amounts of each account do not change. Table 10-6 illustrates the new balance sheet.

Thus, for a stock dividend, an amount equal to the market value of the stock dividend is transferred from retained earnings to the capital share accounts. When stock is split, only the number of shares changes, and the per share value decreases proportionately. Despite this dissimilarity in accounting treatment, remember that no real economic difference exists between a split and a dividend.

Rationale for a Stock Dividend or Split

Although stock dividends and splits occur far less frequently than cash dividends, a significant number of companies choose to use these share distributions either with or in lieu of cash dividends. Since our earlier conclusion that no economic benefit results, how do corporations justify these distributions?

Proponents of stock dividends and splits frequently maintain that shareholders receive a key benefit because the price of the share will not fall precisely in proportion to the share increase. For a two-for-one split, the price of the share might not decrease a full 50 percent, and the shareholder is left with a higher total value. There are two reasons for this disequilibrium. First, many financial executives believe that an optimal price range exists. Within this range, the total market value of the common shareholders is thought to be maximized. As the price exceeds this

Table 10-4 L. Bernard Corporation Balance Sheet Before Stock Dividend

Common Equity:	
Common shares (1,000,000 shares outstanding)	$10,000,000
Retained earnings	15,000,000
Total equity	$25,000,000

Table 10-5 L. Bernard Corporation Balance Sheet After Stock Dividend

Common Equity:	
Common shares (1,150,000 shares outstanding)	$12,100,000
Retained earnings	12,900,000
Total equity	$25,000,000

Table 10-6 L. Bernard Corporation Balance Sheet After Stock Split

Common Equity:	
Common shares (2,000,000 shares outstanding)	$10,000,000
Retained earnings	15,000,000
Total equity	$25,000,000

range, fewer investors can purchase the shares, thereby restraining the demand. Consequently, downward pressure is placed on its price. The second explanation relates to the informational content of the dividend-split announcement. Stock dividends and splits have generally been associated with companies with growing earnings. The announcement of a stock dividend or split has therefore been perceived as favourable news. The empirical evidence, however, fails to verify these conclusions. Most studies indicate that investors are perceptive in identifying the true meaning of a share distribution. If the stock dividend or split is not accompanied by a positive trend in earnings and increases in cash dividends, price increases surrounding the stock dividend or split are insignificant.[27] Therefore, we should be suspicious of the assertion that a stock dividend or split can help increase the investors' worth.

A second reason cited for stock dividends or splits is the conservation of corporate cash. If a company is encountering cash problems, it may substitute a stock dividend for a cash dividend. However, as before, investors will probably look beyond the dividend to ascertain the underlying reason for conserving cash. If the stock dividend is an effort to conserve cash for attractive investment opportunities, the shareholder may bid up the share price. If the move to conserve cash relates to financial difficulties within the firm, the market price will most likely react adversely.

Share Repurchases

OBJECTIVE 7

If management intends to repurchase a block of the firm's outstanding shares, it should make this information public. All investors should be given the opportunity to work with complete information. They should be told the purpose of the repurchase, as well as the method to be used to acquire the shares.

As to the method of **share repurchase**, three options are available. First, the shares could be bought in the open market. Here the firm acquires the shares through a stockbroker at the going market price. This approach may place an upward pressure on the share price until the shares are acquired. Also, commissions must be paid to the stockbrokers as a fee for their services.

Share repurchase
The repurchase of common shares by the issuing firm for any of a variety of reasons resulting in a reduction of shares outstanding.

The second method is to make a tender offer to the firm's shareholders. A **tender offer** is a formal offer by the company to buy a specified number of shares at a predetermined and stated price. The tender price is set above the current market price in order to attract sellers. A tender offer is best when a relatively large number of shares are to be bought, since the company's intentions are clearly known and each shareholder has the opportunity to sell the shares at the tendered price.

Tender offer
The formal offer by the company to buy a specified number of shares at a predetermined and stated price.

The third and final method for repurchasing shares entails the purchase of the shares from one or more major shareholders. These purchases are made on a negotiated basis. Care should be taken to ensure a fair and equitable price. Otherwise the remaining shareholders may be hurt as a result of the sale.

A number of reasons have been given for repurchasing shares. Examples of such benefits include:

1. A means for providing an "internal" investment opportunity.
2. An approach for modifying the firm's capital structure.
3. A favourable impact upon earnings per share.
4. The elimination of minority ownership groups of shareholders.
5. Earnings per share is not diluted as occurs with mergers and options.
6. The reduction in the firm's costs associated with servicing small shareholders.

[27]See James A. Millar and Bruce D. Fielitz, "Stock Split and Stock-Dividend Decisions," *Financial Management* (Winter 1973), pp. 35–45; and Eugene Fama, Lawrence Fisher, Michael Jensen, and Richard Roll, "The Adjustment of Stock Prices to New Information," *International Economic Review* (February 1969), pp. 1–21.

Share Repurchase as a Dividend Decision

Clearly, the payment of a cash dividend is the conventional method for distributing a firm's profits to its owners. However, it need not be the only way. Another approach is to repurchase the firm's shares. The concept may best be explained by an example.

EXAMPLE

The Telink Inc. is planning to pay $4 million ($4 per share) in dividends to its common shareholders. The following earnings and market price information is provided for Telink's:

Net income	$7,500,000
Number of shares	1,000,000
Earnings per share	$7.50
Market price per share after dividend payment	$60
Price/earnings ratio	8

In a recent meeting several board members, who are also major shareholders, questioned the need for a dividend payment. They maintain that they do not need the income, so why not allow the firm to retain the funds for future investments? In response, management contends that the available investments are not sufficiently profitable to justify retention of the income. That is, the investors' required rates of return exceed the expected rates of return that could be earned with the additional $4 million in investments.

Since management opposes the idea of retaining the profits for investment purposes, one of the firm's directors has suggested that the $4 million be used to repurchase the company's shares. In this way, the value of the shares should increase. This result may be demonstrated as follows:

1. **Assume that shares are repurchased by the firm at the $60 market price (ex-dividend price) plus the contemplated $4 dividend per share, or for $64 per share.**
2. **Given a $64 price, 62,500 shares would be repurchased ($4 million/$64 price).**
3. **If net income is not reduced, but the number of shares declines as a result of the share repurchase, earnings per share would increase from $7.50 to $8, computed as follows:**

$$\text{Earnings per share} = \frac{\text{net income}}{\text{outstanding shares}}$$

$$\text{(before repurchase)} = \frac{\$7{,}500{,}000}{1{,}000{,}000}$$

$$= \$7.50$$

$$\text{(after repurchase)} = \frac{\$7{,}500{,}000}{(1{,}000{,}000 - 62{,}500)}$$

$$= \$8$$

4. **Assuming that the price/earnings ratio remains at 8, the new price after the repurchase would be $64, up from $60, where the increase exactly equals the amount of the dividend forgone.**

In this example, Telink's shareholders are essentially provided the same value, whether a dividend is paid or shares are repurchased. If management pays a dividend, the investor will have a share valued at $60 plus $4 received from the dividend. On the other hand, if shares are repurchased in lieu of the dividend, the share will be worth $64. These results were based upon assuming (1) the share is being repurchased at the exact $64 price, (2) the $7,500,000 net income is unaffected by the repurchase, and (3) the price/earnings ratio of 8 does not change after the repurchase. Given these assumptions, however, the share repurchase serves as a perfect substitute for the dividend payment to the shareholders.

The Investor's Choice

Given the choice between a share repurchase and a dividend payment, which would an investor prefer? In perfect markets, where there are no taxes, no commissions when buying and selling shares, and no informational content assigned to a dividend, the investor would be indifferent with regard to the choices. The investor could create a dividend stream by selling shares when income is needed.

If market imperfections exist, the investor must be able to discern the tax consequences of a share repurchase. A firm's repurchase of its shares through either the tender offer or the negotiated basis has tax implications on the investor. Revenue Canada has asserted that the firm will be deemed to pay a dividend on shares either repurchased through a tender offer or negotiated.[28] The amount of the dividend is equal to the difference between the purchase price of the share and the paid-up capital (i.e., the issuing price of the shares) for tax purposes. The consequences of these tax regulations to shareholders is the loss of any capital gains and their preferential tax treatment; each shareholder is required to report any dividends received. The tax consequences of these two procedures of share repurchase differ from the tax consequences which result if a firm uses the procedure of repurchasing its shares on the open market. If a firm repurchases its shares on the open market in the normal manner, a shareholder will incur a capital gain and benefit from the preferential tax treatment.[29] However, the open market repurchase may cause the firm to pay too high a price for the repurchased share, which is to the detriment of the remaining shareholders. If a relatively large number of shares are being bought, the price may be bid up too high, only to fall after the repurchase operation. Second, as a result of the repurchase, the market may perceive the riskiness of the corporation as increasing, which would lower the price/earnings ratio and the value of the share.

Financing or Investment Decision

The repurchase of shares when the firm has excess cash may be regarded as a dividend decision. However, a share repurchase may also be viewed as a financing decision. By issuing debt and then repurchasing shares, a firm can immediately alter its debt-equity mix toward a higher proportion of debt. Rather than choosing to distribute cash to the shareholders, management is using a share repurchase as a means to change the corporation's capital structure.

In addition to dividend and financing decisions, many managers consider a share repurchase an investment decision. When equity prices are depressed in the marketplace, management may view the firm's own share as being materially undervalued and representing a good investment opportunity. While the firm's management may be wise to repurchase shares at unusually low prices, this decision cannot and should not be viewed in the context of an investment decision.

---

[28]Canadian Income Tax Act with Regulations, *Deemed Dividend*, 65th ed. (North York, ON: CCH Canadian Limited, 1995) 84(3).

[29]Canadian Income Tax Act with Regulations, *Deemed Dividend*, 65th ed. (North York, ON: CCH Canadian Limited, 1995) 84(6).

Buying its own shares cannot provide expected returns as other investments do. No company can survive, much less prosper, by investing only in its own shares.

SUMMARY

In determining the firm's dividend policy, the key issue is the dividend payout ratio (the percentage of the earnings paid out in dividends). This decision has an immediate impact upon the firm's financial mix. As the dividend payment is increased, less funds are available internally for financing investments. Consequently, if additional equity capital is needed, the company has to issue new common shares. Keeping this interaction between the level of dividends and financing in mind, management has to determine the best dividend policy for the company's investors. However, selection of the most beneficial dividend payment is not easily accomplished. Management cannot apply an equation to resolve the question. We simply have been unable to disentangle the relationship between dividend policy and share price.

In its simplest form, the dividend payment is a residual factor. In this context, the dividend equals the remaining internal capital after financing the equity portion of investments. However, this single criterion fails to recognize (1) the tax benefit of capital gains, (2) agency costs, (3) the clientele effect, and (4) the informational content of a given policy. Furthermore, other factors which are considered include the firm's liquidity position, the accessibility to capital markets, inflation, legal restrictions, the stability of earnings, and the desire of investors to maintain control of the company.

Given the firm's investment opportunities, and the imperfections in the market, the financial manager should probably apply the residual dividend theory over the long term. In essence, the firm's investment opportunities are to be projected throughout a multiple-year planning horizon. Given these investment needs, the target debt mix, and the anticipated earnings, and the amount of money available to pay dividends for the planning period may be determined. The dividend payments should then be made so that large and unexpected changes in the dividend per share are avoided.

Stock dividends and stock splits have been used by corporations either in lieu of or to supplement cash dividends. At the present, no empirical evidence conclusively identifies a relationship between stock dividends and splits and the market price of the share. Yet a stock dividend or split could conceivably be used to keep the share price within an optimal trading range. Also, if the investors perceived that the stock dividend contained favourable information about the firm's operations, the price of the share could increase.

As an alternative to paying a dividend, management can repurchase shares. This latter approach may have a tax advantage for the shareholder; however, investors may still prefer dividends to a share repurchase.

STUDY QUESTIONS

10-1. What is meant by the term *dividend payout ratio*?

10-2. Explain the tradeoff between retaining internally generated funds and paying cash dividends.

10-3. **a.** What are the assumptions of a perfect market?
b. What effect does dividend policy have on the share price in a perfect market?

10-4. What is the impact of flotation costs on the financing decision?

10-5. **a.** What is the residual dividend theory?
b. Why is this theory operational only in the long term?

10-6. Why might investors prefer capital gains to the same amount of dividend income?

10-7. What legal restrictions may limit the amount of dividends to be paid?

10-8. How does a firm's liquidity position affect the payment of dividends?

10-9. How can ownership control constrain the growth of a firm?

10-10. **a.** Why is a stable dollar dividend policy popular from the viewpoint of the corporation?
b. Is it also popular with investors? Why?

10-11. Explain declaration date, date of record, and ex-dividend date.

10-12. What are the advantages of a stock split or dividend over a cash dividend?

10-13. Why would a firm repurchase its own shares?

SELF-TEST PROBLEMS

ST-1. *(Dividend Growth Rate)* Schulz Inc. maintains a constant dividend payout ratio of 35 percent. Earnings per share last year were $8.20 and are expected to grow indefinitely at a rate of 12 percent. What will be the dividend per share this year? In five years?

ST-2. *(Residual Dividend Theory)* Britton Corporation is considering four investment opportunities. The required investment outlays and expected rates of return for these investments are shown below. The firm's cost of capital is 14 percent. The investments are to be financed by 40 percent debt and 60 percent common equity. Internally generated funds totalling $750,000 are available for reinvestment.

a. Which investments should be accepted? According to the residual dividend theory, what amount should be paid out in dividends?

b. How would your answer change if the cost of capital were 10 percent?

Investment	Investment Cost	Internal Rates of Return
A	$275,000	17.50%
B	325,000	15.72
C	550,000	14.25
D	400,000	11.65

ST-3. *(Stock Split)* The debt and equity section of the Robson Corporation balance sheet is shown below. The current market price of the common shares is $20. Reconstruct the financial statement assuming that (a) a 15 percent stock dividend is issued and (b) a two-for-one stock split is declared.

Robson Corporation	
Debt	$1,800,000
Common equity	
Common shares (100,000 shares)	200,000
Contributed surplus	400,000
Retained earnings	900,000
	$3,300,000

STUDY PROBLEMS (SET A)

10-1A. *(Flotation Costs and Issue Size)* Your firm needs to raise $10 million. Assuming that after-tax issuing costs are expected to be $15 per share and that the market price of the share is $120, how many shares would have to be issued? What is the dollar size of the issue?

10-2A. *(Flotation Costs and Issue Size)* If after-tax issuing costs for a common share issue are 18 percent, how large must the issue be so that the firm will net $5,800,000? If the shares sell for $85 each, how many shares must be issued?

10-3A. *(Residual Dividend Theory)* Terra Cotta finances new investments by 40 percent debt and 60 percent equity. The firm needs $640,000 for financing new investments. If retained earnings equal $400,000, how much money will be available for dividends in accordance with the residual dividend theory?

10-4A. *(Stock Dividend)* RCB has 2 million common shares outstanding. Net income is $550,000, and the P/E ratio for the share is ten. Management is planning a 20 percent stock dividend. What will be the price of the share after the stock dividend? If an investor owns 100 shares prior to the stock dividend, does the total value of his or her shares change? Explain.

10-5A. *(Dividends in Perfect Markets)* The management of Harris, Inc., is considering two dividend policies for the years 1996 and 1997, one and two years away. In 1998 the management is planning to liquidate the firm. One plan would pay a dividend of $2.50 in 1996 and 1997 and a liquidating dividend of $45.75 in 1998. The alternative would be to pay out $4.25 in dividends in 1996, $4.75 in dividends in 1997, and a final dividend of $40.66 in 1998. The required rate of return for the common shareholders is 18 percent. Management is concerned about the effect of the two dividend streams on the value of the common share.

- **a.** Assuming perfect markets, what would be the effect?
- **b.** What factors in the real world might change your conclusion reached in part (a)?

10-6A. *(Long-Term Residual Dividend Policy)* Stetson Manufacturing Inc. has projected its investment opportunities over a five-year planning horizon. The cost of each year's investment and the amount of internal funds available for reinvestment for that year are given below. The firm's debt-equity mix is 35 percent debt and 65 percent equity. There are currently 100,000 common shares outstanding.

- **a.** What would be the dividend each year if the residual dividend theory were used on a year-to-year basis?
- **b.** What target stable dividend can Stetson establish by using the long-term residual dividend theory over the future planning horizon?
- **c.** Why might a residual dividend policy applied to the five years as opposed to individual years be preferable?

Year	Cost of Investments	Internal Funds Available for Reinvestment or for Dividends
1	$350,000	$250,000
2	475,000	450,000
3	200,000	600,000
4	980,000	650,000
5	600,000	390,000

10-7A. *(Stock Split)* You own 5 percent of the Trexco Corporation's common shares, which most recently sold for $98 prior to a two-for-one stock split announcement. Before the split there are 25,000 common shares outstanding.

- **a.** Relative to now, what will be your financial position after the stock split? (Assume the share price falls proportionately.)
- **b.** The executive vice-president in charge of finance believes the price will only fall 40 percent after the split because she feels the price is above the optimal price range. If she is correct, what will be your net gain?

10-8A. *(Dividend Policies)* The earnings for Crystal Cargo Inc. have been predicted for the next five years and are listed below. There are 1 million shares outstanding. Determine the yearly dividend per share to be paid if the following policies are enacted:

- **a.** Constant dividend payout ratio of 50 percent.
- **b.** Stable dollar dividend targeted at 50 percent of the earnings over the five-year period.
- **c.** Small, regular dividend of $.50 per share plus a year-end extra when the profits in any year exceed $1,500,000. The year-end extra dividend will equal 50 percent of profits exceeding $1,500,000.

Year	Profits After Taxes
1	$1,400,000
2	2,000,000
3	1,860,000
4	900,000
5	2,800,000

10-9A. *(Repurchase of Shares)* The Dunn Corporation is planning to pay dividends of $500,000. There are 250,000 shares outstanding, with an earnings per share of $5. The share should sell for $50 after the ex-dividend date. If instead of paying a dividend, management decides to repurchase shares,

- **a.** What should be the repurchase price?
- **b.** How many shares should be repurchased?
- **c.** What if the repurchase price is set below or above your suggested price in part (a)?
- **d.** If you own 100 shares, would you prefer that the company pay the dividend or repurchase shares?

10-10A. *(Flotation Costs and Issue Size)* D. Butler Inc. needs to raise $14 million. Assuming that after-tax issuing costs are expected to be 10 percent of the market price and that the market price of the share is $95, how many shares would have to be issued? What is the dollar size of the issue?

10-11A. *(Residual Dividend Theory)* Terra Cotta finances new investments with 70 percent debt and the rest in equity. The firm needs $1.2 million for financing a new acquisition. If retained earnings are $450,000, how much money will be available for dividends in accordance with the residual dividend theory?

10-12A. *(Stock Split)* You own 20 percent of Rainy Corporation which recently sold common shares for $86 prior to a two-for-one stock-split announcement. Before the split there are 80,000 common shares outstanding.

- **a.** Relative to now, what will be your financial position after the stock split? (Assume the share price falls proportionately.)
- **b.** The executive vice-president in charge of finance believes the price will fall only 45 percent after the split, because she feels the price is above the optimal price range. If she is correct, what will be your net gain?

STUDY PROBLEMS (SET B)

10-1B. *(Flotation Costs and Issue Size)* Your firm needs to raise $12 million. Assuming that after-tax issuing costs are expected to be $17 per share and that the market price of the share is $115, how many shares would have to be issued? What is the dollar size of the issue?

10-2B. *(Flotation Costs and Issue Size)* If after-tax issuing costs for a common share issue are 14 percent, how large must the issue be so that the firm will net $6,100,000? If the share sells for $76 per share, how many shares must be issued?

10-3B. *(Residual Dividend Theory)* Steven Miller finances new investments by 35 percent debt and 65 percent equity. The firm needs $650,000 for financing new investments. If retained earnings equal $375,000, how much money will be available for dividends in accordance with the residual-dividend theory?

10-4B. *(Stock Dividend)* DCA has 2.5 million common shares outstanding. Net income is $600,000, and the P/E ratio for the share is 10. Management is planning an 18 percent stock dividend. What will be the price of the share after the stock dividend? If an investor owns 120 shares prior to the stock dividend, does the total value of his or her shares change? Explain.

10-5B. *(Dividends in Perfect Markets)* The management of Montford Inc. is considering two dividend policies for the years 1997 and 1998, one and two years away. In 1999 the management is planning to liquidate the firm. One plan would pay a dividend of $2.55 in 1997 and 1998 and a liquidating dividend of $45.60 in 1999. The alternative would be to pay out $4.35 in dividends in 1997, $4.70 in dividends in 1998, and a final dividend of $40.62 in 1999. The required rate of return for the common shareholders is 17 percent. Management is concerned about the effect of the two dividend streams on the value of the common shares.

a. Assuming perfect markets, what would be the effect?

b. What factors in the real world might change your conclusion reached in part (a)?

10-6B. *(Long-Term Residual Dividend Policy)* Wells Manufacturing Inc. has projected its investment opportunities over a five-year planning horizon. The cost of each year's investment and the amount of internal funds available for reinvestment for that year are given below. The firm's debt-equity mix is 40 percent debt and 60 percent equity. There are currently 125,000 common shares outstanding.

a. What would be the dividend each year if the residual dividend theory were used on a year-to-year basis?

b. What target stable dividend can Wells establish by using the long-term residual dividend theory over the future planning horizon?

c. Why might a residual dividend policy applied to the five years as opposed to individual years be preferable?

Year	Cost of Investments	Internal Funds Available for Reinvestment or for Dividends
1	$360,000	$225,000
2	450,000	440,000
3	230,000	600,000
4	890,000	650,000
5	600,000	400,000

10-7B. *(Stock Split)* You own 8 percent of the Standlee Corporation's common shares, which most recently sold for $98 per share prior to a two-for-one stock split announcement. Before the split there are 30,000 common shares outstanding.

a. Relative to now, what will be your financial position after the stock split? (Assume the share price falls proportionately.)

b. The executive vice-president in charge of finance believes the price will fall only 45 percent after the split, because she feels the price is above the optimal price range. If she is correct, what will be your net gain?

10-8B. *(Dividend Policies)* The earnings for Carlson Cargo Inc. have been predicted for the next five years and are listed below. There are 1 million shares outstanding. Determine the yearly dividend per share to be paid if the following policies are enacted:

a. Constant dividend payout ratio of 40 percent.

b. Stable dollar dividend targeted at 40 percent of the earnings over the five-year period.

c. Small, regular dividend of $.50 per share plus a year-end extra when the profits in any year exceed $1,500,000. The year-end extra dividend will equal 50 percent of profits exceeding $1,500,000.

EXAMPLE

Suppose Shark Associates Inc. have obtained a $300,000 line of credit at a nominal interest rate of 12 percent but estimates that it will actually need $200,000 in the near future. The bank manager has offered Shark Associates the choice among five independent scenarios that Shark should consider. As the financial analyst for the company, respond to the bank manager by calculating the EAR for each scenario.

Scenario I

Shark must maintain a 10 percent compensating balance but does not have a deposit at the lending bank.

$$EAR = \left[1 + \frac{0.12}{(1 - 0.10)}\right]^1 - 1 = 13.33\%$$

One can also solve this problem as follows: In order to net $200,000 Shark must borrow $222,222.22 [$200,000 / (1 – .10)]. The same EAR result is obtained if we divide the interest cost of $26,666.67 ($222,222.22 × .12) by the funds used by Shark.

$$EAR = \left[1 + \frac{\$26{,}666.67}{(\$222{,}222.22 - \$22{,}222.22)}\right]^1 - 1 = 13.33\%$$

Scenario II

Instead of the 10 percent compensating balance, interest will be compounded daily. Using the EAR formula,

$$EAR = \left[1 + \frac{0.12}{365}\right]^{365} - 1 = 12.75\%$$

Note: Suppose that in addition to the 10 percent compensating balance, interest was compounded, say semiannually. What is the EAR?

$$EAR = \left[1 + \frac{0.1333}{2}\right]^2 - 1 = 13.78\%$$

The EAR has increased because interest is being compounded twice a year.

Scenario III

The loan will be made on a discounted basis, which means that interest will have to be paid in advance, as opposed to the end of the period.

Similar to the calculation in Scenario I, the EAR is determined as follows:

$$EAR = \left[1 + \frac{0.12}{(1 - 0.12)}\right]^1 - 1 + 13.64\%$$

Again, one can also solve this problem as follows: In order to net $200,000 Shark must borrow $227,272.73, in order to cover the interest cost of $27,272.73 ($227,272.73 × .12). The EAR result is obtained through the following equation:

$$EAR = \left[1 + \frac{\$27{,}272.73}{(\$227{,}272.73 - \$27{,}272.73)}\right]^1 - 1 = 13.64\%$$

Scenario IV

The compensating balance requirement is 5 percent; however, the loan will be made on a discounted basis. The EAR is solved through the following equation:

$$EAR = \left[1 + \frac{0.12}{(1 - 0.12 - 0.05)}\right]^1 - 1 = 14.46\%$$

Alternatively, Shark must borrow $240,963.85 [$200,000/(1 – .05 – .12)] in order to cover the interest cost of $28,915.66 ($240,963.85 × .12). The EAR result is obtained through the following equation:

$$EAR = \left[1 + \frac{\$28{,}915.66}{(\$240{,}963.85 - \$28{,}915.66 - \$12{,}048.19)}\right]^1 - 1 = 14.46\%$$

Scenario V

The bank has analyzed the timing of the $200,000 loan request and agrees with Shark that $50,000 will be required in the first 30 days and $150,000 will be required for the next 61 days. Accordingly, the line of credit is for 91 days only. In addition to the 12 percent interest on the loan, a commitment fee equal to half of 1 percent on the unused portion of the line will apply.

This scenario can be solved using the following four-step procedure:

1. Calculate the interest:

First 30 days: $50,000 × .12 × 30/365	=	$ 493.15
Next 61 days: $150,000 × .12 × 61/365	=	3,008.22
Total	=	$3,501.37

2. Calculate the commitment fee:

First 30 days:($300,000 − $50,000) × .005 × 30/365	=	$ 102.74
Next 61 days: ($300,000 − $150,000) X .005 × 61/365	=	125.34
Total	=	$ 228.08
Total Cost ($3,501.37 + $228.08)	=	$3,729.45

3. Calculate the weighted average loan amount:

($50,000 × 30/91) + ($150,000 × 61/91) = $117,032.97

4. Calculate the EAR:

$$EAR = \left[1 + \frac{\$3{,}729.45}{\$117{,}032.97}\right]^{365/91} - 1 = 13.41\%$$

To summarize:

Scenario	Description	EAR
I	10% compensating balance	13.33%
II	Daily compounding	12.75%
	10% compensating balance & semi-annual compounding	13.78%
III	Discounted loan	13.64%
IV	Discounted loan with 5% compensating balance	14.46%
V	91-day line of credit with .5% commitment fee	13.41%

Clearly, Scenario II, daily compounding has the lowest EAR.[6]

Transaction Loans

Still another form of unsecured short-term bank credit can be obtained in the form of **transaction loans**. Here the loan is made for a specific purpose.[7] This is the type of loan that most individuals associate with bank credit. The loan is obtained by signing a promissory note.

Transaction loan

A loan in which the proceeds are designated for a specific purpose—for example, a bank loan used to finance the acquisition of a piece of equipment.

Unsecured transaction loans are very similar to a line of credit with regard to cost, term to maturity, and compensating balance requirements. In both instances commercial banks often require that the borrower clean up its short-term loans for a 30- to 45-day period during the year.[8] This means, very simply, that the borrower

[6]We would like to thank Professor Arshad Ahmed from Concordia University for contributing this problem.

[7]Transaction loans are usually made to finance a nonrecurring purchase.

[8]The "clean-up" provision is very appropriate for those firms that have seasonal patterns in their business (for example, hotel or restaurants in certain vacation areas).

must be free of any bank debt for the stated period. The purpose of such a requirement is to ensure that the borrower is not using short-term bank credit to finance a part of its permanent needs for funds.

Unsecured Sources: Commercial Paper

Only the largest and most creditworthy companies are able to use **commercial paper**, which is simply a short-term *promise to pay* that is sold in the market for short-term debt securities.

Commercial paper

Short-term loans by the most creditworthy borrowers that are bought and sold in the market for short-term debt securities.

Credit Terms

The maturity of this credit source is generally six months or less, although some issues carry nine-month maturities. The interest rate on commercial paper is generally slightly lower (0.5 to 1 percent) than the prime rate on commercial bank loans. Furthermore, the interest cost on commercial paper is determined on a discount basis.

New issues of commercial paper are either placed directly (sold by the issuing firm directly to the investing public) or dealer placed. Dealer placement involves the use of a commercial paper dealer, who sells the issue for the issuing firm. Many major companies, such as Marlborough Properties, Sobey Stores and Eaton's issue their commercial paper. The volume of direct versus dealer placements favours direct placements. Dealers are used primarily by industrial firms that either make only infrequent use of the commercial paper market or have difficulty placing the issue because of their small size.

Commercial Paper as a Source of Short-Term Credit

A number of advantages accrue to the user of commercial paper:

1. **Interest rate.** Commercial paper rates are generally lower than rates on bank loans and comparable sources of short-term financing. Figure 11-1 displays historical rates of interest on short-term bank credit and commercial paper.
2. **Compensating balance requirement.** No minimum balance requirements are associated with commercial paper. However, issuing firms usually find it desirable to maintain lines of credit agreements sufficient to back up their short-term financing needs in the event that a new issue of commercial paper cannot be sold or an outstanding issue cannot be repaid when due.
3. **Amount of credit.** Commercial paper offers the firm with very large credit needs a single source for all its short-term financing. Loan restrictions placed on the financial institutions by the regulatory authorities may cause a firm to obtain the necessary funds from a number of financial institutions.
4. **Prestige.** Since it is widely recognized that only the most creditworthy borrowers have access to the commercial paper market, its use signifies a firm's credit status.

Using commercial paper for short-term financing, however, involves a very important risk. That is, the commercial paper market is highly impersonal and denies even the most creditworthy borrower any flexibility in terms of repayment. When bank credit is used, the borrower has someone with whom he or she can work out any temporary difficulties that might be encountered in meeting a loan deadline. This flexibility simply does not exist for the user of commercial paper.

Estimating the Cost of Commercial Paper

The cost of commercial paper can be estimated using the simple effective annual cost of credit equation (EAR). The key points to remember are that commercial paper interest is usually discounted and that if a dealer is used to place the issue, a fee is charged. Even if a dealer is not used, the issuing firm will incur costs associated with preparing and placing the issue, which also must be considered in estimating the cost of credit.

Figure 11-1 Historical Rates of Interest on Selected Financial Instruments

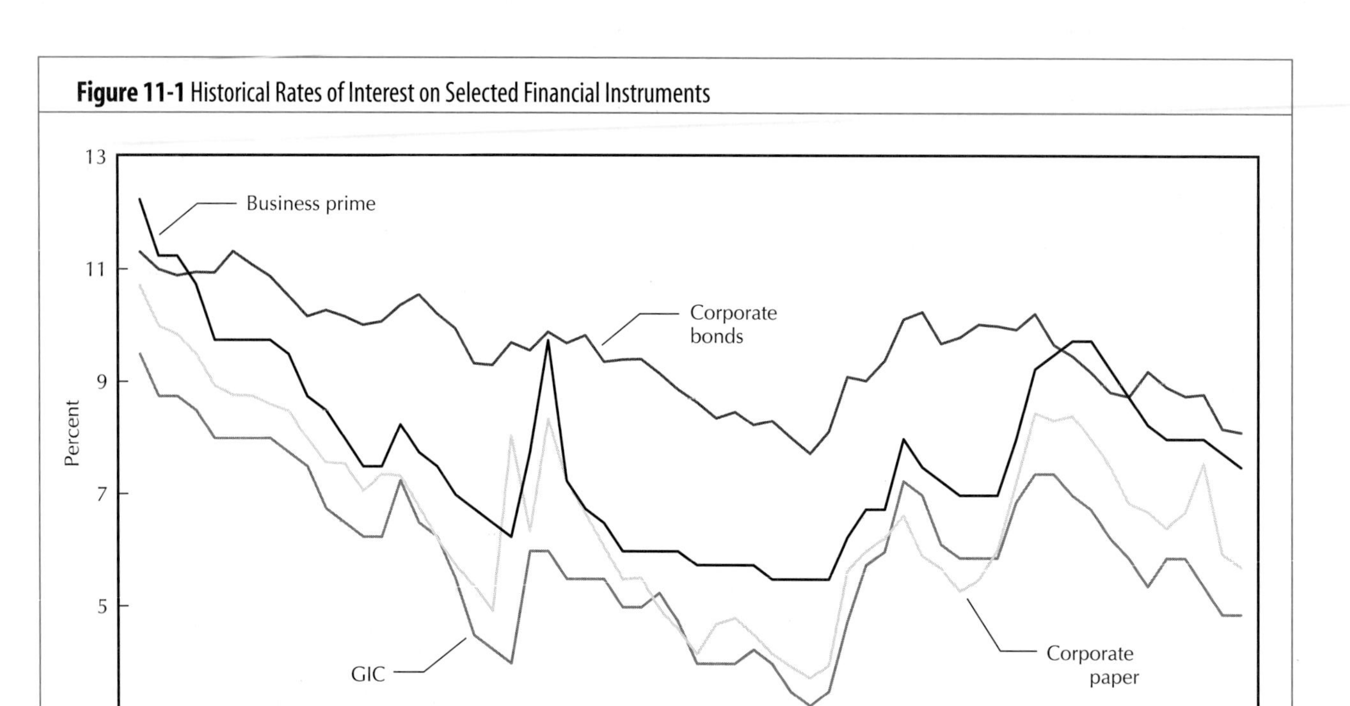

Source: *Bank of Canada Review*, "Financial Market Statistics S," Various Issues (Table F1).

EXAMPLE

The EPG Mfg. Company uses commercial paper regularly to support its needs for short-term financing. The firm plans to sell $100 million in nine-month-maturity paper on which it expects to have to pay discounted interest at an annual rate of 12 percent per annum. In addition, EPG expects to incur a cost of approximately $100,000 in dealer placement fees and other expenses of issuing the paper. The effective annual cost of credit to EPG can be calculated as follows:

$$EAR = \left[1 + \frac{(\$9{,}000{,}000 + \$100{,}000)}{(\$100{,}000{,}000 - \$100{,}000 - \$9{,}000{,}000)}\right]^{365/273} - 1$$

$$= 1.1361 - 1 = 13.61\%$$

where the interest cost is calculated as $100 million $\times$ .12 $\times$ 273/365 = $9 million. Thus, the effective annual cost of credit to EPG is 13.61 percent.

Secured Sources: Accounts Receivable Loans

Secured sources of short-term credit have certain assets of the firm pledged as collateral to secure the loan. Upon default of the loan agreement, the lender has first claim to the pledged assets in addition to its claim as a general creditor of the firm. Hence the secured credit agreement offers an added margin of safety to the lender.

Generally, a firm's receivables are among its most liquid assets. For this reason they are considered by many lenders to be prime collateral for a secured loan. Two basic procedures can be used in arranging for financing based on receivables: pledging and factoring.

Pledging Accounts Receivable

Under the **pledging** arrangement the borrower simply pledges accounts receivable as collateral for a loan obtained from either a commercial bank or a finance company. The amount of the loan is stated as a percent of the face value of the receivables pledged. If the firm provides the lender with a general line on its receivables, then all of the borrower's accounts are pledged as security for the loan. This method of pledging is simple and inexpensive. However, since the lender has no control over the quality of the receivables being pledged, it will set the maximum loan at a relatively low percent of the total face value of the accounts, generally ranging downward from a maximum of around 75 percent.

Still another approach to pledging involves the borrower's presenting specific invoices to the lender as collateral for a loan. This method is somewhat more expensive in that the lender must assess the creditworthiness of each individual account pledged; however, given this added knowledge, the lender will be willing to increase the loan as a percent of the face value of the invoices. In this case the loan might reach as high as 85 or 90 percent of the face value of the pledged receivables.

Pledging

An arrangement whereby the firm obtains a loan from a commercial bank or a finance company using its accounts receivable as collateral.

Credit Terms Accounts receivable loans generally carry an interest rate 2 to 5 percent higher than the bank's prime lending rate. Finance companies charge an even higher rate. In addition, the lender will usually charge a handling fee stated as a percent of the face value of the receivables processed, which may be as much as 1 to 2 percent of the face value.

EXAMPLE

The A. B. Good Company sells electrical supplies to building contractors on terms of net 60. The firm's average monthly sales are \$100,000; thus, its average receivables balance is \$200,000 over the firm's two-month credit terms. The firm pledges all its receivables to a local bank, which in turn advances up to 70 percent of the face value of the receivables at 3 percent over prime and with a 1 percent processing charge on all receivables pledged. A. B. Good follows a practice of borrowing the maximum amount possible, and the current prime rate is 10 percent.

The effective annual cost of using this source of financing for a full year is computed as follows:

$$EAR = \left[1 + \frac{(\$3{,}033 + \$2{,}000)}{\$140{,}000}\right]^{365/60} - 1 = 1.2397 - 1 = 23.97\%$$

where the total dollar cost of the loan consists of both the two-month interest expense [.13 × (2/12) × .70 × \$200,000 = \$3,033] and the annual processing fee (.01 × \$100,000 × 2 months = \$2,000). The amount of credit extended is .70 × \$200,000 = \$140,000. Note that the processing charge applies to all receivables pledged. Thus, the A. B. Good Company pledges \$100,000 each month on which a 1 percent fee must be paid, for a total monthly charge of \$1,000.

The lender, in addition to making advances or loans, may be providing certain credit services to the borrower. For example, the lender may provide billing and collection services. The value of these services should not be considered a part of the cost of credit. In the preceding example, A. B. Good Company may save credit department expenses of \$850 per month by pledging all its accounts and letting the lender provide those services. In this case, the cost of short-term credit is only

$$EAR = \left[1 + \frac{(\$3{,}033 + 2{,}000 - \$1{,}700)}{\$140{,}000}\right]^{365/60} - 1 = 1.1539 - 1 = 15.39\%$$

Advantages and Disadvantages of Pledging The primary advantage of pledging as a source of short-term credit is the flexibility it provides the borrower. Financing is available on a continuous basis. The new accounts created through credit sales provide the collateral for the financing of new production. Furthermore, the lender may provide credit services that eliminate or at least reduce the need for similar services within the firm. The primary disadvantage associated with this method of financing is its cost, which can be relatively high compared with other sources of short-term credit, owing to the level of the interest rate charged on loans and the processing fee on pledged accounts.

Factoring

The sale of a firm's accounts receivable to a financial intermediary known as a factor.

Factor

A firm that acquires the receivables of other firms. A commercial finance company that engages solely in the factoring of receivables is known as an old line factor.

Factoring Accounts Receivable

Factoring receivables involves the outright sale of a firm's accounts to a financial institution called a **factor**. A factor is a financial institution that acquires the receivables of other firms. The factor may be a commercial finance company that engages solely in the factoring of receivables (known as an **old line factor**) or it may be a commercial bank. The factor, in turn, bears the risk of collection and, for a fee, services the accounts. The fee is stated as a percent of the face value of all receivables factored (usually from 1 to 3 percent).

The factor typically does not make payment for factored accounts until the accounts have been collected or the credit terms have been met. Should the firm wish to receive immediate payment for its factored accounts, it can borrow from the factor, using the factored accounts as collateral. The maximum loan the firm can obtain is equal to the face value of its factored accounts less the factor's fee (1 to 3 percent) less a reserve (6 to 10 percent) less the interest on the loan. For example, if $100,000 in receivables is factored, carrying 60-day credit terms, a 2 percent factor's fee, a 6 percent reserve, and interest at 1 percent per month on advances, then the maximum loan or advance the firm can receive is computed as follows:

Face amount of receivables factored	$100,000
Less: Fee (.02 × $100,000)	(2,000)
Reserve (.06 × $100,000)	(6,000)
Interest (.01 × $92,000 × 2 months)	(1,840)
Maximum advance	$ 90,160

Note that interest is discounted and calculated based upon a maximum amount of funds available for advance ($92,000 = $100,000 − $2,000 − $6,000). Thus, the effective annual cost of credit can be calculated as follows:

$$EAR = \left[1 + \frac{(\$1,840 + \$2,000)}{\$90,160}\right]^{365/60} - 1 = .2888 = 28.88\%$$

Inventory loans

Short-term loans that are secured by the pledge of inventories. The type of pledge or security agreement varies and can include floating liens, chattel mortgage agreements, field warehouse financing agreements, and terminal warehouse agreements.

Floating lien

An agreement in which the borrower gives the lender a lien against all its inventories.

Secured Sources: Inventory Loans

Inventory loans provide a second source of security for short-term secured credit. The amount of the loan that can be obtained depends on both the marketability and perishability of the inventory. Some items, such as raw materials (grains, oil, lumber, and chemicals), are excellent sources of collateral, since they can easily be liquidated. Other items, such as work in process inventories, provide very poor collateral because of their lack of marketability.

There are several methods by which inventory can be used to secure short-term financing. These include a floating or blanket lien, chattel mortgage, field warehouse receipt, and terminal warehouse receipt.

Floating Lien Agreement

Under a **floating lien** agreement the borrower gives the lender a lien against all its inventories. This provides the simplest but least secure form of inventory collateral. The borrowing firm maintains full control of the inventories and continues to sell

The Ins and Outs of Factoring

Getting In

Factors are necessarily selective in the companies they agree to take on as clients. Detailed negotiations take place with prospective clients and their financial advisers; two key items of information that the factor will require are:

- Audited accounts for the past year and the current management accounts.
- A list of the customers and a monthly turnover projection for each.

Commencement date. Once agreement has been reached a date will be arranged for the factor to assume responsibility for the sales ledger.

Introducing the service. At the start of the factoring arrangement, it is useful for the client to send a letter to customers explaining the introduction of the factor. Where appropriate, the factor may speak personally to some customers.

Invoices. Once factoring starts, clients will be encouraged to invoice their customers immediately a valid debt has been created, passing copies of invoices to the factor or, in some cases, passing the invoices to the factor for mailing. Clients must ensure that their invoicing procedure is efficient, with the terms and conditions clearly stated so that they will be understood by customers.

Factoring. With recourse and non-recourse factoring, clients may only draw on a financial facility within the credit limits agreed for each customer. These credit limits must be monitored constantly to ensure they are realistic.

Credit cover. In the case of non-recourse factoring, clients must ensure they have credit approval from the factor before delivering the goods or providing the service. They must seek credit cover for new customers before delivery, and should monitor the level of credit cover on existing customers to ensure that it is adequate.

Management accounts. During the course of the year, a factor will ask to see its client's management accounts, usually at about quarterly intervals, in order to monitor the progress of the business.

Day-to-day operation. A number of routine procedures and associated documentation will be employed in the operation of the factoring arrangement, depending on the precise system of the factor concerned. Factors will usually supply a concise set of operating instructions to guide the client.

Getting Out

Companies should not be inhibited from entering into factoring arrangements for fear that they may find it difficult to discontinue the service. Indeed they may wish to discontinue for a variety of reasons. For instance:

- Acquisition by a larger parent who does not wish to factor.
- Change in the pattern of business, resulting for example in a significant reduction in customers.
- Replacement from another source of funding at present being taken from the factor.
- Generation of increased profits rendering factoring unnecessary.

A factoring agreement will almost certainly contain a clause relating to the notice requirement—usually about three months on either side. On deciding to withdraw from factoring, clients will need to address themselves to just three principal matters:

- The availability of funds to finance the buying back of the outstanding debts from the factor.
- The setting up of a sales ledger system and the acquisition of necessary sales ledger administration, credit control and collection staff, to be effective from the agreed handing-over date.
- Credit insurance arrangements to replace the credit cover at present provided by the factor in the case of a non-recourse agreement.

Factors have no wish to hold clients to agreements if their services are no longer needed. Indeed they will provide every assistance to effect a smooth change-over when a client wishes to terminate the arrangement.

Source: Michael Maberly, "Factoring: A Catalyst for Growth and Profits," *Accountancy*, Institute of Chartered Accountants 97 (April 1986), pp. 122–124.

and replace them as it sees fit. Obviously, this lack of control over the collateral greatly dilutes the value of this type of security to the lender. Correspondingly, loans made with floating liens on inventory as collateral are generally limited to a relatively modest fraction of the value of the inventories covered by the lien. In addition, floating liens usually include future as well as existing inventories.

Chattel Mortgage Agreements

Chattel mortgage

A loan agreement in which the lender can increase the security interest by having specific items of inventory identified in the loan agreement. The borrower retains title to the inventory but cannot sell the items without the lender's consent.

The lender can increase its security interest by having specific items of inventory identified (by serial number or otherwise) in the security agreement. Such an arrangement is provided by a **chattel mortgage**. The borrower retains title to the inventory but cannot sell the items without the lender's consent. This type of agreement is costly to implement, because specific items of inventory must be identified. It is used only for major items of inventory such as machine tools or other capital assets.

Field Warehouse Financing Agreements

Field warehouse agreement

A security agreement in which inventories are used as collateral, physically separated from the firm's other inventories, and placed under the control of a third-party field warehousing firm.

Increased lender control over inventories used as loan collateral can be obtained through the use of a **field warehouse agreement.** Here the inventories used as collateral are physically separated from the firm's other inventories and placed under the control of a third-party field warehousing firm. Inventories are not removed from the borrower's premises, but are placed under the control of a third party who is responsible for protecting the security interests of the lender. This arrangement is particularly useful when large, bulky items are used as collateral. For example, a refinery might use a part of its inventory of fuel oil to secure a short-term bank loan. Under a field warehousing agreement the oil reserves would be set aside in specific tanks or storage vessels, which would be controlled (monitored) by a field warehousing concern.

The warehousing firm, upon receipt of the inventory, takes full control of the collateral. This means that the borrower is no longer allowed to use or sell the inventory items without the consent of the lender. The warehousing firm issues a warehouse receipt for the merchandise, which carries title to the goods represented therein. The receipt may be negotiable, in which case title can be transferred through sale of the receipt, or nonnegotiable, whereby title remains with the lender. In a negotiable receipt arrangement the warehouse concern will release the goods to whoever holds the receipt, whereas a nonnegotiable receipt allows the goods to be released only on the written consent of the lender.

The cost of such a loan can be quite high, since the services of the field warehouse company must be paid for by the borrower.

EXAMPLE

The M. M. Richards Company follows a practice of obtaining short-term credit based on its seasonal finished goods inventory. The firm builds up its inventories of outdoor furniture throughout the winter months for sale in spring and summer. Thus, for the two-month period ended March 31, it uses its fall and winter production of furniture as collateral for a short-term bank loan. The bank lends the company up to 70 percent of the value of the inventory at 14 percent interest plus a fixed fee of $2,000 to cover the costs of a field warehousing arrangement. During this period the firm usually has about $200,000 in inventories, which it borrows against. The annual effective annual cost of the short-term credit is therefore

$$EAR = \left[1 + \frac{(\$3{,}222 + \$2{,}000)}{\$140{,}000}\right]^{365/60} - 1 = .2495 = 24.95\%$$

where the financing cost consists of two months' interest ($\$140{,}000 \times .14 \times 60/365 = \$3{,}222$) plus the field warehousing fee of $2,000.

Terminal Warehouse Agreements

The **terminal warehouse agreement** differs from the field warehouse agreement in only one respect. Here the inventories pledged as collateral are transported to a public warehouse that is physically removed from the borrower's premises. The lender has an added degree of safety or security because the inventory is totally removed from the borrower's control. Once again the cost of this type of arrangement is increased because the warehouse firm must be paid by the borrower and the inventory must be transported to and eventually from the public warehouse.

Terminal warehouse agreement

A security agreement in which the inventories pledged as collateral are transported to a public warehouse that is physically removed from the borrower's premises. This is the safest (and a costly) form of financing secured by inventory.

The same warehouse receipt procedure described earlier for field warehouse loans is used. Again, the cost of this type of financing can be quite high.

SUMMARY

Estimating the Cost of Short-Term Credit

In estimating the cost of short-term credit, we use either the approximate cost-of-credit (RATE) equation or the effective annual rate (EAR) equation.

$$RATE = \frac{\text{interest}}{\text{principal} \times \text{time}}$$

or

$$EAR = \left(1 + \frac{i}{m}\right)^m - 1$$

where i is the nominal rate of interest per year and m is the number of compounding periods within the year.

Sources of Short-Term Credit

Three basic factors provide the key considerations in selecting a source of short-term financing: (1) the effective annual cost of credit, (2) the availability of financing in the amount and for the time needed, and (3) the effect of the use of credit from a particular source on the cost and availability of other sources of credit.

The various sources of short-term credit can be categorized into two groups: unsecured and secured. Unsecured credit offers no specific assets as security for the loan agreement. The primary sources include trade credit, lines of credit and unsecured transaction loans from commercial banks, and commercial paper. Secured credit is generally provided to business firms by commercial banks, finance companies, and factors. The most popular sources of security involve the use of accounts receivable and inventories. Loans secured by accounts receivable include pledging agreements in which a firm pledges its receivables as security for a loan, and factoring agreements in which the firm sells the receivables to a factor. A primary difference in these two arrangements relates to the ability of the lender to seek payment from the borrower in the event that the accounts used as collateral become uncollectible. In a pledging arrangement the lender retains the right of recourse in the event of default, whereas in factoring a lender is generally is without recourse.

Loans secured by inventories can be made using one of several types of security arrangements. Among the most widely used are the floating lien, chattel mortgage, field warehouse agreement, and terminal warehouse agreement. The form of agreement used will depend on the type of inventories being pledged as collateral and the degree of control the lender wishes to exercise over the loan collateral.

STUDY QUESTIONS

11-1. What distinguishes short-term, intermediate-term, and long-term debt?

11-2. What considerations should be used in selecting a source of short-term credit? Discuss each.

11-3. How can the formula "interest = principal × rate × time" be used to estimate the annual approximate cost of short-term credit?

11-4. How does compounding influence the calculation of the effective annual cost of short-term credit?

11-5. There are three major sources of unsecured short-term credit. List and discuss the distinguishing characteristics of each.

11-6. What is meant by the following trade credit terms: 2/10, net 30? 4/20, net 60? 3/15, net 45?

11-7. Define the following:

- **a.** Line of credit
- **b.** Commercial paper
- **c.** Compensating balance
- **d.** Prime rate

11-8. List and discuss four advantages of the use of commercial paper.

11-9. What "risk" is involved in the firm's use of commercial paper as a source of short-term credit? Discuss.

11-10. List and discuss the distinguishing features of the principal sources of secured credit based upon accounts receivable.

11-11. List and discuss the distinguishing features of the primary sources of secured credit based upon inventories.

SELF-TEST PROBLEMS

ST-1. *(Analyzing the Cost of a Commercial Paper Offering)* The Marilyn Sales Company is a wholesale machine tool broker which has gone through a recent expansion of its activities resulting in a doubling of its sales. The company has determined that it needs an additional $200 million in short-term funds to finance peak season sales during roughly six months of the year. Marilyn's treasurer has recommended that the firm utilize a commercial paper offering to raise the needed funds. Specifically, he has determined that a $200 million offering would require 10 percent interest (paid in advance or discounted) plus a $125,000 placement fee. The paper would carry a six-month maturity. What is the effective annual cost of credit?

ST-2. *(Analyzing the Cost of Short-Term Credit)* The Samples Mfg. Co. provides specialty steel products to the oil industry. Although the firm's business is highly correlated with the cyclical swings in oil exploration activity, it also experiences some significant seasonality. The firm is currently concerned with the seasonality in its need for funds. The firm needs $500,000 for the two-month July–August period each year, and as a result the company's vice president of finance is currently considering the following three sources of financing:

- **a.** Establish a line of credit with the Bank of Montreal. The bank has agreed to provide Samples with the needed $500,000, carrying an interest rate of 14 percent with interest discounted and a compensating balance of 20 percent of the loan balance. Samples does not have a bank account with the Bank of Montreal and would have to establish one to satisfy the compensating balance requirement.
- **b.** Samples can forgo its trade discounts over the two months of July and August when the funding will be needed. The firm's discount terms are 3/15, net 30, and the firm averages $500,000 in trade credit purchases during July and August.
- **c.** Finally, Samples could enter into a pledging arrangement with a local finance company. The finance company has agreed to extend Samples the needed $500,000, if it pledges $750,000 in receivables. The finance company has offered to advance the

$500,000, with 12 percent annual interest payable at the end of the two-month loan term. In addition, the finance company will charge a one-half of 1 percent fee based on pledged receivables to cover the cost of processing the company's accounts (this fee is paid at the end of the loan period).

Analyze the effective annual cost of each of the alternative sources of credit and select the best one. Note that a total of $500,000 will be needed for a two-month period (July–August) each year.

ST-3. *(Analyzing the Cost of Short-Term Credit)* The treasurer of the Lights-a-Lot Mfg. Company is faced with three alternative bank loans. The firm wishes to select the one that minimizes its cost of credit on a $200,000 loan that it requires in the next ten days. Relevant information for the three loan configurations is found below:

a. An 18 percent rate of interest with interest paid at year-end and no compensating balance requirement.

b. A 16 percent rate of interest but carrying a 20 percent compensating balance requirement. This loan also calls for interest to be paid at year-end.

c. A 14 percent rate of interest that is discounted plus a 20 percent compensating balance requirement.

Analyze the effective annual cost of each of these alternatives. You may assume that the firm would not normally maintain any bank balance that might be used to meet the 20 percent compensating balance requirements of alternatives (b) and (c).

STUDY PROBLEMS (SET A)

11-1A. *(Cost of Trade Credit)* Sage Construction Company purchases $480,000 in doors and windows from Crenshaw Doors under credit terms of 1/15, net 45. Assuming that Sage takes advantage of the cash discount by paying on day 15, answer the following questions:

a. What is Sage's average monthly payables balance? You may assume a 365-day year and that the accounts payable balance includes the gross amount owed (that is, no discount has been taken).

b. If Sage were to decide to pass up the cash discount and extend its payment until the end of the credit period, what would its payable balance become?

c. What is the opportunity cost of not taking the cash discount?

11-2A. *(Estimating the Cost of Bank Credit)* Paymaster Enterprises has arranged to finance its seasonal working-capital needs with a short-term bank loan. The loan will carry a rate of 12 percent per annum with interest paid in advance (discounted). In addition, Paymaster must maintain a minimum demand deposit with the bank of 10 percent of the loan balance throughout the term of the loan. If Paymaster plans to borrow $100,000 for a period of three months, what is the effective annual cost of the bank loan?

11-3A. *(Estimating the Cost of Commercial Paper)* On February 3, 199X, the Western Company plans a commercial paper issue of $20 million. The firm has never used commercial paper before but has been assured by the firm placing the issue that it will have no difficulty raising the funds. The commercial paper will carry a 270-day maturity and will require interest based upon a rate of 11 percent per annum. In addition, the firm will have to pay fees totalling $200,000 in order to bring the issue to market and place it. What is the effective annual cost of the commercial paper issue to Western?

11-4A. *(Cost of Trade Credit)* Calculate the effective annual cost of the following trade credit terms where payment is made on the net due date.

a. 2/10, net 30

b. 3/15, net 30

c. 3/15, net 45

d. 2/15, net 60

11-5A. *(Annual Percentage Rate)* Compute the cost of the trade credit terms in Problem 11-4A using the annual approximate rate method.

11-6A. *(Cost of Short-Term Financing)* The R. Morin Construction Company needs to borrow $100,000 to help finance the cost of a new $150,000 hydraulic crane used in the firm's commercial construction business. The crane will pay for itself in one year and the firm is considering the following alternatives for financing its purchase:

Alternative A—The firm's bank has agreed to lend the $100,000 at a rate of 14 percent. Interest would be discounted, and a 15 percent compensating balance would be required. However, the compensating balance requirement would not be binding on R. Morin, since the firm normally maintains a minimum demand deposit (chequing account) balance of $25,000 in the bank.

Alternative B—The equipment dealer has agreed to finance the equipment with a one-year loan. The $100,000 loan would require payment of principal and interest totalling $116,300.

a. Which alternative should R. Morin select?

b. If the bank's compensating balance requirement were to necessitate idle demand deposits equal to 20 percent of the loan, what effect would this have on the cost of the bank loan alternative?

11-7A. *(Cost of Short-Term Bank Loan)* On July 1, 199X, the Southwest Forging Corporation arranged for a line of credit with the bank. The terms of the agreement called for a $100,000 maximum loan with interest set at 1 percent over prime. In addition, the firm has to maintain a 20 percent compensating balance in its demand deposit throughout the year. The prime rate is currently 12 percent.

a. If Southwest normally maintains a $20,000-to-$30,000 balance in its chequing account with the bank, what is the effective annual cost of credit through the line-of-credit agreement where the maximum loan amount is used for a full year?

b. Recompute the effective annual cost of credit to Southwest if the firm will have to borrow the compensating balance and it borrows the maximum possible under the loan agreement. Again, assume the full amount of the loan is outstanding for a whole year.

11-8A. *(Cost of Commercial Paper)* Tri-Lite Enterprises plans to issue commercial paper for the first time in the firm's 35-year history. The firm plans to issue $500,000 in 180-day maturity notes. The paper will carry a 10.25 percent rate with discounted interest and will cost Tri-Lite $12,000 (paid in advance) to issue.

a. What is the effective annual cost of credit to Tri-Lite?

b. What other factors should the company consider in analyzing whether to issue the commercial paper?

11-9A. *(Cost of Accounts Receivable)* Johnson Enterprises Inc. is involved in the manufacture and sale of electronic components used in small AM-FM radios. The firm needs $300,000 to finance an anticipated expansion in receivables due to increased sales. Johnson's credit terms are net 60, and its average monthly credit sales are $200,000. In general, the firm's customers pay within the credit period; thus, the firm's average accounts receivable balance is $400,000.

Chuck Idol, Johnson's comptroller, approached the firm's bank with a request for a loan for the $300,000 using the firm's accounts receivable as collateral. The bank offered to make the loan at a rate of 2 percent over prime plus a 1 percent processing charge on all receivables pledged ($200,000 per month). Furthermore, the bank agreed to lend up to 75 percent of the face value of the receivables pledged.

a. Estimate the effective annual cost of the receivables loan to Johnson where the firm borrows the $300,000. The prime rate is currently 11 percent.

b. Idol also requested a line of credit for $300,000 from the bank. The bank agreed to grant the necessary line of credit at a rate of 3 percent over prime and required a 15 percent compensating balance. Johnson currently maintains an average demand deposit of $80,000. Estimate the effective annual cost of the line of credit to Johnson.

c. Which source of credit should Johnson select? Why?

11-10A. *(Cost of Factoring)* MDM Inc. is considering factoring its receivables. The firm has credit sales of $400,000 per month and has an average receivables balance of $800,000 with 60-day credit terms. The factor has offered to extend credit equal to 90 percent of the receivables factored less interest on the loan at a rate of 1.5 percent per month. The 10 percent difference in the advance and the face value of all receivables factored consists of a 1 percent factoring fee plus a 9 percent reserve, which the factor maintains. In addition, if MDM, Inc., decides to factor its receivables, it will sell them all, so that it can reduce its credit department costs by $1,500 a month.

a. What is the effective annual cost of borrowing the maximum amount of credit available to MDM Inc. through the factoring agreement?

b. What considerations other than cost should be accounted for by MDM Inc. in determining whether or not to enter the factoring agreement?

11-11A. *(Cost of Secured Short-Term Credit)* The Sean-Janeow Import Co. needs $500,000 for the three-month period ending September 30, 199X. The firm has explored two possible sources of credit.

a. S-J has arranged with its bank for a $500,000 loan secured by accounts receivable. The bank has agreed to advance S-J 80 percent of the value of its pledged receivables at a rate of 11 percent plus a 1 percent fee based on all receivables pledged. S-J's receivables average a total of $1 million year-round.

b. An insurance company has agreed to lend the $500,000 at a rate of 9 percent per annum, using a loan secured by S-J's inventory of salad oil. A field warehouse agreement would be used, which would cost S-J $2,000 a month.

Which source of credit should S-J select? Explain.

11-12A. *(Cost of Short-Term Financing)* You plan to borrow $20,000 from the bank to pay for inventories for a gift shop you have just opened. The bank offers to lend you the money at 10 percent annual interest for the six months the funds will be needed.

a. Calculate the effective annual rate of interest on the loan.

b. In addition, the bank requires you to maintain a 15 percent compensating balance in the bank. Since you are just opening your business, you do not have a demand deposit at the bank that can be used to meet the compensating balance requirement. This means that you will have to put 15 percent of the loan amount from your own personal money (which you had planned to use to help finance the business) in a chequing account. What is the effective annual cost of the loan now?

c. In addition to the compensating balance requirement in (b), you are told that interest will be discounted. What is the effective annual rate of interest on the loan now?

11-13A. *(Cost of Factoring)* A factor has agreed to lend the JVC Corporation working capital on the following terms. JVC's receivables average $100,000 per month and have a 90-day average collection period (note that accounts receivable average $300,000 because of the 90-day average collection period). The factor will charge 12 percent interest on any advance (1 percent per month paid in advance), will charge a 2 percent processing fee on all receivables factored, and will maintain a 20 percent reserve. If JVC undertakes the loan it will reduce its own credit department expenses by $2,000 per month. What is the effective annual rate of interest to JVC on the factoring arrangement? Assume that the maximum advance is taken.

STUDY PROBLEMS (SET B)

11-1B. *(Cost of Trade Credit)* Clearwater Construction Company purchases $600,000 in parts and supplies under credit terms of 2/30, net 60. Assuming that Clearwater takes advantage of the cash discount by paying on day 30, answer the following questions:

a. What is Clearwater's average monthly payables balance? You may assume a 365-day year and that the accounts payable balance includes the gross amount owed (that is, no discount has been taken).

b. If Clearwater were to decide to pass up the cash discount and extend its payment until the end of the credit period, what would its payable balance become?

c. What is the opportunity cost of not taking the cash discount?

11-2B. *(Estimating the Cost of Bank Credit)* Dee's Christmas Trees Inc. is evaluating options for financing its seasonal working-capital needs. A short-term loan from a bank would carry a 14 percent annual interest rate, with interest paid in advance (discounted). If this option is chosen, Dee's would also have to maintain a minimum demand deposit equal to 10 percent of the loan balance, throughout the term of the loan. If Dee's needs to borrow $125,000 for the upcoming three months before Christmas, what is the effective annual cost of the loan?

11-3B. *(Estimating the Cost of Commercial Paper)* Duro Auto Parts would like to exploit a production opportunity overseas, and is seeking additional capital to finance this expansion. The company plans a commercial paper issue of $15 million on February 3, 199X. The firm has never issued commercial paper before, but has been assured by the investment dealer placing the issue that it will have no difficulty raising the funds, and that this method of financing is the least expensive option, even after the $150,000 placement fee. The issue will carry a 270-day maturity and will require interest based on an annual rate of 12 percent. What is the effective annual cost of the commercial paper issue to Duro?

11-4B. *(Cost of Trade Credit)* Calculate the effective annual cost of the following trade credit terms where payment is made on the net due date.

a. 2/10, net 30

b. 3/15, net 30

c. 3/15, net 45

d. 2/15, net 60

11-5B. *(Annual Percentage Rate)* Compute the cost of the trade credit terms in Problem 11-4B using the annual approximate rate method.

11-6B. *(Cost of Short-Term Financing)* Vitra Glass Company needs to borrow $150,000 to help finance the cost of a new $225,000 kiln to be used in the production of glass bottles. The kiln will pay for itself in one year and the firm is considering the following alternatives for financing its purchase:

Alternative A—The firm's bank has agreed to lend the $150,000 at a rate of 15 percent. Interest would be discounted, and a 16 percent compensating balance would be required. However, the compensating balance requirement would not be binding on Vitra, since the firm normally maintains a minimum demand deposit (chequing account) balance of $25,000 in the bank.

Alternative B—The kiln dealer has agreed to finance the equipment with a one-year loan. The $150,000 loan would require payment of principal and interest totaling $163,000.

a. Which alternative should Vitra select?

b. If the bank's compensating balance requirement were to necessitate idle demand deposits equal to 20 percent of the loan, what effect would this have on the cost of the bank loan alternative?

11-7B. *(Cost of Short-Term Bank Loan)* Lola's Ice Cream recently arranged for a line of credit with the bank. The terms of the agreement called for a $100,000 maximum loan with interest set at 2 percent over prime. In addition, Lola's must maintain a 15 percent compensating balance in its demand deposit throughout the year. The prime rate is currently 12 percent.

a. If Lola's normally maintains a $15,000-to-$25,000 balance in its chequing account with the bank, what is the effective annual cost of credit through the line-of-credit agreement where the maximum loan amount is used for a full year?

b. Recompute the effective annual cost of credit to Lola's Ice Cream if the firm has to borrow the compensating balance and it borrows the maximum possible under the loan agreement. Again, assume the full amount of the loan is outstanding for a whole year.

11-8B. *(Cost of Commercial Paper)* Luft Inc. recently acquired production rights to an innovative sailboard design but needs funds to pay for the first production run, which is expected to sell briskly. The firm plans to issue $450,000 in 180-day maturity notes. The paper will carry an 11 percent rate with discounted interest and will cost Luft $13,000 (paid in advance) to issue.

a. What is the effective annual cost of credit to Luft?

b. What other factors should the company consider in analyzing whether to issue the commercial paper?

11-9B. *(Cost of Accounts Receivable)* TLC Enterprises Inc. is a wholesaler of toys and curios. The firm needs $400,000 to finance an anticipated expansion in receivables due to increased sales. TLC's credit terms are net 60, and its average monthly credit sales are $250,000. In general, the TLC's customers pay within the credit period; thus, the firm's average accounts receivable balance is $500,000.

Kelly Leaky, TLC's comptroller, approached the firm's bank with a request for a loan for the $400,000 using the firm's accounts receivable as collateral. The bank offered to make the loan at a rate of 2 percent over prime plus a 1 percent processing charge on all receivables pledged ($250,000 per month). Furthermore, the bank agreed to lend up to 75 percent of the face value of the receivables pledged.

a. Estimate the effective annual cost of the receivables loan to TLC where the firm borrows the $400,000. The prime rate is currently 11 percent.

b. Leaky also requested a line of credit for $400,000 from the bank. The bank agreed to grant the necessary line of credit at a rate of 3 percent over prime and required a 15 percent compensating balance. TLC currently maintains an average demand deposit of $100,000. Estimate the effective annual cost of the line of credit to TLC.

c. Which source of credit should TLC select? Why?

11-10B. *(Cost of Factoring)* To increase profitability, a management consultant has suggested to the Dal Molle Fruit Company that it consider factoring its receivables. The firm has credit sales of $300,000 per month and has an average receivables balance of $600,000 with 60-day credit terms. The factor has offered to extend credit equal to 90 percent of the receivables factored less interest on the loan at a rate of 1.5 percent per month. The 10 percent difference in the advance and the face value of all receivables factored consists of a 1 percent factoring fee plus a 9 percent reserve, which the factor maintains. In addition, if Dal Molle decides to factor its receivables, it will sell them all, so that it can reduce its credit department costs by $1,400 a month.

a. What is the effective annual cost of borrowing the maximum amount of credit available to Dal Molle, through the factoring agreement?

b. What considerations other than cost should be accounted for by Dal Molle, in determining whether or not to enter the factoring agreement?

11-11B. *(Cost of Secured Short-Term Credit)* DST Inc., a producer of inflatable river rafts, needs $400,000 for the three-month summer season, ending September 30, 199X. The firm has explored two possible sources of credit.

a. DST has arranged with its bank for a $400,000 loan secured by accounts receivable. The bank has agreed to advance DST 80 percent of the value of its pledged receivables at a rate of 11 percent plus a 1 percent fee based on all receivables pledged. DST's receivables average a total of $1 million year-round.

b. An insurance company has agreed to lend the $400,000 at a rate of 10 percent per annum, using a loan secured by DST's inventory. A field warehouse agreement would be used, which would cost DST $1,500 a month.

Which source of credit should DST select? Explain.

11-12B. *(Cost of Secured Short-Term Financing)* You are considering a loan of $25,000 to finance inventories for a janitorial supply store that you plan to open. The bank offers to lend you the money at 11 percent annual interest for the six months the funds will be needed.

a. Calculate the effective annual rate of interest on the loan.

b. In addition, the bank requires you to maintain a 15 percent compensating balance in the bank. Since you are just opening your business, you do not have a demand deposit at the bank that can be used to meet the compensating balance requirement. This means that you will have to put 15 percent of the loan amount from your own personal money (which you had planned to use to help finance the business) in a chequing account. What is the effective annual cost of the loan now?

c. In addition to the compensating balance requirement in (b), you are told that interest will be discounted. What is the effective annual rate of interest on the loan now?

11-13B. *(Cost of Financing)* Tanglewood Roofing Supply Inc. has agreed to borrow working capital from a factor on the following terms: Tanglewood's receivables average $150,000 per month and have a 90-day average collection period (note that accounts receivable average $450,000 because of the 90-day average collection period). The factor will charge 13 percent interest on any advance (1 percent per month paid in advance), will charge a 2 percent processing fee on all receivables factored, and will maintain a 15 percent reserve. If Tanglewood undertakes the loan it will reduce its own credit department expenses by $2,000 per month. What is the effective annual rate of interest to Tanglewood on the factoring arrangement? Assume that the maximum advance is taken.

SELF-TEST SOLUTIONS

SS-1. The discounted interest cost of the commercial paper issue is calculated as follows:

$$\text{Interest expense} = .10 \times \$200 \text{ million} \times \frac{6}{12} = \$10 \text{ million}$$

The effective annual cost of credit can now be calculated as follows:

$$EAR = \left(1 + \frac{\$10{,}000{,}000 + \$125{,}000}{\$200{,}000{,}000 - \$125{,}000 - \$10{,}000{,}000}\right)^{12/6} - 1 = 0.1095 = 10.95\%$$

SS-2. a. Interest for two months = .14 × .16667 × \$500,000 = \$11,667

Loan proceeds (for \$500,000 loan) = \$500,000 − (.2 × \$500,000 + \$11,667)
= \$388,333

$$EAR = \left(1 + \frac{\$11{,}667}{\$388{,}333}\right)^{12/2} - 1 = 0.1944 = 19.44\%$$

Note that Samples would actually have to borrow more than the needed \$500,000 in order to cover the compensating balance requirement. However, as we demonstrated earlier, the effective annual cost of credit will not be affected by adjusting the loan amount for interest expense changes accordingly.

b. The estimation of the cost of forgoing trade discounts is generally quite straightforward; however, in this case the firm actually stretches its trade credit for purchases made during July beyond the due date by an additional 30 days. If it is able to do this without penalty, then the firm effectively forgoes a 3 percent discount for not paying within 15 days and does not pay for an additional 45 days (60 days less the discount period of 15 days). Thus, for the July trade credit, Samples' cost is calculated as follows:

$$EAR = \left(1 + \frac{0.03}{0.97}\right)^{365/45} - 1 = 0.2803 = 28.03\%$$

However, for the August trade credit the firm actually pays at the end of the credit period (the 30th day), so that the cost of trade credit becomes

$$EAR = \left(1 + \frac{0.03}{0.97}\right)^{365/15} - 1 = 1.0984 = 109.84\%$$

c. Interest for two months = $.12 \times \frac{2}{12} \times \$500{,}000 = \$10{,}000$

Pledging fee = .005 × \$750,000 = \$3,750

$$EAR = \left(1 + \frac{\$10{,}000 + \$3{,}750}{\$500{,}000}\right)^{12/2} - 1 = .1768 = 17.68\%$$

SS-3. a.

$$EAR = \left[1 + \frac{(.18 \times \$200{,}000)}{\$200{,}000}\right]^{1} - 1 = .18 = 18.00\%$$

b.

$$EAR = \left[1 + \frac{(.16 \times \$250{,}000)}{\$250{,}000 - (.2 \times \$250{,}000)}\right]^{1} - 1 = .20 = 20.00\%$$

c.

$$EAR = \left[1 + \frac{(.14 \times \$303{,}030.30)}{\$303{,}030.30 - (.14 \times \$303{,}030.30) - (.20 \times \$303{,}030.30)}\right]^{1} - 1$$
$$= .2121 = 21.21\%$$

Alternative (a) offers the lower-cost service of financing, although it carries the highest stated rate of interest. The reason for this, of course, is that there is no compensating balance requirement nor is interest discounted for this alternative.

COST OF CAPITAL

CHAPTER 12

LEARNING OBJECTIVES

After reading this chapter you should be able to

1. Explain the concept and purpose of determining a firm's cost of capital.
2. Identify the factors that determine a company's cost of capital.
3. Describe the assumptions made in computing a firm's *weighted average cost of capital*.
4. Calculate a corporation's weighted cost of capital.
5. Explain how PepsiCo calculates its cost of capital.
6. Compute the cost of capital for an individual project when the firm's weighted cost of capital is not appropriate as the discount rate.

INTRODUCTION

In its 1995 Annual Report, *Cambridge Shopping Centres Ltd. outlines the guidelines that it uses to accept acquisition, development, and investment opportunities. These guidelines stipulate that (1) the gross cash flow during the first year of operation be equal to the associated borrowing cost and (2) the internal rate of return of the project be in excess of the company's blended (weighted) cost of capital.*[1] *The comparison of a project's internal rate of return to the company's cost of capital implies that management is incorporating into the cost of capital an assessment of risk which the project's return must compensate for if the project is to be accepted. Cambridge Shopping Centres Ltd. has indicated that the major risk associated with acquisition and development projects is the uncertainty in realizing the forecasted revenues and costs of the projects. This company has identified some of the important components of this uncertainty and formulated guidelines to ensure a project's success. For example, the company indicates that a project will not proceed until commitments are secured from anchor tenants, land acquisition agreements have been completed, a fixed-contract construction is in place, and suitable financing has been arranged.*

In this chapter, we examine how a firm computes the cost of capital in order to accept or reject potential investment opportunities.

[1]Cambridge Shopping Centres Ltd., *Annual Report* (1995), p. 36.

CHAPTER PREVIEW

Having studied risk and rates of return (Chapter 3), the valuation of bonds and common shares (Chapter 5 and Chapter 6), and the various sources of financing, we now examine the cost of capital. As shown in Figure 12-1, the concepts presented in this chapter provide an important link from the firm's financing decision to its investment decision.

The term *cost of capital* may be used interchangeably with the firm's required rate of return, the hurdle rate, the discount rate, and the firm's opportunity cost of funds.

In this chapter, we examine the concepts used in describing a firm's cost of capital, as well as the procedures used to estimate it. These procedures are then extended to determine the cost of capital for a multi-divisional firm, PepsiCo. Finally, we examine how the required rate of return can be used to evaluate individual projects.

OBJECTIVE 1

THE CONCEPT OF THE COST OF CAPITAL

Cost of capital

The cost of capital is the opportunity cost of funds of the firm's investors. The cost of capital is the rate of return that could be earned elsewhere in the capital markets with a similar amount of risk. Thus it is the rate of return the firm must earn on its investments in order to satisfy the required rate of return of the firm's investors.

The **cost of capital** is the opportunity cost of using funds to invest in new projects. In other words, the cost of capital is that rate of return on the firm's total investment which earns the required rates of return needed to satisfy all the sources of long-term financing. If the firm earns exactly its cost of capital on an investment project, we expect the price of its common shares to remain unchanged following the acceptance of the project. By the same reasoning, if the firm earns a rate of return higher than the cost of capital then the excess return will lead to an increase in value of the firm's common shares and consequently, an increase in shareholder wealth.

To illustrate the calculation of the firm's weighted cost of capital, consider the capital structure of the Salinas Corporation found in Table 12-1. The company has three sources of capital: debt, preferred shares, and common shares. Management is considering a $200,000 investment opportunity with an expected internal rate of

Figure 12-1 Cost of Capital: Connecting Investment and Financing Decisions

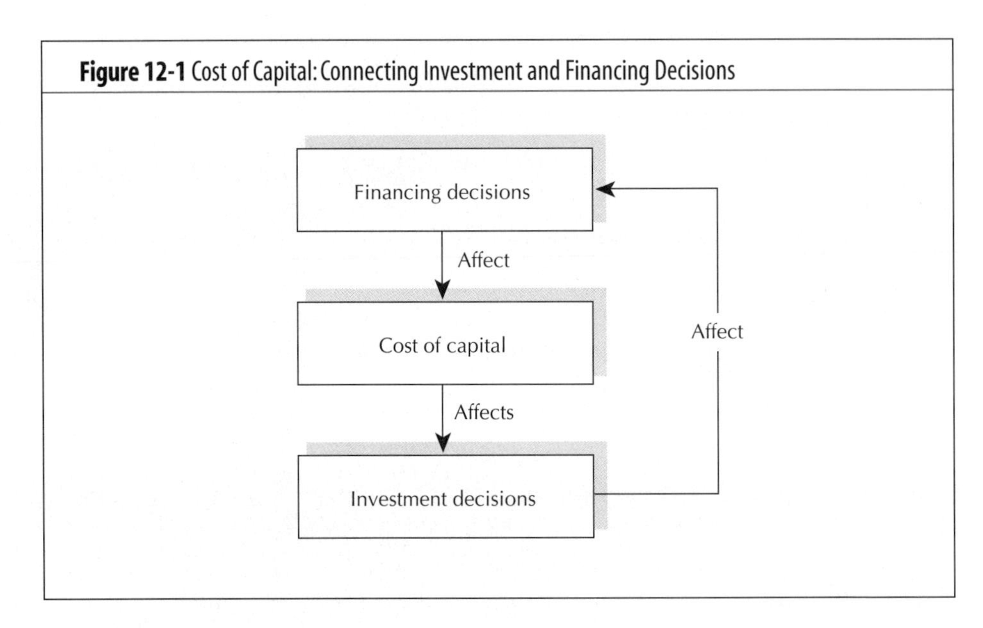

Table 12-1 Salinas Corporation Capital Structure

	Amount	Percentage of Capital Structure
Bonds (18% interest rate)	$ 600,000	30%
Preferred shares (10% dividend rate)	200,000	10%
Common shares	1,200,000	60%
Total liabilities and equity	$2,000,000	100%

return of 15 percent. At the moment, the firm's cost of capital (that is, its required rate of return) for each source of financing is as follows:[2]

	After tax
Cost of debt capital	10%
Cost of preferred shares	12
Cost of common shares	16

Given this information, should the firm make the investment? The creditors and preferred shareholders would probably encourage us to undertake the project. However, since the 15 percent internal rate of return on the investment is less than the common shareholders' required rate of return, shareholders might argue that the investment should be forgone. What decision should the financial manager make?

To answer this question, we must know what percentage of the $200,000 is to be provided by each type of investor. If we intend to maintain the same capital structure mix as reflected in Table 12-1 (30 percent debt, 10 percent preferred shares, and 60 percent common shares), we could compute a weighted cost of capital, where the weights equal the percentage of capital to be financed by each source. Table 12-2 shows that the weighted cost of the individual sources of capital is 13.8 percent. From this calculation we would conclude that an investment offering at least a 13.8 percent return would be acceptable to the company's investors. The investment should be undertaken, since the 15 percent rate of return more than satisfies all investors as indicated by a 13.8 percent weighted cost of capital. Again, the **weighted cost of capital** for a firm is equal to the cost of each source of financ-

Weighted cost of capital

A weighted average of the individual costs of financing, where the weights equal the percentage of financing from each source of financing.

Table 12-2 Salinas Corporation Weighted Cost of Capital

	Weights (Percentage of Financing)	After-tax Cost of Individual Sources	Weighted Cost
Debt	30%	10%	3.0%
Preferred shares	10	12	1.2%
Common shares	60	16	9.6%
	100%	Weighted cost of capital:	13.8%

[2]The underlying theme of the cost of capital tells us that investment and financing decisions should be based on the investor's marginal cost of capital and not on some criterion arbitrarily selected by management. By using the marginal cost of capital which has been determined in the financial markets, management imposes market discipline on its decisions. In addition. this type of approach ensures that market or intrinsic value is added to shareholder wealth.

ing (debt, preferred shares, and common shares) multiplied by the percentage of the financing provided by that source.

In summary, two basic elements are necessary to calculate the weighted cost of capital:

1. Estimates of the required rates of return for each of the firm's sources of capital, and
2. The proportion of each source of capital used by the firm.

In this chapter, we will concern ourselves with determining the first of these elements and take the second element as given.

The Importance of Using a Weighted Cost of Capital

The weighted cost of capital may be fine in theory, but what if a company could borrow the entire amount needed for an investment in a new product line? Is it really necessary to use the weighted cost of capital, or would it be all right to make the decision based simply on the cost of the debt which provides the funding?

To answer these questions, we examine the following example. Management of the Poling Corporation believes it could earn 14 percent from purchasing $500,000 in new equipment, which would allow the firm to expand its business. Although the firm endeavors to maintain a capital structure with equal amounts of debt and equity, the bank is willing to loan the firm the entire $500,000 at an interest rate of 12 percent. The financial manager knows that the firm's earnings per share will increase if the firm earns a rate of return which exceeds the cost of financing or the cost of debt at 12 percent.

Poling's financial manager has also estimated the firm's cost of equity at 18 percent. However, management has decided to make the investment and finance it by borrowing the money from the bank at 12 percent.

The following year, management finds another investment opportunity costing $500,000, but with an expected internal rate of return of 17 percent which is better than the previous year's rate of return of 14 percent. However, this time the bank is unwilling to lend Poling any more money. The loan officer at the bank indicates that Poling has used up all of its debt capacity. The firm must now issue new common shares before the bank will be agreeable to fund any more loans. As a result, management decides to reject the investment since it does not earn the cost of equity of 18 percent.

What is the moral of the Poling example? Intuitively, we can see that Poling's management has made a mistake. The investment made by the firm in the first year has caused the firm to lose an opportunity to make a better investment decision in the second year.

We can conclude that a firm should never use a single cost of financing as the hurdle rate (discount rate) for making investment decisions. In this example the use of debt has caused the firm to implicitly use up some of its debt capacity for future investments and until the firm can complement the use of debt with equity, it will not be able to use more debt in the future.[3] Thus, financial managers should always use the weighted cost of capital and not an individual cost of funds, as the discount rate for investment decisions. We now examine in greater detail the weighted cost of capital.

[3]By financing entirely by debt, the firm has used some of its debt capacity, thereby denying the use of debt at some time in the future. Also, as more debt is used, the common shareholders require a higher rate of return because of the increase in financial risk to their investment. We have discussed the effects of financial risk on return in Chapter 8.

> **PERSPECTIVE IN FINANCE**
>
> *When we compute a firm's weighted cost of capital, we are simply calculating the average cost of money supplied by all investors, those being creditors and shareholders.*

FACTORS DETERMINING THE WEIGHTED COST OF CAPITAL

OBJECTIVE 2

To gain further insight into the meaning of the firm's cost of capital, we need to consider the elements in the business environment that cause a company's weighted cost of capital to be high or low.[4] Looking at Figure 12-2, we see four primary factors: general economic conditions, the marketability of the firm's securities (market conditions), operating and financing conditions within the company, and the amount of financing needed for new investments. These four variables also relate to our discussion in Chapter 3, where we separated an investor's required rate of return into risk-free rate of return and the risk premium. These two aspects of risk are also key ingredients in the firm's cost of capital.

> **PERSPECTIVE IN FINANCE**
>
> *The cost of capital for a firm is driven by the demand for and supply of money in the economy and the riskiness of the firm.*

Factor 1: General Economic Conditions

As discussed in Chapter 2, general economic conditions determine the demand for and supply of capital within the economy, as well as the level of expected inflation. This economic variable is reflected in the riskless rate of return. This rate represents the rate of return on risk-free investments, such as the interest rate on short-term Government of Canada Treasury bills. In principle, as the demand for money in the economy changes relative to the supply, investors alter their required rate of return.[5] For example, if the demand for money increases without an equivalent increase in the supply, lenders will raise their required interest rate. At the same time, if inflation is expected to deteriorate the purchasing power of the dollar, investors require a higher rate of return to compensate for this anticipated loss.[6]

Factor 2: Market Conditions

When an investor purchases a security with significant investment risk, the security must offer the investor an opportunity to earn additional returns to make the investment attractive. Essentially, as risk increases, the investor requires a higher rate of return. This increase is called a **risk premium**. An increase in the investor's required rate of return causes the firm's cost of capital to increase. Remember, we have defined risk as the potential variability of future returns. If the security is not readily marketable when the investor wants to sell, or even if a continuous demand

Risk premium

The additional return expected for assuming risk.

[4]There is an inverse relationship between value and the cost of capital, given that all other conditions remain the same. For example, a decrease in the cost of capital or discount rate which is used to find the present value, causes the value of the cash flow stream to increase.

[5]The cost of capital is, in part, dependent on the demand and supply of money in the economy. The cost of capital is higher in Canada, where the supply of funds is less, than in other countries such as Japan, where the savings rate is higher.

[6]This effect was referenced earlier in Chapter 2 as the Fisher effect.

Figure 12-2 Primary Factors Influencing the Cost of a Particular Source of Capital

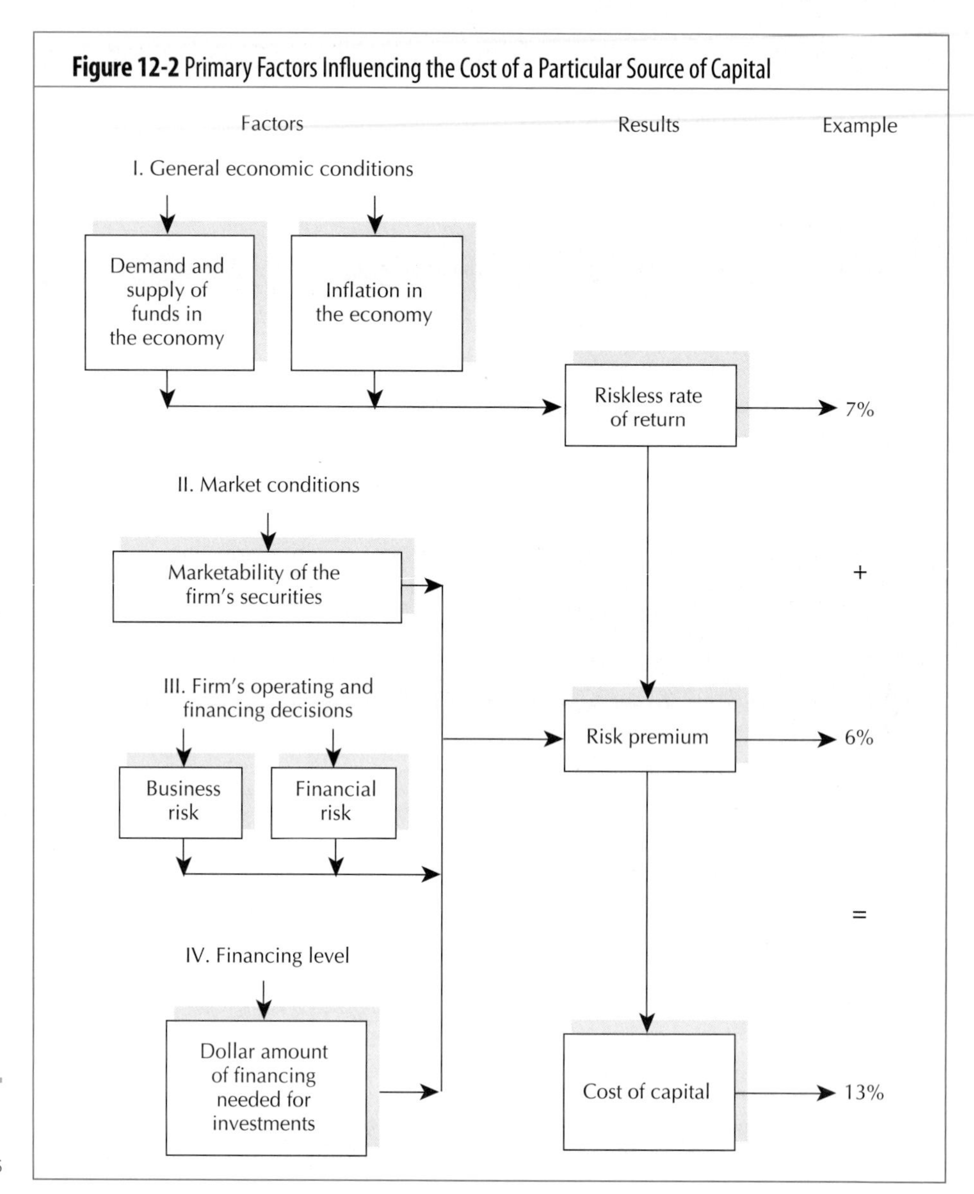

Business risk

The relative dispersion or variability in the firm's expected earnings before interest and taxes (EBIT). The nature of the firm's operations causes its business risk, especially the mix between fixed operating costs and variable costs.

Financial risk

The added variability in earnings available to the firm's common shareholders, and the added chance of insolvency caused by the use of securities, such as debt, that require a fixed payment to the investors regardless of the company's earnings.

for the security exists but the price varies significantly, an investor will require a relatively high rate of return. On the other hand, if a security is readily marketable and its price is reasonably stable, the investor will require a lower rate of return and the company's cost of capital will be lower.

Factor 3: Firm's Operating and Financing Decisions

Risk, or the variability of returns, also results from decisions made within the company. Risk resulting from these decisions is generally divided into two types: business risk and financial risk. **Business risk** is the variability in returns on assets and is affected by the company's investment decisions. **Financial risk** is the increased variability in returns to common shareholders as a result of financing with debt or preferred shares.[7] As business risk and financial risk increase or decrease, the

[7]During the 1980s, some North American firms issued large amounts of debt to finance growth through leveraged buyouts. Consequently, the financial risk of these companies was increased dramatically. However, in computing the firm's weighted cost of capital, we assume that management does make such changes in the firm's capital structure.

investor's required rate of return (and the cost of capital) will move in the same direction.[8]

Factor 4: Amount of Financing

The last factor determining the firm's cost of funds is the level of financing the firm requires. As the financing requirements of the firm become larger, the weighted cost of capital increases for several reasons. For instance, as more securities are issued, additional **flotation costs**, or the cost incurred by the firm from issuing securities, will affect the percentage cost of the funds to the firm. Also, as management approaches the market for large amounts of capital relative to the firm's size, the investors' required rate of return may rise. Suppliers of capital become hesitant to grant relatively large sums without evidence of management's capability to absorb this capital into the business. This is typically "too much too soon." Also, as the size of the issue increases, there is greater difficulty in placing it in the market without reducing the price of the security, which also increases the firm's cost of capital.

Flotation costs

The underwriter's spread and issuing costs associated with the issuance and marketing of new securities.

A Summary Illustration

To summarize, the important variables influencing a corporation's cost of capital include the following:

1. **General economic conditions.** This factor determines the risk-free rate or riskless rate of return.
2. **Marketability of a company's securities.** As the marketability of a security increases, investors' required rates of return decrease, lowering the corporation's cost of capital.
3. **Operating and financial decisions made by management.** If management accepts investments with high levels of risk or if it uses debt or preferred shares extensively, the firm's risk increases. Investors then require a higher rate of return, which causes a higher cost of capital to the company.
4. **Amount of financing needed.** Requests for larger amounts of capital increase the firm's cost of capital.

Figure 12-2 illustrates the cost of capital for a particular source. The risk-free rate, determined by the general economic conditions, is 7 percent. However, owing to the additional risks associated with the security, the firm has to earn an additional 6 percent to satisfy the investors' required rate of return of 13 percent.

ASSUMPTIONS OF THE WEIGHTED COST OF CAPITAL MODEL

OBJECTIVE 3

In a complex business world, difficulties quickly arise in computing a corporation's cost of capital. For this reason, we make several simplifying assumptions.

Constant Business Risk

Business risk is defined as the potential variability of future returns on an investment. The level of business risk within a firm is determined by management's investment policies. As a result, both the investor's required rate of return on a firm's securities and the firm's cost of capital is a function of the firm's current business risk. If this risk level is altered, the corporation's investors will naturally change their required rates of return, which in turn modifies the cost of capital. However, the amount of change in the cost of capital resulting from a given

[8]Both forms of risk, business and financial, were explained and illustrated in Chapter 8.

increase or decrease in business risk is difficult to assess. For this reason, the cost of capital calculation assumes that any investment under consideration will not significantly change the firm's business risk. In other words, *the cost of capital is an appropriate investment criterion only for an investment having a business risk level similar to that of existing assets.*

Constant Financial Risk

Financial risk has been defined as the increased variability in returns on common shares resulting from the increased use of debt and preferred shares financing.[9] Also, financial risk relates to the threat of bankruptcy. As the percentage of debt in the capital structure increases, the possibility that the firm will be unable to pay interest and the principal balance is also increased. As a result, the level of financial risk in a company has an impact upon the investors' required rate of return. As the amount of debt rises, the common shareholders will increase their required rate of return. In other words, *the costs of individual sources of capital are a function of the current financial structure.* For this reason, the data used in computing the cost of capital are appropriate only if management continues to use the same financial mix. If the present capital structure consists of 40 percent debt, 10 percent preferred shares, and 50 percent common shares, this capital structure is assumed to be maintained in the financing of future investments.

Constant Dividend Policy

A third assumption required in estimating the cost of capital relates to the corporation's dividend policy. For ease of computation, we generally assume that a firm's dividends are increasing at a constant annual growth rate. We also assume that this growth is a function of the firm's earning capabilities and not merely the result of paying out a larger percentage of the company's earnings. Thus, it is implicitly assumed that the dividend payout ratio (dividends/net income) is constant.

The assumptions of the weighted cost of capital model are quite restrictive. In practice, the financial manager may need a range of possible cost of capital values rather than a single-point estimate in order to perform an investment analysis. For example, it may be more appropriate to talk in terms of a 10 to 12 percent range as an estimate of the firm's cost of capital, rather than assuming that a precise number can be determined. In this chapter, however, our principal concern will be with calculating a single cost of capital figure.

PERSPECTIVE IN FINANCE

To compute a firm's weighted cost of capital, we must assume that the firm's financial mix will not change, that we will continue to invest in projects of about the same risk as we have done in the past, and that we will not change the percentage of earnings paid out in dividends to the shareholders.

OBJECTIVE 4

COMPUTING THE WEIGHTED COST OF CAPITAL

A firm's weighted cost of capital is a composite of the individual costs of financing, weighted by the percentage of financing provided by each source. Therefore, a firm's weighted cost of capital is a function of (1) the individual costs of capital and (2) the makeup of the capital structure—the percentage of funds provided by debt,

[9] This concept is further explained in Chapter 8.

preferred shares, and common shares.[10] Also, as we noted earlier, the amount of funds needed affects the cost of capital. We will discuss this last consideration, the level of financing, later.

As we explain the procedures for computing a company's cost of capital, it is helpful to remember three basic steps:

1. Calculate the costs of capital for each individual source of financing used by the firm; these costs generally include debt, preferred shares, and common shares.
2. Determine the percentage of each source of financing in terms of their market value.
3. Using the individual costs and the capital structure percentages in the first two steps, compute an overall or weighted cost of capital for the various amounts of financing that might be needed.

The computations are not difficult if we understand our purpose: We want to calculate the firm's overall cost of capital. As a simple exercise, calculate the average age of students in a course where 40 percent are 19 years old, 50 percent are 20 years old, and 10 percent are 21 years old. We can easily find the average age to be 19.7 years by weighting each age by the percentage in each age category [(40%)(19) + (50%)(20) + (10%)(21)]. In a similar way, the weighted cost of capital is estimated by weighing the cost of each individual source by the percentage of financing it provides. If we finance an investment by 40 percent debt at a 10 percent cost and 60 percent common equity at a cost of 18 percent, the weighted cost of capital is 14.8 percent [(.40)(10%) + (.60)(18%) = 14.8%]. Thus, while the details become somewhat involved, the basic approach is relatively simple. Figure 12-3 summarizes the steps of this approach.

Determining Individual Costs of Capital

Companies have created a large variety of financing instruments in an attempt to attract new investors. However, our interest will be restricted to the three basic types of securities: debt, preferred shares, and common shares. In calculating their respective costs, the objective is to determine the rate of return the company must earn on its investments to satisfy investors' required rates of return after allowing for any issuing costs incurred in raising new funds. Also, since the cash flows used in capital-budgeting analysis (net present value, profitability index, and internal rate of return) are on an after-tax basis, the required rates of return should also be expressed on an after-tax basis.

Cost of Debt

The **cost of debt** may be defined as the rate that must be received from an investment in order to achieve the required rate of return for the creditors. In Chapter 5 the required rate of return for debt capital was found by a trial-and-error process or with the use of a financial calculator, where we solved for k_b in the following equation:

Cost of debt

The rate that has to be received from an investment in order to achieve the required rate of return for the creditors. The cost is based on the debtholders' opportunity cost of funds in the capital markets.

Figure 12-3 Computing the Weighted Cost of Capital: The Basic Steps

Remember: The firm's weighted cost of capital is calculated as follows:

1. Compute the cost of capital for each and every source of financing, i.e., each source of debt, preferred shares, and common equity.
2. Determine the percentage of each source of financing in terms of their market value.
3. Calculate a weighted average of the costs of capital using the percentage of each source as the weights.

[10]Examples of how a firm can lower its cost of capital include: offering incentives to customers to prepay for products that will be needed at a later date; using government subsidies from foreign countries relating to the transfer of technology; and using joint ventures in foreign countries to lower the cost of equity.

$$V_b = \frac{\$I_1}{(1+k_b)^1} + \frac{\$I_2}{(1+k_b)^2} + \ldots + \frac{\$I_n}{(1+k_b)^n} + \frac{\$M}{(1+k_b)^n} \qquad (12\text{-}1)$$

If we use the present value tables, the equation would be restated as follows:

$$V_b = \$I_t[PVIFA_{k_b,n}] + \$M[PVIF_{k_b,n}]$$

where V_b = the market price of the debt
$\$I_t$ = the annual dollar interest paid to the investor
$\$M$ = the maturity value of the debt
k_b = the required rate of return of the debt holder
n = the number of years to maturity

BACK TO THE BASICS

When we calculate the bondholder's required rate of return we rely on the observed market price of the firm's bonds to be an accurate reflection of their worth given all available information about the riskiness of those bonds. If the bond price did not fully reflect all available information then our calculated required rate of return would not be an accurate reflection of the bondholder's opportunity cost of funds. When we accept the observed market price of a firm's bonds (or other securities) in calculating required returns, it is captured in ***Axiom 6: Efficient Capital Markets—The Markets Are Quick and Prices Are Right.*** *What we mean here, very simply, is that investors are ever vigilant and quickly act on information that affects the riskiness and consequently the price of a firm's bonds and other securities.*

Assume that an investor is willing to pay $908.32 for an annual coupon bond. The security pays a nominal interest rate of 8 percent on a par value of $1,000 and matures in 20 years. By trial and error, the investor's required rate of return is found to be 9 percent, which is the rate that sets the present value of the future interest payments and the maturity value equal to the price of the bond, or

$$\$908.32 = \sum_{t=1}^{20} \frac{\$80}{(1+.09)^t} + \frac{\$1{,}000}{(1+.09)^{20}}$$

However, if both brokerage commissions and legal or accounting fees are incurred in issuing the security, the company will not receive the full $908.32 market price. Both issuing expenses and interest expenses are considered by Revenue Canada as tax-deductible expenses and can be written off as deductions against taxable income. As a result, the actual cost of these funds to the firm will be less than the investor's 9 percent required rate of return. In our example, suppose that management has determined that issuing expenses per $1,000 bond is $58.32. If the combined corporate tax rate is 50 percent for this firm, the after-tax cost of the issuing expenses is (1 – .5)($58.32) or $29.16. If the after-tax issuing costs are deducted from the price of the bond, the net price after-tax of this bond would be ($908.32 – $29.16) or $879.16. In addition, interest expenses are tax deductible to the firm. The firm pays (1 – .5)(8 percent) or 4 percent annual interest on a security that has $1,000 par value and matures in 20 years. As a result, the after-tax interest expense to the firm is $40. The net effect of these changes on Equation (12-1) is as follows:

$$NP_0 = \sum_{t=1}^{n} \frac{\$I_t(1-T)}{(1+K_b)^t} + \frac{\$M}{(1+K_b)^n} \qquad (12\text{-}2)$$

where NP_0 represents the net price (after-tax) received by the firm from the debt, T equals the combined corporate tax rate, K_b equals the after-tax cost of debt, and the remaining variables retain their meaning from Equation (12-1). If in the present example the company nets $879.16 after-tax issuing costs, the equation should read

$$\$879.16 = \sum_{t=1}^{20} \frac{\$40}{(1+K_b)^t} + \frac{\$1{,}000}{(1+K_b)^{20}}$$

$$= \$40[PVIFA_{K_b,20}] + \$1{,}000[PVIF_{K_b,20}]$$

Solving for K_b in Equation (12-2) is achieved by trial and error when using the present-value tables or by using a financial calculator. If 5 percent is selected as a trial discount rate, a present value of $875.38 results. A choice of 4 percent as a trial discount rate gives a present value of $1,000.00. With this information, we may conclude that the after-tax cost of debt is between 4 and 5 percent. Therefore, we may approximate it by interpolating between these two rates. The computation is shown as follows:

Rate	Value	Differences in Values	
4%	$1,000.00	$120.84	$124.62
K_b	879.16 net price		
5%	$875.38		

Solving for K_b by interpolation,

$K_b = .04 + (0.97)\,(.05 - .04) = .0497 = 4.97\%$

This problem can also be solved by the following approach. Since the company will not receive the full $908.32 market price, the effective cost of these funds is larger than the investor's 9 percent required rate of return. To adjust for this difference, we would simply use the market price less issuing costs in place of the market price in Equation (12-1). Thus, the equation becomes

$$P_0 - IC = \sum_{t=1}^{n} \frac{\$I_t}{(1 + k_b)^t} + \frac{\$M}{(1 + k_b)^n} \qquad (12\text{-}3)$$

where P_0 represents the market price of the debt, IC represents the issuing costs of the new debt, k_b equals the before-tax cost of debt, and the remaining variables retain their meaning from Equation (12-1). If in the present example the company nets $850 after issuing costs, the equation should read

$$\$908.32 - \$58.32 = \$850 = \sum_{t=1}^{20} \frac{\$80}{(1 + k_b)^t} + \frac{\$1{,}000}{(1 + k_b)^{20}}$$

$$= \$80(PVIFA_{k_b,20}) + \$1{,}000(PVIF_{k_b,20})$$

Solving for k_b in Equation (12-3) can be achieved by trial and error or by using a financial calculator. If 10 percent is selected as a trial discount rate, a present value of $830.12 results. We know that the rate is above 9 percent because a 9 percent rate gave us a $908.32 value. We need the discount rate that gives us an $850 value. With this information, we may conclude that the before-tax cost of the debt capital is between 9 and 10 percent; therefore, we may approximate it by interpolating between these two rates. The computation may be shown as follows:[11]

[11]An alternative to the estimation procedure for interpolation is to use the following approximation formula for yield:

$$= \frac{\$\text{Coupon} + \dfrac{\text{Face Value} - \text{Price}}{\text{Maturity}}}{\dfrac{\text{Price} + \text{Face Value}}{2}}$$

The approximation formula for yield can be modified to incorporate the effect of after-tax issuing costs as follows:

$$= \frac{\$\text{Coupon} + \dfrac{\text{Face Value} - \text{Net Price}}{\text{Maturity}}}{\dfrac{\text{Net Price} + \text{Face Value}}{2}}$$

where, Net Price = Market Price – After-tax Issuing Costs

Rate	Value	Differences in Values	
9%	$908.32	$58.32	$78.20
k_b	850.00 proceeds		
10%	$830.12		

Solving for k_b by interpolation,

$$k_b = 0.09 + \left(\frac{\$58.32}{\$78.20}\right)(0.10 - 0.09) = 0.0975 = 9.75\%$$

We could also compute the before-tax cost of debt by using a financial calculator where $N = 20$, PV = 850 (–850 on some calculators), and $FV = 1{,}000$. Then press "*I/Y*" to solve for k_b. The result would be 9.79 percent (a slight rounding difference).

Thus, the company's cost of debt, before recognizing the tax deductibility of interest, is 9.75 percent. However, we want to know the after-tax cost of the debt, not the before-tax cost. Since interest is a tax-deductible expense, for every $1 we pay in interest, we lower the firm's tax liability by $1 times the tax rate. If our company has a combined corporate tax rate of 50 percent, then a dollar in interest means that we save $.50 in taxes. Applying the same logic to our cost of debt, we may correctly conclude that the after-tax interest cost is found by multiplying the before-tax interest rate by [1 – (tax rate)]. If T is the company's combined corporate tax rate and k_b is the before-tax cost of debt, the after-tax cost of new debt financing, K_b, is found as follows:

$$\text{after-tax cost of debt} = K_b = k_b(1 - T) \quad (12\text{-}4)$$

If in the present example the corporation's tax rate is 50 percent, then the after-tax cost of debt is 4.88 percent:

$$\text{after-tax cost of debt} = K_b = 9.75\%(1 - .5) = 4.88\%$$

In summary, the firm must earn 4.88 percent on its borrowed capital after the payment of taxes.[12] In doing so, the investors will earn a 9 percent rate of return (their required rate) on the $908.32 investment, the market price of the bond.

BACK TO THE BASICS

The tax deductibility of interest expense favours the use of debt financing. This is an example of ***Axiom 8: Taxes Bias Business Decisions.*** *The tax deductibility of interest, other things remaining constant, serves to encourage firms to use more debt in their capital structure than they would otherwise.*

Cost of Preferred Shares

Cost of preferred shares

The rate of return the firm must earn on the preferred shareholders' investment to satisfy their required rate of return. The cost is based on the preferred shareholders' opportunity cost of debt in the capital markets.

Determining the **cost of preferred shares** follows the same logic as the cost of debt computations. The objective is to find the rate of return that must be earned on the preferred shareholders' investment to satisfy their required rate of return.

In Chapter 6 the value of a preferred share, P_0, that is nonmaturing and promising a constant dividend per period was defined as follows:

$$P_0 = \frac{\text{dividend}}{\text{required rate of return for a preferred shareholder}} = \frac{D}{k_p} \quad (12\text{-}5)$$

From this equation, the required rate of return, k_p, is defined as

$$k_p = \frac{\text{dividend}}{\text{market price}} = \frac{D}{P_0} \quad (12\text{-}6)$$

[12]The slight difference in the after-tax cost of debt between both approaches represents a rounding error.

If, for example, a preferred share pays $1.50 in annual dividends and sells for $15, the investors' required rate of return is 10 percent:

$$k_p = \frac{\$1.50}{\$15} = 10\%$$

Yet even if these preferred shareholders have a 10 percent required rate of return, the effective cost of this capital will be greater owing to the issuing costs incurred in issuing the security. In our example, a firm will net $13.50 per share, the market price less the after-tax issuing costs, rather than the full $15 market price. The cost of preferred shares, K_p, should be calculated using the net price received by the company. Therefore

$$K_p = \frac{\text{dividend}}{\text{net price}} = \frac{D}{NP_0} \tag{12-7}$$

For the preceding example, the cost would be

$$K_p = \frac{\$1.50}{\$13.50} = 11.11\%$$

Since preferred share dividends are not tax-deductible, no adjustment for taxes is required. Thus, the firm must earn the cost of preferred capital after taxes have been paid, which for the preceding example was 11.11 percent.

Cost of Common Shares

While debt and preferred shares must be issued to receive any new money from these sources, common shareholders can provide additional capital in one of two ways. First, new common shares may be issued. Second, the earnings available to common shareholders can be retained, in whole or in part, within the company and used to finance future investments. Retained earnings represent the largest source of capital for most North American corporations. As much as 70 percent of a company's financing in any year may come from the profits retained within the business. To distinguish between these two sources, we will use the term **internal common equity** to designate the profits retained within the business for investment purposes, and **external common equity** to represent a new issue of common shares.

Internal common equity
Profits retained within the business for investment purposes.

External common equity
A new issue of common shares.

Cost of Internal Common Equity When managers are considering the retention of earnings as a means for financing an investment, they are serving in a fiduciary capacity. That is, the shareholders have entrusted the company assets to management. If the company's objective is to maximize the wealth of its common shareholders, management should retain the profits only if the company's investments within the firm are at least as attractive as the shareholders' next best investment opportunity.[13] Otherwise the profits should be paid out in dividends, thus permitting the investor to invest more profitably elsewhere.

How can management know the shareholders' alternative investment opportunities? Certainly identifying those specific investments is not feasible. However, the common shareholders' required rate of return should be a function of competing investment opportunities. If the only other investment alternative of similar risk has a 12 percent return, one would expect a rational investor to set a minimum acceptable return on the investment at 12 percent. In other words, *the common shareholders' required rate of return should be equal to the expected rate of the best competing investment available.* Thus, if the common shareholders' required rate of return is used as a minimum return for investments financed by common share

[13]Other factors may justify management's not adhering completely to this principle. These issues were examined when we discussed dividend policy in Chapter 10.

investors, management may be assured that its investment policies are acceptable to the common shareholder.

Cost of common equity

The rate of return the firm must earn in order for the common shareholders to receive their required rate of return. The rate is based on the opportunity cost of funds for the common shareholders in the capital markets.

The cost of **common equity** is the opportunity cost of the common shareholder's funds. The implication of the cost of common equity being an opportunity cost is that management should not invest the common shareholder funds if they cannot earn a rate of return at least comparable to what the shareholders could earn elsewhere on another investment of similar risk. This type of rate of return is also known as the common shareholders' required rate of return. Hence, the cost of common equity is the rate of return the firm must earn to satisfy the common shareholders' required rate of return. This concept should be familiar since it was also used to describe the cost of debt and the cost of preferred shares.

The cost of equity is more difficult to measure than the cost of debt or preferred shares. To estimate a firm's cost of equity, we have to make assumptions about such things as the firm's future growth prospects and the additional rates of return that will be required by shareholders for assuming more risk—assumptions that were not necessary in computing the cost of debt or the cost of preferred shares. The accuracy of our estimate, therefore, will significantly depend on the correctness of our assumptions. For this reason, we typically use several approaches to calculate the cost of equity, so as to check the reasonableness of our answer. Three of these techniques include: (1) the dividend-growth model, (2) the capital asset pricing model, and (3) the risk-premium approach.

The Dividend-Growth Model In Chapter 6, the value of a common share was found to be equal to the present value of the expected future dividends, discounted at the common shareholders' required rate of return. Since the shares have no maturity date, these dividends extend to infinity. For an investor with a required rate of return of k_c, the value of a common share, P_0, promising dividends of D_t in year t would be

$$P_0 = \frac{D_1}{(1 + k_c)^1} + \frac{D_2}{(1 + k_c)^2} + \ldots + \frac{D_n}{(1 + k_c)^n} + \ldots + \frac{D_\infty}{(1 + k_c)^\infty} \quad (12\text{-}8)$$

Since the market price of the security, P_0, is known, the required rate of return of an investor purchasing the security at this price can be determined by estimating future dividends, D_t, and solving for k_c using Equation (12-8). Furthermore, if the dividends are increasing at a constant annual rate of growth[14] (g) that is less than k_c (the required rate), then k_c may be measured as follows:[15]

$$k_c = \left(\frac{\text{dividend in year 1}}{\text{market price}}\right) + \left(\begin{array}{c}\text{annual growth rate}\\ \text{in dividends}\end{array}\right) = \frac{D_1}{P_0} + g \quad (12\text{-}9)$$

To convert from the common investor's required rate of return in Equation (12-9) to the cost of internal common funds, no adjustment is required for taxes. Dividends paid to the firm's common shareholders are not tax-deductible; therefore, the cost is already on an after-tax basis. Also, issuing costs are not involved in computing the cost of internal common equity since the funds are already within the business. Thus, the investor's required rate of return, k_c, is the same as the cost of internal common equity, K_c.

EXAMPLE

To demonstrate the computation, the Talbot Corporation's common shareholders recently received a $2 dividend per share, and they expect dividends to grow at an annual rate of 10 percent. If the market price of the security is $50, the investor's required rate of return is

[14]In the dividend-growth model, growth comes from the retention of earnings and is determined as follows: growth = percentage of earnings retained in the firm × return on equity.

[15]For additional explanation, see Chapter 6.

$$k_c = K_c = \frac{D_1}{P_0} + g \qquad (12\text{-}10)$$

$$= \frac{\$2(1 + .10)}{\$50} + .10$$

$$= \frac{\$2.20}{\$50} + .10 = .144$$

$$= 14.4\%$$

Note that the forthcoming dividend, D_1, is estimated by taking the past dividend, \$2, and increasing it by 10 percent, the expected growth rate. That is, $D_1 = D_0\,(1.10) = \$2(1.10) = \$2.20$.

BACK TO THE BASICS

The dividend-growth model for common stock is based on two of our axioms of finance: Axiom 2: The Time Value of Money—A Dollar Received Today Is Worth More Than a Dollar Received in the Future, and Axiom 3: Cash—Not Profits—Is King: Measuring the Timing of Costs and Benefits.

The dividend-growth model has been a relatively popular approach for calculating the cost of equity. The primary difficulty, as you might expect, is estimating the expected growth rate in future dividends. The following methods can be used to calculate the growth rate:

1. **Compounded growth rates determined from DPS or EPS data**. For example, the use of data points such as the DPS today versus the DPS five years ago.
2. **Average compounded growth rates from DPS or EPS data**. For example, the use of an average of compounded growth rates over a five-year period.
3. **Analyst forecasts**. For example, investment advisory services such as Value Line or even services that collect and publish the forecasts of a large number of analysts. For instance, Institutional Broker's Estimate System (IBES) publishes earnings per share forecasts made by about 2,000 analysts on a like number of shares. Growth estimates are generally available only for about five years, and not for the indefinite future, as required by the constant-growth model. Also, the analysts usually state their forecasts in terms of earnings rather than dividends, which does not meet the strict requirements of the dividend-growth model. Even so, the earnings information is helpful, since dividend growth in the long run is dependent on earnings. Nevertheless, the analysts' forecasts are helpful because they provide direct measures of the expectations that determine values in the market.
4. **Management forecasts for sustainable growth**. Again, this type of method makes use of inside information in which managers can provide direct measures of the expectations that determine values in the market.
5. **Percentage of earnings retained in the firm times return on equity**. This method requires that we assume new shares will not be issued and that book value figures are approximations.

The use of analysts' forecasts in conjunction with the dividend-growth model to compute required rates of return for the Standard and Poor's 500 shares has been studied by Harris.[16] Computing an average of the analysts' forecasts of five-year growth rates in EPS, Harris used this average as a proxy for the growth rate in dividends. Then, using the dividend-growth model, Equation (12-9), he estimated an average cost of equity for the S&P 500 shares. He next compared these required

[16]Robert Harris, "Using Analysts' Forecasts to Estimate Shareholder Required Returns," *Financial Management* (Spring 1986), pp. 58–67.

Table 12-3 Required Rates of Return and Risk Premiums

	Government Bond Yield	S&P 500 Required Return	Risk Premium
1982			
Quarter 1	14.27	20.81	6.54
Quarter 2	13.74	20.68	6.94
Quarter 3	12.94	20.23	7.29
Quarter 4	10.72	18.58	7.86
Average	12.92	20.08	7.16
1983			
Quarter 1	10.87	18.07	7.20
Quarter 2	10.80	17.76	6.96
Quarter 3	11.79	17.90	6.11
Quarter 4	11.90	17.81	5.91
Average	11.34	17.88	6.54
1984			
Quarter 1	12.09	17.22	5.13
Quarter 2	13.21	17.42	4.21
Quarter 3	12.83	17.34	4.51
Quarter 4	11.78	17.05	5.27
Average	12.48	17.26	4.78
1982–1984 Average	12.25	18.41	6.16

Source: Robert Harris, "Using Analysts' Forecasts to Estimate Shareholder Required Returns," *Financial Management* (Spring 1986), p. 62.

rates with the yields on U.S. Treasury bonds to see how much risk premium common shareholders were expecting. The analysis was conducted for each quarter from 1982 through 1984. Results of the Harris study are presented in Table 12-3. The findings suggest that common shareholders have required a return of between 17.26 and 20.08 percent on average each year for 1982 through 1984. For the three-year period, the average required rate of return was 18.41 percent. The average risk premiums of common shareholders each year, which are shown in the last column of Table 12-3, ranged from 4.28 percent to 7.16 percent, for an average of 6.16 percent. However, we should remember that these returns apply only for equity investments of average riskiness. As we well know, the risk of individual securities will differ from the average, as will the shareholders' required returns.

Although the results in Table 12-3 look reasonable, the same computations for individual shares may not be as plausible, largely because of measurement errors that occur when only one or a few shares are analyzed. Moreover, the constant growth assumption of the model may be inconsistent with reality.

The CAPM Approach Drawing from Chapter 6, we can estimate the cost of equity using the capital asset pricing model (CAPM). Remember that investors should require a rate of return that at least equals the risk-free rate plus a risk premium appropriate for the level of systematic risk associated with the particular security. Using the CAPM, we may represent the equity required rate of return (cost of internal equity) as follows:

$$k_c = K_c = k_f + \beta(k_m - k_f) \tag{12-11}$$

where k_c, K_c = the required rate of return of the equity shareholders, and also the cost of internal equity capital

k_f = the risk-free rate

β = beta, or the measure of a share's systematic risk

k_m = the expected rate of return for the market as a whole—that is, the expected return for the "average security"

For example, assume the risk-free rate is 7 percent, the expected return in the market is 16 percent, and the beta for Talbot Corporation's common shares is .82. Then the cost of internal equity would be estimated as follows:

$$K_c = \underset{k_f}{7\%} + \underset{\beta}{.82}(\underset{k_m}{16\%} - \underset{k_f}{7\%})$$
$$= 14.4\%$$

Although using the CAPM appears relatively easy, its application is not entirely straightforward, particularly in the corporate setting. In estimating the risk-free rate, the market rate, and the security's beta, the goal is to describe the expectations in the minds of the investors, since it is these expectations that determine how assets are valued. Such a task is difficult. However, as indicated earlier, financial service companies now help provide information about investor expectations.

BACK TO THE BASICS

The capital asset pricing model (CAPM) is a formal representation of ***Axiom 1: The Risk-Return Tradeoff—We Won't Take on Additional Risk Unless We Expect to Be Compensated with Additional Return.*** *This model indicates, in the form of an equation, that additional returns are needed to compensate for additional risk. The added risk is measured in terms of systematic or nondiversifiable risk, and the additional return is calculated using the beta coefficient and the market risk premium (the difference between the expected rate of return on the market portfolio of all risky securities and the risk-free rate).*

The Risk-Premium Approach Since we know that common shareholders will demand a return premium above the bondholder's required rate of return, we may state the cost of equity as follows:

$$K_c = K_b + RP_c \tag{12-12}$$

where, as before, K_c and K_b represent the cost of internal common equity and debt, respectively. RP_c is the additional return premium common shareholders expect for assuming greater risk than bondholders. Since we can compute the cost of debt with some degree of confidence, the key to estimating the cost of equity is in knowing RP_c, the risk premium.

We again are in some difficulty, because we have no direct means of computing RP_c. We can only draw from our experience, which tells us that the risk premium of a firm's common shares relative to its own bonds has for the most part been between 3 and 5 percent. In times when interest rates are historically high, the premium is usually low. In years when interest rates are at historical lows, the premium has been higher. Using an average premium of 4 percent, we would approximate the cost of equity capital as follows:

$$K_c = K_b + 4\%$$

For a firm with A++ rated bonds that have a cost of 9 percent, the cost of equity would be estimated to be 13 percent (9 percent cost of debt plus the 4 percent average premium); a more risky company, with bonds that are rated B++ with a 13 percent cost, could expect its cost of equity to approximate 17 percent (13 percent + 4 percent).

The risk-premium approach is somewhat similar in concept to CAPM in that both recognize that common shareholders require a risk premium. The differences between the two approaches come from using different beginning points (CAPM uses the risk-free rate and the risk-premium approach uses the firm's cost of debt) and in how the risk premium is estimated. Both estimates of the risk premium involve subjectivity; however, CAPM has a more developed conceptual basis. Even so, the risk-premium approach is at times the best we can do, especially when the dividend-growth model and the CAPM give unreasonable estimates. Even if the dividend-growth and CAPM approaches are thought to fit the situation, the risk-premium technique gives us a good way to verify the reasonableness of our results.

Cost of New Common Shares If internal common equity does not provide all the equity capital needed for new investments, the firm may need to issue new common shares. Again, this capital should not be acquired from the investors unless the expected returns on the prospective investments exceed a rate sufficient to satisfy the shareholders' required rate of return. Since the required rate of return of common shareholders was measured using Equation (12-10), the only adjustment necessary is to consider the potential issuing costs incurred from issuing the shares. We recognize the issuing costs by reducing the market price of the shares by the after-tax issuing costs.[17] Thus, the cost of new common shares, K_{nc}, is

$$K_{nc} = \frac{D_1}{NP_0} + g \tag{12-13}$$

where NP_0 equals the net proceeds after-tax per share received by the company. Suppose that the after-tax issuing costs are 15 percent of the market price, the cost of capital for the new common shares, or external common, would be 15.18 percent, calculated as follows:

$$\begin{aligned} K_{nc} &= \frac{\$2.20}{\$50 - .15(\$50)} + .10 \\ &= \frac{\$2.20}{\$42.50} + .10 = .1518 \\ &= 15.18\% \end{aligned}$$

In this example, if management achieves a 15.18 percent return on the net capital received from common shareholders, it will satisfy the investors' required rate of return of 14.4 percent, as determined earlier by Equation (12-11).

Selection of Capital Structure Weights

The individual costs of capital will be different for each source of capital in the firm's capital structure. To use these costs of capital in our investment analyses, we must compute a composite or weighted average cost of capital.

The weights for computing this overall cost should reflect the market values of the different sources in the corporation's financing mix. For instance, the balance sheet may indicate that creditors have financed 30 percent of the investments and that the common shareholders have provided the remaining 70 percent. These proportions reflect an application of accounting principles whereby debt and common shares are carried at their book values on the balance sheet. Remember that the book values of these financial instruments represent the historical costs of these instruments when they were issued. In finance, the value of an asset is reported in terms of its present value. Thus, we must determine a more accurate representation of the proportions of debt and common shares to be used in the firm's financing mix. As a result, we estimate the proportions of the financing mix by using the market values of debt and common shares. Hence, the firm's weighted cost of capital reflects the market values of the different sources of financing.

[17]For another approach to adjusting for flotation costs, see John R. Ezzell and R. Burr Porter, "Flotation Costs and the Weighted Average Cost of Capital," *Journal of Financial and Quantitative Analysis* 11 (September 1976), pp. 403–413.

Several choices are available for selecting the financing weights for the weighted cost of capital. Theoretically, the actual mix to be used in financing the proposed investments should be used as the weights. But this approach presents a problem. The costs of capital for individual sources depend upon the firm's financial risk, which in turn is affected by its financial mix.[18] If management alters the present financial structure, the individual costs will change, making it more difficult to compute the cost of capital. Thus, we will assume that the company's financial mix is relatively stable and that these weights will closely approximate future financing. Although this assumption may not be strictly met in any particular year, firms frequently have an **optimal capital structure** (desired debt-equity mix), which is maintained over the long term. The target financing mix provides the appropriate weights to be used in calculating the firm's weighted cost of capital.

Optimal capital structure

The unique capital structure that minimizes the firm's weighted cost of capital.

EXAMPLE

The Ash Company's current financing mix is contained in Table 12-4. The firm's chief financial officer, Tony Ash, does not want to alter the firm's financial risk, instead he chooses to maintain the same relative mix of capital in financing future investments. Thus, in computing Ash's weighted cost of capital we make the following assumptions: (1) a constant financial mix and (2) the weights represent market values of the different sources of financing.

Computing the Weighted Cost of Capital

Let's now compute the weighted cost of capital for a firm, which is simply the weighted average of the individual costs, given the firm's financial mix. This calculation is best illustrated with our example.

From Table 12-4, we have estimated Ash's financial mix for the purpose of computing the firm's weighted or overall cost of capital. Let's further assume that management has computed the individual costs of capital for the firm, shown in Table 12-5. For the time being, we will restrict equity financing to the retained earnings available for reinvestment, or internally generated common equity. We are assuming Ash will not issue any new common shares. Therefore, the cost of new common shares is not relevant. Table 12-6 combines the weights from Table 12-4 and the individual cost for each security from Table 12-5 into a single weighted cost of capital. Given that the assumptions of the weighted cost of capital concept are met and that common equity requirements can be satisfied internally, the company's

Table 12-4 Ash Inc. Capital Structure

Investor Group	Amount of Funds Raised ($)[a]	Percentage of Total
Bonds	$1,750,000	35%
Preferred shares	250,000	5
Common shares	3,000,000	60
Total new financing	$5,000,000	100%

[a]Funds raised represent market values.

[18]During the 1980s, many firms increased their use of debt. Relatively easy access to credit was used to fund office buildings, shopping centres, and housing developments. Many North American firms shifted financing from common shares to debt during the eighties. In addition, the increased debt caused interest payments to absorb a higher proportion of the firm's cash flows. Such large increases in debt have significantly altered the capital structure of most firms and impacted the level of financial risk.

Table 12-5 After-Tax Component Costs of Capital for Ash Inc.

Investor Group	After-Tax Component Costs
Bonds	7%
Preferred shares	13
Common equity	
Internally generated common	16
New common	18

weighted cost of capital is 12.7 percent. This rate is the firm's minimum acceptable rate of return for new investments.

We have now observed the process for measuring a firm's weighted cost of capital. We cannot overstate the importance of this calculation with respect to evaluating investment proposals. However, the limiting effects of our assumptions, such as the requirement of constant business and finance risk, reduces the level of confidence which we can place in the measurement techniques. At best, we can only approximate the firm's weighted cost of capital. Our approximation can be improved upon by testing our results over a reasonable range of values, as is done in sensitivity analyses.

We now examine the effects of the level of financing on the firm's cost of capital. In other words, as the firm continues to raise more capital for investment purposes, how is the cost of capital affected?

Level of Financing and the Weighted Cost of Capital

In the previous example, we assumed that no new common shares were to be issued. If new common shares are to be issued, the firm's weighted cost of capital will increase, because external equity capital has a higher cost than internal equity due to the issuing costs.

Generally the firm should use its cheapest sources of funds first, while maintaining its desired debt-equity mix. In other words, since internally generated funds cost less, they should be blended with debt until fully exhausted. Beyond this point, the weighted cost of capital increases, since the firm has to rely on new shares for its equity financing. This basic concept is best explained through an example.[19]

Table 12-6 Weighted Cost of Capital for Ash Inc. If Only Internal Common Is Used

(1) Investor Group	(2) Weight[a]	(3) Individual After-Tax Costs[b]	(4) Weighted Costs
Bonds	35%	7%	2.45%
Preferred shares	5	13	0.65
Common equity			
(internally generated)	60	16	9.60
Weighted cost of capital (K_0)			12.70%

[a]Taken from the desired financing mix presented in Table 12-4.
[b]Taken from Table 12-5.

[19]Although we assume a constant debt-equity mix, we have no reason or need to assume a constant mix between internally generated equity and new common shares. Rather, we assume that we first use internal equity funds until exhausted; if more equity is needed, we issue new common shares.

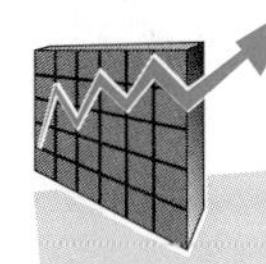

Financing Corporate Canada in the 1990's

Looking ahead, Canadian companies must adapt not only to the central bankers' brave new world of price stability, but also global economic growth patterns that will differ markedly from those of the 1980's. Very slow growth in domestic demand is likely. Beyond the global slowdown of the early nineties, rapid growth will be largely found in unfamiliar, difficult markets. Slowdowns in demand always expose weak balance sheets and penalize those who have financed too aggressively. This is a material reality in an environment where the Bank of Canada will likely continue to pursue its goal of price stability within a moderate range.

Industrial Canada's Balance Sheet: 1978, 1988, and 1998

What does all this mean for the financial management of Corporate Canada? Comparing the aggregate balance sheet of large Canadian industrial corporations in 1978 with what it might look like in 1998 is instructive.

There has been a major reduction in working capital ratios during the last decade. Companies now routinely run their operations with less cash, lower receivables, and especially lower inventories. This trend will likely continue. Corporate treasurers, like their counterparts in physical operations, now have a bewildering array of new tools at their disposal. 'Just-in-time' delivery relates not only to assembly line parts, but also to cash in the payment stream.

Continued Financial Innovation: Implications

Securitization of receivables is now a hot topic. Commodity-linked securities, interest rate swaps and liquid futures and options markets are providing the capacity to restructure and transform the nature of the risks that the corporation must accept. These financial products are permitting corporate treasurers to manipulate the exposure of their companies to the economic factors intrinsic to their business, choosing which risks to accept and which to hedge. The financial markets are serving up a growing menu of these products at increasingly attractive prices.

Accounting standards are coming under great stress in this environment. The visible balance sheet does not necessarily give an accurate picture of a company's position when swaps, hedges, futures, and options are routinely employed to change one form of risk into another. Regulators have become well aware of the degree to which companies can sidestep one country's taxes and regulations by nimbly shifting operations and assets to another. The result is already visible in heightened international coordination of tax and regulatory policies, but as usual, the bureaucrats are well behind the corporate innovators.

Fixed Assets

Fixed asset investment over the next decade is likely to be dominated by continuing strength in machinery and equipment purchases. As demand growth shifts even further toward export markets, and the pace of technological change continues to be very rapid, there must be a continuing focus on change and renewal. At the same time, tougher environmental regulation will accelerate obsolescence.

Investments in Affiliates

We predict 'investment in affiliates' will be the fastest-growing asset category of Canadian corporate balance sheets. Rising competition from abroad and greater business opportunities in foreign markets suggest that cross-border affiliations and associations will be useful and important. Marketing in one country, production in others, financial direction in a third: all this is becoming familiar. Since 1978, the share of corporate assets described as 'long-term equity investment in affiliated Canadian companies' has doubled, and the share of 'long-term equity investment in affiliated foreign companies' has tripled.

The Liability Side of the Balance Sheet

On the liability side, short-term bank loans are likely to continue to diminish in importance, as the working capital they support dwindles down. Thus while short-term bank financing will continue to play an important role during cyclical downturns, its secular trend continues to be downward. In the 'permanent finance' category, the key question is whether debt or equity will play an increased role over time. While profits decline and investment outlays continue to grow in the short term, corporations are financing with debt, causing the debt/equity ratio to rise in the short term.

Debt/Equity Ratios Unlikely To Trend Upwards

Our longer term view is that debt/equity ratios will not rise significantly during the 1990's. The average 1978 debt/equity ratio for (all, not just industrial) Canadian non-financial corporations was 0.86. In the midst of the 1982 recession, it climbed to 1.09. In the extended subsequent recovery, it was reduced back below 0.9. By 1991, given our forecast of slowing world growth, falling commodity prices and domestic disinflation, the ratio will likely have climbed to over 1.0 again. Longer term, we believe the corporate sector will want to make another assault on even lower debt ratios. We think that the ratio will be below 0.9 again in the late 1990's.

Major Upturn in Capital Availability Likely

Canada could become a major capital exporter before the end of the decade. Today, an excessive amount of domestic saving has to be devoted to financing the fiscal deficit. A further big chunk has been going to finance the house buying ambitions of the baby-boom generation. But these pressures will likely fall off over the next few years. Meanwhile, personal saving itself should increase at a healthy pace encouraged by major shifts in tax structure. The GST represents a shift from taxing personal income to taxing consumption.

Also, tax sheltered contribution rates to money-purchase pension plans and Registered Retirement Savings Plans (RRSPs) are soon to rise. And contractual saving, such as the rollover of investment income in trusteed pension plans and RRSPs, continues to rise.

Much of New Canadian Financial Capital Will Be Risk-Seeking

Much of the large amounts of institutional money that will be available for investment, as well as some of the rapidly growing non-institutionally managed pools of capital based on self-directed RRSPs and closed-end mutual funds will be risk capital. One impetus is the trend toward relaxing investment restrictions on institutional investors through the adoption of 'prudent portfolio' pension investment rules. Also financial institutions and financial intermediaries, faced with intense competition, lower margins and the greater savings inflows, will be increasingly willing to consider higher-risk investment alternatives to achieve satisfactory returns.

For these reasons, we are likely to see very heavy participation by Canadian savers in equity finance for Canadian companies, more than they will in fact need to finance domestic operations. Fortunately, the government has recognized the desirability of encouraging international diversification by Canadian pension funds, permitting them to move over the next five years to place up to 20%, instead of only 10% of their portfolios in foreign assets.

Foreign Portfolio Investors Will Be a Factor Too

Canadian corporations have enjoyed relatively easy access to foreign capital markets. Earlier, we focused on the increasing role foreign direct investment will play in shaping Canadian corporate balance sheets. But it is wrong to ignore the role foreign portfolio investors could play. Currency hedging is increasingly cost-effective. Complex international swaps are now increasingly matter-of-course. So large corporations will increasingly access global markets for their financing in the 1990's.

Unregulated off-shore markets will from time to time provide attractive financing opportunities to those Canadian corporations with the credit quality to access them. Liquid markets in interest rate and currency swaps will facilitate financing abroad. In the 1980's, international markets provided very attractive sources of debt financing to Canadian corporations. In the 1990's, this availability of funding should extend to equities, a process that has already begun with the privatization initiatives in the U.K. and Europe.

A New Canadian Competitive Advantage: Risk Finance

We conclude that Canadian corporations will not be without some important competitive advantages as they face the demanding 1990's. They may not be world leaders in R&D or have special entries into some emerging new growth markets. But they do have access to a well-educated labour force, unbounded natural resources, strong infrastructure, and good international linkages to export markets. This article points to yet another emerging strength Canadian corporations have: good access to a broad range of risk finance. May they use it productively.

Source: An excerpt from "Financing Corporate Canada In The 1990's," by John Grant, Mary Webb and Peter Hendrick in *Canadian Investment Review* 3 (Spring 1990), p. 9.

EXAMPLE

The Crisp Corporation, an independent oil company, is contemplating three major capital investments this year. The first proposal is the acquisition of equipment used to examine geological formations. This new equipment should improve the success ratio in discovering productive oil and gas reserves. The second proposal requires an investment into water flooding equipment. This process involves the injection of large amounts of water into underground reserves, which permits a more efficient recovery of the minerals. The third proposal involves an investment into new and advanced drilling equipment which will offer cost savings in drilling for oil and gas. The costs and expected returns for these three possible investments are shown in Table 12-7. Management must decide which of these projects should be accepted.

If any of the proposed projects are accepted, the financing will consist of 50 percent debt and 50 percent common equity. Based upon the anticipated profits this year, the company should have $1,500,000 in profits for reinvestment (internally generated common equity). The costs of capital for each source of financing have been computed and are presented in Table 12-8.

The weighted cost of capital, K_0, would be calculated as follows:

$$K_0 = \left[\begin{pmatrix}\text{percentage}\\ \text{of debt}\\ \text{financing}\end{pmatrix} \times \begin{pmatrix}\text{after-tax}\\ \text{cost of}\\ \text{debt}\end{pmatrix}\right] + \left[\begin{pmatrix}\text{percentage of}\\ \text{common equity}\\ \text{financing}\end{pmatrix} \times \begin{pmatrix}\text{cost of}\\ \text{common}\\ \text{equity}\end{pmatrix}\right] \quad (12\text{-}14)$$

If only internally generated common equity is utilized, the weighted cost of capital is 10 percent.

$$K_0 = [(50\%)(6\%) + (50\%)(14\%)] = 10 \text{ percent}$$

However, when new common shares are used rather than internally generated common equity, the weighted cost of capital is 12 percent:

$$K_0 = [(50\%)(6\%) + (50\%)(18\%)] = 12 \text{ percent}$$

Which weighted cost of capital should be used in evaluating the three investments?

To answer this question, we must first rank the projects in descending order by their respective internal rates of return. Second, we must calculate the level of total financing at which point our internally generated common equity is exhausted. In the Crisp Corporation example, $3 million in total new investments may be

Table 12-7 Crisp Corporation's Investment Opportunities

	Investment Cost	Expected Internal Rate of Return
Geological equipment	$1,500,000	14%
Water flooding equipment	2,000,000	18
Drilling equipment	2,500,000	11

financed with debt and internally generated common equity, without having to change the present financial mix of 50 percent debt and 50 percent common equity and without having to issue common shares. The $3 million is determined by solving the following equation:

$$\text{internally generated common equity financing available} = \text{percentage of common equity financing} \times \text{total financing from all sources} \qquad (12\text{-}15)$$

For the Crisp Corporation,

$\$1,500,000 = 50\% \times \text{total financing}$

which indicates that if Crisp has $1,500,000 in internally generated common equity, and management maintains a 50 percent debt ratio, Crisp will be able to finance $3 million of total investments without issuing new common shares. Equation (12-15) may be changed to solve directly for the amount of total financing:

$$\text{total financing from all sources} = \frac{\text{internally generated common equity}}{\text{percentage of common equity financing}} \qquad (12\text{-}16)$$

$$= \frac{\$1,500,000}{.50}$$

$$= \$3,000,000$$

Therefore, for a total investment level of $3 million or less, the firm's weighted cost of capital is expected to be 10 percent. Beyond this level of total financing our internally generated common equity is totally exhausted, and the weighted cost of capital increases to 12 percent. This reflects the increased cost of the common shares to be issued beyond the $3 million in total financing from both debt and common equity.

Figure 12-4 illustrates the relationship between the weighted costs of capital and the amount of financing being sought. The graph depicts the firm's **weighted marginal cost of capital**. The term *marginal* is used because the computed cost of capital shows the weighted cost of each additional dollar of financing. This marginal cost of capital represents the appropriate criterion for making investment decisions. Thus, the firm should continue to invest up to the point where the marginal rate of return earned on the new investment (IRR) equals the marginal cost of new capital. Figure 12-5 illustrates this comparison in which the firm's optimal investment schedule or capital budget is found to be $3,500,000. The company should invest in water flooding machinery and the geological equipment. However, since the weighted marginal cost of capital is greater than the expected internal rate of return of the drilling equipment, this investment should be rejected.

Weighted marginal cost of capital

The composite or average cost of all sources of financing for each additional dollar of financing. The marginal cost of capital represents the appropriate criterion for making investment decisions.

Table 12-8 Crisp Corporation's Individual Costs of Capital

Source	Cost
Debt (after-tax)	6%
Internally generated common equity ($1,500,000)	14
New common shares	18

Figure 12-4 Crisp Corporation's Weighted Marginal Cost of Capital

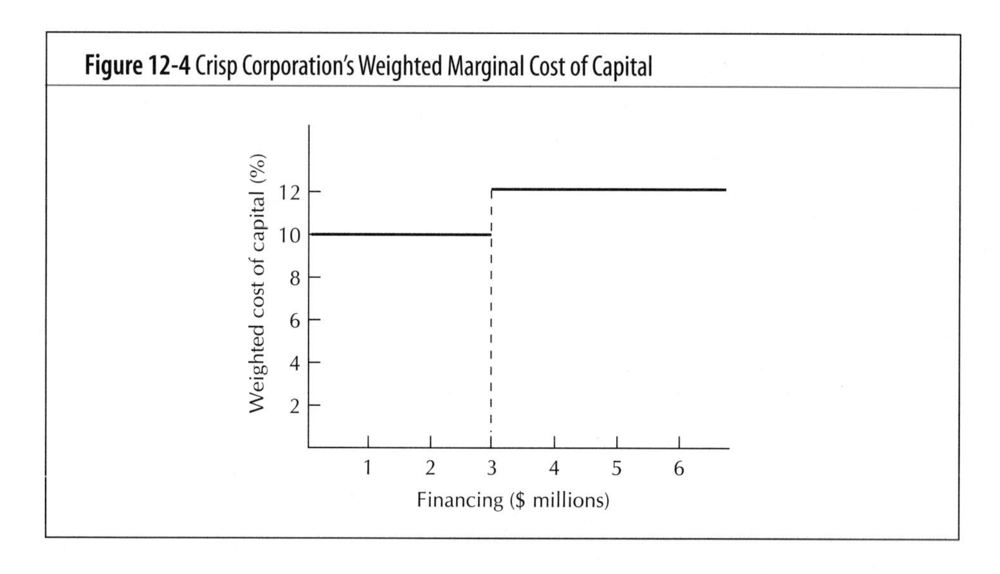

The Effect of New Financing on the Marginal Cost of Capital

Thus far, we have considered only the effect of increases of issuing new common shares on the weighted marginal cost of capital. Similar effects will occur as the cost of any source of financing increases. If the 6 percent cost of debt capital for Crisp Corporation increased to 8 percent after the firm issued $2 million in bonds, an increase in the weighted marginal cost of capital would have occurred at the $4 million financing level from all sources. This break in the marginal cost of capital curve is determined as

$$\begin{aligned}\text{total financing available with the lower-cost debt} &= \frac{\text{total debt available at lower cost}}{\text{percentage of debt financing}} \qquad (12\text{-}17)\\ &= \frac{\$2{,}000{,}000}{.5}\\ &= \$4{,}000{,}000\end{aligned}$$

As a general rule, changes in the weighted marginal cost of capital will take place when the cost of an individual source increases. The break in the marginal cost of capital curve will occur at the dollar financing level in which

Figure 12-5 Crisp Corporation: Comparison of Investment Returns and the Weighted Marginal Cost of Capital

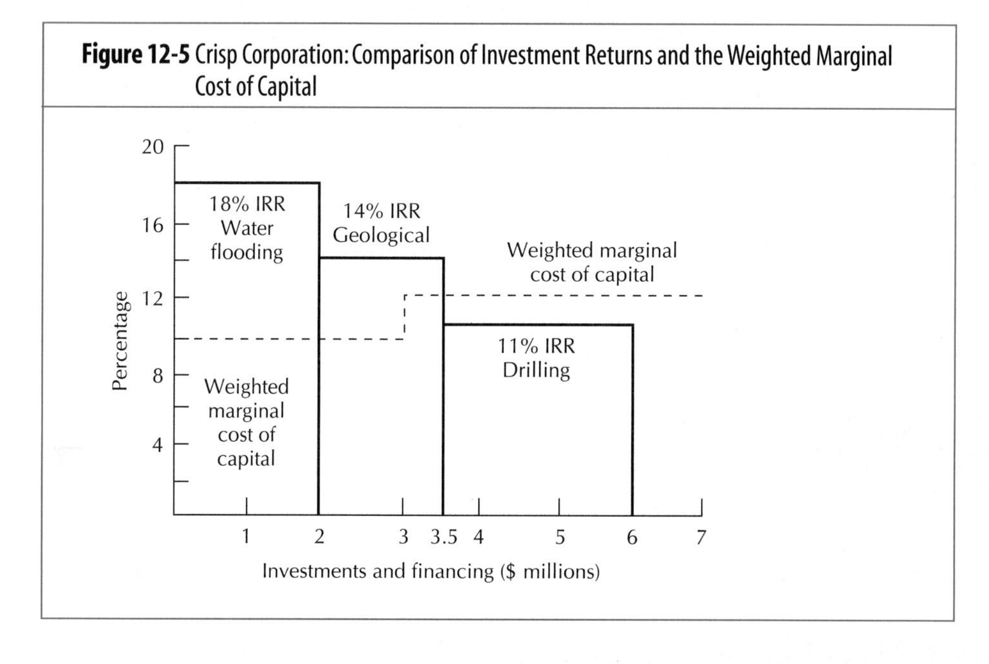

$$\frac{\text{total financing}}{\text{from all sources}} = \frac{\text{maximum amount of a lower-cost source of capital}}{\text{percentage financing provided by the source}} \tag{12-18}$$

Summary of Computations

1. Determine the percentage of financing to be used from each source of capital (debt, preferred shares, and common equity).
2. Compute the points on the marginal cost of capital curve where the weighted cost will increase.
3. Calculate the costs of each individual source of capital.
4. Compute the weighted cost of capital for the company, which will be different as the amount of financing increases.
5. Construct a graph that compares the internal rates of return for prospective investment opportunities with the marginal weighted cost of capital, which will indicate which investments should be accepted.

Now that we have examined the concepts and methods for computing a firm's cost of capital, we thought it might be interesting to see a real-world example. So we asked the chief financial officer of PepsiCo if the company would share with us how they calculate their cost of capital. Agreeing to do so, the company's Director of Capital Planning worked with us in developing a presentation of the company's cost-of-capital methodology. In the next section, we present this methodology.

OBJECTIVE 5

IF IT'S NOT THE WEIGHTED COST OF CAPITAL, IT'SNOT A PEPSI

So far, we have dealt largely in the world of theory and concept, with some hypothetical examples to help us better understand how we might calculate a firm's cost of capital. But we might ask, Do companies actually do what we are talking about? Do they calculate and use the cost of capital in their decision-making process? To be honest, many do not; but many others do. PepsiCo, Inc., is one firm that does. PepsiCo's chief financial officer believes strongly that strategies within the company's major divisions should be evaluated based on their contribution to shareholder value. If a strategy cannot be expected to increase shareholder value, then it should not be undertaken. At PepsiCo, and at any firm for that matter, a proposed strategy to increase shareholder value requires earning a rate of return that exceeds the cost of capital. Thus estimating the firm's cost of capital is a critical part of the evaluation process. In the words of PepsiCo's management:

> An investment's present value is determined by discounting its future cash flows by the appropriate cost of capital. If the investment's risk approximates PepsiCo's overall risk, the company's cost of capital is the appropriate discount rate. However, PepsiCo's business segments have different levels of risk associated with their investment opportunities. Using the company's overall cost of capital to evaluate each division's projects can lead to the rejection of good investments in low-risk divisions and the acceptance of poor projects in high-risk divisions. To avoid this problem, PepsiCo calculates the divisional cost of capital for its snack, beverage, and restaurant divisions. These figures provide better estimates of divisional risk and are used to enhance the company's investment process.

A description of how PepsiCo's managers calculate their weighted cost of capital—mostly in their own words—follows.[20]

[20] This presentation of PepsiCo's cost of capital was provided by John S. Waldek, Director for Capital Planning, who works as part of the team that computes PepsiCo's cost of capital.

PepsiCo's Weighted Average Cost of Capital: The Basic Computation

PepsiCo's policy is to estimate the weighted average cost of capital (K_0) for each of the major divisions, which include beverages, snack foods, and restaurants. The equation used to compute the weighted average cost of capital for the three divisions may be represented as follows:

$$K_0 = \left[\left(\begin{array}{c}\text{percentage}\\\text{of debt}\\\text{financing}\end{array}\right) \times \left(\begin{array}{c}\text{after-tax}\\\text{cost of}\\\text{debt}\end{array}\right)\right] + \left[\left(\begin{array}{c}\text{percentage of}\\\text{common equity}\\\text{financing}\end{array}\right) \times \left(\begin{array}{c}\text{cost of}\\\text{common}\\\text{equity}\end{array}\right)\right] \quad (12\text{-}14)$$

Once the weighted cost of capital is calculated for each division, the company's overall weighted cost of capital is determined to be the weighted average of the divisional cost of capital. The process used by PepsiCo for calculating the weighted cost of capital is as follows:

1. Estimate the cost of equity and the cost of debt for each division.
2. Determine the optimal capital structure—debt-to-equity mix—for each division.
3. Weight the individual costs of capital (debt and equity) by the optimal capital structure to determine a weighted cost of capital for each division.
4. Calculate a cost of capital for the firm as a whole.

Let's look at each of these steps in turn.

Calculating the Individual Costs of Capital

PepsiCo recognizes two types of capital when computing its cost of capital: common equity and debt. We will first explain the process for computing the cost of equity.

Calculating the Cost of Common Equity

PepsiCo's cost of common equity, K_c, is based on the capital asset pricing model (CAPM), represented earlier in Equation (12-11) as follows:

$$K_c = k_f + \beta(k_m - k_f) \quad (12\text{-}11)$$

So estimating a division's cost of equity involves determining (1) the risk-free rate, (2) the market expected return, and (3) the beta for the division.

Estimating the Risk-Free Rate Thirty-year Treasury bonds are used as the measure for the risk-free rate for two reasons. First, these bonds are essentially free of default risk. Second, they capture the long-term inflation expectations of investors associated with investments in long-term assets. At the time PepsiCo was calculating its cost of capital for this presentation, the rate on thirty-year Treasury bonds was 7.28 percent.

Estimating the Market's Expected Return The expected market return equals the expected return on a well-diversified portfolio of common stocks. The ALCAR Group, a consulting firm in Chicago, provides PepsiCo with quarterly estimates of this return. The market return, according to the ALCAR Group, was 11.48 percent at the time PepsiCo was computing its cost of capital.

Estimating the Beta for a Division The betas for PepsiCo's three business divisions, or segments as they are called, cannot be calculated directly because the segments are not publicly traded in the stock market. As a result, PepsiCo uses two methods to estimate these divisional betas: (1) single-industry or -division firms, and (2) betas based on the accounting data of each business segment. Both of these techniques will be described in Chapter 15 when we discuss project betas. An average of the two betas is then used in computing the cost of equity.

Table 12-9 Single-industry or -division Firms for PepsiCo's Business Segment

Restaurant	Snack Food	Beverage
Bob Evans	Borden	Anheuser Busch
Carl Karcher	Campbell's Soup	A & W Brands
Dairy Queen	CPC International	Coca-Cola
JB's Restaurants	General Mills	Coca-Cola Enterprises
Luby's Cafeterias	Gerber	
Mc Donald's	Heinz	
National Pizza	Hershey	
Piccadilly	Kellogg	
Ryan's Family	Quaker Oats	
Vicorp	Ralston Purina	
Wendy's	Sara Lee	

Pure-play betas. The process for estimating divisional betas based on peer groups may be outlined as follows:

1. Select a group of single-industry or -division firms that are publicly traded firms for each of PepsiCo's three divisions (segments) and acquire estimates of these firm's betas. The group consists of companies that are considered to have similar business risk to that of the respective PepsiCo business segments. The firms making up the pure-play for the three business segments (restaurants, snack foods, and beverages) are shown in Table 12-9.
2. Determine the betas for each of the single-industry or -division firms, as reported by *Value Line.*[21]
3. Adjust the betas of the single-industry or -division firms to account for their different capital structures (debt-to-equity mixes). While the single-industry or -division firms may have similar business risk, they have different financial risk; that is, they use different amounts of debt in their capital structures—some more and some less. The more debt a firm uses in its financing mix, the greater the firm's systematic risk will be—and the higher its beta. Thus, to make the betas of the single-industry or -division firms comparable with PepsiCo's business segments, the beta for each single-industry or -division firm must be adjusted to eliminate the differences attributable to unequal financial risk. This adjustment process was performed on the single-industry or -division firms, with the results shown in Table 12-10. In this table, we see the betas for each company, where the average pure-play betas are as follows.[22]

Restaurants	1.01
Snack foods	1.03
Beverages	1.11

Accounting betas. Determining betas for individual stocks is an inexact science at best. Therefore, to increase their confidence in the pure-play beta estimates,

[21] *Value Line* is an investment service group that publishes betas for a large group of companies.

[22] We are not interested in delving into the details of how betas are adjusted for different levels of financial risk. Just trust us or see J. Fred Weston and Thomas E. Copeland, *Managerial Finance* (Fort Worth: The Dryden Press, 1992), pp. 419–420 for an explanation.

management uses a secondary approach—accounting betas—to estimate the systematic risk for PepsiCo's divisions. As with the pure-play betas, accounting betas measure the relative risk of a business versus the market as a whole. However, while the pure-play betas are based on market rates of return, the accounting betas are calculated using accounting data taken from each segment's financial statements. The technique for measuring accounting betas is beyond the scope of our study. However, the results of these calculations as shown in Table 12-10 are as follows:

Accounting betas:	
Restaurants	1.34
Snack foods	1.00
Beverages	1.04

The final beta estimates for PepsiCo's divisions are obtained by taking the average of the pure-play betas and the accounting betas. These averages are shown in the last row of Table 12-10, and are as follows:

PepsiCo business segment betas:	
Restaurants	1.17
Snack foods	1.02
Beverages	1.07

Given the above calculations, the cost of common equity, K_c, for each division can now be determined as follows:

Table 12-10 Estimates of the Pure-Play Betas for PepsiCo Inc. Business Segments

Restaurant		Snack Food		Beverage	
Bob Evans	1.08	Borden	1.00	Anheuser Busch	1.05
Carl Karcher	0.71	Campbell's Soup	1.06	A & W Brands	1.69
Dairy Queen	1.11	CPC International	1.04	Coca-Cola	1.14
JB's Restaurants	0.54	General Mills	1.08	Coca-Cola Enterprises	0.56
Luby's Cafeterias	1.10	Gerber	1.36		
Mc Donald's	0.97	Heinz	0.98		
National Pizza	0.97	Hershey	1.08		
Piccadilly	0.76	Kellogg	1.09		
Ryan's Family	1.40	Quaker Oats	0.86		
Vicorp	1.29	Ralston Purina	0.74		
Wendy's	1.26	Sara Lee	1.04		
Average pure-play betas for restaurants	1.01	Average pure-play betas for snacks	1.03	Average pure-play betas for beverages	1.11
Accounting beta—restaurant segment	1.34	Accounting beta—snack segment	1.00	Accounting beta—beverage segment	1.04
Beta estimate for restaurant segment	1.17	Beta estimate for snack segment	1.02	Beta estimate for beverage segment	1.07

	Risk-Free Rate	+	Beta	(Expected Market Return	−	Risk-Free Rate)	=	Cost of Equity
Restaurants	7.28%	+	1.17	(11.48%	−	7.28%)	=	12.20%
Snack foods	7.28%	+	1.02	(11.48%	−	7.28%)	=	11.56%
Beverages	7.28%	+	1.07	(11.48%	−	7.28%)	=	11.77%

Calculating the Cost of Debt

The after-tax cost of debt is equal to:

$$K_b = k_b(1 - T) \tag{12-4}$$

where k_b = before-tax cost of debt
T = marginal tax rate

The cost of debt for each division will differ depending on the characteristics of the industry and the division's target leverage. Companies with greater risk pay a higher interest rate on debt than do less risky firms. That is, the greater the risk, the greater will be an investor's added required rate of return above the risk-free rate. This return premium is called the **spread to Treasuries**. For example, given the business risks involved and the higher target leverage, restaurants are assigned a higher spread to Treasuries than beverage and snack foods.

Spread to Treasuries

The difference between the interest rate on corporate debt and a Treasury security (If a company's bonds yield 11 percent compared with 7 percent for a Treasury security, the spread to Treasury is 4 percent).

The spread to Treasuries for these segments are estimated and added to the thirty-year Treasury bond to determine the cost of debt. This cost is then adjusted to an after-tax cost, where the tax rate used is 38 percent. The results of these computations are as follows:

	Treasury Rate	+	Spread to Treasuries	=	Pretax Cost of Debt	x	(1 – Tax rate)	=	After-Tax Cost of Debt
Restaurants	7.28%	+	1.65%	=	8.93%	x	0.62		5.54%
Snack foods	7.28%	+	1.15%	=	8.43%	x	0.62		5.23%
Beverages	7.28%	+	1.23%	=	8.51%	x	0.62		5.28%

Optimal Capital Structure

The optimal capital structure for each business segment is used to weight the costs of debt and equity to determine the division's weighted cost of capital. In this case, management set the optimal debt-to-total assets for each division as follows:

Restaurants	30%
Snack foods	20%
Beverages	26%

PepsiCo's Weighted Cost of Capital: The Final Results

Using all the information provided to this point, we can now calculate the weighted cost of capital for each of PepsiCo's divisions as follows:

$$\begin{pmatrix}\text{weighted cost}\\ \text{of capital}\end{pmatrix} = \begin{pmatrix}\text{cost}\\ \text{of equity}\end{pmatrix}\begin{pmatrix}\text{percent}\\ \text{of equity}\end{pmatrix} + \begin{pmatrix}\text{after-tax}\\ \text{cost of debt}\end{pmatrix}\begin{pmatrix}\text{percent}\\ \text{of debt}\end{pmatrix} \tag{12-19}$$

Using the above equation, we can find the weighted cost of capital for each of the divisions as follows:

Restaurants	(12.20%) (0.70)	+	(5.54%) (0.30)	=	10.20%
Snack foods	(11.56%) (0.80)	+	(5.23%) (0.20)	=	10.29%
Beverages	(11.77%) (0.74)	+	(5.28%) (0.26)	=	10.08%

Thus the weighted cost of capital for each of the divisions is slightly above ten percent. The firm then takes these divisional costs of capital and calculates the overall cost of capital for PepsiCo, where the overall cost of capital for PepsiCo is a weighted average of the divisional costs, with the weights being equal to the value of each division as a percent of the total value of PepsiCo. For instance, assume that the percentage of PepsiCo value attributable to each division is as follows:

Restaurants	25%
Snack foods	30%
Beverages	45%

Given the cost of capital for each division, as computed above, and the relative value of each division, the weighted cost of capital for PepsiCo would be found as follows:

$$\begin{aligned}\text{PepsiCo weighted cost of capital} &= \begin{pmatrix}\text{restaurant}\\ \text{percentage value}\end{pmatrix}\begin{pmatrix}\text{restaurant}\\ \text{cost of capital}\end{pmatrix} \\ &+ \begin{pmatrix}\text{snack foods}\\ \text{percentage value}\end{pmatrix}\begin{pmatrix}\text{snack foods}\\ \text{cost of capital}\end{pmatrix} \\ &+ \begin{pmatrix}\text{beverages}\\ \text{percentage value}\end{pmatrix}\begin{pmatrix}\text{beverages}\\ \text{cost of capital}\end{pmatrix} \\ &= (0.25)(10.20\%) + (0.30)(10.29\%) + (0.45)(10.08\%) \\ &= 10.18\%\end{aligned} \tag{12-20}$$

We have now observed firsthand how a firm, in this case PepsiCo, Inc., computes its cost of capital. While it has been somewhat involved, we have had the opportunity to "sit at the feet" of one of the best managed and most highly respected companies in the world. We will now change our perspective from measuring the cost of capital for an *entire firm or division* to asking how we would determine the cost of capital for an *individual project* we are considering as a prospective investment.

REQUIRED RATE OF RETURN FOR INDIVIDUAL PROJECTS: AN ALTERNATIVE APPROACH

OBJECTIVE 6

Two basic assumptions were made in the preceding sections in computing the firm's weighted cost of capital as a cutoff rate for new capital investments. First, the riskiness of the project being evaluated is assumed to be similar to the riskiness of the company's existing assets. In other words, acceptance of the investment will not alter the firm's overall business risk. Second, future investments are assumed to be financed in the same proportions of debt, preferred shares, and common shares as past investments. Frequently these two assumptions are not met in practice.

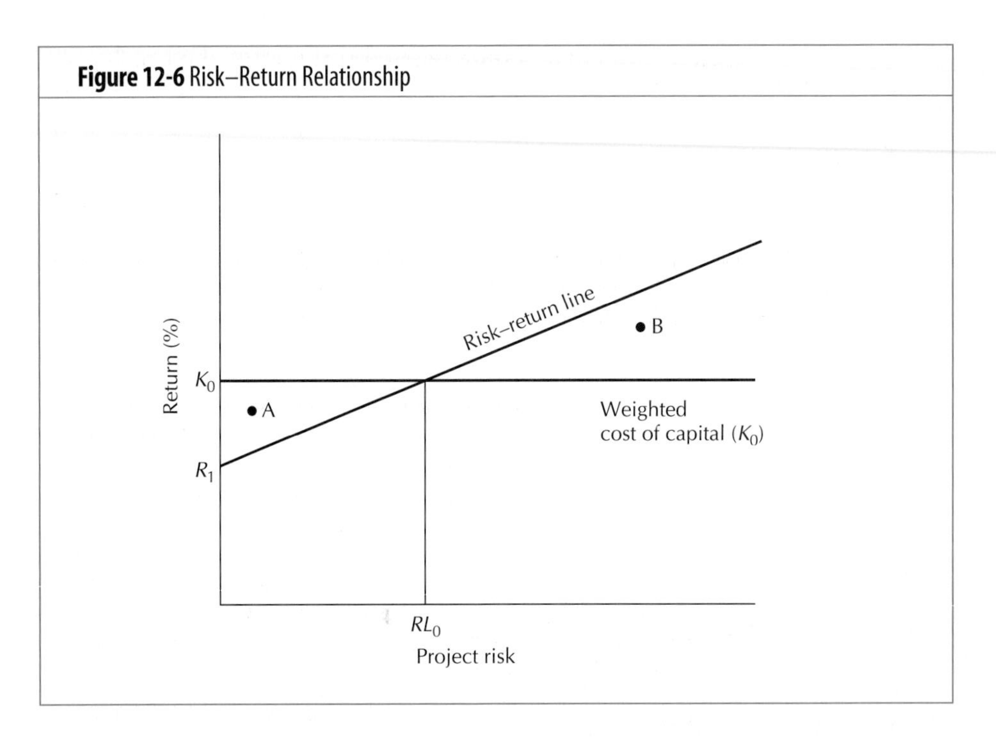

Figure 12-6 Risk–Return Relationship

Firms often do make investments that have risk characteristics different from the firm's existing assets. The firm may also finance a given project with all debt, and then rely solely on equity for the next investment. Also, given the nature of an investment, the firm may be able to use more debt than what was possible for prior investments.

Figure 12-6 illustrates the problem that occurs from using the weighted cost of capital for projects having different levels of risk. The line which represents the firm's weighted cost of capital, K_0, does not allow for varying levels of project risk and is appropriate only for projects having a level of risk at point RL_0. The same figure illustrates the **risk-return line** which describes an increasing cost of capital with increasing levels of risk. In other words, we should use different costs of capital (required rates of return) for investments having different levels of risk. Investments A and B in Figure 12-6 illustrate any incorrect decisions that might result from the use of the firm's cost of capital as the cutoff rate for project acceptance where the risk inherent in A is below the firm's risk level of RL_0 and project B's risk exceeds RL_0. Investment A would be rejected and investment B accepted if the required rate being used were the weighted marginal cost of capital. However, if the risk-return line accurately measures the market risk-return relationship, investment A should be accepted and investment B rejected. Although the expected return for investment A is below the firm's weighted marginal cost of capital, the reduction in risk sufficiently justifies accepting the investment's lower expected return.

Risk-return line

The relation between the appropriate required rates of return for investments having different amounts of risk.

In determining a fair or appropriate rate of return for a given level of project risk, we will examine the different types of risk. We could think of risk as the total variability in returns for the investment, without regard for how it relates to the firm's other investments. This type of risk is called *project risk* or *risk in isolation.* We could alternatively think of the risk in terms of its contribution to the riskiness of the firm's portfolio of assets. Taking this contribution-to-firm approach, we would measure risk as the standard deviation of the returns for all the firm's assets taken together including the new project. A third option, and the one we choose to take is to see risk in terms of risk to shareholders. We assume that the firm's investors hold a diversified portfolio of investments and that we want our measure of risk to indicate how much incremental risk from the project is added to the investors' (not the firm's) diversified portfolio. In this situation, the capital asset pricing model (CAPM) provides us with a technique to evaluate a project. Using the CAPM, the required rate of return for the jth project may be expressed as follows:[23]

$$K_j = k_f + (k_m - k_f)\beta_j \quad (12\text{-}21)$$

where β_j (beta) identifies the volatility of the jth project returns relative to the investor's widely diversified portfolio. Thus, β_j represents the effect of the jth project upon the riskiness of the investor's portfolio, which is also the relevant risk used in the market to value the asset. As in Chapter 3, k_m and k_f represent the expected return for the diversified portfolio and for a risk-free asset, respectively.[24]

EXAMPLE

To illustrate the use of the capital asset pricing model for determining the required rate of return for a project, suppose that a firm is considering a capital investment with an expected return of 16 percent. Management believes that the riskiness of the project should be analyzed in terms of its contribution to the risk of a diversified investor's portfolio. The expected return for a diversified portfolio of assets, k_m, is 14 percent, and the risk-free rate, k_f, is 6 percent. Management has estimated that the beta for the project is 1.12 which indicates that a 1 percent change in the market portfolio's risk premium (that is, $k_m - k_f$) will produce an expected 1.12 percent change in the investment's risk premium ($K_j - k_f$). We may therefore estimate the appropriate required rate of return for the project, as follows:

$$\begin{aligned} K_j &= k_f + (k_m - k_f)\beta_j \quad (12\text{-}21) \\ &= .06 + (.14 - .06)1.12 = .1496 \\ &= 14.96\% \end{aligned}$$

Since the 16 percent expected return for the investment exceeds the required rate of return of 14.96 percent, the investment should be made.

While conceptually attractive, the use of the capital asset pricing model in calculating a project required rate of return is difficult, owing to measurement problems. The primary difficulty lies in measuring the project's beta, or the systematic risk of the project. One approach to circumvent this problem is to use the beta of a "single-industry or -division" firm as the project's beta or alternating by using selected accounting data for this purpose. In addition, the model maintains that the relevant risk is limited to the portion of the risk that the investor cannot eliminate through diversification. For this premise to hold, bankruptcy costs are assumed to equal zero if the firm falls. In reality, bankruptcy costs are generally significant. If a project increases significantly the probability of firm bankruptcy, total variability of project returns is the appropriate risk measure.

---

[23]For further explanation, see Ezra Solomon, "Measuring a Company's Cost of Capital," *Journal of Business* (October 1955), pp. 95–117; Steward C. Myers, "Interactions of Corporate Financing and Investment Decision—Implications for Capital Budgeting," *Journal of Finance* (March 1974), pp. 1–25; Richard S. Bower and Jeffrey M. Jenks, "Divisional Screening Rates," *Financial Management* (Autumn 1975), pp. 42–49; Donald I. Tuttle and Robert H. Litzenberger, "Leverage Diversification and Capital Market Effects on a Risk-Adjusted Capital Budgeting Framework," *Journal of Finance* (June 1968), pp. 427–443; and John D. Martin and David F. Scott, Jr., "Debt Capacity and the Capital Budgeting Decision," *Financial Management* (Summer 1976), pp. 7–14.

[24]For our purposes, we shall assume that the firm finances all its investments with equity. Thus, we need not be concerned with the problems that arise from having to adjust for different financing mixes. This issue is simply beyond the scope of our studies; however to see how it is done, refer to Thomas Conine and Maurry Tamarkin, "Divisional Cost of Capital Estimation: Adjusting For Financial Leverage," *Financial Management* (Spring 1985), pp. 54–58.

SUMMARY

The Concept of the Cost of Capital

The cost of capital is an important concept within financial management. In making an investment, the cost of capital is the rate of return that must be achieved on the company's projects in order to satisfy the investor's required rate of return. If the rate of return from the firm's investment equals the cost of capital, the price of the shares should remain unchanged. In other words, the firm's cost of capital may be defined as the rate of return from an investment that will leave the company's share price unchanged. Therefore, the cost of capital, if certain assumptions are met, represents the minimum acceptable rate of return for the firm's new investments.

Factors Determining the Firm's Cost of Capital

The factors that affect a firm's cost of capital include:

- General economic conditions, as reflected in the demand and supply of funds in the economy and any inflationary pressures that exist.
- The marketability of the firm's securities.
- The firm's operating and financial risks.
- The dollar amount of financing needed for future investments.

Assumptions of the Weighted Cost of Capital Model

The principal assumptions made when computing a firm's weighted cost of capital are:

- Constant business risk across new investments.
- Constant financial risk (ie., the firm maintains the same debt-to-equity ratio in financing new investments).
- Constant dividend payout ratio (dividends divided by earnings).

Computing the Weighted Cost of Capital

To compute the firm's weighted cost of capital:

- Calculate individual costs of capital.
  a. The cost of debt is equal to the effective interest rate on new debt net of issuing costs and adjusted for the tax deductibility of the interest expense.
  b. The cost of preferred shares is equal to the dividend yield on new preferred shares, net of any issuing costs.
  c. The cost of internally generated common equity is equal to the dividend yield on the common shares plus the anticipated percentage increase in dividends (and in the price of the shares) during the forthcoming year).
  d. The cost of new common shares is the same as the cost of internally generated common, but should recognize the effect of flotation costs.
  e. We may also compute the cost of equity by using the CAPM or the risk-premium technique.
- To find the weighted cost of capital, the individual costs of financing are weighted by the percentage of financing provided by each source.
- Because the amount of financing has an effect upon the firm's weighted cost of capital, the expected return from an investment must be compared with the marginal cost of financing the project. If the cost of capital increases as the level of financing increases, we should use the marginal cost of capital, and not the average cost of all the funds raised. Following the basic economic principle of marginal analysis, investments should be made to the point where marginal revenue (internal rate of return) equals the marginal cost of capital.

If it's Not the Weighted Cost of Capital, It's Not a Pepsi

PepsiCo calculates its cost of capital for each of its three major divisions—beverages, snack foods and restaurants. In the chapter, we present the approach and procedures used by the PepsiCo management in computing the firm's weighted cost of capital.

Required Rate of Return for Individual Projects: An Alternative Approach

Owing to the limited assumptions associated with an overall cost of capital for the firm, a single hurdle rate is not generally appropriate for all the investments a firm will analyze. In particular, if the risk associated with a specific investment is significantly different from the firm's existing assets, the weighted cost of capital should not be employed. In this context, a minimum acceptable rate of return that recognizes the different risk levels has to be used. To implement such an approach, financial executives must identify the appropriate measure of risk; generally an analysis of systematic risk through the CAPM model is the best choice.

STUDY QUESTIONS

12-1. Define the term *cost of capital.*

12-2. Why do we calculate a firm's cost of capital?

12-3. In computing the cost of capital, which sources of capital do we consider?

12-4. In general, what factors determine a firm's cost of capital? In answering this question, identify the factors that are within management's control and those that are not.

12-5. What limitations exist in using the firm's cost of capital as an investment hurdle rate?

12-6. How does a firm's tax rate affect its cost of capital? What is the effect of the issuing costs associated with a new security issue?

12-7. **a.** Distinguish between internal common equity and new common shares.
b. Why is a cost associated with internal common equity?
c. Compare approaches that could be used in computing the cost of common equity.

12-8. Define the expression *marginal cost of capital.* Why is the marginal cost of capital an appropriate investment criterion?

12.9. Describe the approach taken by PepsiCo, Inc., in computing the weighted cost of capital.

12-10. How may we avoid the limitation of the weighted cost of capital approach when it requires that we assume business risk is constant?

SELF-TEST PROBLEMS

ST-1. *(Individual Costs of Capital)* Compute the cost for the following sources of financing:

a. A $1,000 par value bond with a market price of $970 and a coupon interest rate of 10 percent. After-tax issuing costs for a new issue would be approximately 5 percent. The bonds mature in ten years and the combined corporate tax rate is 44 percent.

b. A preferred share selling for $100 with an annual dividend payment of $8. If the company sells a new issue, the after-tax issuing cost will be $9 per share.

c. Internally generated common equity totalling $4.8 million. The company's common share sells for $75 and the dividends per share were $9.80 last year. These dividends are expected to increase at an 8-percent growth rate into the indefinite future.

d. New common shares where the most recent dividend was $9.80. The company's dividends per share should continue to increase at an 8 percent growth rate into the indefinite future. The market price of the share is currently $75; however, after-tax issuing of $6 per share is expected if the new shares are issued.

ST-2. *(Level of Financing)* The Argue Company has the following capital structure mix:

Debt	30%
Preferred shares	15
Common shares	55
	100%

Assuming that management intends to maintain the above financial structure, what amount of total investments may be financed if the firm (a) restricts its common financing to the $200,000 that is available from internally generated common equity, or (b) uses $200,000 of internally generated common equity plus $300,000 in new common shares?

ST-3. *(Marginal Cost-of-Capital Curve)* The Zenor Corporation is considering three investments. The costs and expected returns of these projects are shown below:

Investment	Investment Cost	Internal Rate of Return
A	$165,000	17%
B	200,000	13
C	125,000	12

The firm would finance the projects by 40 percent debt and 60 percent common equity. The after-tax cost of debt is 7 percent. Internally generated common totaling $180,000 is available, and the common shareholders' required rate of return is 19 percent. If new shares are issued, the after-tax cost will be 22 percent.

a. Construct a weighted marginal cost of capital curve.

b. Which projects should be accepted?

ST-4. *(Weighted Cost of Capital)* David Osborn is the new vice-president-finance for Jozell Brister, Inc. He is preparing his recommendations for the firm's capital budget. With the information provided below, prepare a graph comparing the company's weighted cost of capital and the prospective investment returns. Which investments should be made?

Investment	Investment Cost	Internal Rate of Return
A	$200,000	18%
B	125,000	16
C	150,000	12
D	275,000	10

The firm's capital structure consists of $2 million in debt, $500,000 in preferred shares, and $2.5 million in common equity. This capital mix is to be maintained for future investments.

The cost of debt (before-tax) is 14.14 percent.

The company's preferred shares each sell for $95 and pay a 14 percent dividend rate on a par value of $100. A new offering of this share would entail underwriting costs and a price discount after-tax of 8 percent of the present market price.

The common equity portion of the investments will be financed first by profits retained within the company of $150,000. If additional common financing is needed, new common shares can be issued at the $30 current price less after-tax issuing costs of $6 per share. Management expects to pay a dividend at the end of this year of $2.50, and dividends should increase at an annual rate of 9 percent thereafter. The firm's combined corporate tax rate is 44 percent.

ST-5. *(Individual Project—Required Return)* Scudder Corporation is evaluating three investments that have the expected returns and betas listed below. Managers want to determine the required rates of return of the projects using the capital asset pricing model. The expected return of a diversified portfolio is 15 percent. The rate on Government of Canada bonds is 8 percent. Which investments should they make?

Investment	Expected Return	Beta
A	18.8%	1.1
B	13.5	.90
C	15.0	.80

STUDY PROBLEMS (SET A)

12-1A. *(Individual or Component Costs of Capital)* Suppose that a firm has a combined corporate tax rate of 44 percent. Compute the cost for the following sources of financing:

- **a.** A bond that has a $1,000 par value (face value) and a contract or coupon interest rate of 11 percent. A new issue would have an issuing cost of 5 percent of the $1,125 market value. The bonds mature in ten years.
- **b.** A new common share issue that paid a $3.27 dividend last year. The earnings per share have grown at a rate of 7 percent per year. This growth rate is expected to continue into the foreseeable future. The company maintains a constant dividend/earnings ratio of 30 percent. The price of this share is now $43.00, but 5 percent after-tax issuing costs are anticipated.
- **c.** Internal common equity where the current market price of the common share is $43. The expected dividend this coming year should be $3.50, increasing thereafter at a 7 percent annual growth rate.
- **d.** A preferred share issue paying a 9 percent dividend on a $150 par value. If a new issue is offered, after-tax issuing costs will be 12 percent of the current price of $175.
- **e.** A bond selling to yield 12 percent after issuing costs, but prior to adjusting for the combined corporate tax rate of 44 percent. In other words, 12 percent is the rate that equates the net proceeds from the bond with the present value of the future cash flows (principal and interest).

12-2A. *(Level of Financing)* The Mathews Company has the following capital structure mix:

Debt	$525,000
Preferred shares	225,000
Common shares	450,000

Using the capital structure mix, compute the total investment amount if the company uses

- **a.** $700,000 of debt
- **b.** $67,500 of preferred shares
- **c.** $300,000 of retained earnings only, or
- **d.** $100,000 of retained earnings plus $600,000 of new common shares.

12-3A. *(Individual or Component Costs of Capital)* Suppose that a firm has a combined corporate tax rate of 44 percent. Compute the cost for the following sources of financing:

- **a.** A bond selling to yield 8 percent after issuing cost, but prior to adjusting for the combined corporate tax rate of 44 percent. In other words, 8 percent is the rate that equates the net proceeds from the bond with the present value of the future cash flows (principal and interest).

b. A new common share issue has recently paid a $1.05 dividend. The earnings per share have grown at a rate of 5 percent per year. This growth rate is expected to continue into the foreseeable future. The company maintains a constant dividend/earnings ratio of 40 percent. The price of this share is now $25, but 9 percent after-tax issuing costs are anticipated.

c. A bond that has a $1,000 par value (face value) and a contract or coupon interest rate of 12 percent. A new issue would net the company 90 percent of the $1,150 market value. The bonds mature in 20 years.

d. A preferred share issue paying a 7 percent dividend on a $100 par value. If a new issue is offered, the company can expect to receive a net price of $85 per share.

e. Internal common equity where the current market price of the common share is $25. The expected dividend this forthcoming year should be $1.10, increasing thereafter at a 5 percent annual growth rate.

12-4A. *(Cost of Equity)* Salte Corporation is issuing new common shares at a market price of $27. Dividends last year were $1.45 and are expected to grow at an annual rate of 6 percent forever. After-tax issuing costs will be 6 percent of market price. What is Salte's cost of equity?

12-5A. *(Cost of Preferred Shares)* The preferred shares of Walter Industries sell for $36 and pay $2.50 in dividends. The net price received from the security after tax issuing costs are deducted is $32.50. What is the cost of capital for the preferred shares?

12-6A. *(Cost of Debt)* Belton is issuing a $1,000 par value bond that pays 7 percent annual interest and matures in 15 years. Investors are willing to pay $958 for the bond. Issuing costs will be 11 percent of market value. The company is in a 44 percent tax bracket. What will be the firm's after-tax cost of debt on the bond?

12-7A. *(Cost of Debt)* The Zephyr Corporation is contemplating a new investment to be financed 33 percent from debt. The firm could sell new $1,000 par value bonds at a net price of $945. The coupon interest rate is 12 percent, and the bonds would mature in 15 years. If the company is in a 44 percent tax bracket, what is the after-tax cost of capital to Zephyr for bonds?

12-8A. *(Cost of Preferred Shares)* Your firm is planning to issue preferred shares. The share sells for $115; however, if new shares are issued, the company would receive a net price of $98. The par value of the share is $100 and the dividend rate is 14 percent. What is the cost of capital for the shares to your firm?

12-9A. *(Cost of Internal Equity)* Pathos Co.'s common shares are currently selling for $21.50. Dividends paid last year were $.70. After-tax issuing costs on new shares will be 10 percent of market price. The dividends and earnings per share are projected to have an annual growth rate of 15 percent. What is the cost of internal common equity for Pathos?

12-10A. *(Cost of Debt)* Sincere Stationery Corporation needs to raise $500,000 to improve its manufacturing plant. It has decided to issue a $1,000 par value bond with a 14 percent annual coupon rate and a 10-year maturity. If the investors require a 9 percent rate of return

a. Compute the market value of the bonds.

b. What will the net price be if issuing costs are 10.5 percent of the market price and the combined corporate tax rate is 44 percent?

c. How many bonds will the firm have to issue to receive the needed funds?

d. What is the firm's after-tax cost of debt?

12-11A. *(Cost of Debt)*

a. Rework problem 12-10A assuming a 10 percent coupon rate. What effect does changing the coupon rate have on the firm's after-tax cost of capital?

b. Why is there a change?

12-12A. *(Cost of Equity)* The common share for the Bestsold Corporation sells for $58. If a new issue is sold, the after-tax issuing cost is estimated to be 8 percent. The company pays 50 percent of its earnings in dividends, and a $4 dividend was recently paid. Earnings per share five years ago were $5. Earnings are expected to continue to grow at the same annual rate in the future as during the past five years. The firm's combined corporate tax rate is 44 percent. Calculate the cost of (a) internal common and (b) external common.

12-13A. *(Individual Project—Required Return)* Welton Corporation is evaluating two investments that have the expected returns and betas listed below. Managers want to determine the required rates of return of the projects using the capital asset pricing model. The expected return of a diversified portfolio is 16 percent. The rate on Government of Canada bonds is 7 percent. Which investments should they make?

Investment	Expected Return	Beta
A	15.7%	0.85
B	17.5	1.05

12-14A. *(Weighted Cost of Capital)* The capital structure for the Carion Corporation is provided below. The company plans to maintain its debt structure in the future. If the firm has a 5.5 percent cost of debt, a 13.5 percent cost of preferred shares, and an 18 percent cost of common shares, what is the firm's weighted cost of capital?

Capital Structure ($000)	
Bonds	$1,083
Preferred shares	268
Common shares	3,681
	$5,032

12-15A. *(Level of Financing)* Using the capital structure mix in problem 12-14A, compute the total investment amount if the company uses

a. $200,000 of debt
b. $40,000 of preferred shares
c. $100,000 of retained earnings only, or
d. $100,000 of retained earnings plus $50,000 of new common shares.

12-16A. *(Individual Project—Required Return)* Hastings Inc. is analyzing several investments. The expected returns and standard deviations of each project are tabulated below. The firm's cost of capital is 16.5 percent, and the standard deviation of the average project is 7 percent. The current rate on long-term Government of Canada bonds is 8.5 percent. Which investments should the firm accept?

Project	Investment's Expected Return	Standard Deviations of Returns
A	18.0%	7.0%
B	13.8	6.0
C	15.3	7.8
D	11.4	5.0

12-17A. *(Weighted Cost of Capital)* Calvert Inc.'s capital structure is provided below. Issuing costs would be (a) 15 percent of market value for a new bond issue, (b) after-tax issuing costs of $1.21 per share for common shares, and (c) after-tax issuing costs of $2.01 per share for preferred shares. The dividends for common shares were $2.50 last year and are projected to have an annual growth rate of 6 percent. The firm is in a 44 percent tax bracket. What is the weighted cost of capital if the firm finances in the proportions shown below? Market prices are $1,035 for bonds, $19 for preferred shares, and $35 for common shares. There will be $500,000 of internal common available.

Calvert Inc. Balance Sheet

Type of Financing		Percentage of Future Financing
Bonds (8%, $1,000 par, 16-year maturity)		38%
Preferred shares (5,000 shares outstanding, $50 par, $1.50 dividend)		15
Common shares		47
	Total	100%

12-18A. *(Weighted Cost of Capital)* The Bach's Candy Corporation has determined the marginal costs of capital for debt after-taxes, preferred shares, and common equity as follows:

Source	Amount of Capital	Cost
Debt	$0–$175,000	4.8%
	$175,001–$300,000	5.5
	Over $300,000	6.0
Preferred shares	$0–50,000	10.0
	$50,001–$75,000	12.0
	Over $75,000	13.0
Common shares	$0–$400,000[a]	15.0
	$400,001–$750,000	18.0
	Over $750,000	22.0

[a] $400,000 is available from internally generated common equity.

The firm maintains a capital mix of 45 percent debt, 5 percent preferred shares, and 50 percent common equity. Construct Bach's weighted marginal cost of capital curve.

12-19A. *(Marginal Cost of Capital)* Mary Basett Inc. a national advertising firm, is analyzing the following investment opportunities.

Investment	Investment Cost	Internal Rate of Return
A	$ 50,000	14.5%
B	$200,000	17.9
C	$325,000	15.6
D	$125,000	12.4
E	$400,000	10.9
F	$ 75,000	13.8

The information needed to calculate the firm's weighted marginal cost of capital is presented below. Construct Basset's weighted marginal cost of capital curve and decide which investments should be accepted.

Source	Percentage	Amount of Capital	After-tax Cost
Debt	40%	$0–$300,000	4.5%
		Over $300,000	6.0
Preferred shares	8	$0–50,000	9.5
		$50,001–$100,000	10.5
		Over $100,000	11.0
Common shares	52	$0–$520,000	16.0
		Over $520,000	18.0

12-20A. *(Weighted Cost of Capital)* Blacktop Chemical Co. is considering five investments. The cost of each of these investments is shown below. Retained earnings of $650,000 will be available for investment purposes, and management can issue the following securities:

1. **Bonds.** $270,000 can be issued at an after-issuing, before-tax cost of 8.5 percent. Over $270,000 the cost will be 9.75 percent.
2. **Preferred shares.** The shares can be issued at the prevailing market price. The firm incurs issuing costs of $1.55 per share up to an issue size of $90,000 and thereafter $2.80 per share.
3. **Common shares.** The shares will be issued at the market price. For an issue of $250,000 issuing costs will be $1 per share. For any additional issue of common shares, the issuing costs will be $1.75 per share.

The firm's combined tax rate is 44 percent. The firm's common share paid a dividend of $1.80 last year and is expected to grow at an annual rate of 9 percent. Market prices are $975 for bonds, $39 for preferred shares and $23 for common shares. Determine which projects should be accepted, based upon a comparison of the internal rate of return (IRR) of the investments and the weighted marginal cost of capital. The firm's capital structure at market value is shown below, and this mix is to be used for future investments.

Investment	Investment Cost	Internal Rate of Return
A	200,000	16.0%
B	650,000	12.0
C	115,000	9.0
D	875,000	10.0
E	180,000	15.0
Total	$2,020,000	

Capital Structure	Amount of Capital	Percentage of Financing
Bonds (9%, $1000, 18-year maturity)	$3,000,000	43%
Preferred shares (10%, $45 par, 30,000 shares outstanding)	1,350,000	20
Common shares	2,600,000	37
Total	$6,950,000	100%

12-21A. *(Weighted Cost of Capital)* Heard Ski Inc. is a regional manufacturer of ski equipment. The firm's target financing mix appears as follows:

	Percentage
Debt	30%
Preferred shares	10
Common shares	60%
Total	100%

The firm's management is analyzing the following investment opportunities.

Investment	Investment Cost	Internal Rate of Return
A	$175,000	16.0%
B	$100,000	14.0
C	$125,000	12.0
D	$200,000	10.0
E	$250,000	9.0
F	$150,000	8.0

Using the accept-reject criterion, Paul Heard, president of the firm, has compiled the necessary data for computing the firm's weighted marginal cost of capital. The cost information indicates the following:

1. Debt can be raised at the following after-issuing, before-tax costs:

Amount	Cost
$0–$150,000	8.0%
$150,001–$225,000	9.0
Over $225,000	10.5

2. Preferred shares can be issued paying an annual dividend of $8.50. The par value of the stock is $100. Also, the market price of a preferred share is $100. If new shares were issued, the company would receive a net price of $80 on the first $75,000. Thereafter, the net amount received would be reduced to $75.
3. Common equity is provided first by internally generated funds. Profits for the year that should be available for reinvestment purposes are projected at $150,000. Additional common shares can be issued at the current $72 market price less 15 percent in issuing costs. However, a 20 percent issuing cost is expected if more than $225,000 in new common shares is issued. The dividend per share was $2.75 last year, and the long-term growth rate for dividends is 9 percent.

a. Given that the firm's combined tax rate is 44 percent, compute the firm's weighted marginal cost of capital at a financing level up to $1 million.

b. Construct a graph that presents the firm's weighted marginal cost of capital relative to the amount of financing.

c. What is the appropriate size of the capital budget, and which projects should be accepted?

INTEGRATIVE PROBLEM

Three-Six Systems Inc. is starting its capital planning process for 1997. Bill Reichenstein, the firm's CFO, calculates the weighted cost of capital each year to be used as the firm's discount rate in its net present value analysis. You have been asked to help Reichenstein compute the firm's cost of capital. In fact, he has asked you to gather the data you believe to be necessary to do the job. Your efforts have resulted in the following information:

1. The company's bonds currently are rated as AA bonds and have a market value of $850 on a $1,000 par value. The bond's coupon interest rate is 7 percent. The bonds mature in 12 years. A new issue would have an after-tax issuing cost of 4 percent of the market value.
2. The company has preferred shares that pay a 9 percent dividend rate on a $100 par value. If a new issue is offered, the after-tax issuing costs would be 10 percent of the current price of $90.
3. The firm's common shares paid a dividend of $3.50 last year. The earnings per share have grown at a rate of 7 percent per year. This growth rate is expected to continue into the foreseeable future. The company maintains a constant dividend/earnings ratio of 25 percent. The shares sell for $55, but there would be a 12 percent after-tax issuing cost on a new issue.
4. The existing short-term bank notes have an interest rate of 9 percent, but the notes are due to be renewed at an 8 percent interest rate next month.
5. The firm's current capital structure, as presented in the financial statements, is as follows:

Short-term bank notes	$150,000
Bonds (650 bonds outstanding)	650,000
Preferred shares (1,000 shares outstanding)	100,000
Common shares (10,000 shares)	300,000

6. There should be $300,000 of retained earnings available for reinvestment purposes in the coming year.
7. The firm's combined tax rate is 50 percent.
8. Mike Robinson, the vice president for operations mentioned that he uses four companies for peer-group comparisons. These four firms, along with their betas that have been adjusted for financial leverage effects, are as follows:

	Beta
Souther Electronics Inc.	1.10
Electronics Arts Inc.	1.20
Adaptec Inc.	1.30
Banyan Systems Inc.	0.90

In addition to the company data, you have learned that Government of Canada bonds currently sell to yield the investors a return of 5.5 percent. The spread above Government of Canada bonds for AA bonds has averaged 4 percent. Finally, you learn from a recent publication by an investment dealer that the market return premium of common shares above risk-free securities ($k_m - k_f$) has averaged 6.5 percent.

Assignment:

a. Calculate the after-tax cost of the short-term notes.

b. Compute the after-tax cost of the bonds by using two different methods: (1) by finding the rate of return that the bondholders will earn on the bonds if held to maturity, and (2) by using the spread over Government of Canada bonds. Calculate the average of the cost of bonds according to these two methods.

c. What is the cost of the firm's preferred shares?

d. Estimate the cost of internal common equity, first based on the dividend-growth model and then by using the capital asset pricing model to estimate this firm's beta, Robinson uses the average beta of peer-group firms. What is the average of these two estimates?

e. Compute the cost of new common shares according to the dividend-growth model.

f. Determine the market value weights to be used in computing the firm's weighted cost of capital.

g. Calculate the weighted cost of capital for Three-Six Systems, Inc. assuming (1) only internal common is used in financing the equity portion of new investments, and then (2) new common shares are issued to finance the firm's investments.

h. Draw a graph showing the marginal weighted cost of capital for the firm.

STUDY PROBLEMS (SET B)

12-1B. *(Individual or Component Costs of Capital)* Suppose that a firm has a combined corporate tax rate of 44 percent. Compute the cost for the following sources of financing:

a. A bond that has a $1,000 par value (face value) and a contract or coupon interest rate of 12 percent. A new issue would have an issuing cost of 6 percent of the $1,125 market value. The bonds mature in 10 years.

b. New common shares will be issued with an expected dividend of $3.25. The earnings per share have grown at a rate of 7 percent per year. This growth rate is expected to continue into the foreseeable future. The company maintains a constant dividend/earnings ratio of 30 percent. The price of this share is now $43.50, but 5 percent after-tax issuing costs are anticipated.

c. Internal common equity where the current market price of the common share is $43.50. The expected dividend this coming year should be $3.25, increasing thereafter at a 7 percent annual growth rate.

d. A preferred share issue paying a 10 percent dividend on a $125 par value. If a new issue is offered, after-tax issuing costs will be 12 percent of the current price of $150.

e. A bond selling to yield 13 percent after issuing costs, but prior to adjusting for the combined corporate tax rate of 44 percent. In other words, 13 percent is the rate that equates the net proceeds from the bond with the present value of the future cash flows (principal and interest).

12-2B. *(Level of Financing)* The Collier Company has the following capital structure mix:

Debt	$600,000
Preferred shares	200,000
Common shares	400,000

Using the capital structure mix, compute the total investment amount if the company uses

a. $500,000 of debt

b. $50,000 of preferred shares

c. $275,000 of retained earnings only, or

d. $125,000 of retained earnings plus $600,000 of new common shares

12-3B. *(Individual or Component Costs of Capital)* Suppose that a firm has a combined corporate tax rate of 44 percent. Compute the cost for the following sources of financing:

a. A bond selling to yield 9 percent after issuing cost, but prior to adjusting for the combined corporate tax rate of 44 percent. In other words, 9 percent is the rate that equates the net proceeds from the bond with the present value of the future cash flows (principal and interest).

b. A new common share issue has recently paid a $1.25 dividend. The earnings per share have grown at a rate of 6 percent per year. This growth rate is expected to continue into the foreseeable future. The company maintains a constant dividend/earnings ratio of 40 percent. The price of this share is now \$30, but 9 percent after-tax issuing costs are anticipated.

c. A bond that has a \$1,000 par value (face value) and a contract or coupon interest rate of 13 percent. A new issue would net the company 90 percent of the \$1,125 market value. The bonds mature in 20 years.

d. A preferred share issue paying a 7 percent dividend on a \$125 par value. If a new issue is offered, the company can expect to receive a net price of \$90 per share.

e. Internal common equity where the current market price of the common share is \$30. The expected dividend this forthcoming year should be \$1.325, increasing thereafter at a 6 percent annual growth rate.

12-4B. (*Cost of Equity*) Falon Corporation is issuing new common shares at a market price of \$28. Dividends last year were \$1.30 and are expected to grow at an annual rate of 7 percent forever. After-tax issuing costs will be 6 percent of market price. What is Falon's cost of equity?

12-5B. *(Cost of Preferred Shares)* The preferred shares of Gator Industries sell for \$35 and pay \$2.75 in dividends. The net price received from the security when after-tax issuing costs are deducted, is \$32.50. What is the cost of capital for the preferred shares?

12-6B. *(Cost of Debt)* Temple is issuing a \$1,000 par value bond that pays 8 percent annual interest and matures in 15 years. Investors are willing to pay \$950 for the bond. Issuing costs will be 11 percent of market value. The company is in a 44 percent tax bracket. What will be the firm's after-tax cost of debt on the bond?

12-7B. *(Cost of Debt)* The Walgren Corporation is contemplating a new investment to be financed 33 percent from debt. The firm could sell new \$1,000 par value bonds at a net price of \$950. The coupon interest rate is 13 percent, and the bonds would mature in 15 years. If the company is in a 44 percent tax bracket, what is the after-tax cost of capital to Walgren for bonds?

12-8B. *(Cost of Preferred Shares)* Your firm is planning to issue preferred shares. The share sells for \$120; however, if new shares are issued, the company would receive a net price of only \$97. The par value of the share is \$100 and the dividend rate is 13 percent. What is the cost of capital for the shares to your firm?

12-9B. *(Cost of Internal Equity)* Oxford Inc.'s common share is currently selling for \$22.50. Dividends paid last year were \$80. After-tax issuing costs on new shares will be 10 percent of market price. The dividends and earnings per share are projected to have an annual growth rate of 16 percent. What is the cost of internal common equity for Oxford?

12-10B. *(Cost of Debt)* Gillian Stationery Corporation needs to raise \$600,000 to improve its manufacturing plant. It has decided to issue a \$1,000 par value bond with a 15 percent annual coupon rate and a ten-year maturity. If the investors require a 10 percent rate of return

a. Compute the market value of the bonds.

b. What will the net price be if issuing costs are 11.5 percent of the market price and the firm's combined corporate tax rate is 44 percent?

c. How many bonds will the firm have to issue to receive the needed funds?

d. What is the firm's after-tax cost of debt?

12-11B. *(Cost of Debt)*

a. Rework Problem 12-10B assuming a 10 percent coupon rate. What effect does changing the coupon rate have on the firm's after-tax cost of capital?

b. Why is there a change?

12-12B. *(Cost of Equity)* The common shares for the Hetterbrand Corporation sell for \$60. If a new issue is sold, the after-tax issuing cost is estimated to be 9 percent. The company pays 50 percent of its earnings in dividends, and a \$4.50 dividend was recently paid. Earnings per share five years ago were \$5. Earnings are expected to continue to grow at the same annual rate in the future as during the past five years. The firm's combined corporate tax rate is 44 percent. Calculate the cost of (a) internal common and (b) external common.

12-13B. *(Individual Project—Required Return)* Dellington Corporation is evaluating two investments that have the expected returns and betas listed below. Managers want to determine the required rates of return of the projects using the capital asset pricing model. The expected return of a diversified portfolio is 16 percent. The rate on Government of Canada bonds is 7 percent. Which investments should they make?

Investment	Expected Return	Beta
A	20.0%	1.25
B	16.2	.95

12-14B. *(Weighted Cost of Capital)* The capital structure for the Bias Corporation is provided below. The company plans to maintain its debt structure in the future. If the firm has a 6 percent cost of debt, a 13.5 percent cost of preferred shares, and a 19 percent cost of common shares, what is the firm's weighted cost of capital?

Capital Structure ($000)	
Bonds	$1,100
Preferred shares	250
Common shares	3,700
	$5,050

12-15B. *(Level of Financing)* Using the capital structure mix in problem 12-14B, compute the total investment amount if the company uses

- **a.** $150,000 of debt
- **b.** $60,000 of preferred shares
- **c.** $120,000 of retained earnings only, or
- **d.** $120,000 of retained earnings plus $60,000 of new common shares

12-16B. *(Individual Project—Required Return)* Badger Inc. is analyzing several investments. The expected returns and standard deviations of each project are tabulated below. The firm's cost of capital is 16.5 percent, and the standard deviation of the average project is 7 percent. The current rate on long-term Government of Canada bonds is 8.5 percent. Which investments should the firm accept?

Project	Investment's Expected Return	Standard Deviations of Returns
A	20.0%	8.0%
B	16.2	6.8
C	15.0	7.0
D	11.2	6.1

12-17B. *(Weighted Cost of Capital)* R. Stewart Inc.'s capital structure is provided below. Issuing costs would be (a) 13 percent of market value for a new bond issue, (b) after-tax issuing costs of $1.25 per share for common shares, and (c) after-tax issuing costs of $2.50 per share for preferred shares. The dividends for common shares were $3.25 last year and are projected to have an annual growth rate of 6 percent. The firm is in a 44 percent tax bracket. What is the weighted cost of capital if the firm finances in the proportions shown below? Market prices are $1,040 for bonds, $18 for preferred shares, and $30 for common shares. There will be $250,000 of internal common.

R. Stewart Inc. Balance Sheet

Type of Financing		Percentage of Future Financing
Bonds (8%, $1,000 par, 16-year maturity)		38%
Preferred shares (5,000 shares outstanding, $50 par, $1.50 dividend)		15
Common shares		47
	Total	100%

12-18B. *(Weighted Cost of Capital)* The Hun Sen Corporation has determined the marginal costs of capital for debt after taxes, preferred shares, and common equity as follows:

Source	Amount of Capital	Cost
Debt	\$0–\$200,000	6.0%
	\$200,001–\$350,000	7.5
	Over \$350,000	9.0
Preferred shares	\$0–70,000	11.0
	\$70,001–\$100,000	13.0
	Over \$100,000	14.0
Common shares	\$0–\$500,000[a]	16.0
	\$500,001–\$750,000	20.0
	Over \$750,000	22.0

[a]\$500,000 is available from internally generated common equity.

The firm maintains a capital mix of 50 percent debt, 10 percent preferred shares, and 40 percent common equity. Construct Hun Sen's weighted marginal cost of capital curve.

12-19B. *(Marginal Cost of Capital)* TNC Inc. a national advertising firm, is analyzing the following investment opportunities.

Investment	Investment Cost	Internal Rate of Return
A	\$100,000	17.0%
B	\$150,000	16.0
C	\$325,000	15.0
D	\$175,000	9.0
E	\$420,000	14.0
F	\$100,000	12.0

The information needed to calculate the firm's weighted marginal cost of capital is presented below. Construct TNC's weighted marginal cost of capital curve and decide which investments should be accepted.

Source	Percentage	Amount of Capital	After-Tax Cost
Debt	45%	\$0–\$300,000	7.0%
		Over \$300,000	8.0
Preferred shares	10	\$0–\$50,000	10.0
		\$50,001–\$100,000	11.0
		Over \$100,000	13.0
Common shares	45	\$0–\$520,000	18.0
		Over \$520,000	20.0

12-20B. *(Weighted Cost of Capital)* The Hannitin Co. is considering five investments. The cost of each of these investments is shown below. Retained earnings of \$850,000 will be available for investment purposes, and management can issue the following securities:

1. **Bonds.** \$350,000 can be issued at an after-issuing, before-tax cost of 9.0 percent. Over \$350,000 the cost will be 10.75 percent.
2. **Preferred shares.** The shares can be issued at the prevailing market price. The firm incurs issuing costs of \$1.75 per share up to an issue size of \$100,000 and thereafter \$2.40 per share.

3. **Common shares.** The shares will be issued at the market price. For an issue of $300,000 issuing costs will be $2 per share. For any additional issue of common shares, the issuing costs will be $2.75 per share.

The firm's combined tax rate is 44 percent. The firm's common share paid a dividend of $2.50 last year and is expected to grow at an annual rate of 10 percent. Market prices are $975 for bonds, $49 for preferred shares and $43 for common shares. Determine which projects should be accepted, based upon a comparison of the internal rate of return (IRR) of the investments and the weighted marginal cost of capital. The firm's capital structure at market value is shown below, and this mix is to be used for future investments.

Investment	Investment Cost	Internal Rate of Return
A	$ 250,000	21.0%
B	$ 700,000	17.0
C	$ 100,000	12.0
D	$ 500,000	13.0
E	$ 250,000	17.0
Total	$1,800,000	

Capital Structure	Amount of Capital	Percentage of Financing
Bonds (9%, $1,000, 18-year maturity)	$2,800,000	35%
Preferred shares (10%, $45 par, 30,000 shares outstanding)	1,200,000	15
Common shares	4,000,000	50
Total	$8,000,000	100%

12-21B. *(Weighted Cost of Capital)* Dalton Ski Inc. is a regional manufacturer of ski equipment. The firm's target financing mix appears as follows:

	Percentage
Debt	30%
Preferred shares	10
Common shares	60
Total	100%

The firm's management is analyzing the following investment opportunities.

Investment	Investment Cost	Internal Rate of Return
A	$175,000	16.0%
B	$100,000	14.0
C	$125,000	12.0
D	$200,000	10.0
E	$250,000	9.0
F	$150,000	8.0

Using the accept-reject criterion, Carter Dalton, president of the firm, has compiled the necessary data for computing the firm's weighted marginal cost of capital. The cost information indicates the following:

1. Debt can be raised at the following after-issuing, before-tax costs:

Amount	Cost
$0–$200,000	9.0%
$200,001–$300,000	10.0
Over $300,000	11.0

2. Preferred shares can be issued paying an annual dividend of $10. The par value of the stock is $100. Also, the market price of a preferred share is $100. If new shares were issued, the company would receive a net price of $90 on the first $100,000. Thereafter, the net amount received would be reduced to $85.
3. Common equity is provided first by internally generated funds. Profits for the year that should be available for reinvestment purposes are projected at $200,000. Additional common shares can be issued at the current $80 market price less 15 percent in issuing costs. However, a 20 percent issuing cost is expected if more than $250,000 in common shares is issued. The dividend per share was $4.00 last year, and the long-term growth rate for dividends is 12 percent.

a. Given that the firm's combined tax rate is 44 percent, compute the firm's weighted marginal cost of capital at a financing level up to $1 million.

b. Construct a graph that presents the firm's weighted marginal cost of capital relative to the amount of financing.

c. What is the appropriate size of the capital budget, and which projects should be accepted?

SELF-TEST SOLUTIONS

The following notations are used in this group of problems:

k_b = the before-tax cost of debt
K_b = the after-tax cost of debt
K_p = the after-tax cost of preferred shares
K_c = the after-tax cost of internal common equity
K_{nc} = the after-tax cost of new common shares
T = the combined corporate tax rate
D_t = the dollar dividend per share, where D_0 is the most recently paid dividend and D_1 is the forthcoming dividend
P_0 = the value (present value) of a security
NP_0 = the value of a security less any issuing costs incurred in issuing the security

SS-1. a.

$$\$921.50 = \sum_{t=1}^{10} \frac{\$100}{(1+k_b)^t} + \frac{\$1{,}000}{(1+k_b)^{10}}$$

Rate	*Value*		
11%	$940.90	$19.40 (11% to k_b%)	$53.90 (11% to 12%)
k_b%	$921.50		
12%	$887.00		

$$k_b = 0.11 + \left(\frac{\$19.40}{\$53.90}\right)0.01 = 11.36\%$$

$$K_b = 11.36\%\ (1 - 0.44) = 6.36\%$$

b. $K_p = \frac{D}{NP_0}$

$$K_p = \frac{\$8}{\$100 - \$9} = 8.79\%$$

c. $K_c = \frac{D_1}{P_0} + g$

$$K_c = \frac{\$9.80(1 + .08)}{\$75} + 0.08 = 22.0\%$$

d. $K_{nc} = \frac{D_1}{NP_0} + g$

$$K_{nc} = \frac{\$9.80(1 + .08)}{\$75 - \$6} + 0.08 = 23.0\%$$

SS-2. Dollar breaks = $\frac{\text{amount of financing at a given cost}}{\text{percentage of funds provided by the specific source}}$

a. $\frac{\$100{,}000}{0.30} = \$333{,}333.33$

b. $\frac{\$150{,}000}{0.30} = \$500{,}000$

c. $\frac{\$40{,}000}{0.15} = 266{,}667.67$

d. $\frac{\$90{,}000}{0.15} = \$600{,}000$

e. $\frac{\$200{,}000}{0.55} = \$363{,}636.36$

f. $\frac{\$500{,}000}{0.55} = \$909{,}090.91$

SS-3. a. The increase (break) in the weighted marginal cost of capital curve will occur resulting from the cost of debt:

$$\frac{\$120{,}000}{.40} = \$300{,}000$$

Increase from the cost of common:

$$\frac{\$180{,}000}{.60} = \$300{,}000$$

Weighted cost of capital (K_0) for

$0–$300,000 Total Financing

	Weights	Weighted Costs	Costs
Debt	40%	7%	2.80%
Common shares	60	19	11.40%
			K_0 = 14.20%

Over $300,000 Total Financing

	Weights	Weighted Costs	Costs
Debt	40%	11%	4.40%
Common shares	60	22	13.20%
			K_0 = 17.60%

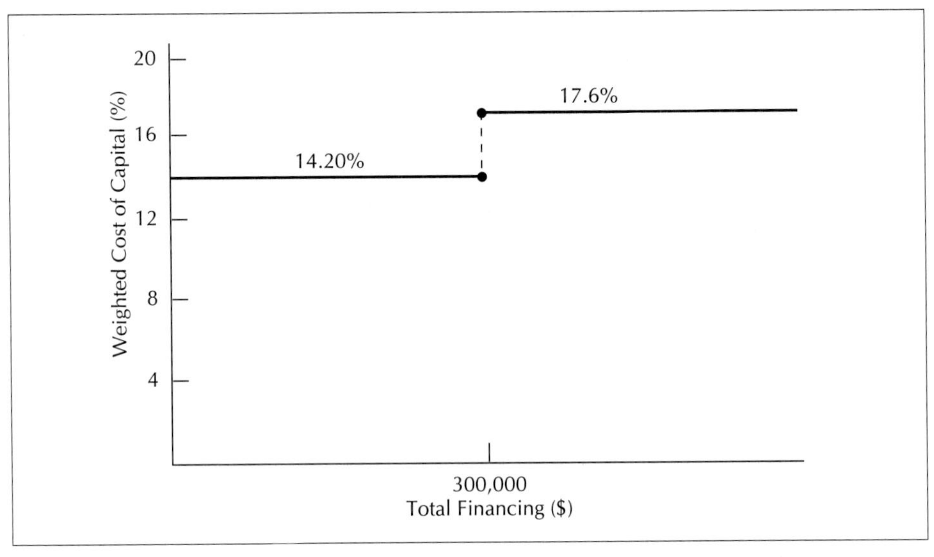

b. Only project A should be accepted.

SS-4. (1) Compute weights

	Capital Structure	Capital Mix (Weights)
Debt	$2,000,000	40%
Preferred shares	500,000	10
Common shares	2,500,000	50
	$5,000,000	100%

(2) Compute individual costs

Debt:

$0–$120,000: 14.14% (1 – .44) = 7.92%

over $120,000: 17.68% (1 – .44) = 9.90%

Preferred Shares:

$0–$50,000: $\frac{\$14}{\$95(1-.08)} = \frac{\$14}{\$87.40} = 16.02\%$

over $50,000: $\frac{\$14}{\$95(1-.11)} = \frac{\$14}{\$84.55} = 16.56\%$

Common Shares:

$0–$150,000: $\frac{\$2.50}{\$30} + .09 = .1733 = 17.33\%$

over $150,000: $\frac{\$2.50}{\$27} + .09 = .1826 = 18.26\%$

(3) Calculate the increase (break) in the weighted marginal cost of capital curve caused by an increase in the cost of common shares:

$$\textbf{Debt} = \frac{\$120{,}000}{.40} = \$300{,}000$$

$$\textbf{Preferred Shares} = \frac{\$50{,}000}{.10} = \$500{,}000$$

$$\textbf{Common Shares} = \frac{\$150{,}000}{.50} = \$300{,}000$$

(4) Construct the weighted cost of capital curve

$0-$300,000 Total Financing

	Weights	Weighted Costs	Costs
Debt	40%	7.92%	3.17%
Preferred shares	10	16.02	1.60%
Common shares	50	17.33	8.67%
			K_0 = 13.44%

At Least $300,001 But Not More Than $500,000

	Weights	Weighted Costs	Costs
Debt	40%	9.90%	3.96%
Preferred shares	10	16.02	1.60%
Common shares	50	18.26	9.13%
			K_0 = 14.69%

More than $500,000

	Weights	Weighted Costs	Costs
Debt	40%	9.90%	3.96%
Preferred shares	10	16.56	1.66%
Common shares	50	18.26	9.13%
			K_0 = 14.75%

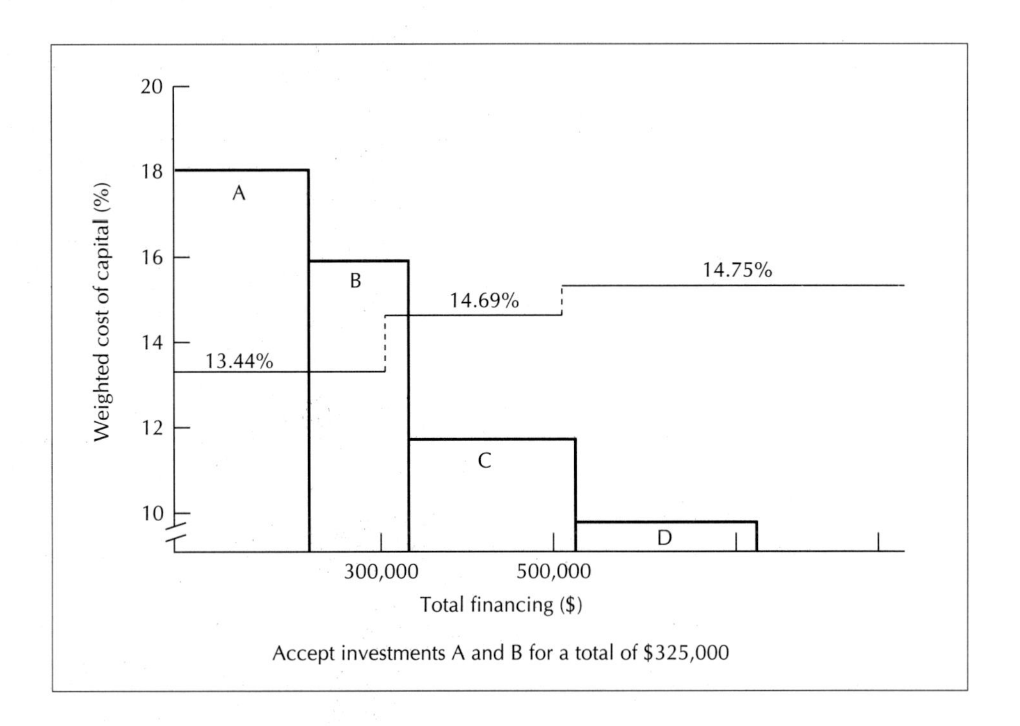

SS-5.

Project	Rate of Return	Required Return	Expected Decision
A	0.08 + (0.15 − 0.08)(1.1) = 15.7%	18.8%	Accept
B	0.08 + (0.15 − 0.08)(0.90) = 14.3%	13.5%	Reject
C	0.08 + (0.15 − 0.08)(0.80) = 13.6%	15.0%	Accept

CAPITAL-BUDGETING TECHNIQUES

CHAPTER 13

LEARNING OBJECTIVES

After reading this chapter you should be able to

1. Explain the difficulty encountered in finding profitable projects in competitive markets and the importance of the search.
2. Determine the acceptability of investment proposals by nondiscounted cash flow criteria.
3. Determine the acceptability of investment proposals by discounted cash flow criteria.
4. Explain the importance of ethical considerations in capital budgeting decisions.
5. Explain the trends in the use of different capital-budgeting criteria.

INTRODUCTION

In 1988, the Ford Motor Company made a decision to reenter the minivan market with a new challenger to Chrysler's Caravan and Voyager. Over the past decade, a number of challengers, including the Ford Aerostar, GM's APV, and a number of Japanese models, had entered the ring against the Chrysler minivan all with the same result. The Chrysler minivan had scored a knockout against all comers and had continued to dominate the minivan market by a wide margin.

Given the history of challengers to Chrysler's minivan, it was not an easy decision to challenge the champ. Moreover, the stakes involved were so large that the outcome of this decision would have a major effect on Ford's future value. To challenge Chrysler's dominance in this market, Ford committed $1.5 billion, with roughly $500 million going toward design, engineering, and testing. The result of this was the Ford Windstar, which Ford unveiled in 1994. Whether this decision was a good one only time will tell, but early reports indicate that Ford has come up with a winner.

In this chapter we will look at the process of decision making with respect to investment in fixed assets—that is, should a proposed project be accepted or should it be rejected? We will refer to this process as ***capital budgeting.*** *Typically, these investments involve rather large cash outlays at the outset and commit the firm to a particular course of action over a relatively long period. Thus, if a capital budgeting decision is incorrect, reversing it tends to be costly. In evaluating capital investment proposals, we compare the costs and benefits of each in a number of ways. Some of these methods*

Capital budgeting

The decision-making process with respect to investment in fixed assets. Specifically, it involves measuring the incremental cash flows associated with investment proposals and evaluating those proposed investments.

take into account the time value of money, two do not; however, each of these methods is used frequently in the real world. As you will see, our preferred method of analysis will be the net present value (NPV) method, which compares the present value of inflows and outflows.

CHAPTER PREVIEW

In the next four chapters, we examine the capital-budgeting decision and how it is applied in a business environment. We begin by examining the purpose and the importance of the capital-budgeting process. Next, several capital-budgeting criteria are presented, followed by a look at capital budgeting in practice. Keep in mind that we are developing a framework for decision making on capital investments.

OBJECTIVE 1

FINDING PROFITABLE PROJECTS

Without question, it is easier to *evaluate* profitable projects that it is to *find* them. In competitive markets, generating ideas for profitable projects is extremely difficult.

Moreover, industries that generate large profits attract new entrants. The additional competition and added capacity can result in profits being driven down. For this reason, a firm must have a systematic strategy for generating capital-budgeting projects. Without this flow of new ideas, the firm cannot grow or even survive for long, being forced to live off profits from existing projects with limited lives. So where do these ideas come from for new products, for ways to improve existing products, or for ways to make existing products more profitable? The answer is from inside the firm—from everywhere inside the firm.

BACK TO THE BASICS

The fact that profitable projects are difficult to find relates directly to ***Axiom 5: The Curse of Competitive Markets—Why It's Hard to Find Exceptionally Profitable Projects.*** *The prerequisite of finding a successful investment within a competitive market means that a firm must reduce the competition by creating barriers to entry either through product differentiation or cost advantages.*

Typically, a firm has a research and development department that searches for ways of improving on existing products or finding new products. Ideas may come from within the R&D department or be based on referral ideas from executives, sales personnel, or anyone in the firm. For example, at Ford Motor Company prior to the 1980s, ideas for product improvement had typically been generated in Ford's R&D department. Unfortunately, this strategy was not enough to keep Ford from losing much of its market share to the Japanese. In an attempt to cut costs and improve quality, Ford moved from strict reliance on an R&D department to seeking the input of employees at all levels for new ideas. Bonuses are now provided to workers for their cost-cutting suggestions, and assembly line personnel who can see the production process from a hands-on point of view are now brought into the hunt for new projects. The effect on Ford has been positive and significant. Although not all suggested projects prove to be profitable, many new ideas generated from within the firm turn out to be good ones. The best way to evaluate new investment proposals is the topic of the remainder of this chapter.

BASIC FINANCIAL MANAGEMENT IN PRACTICE

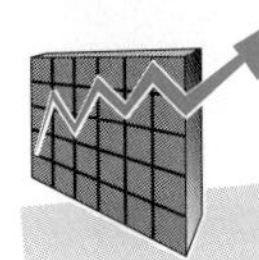

Finding Profitable Projects in Competitive Markets—Creating Them by Developing a Cost Advantage

Federal Express has been able to win a strong competitive position by restructuring the traditional production-cost chain in its industry. Federal Express innovatively redefined the production-cost chain for rapid delivery of small parcels. Traditional firms like Emery and Airborne Express operated by collecting freight packages of varying sizes, shipping them to their destination points via air freight and commercial airlines, and then delivering them to the addressee. Federal Express opted to focus only on the market for overnight delivery of small packages and documents. These were collected at local drop points during the late afternoon hours, flown on company-owned planes during early evening hours to a central hub in Memphis, where—from 11 p.m. to 3 a.m. each night—all parcels were sorted and then reloaded on company planes and flown during the early morning hours to their destination points, where they were delivered the next morning by company personnel using company trucks. The cost structure so achieved by Federal Express was low enough to permit it to guarantee overnight delivery of a small parcel anywhere in North America.

Source: Arthur A. Thompson, Jr., *Economics of the Firm: Theory and Practice* (Englewood Cliffs, NJ: Prentice-Hall, 1989), p. 451. Based on information in Michael E. Porter, *Competitive Advantage* (New York: Free Press, 1985), p. 109.

NONDISCOUNTED CASH FLOW CRITERIA FOR CAPITAL-BUDGETING DECISIONS

OBJECTIVE 2

We are now ready to consider the interpretation of cash flows. Cash flows represent the benefits generated from accepting a capital-budgeting proposal. In the remainder of this chapter, we will assume a given cash flow is generated by a project and will determine whether or not that project should be accepted or rejected.

We will consider five commonly used criteria for determining acceptability of investment proposals. The first two are the least sophisticated in that they do not incorporate the time value of money into their calculations; the final three do take it into account.

Nondiscounted Cash Flow Criterion 1: Payback Period

The **payback period** is the number of years needed to recover the initial cash outlay. As this criterion measures how quickly the project will return its original investment, it deals with cash flows rather than accounting profits. However, this criterion ignores the time value of money since it does not discount these cash flows back to the present. The accept-reject criterion involves whether or not the project's payback period is less than or equal to the firm's maximum desired payback period. For example, if a firm's maximum desired payback period is three years and an

Payback period

A capital-budgeting criterion defined as the number of years required to recover the initial cash investment.

investment proposal requires an initial cash outlay of $10,000 and yields the following set of annual cash flows, what is its payback period? Should the project be accepted?

Year	Cash Flow After Tax
1	$2,000
2	4,000
3	3,000
4	3,000
5	1,000

In this case, after three years the firm will have recaptured $9,000 on an initial investment of $10,000, leaving $1,000 of the initial investment to be recouped. During the fourth year a total of $3,000 will be returned from this investment, and, assuming it will flow into the firm at a constant rate over the year, it will take one-third of the year ($1,000/$3,000) to recapture the remaining $1,000. Thus, the payback period on this project is three and a third years, which is more than the desired payback period. Using the payback period criterion, the firm would reject this project.

Although the payback period is used frequently, it does have some rather obvious drawbacks, which can best be demonstrated through the use of an example. Consider two investment projects, A and B, which involve an initial cash outlay of $10,000 each and produce the annual cash flows shown in Table 13-1. Both projects have a payback period of two years; therefore, in terms of the payback period criterion both are equally acceptable.

However, if we had our choice, it is clear we would select A over B—for the following reasons. First, regardless of what happens after the payback period, project A returns our initial investment to us earlier within the payback period. Thus, if there is a time value of money, the cash flows occurring within the payback period should not be weighted equally, as they are. The payback method does not incorporate any degree of riskiness in determining the time value of cash flows since financing costs are ignored. In addition, all cash flows that occur after the payback period are ignored. This violates the principle that investors desire more in the way of benefits rather than less—a principle that is hard to deny, especially when we are talking about money. Finally, the payback method does not answer the basic question of whether the project is creating an absolute amount of value. In fact, the criterion used to make any decision is relative to some standard, since the payback method requires that we arbitrarily or subjectively select another investment as a standard for comparison purposes.

Discounted payback period

A variation of the payback period defined as the number of years required to recover the initial cash outlay from the discounted net cash flows.

To deal with the criticism that the payback period ignores the time value of money, some firms use the **discounted payback period** approach. The discounted payback period method is similar to the traditional payback period except that it uses discounted net cash flows rather than actual nondiscounted net cash flows in calculating the payback period. The discounted payback period is defined as the

Table 13-1 Payback Period Example Projects

	A	B
Initial cash outlay	–$10,000	–$10,000
Annual net cash inflows:		
Year 1	6,000	5,000
2	4,000	5,000
3	3,000	0
4	2,000	0
5	1,000	0

number of years needed to recover the initial cash outlay from the *discounted net cash flows.* The accept-reject criterion then becomes whether the project's discounted payback period is less than or equal to the firm's maximum desired discounted payback period. Using the assumption that the required rate of return on projects A and B illustrated in Table 13-1 is 17 percent, the discounted cash flows from these projects are given in Table 13-2. The discounted payback period for project A is 3.07 years, calculated as follows:

$$\text{Discounted payback period}_A = 3.0 + \frac{\$74}{\$1{,}068} = 3.07 \text{ years.}$$

If project A's discounted payback period was less than the firm's maximum desired payback period, then Project A would be accepted. Project B, on the other hand, does not have a discounted payback period because it never fully recovers the project's initial cash outlay, and thus should be rejected. Thus, while the discounted payback period is superior to the traditional payback period, in that it accounts for the time value of money in its calculations, its use is limited by the arbitrariness of the process used to select the maximum desired payback period. Moreover, as we will soon see, the net present value criterion is theoretically superior and no more difficult to calculate.

Although these deficiencies limit the value of the payback period as a tool for investment evaluation, the payback period method does have several positive features. First, it deals with cash flows, as opposed to accounting profits, and therefore focuses on the true timing of the project's benefits and costs, even though these cash flows are not adjusted for the time value of money. Second, it is easy to visualize, quickly understood, and easy to calculate. Finally, while the payback period method does have serious deficiencies, it is often used as a rough screening device to eliminate projects whose returns do not materialize until later years. This approach emphasizes the earliest returns, which in all likelihood are less uncertain, and provides for the liquidity needs of the firm. It should also be noted that the deficiency of not accounting for the time value of money can be overcome by calculating a discounted payback period in which cash flows are discounted at the project's cost of capital. Although the advantages of the payback period and discounted payback methods are certainly significant, their disadvantages severely limit their value as discriminating capital-budgeting criteria.

Table 13-2 Discounted Payback Period Example Using a 17 Percent Required Rate of Return

Project A

Year	Nondiscounted Cash Flows	$PVIF_{17\%,n}$	Discounted Cash Flows	Cumulative Discounted Cash Flows
0	−$10,000	1.0000	−$10,000 .	−$10,000
1	6,000	0.8547	5,128	− 4,872
2	4,000	0.7305	2,922	− 1,950
3	3,000	0.6244	1,873	− 77
4	2,000	0.5337	1,067	990
5	1,000	0.4561	456	1,446

Project B

Year	Nondiscounted Cash Flows	$PVIF_{17\%,n}$	Discounted Cash Flows	Cumulative Discounted Cash Flows
0	−$10,000	1.0000	−$10,000	−$10,000
1	5,000	0.8547	4,275	− 5,725
2	5,000	0.7305	3,653	− 2,072
3	0	0.6244	0	− 2,072
4	0	0.5337	0	− 2,072
5	0	0.4561	0	− 2,072

Nondiscounted Cash Flow Criterion 2: Accounting Rate of Return

Accounting rate of return (AROR)

A capital-budgeting criterion that relates the returns generated by the project, as measured by average accounting profits after tax, to the average dollar size of the investment required.

The **accounting rate of return (AROR)** compares the average after-tax profits with the average dollar size of the investment.[1] The average profits figure is determined by adding up the after-tax profits generated by the investment over its life and dividing by the number of years. The average investment is determined by adding the initial outlay and the project's actual expected salvage value and dividing by two. This computation attempts to calculate the average value of an investment by simply averaging the initial and liquidation values. Thus, the accounting rate of return for an investment with an expected life of n years can be calculated as follows:

$$AROR = \frac{\sum_{t=1}^{n} (\text{accounting profit after tax}_t)/n}{(\text{initial outlay} + \text{expected salvage value})/2} \tag{13-1}$$

The accept-reject criterion associated with the accounting rate of return compares the calculated return with a minimum acceptable AROR level. If the AROR is greater than this minimum acceptable level, the project is accepted; otherwise it is rejected.

Consider an investment in new machinery that requires an initial outlay of $20,000 and has an expected salvage value of zero after five years. Assume that this machine, if acquired, will result in an increase in after-tax profits of $800 each year for five years. In this case the average accounting profit is $800, while the average investment is ($20,000 + 0)/2, or $10,000. For this example, the AROR would be $800/$10,000, or 8 percent.

This technique seems straightforward enough, but its limitations detract significantly from its value as a discriminating capital-budgeting criterion. To examine these limitations, let us first determine the AROR of three proposals, each with an expected life of five years. Assume that the initial outlay associated with each project is $10,000 and it will have an expected salvage value of zero in five years. The minimum acceptable AROR for this firm is 8 percent, and the annual accounting profits from the three proposals are given in Table 13-3. In each case, the average annual accounting profit is $500 and the average investment is $5,000—that is, ($10,000 + 0)/2. Therefore, the AROR is 10 percent for each project, which indicates that the AROR method does not do an adequate job of discriminating among these projects.

A casual examination leads us to the conclusion that project B is the best, as it yields its returns earlier than either project A or C. However, the AROR technique gives equal weight to all returns within the project's life without any regard for the time value of money. In other words, the AROR technique does not incorporate any degree of riskiness in determining the time value of cash flows since financing costs have been ignored. The second major disadvantage associated with the AROR method is that it deals with accounting profit figures rather than cash flows. For this reason it does not truly recognize the proper timing of the benefits. Finally, the AROR technique does not answer the basic question of whether the project is creating an absolute amount of value. In fact, the criterion used to make any decision is relative to some standard, since the AROR technique requires that we arbitrarily or subjectively select another investment as a standard for comparison purposes.

Table 13-3 Annual Accounting Profits After Tax

Year	A	B	C
1	$ 0	$500	$ 0
2	1,000	500	0
3	500	500	0
4	500	500	0
5	500	500	2,500

[1] Also known as the average rate of return method.

Despite the criticisms, the accounting rate of return has been a relatively popular tool for capital-budgeting analysis, primarily because it involves familiar terms that are easily accessible. Also it is easily understood. The AROR provides a measure of accounting profits per average dollar invested, and the intuitive appeal of this measurement has kept the method alive over the years.[2] For our purposes, the AROR is inadequate, as it does not treat cash flows and does not take account of the time value of money.

BACK TO THE BASICS

The final three capital-budgeting criteria all incorporate ***Axiom 2: The Time Value of Money—A Dollar Received Today Is Worth More Than a Dollar Received in the Future*** *in their calculations. If we are at all to make rational business decisions we must recognize that money has a time value. In examining the following three capital-budgeting techniques you will notice that this axiom is the driving force behind each of them.*

DISCOUNTED CASH FLOW CRITERIA FOR CAPITAL-BUDGETING DECISIONS

OBJECTIVE 3

The final three capital-budgeting criteria to be examined base decisions on the investment's cash flows after adjusting for the time value of money. For the time being, the problem of incorporating risk into the capital-budgeting decision is ignored. This issue will be examined in Chapter 15. In addition, we will assume that the appropriate discount rate, required rate of return, or cost of capital is given. The determination of this rate was the topic of Chapter 12.

We will examine three discounted cash flow capital-budgeting techniques—net present value, profitability index, and internal rate of return.

Net Present Value

The **net present value** (**NPV**) of an investment proposal is equal to the present value of its annual net cash flows after tax less the investment's initial outlay. The net present value can be expressed as follows:

Net present value (NPV)

A capital-budgeting technique defined as the present value of the future net cash flows after tax less the project's initial outlay.

$$NPV = \sum_{t=1}^{n} \frac{CFAT_t}{(1+k)^t} - IO \tag{13-2}$$

where $CFAT_t$ = the annual cash flow after tax in time period t (this can take on either positive or negative values)

k = the appropriate discount rate, that is, the required rate of return or cost of capital[3]

IO = the initial cash outlay

n = the project's expected life

The project's net present value gives a measurement of the net value of an investment proposal in terms of today's dollars. Since all cash flows are discounted back to the present, comparing the difference between the present value of the annual cash flows and the investment outlay does not violate the time value of money assumption. The difference between the present value of the annual cash flows and

[2]It should be pointed out that the AROR will assess the rate of return based upon accounting information; thus, it is possible to analyze the ultimate effect of an alternative on the financial statements.

[3]The required rate of return or cost of capital is the rate of return necessary to justify raising funds to finance the project or, alternatively, the rate of return necessary to maintain the firm's current market price per share. These terms have been defined in Chapter 12.

the initial outlay determines the net present value of accepting the investment proposal in terms of today's dollars. Whenever the project's NPV is greater than or equal to zero, we will accept the project; and whenever there is a negative value associated with the acceptance of a project, we will reject the project. If the project's net present value is zero, then it returns the required rate of return and should be accepted. This accept-reject criterion is illustrated below:

$NPV \geq 0.0$ Accept
$NPV < 0.0$ Reject

Thus, the decision criterion not only provides us with a basis to accept or reject projects but also provides an absolute amount of value which can be measured in terms of shareholder wealth. The following example illustrates the use of the net present value capital-budgeting criterion.

Calculator Solution

DATA INPUT	FUNCTION KEY
40,000	+/− ENTER
	↓
15,000	ENTER ↓
1	ENTER ↓
14,000	ENTER ↓
1	ENTER ↓
13,000	ENTER ↓
1	ENTER ↓
12,000	ENTER ↓
1	ENTER ↓
11,000	ENTER ↓
1	ENTER ↓
	NPV
12	ENTER ↓

FUNCTION KEY	ANSWER
CPT	7,674.63**

*Note: If you are using a TI BA II Plus calculator, you must first get the cash flow worksheet by pressing both CF and 2nd CLRWORK. To solve for the NPV, press the NPV button. For further explanation see the student *Study Guide.*

** The difference between the two answers represents rounding error.

EXAMPLE

A firm is considering new machinery, for which the cash flows after-tax are shown in Table 13-4. If the firm has a 12 percent required rate of return, the present value of the cash flow after-tax is $47,675, as calculated in Table 13-5. Furthermore, the net present value of the new machinery is $7,675. Since this value is greater than zero, the net present value criterion indicates that the project should be accepted.

Note that the worth of the net present value calculation is a function of the accuracy of cash flow predictions. Before the NPV criterion can be reasonably applied, incremental costs and benefits must first be estimated, including the initial outlay, the differential flows over the project's life, and the terminal cash flow.

In comparing the NPV criterion with those that we have already examined, we find it far superior. First of all, it deals with cash flows rather than accounting profits. In this regard it is sensitive to the true timing of the benefits resulting from the project. Moreover, recognizing the time value of money allows comparison of the benefits and costs in a logical manner. Finally, since projects are accepted only if a positive net present value is associated with them, the acceptance of a project using this criterion will maximize the value of the firm. In other words, this decision criterion implies sustainable growth of shareholders' wealth, which is consistent with our goal in financial management.

The disadvantage of the NPV method stems from the need for detailed, long-term forecasts of the incremental cash flows accruing from the project's acceptance. In spite of this drawback, the net present value is the most theoretically correct criterion that we will examine. The following example provides an additional illustration of its application.

EXAMPLE

A firm is considering the purchase of a new computer system, which will cost $30,000 initially, to aid in credit billing and inventory management. The incremental cash flow after tax resulting from this project is provided in Table 13-6.

Table 13-4 NPV Illustration of Investment in New Machinery

	Cash Flow After Tax
Initial outlay	−$40,000
Inflow year 1	15,000
Inflow year 2	14,000
Inflow year 3	13,000
Inflow year 4	12,000
Inflow year 5	11,000

Table 13-5 Calculation for NPV Illustration of Investment in New Machinery

	Cash Flow After-tax	Factor at 12 Percent	Present Value
Inflow year 1	$15,000	.8929	$13,394
Inflow year 2	14,000	.7972	11,161
Inflow year 3	13,000	.7118	9,253
Inflow year 4	12,000	.6355	7,626
Inflow year 5	11,000	.5674	6,241
Present value of cash flows			$47,675
Investment initial outlay			−40,000
Net present value			$7,675

The required rate of return demanded by the firm is 10 percent. In order to determine the system's net present value, the three-year $15,000 cash flow annuity is first discounted back to the present at 10 percent. From Appendix D in the back of this book, we find that $PVIFA_{10\%,3yr}$ is 2.4869. Thus, the present value of this $15,000 annuity is $37,304.

Because the cash inflows have been discounted back to the present, they can now be compared with the initial outlay, since both of the flows are now stated in terms of today's dollars. Subtracting the initial outlay ($30,000) from the present value of the cash inflows ($37,304), we find that the system's net present value is $7,304. Since the NPV on this project is positive, the project should be accepted.

Profitability Index (Benefit/Cost Ratio)

The **profitability index (PI)**, or **benefit/cost ratio**, is the ratio of the present value of the future net cash flows to the initial outlay. While the net present value investment criterion gives a measure of the absolute dollar desirability of a project, the profitability index provides a relative measure of an investment proposal's desirability—that is, the ratio of the present value of its future net benefits to its initial cost. The profitability index can be expressed as follows:

$$\text{Profitability Index} = \frac{\sum_{t=1}^{n} \frac{CFAT_t}{(1+k)^t}}{IO} \tag{13-3}$$

where $CFAT_t$ = the annual cash flow after tax in time period t (this can take on either positive or negative values)

k = the appropriate discount rate; that is, the required rate of return or the cost of capital

IO = the initial cash outlay

n = the project's expected life

The decision criterion with respect to the profitability index is to accept the project if the PI is greater than or equal to 1.00, and to reject the project if the PI is less than 1.00.

Profitability index (PI) or benefit/cost ratio

A capital-budgeting criterion defined as the ratio of the present value of the future net cash flows to the initial outlay.

Table 13-6 NPV Example Problem of a Computer System

	Cash Flow After Tax
Initial outlay	−$30,000
Year 1	15,000
Year 2	15,000
Year 3	15,000

PI ≥ 1.0	Accept
PI < 1.0	Reject

Looking closely at this criterion, we see that it yields the same accept-reject decision as does the net present value criterion. Whenever the present value of the project's net cash flow is greater than its initial cash outlay, the project's net present value will be positive, signalling a decision to accept the project. When this is true, the project's profitability index will also be greater than 1, as the present value of the net cash flows (the PI's numerator) is greater than its initial outlay (the PI's denominator). While these two decision criteria will always yield the same decision, they will not necessarily rank acceptable projects in the same order. This problem of conflicting ranking will be dealt with at a later point.

Since the net present value and profitability index criteria are essentially the same, they have the same advantages over the other criteria examined. Both employ cash flows, recognize the timing of the cash flows, and are consistent with the goal of maximization of shareholders' wealth. The major disadvantage of both the profitability index and net present value criteria is caused by the use of uncertain cash flows which must be determined by forecasting.

EXAMPLE

A firm with a 10 percent required rate of return is considering investing in a new machine with an expected life of six years. The cash flows after tax resulting from this investment are given in Table 13-7. Discounting the project's future net cash flows back to the present yields a present value of $53,683; dividing this value by the initial outlay of $50,000 gives a profitability index of 1.0737, as shown in Table 13-8. This tells us that the present value of the future benefits accruing from this project is 1.0737 times the level of the initial outlay. Since the profitability index is greater than 1.0, the project should be accepted.

Internal Rate of Return

Internal rate of return (IRR)

A capital-budgeting technique that reflects the rate of return a project earns. Mathematically, the IRR is the discount rate that equates the present value of the inflows with the present value of the outflows.

The internal rate of return (IRR) attempts to answer this question: What rate of return is required so that the project's NPV will be equal to zero? For computational purposes, the **internal rate of return** is defined as the discount rate that equates the present value of the project's future net cash flows with the project's initial cash outlay. Mathematically, the internal rate of return is defined as the value IRR in the following equations:

$$NPV = 0 = \sum_{t=1}^{n} \frac{CFAT_t}{(1 + IRR)^t} - IO \quad (13\text{-}4a)$$

$$IO = \sum_{t=1}^{n} \frac{CFAT_t}{(1 + IRR)^t} \quad (13\text{-}4b)$$

Table 13-7 Profitability Index (PI) Illustration of Investment in New Machinery

	Cash Flow After Tax
Initial outlay	−$50,000
Inflow year 1	15,000
Inflow year 2	8,000
Inflow year 3	10,000
Inflow year 4	12,000
Inflow year 5	14,000
Inflow year 6	16,000

Table 13-8 Calculation for PI Illustration of Investment in New Machinery

	Cash Flow After tax	Factor at 10 Percent	Present Value
Initial outlay	−$50,000	1.0	−$50,000
Inflow year 1	15,000	0.9091	13,637
Inflow year 2	8,000	0.8265	6,612
Inflow year 3	10,000	0.7513	7,513
Inflow year 4	12,000	0.6830	8,196
Inflow year 5	14,000	0.6209	8,693
Inflow year 6	16,000	0.5645	9,032

$$\text{Profitability Index} = \frac{\sum_{t=1}^{n} \frac{CFAT_t}{(1+k)^t}}{IO}$$

$$= \frac{\$13{,}637 + \$6{,}612 + \$7{,}513 + \$8{,}196 + \$8{,}693 + \$9{,}032}{\$50{,}000}$$

$$= \frac{\$53{,}683}{\$50{,}000}$$

$$= 1.0737$$

where $CFAT_t$ = the annual cash flow after tax in time period t (this can take on either a positive or negative value)

IO = the initial cash outlay

n = the project's expected life

IRR = the project's internal rate of return

In effect, the IRR is analogous to the concept of the yield to maturity for bonds, which was examined in Chapter 5. In other words, a project's internal rate of return is simply the rate of return that the project earns. The decision criterion associated with the internal rate of return is to accept the project if the internal rate of return is greater than or equal to the required rate of return. We reject the project if its internal rate of return is less than this required rate of return. This accept-reject criterion is illustrated below:

IRR $\geq$ required rate of return Accept
IRR $<$ required rate of return Reject

If the internal rate of return on a project is equal to the shareholders' required rate of return, then the project should be accepted. This is because the firm is earning the rate that its shareholders are requiring. However, the acceptance of a project with an internal rate of return below the investors' required rate of return will decrease the firm's stock price.

If the NPV is positive, the IRR must be greater than the required rate of return, k. Thus, all the discounted cash flow criteria are consistent and will give similar accept-reject decisions. In addition, since the internal rate of return is another discounted cash flow criterion, it exhibits the same general advantages and disadvantages as both the net present value and profitability index, but is also tedious to calculate if a financial calculator is not available.[4]

An important disadvantage of the IRR relative to the NPV deals with the implied reinvestment rate assumptions made by these two methods. The NPV method implicitly assumes that cash flows received over the life of the project are reinvested back in projects that earn the required rate of return. For example, if we have a mining project with a ten-year expected life that produces a $100,000 cash flow at

[4]In general, the IRR must be calculated by means of iteration since there is no explicit solution except in some special cases such as even cash flows.

the end of the second year, the NPV technique assumes that this $100,000 is reinvested during the period extending from three to ten years at the required rate of return. The use of the IRR, on the other hand, implies that cash flows over the life of the project can be reinvested at the IRR. Thus, if the mining project we just looked at has a 40 percent IRR, the use of the IRR implies that the $100,000 cash flow that is received at the end of year 2 could be reinvested at 40 percent over the remaining life of the project. In effect, *the NPV method implicitly assumes that cash flows over the life of the project can be reinvested at the firm's required rate of return; whereas, the use of the IRR method implies that these cash flows could be reinvested at the IRR.* The NPV method makes a more conservative assumption because cash flows are expected to be reinvested at the investors' required rate of return and implies sustainable growth of shareholder's wealth. In other words, cash flows earned from the project will be either (1) returned in the form of dividends to shareholders who demand the required rate of return on their investment, or (2) reinvested in investment projects that earn a rate of return that is greater than or equal to the investors' required rate of return. If these cash flows are invested in a new project, then they are simply substituting for external funding on which the required rate of return is demanded. Thus, the opportunity cost of these funds is the required rate of return. The bottom line to all this is that the NPV method makes the best reinvestment rate assumption and, as such is superior to the IRR method. Why should we care which method is used if both methods give similar accept-reject decisions? The answer, as we will see in the next chapter, is that while they may give the same accept-reject decision, they may rank projects differently in terms of desirability.

Computing the IRR with a Financial Calculator

An internal rate of return problem can be solved with a few keystrokes on today's calculators. In Chapter 4, whenever we were determining time value of money problems for *i*, we were really solving for the internal rate of return. For instance, in Chapter 4, when we solved for the rate that $100 must be compounded annually for it to grow to $179.10 in ten years, we were actually solving for the problem's internal rate of return. Thus, with financial calculators we need only input the initial outlay, the cash flows and their timing, and then use the function key **I/Y** or the **IRR** button to calculate the internal rate of return. On some calculators it is necessary to input the compute key, **CPT**, before inputting the function key to be calculated.

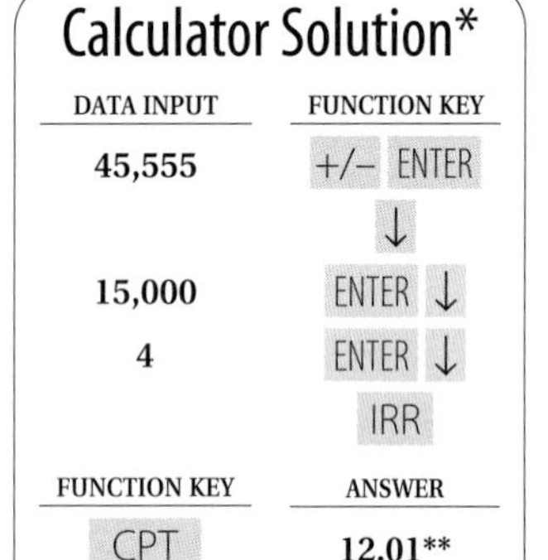

*Note: If you are using a TI BA II Plus calculator, you must first get the cash flow worksheet by pressing both CF and 2nd CLRWORK. To solve for IRR, press the IRR button.
For further explanation see the student *Study Guide.*

**The difference between the two answers represents rounding error.

Computing the IRR for Even Cash Flows

In this section, however, we are going to put our calculators aside and examine the mathematical process of calculating internal rates of return for a better understanding of internal rates of return.

The calculation of a project's internal rate of return can be either very simple or relatively complicated. As an example of straightforward solution, assume that a firm with a required rate of return of 10 percent is considering a project that involves an initial outlay of $45,555. If the investment is taken, the cash flow after tax is expected to be $15,000 per annum over the project's four-year life.[5] In this case, the internal rate of return is equal to IRR in the following equation:

$$\$45{,}555 = \frac{\$15{,}000}{(1 + IRR)^1} + \frac{\$15{,}000}{(1 + IRR)^2} + \frac{\$15{,}000}{(1 + IRR)^3} + \frac{\$15{,}000}{(1 + IRR)^4}$$

From our discussion of the present value of an annuity in Chapter 4, we know that this equation can be reduced to

$$\$45{,}555 = \$15{,}000\left[\sum_{t=1}^{4} \frac{1}{(1 + IRR)^t}\right]$$

Appendix D at the back of the book gives table values for the $PVIFA_{i,n}$ for various combinations of *i* and *n*, which further reduces this equation to

[5]Even cash flows arise in many situations, such as in long-term leasing or loan amortization.

$\$45{,}555 = \$15{,}000\ (PVIFA_{IRR,n})$

Dividing both sides by $15,000, this becomes

$3.0370 = PVIFA_{IRR,4}$

Hence, we are looking for a $PVIFA_{i,n}$ of 3.037 in the four-year row of Appendix D. This value occurs when i equals 12 percent, which means that 12 percent is the internal rate of return for the investment. Therefore, since 12 percent is greater than the 10 percent required return, the project should be accepted.

Computing the IRR for Uneven Cash Flows

Unfortunately, while solving for the IRR is quite easy when using a financial calculator or spreadsheet, it can be solved directly in the tables only when the future after-tax net cash flows are in the form of an annuity or a single payment. When a financial calculator is not available and these flows are in the form of an uneven series of flows, a trial-and-error approach is necessary. To do this, we first determine the present value of the future after-tax net cash flows using an arbitrary discount rate. If the present value of the future cash flows at this discount rate is larger than the initial outlay, the rate is increased; if it is smaller than the initial outlay, the discount rate is lowered, and the process begins again. This search routine is continued until the present value of the future cash flow after tax is equal to the initial outlay. The interest rate that creates this situation is the internal rate of return.

To illustrate the procedure, consider an investment proposal that requires an initial outlay of $3,817 and returns $1,000 at the end of year 1, $2,000 at the end of year 2, and $3,000 at the end of year 3. In this case, the internal rate of return must be determined using trial and error. This process is presented in Table 13-9, in which an arbitrarily selected discount rate of 15 percent was chosen to begin the process. The trial-and-error technique slowly centres in on the project's internal rate of return of 22 percent. The project's internal rate of return is then compared with the firm's required rate of return, and if the IRR is the larger, the project is accepted.

EXAMPLE

A firm with a required rate of return of 10 percent is considering three investment proposals. Given the information in Table 13-10, management plans to calculate the internal rate of return for each project and determine which projects should be accepted.

Because project A is an annuity, we can easily calculate its internal rate of return by determining the $PVIFA_{i,n}$, necessary to equate the present value of the future cash flows with the initial outlay. This computation is done as follows:

$$IO = \sum_{t=1}^{n} \frac{CFAT_t}{(1 + IRR)^t}$$

$$\$10{,}000 = \sum_{n=1}^{4} \frac{\$3{,}362}{(1 + IRR)^t}$$

$$\$10{,}000 = \$3{,}362\ (PVIFA_{i,4})$$

$$2.9744 = PVIFA_{i,4}$$

We are looking in the four-year row of Appendix D, for a $PVIFA_{i,n}$ of 2.9744, which occurs in the i = 13 percent column. Thus, 13 percent is the internal rate of return. Since this rate is greater than the firm's required rate of return of 10 percent, the project should be accepted.

Project B involves a single future cash flow of $13,605, resulting from an initial outlay of $10,000; thus, its internal rate of return can be determined directly from the present value table in Appendix B as follows:

Calculator Solution*

DATA INPUT	FUNCTION KEY
3,817	+/– ENTER ↓
1,000	ENTER ↓
1	ENTER ↓
2,000	ENTER ↓
1	ENTER ↓
3,000	ENTER ↓
1	ENTER ↓
	IRR

FUNCTION KEY	ANSWER
CPT	21.98

*Note: If you are using a TI BA II Plus calculator, you must first get the cash flow worksheet by pressing both CF and 2nd CLRWORK. To solve for IRR, press the IRR button.
For further explanation see the student *Study Guide.*

Table 13-9 Computing the IRR for Uneven Cash Flows Without a Financial Calculator

Initial outlay	–$3,817
Flow year 1	1,000
Flow year 2	2,000
Flow year 3	3,000

Solution: Pick an arbitrary discount rate and use it to determine the present value of the inflows. Compare the present value of the inflows with the initial outlay; if they are equal, you have determined the IRR. Otherwise, raise the discount rate if the present value of the inflows is larger than the initial outlay or lower the discount rate if the present value of the inflows is less than the initial outlay. Repeat this process until the IRR is found.

1. Try $i = 15$ percent:

	Net Cash Flows	Present Value Factor at 15 Percent	Present Value
Inflow year 1	$1000	.8696	$ 870
Inflow year 2	2000	.7561	1,512
Inflow year 3	3000	.6575	1,973
Present value of inflows			$4,355
Initial outlay			–$3,817

2. Try $i = 20$ percent:

	Net Cash Flows	Present Value Factor at 20 Percent	Present Value
Inflow year 1	$1000	.8333	$ 833
Inflow year 2	2000	.6944	1,389
Inflow year 3	3000	.5787	1,736
Present value of inflows			$3,958
Initial outlay			–$3,817

3. Try $i = 22$ percent:

	Net Cash Flows	Present Value Factor at 22 Percent	Present Value
Inflow year 1	$1000	.8197	$ 820
Inflow year 2	2000	.6719	1,344
Inflow year 3	3000	.5507	1,652
Present value of inflows			$3,816
Initial outlay			–$3,817

Table 13-10 Three IRR Investment Proposal Examples

	A	B	C
Initial outlay	– $10,000	– $10,000	– $10,000
Inflow year 1	3,362	0	1,000
Inflow year 2	3,362	0	3,000
Inflow year 3	3,362	0	6,000
Inflow year 4	3,362	13,605	7,000

$$IO = \sum_{t=1}^{n} \frac{CFAT_t}{(1 + IRR)^t}$$

$$\$10,000 = \frac{\$13,605}{(1 + IRR)^4}$$

$$\$10,000 = \$13,605\ (PVIF_{i,4yr})$$
$$0.7350 = PVIF_{i,4yr}$$

This tells us that we should look for a $PVIF_{i,4yr}$ of 0.735 in the four-year row of Appendix B, which occurs in the $i = 8$ percent column. We may therefore conclude that 8 percent is the internal rate of return. Since this rate is less than the firm's required rate of return of 10 percent, project B should be rejected.

The uneven nature of the future cash flows associated with project C necessitates the use of the trial-and-error method. The internal rate of return for project C is equal to the value of IRR in the following equation:

$$\$10,000 = \frac{\$1000}{(1 + IRR)^1} + \frac{\$3000}{(1 + IRR)^2} + \frac{\$6000}{(1 + IRR)^3} + \frac{\$7000}{(1 + IRR)^4}$$

Arbitrarily selecting a discount rate of 15 percent and substituting it into Equation (13-5) for IRR reduces the right-hand side of the equation to \$11,086, as shown in Table 13-11. Therefore, since the present value of the future cash flows is larger than the initial outlay, we must raise the discount rate in order to find the project's internal rate of return. Substituting 20 percent for the discount rate, the right-hand side of Equation (13-5) now becomes \$9,764. As this is less than the initial outlay of \$10,000, we must now decrease the discount rate. In other words, we know that the internal rate of return for this project is between 15 and 20 percent. Since the present value of the future flows discounted back to the present at 20 percent was only \$236 too low, a discount rate of 19 percent is selected. As shown in Table 13-11, a discount rate of 19 percent reduces the present value of the future inflows down to \$10,010, which is approximately the same as the initial outlay. Consequently, project C's internal rate of return is

Figure 13-1 Net Present Value Profile

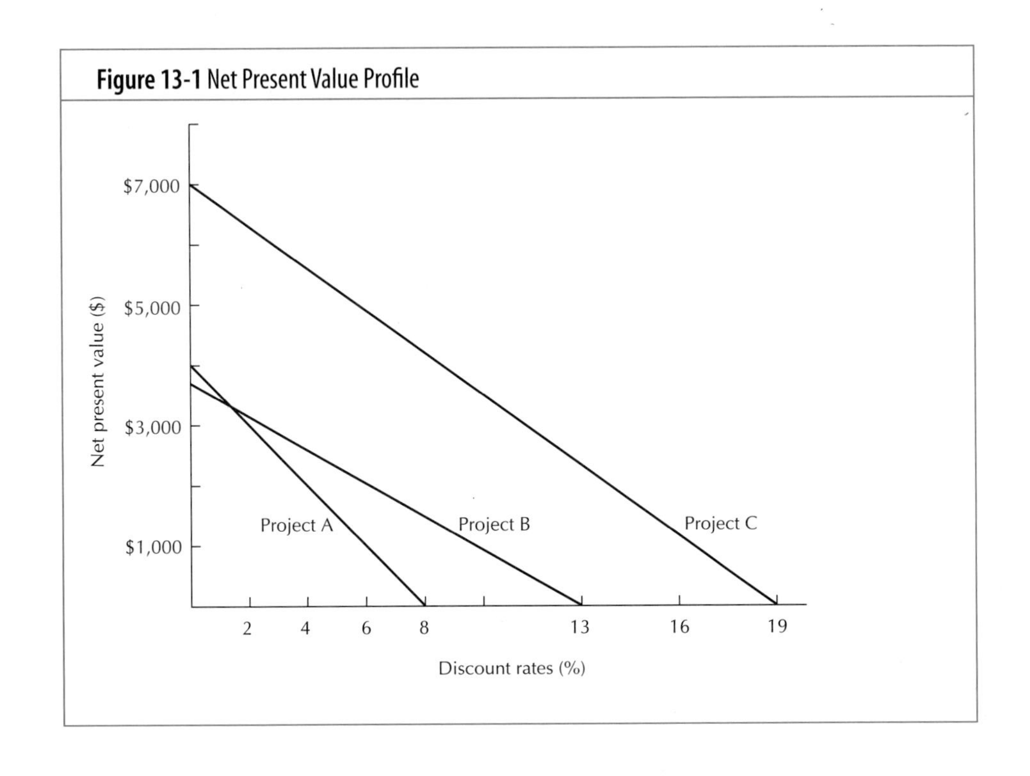

Calculator Solution*

DATA INPUT	FUNCTION KEY
10,000	+/− ENTER
	↓
1,000	ENTER ↓
1	ENTER ↓
3,000	ENTER ↓
1	ENTER ↓
6,000	ENTER ↓
1	ENTER ↓
7,000	ENTER ↓
1	ENTER ↓
	IRR

FUNCTION KEY	ANSWER
CPT	19.04

*Note: If you are using a TI BA II Plus calculator, you must first get the cash flow worksheet by pressing both CF and 2nd CLRWORK. To solve for IRR, press the IRR button.
For further explanation see the student *Study Guide*.

Table 13-11 Computing the IRR for Project C

1. Try $i = 5$ 15 percent:

	Net Cash Flows	Present Value Factor at 15 Percent	Present Value
Inflow year 1	$1,000	.8696	$ 870
Inflow year 2	3,000	.7561	2,268
Inflow year 3	6,000	.6575	3,945
Inflow year 4	7,000	.5718	4,003
Present value of inflows			$11,086
Initial outlay			−$10,000

2. Try $i =$ 20 percent:

	Net Cash Flows	Present Value Factor at 20 Percent	Present Value
Inflow year 1	$1,000	.8333	$ 833
Inflow year 2	3,000	.6944	2,083
Inflow year 3	6,000	.5787	3,472
Inflow year 4	7,000	.4823	3,376
Present value of inflows			$ 9,764
Initial outlay			−$10,000

3. Try $i =$ 19 percent:

	Net Cash Flows	Present Value Factor at 19 Percent	Present Value
Inflow year 1	$1,000	.8403	$ 840
Inflow year 2	3,000	.7062	2,119
Inflow year 3	6,000	.5934	3,560
Inflow year 4	7,000	.4987	3,491
Present value of inflows			$10,010
Initial outlay			−$10,000

approximately 19 percent.[6] Since the internal rate of return is greater than the firm's required rate of return of 10 percent, this investment should be accepted.

The results from the above example can be shown graphically using a net present value profile. This type of figure is constructed by plotting net present values versus discount rates for a particular project. In Figure 13-1, we have plotted the net present values versus the discount rates for projects A, B, and C. The steepness of each graph reflects the project's sensitivity to various discount rates. In this example, the graph of project B shows the greatest steepness (greatest slope) as compared to the plots of the other projects. Thus, project B shows the greatest sensitivity to different discount rates as compared to projects A and C.

[6]If desired, the actual rate can be more precisely approximated through interpolation as follows:

	Discount Rate	Present Value
	19 %	$10,010
	20	9,764
Difference	1 %	$ 246

$$\text{Proportion \$10 is of \$246} = \frac{\$10}{\$246} = 0.407$$

$$19\% + .0407\ (1\%) = 19.0407\%$$

Complications with the IRR: Multiple Rates of Return

While any project can have only one NPV and one PI, a single project under certain circumstances can have more than one IRR. The reason for this can be traced to the calculations involved in determining the IRR. Equation (13-4b) states that the IRR is the discount rate that equates the present value of the project's future net cash flows with the project's initial outlay:

$$IO = \sum_{t=1}^{n} \frac{CFAT_t}{(1 + IRR)^t} \tag{13-4b}$$

However, since Equation (13-4b) is a polynomial of a degree n, it has n solutions. Now if the initial outlay (IO) is the only negative cash flow and all the annual cash flows after tax ($CFAT_t$) are positive, then all but one of these n solutions is either a negative or imaginary number and there is no problem. But problems occur when there are sign reversals in the cash flow stream; in fact there can be as many solutions as there are sign reversals. Thus, a normal pattern with a negative initial outlay and positive annual after-tax cash flows (–, +, +, +,..., +) after that has only one sign reversal, hence only one positive IRR. However, a pattern with more than one sign reversal can have more than one IRR. Consider, for example, the following pattern of cash flows.[7]

	Cash Flow After Tax
Initial outlay	–$ 1,600
Year 1	+$10,000
Year 2	–$10,000

In this pattern of cash flows there are two sign reversals, from –$1,600 to +$10,000 and then from +$10,000 to –$10,000, so there can be as many as two positive IRRs that will make the present value of the future cash flows equal to the initial outlay. In fact two internal rates of return solve this problem, 25 and 400 percent. Graphically what we are solving for is the discount rate that makes the project's NPV equal to zero; as Figure 13-2 illustrates, this occurs twice.

Figure 13-2 Multiple IRRs

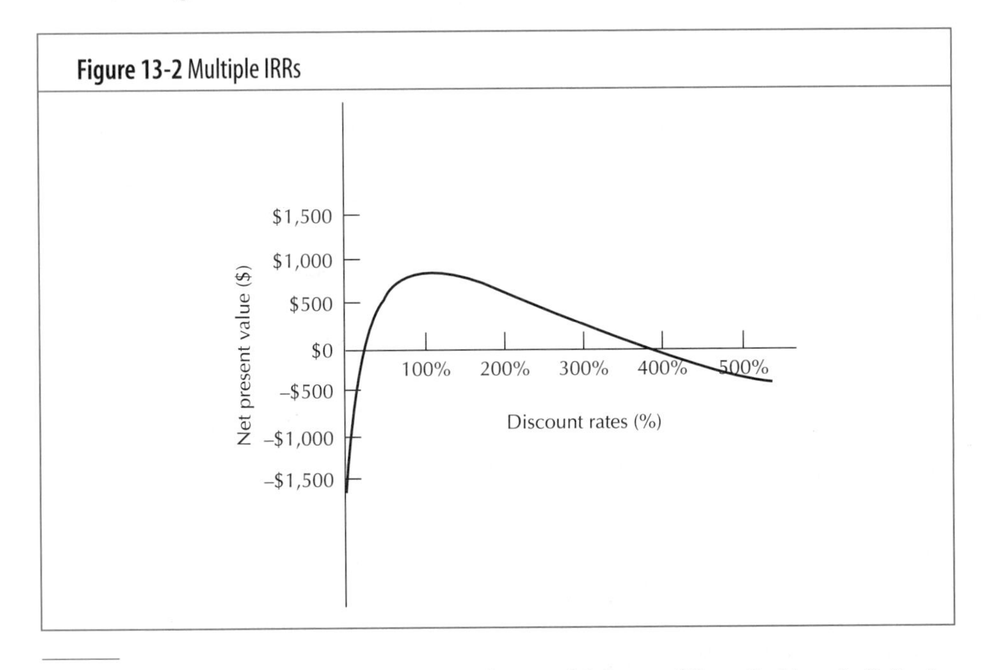

[7]This example is taken from James H. Lorie and Leonard J. Savage, "Three Problems in Rationing Capital," *Journal of Business* 28 (October 1955), pp. 229–239.

Which solution is correct? The answer is that neither solution is valid. While each fits the definition of IRR, neither provides any insight into the true project returns. What has happened is that the firm has borrowed or received $10,000 from the project at the end of year 1 and must return this sum at the end of year 2. The real question is this: What is the use of this $10,000 worth for one year? The answer depends upon what the investment opportunities are like to the firm. If they are 30 percent, then the value of receiving $10,000 at the end of year 1 and returning it at the end of year 2 is $3,000 received at the end of year 2, since this is what the firm would be left with. The internal rate of return on this (an initial outlay of $1,600 and a positive inflow of $3,000 at the end of year 2) is 37 percent. Thus, neither of the internal rates of return, 25 or 400 percent, was valid. In summary, when there is more than one sign reversal in the cash flow stream, the possibility of multiple IRRs exists, and the normal interpretation of the IRR loses its meaning.

Modified Internal Rate of Return

Modified internal rate of return (MIRR)

A variation of the IRR capital-budgeting criterion defined as the discount rate that equates the present value of the project's annual cash outlays with the present value of the project's terminal value, where the terminal value is defined as the sum of the future value of the project's cash inflows compounded to the project's termination at the project's required rate of return.

The primary drawback of the internal rate of return relative to the net present value method is the reinvestment rate assumption made by the internal rate of return. Recently, a new technique, the **modified internal rate of return (MIRR)**, has gained popularity as an alternative to the IRR method because it allows the decision-maker to directly specify the appropriate reinvestment rate. As a result the MIRR provides the decision maker with the intuitive appeal of the IRR coupled with an improved reinvestment rate assumption.

The driving force behind the MIRR is the assumption that all cash inflows over the life of the project are reinvested at the required rate of return until the termination of the project. Thus, to calculate the MIRR, we take all the annual cash inflows after tax, $CIAT_t$'s and find their future value at the end of the project's life, compounded at the required rate of return. We will call this project's *terminal value*, or *TV*. We then calculate the present value of the project's cash *outflows*. We do this by discounting all cash *outflows*, $COAT_t$, back to present at the required rate of return. If the initial outlay is the only cash *outflow*, then the initial outlay is the present value of the cash *outflows*. The MIRR is the discount rate that equates the present value of the cash *outflows* with the present value of the project's *terminal value*.[8] Mathematically, the modified internal rate of return is defined as the value of MIRR in the following equation

$$PV_{outflows} = PV_{inflows} \tag{13-5}$$

$$\sum_{t=0}^{n} \frac{COAT_t}{(1+k)^t} = \frac{\sum_{t=1}^{n} CIAT_t (1+k)^{n-t}}{(1+MIRR)^n} \tag{13-6}$$

To facilitate the presentation, we define the terminal value (TV) as:

$$TV = \sum_{t=1}^{n} CIAT_t(1+k)^{n-t} \tag{13-7}$$

and solve for MIRR as follows

$$PV_{outflows} = \frac{TV}{(1+MIRR)^n} \tag{13-8}$$

where $COAT_t$ = the annual cash outflow after tax in time period t
$CIAT_t$ = the annual cash inflow after tax in time period t

[8]You will noticed that we differentiate between annual cash *in*flows and annual cash *out*flows, compounding all the inflows to the end of the project and bringing all the outflows back to present as part of the present value of the costs. While there are alternative definitions of the MIRR, this is the most widely accepted definition. For an excellent discussion of the MIRR, see William R. McDaniel, Daniel E. McCarty, and Kenneth A. Jessell, "Discounted Cash Flow with Explicit Reinvestment Rates: Tutorial and Extention," *The Financial Review* (August 1988), pp. 369–385.

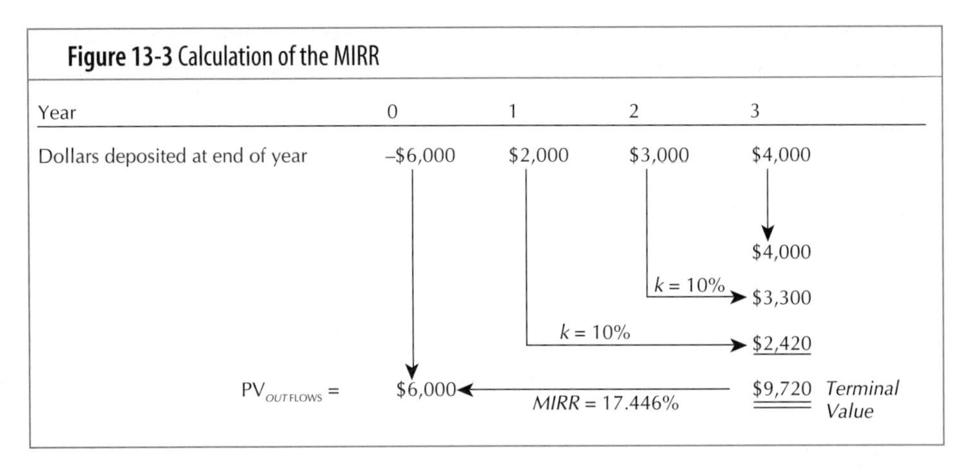

TV = the terminal value of the CIAT's compounded at the required rate of return to the end of the project
n = the project's expected life
$MIRR$ = the project's modified internal rate of return
k = the appropriate discount rate; that is, the required rate of return or cost of capital

EXAMPLE

Let's look at an example of a project with a three-year life and a required rate of return of 10 percent assuming the following cash flows are associated with it:

Cash Flows After Tax		Cash Flows After Tax	
Initial outlay	–$6,000	Year 2	$3,000
Year 1	2,000	Year 3	4,000

The calculation of the MIRR can be viewed as a three-step process, which is also shown graphically in Figure 13-3.

Step 1: Determine the present value of the project's cash *out*flows. In this case, the only *out*flow is the initial outlay of $6,000, which is already at the present, thus it becomes the present value of the cash *out*flows.

Step 2: Determine the terminal value of the project's cash *in*flows. To do this, we merely use the project's required rate of return to calculate the future value of the project's three cash *in*flows at the termination of the project. In this case, the *terminal value* becomes $9,720.

Step 3: Determine the discount rate that equates the present value of the *terminal value* and the present value of the project's cash *out*flows. Thus the MIRR is calculated to be 17.446 percent.

For our example, the calculations are as follows:

$$\$6{,}000 = \sum_{t=1}^{3} CIAT_t(1+k)^{n-t}/(1+MIRR)^n$$

$$\$6{,}000 = \frac{\$2{,}000(1+.10)^2 + \$3{,}000(1+.10)^1 + \$4{,}000(1+.10)^0}{(1+MIRR)^3}$$

$$\$6{,}000 = \frac{\$2{,}420 + \$3{,}300 + \$4{,}000}{(1+MIRR)^3}$$

$$\$6{,}000 = \frac{\$9{,}720}{(1+MIRR)^3}$$

$$MIRR = 17.446\%$$

Thus the MIRR for this project (17.446 percent) is less than its IRR which comes out to 20.614 percent. In this case, it only makes sense that the IRR should be greater than the MIRR, because the IRR allows intermediate cash *in*flows to grow at the IRR rather than the required rate of return.

In terms of decision rules, if the project's MIRR is greater than or equal to the project's required rate of return, the project should be accepted; if not, it should be rejected:

$MIRR$ ≥ required rate of return: Accept
$MIRR$ < required rate of return: Reject

Because of the frequent use of the IRR in the real world as a decision-making tool, and its limiting reinvestment rate assumption, the MIRR has become increasingly popular as an alternative decision-making tool.

OBJECTIVE 4

ETHICS IN CAPITAL BUDGETING

Although it may not seem obvious, ethics has a role in capital budgeting. Beech-Nut provides an example of how these rules have been violated in the past and what the consequences can be. No doubt this project appeared to have a positive net present value associated with it, but in fact, it cost Beech-Nut tremendously. The Ethics in Financial Management insert, "Bad Apple for Baby," provides a narrative of what took place.

BACK TO THE BASICS

Ethics and ethical considerations continually crop up when capital-budgeting decisions are being made. This brings us back to the fundamental ***Axiom 10: Ethical Behaviour Is Doing the Right Thing, and Ethical Dilemmas Are Everywhere in Finance.*** *As "Bad Apple for Baby" points out, the most damaging event a business can experience is a loss of the public's confidence in its ethical standards. In making capital-budgeting decisions we must be aware of this, and that ethical behaviour is doing the right thing and it is the right thing to do.*

OBJECTIVE 5

A GLANCE AT ACTUAL CAPITAL-BUDGETING PRACTICES

During the past forty years, the popularity of capital-budgeting techniques has shifted rather dramatically. In the 1950s the payback period and AROR methods dominated capital budgeting, but through the 1970s and 1980s the discounted cash flow decision techniques slowly displaced the nondiscounted techniques. Table 13-12 provides the results of surveys on capital-budgeting practices of North American firms showing the popularity of the internal rate of return and the net present values methods.

Interestingly, while a large majority of firms use the NPV and IRR as their primary techniques, a large majority of firms also use the payback period as a secondary decision method for capital budgeting. In a sense they are using the payback period to control for risk. The logic behind this is that since the payback period dramatically emphasizes early cash flows, which are presumably more certain—that is, have less risk—than cash flows that occur later in a project's life, its use will lead to projects with more certain cash flows.

Bad Apple for Baby

It's a widely held, but hard to prove, belief that a company gains because it is perceived as more socially responsive than its competitors. Over the years, the three major manufacturers of baby food—Gerber Products, Beech-Nut Nutrition and H. J. Heinz—had, with almost equal success, gone out of their way to build an image of respectability.

Theirs is an almost perfect zero-sum business. They know, at any given time, how many babies are being born. They all pay roughly the same price for their commodities, and their manufacturing and distribution costs are almost identical. So how does one company gain a market share edge over another, especially in a stagnant or declining market?

The answer for Beech-Nut was to sell a cheaper, adulterated product. Beginning in 1977, the company began buying a chemical concoction, made up mostly of sugar and water, and labelling it as apple juice. Sales of that product brought Beech-Nut an estimated $60 million between 1977 and 1982, while reducing material costs about $250,000 annually.

When various investigators tried to do something about it, the company stonewalled. Among other things, they shipped the bogus juice out of a plant in New York to Puerto Rico, to put it beyond the jurisdiction of federal investigators, and they even offered the juice as a give-away to reduce their stocks after they were finally forced to discontinue selling it.

In the end, the company pleaded guilty to 215 counts of introducing adulterated food into commerce and violating the Federal Food Drug and Cosmetic Act. The FDA fined Beech-Nut $2 million.

In addition, Beech-Nut's president, Neils Hoyvald, and its vice president of operations, John Lavery, were found guilty of similar charges. Each faces a year and one day in jail and a $100,000 fine. Both are now out on appeal on a jurisdiction technicality.

Why did they do it? The Fort Washington, Pa.-based company will not comment. But perhaps some portion of motive can be inferred from a report Hoyvald wrote to Nestle, the company which had acquired Beech-Nut in the midst of his coverup. "It is our feeling that we can report safely now that the apple juice recall has been completed. If the recall had been effectuated in early June [when the FDA had first ordered it], over 700,000 cases in inventory would have been affected...due to our many delays, we were only faced with having to destroy 20,000 cases."

One thing is clear: Two executives of a company with an excellent reputation breached a trust and did their company harm.

Since 1987, when the case was brought to a close, Beech-Nut's share of the overall baby food market has fallen from 19.1% to 15.8%. So what was gained in the past has been lost in the present, and perhaps for the future as well.

Source: Stephen Kindel, "Bad Apple for Baby," *Financial World*, June 27, 1989, p. 48.

Table 13-12 Surveys of Capital-Budgeting Practices

Capital-Budgeting Technique	Percent of Respondents Using Each Technique Bierman 1992	Blazouske, Cralin, and Kim 1985
Primary method:		
NPV	63%	25%
IRR	88	40
Payback	24	19
AROR	15	9
Secondary method:		
NPV	22%	11%
IRR	11	13
Payback	59	44
AROR	18	9

Source: H. Bierman, Jr., "Capital Budgeting in 1992: A Survey," *Financial Management* (Autumn 1993), p. 24; J. Blazouske, I. Carlin, and S. Kim, "Current Capital Budgeting Practices in Canada," *CMA The Management Accounting Magazine* (March 1988), pp. 51–54.

A reliance on the payback period came out even more dramatically in a recent study of the capital-budgeting practices of 12 large manufacturing firms.[9] Information for this study was gathered from interviews over one to three days in addition to an examination of the records of about 400 projects. This study revealed a number of things of interest. First, firms were typically found to categorize capital investments as mandatory (regulations and contracts, capitalized maintenance, replacement of antiquated equipment, product quality) or discretionary (expanded markets, new businesses, cost cutting), with the decision-making process being different for mandatory and discretionary projects. Second, it was found that the decision-making process was different for projects of differing size. In fact, approval authority tended to rest in different locations, depending upon the size of the project. Table 13-13 provides the typical levels of approval authority.

The study also showed that while the discounted cash flow methods are used at most firms, the simple payback criterion was the measure relied upon primarily in one-third of the firms examined. The use of the payback period seemed to be even more common for smaller projects, with firms severely simplifying the discounted cash flow analysis and/or relying primarily on the payback period. Thus, while discounted cash flow decision techniques have become more widely accepted, their use depends to an extent upon the size of the project and where within the firm the decision is being made.

Table 13-13 Project Size and Decision-Making Authority

Project Size	Typical Boundaries	Primary Decision Site
Very small	Up to $100,000	Plant
Small	$100,000 to $1,000,000	Division
Medium	$1 million to $10 million	Corporate investment committee
Large	Over $10 million	CEO & board

[9] Marc Ross, "Capital Budgeting Practices of Twelve Large Manufacturers," *Financial Management* 15 (Winter 1986), pp. 15–22.

SUMMARY

Before a profitable project can be adopted, it must be identified or found. Unfortunately, coming up with ideas for new products, for ways to improve existing products, or for ways to make existing products more profitable is extremely difficult. In general, the best source of ideas for these new, potentially profitable products is from within the firm.

The process of capital-budgeting involves decision making with respect to investment in fixed assets. We examine five commonly used criteria for determining the acceptance or rejection of capital-budgeting proposals. The first two methods, the payback period and the accounting rate of return are nondiscounted and do not incorporate the time value of money into their calculations. However, the payback method can be modified to recognize the time value of money. The discounted methods, net present value, profitability index, and internal rate of return, do account for the time value of money. These methods are summarized in Table 13-14.

Ethics and ethical decisions continuously crop up in capital budgeting. Just as with all other areas of finance, violating ethical considerations causes a loss of public confidence, which can have a significant negative effect on shareholder wealth.

TABLE 13-14 Capital-Budgeting Criteria

Nondiscounted Cash Flow Methods

1. Payback period = number of years required to recapture the initial investment
   Accept if payback $\leq$ maximum acceptable payback period
   Reject if payback $>$ maximum acceptable payback period

Advantages:
1. Uses cash flows.
2. Is easy to calculate and understand.
3. May be used as rough screening device.

Disadvantages:
1. Ignores the time value of money.
2. Ignores cash flows occurring after the payback period.

2. Accounting rate of return =
$$\frac{\sum_{t=1}^{n} (\text{accounting profit after tax}_t)/n}{(\text{initial investment} + \text{expected salvage value})/2}$$

where n = project's expected life

Accept if AROR $\geq$ minimum acceptable rate of return
Reject if AROR $<$ minimum acceptable rate of return

Advantages:
1. Involves familiar, easily accessible terms.
2. Is easy to calculate and understand.

Disadvantages:
1. Ignores the time value of money.
2. Uses accounting profits rather than cash flows.

Discounted Cash Flow Methods

3. Net present value = present value of the annual cash flows after tax less the investment's initial outlay

$$NPV = \sum_{t=1}^{n} \frac{CFAT_t}{(1+k)^t} - IO$$

where,
$CFAT_t$ = the annual after-tax cash flow in time period t (this can take on either positive or negative values)
k = the appropriate discount rate, that is, the required rate of return or the cost of capital[a]
IO = the initial cash outlay
n = the project's expected life

Accept if NPV ≥ 0.0
Reject if NPV < 0.0

Advantages:
1. Uses cash flows.
2. Recognizes the time value of money.
3. Is consistent with the firm goal of shareholder wealth maximization.

Disadvantages:
1. Requires detailed long-term forecasts of the incremental benefits and costs.

TABLE 13-14 (continued)

Discounted Cash Flow Methods (continued)

4. Profitability index = the ratio of the present value of the future net cash flows to the initial outlay

$$\text{Profitability Index} = \sum_{t=1}^{n} \frac{CFAT_t}{(1+k)^t} \backslash IO$$

Accept if PI $\geq$ 1.0
Reject if PI $<$ 1.0

Advantages:
1. Uses cash flows.
2. Recognizes the time value of money.
3. Is consistent with the firm goal of shareholder wealth maximization.

Disadvantages:
1. Requires detailed long-term forecasts of the incremental benefits and costs.

5. Internal rate of return = the discount rate that equates the present value of the project's future net cash flows with the project's initial outlay

$$IO = \sum_{t=1}^{n} \frac{CFAT_t}{(1+IRR)^t}$$

where IRR = the project's internal rate of return

Accept if IRR $\geq$ required rate of return
Reject if IRR $<$ required rate of return

Advantages:
1. Uses cash flows.
2. Recognizes the time value of money.

Disadvantages:
1. Requires detailed long-term forecasts of the incremental benefits and costs.
2. Can involve tedious calculations.
3. Possibility of multiple IRRs.

6. Modified internal rate of return = the discount rate that equates the present value of the project's cash *outflows* with the present value of the project's *terminal value*.

$$\sum_{t=1}^{n} \frac{COAT_t}{(1+k)^t} = \frac{\sum_{t=1}^{n} CIAT_t(1+k)^{n-t}}{(1+MIRR)^n}$$

Given that

$$TV = \sum_{t=1}^{n} CIAT_t(1+k)^{n-t}$$

$$PV_{outflows} = \frac{TV}{(1+MIRR)^n}$$

where $COAT_t$ = the annual cash *outflow* after tax in time period t
$CIAT_t$ = the annual cash *inflow* after tax in time period t
TV = the terminal value of $CIAT$'s compounded at the required rate of return to the end of the project

Accept if $MIRR \geq$ the required rate of return
Reject if $MIRR <$ the required rate of return

Advantages:
1. Uses cash flows.
2. Recognizes the time value of money.
3. In general, is consistend with goal of maximization of shareholder wealth.

Disadvantages:
1. Requires detailed long-term forecasts of the incremental benefits and costs.

[a]The cost of capital was discussed in Chapter 12.

STUDY QUESTIONS

13-1. Why is the capital-budgeting decision such an important process? Why are capital-budgeting errors so costly?

31-2. What are the criticisms of the use of the payback period as a capital-budgeting technique? What are its advantages? Why is it so frequently used?

13-3. In some countries, expropriation of foreign investments is a common practice. If you were considering an investment in one of those countries, would the use of the payback period criterion seem more reasonable than it otherwise might? Why?

13-4. What are the criticisms of the use of the accounting rate of return as a capital-budgeting technique? What are its advantages?

13-5. Briefly compare and contrast the three discounted cash flow criteria. What are the advantages and disadvantages of the use of each of these methods?

13-6. What is the advantage of using the MIRR as opposed to the IRR decision criteria?

SELF-TEST PROBLEM

ST-1. The B. Pari Corporation is considering signing a one-year contract with one of two computer-based marketing firms. While one is more expensive, it offers a more extensive program and thus will provide higher cash flows after tax. Assume these two options are mutually exclusive and that the required rate of return is 12 percent. Given the following cash flows after tax:

Year	Option A	Option B
0	−$50,000	−$100,000
1	70,000	130,000

a. Calculate the net present value.
b. Calculate the profitability index.
c. Calculate the internal rate of return.

STUDY PROBLEMS (SET A)

13-1A. (*AROR Calculation*) Two mutually exclusive projects are being evaluated using the accounting rate of return. Each project has an initial cost of $20,000 and a salvage value of $4,000 after six years. Given the following information:

	Annual Accounting Profits After Tax	
Year	Project A	Project B
1	$ 2,000	$10,000
2	2,000	10,000
3	2,000	10,000
4	13,000	5,000
5	13,000	5,000
6	14,000	5,000

a. Determine each project's AROR.
b. Which project should be selected, using this criterion? Would you support that recommendation? Why or why not?

13-2A. (*Payback Period Calculations*) You are considering three independent projects, project A, project B, and project C. The required rate of return is 10 percent on each. Given the following cash flow information, calculate the payback and discounted payback period for each.

Year	Project A	Project B	Project C
0	–$1,000	–$10,000	–$5,000
1	600	5,000	1,000
2	300	3,000	1,000
3	200	3,000	2,000
4	100	3,000	2,000
5	500	3,000	2,000

If you require a three-year payback for both the traditional and discounted payback period methods before an investment can be accepted, which projects would be accepted under each criterion?

13-3A. (*IRR Calculation*) Determine the internal rate of return on the following projects:

a. An initial outlay of $10,000 resulting in a cash flow after tax of $1,993 at the end of each year for the next 10 years.

b. An initial outlay of $10,000 resulting in a cash flow after tax of $2,054 at the end of each year for the next 20 years.

c. An initial outlay of $10,000 resulting in a cash flow after tax of $1,193 at the end of each year for the next 12 years.

d. An initial outlay of $10,000 resulting in a cash flow after tax of $2,843 at the end of each year for the next 5 years.

13-4A. (*IRR Calculation*) Determine the internal rate of return on the following projects:

a. An initial outlay of $10,000 resulting in a single cash flow after tax of $17,182 after 8 years.

b. An initial outlay of $10,000 resulting in a single cash flow after tax of $48,077 after 10 years.

c. An initial outlay of $10,000 resulting in a single cash flow after tax of $114,943 after 20 years.

d. An initial outlay of $10,000 resulting in a single cash flow after tax of $13,680 after 3 years.

13-5A. (*IRR Calculation*) Determine the internal rate of return to the nearest percent on the following projects:

a. An initial outlay of $10,000 resulting in a cash flow after tax of $2,000 at the end of year 1, $5,000 at the end of year 2, and $8,000 at the end of year 3.

b. An initial outlay of $10,000 resulting in a cash flow after tax of $8,000 at the end of year 1, $5,000 at the end of year 2, and $2,000 at the end of year 3.

c. An initial outlay of $10,000 resulting in a cash flow after tax of $2,000 at the end of years 1 through 5 and $5,000 at the end of year 6.

13-6A. (*Cash Flow—Capital-Budgeting Calculation*) The R. Boisjoly Chemical Corporation is considering the purchase of a new machine that costs $80,000. The expected life of this machine is three years, after which it will be discarded, having no value. The new machine is expected to produce cash flows after tax of $55,000 per year. The firm's required rate of return on this type of project is 10 percent. Given the above information, determine the following:

a. The net present value.

b. The profitability index.

c. The internal rate of return.

13-7A. (*Cash Flow—Capital-Budgeting Calculation*) The Mad Dog Hansen Electronic Components Corporation is considering the purchase of a new cash register at a cost of $40,000. This cash register is expected to have a life of six years, at which time the machine will be discarded, having no value. Material efficiencies resulting from the purchase would result in savings after tax of $9,000 per year. The firm has a required rate of return of 20 percent, calculate

a. The net present value.
b. The profitability index.
c. The internal rate of return.

13-8A. (*NPV, PI, and IRR Calculations*) Brownsville Girl Inc. is considering a major expansion of its product line and has estimated the following cash flows after tax associated with such an expansion. The initial outlay associated with the expansion would be $1,950,000 and the project would generate incremental cash flows after tax of $450,000 per year for six years. The appropriate required rate of return is 9 percent.

a. Calculate the net present value.
b. Calculate the profitability index.
c. Calculate the internal rate of return.
d. Should this project be accepted?

13-9A. (*Net Present Value, Profitability Index, and Internal Rate of Return Calculations*) You are considering two independent projects, project A and project B. The initial cash outlay associated with project A is $50,000 and the initial cash outlay associated with project B is $70,000. The required rate of return is 12 percent. The expected annual cash flows after tax from each project are as follows:

Year	Project A	Project B
0	–$50,000	–$70,000
1	12,000	13,000
2	12,000	13,000
3	12,000	13,000
4	12,000	13,000
5	12,000	13,000
6	12,000	13,000

Calculate the NPV, PI and IRR for each project and indicate if the project should be accepted.

13-10. (*NPV with Varying Required Rates of Return*) Big Steve's, makers of swizzle sticks, is considering the purchase of a new plastic stamping machine. This investment requires an initial outlay of $100,000 and will generate cash flows after tax of $18,000 per year for 10 years. For each of the listed required rates of return, determine the project's net present value.

a. The required rate of return is 10 percent.
b. The required rate of return is 15 percent.
c. Would the project be accepted under part (a) or (b)?
d. What is this project's internal rate of return?

13-11A. (*Internal Rate of Return Calculations*) Given the following cash flows after tax, determine the internal rate of return for projects A, B, and C.

	Project A	Project B	Project C
Initial Investment:	$50,000	$100,000	$450,000
Cash Inflows:			
Year 1	$10,000	25,000	200,000
Year 2	15,000	25,000	200,000
Year 3	20,000	25,000	200,000
Year 4	25,000	25,000	–
Year 5	30,000	25,000	–

13-12A. (*NPV with Varying Rates of Return*) Dowling Sportswear is considering building a new factory to produce aluminum baseball bats. This project would require an initial cash outlay of $5,000,000 and will generate annual cash flows after tax of $1 million per year for eight years. Calculate the project's NPV given:

a. A required rate of return of 9 percent.
b. A required rate of return of 11 percent.

c. A required rate of return of 13 percent.
d. A required rate of return of 15 percent.

13-13A. (*MIRR Calculation*) Emily's Soccer Mania is considering building a new plant. This project would require an initial cash outlay of $10 million and will generate annual cash flows after tax of $3 million per year for ten years. Calculate the project's MIRR, given:

a. A required rate of return of 10 percent.
b. A required rate of return of 12 percent.
c. A required rate of return of 14 percent.

INTEGRATIVE PROBLEM

Your first assignment in your new position as assistant financial analyst at Cale Products is to evaluate two new capital-budgeting proposals. Since this is your first assignment, you have been asked not only to provide a recommendation, but also to respond to a number of questions aimed at judging your understanding of the capital-budgeting process. This is a standard procedure for all new financial analysts at Cale, and will serve to determine whether you are moved directly into the capital-budgeting analysis department or are provided with remedial training. The memorandum you received outlining your assignment follows:

TO: The New Financial Analysts

FROM: Mr. V. Morrison, CEO, Cale Products

RE: Capital-Budgeting Analysis

Provide an evaluation of two proposed projects, both with five-year expected lives and identical initial outlays of $110,000. Both these projects involve additions to Cale's highly successful Avalon product line, and as a result, the required rate of return on both projets has been established at 12 percent. The expected cash flows after tax from each project are as follows:

	Project A	Project B
Initial Outlay	–$110,000	–$110,000
Year 1	20,000	40,000
Year 2	30,000	40,000
Year 3	40,000	40,000
Year 4	50,000	40,000
Year 5	70,000	40,000

In evaluating these projects, please respond to the following questions:

a. Why is the capital-budgeting process so important?
b. Why is it difficult to find exceptionally profitable projects?
c. What is the payback period on each project? If Cale's imposes a three-year maximum acceptable payback period, which of these projects should be accepted?
d. What are the criticisms of the payback period?
e. What are the discounted payback periods for each of these projects? If Cale requires a three-year maximum acceptable discounted payback period on new projects, which of these projects should be accepted?
f. What are the drawbacks or deficiencies of the discounted payback period? Do you feel either the payback or discounted payback period should be used to determine whether or not these projects should be accepted? Why or why not?
g. Determine the net present value for each of these projects. Should they be accepted?
h. Describe the logic behind the net present value.
i. Determine the profitability index for each of these projects. Should they be accepted?
j. Would you expect the net present value and profitability index methods to give consistent accept-reject decisions? Why or why not?

k. What would happen to the net present value and profitability index for each project if the required rate of return increased? If the required rate of return decreased?

l. Determine the internal rate of return for each project. Should they be accepted?

m. How does a change in the required rate of return affect the project's internal rate of return?

n. What reinvestment rate assumptions are implicitly made by the net present value and internal rate of return methods? Which one is better?

STUDY PROBLEMS (SET B)

13-1B. (*AROR Calculation*) Two mutually exclusive projects are being evaluated using the accounting rate of return. Each project has an initial cost of $35,000 and a salvage value of $7,000 after six years. Given the following information:

Annual Accounting Profits After Tax

Year	Project A	Project B
1	$ 2,500	$10,500
2	2,500	10,500
3	2,500	10,500
4	13,000	5,000
5	13,000	5,000
6	14,000	5,000

a. Determine each project's AROR.

b. Which project should be selected, using this criterion? Would you support that recommendation? Why or why not?

13-2B. (*Payback Period Calculations*) You are considering three independent projects, project A, project B, and project C. The required rate of return is 12 percent on each. Given the following cash flow information calculate the payback and discounted payback period for each.

Year	Project A	Project B	Project C
0	−$ 900	−$ 9,000	−$7,000
1	600	5,000	2,000
2	300	3,000	2,000
3	200	3,000	2,000
4	100	3,000	2,000
5	500	3,000	2,000

If you require a three-year payback for both the traditional and discounted payback period methods before an investment can be accepted, which projects would be accepted under each criterion?

13-3B. (*IRR Calculation*) Determine the internal rate of return on the following projects:

a. An initial outlay of $10,000 resulting in a single cash flow after tax of $19,926 after 8 years.

b. An initial outlay of $10,000 resulting in a single cash flow after tax of $20,122 after 12 years.

c. An initial outlay of $10,000 resulting in a single cash flow after tax of $121,000 after 22 years.

d. An initial outlay of $10,000 resulting in a single cash flow after tax of $19,254 after 5 years.

13-4B. (*IRR Calculation*) Determine the internal rate of return on the following projects:

a. An initial outlay of $10,000 resulting in a cash flow after tax of $2,146 at the end of each year for the next 10 years.

b. An initial outlay of $10,000 resulting in a cash flow after tax of $1,960 at the end of each year for the next 20 years.

c. An initial outlay of $10,000 resulting in a cash flow after tax of $1,396 at the end of each year for the next 12 years.

d. An initial outlay of $10,000 resulting in a cash flow after tax of $3,197 at the end of each year for the next 5 years.

13-5B. (*IRR Calculation*) Determine the internal rate of return to the nearest percent on the following projects:

a. An initial outlay of $10,000 resulting in a cash flow after tax of $3,000 at the end of year 1, $5,000 at the end of year 2, and $7,500 at the end of year 3.

b. An initial outlay of $12,000 resulting in a cash flow after tax of $9,000 at the end of year 1, $6,000 at the end of year 2, and $2,000 at the end of year 3.

c. An initial outlay of $8,000 resulting in a cash flow after tax of $2,000 at the end of years 1 through 5 and $5,000 at the end of year 6.

13-6B. (*Cash Flow—Capital-Budgeting Calculation*) The R. J. Water, Inc., is considering the purchase of a new bottling machine that costs $150,000. The expected life of this machine is seven years, after which it will be discarded, having no value. The new machine is expected to produce cash flows after tax of $35,000 per year. The firm's required rate of return on this type of project is 13 percent. Given the above information, determine the following:

a. The net present value.

b. The profitability index.

c. The internal rate of return.

13-7B. (*Cash Flow—Capital-Budgeting Calculation*) The P. J. Corporation is considering the purchase of a new wrapping machine at a cost of $160,000. This new machine is expected to have a life of six years, at which time the machine will be discarded, having no value. Material efficiencies resulting from the purchase would result in savings after tax of $25,000 per year. The firm has a required rate of return on this type of project of 15 percent. Calculate

a. The net present value.

b. The profitability index.

c. The internal rate of return.

13-8B. (*NPV, PI, and IRR Calculations*) Anderson, Inc., is considering a major expansion of its product line and has estimated the following cash flows after tax associated with such an expansion. The initial outlay associated with the expansion would be $2,500,000 and the project would generate incremental cash flows after tax of $750,000 per year for six years. The appropriate required rate of return is 11 percent.

a. Calculate the net present value.

b. Calculate the profitability index.

c. Calculate the internal rate of return.

d. Should this project be accepted?

13-9B. (*Net Present Value, Profitability Index, and Internal Rate of Return Calculations*) You are considering two independent projects, project A and project B. The initial cash outlay associated with project A is $45,000 and the initial cash outlay associated with project B is $70,000. The required rate of return is 12 percent. The expected annual cash flows after tax from each project are as follows:

Year	Project A	Project B
0	–$45,000	–$70,000
1	12,000	14,000
2	12,000	14,000
3	12,000	14,000
4	12,000	14,000
5	12,000	14,000
6	12,000	14,000

Calculate the NPV, PI and IRR for each project and indicate if the project should be accepted.

13-10B. (*Internal Rate of Return Calculations*) Given the following cash flows after tax, determine the internal rate of return for projects A, B, and C.

	Project A	Project B	Project C
Initial Investment:	$75,000	$95,000	$395,000
Cash Inflows:			
Year 1	$10,000	25,000	150,000
Year 2	10,000	25,000	150,000
Year 3	30,000	25,000	150,000
Year 4	25,000	25,000	–
Year 5	30,000	25,000	–

13-11B. (*NPV with Varying Rates of Return*) Mo-Lee's Sportswear is considering building a new factory to produce soccer equipment. This project would require an initial cash outlay of $10,000,000 and will generate annual cash flows after tax of $2.5 million per year for eight years. Calculate the project's NPV given:

a. A required rate of return of 9 percent.
b. A required rate of return of 11 percent.
c. A required rate of return of 13 percent.
d. A required rate of return of 15 percent.

13-12B. (*MIRR Calculation*) Artie's Soccer Stuff is considering building a new plant. This project would require an initial cash outlay of $8 million and will generate annual cash flows after tax of $2 million per year for eight years.

Calculate the project's MIRR, given:

a. A required rate of return of 10 percent.
b. A required rate of return of 12 percent.
c. A required rate of return of 14 percent.

SELF-TEST SOLUTION

SS-1. a.

$$NPV_A = \frac{\$70{,}000}{(1+.12)^1} - \$50{,}000$$
$$= \$70{,}000(.8929) - \$50{,}000$$
$$= \$62{,}503 - \$50{,}000$$
$$= \$12{,}503$$

$$NPV_B = \frac{\$130{,}000}{(1+.12)^1} - \$100{,}000$$
$$= \$130{,}000(.8929) - \$100{,}000$$
$$= \$116{,}077 - \$100{,}000$$
$$= \$16{,}077$$

b.

$$\text{Profitability Index (A)} = \frac{\$62{,}510}{\$50{,}000}$$
$$= 1.2502$$

$$\text{Profitability Index (B)} = \frac{\$116{,}090}{\$100{,}000}$$
$$= 1.1609$$

c.

$$\$50{,}000 = \$70{,}000(PVIF_{i,1yr})$$
$$.7143 = PVIF_{i,1yr}$$

Looking for the value of $PVIF_{i,1yr}$ in Appendix B at the back of the book, a value of .7143 is found in the 40 percent column. Thus, the IRR is 40 percent.

$$\$100{,}000 = \$130{,}000(PVIF_{i,1yr})$$
$$.7692 = PVIF_{i,1yr}$$

Looking for a value of $PVIF_{i,1yr}$ in Appendix B, a value of .7692 is found in the 40 percent column. Thus the IRR is 30 percent.

CASH FLOWS AND OTHER TOPICS IN CAPITAL BUDGETING

CHAPTER 14

LEARNING OBJECTIVES

After reading this chapter you should be able to

1. Identify guidelines by which we measure cash flows.
2. Explain how the capital-budgeting process is used to make decisions on expansion projects.
3. Explain how the capital-budgeting process is used to make decisions on replacement projects.
4. Explain how the capital-budgeting decision process changes when a limit is placed on the dollar size of the capital budget.

INTRODUCTION

In 1994, the Ford Motor Company introduced the Ford Windstar to its lineup of cars, trucks, and minivans. In the introduction to the previous chapter we discussed the importance of this $1.5 billion investment by Ford that was targeted directly at Chrysler's Caravan/Voyager, which had been dominating the minivan market since its beginnings. Although this capital-budgeting decision may, on the surface seem like a relatively simple decision, the forecasting of the expected cash flows associated with the Windstar were, in fact, quite complicated.

To begin with, Ford was introducing a product that was to compete directly with some of its own products, the Ford Aerostar and the Mercury Villager. Thus, some of the sales of the Windstar would be cannibalizing sales of other Ford products. In addition, Chrysler had been in the process of a major redesign since its introduction. Given that fact, Ford was confronted with the possibility of having its hands full trying to increase its market share in the minivan market above its current level. From Ford's point of view, increasing the market share might not have been the objective. Simply preventing Ford from losing market share may have been all that Ford was looking for from the Windstar.

In fact, in many very competitive markets, the evolution and introduction of new products may serve more to preserve market share than to expand it. Certainly, that's the case in the computer market in which Apple, Dell, Compaq, and IBM are introducing upgraded models that continually render current models obsolete.

Does competing with yourself just to maintain market share make sense, or is it negative thinking that should be avoided? The answer to this deals with how we esti-

mate a project's future cash flows. In this chapter, we try to gain an understanding of what a relevant cash flow is. We evaluate projects relative to their base case—that is, what would have happened if the project had not been carried out. In the case of the Ford Windstar, we ask what would have happened to the Ford Aerostar and the Mercury Villager sales that were expected to be captured by the Windstar if the Windstar had not been introduced. Would they have been lost to a new generation of the Caravan/Voyager? It is questions like these, all leading us to an understanding of what are and are not relevant cash flows, that are addressed in this chapter.

CHAPTER PREVIEW

In this chapter, we continue our discussion of the decision rules which are used to decide whether or not to invest in new projects. First, we will examine what is a relevant cash flow and how to calculate the relevant cash flow. We then examine how the capital-budgeting process is used to make decisions on whether to accept or to reject expansion and replacement projects for investment purposes. Finally, we examine some of the problems that are encountered in the capital budgeting decision. For example, which projects to choose if there are constraints on the number of projects that can be accepted or the total budget is limited?

OBJECTIVE 1

GUIDELINES TO CAPITAL BUDGETING

Capital budgeting is a decision-making process for investment in fixed assets; specifically, it involves measuring the incremental cash flows associated with investment proposals and evaluating the attractiveness of these cash flows relative to the project's cost.[1] Typically these investments involve rather large cash outlays at the outset and commit the firm to a particular course of action over a relatively long time period. Thus, if a capital-budgeting decision is incorrect, reversing it tends to be costly.

For example, about 30 years ago the Ford Motor Company's decision to produce the Edsel entailed an outlay of $250 million to bring the car to market and losses of approximately $200 million during the 2.5 years it was produced—in all, a $450 million mistake.[2] This type of decision is costly to reverse. Fortunately for Ford, it was able to convert Edsel production facilities to produce the Mustang, thereby avoiding an even larger loss.

PERSPECTIVE IN FINANCE

This chapter deals with decision rules for deciding when to invest in new projects. Before we can develop these rules we must be able to measure the benefits and costs of the new project by calculating cash flow. In reading over the upcoming section try not to look upon calculating cash flow items in a checklist, but instead try to think about what cash flows are and what creates them.

[1]The only relevant measure of a project's initial outlay is the market value of its assets. Historical costs measure book value and are not an appropriate measure of value for the decision making with respect to maximizing shareholder wealth.

[2]"The Edsel Dies, and Ford Regroups Survivors," *Business Week*, November 28, 1959, p. 27.

In measuring cash flows, we will be interested in only the incremental cash flow after tax that can be attributed to the project being evaluated.[3] A project proposal will present a forecast of expected cash inflows and cash outflows which must increase shareholder wealth for the project to be accepted. Obviously, the worth of our decision depends upon the accuracy of our cash-flow estimates.

The measurement of cash flows will be affected by the type of project.[4] The majority of projects encountered in a firm's capital budget are classified as either of two types: a replacement project or an expansion project. A replacement project involves the acquisition of a new asset for the purposes of either maintenance or cost reduction. The net cash flows of a replacement project will be measured in terms of the amount of net savings after taxes that accrues to the firm. An expansion project involves the acquisition of a new asset for the purpose of increasing production capacity. The net cash flows of an expansion project will be measured in terms of the amount of net operating income that accrues to the firm. Also encountered within the firm's capital budget are a small minority of projects called "soft projects." This type of project is difficult to evaluate by capital-budgeting techniques because the measurement of either the expected cash inflow or the cash outflow is imprecise; for example, the benefits to society of installing pollution control devices in factories.

Video Case 4 

To evaluate investment proposals, we must first set guidelines by which we measure the value of each proposal.

Cash Flows Rather than Accounting Profits

We will use cash flows, not accounting profits, as our measurement tool. The firm receives and is able to reinvest cash flows, whereas accounting profits are shown when they are earned rather than when the money is actually in hand. Unfortunately, a firm's accounting profits and cash flows may not be timed to occur together. For example, capital expenses such as vehicles, plants and equipment, are depreciated over a number of years, with their annual depreciation subtracted from profit. Cash flow correctly reflects the timing of benefits and costs; that is, when the money is received, when it can be reinvested, and when it must be paid out.

BACK TO THE BASICS

If we are to make intelligent capital-budgeting decisions we must accurately measure the timing of the benefits and costs, that is, when we receive money and when it leaves our hands. ***Axiom 3: Cash—Not Profits—Is King: Measuring the Timing of Costs and Benefits*** *speaks directly to this. Remember, it is cash inflows that can be reinvested and cash outflows that involve paying out money.*

Think Incrementally

Interestingly, calculating cash flow from a project may not be enough. Decision makers must ask: what new cash flows will the company as a whole receive if it takes on a given project? In fact, we may find that not all cash flows a firm expects

[3]Good capital budgeting decisions consider all relevant incremental cash flows. This decision process includes the consequences of not taking on the project. It also includes the value of opening up new options for the future. Unfortunately, while talking about this is easy, implementing it is not.

[4]Examples of capital investments include the acquisition of another company, building a plant, introducing a new product, and purchasing equipment. One of the largest capital budgeting projects ever undertaken was the tunnel built under the English Channel by the Eurotunnel Company. The original estimated cost of this project in 1987 when it was started was approximately $8 billion.

from an investment proposal are incremental in nature. But, in measuring cash flows, the trick is to think incrementally. In doing so, we will see that only incremental cash flows after tax matter. There are a number of things a firm must consider in its decision-making process.

BACK TO THE BASICS

In order to measure the true effects of our decisions we will analyze the benefits and costs of projects on an incremental basis, which relates directly to ***Axiom 4: Incremental Cash Flows—It's Only What Changes That Counts.*** *In fact, we will compare the cash flows with and without the project taken on in order to measure the incremental effect.*

Incremental cash flows

The cash flows that result from the acceptance of a capital-budgeting project.

Beware of Cash Flows Diverted from Existing Products

Assume for a moment that we are managers of a firm considering a new product line that might compete with one of our existing products and possibly reduce its sales. In determining cash flows associated with the proposed project, we should consider only the incremental sales brought to the company as whole. New product sales achieved at the cost of losing sales of other products in our line are not considered a benefit of adopting the new product.

Remember that we are only interested in sales dollars to the firm if this project is accepted as opposed to what the sales dollars would be if the project is rejected. Just moving sales from one product line to a new product line does not bring anything new into the company, but if sales are captured from our competitors or if sales that would have been lost to new competing products are retained then these are relevant incremental cash flows. In each case these are the incremental cash flows to the firm—looking at the firm as a whole with the new product versus without the new product.

Look for Incidental or Synergistic Effects

Although in some cases a new project may take sales away from a firm's current projects, in other cases a new effort may actually bring new sales to the existing line. For example, an airline may open a new route to a North American city. The new route not only means new tickets on this new route, but also means increasing ticket sales on connecting routes. If managers were to look only at the revenue from ticket sales from the new route, they would miss the incremental cash flow to the firm as a whole that results from taking on the new route. This is called an incidental or synergistic effect. As a result, a manager when analyzing the firm's bottom line must consider any cash flow to any part of the company that may result from the new project under consideration.

Include Working-Capital Requirements

Many times a new project will involve additional investment in working capital. This may take the form of new inventory to stock a sales outlet, additional investment in accounts receivable resulting from additional credit sales, or increased investment in cash to operate cash registers, and more. Working-capital requirements are considered a cash flow even though they do not leave the company. Why is investment in inventory considered a cash outflow when the goods are still in the store? Because the firm does not have access to the inventory's cash value, the firm cannot use the money for other investments. Generally, working-capital requirements are tied up over the life of the project. When the project terminates there is usually an offsetting cash inflow as the working capital is recovered.

Consider Incremental Expenses

Just as cash inflows from a new project are measured on an incremental basis, expenses should also be measured on an incremental basis. For example, the introduction of a new product line requires the training of sales staff and the cash flow after tax associated with the training program must be considered a cash flow and charged to the project. If accepting a new project dictates that a production facility be re-engineered, the cash flow after tax associated with the capital investment should be charged against the project. Again, any incremental cash flow after tax affecting the company as a whole is a relevant cash flow—whether it is flowing in or flowing out.

Sunk Costs Are Not Incremental Cash Flows

Only cash flows that are affected by the decision making at the moment are relevant cash flows in capital budgeting. The manager must ask two questions: (1) Will this cash flow take place if the project is accepted? (2) Will this cash flow take place if the project is rejected? If the answers are yes to the first question and no to the second question, the cash flow is incremental. For example, let's assume you are considering the introduction of a new taste treat called Puddin' in a Shoe. You would like to do some test marketing before production. If you are considering the decision to test market and have not yet done so, the costs associated with the test marketing are relevant cash flows. On the other hand, if you have already test marketed, the cash flows involved in the test marketing are no longer relevant in project evaluation. It is a matter of timing. Regardless of what you may decide about future production, the cash flows allocated to marketing have already taken place. As a rule, any cash flows that are not part of the accept-or-reject decision should not be included in the capital-budgeting analysis.

Account for Opportunity Costs

Now we will focus on the cash flows that are lost because assets already in place are now being used for a new project. This is the opportunity cost of doing business. For example, a product may use valuable floor space in a production facility. Although the cash flow is not obvious, the real question remains: What else could be done with this space? The space could have been rented out, or another product could have been stored there. The key point is that the opportunity costs have occurred because existing assets are now used for the new project and these opportunity costs must be included in the net cash flow of the new project.

Cash Flows from Changes in Overhead

Although we certainly want to include any incremental cash flows which are caused by changes from overhead expenses such as utilities and salaries, we also want to make sure that these are truly incremental cash flows. Many times, overhead expenses—heat, light, rent—would occur whether or not a given project is accepted. In many cases, expenses cannot be allocated to a specific project. Thus, we must determine whether or not these expenses are incremental cash flows and relevant to capital budgeting.

Ignore Interest Payments and Financing Flows

In evaluating new projects and determining cash flows, we must separate the investment decision from the financing decision. Interest payments and other financing cash flows that might result from raising funds to finance a project should not be considered incremental cash flows. If accepting a project means we have to raise new funds by issuing new bonds, interest charges associated with raising funds are not a relevant cash outflow. When we discount the incremental cash flows back to the present at the required rate of return, we are implicitly accounting for the cost of raising funds to finance the new project. In essence, the required rate of return reflects the cost of the funds needed to support the project. Managers must first determine the desirability of the project and then decide how best to finance it.

PERSPECTIVE IN FINANCE

In examining the discounted cash flow criteria for capital-budgeting decisions, we showed that these criteria were superior to nondiscounted cash flow criteria because they used the concept of cash flows and accounted for the time value of money. As a result, we now apply these criteria to capital-budgeting decisions for expansion and replacement projects.

OBJECTIVE 2

CAPITAL BUDGETING FOR EXPANSION PROJECTS

An expansion project for manufacturing and processing firms generally requires the acquisition of new assets for either initiating a production line or increasing the existing production capacity. The decision criteria for evaluating expansion projects are dependent upon how accurately we measure the net cash flows of the project. To facilitate the measurement of cash flows during the project's life, we arbitrarily divide the project into three periods: the initial period when the project set-up costs are incurred, the operating period when the project incurs both revenues and expenses for manufacturing the product, and the final period when the project is terminated. As a result of dividing the project into three periods, we can now separate the net cash flows of an expansion project into one of three categories: (1) the initial outlay, (2) the incremental net cash flows over the project's operating life, (3) the terminal cash flow. In introducing these cash flow categories, we first determine what factors influence each category of cash flow and then determine the value of the project's cash flows.

EXAMPLE

The managing director of an advertising agency in Toronto is presently in the process of evaluating whether to accept or reject a new account. The account's expenses include an initial outlay of cash which has been estimated at $100,000. The expenses for the initial outlay include $25,000 in fees for marketing tests on a new product for which results have been obtained one month ago. The managing director estimates that the account will generate after-tax net cash flows of $35,000 over the next five years. An additional $5,000 in after-tax expenses will be required to close the account after five years. It is company policy to use a 16 percent required rate of return for projects of this risk. Should the managing director accept or reject this account?

$NPV = PV$(incremental net cash flows over the project's operating life)
$\quad + PV$(terminal cash flow) – initial outlay

$NPV = \$35{,}000(PVIFA_{16\%,5yrs}) + (-\$5{,}000)(PVIF_{16\%,5yrs}) - (\$100{,}000 - \$25{,}000)$

$NPV = \$35{,}000(3.2743) - \$5{,}000(.4761) - \$75{,}000$

$NPV = \$41{,}981$

The managing director should accept the account because the NPV is positive. It should be noted that the marketing expense of $25,000 is considered a sunk cost and should not be included in the project's initial outlay.

Table 14-1 Summary of Calculation of Initial Outlay Incremental Cash Flow After Tax

1. Capital cost of asset
2. Opportunity costs
3. Incremental nonexpense outlays incurred (for example, working-capital investments)
4. Incremental cash expenses on an after-tax basis (for example, training expenses)
5. In a replacement decision, the cash flow associated with the sale of the old machine

Initial Outlay

The most important component of the **initial outlay** for an expansion project is the immediate cash outflow required to purchase an asset and to put it into operating order. Revenue Canada designates the purchase cost of an asset to the firm as the asset's capital cost.[5] The capital cost of an asset also includes any costs associated with acquiring or putting the asset into working order, such as shipping fees, installation costs, special engineering costs, and any incremental administrative costs such as accounting expenses.

Initial outlay

The immediate cash outflow necessary to purchase the asset and put it in operating order.

The initial outlay must also include any opportunity costs that are associated with the implementation of the project. For example, the firm should include the fair market value net of any tax effects of land that is owned by the firm but is being used for the project. Furthermore, any lost revenues (e.g., parking fees) from this land should also be included as an opportunity cost to the firm. The initial outlay also includes any nonexpense cash outlays. If we are considering a new sales outlet, there might be incremental cash flows associated with the investment in working capital such as any increases in inventory or cash which are necessary to operate the sales outlet. While these cash flows are not included in the cost of the asset or even expensed on the books, they must be included in our analysis. In addition, the initial outlay will include investment tax credits and any cash expenses on an after-tax basis, such as any training expenses associated with operating the new asset.

Obviously, determining the initial outlay is a complex calculation. Table 14-1 summarizes some of the more common calculations involved in determining the initial outlay. This list is by no means exhaustive, but it should help simplify the calculations involved in the example that follows.[6]

PERSPECTIVE IN FINANCE

At this point we should note that the incremental nature of the cash flow is of great importance. In many cases if the project is not accepted, then "status quo" for the firm will simply not continue. In estimating incremental cash flows we must be realistic in choosing our base case. The reading in Basic Financial Management in Practice, "Using the Right Base Case," deals with precisely this question.

[5]The Income Tax Act, section 13(7.1), refers to the deemed capital cost of depreciable property as the capital cost of the asset net of any investment tax credit or governmental assistance.

[6]The evaluation of new projects should also look at the strategic options provided by a new investment. In light of the opening up of borders in Europe in 1992, many companies based their investment decision in part on the access to the marketing and distribution channels.

EXAMPLE

In order to clarify the calculation of the initial outlay, consider an example of a manufacturing company in the electronic components sector. Assume that this company has a combined corporate tax rate of 45 percent with a 15 percent required rate of return or cost of capital. Management is considering the expansion of a manufacturing line by purchasing a $500,000 hand-operated assembly machine. To put the new machine in running order, it is necessary to pay shipping charges of $30,000 and installation charges of $30,000. The new machine will require an increase in work-in-progress inventory of $50,000 and its expected useful life is ten years.

The capital cost of the new machine would be the $500,000 purchase cost plus the $30,000 shipping and the $30,000 installation fees, for a total of $560,000. An incremental outflow occurs with the increased investment in inventory. The increase in work-in-process inventory of $50,000 must also be considered part of the initial outlay, with an offsetting inflow of $50,000 corresponding to the recovery of this inventory occurring at the termination of the project. In effect, the firm invests $50,000 in inventory now, resulting in an initial cash outlay, and liquidates this inventory in ten years, resulting in a cash inflow at the end of the project. The total outlays associated with the new machine are $560,000 for its capital cost and $50,000 in investment in inventory, for a total of $610,000. Thus, the net initial outlay associated with this project is $610,000. These calculations are summarized in Table 14-2.

Incremental Net Cash Flows over the Project's Operating Life

In considering an investment proposal, we must determine the net cash flows generated from a project during the operating period. Again, we emphasize that the cash flows being measured must be incremental. As a result, the incremental net cash flow occurring over the life of an expansion project includes the following items: increases in operating revenues, labour or material savings, and reductions in selling expenses. Incremental increases in any overhead items such as utilities, heat, light, and executive salaries must be included in the determination of incremental operating expenses.

Another important consideration in determining the net operating cash flows is the effect of depreciation for tax purposes on the firm's taxes payable. For the purposes of accounting, depreciation and depletion are recognized on the balance sheet as an allocation of the cost of a depreciable asset or a natural resource over its useful life. In other words, accounting depreciation or depletion expenses are non-cash expenses and are generally included in operating expenses. For the purposes of finance, we are only interested in determining cash expenses. As a result, non-

Table 14-2 Calculation of Initial Outlay for Example Problem

Outflows:	
Purchase price	$500,000
Shipping fee	30,000
Installation fee	30,000
Capital cost of machine	$560,000
Increased investment in inventory	50,000
Total outflows	$610,000
Inflows:	
None	– 0
Net initial outlay	$610,000

I. Patrick Barwise, Paul Marsh, and Robin Wensley on Using the Right Base Case

Finance theory assumes that a project will be evaluated against its base case, that is, what will happen if the project is not carried out. Managers tend to explore fully the implications of adopting the project but usually spend less time considering the likely outcome of not making the investment. Yet unless the base case is realistic, the incremental cash flows—the difference between the "with" and the "without" scenarios—will mislead.

Often companies implicitly assume that the base case is simply a continuation of the status quo, but this assumption ignores market trends and competitor behavior. It also neglects the impact of changes the company might make anyway, like improving operations management.

Using the wrong base case is typical of product launches in which the new product will likely erode the market for the company's existing product line. Take Apple Computer's introduction of the Macintosh SE. The new PC had obvious implications for sales of earlier generation Macintoshes. To analyze the incremental cash flows arising from the new product, Apple would have needed to count the lost contribution from sales of its existing products as a cost of the launch.

Wrongly applied, however, this approach would equate the without case to the status quo: it would assume that without the SE, sales of existing Macintoshes would continue at their current level. In the competitive PC market, however, nothing stands still. Competitors like IBM would likely innovate and take share away from the earlier generation Macintoshes—which a more realistic base case would have reflected. Sales of existing products would decline even in the base case.

Consider investments in the marketing of existing brands through promotions, media budgets, and the like. They are often sold as if they were likely to lead to ever-increasing market share. But competitors will also be promoting their brands, and market shares across the board still have to add up to 100%. Still, such an investment is not necessarily wasted. It may just need a more realistic justification: although the investment is unlikely to increase sales above existing levels, it may prevent sales from falling. Marketers who like positive thinking may not like this defensive argument, but it is the only argument that makes economic sense in a mature market.

In situations like this, when the investment is needed just to maintain market share, the returns may be high in comparison with the base case, but the company's reported profits may still go down. Senior managers are naturally puzzled at apparently netting only 5% on a project that had promised a 35% return.[7] Without the investment, however, the profit picture would have looked even worse, especially in the longer term.

Some projects disappoint for other reasons. Sometimes the original proposals are overoptimistic, partly because the base case is implicit or defined incorrectly. That is, if managers are convinced that the investment is sound and are frustrated because the figures fail to confirm their intuition, they may over-inflate projections of sales or earnings. But misstating the base case and then having to make unrealistic projections are unlikely to cancel each other out; they merely cloud the analysis.

Source: Reprinted by permission of the *Harvard Business Review*. An excerpt from "Must Finance and Strategy Clash?" by I. Patrick Barwise, Paul Marsh, and Robin Wensley (September–October, 1989).

---

[7]Joseph L. Bower, *Managing the Resource Allocation Process* (Boston: Harvard Business School Press, 1986), p. 13.

cash expenses such as the accounting expenses of depreciation and depletion must be added back to operating expenses in order to obtain real flows of cash from operations.

PERSPECTIVE IN FINANCE

The analysis of an investment proposal for tax purposes requires that we determine the effect of depreciation on taxable income. For purposes of taxes, depreciation expenses are called capital cost allowances and they decrease the firm's taxable income. In other words, the effect of capital cost allowance on taxable income causes a decrease in the firm's taxes which in turn increases the firm's profits. However, for the purposes of finance, the analysis of projects requires that we maximize shareholders' wealth. In fact, projects are chosen on the basis of how much economic wealth is transferred from the project to the firm. Because the emphasis in financial decision making is to increase the shareholder's economic wealth, management will choose a project on the basis of how much economic wealth it generates and not on the basis of the amount of tax savings generated from the project. Nevertheless, we must bear in mind that although the capital cost allowance is not a cash flow item, the permitted write-off in terms of a capital cost allowance does have a tax effect on the project's cash flows. As a result, our decision-making process should account for the tax effects of capital cost allowance, but only to the extent that it provides a more accurate basis to evaluate the project's cash flows.

Depreciation Methods for Tax Purposes

Revenue Canada permits a taxpayer to write off the value of depreciable assets by using the capital cost allowance system. For the purposes of taxes, the portion of the asset's value being written off is called a *capital cost allowance*. In other words, capital cost allowances are the equivalent of depreciation expenses for accounting purposes. To understand how the capital cost allowance system functions, we first define some of the system's terms such as asset classes and undepreciated capital cost.

Revenue Canada has classified all depreciable assets into groups, called asset classes, which specify both the method of depreciation and the rate at which the asset can be written off. The majority of depreciable assets belong to asset classes which require the use of the declining balance method to write off the asset's value.[8] Some of the major asset classes and their depreciation methods are shown in Table 14-3.

A small number of depreciable assets are classified in asset classes which use methods other than the declining balance method. For example, asset class 14 contains limited life intangibles such as patents, franchises, concessions, or licenses. Revenue Canada requires that these assets be written off over their legal life using a straight-line method of depreciation. Other examples of depreciation methods include asset classes 24, 27, and 34, which permit the asset to be fully written off over three years. The capital cost allowances are determined from the following capital cost allowance rates: 25 percent in the first year, 50 percent in the second year, and 25 percent in the third year.

For tax purposes, Revenue Canada defines undepreciated capital cost (UCC) of an asset, as the capital cost of the asset net of any capital cost allowances claimed by the firm. The UCC of the asset class is the sum of UCCs for each asset in the class net of any capital cost allowances for assets in the asset class.

Revenue Canada allows a taxpayer to claim any amount of capital cost allowance up to a maximum which is determined from the capital cost allowance

[8]The declining balance method of depreciation encourages more noncash expenses to occur in the earlier years of the project. As a result, the declining balance method is used to minimize taxable income and thereby postpone taxes. This advantage is somewhat reduced for asset classes which require that the half-net amount rule be applied to the asset's UCC.

Table 14-3 Depreciation Methods of the Major Asset Classes

Class	Summary of Items	Method and CCA Rate
1	Buildings or other structures including components parts acquired by taxpayer after 1987	Declining balance method at 4 percent
3	Buildings or other structures including components parts acquired by taxpayer before 1987	Declining balance method at 5 percent
8	Kilns, vats, electrical equipment acquired after May 25, 1976 or equipment not included in another class	Declining balance method at 20 percent
9	Aircraft acquired May 25, 1976	Declining balance method at 25 percent
10	General purpose electronic data-processing equipment, systems software, electronic communications control equipment, and automotive equipment	Declining balance method at 30 percent
12	Books of a lending library, chinaware, cutlery, linen, uniforms, computer software acquired after May 25, 1976	100 percent
13	Leasehold interests	Straight-line method over unexpired term of lease (i.e., 5yr < unexpired term < 40yr)
14	Patents acquired before 1993, franchises, concessions, or licenses for a limited period	Remaining legal life
29	Property acquired by a taxpayer before 1988 to be used directly or indirectly by him or her in Canada primarily in manufacturing or processing of goods for sale or lease	25 percent 1st year 50 percent 2nd year 25 percent 3rd year
39	Property acquired by a taxpayer after February 1992 that is not included in class 29	Declining balance method at 25 percent
44	Patents acquired after 1993	Declining balance method at 25 percent

rate of each asset class and the UCC of the asset at the beginning of the year. The amount of capital cost that is written off in a year is known as the capital cost allowance for tax purposes. A taxpayer is only permitted to claim a capital cost allowance on assets that are in productive use. For the purposes of capital budgeting, we are only interested in the assets purchased for the project. As a result, we determine the amount of capital cost allowance (CCA) which can be claimed for an asset as follows:

$$CCA = (UCC_n)(d) \tag{14-1}$$

where CCA = amount of capital cost allowance to be written off for the asset in year n

UCC_n = undepreciated capital cost of an asset at the beginning of year n

d = capital cost allowance rate

For most assets, Revenue Canada requests that in the first year of productive use the half-net amount rule be used to determine the amount of capital cost allowance. This rule requires that 50 percent of the net acquisitions amount, the difference between any cost of purchase and any proceeds from disposition of assets within the asset class during the year, be added to the undepreciated capital cost at the beginning of the first year. The remaining 50 percent of the net amount is added to the undepreciated capital cost at the end of the first year. No half-net amount rule is applied to asset class 14.

To illustrate the declining balance method of depreciation, consider the firm which is purchasing the hand-operated assembly machine. This newly installed machine has a capital cost of $560,000. The assembly machine is one of several assets in asset class 39 which uses a declining balance method of depreciation at a CCA rate of 30 percent. In the machine's first year of productive use, the maximum amount of CCA that can be claimed by the firm is $84,000, the product of the capital cost allowance rate of 30 percent and 50 percent of the net acquisitions amount (.50 × $560,000) or $280,000.

For purposes of finance, it is easier to think of the amount of depreciation expense for tax purposes in terms of tax shields. A tax shield is determined as follows:

$$\text{tax shield} = (CCA)(T) \tag{14-2}$$

where CCA = capital cost allowance
T = the firm's combined tax rate

In the first year, the firm inherits a tax shield of $37,800, the product of the CCA ($84,000) and the combined corporate tax rate (45%). The CCAs and the tax shields for the 10 years of economic life are shown in Table 14-4.

In the first year of the project, the CCA from the depreciation of the new machine will be based on the UCC which is adjusted for the half-net amount rule. Thus, the UCC at the beginning of the first year is $280,000, the product of the UCC ($560,000) and the adjustment for the half-net amount rule (50 percent). The CCA is $84,000, the product of the UCC ($280,000) and CCA rate (30 percent). At the end of the first year, the UCC is $476,000, the sum of UCC at the beginning of the year less the CCA ($560,000 – $84,000) and the other half of the adjustment for half-net amount rule ($280,000). The UCC at the beginning of the second year is $476,000, the UCC at the end of the first year. The CCA for the second year is $142,800, the product of the UCC at the beginning of the second year ($476,000) and the CCA rate (30 percent). The UCC at the end of the second year is $333,200, the difference between the UCC at the beginning of the year ($476,000) and the CCA ($142,800). This procedure for the second year is repeated to determine the annual UCC and the CCA for the useful life of the asset.

We now examine how the firm uses tax shields to determine the net operating cash flow for capital budgeting purposes. For an expansion project the net operating cash flow is the sum of the project's net operating income and the tax shields generated from the depreciation of the project's assets. We summarize the calculation for net operating cash flow as follows:[9]

	Incremental operating revenues
Deduct:	Incremental operating expenses (includes accounting depreciation and depletion expenses)
Add:	Accounting depreciation and depletion expenses
	Net operating revenue
Deduct:	Tax on net operating revenue
	Net operating income
Add:	Tax shields from depreciation & depletion
	Net operating cash flow after tax

In order to finance an investment project, a firm will issue securities and incur financing costs such as dividend payments, interest payments or sinking fund pay-

[9]This relationship can also be described as follows:
Net Operating Cash Flows = (Net Operating Revenue)$(1 - T)$ + (CCA)(T)
where T is the combined corporate tax rate and CCA is the capital cost allowances.

Table 14-4 Schedule for Capital Cost Allowances and Tax Shields

Year	Undepreciated Capital Cost Beginning	Capital Cost Allowance	Tax Shields	Undepreciated Capital Cost End
1	$280,000[a]	$ 84,000	$37,800	$476,000
2	476,000	142,800	64,260	333,200
3	333,200	99,960	44,982	233,240
4	233,240	69,972	31,487	163,268
5	163,268	48,980	22,041	114,288
6	114,288	34,286	15,429	80,001
7	80,001	24,000	10,800	56,001
8	56,001	16,800	7,560	39,201
9	39,201	11,760	5,292	27,441
10	27,441	8,232	3,704	19,208

[a]$280,000 represents the effect of the half-net amount rule.

Table 14-5 Summary of Calculation of Incremental Net Cash Flows on an After-Tax Basis

1. Added revenue offset by increased expenses
2. Labour and material savings
3. Increases in overhead incurred
4. Tax shields from depreciation and depletion expenses
5. Do not include interest expenses if the project is financed by issuing debt, as this is accounted for in the required rate of return

ments. These type of financing costs should not be included in net operating cash flows since the cost of funds needed to support the project are implicitly accounted for by discounting the project back to the present using the required rate of return. We define the required rate of return for a project as the rate the firm must earn to justify raising funds to finance it. Table 14-5 lists some of the factors that might be involved in determining a project's incremental cash flows.

EXAMPLE

Suppose that the new hand-operated assembly machine from the previous example requires one full-time operator who will be paid an annual salary of $35,000 with fringe benefits of $5,000 per year. The operation of this machine is expected to require variable overtime of $10,000 per year. In addition, the cost of defects is expected to increase by $5,000 per year, whereas the maintenance cost is expected to increase by $10,000 annually. For accounting purposes, the hand-operated assembly machine will be depreciated over 10 years using the straight-line method. Finally, management estimates that the annual operating revenue will increase by $275,000. Table 14-6 shows the calculation to determine the net operating income for the operating period.

The annual net operating income is $115,500, the difference between the annual net operating revenue ($210,000) and taxes ($94,500). Note that the annual net operating revenue is the difference between incremental operating revenues and incremental operating expenses net of the accounting depreciation expense. The adjustment of adding back accounting depreciation expense is required since depreciation is a non-cash expense. Table 14-7 shows the net operating cash flows after taxes over the useful life of the asset.

Thus, the net operating cash flow after tax for the first year of operation is $153,300, the sum of the net operating income after taxes ($115,500) and the

Table 14-6 Calculation of the Annual Net Operating Income for an Expansion Project

Revenue:	
Incremental operating revenue	$275,000
Costs:	
Salary	−35,000
Variable overtime	−10,000
Fringe benefits	−5,000
Defects	−5,000
Maintenance expense	−10,000
Accounting depreciation	−56,000
Incremental operating expenses (including accounting depreciation)	121,000
Add: Accounting depreciation	56,000
Net operating revenue before taxes	$210,000
Taxes (@45%)	−94,500
Net operating income	$115,500

Table 14-7 Present Value of Net Operating Cash Flow After Taxes Over the Useful Life of the Asset

Year	Net Operating Income	Tax Shield	Cash Flow After Tax	Present Value of Cash Flow After tax
1	$115,500	$37,800	$153,300	$133,304
2	115,500	64,260	179,760	135,924
3	115,500	44,982	160,482	105,520
4	115,500	31,487	146,987	84,040
5	115,500	22,041	137,541	68,382
6	115,500	15,429	130,929	56,604
7	115,500	10,800	126,300	47,481
8	115,500	7,560	123,060	40,229
9	115,500	5,292	120,792	34,337
10	115,500	3,704	119,204	29,465

tax shield from depreciation ($37,800) of the new machine. Remember that the cash flows after tax over the useful life of the asset should be converted into today's dollars. Table 14-7 shows these cash flows discounted at 15 percent, the project's required rate of return.

Terminal Cash Flow

The calculation of the terminal cash flow must be representative of the economic value of the project when it is terminated by the firm. Cash flows associated with the project's termination generally include the salvage value of the project, any tax effects caused by the salvage value, taxable gains associated with its sale, and any expenses recovered from the initial outlay such as working-capital investment.

Generally, the most important terminal cash flow that a manager will add to the cash flows of the project in its last year is the salvage value of the asset. However, the time value of this cash flow must be determined by discounting the salvage value (SV) by the project's required rate of return.

$$PV(SV_n) = \left[\frac{SV_n}{(1+k)^n}\right] \tag{14-3}$$

where SV_n = salvage value of the asset in year n
n = year in which the asset is salvaged
k = required rate of return on the project

In most cases, a firm will sell an asset in order to terminate a project, but the sale will not terminate the asset class. In these cases, Revenue Canada requires that for tax purposes the firm subtract the lesser of the salvage value or the original capital cost of the asset from the undepreciated capital cost of the asset class.

When a project is terminated, the sale of assets will create one of two possible cases for tax purposes: either the asset class is terminated or the asset class is kept open. In the majority of capital-budgeting projects, the sale of an asset does not terminate the asset class. For those cases in which the asset class is terminated by the sale of an asset we refer you to Appendix 14A for the treatment of the terminal cash flow. In the case in which the asset class is not terminated, Revenue Canada requires that for tax purposes the firm subtract the lesser of the salvage value or the original capital cost of the asset from the undepreciated capital cost of the asset class. However, for purposes of capital budgeting, the sale of the asset causes the firm to forgo any future tax shields which it would inherit by continuing to own the asset. The sale of the asset when the project ends requires that the firm determine the value of all tax shields which it forgoes from the salvage year to time infinity. Normally, the amount of tax shields inherited in one year is determined as follows:

$$\text{Annual tax shield} = (UCC_n)(d)(T) \tag{14-4}$$

To determine the value of the tax shields from the year of salvage to time infinity, we apply the principles used to value a perpetuity. In this case, we are determining the value of a series of tax shields which extend from period n to infinity. The value of these lost tax shields to the project at $t = n$ is determined as follows:

$$\text{Value of lost tax shields} = \frac{\left(UCC_n - \text{lesser of} \begin{Bmatrix} SV_n \\ \text{or} \\ C_0 \end{Bmatrix}\right)(d)(T)}{(k+d)} \tag{14-5}$$

In equation 14-5, the values of UCC_n and SV_n represent the values of the asset to be sold. The appropriate discount rate for the valuation of these lost tax shields is the project's required rate of return which is adjusted by the CCA rate. This discount rate reflects the appropriate risk level for the project's tax shields. It should also be noted that the *UCC* of the asset class would be adjusted by subtracting the lesser of the salvage value or the original capital cost.

In this case, the sale of an asset at a value that is greater than the original capital cost and the *UCC* balance of the asset will create both a capital gain and recapture CCAs. For the purposes of capital budgeting, three-quarters of the capital gains is reported as a terminal cash flow.

In addition to the cash flows from the salvage value, there may be cash flows associated with the termination of the project. For example, at the close of a strip-mining operation, the mine must be refilled in an ecologically acceptable manner. Finally, any working-capital outlay required at the initiation of the project—for example, increased inventory needed for the operation of a new plant—will be recovered at the termination of the project. In effect, the increased inventory required by the project can be liquidated when the project expires. For example, the amount of inventory expense recovered would be determined as follows:

$$PV(\text{recovery of working capital}) = \frac{(\text{inventory expense})}{(1+k)^n} \tag{14-6}$$

Table 14-8 Summary of Calculation of Terminal Cash Flow on After-Tax Basis

1. The salvage value of the project
2. Any tax effects associated with the termination of the project
3. Cash outlays associated with the project's termination
4. Recapture of nonexpense outlays that occurred at the project's initiation (for example, working capital investment)

Table 14-8 provides a sample list of some of the factors that might affect a project's terminal cash flow.

EXAMPLE

The hand-held assembly machine is expected to have a salvage value of \$50,000 at the end of ten years of useful life. The sale of the hand-held assembly machine is not expected to terminate the asset class for tax purposes. In addition, the firm will recapture the inventory expense at the end of the tenth year. Determine the firm's terminal cash flow.

To solve for the terminal cash flow of this project, we first determine the present value of the salvage value.

$$PV(SV_n) = \frac{SV_n}{(1 + k)^n} \tag{14-3}$$
$$= \frac{50{,}000}{(1.15)^{10}}$$
$$= 12{,}360$$

The second step is to determine the value of the lost tax shields in today's dollars to the project because the asset is being salvaged.

$$PV(\text{lost tax shields}) = \frac{\frac{(UCC_n - SV_n)(d)(T)}{(k + d)}}{(1 + k)^n} \tag{14-7}$$
$$= \frac{\frac{(19{,}208 - 50{,}000)(.30)(.45)}{(.15 + .30)}}{(1 + .15)^{10}}$$
$$= -2{,}283$$

Finally, we determine the amount of recovered inventory expense which will be claimed by the firm when terminating the project.

$$PV(\text{recovery of working capital}) = \frac{\text{inventory expenses}}{(1 + k)^n} \tag{14-6}$$
$$= \frac{50{,}000}{(1.15)^{10}}$$
$$= 12{,}360$$

Thus, the terminal cash flow is \$90,762 and its present value is \$22,437. As shown in Figure 14-1, the cash inflow in year ten is the sum of the terminal cash flow (\$90,762) and the net operating cash flow after tax (\$119,204).

Determining Net Present Value

The project's net present value (NPV) can now be determined from the present value of each of the project's annual net cash flows. The calculation would require that the net operating cash flows over the life of the project be added to the terminal

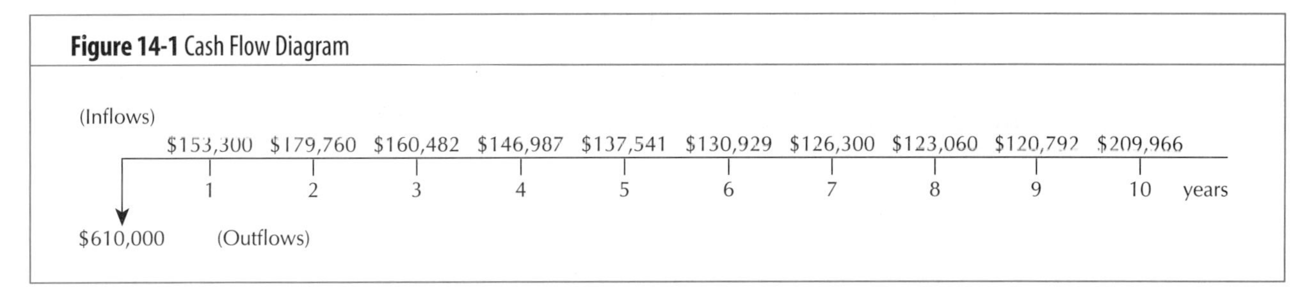

Figure 14-1 Cash Flow Diagram

cash flow and then subtracted from the initial outlay. The following equation shows this procedure:

$$NPV = PV(\text{net operating cash flows}) + PV(\text{terminal cash flow}) - IO \tag{14-8}$$

This equation can be rewritten as follows:

$$NPV = \sum_{t=1}^{n} \frac{CFAT_t}{(1 + k)^t} - IO \tag{14-9}$$

The firm should accept the project if the NPV is positive, reject the project if NPV is negative and be indifferent to accepting or rejecting the project if NPV is zero.

A simple tool which effectively illustrates the cash flows of this project is the cash flow diagram in Figure 14-1.

The NPV of this project is determined by discounting the annual net cash flow at the project's required rate of return over the ten years of project life. The equation for NPV is shown as follows:

$$\begin{aligned} NPV = {} & \frac{\$153{,}300}{(1.15)^1} + \frac{\$179{,}760}{(1.15)^2} + \frac{\$160{,}482}{(1.15)^3} + \frac{\$146{,}987}{(1.15)^4} + \\ & \frac{\$137{,}541}{(1.15)^5} + \frac{\$130{,}929}{(1.15)^6} + \frac{\$126{,}300}{(1.15)^7} + \\ & \frac{\$123{,}060}{(1.15)^8} + \frac{\$120{,}792}{(1.15)^9} + \frac{\$209{,}966}{(1.15)^{10}} - \$610{,}000 \end{aligned}$$

$$NPV = \$147{,}722$$

Thus, the firm should accept the project of purchasing the hand-held assembly machine which has a positive NPV.

BACK TO THE BASICS

In this chapter, it is easy to get caught up in the calculations and forget that before the calculations can be made, someone has to come up with the idea for the project. In some of the example problems, you may see projects that appear to be profitable. Unfortunately, as we learned in ***Axiom 5: The Curse of Competitive Markets—Why It's Hard to Find Exceptionally Profitable Projects****, it is unusual to find projects with dramatically high returns because of the very competitive nature of the business. Thus, keep in mind that capital budgeting not only involves the estimation and evaluation of the project's cash flows, but it also includes the process of coming up with the idea for the project in the first place.*

Determining the Net Present Value: The Annuity Approach

The net present value of a project can also be determined by applying principles which we used to value annuities and perpetuities. This approach separates the project's cash flows into components such as the project's annual net operating incomes and tax shields from depreciation, which can be valued using the same principles used to value either an annuity or a perpetuity of cash flows. The present value of each of these cash flows' components is then used to determine the net present value of the project.

First we examine the annuity which represents the annual net operating incomes which occur over the life of the project. The present value of this annuity is determined as follows:

$$PV(\text{net operating income}) = (\text{annual net operating income})(PVIFA_{k,n}) \tag{14-10}$$

The second component of the project's cash flows represents the tax shields from depreciation of the newly purchased asset. Most assets are written off for tax purposes using the declining balance method of depreciation. This particular method depreciates the capital cost (C_0) over an infinite number of years. An infinite series of tax shields which decrease at a constant rate can be valued as a perpetuity.[10] The present value of the tax shields from the purchase of an asset is determined as follows:

$$PV(\text{tax shields of purchased asset}) = \left[\frac{(C_0)(d)(T)}{(k+d)}\right] \tag{14-11}$$

However, Revenue Canada requires that the undepreciated capital cost of a new asset in asset class 39 be adjusted for the half-net amount rule. As a result of this adjustment, the present value of the adjusted tax shields from the purchase of an asset is determined as follows:

$$PV\left(\begin{array}{c}\text{adjusted tax}\\ \text{shields of purchased asset}\end{array}\right) = \left[\frac{(C_0)(d)(T)}{(k+d)}\right]\left[\frac{1+(.5)(k)}{(1+k)}\right] \tag{14-12}$$

The firm is only permitted to use for capital-budgeting purposes those tax shields which occur over the useful life of the asset. Once the asset is sold, any future tax shield generated by the asset cannot be used by this firm for capital-budgeting purposes. These future tax shields represent a series of an infinite number of tax shields which decrease at a constant rate. Again, this type of series is valued as a

---

[10]In the case of a declining balance method of depreciation whereby the capital cost of the asset is written off at a constant CCA rate (d), we can express the tax shield (TS) in any year in terms of the tax shield paid at the end of the previous year, $(C_0)(T)(d)(1-d)^{-1}$. For example, the expected tax shield one year hence is simply $(C_0)(T)(d)(1-d)^{-1}(1+g)$ where $g=-d$. Likewise, the tax shield at the end of t years is $(C_0)(T)(d)(1-d)^{-1}(1+g)^t$. Using this notation, the valuation equation of tax shields inherited by the firm can be rewritten as follows:

$$V_{TS} = \frac{(C_0)(T)(d)(1-d)^{-1}(1-d)^1}{(1+k)^1} + \frac{(C_0)(T)(d)(1-d)^{-1}(1-d)^2}{(1+k)^2} + \ldots$$
$$\frac{(C_0)(T)(d)(1-d)^{-1}(1-d)^n}{(1+k)^n} + \ldots \frac{(C_0)(T)(d)(1-d)^{-1}(1-d)^\infty}{(1+k)^\infty}$$

If both sides of the equation are multiplied by $(1+k)/(1-d)$ and then the equation is subtracted from the product, the result is

$$\frac{V_{TS}(1+k)}{(1-d)} - V_{TS} = (C_0)(T)(d)(1-d)^{-1} - \frac{(C_0)(T)(d)(1-d)^{-1}(1-d)^\infty}{(1+k)^\infty}$$

If $k > g$ where $g = -d$, which normally should hold, $[(C_0)(T)(d)(1-d)^{-1}(1+g)/(1+k)]$ approaches zero. As a result,

$$\frac{V_{TS}(1+k)}{(1-d)} - V_{TS} = (C_0)(T)(d)(1-d)^{-1}$$

$$V_{TS}\left(\frac{1+k}{1-d}\right) - V_{TS}\left(\frac{1-d}{1-d}\right) = (C_0)(T)(d)(1-d)^{-1}$$

$$V_{TS}\left[\frac{(1+k)-(1-d)}{1-d}\right] = (C_0)(T)(d)(1-d)^{-1}$$

$$V_{TS}(k+d) = (C_0)(T)(d)(1-d)^{-1}(1-d)$$

$$V_{TS} = \frac{(C_0)(T)(d)}{(k+d)}$$

perpetuity and its value at $t = n$ is determined as follows when the asset class is kept open:

$$\text{(Value of lost tax shields at } t = n) = \left[\frac{(SV_n)(d)(T)}{(k + d)}\right] \qquad (14\text{-}13)$$

The value of this series of tax shields has been determined when the asset is salvaged and does not represent their present value.

In order to adjust for their present value we discount this series of tax shields by the project's required rate of return as follows:

$$PV(\text{lost tax shields}) = \left[\frac{(SV_n)(d)(T)}{(k + d)}\right]\left[\frac{1}{(1 + k)^n}\right] \qquad (14\text{-}14)$$

For capital budgeting purposes, we define the inherited tax shields from the purchased asset at $t = 0$, as the difference between the present value of the tax shields from purchasing the asset and the present value of the tax shields forgone by the sale of the asset. The present value of tax shields over the useful life of the asset is determined as follows:

$$PV\left(\begin{array}{c}\text{inherited tax}\\ \text{shields of purchased asset}\end{array}\right) =$$

$$\left[\frac{(C_0)(d)(T)}{(k + d)}\right]\left[\frac{1 + (.5)(k)}{(1 + k)}\right] - \left[\frac{(SV_n)(d)(T)}{(k + d)}\right]\left[\frac{1}{(1 + k)^n}\right] \qquad (14\text{-}15)$$

In most cases, the firm will terminate a project, but will not close an asset class for tax purposes. As a result, we determine the present value of the tax shields over the useful life of the asset by using the salvage value.

The last step in our procedure is to include all other terms, such as the present value of the salvage value and any recaptured expense from the initial outlay, in the NPV equation. The NPV equation for the project is determined as follows:

$$\begin{aligned} NPV = \; & PV(\text{project's net operating income}) + \\ & \text{PV(inherited tax shields of purchased asset)} + \\ & \text{PV(salvage value)} + \\ & \text{PV(recovery of working capital)} - IO \end{aligned} \qquad (14\text{-}16)$$

EXAMPLE

Using the annuity approach, we will now redo the previous example in which a firm is considering the purchase of a hand-held assembly machine.

***1st Step:* The annual net operating income for the project has been determined to be \$115,500 as shown in Table 14-6. The present value of the series of annual net operating incomes over the ten years of useful life of the asset is determined as follows:**

$$\begin{aligned} PV(\text{net operating income}) &= (\text{annual net operating income})(PVIFA_{15\%,10yr}) \\ &= (115{,}500)(5.0188) \\ &= 579{,}671 \end{aligned}$$

Thus, the present value of the net operating incomes over the ten years of the project is \$579,671.

***2nd Step:* The present value of the tax shields over the useful life of the asset is determined from the difference between the present value of the tax shields from purchasing the asset and the present value of the tax shields forgone by the sale of the asset at $t = 0$. Given a capital cost (C_0) of \$560,000, a combined corporate tax rate (T) of 45%, a capital cost allowance rate (d) of 30% and the project's required rate of return, k, of 15%, the present value of the adjusted tax shields from purchasing the asset is determined as follows:**

$$PV\left(\begin{array}{c}\text{adjusted tax}\\ \text{shields of purchased asset}\end{array}\right) = \left[\frac{(C_0)(d)(T)}{(k + d)}\right]\left[\frac{1 + (.5)(k)}{(1 + k)}\right] \qquad (14\text{-}12)$$

$$= \left[\frac{(560{,}000)(.30)(.45)}{(.15 + .30)}\right]\left[\frac{1 + (.5)(.15)}{(1 + .15)}\right]$$

$$= [168{,}000][.9348]$$
$$= 157{,}046$$

Thus, the present value of the tax shields from purchasing the asset is $157,046.

Given the salvage value of $50,000, the present value of the tax shields forgone by the sale of the asset at $t = 0$, when the asset class is kept open, is determined as follows:

$$PV(\text{lost tax shields}) = \left[\frac{(SV_n)(d)(T)}{(k + d)}\right]\left[\frac{1}{(1 + k)^n}\right] \tag{14-14}$$

$$= \left[\frac{(50{,}000)(.30)(.45)}{(.15 + .30)}\right]\left[\frac{1}{(1 + .15)^{10}}\right]$$

$$= [15{,}000][.2472]$$
$$= 3{,}708$$

Thus, the present value of the tax shields forgone by the sale of the asset at $t = 0$ is $3,708.

We can now determine the present value of the tax shields over the useful life of the asset as follows:

$$PV\left(\begin{matrix}\text{inherited tax}\\ \text{shields of purchased asset}\end{matrix}\right) = \left[\frac{(C_0)(d)(T)}{(k + d)}\right]\left[\frac{1 + (.5)(k)}{(1 + k)}\right] - \left[\frac{(SV_n)(d)(T)}{(k + d)}\right]\left[\frac{1}{(1 + k)^n}\right] \tag{14-15}$$

$$= 157{,}046 - 3{,}708$$
$$= 153{,}338$$

Thus, the present value of the tax shields over the useful life of the asset is $153,338.

***3rd Step:* The present value of the salvage value is determined as follows:**

$$PV(\text{salvage value}) = (SV_n)(PVIF_{15\%,10\text{yr}})$$
$$= (50{,}000)(.2472)$$
$$= 12{,}360$$

Thus, the present value of the salvage value is $12,360.

***4th Step:* The present value of any recaptured expense from the initial outlay is determined as follows:**

$$PV(\text{recovery of working capital}) = (\text{inventory expenses})(PVIF_{15\%,10\text{yr}})$$
$$= (50{,}000)(.2472)$$
$$= 12{,}360$$

Thus, the present value of any recaptured expense from the initial outlay is $12,360.

***5th Step:* The NPV equation for this problem is determined as follows:**

$$\begin{aligned} NPV = {} & PV(\text{project's net operating income}) + \\ & PV(\text{inherited tax shields of purchased asset}) + \\ & PV(\text{salvage value}) + \\ & PV(\text{recovery of working capital}) - IO \end{aligned}$$

$$\begin{aligned} NPV = {} & (\$115{,}500)(5.0188) + [(168{,}000)(.9348)] - [(15{,}000)(.2472)] + \\ & (\$50{,}000)(.2472) + (\$50{,}000)(.2472) - \$610{,}000 \end{aligned}$$

$$NPV = \$147{,}729$$

Given the asset class is kept open, the NPV of this project is $147,729. Since the NPV is positive, the firm should accept the project. The two approaches provide a similar estimate for NPV given rounding error; however, the annuity approach makes use of valuation principles seen in Chapter 4.

CAPITAL BUDGETING FOR REPLACEMENT PROJECTS

OBJECTIVE 3

We now examine replacement projects which involve the acquisition of a new asset to replace an older, less efficient asset. To evaluate whether to accept or reject a replacement project, we will use the annuity approach to determine the NPV of the project. However, the replacement project differs from the expansion project because the net cash flows of a replacement project will be measured in terms of the amount of net incremental savings after taxes that accrues to the firm. In addition to the cash flows which we described for the initial outlay of an expansion project, we also consider the cash flow associated with the selling price of the old asset and the tax shields lost from the sale of the old asset.

Although the old asset is sold, the asset class remains open during the first year of the replacement project. We use the concept of incremental capital cost to adjust for the tax shields lost because of the sale of the old machine. The incremental capital cost of the new machine is calculated as the difference between the capital cost of the new machine and the current salvage value of the old machine.

$$\Delta C_0 = C_{0,\text{new}} - SV_{0,\text{old}} \tag{14-17}$$

Table 14-1 provides a summary of the considerations which are used in determining the initial outlay of a replacement project.

The incremental net operating cash flows occurring over the life of a replacement project also include items such as increases in operating revenues, labour or material savings, and reductions in selling expenses. The net operating cash flow over the life of the project is the sum of the project's annual net operating incomes and the tax shields generated from the depreciation of the project's assets. The annual net operating cash flow is determined as follows:

	Incremental operating savings
Deduct:	Incremental operating expenses (includes accounting depreciation and depletion expenses)
Add:	Depreciation and depletion expenses
	Net operating savings before taxes
Deduct:	Tax on net operating savings
	Net operating savings after taxes
Add:	Tax shields from depreciation and depletion
	Net operating cash flow

Table 14-9 summarizes some of the factors used in determining the net operating cash flows of a replacement project.

The series of annual net operating savings after tax can be valued as an annuity. The present value of this annuity is determined as follows:

$$PV\left(\begin{array}{c}\text{net operating}\\ \text{savings after taxes}\end{array}\right) = \left(\begin{array}{c}\text{annual net}\\ \text{operating savings after taxes}\end{array}\right)\left(PVIFA_{k,n}\right) \tag{14-18}$$

The approach of valuing tax shields inherited by the firm over the life of the replacement project is similar to that of the expansion project. However, we must treat the salvage value of the new asset as an incremental cash flow by considering the salvage value of the old asset if it were not replaced. This consideration is an example of an opportunity cost for a replacement project. As a result, the incremental salvage value (ΔSV_n) for the new asset is determined as follows:

$$\Delta SV_n = SV_{n,\text{new}} - SV_{n,\text{old}} \tag{14-19}$$

We now examine the adjustments which have to be made to the tax shields because of the changes to both the capital cost and salvage value of the new asset. The incremental capital cost and the incremental salvage value of an asset affect the present value of the tax shields over the life of a project when the asset is kept open. We determine the present value of these tax shields over the life of the project as follows:

$$PV\left(\begin{array}{c}\text{inherited tax shields}\\ \text{of purchased asset}\end{array}\right) = \left[\frac{(\Delta C_0)(d)(T)}{(k+d)}\right]\left[\frac{1+(.5)(k)}{(1+k)}\right] - \left[\frac{(\Delta SV_n)(d)(T)}{(k+d)}\right]\left[\frac{1}{(1+k)^n}\right] \quad (14\text{-}20)$$

Finally, we include the recovery of working capital from the initial outlay such as inventory expenses, in the NPV equation. The NPV equation for a replacement project is determined as follows:

$$\begin{aligned} NPV = {} & PV(\text{Project's net operating savings after taxes}) + \\ & \text{PV(Inherited tax shields of purchased asset)} + \\ & \text{PV(Incremental salvage value)} + \\ & \text{PV(Recovery of working capital)} - IO \end{aligned} \quad (14\text{-}21)$$

EXAMPLE

In order to clarify the annuity approach for determining the net present value of a replacement project, we consider an example of a company that has a combined corporate tax rate of 45 percent. This company is considering the purchase of a $30,000 machine which will replace an existing machine that has been used for gas and oil drilling. Both of these machines are classified in asset class 28 for tax purposes. Revenue Canada requires that assets in this asset class be written off using the declining balance method of depreciation at a CCA rate of 30 percent. Five years ago, the existing machine had a capital cost of $20,000. However, the existing machine can now be sold to a scrap dealer for $15,000. The existing machine is expected to have a salvage value of zero in five years, while the new machine is expected to be salvaged for $8,000 in five years. To put the new machine in running order, it is necessary to pay shipping charges of $2,000 and installation charges of $3,000. Because the new machine will work faster than the existing one, it will require an increase in work-in-progress inventory of $5,000.

Table 14-9 Summary of Calculation of Incremental After-Tax Cash Flows

A. Initial Outlay
   1. Capital cost of asset
   2. Opportunity costs
   3. Incremental nonexpense outlays incurred (for example, working-capital investments)
   4. Incremental cash expenses on an after-tax basis (for example, training expenses)
   5. In a replacement decision, the cash flow associated with the sale of the old machine

B. Incremental Cash Flows over the Project's Life
   1. Added revenue offset by increased expenses
   2. Labour and material savings
   3. Increases in overhead incurred
   4. Tax shields from depreciation and depletion expenses
   5. Do not include interest expenses if the project is financed by issuing debt, as this is accounted for in the required rate of return

C. Terminal Cash Flow
   1. The salvage value of the project
   2. Any tax effects associated with the termination of the project
   3. Cash outlays associated with the project's termination
   4. Recapture of nonexpense outlays that occurred at the project's initiation (for example, working capital investment)

The purchase of the new machine is expected to reduce salaries by $10,000 per year and fringe benefits by $1,000 annually. These benefits accrue to the firm because it will take only one person to operate the new machine, whereas the old machine requires two operators. In addition, the cost of defects will fall from $8,000 per year to $3,000. However, maintenance expenses will increase by $4,000 annually.

The sale of the new machine in five years is not expected to terminate the asset class for tax purposes. The firm's required rate of return on this type of project is 12 percent. Determine whether the firm should accept or reject this replacement project.

We now solve the NPV for this project using the annuity approach:

Step 1: To determine the initial outlay we first add to the capital cost of the machine ($30,000), any costs associated with acquiring or putting the asset in working order, such as the installation charge ($3,000) and the shipping charge ($2,000). Since the investment in working capital also increases, as more money will be immediately tied up in inventory ($5,000), this amount should be added as well. Note, the inventory investment will be recovered at the end of the project so that this amount will be recovered at the end of the project. Finally, since this is a replacement project, the current salvage value of the old machine ($15,000) is subtracted from the initial outlay. These calculations are summarized in Table 14-10.

Step 2: We now determine the present value of the net operating savings after taxes. The annual net operating savings after taxes is the difference between the net operating savings before taxes and taxes payable on these savings. Table 14-11 shows the calculation for the annual net operating savings which occur over the life of the project.

The present value of the series of annual net operating savings after taxes is determined as follows:

$$
\begin{aligned}
PV\left(\begin{matrix}\text{net operating}\\ \text{savings after taxes}\end{matrix}\right) &= \left(\begin{matrix}\text{annual net}\\ \text{operating savings after taxes}\end{matrix}\right)\left(PVIFA_{12\%,5yr}\right)\\
&= (6{,}600)(3.6048)\\
&= 23{,}792
\end{aligned}
$$

Thus the present value of the net operating savings after taxes is $23,792.

Step 3: We now determine the present value of the incremental salvage value. In our example, the new machine is expected to salvage for $8,000 at the end of five years of useful life. If the existing machine is not replaced by the new machine, the existing machine is expected to have no salvage value at the end of five years of additional useful life. As a result, we determine the present value of the incremental salvage value as follows:

Table 14-10 Calculation of Initial Outlay for Example Problem

Outflows:	
Purchase price	$30,000
Shipping fee	2,000
Installation fee	3,000
Capital cost of machine	$ 35,000
Increased investment in inventory	5,000
Total outflows	$ 40,000
Inflows:	
Salvage value of old machine	−15,000
Net initial outlay	$ 25,000

Note: Assume that the asset class remained open with the purchase of the new machine.

Table 14-11 Schedule for the Annual Net Operating Savings After Tax

Savings:	
Reduced salary	$10,000
Reduced fringe benefits	1,000
Reduced defects	5,000
Costs:	
Increased maintenance costs	−4,000
Net operating savings before taxes	$12,000
Taxes (@45%)	−5,400
Net operating savings after taxes	$ 6,600

$$PV(\text{incremental salvage value}) = \frac{(SV_{\text{new}} - SV_{\text{old}})}{(1+k)^n} \quad (14\text{-}22)$$

$$= \frac{(8{,}000 - 0)}{(1.12)^5}$$

$$= 4{,}539$$

Thus, the present value of the incremental salvage value is $4,539.

Step 4: The present value of the tax shields occurring over the useful life of the asset is affected by both the incremental capital cost and the incremental salvage value of the new machine when the asset class is kept open. In this example, the incremental capital cost is $20,000 and the incremental salvage value is $8,000. As a result, the present value of the tax shields occurring over the life of the asset is determined as follows:

$$PV\left(\begin{array}{c}\text{inherited tax shields}\\ \text{of purchased asset}\end{array}\right) = \left[\frac{(\Delta C_0)(d)(T)}{(k+d)}\right]\left[\frac{1+(.5)(k)}{(1+k)}\right] - \left[\frac{(\Delta SV_n)(d)(T)}{(k+d)}\right]\left[\frac{1}{(1+k)^n}\right] \quad (14\text{-}20)$$

$$= \left[\frac{(20{,}000)(.30)(.45)}{(.30+.12)}\right]\left[\frac{1+(.5)(.12)}{(1+.12)}\right] - \left[\frac{(8{,}000)(.30)(.45)}{(.30+.12)}\right]\left[\frac{1}{(1+.12)^5}\right]$$

$$= [6{,}429][.9464] - [2{,}571][.5674]$$

$$= 6{,}084 - 1{,}459$$

$$= 4{,}625$$

Thus the present value of the tax shields occurring over the useful life of the asset is $4,625.

Step 5: We also must determine the present value of any recovery of working capital. In this example the initial investment in inventory expense of $5,000 must be recaptured by the firm. We determine the present value of any recovery of working capital as follows:

$$PV\left(\begin{array}{c}\text{recovery of}\\ \text{working capital}\end{array}\right) = \left(\text{inventory expenses}\right)\left(PVIF_{12\%,5\text{yr}}\right)$$

$$= (5{,}000)(0.5674)$$

$$= 2{,}837$$

Thus, the present value of the recovery of working capital from the initial outlay is $2,837.

Step 6: The NPV equation for this replacement project is determined as follows:

$$
\begin{aligned}
NPV &= PV(\text{project's net operating savings after taxes}) + \\
&\quad PV(\text{inherited tax shields of purchased asset}) + \\
&\quad PV(\text{incremental salvage value}) + \\
&\quad PV(\text{recovery of working capital}) - IO \\
&= 23{,}792 + 4{,}625 + 4{,}539 + 2{,}837 - 25{,}000 \\
&= 10{,}793
\end{aligned}
\qquad (14\text{-}23)
$$

Thus, the NPV of this replacement project is $10,793. The firm should accept this replacement project since the NPV is positive.

In this section we focused solely on the decision-making process that is required for a single capital-budgeting project. Sometimes, financial executives may have to choose among a number of acceptable projects. At other times the number of projects that can be accepted, or the total budget, is limited; that is, capital rationing is imposed. In these cases, executives may assign different rankings which depend upon the discounted cash flow criterion being used. In the next section we examine the reasons for these differences. We explore the rationale behind capital rationing and project ranking and the ways in which they affect capital-budgeting decisions.

CAPITAL RATIONING

OBJECTIVE 4

The use of our capital-budgeting rules implies that the size of the capital budget is determined by the availability of acceptable investment proposals. However, a firm may place a limit on the dollar size of the capital budget. This situation is called **capital rationing**.

Capital rationing

The placing of a limit by the firm on the dollar size of the capital budget.

Using the internal rate of return as the firm's decision rule, the firm accepts all projects with an internal rate of return greater than the firm's required rate of return. This rule is illustrated in Figure 14-2, where projects A through E would be chosen. However, when capital rationing is imposed, the dollar size of the total investment is limited by the budget constraint. In Figure 14-2 the budget constraint of $X precludes the acceptance of an attractive investment, project E. This situation obviously contradicts prior decision rules. Moreover, the solution of choosing the projects with the highest internal rate of return is complicated by the fact that some projects may be indivisible; for example, it is meaningless to recommend that half a computer be acquired.

PERSPECTIVE IN FINANCE

It is always somewhat uncomfortable to deal with problems associated with capital rationing because, under capital rationing, projects with positive net present values are rejected. This is a situation that violates the goal of the firm of shareholder wealth maximization. However, in the real world capital rationing does exist, and managers must deal with it. When firms impose capital constraints, they are recognizing that they do not have the ability to profitably handle more than a certain number or dollar value of new projects. In this case, perhaps a better approach would be to recalculate the incremental cash flows associated with the marginal project.

Rationale for Capital Rationing

We will first ask why capital rationing exists and whether or not it is rational. In general, three principal reasons are given for imposing a capital-rationing constraint. First, management may feel that market conditions are temporarily adverse. In the period surrounding the stock market crash of 1987, this reason was frequently given. At that time interest rates were high, and share prices were depressed. Second, there may be a shortage of qualified managers to direct new projects; this circumstance can occur when projects are of a highly technical nature and qualified managers are simply not available. Third, there may be intangible considerations. For example, the management may simply fear debt, wishing to avoid interest payments at any cost. Or perhaps the issuing of common shares may be limited to preserve the current owners' strict voting control over the company or to maintain a stable dividend policy.

Despite strong evidence that capital rationing exists in practice, the question remains as to its effect on the firm. In brief, the effect is negative, and its degree depends upon the severity of the rationing. If the rationing is minor and brief, then the firm's share price will not suffer to any great extent. In this case, any capital rationing can probably be excused, although it should be noted that any capital rationing which rejects projects with positive net present values is contrary to the firm's goal of maximization of shareholders' wealth. If capital rationing is a result of the firm's decision to limit dramatically the number of new projects or to limit total investment to internally generated funds, then this policy will eventually have a significantly negative effect on the firm's share price. For example, a lower share price will eventually result from lost competitive advantage if, owing to a decision to limit its capital budget arbitrarily, a firm fails to upgrade its products and manufacturing process.

Capital Rationing and Project Selection

If the firm decides to impose a capital constraint on investment projects, it must set an appropriate decision criterion to select the set of projects with the highest net present value subject to the capital constraint. This guideline may preclude merely taking the highest-ranked projects in terms of the profitability index or the internal rate of return. If the projects shown in Figure 14-2 are divisible, the last project accepted may be only partially accepted. Although partial acceptances may be possible in some cases, the indivisibility of most capital investments prevents it. If a project is a sales outlet or a truck, it may be meaningless to purchase half a sales outlet or half a truck.

To illustrate this procedure, consider a firm with a budget constraint of $1 million and five indivisible projects available to it, as given in Table 14-12. If the highest-ranked projects were taken, projects A and B would be taken first. At that point there would not be enough funds available to take project C; hence, projects D and E would be taken. However, a higher total net present value is provided by the combination of projects A and C. Thus projects A and C should be selected from the set of projects available. This illustrates our guideline: to select the set of projects that maximizes the firm's net present value.

Figure 14-2 Projects Ranked by IRR

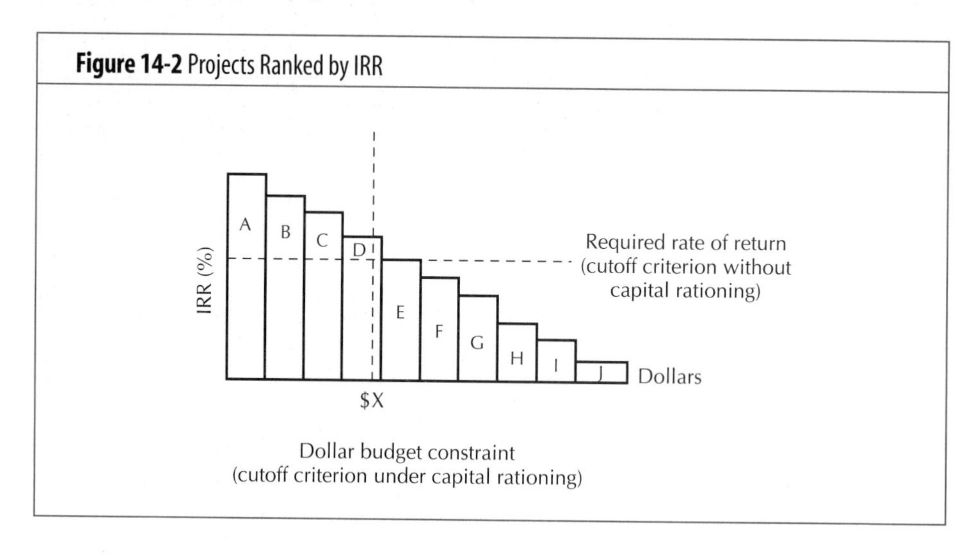

Table 14-12 Capital-Rationing Example of Five Indivisible Projects

Project	Initial Outlay	Profitability Index	Net Present Value
A	$200,000	2.4	$280,000
B	200,000	2.3	260,000
C	800,000	1.7	560,000
D	300,000	1.3	90,000
E	300,000	1.2	60,000

Project Ranking

In the past, we have proposed that all projects with a positive net present value, a profitability index greater than 1.0, or an internal rate of return greater than the required rate of return be accepted, assuming there is no capital rationing. However, this acceptance is not always possible. In some cases, in which two projects are judged acceptable by the discounted cash flow criteria, it may be necessary to select only one of them, as they are mutually exclusive. **Mutually exclusive projects** occur when a set of investment proposals perform essentially the same task; acceptance of one will necessarily mean rejection of the others. For example, a company considering the installation of a computer system may evaluate three or four systems, all of which may have positive net present values; but the acceptance of one system will automatically mean rejection of the others. In general, to deal with mutually exclusive projects, we will simply rank them by means of the discounted cash flow criteria and select the project with the highest ranking. On occasion, however, problems of conflicting ranking may arise.

Mutually exclusive projects

A set of projects that perform essentially the same task, so that acceptance of one will necessarily mean rejection of the others.

Problems in Project Ranking

There are three general types of ranking problems—the size disparity problem, the time disparity problem, and the unequal lives problem. Each involves the possibility of conflict in the ranks yielded by the various discounted cash flow capital-budgeting criteria. As noted in the previous chapter, when one discounted cash flow criterion gives an accept signal, they will all give an accept signal, but they will not necessarily rank all projects in the same order. In most cases this disparity is not critical; however, for mutually exclusive projects the ranking order is important.

Size Disparity The size disparity problem occurs when mutually exclusive projects of unequal size are examined. This problem is most easily illustrated with the use of an example.

EXAMPLE

Suppose that a firm with a cost of capital of 10 percent is considering two mutually exclusive projects, A and B. Project A involves a $200 initial outlay and cash inflow of $300 at the end of one year, while project B involves an initial outlay of $1,500 and a cash inflow of $1,900 at the end of one year. The net present value, profitability index, and internal rate of return for these projects are given in Figure 14-3.

In this case, if the net present value criterion is used, project B should be accepted, while if the profitability index or the internal rate of return criterion is used, project A should be chosen. The question now becomes: Which project is better? The answer depends on whether or not capital rationing exists. Without capital rationing, project B is better because it provides the largest increase in shareholder wealth; that is, it has a larger net present value. If there is a capital constraint, the problem then focuses on what can be done with the additional $1,300 that is freed if project A is chosen (costing $200, as opposed to

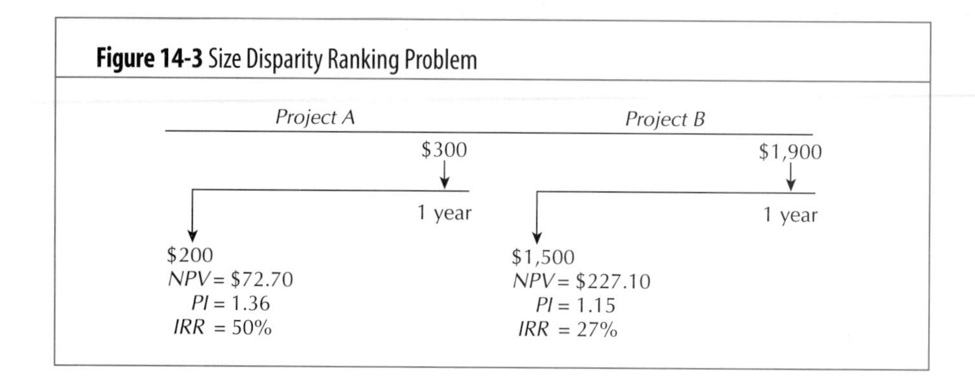

Figure 14-3 Size Disparity Ranking Problem

$1,500). If the firm can earn more on project A plus the project financed with the additional $1,300 than it can on project B, then project A and the marginal project should be accepted. In effect, we are attempting to select the set of projects that maximize the firm's NPV. Thus, if the marginal project has a net present value greater than $154.40, selecting it plus project A with a net present value of $72.70 will provide a net present value greater than $227.10, the net present value for project B.

In summary, whenever the size disparity problem results in conflicting rankings between mutually exclusive projects, the project with the largest net present value will be selected, provided there is no capital rationing. When capital rationing exists, the firm should select the set of projects with the largest net present value.

Time Disparity The *time disparity problem* and the conflicting rankings that accompany it result from the differing reinvestment assumptions made by the net present value and internal rate of return decision criteria. The NPV criterion assumes that cash flows over the life of the project can be reinvested at the required rate of return or cost of capital, while the IRR criterion implicitly assumes that the cash flows over the life of the project can be re-invested at the internal rate of return. Again, this problem may be illustrated through the use of an example.

EXAMPLE

Suppose a firm with a required rate of return or cost of capital of 10 percent and with no capital constraints is considering the two mutually exclusive projects illustrated in Figure 14-4. The net present value and profitability index indicate that project A is the better of the two, while the internal rate of return indicates that project B is better. Project B receives its cash flows after tax earlier than project A, and the different assumptions made as to how these flows can be reinvested result in the difference in rankings. Which criterion should be followed depends upon which reinvestment assumption is used. The net present value criterion is preferred in this case because it makes the most acceptable assumption for the wealth-maximizing firm. It is certainly the most conservative assumption that can be made, since the required rate of return is the lowest possible reinvestment rate. Moreover, as we have already noted, the net present value maximizes the value of the firm and the shareholders' wealth.

Unequal Lives The final ranking problem to be examined centres on the question of whether or not it is appropriate to compare mutually exclusive projects with different life spans.

Figure 14-4 Time Disparity Ranking Problem

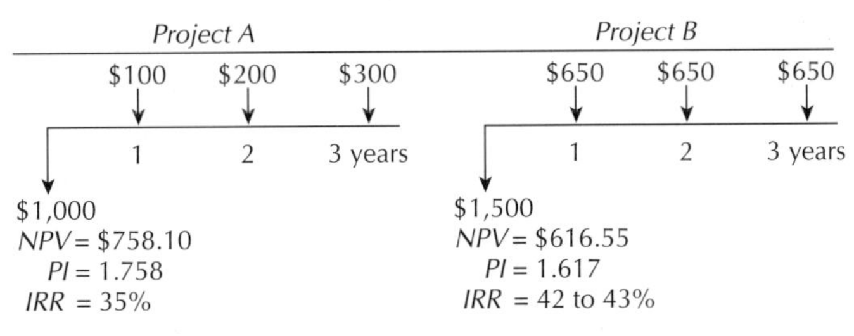

EXAMPLE

Suppose a firm with a 10 percent required rate of return is faced with the problem of replacing an aging machine and is considering two replacement machines, one with a three-year life and one with a six-year life. The relevant cash flow information for these projects is given in Figure 14-5.

Examining the discounted cash flow criteria, we find that the net present value and profitability index criteria indicate that project B is the better project, while the internal rate of return favours project A. This inconsistency in ranking is caused by the different life spans of the projects being compared. In this case the decision is a difficult one because the projects are not comparable.

The problem of incomparability of projects with different lives arises because future profitable investment proposals may be rejected without being included in the analysis. This can easily be seen in a replacement problem such as the present example, in which two mutually exclusive machines with different lives are being considered. In this case a comparison of the net present values alone on each of these projects would be misleading. If the project with the shorter life were taken, at its termination the firm could replace the machine and receive additional benefits, while acceptance of the project with the longer life would exclude this possibility, a possibility that is not included in the analysis.

The key question thus becomes: Does today's investment decision include all future profitable investment proposals in its analysis? If not, the projects are not comparable. In this case, if project B is taken, then the project that could have been taken after three years when project A terminates is automatically rejected without being included in the analysis. Thus, acceptance of project B not only forces rejection of project A, but also forces rejection of any replacement machine that might have been considered for years 4 through 6 without including this replacement machine in the analysis.

Figure 14-5 Unequal Lives Ranking Problem

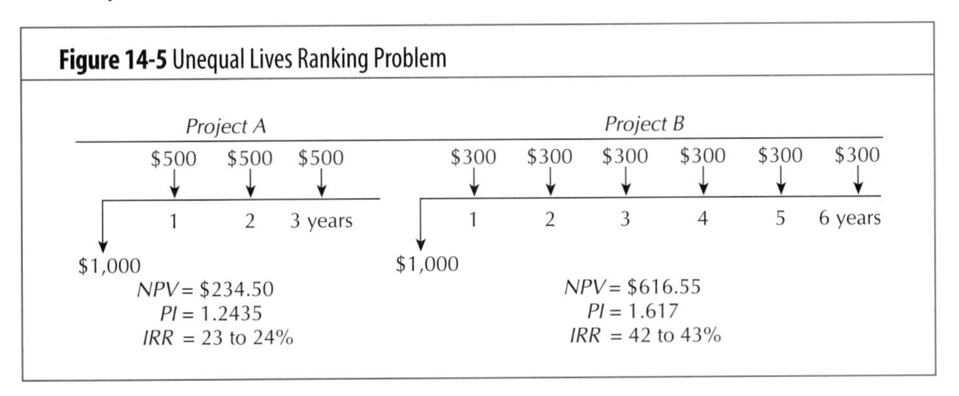

There are several methods to deal with this situation. The first option is merely to assume that the cash inflows from the shorter-lived investment will be reinvested at the cost of capital until the termination of the longer-lived asset. While this approach is the simplest, merely calculating the net present value, it actually ignores the problem at hand—that of allowing for participation in another replacement opportunity with a positive net present value. The proper solution thus becomes the projection of reinvestment opportunities into the future—that is, making assumptions about possible future investment opportunities.

Unfortunately, while the first method is too simplistic to be of any value, the second is extremely difficult, requiring extensive cash flow forecasts. The final technique for confronting the problem is to assume that reinvestment opportunities in the future will be similar to the current ones. The two most common ways of doing this are by creating a replacement chain to equalize life spans or calculating the project's equivalent annual annuity (EAA). Using a replacement chain, the present example would call for the creation of a two-chain cycle for project A—that is, we assume that project A can be replaced with a similar investment at the end of three years. Thus, project A would be viewed as two A projects occurring back to back, as illustrated in Figure 14-6. The net present value on this replacement chain is \$426.50, which is comparable with project B's net present value. Therefore, project A should be accepted because the net present value of its replacement chain is greater than the net present value of project B.

Unequal Lives—The Equivalent Annual Annuity (EAA) Approach One problem with replacement chains is that depending upon the life of each project, it can be quite difficult to come up with equivalent lives. For example, if the two projects had 7- and 13-year lives, a replacement chain of 91 years would be needed to establish equivalent lives. In this case, it is easier to determine the project's **equivalent annual annuity (EAA)**. A project's EAA is simply an annuity cash flow that yields the same present value as the project's NPV. To calculate a project's EAA we need only calculate a project's NPV and then divide that number by the $PVIFA_{k,n}$ to determine the dollar value of an n-year annuity that would produce the same NPV as the project. This can be done in two steps as follows:

Equivalent annual annuity

An annual cash flow that yields the same present value as the project's NPV. The equivalent annual annuity is calculated by dividing the projet's NPV by the appropriate PVIFA.

Step 1: Calculate the project's NPV. In Figure 14-6, we determined that project A had an NPV of 234.50 and project B had an NPV of \$306.50.

Step 2: Calculate the EAA. The EAA is determined by dividing each project's NPV by the $PVIFA_{k,n}$ where k is the required rate of return and n is the project's life. This determines the level of an annuity cash flow that would produce the same NPV as the project. For project A the $PVIFA_{k,n}$ is equal to 2.4869 and the $PVIFA_{k,n}$ for project B is equal to 4.3553. Dividing each project's NPV by the appropriate $PVIFA_{k,n}$, we determine the EAA for each project:

Figure 14-6 Replacement Chain Illustration: Two A Projects

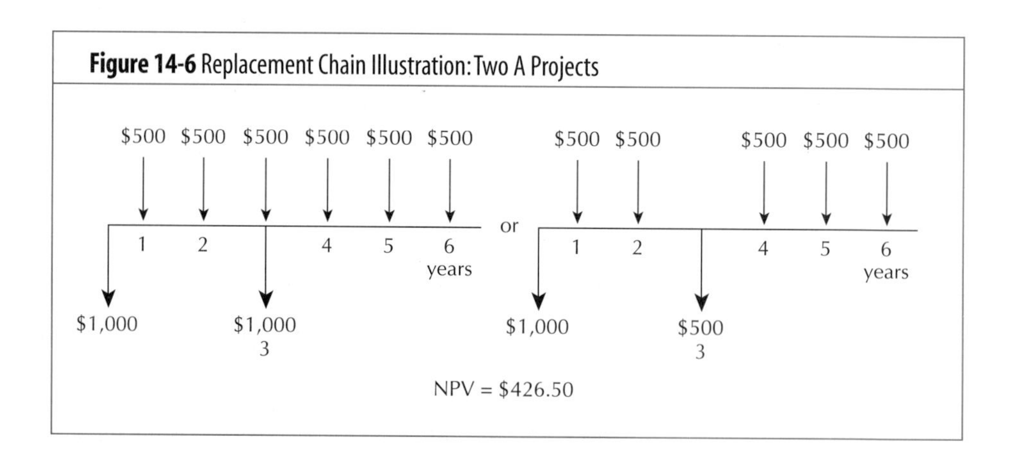

$$EAA_A = \frac{NPV_A}{PVIFA_{k,n}}$$

$$= \frac{\$234.50}{2.4869}$$

$$= \$94.29$$

$$EAA_B = \frac{NPV_B}{PVIFA_{k,n}}$$

$$= \frac{\$306.29}{4.3553}$$

$$= \$70.33$$

How do we interpret the EAA? For a project with an n-year life, it tells us what the dollar value is of an n-year annual annuity that would provide the same NPV as the project. Thus, for project A it means that a three-year annuity of $94.29 given a discount rate of 10 percent, would produce an NPV the same as project A's NPV which is $234.50. We can now compare the EAAs directly to determine which project is better. We can do this because we now have found the level of annual annuity that produces an NPV equivalent to the project's NPV. Since they are both annual annuities, they are comparable. An easy way to see this is to use the EAA's to create infinite life replacement chains. To do this we need only calculate the present value of an infinite stream or perpetuity of equivalent annual annuities. This is done by using the present value of an annuity formula, that is, simply dividing the equivalent annual annuity by the appropriate discount rate. In this case we find:

$$NPV_A = \frac{\$94.29}{.10} = \$942.90$$

$$NPV_B = \frac{\$70.33}{.10} = \$703.30$$

Here we calculated the present value of an infinite life replacement chain. Since the EAA method provides the same results as the infinite life replacement chain, it really doesn't matter which method you prefer to use.

SUMMARY

Guidelines for Capital Budgeting

In this chapter, we examined the measurement of incremental cash flows associated with a firm's investment proposals which are used to evaluate these proposals. Relying on **Axiom 3: Cash—Not Profits—Is King: Measuring the Timing of Costs and Benefits** and **Axiom 4: Incremental Cash Flows—It's Only What Changes That Counts**, we focused only on the incremental or different cash flows after tax attributed to the investment proposal. Care was taken to account for cash flows diverted from existing products, look for incidental or synergistic effects, considering working capital requirements, consider incremental expenses, ignore sunk costs, account for opportunity costs, examine overhead costs carefully, and ignore interest payments and financing flows.

The cash flows of both expansion and replacement projects fall into one of three categories: (1) the initial outlay, (2) the incremental cash flows over the project's life, and (3) the terminal cash flow. A summary of factors which affect the measurement of these cash flows was presented in Table 14-9.

The availability of funds creates several complications in the capital-budgeting process. Capital rationing can cause the rejection of projects with positive NPVs.

Although capital rationing poses an obstacle to the goal of maximization of shareholder's wealth, financial executives must deal with capital rationing by selecting new projects which maximize the total NPV of the projects. We also examine the problems associated with the evaluation of mutually exclusive projects. Mutually exclusive projects occur when a set of investment proposals perform essentially the same task. Mutually exclusively projects are ranked by means of the discounted cash flow criteria and a project with the highest ranking is selected. Conflicting rankings may arise because of the size disparity problem, the time disparity problem, and unequal lives. The problem of incomparability of projects with different lives is not simply a result of the different lives; rather, it arises because future profitable investment proposals may be rejected without being included in the analysis. Replacement chains and equivalent annual annuities are presented as possible solutions to this problem.

STUDY QUESTIONS

14-1. Why do we focus on cash flows rather than accounting profits in making our capital-budgeting decisions? Why are we interested only in incremental cash flows rather than cash flows?

14-2. If a project requires additional investment in working capital, how should this be treated in calculating cash flows?

14-3. How do sunk costs affect the determination of cash flows associated within an investment proposal?

14-4. If depreciation is not a cash flow expense, does it affect the level of cash flows from a project in any way?

14-5. Compare and contrast the cash flows of an expansion project and of a replacement project.

14-6. Explain how capital cost allowances influence cash flows for capital-budgeting purposes.

14-7. Explain the difference between terminal loss and recaptured depreciation.

14-8. Often government action to push or slow down the economy is implemented by altering the capital-budgeting decision-making process of the firm. How might the following changes affect capital-budgeting decisions?

- **a.** A movement by the Bank of Canada to reduce interest rates
- **b.** A change in the tax laws which increases the capital cost allowance rate
- **c.** A change in tax laws that prohibits installation costs to be included in the calculation of the capital cost

14-9. What are mutually exclusive projects? Why might the existence of mutually exclusive projects cause problems in the implementation of the discounted cash flow capital-budgeting criteria?

14-10. Why do firms practice capital rationing for the purposes of capital budgeting?

14-11. How should managers compare two mutually exclusive projects of unequal size? Would your approach change if capital rationing existed?

14-12. What causes the time disparity ranking problem? What reinvestment rate assumptions are associated with the net present value and internal rate of return capital-budgeting criteria?

14-13. When might two mutually exclusive projects having unequal lives be incomparable? How should managers deal with this problem?

SELF-TEST PROBLEMS

ST-1. The Scotty Gator Corporation of Meadville, maker of Scotty's electronic components, is considering the replacement of one of its existing hand-operated assembly machines with a new fully automated machine. This replacement project would mean the elimination of one employee, generating salary and benefit savings. Given the following information, determine the cash flows after tax associated with this replacement project.

Existing situation:	One full-time machine operator—salary and benefits, $25,000 per year
	Cost of maintenance—$2,000 per year
	Cost of defects—$6,000
	Capital cost of old machine—$50,000
	Depreciation method—declining balance method
	CCA rate—30 percent
	Expected life—4 years
	Machine's age—two years old
	Expected salvage value in two years—0
	Current salvage value—$5,000
	Combined corporate tax rate—45 percent
Proposed situation:	Fully automated machine
	Cost of machine—$60,000
	Installation fee—$3,000
	Shipping fee—$3,000
	Cost of maintenance—$3,000 per year
	Cost of defects—$3,000 per year
	Expected life—two years
	Expected salvage value—$20,000
	Depreciation method—declining balance method
	CCA rate—30 percent
	Required rate of return—15 percent

Assume that the asset class is kept open after the sale of the fully automated machine.

ST-2. Given the cash flow information in problem ST-1 and a required rate of return of 15 percent, complete the following for the new, fully automated machine:

a. Net present value
b. Profitability index
c. Internal rate of return

Should this project be accepted?

ST-3. The N. Sen Corp. is considering two mutually exclusive projects. Both require an initial outlay of $25,000 and will operate for five years. The probability distributions associated with each project for years 1 through 5 are given below:

Net Cash Flow After Tax, Years 1–5

Project A		Project B	
Probability	Cash Flow After Tax	Probability	Cash Flow After Tax
.20	$10,000	.20	$ 6,000
.60	15,000	.60	18,000
.20	20,000	.20	30,000

Since project B is the riskier of the two projects, the management of N. Sen has decided to apply a required rate of return of 18 percent to its evaluation, but only a 12 percent required rate of return to project A.

a. Determine the expected value of each project's net cash flows after tax.

b. Determine each project's risk-adjusted net present value.

ST-4. G. Norohna and Co. is considering two mutually exclusive projects. The expected values for each project's net cash flows after tax are given below.

Year	Project A	Project B
0	–$300,000	–$300,000
1	100,000	200,000
2	200,000	200,000
3	200,000	200,000
4	300,000	300,000
5	300,000	400,000

The company has decided to evaluate these projects using the certainty equivalent method. The certainty equivalent coefficients for each project's net cash flows after tax are given below.

Year	Project A	Project B
0	1.00	1.00
1	.95	.90
2	.90	.80
3	.85	.70
4	.80	.60
5	.75	.50

Given that this company's normal required rate of return is 15 percent and the after-tax risk-free rate is 8 percent, which project should be selected?

STUDY PROBLEMS (SET A)

14-1A. *(Capital Gains)* The J. Harris Corporation is considering the sale of its only assembly machine in asset class 39. This machine had a capital cost of $30,000 three years ago. The expected life of the machine when purchased was ten years after which it is expected to have a salvage value of zero. Revenue Canada requires that assets in this asset class be written off using a declining balance method at a CCA rate of 30 percent. This assembly machine would now sell for $35,000. Assume that the firm has a 45 percent combined corporate tax rate.

a. What would be the taxes associated with this sale?

b. If the old machine were sold for $25,000, what would be the taxes associated with this sale?

c. If the old machine were sold for $15,000, what would be the taxes associated with this sale?

d. If the old machine were sold for $12,000, what would be the taxes associated with this sale?

14-2A. *(Expansion Project—Capital-Budgeting Calculation)* The R. Boisjoly Chemical Corporation is considering the purchase of a new machine that has a capital cost of $80,000. The new machine is classified in asset class 8 for tax purposes. Revenue Canada requires that assets in this asset class be written off using the declining balance method at a CCA rate of 20 percent. The expected life of this machine is two years, after which it can be sold for $50,000. The asset class will be kept open after the sale of the asset. This new machine will produce net operating revenues (excluding accounting depreciation) of $55,000 per year. The firm's combined corporate tax rate is 45 percent, and its required rate of return is 10 percent. Given the above information, determine the following:

- **a.** The net present value
- **b.** The profitability index
- **c.** The internal rate of return

14-3A. *(Expansion Project—Capital-Budgeting Calculation)* The C. Duncan Chemical Corporation is considering the purchase of a new machine that has a capital cost of $200,000. The new machine is classified in asset class 39 for tax purposes. Revenue Canada requires that assets in this asset class be written off using the declining balance method at a CCA rate of 30 percent. The expected life of this machine is two years, after which it can be sold for $50,000. The asset class will remain open after the sale of the asset. This new machine will produce net operating revenues (excluding accounting depreciation) of $90,000 per year. The firm's combined corporate tax rate is 45 percent, and the firm's required rate of return is 15 percent. Find the following:

- **a.** The net present value
- **b.** The profitability index
- **c.** The internal rate of return

14-4A. *(Expansion Project—Capital-Budgeting Calculation)* The Woodchuck Corporation is considering the purchase of a new machine that has a capital cost of $450,000. The new machine is classified in asset class 8 for tax purposes. Revenue Canada requires that assets in this asset class be written off using the declining balance method at a CCA rate of 20 percent. The expected life of this machine is four years, after which it can be sold for $150,000. The asset class will be kept open after the sale of the asset. This new machine will produce net operating revenues (excluding accounting depreciation) of $115,000 per year. The firm's combined corporate tax rate is 45 percent, and the firm's required rate of return is 10 percent. Given the above information, determine the following:

- **a.** The net present value
- **b.** The profitability index
- **c.** The internal rate of return

14-5A. *(Expansion Project—Capital-Budgeting Calculation)* The Labelit Corporation is considering the purchase of a new printing machine that costs $450,000. In order to get the automated machine in running order, there would be a $5,000 shipping fee and a $10,000 installation charge. In addition, the new machine will require an investment in raw materials and goods-in-process inventories of $5,000. This increase in inventory cost of $5,000 will be recovered at the end of the five-year project. The new machine is classified in asset class 8 for tax purposes. Revenue Canada requires that assets in this asset class be written off using the declining balance method at a CCA rate of 20 percent. The expected life of this machine is five years, after which it can be sold for $150,000. The asset class will be kept open after the sale of the asset. This new machine will produce incremental operating revenues of $325,000 and incremental operating expenses of $210,000 (including accounting depreciation of $93,000) per year. The firm's combined corporate tax rate is 45 percent, and the firm's required rate of return is 10 percent. Given the above information, determine the following:

- **a.** The net present value
- **b.** The profitability index
- **c.** The internal rate of return

14-6A. *(Replacement Project—Capital-Budgeting Calculation)* The Mad Dog Hansen Electronic Components Corporation is considering the replacement of its existing machine with a new automated machine. The existing machine was purchased and installed at a capital cost of $30,000 three years ago. The existing machine when purchased had an expected life of six years, at which time the machine was expected to have a salvage value of zero. Both machines belong to asset class 8. Revenue Canada requires that assets in this asset class be written off using the declining balance method at a CCA rate of 20 percent. The new automated machine being considered has a capital cost of $80,000 and an expected life of three years. The new machine is expected to salvage for $4,000 in three years and the asset class will be kept open. Material efficiencies resulting from the replacement would result in savings of $30,000 per year before depreciation taxes. Currently, the old machine could be sold for $15,000. The firm's combined corporate tax rate is 45 percent, and the firm has a required rate of return of 20 percent. Calculate the following:

- **a.** The net present value
- **b.** The profitability index
- **c.** The internal rate of return

14-7A. *(Replacement Project—Capital-Budgeting Calculation)* The L. Knutson Company, a manufacturer of electronic components in the 45 percent combined corporate tax-rate bracket, is considering the purchase of a new, fully automated machine to replace an older, manually operated one. Both machines belong to asset class 8. Revenue Canada requires that these types of machines be written off using a declining balance method at a CCA rate of 20 percent.

The old machine had an expected life of ten years, at which time the salvage value was expected to be zero. After five years of productive use the old machine can now be sold for $25,000. It took one employee to operate the old machine, and he earned $15,000 per year in salary and $2,000 per year in fringe benefits. The annual costs of maintenance and defects associated with the old machine were $7,000 and $3,000, respectively. Ten years ago, the old machine had a capital cost of $40,000. The replacement machine being considered has a purchase price of $50,000. The salvage value of this new machine is expected to be $10,000 after five years, and the asset class will continue to remain open. In order to get the automated machine in running order, there would be a $3,000 shipping fee and a $2,000 installation charge. In addition, because the new machine would work faster than the old one, investment in raw materials and goods-in-process inventories would need to be increased by a total of $5,000. This increase in inventory cost of $5,000 would be recovered at the end of the five-year project. The annual costs of maintenance and defects on the new machine would be $2,000 and $4,000, respectively. The acquisition of the new machine would cause the firm to incur an additional $5,000 in administrative costs in the form of accounting costs. Finally, in order to purchase the new machine, it appears the firm would have to borrow an additional $20,000 at 10 percent interest from its local bank, resulting in additional interest payments of $2,000 per year. The required rate of return on projects of this kind is 20 percent.

- **a.** What is the project's initial outlay?
- **b.** What are the differential cash flows after tax over the project's life?
- **c.** What is the terminal cash flow?
- **d.** Draw a cash flow diagram for this project.
- **e.** What is its net present value?
- **f.** What is its profitability index?
- **g.** What is its internal rate of return?
- **h.** Should the project be accepted? Why or why not?

14-8A. *(Cash Flow Calculations)* The Winky Corporation, maker of electronic components, is considering the replacement of a hand-operated machine used in the manufacture of electronic components with a new, fully automated machine. Given the following information, determine the after-tax cash flows associated with this replacement project. Assume that both machines belong to asset class 39 for tax purposes.

Existing situation:	Two full-time machine operators—salaries $10,000 each per year
	Cost of maintenance—$5,000 per year
	Cost of defects—$5,000
	Capital cost of old machine—$30,000
	Expected life—10 years
	Machine's age—7 years old
	Expected salvage value—$0
	Depreciation method—declining balance method
	CCA rate—30 percent
	Current salvage value—$10,000
	Combined corporate tax rate—45 percent

Proposed situation:	Fully automated machine
	Cost of machine—$55,000
	Installation fee—$5,000
	Cost of maintenance—$6,000 per year
	Cost of defects—$2,000 per year
	Expected life—3 years
	Expected salvage value—$0
	Depreciation method—declining balance method
	CCA rate—30 percent
	Required rate of return—15 percent

Assume the asset class will be kept open with the sale of the automated machine after three years.

14-9A. *(Capital-Budgeting Calculation)* Given the cash flow information in problem 14-8A and a required rate of return of 15 percent, compute the following for the automated machine:

a. Net present value
b. Profitability index
c. Internal rate of return

Should this project be accepted?

14-10A. *(Capital Rationing)* The Cowboy Hat Company of Dawson Creek is considering seven capital investment proposals, for which the funds available are limited to a maximum of $12 million. The projects are independent and have the following costs and profitability indexes associated with them:

Project	Cost	Profitability Index
A	$4,000,000	1.18
B	3,000,000	1.08
C	5,000,000	1.33
D	6,000,000	1.31
E	4,000,000	1.19
F	6,000,000	1.20
G	4,000,000	1.18

a. Under strict capital rationing, which projects should be selected?
b. Is this an optimal strategy?

14-11A. *(Unequal Lives Ranking Problem)* The B. T. Knight Corporation is considering two mutually exclusive pieces of machinery that perform the same task. The two alternatives available provide the following set of cash flows after tax:

Year	Equipment A	Equipment B
0	−$20,000	−$20,000
1	12,590	6,625
2	12,590	6,625
3	12,590	6,625
4		6,625
5		6,625
6		6,625
7		6,625
8		6,625
9		6,625

Equipment A has an expected life of three years, while equipment B has an expected life of nine years. Assume a required rate of return of 15 percent.

- **a.** Calculate each project's payback period.
- **b.** Calculate each project's net present value.
- **c.** Calculate each project's internal rate of return.
- **d.** Are these projects comparable?
- **e.** Compare these projects using replacement chains. Which project should be selected? Support your recommendation.

14-12A. *(Time Disparity Ranking Problem)* The R. Rogers Corporation is considering two mutually exclusive projects. The cash flows after tax associated with those projects are as follows:

Year	Project A	Project B
0	–$50,000	–$50,000
1	15,625	0
2	15,625	0
3	15,625	0
4	15,625	0
5	15,625	$100,000

The required rate of return on these projects is 10 percent.

- **a.** What is each project's payback period?
- **b.** What is each project's net present value?
- **c.** What is each project's internal rate of return?
- **d.** What has caused the ranking conflict?
- **e.** Which project should be accepted? Why?

14-13A. *(Size Disparity Ranking Problem)* The D. Dorner Farms Corporation is considering purchasing one of two fertilizer-herbicides for the upcoming year. The more expensive of the two is the better and will produce a higher yield. Assume these projects are mutually exclusive and that the required rate of return is 10 percent. Given the following cash flows after tax:

Year	Project A	Project B
0	–$500	–$5,000
1	700	6,000

- **a.** Calculate the net present value.
- **b.** Calculate the profitability index.
- **c.** Calculate the internal rate of return.
- **d.** If there is no capital-rationing constraint, which project should be selected? If there is a capital-rationing constraint, how should the decision be made?

14-14A. *(Cash Flow—Capital-Budgeting Calculation)* The Beamer Corp. is considering the expansion of its highly technical construction facilities. The construction will take a total of four years until it is completed and ready for operation. The following data and assumptions describe the proposed expansion:

1. In order to make this expansion feasible, installation expenditures of $200,000 must be made immediately to ensure that the construction facilities are competitively efficient ($t = 0$).
2. At the end of the first year the land will be purchased and construction on stage 1 of the facilities will begin, involving a cash outflow of $150,000 for the land and $300,000 for the construction facilities. For tax purposes, the construction facilities belong to asset class 1. Revenue Canada requires that assets in this asset class be written off using a declining balance method at a CCA rate of 4 percent.
3. Stage 2 of the construction will involve a $300,000 cash outflow at the end of year 2. This cash outlay is for the purchase of a processing machine. For tax purposes, the

processing machine belongs to asset class 39. Revenue Canada requires that assets in this asset class be written off using a declining balance method at a CCA rate of 30 percent.

4. At the end of year 3, when production begins, inventory will be increased by $50,000.
5. At the end of year four, the first sales revenue from operations of the new plant is estimated at $800,000 and continues at that level for ten years (with the final flow from sales occurring at the end of year 13).
6. Annual operating costs on these sales include $100,000 of fixed operating costs per year and variable operating costs equal to 40 percent of sales.
7. When the plant is closed, it is expected that the processing machine will salvage at $30,000 and asset class 39 will be kept open. The construction facilities are expected to salvage for $50,000 and asset class 1 will also be kept open.
8. When the plant is closed, the land will be sold for $200,000 ($t = 13$).
9. The company is in the 45-percent combined corporate tax-rate bracket. Given a 12 percent required rate of return, what is the NPV of this project? Should it be accepted?

14-15A. *(Cash Flow—Capital-Budgeting Calculation)* The Steel Mill Corporation of Asbury Park must replace its executive jet and is considering two mutually exclusive planes as replacements. As far as it is concerned, both planes are identical and each will provide annual benefits, before maintenance, taxes, and depreciation, of $40,000 per year for the life of the plane; however, the costs on each are decidedly different. The first is a Steel Mills and costs $94,000. It has an expected life of seven years and will require major maintenance at the end of year 4. The annual maintenance costs are $8,000 for the first three years, $25,000 in year 4, and $10,000 for years 5 through 7. Its salvage value at the end of year 7 is $24,000.

The other plane is a Honeybilt Jet costing $100,000 and also having an expected life of seven years. Its maintenance expenses are expected to be $9,000 the first four years, rising to $18,000 annually for years 5 through 7. In years 3 and 5 the jet engine must be overhauled at an expense of $15,000 each time; this overhaul expense is in excess of the maintenance expenses. Finally, at the end of year 7 this jet will have a salvage value of $30,000. Both planes being considered belong to asset class 9. Revenue Canada requires that assets in this asset class 9 be written off using the declining balance method at a CCA rate of 25 percent. At the end of year 7, the asset class will remain open. Given a 45 percent combined corporate tax rate and a 10 percent required rate of return, what is the NPV on each jet? Which should be taken?

INTEGRATIVE PROBLEM

It's been two months since you took a position as an assistant financial analyst at Cale Products. While your boss has been pleased with your work, he is still a bit hesitant about unleashing you without supervision. Your next assignment involves both the calculation of cash flows associated with a new investment under consideration and the evaluation of several mutually exclusive projects. Given your lack of tenure at Cale, you have been asked not only to provide a recommendation, but also to respond to a number of questions aimed at judging your understanding of the capital-budgeting process. The memorandum you received outlining your assignment follows:

TO: The Financial Analyst
FROM: Mr. V. Morrison, CEO, Cale Products
RE: Cash Flow Analysis and Capital Rationing

We are currently considering the purchase of a new fully automated machine to replace an older, manually operated one. The machine being replaced, now five years old, originally had an expected life of ten years after which it could be sold for $10,000. Its salvage value now is estimated at $40,000. This type of machine belongs to asset class 8, which requires the business to write off the asset value using a declining balance method at a CCA rate of 20%. The old machine took one operator, who earned $15,000 per year in salary and $2,000 per year in fringe benefits. Since the machine is fully automated, this worker will no longer be needed. The annual costs of maintenance and defects associated with the old machine were $7,000 and $3,000 respectively. The replacement machine being considered has a purchase price of

$50,000 and an expected salvage value after five years of $10,000. To get the automated machine in running order, there would be a $3,000 shipping fee and a $2,000 installation charge. In addition, because the new machine would work faster than the old one, investment in raw materials and work-in-process inventories would need to be increased by a total of $5,000. The annual costs of maintenance and defects on the new machine would be $2,000 and $4,000, respectively. The new machine also requires maintenance workers to be specially trained; fortunately, a similar machine was purchased three months ago, and at that time the maintenance workers went through the $5,000 training program needed to familiarize themselves with the new equipment. Cale's management is uncertain whether or not to charge half of this training fee toward the new project. Finally, to purchase the new machine, it appears the firm would have to borrow an additional $20,000 at 10 percent from its local bank, resulting in additional interest payments of $2,000 per year. The required rate of return on projects of this kind is 20 percent and Cale has a combined tax rate of 50 percent.

a. Should Cale's focus on cash flows or accounting profits in making our capital-budgeting decisions? Should we be interested in incremental cash flows, incremental profits, total cash flows or total profits?

b. How does depreciation affect cash flows?

c. How do sunk costs affect the determination of cash flows?

d. What is the project's initial outlay?

e. What are the incremental cash flows over the project's life?

f. What is the terminal cash flow?

g. Draw a cash flow diagram for this project.

h. What is its net present value?

i. What is its internal rate of return?

j. Should the project be accepted? Why or why not?

You have also been asked for your views on three unrelated sets of projects. Each set of projects involves two mutually exclusive projets. These projects follow:

k. Cale is considering two investments with one-year lives. The more expensive of the two is the better and will produce more savings. Assume these projects are mutually exclusive and that the required rate of return is 10 percent. Given the following net cash flows after tax:

Year	Project A	Project B
0	−$195,000	−$1,200,000
1	240,000	1,650,000

1. Calculate the net present value.
2. Calculate the profitability index.
3. Calculate the internal rate of return.
4. If there is no capital-rationing constraint, which project should be selected? If there is a capital-rationing constraint, how should the decision be made?

l. Cale is considering two additional mutually exclusive projects. The cash flows associated with these projects are as follows:

Year	Project A	Project B
0	−$100,000	−$100,000
1	32,000	0
2	32,000	0
3	32,000	0
4	32,000	0
5	32,000	$200,000

The required rate of return on these projects is 11 percent.

1. What is each project's payback period?
2. What is each project's net present value?
3. What is each project's internal rate of return?

4. What caused the ranking conflict?
5. Which project should be accepted? Why?

m. The final two mutually exclusive projects that Cale is considering involve mutually exclusive pieces of machinery that perform the same task. The two alternatives available provide the following set of cash flows after tax:

Year	Project A	Project B
0	–$100,000	–$100,000
1	65,000	32,500
2	65,000	32,500
3	65,000	32,500
4	0	32,500
5	0	32,500
6	0	32,500
7	0	32,500
8	0	32,500
9	0	32,500

Project A has an expected life of three years; whereas, project B has an expected life of nine years. Assume a required rate of return of 14 percent.

1. Calculate each project's payback period.
2. What is each project's net present value?
3. What is each project's internal rate of return?
4. Are these projects comparable?
5. Compare these projects using replacement chains and EAAs. Which project should be selected? Support your recommendation.

STUDY PROBLEMS (SET B)

14-1B. *(Capital Gains Tax)* The R. T. Kleinman Corporation is considering the sale of its only assembly machine in asset class 39. This machine had a capital cost of $40,000 three years ago. The expected life of the machine when purchased was 10 years, after which it is expected to have a salvage value of zero. Revenue Canada requires that assets in this asset class be written off using a declining balance method at a CCA rate of 30 percent. This assembly machine would now sell for $45,000. Assume that the firm has a 45 percent combined corporate tax rate.

a. What would be the taxes associated with this sale?

b. If the old machine were sold for $30,000, what would be the taxes associated with this sale?

c. If the old machine were sold for $15,000, what would be the taxes associated with this sale?

14-2B. *(Expansion Project—Capital-Budgeting Calculation)* The Kensinger Corporation is considering the purchase of a new machine that has a capital cost of $300,000. The new machine is classified in asset class 39 for tax purposes. Revenue Canada requires that assets in this asset class be written off using the declining balance method at a CCA rate of 30 percent. The expected life of this machine is four years, after which it can be sold for $75,000. The asset class will remain open after its sale. Because of reductions in defects and material savings, the new machine will produce cash benefits of $110,000 per year before depreciation and taxes. The firm's combined corporate tax rate is 45 percent, and the firm's required rate of return is 15 percent. Find:

a. The net present value

b. The profitability index

c. The internal rate of return.

14-3B. *(Expansion Project—Capital-Budgeting Calculation)* The Boiling Chemical Corporation is considering the purchase of a new machine that has a capital cost of $100,000. The new machine is classified in asset class 8 for tax purposes. Revenue Canada requires that assets in this asset class be written off using the declining balance method at a CCA rate of 20 percent. The expected life of this machine is five years, after which it can be sold for $80,000. The asset class will be kept open after the sale of the asset. As a result of reductions in defects and material savings, the new machine will produce cash benefits of $85,000 per year before depreciation and taxes. The firm's combined corporate tax rate is 45 percent, and the firm's required rate of return is 10 percent. Given the above information, determine the following:

- **a.** The net present value
- **b.** The profitability index
- **c.** The internal rate of return

14-4B. *(Expansion Project—Capital-Budgeting Calculation)* The BJ Manufacturing Corporation is considering the purchase of a new machine that has a capital cost of $500,000. The new machine is classified in asset class 8 for tax purposes. Revenue Canada requires that assets in this asset class be written off using the declining balance method at a CCA rate of 20 percent. The expected life of this machine is four years, after which it can be sold for $110,000. The asset class will be kept open after its sale. This new machine will produce net operating revenues (excluding accounting depreciation) of $145,000 per year. The firm's combined corporate tax rate is 45 percent, and the firm's required rate of return is 10 percent. Given the above information, determine the following:

- **a.** The net present value
- **b.** The profitability index
- **c.** The internal rate of return

14-5B. *(Expansion Project—Capital-Budgeting Calculation)* The Printit Corporation is considering the purchase of a new printing machine that costs $400,000. In order to get the automated machine in running order, there would be a $5,000 shipping fee and a $10,000 installation charge. In addition, the new machine will require an investment in raw materials and goods-in-process inventories of $5,000. This increase in inventory cost of $5,000 will be recovered at the end of the five-year project. The new machine is classified in asset class 8 for tax purposes. Revenue Canada requires that assets in this asset class be written off using the declining balance method at a CCA rate of 20 percent. The expected life of this machine is five years, after which it can be sold for $100,000. The asset class will be kept open after the sale of the asset. This new machine will produce incremental operating revenues of $365,000 and incremental operating expenses of $240,000 (including accounting depreciation of $83,000) per year. The firm's combined corporate tax rate is 45 percent, and the firm's required rate of return is 10 percent. Given the above information, determine the following:

- **a.** The net present value
- **b.** The profitability index
- **c.** The internal rate of return

14-6B. *(Replacement Project—Capital-Budgeting Calculation)* The G. Rod Electronic Components Corporation is considering the replacement of its existing machine with a new automated machine. The existing machine was purchased and installed at a capital cost of $25,000 three years ago. The existing machine when purchased had an expected life of six years, at which time the machine was expected to have a salvage value of zero. Both machines belong to asset class 8. Revenue Canada requires that assets in this asset class be written off using the declining balance method at a CCA rate of 20 percent. The new automated machine being considered has a capital cost of $75,000 and an expected life of three years. The new machine is expected to salvage for $4,000 in three years and at that time the asset class will be kept open. Material efficiencies resulting from the replacement would result in savings of $40,000 per year before taxes. Currently, the old machine could be sold for $15,000. The firm's combined corporate tax rate is 45 percent, and the firm has a required rate of return of 20 percent. Calculate the following:

- **a.** The net present value
- **b.** The profitability index
- **c.** The internal rate of return

14-7B. *(Comprehensive Cash Flow—Capital-Budgeting Calculation)* The L. Bellich Company, a manufacturer of electronic components in the 45-percent combined corporate tax-rate bracket, is considering the purchase of a new, fully automated machine to replace an older, manually operated one. Both machines belong to asset class 8. Revenue Canada requires that these types of machines be written off using a declining balance method at a CCA rate of 20 percent.

The old machine had an expected life of ten years, at which time the salvage value was expected to be zero. After five years of productive use, the old machine can now be sold for $30,000. It took one person to operate the old machine, and he earned $14,000 per year in salary and $6,000 per year in fringe benefits. The annual costs of maintenance and defects associated with the old machine were $6,000 and $2,000, respectively. Ten years ago, the old machine had a capital cost of $37,500. The replacement machine being considered has a purchase price of $40,000. The salvage value of the new machine is expected to be $15,000 after five years, and the asset class will continue to remain open. In order to get the automated machine in running order, there would be a $6,000 shipping fee and a $4,000 installation charge. In addition, because the new machine would work faster than the old one, investment in raw materials and goods-in-process inventories would need to be increased by a total of $7,000. This increase in inventory costs will be recovered at the end of the five-year project. The annual costs of maintenance and defects on the new machine would be $1,000 and $1,000, respectively. The acquisition of the new machine requires an additional $5,000 in administrative costs in the form of accounting costs. Finally, in order to purchase the new machine, it appears the firm would have to borrow an additional $25,000 at 10 percent interest from its local bank, resulting in additional interest payments of $2,500 per year. The required rate of return on projects of this kind is 20 percent.

a. What is the project's initial outlay?

b. What are the differential cash flows after tax over the project's life?

c. What is the terminal cash flow?

d. Draw a cash flow diagram for this project.

e. What is its net present value?

f. What is its profitability index?

g. What is its internal rate of return?

h. Should the project be accepted? Why or why not?

14-8B. *(Cash Flow Calculations)* The J. R. Woolridge Corporation, maker of electronic components, is considering the replacement of a hand-operated machine used in the manufacture of electronic components with a new, fully automated machine. Given the following information, determine the cash flows after tax associated with this replacement project. Assume that both machines belong to asset class 39 for tax purposes.

Existing situation:	Two full-time machine operators' salaries—$12,000 each per year
	Cost of maintenance—$6,000 per year
	Cost of defects—$5,000
	Capital cost of old machine—$40,000
	Expected life—10 years
	Machine's age—7 years old
	Expected salvage value—$0
	Depreciation method—declining balance method
	CCA rate—30 percent
	Current salvage value—$10,000
	Combined corporate tax rate—45 percent
Proposed situation:	Fully automated machine
	Cost of machine—$55,000
	Installation fee—$5,000
	Cost of maintenance—$6,000 per year
	Cost of defects—$2,000 per year
	Expected life—3 years
	Expected salvage value—$0
	Depreciation method—declining balance method
	CCA rate —30 percent
	Required rate of return—15 percent

Assume the asset class will be kept open with the sale of the automated machine after three years.

14-9B. (*Capital-Budgeting Calculation*) Given the cash flow information in problem 14-8B and a required rate of return of 17 percent, compute the following for the automated machine:

a. Net present value
b. Profitability index
c. Internal rate of return

Should this project be accepted?

14-10B. *(Size Disparity Ranking Problem)* The Unk's Farms Corporation is considering purchasing one of two fertilizer-herbicides for the upcoming year. The more expensive of the two is the better and will produce a higher yield. Assume these projects are mutually exclusive and that the required rate of return is 10 percent. Given the following net cash flows after tax:

Year	Project A	Project B
0	−$650	−$4,000
1	800	5,500

a. Calculate the net present value.
b. Calculate the profitability index.
c. Calculate the internal rate of return.
d. If there is no capital-rationing constraint, which project should be selected? If there is a capital-rationing constraint, how should the decision be made?

14-11B. *(Time Disparity Ranking Problem)* The Z. Bello Corporation is considering two mutually exclusive projects. The cash flows after tax associated with those projects are as follows:

Year	Project A	Project B
0	−$50,000	−$50,000
1	16,000	0
2	16,000	0
3	16,000	0
4	16,000	0
5	16,000	$100,000

The required rate of return on these projects is 11 percent.

a. What is each project's payback period?
b. What is each project's net present value?
c. What is each project's internal rate of return?
d. What has caused the ranking conflict?
e. Which project should be accepted? Why?

14-12B. *(Unequal Lives Ranking Problem)* The Battling Bishops Corporation is considering two mutually exclusive pieces of machinery that perform the same task. The two alternatives available provide the following set of cash flows after tax:

Year	Equipment A	Equipment B
0	–$20,000	–$20,000
1	13,000	6,500
2	13,000	6,500
3	13,000	6,500
4		6,500
5		6,500
6		6,500
7		6,500
8		6,500
9		6,500

Equipment A has an expected life of three years, while equipment B has an expected life of nine years. Assume a required rate of return of 14 percent.

a. Calculate each project's payback period.

b. Calculate each project's net present value.

c. Calculate each project's internal rate of return.

d. Are these projects comparable?

e. Compare these projects using replacement chains. Which project should be selected? Support your recommendation.

14-13B. *(Capital Rationing)* The Taco Toast Company is considering seven capital investment proposals, for which the funds available are limited to a maximum of $12 million. The projects are independent and have the following costs and profitability indexes associated with them:

Project	Cost	Profitability Index
A	$4,000,000	1.18
B	3,000,000	1.08
C	5,000,000	1.33
D	6,000,000	1.31
E	4,000,000	1.19
F	6,000,000	1.20
G	4,000,000	1.18

a. Under strict capital rationing, which projects should be selected?

b. Is this an optimal strategy?

14-14B. *(Cash Flow—Capital-Budgeting Calculation)* The Hokie Tech Corp. is considering expanding its highly technical construction facilities. The construction will take a total of four years until it is completed and ready for operation. The following data and assumptions describe the proposed expansion:

1. In order to make this expansion feasible, installation expenditures of $175,000 must be made immediately to ensure that the construction facilities are competitively efficient ($t = 0$).
2. At the end of the first year the land will be purchased and construction on stage 1 of the facilities will begin, involving a cash outflow of $100,000 for the land and $300,000 for the construction facilities. For tax purposes, the construction facilities

belong to asset class 1. Revenue Canada requires that assets in this asset class be written off using a declining balance method at a CCA rate of 4 percent.

3. Stage 2 of the construction will involve a $300,000 cash outflow at the end of year 2. This cash outlay is for the purchase of a processing machine. For tax purposes, the processing machine belongs to asset class 39. Revenue Canada requires that assets in this asset class be written off using a declining balance method at a CCA rate of 30 percent.
4. At the end of year 3, when production begins, inventory will be increased by $50,000.
5. At the end of year four, the first sales revenue from operations of the new plant is estimated at $800,000 and continues at that level for ten years (with the final flow from sales occurring at the end of year 13).
6. Annual operating costs on these sales include $100,000 of fixed operating costs per year and variable operating costs equal to 35 percent of sales.
7. When the plant is closed, it is expected that the processing machine will salvage at $35,000 and asset class 39 will be kept open. The construction facilities are expected to salvage for $50,000 and asset class 1 will also be kept open.
8. When the plant is closed, the land will be sold for $225,000 ($t = 13$).
9. The company is in the 45-percent combined corporate tax-rate bracket. Given a 12 percent required rate of return, what is the NPV of this project? Should it be accepted?

14-15B. *(Cash Flow—Capital-Budgeting Calculation)* The M. Jose Corporation of Ontario must replace its executive jet and is considering two mutually exclusive planes as replacements. As far as it is concerned, both planes are identical and each will provide annual benefits, before maintenance, taxes, and depreciation, of $50,000 per year for the life of the plane; however, the costs on each are decidedly different. The first is a Lays Jet and costs $94,000. It has an expected life of seven years and will require major maintenance at the end of year 4. The annual maintenance costs are $7,500 for the first three years, $25,000 in year 4, and $12,000 for years 5 through 7. Its salvage value at the end of year 7 is $24,000.

The other plane is a Honeybilt Jet costing $100,000 and also having an expected life of seven years. Its maintenance expenses are expected to be $9,000 the first four years, rising to $18,000 annually for years 5 through 7. In years 3 and 5 the jet engine must be overhauled at an expense of $13,000 each time; this overhaul expense is in excess of the maintenance expenses. Finally, at the end of year 7 this jet will have a salvage value of $30,000. Both planes being considered belong to asset class 9. Revenue Canada requires that assets in asset class 9 be written off using the declining balance method at a CCA rate of 25 percent. At the end of year 7, the asset class will remain open. Given a 45 percent combined corporate tax rate and a 10 percent required rate of return, what is the NPV on each jet? Which should be taken?

CASE PROBLEM

HARDING PLASTIC MOLDING COMPANY

Capital Budgeting: Ranking Problems

On January 11, 1996, the finance committee of Harding Plastic Molding Company (HPMC) met to consider eight capital-budgeting projects. Present at the meeting were Robert L. Harding, president and founder, Susan Jorgensen, comptroller, and Chris Woelk, head of research and development. Over the past five years, this committee has met every month to consider and make final judgment on all proposed capital outlays brought up for review during the period.

Harding Plastic Molding Company was founded in 1968 by Robert L. Harding to produce plastic parts and molding for the Oshawa automakers. For the first ten years of operations, HPMC worked solely as a subcontractor for the automakers, but since then has made strong efforts to diversify in an attempt to avoid the cyclical problems faced by the auto industry. By 1996, this diversification attempt had led HPMC into the production of over 1,000 different items, including kitchen utensils, camera housings, and phonographic and recording equipment. It also led to an increase in sales of 800 percent during the 1978–1996 period. As this dramatic increase in sales was parallelled by a corresponding increase in production volume, HPMC was forced, in late 1994, to expand production facilities. This plant and equipment expansion involved capital expenditures of approximately $10.5 million and resulted in an increase of production capacity of about 40 percent. Because of this increased production capacity, HPMC has made a concerted effort to attract new business and consequently has recently entered into contracts with a large toy firm and a major discount department store chain. While non-auto-related business has grown significantly, it still only represents 32 percent of HPMC's overall business. Thus, HPMC has continued to solicit nonautomotive business, and as a result of this effort and its internal research and development, the firm has four sets of mutually exclusive projects to consider at this month's finance committee meeting.

Over the past ten years, HPMC's capital-budgeting approach has evolved into a somewhat elaborate procedure in which new proposals are categorized into three areas: profit, research and development, and safety. Projects falling into the profit or research and development areas are evaluated using present value techniques, assuming a 10 percent opportunity rate; those falling into the safety classification are evaluated in a more subjective framework. Although research and development projects have to receive favourable results from the present value criteria, there is also a total dollar limit assigned to projects of this category, typically running about $750,000 per year. This limitation was imposed by Harding primarily because of the limited availability of quality researchers in the plastics industry. Harding felt that if more funds than this were allocated, "we simply couldn't find the manpower to administer them properly." The benefits derived from safety projects, on the other hand, are not in terms of cash flows after tax; hence, present value methods are not used at all in their evaluation. The subjective approach used to evaluate safety projects is a result of the pragmatically difficult task of quantifying the benefits from these projects into dollar terms. Thus, these projects are subjectively evaluated by a management-worker committee with a limited budget. All eight projects to be evaluated in January are classified as profit projects.

The first set of projects listed on the meeting's agenda for examination involve the utilization of HPMC's precision equipment. Project A calls for the production of vacuum containers for thermos bottles produced for a large discount hardware chain. The containers would be manufactured in five different size and colour combinations. This project would be carried out over a three-year period, for which HPMC would be guaranteed a minimum return plus a percentage of the sales. Project B involves the manufacture of inexpensive photographic equipment for a national photography outlet. Although HPMC currently has excess plant capacity, each of these projects would utilize precision equipment of which the excess capacity is limited. Thus, adopting either project would tie up all precision facilities. In addition, the purchase of new equipment would be both prohibitively expensive and involve a time delay of approximately two years, thus making these projects mutually exclusive. (The cash flows after tax associated with these two projects are given in Exhibit 1.)

The second set of projects involves the renting of computer facilities over a one-year period to aid in customer billing and perhaps inventory control. Project C entails the evaluation of a customer billing system proposed by Advanced Computer Corporation. Under this system all the bookkeeping and billing presently being done by HPMC's accounting department would be done by Advanced. In addition to saving costs involved in bookkeeping, Advanced would provide a more efficient billing system and do a credit analysis of delinquent customers, which could be used in the future for in-depth credit analysis. Project D is proposed by International Computer Corporation and includes a billing system similar to that offered by Advanced and, in addition, an inventory control system that will keep track of all raw materials and parts in stock and reorder when necessary, thereby reducing the likelihood of material stockouts, which have become more and more frequent over the past three years. (The cash flows after tax for these projects are given in Exhibit 2.)

The third decision that faces the financial directors of HPMC involves a newly developed and patented process for molding hard plastics. HPMC can either manufacture and market the equipment necessary to mold such plastics or it can sell the patent rights to Polyplastics Incorporated, the world's largest producer of plastics products. (The cash flows after tax for projects E and F are shown in Exhibit 3.) At present, the process has not been fully tested, and if HPMC is going to market it itself, it will be necessary to complete this testing and begin production of plant facilities immediately. On the other hand, the selling of these patent rights to Polyplastics would involve only minor testing and refinements, which could be completed within the year. Thus, a decision as to the proper course of action is necessary immediately.

The final set of projects up for consideration revolve around the replacement of some of the machinery. HPMC can go in one of two directions. Project G suggests the purchase and installation of moderately priced, extremely efficient equipment with an expected life of five years; project H advocates the purchase of a similarly priced, although less efficient, machine with a life expectancy of ten years. (The cash flows after tax for these alternatives are shown in Exhibit 4.)

As the meeting opened, debate immediately centered on the most appropriate method for evaluating all the projects. Harding suggested that as the projects to be considered were mutually exclusive, perhaps their usual capital-budgeting criterion of net present value was inappropriate. He felt that, in examining these projects, perhaps they should be more concerned with relative profitability or some measure of yield. Both Jorgensen and Woelk agreed with Harding's point of view, with Jorgensen advocating a profitability index approach and Woelk preferring the use of the internal rate of return. Jorgensen argued that the use of the profitability index would provide a benefit-cost ratio, directly implying relative profitability. Thus, they merely need to rank these projects and select those with the highest profitability index. Woelk agreed with Jorgensen's point of view, but suggested that the calculation of an internal rate of return would also give a measure of profitability and perhaps be somewhat easier to interpret. To settle the issue, Harding suggested that they calculate all three measures, as they would undoubtedly yield the same ranking.

From here the discussion turned to an appropriate approach to the problem of differing lives among mutually exclusive projects E and F, and G and H. Woelk argued that there really was not a problem here at all, that as all the cash flows after tax from these projects can be determined, any of the discounted cash flow methods of capital budgeting will work well. Jorgensen argued that although this was true, some compensation should be made for the fact that the projects being considered did not have equal lives.

Exhibit 1 Harding Plastic Molding Company

Cash Flows After Tax

Year	Project A	Project B
0	–$ 75,000	–$75,000
1	10,000	43,000
2	30,000	43,000
3	100,000	43,000

Exhibit 2 Harding Plastic Molding Company

Cash Flows After Tax

Year	Project C	Project D
0	–$ 8,000	–$20,000
1	11,000	25,000

Exhibit 3 Harding Plastic Molding Company

Cash Flows After Tax

Year	Project E	Project F
0	–$ 30,000	–$271,500
1	210,000	100,000
2		100,000
3		100,000
4		100,000
5		100,000
6		100,000
7		100,000
8		100,000
9		100,000

Exhibit 4 Harding Plastic Molding Company

Cash Flows After Tax

Year	Project G	Project H
0–	$500,000	–$500,000
1	225,000	150,000
2	225,000	150,000
3	225,000	150,000
4	225,000	150,000
5	225,000	150,000
6		150,000
7		150,000
8		150,000
9		150,000
10		150,000

Questions

1. Was Harding correct in stating that the NPV, PI, and IRR necessarily will yield the same ranking order? Under what situations might the NPV, PI, and IRR methods provide different rankings? Why is it possible?
2. What are the NPV, PI, and IRR for projects A and B? What has caused the ranking conflicts? Should project A or B be chosen? Might your answer change if project B is a typical project in the plastic molding industry? For example, if projects for HPMC generally yield approximately 12 percent, is it logical to assume that the IRR for project B of approximately 33 percent is a correct calculation for ranking purposes? (*Hint:* Examine the reinvestment rate assumption.)
3. What are the NPV, PI, and IRR for projects C and D? Should project C or D be chosen? Does your answer change if these projects are considered under a capital constraint? What return on the marginal $12,000 not employed in project C is necessary to make one indifferent to choosing one project over the other under a capital-rationing situation?
4. What are the NPV, PI, and IRR for projects E and F? Are these projects comparable even though they have unequal lives? Why? Which project should be chosen? Assume that these projects are not considered under a capital constraint.
5. What are the NPV, PI, and IRR for projects G and H? Are these projects comparable even though they have unequal lives? Why? Which project should be chosen? Assume that these projects are not considered under a capital constraint.

SELF-TEST SOLUTIONS

SS-1: *STEP 1*: First calculate the initial cash outlay.

Initial Outlay

Outflows:	
Cost of machine	$ 60,000
Installation fee	3,000
Shipping fee	3,000
Inflows:	
Salvage value—old machine	–5,000
Initial cash outlay	$ 61,000

STEP 2: Calculate the incremental cash flows after tax over the project's life.

Schedule for Annual Net Operating Cash Flows After Tax

	Year 1	Year 2
Savings:		
Reduced salary	$ 25,000	$ 25,000
Reduced defects	3,000	3,000
Costs:		
Increased maintenance	–1,000	–1,000
Net savings before taxes	$ 27,000	$ 27,000
Taxes (@45%)	–12,150	–12,150
Net savings after taxes	$ 14,850	$ 14,850
Tax shields from depreciation	4,118	7,000
Net cash flows after tax	$ 18,968	$ 21,850

Schedule for Capital Cost Allowances and Tax Shields

Year	UCC (Beg)	Capital Cost Allowance	Tax Shields	UCC (End)
1	1$30,500	$ 9,150	$4,118	$51,850
11	51,850	15,555	7,000	36,295

STEP 3: Calculate the terminal cash flow.

Salvage value—new machine	$20,000
Tax shield adjustment	4,889[a]
Terminal cash flow	$24,889

[a] $\left[\frac{(\$36,295 - \$20,000)(.30)(.45)}{(.30 + .15)}\right] = \$4,889$

	Year 0	1	2
Initial outlay	–$61,000		
Net operating cash flow		$18,968	$21,850
Salvage value			20,000
Tax shield adjustment			4,889
Net annual cash flow	–$61,000	$18,968	$46,739

SS-2: a. 1st Approach:

$$NPV = \frac{\$18{,}968}{(1.15)^1} + \frac{\$46{,}739}{(1.15)^2} - \$61{,}000$$
$$= \$16{,}494 + \$35{,}341 - \$61{,}000$$
$$= -\$9{,}165$$

2nd Approach:

$$NPV = PV(\text{project's net savings after taxes}) + PV(\text{tax shields from depreciation}) + PV(\text{salvage value}) - IO$$
$$= (\$14{,}850)(1.6257) + \left[\frac{(61{,}000)(.30)(.45)}{(.30 + .15)}\right]\left[\frac{1.075}{1.15}\right] - \left[\frac{(\$20{,}000)(.30)(.45)}{(.30 + .15)}\right]\left[\frac{1}{(1.15)^2}\right] + (\$20{,}000)(.7561) - \$61{,}000$$
$$= \$24{,}142 + \$17{,}107 - \$4{,}537 + \$15{,}122 - \$61{,}000$$
$$= -\$9{,}166$$

b.

$$\text{Profitability Index} = \frac{\sum_{t=1}^{n} \frac{CFAT_t}{(1+K)^t}}{IO}$$
$$= \frac{\$51{,}832}{\$61{,}000}$$
$$= 0.8497$$

c. Through linear interpolation, upper and lower boundaries for IRR can be estimated by substitution in the following equations

$$NPV = \sum_{t=1}^{n} \frac{CFAT_t}{(1+IRR)^t} - IO$$

$$NPV = \frac{\$18{,}968}{(1+IRR)^1} + \frac{\$46{,}738}{(1+IRR)^2} - \$61{,}000$$

Using a calculator, we find that the value of IRR is 4.45 percent. As a result, the project's internal rate of return is less than the firm's required rate of return of 15 percent, and the project should not be accepted.

Thus, the project should be rejected since the NPV is negative, the PI is less than 1.0, and the IRR is less than the required rate of return of 15 percent.

SS-3. a.

$$\overline{X} = \sum_{i=1}^{N} X_i P(X_i)$$

$$\begin{aligned}\overline{X}_A &= 0.20(\$10{,}000) + 0.60(\$15{,}000) + 0.20(\$20{,}000)\\ &= \$2{,}000 + \$9{,}000 + \$4{,}000\\ &= \$15{,}000\end{aligned}$$

$$\begin{aligned}\overline{X}_B &= 0.20(\$6{,}000) + 0.60(\$18{,}000) + 0.20(\$30{,}000)\\ &= \$1{,}200 + \$10{,}800 + \$6{,}000\\ &= \$18{,}000\end{aligned}$$

b.

$$NPV = \sum_{t=1}^{n} \frac{CFAT_t}{(1+k^*)^t} - IO$$

$$\begin{aligned}NPV_A &= \$15{,}000(3.6048) - \$25{,}000\\ &= \$54{,}072 - \$25{,}000\\ &= \$29{,}072\end{aligned}$$

$$\begin{aligned}NPV_B &= \$18{,}000(3.1272) - \$25{,}000\\ &= \$56{,}289.60 - \$25{,}000\\ &= \$31{,}289.60\end{aligned}$$

SS-4. Project A:

Year	(A) Expected Cash Flow After Tax	(B)	(A · B) (Expected Cash Flow) × (α_t)	Present Value Factor at 8%	Present Value
0	–\$300,000	1.00	–\$300,000	1.000	–\$300,000.00
1	100,000	.95	95,000	0.9259	87,960.50
2	200,000	.90	180,000	0.8573	154,314.00
3	200,000	.85	170,000	0.7938	134,946.00
4	300,000	.80	240,000	0.7350	176,400.00
5	300,000	.75	225,000	0.6806	153,135.00
					NPV_A = \$406,755.50

Project B:

Year	(A) Expected Cash Flow After Tax	(B)	(A · B) (Expected Cash Flow) × (α_t)	Present Value Factor at 8%	Present Value
0	–\$300,000	1.00	–\$300,000	1.000	–\$300,000
1	200,000	.90	180,000	0.9259	166,662
2	200,000	.80	160,000	0.8573	137,168
3	200,000	.70	140,000	0.7938	111,132
4	300,000	.60	180,000	0.7350	132,300
5	400,000	.50	200,000	0.6806	136,120
					NPV_B = \$383,382

Thus, project A should be selected since it has the higher NPV.

APPENDIX 14A

THE TERMINAL CASH FLOW WHEN AN ASSET CLASS IS TERMINATED

As discussed earlier, the termination of a project triggers cash flows such as the salvage value, cash outlays associated with the project's termination and any recovery of nonexpense outlays that occurred at the project's initiation. In addition to these considerations, the sale of an asset(s) can terminate both the project and the asset class. For this case, we make the following adjustments.

First, equation (14-15) must be adjusted to reflect the termination of the asset class. The salvage value in equation (14-15) is now substituted with the UCC_n of the asset class. As a result, we now use the following equation:

$$PV\left(\begin{array}{c}\text{inherited tax}\\ \text{shields of purchased asset}\end{array}\right) = \left[\frac{(C_0)(d)(T)}{(k+d)}\right]\left[\frac{1+(.5)(k)}{(1+k)^n}\right] - \left[\frac{(UCC_n)(d)(T)}{(k+d)}\right]\left[\frac{1}{(1+k)^n}\right] \quad (14A\text{-}15)$$

where

$$UCC_n = \left[C_0\right]\left[1-\left(\frac{d}{2}\right)\right]\left[\left(1-d\right)^{(n-1)}\right] \quad (14A\text{-}1)$$

We also make adjustments to account for either recaptured depreciation or a terminal loss because the asset class is being terminated[11]. These adjustments can be summarized in three scenarios depending on whether the UCC_n of the asset class is greater, smaller or equal to the lesser of the salvage value or the capital cost of the asset.

Scenario (i) $SV_n > UCC_n$

If the salvage value is greater than the undepreciated capital cost of the asset class, the firm has written off the asset too rapidly. In this situation, Revenue Canada requires that the firm report a recaptured depreciation because the asset class is being terminated.

The amount of recaptured depreciation to be reported is determined from the difference between the undepreciated capital cost of the asset class and the lesser of either the salvage value (SV_n) or the original capital cost (C_0) of the asset to be salvaged.

$$\text{Recaptured depreciation} = UCC_n - \text{lesser of:}\left\{\begin{array}{l}C_0\\ SV_n\end{array}\right. \quad (14A\text{-}2)$$

The recaptured depreciation is added to the firm's taxable income since both the salvage value and the original capital cost are greater than the UCC_n of the asset class. If the firm has written off the asset too rapidly for the purposes of capital budgeting, the firm is required to subtract the portion of tax shields which the firm is not entitled to claim from the cash flows of the project. The value of these tax shields is discounted by the project's required rate of return to determine their present value. The present value of the tax shields from terminating the asset class under this scenario is determined as follows:

[11] It should be noted that Revenue Canada requires that recaptured depreciation be included as a source in calculating taxable income [ITA, 13(1)]; whereas, a terminal loss should be included as a deduction in calculating taxable income [ITA, 20(16)]. For the purposes of capital budgeting, we make these adjustments to the terminal cash flow in terms of tax shields. Some tax practitioners make these adjustments by multiplying either the deduction or source by the corporate tax rate; however, this approximation fails to consider the concept of tax shields or the concept of cash flow.

$$PV\left(\begin{array}{c}\text{tax shields from}\\ \text{recaptured depreciation}\end{array}\right) = \left[\frac{(\text{recaptured depreciation})(d)(T)}{(1+k)^n}\right] \quad (14A\text{-}3)$$

Since the lesser of either the salvage value or the capital cost is greater than UCC_n of the asset class, the amount will be negative, indicating that these tax shields will be subtracted from the cash flows of the project.

Scenario (ii) $SV_n < UCC_n$

In this scenario, the firm has written off the asset too slowly such that the salvage value is less than the undepreciated capital cost of the asset class. In this situation, Revenue Canada allows the firm to report a terminal loss because the asset class is being terminated.

The amount of terminal loss to be reclaimed is determined from the difference between the undepreciated capital cost (UCC_n) of the asset class and the salvage value (SV_n) of the asset to be salvaged.

$$\text{Terminal loss} = UCC_n - SV_n \quad (14A\text{-}4)$$

The terminal loss is subtracted from the firm's taxable income since the salvage value is less than the UCC_n of the asset class. If the firm has written off the asset too slowly for purposes of capital budgeting, the firm would add the portion of tax shields which the firm should have used to the cash flows of the project. The value of these tax shields is discounted by the project's required rate of return to determine their present value. The present value of the tax shields from terminating the asset class under this scenario is determined as follows:

$$PV(\text{tax shields from terminal loss}) = \left[\frac{(UCC_n - SV_n)(d)(T)}{(1+k)^n}\right] \quad (14A\text{-}5)$$

Since the salvage value is less than UCC_n, the amount will be positive, which indicates that these tax shields will be added to the cash flows of the project.

Scenario (iii) $UCC_n = SV_n$

In this scenario, the salvage value of the asset equals the UCC_n of the asset class and no adjustment to the tax shields is required.

The sale of the assets when terminating a project may also cause the firm to incur a capital gain which must be included in the terminal cash flow. Revenue Canada requires that corporations report three quarters of the capital gain as taxable income.

As result of these adjustments to the terminal cash flow, equation (14-16) is now restated as follows:[12]

$$\begin{aligned} NPV = {} & PV(\text{project's net operating income}) + \\ & PV(\text{inherited tax shields of purchased asset}) + \\ & PV(\text{salvage value}) + \\ & PV(\text{adjustment for the termination of the asset class}) + \quad (14A\text{-}6) \\ & PV(\text{recovery of working capital}) - IO. \end{aligned}$$

[12] Although these adjustments have been made for the terminal cash flow of an expansion project, these adjustments also hold for a replacement project. In this case, we consider the adjustments in terms of the incremental C_0 and the incremental UCC_n.

CAPITAL BUDGETING AND RISK ANALYSIS

CHAPTER 15

LEARNING OBJECTIVES

After reading this chapter you should be able to

1. Explain what the appropriate measure of risk is for capital-budgeting purposes.
2. Determine the acceptability of a new project using both the certainty equivalent and risk-adjusted discount rate methods of adjusting for risk.
3. Explain the use of simulation and probability trees for imitating the performance of a project under evaluation.
4. Explain the effect of inflation on the capital-budgeting decision.

INTRODUCTION

As international competition increases and technology changes at an ever quickening pace, risk and uncertainty play an increasingly important role in business decisions. In this chapter, we examine problems in measuring risk and approaches for dealing with it as it affects business decisions.

We look at problems similar to ones faced by the Chief Financial Officer at Merck & Co., the giant pharmaceutical firm. It takes an average of ten years and an investment of $359 million to bring a new drug to market. Quite an investment, particularly when you consider the uncertainty involving what the competition will be like in ten years and what constraints surrounding pricing might be in place as a result of health care reform. Moreover, not all new drugs introduced are profitable. In fact, only three out of ten new drugs that Merck introduces produce a positive net value. How should Merck evaluate projects with uncertain returns that stretch out well into the future? Certainly, they should not treat all projects in the same way, but how should they ensure that the decisions they make correctly reflect a project's uncertainty? Complicating Merck's task is the question of what is the appropriate measure of risk for a new project.

Axiom 1: The Risk-Return Tradeoff *states that investors demand a higher return for taking on additional risk; in this chapter, we modify our capital-budgeting criteria to allow for different levels of risk for different projects. In fact, in so doing we examine the experience of Merck and try to gain an understanding of how Merck deals with the risk and uncertainty that surround all its capital-budgeting decisions.*

CHAPTER PREVIEW

In this chapter, we examine how risk affects the decision rules used to invest in new projects. In our previous discussion of capital-budgeting techniques, we implicitly assumed that the level of risk associated with each investment was the same. In this chapter, we examine the capital-budgeting decision without this assumption. We begin with a discussion of what measure of risk is relevant in capital-budgeting decisions. We then consider various methods of incorporating risk into the capital-budgeting decision. Finally, we examine other approaches to evaluate risk in capital budgeting.

OBJECTIVE 1

RISK AND THE CAPITAL-BUDGETING DECISION

Up to this point we have ignored risk in the capital-budgeting decision; in fact, we have discounted expected cash flows back to the present and ignored any uncertainty that is associated with the discount factor.[1] In reality the future cash flows associated with the introduction of a new sales outlet or a new product are estimates of what is expected to happen in the future, not necessarily what will happen in the future.[2] In effect, the cash flows we have discounted back to the present have only been our best estimate of the expected future cash flows.[3] A cash flow diagram based on the possible outcomes of an investment proposal rather than the expected values of these outcomes appears in Figure 15-1.

In this section we will assume that risk exists; consequently, we do not know beforehand what cash flows will actually result from a new project. However, we do have expectations concerning the possible outcomes and are able to assign probabilities to these outcomes. Stated another way, although we do not know what the cash flows resulting from the acceptance of a new project will be, we can formulate the probability distributions from which the flows will be drawn.

Chapter 3 indicated that risk occurs if there is some question as to the future outcome of an event. We will now proceed with an examination of the logic behind this definition. Again, risk is defined as the potential variability in future cash flows.

The fact that variability reflects risk can easily be shown with a coin toss. Consider the possibility of flipping a coin—heads you win, tails you lose—for 25 cents with your finance professor. Most likely, you would be willing to take on this game because the utility gained from winning 25 cents is about equal to the utility lost if you lose 25 cents. Conversely, if the flip is for $1,000 you may be willing to play only if you are offered more than $1,000 if you win—say, you win $1,500 if it turns out heads and lose $1,000 if it turns out tails. In each case, the probability of winning and losing is the same; that is, there is an equal chance that the coin will land heads or tails. In each case, however, the width of the dispersion changes, which is why the second coin toss is more risky and why you may not take the chance unless the

[1]It is assumed that all cash flows mentioned here are after taxes.

[2]The capital budgeting process is dynamic and must constantly adjust to new forecasts of future cash flows. As the project moves forward, some uncertainty is resolved leading to estimates of future cash flows and subsequent cash re-evaluations of the project. A good example of a project bouncing back and forth between acceptable and unacceptable is the Quebec Hydro electric power project in Northern Quebec. While much of the uncertainty surrounding the decision to go or not to go with this project is centred on environmental concerns and demand forecasts, it still provides a good example of re-evaluating and reforecasting distribution of future events.

[3]Sometimes the risk associated with a project can be in the estimation of the initial outlay. The Eurotunnel project was initiated in 1987 at a cost of $8 billion. By April of 1990 the cost overruns had pushed the estimate to $12.33 billion and at one point in early 1990 had caused a freeze on bank credit, putting the project in jeopardy.

Figure 15-1 Cash Flow Diagram Based on Possible Outcomes

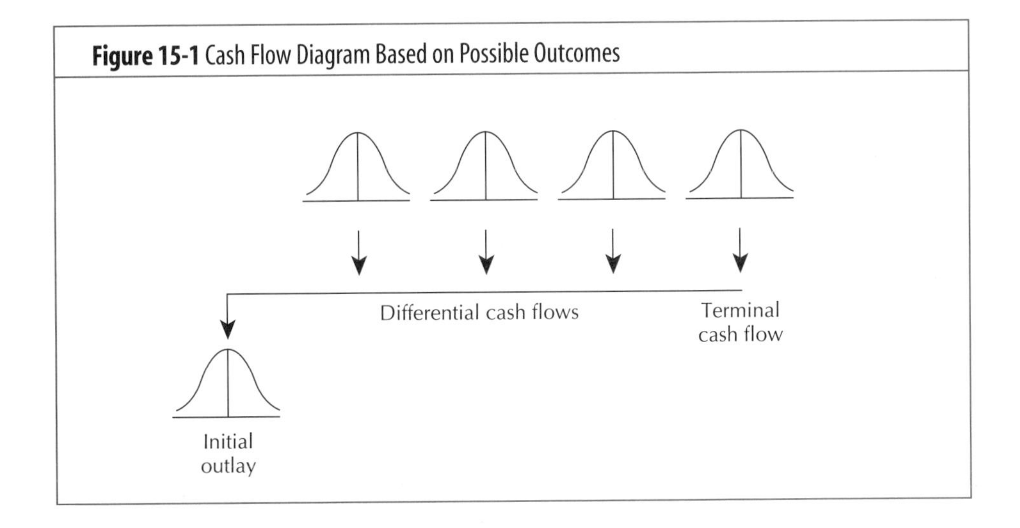

payoffs are altered. It should be noted that only the dispersion changes; the probability of winning or losing is the same in each case. Thus, the potential variability in the future reflects the risk.

In the remainder of this chapter we assume that although future cash flows are not known with certainty, the probability distribution from which they come is known. Since the dispersion of possible outcomes reflects risk, we are prepared to use a measure of dispersion or variability later in the chapter when we quantify risk.

In the pages that follow, there are only two basic issues that we address: (1) What is risk in terms of capital-budgeting decisions? and (2) How should risk be incorporated into the capital-budgeting process?

BACK TO THE BASICS

*Our discussion of capital budgeting and risk is based on **Axiom 9: All Risk Is Not Equal—Some Risk Can Be Diversified Away, and Some Cannot.** Axiom 9 describes how difficult it is to measure a project's risk as a result of diversification. The difficulty occurs because diversification takes place both within the firm and within the shareholder's portfolio. Diversification occurs within the firm if the new project is just one of many projects; whereas, diversification occurs within the shareholder's portfolio if the company's stock is just one of many stocks that the shareholder holds.*

What Measure of Risk is Relevant in Capital Budgeting?

Before we begin our discussion of how to adjust for risk, it is important to determine just what type of risk we are to adjust for. In capital budgeting, a project's risk can be looked at in three levels. First there is **project standing alone risk**, which is a project's risk that ignores the fact that much of this risk will be diversified away as the project is combined with the firm's other projects and assets. Second, we have the project's **contribution-to-firm risk**. This type of risk represents the amount of risk that the project contributes to the firm as a whole. This measure considers the fact that some of the project's risk will be diversified away as the project is combined with the firm's other projects and assets, but ignores the effect of any diversification by the firm's shareholders. Finally, there is **systematic risk**, which is the risk of the project from the point of view of a well-diversified shareholder. This measure considers the fact that some of a project's risk will be diversified away by shareholders as they combine this security with other securities in their portfolio. Figure 15-2 illustrates the three levels of risk.

Project standing alone risk

The risk of a project standing alone is measured by the variability of the project's expected returns. This type of risk considers the risk of a project, ignoring the fact that it is only one of many projects within the firm, and the firm's securities are but some of many securities within the shareholder's portfolio.

Project's contribution-to-firm risk

The amount of risk that a project contributes to the firm as a whole. This type of risk considers the effects of diversification among different projects within the firm, but ignores the effects of shareholders' diversification of their portfolios.

Systematic risk

The risk of a project measured from the point of view of a well-diversified shareholder. This type of risk considers that the project is only one of many projects within the firm, and the firm's securities are but some of many securities within the shareholder's portfolio.

Figure 15-2 Looking at Three Measures of a Project's Risk

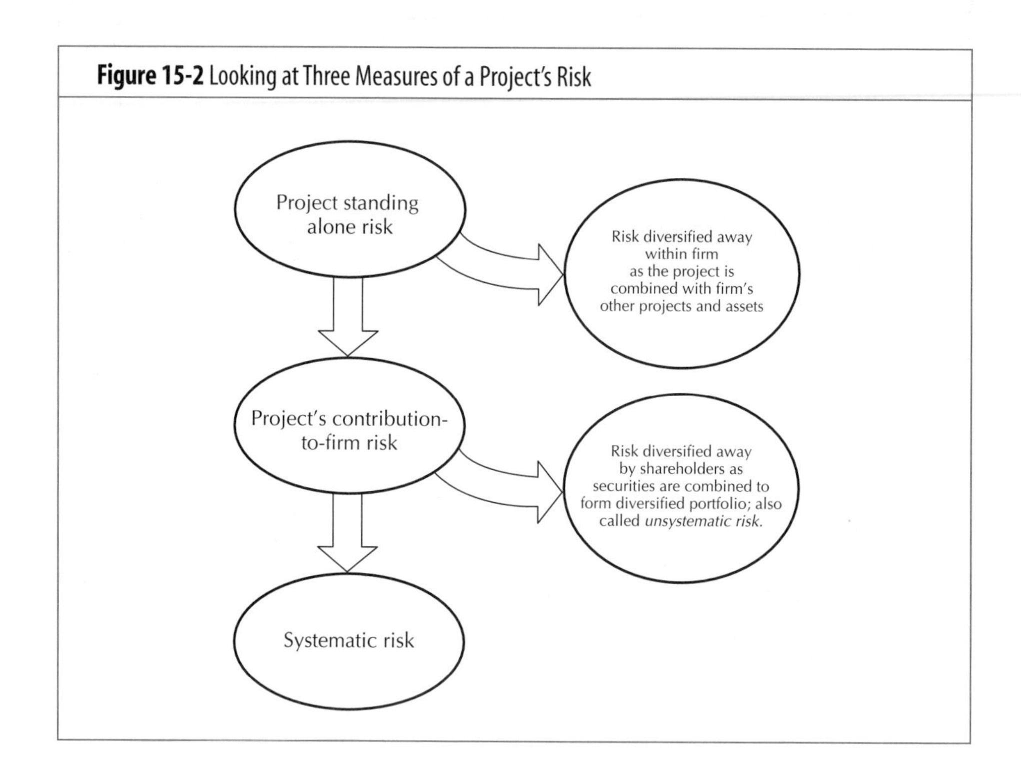

Should we be interested in total project risk? The answer is no. Perhaps the easiest way to understand why not is to look at an example. Let's take the case of research and development projects at Johnson & Johnson. Each year Johnson & Johnson takes on hundreds of new R&D projects, knowing that they only have about a 10 percent probability of being successful. If they are successful, the profits can be enormous; if they fail, the investment is lost. If we examined these R&D projects individually and measured their total project risk, we would have to judge them to be enormously risky. However, if we consider the effect of the diversification that comes about from taking on several hundred independent R&D projects a year, all with about a 10 percent chance of success, we can see that these R&D projects do not add much in the way of risk to Johnson & Johnson. In short, since much of a project's risk is diversified away within the firm, total project risk is an inappropriate measure of the meaningful level of risk of a capital-budgeting project.

Should we be interested in the project's contribution to the risk of the firm? Once again, the answer is no, provided investors are well-diversified and there is no chance of bankruptcy. From our earlier discussion of risk in Chapter 3, we saw that as a shareholder combined a security with other securities to form a diversified portfolio, much of the risk of this security would be diversified away. Thus, all that affects the shareholders is the systematic risk of the project, and as such is all that is theoretically relevant for capital budgeting.

Measuring Risk for Capital Budgeting Purposes and a Dose of Reality—Is Systematic Risk All There Is?

According to the CAPM, systematic risk is the only relevant risk for capital-budgeting purposes. The use of systematic risk in evaluating capital-budgeting decisions is complicated by the fact that the firm may have undiversified shareholders, including owners of small corporations. For these types of investors, the relevant measure of risk is the project's contribution to firm risk.

The possibility of bankruptcy also affects our view of what measure of risk is relevant. In the real world there is a cost associated with bankruptcy. First, if the firm fails, its assets in general cannot be sold for their true economic value. Moreover, the amount of money actually available for distribution to shareholders is further reduced by liquidation and legal fees that must be paid. Finally, the opportunity cost associated with the delays related to the legal process further reduces the funds available to the shareholder. Since costs are associated with bankruptcy, any reduction in the probability of a firm's bankruptcy has a very real value associated with it.

Indirect costs of bankruptcy also affect other areas of the firm including production, sales, and the quality and efficiency of management. For example, firms with a higher probability of bankruptcy may have a more difficult time in recruiting and retaining quality managers because jobs with that firm are viewed as being less than secure. Suppliers also may be less willing to sell on credit. Finally, customers may lose confidence and fear that the firm may not be around to honour the warranty or to supply the spare parts for the product in the future. In other words, as the probability of bankruptcy increases, the eventual bankruptcy may become self-fulfilling as potential customers and suppliers flee. Thus, the influence that a project has on possible bankruptcy costs which are incurred by a firm causes the project's contribution to the firm's risk to be a relevant measure of risk.

Finally, problems in measuring a project's systematic risk make its implementation extremely difficult. As we will see later on in this chapter, it is much easier talking about a project's systematic risk than measuring it.

Given all of these considerations, what do we use? The answer is that we will address both measures. We know in theory that systematic risk is correct. We also know that bankruptcy costs and undiversified shareholders violate the assumptions of the theory, which brings us back to the concept of a project's contribution to the firm's risk. Still, the concept of systematic risk holds value for capital budgeting decisions because shareholders are being compensated for taking on this type of risk. As such, we will concern ourselves with both the project's contribution to the firm's risk and the project's systematic risk without making any specific allocation of importance between the two for capital-budgeting purposes.

BACK TO THE BASICS

All the methods used to compensate for risk in capital budgeting find their roots in ***Axiom 1: The Risk-Return Tradeoff—We Won't Take on Additional Risk Unless We Expect to Be Compensated with Additional Return.*** *In fact, the risk-adjusted discount method puts this concept directly into play.*

METHODS FOR INCORPORATING RISK INTO CAPITAL BUDGETING

OBJECTIVE 2

In the preceding chapters we ignored any risk differences between projects. This assumption simplifies matters but it is not valid; different investment projects do, in fact, contain different levels of risk. We will now look at several methods for incorporating risk into the capital-budgeting process. The first technique, the certainty equivalent approach, attempts to incorporate the manager's utility function into the analysis. The second technique, the risk-adjusted discount rate, is based on the notion that investors require higher rates of return on more risky projects.

Certainty equivalent approach

A technique for incorporating risk into the capital-budgeting decision in which the decision maker substitutes a set of equivalent riskless cash flows for the expected cash flows and then discounts these cash flows back to the present.

Certainty Equivalent Approach

The **certainty equivalent approach** involves a direct attempt to allow the decision maker to incorporate utility function into the analysis. The financial executive is allowed to substitute the certain dollar amount that he or she feels is equivalent to the expected but risky cash flow offered by the investment for that risky cash flow in the capital-budgeting analysis. In effect, a set of riskless cash flows is substituted for the original risky cash flows.

Certainty equivalents

The amount of cash a person would require with certainty to make him or her indifferent between this certain sum and a particular risky or uncertain sum.

To illustrate the concept of a **certainty equivalent,** let us look at a simple coin toss. Assume you can play the game only once and if it comes out heads, you win $10,000, and if it comes out tails you win nothing. Obviously, you have a 50 percent chance of winning $10,000 and a 50 percent chance of winning nothing, with an expected value of $5,000. Thus, $5,000 is your uncertain expected value outcome. The certainty equivalent then becomes the amount you would demand in order to make you indifferent with regard to playing and not playing the game. If you are indifferent with respect to receiving $3,000 for certain and not playing the game, then $3,000 is the certainty equivalent.

In order to simplify future calculations and problems, let us define certainty equivalent coefficients (α_t) that represent the ratio of the certain outcome to the risky outcome, between which the financial manager is indifferent. In equation form, α_t can be represented as follows:

$$\alpha_t = \frac{\text{certain cash flow after tax}_t}{\text{risky cash flow after tax}_t} \tag{15-1}$$

Thus, the alphas can vary between 0, in the case of extreme risk, and 1, in the case of certainty. To obtain the value of the equivalent certain cash flow, we need only multiply the risky cash flow and the α_t. When this is done, we are indifferent with respect to the certain and the risky cash flows. In the preceding example of the simple coin toss, the certain cash flow was $3,000, while the risky cash flow was $5,000 (the expected value of the coin toss); thus, the certainty equivalent coefficient is 3,000/5,000 = .6. In summary, by multiplying the certainty equivalent coefficient (α_t) times the expected but risky cash flow, we can determine an equivalent certain cash flow.

Once this risk is taken out of the project's cash flows, those cash flows are discounted back to the present at the risk-free rate of interest, and the project's net present value or profitability index is determined. If the internal rate of return is calculated, it is then compared with the risk-free rate of interest rather than the firm's required rate of return in determining whether or not it should be accepted or rejected. The certainty equivalent method is based on the following equation:

$$NPV = \sum_{t=1}^{n} \frac{\alpha_t CFAT_t}{(1 + k_f)^t} - IO \tag{15-2}$$

where α_t = the certainty equivalent coefficient in time period t
$CFAT_t$ = the annual expected cash flow after tax in time period t
IO = the initial cash outlay
n = the project's expected life
k_f = the risk-free interest rate[4]

The certainty equivalent method is described by the following steps:

Step 1: Risk is removed from the cash flows by substituting equivalent certain cash flows for the risky cash flows. If the certainty equivalent coefficient (α_t) is given, this is done by multiplying each risky cash flow by the appropriate α_t value.

Step 2: These riskless cash flows are then discounted back to the present at the riskless rate of interest.

Step 3: The normal capital-budgeting criteria are then applied, except in the case of the internal rate of return criterion, where the project's internal rate of return is compared with the risk-free rate of interest rather than the firm's required rate of return.

EXAMPLE

A firm with a 10 percent required rate of return is considering building new research facilities with an expected life of five years. The initial outlay associated with this project involves a certain cash outflow of $120,000. The expected cash inflows and certainty equivalent coefficients, α_t, are as follows:

[4]Note that the risk-free interest rate is used to discount back the cash flows to the present and the risk of the cash flows is obtained b y the estimation of the certainty equivalent coefficient.

Year	Expected Cash Flow After Tax	Certainty Equivalent Coefficient, α_t
1	$10,000	.95
2	20,000	.90
3	40,000	.85
4	80,000	.75
5	80,000	.65

The risk-free rate of interest is 6 percent. What is the project's net present value?

To determine the net present value of this project using the certainty equivalent approach, we must first remove the risk from the future cash flows. We do so by multiplying each expected cash flow by the corresponding certainty equivalent coefficient, α_t.

Expected Cash Flow After Tax	Certainty Equivalent Coefficient, α_t	$\alpha_t \times$ (Expected Cash Flow After Tax) = Equivalent Riskless Cash Flow After Tax
$10,000	.95	$ 9,500
20,000	.90	18,000
40,000	.85	34,000
80,000	.75	60,000
80,000	.65	52,000

The equivalent riskless cash flows are then discounted back to the present at the riskless interest rate, not the firm's required rate of return. The required rate of return would be used if this project had the same level of risk as a typical project for this firm. However, these equivalent cash flows have no risk at all; hence, the appropriate discount rate is the riskless rate of interest. The equivalent riskless cash flows can be discounted back to present at the riskless rate of interest, 6 percent, as follows:

Year	Equivalent Riskless Cash Flow After Tax	Present Value Factor at 6 Percent	Present Value
1	$ 9,500	.9434	$ 8,962.30
2	18,000	.8900	16,020.00
3	34,000	.8396	28,546.40
4	60,000	.7921	47,526.00
5	52,000	.7473	38,859.60

$$NPV = -\$120,000 + \$8,962.30 + \$16,020.00 + \$28,546.40 + \$47,526.00 + \$38,859.60$$
$$= \$19,914.30$$

Applying the normal capital-budgeting decision criteria, we find that the project should be accepted, as its net present value is greater than zero.

Risk-Adjusted Discount Rates

The use of risk-adjusted discount rates is based on the assumption that investors demand higher returns for more risky projects. This assumption is also important to the CAPM. Figure 15-3 illustrates the relationship between risk and return.[5]

[5]Note that this relationship does not represent a continuous distribution but a discrete relationship between risk and return. In a continuous distribution, the probability of a range of outputs is given by the area under the curve over that range. However, for a continuous distribution, the probability of a return taking on a particular value, say 15 percent, is zero.

The required rate of return on any investment should include compensation for delaying consumption. As a result, the required rate of return is equal to the risk-free rate of return, plus compensation for any risk taken on. If the risk associated with the investment is greater than the risk involved in a typical endeavour, the discount rate is adjusted upward to compensate for this added risk. Once the firm determines the appropriate required rate of return for a project with a given level of risk, the cash flows are discounted back to the present at the **risk-adjusted discount rate**. Then the normal capital-budgeting criteria are applied, except in the case of the internal rate of return. In this particular case the hurdle rate becomes the risk-adjusted discount rate. Expressed mathematically, the net present value using the risk-adjusted discount rate becomes[6]

Risk-adjusted discount rate

A method for incorporating the project's level of risk into the capital-budgeting process, in which the discount rate is adjusted upward to compensate for higher than normal risk or downward to adjust for lower than normal risk.

$$NPV = \sum_{t=1}^{n} \frac{CFAT_t}{(1 + k^*)^t} - IO \tag{15-3}$$

where $CFAT_t$ = the annual expected cash flow after tax in time period t
IO = the initial cash outlay
k^* = the risk-adjusted discount rate
n = the project's expected life

The logic behind the risk-adjusted discount rate stems from the idea that if the level of risk in a project is different from that in the typical firm project, then management must incorporate the shareholders' probable reaction to this new endeavour into the decision-making process. If the project has more risk than a typical project, then a higher required rate of return should apply. Otherwise, marginal projects will lower the firm's share price—that is, reduce shareholders' wealth. This will occur as the market raises its required rate of return on the firm to reflect the addition of a more risky project, while the incremental cash flows resulting from the acceptance of the new project are not large enough to fully offset this change. By the same logic, if the project has less than normal risk, a reduction in the required rate of return is appropriate. Thus, the risk-adjusted discount method attempts to apply more stringent standards—that is, require a higher rate of return—to projects that will increase the firm's risk level.

Figure 15-3 Risk–Return Relationship

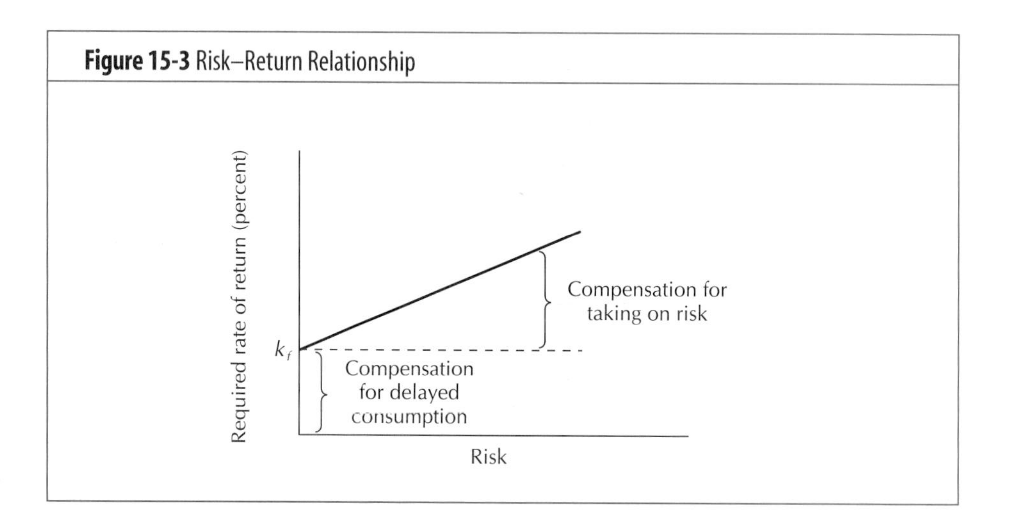

[6]Cash outflows as well as cash inflows, when not known with certainty, should be adjusted for risk. For example, the Eurotunnel project's initial outlay was spread over 5 years and had a good deal of uncertainty concerning what the actual cost would be. These cash outflows should also have been adjusted for uncertainty when evaluating the project.

EXAMPLE

A toy manufacturer is considering the introduction of a line of fishing equipment with an expected life of five years. In the past, this firm has been quite conservative in its investment in new products, sticking primarily to standard toys. In this context, the introduction of a line of fishing equipment is considered an abnormally risky project. Management feels that the normal required rate of return for the firm of 10 percent is not sufficient. Instead, the minimum acceptable rate of return on this project should be 15 percent. The initial outlay would be $110,000, and the expected cash flows after tax from this project are as given below:

Year	Expected Cash Flow After Tax
1	$30,000
2	30,000
3	30,000
4	30,000
5	30,000

Discounting this annuity back to the present at 15 percent yields a present value of the future cash flows of $100,560. Since the initial outlay on this project is $110,000, the net present value becomes –$9,440, and the project should be rejected. If the normal required rate of return of 10 percent had been used as the discount rate, the project would have been accepted with a net present value of $3,730.

In practice, when the risk-adjusted discount rate is used, projects are generally grouped according to purpose, or risk class; then the discount rate preassigned to that purpose or risk class is used.[7] For example, a firm with a required rate of return of 12 percent might use the following rate-of-return categorization:

Project	Required Rate of Return
Replacement decision	12%
Modification or expansion of existing product line	15
Project unrelated to current operations	18
Research and development operations	25

The purpose of this categorization of projects is to make their evaluation easier, but it also makes the evaluation less meaningful since the calculations for required rate of return are made on an arbitrary basis. The tradeoffs involved in the classification above are obvious: time and effort are minimized, but only at the cost of precision.

Certainty Equivalent versus Risk-Adjusted Discount Rate Methods

The primary difference between the certainty equivalent approach and the risk-adjusted discount rate approach involves the point at which the adjustment for

[7]Many firms actually calculate different risk-adjusted discount rates for different company divisions. Again, this procedure is far from exact, but it does recognize that different divisions have different costs of capital and different projects being considered by each division have different levels of risk.

risk is incorporated into the calculations. The certainty equivalent penalizes or adjusts downward the value of the expected annual cash flows, $CFAT_t$, which results in a lower net present value for a risky project. The risk-adjusted discount rate, on the other hand, leaves the cash flows at their expected value and adjusts the required rate of return, k, upward to compensate for added risk. In either case the project's net present value is being adjusted downward to compensate for additional risk. The computational differences are illustrated in Table 15-1.

In addition to the difference in point of adjustment for risk, the risk-adjusted discount rate makes the implicit assumption that risk becomes greater as we move further out in time. While this is not necessarily a good or bad assumption, we should be aware of it and understand it. Let's look at an example in which the risk-adjusted discount rate is used and then determine what certainty equivalent coefficients, α_t, would be necessary to arrive at the same solution.

EXAMPLE

Assume that a firm with a required rate of return of 10 percent is considering introducing a new product. This product has an initial outlay of $800,000, an expected life of 15 years, and cash flows after tax of $100,000 each year during its life. Because of the increased risk associated with this project, management is requiring a 15 percent rate of return. Let us also assume that the risk-free rate of return is 6 percent.

If the firm chose to use the certainty equivalent method, the certainty equivalent cash flows after tax would be discounted back to the present at 6 percent, the risk-free rate of interest. The present value of the $100,000 cash flow after tax occurring at the end of the first year discounted back to present at 15 percent is $87,000. The present value of this $100,000 flow discounted back to the present at the risk-free rate of 6 percent is $94,300. Thus, if the certainty equivalent approach were used, a certainty equivalent coefficient, α_1, of .9226 would be necessary to produce a present value of $87,000. In other words, the same results can be obtained in the first year by using the risk-adjusted discount rate and adjusting the discount rate up to 15 percent or by using the certainty equivalent approach and adjusting the expected cash flows by a certainty equivalent coefficient of .9226.

Under the risk-adjusted discount rate, the present value of the $100,000 cash flow occurring at the end of the second year becomes $75,600, and to produce an identical present value under the certainty equivalent approach a certainty equivalent coefficient of .8494 would be needed. Following this through for the life of the project yields the certainty equivalent coefficients given in Table 15-2.

Table 15-1 Computational Steps in the Certainty Equivalent and Risk-Adjusted Discount Rate Methods

Certainty Equivalent	Risk-Adjusted Discount Rate
STEP 1: Adjust the expected cash flows after tax, $CFAT_t$, downward for risk by multiplying them by the corresponding certainty equivalent, risk coefficient, α_t.	STEP 1: Adjust the discount rate upward for risk.
STEP 2: Discount the certainty equivalent, riskless, cash flows after tax back to the present using the risk-free rate of interest.	STEP 2: Discount the expected cash flows after tax back to present using the risk-adjusted discount rate.
STEP 3: Apply the normal decision criteria, except in the case of the internal rate of return, where the risk-free rate of interest replaces the required rate of return as the hurdle rate.	STEP 3: Apply the normal decision criteria, except in the case of the internal rate of return, where the risk-adjusted discount rate replaces the required rate of return as the hurdle rate.

Bennett Stewart on the Calculation of a Business Risk Index Used in Setting a Firm's Required Rate of Return

The calculation of the Business Risk Index focuses on a series of ratios grouped into four Risk Factors. The four Risk Factors found to be most important are:

1. Operating Risk
2. The Risk in Achieving Profitable Growth
3. Asset Management
4. Size and Diversity of Operations

Each of these is discussed in more detail in the following sections.

Operating Risk

The first set of ratios measures the risk in achieving a predictable level of operating profitability over time. It gauges investors' confidence (or lack thereof) in predicting future rates of return on capital already employed. When a unit's profitability is judged to be more unpredictable than that of other companies in the industry, its risk index, and thus its required return, are adjusted upward.

Risk in Achieving Profitable Growth

Our research demonstrates that unrealized investment opportunities create an additional risk for investors. As the perceived value of future investments changes, this leads to dramatic movements in stock prices and great risk for investors. For example, as IBM's success in personal computers became apparent, investors marked down the likely value from Digital Equipment Corporation's (DEC's) participation in the office of the future. The result: DEC's stock price fell by nearly $30. Similar dramatic stock price movements have been experienced by Apple Computer, Genentech, and Peoplexpress. The common attributes are rapid growth, great profit potential, and high P/E multiples.

It is precisely these attributes that our statistical research revealed as conveying high business risk. Accordingly, a single measure of risk is derived by weighting the unit's rate of return and growth rate. A business unit whose combined score is higher than that of its business peers will be deemed to have a higher price-earnings ratio and more risk in achieving continued profitable growth.

Asset Quality

The Risk Factor that measures the quality of a company's assets has four components: Working Capital Management, Plant Intensity, Plant Newness, and Useful Plant Life. Maintaining low and stable levels of working capital—inventory and receivables, less payables and accruals—relative to competitors demonstrates a superior budgeting and control capability. Managing working capital effectively also provides an additional source of cash during recessions and, in expansionary cycles, minimizes the need for external capital to grow. This is viewed positively by investors and results in a reduction in the company's BRI.

The extent to which a company is more plant-intensive than other companies in its business suggests that management has established a dominant position through a low-cost production capability. The fact that management is willing to commit a large amount of fixed capital is an indication of management confidence that results in a reduction in the Risk Index.

Also, the newer the plant, the less is the perceived risk in a company's assets. Older plants, because of their higher operating cost, generally are the first to be shut down in a recession, only to be reopened in expansionary times. As a result, companies operating older, less efficient plants will experience more volatile returns over a business cycle than companies with new plants. Thus, they are penalized with a higher risk index.

A related characteristic is useful plant life. Having to replace important assets frequently exposes the company to risk, and is a sign of rapid and often unpredictable technological change. In contrast, longer-lived assets are viewed by investors as providing operating stability and, hence, serve to reduce risk.

Size and Diversity

Absolute size of assets and the diversification of earnings sources tend to stabilize a company's performance. The size of the company often suggests the degree of market power it has in its industry, the length of its track record, all factors generally associated with lower business risk. Also, geographic diversification is viewed as mitigating overall business risk. The reason is that a company's investment in other countries reduces investors' exposure to business cycles in the United States.

Source: Bennett Stewart, "Commentary: A Framework for Setting Required Rates of Return by Line of Business," in *Six Roundtable Discussions of Corporate Finance with Joel Stern*, ed. Donald H. Chew, Jr. (New York: Quorum Books, 1986).

Table 15-2 Certainty Equivalent Coefficients Yielding the Same Results as the Risk-Adjusted Discount Rate of 15 Percent in the Illustrative Example

Year	1	2	3	4	5	6	7	8	9	10
α_t:	.9226	.8494	.7833	.7222	.6653	.6128	.5654	.5215	.4797	.4427

What does this analysis suggest? It indicates that if the risk-adjusted discount rate method is used, we are adjusting downward the value of future cash flows that occur further in the future more severely than earlier cash flows. We can easily see this by comparing Equations (15-2) and (15-3). The net present value using the risk-adjusted discount rate is expressed as

$$NPV = \sum_{t=1}^{n} \frac{CFAT_t}{(1 + k^*)^t} - IO \qquad (15\text{-}3)$$

and the certainty equivalent net present value is measured by

$$NPV = \sum_{t=1}^{n} \frac{\alpha_t CFAT_t}{(1 + k_f)^t} - IO \qquad (15\text{-}2)$$

Thus, for the net present values under each approach to be equivalent, the following must be true:

$$\frac{CFAT_t}{(1 + k^*)^t} = \frac{\alpha_t \, CFAT_t}{(1 + k_f)^t} \qquad (15\text{-}4)$$

Solving for α_t yields

$$\alpha_t = \frac{(1 + k_f)^t}{(1 + k^*)^t} = \left(\frac{1 + k_f}{1 + k^*}\right)^t \qquad (15\text{-}5)$$

Thus, since k_f and k^* are constants and k^* is greater than k_f, the value of α_t, or $[(1 + k_f)/(1 + k^*)]^t$, must decrease as t increases for the present values under each approach to be equivalent. This is exactly what we concluded from Table 15-2.

In summary, the use of the risk-adjusted discount rate assumes that risk increases over time and that cash flows occurring further in the future should be more severely penalized.

PERSPECTIVE IN FINANCE

If performed properly either of these methods can do a good job of adjusting for risk. However, by far, the most popular method of risk adjustment is the risk-adjusted discount rate. The reason for the popularity of the risk-adjusted discount rate over the certainty equivalent approach is purely and simply its ease of implementation.

The Risk-Adjusted Discount Rate and Measuring a Project's Systematic Risk

When we initially talked about systematic risk or beta, we were talking about measuring it for the entire firm. As you recall, while we could estimate a firm's beta using historical data, we did not have complete confidence in our results. As we will see, estimating the appropriate level of systematic risk for a single project is even more fraught with difficulties. To truly understand what it is that we are trying to do and the difficulties that we will encounter let us step back a bit and examine systematic risk and the risk adjustment for a project.

What we are trying to do is to use the CAPM to determine the level of risk and the appropriate risk-return tradeoffs for a particular project. We will then take the expected return on this project and compare it with the risk-return tradeoffs suggested by the CAPM to determine whether or not the project should be accepted. If the project appears to be a typical one for the firm, using the CAPM to determine the appropriate risk-return tradeoffs and then judging the project against them may be a warranted approach. But if the project is not a typical project, what do we do? Historical data generally do not exist for a new project. In fact, for some capital investments (for example, a truck or a new building) historical data would not have much meaning. What we need to do is make the best out of a bad situation. We either (1) use historical accounting data, if available, to substitute for historical price data in estimating systematic risk, or (2) attempt to find a substitute firm in the same industry as the capital-budgeting project and use the substitute firm's estimated systematic risk as a proxy for the project's systematic risk.

Beta Estimation Using Accounting Data

When we are dealing with a project that is identical to the firm's other project, we need only estimate the level of systematic risk for the firm and use that estimate as a proxy for the project's risk. Unfortunately, when projects are not typical of the firm this approach does not work. For example, when R. J. Reynolds introduces a new food through one of its food products divisions, this new product most likely carries with it a different level of systematic risk than is typical for Reynolds as a whole.

To get a better approximation of the systematic risk level on this project we will estimate the level of systematic risk for the food division and use that as a proxy for the project's systematic risk. Unfortunately, historical stock price data are available only for the company as a whole, and as you recall historical stock return data are generally used to estimate a firm's beta. Thus, we are forced to use accounting return data rather than historical stock return data for the division to estimate the division's systematic risk. To estimate a project's beta using accounting data we need only run a time series regression of the division's return on assets (net income/total assets) on the market index (the TSE 300 Index). The regression coefficient from this equation would be the project's accounting beta and would serve as an approximation for the project's true beta or measure of systematic risk. Alternatively, a multiple regression model based on accounting data could be developed to explain betas. The results of this model could then be applied to firms which are not publicly traded to estimate their betas.

How good is the accounting beta technique? It certainly is not as good as a direct calculation of the beta. In fact, one study has indicated that the correlation between the accounting beta for earnings and the market beta was in the range of 0.4 and 0.7; however, better luck has been experienced with multiple regression models used to predict betas.[8] Unfortunately, in many cases there may not be any realistic alternative to the calculation of the accounting beta. Owing to the importance of adjusting for a project's risk, the accounting beta method is a preferred alternative to doing nothing.

Pure play method

A method of estimating a project's beta that attempts to identify a publicly traded firm that is engaged solely in the same business as the project and uses that beta as a proxy for the project's beta.

The Pure Play Method for Estimating a Project's Beta

Whereas the accounting beta method attempts to directly estimate a project or division's beta, the **pure play method** attempts to identify publicly traded firms

[8]For further information on the market-determined and accounting-determined risk measures, refer to W. Beaver, P. Kettler, and M. Scholes, "The Association Between Market Determined and Accounting Determined Risk Measures," *The Accounting Review* 45 (Oct. 1970), pp. 654-682.

that are engaged solely in the same business as the project or division. Once the proxy or pure play firm is identified, its systematic risk is determined and then used as a proxy for the project or division's level of systematic risk. What we are doing is looking for a publicly traded firm on the outside that looks like our project and using that firm's required rate of return to judge our project. In doing so we are presuming that the systematic risk and the capital structure of the proxy firm are identical to those of the project.[9]

In using the pure play method it should be noted that a firm's capital structure is reflected in its beta. When the capital structure of the proxy firm is different from that of the project's firm, some adjustment must be made for this difference. Although not a perfect approach, it does provide some insights as to the level of systematic risk a project might have. In Chapter 12 we saw how PepsiCo uses the pure play method to calculate a division's beta and required rate of return.

OBJECTIVE 3

OTHER APPROACHES TO EVALUATING RISK IN CAPITAL BUDGETING

Simulation

Simulation

The process of imitating the performance of an investment project under evaluation using a computer. This process is accomplished by randomly selecting observations from each of the distributions that affect the outcome of the project, combining those observations to determine the final output of the project, and continuing with this process until a representative record of the project's probable outcome is assembled.

Another method for evaluating the effects of risk on an investment decision is through the use of **simulation**. Whereas the certainty equivalent and risk-adjusted discount rate approaches provided us with a single value for the risk-adjusted net present value, a simulation approach gives us a probability distribution for the investment's net present value or internal rate of return. Simulation imitates the performance of the project under evaluation. This is done by randomly selecting observations from each of the assumed probability distributions for factors that affect the outcome of the project, combining those observations to determine the final output of the project, and continuing with this process until a representative record of the project's probable outcome is assembled.

The easiest way to develop an understanding of the computer simulation process is to follow through an example simulation for an investment project evaluation. Suppose a chemical producer is considering an extension to its processing plant. The simulation process is portrayed in Figure 15-4. First the probability distributions are determined for all the factors that affect the project's returns; in this case, let us assume there are nine such variables:

1. Market size
2. Selling price
3. Market growth rate
4. Share of market (which results in physical sales volume)
5. Investment required
6. Residual value of investment
7. Operating costs
8. Fixed costs
9. Useful life of facilities

Then the computer randomly selects one observation from each of the probability distributions, according to its chance of actually occurring in the future. These nine observations are combined, and a net present value or internal rate of return figure is calculated. This process is repeated as many times as desired, until a representative distribution of possible future outcomes is assembled. Thus, the inputs to a simulation include all the principal factors affecting the project's profitability, and

[9]While portfolio betas tend to be stable, individual betas are not necessarily stable and not always meaningful. In fact, many individual securities have coefficients of determination of 10 percent or less, indicating that the characteristic line with a slope of beta explains only 10 percent or less of the variability of the security's returns.

the simulation output is a probability distribution of net present values or internal rates of return for the project. The decision maker bases the decision on the full range of possible outcomes. The project is accepted if the decision maker feels that enough of the distribution lies above the normal cutoff criteria ($NPV \geq 0$, $IRR \geq$ required rate of return).

Figure 15-4 Capital-Budgeting Simulation

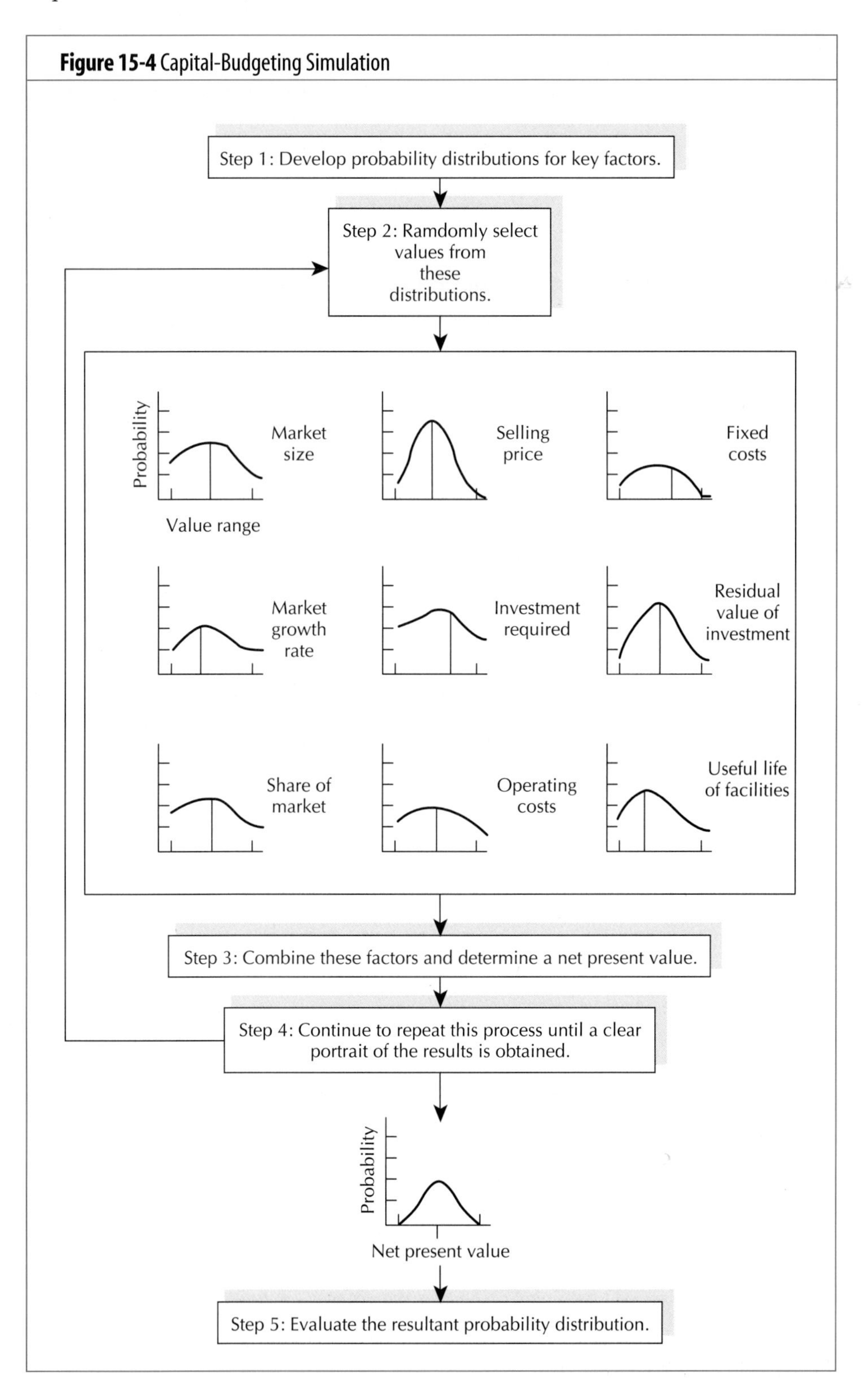

Suppose that the output from the simulation of a chemical producer's project is as given in Figure 15-5. This output provides the decision maker with the probability of different outcomes occurring in addition to the range of possible outcomes. Sometimes called **scenario analysis**, this examination identifies the range of possible outcomes under the worst, best, and most likely case. The firm's management will examine the distribution to determine the project's level of risk, and then make the appropriate adjustment.

Scenario analysis

Simulation analysis that focuses on an examination of the range of possible outcomes.

You'll notice that while the simulation approach helps us to determine the amount of total risk that a project has, it does not differentiate between systematic and unsystematic risk. Since systematic risk cannot be diversified away for free, the simulation approach does not provide a complete method of assessment. However, it does provide important insights as to the total risk level of a given investment project. Now we will look briefly at how the simulation approach can be used to perform sensitivity analysis.

Sensitivity Analysis Through the Simulation Approach

Sensitivity analysis involves determining how the distribution of possible net present values or internal rates of return for a particular project is affected by a change in one particular input variable. This is done by changing the value of one input variable while holding all other input variables constant. The distribution of possible net present values or internal rates of return that is generated is then compared with the distribution of possible returns generated before the change was made to determine the effect of the change. For this reason, sensitivity analysis is commonly called *"What if?" Analysis.*

Sensitivity analysis

The process of determining how the distribution of possible returns for a particular project is affected by a change in one particular input variable.

For example, the chemical producer that was considering a possible expansion to its plant may wish to determine the effect of a more pessimistic forecast of the anticipated market growth rate. After the more pessimistic forecast replaces the original forecast in the model, the simulation is rerun. The two outputs are then compared to determine how sensitive the results are to the estimate of the revised market growth rate.

By modifying assumptions made about the values and ranges of the input factors and rerunning the simulation, management can determine how sensitive the outcome of the project is to these changes. If the output appears to be highly sensitive to one or two of the input factors, financial managers may then wish to spend additional time refining those input estimates to make sure they are accurate.

Probability Trees

A **probability tree** is a graphic exposition of the sequence of possible outcomes; it presents the decision maker with a schematic representation of the problem in which all possible outcomes are illustrated. Moreover, the computations and

Probability tree

A schematic representation of a problem in which all possible outcomes are graphically displayed.

Figure 15-5 Output from Simulation

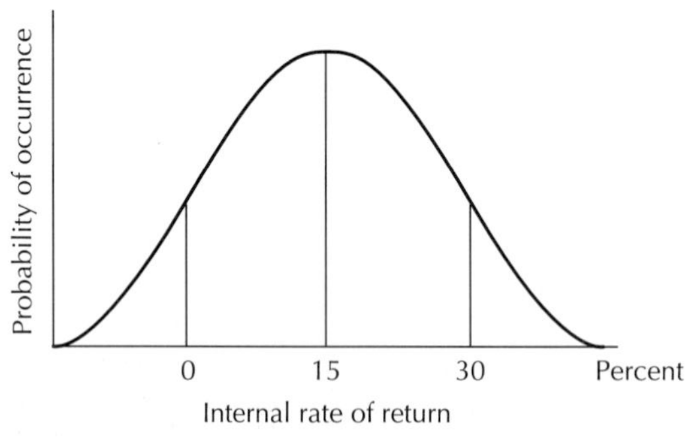

Probability of occurrence

0 15 30 Percent

Internal rate of return

results of the computations are shown directly on the tree, so that the information can be easily understood.

To illustrate the use of a probability tree, suppose a firm is considering an investment proposal that requires an initial outlay of $1 million and will yield resultant cash flows for the next two years. Table 15-3 shows that we have assumed that there are three possible outcomes during the first year. Figure 15-6 illustrates each of these three possible alternatives on the probability tree as one of the three possible branches.

The second step in the probability tree is to continue drawing branches in a similar manner so that each of the possible outcomes during the second year is represented by a new branch. For example, if outcome 1 occurs in year 1, a 20 percent chance of a $300,000 cash flow and an 80 percent chance of a $600,000 cash flow in year 2 have been projected. Two branches would be sent out from the outcome 1 node, reflecting these two possible outcomes. The cash flows that occur if outcome 1 takes place and the probabilities associated with them are called *conditional outcomes* and *conditional probabilities* because they can occur only if outcome 1 occurs during the first year. Finally, to determine the probability of the sequence of a $600,000 flow in year 1 and a $300,000 outcome in year 2, the probability of the $600,000 flow (.5) is multiplied by the conditional probability of the second flow (.2), telling us that this sequence has a 10 percent chance of occurring; this probability is called its **joint probability**. Letting the values in Table 15-4 represent the conditional outcomes and their respective conditional probabilities, we can complete the probability tree, as shown in Figure 15-7.

Joint probability

The probability of two different sequential outcomes occurring.

The financial executive, by examining the probability tree, is provided with the expected internal rate of return for the investment, the range of possible outcomes, and a listing of each possible outcome with the probability associated with it. In this case, the expected internal rate of return is 14.74 percent, and there is a 10 percent chance of incurring the worst possible outcome with an internal rate of return of –7.55 percent. There is a 2 percent probability of achieving the most favourable outcome, an internal rate of return of 37.98 percent.

Decision making with probability trees does not mean simply the acceptance of any project with an internal rate of return greater than the firm's required rate of

Table 15-3 Possible Outcomes in Year 1

		Probability	
	.5	.3	.2
	Outcome 1	Outcome 2	Outcome 3
Cash flow after tax	$600,000	$700,000	$800,000

Figure 15-6 First Stage of a Probability Tree Diagram

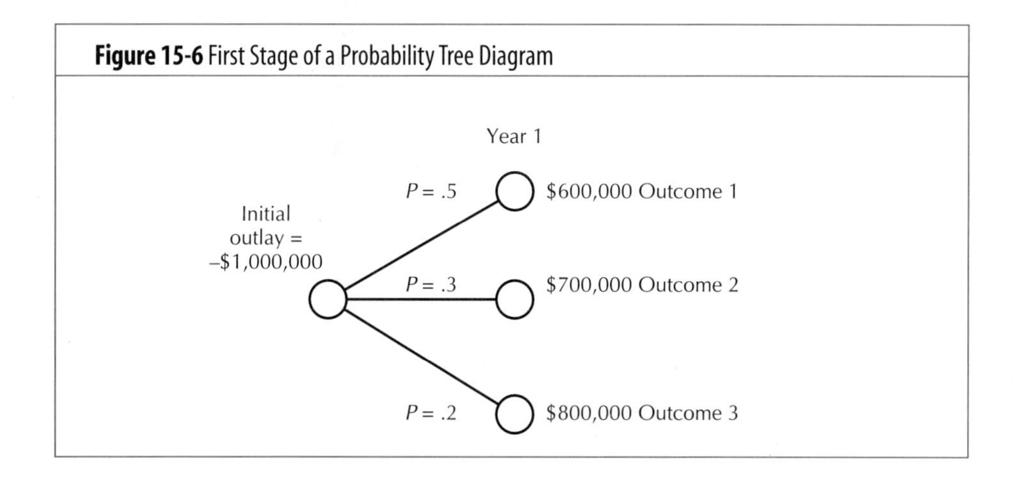

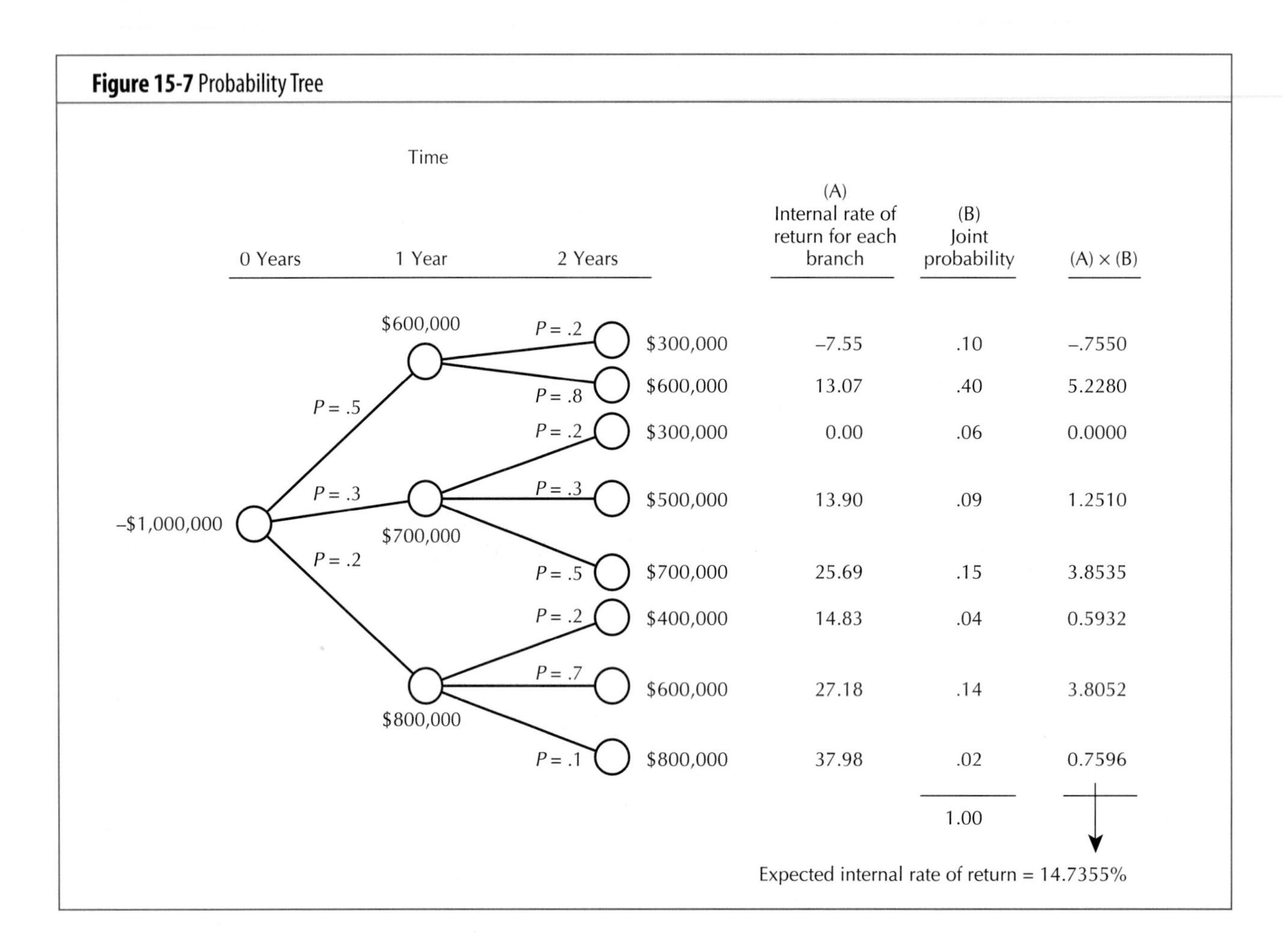

Figure 15-7 Probability Tree

return, because the project's required rate of return has not yet been adjusted for risk. As a result, the financial decision maker must examine the entire distribution of possible internal rates of return and then, based on that examination, decide, given her or his aversion to risk, if enough of this distribution is above the appropriate (risk-adjusted) required rate of return to warrant acceptance of the project. Thus, the probability tree allows the executive to quickly visualize the possible future events, their probabilities, and their outcomes. In addition, the calculation of the expected internal rate of return and enumeration of the distribution should aid the financial executive in determining the risk level of the project.

Table 15-4 Conditional Outcomes and Probabilities for Year 2

	If		If		If	
	Outcome 1		Outcome 2		Outcome 3	
Year 1	$CFAT_1$ = $600,000		$CFAT_1$ = $700,000		$CFAT_1$ = $800,000	
	Then		Then		Then	
Year 2	$CFAT_2$	Probability	$CFAT_2$	Probability	$CFAT_2$	Probability
	$300,000	.2	$300,000	.2	$400,000	.2
	600,000	.8	500,000	.3	600,000	.7
			700,000	.5	800,000	.1

OTHER SOURCES AND MEASURES OF RISK

Time Dependence of Cash Flows

Up to this point, in all approaches other than the probability tree we have assumed that the cash flow in one period is independent of the cash flow in the previous period. While this assumption is appealing because of its simplifying nature, in many cases it is also invalid. For example, if a new product is introduced and the initial public reaction is poor, resulting in low initial cash flows, then cash flows in future periods are likely to be low. An extreme example of this is Ford's experience with the Edsel. Poor consumer acceptance and sales in the first year were followed by even poorer results in the second year. If the Edsel had been received favourably during its first year, it quite likely would have done well in the second year. The end effect of time dependence of cash flows is to increase the risk of the project over time. That is, since large cash flows in the first period lead to large cash flows in the second period, and low cash flows in the first period lead to low cash flows in the second period, the probability distribution of possible net present values tends to be wider than if the cash flows were not dependent over time. The greater the degree of correlation between flows over time, the greater will be the dispersion of the probability distribution.

Skewness

In all previous approaches other than simulation and probability trees, we have assumed that the distributions of net present values for projects being evaluated are normally distributed. This assumption is not always valid. When it is true, the standard deviation provides an adequate measure of the distribution's dispersion; however, when the distribution is not normally distributed, reliance on the standard deviation can be misleading.

A distribution that is not symmetric is said to be *skewed.* A **skewed distribution** has either a longer "tail" to the right or to the left. For example, if a distribution is skewed to the right, most values will be clustered around the left end, and the distribution will apear to have a long tail on the right end of the range of values. Graphic illustrations of skewed distributions are presented in Figure 15-8.

Skewed distribution

A distribution that has a longer "tail" to the right or the left.

The difficulty associated with skewed distributions arises from the fact that the use of the expected value and standard deviation alone may not be enough to differentiate properly between two distributions. For example, the two distributions shown in Figure 15-9 have the same expected value and the same standard deviation; however, distribution A is skewed to the left while B is skewed to the right. If a financial manager were given a choice between these two distributions, assuming they are mutually exclusive, he or she would most likely choose distribution B, because it involves less chance of a negative net present value. Thus, skewness can affect the level of risk and the desirability of a distribution.

INFLATION AND CAPITAL-BUDGETING DECISIONS

OBJECTIVE 4

It is always important to consider the effect of inflation in capital-budgeting decisions. While every project is affected a little differently by inflation, there are four general ways in which inflation can affect capital-budgeting decisions:

1. Increased inflation will cause the required rate of return on the project to rise. As cash flows received in the future will buy less with increased inflation, investors will demand a higher required rate of return on funds invested. This is called the

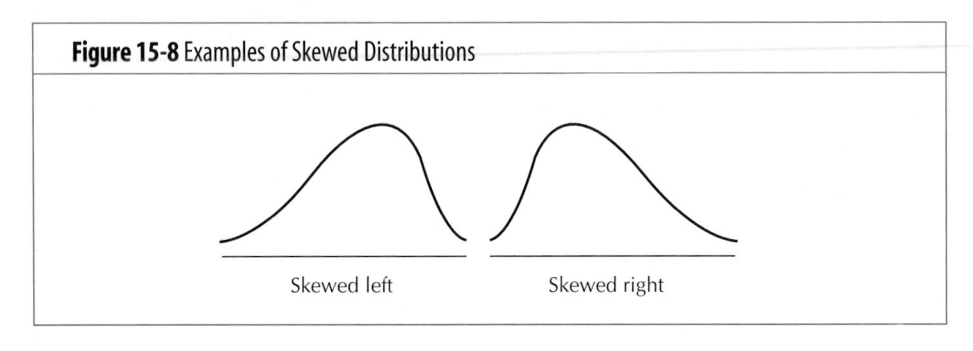

Figure 15-8 Examples of Skewed Distributions

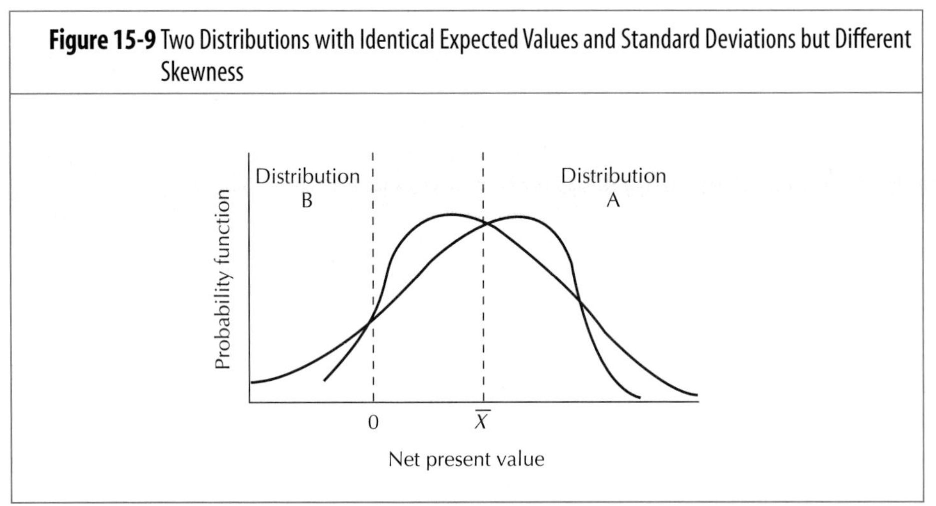

Figure 15-9 Two Distributions with Identical Expected Values and Standard Deviations but Different Skewness

Fisher Effect and was referred to in Chapter 2. It states that the required rate of return on a project is

$$k_j = k_j^* + \rho \tag{15-6}$$

where k_j is the required rate of return in nominal terms, k_j^* is the required rate of return in real terms, and ρ is the weighted average anticipated inflation rate over the life of the project.[10]

2. Both anticipated cash inflows and outflows could be affected by inflation. Inflation may well affect materials, wages, sales, and administrative expenses, while also affecting the selling price of the item being produced. It is obviously impossible to generalize on exactly how it will affect these items for all projects, but it is safe to assume it could have a significant impact on both cash inflows and outflows.
3. The salvage value of the project could also be affected by inflation. Again, it is impossible to generalize on exactly what the relationship will be on all capital-budgeting projects.
4. The fact that capital cost allowance (CCA) does not change with inflation also distorts capital-budgeting decisions. Since CCA charges are based upon original costs rather than replacement costs, they are not affected by inflation. As such, if inflation increases while CCA remains constant, a smaller percentage of the cash inflows is sheltered by CCAs and thus a larger percentage is taxed. The end result of this is that the real cash flows decline over time as taxes increase faster than inflation because everything but CCA is free to rise with inflation.

[10]Actually $(1 + k_j) = (1 + k_j^*)(1 + \rho)$; thus, $k_j = k_j^* + \rho + k_j^*\rho$. However, since $k_j^*\rho$ is assumed to be extremely small and inconsequential, it is generally dropped from consideration.

How can financial executives adjust capital-budgeting decisions for inflation? They must anticipate and include any effects of expected inflation in the cash flow estimates and also include it in the estimate of the required rate of return. Since it does affect the magnitude of the cash flows, it must be allowed for in order to produce an appropriate capital-budgeting decision.

SUMMARY

In this chapter, we examined the problem of incorporating risk into the capital-budgeting decision. First we explored just what type of risk to adjust for: project standing alone risk, the project's contribution to firm risk and the project's systematic risk. In theory, systematic risk is the appropriate risk measure, but bankruptcy costs and the issue of undiversified shareholders also give weight to considering a project's contribution to firm risk as the appropriate risk measure. Both measures of risk have merit and we avoid making any specific allocation of importance between the two in capital budgeting.

Two commonly used methods for incorporating risk into capital budgeting are (1) the certainty equivalent method and (2) risk-adjusted discount rates. The certainty equivalent approach involves a direct attempt to incorporate the decision maker's utility function into the analysis. Under this method, cash flows are adjusted downward by multiplying them by certainty equivalent coefficients, α_t, which transform the risky cash flows into equivalent certain cash flows in terms of desirability. A project's net present value using the certainty equivalent method for adjusting for risk becomes

$$NPV = \sum_{t=1}^{n} \frac{\alpha_t CFAT_t}{(1 + k_f)^t} - IO$$

The risk-adjusted discount involves an upward adjustment of the discount rate to compensate for risk. This method is based on the concept that investors demand higher returns for riskier projects. A project's net present value using the risk-adjusted discount method for adjusting for risk becomes

$$NPV = \sum_{t=1}^{n} \frac{CFAT_t}{(1 + k^*)^t} - IO$$

The simulation and probability tree methods are used to provide information as to the location and shape of the distribution of possible outcomes. Decisions could be based directly on these methods, or they could be used to determine input into either certainty equivalent or risk-adjusted discount method approaches.

Inflation does affect the capital budgeting decision through the following variables: the required rate of return on a project, the anticipated cash inflows and outflows, and the salvage value of the project.

STUDY QUESTIONS

15-1. In Chapter 13 we examined the payback period capital-budgeting criterion. Often this capital-budgeting criterion is used as a risk-screening device. Explain the rationale behind its use.

15-2. The use of the risk-adjusted discount rate assumes that risk increases over time. Justify this assumption.

15-3. What are the similarities and differences between the risk-adjusted discount rate and certainty equivalent methods for incorporating risk into the capital-budgeting decision?

15-4. What is the value of using the probability tree technique for evaluating capital-budgeting projects?

15-5. Explain how simulation works. What is the value in using a simulation approach?

15-6. What does time dependence of cash flows mean? Why might cash flows be time dependent? Give some examples.

15-7. What does skewness mean? If a distribution is skewed, how does this affect the significance of its standard deviation and mean?

SELF-TEST PROBLEMS

ST-1. The N. Sen Corp. is considering two mutually exclusive projects. Both require an initial outlay of $25,000 and will operate for five years. The probability distributions associated with each project for years 1 through 5 are given below:

Net Cash Flow After Tax, Years 1–5

Project A		Project B	
Probability	Cash Flow After Tax	Probability	Cash Flow After Tax
.20	$10,000	.20	$ 6,000
.60	15,000	.60	18,000
.20	20,000	.20	30,000

Since project B is the riskier of the two projects, the management of N. Sen has decided to apply a required rate of return of 18 percent to its evaluation, but only a 12 percent required rate of return to project A.

a. Determine the expected value of each project's net cash flows after tax.

b. Determine each project's risk-adjusted net present value.

ST-2. G. Norohna and Co. is considering two mutually exclusive projects. The expected values for each project's net cash flows after tax are given below.

Year	Project A	Project B
0	–$300,000	–$300,000
1	100,000	200,000
2	200,000	200,000
3	200,000	200,000
4	300,000	300,000
5	300,000	400,000

The company has decided to evaluate these projects using the certainty equivalent method. The certainty equivalent coefficients for each project's net cash flows after tax are given below.

Year	Project A	Project B
0	1.00	1.00
1 .	95	.90
2 .	90	.80
3	.85	.70
4	.80	.60
5	.75	.50

Given that this company's normal required rate of return is 15 percent and the after-tax risk-free rate is 8 percent, which project should be selected?

STUDY PROBLEMS (SET A)

15-1A. *(Risk-Adjusted NPV)* The MCAT Corporation is considering two mutually exclusive projects. Both require an initial outlay of $10,000 and will operate for five years. For years 1 through 5, the probability distributions associated with annual net cash flow after tax for each project are given as follows:

Probability Distribution for Net Cash Flows After Tax, Years 1-5

Project A		Project B	
Probability	Cash Flow	Probability	Cash Flow
.15	$ 4,000	.15	$ 2,000
.70	5,000	.70	6,000
.15	6,000	.15	10,000

Because project B is the riskier of the two projects, the management of MCAT Corporation has decided to apply a required rate of return of 15 percent to its evaluation but only a 12 percent required rate of return to project A.

a. Determine the expected value of each project's cash flows.

b. Determine each project's risk-adjusted net present value.

c. What other factors might be considered in deciding between these two projects?

15-2A. *(Risk-Adjusted NPV)* The Goblu Corporation is evaluating two mutually exclusive projects, both of which require an initial outlay of $100,000. Each project has an expected life of five years. The probability distributions associated with the annual net cash flows after tax from each project are given below:

Net Cash Flow After Tax, Years 1–5

Project A		Project B	
Probability	Cash Flow After Tax	Probability	Cash Flow After Tax
.10	$35,000	.10	$10,000
.40	40,000	.20	30,000
.40	45,000	.40	45,000
.10	50,000	.20	60,000
		.10	80,000

The normal required rate of return for Goblu is 10 percent, but since these projects are riskier than most, they are requiring a higher-than-normal rate of return on them. On project A they are requiring a 12 percent, and on project B a 13 percent rate of return.

a. Determine the expected value for each project's net cash flows after tax.

b. Determine each project's risk-adjusted net present value.

c. What other factors might be considered in deciding between these projects?

15-3A. *(Certainty Equivalents)* The V. Coles Corp. is considering two mutually exclusive projects. The expected values for each project's net cash flows after tax are given below:

Year	Project A	Project B
0	–$1,000,000	–$1,000,000
1	500,000	500,000
2	700,000	600,000
3	600,000	700,000
4	500,000	800,000

The management has decided to evaluate these projects using the certainty equivalent method. The certainty equivalent coefficients for each project's net cash flows after tax are given below:

Year	Project A	Project B
0	1.00	1.00
1	.95	.90
2	.90	.70
3	.80	.60
4	.70	.50

Given that this company's normal required rate of return is 15 percent and the risk-free rate is 5 percent, which project should be selected?

15-4A. *(Certainty Equivalents)* Neustal Inc. has decided to use the certainty equivalent method in determining whether or not a new investment should be made. The expected net cash flows after tax associated with this investment and the estimated certainty equivalent coefficients are as follows:

Year	Expected Values for Cash Flows After Tax	Certainty Equivalent Coefficients
0	−$90,000	1.00
1	25,000	0.95
2	30,000	0.90
3	30,000	0.83
4	25,000	0.75
5	20,000	0.65

Given that Neustal's normal required rate of return is 18 percent and that the risk-free rate is 7 percent, should this project be accepted?

15-5A. *(Risk-Adjusted Discount Rates and Risk Classes)* The G. Wolfe Corporation is examining two capital-budgeting projects with five-year lives. The first, project A, is a replacement project; the second, project B, is a project unrelated to current operations. The G. Wolfe Corporation uses the risk-adjusted discount rate method and groups projects according to purpose and then uses a required rate of return or discount rate that has been preassigned to that purpose or risk class. The expected net cash flows after tax for these projects are given below:

	Project A	Project B
Initial Investment:	$250,000	$400,000
Cash Inflows:		
Year 1	$ 30,000	$135,000
Year 2	40,000	135,000
Year 3	50,000	135,000
Year 4	90,000	135,000
Year 5	130,000	135,000

The purpose/risk classes and preassigned required rates of return are as follows:

Purpose	Required Rate of Return
Replacement decision	12%
Modification or expansion of existing product line	15
Project unrelated to current operations	18
Research and development operations	20

Determine the project's risk-adjusted net present value.

15-6A. *(Certainty Equivalents)* "Nacho" Nachtmann Company uses the certainty equivalent approach when it evaluates risky investments. The company presently has two mutually exclusive investment proposals, with an expected life of four years each, to choose from with money it received from the sale of part of its toy division to another company. The expected net cash flows after tax are given below:

Year	Project A	Project B
0	−$50,000	−$50,000
1	15,000	20,000
2	15,000	25,000
3	15,000	25,000
4	45,000	30,000

The certainty equivalent coefficients for the net cash flows after tax are as follows:

Year	Project A	Project B
0	1.00	1.00
1	.95	.90
2	.85	.85
3	.80	.80
4	.70	.75

Which of the two investment proposals should be chosen, given that the risk-free rate of return is 6 percent?

15-7A. *(Probability Trees)* The M. Solt Corporation is evaluating an investment proposal with an expected life of two years. This project will require an initial outlay of $1,200,000. The resultant possible net cash flows after tax are given below:

Possible Outcomes in Year 1			
		Probability	
	.6	.3	.1
	Outcome 1	Outcome 2	Outcome 3
Net cash flow after tax =	$700,000	$850,000	$1,000,000

Conditional Outcomes and Probabilities for Year 2

If $CFAT_1 = \$700,000$		If $CFAT_1 = \$850,000$		If $CFAT_1 = \$1,000,000$	
$CFAT_2$	Probability	$CFAT_2$	Probability	$CFAT_2$	Probability
\$ 300,000	.3	\$ 400,000	.2	\$ 600,000	.1
700,000	.6	700,000	.5	900,000	.5
1,100,000	.1	1,100,000	.2	1,100,000	.4
		1,300,000	.1		

a. Construct a probability tree representing the possible outcomes.
b. Determine the joint probability of each possible sequence of events taking place.
c. What is the expected IRR of this project?
d. What is the range of possible IRRs for this project?

15-8A. *(Probability Trees)* V. Janjigian Inc. is considering expanding its operations into computer-based basketball games. Janjigian feels that there is a three-year life associated with this project, and it will initially involve an investment of \$100,000. It also feels there is a 60 percent chance of success and a cash flow after tax of \$100,000 in year 1 and a 40 percent chance of "failure" and a \$10,000 cash flow after tax in year 1. If the project "fails" in year 1, there is a 60 percent chance that it will produce net cash flows after tax of only \$10,000 in years 2 and 3. There is also a 40 percent chance that it will really fail and Janjigian will earn nothing in year 2 and get out of this line of business, with the project terminating and no net cash flow after tax occurring in year 3. If, on the other hand, this project succeeds in the first year, then net cash flows after tax in the second year are expected to be \$200,000, \$175,000, or \$150,000 with probabilities of .30, .50, and .20, respectively. Finally, if the project succeeds in the third and final year of operation, the net cash flows after tax are expected to be either \$30,000 more or \$20,000 less than they were in year 2, with an equal chance of occurrence.
a. Construct a probability tree representing the possible outcomes.
b. Determine the joint probability of each possible sequence of events.
c. What is the expected IRR?
d. What is the range of possible IRRs for this project?

15-9A. *(Probability Trees)* The E. Swank Corporation is considering an investment project with an expected life of two years. The initial outlay on this project would be \$600,000, and the resultant possible net cash flows after tax are given below:

Possible Outcomes in Year 1

	Probability		
	.4	.4	.2
	Outcome 1	Outcome 2	Outcome 3
Net cash flow after tax =	\$300,000	\$350,000	\$450,000

Conditional Outcomes and Probabilities for Year 2

If $CFAT_1 = \$300,000$		If $CFAT_1 = \$350,000$		If $CFAT_1 = \$450,000$	
$CFAT_2$	Probability	$CFAT_2$	Probability	$CFAT_2$	Probability
\$200,000	.3	\$250,000	.2	\$ 300,000	.2
300,000	.7	450,000	.5	500,000	.5
		650,000	.3	700,000	.2
				1,000,000	.1

a. Construct a probability tree representing the possible outcomes

b. Determine the joint probability of each possible sequence of events taking place.
c. What is the expected IRR of this project?
d. What is the range of possible IRRs for this project?

INTEGRATIVE PROBLEM

It's been four months since you took a position as an assistant analyst at Cale Products. During that time, you have been promoted and now are working as a special assistant for capital budgeting to the CEO. Your latest assignment involves the analysis of several risky projects. Since this is your first assignment dealing with risk analysis, you have been asked not only to provide a recommendation on the projects in question, but also to respond to a number of questions aimed at judging your understanding of analysis and capital budgeting. The memorandum you received outlining your assignment follows:

Provide a written response to the following questions:

a. In capital budgeting, risk can be measured from three perspectives. What are those three measures of a project's risk?
b. According to the CAPM, which measurement of a project's risk is relevant? What complications does reality introduce into the CAPM view of risk and what does that mean for our view of the relevant measure of a project's risk?
c. What are the similarities and differences between the risk-adjusted discount rate and certainty equivalent methods for incorporating risk into the capital-budgeting decision?
d. Why might we use the probability tree technique for evaluating capital-budgeting projects?
e. Explain how simulation works. What is the value in using a simulation approach?
f. What is sensitivity analysis and what is its purpose?
g. What does time dependence of cash flows mean? Why might cash flows be time-dependent? Give some examples?
h. Cale Products is using the certainty equivalent approach to evaluate two mutually exclusive investment proposals with an expected life of four years. The expected net cash flows after tax are given below:

Year	Project A	Project B
0	−$150,000	−$200,000
1	40,000	50,000
2	40,000	60,000
3	40,000	60,000
4	100,000	50,000

The certainty equivalent coefficients for the net cash flows after tax are as follows:

Year	Project A	Project B
0	1.00	1.00
1	0.90	0.95
2	0.85	0.85
3	0.80	0.80
4	0.70	0.75

Which of the two investment proposals should be chosen, given that the after-tax risk free rate of return is 7 percent?

i. Cale Products is considering an additional investment project with an expected life of two years and would like some insights on the level of risk this project has using the probability tree method. The initial outlay on this project would be $600,000 and the resultant possible cash flows are given below:

Possible Outcomes in Year 1

	Probability		
	.4	.4	.2
	Outcome 1	Outcome 2	Outcome 3
Cash flow =	$300,000	$350,000	$450,000

Conditional Outcomes and Probabilities for Year 2

If		If		If	
$CFAT_1$ = $300,000		$CFAT_1$ = $350,000		$CFAT_1$ = $450,000	
$CFAT_2$	Probability	$CFAT_2$	Probability	$CFAT_2$	Probability
$200,000	.3	$250,000	.2	$ 300,000	.2
300,000	.7	450,000	.5	500,000	.5
		650,000	.3	700,000	.2
				1,000,000	.1

1. Construct a probability tree representing the possible outcomes.
2. Determine the joint probability of each possible sequence of events taking place.
3. What is the expected IRR of this project?
4. What is the range of possible IRRs for this project?

STUDY PROBLEMS (SET B)

15-1B. *(Risk-Adjusted NPV)* The Cake-O-Las Corporation is considering two mutually exclusive projects. Each of these projects requires an initial outlay of $10,000 and will operate for five years. The probability distributions associated with each project for years 1 through 5 are given as follows:

Net Cash Flow After Tax, Years 1–5

Project A		Project B	
Probability	Net Cash Flow After Tax	Probability	Net Cash Flow After Tax
.20	$5,000	.20	$ 3,000
.60	6,000	.60	7,000
.20	7,000	.20	11,000

Since project B is the riskier of the two projects, the management of Cake-O-Las Corporation has decided to apply a required rate of return of 18 percent to its evaluation but only a 13 percent required rate of return to project A.

a. Determine the expected value of each project's net cash flow after tax.
b. Determine each project's risk-adjusted net present value.
c. What other factors might be considered in deciding between these two projects?

15-2B. *(Risk-Adjusted NPV)* The Dorf Corporation is evaluating two mutually exclusive projects, both of which require an initial outlay of $125,000. Each project has an expected life of five years. The probability distributions associated with the annual net cash flows from each project are given below:

Net Cash Flow After Tax, Years 1–5			
Project A		Project B	
Probability	Cash Flow After Tax	Probability	Cash Flow After Tax
.10	$40,000	.10	$20,000
.40	45,000	.20	40,000
.40	50,000	.40	55,000
.10	55,000	.20	70,000
		.10	90,000

The normal required rate of return for Dorf is 10 percent, but since these projects are riskier than most, Dorf is requiring a higher-than-normal rate of return on them. It is requiring an 11 percent rate of return on project A and a 13 percent rate of return on project B.

a. Determine the expected value for each project's net cash flows after tax.

b. Determine each project's risk-adjusted net present value.

c. What other factors might be considered in deciding between these projects?

15-3B. *(Certainty Equivalents)* The Temco Corp. is considering two mutually exclusive projects. The expected values for each project's net cash flows after tax are given below:

Year	Project A	Project B
0	–$1,000,000	–$1,000,000
1	600,000	600,000
2	750,000	650,000
3	600,000	700,000
4	550,000	750,000

Temco has decided to evaluate these projects using the certainty equivalent method. The certainty equivalent coefficients for each project's net cash flows after tax are given below:

Year	Project A	Project B
0	1.00	1.00
1	.90	.95
2	.90	.75
3	.75	.60
4	.65	.60

Given that this company's normal required rate of return is 15 percent and the risk-free rate is 5 percent, which project should be selected?

15-4B. *(Certainty Equivalents)* Perumperal Inc. has decided to use the certainty equivalent method in determining whether or not a new investment should be made. The expected net cash flows after tax associated with this investment and the estimated certainty equivalent coefficients are as follows:

Year	Expected Values for Cash Flows After Tax	Certainty Equivalent Coefficients
0	–$100,000	1.00
1	30,000	0.95
2	25,000	0.90
3	30,000	0.83
4	20,000	0.75
5	25,000	0.65

Given that Perumperal's normal required rate of return is 18 percent and that the risk-free rate is 8 percent, should this project be accepted?

15-5B. *(Risk-Adjusted Discount Rates and Risk Classes)* The Kick 'n' MacDonald Corporation is examining two capital-budgeting projects with five-year lives. The first, project A, is a replacement project; the second, project B, is a project unrelated to current operations. The Kick 'n' MacDonald Corporation uses the risk-adjusted discount rate method and groups projects according to purpose and then uses a required rate of return or discount rate that has been preassigned to that purpose or risk class. The expected net cash flows after tax for these projects are given below:

	Project A	Project B
Initial Investment:	$300,000	$450,000
Cash Inflows:		
Year 1	$ 30,000	$130,000
Year 2	40,000	130,000
Year 3	50,000	130,000
Year 4	80,000	130,000
Year 5	120,000	130,000

The purpose/risk classes and preassigned required rates of return are as follows:

Purpose	Required Rate of Return
Replacement decision	13%
Modification or expansion of existing product line	16
Project unrelated to current operations	18
Research and development operations	20

Determine the project's risk-adjusted net present value.

15-6B. *(Certainty Equivalents)* The M. Jose Company uses the certainty equivalent approach when it evaluates risky investments. The company presently has two mutually exclusive investment proposals, with an expected life of four years each, to choose from with money it received from the sale of part of its toy division to another company. The expected net cash flows after tax are given below:

Year	Project A	Project B
0	−$75,000	−$75,000
1	20,000	25,000
2	20,000	30,000
3	15,000	30,000
4	50,000	25,000

The certainty equivalent coefficients for the net cash flows after tax are as follows:

Year	Project A	Project B
0	1.00	1.00
1	.90	.95
2	.85	.85
3	.80	.80
4	.70	.75

Which of the two investment proposals should be chosen, given that the risk-free rate of return is 7 percent?

15-7B. *(Probability Trees)* The Buckeye Corporation is evaluating an investment proposal with an expected life of two years. This project will require an initial outlay of $1,300,000. The resultant possible net cash flows after tax are given below:

Possible Outcomes in Year 1

	Probability		
	.6	.3	.1
	Outcome 1	Outcome 2	Outcome 3
Net cash flow after tax =	$750,000	$900,000	$1,500,000

Conditional Outcomes and Probabilities for Year 2

If		If		If	
$CFAT_1 = \$750,000$		$CFAT_1 = \$900,000$		$CFAT_1 = \$1,500,000$	
$CFAT_2$	Probability	$CFAT_2$	Probability	$CFAT_2$	Probability
$ 300,000	.10	$ 400,000	.2	$ 600,000	.3
700,000	.50	700,000	.5	900,000	.6
1,100,000	.40	900,000	.2	1,100,000	.1
		1,300,000	.1		

a. Construct a probability tree representing the possible outcomes.

b. Determine the joint probability of each possible sequence of events taking place.

c. What is the expected IRR of this project?

d. What is the range of possible IRRs for this project?

15-8B. *(Probability Trees)* The OSUCOWBOYS Corporation is considering an investment project with an expected life of two years. The initial outlay on this project would be $700,000, and the resultant possible net cash flows after tax are given as follows:

Possible Outcomes in Year 1

	Probability		
	.4	.4	.2
	Outcome 1	Outcome 2	Outcome 3
Net cash flow after tax =	$325,000	$350,000	$475,000

Conditional Outcomes and Probabilities for Year 2

If		If		If	
$CFAT_1$ = $325,000		$CFAT_1$ = $350,000		$CFAT_1$ = $475,000	
$CFAT_2$	Probability	$CFAT_2$	Probability	$CFAT_2$	Probability
$ 200,000	.3	$ 250,000	.2	$ 300,000	.2
350,000	.7	400,000	.5	550,000	.5
		650,000	.3	600,000	.2
				1,000,000	.1

a. Construct a probability tree representing the possible outcomes.

b. Determine the joint probability of each possible sequence of events taking place.

c. What is the expected IRR of this project?

d. What is the range of possible IRRs for this project?

15-9B. *(Probability Trees)* Mac's Buffaloes Inc. is considering expanding its operations into computer-based basketball games. Mac's Buffaloes feels that there is a three-year life associated with this project, and it will initially involve an investment of $120,000. It also feels there is a 70 percent chance of success and a net cash flow after tax of $100,000 in year 1 and a 30 percent chance of "failure" and a $10,000 net cash flow after tax in year 1. If the project "fails" in year 1, there is a 60 percent chance that it will produce net cash flows after tax of only $10,000 in years 2 and 3. There is also a 40 percent chance that it will really fail and Mac's Buffaloes will earn nothing in year 2 and get out of this line of business, with the project terminating and no net cash flow after tax occurring in year 3. If, on the other hand, this project succeeds in the first year, then net cash flows after tax in the second year are expected to be $225,000, $180,000, or $140,000 with probabilities of .30, .50, and .20, respectively. Finally, if the project succeeds in the third and final year of operation, the net cash flows after tax are expected to be either $30,000 more or $20,000 less than they were in year 2, with an equal chance of occurrence.

a. Construct a probability tree representing the possible outcomes.

b. Determine the joint probability of each possible sequence of events.

c. What is the expected IRR?

d. What is the range of possible IRRs for this project?

SELF-TEST SOLUTIONS

SS-1. a.

$$\bar{X} = \sum_{i=1}^{N} X_i P(X_i)$$

$$\begin{aligned}\bar{X}_A &= 0.20(\$10{,}000) + 0.60(\$15{,}000) + 0.20(\$20{,}000)\\ &= \$2{,}000 + \$9{,}000 + \$4{,}000\\ &= \$15{,}000\end{aligned}$$

$$\begin{aligned}\bar{X}_B &= 0.20(\$6{,}000) + 0.60(\$18{,}000) + 0.20(\$30{,}000)\\ &= \$1{,}200 + \$10{,}800 + \$6{,}000\\ &= \$18{,}000\end{aligned}$$

b.

$$NPV = \sum_{t=1}^{n} \frac{CFAT_t}{(1 + k^*)^t} - IO$$

$$
\begin{aligned}
NPV_A &= \$15{,}000(3.6048) - \$25{,}000 \\
&= \$54{,}072 - \$25{,}000 \\
&= \$29{,}072
\end{aligned}
$$

$$
\begin{aligned}
NPV_B &= \$18{,}000(3.1272) - \$25{,}000 \\
&= \$56{,}289.60 - \$25{,}000 \\
&= \$31{,}289.60
\end{aligned}
$$

SS-2. Project A:

Year	(A) Expected Cash Flow After Tax	(B)	(A × B) (Expected Cash Flow) × (α_t)	Present Value Factor at 8%	Present Value
0	−$300,000	1.00	−$300,000	1.0000	−$300,000.00
1	100,000	.95	95,000	0.9259	87,960.50
2	200,000	.90	180,000	0.8573	154,314.00
3	200,000	.85	170,000	0.7938	134,946.00
4	300,000	.80	240,000	0.7350	176,400.00
5	300,000	.75	225,000	0.6806	153,135.00
				$NPV_A =$	$406,755.50

Project B:

Year	(A) Expected Cash Flow After Tax	(B)	(A × B) (Expected Cash Flow) × (α_t)	Present Value Factor at 8%	Present Value
0	−$300,000	1.00	−$300,000	1.0000	−$300,000
1	200,000	.90	180,000	0.9259	166,662
2	200,000	.80	160,000	0.8573	137,168
3	200,000	.70	140,000	0.7938	111,132
4	300,000	.60	180,000	0.7350	132,300
5	400,000	.50	200,000	0.6806	136,120
				$NPV_B =$	$383,382

Thus, project A should be selected since it has the higher NPV.

APPLICATIONS OF CAPITAL BUDGETING

CHAPTER 16

LEARNING OBJECTIVES

After reading this chapter you should be able to

1. Explain the bond-refunding decision.
2. Explain the factors used to estimate a firm's value.
3. Explain the steps used to determine a firm's acquisition value.
4. Explain the various types of leases.
5. Explain the steps used to determine whether to lease or to purchase an asset.

INTRODUCTION

In 1995, several Canadian companies were involved in some very large corporate takeovers. Montreal-based Seagram Co. sold its stake in E. I. du Pont de Nemours & Co. for $12.2 billion and then purchased the entertainment giant MCA Inc. for US$ 5.7 billion. Wallace McCain, formerly the chief executive of McCain Foods Ltd. led a successful takeover bid of Maple Leaf Foods Inc. for $1.2 billion. And Alberta Energy Co. Ltd. announced a $1.1 billion friendly merger with Toronto's Conwest Exploration Co. Ltd.[1] During 1995, 14 deals worth more than $1 billion were reported in Canada. Crosbie & Co., the Toronto-based merchant bank, has indicated that the total value of these types of transactions in 1995 was worth $77.4 billion.[2]

In each of these corporate takeovers, the value of the acquisition must be determined. Financial analysts and business valuators use any of several valuation methods to obtain an estimate of the worth of the assets. However, the actual value of the acquisition will be arrived at through the negotiation process. In this chapter, we examine one approach which is based on valuing the benefits and costs of the acquisition. The approach is the same as the one used in the capital-budgeting decisions concerning expansion and replacement projects. In fact, the approach of comparing benefits and costs is applied to many business decisions. As we will see, the cost/benefit analysis can be applied to the bond-refunding decision, as well as the decision to lease or to purchase an asset.

[1]Adapted from: Susan Noakes, "Mergers big news again," *The Financial Post,* December 23, 1995, p. 5.

[2]Adapted from: Terry Weber, "Big deals in '95 lift M & A's value by 60%," *The Financial Post,* January 6, 1996, p. 5.

CHAPTER PREVIEW

The final chapter of this section examines how the capital-budgeting decision can be applied to several important business decisions that face financial executives. We begin by examining the decision of whether to refund a bond issue. Then, we apply the capital-budgeting process to determine the value of the firm. Finally, we apply the capital-budgeting process to the decision of either purchasing or leasing an asset.

BACK TO THE BASICS

In applying the capital-budgeting process to business decisions, keep in mind that we rely on the following axioms:
Axiom 1: The Risk-Return Tradeoff
Axiom 2: The Time Value of Money
Axiom 3: Cash—Not Profits—Is King
Axiom 4: Incremental Cash Flows
It should be noted, however, that in each example the required rate of return of the project, as well as the type of cash flow measured, will differ. Nevertheless, in each example the process of measuring the incremental cash flows associated with the investment proposals and evaluating those proposed investments remains the same.

OBJECTIVE 1

REFUNDING OF BONDS

Because bonds have a maturity date, their retirement is a crucial matter. Bonds can be retired at maturity, at which time the bondholder receives the par value of the bond, or they can be retired prior to maturity. Early redemption is generally accomplished through the use of a call provision or a sinking fund.

The Use of the Call Provision

Call provision

A provision that entitles the corporation to repurchase its bonds or preferred shares from their holders at stated prices over specified periods.

Call premium

The difference between the call price and the security's par value.

Sinking fund

A required annual payment that allows for the periodic retirement of debt.

A **call provision** entitles the corporation to repurchase or "call" the bonds from their holders at stated prices over specified periods. This feature provides the firm with the flexibility to recall its debt and replace it with lower-cost debt if interest rates fall. The terms of the call provision are provided in the indenture and generally state the call price above the bond's par value. The difference between the call price and the par value is referred to as the **call premium**. The size of this premium changes over time, becoming smaller as the date of call approaches the bond's scheduled maturity. It is also common to prohibit calling the bond during its first years. Obviously, a call provision works to the disadvantage of the bondholder, who for this reason is generally compensated by a higher rate of return on the bond.

The Use of the Sinking Fund Provision

A bond covenant may have a provision for a **sinking fund** which requires the periodic repayment of debt, thus reducing the total amount of debt outstanding.[3] When a sinking fund is set up, the firm makes annual payments to the trustee, who can then purchase the bonds in the capital markets or use the call provision. The

[3]Serial Bonds are sinking fund bonds that must be redeemed at par. They are more valuable than the pure sinking bonds since they cannot be repurchased in the marketplace by the borrower.

advantage is that the annual retirement of debt through a sinking fund reduces the amount needed to retire the remaining debt at maturity. Otherwise the firm would face a major payment at maturity. If the firm were experiencing temporary financial problems when the debt matured, both the repayment of the principal and the firm's future could be jeopardized. The use of a sinking fund and its periodic retirement of debt eliminates this potential danger.

PERSPECTIVE IN FINANCE

The bond-refunding decision is actually nothing more than a capital-budgeting decision. While a casual glance at the calculations may appear intimidating, all that the financial manager really does is determine the present value of the annual net cash benefits from the refund and subtract out the initial outlay.

The Bond-Refunding Decision

The **bond-refunding decision**—that is, whether or not to call an entire issue of bonds—is similar to the capital-budgeting decision. The present value of the stream of benefits from the refunding decision is compared with the present value of its costs. If the benefits outweigh the costs, the refunding is done.

Bond-refunding decision

The choice of whether or not to call an entire issue of bonds.

The costs of refunding a bond issue include issuing costs (e.g., printing and advertising costs, filing fees, legal and accounting fees, registration and transfer fees), the cost of recalling the old issue and any interest expenses occurring during the bond overlap period. An overlap period normally occurs because firms wish to obtain the funds from the new issue before calling the old bonds. During this overlap period, there is a risk of a rise in interest rates or a drying up of funds in the capital markets. Thus, the cost associated with the additional interest payments can be viewed as the cost of eliminating this risk. The call premium is treated as an expense that occurs during the year of refunding although it is not tax deductible. Revenue Canada requires that the issuing costs of securities be amortized over five years. The amount of issuing costs is the *lesser* of (i) the amount of issuing costs occurring from the amortization; or (ii) the amount by which the unamortized balance exceeds the sum of the amortized amounts. This regulation is applied to securities issued after 1987. As a result, the present value of all tax shields from the issuing costs must be subtracted from the flotation cost to determine the appropriate cost level. The benefits of refunding bonds include any savings from interest expenses and any return from the investment of funds during the overlap period. The interest savings are achieved by replacing older, higher-cost debt with less expensive debt as interest rates drop.

Although the calculations associated with the bond-refunding decision appear to be quite complex, we are really just comparing a project's benefits to its costs. This type of cost/benefit analysis enables the firm to decide whether to reject or to accept the refunding of an old bond issue with a new bond issue that has a lower coupon rate. Since both annual interest costs and the annual tax benefits from the flotation costs of the new issue are constant over the maturity of the new bond, we use our knowledge of annuities to determine the present value of these costs to the firm. The bond-refunding project requires that we use the after-tax cost of borrowing on the new bonds to discount the annual benefits and costs. The after-tax cost of borrowing on the new bonds reflects the firm's default risk, that is the risk of the firm defaulting on interest or principal payments. All other costs or benefits associated with bond refunding, are known with certainty and do not present the firm with any additional risk. Finally, we will decide whether to reject or accept the bond-refunding project on the basis of whether or not the project shows a net benefit to the firm. A net benefit to the firm occurs if the project shows a greater amount of benefits from the interest savings and the return on investment during the overlap period as compared to the costs of bond refunding. We have used principles of a cost/benefit analysis when examining the net present value technique;

this measure of value enables financial managers to evaluate whether to accept or reject investment decisions made by the firm. The following example illustrates the calculations for a bond-refunding decision.[4]

EXAMPLE

Suppose that interest rates have fallen to a level such that a firm is considering the decision of refunding a $125 million, 10 percent debenture issue that has 20 years left to maturity. These old debentures contain a call provision and can be called at a premium of 5 percent on par value. The old debenture issue can be replaced with a $125 million issue of 7.5 percent, 20-year debentures. The firm will incur $1,750,000 in issuing costs for the new issue. The overlap period during which both issues will be outstanding is expected to be one month. Finally, the firm expects a nominal return of 6.50 percent; interest is compounded monthly on short-term investments. Assume that the old bonds were issued before 1987 such that tax shields from flotation costs were expensed in the year of issue and that the firm belongs to a tax bracket of 45 percent.

The procedure for deciding whether to accept or reject the bond-refunding project involves determining the cash flows from the project's costs and benefits. The appropriate discount factor for a bond-refunding project is the after-tax cost of debt for the new bonds which equals 4.125% (7.5% × (1 − .45)). The costs of the bond-refunding decision include: (1) the cost of the call premium of the old bond issue, $6,250,000; (2) the net flotation costs of the new issue, $701,298; and, (3) the overlap interest costs after tax of the old debenture issue, $572,917. To determine the net flotation costs, we make use of our knowledge of the time value of money by treating the tax shields obtained from the flotation costs as an annuity and calculating the present value of this annuity. The benefits of the bond-refunding decision include the interest savings caused by the lower coupon rate of the new bond issue, $23,101,891 and the return on investment which occurs during the overlap period, $372,625. Again, we determine the present value of the after-tax interest savings by treating the interest savings as an annuity and determining the present value of this annuity. Table 16-1 shows these calculations. A comparison of the benefits to the costs of the bond-refunding decision shows a net benefit (net present value) of $15,950,301. Since the net benefit (net present value) is positive, the bond-refunding project should be accepted.

A typical complication in the analysis of the bond-refunding decision involves a difference in the maturities of the new and old bonds. The old bonds, which have been outstanding for some length of time, may have a shorter maturity than the proposed new bonds that are intended to replace them. Actually, this complication is quite easy to accommodate. The only alteration in the analysis is that only the net benefits up to the maturity of the old bonds are considered; after the old bonds terminate, the analysis terminates.

It should also be noted that the procedure to decide whether a preferred share issue should be refunded is the same as the bond-refunding decision, except for the treatment of dividends, which are not tax deductible. Although preferred shares do not have a finite maturity, the present value of the dividend savings is calculated as the valuation of a perpetuity.

[4]The subject of the appropriate discount rate to be used in discounting the benefits of a bond refund back to present has received considerable attention. See, for example, Thomas H. Mayor and Kenneth G. McCoin, "The Rate of Discounting in Bond Refunding," *Financial Management* 3 (Autumn 1974), 54–58; and Aharon R. Ofer and Robert A. Taggart, Jr., "Bond Refunding: A Clarifying Analysis," *Journal of Finance* 32 (March 1977), pp. 21–30. Ofer and Taggart show that the relevant discount rate is the after-tax cost of the refunding bonds when new bonds are used to replace the existing bonds.

Table 16-1 Calculations Illustrating the Bond-Refunding Decision

Step 1:

Calculate the costs.

(a) Determine the cost of the call premium:

($125,000,000/$1,000) × $50 = $6,250,000

(b) Determine the flotation costs:

Annual flotation costs = $1,750,000/5 = $350,000

Annual tax shield = ($350,000) × (0.45) = $157,500 per year

Present value of all tax shields = $157,500 × $(PVIFA_{4.125\%, 5})$

= $157,500 × (4.4362)

= $698,702

Net flotation costs = flotation costs – present value of all tax shields

= $1,750,000 − $698,702

= $1,051,298

(c) Determine overlap interest costs:

After-tax interest rate of old debenture issue = 10% × (1 − 0.45)

= 5.50%

Interest cost = 0.055 × (1/12) × ($125,000,000)

= $572,917

(d) Determine total costs:

Total costs = Call premium + net flotation costs + overlap interest costs

= $6,250,000 + $1,051,298 + $572,917

= $7,874,215

Step 2:

Calculate the cash benefit from eliminating the old bonds through refunding.

(a) Determine present value of the interest savings:

Present value of interest savings = ($125,000,000) × (0.10 − 0.075) × (1 − 0.45) × $(PVIFA_{4.125\%, 20})$

= ($1,718,750) × (13.4411)

= $23,101,891

(b) Determine the investment benefits during overlap period:

Investment benefits = ($125,000,000) × (0.00542)[a] × (1 − 0.45)

= $372,625

(c) Determine total benefits:

Total benefits = $23,101,891 + $372,625

= $23,474,516

Step 3:

Calculate the refunding decision's net present value.

(a) Present value of the benefits (from Step 2)	$23,474,516
(b) Less present value of the costs (from Step 1)	7,874,215
(c) Equals net present value	$15,600,301

[a]Monthly compounding: $[1 + (.065/12)]^1 - 1 = 0.542\%$

DETERMINING A FIRM'S ACQUISITION VALUE

OBJECTIVE 2

One of the first problems in analyzing a potential acquisition involves placing a value on the acquired firm. This task is not easy. The value of a firm depends not only upon its cash flow generation capabilities, but also upon the operating and financial characteristics of the acquiring firm. As a result, no single dollar value exists for a company. Instead, a range of values is determined that would be economically justifiable to the prospective acquirer. The final price within this range is then negotiated by the two managements.

To determine an acceptable price for a corporation, a number of factors are carefully evaluated. We know that the objective of the acquiring firm is the maximization of shareholders' wealth (share price). However, quantifying the relevant variables for this purpose is difficult at best. For instance, the primary reason for a merger might be to acquire managerial talent, or to complement a strong sales staff with an excellent production department. This potential synergistic effect is difficult to measure using the historical data of the companies involved. Even so, several quantitative variables are frequently used in an effort to estimate a firm's value. These factors include (1) book value, (2) appraisal value, (3) market price of the firm's common shares, and (4) expected cash flows.

Book Value

Book value

The depreciated value of a company's assets (original cost less accumulated depreciation) less the firm's outstanding liabilities.

The **book value** of a firm's net worth is the balance sheet amount of the assets less its outstanding liabilities, or in other words, the owners' equity. For example, if a firm's historical cost less accumulated depreciation is $10 million and the firm's debt totals $4 million, the aggregate book value is $6 million. Furthermore, if 100,000 common shares are outstanding, the book value per share is $60 ($6 million/100,000 shares).

Book value does not measure the true market value of a company's net worth because it is based on the historical cost of the firm's assets. Seldom do such costs bear a relationship to the value of the organization or its ability to produce earnings.

Although the book value of an enterprise is clearly not the most important factor, it should not be overlooked. It can be used as a starting point to be compared with other analyses. Also, a study of the firm's working capital is particularly important to acquisitions involving a business consisting primarily of liquid assets, such as financial institutions. Furthermore, in industries where the ability to generate earnings requires large investments in such items as steel, cement, and petroleum, the book value could be a critical factor, especially where plant and equipment are relatively new.

Appraisal Value

Appraisal value

The value of a company as stated by an independent appraisal firm. Although the techniques used by appraisers vary widely, the appraisal value is often related to the replacement cost of the asset.

An **appraisal value** of a company may be acquired from an independent appraisal firm. The techniques used by appraisers vary widely; however, this value is often closely tied to replacement cost. This method of analysis is not adequate by itself, since the value of individual assets may have little relation to the firm's overall ability to generate earnings, and thus the going-concern value of the firm. However, the appraised value of an enterprise may be beneficial when used in conjunction with other valuation methods. Also, the appraised value may be an important factor in special situations, such as in financial companies, natural resource enterprises, or organizations that have been operating at a loss.[5]

The use of appraisal values does yield several additional advantages. The value determined by independent appraisers may permit the reduction of accounting goodwill by increasing the recognized worth of specific assets. Goodwill results when the purchase price of a firm exceeds the value of the individual assets. Consider a company having a book value of $60,000 that is purchased for $100,000 (the $40,000 difference is goodwill). The $60,000 book value consists of $20,000 in working capital and $40,000 in fixed assets. However, an appraisal might suggest that the current values of these assets are $25,000 and $55,000, respectively. The $15,000 increase ($55,000 – $40,000) in fixed assets permits the acquiring firm to record a larger depreciation expense than would otherwise be possible, thereby reducing taxes.

---

[5]The assets of a financial company and a natural resources firm largely consist of securities and natural reserves, respectively. The value of these individual assets has a direct bearing on the firm's earning capacity. Also, a company operating at a loss may only be worth its liquidation value, which would approximate the appraisal value.

A second reason for an appraisal is to provide a test of the reasonableness of results obtained through methods based upon the going-concern concept. Third, the appraiser may uncover strengths and weaknesses that otherwise might not be recognized, such as in the valuation of patents, secret processes, and partially completed R&D expenditures.

Thus, the appraisal procedure is generally worthwhile if performed with additional evaluation processes. In specific instances, it may be an important instrument for valuing a corporation.

Stock Market Value

The **stock market value**, as expressed by stock market quotations, is another approach to estimate the net worth of a business. If the stock is listed on a major securities exchange, such as the Toronto Stock Exchange, and is widely traded, an approximate value can be established on the basis of the market value. The justification is based upon the fact that the market quotations indicate the consensus of investors as to a firm's cash flow potential and the corresponding risk.

Stock market value

The value observed in the marketplace in which buyers and sellers negotiate a mutually acceptable price for the asset.

The market-value approach is the one most frequently used in valuing large corporations. However, this value can change abruptly. Analytical factors compete with purely speculative influences and are subject to people's sentiments and personal decisions. Thus,

> the market is not a weighing machine, on which the value of each issue is recorded by an exact and impersonal mechanism, in accordance with the specific qualities. Rather, should we say that the market is a voting machine, whereon countless individuals register choices which are the product partly of reason and partly of emotion.[6]

In short, the market-value approach is probably the one most widely used for determining the worth of a firm; a 10 to 20 percent premium above the market price is often required as an inducement for the current owners to sell their shares.[7] Even so, executives who place their entire reliance upon this method are subject to an inherent danger of market psychology.

The "Chop-Shop" Value

The **"chop-shop"** approach to valuation was first proposed by Dean Lebaron and Lawrence Speidell of Batterymarch Financial Management. Specifically, it attempts to identify multi-industry companies that are undervalued and would be worth more if separated into their parts. This approach conceptualizes the practice of attempting to buy assets below their replacement cost.

Chop-shop value

The value obtained by valuing a company's various business segments. This approach attempts to identify companies from single industries and computes the average valuation ratios.

Any time we confront a technique that suggests that stocks may be inefficiently priced, we must be a bit skeptical. In the case of multi-industry firms, the inefficiency in pricing may be brought on by the high cost of obtaining information. Alternatively, these firms may be worth more if split up because of agency problems. Shareholders of multi-industry companies may feel they have less control of the firm's managers, since additional layers of management may have developed within multi-industry firms. These agency costs may take the form of increased expenditures necessary to monitor the managers, costs associated with organizational change and opportunity costs associated with poor decisions made as a

[6]Benjamin Graham, David L. Dodd, and Sidney Cottle, *Security Analysis*, 4th ed. (New York: McGraw-Hill, 1962), p. 42.

[7]The price paid for a firm in an acquisition may exceed the current market price for the following reasons: (1) the increased efficiencies provided by the combination; (2) an inducement for owners to sell the company; (3) the benefits provided from being able to control the acquired company; and (4) the possible tax advantages to the buyer.

result of the manager acting in his or her own best interests rather than the best interest of the shareholders.

The "chop-shop" approach attempts to value companies by their various business segments. As it is implemented by Batterymarch, it first attempts to find "pure-play" companies—that is, companies in a single industry—from which it computes average "valuation ratios." The ratios frequently used compare total capitalization (debt plus equity) to total sales, to assets, and to income. In effect, these ratios represent the average value of a dollar of sales, a dollar of assets, and a dollar of income for a particular industry based upon the average of all pure companies in that industry. Assuming that these ratios hold for the various business segments of a multi-industry firm, the firm can then be valued by its parts.

For the chop-shop valuation technique to be feasible, we must naturally have information about the various business segments within the firm. This requirement is fulfilled, at least in part, by the reporting rules set forth by the Accounting Standards Committee of the CICA. This standard (*CICA Handbook*, 1700.07) requires that firms disclose segmented information which describes the extent of their total operations, first by industry and second by geographic area. Of course, we know that not all firms in the same industry are in fact the same—some simply have more potential growth or earnings ability than others. This methodology should be viewed cautiously since financial managers are reluctant to release information which competitors can use for their benefit.

The "chop-shop" approach is actually a three-step process:

OBJECTIVE 3

1. Identify the firm's various business segments and calculate the average capitalization ratios for firms in those industries.
2. Calculate a "theoretical" market value based upon each of the average capitalization ratios.
3. Average the "theoretical" market values to determine the "chop-shop" value of the firm.

EXAMPLE

To illustrate the chop-shop approach, consider Cavos Inc. with common shares currently trading at a total market price of $13 million. For Cavos, the accounting data set forth four business segments: industrial specialties, basic chemicals, consumer specialties, and basic plastics. Data for these four segments are as follows:

Business Segment	Segment Sales ($000)	Segment Assets ($000)	Segment Income ($000)
Industrial specialties	$2,765	$2,206	$186
Basic chemicals	5,237	4,762	165
Consumer specialties	2,029	1,645	226
Basic plastics	1,506	1,079	60
Total	$11,537	$9,692	$637

The three steps for valuing Cavos would be

***Step 1:* We first identify "pure" companies, those being firms that operate solely in one of the above industries; we then calculate the average capitalization ratios for those firms. This could easily be done using a computer database, such as the TSE/Western Database for TSE 300 Index returns, which provides detailed financial information on most publicly traded firms. Assume the average capitalization ratios for the four business segments in which Cavos is active have been determined and are as shown in Table 16-2.**

Step 2: Once we have calculated the average market capitalization ratios for the various market segments, we need only multiply Cavos' segment values (that is, segment sales, segment assets, and segment income) times the capitalization ratios to determine the theoretical market values. This is done in Table 16-3.

Step 3: Finally, the theoretical values must be averaged to calculate the "chop-shop" value of the firm. The average of the three theoretical values in Table 16-3 is $18,913,000, computed as follows:

$$\frac{\text{value based on sales} + \text{value based on assets} + \text{value based on income}}{3}$$

or

$$\frac{(\$23{,}518{,}500 + \$21{,}087{,}700 + \$12{,}132{,}800)}{3}$$

$$= \$18{,}913{,}000$$

Hence, Cavos, Inc., is selling for significantly less than its "chop-shop" value, $13 million compared to $18.9 million.

The major limitation of a valuation model, such as the "chop-shop" approach, is that it is not derived from any theoretical basis. What it does is assume that average industry capitalization relationships—in this case, ratios of capitalization to sales, assets, and operating income—hold for all firms or conglomerate subsidiaries in that particular industry. Of course, this is frequently not the case. It is easy to identify specific companies that simply produce superior products and whose future earnings growth is, as a result, brighter. These companies, because of their expected future growth, should have their sales, assets, and operating income valued higher. This only makes sense, because as we know the valuation of any asset is based upon the market's expectations.

The "chop-shop" valuation approach reflects the idea that the replacement value of a firm's assets may exceed the value placed upon the firm as a whole in the market. This approach attempts to value the multi-industry firm by its parts. Moreover, as we will see as we explore the cash flow approach to valuation, there simply is no way to estimate the value of a takeover candidate with complete confidence. Thus, this method may provide the decision maker with additional information.

The Cash Flow Value

Our last look at valuation models should be familiar to us, given our prior work in finding the present value of cash flows, as we did in our studies in capital budgeting. Using the cash flow approach to determine the firm's acquisition value requires that we estimate the incremental net cash flows available to the bidding firm as a result of the merger or acquisition. The present value of these cash flows represents the maximum amount that should be paid for the target firm. If initial outlay is subtracted out, we determine the net present value of the acquisition. This

Table 16-2 Average Capitalization Ratios for Industries in Which Cavos Inc. Is Active

Business Segment	Capitalization/ Sales	Capitalization/ Assets	Capitalization/ Operating Income
Industrial specialties	0.61	1.07	21.49
Basic chemicals	2.29	2.43	17.45
Consumer specialties	3.58	2.92	19.26
Basic plastics	1.71	2.18	15.06

Table 16-3 Calculation of the "Theoretical Values" for Cavos Inc. Using Market Capitalization Ratios

Value Based on Market Capitalization/Sales

Business Segment	(A) Market Capitalization/ Sales	(B) Segment Sales ($000)	(A) × (B) Theoretical Market Value ($000)
Industrial specialties	0.61	$2,765	$ 1,686.7
Basic chemicals	2.29	5,237	11,992.7
Consumer specialties	3.58	2,029	7,263.8
Basic plastics	1.71	1,506	2,575.3
Total			$23,518.5

Value Based on Market Capitalization/Assets

Business Segment	(A) Market Capitalization/ Assets	(B) Segment Assets ($000)	(A) × (B) Theoretical Market Value ($000)
Industrial specialties	1.07	$2,206	$ 2,360.4
Basic chemicals	2.43	4,762	11,571.7
Consumer specialties	2.92	1,645	4,803.4
Basic plastics	2.18	1,079	2,352.2
Total			$21,087.7

Value Based on Market Capitalization/Income

Business Segment	(A) Market Capitalization/ Income	(B) Segment Income ($000)	(A) × (B) Theoretical Market Value ($000)
Industrial specialties	21.49	$186	$3,997.1
Basic chemicals	17.45	165	2,879.3
Consumer specialties	19.26	226	4,352.8
Basic plastics	15.06	60	903.6
Total			$12,132.8

approach is similar to that of the capital-budgeting problem, but the approach differs in the estimation of the initial outlay.

Finding the present value of the cash flows for the acquisition value of a firm involves a five-step process:

1. Estimate the incremental cash flows available from the target firm. This estimation of incremental after-tax cash flows includes all synergistic cash flows (including those to both the bidding and target firms) created as a result of the acquisition. It should also be noted that interest expenses are not included in these cash flows, as they are accounted for in the required rate of return.
2. Estimate the risk-adjusted discount rate associated with cash flows from the target firm. The target firm's, not the bidding firm's, required rate of return is appropriate here. If there is any anticipated change in financing associated with the target firm as a result of the acquisition, this change should be taken into account.
3. Calculate the present value of the incremental cash flows from the target firm.

4. Estimate the initial outlay associated with the acquisition. The initial outlay is defined here as the market value of all securities and cash paid out plus the market value of all debt liabilities assumed.
5. Calculate the net present value of the acquisition by subtracting the initial outlay from the present value of the incremental cash flows from the target firm.

Financial executives have difficulty in accurately estimating the synergistic gains which are obtained from the merger of two firms. This type of estimation also increases the uncertainty of the incremental cash flows after tax. For example, it is very difficult to estimate the gains that might be expected from any reduction in bankruptcy costs, increased market power, or reduction in agency costs that might occur. Still, it is imperative to estimate these gains if we are to place a value on the target firm. Once the required rate of return is determined, the present value of the incremental cash flows from acquiring the target firm can then be calculated.

The final step then becomes the calculation of the initial outlay associated with the acquisition.

EXAMPLE

Let's look at the example of Tabbypaw Pie Inc., which is being considered as a possible takeover target by ALF Inc. Currently Tabbypaw Pie uses 30 percent debt in its capital structure, but ALF plans on increasing the debt ratio to 40 percent (we will assume that only debt and equity are used) once the acquisition is completed. The after-tax cost of debt capital for Tabbypaw Pie is estimated to be 7 percent, and we will assume that this rate does not change as Tabbypaw's capital structure changes. The cost of equity after the acquisition is expected to be 20.8 percent. The current market value of Tabbypaw's debt outstanding is $110 million, all of which will be assumed by ALF. Also, let's assume that ALF intends to pay $260 million in cash and common shares for all of Tabbypaw Pie's shares in addition to assuming all of Tabbypaw's debt. Currently, the market price of Tabbypaw Pie's common shares is $210 million.

Step 1**: Estimate the incremental cash flows from the target firm, including the synergistic flows, such as any possible flows from the firm's use of tax credits. One of the firm's financial analysts has estimated that the firm will be eligible to use the following accounting depreciation expenses and tax shields:**

Year	1996	1997	1998	1999	2000 and thereafter
Accounting Depreciation	$39 M	$40 M	$41 M	$42 M	$43 M
Tax Shields	60	42	29	21	7

The estimation for Tabbypaw's cash flows after tax is provided in Table 16-4. Here we are assuming that any cash flows after 2000 will be constant at $85 million. Also, we subtract any funds that must be reinvested in the firm in the form of capital expenditures that are required to support the firm's increasing profits.

Step 2: **Determine an appropriate risk-adjusted discount rate for evaluating Tabbypaw. Here we will use the weighted cost of capital (WCC) for Tabbypaw as our discount rate, where the weighted cost of capital is calculated as**

$$WCC = W_d K_b + W_c K_c \quad (16\text{-}1)$$

where W_d, W_b = the percentage of funds provided by debt and common, respectively

K_b, K_c = the cost of debt and common, respectively

Table 16-4 Estimated Incremental Cash Flows from Tabbypaw Pie Inc. ($ millions)

	1996	1997	1998	1999	2000 and thereafter
Incremental operating revenues	$496	$536	$606	$670	$731
Incremental operating expenses (including accounting depreciation)	–421	–455	–517	–577	–632
+ Accounting depreciation	39	40	41	42	43
Net operating revenue before taxes	114	121	130	135	142
Taxes (@45%)	– 51	– 54	– 59	– 61	– 64
Net operating income	63	67	71	74	78
Tax shields	60	42	29	21	7
Net after-tax cash flow	$123	$109	$100	$ 95	$ 85

For Tabbypaw,

$$WCC = (.40)(.07) + (.60)(.208) = .1528, \text{ or } 15.28\%$$

Step 3: **Next we must calculate the present value of the incremental cash flows expected from the target firm, as given in Table 16-4. Assuming that cash flows do not change after 2000, but continue at the 2000 level in perpetuity, and discounting these cash flows at 15.28 percent, we get**

$$\begin{pmatrix}\text{present value}\\ \text{of all}\\ \text{cash flows}\end{pmatrix} = \begin{pmatrix}\text{present value}\\ \text{of 1996–1999}\\ \text{cash flows}\end{pmatrix} + \begin{pmatrix}\text{present value of cash}\\ \text{flows of 2000}\\ \text{and thereafter}\end{pmatrix} \qquad (16\text{-}2)$$

where the present value of cash flows for the first four years of 1996 through 1999 would be $307.408 million, determined as follows:

$$PV(\text{cash flows 1996–1999}) = \frac{\$123}{(1.1528)^1} + \frac{\$109}{(1.1528)^2} + \frac{\$100}{(1.1528)^3} + \frac{\$95}{(1.1528)^4}$$

and the present value of the $85 million cash flow stream,[8] beginning in 2000 is computed to be $314.978

$$PV(\text{cash flows 2000, thereafter}) = \frac{\left(\dfrac{\$85}{.1528}\right)}{(1.1528)^4} = \$314.978$$

Thus, the present value of the cash inflows associated with the acquisition of Tabbypaw Pie by ALF is $622.759 million, or $622,759,000; that is, the sum of $307,781,000 and $314,978,000.

Step 4: **Next, we estimate the initial outlay associated with the acquisition. As already noted, the initial outlay is defined as the market value of all securities and cash paid out plus the market value of all debt liabilities assumed. In this case, the market value of the debt obligations is $110 million. This amount, along with the acquisition price of $260 million, comprise the initial outlay of $370 million used to calculate the acquisition's initial outlay.**

---

[8]Remember that we find the present value of an infinite stream of cash flows, where the amount is constant in each year, as follows:

$$\text{value} = \frac{\text{annual cash flow}}{\text{required rate of return}}$$

Since the cash flows do not begin until the fifth year, our equation is finding the value at the end of the fourth year; thus, we must discount the value back four years.

Step 5: **Finally, the net present value of the acquisition is calculated by subtracting the initial outlay (calculated in step 4) from the present value of the incremental cash flows from the target firm (calculated in step 3):**

$$
\begin{aligned}
NPV\text{acquisition} &= PV\text{inflows} - \text{initial outlay} && (16\text{-}3)\\
&= \$622{,}759{,}000 - \$370{,}000{,}000\\
&= \$252{,}759{,}000
\end{aligned}
$$

Thus, this acquisition should be undertaken since it has a positive net present value. In fact, ALF could pay up to $622,759,000 million for Tabbypaw Pie.

PERSPECTIVE IN FINANCE

Leasing provides an alternative to buying an asset in order to acquire its services. Although some leases involve maturities of more than ten years, most do not. Thus, lease financing is classified as a source of intermediate term credit. Today, virtually any type of asset can be acquired through a lease agreement.

LEASING VERSUS PURCHASING

OBJECTIVE 4

We begin our discussion by defining the major types of lease arrangements. Next we briefly review the history and describe the present accounting practices for leasing. We examine the lease versus purchase decision, and we conclude by investigating the potential benefits of leasing.

Types of Lease Arrangements

There are four major types of lease agreements: direct leasing, net and net-net leasing, sale and leaseback, and leveraged leasing. Most lease agreements fall into one of these categories. However, the particular lease agreement can take one of two forms. (1) The **financial lease** constitutes a noncancellable contractual commitment on the part of the firm leasing the asset (the lessee) to make a series of payments to the firm that actually owns the asset (the lessor) for use of the asset. (2) The **operating lease** differs from the financial lease only with respect to its cancellability. An operating lease can be cancelled after proper notice to the lessor any time during its term. Thus, operating leases are by their very nature sources of short-term financing. The balance of this chapter is concerned with the financial lease, which provides the firm with a form of intermediate-term financing most comparable to debt financing.

Financial lease

A noncancellable contractual commitment on the part of the firm leasing the asset (the lessee) to make a series of payments to the firm that actually owns the asset (the lessor) for the use of the asset.

Operating lease

A contractual commitment on the part of the firm leasing the asset (the lessee) to make a series of payments to the firm that actually owns the asset (the lessor) for the use of the asset. An operating lease differs from a financial lease in that it can be cancelled at any time after proper notice has been given to the lessor.

Direct Leasing

In a **direct lease** the firm acquires the services of an asset it did not previously own. Direct leasing is available through a number of financial institutions, including manufacturers, banks, finance companies, independent leasing companies, and special purpose leasing companies.[9] In the lease arrangement, the lessor purchases the asset and leases it to the lessee. In the case of the manufacturer lessor, however, the acquisition step is not necessary.

Direct lease

In this type of leasing arrangement, the lessor purchases the asset and leases it to the lessee.

[9]Many leasing companies specialize in the leasing of a single type of asset. For example, a number of firms lease computers exclusively, and others lease only automobiles.

Sale and Leaseback

Sale and leaseback

An arrangement arising when a firm sells land, buildings, or equipment that it already owns and simultaneously enters into an agreement to lease the property back for a specified period under specific terms.

A **sale and leaseback** arrangement arises when a firm sells land, buildings, or equipment that it already owns to a lessor and simultaneously enters into an agreement to lease the property back for a specified period under specific terms. The lessor involved in the sale and leaseback varies with the nature of the property involved and the lease period. When land is involved and the corresponding lease is long term, the lessor is generally a life insurance company. If the property consists of machinery and equipment, then the maturity of the lease will probably be intermediate term and the lessor could be an insurance company, commercial bank, or leasing company.

The lessee firm receives cash in the amount of the sales price of the assets sold and the use of the asset over the term of the lease. In return, the lessee must make periodic rental payments through the term of the lease and give up any salvage or residual value to the lessor.

Net and Net-Net Leases

Net and net-net leases

In a net lease agreement the lessee assumes the risk and burden of ownership over the term of the lease. This type of agreement means that the lessee must pay insurance and taxes on the asset as well as maintain the operating condition of the asset. In a net-net lease the lessee must, in addition to the requirements of a net lease, return the asset to the lessor at the end of the lease still worth a pre-established value.

In the jargon of the leasing industry, a financial lease can take one of two basic forms: a net lease or a net-net lease. In a **net lease** agreement the lessee firm assumes the risk and burden of ownership over the term of the lease. That is, the lessee must maintain the asset, as well as pay insurance and taxes on the asset. A **net-net lease** requires that the lessee meet all the requirements of a net lease as well as return the asset, still worth a pre-established value, to the lessor at the end of the lease term.

Leveraged Leasing

Leveraged lease

A leasing arrangement in which the lessor will generally supply equity funds of 20 to 30 percent of the purchase price and borrow the remainder from a third-party lender.

In the leasing arrangements discussed thus far, only two participants have been identified: the lessor and lessee. In a **leveraged lease** a third participant is added. The added party is the lender who helps finance the acquisition of the asset to be leased. From the point of view of the lessee there is no difference in a leveraged lease, direct lease, or sale and leaseback arrangement. However, with a leveraged lease, the method of financing used by the lessor in acquiring the asset receives special consideration. The lessor will generally supply equity funds to 20 to 30 percent of the purchase price and borrow the remainder from a third-party lender, which may be a commercial bank or insurance company. In some arrangements the lessor firm sells bonds, which are guaranteed by the lessee. This guarantee serves to reduce the risk and thus the cost of the debt. The majority of financial leases are leveraged leases.

OBJECTIVE 5

The Lease Versus Purchase Decision

The lease versus purchase decision is a hybrid capital-budgeting problem that forces the analyst to consider the consequences of alternative forms of financing on the investment decision. When we discussed the cost of capital in Chapter 12 and capital budgeting in Chapters 13–15, we assumed that all new financing would be undertaken in accordance with the firm's optimal capital structure. When analyzing an asset that is to be leased, the analysis must be altered to consider financing through leasing as opposed to the use of the more traditional debt and equity sources of funds. Thus, the lease versus purchase decision requires a standard capital-budgeting type of analysis, as well as an analysis of two alternative financing packages. The lease-purchase decision involves the analysis of two basic issues:

1. Should the asset be purchased using the firm's optimal financing mix?
2. Should the asset be financed using a financial lease?

The answer to the first question can be obtained through an analysis of the project's net present value (*NPV*) following the method laid out in Chapters 13 and 14. However, regardless of whether the asset should or should not be purchased, it may be advantageous for the firm to lease it. That is, the cost savings accruing through leasing might be great enough to offset a negative net present value resulting from the purchase of an asset. For example, the Alpha Mfg. Co. is considering the acquisition of a new computer-based inventory and payroll system. The computed net present value of the new system based upon normal purchase financing is –$40, indicating that acquisition of the system through purchasing or ownership is not warranted. However, an analysis of the cost savings resulting from leasing the system (referred to here as the net advantage of leasing—*NAL*) indicates that the lease alternative will produce a present value cost saving of $60 over normal purchase financing. Therefore, the net present value of the system if leased is $20 (the net present value if leased equals the *NPV* of a purchase plus the net advantage of leasing, or – $40 + $60). Thus, the system's services should be acquired via the lease agreement.

In the pages that follow we will (1) review briefly the concept of a project's net present value, which we will refer to as the net present value of purchase, or *NPV(P)*; (2) introduce a model for estimating the net present value advantage of leasing over normal purchase financing, which we will refer to as the net advantage of lease financing, or *NAL*; (3) present a flow chart that can be used in performing lease-purchase analyses based upon *NPV(P)* and *NAL*; and (4) provide a comprehensive example of a lease-purchase analysis.

The Lease-Purchase Algorithm

Answers to both questions posed above can be obtained using the two equations found in Table 16-5. The first equation is simply the net present value of purchasing the proposed project, discussed in Chapters 13 and 14. The second equation determines the net present value advantage of leasing (*NAL*). *NAL* represents an accumulation of the cash flows (both inflows and outflows) associated with leasing as opposed to purchasing the asset. Specifically, through leasing the firm avoids certain operating expenses, O_t, but incurs the after-tax rental expense, $R_t(1 - T)$. By leasing, furthermore, the firm loses the tax-deductible expense associated with depreciation, $T(CCA_t)$. Finally, the firm does not receive the salvage value from the asset, SV_n, if it is leased, but it does not have to make the initial cash outlay to purchase the asset, *IO*. Thus, *NAL* reflects the cost savings associated with leasing, net of the opportunity costs of not purchasing.

Note that the after-tax cost of new debt is used to discount the *NAL* cash flows other than the salvage value, SV_n. This is justified on the basis that the affected cash flows are very nearly riskless and certainly no more risky than the interest and principal accruing to the firm's creditors (which underlie the rate of interest charged to the firm for its debt).[10] Since SV_n is not a risk-free cash flow but depends upon the market price for the leased asset in year *n*, a rate higher than k_b, the interest rate before taxes, is appropriate. Since the salvage value of the leased asset was discounted using the weighted cost of capital when determining *NPV(P)*, we again use this rate here when calculating *NAL*.

---

[10]The argument for using the firm's borrowing rate to discount these tax shelters goes as follows: The tax shields are relatively free of risk in that their source (depreciation, interest, rental payments) can be estimated with a high degree of certainty. There are, however, two sources of uncertainty with regard to these tax shelters: (1) the possibility of a change in the firm's tax rate and (2) the possibility that the firm might become bankrupt at some future date. If we attach a very low probability to the likelihood of a reduction in the tax rate, then the prime risk associated with these tax shelters is the possibility of bankruptcy wherein they would be lost forever (certainly, all tax shelters after the date of bankruptcy would be lost). We now note that the firm's creditors also faced the risk of the firm's bankruptcy when they lent the firm funds at the rate k_b. If this rate, k_b, reflects the market's assessment of the firm's bankruptcy potential as well as the time value of money, then it offers an appropriate rate for discounting the interest shelters generated by the firm. Note also that the O_t cash flows are generally estimated with a high degree of certainty (in the case where they represent insurance premiums they may be contractually set) such that K_b, the interest rate after taxes, is appropriate as a discount rate here also.

Table 16-5 The Lease-Purchase Model

Equation One—Net present value of purchase [$NPV(P)$]:

$$NPV(P) = \sum_{t=1}^{n} \frac{CFAT_t}{(1+K)^t} - IO \quad (16\text{-}4)$$

where $CFAT_t$ = the annual cash flow after tax in period t resulting from the asset's purchase (note that $CFAT_n$ also includes any salvage value expected from the project and the tax effects which occur from the termination of the asset class).

K = the firm's weighted cost of capital applicable to the project being analyzed and the particular mix of financing used to acquire the project.

IO = the initial cash outlay required to purchase the asset in period zero (now).

n = the productive life of the project.

Equation Two—Net Advantage of Leasing (NAL):

$$NAL = \sum_{t=1}^{n} \frac{O_t(1-T) - R_t(1-T) - I_t(T) - T(CCA_t)}{(1+K_b)^t} - \frac{SV_n}{(1+K_s)^n} - \frac{(UCC_n - SV_n)(d)(T)}{(1+K_s)^n} + IO \quad (16\text{-}5)$$

where O_t = any net operating income incurred in period t that is incurred only where the asset is purchased. Most often this consists of maintenance expenses and insurance that would be paid by the lessor

R_t = the annual rental for period t

T = the combined tax rate on corporate income

CCA_t = capital cost allowance (depreciation) in period t for the asset

SV_n = the salvage value of the asset expected in year n

K_s = the discount rate used to find the present value of SV_n. This rate should reflect the risk inherent in the estimated SV_n. For simplicity the weighted cost of capital (K) is often used as a proxy for this rate. Also, note that this rate is the same one used to discount the salvage value in $NPV(P)$.

UCC_n = the undepreciated capital cost of the asset class in year n

IO = the purchase price of the asset, which is not paid by the firm in the event the asset is leased

K_b = the after-tax rate of interest on borrowed funds (i.e., $K_b = k_b(1-T)$ where k_b is the before-tax borrowing rate for the firm). This rate is used to discount the relatively certain cash flow after-tax savings that accrue through the leasing of the asset.

Note: This analysis makes the implicit assumption that a dollar of lease financing is equivalent to a dollar of loan. This form of equivalence is only one of several that might be used. The interested reader is referred to A. H. Ofer, "The Evaluation of the Lease Versus Purchase Alternatives," *Financial Management* 5 (Summer 1976), pp. 67–74.

Figure 16-1 contains a flow chart that can be used in performing lease purchase analyses. The analyst first calculates *NPV(P)*. If the project's net present value is positive, then the left-hand branch of Figure 16-1 should be followed. Tracing through the left branch we now compute *NAL*. If *NAL* is positive, the lease alternative offers a positive present-value cost advantage over normal purchase financing in which case the asset should be leased. Should *NAL* be negative, then the purchase alternative should be selected. Return to the top of Figure 16-1 once again. This time we assume that *NPV(P)* is negative and the analyst's attention is directed to the right-hand side of the flow chart. The only hope for the project's acceptance at this point is a favourable set of lease terms. In this circumstance the project would be acceptable and thus leased only if *NAL* were large enough to offset the negative *NPV(P)* [that is, where *NAL* was greater than the absolute value of *NPV(P)* or, equivalently, where $NAL + NPV(P) \geq 0$].[11]

[11]That is, $NAL = NPV(L) - NPV(P)$, where $NPV(L)$ is the net present value of the asset if leased. Thus, the sum of $NPV(P)$ and NAL is the net present value of the asset if leased. See Lawrence D. Schall, "The Lease-or-Buy and Asset Acquisition Decisions," *Journal of Finance* 29 (September 1974), pp. 1203–1214, for a development of net present value of leasing and purchasing equations.

Figure 16-1 Lease-Purchase Analysis

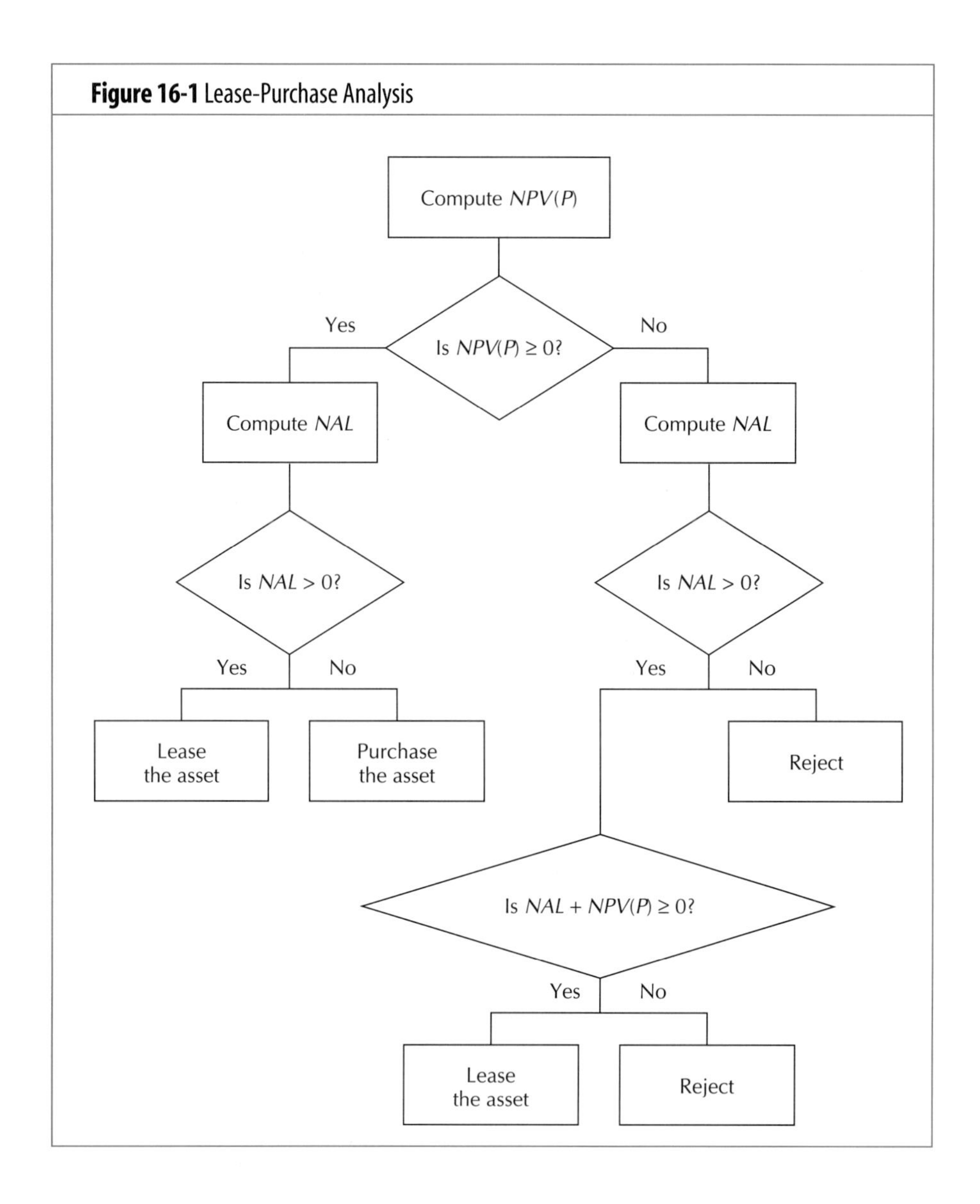

Case Problem in Lease-Purchase Analysis

The Waynesboro Plastic Molding Company (WPM) is now deciding whether to purchase an automatic casting machine. The machine will cost $15,000 and for tax purposes will be depreciated at the CCA rate of 30 percent using a declining balance method of depreciation. The machine's expected useful life is five years and it is expected to have a salvage value of $5,000 at the end of the five years.[12] The asset class will be kept open after the machine is salvaged. For accounting purposes, the firm will depreciate the machine over a five-year period using a straight-line depreciation method. Over the next five years, the project is expected to generate annual operating revenues of $16,000 per year and annual operating expenses of $4,000 (including accounting depreciation). The annual payments for leasing an automatic casting machine are $3,800. WPM has a combined corporate tax rate of 50 percent and a target debt ratio of 40 percent for projects of this type. Finally, WPM esti-

[12]The problem example is a modification of the well-known example from R. W. Johnson and W. G. Lewellen, "Analysis of the Lease-or-Buy Decision," *Journal of Finance* 27 (September 1972), pp. 815–823.

mates that its after-tax cost of capital for projects of similar risk is 12 percent and it can borrow funds at a before-tax rate of 8 percent.

Computing* NPV(P)*—Should the Asset Be Purchased? The first step in analyzing the lease-purchase problem involves computing the net present value under the purchase alternative. The relevant cash flow computations are presented in Table 16-6.

The annual cash flows after tax are discounted at the weighted cost of capital of 12 percent. Table 16-7 shows that the *NPV* of the project is the sum of the discounted cash flows, the present value of the salvage value, the lost tax shields of the project and the initial cash outlay.

The second question concerns whether the asset should be leased. This can be answered by considering the net advantage to leasing (*NAL*).

Table 16-6 Computing Project Annual Cash Flows After Tax ($CFAT_t$) Associated with Asset Purchase

Year:	1	2	3	4	5
Operating revenue	$16,000	$16,000	$16,000	$16,000	$16,000
Operating expense (including accounting depreciation)	($ 4,000)	(4,000)	(4,000)	(4,000)	(4,000)
Accounting depreciation	3,000	3,000	3,000	3,000	3,000
Net operating revenue	15,000	15,000	15,000	15,000	15,000
Taxes	(7,500)	(7,500)	(7,500)	(7,500)	(7,500)
Net operating income	7,500	7,500	7,500	7,500	7,500
Tax shields[a]	1,125	1,913	1,339	937	656
Cash flows after tax	$ 8,625	9,413	8,839	8,437	8,156

[a] The schedule for tax shields is determined as follows:

Year	*UCC*(beg)	*CCA*	Tax Shields	*UCC*(end)
1	$ 7,500.00	$2,250.00	$1,125.00	$12,750.00
2	12,750.00	3,825.00	1,912.50	8,925.00
3	8,925.00	2,677.50	1,338.75	6,247.50
4	6,247.50	1,874.25	937.13	4,373.25
5	4,373.25	1,311.98	655.99	3,061.28

Table 16-7 Calculating *NPV*(*P*)

Year t	Annual Cash Flow $CFAT_t$	Discount Factor for 12 Percent	Present Value
1	$8,625	.8929	$7,701
2	9,413	.7972	7,504
3	8,839	.7118	6,292
4	8,437	.6355	5,362
5	8,156	.5674	4,628
5 (Salvage—SV_n)	5,000	.5674	2,837
5 Lost tax shields	−693[a]	.5674	−393
Initial cash flow			−15,000
		NPV(*P*)	$18,931

[a]lost tax shields $= \dfrac{(UCC_n - SV_n)(d)(T)}{(K + d)} = \dfrac{(\$3,061 - \$5,000)(.30)(.50)}{(.12 + .30)} = -\$693$

The project's *NPV*(*P*) is a positive $18,931, indicating that the asset should be acquired.

The calculation to determine *NAL* involves solving Equation (16-5) presented earlier in Table 16-5. To do this, we first estimate all those cash flows that are to be discounted at the firm's after-tax cost of debt, K_b. These include $O_t(1 - T)$, $R_t(1 - T)$, $I_t(T)$ and $CCA_t(T)$.

The operating expenses associated with the asset that will be paid by the lessor if we lease—that is, the O_t—generally consist of certain maintenance expenses and insurance. WPM estimates them to be $1,000 per year over the life of the project. The annual rental or lease payments, R_t, are given and equal $3,800.

The interest tax shelter lost because the asset is leased and not purchased must now be estimated. This tax shelter is lost because the firm does not borrow any money if it enters into the lease agreement. Table 16-8 contains the principal and interest components for a $15,000 loan. Note that the interest column supplies the needed information for the interest tax shelter that is lost if the asset is leased, I_t.[13]

The next steps in calculating *NAL* involve the determination of the present value of the salvage value and the present value of any lost tax shields. However, in the case of leasing, the lessee does not have ownership of the asset. As a result, the lessee cannot claim the benefits of the salvage value. Thus, the terms for the salvage value are subtracted from the benefits of leasing in Equation (16-5). The computation of *NAL* is shown in Table 16-9.

The resulting *NAL* is a negative (–$784) which indicates that purchasing is preferred to leasing.

Note that the lease payments used in this example were made at the end of each year. In practice, most lease payments are made at the beginning of each year. A series of constant payments made at the beginning of the year can be valued in terms of an annuity due.[14] The effect of paying leasing payments at the beginning of each year causes the computation for *NAL* to change. We examine this effect by redoing the previous example, as follows:

[13]Technically the firm does not lose the interest tax shelter on a $15,000 loan if it leases. In fact, the firm would lose the tax shelter on only that portion of the $15,000 purchase price that it would have financed by borrowing, for example, 40 percent of the $15,000 investment, or $6,000. However, if the firm leases the $15,000 asset, it has effectively used 100 percent levered (nonowner) financing. This means that the leasing of this project uses not only its 40 percent allotment of levered (debt) financing, but an additional 60 percent as well. Thus, by leasing the $15,000 asset the lessee forfeits the interest tax shelter on a 40 percent or $6,000 loan plus an additional 60 percent of the $15,000 purchase price, or an additional $9,000 loan. In total, leasing has caused the firm to forgo the interest tax savings on a loan equal to 100 percent of the leased asset's purchase price. Once again, we note that this analysis presumes $1 of lease financing is equivalent to $1 of loan financing. See Note to Table 16-5.

[14]We have examined the concept of annuity due in Chapter 4. You can easily verify this result as follows: Note first that changing from a regular annuity to an annuity due affects only the first and last annuity payments. In this example, this means that the first lease payment of $1,900 (after tax) is paid immediately such that its present value is $1,900. However, the final lease payment is now made at the beginning of year 4. The present value of the after-tax lease payment in the fifth year is –$243 (–$296 × .8219), the present value of the sum between the tax shelters from depreciation and the after-tax operating expenses paid by the lessor at the end of the year.

Step 1: Solving for Term 1: Present Value of After-Tax Lease Payments.

Year	After-tax Operating Expenses Paid by Lessor		After-tax Rental Expense		Tax Shelter on Loan Interest		Tax Shelter on Deprecia-tion		Total		Discount Factor at 4 Percent		Present Value
t	$O_t(1-T)$	–	$R_t(1-T)$	–	$I_t(T)$	–	$CCA_t(T)$	=	SUM	×	*DF*	=	*PV*
0	$ 0		$1,900		$ 0		$ 0		–$1,900		1.0000		–$1,900
1	500		1,900		600		1,125		–3,125		.9615		–3,005
2	500		1,900		498		1,913		–3,811		.9246		–3,524
3	500		1,900		387		1,339		–3,126		.8890		–2,779
4	500		1,900		268		937		–2,605		.8548		–2,227
5	500		0		140		656		–296		.8219		–243
											Sum of present value		–$13,678

Step 2:

Solving for Term 2 $= \dfrac{-SV_n}{(1+K_s)^n} = \dfrac{-\$5{,}000}{(1+.12)^5} = -\$5{,}000 \times .5674 = -\$2{,}837$

Step 3:

Term 3 $= -\left[\dfrac{(UCC_n - SV_n)(d)(T)}{(K_s + d)}\right]\left[\dfrac{1}{(1+K_s)^n}\right] = -\left[\dfrac{(3{,}061 - 5{,}000)(.30)(.50)}{(.12 + .30)}\right]\left[.5674\right] = \$393$

Step 4: *Term 4* = *I0* = $15,000

Step 5: *Calculate NAL* = *Term* 1 – *Term* 2 – *Term* 3 + *Term* 4

= –$13,678 – $2,837 + $393 + $15,000 = –$1,122

We adjust for the change from a regular annuity to an annuity due by including –$1,900 as the discounted cash flow at time $t = 0$ and changing the discounted cash flow for the fifth year to –$2,687 (–$243 – $2,837 + $393). In this example in which payments are made at the beginning of the year, the *NAL* is

Table 16-8 Term Loan Amortization Schedule

End of Year t	Instalment Payment[a] A_t	Interest[b] I_t	Principal Repayment[c] P_t	Remaining Balance[d] RB_t
0	—	—	—	$15,000.00
1	$3,756.57	$1,200.00	$2,556.57	12,443.43
2	3,756.57	995.47	2,761.10	9,682.33
3	3,756.57	774.59	2,981.98	6,700.35
4	3,756.57	536.03	3,220.54	3,479.81
5	3,756.57	278.38	3,478.19	1.62[e]

[a]The annual instalment payment, A_t, is found as follows:

$$A_t = \frac{\$15{,}000}{\sum_{t=1}^{5} \frac{1}{(1+.08)^t}} = 3{,}756.57$$

[b]Annual interest expense is equal to 8 percent of the outstanding loan balance. Thus, for year 1 the interest expense, I_1 is found as follows:

$I_1 = .08(\$15{,}000) = \$1{,}200$

[c]Principal repayment for year t, P_t, is the difference in the loan payment, A, and interest for the year. Thus, for year 1 we compute:

$P_1 = A_1 - I_1 = \$3{,}756.57 - \$1{,}200 = \$2{,}556.57$

[d]The remaining balance of the end of year 1, RB_1, is the difference in the remaining balance for the previous year, RB_0, and the principal payment in year 1, P_1. Thus, at the end of year 1,

$RB_1 = RB_0 - P_1 = \$15{,}000 - \$2{,}556.57 = \$12{,}443.43$

[e]The $1.62 difference in RB_4 and P_5 is due to rounding error.

Table 16-9 Computing the Net Advantage to Leasing (*NAL*)

Overview: To solve for *NAL* we use Equation (16-5), which was discussed in Table 16-5. This equation contains three terms and is repeated below for convenience.

$$NAL = \sum_{t=1}^{n} \frac{O_t(1-T) - R_t(1-T) - I_t(T) - T(CCA_t)}{(1+K_b)^t} - \frac{SV_n}{(1+K_s)^n} - \frac{(UCC_n \quad SV_n)(d)(T)}{(1+K_s)^n} + IO \quad (16\text{-}5)$$

Step 1:

$$\text{Solving for Term 1} = \sum_{t=1}^{n} \frac{O_t(1-T) - R_t(1-T) - I_t(T) - T(CCA_t)}{(1+K_b)^t}$$

Year	After-tax Operating Expenses Paid by Lessor[a]		After-tax Lease Expense[b]		Tax Shelter on Loan Interest[c]		Tax Shelter on Depreciation[d]		Total		Discount Factor at 4 Percent		Present Value
t	$O_t(1-T)$	−	$R_t(1-T)$	−	$I_t(T)$	−	$CCA_t(T)$	=	SUM	×	*DF*	=	*PV*
1	$500		$1,900		$600		$1,125		–$3,125		.9615		–$3,005
2	500		1,900		498		1,913		–3,811		.9246		–3,524
3	500		1,900		387		1,339		–3,126		.8890		–2,779
4	500		1,900		268		937		–2,605		.8548		–2,227
5	500		1,900		140		656		–2,196		.8219		–1,805
											Sum of present value		–$13,340

Step 2:

$$\text{Solving for Term 2} = \frac{-SV_n}{(1+K_s)^n} = \frac{-\$5{,}000}{(1+.12)^5} = -\$5{,}000 \times .5674^{e} = -\$2{,}837$$

Step 3:

$$\text{Term 3} = -\left[\frac{(UCC_n - SV_n)(d)(T)}{(K_s + d)}\right]\left[\frac{1}{(1+K_s)^n}\right] = -\left[\frac{(3{,}061 - 5{,}000)(.30)(.50)}{(.12 + .30)}\right]\left[.5674\right] = \$393$$

Step 4: *Term* 4 = *IO* = $15,000

Step 5: *Calculate NAL* = *Term* 1 − *Term* 2 − *Term* 3 + *Term* 4
= –$13,340 – $2,837 + $393 + $15,000 = –$784

[a]After-tax lessor-paid operating expenses are found by $O_t(1-T) = \$1{,}000(1-0.5) = \$500$.
[b]After-tax rent expense for year 1 is computed as follows: $R_t(1-T) = \$3{,}800(1-0.5) = \$1{,}900$
[c]The values of interest expense were calculated in Table 16-8 for a $15,000 loan. For year 1 the interest rate tax shelter is 0.5 × $1,200 = $600.
[d]The tax shelter from depreciation is found as follows: $(CCA_t)T = \$2{,}250 \times 0.5 = \$1{,}125$.
[e]K_s was estimated to be the same as the firm's weighted cost of capital, 12 percent.

–$1,122 (– $13,678 + $2,837 + $393 + $15,000). According to our lease or purchase decision-making strategy which was illustrated in Figure 16-1, the asset should not be leased since the *NPV*(*P*) is positive but the *NAL* is negative.

To summarize the procedure used to analyze the lease or purchase decision: first, the project's net present value is computed. This analysis produced a positive *NPV*(*P*) equal to $18,931, which indicated that the asset should be purchased. Upon computing the net advantage to leasing, we found that leasing is the preferred method of financing when lease payments are made at the end of the year. However, the asset should be purchased if the leasing payments are made at the beginning of the year.

Potential Benefits from Leasing[15]

A number of purported advantages have been associated with leasing as opposed to debt financing. These benefits include flexibility and convenience, lack of restric-

[15]The contributions of Paul F. Anderson in the preparation of this discussion are gratefully acknowledged.

tions, avoiding the risk of obsolescence, conservation of working capital, 100 percent financing, tax savings, and availability of credit. However, before we discuss the relative merits of each purported advantage, we review briefly the economic character of leasing and purchasing.

Figure 16-2 illustrates the participants and transactions involved in leasing (the right-hand side of the figure) and purchasing (the left-hand side). In purchasing, the asset is financed via the sale of securities and the purchaser acquires title to the asset (including both the use and salvage value of the asset). In leasing, the lessee acquires the use value of the asset but uses the lessor as an "intermediary" to finance and purchase the asset. The key feature of leasing as opposed to purchasing is the interjection of a financial intermediary (the lessor) into the scheme used to acquire the asset's services. Thus, the basic question that arises in lease-purchase analysis is, why does adding another financial intermediary (the lessor) save the lessee money? Some of the "traditional" answers to this question are discussed below. As you read through each, simply remember that the lessee is "hiring" the lessor to perform the functions associated with ownership that he or she would perform if the asset were purchased. Thus, for the lease to be "cheaper" than owning, the lessor must be able to perform these functions of ownership at a lower cost than the lessee could perform them.

Flexibility and Convenience

A variety of potential benefits are often included under the rubric of flexibility and convenience. It is argued, for example, that leasing provides the firm with flexibility because it allows for piecemeal financing of relatively small asset acquisitions. Debt financing of such acquisitions can be costly and difficult to arrange. Leases, on the other hand, may be arranged more quickly and with less documentation.

Another flexibility argument notes that leasing may allow a division or subsidiary manager to acquire equipment without the approval of the corporate capital-budgeting committee. Depending upon the firm, the manager may be able to avoid the time-consuming process of preparing and presenting a formal acquisition proposal.

A third flexibility advantage relates to the fact that some lease payment schedules may be structured to coincide with the revenues generated by the asset, or they may be timed to match seasonal fluctuations in a given industry. Thus, the firm is able to synchronize its lease payments with its cash cycle—an option rarely available with debt financing.

Arguments for the greater convenience of leasing take many forms. It is sometimes stated that leasing simplifies bookkeeping for tax purposes because it eliminates the need to prepare time-consuming depreciation tables and subsidiary fixed asset schedules. It is also pointed out that the fixed-payment nature of lease rentals allows more accurate forecasting of cash needs. Finally, leasing allows the firm to avoid the "problems" and "headaches" associated with ownership. Executives often note that leasing "keeps the company out of the real estate business." Implicit in this argument is the assumption that the firm's human and material resources may be more profitably allocated to its primary line of business and that it is better to allow the lessor to deal with the obligations associated with ownership.

It is difficult to generalize about the validity of the various arguments for greater flexibility and convenience in leasing. Some companies, under specific conditions, may find leasing advantageous for some of the reasons listed above. In practice, the tradeoffs are likely to be different for every firm. The relevant issue is often that of shifting functions. By leasing a piece of capital equipment, the firm may effectively shift bookkeeping, disposal of used equipment, and other functions to the lessor. The lessee will benefit in these situations if the lessor is able to perform the functions at a lower cost than the lessee and is willing to pass on the savings in a lower lease rate.

The arguments that follow should be viewed in a similar vein. The lessee must estimate how much this greater flexibility and convenience is worth. In many cases,

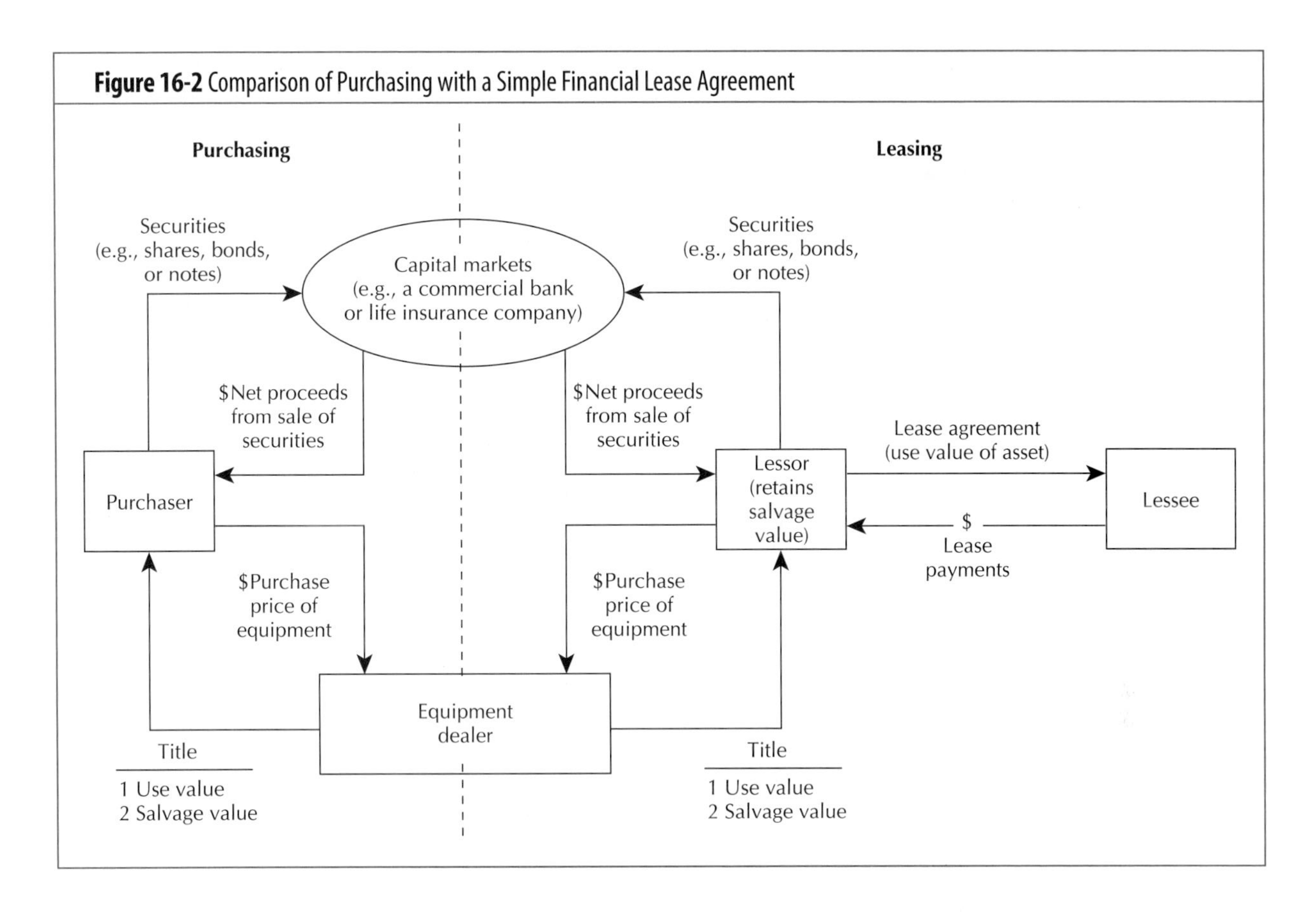

the benefits enjoyed by the firm are not worth the cost. Compounding the problem is the fact that it is often difficult for a lessee firm to quantify such cost-benefit tradeoffs.

Lack of Restrictions

Another suggested advantage relates to the lack of restrictions associated with a lease. Unlike term loan agreements or bond indentures, lease contracts generally do not contain protective covenant restrictions. Furthermore, in calculating financial ratios under existing covenants, it is sometimes possible to exclude lease payments from the firm's debt commitments. Once again, the extent to which lack of restrictions benefits a lessee will depend on the price it must pay. If a lessor views its security position to be superior to that of a lender, it may not require a higher return on the lease to compensate for the lack of restrictions on the lessee. On the other hand, if the prospective lessee is viewed as a marginal credit risk, a higher rate may be charged.

Avoiding the Risk of Obsolescence

Similar reasoning applies to another popular argument for leasing. This argument states that a lease is advantageous because it allows the firm to avoid the risk that the equipment will become obsolete. In actuality, the risk of obsolescence is passed on to the lessee in any financial lease. Since the original cost of the asset is fully amortized over the basic lease term, all of the risk is borne by the lessee. Only in a cancellable operating lease is it sometimes possible to avoid the risk of obsolescence.

A related argument in favour of leasing states that a lessor will generally provide the firm with better and more reliable service in order to maintain the resale value of the asset. The extent to which this is true depends on the lessor's own cost-benefit tradeoff. If the lessor is a manufacturing or a leasing company that specializes in a particular type of equipment, it may be profitable to maintain the equipment's resale value by ensuring that it is properly repaired and maintained. Because of their technical and marketing expertise, these types of lessors may be able to operate successfully in the secondary market for the equipment. On the other hand, bank lessors or independent financial leasing companies would probably find it too expensive to follow this approach.

Conservation of Working Capital

One of the oldest and most widely used arguments in favour of leasing is the assertion that a lease conserves the firm's working capital. Indeed, many managers within the leasing industry consider this to be the number one advantage of leasing.[16] The conservation argument runs as follows: Since a lease does not require an immediate outflow of cash to cover the full purchase price of the asset, funds are retained in the business.

It is clear that a lease does require a lower initial outlay than a cash purchase. However, the cash outlay associated with the purchase option can be reduced or eliminated by borrowing the down payment from another source. This argument leads us directly into the next purported advantage of lease financing.

100 Percent Financing

Another alleged benefit of leasing is embodied in the argument that a lease provides the firm with 100 percent financing. It is pointed out that the borrow-and-buy alternative generally involves a down payment, whereas leasing does not. Given that investors and creditors are reasonably intelligent, however, it is sensible to conclude that they consider similar amounts of lease and debt financing to add equivalent amounts of risk to the firm. Thus, a firm uses up less of its capacity to raise non-equity funds with debt than with leasing. In theory, it could issue a second debt instrument to make up the difference—that is, the down payment.

Tax Savings

The extent to which leasing provides an economic advantage must be weighed against the variables discussed in the *NAL* equation (16-5). For example, a firm which purchases an asset inherits the tax shields from depreciation during the asset's useful life. These deductions cannot be claimed by the lessee firm when leasing the asset. As a result, leasing can offer an economic advantage to the firm only if the tax deductions from the leasing payments are greater than the tax deductions from the capital cost allowances if the asset was purchased.

Ease of Obtaining Credit

Another alleged advantage of leasing is that firms with poor credit ratings are able to obtain assets through leases when they are unable to finance the acquisitions with debt capital. The obvious counterargument is that the firm will certainly face a high lease rate in order to compensate the lessor for bearing the risk of default.

[16]For example, see L. Rochwarger, "The Flexible World of Leasing," *Fortune* 39 (November 1974), pp. 56–59.

SUMMARY

Because bonds have a maturity date, we examined the problem of their retirement. Bonds may be retired at maturity, or they may be retired prior to maturity through the use of a call provision or a sinking fund. A cost-benefit analysis is used to accept or reject the bond-refunding decision.

Determining the value of a firm is a difficult task. The valuation procedure requires projecting the firm's future profitability, as well as considering the effects of joining two businesses into a single operation. In estimating a firm's value, several factors are frequently considered, including (1) the firm's book value, (2) its appraisal value, (3) the stock market value of a firm's common shares, (4) its chop-shop value and the present value of the future free cash flows.

There are four basic types of lease arrangements:

1. Direct lease
2. Sale and leaseback
3. Net and Net-Net leases
4. Leveraged lease

The lease-purchase decision is a hybrid capital-budgeting problem wherein the financial executive must consider both the investment and financing aspects of the decision. The method we recommend for analyzing the lease-versus-purchase choice involves first calculating the net present value of the asset if it were purchased. Then we calculate the net advantage of leasing over purchasing.

Many and varied factors are often claimed as advantages of leasing over the firm's usual debt-equity financing mix. Many of the arguments have been found to be at least partly fallacious. However, a complete lease-purchase analysis using a model similar to the one discussed here should provide a rational basis for determining the true advantages of lease financing.

STUDY QUESTIONS

16-1. Although the bond-refunding decision is analyzed in much the same way as the capital-budgeting decision, one major difference is the discount rate used. What discount rate is used in the refunding decision, and what is the rationale of using this discount rate?

16-2. Why is book value alone not a significant measure of the worth of a company?

16-3. Explain the underlining principles of the chop-shop valuation procedure.

16-4. Compare the net present value approach used in valuing a merger with the same approach in capital budgeting.

16-5. Define each of the following:

- **a.** Financial lease
- **b.** Sale and leaseback arrangement
- **c.** Leveraged leasing
- **d.** Operating lease

16-6. List and discuss each of the potential benefits from lease financing.

SELF-TEST PROBLEMS

ST-1. *(Refunding Decision)* The A. Fields Wildcats Corporation currently has outstanding a 20-year $10 million bond issue with a 9.25 percent interest rate callable at $1,030. Because of a decline in interest rates, Fields would be able to refund the issue with a $10 million issue of 8 percent 20-year bonds. The flotation costs for refunding the issue are $200,000. The firm expects a nominal return of 6.50 percent, compounded monthly. The overlap period during which both issues would be expected to be outstanding is one month. Assume that the old bonds were issued before 1987 such that tax shields from flotation costs were expensed in the year of issue. The firm has a combined corporate tax rate of 45 percent. Should the firm refund the old bond issue?

ST-2. *(Valuation of a firm)* Using the chop-shop approach, assign a value for the Calvert Corporation, where its common shares are currently trading at a total market price of $5 million. For Calvert, the accounting data set forth two business segments: auto sales and auto specialities. Data for the firm's two segments are as follows:

Business Segment	Segment Sales ($000)	Segment Assets ($000)	Segment Income ($000)
Auto sales	$3,000	$1,000	$150
Auto specialities	2,500	3,000	500
Total	$5,500	$4,000	$650

Industry data for "pure-play" firms have been compiled and are summarized as follows:

Business Segment	Capitalization/ Sales ($000)	Capitalization/ Assets ($000)	Capitalization/ Operating Income ($000)
Auto sales	1.40	3.20	18.00
Auto specialities	0.80	0.90	8.00

ST-3. *(Lease Versus Purchase Analysis)* Jensen Trucking Inc. is considering the possibility of leasing a $100,000 truck-servicing facility. This newly developed piece of equipment facilitates the cleaning and servicing of diesel tractors used on long-haul runs. The firm has evaluated the possible purchase of the equipment and found it to have an $8,000 net present value. However, an equipment leasing company has approached Jensen with an offer to lease the equipment for an annual rental charge of $30,457.85 payable at the beginning of each of the next four years. In addition, should Jensen lease the equipment it would receive insurance and maintenance valued at $4,000 per year (assume that this amount would be payable at the beginning of each year if purchased separately from the lease agreement). Additional information pertaining to the lease and purchase alternatives is found in the following table:

Acquisition price	$100,000
Useful life (used in analysis)	4 years
Salvage value (estimated)	$35,000
CCA rate	30%
Borrowing rate	12%
Combined corporate tax rate	45%
Weighted cost of capital (based on a target debt/total asset ratio of 30%)	16%
Accounting method of depreciation	Straight-line
Tax method of depreciation	Declining balance

The firm expects that the asset class will remain open after the truck-servicing facility is salvaged.

a. Calculate the net advantage of leasing (*NAL*) the equipment.

b. Should Jensen lease the equipment?

STUDY PROBLEMS (SET A)

16-1A. *(Refunding Decision)* The Reynolds Corporation currently has outstanding a 15-year \$225 million bond issue with an 11.25 percent interest rate callable at \$1,055. Because of a decline in interest rates, Reynolds would be able to refund the issue with a \$225 million issue of 8.5 percent 15-year bonds. The flotation costs for refunding the issue are \$325,000. The overlap period during which both issues would be expected to be outstanding is one month. The company expects a nominal return of 5.50% compounded monthly. Assume that the old bonds were issued before 1987 such that tax shields from flotation costs were expensed in the year of the issue. The firm has a combined corporate tax rate of 45 percent. Should the firm refund the old bond issue?

16-2A. *(Refunding Decision)* Three years ago the R. Wittman Corporation issued a 30-year \$50 million bond issue with a 7 percent interest rate callable at \$108. During the past three years interest rates have fallen, allowing the Wittman Corporation to replace this bond issue with a 27-year \$50 million bond issue with a coupon rate of 6 percent. Flotation costs will drain an additional \$400,000. The overlap period during which both issues are expected to be outstanding is one month. The firm expects a nominal return of 6.50 percent compounded monthly. Assume that the old bonds were issued before 1987 such that tax shields from flotation costs were expensed in the year of issue. The firm has a combined corporate tax rate of 45 percent. Should the firm refund the old bond issue?

16-3A. *(Lease Versus Purchase Analysis)* Early in the spring of 1996 the Jonesboro Steel Corporation (JSC) decided to purchase a small computer. The computer is designed to handle the inventory, payroll, shipping, and general clerical functions for small manufacturers like JSC. The firm estimates that the computer will cost \$60,000 to purchase and will last four years, at which time it can be salvaged for \$10,000. The firm's combined tax rate is 45 percent, and its weighted cost of capital for projects of this type is estimated to be 12 percent. Over the next four years the management of JSC feels the computer will reduce operating expenses by \$42,000 a year (accounting depreciation expenses included). For accounting purposes, the firm utilizes straight-line depreciation. However, the firm plans to depreciate the asset at a CCA rate of 30 percent using the declining balance method for tax purposes. The asset class is expected to remain open after the computer is salvaged.

JSC is also considering the possibility of leasing the computer. The computer sales firm has offered JSC a four-year lease contract with payments at the beginning of each year of \$18,000. In addition, if JSC leases the computer, the lessor will absorb insurance and maintenance expenses valued at \$2,000 per year. Thus, JSC will save \$2,000 at the beginning of each year it leases the asset (on a before-tax basis).

- **a.** Evaluate the net present value of the computer purchase. Should the computer be acquired via purchase? [*Hint:* Refer to Tables 16-6 and 16-7.]
- **b.** If JSC uses a 40 percent target debt to total assets ratio, evaluate the net present value advantage of leasing. JSC can borrow at a rate of 8 percent with annual installments paid over the next four years. [*Hint:* Recall that the interest tax shelter lost through leasing is based upon a loan equal to the full purchase price of the asset or \$60,000.]
- **c.** Should JSC lease the asset?

16-4A. *(Lease Versus Purchase Analysis)* S. S. Johnson Enterprises (SSJE) is evaluating the acquisition of a heavy-duty forklift with 20,000- to 24,000-pound lift capacity. SSJE can purchase the forklift through the use of its normal financing mix (30 percent debt and 70 percent common equity) or lease it. Pertinent details follow:

Acquisition price of the forklift	\$20,000
Useful life	4 years
Salvage value (estimated)	\$4,000
CCA rate	30%
Annual after-depreciation cash savings from the forklift	\$11,000
Rate of interest on a 4-year instalment loan	10%

Combined corporate tax rate	45%
Annual leasing payments at beginning of the year (4-year lease)	$6,000
Annual operating expenses at the beginning of the year included in the lease	$1,000
Weighted cost of capital	12%
Accounting method of depreciation	Straight-line
Tax method of depreciation	Declining balance

SSJE expects that the asset class will be kept open after the forklift is salvaged.

a. Evaluate whether the forklift acquisition is justified through purchase.

b. Should SSJE lease the asset?

16-5A. *(Chop-Shop Valuation)* Using the chop-shop approach, assign a value for Dabney Inc., whose common shares are currently trading at a total market price of $10 million. For Dabney, the accounting data sets forth three business segments: consumer wholesaling, specialty services, and retirement centres. Data for the firm's three segments are as follows:

Business Segment	Segment Sales ($000)	Segment Assets ($000)	Segment Income ($000)
Consumer wholesaling	$2,000	$1,500	$175
Specialty services	1,000	500	100
Retirement centres	3,500	4,000	600
Total	$6,500	$6,000	$875

Industry data for "pure-play" firms have been compiled and are summarized as follows:

Business Segment	Capitalization/ Sales	Capitalization/ Assets	Capitalization/ Operating Income
Consumer wholesaling	0.80	.70	12.00
Specialty services	1.20	1.00	6.00
Retirement centres	1.20	.70	7.00

16-6A. *(Chop-Shop Valuation)* Using the chop-shop approach, assign a value for Aramus Inc., whose common shares are currently trading at a total market price of $20 million. For Aramus the accounting data set forth three business segments: consumer wholesaling, specialty services, and retirement centres. Data for the firm's three segments are as follows:

Business Segment	Segment Sales ($000)	Segment Assets ($000)	Segment Income ($000)
Consumer wholesaling	$ 3,500	$ 1,000	$ 350
Specialty services	2,000	1,500	250
Retirement centres	6,500	8,500	1,200
Total	$12,000	$11,000	$1,800

Industry data for "pure-play" firms have been compiled and are summarized as follows:

Business Segment	Capitalization/ Sales	Capitalization/ Assets	Capitalization/ Operating Income
Consumer wholesaling	1.00	.80	8.00
Specialty services	.90	.80	10.00
Retirement centres	1.20	1.00	7.00

16-7A. *(Free Cash Flow Valuation)* The Argo Corporation is viewed as a possible takeover target by Hilary Inc. Currently, Argo uses 20 percent debt in its capital structure, but Hilary plans to increase the debt ratio to 30 percent if the acquisition is consummated. The after-tax cost of debt capital for Argo is estimated to be 8 percent, which holds constant. The cost of equity after the acquisition is expected to be 18 percent. The current market value of Argo's debt outstanding is $40 million, all of which will be assumed by Hilary. Hilary intends to pay $250 million in cash and common shares for all of Argo's shares in addition to assuming all of Argo's debt. Currently, the market price of Argo's common shares is $200 million. Selected items from Argo's financial data are as follows:

	1997	1998	1999	2000	Thereafter
Net sales	$200 M	$225 M	$240 M	$250 M	$275 M
Administrative and selling expenses	15	20	27	28	30
Depreciation	10	15	17	20	24
Capital expenditures	12	13	15	17	20

In addition, the cost of goods sold runs 60 percent of sales and the marginal tax rate is 45 percent. For tax purposes, the tax shields are $30, $21, $15, $10, $4 for the years 1997, 1998, 1999, 2000, and thereafter, respectively. Compute the net present value of the acquisition.

16-8A. *(Free Cash Flow Valuation)* The Whatever Corporation is viewed as a possible takeover target by Big Boy Inc. Currently, Whatever uses 25 percent debt in its capital structure, but Big Boy plans to increase the debt ratio to 40 percent if the acquisition is consummated. The after-tax cost of debt capital for Whatever is estimated to be 10 percent, which holds constant. The cost of equity after the acquisition is expected to be 20 percent. The current market value of Whatever's debt outstanding is $30 million, all of which will be assumed by Big Boy. Big Boy intends to pay $150 million in cash and common shares for all of Whatever's shares in addition to assuming all of Whatever's debt. Currently, the market price of Whatever's common shares is $200 million. Selected items from Whatever's financial data are as follows:

	1997	1998	1999	2000	Thereafter
Net sales	$300 M	$330 M	$375 M	$400 M	$425 M
Administrative and selling expenses	40	50	58	62	65
Depreciation	25	30	35	38	40
Capital expenditures	30	37	45	48	50

In addition, the cost of goods sold runs 60 percent of sales and the marginal tax rate is 45 percent. For tax purposes, the tax shields are $51, $36, $25, $18, $6 for the years 1997, 1998, 1999, 2000, and thereafter, respectively. Compute the net present value of the acquisition.

STUDY PROBLEMS (SET B)

16-1B. *(Refunding Decision)* The Hook'ems Corporation currently has outstanding a 20-year $12 million bond issue with a 9.5 percent interest rate callable at $103. Because of a decline in interest rates, Hook'ems would be able to refund the issue with a $12 million issue of 8 percent 20-year bonds. The flotation costs for refunding the issue are $250,000. The overlap period during which both issues would be expected to be outstanding is one month. The firm expects a nominal return of 7.50 percent compounded monthly. Assume that the old bonds were issued before 1987 such that tax shields from flotation costs were expensed in the year of issue. The firm has a combined corporate tax rate of 45 percent. Should the firm refund the old bond issue?

16-2B. *(Refunding Decision)* Three years ago the Boulder Corporation issued a 30-year $45 million bond issue with a 6.75 percent interest rate callable at $108. During the past three years interest rates have fallen, allowing the Wittman Corporation to replace this bond issue with a 27-year $50 million bond issue with a coupon rate of 6 percent. Flotation costs will drain an additional $350,000. The overlap period during which both issues are expected to be outstanding is one month. The firm expects a nominal return of 5.50 percent compounded monthly. Assume that the old bonds were issued before 1987 such that tax shields from flotation costs were expensed in the year of issue. The firm has a combined corporate tax rate of 45 percent.

16-3B. *(Chop-Shop Valuation)* Using the chop-shop approach, assign a value for Wrongway Inc., whose common shares are currently trading at a total market price of $10 million. For Wrongway the accounting data sets forth three business segments: consumer wholesaling, specialty services, and retirement centres. Data for the firm's three segments are as follows:

Business Segment	Segment/ Sales ($000)	Segment/ Assets ($000)	Segment/ Income ($000)
Consumer wholesaling	$2,200	$ 650	$200
Specialty services	1,000	700	150
Retirement centres	3,500	5,000	500
Total	$6,700	$6,300	$850

Industry data for "pure-play" firms have been compiled and are summarized as follows:

Business Segment	Capitalization/ Sales	Capitalization/ Assets	Capitalization/ Operating Income
Consumer wholesaling	0.80	1.00	8.00
Specialty services	1.20	.90	10.00
Retirement centres	1.20	1.10	12.00

16-4B. *(Free Cash Flow Valuation)* The Brown Corporation is viewed as a possible takeover target by Cicron Inc. Currently, Brown uses 20 percent debt in its capital structure, but Cicron plans to increase the debt ratio to 25 percent if the acquisition is consummated. The after-tax cost of debt capital for Brown is estimated to be 8 percent, which should not change. The cost of equity after the acquisition is expected to be 22 percent. The current market value of Brown's debt outstanding is $75 million, all of which will be assumed by Cicron. Cicron intends to pay $225 million in cash and common shares for all of Brown's shares in addition to assuming all of Brown's debt. Currently, the market price of Brown's common shares is $200 million. Selected items from Brown's financial data are as follows:

	1997	1998	1999	2000	Thereafter
Net sales	$260 M	$265 M	$280 M	$290 M	$300 M
Administrative and selling expenses	25	25	25	30	30
Depreciation	15	17	18	23	30
Capital expenditures	22	18	18	20	22

In addition, the cost of goods sold runs 60 percent of sales and the marginal tax rate is 45 percent. For tax purposes, the tax shields are $32, $22, $16, $11, $4 for the years 1997, 1998, 1999, 2000, and thereafter, respectively. Compute the net present value of the acquisition.

16-5B. *(Free Cash Flow Valuation)* The Little Corporation is viewed as a possible takeover target by Big, Inc. Currently, Little uses 20 percent debt in its capital structure, but Big plans to increase the debt ratio to 50 percent if the acquisition is consummated. The after-tax cost of debt capital for Little is estimated to be 15 percent, which holds constant. The cost of equity after the acquisition is expected to be 25 percent. The current market value of Little's debt outstanding is $12 million, all of which will be assumed by Big. Big intends to pay $15 million in cash and common shares for all of Little's shares in addition to assuming all of Little's debt. Currently, the market price of Little's common shares is $35 million. Selected items from Little's financial data are as follows:

	1997	1998	1999	2000	Thereafter
Net sales	$200 M	$220 M	$245 M	$275 M	$300 M
Administrative and selling expenses	30	35	38	40	45
Depreciation	18	20	22	25	30
Capital expenditures	20	22	25	28	30

In addition, the cost of goods sold runs 70 percent of sales and the marginal tax rate is 45 percent. For tax purposes, the tax shields are $32, $22, $16, $11, $4 for the years 1997, 1998, 1999, 2000, and thereafter, respectively. Compute the net present value of the acquisition.

16-6B. *(Lease Versus Purchase Analysis)* Early in the spring of 1996 Lubin Landscaping Inc. decided to purchase a truck-mounted lawn fertilizer tank and spray unit. The truck would replace its hand-held fertilizer tanks, providing substantial reductions in labour expense. The firm estimates that the truck will cost $65,000 to purchase and will last four years, at which time it can be salvaged for $8,000. The firm's combined tax rate is 45 percent, and its weighted cost of capital for projects of this type is estimated to be 14 percent. Over the next four years the management of Lubin feels the truck will reduce operating expenses by $45,250 a year (accounting depreciation included). For accounting purposes, the firm utilizes straight-line depreciation. However, the firm plans to depreciate the asset at a CCA rate of 30 percent using the declining balance method for tax purposes. The asset class is expected to remain open after the equipment is salvaged.

Lubin is also considering the possibility of leasing the truck. The truck sales firm has offered Lubin a four-year lease contract with payments at the beginning of the year of $20,000. In addition, if Lubin leases the truck, the lessor will absorb insurance and maintenance expenses valued at $2,250 per year. Thus, Lubin will save $2,250 at the beginning of each year if it leases the asset (on a before-tax basis).

a. Evaluate the net present value of the truck purchase. Should the truck be acquired via purchase? [*Hint:* Refer to Tables 16-6 and 16-7.]

b. If Lubin uses a 40 percent target debt to total assets ratio, evaluate the net present value advantage of leasing. Lubin can borrow at a rate of 10 percent with annual installments paid over the next four years. [*Hint:* Recall that the interest tax shelter lost through leasing is based upon a loan equal to the full purchase price of the asset or $65,000.]

c. Should Lubin lease the asset?

16-7B. *(Lease Versus Purchase Analysis)* Music Live Inc. a carnival-operating firm, is considering the acquisition of a new German-made carousel, with a passenger capacity of 30. Music Live can purchase the carousel through the use of its normal financing mix (30 percent debt and 70 percent common equity) or lease it. Pertinent details follow:

Acquisition price of the carousel	\$25,000
Useful life	4 years
Salvage value (estimated)	\$5,000
CCA rate	30%
Annual after depreciation cash savings from the carousel	\$13,250
Rate of interest on a 4-year instalment loan	11%
Combined corporate tax rate	45%
Annual leasing payments at beginning of the year (4-year lease)	\$7,000
Annual operating expenses at the beginning of the year included in the lease	\$1,250
Weighted cost of capital	13%
Accounting method of depreciation	Straight-line
Tax method of depreciation	Declining balance

The firm expects that the asset class will be kept open after the carousel is salvaged.

a. Evaluate whether the carousel acquisition is justified through purchase.

b. Should Music Live lease the asset?

SELF-TEST SOLUTIONS

SS-1.

Calculations Illustrating the Bond-Refunding Decision.

Step 1:

Calculate the costs.

(a) Determine the cost of the call premium:

($10,000,000/$1,000) × \$30 = \$300,000

(b) Determine the flotation costs:

Annual flotation costs = \$200,000/5 = \$40,000

Annual tax shield = ($40,000) × (0.45) = \$18,000 per year

Present value of all tax shields = \$18,000 × ($PVIFA_{4.4\%, 5}$)

= \$18,000 × (4.4022)

= \$79,240

Net flotation costs = flotation costs – present value of all tax shields

= \$200,000 − \$79,240

= \$120,760

(c) Determine overlap interest costs:

After-tax interest rate of old debenture issue = 9.25% × (1 − 0.45)

= 5.0875%

Interest cost = 0.050875 × (1/12) × ($10,000,000)

= \$42,395

(d) Determine total costs:

Total costs = Call premium + net flotation costs + overlap interest costs

= \$300,000 + \$120,760 + \$42,395

= \$463,155

Step 2:

Calculate the cash benefit from eliminating the old bonds through refunding.

(a) Determine present value of the interest savings:

$$\text{Present value of interest savings} = (\$10{,}000{,}000) \times (0.0925 - 0.08) \times (1 - 0.45) \times (PVIFA_{4.4\%,\,20})$$
$$= (\$68{,}750) \times (13.1214)$$
$$= \$902{,}096$$

(b) Determine the investment benefits during overlap period:

$$\text{Investment benefits} = (\$10{,}000{,}000) \times (0.00542)^{a} \times (1 - 0.45)$$
$$= \$29{,}810$$

(c) Determine total benefits:

$$\text{Total benefits} = \$902{,}096 + \$29{,}810$$
$$= \$931{,}906$$

Step 3:

Calculate the refunding decision's net present value.

(a) Present value of the benefits (from Step 2)	\$931,906
(b) Less present value of the costs (from Step 1)	463,155
(c) Equals net present value	\$468,751

[a]Monthly compounding: $[1 + (.065/12)]^{1} - 1 = 0.542\%$

SS-2.

Calvert Corporation:

Business Segment	Market Capitalization/ Sales	Segment Sales	Theoretical Values (\$000)
Auto sales	1.40	\$3,000	\$4,200
Auto specialities	0.80	2,500	2,000
			\$6,200

Business Segment	Market Capitalization/ Assets	Segment Assets	Theoretical Values (\$000)
Auto sales	3.20	\$1,000	\$3,200
Auto specialities	0.90	3,000	2,700
			\$5,900

Business Segment	Market Capitalization/ Income	Segment Income	Theoretical Values (\$000)
Auto sales	18.00	\$150	\$2,700
Auto specialities	8.00	500	4,000
			\$6,700

Theoretical value based on:	
Sales	\$6,200
Assets	5,900
Income	6,700
Average value	\$6,267

Based on this valuation, Calvert is currently selling for significantly less than its "chop-shop" valuation—\$5 million is \$6,267,000.

SS-3. **a.** *NAL* = –$22,346.

b. The *NAL* is negative, indicating that a purchase would offer cost savings over a lease and therefore a purchase should be used.

Year	After-tax Operating Expenses Paid by Lessor[a]		After-tax Rental Expense[b]		Tax Shelter on Loan Interest[c]		Tax Shelter on Depreciation[d]		Total		Discount Factor at 6.6 Percent		Present Value
t	$O_t(1-T)$	–	$R_t(1-T)$	–	$I_t(T)$	–	$CCA_t(T)$	=	SUM	×	*DF*	=	*PV*
0	$2,200		$16,752		0		0		–$14,552		1.0000		–$14,552
1	2,200		16,752		6,600		6,750		–27,902		.9381		–26,175
2	2,200		16,752		5,395		11,475		–31,422		.8800		–27,651
3	2,200		16,752		4,013		8,033		–26,598		.8255		–21,957
4	0		0		2,429		5,623		–8,052		.7744		–6,235
											Total		–$96,570
					Salvage value								–27,104
					Lost tax shields								–1,328
					Plus: Initial outlay								100,000
					NAL(–$96,570 – $27,104 + $1,328 + $100,000)								–$22,346

[a] $4,000 (1 – 0.45) = $2,200. For simplicity we assume that the tax shields on expenses paid at the beginning of the year are realized immediately.

[b] $16,752 = $30,458 (1 – 0.45)

[c] Based upon a $100,000 loan with four end-of-year instalments and a 12 percent rate of interest. The loan payments equal $32,923.

[d] The machine is depreciated at a CCA rate of 30 percent using a declining balance method. The following schedule shows the tax shields over four years:

Schedule of Tax Shields

Year	*UCC*(beg)	*CCA*	Tax Shields	*UCC*(end)
1	50,000	15,000	6,750	85,000
2	85,000	25,500	11,475	59,500
3	59,500	17,850	8,033	41,650
4	41,650	12,495	5,623	29,155

Introduction to Working-Capital Management

CHAPTER 17

LEARNING OBJECTIVES

After reading this chapter you should be able to

1. Define net working capital.
2. Describe the risk-return tradeoff involved in managing a firm's working capital.
3. Describe the hedging principle or principle of self-liquidating debt and the relevance of permanent and temporary sources of financing.

INTRODUCTION

Data from Statistics Canada indicate that in 1994, Canadian nonfinancial industries invested on average 28 cents in working capital for each $1 of sales of goods and services. Fortune *magazine reported that U.S. companies, on average, invested more than 15 cents in working capital for each $1 of sales. In 1990, American Standard fit into this mould very well with over $735 million invested in working capital. By 1993, American Standard had revenues totalling $4.2 billion but had reduced its net working capital roughly in half.*

In 1990, American Standard had three primary product lines: plumbing supplies, air conditioners, and brakes for trucks and buses. The firm faced static sales and huge interest payments (the result of a $3.1 billion junk bond issue used to stave off a hostile takeover attempt by Black & Decker in 1989). To improve the firm's operating performance, its chairman, Emmanuel Kampouris, introduced a strategy aimed at reducing the firm's $735 million in net working capital to zero by 1996. This would be feasible if the company can reduce its inventories so low that they can be financed without borrowing. The idea is to deliver goods and bill customers more rapidly so that customer payments are sufficient to pay for minimal stocks of inventories. Kampouris sought to accomplish this ambitious goal through implementation of a lean manufacturing system known as a demand flow technology. Under this system, plants manufacture products as a customers order them. Suppliers deliver straight to the assembly line, *thus reducing stocks of parts, and plants ship the products as soon as they are completed. The system dramatically reduces inventories of both parts and finished goods. To date, American Standard invests only 5 cents out of each sales dollar in working capital compared to the norm of 15 cents. By saving interest payments on supplies, the company has increased its cash flow by $60 million a year.*

CHAPTER PREVIEW

Working capital

The firm's total investment in current assets or assets which the firm expects to be converted into cash within a year or less.

In this chapter, we examine how a firm manages its investment in working capital. Traditionally, working capital is defined as the firm's total investment in current assets. Net working capital, on the other hand, is the difference between the firm's current assets and its current liabilities. We begin by showing the importance of properly managing current assets and current liabilities. Then, we examine the risk-return tradeoff that occurs between liquidity and profitability when a firm manages its working capital. Finally, we examine how the hedging principle of working-capital management can be used to determine the appropriate level of working capital for a firm.

OBJECTIVE 1

THE IMPORTANCE OF WORKING-CAPITAL MANAGEMENT

In finance, we refer to the term, **net working capital**, as the difference in the firm's current assets and its current liabilities:

net working capital = current assets − current liabilities (17-1)

Net working capital

The difference between the firm's current assets and its current liabilities.

Current assets comprise all assets that the firm expects to convert into cash within a year and includes cash, marketable securities, accounts receivable and inventories. Current liabilities include all current obligations that are expected to be eliminated within a year either through the use of current assets or by the creation of other current liabilities. Examples of current liabilities include: accounts payable, wages or salaries earned, property taxes, payroll taxes, and accrued income taxes. Thus, in managing the firm's net working capital, we are concerned with managing the firm's *liquidity*. This entails examining two related aspects of the firm's operations:

1. Managing the firm's current assets.
2. Managing the firm's short-term or current liabilities.

To understand the management of the firm's current assets and liabilities, we examine the elements found on both the right-hand and the left-hand sides of the balance sheet. The elements on the right-hand side represent the claims on the firm by creditors and owners and include the short-term liabilities, long-term debt, and shareholders' equity. The elements on the left-hand side represent the firm's cash-generating ability and include current assets, capital assets, and intangible assets. Thus, the claims on the right-hand side of the balance sheet reflect the method chosen by the firm to finance its assets or the firm's ability to generate cash. Furthermore, the maturity of the claims should follow the cash-generating characteristics of the assets being financed. For example, an asset that is expected to provide cash flows over an extended period of time such as five years should be financed with a claim that has a pattern of similar cash flow requirements. This financial management practice should also be followed when financing current assets by short-term financing claims.

The importance of managing both current assets and current liabilities cannot be overstated. As we will see, for many firms current assets represent over half of the total assets. Moreover, surveys of both small business owners and financial managers indicate that the majority of their time is taken by the management of the day-to-day operations of the firm. This is largely the management of current assets and liabilities. Finally, for smaller firms, the management of current assets and current liabilities takes on even greater importance. For smaller firms, access to capital markets, and the long-term sources of financing they supply, is limited. As such, smaller firms are forced to rely more heavily on short-term sources of financing, such as trade credit, accounts receivable, and inventory loans.

MANAGING CURRENT ASSETS AND LIABILITIES

Generally speaking, the greater the firm's investment in current assets, the greater its liquidity. A firm may choose to increase its liquidity by investing additional funds in cash or marketable securities. However, this type of investment involves a tradeoff because current assets earn little or no return. As a result, the firm will reduce its risk of insolvency only by reducing its overall return on invested funds, and vice versa.

Working-Capital Management and the Risk-Return Tradeoff

OBJECTIVE 2

The *risk-return* involved in managing a firm's *working capital* involves a tradeoff between a firm's liquidity and its profitability. By maintaining a large investment in current assets like cash, marketable securities or inventory a firm reduces the chance of production stoppages, lost sales from inventory shortages, and the inability to pay bills on time. However, a firm would be expected to earn a lower return on an investment in cash or inventory as compared to an investment in its fixed assets. The additional investment in working capital will cause the firm's return on investment to decrease because the profits are not maintained in the same proportion as the increase in assets.

BACK TO THE BASICS

Many of the working-capital decisions made by financial managers involve risk-return tradeoffs between liquidity and profitability. The principles that guide these decisions are the same ones set out in ***Axiom 1: The Risk-Return Tradeoff—We Won't Take on Additional Risk Unless We Expect to Be Compensated with Additional Return.*** *The firm which holds more current assets and uses more of both short- and long-term financing, will have less risk since it has greater liquidity, but will earn less of a return.*

The firm's use of current liabilities versus long-term debt also involves a risk-return tradeoff. Other things remaining the same, the greater the firm's reliance on short-term debt or current liabilities in financing its asset investments, the greater the risk of insolvency. *The use of current liabilities or short-term debt as opposed to long-term debt subjects the firm to a greater risk of insolvency for two reasons.*[1] *First, short-term debt, due to its maturity, must be repaid or rolled over more often, and so it increases the possibility that the firm's financial condition might deteriorate to a level at which the needed funds might not be available.*[2]

A second disadvantage of short-term debt is the uncertainty of interest costs from year to year. For example, a firm borrowing during a six-month period each

[1]Most current liabilities are financed with working-capital loans from banks. Many capital loans have provisions that require the borrower to be debt-free over some period of the lending cycle.

[2]The dangers of such a policy are readily apparent in the experiences of firms that have experienced insolvency. For example, Olympia and York Developments Ltd. had to retire its program of selling commercial paper during the first quarter of 1992 because the company was experiencing a liquidity crisis.

year to finance a seasonal expansion in current assets might incur a different rate of interest each year. This rate reflects the current rate of interest at the time of the loan, as well as the lender's perception of the firm's riskiness. If fixed rate long-term debt were used, the interest cost would be known for the entire period of the loan agreement.

On the other hand, the use of current liabilities offers some very real advantages, in that they can be less costly than long-term financing and they provide the firm with a flexible means of financing its fluctuating needs for assets. Current liabilities offer the firm a flexible source of financing. They can be used to match the timing of a firm's needs for short-term financing. If, for example, a firm needs funds for a three-month period during each year to finance a seasonal expansion in inventories, then a three-month loan can provide substantial cost savings over a long-term loan (even if the interest rate on short-term financing should be higher). The fact that the use of long-term debt in this situation involves borrowing for the entire year rather than for the period when the funds are needed increases the amount of interest the firm must pay.

This brings us to the second advantage generally associated with the use of short-term financing. In general, interest rates on short-term debt are lower than on long-term debt for a given borrower. Figure 17-1 shows the relationship between interest rates and loan maturity which is known as the **term structure of interest rates.**[3] An alternative to the use of short-term debt is long-term debt which can reduce a firm's risk of insolvency but at the expense of a reduction in its return on invested funds. Once again, we see that the risk-return tradeoff involves an increased risk of insolvency versus increased profitability. It should be noted that this term structure reflects the rates of interest applicable to a given borrower at a particular point in time; it would not, for example, describe the rates of interest available to another borrower or even those applicable to the same borrower at a different point in time.

Figure 17-1 The Term Structure of Interest Rates

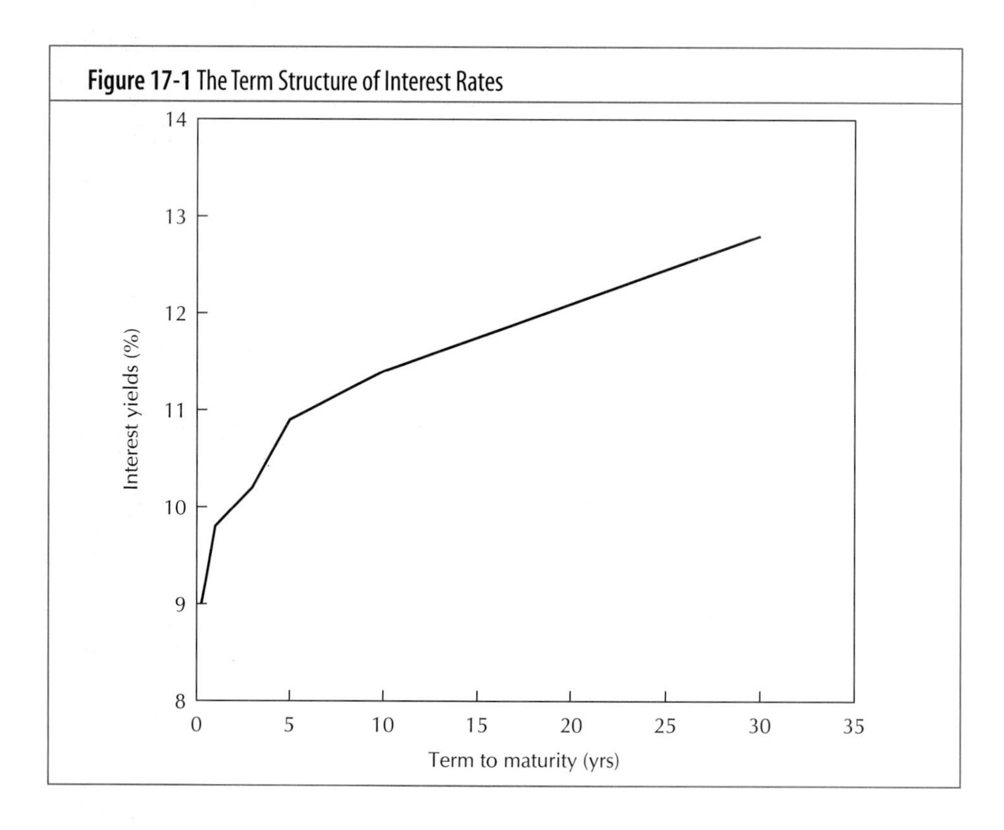

[3]The concept of term structure of interest rates was examined in detail in Chapter 2.

How Firms Manage Their Working Capital

So how do firms located in Canada manage their working capital? The following table provides 1992 data for the aggregate percent of current assets and current liabilities for two industries: Motor vehicles, motor vehicles parts and accessories manufacturing, as well as newspaper publishing and printing. The median for current assets in medium-sized firms in the motor vehicles, motor vehicles parts and accessories manufacturing industry is greater than the median for medium-sized firms in the newspaper publishing and printing industry. Similarly, the median for the current ratio of medium-sized firms in the motor vehicles, motor vehicles parts and accessories manufacturing industry is greater than the median for the current ratio of medium-sized firms in the newspaper publishing and printing industry. These results are consistent with the following trends in working-capital management. First, current assets are a major component of a firm's investment and can constitute as much as 65 percent of the firm's assets. Second, the current ratios vary both across the industries and over time. The message then is this. Working-capital management is extremely important to the firm's financial well-being and deserves serious consideration!

Source: Statistics Canada, *Financial Performance Indicators for Canadian Business* Catalogue No. 61F0058XPE (Ottawa: Minister of Industry, 1996).

	Motor Vehicle, Motor Vehicle Parts, and Accessories Manufacturing		Newspaper Publishing and Printing	
	Small Firms with Revenue between $500,000 and $5 Million	Medium Firms with Revenue between $5 Million and $25 Million	Small Firms with Revenue between $500,000 and $5 Million	Medium Firms with Revenue between $5 Million and $25 Million
Current assets (% of total assets)	65	63	43	37
Current liabilities (% of total assets)	53	54	25	60
Current ratio	0.89	1.28	1.77	0.89

OBJECTIVE 3

APPROPRIATE LEVEL OF WORKING CAPITAL

Hedging principle (principle of self-liquidating debt)

A working-capital management policy which states that the cash flow generating characteristics of a firm's investments should be matched with the cash flow requirements of the firm's sources of financing. Very simply, short-lived assets should be financed with short-term sources of financing and long-lived assets should be financed with long-term sources of financing.

Managing the firm's net working capital (that is, its liquidity) has been shown to involve simultaneous and interrelated decisions regarding investment in current assets and use of current liabilities. Fortunately, a guiding principle exists that can be used as a benchmark for the firm's working-capital policies: the **hedging principle**, or **principle of self-liquidating debt**. This principle provides a guide to the maintenance of a level of liquidity sufficient for the firm to meet its maturing obligations on time.

PERSPECTIVE IN FINANCE

In Chapter 9 we discussed the firm's financing decision in terms of the choice between debt and equity sources of financing. However, there is yet another critical dimension of the firm's financing decision. This relates to the maturity structure of the firm's debt. How should the decision be made as to whether to use short-term or current debt or longer maturity debt? This is one of the fundamental questions addressed in this chapter and one that is critically important to the financial success of the firm. Basically, the hedging principle is one possible rule of thumb for guiding a firm's debt maturity financing decisions. This principle states that financing maturity should follow the cash-generating characteristics of the asset being financed. For example, an asset that is expected to provide cash flows over an extended period of time such as five years should, in accordance with the hedging principle, be financed with debt instruments that have a pattern of similar cash flow requirements. Note that if the hedging principle is followed, the firm's debt will "self-liquidate" since the assets being financed will generally provide sufficient cash to retire the debt as it comes due.

Hedging Principle

Very simply, the hedging principle involves matching the cash-flow-generating characteristics of an asset with the maturity of the source of financing used to finance its acquisition. For example, a seasonal expansion in inventories, according to the hedging principle, should be financed with a short-term loan or current liability. The rationale underlying the rule is straightforward. Funds are needed for a limited period of time, and when that time has passed, the cash needed to repay the loan will be generated by the sale of the extra inventory items. Obtaining the needed funds from a long-term source (longer than one year) would mean that the firm would still have the funds after the inventories they helped finance had been sold. In this case the firm would have "excess" liquidity, which it either holds in cash or invests in low-yield marketable securities until the seasonal increase in inventories occurs again and the funds are needed.

Consider a second example in which a firm purchases a new conveyor belt system, which is expected to produce cash savings to the firm by eliminating the need for two labourers and, consequently, their salaries. This amounts to an annual savings of $14,000, while the conveyor belt costs $150,000 to install and will last 20 years. If the firm chooses to finance this asset with a one-year note, then it will not be able to repay the loan from the cash flow generated by the asset. In accordance with the hedging principle, the firm should finance the asset with a source of financing that more nearly matches the expected life and cash-generating characteristics of the asset.[4] In this case, a 15- to 20-year loan would be more appropriate.

[4]The maturity-matching concept is roughly equivalent to an immunization strategy based upon the concept of duration. Immunization provides protection from interest rate exposure, while the maturity-matching concept as it is applied to working capital management is intended to match up operating cash flows with cash flow requirements of the firm's financing sources. Maturity matching does not insure the firm against the risks of uncertain cash flow shortfalls but attempts simply to match up the expected cash flow producing capacity of the firm's assets with the cash flow requirements of the assets being financed.

Permanent and Temporary Assets

The notion of maturity matching in the hedging principle can be most easily understood when we think in terms of the distinction between permanent and temporary investments in assets as opposed to the more traditional fixed and current asset categories. A **permanent asset investment** in an asset is an investment that the firm expects to hold for a period longer than one year. Note that we are referring to the period of time the firm plans to hold an investment, and not the useful life of the asset. For example, permanent investments are made in the firm's minimum level of current assets, as well as in its fixed assets. **Temporary asset investments**, on the other hand, are comprised of current assets that will be liquidated and not replaced within the current year. Thus, some part of the firm's current assets is permanent and the remainder is temporary. For example, a seasonal increase in level of inventories is a temporary investment; the buildup in inventories will be eliminated when it is no longer needed.

Permanent asset investment

An investment in an asset that the firm expects to hold for the foreseeable future. Permanent investments can be made in fixed assets and current assets. For example, the minimum level of inventory the firm plans to hold for the foreseeable future is a permanent investment.

Temporary asset investment

Investments in assets that the firm plans to liquidate within a period no longer than one year. Although temporary investments can be made in fixed assets, this is not the usual case. Temporary investments generally are made in inventories and receivables.

Spontaneous, Temporary, and Permanent Sources of Financing

Since total assets must always equal the sum of spontaneous, temporary, and permanent sources of financing, the hedging approach provides the financial manager with the basis for determining the sources of financing to use at any point in time.

Now, what constitutes a temporary, permanent, and spontaneous source of financing? Temporary sources of financing consist of current liabilities. Short-term notes payable constitute the most common example of a temporary source of financing. Examples of notes payable include unsecured bank loans, commercial paper, and loans secured by accounts receivable and inventories. Permanent sources of financing include intermediate-term loans, long-term debt, preferred shares, and common equity.

Trade credit and other types of accounts payable are spontaneous sources of financing which expand or contract as a result of the level of a firm's business activity. For example, as the firm acquires materials for its inventories, the firm's suppliers will provide trade credit to the firm on demand. Once trade credit is granted, this source of financing is said to occur spontaneously with increasing or decreasing levels of business activity. Trade credit appears on the firm's balance sheet as accounts payable, and the size of the accounts payable balance varies directly with the firm's purchases of inventory items. In turn, inventory purchases are related to anticipated sales. Thus, part of the financing needed by the firm is spontaneously provided in the form of trade credit.

Trade credit

Credit made available by a firm's suppliers in conjunction with the acquisition of materials. Trade credit appears on the balance sheet as accounts payable.

In addition to trade credit, wages and salaries payable, accrued interest, and accrued taxes also provide valuable sources of spontaneous financing. These expenses accrue throughout the period until they are paid. For example, if a firm has a wage expense of $10,000 a week and pays its employees monthly, then its employees effectively provide financing equal to $10,000 by the end of the first week following a payday, $20,000 by the end of the second week, and so forth. Since these expenses generally arise in direct conjunction with the firm's ongoing level of operations, they too are referred to as *spontaneous.*

Hedging Principle: Graphic Illustration

The hedging principle can now be stated very succinctly—Asset needs of the firm not financed by spontaneous sources should be financed in accordance with this rule: Permanent asset investments are financed with permanent sources, and temporary investments are financed with temporary sources.

The hedging principle is depicted in Figure 17-2. Total assets are broken down into temporary and permanent asset investment categories. The firm's permanent

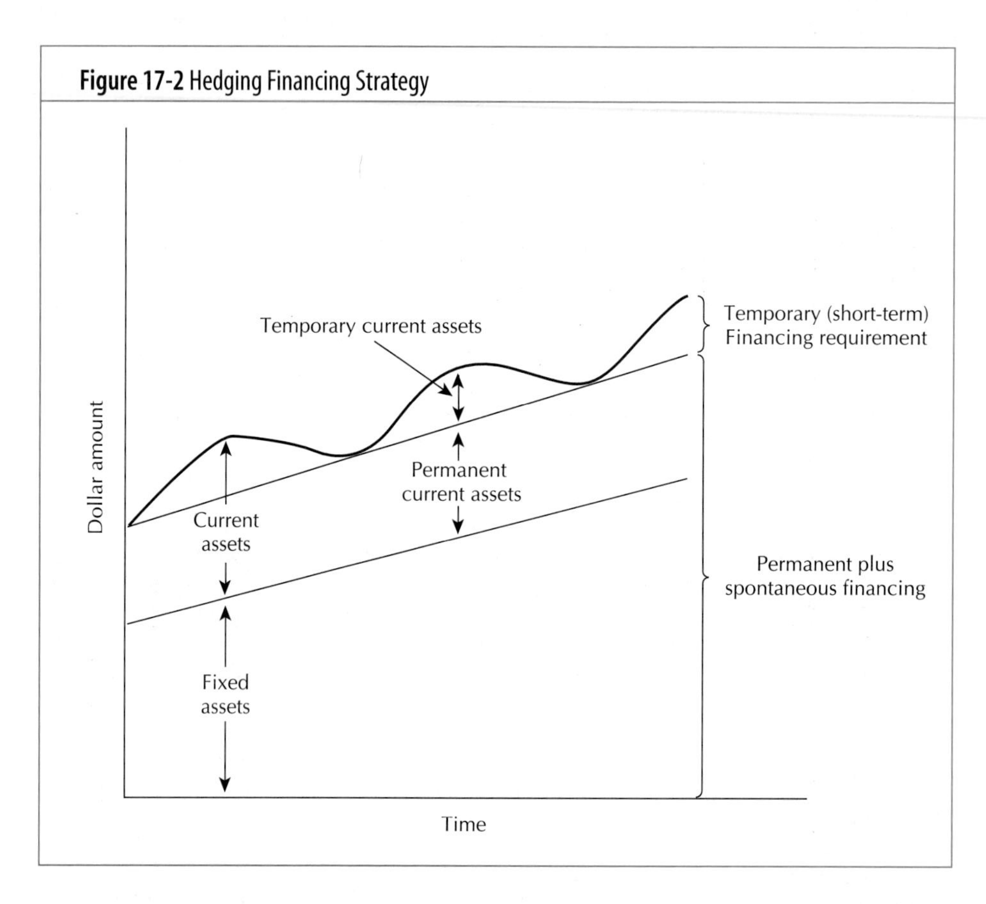

investment in assets is financed by the use of permanent sources of financing (intermediate- and long-term debt, preferred shares, and common equity) or spontaneous sources (trade credit and other accounts payable).[5] Its temporary investment in assets is financed with temporary (short-term) debt.

Modifications to the Hedging Principle

Figures 17-3 and 17-4 depict two modifications of the strict hedging approach to working-capital management. In Figure 17-3 the firm follows a more cautious plan whereby permanent sources of financing exceed permanent assets in the trough of the asset cycle so that excess cash is available (which must be invested in marketable securities).[6] Note that the firm actually has excess liquidity during the low ebb of its asset cycle, and thus faces a lower risk of being caught short of cash than a firm that follows the pure hedging approach. However, the firm also increases its investment in relatively low-yielding assets so that its return on investment is diminished.

In contrast, Figure 17-4 depicts a firm that continually finances a part of its permanent asset needs with temporary or short-term funds and thus follows a more aggressive strategy in managing its working capital. Even when its investment in asset needs is lowest, the firm must still rely on short-term or temporary financing. Such a firm would be subject to increased risks of a cash shortfall in that it must depend on a continual rollover or replacement of its short-term debt with more short-term debt. The benefit derived from following such a policy relates to the possible savings resulting from the use of lower-cost short-term debt.

[5]For illustration purposes, spontaneous sources of financing are treated as if their amount were fixed. In practice, of course, spontaneous sources of financing fluctuate with the firm's purchases and its expenditures for wages, salaries, taxes, and other items that are paid on a delayed basis.

[6]Marketable securities and cash management are discussed in Chapter 18.

Figure 17-3 Conservative Financing Strategy: Spontaneous Plus Permanent Sources of Financing Exceed Permanent Assets

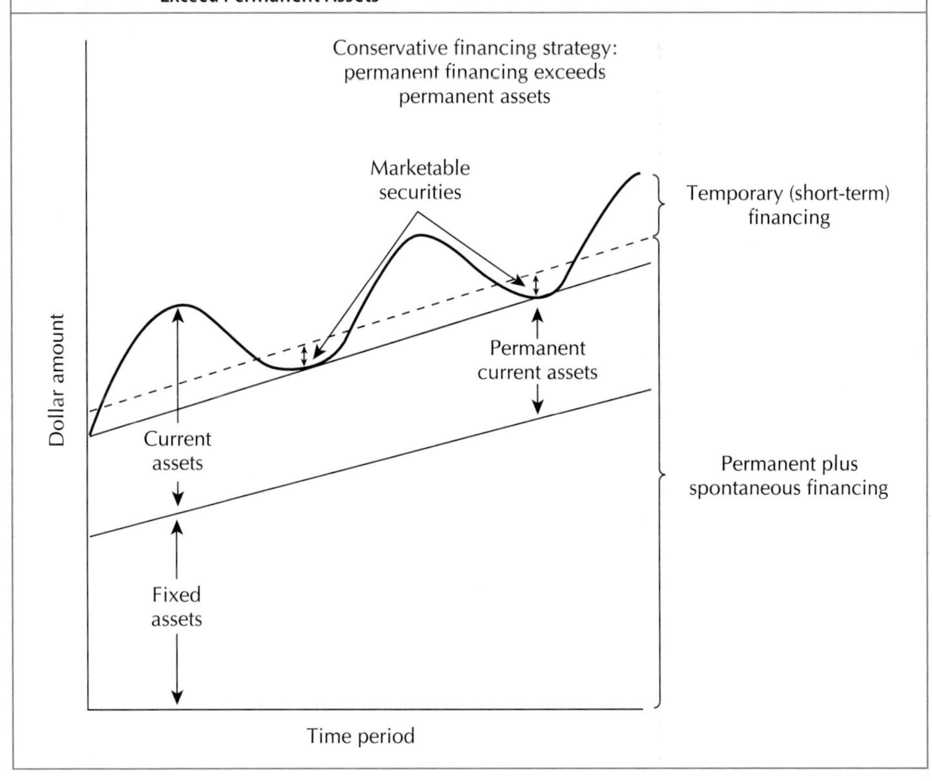

Figure 17-4 Aggressive Financing Strategy: Permanent Reliance on Short-Term Financing

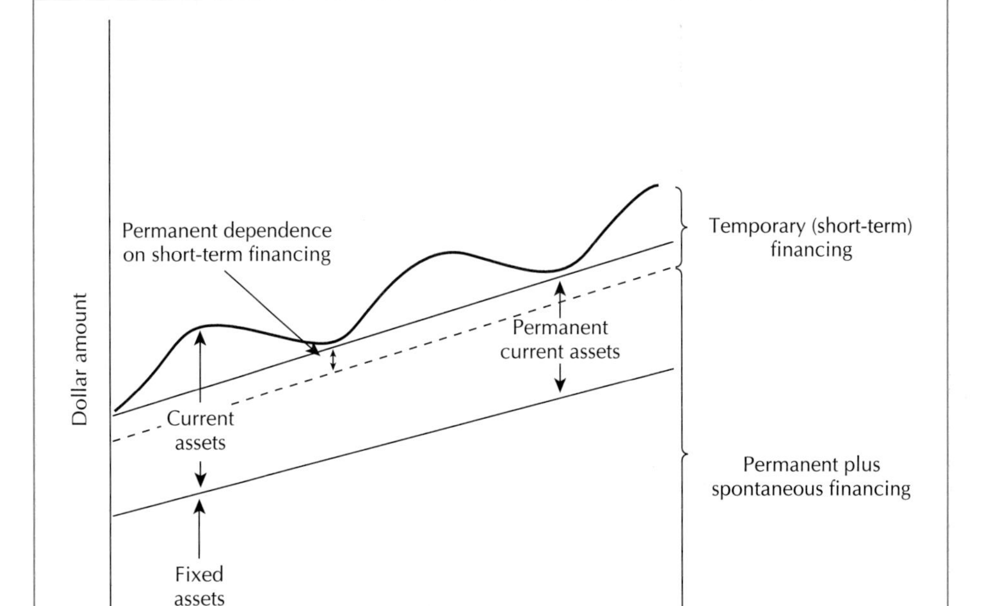

Most firms will not exclusively follow any one of the three strategies outlined here in determining their reliance on short-term credit. Instead, a firm will at times find itself overly reliant on permanent financing and thus holding excess cash; at other times it may have to rely on short-term financing throughout an entire operating cycle. The hedging approach does, however, provide an important benchmark that can be used to guide decisions regarding the appropriate use of short-term credit.

BASIC FINANCIAL MANAGEMENT IN PRACTICE

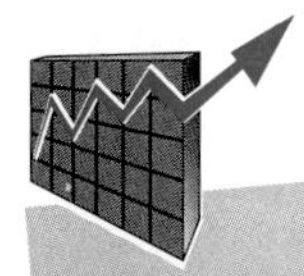

Hedging Interest Rate Risk

Every form of business activity contains risk; it can never be completely eliminated. In finance, we classify the risk to which a firm is exposed as either commercial (business) or financial risk. We include interest rate risk and foreign exchange risk as components of financial risk. In this excerpt from the article entitled, "Hedging: A User's Manual," we examine how hedging techniques are used to reduce interest rate risk.

Market interest rates are subject to change. When interest rates are on the rise, those who have borrowed funds at a variable rate pay more interest than they expected and are in a weaker position than competitors with fixed-rate financing. But fixed-rate financing can become a disadvantage if interest rates start to fall. As lenders, it's preferable to make variable-rate investments when rates are increasing and fixed-rate investments when rates are decreasing.

A Strategy to Manage Financial Risk

Finding ways to protect an entity against financial risks demands strong technical knowledge, serious planning and an ability to anticipate events. Clairvoyants excepted, those determined to establish an effective hedging strategy must take three steps: define the objectives, measure exposure to financial risk and apply the strategy.

Defining the objectives—The objectives of a hedging strategy should be defined in terms of amount, duration and nature of coverage. What's the proportion of the overall value of the financial instrument that will be hedged and what level of risk is acceptable? Is complete or partial coverage needed or can the company opt instead for a form of self-insurance? How long are the financial markets expected to be unstable?

Following that, the type of coverage must be settled. When will it take effect? Will the value of the financial instrument be guaranteed at a given date or will the coverage be continuous, ensuring that the value of the instrument will always exceed a predetermined minimum value? Will it be used to freeze a value or a profit, or to curb losses while making it possible to take advantage of a favourable market swing?

Measuring the exposure—Start by calculating the difference between the market value of the assets whose value is influenced by the risk factor identified (the exchange rate for example) and the market value of the liabilities whose value is influenced by this same risk factor. Possible variations of the risk factor are then applied to the difference measured in order to estimate the gains or losses that could result from the exposure.

Choosing the Instrument

Hedging against interest rate risk—Once the exposures to interest rate risk have been properly described (in terms of amounts, maturities and cash flows), a selection can be made from a variety of instruments. Despite their numbers, these instruments are used mainly to perform three functions: to freeze a rate, to change the structure of a rate or to guarantee a rate.

Freezing an interest rate—Forward rate agreements (FRAs) can be used for this purpose and involve a simple agreement between the company that intends to lend or borrow and the financial institution which guarantees that, as long as the financial transaction takes place within a specified period, the interest rate applicable to it will be the one negotiated when the agreement was signed.

Suppose a company plans to borrow $20 million for a three-month period six months from now. The current interest rate is 10%, which is acceptable to the company, but an increase in interest rates is anticipated. The current market rate for six-month FRAs is 10.5%. The agreement is therefore signed with Bank Y at 10.5%. Six months later, the company receives a loan from Bank Z (note that the loan does not have to be negotiated with the bank that sold the FRA) at the current market rate, which is 10.75%. When the loan matures, the company will reimburse Bank Z the amount of the loan plus $536,027 in interest (10.75% × $20 million × 91/365). The company will pay $12,466 (0.25% × $20 million × 91/365) more than the amount that would have been paid at the FRA rate, and, consequently, will receive $12,466 from Bank Y.

FRAs do not govern the cash flows that will result from hedged loans. Only the effect of a rate differential will change hands between the parties. The rates are determined using a reference rate such as a LIBOR (the London Interbank Offered Rate)

Forward rate agreements are generally used for short-term cash management purposes, and the resulting cash payments are usually made at the end of the hedged period.

Changing the structure of an interest rate—Interest rate swaps offer borrowers the possibility of reducing the cost of a loan below the minimum normally available to them on the market. A lender can also choose a swap to obtain a better return on investment.

Swaps exist because two entities intending to lend or borrow are rarely entitled to the same conditions on the market. Swaps enable the two to exchange interest expenses or interest income related to a specific financial instrument without exchanging the principal or the related responsibility.

Suppose that Company A, which is well established in its industry, is able to borrow at preferred, variable or fixed rates. This company has a privileged relationship with lesser-known Company B which has a higher borrowing cost. Company A wishes to borrow, for the short term and at a variable rate, an amount approximately equivalent to that which Company B wishes to borrow for the long term and at a fixed rate. When Company A borrows at a variable rate, it pays LIBOR plus 1% and when it borrows at a fixed rate, it pays 10%. When Company B borrows at a variable rate, it pays LIBOR plus 2%, whereas a fixed-rate loan would cost it 14%.

In the absence of any agreement, Company A would pay LIBOR + 1% to obtain variable-rate financing and B would pay 14% for its fixed-rate financing. For mutual benefit, A and B then enter into an agreement to exchange interest rates: A will borrow at a fixed rate (10%), whereas B will borrow at a variable rate (LIBOR + 2%). A and B will exchange their interest flows and, for each party to benefit, B will reimburse A its 10% interest on the fixed-rate loan, to which it will add the 2% premium it is required to pay on the variable-rate loan. A, for its part, will pay B the LIBOR. In this way, A will pay as if it were indebted at a variable rate; that is, 1% less than the best rate it was offered (LIBOR + 1%). B will pay as if it were indebted at a fixed rate; that is, 2% less than what it would have obtained on the market. While the swap can, as in this case, be arranged by mutual agreement, in most cases transactions are arranged through banks holding important swap portfolios.

Guaranteeing an interest rate—Here, the objective is to limit the risk of rate fluctuations by setting a maximum (cap) on the rate increase that a borrower is willing to accept, or by determining a minimum level (floor) of interest yield that an investor is prepared to accept.

In both cases, the services of an entity willing to take on the role of an insurer are required. For a fee, this entity will assume (for the borrower) the risk that the rate will increase, or (for the investor) the risk that the rate will decrease. Of course, the higher the cap or the lower the floor, the lower the fee. Again, banks are the most active participants in this coverage market.

The same applies to cases where a company gives up to the bank a portion of the potential profit it could obtain from a drop (as a borrower) or increase (as an investor) in interest rates, by determining a maximum and minimum rate (a collar) that will apply.

Take the case of a company that wants to be covered against increases to the cost of its one-year, variable-rate $10 million loan. The company and the bank enter into a collar agreement, which provides that the total interest cost for the loan will not exceed 10%. The company, which currently pays 9.5% interest on this debt, agrees to hand over to the bank any interest savings it will realize if the real interest rate falls below 8%. Given the volatility of interest rates at the time of the agreement, and considering the cap requested and the floor conceded, the bank decides to charge $21,000 for this transaction. As it turns out, conditions are such that the company has to pay an average interest rate of only 7.75% to its creditor. In addition to the $21,000 in fees that it was charged, it will have to pay $25,000 to the bank [(8 − 7.75%) × 10 million] at the end of the agreement.

Source: Adapted from: Sylvie Léger and Jacques Fortin, "Hedging: A User's Manual," *CAmagazine* (April 1994), pp. 20–24.

SUMMARY

Working-capital management involves managing the firm's liquidity, which, in turn, involves managing (1) the firm's investment in current assets and (2) its use of current liabilities. Each of these problems involves risk-return tradeoffs. Investing in current assets reduces the firm's risk of insolvency at the expense of lowering its overall rate of return on its investment in assets. Furthermore, the use of long-term sources of financing enhances the firm's liquidity while reducing its rate of return on assets.

STUDY QUESTIONS

17-1. Define and contrast the terms *working capital* and *net working capital.*

17-2. Discuss the risk-return relationship involved in the firm's asset investment decisions as it pertains to working-capital management.

17-3. What advantages and disadvantages are generally associated with the use of short-term debt? Discuss.

17-4. Explain what is meant by the statement "The use of current liabilities as opposed to long-term debt subjects the firm to a greater risk of illiquidity."

17-5. Define the hedging principle. How can this principle be used in the management of working capital?

17-6. Define the following terms:

- **a.** Permanent asset investments
- **b.** Temporary asset investments
- **c.** Permanent sources of financing
- **d.** Temporary sources of financing
- **e.** Spontaneous sources of financing

SELF-TEST PROBLEMS

ST-1. (*Investing in Current Assets and the Return on Common Equity*) Walker Enterprises is presently evaluating its investment in working capital for the coming year. Specifically, the firm is analyzing its level of investment in current assets. The firm projects that it will have sales of $4,000,000 next year, and its fixed assets are projected to total $1,500,000. The firm pays interest at a rate of 10 percent on its short- and long-term debt (which is managed by the firm so as to equal a target of 40 percent of assets). The firm projects its earnings before interest and taxes to be 15 percent of sales and faces a 30 percent tax rate.

- **a.** If Walker decides to follow a working-capital strategy calling for current assets equal to 40 percent of sales, what will be the firm's return on common equity?
- **b.** Answer (a) for a current asset to sales ratio of 60 percent.
- **c.** Throughout our analysis we have assumed that the rate of return earned by the firm on sales is independent of its investment in current assets. Is this a valid assumption?

ST-2. (*Using Marketable Securities to Increase Liquidity*) The balance sheet for the Simplex Mfg. Co. is presented below for the year ended December 31, 1995:

Simplex Mfg. Co. Balance Sheet December 31, 1995			
Cash	$10,000		
Accounts receivable	50,000		
Inventories	40,000		
Total current assets		$100,000	
Net fixed assets		100,000	
Total			$200,000
Current liabilities		$ 60,000	
Long-term liabilities		40,000	
Common equity		100,000	
Total			$200,000

During 1995 the firm earned net income after taxes of $20,000 based on net sales of $400,000.

a. Calculate Simplex's current ratio, net working-capital position, and return on total assets ratio (net income/total assets) using the above information.

b. The vice-president for finance at Simplex is considering a plan for enhancing the firm's liquidity. The plan involves raising $20,000 in common equity and investing in marketable securities that will earn 8 percent before taxes and 4.8 percent after taxes. Calculate Simplex's current ratio, net working capital, and return on total asset ratio after the plan has been implemented. [*Hint:* Net income will now become $20,000, plus .048 times $20,000, or $20,960.]

c. Will the plan proposed in (b) enhance firm liquidity? Explain.

d. What effect does the plan proposed in (b) have on firm profitability? Explain.

ST-3. (*Using Long-Term Debt to Increase Liquidity*) On April 30, 1996, the Jamax Sales Company had the following balance sheet and income statement for the year just ended:

Jamax Sales Company Balance Sheet April 30, 1996			
Current assets		$100,000	
Net fixed assets		200,000	
Total			$300,000
Accounts payable	$30,000		
Notes payable (14%)[a]	40,000		
Total		$ 70,000	
Long-term debt (10%)		100,000	
Common equity		130,000	
Total			$300,000

Partial Income Statement for the year ended April 30, 1996	
Net operating income	$72,800
Less: Interest expense[b]	12,800
Earnings before taxes	$60,000
Less: Taxes (50%)	30,000
Net income	$30,000

[a]The short-term notes are outstanding during the latter half of the firm's fiscal year in response to the firm's seasonal need for funds.
[b]Total interest expense for the year consists of 14% interest on the firm's $40,000 note for a six-month period (.14 × $40,000 × 1/2), $2,800, plus 10% of the firm's $100,000 long-term note for a full year (.10 × $100,000), or $10,000. Thus, total interest expense for the year is $2,800, plus $10,000 or $12,800.

a. Calculate Jamax's current ratio, net working capital, and return on total assets.

b. The treasurer of Jamax was recently advised by the firm's investment dealer that its current ratio was considered below par. In fact, a current ratio of 2.0 was considered to be a sign of a healthy liquidity position. In response to this news the treasurer devised a plan wherein the firm would issue $40,000 in 13 percent long-term debt and pay off its short-term notes payable. This long-term note would be outstanding all year long, and when the funds were not needed to finance the firm's seasonal asset needs, they would be invested in marketable securities earning 8 percent before taxes. If the plan had been in effect last year, other things being the same, what would have been the firm's current ratio, net working capital, and return on total assets ratio? [*Hint:* The firm's net income with the change would have been $29,600.]

c. Did Jamax's liquidity improve to the desired level (based on a desired current ratio of 2.0)?

d. How was the firm's profitability in relation to total investment affected by the change in financial policy?

STUDY PROBLEMS (SET A)

17-1A. (*Investing in Current Assets and the Return on Common Equity*) In June the MacMinn Company began planning for the coming year. A primary concern is its working-capital management policy. In particular, the firm's management is considering the effects of its investment in current assets on the return earned on common shareholders' equity. The firm projects that it will have sales of $8,000,000 next year, and its fixed assets are projected to total $2,000,000. It pays interest at a rate of 12 percent on its short- and long-term debt (which is managed by the firm so as to equal a target of 30 percent of assets). Finally, the firm projects its earnings before interest and taxes to be 15 percent of sales and faces a 30 percent tax rate.

a. If MacMinn decides to follow a working-capital strategy calling for current assets equal to 40 percent of sales, what will be the firm's return on common equity?

b. Answer (a) for a current-asset-to-sales ratio of 60 percent.

c. Throughout our analysis we have assumed that the rate of return earned by the firm on sales is independent of its investment in current assets. Is this a valid assumption?

17-2A. (*Managing Firm Liquidity*) On September 30, 1995, the balance sheet and income statement for Trecor Mfg. Company appeared as follows:

Trecor Mfg. Company Balance Sheet
September 30, 1995

Current assets	$ 500,000
Net fixed assets	500,000
Total	$1,000,000
Accounts payable	$ 100,000
Notes payable (17%)[a]	400,000
Total	$ 500,000
Long-term debt (12%)	100,000
Common equity	400,000
Total	$1,000,000

[a]Short-term notes are used to finance a three-month seasonal expansion in Trecor's asset needs. This period is the same for every year and extends from July through September with the note being due October 1.

Trecor Mfg. Company Income Statement September 30, 1995	
Net operating income	$195,666
Less: Interest expense[b]	(29,000)
Earnings before taxes	166,666
Less: Taxes (40%)	(66,666)
Net income	$100,000

[b]Interest expense was calculated as follows:

Notes payable (.17 × $400,000 × 1/4 year) =	$17,000
Long-term debt (.12 × $100,000 × 1 year) =	12,000
Total =	$29,000

a. Calculate the current ratio, net working capital, return on total assets, and return on common equity ratio for Trecor.

b. Assume that you have just been hired as financial vice-president of Trecor. One of your first duties is to assess the firm's liquidity position. Based upon your analysis, you plan to issue $400,000 in common shares and use the proceeds to retire the firm's notes payable. Recalculate the ratios from (a) and assess the change in the firm's liquidity.

c. Given your actions in (b), assume now that in the future you will finance your three-month seasonal need for $400,000 using a long-term bond issue that will carry an interest cost of 15 percent. (Note that during the time the funds are needed your current assets increase by $400,000 because of increased inventories and receivables.) In addition, you estimate that during the nine months you do not need the funds, they can be invested in marketable securities to earn a rate of 10 percent. Recalculate the financial ratios from (a) for 1996 where all revenues and nonfinance expenses are expected to be the same as in 1995. Analyze the results of your plan.

17-3A. (*Managing Firm Liquidity*) Re-analyze (c) of problem 17-2A assuming that a three-month short-term note is used as opposed to the bonds. The note carries a rate of 15 percent per annum.

17-4A. (*The Hedging Principle*) H. O. Hielregal Inc. estimates that its current assets are about 25 percent of sales. The firm's current balance sheet is presented below:

H. O. Hielregal Balance Sheet December 31, 1995 ($ millions)			
Current assets	$2.0	Trade credit and	
Fixed assets	2.8	accounts payable	$ .8
		Long-term debt	1.0
Total	$4.8	Common equity	3.0
		Total	$4.8

Hielregal pays out all of its net income in cash dividends to its shareholders. Trade credit and accounts payable equal 10 percent of the firm's sales.

a. Based on the following five-year sales forecast, prepare five end-of-year pro forma

Year	Predicted Sales ($ Millions)
19X4	10
19X5	11
19X6	13
19X7	14
19X8	15

balance sheets that indicate "additional financing needed" for each year as a balancing account. Fixed assets are expected to increase by $0.2 million each year.

b. Develop a financing policy for Hielregal using your answer to (a) above that is consistent with the following goals:

1. A minimum current ratio of 2.0 and a maximum of 3.0.
2. A debt-to-total-assets ratio of 35–45 percent. You may issue new common shares to raise equity funds.

STUDY PROBLEMS (SET B)

17-1B. (*Investing in Current Assets and the Return on Common Equity*) The managers of Tharp's Tarps Inc. are considering a possible change in their working-capital management policy. Specifically, they are concerned with the effect of their investment in current assets on the return earned on common shareholders' equity. They expect sales of $7,000,000 next year, and their fixed assets are projected to total $2,000,000. The firm pays 11 percent interest on both short- and long-term debt (which is managed by the firm so as to equal a target of 30 percent of assets) and faces a 30 percent tax rate. Finally, the firm projects its earnings before interest and taxes to be 15 percent of sales.

a. If management follows a working-capital strategy calling for current assets equal to 50 percent of sales, what will be the firm's return on common equity?

b. Answer (a) for a current-asset-to-sales ratio of 40 percent.

c. Is it reasonable to assume, as we did, that the rate of return earned by the firm on sales is independent of its investment in current assets?

17-2B. (*Managing Firm Liquidity*) As of September 30, 1995, the balance sheet and income statement of No-Soy Foods Inc. appeared as follows:

No-Soy Foods Inc. Balance Sheet
September 30, 1995

Current assets	$ 625,000
Net fixed assets	400,000
Total	$1,025,000
Accounts payable	$ 125,000
Notes payable (17%)[a]	400,000
Total	$ 525,000
Long-term debt (12%)	100,000
Common equity	400,000
Total	$1,025,000

[a]Short-term notes are used to finance a three-month seasonal expansion in No-Soy Foods asset needs. This period is the same for every year and extends from July through September with the note being due October 1.

No-Soy Foods Inc. Income Statement
September 30, 1995

Net operating income	$195,666
Less: Interest expense[b]	(29,000)
Earnings before taxes	166,666
Less: Taxes (40%)	(66,666)
Net income	$100,000

[b]Interest expense was calculated as follows:

Notes payable (.17 × $400,000 × 1/4 year)	=	$17,000
Long-term debt (.12 × $100,000 × 1 year)	=	12,000
Total	=	$29,000

a. Calculate the current ratio, net working capital, return on total assets, and return on common equity ratio for No-Soy.

b. Assume that you have just been hired as financial vice-president of No-Soy. One of your first duties is to assess the firm's liquidity position. Based upon your analysis, you plan to issue $400,000 in common shares and use the proceeds to retire the firm's notes payable. Recalculate the ratios from (a) and assess the change in the firm's liquidity.

c. Given your actions in (b), assume now that in the future you will finance your three-month seasonal need for $450,000 using a long-term bond issue that will carry an interest cost of 15 percent. (Note that during the time the funds are needed your current assets increase by $450,000 because of increased inventories and receivables.) In addition, you estimate that during the nine months you do not need the funds, they can be invested in marketable securities to earn a rate of 10 percent. Recalculate the financial ratios from (a) for 1996 where all revenues and nonfinance expenses are expected to be the same as in 1995. Analyze the results of your plan.

17-3B. *Managing Firm Liquidity*) Re-analyze (c) of Problem 17-2B assuming that a three-month short-term note is used as opposed to the bonds. The note carries a rate of 15 percent per annum.

17-4B. *The Hedging Principle*) B. A. Freeman Inc. successfully sells bail bond franchises throughout the country. Its founder, Mr. Freeman, estimates current assets to be 30 percent of sales. The firm's current balance sheet is as follows:

B. A. Freeman Balance Sheet
December 31, 1995 ($ millions)

Current assets	$2.0	Trade credit and	
Fixed assets	2.8	accounts payable	$ .8
		Long-term debt	1.0
Total	$4.8	Common equity	3.0
		Total	$4.8

Freeman pays out all of its net income in cash dividends to its shareholders. Trade credit and accounts payable equal 10 percent of the firm's sales.

a. Based on the following five-year sales forecast, prepare five end-of-year pro forma balance sheets that indicate "additional financing needed" for each year as a balancing account. Fixed assets are expected to increase by $.3 million each year.

Year	Predicted Sales ($ Millions)
19X2	10
19X3	11
19X4	13
19X5	14
19X6	15

b. Develop a financing policy for Freeman using your answer to (a) that is consistent with the following goals:

1. A minimum current ratio of 2.5 and a maximum of 3.5.
2. A debt-to-total-assets ratio of 35–45 percent. You may issue new common shares to raise equity funds.

SELF-TEST SOLUTIONS

SS-1. a. Walker's pro forma balance sheet for next year will appear as follows:

Current assets	$1,600,000	Debt (40% of assets)	$1,240,000
Fixed assets	1,500,000	Owner's equity	1,860,000
Total	$3,100,000	Total	$3,100,000

Projected earnings are calculated as follows:

Sales	$4,000,000
EBIT (15%)	600,000
Interest	(124,000)
EBT	$ 476,000
Taxes	(142,800)
Net income	$ 333,200

Thus, the firm's return on common equity is

$$\frac{\$333,200}{\$1,860,000} = 17.9\%$$

b. Under this scenario, Walker's pro forma balance sheet for next year will appear as follows:

Current assets	$2,400,000	Debt (40% of assets)	$1,560,000
Fixed assets	1,500,000	Owner's equity	2,340,000
Total	$3,900,000	Total	$3,900,000

Projected earnings are calculated as follows:

Sales	$4,000,000
EBIT (15%)	600,000
Interest	(156,000)
EBT	$ 444,000
Taxes	(133,200)
Net income	$ 310,800

Thus, the firm's return on common equity is

$$\frac{\$310,800}{\$2,340,000} = 13.3 \text{ percent}$$

c. One would expect that the firm would earn some positive return from investing in current assets. For example, the firm could hold some of its current assets in marketable securities earning some positive rate of return. Note, however, that as long as these current assets earn a rate of return less than the cost of financing the investment, the rate of return on the common shareholders' equity will decrease with increased investment in current assets.

SS-2. a.

$$\text{current ratio} = \frac{\text{current assets}}{\text{current liabilities}} = \frac{\$100,000}{\$60,000} = 1.67\times$$

$$\text{net working capital} = \text{current assets} - \text{current liabilities} = \$100,000 - \$60,000 = \$40,000$$

$$\text{return on total assets} = \frac{\text{net income}}{\text{total assets}}$$

$$= \frac{\$20{,}000}{\$200{,}000} = 10\%$$

b.

$$\text{current ratio} = \frac{\$120{,}000}{\$60{,}000} = 2\times$$

$$\text{net working capital} = \$120{,}000 - \$60{,}000 = \$60{,}000$$

$$\text{return on total assets} = \frac{\$20{,}960}{220{,}000} = 9.52\%$$

c. Yes, the firm's liquidity position as measured by the current ratio and the amount of net working capital has definitely improved. However, as we see in the answer to (b), profitability has been adversely affected.

d. Simplex's return on total assets declined from 10 to 9.52 percent as a result of the new financing plan. This occurred because it was earning 10 percent after taxes on its $200,000 investment and it invested $20,000 in marketable securities earning only 4.8 percent after taxes. The result was a decline in firm profitability in relation to assets.

SS-3. a.

$$\text{current ratio} = \frac{\text{current assets}}{\text{current liabilities}}$$

$$= \frac{\$100{,}000}{\$70{,}000} = 1.43\times$$

$$\text{net working capital} = \text{current assets} - \text{current liabilities}$$

$$= \$100{,}000 - \$70{,}000$$

$$= \underline{\$30{,}000}$$

$$\text{return on total assets} = \frac{\text{net income}}{\text{total assets}}$$

$$= \frac{\$30{,}000}{\$300{,}000}$$

$$= 10\%$$

b.

$$\text{current ratio} = \frac{\$100{,}000}{\$30{,}000} = 3.33\times$$

$$\text{net working capital} = \$100{,}000 - \$30{,}000 = \$70{,}000$$

$$\text{return on total assets} = \frac{\$29{,}600}{\$300{,}000} = 9.87\%$$

c. Yes, the current ratio of 3.33 is now well above the 2.0 standard when the funds are needed. Furthermore, if we include the $40,000 in marketable securities in current assets, the current ratio rises to 4.67 ($100,000 + $40,000/$30,000 = 4.67) during those times when the funds are seasonally idle.

d. Firm profitability declined slightly because the firm used a permanent source of financing to replace a more flexible short-term source. The firm actually saves interest expense through the plan during the six-month period when the funds are needed (13 percent on long-term funds versus 14 percent on short-term notes); however, during the six months when the funds are not needed, the firm can only earn a before-tax return of 8 percent, while the funds cost 13 percent. In this case, the added liquidity of the plan appears to overshadow the modest loss in expected profitability.

CASH AND MARKETABLE SECURITIES MANAGEMENT

CHAPTER 18

LEARNING OBJECTIVES

After reading this chapter you should be able to

1. Explain why a firm holds cash.
2. Explain why variations in liquid asset holdings occur.
3. Explain various cash management objectives and decisions.
4. Explain the different mechanisms for managing the firm's cash collection and disbursement procedures.
5. Evaluate quantitatively the expected costs and benefits of cash management services.
6. Determine a prudent composition for the firm's marketable securities portfolio.
7. Explain the logic behind some actual cash management practices.

INTRODUCTION

The stability of Canada's financial system is due in part to its payments system. Many consider Canada's payments system to be the finest in the world because of its efficiency and reliability. Most Canadians can deposit a cheque into their bank account and receive immediate credit, even if the cheque itself may take a day or more to return to the bank on which it was drawn.[1] Irrespective of where your bank account is located in Canada, the Canadian banking system gives same-day value. This feature contrasts with other banking systems in the world in which the value to a depositor's account is deferred to accommodate delays in the clearing system or to compensate the bank.

The efficiency of Canada's payments system is demonstrated by the volume of cheques it processes, approximately 6 million cheques in a day. This system also processes another 2 million electronic payment items that are received through sources such as point-of-sales terminals, magnetic tapes, and automated banking machines.[2] The Canadian payments system is now capable of providing same-day value for cheques that have values in excess of $50,000.

In this chapter we examine how the Canadian payments system reduces the float that is incurred in a company's cash collection system. We also examine the various techniques that a company can use to disburse its payments.

---

[1]Source: Adapted from: Bruce McDougall, "Hidden Highways: Canada's Payments Clearing And Settlement System," *Canadian Banker* (January/February 1994), pp. 27–30.

[2]Source: Adapted from: Bruce McDougall, "Beyond Paper: The Future Of Canada's Payments System," *Canadian Banker* (May/June 1994), pp. 22–26.

CHAPTER PREVIEW

In this chapter, we formulate financial policies for the management of cash and marketable securities. We examine three major areas: (1) techniques available to management for favourably influencing cash receipts and disbursement patterns; (2) sensible investment possibilities that enable the company productively to employ excess cash balances; and (3) some straight models that can assist financial officers in deciding how much cash to hold. We begin with a review of the basic cash flow process and the motives that economic units have for holding cash balances. In recent years, the financial function of cash management has grown in stature. Now trade associations and professional certifications exist just for this function, and cash management is usually part of treasury management within the company.

PERSPECTIVE IN FINANCE

It is useful to think of the firm's cash balance as a reservoir that rises with cash inflows and falls with cash outflows. Any nonfinancial firm, such as Northern Telecom, desires to minimize its cash balances consistent with meeting its financial obligations in a timely manner.

Holding too much cash—what analysts tend to call "excess cash"—results in a loss of profitability to the firm. The auto manufacturer, for example, is not in business to build up its cash reservoir. Rather, it wants to manage its cash balance in order to maximize its financial returns—this will enhance shareholder wealth.

OBJECTIVE 1

WHY A COMPANY HOLDS CASH

Cash

Currency, coins, and demand deposit accounts.

Marketable securities

Security investments (financial assets) the firm can quickly convert to cash balances. Also known as near cash or near-cash assets.

Liquid assets

The sum of cash and marketable securities.

Cash is the currency and coins that a firm has on hand in petty cash drawers, in cash registers, or in chequing accounts at the various commercial banks in which its demand deposits are maintained. **Marketable securities** are those security investments the firm can quickly convert into cash balances. Most firms in Canada tend to hold marketable securities with very short maturity periods—less than one year. No law, of course, dictates that instruments with longer terms to maturity cannot be included. Marketable securities are also referred to as near cash or near-cash assets because they can be converted into cash in a short period of time. Taken together, cash and near cash are known as **liquid assets**.

A thorough understanding of why and how a firm holds cash requires an accurate conception of how cash flows into and through the enterprise. Figure 18-1 depicts the process of cash generation and disposition in a typical manufacturing setting. The arrows designate the direction of the flow—that is, whether the cash balance increases or decreases.

Cash Flow Process

The firm experiences irregular increases in its cash holdings from several external sources. Funds can be obtained in the financial markets from the sale of securities, such as bonds, preferred shares, and common shares, or the firm can enter into nonmarketable debt contracts with lenders such as commercial banks. These irregular cash inflows do not occur on a daily basis. They tend to be episodic; the financing arrangements that give rise to them are effected at wide intervals. The reason is that external financing contracts usually involve huge sums of money stemming

Figure 18-1 The Cash Generation and Disposition Process

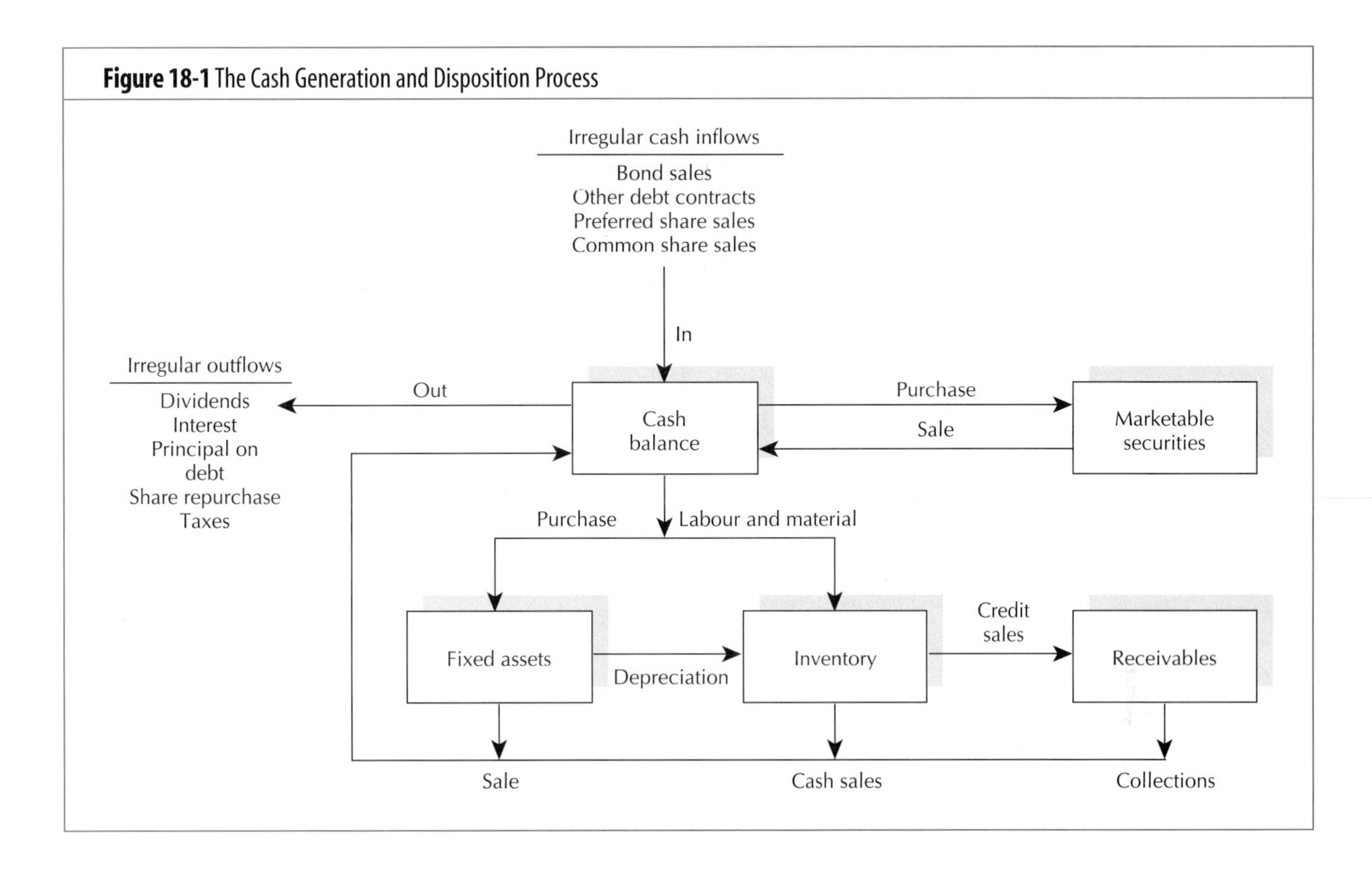

from a major need identified by the company's management, and these needs do not occur every day. For example, a new product might be at the development stage, or a plant expansion might be required to provide added productive capacity.

In most organizations the financial manager responsible for cash management also controls the transactions that affect the firm's investment in marketable securities. As excess cash becomes temporarily available, marketable securities are purchased. When cash is in short supply, a portion of the marketable securities portfolio is liquidated.

Whereas the irregular cash inflows are from external sources, the other main sources of cash to the firm arise from internal operations and occur on a more regular basis. Over long periods, the largest receipts come from accounts receivable collections and to a lesser extent from direct cash sales of finished goods. Many manufacturing concerns also generate cash on a regular basis through the liquidation of scrap or obsolete inventory. In the automobile industry, large and costly machines called chip crushers grind waste metal into fine scrap which brings considerable revenue to the major producers. At various times fixed assets may also be sold, thereby generating some cash inflow. This is not a large source of funds except in unusual situations where, for instance, a complete plant renovation may be taking place.

Apart from the investment of excess cash in near-cash assets, the cash balance decreases for three key reasons. First, on an irregular basis withdrawals are made to (1) pay cash dividends on preferred and common shares, (2) meet interest requirements on debt contracts, (3) repay the principal borrowed from creditors, (4) buy the firm's own shares in the financial markets for use in executive compensation plans or as an alternative to paying a cash dividend, and (5) pay tax bills. Again, by an "irregular basis" we mean items not occurring on a daily or frequent schedule. Second, the company's capital expenditure program designates that fixed assets be acquired at various intervals. Third, inventories are purchased on a regular basis to ensure a steady flow of finished goods off the production line. Note that the arrow linking the investment in fixed assets with the inventory account is labelled *depreciation.* This indicates that a portion of the cost of fixed assets is charged against the products coming off the assembly line. This cost is subsequently recovered through the sale of the finished goods inventory, since the product selling price will be set by management to cover all the costs of production, including depreciation.

The variety of influences that constantly affect the cash balance held by the firm can be synthesized in terms of the classic motives for holding cash, as identified in the literature of economic theory.

Motives for Holding Cash

In a classic economic treatise John Maynard Keynes segmented the firm's, or any economic unit's, demand for cash into three categories: (1) the transactions motive, (2) the precautionary motive, and (3) the speculative motive.[3]

Transactions Motive

Balances held for transactions purposes allow the firm to dispense with cash needs that arise in the ordinary course of doing business. In Figure 18-1, transactions balances would be used to meet the irregular outflows as well as the planned acquisition of fixed assets and inventories.

The relative amount of transactions cash held is significantly affected by the industry in which the firm operates. If revenues can be forecast to fall within a tight range of outcomes, then the ratio of cash and near cash to total assets will be less for the firm than if the prospective cash inflows might be expected to vary over a wide range. It is well known that utilities can forecast cash receipts quite accurately, because of stable demand for their services. This enables the firm to stagger its billings throughout the month and to time them to coincide with planned expenditures. Inflows and outflows of cash are thus synchronized. We would expect the cash holdings of utility firms relative to sales or assets to be less than those associated with a major clothing retailer which has a more variable demand.

Firms competing in the same industry may experience notably different strains on their cash balances. Transactions balances for the juice processing industry are influenced by a seasonal factor. During the summer and autumn months, most juice processors within Canada will incur a large part of their operating expenses for raw materials and labour necessary for picking, grading, and processing of fruits. As a result, juice processors in Canada will suffer sizable cash drains throughout the summer and autumn seasons. Cash balances for juice processors in Canada will therefore be expected to be relatively higher during the summer and autumn seasons than those of juice processors operating in more southerly areas of the United States where processing may occur on a continual basis.

The Precautionary Motive

Precautionary balances are a buffer stock of liquid assets. This motive for holding cash relates to the maintenance of balances to be used to satisfy possible, but as yet indefinite, needs.

In our discussion of transactions balances, we saw that cash flow predictability could affect a firm's cash holdings through synchronization of receipts and dis-

---

[3]John Maynard Keynes, *The General Theory of Employment, Interest, and Money* (New York: Harcourt Brace Jovanovich, 1936).

bursements. Cash flow predictability also has a material influence on the firm's demand for cash through the precautionary motive. The airline industry provides a typical illustration. Air passenger carriers are plagued with a high degree of cash flow uncertainty. The weather, rising fuel costs, and continual strikes by operating personnel make cash forecasting difficult for any airline. The upshot of this problem is that because of all the things that might happen, the minimum cash balances desired by the management of the air carriers tend to be large.

In addition to cash flow predictability, the precautionary motive for holding cash is affected by access to external funds.[4] Especially important are cash sources that can be tapped on short notice. Good banking relationships and established lines of credit can reduce the need to keep cash on hand. This unused borrowing power obviates somewhat the need to invest in precautionary balances.

In actual business practice, the precautionary motive is met to a large extent by the holding of a portfolio of liquid assets, not just cash. Notice in Figure 18-1 the two-way flow of funds between the company's holdings of cash and marketable securities. In large corporate organizations, funds may flow either into or out of the marketable securities portfolio on a daily basis. In this type of organization, the precautionary motive will be met by investment in marketable securities because some actual rate of return can be earned on the near-cash assets as compared to a zero rate of return available on cash holdings.

The Speculative Motive

Cash is held for speculative purposes in order to take advantage of potential profit-making situations.[5] Construction firms that build private dwellings will at times accumulate cash in anticipation of a significant drop in lumber costs. If the price of building supplies does drop, the companies that built up their cash balances stand to profit by purchasing materials in large quantities. This will reduce their cost of goods sold and increase their net profit margin. Generally, the speculative motive is the least important component of a firm's preference for liquidity. The transactions and precautionary motives account for most of the reasons why a company holds cash balances.

VARIATIONS IN LIQUID ASSET HOLDINGS

OBJECTIVE 2

Decisions that concern the amounts of liquid assets to hold rest with the financial executive for cash management. A number of factors that can be expected to influence the financial executive's investment in cash or liquid assets have been reviewed. Not all these factors affect every firm. Moreover, factors that do affect many companies do so in differing degrees. Since cash management choices have different risk-bearing preferences, we might expect that holdings of liquid assets among industries and firms show considerable variation.

In the fourth quarter of 1995, the average of the ratio for holdings of cash and deposits to total assets is shown in Table 18-1 to be 4.5 percent for all industries and 4.1 for the nonfinancial industries. However, for the telecommunication carriers, postal and courier services sector or the petroleum and natural gas sector, the industry sector averages are 1.2 percent and 2.2 percent, respectively. These averages are substantially less than the average reported for all industries. The differ-

[4]According to Keynes, the precautionary motive for holding cash would be greatly increased in the absence of financial markets.

[5]Likewise Keynes indicated that the speculative motive for holding cash will increase with the existence of financial markets.

Table 18-1 Cash and Deposits as a Percentage of Total Assets for Major Industries

Industries	4th Quarter 1995	4th Quarter 1994
All industries	4.5	4.4
Nonfinancial industries	4.1	3.9
Telecommunication carriers and postal and courier services	1.2	0.5
Petroleum and natural gas	2.2	1.8
Building materials and construction	6.0	6.2
Motor vehicles, parts and accessories, and tires	4.5	3.4
Other transportation equipment	7.6	6.2
Household appliances and electrical products	3.9	4.0
Printing, publishing and broadcasting	2.5	2.1
Accommodation, food and beverage, educational, health and recreational services	8.1	7.1
Torstar	9.6 (1994)	3.0 (1993)
Reader's Digest	20.8 (1994)	19.3 (1993)

Source: Statistics Canada, "Quarterly Financial Statistics for Enterprises," Cat. No. 61-008, Fourth Quarter, 1995. Reproduced with the authority of the Minister of Industry, 1996; Torstar Corporation, "Annual Report" (1994); Reader's Digest Association Inc., "Annual Report" (1994).

ence in averages may be explained by the fact that both these industries operate in an environment in which cash flows are highly predictable relative to other industries. Table 18-1 suggests that in industry sectors which compete in a more risky global arena, such as the accommodation, food and beverage, educational, health and recreational services sector or the retailing of motor vehicles, parts and accessories, and tires sector, there is considerable variation in the holding of cash and deposits. This type of variation can be attributed to factors such as the state of the economy, management credit policies and management's risk preferences. For example, in the early nineties the North American economy fell into a period of recession. During this period most consumers postponed the purchases of large ticket items such as cars and trucks. The difficulty of accurately predicting sales demand in the retailing of motor vehicles caused many retailers to carry more of their assets as liquid assets. Companies continually adjust their liquid asset positions to meet operating needs and to respond to the real investment requirements of strategic plans.

Table 18-1 also suggests that it takes more than a knowledge of the industry class in which a firm operates to understand its cash management policies. For example, in 1994 Torstar Corporation and Reader's Digest Association Inc. reported liquid assets of 3.0 percent and 20.8 percent, respectively. The difference in the amounts suggests that each company perceives their liquidity needs differently. Different corporate strategies and different risk-bearing preferences with respect to the chances of running out of cash cause variations in the levels of liquidity investment over a period of time.

The examples given above showing liquid asset holdings of some specific industries and firms are only snapshots at a fixed point in time. In fact, assets are acquired, wasted, and sold every day. We must view the management of cash and liquid assets as a dynamic process. The flow of cash as depicted in Figure 18-1 never ceases. The cash inflows and outflows affecting the size of the firm's cash reservoir occur simultaneously. To ensure effective management of the firm's cash flows, managers must define a cash management system that has clearly defined objectives.

PERSPECTIVE IN FINANCE

Any company can benefit from a properly designed cash management system. If you identify what you believe to be a superbly run business organization, the odds are that firm has in place a sound cash management system. Before we explore several cash management techniques, it is necessary to introduce (1) the risk-return tradeoff, (2) the objectives, and (3) the decisions that comprise the centre of the cash management process. Keep in mind that the billion-dollar company will save millions each year by grasping these concepts—while the small or mid-sized organization may actually enhance its overall chances of survival. The following section provides the rationale and structure for knowing about all of the techniques and financial instruments discussed in the remainder of the chapter.

CASH MANAGEMENT OBJECTIVES AND DECISIONS

OBJECTIVE 3

The Risk-Return Tradeoff

A companywide cash management program must be concerned with minimizing the firm's risk of **insolvency**. In the context of cash management, the term insolvency describes the situation where the firm is unable to meet its maturing liabilities on time. In such a case the company is **technically insolvent** in that it lacks the necessary liquidity to make prompt payment on its current debt obligations. A firm could avoid this problem by carrying large cash balances to pay the bills that come due. Production, after all, would soon come to a halt should payments for raw material purchases be continually late or omitted entirely. The firm's suppliers would just cut off further shipments. In fact, fear of irritating a key supplier by being past due on the payment of a trade payable does cause some financial managers to invest in too much liquidity.

Insolvency

The firm is unable to meet its maturing liabilities on time. Examples are an inability to meet interest payments or to repay debt at maturity. In this case, the firm is said to be "technically insolvent."

The management of the company's cash position, however, is one of those problem areas in which you are criticized if you don't and criticized if you do. True, the production process will eventually be halted should too little cash be available to pay bills. If excessive cash balances are carried, the value of the enterprise in the financial marketplace will be suppressed because of the large cost of income forgone. The explicit return earned on idle cash balances is zero.

The financial manager must strike an acceptable balance between holding either too much cash or too little cash. This is the focal point of the risk-return tradeoff. A large cash investment minimizes the chances of insolvency, but penalizes company profitability. A small cash investment frees excess balances for investment in both marketable securities and longer-lived assets; this enhances company profitability and the value of the firm's common shares, but increases the chances of running out of cash.

BACK TO THE BASICS

The dilemma faced by the financial manager is a clear application of ***Axiom 1: The Risk-Return Tradeoff—We Won't Take on Additional Risk Unless We Expect to Be Compensated with Additional Return.*** *The return obtained from investing cash into marketable securities is offset by risk from both the firm not having enough cash on hand and the level of risk inherent in the marketable securities.*

The Objectives

The risk-return tradeoff can be reduced to two prime objectives for the firm's cash management system:

1. Enough cash must be on hand to meet the disbursal needs that arise in the course of doing business.
2. Investment in idle cash balances must be reduced to a minimum.

Evaluation of these operational objectives, and a conscious attempt on the part of management to meet them, gives rise to the need for some typical cash management decisions.

The Decisions

Two conditions would allow the firm to operate for extended periods with cash balances near or at a level of zero: (1) a completely accurate forecast of net cash flows over the planning horizon and (2) perfect synchronization of cash receipts and disbursements.

Cash flow forecasting is the initial step in any effective cash management program. This is usually accomplished by the finance function's evaluation of data supplied by the marketing and production functions in the company. The cash budget is used to forecast the cash flows over the planning period. Any net cash flows reported in the formal cash budget are estimates, subject to considerable variation. A totally accurate cash flow projection is an ideal, not a reality.

Our discussion of the cash flow process depicted in Figure 18-1 showed that inflows and outflows are not synchronized. Some inflows and outflows are irregular; others are more continual. Some finished goods are sold directly for cash, but more likely the sales will be on account. The receivables, then, will have to be collected before a cash inflow is realized. Raw materials have to be purchased, but several suppliers are probably used, and each may have its own payment date. Further, no law of doing business fixes receivable collections to coincide with raw material payments dates. So the second criterion that would permit a firm to operate with extremely low cash balances is not met in actual practice either.

Given that the firm will, as a matter of necessity, invest in some cash balances, certain types of decisions related to the size of those balances dominate the cash management process. These include decisions that answer the following questions:

1. What can be done to speed up cash collections and slow down or better control cash outflows?
2. What should be the composition of a marketable securities portfolio?
3. How should investment in liquid assets be split between actual cash holdings and marketable securities?

The remainder of this chapter dwells on the first two of these three questions. The third is explored in Appendix 18A.

PERSPECTIVE IN FINANCE

Although the sheer number of cash collection and payment techniques is large, the concepts upon which those techniques rest are quite simple. Controlling the cash inflow and outflow is a major theme of treasury management. But, within the confines of ethical management, the cash manager is always thinking (1) How can I speed up the firm's cash receipts? and (2) How can I slow down the firm's cash payments and not irritate too many important constituencies, such as suppliers?

The critical point is that cash saved becomes available for investment elsewhere in the company's operations, and at a positive rate of return this will increase total profitability. Grasping the elements of cash management requires that you understand the concept of cash float. We address the concept of float and float reduction early in the discussion on collection and disbursement procedures.

COLLECTION AND DISBURSEMENT PROCEDURES

OBJECTIVE 4

The efficiency of the firm's cash management program can be enhanced by knowledge and use of various procedures aimed at (1) accelerating cash receipts and (2) improving the methods used to disburse cash. We will see that greater opportunity for corporate profit improvement lies with the cash receipts side of the funds flow process, although it would be unwise to ignore opportunities for favourably affecting cash-disbursement practices.

Managing the Cash Inflow[6]

A **float** refers to the time delay that a firm incurs from the processing of payments within the payment cycle. The reduction of float lies at the centre of the many approaches employed to speed up cash receipts. Float (or total float) has four elements, as follows:

Float

The length of time from when a cheque is written until the actual recipient can draw upon or use the "goods funds."

1. *Mail float* represents the period of time that elapses between the events of a customer mailing a remittance cheque and a firm beginning to process the cheque.
2. *Processing float* represents the period of time that elapses between the events of a firm beginning to process a remittance cheque and the firm depositing the cheque with the bank.
3. *Transit or banking float* represents the period of time that is required for a deposited cheque to clear through the commercial banking system and become usable funds to the company. The introduction of same-day settlement by the Canadian Payments Association (CPA) Act enables credit for cheques deposited on the day received so that the depositor can use the money immediately.[7]
4. *Disbursing float* represents the period of time that funds are available in the company's bank account before its payment cheque has cleared through the banking system. Typically, funds available in the firm's banks exceed the balances indicated on its own books (ledgers).

We will use the term *float* to refer to the total time delay caused by the mail, processing and transit floats. Float reduction can yield considerable benefits in terms of usable funds that are released for company use and returns produced on such freed-up balances. As an example, for the year 1994 Northern Telecom reported total revenues of $9.247 billion. The amount of usable funds that would be released if Northern Telecom could achieve a one-day reduction in float can be approximated by dividing annual revenues (sales) by the number of days in a year.[8] In this case one day's freed-up balances would be

Annual revenues/365 days
= $9,247,000,000/365
= $25,334,247

[6]The discussions on cash receipt and disbursement procedures draw heavily from the following sources: S. Sarpkaya, *Modern Cash Management*, 2nd ed. (Don Mills, ON: CCH Canadian Ltd., 1991) and from the management of the Canadian Imperial Bank of Commerce.

[7]The Bank of Canada and the Canadian Payments Association provide cheque-clearing facilities for deposit-taking institutions. The cheque-clearing system is referred to as the "Canadian Payments System" in Canada. This term will be used several times in this chapter. The objective of the Canadian Payments System is to ensure that the proper amount of funds is available in a customer's chequing account for the collecting bank, trust, or Caisse Populaire. The Bank of Canada will only then transfer the appropriate credit to the collecting bank's account. A computerized network between clearing centres situated in major regions of the country and the Bank of Canada reduces the transit float by providing same-day settlement of cheques.

[8]Frederick W. Searby, "Use Your Hidden Cash Resources," *Harvard Business Review* 46 (March–April 1968), pp. 71–80.

If these released funds, which represent one day's sales, of approximately \$25.3 million could be invested to return 8 percent a year, then the annual value of the one-day float reduction would be

(sales per day) × (assumed yield) = \$25,334,247 × .08 = \$2,026,740

It is clear that effective cash management can yield impressive opportunities for profit improvement. As a result, any consumer-directed delivery product, such as automated banking machines, debit cards, credit cards, smart cards and home banking facilitates the payment of bills by the consumer and reduces the mail and processing floats incurred by a company.

Figure 18-2 illustrates an elementary, but typical, cash collection system for a hypothetical firm. It also shows the origin of mail float, processing float, and transit float. In this system the customer places with Canada Post his or her remittance cheque, which is then delivered to the firm's headquarters. The mail float is the time delay caused by the procedures involved in mailing a remittance cheque. On the arrival of cheques at the firm's headquarters (or local collection centre), general accounting personnel must go through the bookkeeping procedures needed to prepare them for local deposit. The cheques are then deposited with the firm's bank. The processing float is the time delay caused by procedures involved in the preparation and the deposit of the cheques. Once the firm deposits the remittance cheques with a branch bank, the Canadian banking system provides the firm with immediate credit for the deposit.[9] In fact, all cheques, money orders, drafts, travellers' cheques, and other electronically transacted payments are credited immediately at the clearing point which in this case is the bank where the cheque was

Figure 18-2 An Ordinary Cash-Gathering System

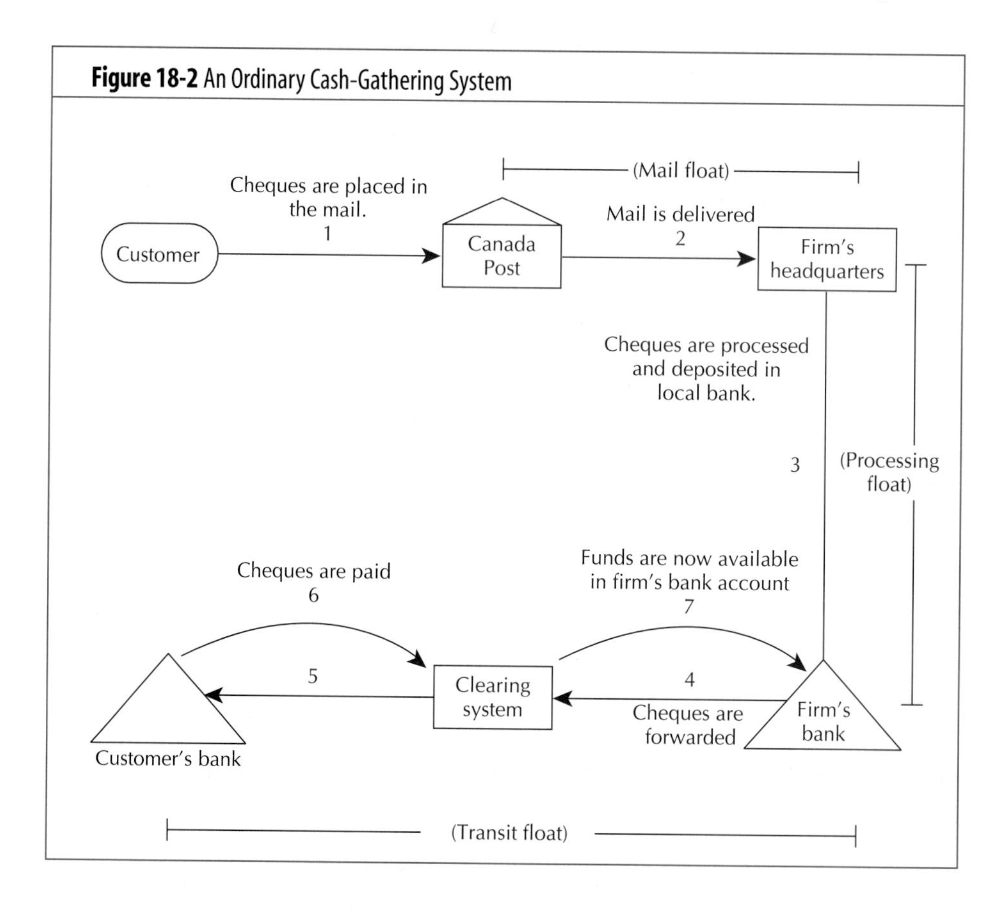

[9]Adapted from: Bruce McDougall, "Beyond Paper: The Future Of Canada's Payments System," *Canadian Banker* (May/June 1994), pp. 22–26.

deposited; same-day value for items deposited is a standard banking practice in Canada. Although, the bank will receive the cheque no later than two days after the cheque was deposited, the related settlement entries on the books of the Bank of Canada occur the following day through the Automated Clearing Settlement System. This system is operated by the Canadian Payment Association who communicate the clearing balances of the various financial institutions to the Bank of Canada. In general, final settlement of the transaction occurs on the next business day. The transit float is the time delay caused by the cheque-clearing system. However, the policy of same-day value obtained at major Canadian banks, credit unions, and trust and loan companies has substantially reduced this type of float.

The Lock-Box Arrangement

An example of a cash collection system which has the objective of releasing funds caused by the float, is the lock-box system. The lock-box system is the most widely used commercial banking service for expediting cash gathering. Such a system speeds up the conversion of receipts into usable funds by reducing both mail and processing float. For large corporations that receive cheques from all parts of the country, float reductions of two to four days are not unusual.

The lock-box arrangement shown in Figure 18-3 is based on a simple procedure. The firm's customers are instructed to mail their remittance cheques not to company headquarters or regional offices, but to a numbered Post Office box. The bank that is providing the lock-box service is authorized to open the box, collect the mail, process the cheques, and deposit the cheques directly into the company's account.

Typically a large bank will collect payments from the lock box at one- to two-hour intervals, 365 days of the year. During peak business hours, the bank may increase the frequency of pick-up of mail.

Once the mail is received at the bank, the cheques will be examined, totalled, photocopied, and microfilmed. A deposit form is then prepared by the bank, and

Figure 18-3 A Simple Lock-Box System

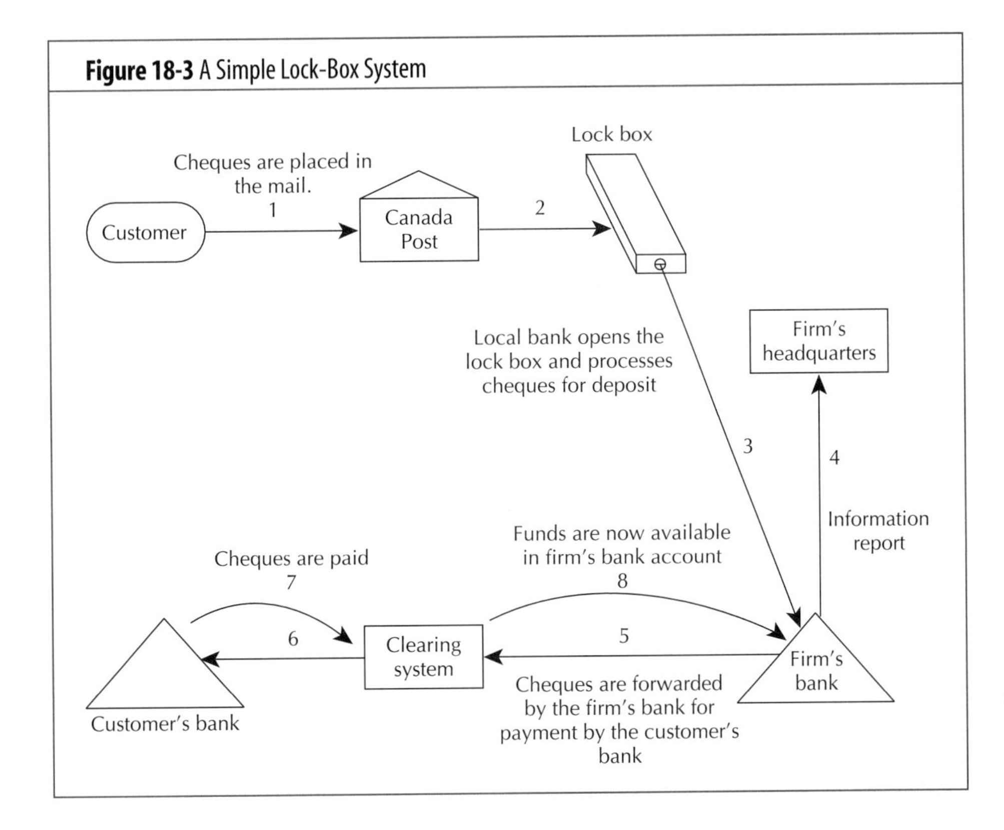

each batch of processed cheques is forwarded to the collection department for clearance. Funds deposited in this manner are usually available for company use in one business day or less.

The bank can notify the firm via some type of telecommunications system that the same-day deposits have been made according to their specified amounts. At the conclusion of each day all cheque photocopies, invoices, deposit slips, and any other documents included with the remittances are mailed to the firm.

Since a firm receives cheques from all over the country, it will have to use several lock boxes to take full advantage of a reduction in mail float. The firm's major bank should be able to offer as a service a detailed lock-box study, analyzing the company's receipt patterns to determine the proper number and location of lock-box receiving points.[10]

Previously in this chapter we calculated the 1994 sales per day for Northern Telecom to be $25.3 million and assumed that the firm could invest its excess cash in marketable securities to yield 8 percent annually. If Northern Telecom could speed up its cash collections by four days (Table 18-2), the results would be startling. The gross annual savings to Northern Telecom (apart from operating the lock-box system) would amount to $8.1 million, as follows:

Video Case 5 

(sales per day) × (days of float reduction) × (assumed yield)
= $25,334,247 × (4) × .08 = $8,106,959

As you might guess, the prospects for generating revenues of this magnitude are important not only to the firms involved, but also to commercial banks that offer lock-box services.

In summary, the benefits of a lock-box arrangement are these:

1. *Increased working cash.* The time required for converting receivables into available funds is reduced. This frees up cash for use elsewhere in the enterprise.
2. *Elimination of clerical functions.* The bank takes over the tasks of receiving, endorsing, totalling, and depositing cheques. With less handling of receipts by employees, better audit control is achieved and the chance of documents becoming lost is reduced.
3. *Early knowledge of dishonoured cheques.* Should a customer's cheque be uncollectible because of lack of funds, it is returned, usually by special handling, to the firm.

These benefits are not free. Usually, the bank levies a charge for each cheque processed through the system. The benefits derived from the acceleration of receipts must exceed the incremental costs of the lock-box system, or the firm would be better off without it. Companies that find the average size of their remittances to be quite small, for instance, might avoid a lock-box plan. One major bank has pointed out that for companies with less than $500,000 in average monthly sales or with customer remittance cheques averaging less than $1,000, the lock-box approach would probably not yield a great enough benefit to offset its costs. Other examples of cash-collection techniques include: **preauthorized payments; depository** (ordinary) **transfer cheques; automated depository transfer cheques;** and **electronic data interchange** (EDI). Table 18-3 summarizes what the objective is for each of these techniques and how each of these techniques is accomplished. Many of these techniques will be used in conjunction with what is known as concentration banking. A **concentration bank** enables a firm to maintain a major disbursing account. Centralizing the firm's pool of cash provides the following benefits:

Preauthorized payment or preauthorized cheque (PAC)

A cheque that resembles the ordinary cheque but does not contain or require the signature of the person on whose account it is being drawn. A PAC is created only with the individual's legal authorization. The PAC system is advantageous when the firm regularly receives a large volume of payments of a fixed amount from the same customer over a long period.

Depository transfer cheques

A nonnegotiable instrument that provides the firm with a means to move funds from local bank accounts to concentration bank accounts.

Automated depository transfer cheque system

A cash management tool that moves funds from local bank accounts to concentration accounts electronically, and thereby eliminates the mail float from the branch bank to the concentration bank.

Electronic data interchange (EDI)

The electronic transfer of funds as well as remittance data between organizations.

Concentration bank

A bank where the firm maintains a major disbursing account.

[10]Several quantitative studies have examined the problem of identifying the optimal locations and numbers of lock boxes, these include: Batlin, C. A., and Susan Hinko. "Lockbox Management and Value Maximization," *Financial Management* 10 (Winter 1981), pp. 39–44; Ferguson, Daniel M. "Optimize Your Firm's Lockbox Selection System," *Financial Executive* 51 (April 1983), pp. 8–12, 14–15, 18–19; and Nauss, Robert M., and Robert E. Markland, "Solving Lock Box Location Problems," *Financial Management* 8 (Spring 1979), pp. 21–27.

Table 18-2 Comparison of Ordinary Cash-Gathering System with Simple Lock-Box System

Step Numbers	Ordinary System	and Time	Advantage of Lock Box
1	Customer writes cheque and places it in the mail	1 Day	
2	Mail is delivered to firm's headquarters	2 Days	Mail will not have to travel as far. Result: save 1 day
3	Accounting personnel process the cheques and deposit them in the firm's local bank	2 Days	Bank personnel prepare cheques for deposit. Result: save 2 days
4 and 5	Cheques are forwarded for payment through the cheque clearing system	1 Day	As the lock boxes are located near the clearing centres of the Canadian payments system, and in Canada the standard banking practice is same-day value for items deposited, the transit float will be reduced. Result: save 1 day
6 and 7	The firm receives notice from its bank that the cheques have cleared and the funds are now "good"	1 Day	
	Total working days	7 Days	Overall result: Save 4 working days

1. *Lower levels of excess cash.* Desired cash balance target levels are set for each branch bank. These target levels consider both compensating balance requirements and necessary working levels of cash. Cash in excess of the target levels can be transferred regularly to concentration banks for deployment by the firm's top-level management.
2. *Better control.* With more cash in fewer accounts, stricter control over available cash is achieved. Quite simply, there are fewer problems. The concentration banks can prepare sophisticated reports that detail corporate-wide movements of funds into and out of the central cash pool.
3. *More efficient investments in near-cash assets.* The coupling of information from the firm's cash forecast with data on available funds supplied by the concentration bank allows the firm to react quickly when making decisions on transferring cash to the marketable securities portfolio.

Table 18-3 Features of Selected Cash Collection Techniques: A Summary

Technique	Objective	How Accomplished
1. Lock-box system	Reduce (1) mail float, (2) processing float, (3) transit float.	Strategic location of lock boxes to reduce mail float and transit float. Firm's commercial bank has access to lock box to reduce processing float. Benefits include (1) increased working cash; (2) elimination of clerical functions; and (3) early knowledge of dishonoured cheques.
2. Preauthorized payments	Reduce (1) mail float and (2) processing float.	The firm's customers give authorization for cheques or electronic debits to be charged to their respective demand deposit accounts. Benefits include (1) highly predictable cash flows; (2) reduced expenses; (3) customer preference; and (4) increased working cash.
3. Depository (ordinary) transfer cheques	Eliminate excess funds in regional banks.	Used in conjunction with concentration banking whereby the firm utilizes several collection centres. The transfer cheque authorizes movement of funds from a local bank to the concentration bank. Benefits include (1) lower levels of excess cash; and (2) better control.
4. Automated depository transfer cheques	Eliminate the mail float associated with the ordinary transfer cheque.	Telecommunications company transmits deposit data to the firm's concentration bank. Benefits include (1) lower levels of excess cash; (2) better control; and (3) more efficient investments in near-cash assets.
5. Electronic data interchange	Enables funds transfers as well as remittance data between organizations. This eliminates the mail and transit float in that only "good funds" are transferred.	The buyer's bank receives and processes the payment from the buyer's transmission. These payments are then forwarded to the destination bank. Benefits include (1) lower levels of excess cash; and (2) better control.

Zero balance accounts

A cash management tool that permits centralized control over cash outflows but also maintains divisional disbursing authority.

Payable-through drafts

A payment mechanism that substitutes for regular cheques in that drafts are not drawn on a bank, but instead are drawn on and authorized by the firm against its demand deposit account. The purpose is to maintain control over field-authorized payments.

Remote disbursing

A cash management service specifically designed to extend the disbursing float.

Later in this chapter a straightforward method for assessing the desirability of a specific cash management service, such as the lock-box arrangement, will be illustrated.

BACK TO THE BASICS

These collection and disbursement procedures are an illustration of ***Axiom 2: The Time Value of Money—A Dollar Received Today Is Worth More Than a Dollar Received in the Future.*** *The faster the firm can take possession of the money to which it is entitled, the sooner the firm can put the money to work generating a return. Similarly, the longer the firm can hold onto the liquid assets in its possession, the greater is the return the firm can receive on such funds.*

Managing the Cash Outflow

Significant techniques and systems for improving the firm's management of cash disbursements include (1) **zero balance accounts**, (2) **payable-through drafts**, and (3) **remote disbursing.** The first two offer markedly better control over companywide payments, and as a secondary benefit they may increase the disbursement float. Table 18-4 summarizes what the objective is for each of these techniques and how each of these techniques is accomplished. In Canada, payment systems which aim at increasing the disbursement float are not effective because both the Canadi-

Table 18-4 Features of Selected Cash Disbursal Techniques: A Summary

Technique	Objective	How Accomplished
1. Zero balance accounts	(1) Achieve better control over cash payments, (2) reduce excess cash balances held in regional banks, and (3) possibly increase disbursing float.	Establish zero balance accounts for all of the firm's disbursing units. These accounts are all in the same concentration bank. Cheques are drawn against these accounts, with the balance in each account never exceeding $0. Divisional disbursing authority is thereby maintained at the local level of management.
2. Payable-through drafts	Achieve effective central office control over field-authorized payments.	Field office issues drafts rather than cheques to settle up payables.
3. Remote disbursing	Extend disbursing float.	Write cheques against demand deposit accounts held in branch banks.

an payments system and the computerized branch bank system ensure same-day processing of payments. However in the United States there are systems, such as remote disbursing, which aim solely at increasing the disbursement float.

PERSPECTIVE IN FINANCE

Our previous work in Chapter 8 presented the popular breakeven model used by financial executives, accountants, and economists. The benefit to the firm of a given cash management service can be assessed in a similar manner. More complicated methods can be presented (some that involve use of an appropriate company discount rate), but the model below is used by managers and is easily explained to them. The important point is: Cash management services are not free.

EVALUATING COSTS OF CASH MANAGEMENT SERVICES

OBJECTIVE 5

A form of breakeven analysis can help the financial manager decide whether a particular collection or disbursement service will provide an economic benefit to the firm. The evaluation process involves a very basic relationship in microeconomics:

$$\text{added costs} = \text{added benefits} \tag{18-1}$$

If Equation (18-1) holds exactly, then the firm is no better or worse off for having adopted the given service. We will illustrate this procedure in terms of the desirability of installing an additional lock box. Equation (18-1) can be restated on a per-unit basis as follows:

$$P = (D)(S)(i) \tag{18-2}$$

where P = increases in per-cheque processing cost if the new system is adopted
D = days saved in the collection process (float reduction)
S = average cheque size in dollars
i = the daily, before-tax opportunity cost (rate of return) of carrying cash

Assume now that cheque processing cost, P, will rise by \$0.18 a cheque if the lock box is used. The firm has determined that the average cheque size, S, that will be mailed to the lock-box location will be \$900. If funds are freed by use of the lock box, they will be invested in marketable securities to yield an annual before-tax return of 8 percent. With these data it is possible to determine the reduction in cheque collection time, D, that is required to justify use of the lock box. That level of D is found to be

$$\$0.18 = (D)(\$900)\left(\frac{.08}{365}\right)$$

$$1.096 \text{ days} = D$$

Thus, the lock box is justified if the firm can speed up its collections by more than 1.096 days. This same style of analysis can be adapted to analyze the other tools of cash management.

PERSPECTIVE IN FINANCE

Designing the marketable securities portfolio is one of the more pleasant tasks in financial management. This is because it is usually done with excess cash that is available for short periods of time. The firm typically has excess cash when operations are going well; some firms never get the opportunity to design such a near-cash portfolio or to spend much time even thinking about it.

We will review the major securities that make up the portfolios designed by cash managers. Note that the firm's main line of business activity will have a key impact on how much risk is assumed in the portfolio.

If the firm is in the business of manufacturing personal computers, it is likely management will feel that business itself is risky enough. Thus, not much additional risk will be embedded in the marketable securities portfolio.

In addition, observe how critical the concept of liquidity is to this aspect of cash management. Most large organizations will transfer funds into and out of the portfolio several times a day. Ready convertibility into cash, therefore, is a prime determinant of the final composition.

OBJECTIVE 6

COMPOSITION OF MARKETABLE SECURITIES PORTFOLIO

Once the design of the firm's cash receipts and payments system has been determined, the financial manager faces the task of selecting appropriate financial assets for inclusion in the firm's marketable securities portfolio.

General Selection Criteria

Certain criteria can provide the financial manager with a useful framework for selecting a proper marketable securities mix. These considerations include evaluation of (1) financial risk, (2) interest rate risk, (3) liquidity, (4) taxability, and (5) yields among different financial assets. From the investor's point of view, we will briefly delineate these criteria.

Financial Risk

Financial risk here refers to the uncertainty of expected returns from a security attributable to possible changes in the financial capacity of the security issuer to make future payments to the security owner. If the chance of default on the terms of the instrument is high (low), then the financial risk is said to be high (low). It is

The Canadian Payments System

We have seen that understanding the concept of float is central to understanding efficient cash management activities. The transit float has been effectively reduced by same-day settlement which is offered by the Canadian payments system. The following discussion emphasizes two important aspects of the Canadian payments system.

The Integrity of the Payments System

There are two fundamental determinants of the integrity of the Canadian payments system or, indeed, of any national payments system. First, withdrawals from the kinds of accounts used for making payments must be subject to stringent security precautions, or the payment orders drawn on those accounts will inevitably become suspect and cease to be accepted in payment. Second, the institutions that participate in the clearing and settlement operations that are at the heart of the national payments system must be carefully regulated and supervised. This is necessary in order to prevent failures as far as possible but also to preserve the basis of mutual trust that underlies the efficient operation of the system. This is why the running of the clearing system and the planning of the national payments system were entrusted by the CPA Act of 1980 to a particular type of institution. They are the ones that offer such accounts as a major part of their business and that are regulated and supervised under federal or provincial laws that also provide for the insurance or the guarantee of the deposits and the inspection of the institutions themselves. Direct access to clearing and settlement facilities must then clearly be, in a Canadian context, under the control of the member institutions of the Canadian Payments Association if the safeguards established by Parliament in the Association's act are to have any meaning or effect.

The first aspect of the integrity of the Canadian payments system—access to accounts at deposit-taking institutions—was addressed by the CPA in 1985 and in early 1986 when it set out a framework for the development of the Canadian payments system. The main parameters of that framework are that, unless deposit-taking institutions can satisfy themselves as to the authenticity of any order to withdraw or pay money out of the accounts that are held with them, they cannot make good on their undertakings to their customers or their obligations under the law. These require the institutions to safeguard the money entrusted to them, to make payments to third parties with that money only when properly authorized to do so, and to protect customers' financial affairs from improper scrutiny. Deposit-taking institutions are under an obligation to their customers to ensure the scrutiny and confidentiality of their accounts, and they are not necessarily relieved of that responsibility if their customers themselves deliberately or accidentally disregard the appropriate security measures. Therefore deposit-taking institutions should have the means to discharge these responsibilities. More specifically, in an electronic world of virtually instantaneous transactions, the issue and validation of any customer identification device that unlocks access to the account must remain strictly confidential between the deposit-taking institution and its customer, and both must ensure that it remains fully protected. Only then can deposit-taking institutions be reasonably certain that the person making a withdrawal or a payment from a terminal is actually the customer to whom they are responsible for the account, and the customer, be reasonably certain that their funds are not at risk of unauthorized use. The experience that member institutions have gained with credit card transactions suggests that transaction-by-transaction authorization can play a significant role in reducing fraud.

The second aspect of the integrity of the Canadian payments system—access to clearing and settlement facilities—has been tackled more recently by the Association. There have hitherto been no clear guidelines to determine what should and should not be accepted for processing in the clearing and settlement system. The Association has now defined items acceptable for clearing to mean basically those items that are drawn on CPA member institutions and on those institutions that are eligible for membership. In this way, the participants in the clearing system and the public at large can have the confidence that the same requirement for insurance or guarantee of the funds lies behind the payments passing through the system as applies to deposits with member institutions. At the same time, the Association certainly intends to continue to operate the clearing system for the public benefit. It is clearly in the public interest to properly identify and define as acceptable for clearing a number of other types of payment items, such as government cheques, grain tickets and postal money orders. Such items will continue to be processed through the national clearing and settlement system, and so will new ones as they are developed, provided that they meet the same kinds of criteria.

Source: Serge Vachon, "Evolution 1987," *Bank of Canada Review* (October 1987), pp. 9–14. Opening remarks to the Canadian Payments System Conference on 21–22 September, 1987 by Serge Vachon, Chairman of the Canadian Payments Association and Advisor to the Governor of the Bank of Canada.

clear that the financial risk associated with holding commercial paper, which we will see shortly is nothing more than a corporate IOU, exceeds that of holding securities issued by the Government of Canada.

In both financial practice and research, when estimates of risk-free returns are desired, the yields available on treasury bills are consulted and the safety of other financial instruments is weighed against them. Because the marketable securities portfolio is designed to provide a return on funds that would otherwise be tied up in idle cash held for transactions or precautionary purposes, the financial manager is not usually willing to assume much financial risk in the hope of greater return.

Interest Rate Risk[11]

Interest rate risk refers to the uncertainty that envelops the expected returns from a financial instrument attributable to changes in interest rates. Of particular concern to the corporate treasurer is the price volatility associated with instruments that have long, as opposed to short, terms to maturity.

To illustrate that bond price changes are an increasing function of maturity for any given difference between the coupon rate and the yield-to-maturity on the security, we assume that a financial manager is weighing the merits of investing temporarily available corporate cash in a new offering of Government of Canada bonds that will mature in either (1) three years or (2) 20 years from the date of issue. The purchase price of the three-year bonds or 20-year bonds is at their par value of $1,000 per security. For the purposes of this illustration, we assume that the term structure for yields on interest rates is flat such that the maturity value of either class of security is equal to par, $1,000, and the coupon rate (stated interest rate) is set at 7 percent, compounded annually.

If after one year from the date of purchase prevailing interest rates rise to 9 percent, the market prices of these currently outstanding Government bonds will fall to bring their yields to maturity in line with what investors could obtain by buying a new issue of a given instrument. The market prices of both the three-year and 20-year obligations will decline. The price of the 20-year instrument will decline by a greater dollar amount, however, than that of the three-year instrument.

One year from the date of issue the price obtainable in the marketplace for the original 20-year instrument, which now has 19 years to go to maturity, can be found by computing P as follows:

$$P = \sum_{t=1}^{19} \frac{\$70}{(1 + .09)^t} + \frac{\$1{,}000}{(1 + .09)^{19}} = \$821.01$$

In the previous expression t is the year in which the particular return, either interest or principal amount, is received; $70 is the annual interest payment; and $1,000 is the contractual maturity value of the bond. The rise in interest rates has forced the market price of the bond down to $821.01.

Now, what will happen to the price of the note that has two years remaining to maturity? In a similar manner, we can compute its price, P:

$$P = \sum_{t=1}^{2} \frac{\$70}{(1 + .09)^t} + \frac{\$1{,}000}{(1 + .09)^{2}} = \$964.84$$

The market price of the shorter-term note will decline to $964.84. It is evident from Table 18-5 that the market value of the shorter-term security was penalized much less by the given rise in the general level of interest rates.

If we extended the illustration, we would see that, in terms of market price, a one-year security would be affected less than a two-year security, a three-month security less than a six-month security, and so on. Equity securities would exhibit the largest price changes because of their infinite maturity periods. To hedge against

[11]The computations in this discussion assume some knowledge of basic interest calculations. The concept can be grasped without the mathematical example, however. The mathematics of finance was covered in Chapter 4.

Table 18-5 Market Price Effect of a Rise in Interest Rates

Item	Three-year Instrument	Twenty-year Instrument
Original price	$1,000.00	$1,000.00
Price after one year	964.84	821.01
Decline in price	$ 35.16	$ 178.99

the price volatility caused by interest rate risk, the firm's marketable securities portfolio will tend to be composed of instruments that mature over short periods.

Liquidity

In the present context of managing the marketable securities portfolio, *liquidity* refers to ability to transform a security into cash. Should an unforeseen event require that a significant amount of cash be immediately available, then a sizable portion of the portfolio might have to be sold. The financial manager will want the cash quickly and will not want to accept a large price concession in order to convert the securities.[12] Thus, in the formulation of preferences for the inclusion of particular instruments in the portfolio, the manager must consider (1) the time period needed to sell the security and (2) the likelihood that the security can be sold at or near its prevailing market price. The latter element, here, means that "thin" markets, where relatively few transactions take place or where trades are accomplished only with large price changes between transactions, will be avoided.

Taxability

The tax treatment of a firm's income which is received from its security investments, does not affect the ultimate mix of the marketable securities portfolio as much as the criteria mentioned earlier. This is because the interest income from most instruments suitable for inclusion in the portfolio is taxable at both the federal and provincial levels. Still, the taxability of interest income and capital gains is seriously evaluated by some corporate treasurers.

Capital gains enjoy a preferential tax treatment as compared to interest income. Under such circumstances, bonds selling at a discount from their face value may be attractive investments to tax-paying firms. Should a high level of interest rates currently exist, the market prices of debt issues that were issued in the past at low coupon rates will be depressed. This, as we said previously, brings their yield to maturity up to that obtainable on a new issue. Part of the yield to maturity on a bond selling at a discount is a capital gain, or the difference between the purchase price and the maturity value. Hence, the return after tax that is earned could be higher than that derived from a comparable issue carrying a higher coupon but selling at par. We say *could be higher,* as the marketplace is rather efficient and recognizes this feature of taxability; consequently, bonds sold at a discount will provide a lower yield than issues that have similar risk characteristics but larger coupons. For short periods, though, a firm might find a favourable yield advantage by purchasing bonds sold at a discount.

Yields

The final selection criterion that we mention is a significant one—the yields that are available on the different financial assets suitable for inclusion in the near-cash

[12]The high demand by corporate treasurers for short-term instruments will generally keep these instruments more liquid than long-term maturity instruments.

portfolio. By now it is probably obvious that the factors of (1) financial risk, (2) interest rate risk, (3) liquidity, and (4) taxability all influence the available yields on financial instruments. The yield criterion involves an evaluation of the risks and benefits inherent in all of these factors. If a given risk is assumed, such as lack of liquidity, a higher yield may be expected on the non-liquid instruments.

Figure 18-4 summarizes our framework for designing the firm's marketable securities portfolio. The four basic considerations are shown to influence the yields available on securities. The financial manager must focus on the risk-return tradeoffs identified through analysis. Coming to grips with these tradeoffs will enable the financial manager to determine the proper marketable securities mix for the company. Let us look now at the marketable securities prominent in firms' near-cash portfolios.

Marketable Security Alternatives[13]

Table 18-6 shows that the Government of Canada Treasury bills are the best-known and most popular short-term investment. The Government of Canada **Treasury bill** is a non-interest-bearing, discount security that the Bank of Canada sells in order to raise money for the Government of Canada. Since Treasury bills are sold on a discount basis, the investor does not receive actual interest payments. The actual dollar return is the difference between the purchase price and the face (par) value of the bill less any transaction costs. However, for comparison purposes with other securities, the return of a Treasury bill can be expressed on a true yield basis. The true yield is the annualized rate of interest which is compounded to the frequency of the bill's term to maturity. For example, suppose three-month Treasury bills are issued on a true yield basis of 6.93 percent. This yield implies that the interest rate on a quarterly basis is (6.93 × 1/4) or 1.73 percent. As a result, a $1,000 Treasury bill has a present value over a period of one quarter of ($1,000/(1+.0173)) or $982.99. In fact, the Government of Canada will be obliged to pay $1,000 after the quarter has expired for every $928.99 of investment in three-month Treasury bills by the investor, today.

Treasury bill

Direct and guaranteed debt obligations of the Canadian government sold on a regular basis by the Bank of Canada.

At present, bills are regularly offered with maturities of three months, six months, and one year. The three-month and six-month bills are auctioned weekly by the Bank of Canada, and the one-year bills are offered biweekly. Every Tuesday the Bank of Canada invites both the banks and investment dealers to submit bids to buy Treasury bills. The allotment of Treasury bills is distributed according to the highest price submitted by bidders.

In the past, this type of auction process enabled the Bank of Canada to set their weekly bank rate at one quarter of one percentage point above the latest average rate for three-month Treasury bills. However in 1996, the Bank of Canada modified their policy of adjusting the bank rate on a weekly basis to a policy in which the bank rate is adjusted when the Bank of Canada, reacting to market pressures, feels

Figure 18-4 Designing the Marketable Securities Portfolio

Considerations →	Influence →	Focus Upon →	Determine
Financial risk Interest rate risk Liquidity Taxability	Yields	Risk vs. return preferences	Marketable securities mix

[13]The discussions on market security alternatives draw heavily from the following sources: S. Sarpkaya, *Modern Cash Management*, 2nd ed. (Don Mills, ON: CCH Canadian Ltd., 1991) and from G.F. Boreham, *Money, Banking, and Finance* (Toronto: Holt, Rinehart and Winston of Canada, Ltd., 1988).

Table 18-6 Volume of Selected Money Market Instruments, 1981–1995 (millions of dollars)

End of Year	Government of Canada Treasury Bills	Government of Canada Bonds Less than 3 Years	Provincial, Municipal Treasury Bills and Notes	Bankers' Acceptances	Commercial Paper	Sales Paper and Consumer Loan Company Paper	Term and Notice Deposits and GICs
1981	20,700	16,594	2,067	6,591	9,314	3,501	55,081
1982	25,725	19,037	3,903	12,647	7,696	1,821	55,618
1983	39,025	17,952	5,297	13,954	9,915	2,372	51,748
1984	49,675	20,061	6,950	13,982	11,341	2,975	54,511
1985	59,400	19,744	7,418	17,007	10,125	4,049	52,879
1986	69,700	23,956	9,943	24,896	10,298	6,102	54,616
1987	74,200	26,701	11,445	31,115	12,017	7,335	62,965
1988	95,100	31,734	10,255	40,191	16,377	8,729	68,375
1989	120,550	36,149	11,921	43,666	19,029	9,898	80,750
1990	135,400	42,216	14,164	44,109	21,851	9,004	90,340
1991	147,600	48,228	13,950	36,151	25,199	6,464	92,450
1992	159,450	54,616	17,670	21,970	25,036	5,971	105,035
1993	165,900	64,470	16,171	26,171	25,246	5,996	93,485
1994	159,550	70,169	17,433	26,607	23,422	8,390	100,087
1995	160,100	83,495	17,067	30,701	21,597	9,257	100,132

Source: *Bank of Canada Review*, various issues (1981–1995).

a change in interest rates is warranted. In other words, this new policy for adjusting the bank rate is intended to act as clearer signal to all Canadians of changes in monetary policy by the Bank of Canada, as well as a signal to indicate where the Bank of Canada wants other interest rates to move. This type of policy is consistent with the theory that suggests Canadian interest rates are determined by demand and supply forces of the money market, not the Bank of Canada.

Treasury bills are marketed by the Bank of Canada only in bearer form. They are purchased, therefore, without the investor's name on them. This attribute makes them easily transferable from one investor to the next. Of prime importance to the corporate treasurer is the fact that a very active secondary market exists for bills. After a Treasury bill has been acquired by the firm, should the need arise to turn it into cash, a group of securities dealers stand ready to purchase it. This highly developed secondary market for Treasury bills not only makes them extremely liquid, but also allows the firm to buy Treasury bills with maturities of a week or even less. This type of flexibility in maturity of Treasury bills enables a cash manager to plan for the payments on projects and the distribution of dividend payments.

As Treasury bills have the full financial backing of the Government of Canada, they are, for all practical purposes, risk-free. This negligible financial risk and high degree of liquidity makes the yields lower than those obtainable on other marketable securities.

Since a Treasury bill is a non-interest-bearing security which is sold at a discount, Revenue Canada requires that for tax purposes the discount be treated as business income for corporations and as interest income for individuals.

The provinces also issue Treasury bills which are sold on a discount basis in denominations of $100,000. Both the yield and liquidity of these types of short-term obligations are determined by the credit standing of the government. In this case, the credit standing of the government is a reflection of debt burden, prospects of local economy, taxing power, and per capita income of the residents.

Federal, provincial and municipal governments of Canada issue short-term bonds and paper in the money market. **Short-term bonds** of the Government of

Canada are direct and guaranteed obligations which have a maturity of less than three years. These favourable features make for an active secondary market with the most recent issues being more liquid. Government of Canada bonds provide semiannual interest coupons and are generally issued in the following denominations: $1,000; $5,000; $25,000; $100,000; and $1,000,000. Generally, the provincial and municipal types of governments use the money market to cover short-term lending programs or to meet current expenditures in anticipation of tax revenues. Provincial and municipal short-term obligations provide higher returns than similar federal government issues. However, provincial and municipal government obligations are not as liquid as federal debt obligations because the issues are generally small in size or are private placements such that only a few investment dealers are required to be involved in the underwriting of the issue.

Short-term bonds

Direct and guaranteed debt obligations which have a maturity of less than three years.

The chartered banks of Canada issue several types of short-term investment securities to their customers. Examples of *bank-issued securities* include bearer term deposit notes, deposit receipts, Eurodollar negotiable certificates of deposit, interbank deposit receipts, swapped deposits, special call loans, corporate certificates of deposit, and banker's acceptances. Each of these instruments is customized to either the specific needs of their clients or the needs of the banks themselves. For example, the bearer term deposit notes are instruments that offer money market dealers a prime credit rating with a competitive rate of return for activities such as short selling; whereas, the special day loans provide chartered banks with a means to finance investment dealers. The most important of these instruments in terms of dollar volume are banker's acceptances and corporate certificates of deposit.

The role of bankers' acceptances in Canada's financial system has increased to such an extent that this instrument is the underlying security for financial interest futures which are sold on the Montreal Futures Exchange. Generally, a **bankers' acceptance** is a draft (order to pay) that has been accepted by a specific bank. Once accepted, the draft becomes a negotiable financial instrument that can be traded in the marketplace.

Bankers' acceptances

A draft (order to pay) drawn on a specific bank by a seller of goods in order to obtain payment for goods that have been shipped (sold) to a customer. The customer maintains an account with that specific bank.

Any chartered bank will issue banker's acceptances once a customer's line of credit is established. A well-developed secondary market exists for banker's acceptances. Furthermore, the credit quality of these financial instruments is enhanced since both the names of the accepting bank and the issuer appear on the draft.[14] Hence, the banker's acceptances of chartered banks are a very safe investment, making the yield advantage over Treasury bills worth looking at from the firm's vantage point.

Certificate of deposit

A nonmarketable receipt for funds that have been deposited with a bank for a fixed time period. The corporate CD is issued at par in blocks of $100,000 and the owner receives the full amount deposited plus earned interest.

A corporate **certificate of deposit** (CD), is a nonmarketable term note issued at par, generally in blocks of $100,000. When the certificate matures, the owner receives the full amount deposited plus the earned interest. Withdrawal of the investment funds before the maturity date causes the investor to incur an interest penalty. To accommodate the individual investor with limited funds, the banks offer certificates of deposit with smaller denominations.

Commercial paper

Short-term unsecured promissory notes sold by large businesses in order to raise cash. Unlike most other money market instruments, commercial paper has no developed secondary market.

Commercial paper refers to short-term, unsecured promissory notes sold by large businesses to raise cash. Because they are unsecured, the issuing side of the market is dominated by large corporations, which typically maintain sound credit ratings. The issuing (borrowing) firm can sell the paper to a dealer who will in turn sell it to the investing public; if the firm's reputation is solid, the paper can be sold directly to the ultimate investor. Commercial paper in past years consistently provided a yield advantage over other near-cash assets of comparable maturity because of the lack of marketability and as result a limited secondary market.

Repurchase agreements (repos) are legal contracts that involve the actual sale of securities by a money market dealer to the lender, with a commitment on the

[14]Bankers' acceptances are referred to among practitioners as a form of "two-name paper." This stems from the fact that should the bank default on the instrument, the holder (investor) can seek payment from the drawer of the acceptance.

part of the money market dealer to repurchase the securities at the contract price plus a stated interest charge. The securities sold to the lender are Government of Canada Treasury bills or other government securities.

Repurchase agreements (repos)

Legal contracts that involve the sale of the short-term securities by a borrower to a lender of funds. The borrower commits to repurchase the securities at a later date at the contract price plus a stated interest charge.

The motivation to buy repurchase agreements rather than a given marketable security is twofold. First, the original maturities of the instruments being sold can, in effect, be adjusted to suit the particular needs of the investing corporation. Funds available for very short time periods, such as one or two days, can be productively employed. The second reason is closely related to the first. The firm could, of course, buy a Treasury bill and then resell it in the market in a few days when cash was required. The drawback here would be the risk involved in liquidating the bill at a price equal to its earlier cost to the firm. The purchase of a repurchase agreement removes this risk. The contract price of the securities that make up the arrangement is fixed for the duration of the transaction. The corporation that buys a repurchase agreement, then, is protected against market price fluctuations throughout the contract period. This makes it a sound alternative investment for funds that are freed up for only very short periods. These types of features enable firms which have cash positions that cannot be accurately forecasted to benefit from repurchase agreements' liquidity and a competitive rate of return on surplus cash holdings.

The maturities may be for a specified time period or may have no fixed maturity date. In the latter case either lender or borrower may terminate the contract without advance notice. Since the interest rates are set by direct negotiation between lender and borrower, no regular published series of yields is available for direct comparison with the other short-term investments. The rates available on repurchase agreements, however, are closely related to, but generally less than, Treasury bill rates of comparable maturities.

Finally, investors can invest in **money market mutual funds** which are composed of a diversified portfolio of short-term, high-grade debt instruments such as Treasury bills, certificates of deposit, and commercial paper. However, there are some funds which will accept more interest rate risk in their portfolios and acquire some corporate bonds.

Money market mutual funds

Investment companies that purchase a diversified array of short-term, high-grade (money market) debt instruments.

The money market funds sell their shares to raise cash, and by pooling the funds of large numbers of small savers, they can build their liquid-asset portfolios. Many of these funds allow the investor to start an account with as little as $1,000. This small initial investment, coupled with the fact that some liquid-asset funds permit subsequent investments in amounts as small as $100, makes this type of outlet for excess cash suited to the small firm. Furthermore, the management of a small enterprise may not be highly versed in the details of short-term investments. By purchasing shares in a liquid-asset fund, the investor is also buying managerial expertise.

Money market mutual funds offer the investing firm a high degree of liquidity. An investor can liquidate some of their shares of the fund in order to obtain cash quickly. However, some fund managers allow both retail clients and businesses to write cheques on their money funds. This type of fund is ideal for large withdrawals since servicing charges are expensive as compared to normal chequing accounts with the banks.[15]

Table 18-7 provides a summary of the features of selected money market instruments. Each instrument is described in terms of its denomination, maturity, interest, form, liquidity and taxability characteristics.

The Yield Structure of Marketable Securities

What type of return can the financial manager expect on a marketable securities portfolio? This is a reasonable question. Some insight can be obtained by looking at

[15]Olev Edur, "Cheque Funds Benefits Funds," *The Globe and Mail*, August 15, 1991, p. B21.

Table 18-7 Features of Selected Money Market Instruments

Instrument	Denominations	Maturities	Basis	Form	Liquidity	Taxability
Government of Canada Treasury bills—direct obligations of the government; provinces also issue Treasury bills	$ 1,000 5,000 25,000 100,000 1,000,000	91 days 182 days 365 days	Discount	Bearer	Excellent secondary market	Taxed federally and provincially
Government of Canada short-term bonds	Wide variation	Less than 3 years	Interest bearing	Bearer or registered	Excellent secondary market	Taxed federally and provincially
Bankers' acceptances—drafts accepted for future payment by chartered banks	Multiples of $100,000	Predominantly from 30 to 90 days	Discount	Bearer	Excellent secondary market	Taxed federally and provincially
Corporate certificates of deposit—marketable receipts for funds deposited in a bank for a fixed time period	$100,000 or more	Few days to one year	Accrued interest	Registered	Subject to interest penalty	Taxed federally and provincially
Commercial paper short-term unsecured promissory notes	$25,000 or more; $1,000 and $5,000 multiples above the initial offering size are sometimes available	3 to 270 days	Discount	Bearer	Limited secondary market	Taxed federally and provincially
Repurchase agreements—legal contracts between a borrower (security seller) and lender (security buyer). The borrower will repurchase at the contract price plus an interest charge	Typical sizes are $100,000 or more	According to terms of contract	Not applicable	Not applicable	Fixed by the agreement; that is, borrower will repurchase	Taxed federally and provincially
Money market mutual funds—holders of diversified portfolios of short-term, high grade instruments	Some require an initial investment as small as $1,000	Your shares can be sold at any time; cheque-service	Net asset value	Registered	Good; provided by the fund itself	Taxed federally and provincially

the past, although we must realize that future returns are not guided by past experience. It is also useful to have some understanding of how the returns on one type of instrument stack up against another. The behaviour of yields on short-term debt instruments over the 1970–1995 period is shown in Table 18-8. An examination of the data in that table permits the following generalizations:

1. The returns from the various instruments are highly correlated in the positive direction over time. That is, the yields tend to rise and fall together.
2. The yields are quite volatile over time. For example, notice the large increases in returns from 1972 to 1974, 1978 to 1979, 1980 to 1981, and 1988 to 1989. The financial manager, then, cannot plan on any given level of returns prevailing

Table 18-8 Annual Yields (Percent) on Selected Marketable Securities

Year	Non-Chequable Savings Deposits	T-Bills 3 Months	Commercial Paper 90-Day	Bankers' Acceptances 30-Day
1970	6.17	6.12	7.34	7.28
1971	4.54	3.62	4.51	4.41
1972	4.00	3.55	5.10	4.89
1973	5.44	5.39	7.45	6.97
1974	8.50	7.78	10.51	10.16
1975	7.00	7.37	7.94	7.71
1976	7.83	8.89	9.17	9.18
1977	6.00	7.35	7.48	7.46
1978	7.08	8.58	8.83	8.52
1979	10.14	11.57	12.07	11.77
1980	11.17	12.68	13.15	13.03
1981	15.42	17.78	18.33	18.52
1982	11.50	13.83	14.15	14.22
1983	6.85	9.32	9.45	9.34
1984	7.69	11.10	11.19	10.90
1985	6.08	9.46	9.56	9.43
1986	6.02	8.99	9.16	9.16
1987	4.81	8.17	8.39	8.16
1988	5.69	9.42	9.66	9.40
1989	8.08	12.02	12.21	12.08
1990	8.77	12.81	13.03	13.04
1991	4.48	8.83	8.90	8.91
1992	2.27	6.51	6.73	6.73
1993	0.77	4.93	4.97	4.71
1994	0.50	5.52	5.66	5.25
1995	0.50	7.05	7.22	7.07

Source: Statistics Canada, *Canadian Economic Observer*, Cat. No. 11-210, 1994/1995 and various issues of *Bank of Canada Review* (1970–1996), Table F1.

over a long time period. Also, observe the relatively high yields during 1974 that led to the formation of a large number of money market mutual funds. High yields prevailed again in 1979–1982 and aided their growth.

The discussion in this chapter on designing the firm's marketable securities portfolio touched upon the essential elements of several near-cash assets. At times it is difficult to sort out the distinguishing features among these short-term investments. To alleviate that problem, Table 18-7 draws together their principal characteristics.

A GLANCE AT ACTUAL CASH MANAGEMENT PRACTICES

OBJECTIVE 7

We have dealt at length in this chapter with (1) the collection and disbursement procedures the firm can use to manage its cash balances more effectively and (2) the composition of the firm's marketable securities portfolio. It is of interest to relate our discussion to the findings of some studies that focus on corporate cash management practices.

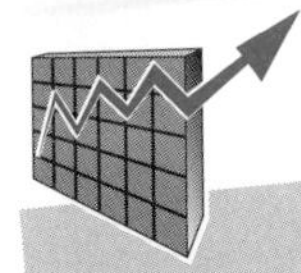

Investment Practices

A study published by Frankle and Collins provided some insights on the investment tendencies of cash managers. The results differ somewhat from those reported in earlier studies. The researchers sampled the *Fortune 1000* list of large industrial firms and achieved a response rate of about 22 percent.

In the Frankle and Collins study the respondents were asked to describe their own approach to cash management—from "aggressive" to "passive." Only those ultimately classified as aggressive or moderate are commented upon here.

One portion of the study asked the managers to identify the percentage of their short-term portfolios invested in the typical marketable securities. The five most popular instruments and the respective percentages are shown below.

Unlike earlier studies, this piece of research indicated that Eurodollar CDs are preferred by aggressive cash managers with commercial paper being ranked second and domestic CDs ranked third. The moderate managers also liked Eurodollar CDs, with repurchase agreements ranked second and commercial paper ranked third.

Another question in the study asked the managers to indicate which investment attributes shaped their ultimate investment decisions. The aggressive managers ranked rate of return first, risk of default second, and maturity third. The moderate managers ranked risk of default first in importance, rate of return second, and maturity third.

Compared to earlier work on cash-management practices, this study suggests that those responsible for the investment of excess corporate cash are assuming even a bit more risk in their portfolios in exchange for higher expected returns. This may be due to the increased emphasis placed by corporate management on the cash-management function and the growing sophistication of those who practice the trade.

Source: Alan W. Frankle and J. Markham Collins, "Investment Practices of the Domestic Cash Manager," *Journal of Cash Management* 7 (May/June 1987), pp. 50–53.

Type of Security	Aggressive Manager	Moderate Manager
Treasury securities	5.55%	8.82%
Commercial paper	18.16%	17.53%
Domestic CDs	11.45%	6.87%
Eurodollar CDs	36.77%	24.28%
Repurchase agreements	11.27%	18.99%

Surveys conducted by (1) Gitman, Moses, and White and (2) Mathur and Loy centred on both the cash management services offered by commercial banks and the investment of excess cash in the marketable securities portfolio.[16] Gitman, Moses, and White surveyed 300 of the 1,000 largest industrial firms in the United States ranked by total sales dollars for 1975 (the *Fortune 1000* list). They received 98 responses to their questionnaire, for a response rate of 32.7 percent. With regard to speeding up the firm's cash receipts, it was found that 80 percent of the respon-

[16]Lawrence J. Gitman, Edward A. Moses, and I. Thomas White, "An Assessment of Corporate Cash Management Practices," *Financial Management* 8 (Spring 1979), pp. 32-41; and Ike Mathur and David Loy, "Corporate-Banking Cash Management Relationships: Survey Results," *Journal of Cash Management* 3 (October–November 1983), pp. 35–46.

dents used lock-box systems and 61 percent used selected disbursing points (remote disbursing) to slow down or better control cash payments.

The Mathur and Loy study centred on the 200 largest industrial firms from the *Fortune 500* list for 1979. Of these 200 firms, 160 were sampled. A response rate of 46 percent was achieved. This study showed that almost all of the respondents used lock-box services (96 percent). In addition, concentration banking was employed by 80 percent of the survey participants.

These same two studies touched upon the management of the marketable securities portfolio. Gitman, Moses, and White found that the most favoured security for the investment of excess cash was commercial paper. It was followed in popularity by repurchase agreements, treasury securities, and CDs. Mathur and Loy also found that commercial paper was the most favoured marketable security investment. Furthermore, repurchase agreements and CDs ranked high on respondents' most favoured lists. Mathur and Loy also reported that, on average, the firms invest in 3.6 different types of marketable securities.

We all know that commercial paper, CDs, and repurchase agreements are more risky investments than treasury securities. It seems that those who are responsible for making marketable security investment decisions in large firms do not consider the risk differences among the traditional money market instruments to be very significant. They are willing to trade the greater riskiness inherent in the "non-treasury" instruments for their higher expected yield.

SUMMARY

Managing Current Assets

Firms hold cash to satisfy transactions, precautionary, and speculative needs for liquidity. Because cash balances provide no direct return, the precautionary motive for investing in cash is satisfied in part by holdings of marketable securities.

Cash Management Objectives and Decisions

The financial manager must (1) make sure that enough cash is on hand to meet the payment needs that arise in the course of doing business and (2) attempt to reduce the firm's idle-cash balances to a minimum.

Collection and Disbursement Procedures

A firm can benefit considerably by reducing its float through the use of (1) lock-box arrangements, (2) preauthorized payments, (3) special forms of depository transfer cheques, and (4) wire transfers. Lock-box systems and preauthorized cheques serve to reduce mail and processing float. Depository transfer cheques and wire transfers move funds between banks; they are often used in conjunction with concentration banking. Both the lock-box and preauthorized cheque systems can be employed as part of the firm's concentration banking setup to speed receipts to regional collection centres. The firm can delay and favourably affect the control of its cash disbursements through the use of (1) zero balance accounts, (2) payable-through drafts, and (3) remote disbursing. Zero balance accounts allow the company to maintain central office control over payments while permitting the firm's several divisions to maintain their own disbursing authority. Because key disbursing accounts are located in one major concentration bank, rather than in multiple banks across the country, excess cash balances that tend to build up in the outlying banks are avoided. Payable-through drafts are legal instruments that look like cheques but are drawn on and paid by the issuing firm rather than its bank. The bank serves as a collection point for the drafts. Effective central office control over field-authorized payments is the main reason such a system is used; it is not used as a major vehicle for extending the disbursing float. Remote disbursing, however, is used to increase disbursing float. Remote disbursing refers to the process of writing payment cheques on banks located in cities distant from the one where the cheque is originated. Before any of these collection and disbursement procedures is initiated by the firm, a careful analysis should be undertaken to see if the expected benefits outweigh the expected costs.

Marketable Securities Portfolio

The factors of (1) financial risk, (2) interest rate risk, (3) liquidity, and (4) taxability affect the yields available on marketable securities. By considering these four factors simultaneously with returns desired from the portfolio, the financial manager can design the mix of near-cash assets most suitable for a firm.

Several marketable securities were evaluated in terms of their features; for example, Treasury bills and short-term government bonds are extremely safe investments. Higher yields are obtainable on bankers' acceptances, CDs, and commercial paper in exchange for greater risk assumption. The firm can hedge against price fluctuations through the use of repurchase agreements. Money market mutual funds, a recent phenomenon of our financial market system, are particularly well suited for the short-term investing needs of small firms.

STUDY QUESTIONS

18-1. What is meant by the cash flow process?

18-2. Identify the principal (classical) motives for holding cash and near-cash assets. Explain the purpose of each motive.

18-3. Define each element of the total float.

18-4. Distinguish between depository transfer cheques and automated depository transfer cheques (ADTC).

18-5. How has the Canadian Payments Association policy of same-day value affected the transit float?

18-6. What are the two major objectives of the firm's cash-management system?

18-7. What three decisions dominate the cash management process?

18-8. Within the context of cash management, what are the key elements of (total) float? Briefly define each element.

18-9. Distinguish between financial risk and interest rate risk as these terms are commonly used in discussions of cash management.

18-10. What is meant when we say, "A money market instrument is highly liquid"?

18-11. Which money market instrument is generally conceded to have no secondary market?

18-12. Your firm invests in only three different classes of marketable securities: commercial paper, Treasury bills, and certificates of deposit. Recently, yields on these money market instruments of three months' maturity were quoted at 6.10, 6.25, and 5.90 percent. Match the available yields with the types of instruments your firm purchases.

18-13. Explain why money market mutual funds can be advantageous to small businesses in managing their cash.

18-14. What two key factors might induce a firm to invest in repurchase agreements rather than a specific security of the money market?

SELF-TEST PROBLEMS

ST-1. *(Costs of Services)* Creative Fashion Designs is evaluating a lock-box system as a cash receipts acceleration device. In a typical year this firm receives remittances totalling $7 million by cheque. The firm will record and process 4,000 cheques over the same time period. The National Bank has informed the management of Creative Fashion Designs that it will process cheques and associated documents through the lock-box system for a unit cost of $.25 per cheque. Creative Fashion Designs' financial manager has projected that cash freed by adoption of the system can be invested in a portfolio of near-cash assets that will yield an annual before-tax return of 8 percent. Creative Fashion Designs' financial analysts use a 365-day year in their procedures.

- **a.** What reduction in cheque collection time is necessary for Creative Fashion Designs to be neither better nor worse off for having adopted the proposed system?
- **b.** How would your solution to (a) be affected if Creative Fashion Designs could invest the freed balances only at an expected annual return of 5.5 percent?
- **c.** What is the logical explanation for the differences in your answers to (a) and (b) above?

ST-2. *(Cash Receipts Acceleration System)* Artie Kay's Komputer Shops is a large, national distributor and retailer of microcomputers, personal computers, and related software. The company has its central offices in Oshawa, not far from the Ford Motor Company executive offices and headquarters. Only recently has Artie Kay's begun to pay serious attention to its cash management procedures. Last week the firm received a proposal from the Royal Bank. The objective of the proposal is to speed up the firm's cash collections.

Artie Kay's now uses a centralized billing procedure. All cheques are mailed to the Oshawa headquarters office for processing and eventual deposit. Remittance cheques now take an average of five business days to reach the Oshawa office. The in-house processing at Artie Kay's is quite slow. Once in Oshawa, another three days are needed to process the cheques for deposit with the Royal Bank.

The daily cash remittances of Artie Kay's average $200,000. The average cheque size is $800. The firm currently earns 10.6 percent on its marketable securities portfolio and expects this rate to continue to be available.

The cash acceleration plan suggested by managers of the Royal Bank involves both a lock-box system and concentration banking. The Royal Bank would be the firm's only concentration bank. Lock boxes would be established in (1) Edmonton, (2) Winnipeg, (3) Montreal, and (4) Moncton. This would reduce mail float by 2.0 days. Processing float would be reduced to a level of 0.5 days. Funds would then be transferred twice each business day by means of automated depository transfer cheques from branch banks of the Royal Bank in Edmonton, Winnipeg, Montreal, and Moncton. Each depository transfer cheque (ADTC) costs $20. These transfers will occur all 270 business days of the year. Each cheque processed through the lock-box system will cost Artie Kay's $.25.

- **a.** What amount of cash balances will be freed if Artie Kay's adopts the system proposed by the Royal Bank?
- **b.** What is the opportunity cost of maintaining the current banking arrangement?
- **c.** What is the projected annual cost of operating the proposed system?
- **d.** Should Artie Kay's adopt the new system? Compute the net annual gain or loss associated with adopting the system.

ST-3. *(Buying and Selling Marketable Securities)* Mountaineer Outfitters has $2 million in excess cash that it might invest in marketable securities. In order to buy and sell the securities, however, the firm must pay a transactions fee of $45,000.

- **a.** Would you recommend purchasing the securities if they yield 12 percent annually and are held for
  1. One month?
  2. Two months?
  3. Three months?
  4. Six months?
  5. One year?
- **b.** What minimum required yield would the securities have to return for the firm to hold them for three months (what is the breakeven yield for a three-month holding period)?

STUDY PROBLEMS (SET A)

18-1A. *(Lock-Box System)* Marino Rug Co. specializes in the manufacture of a wide variety of carpet and tiles. All of the firm's output is shipped to 12 warehouses which are located near the major cities across Canada. The Bank of Montreal is Marino Rug's lead bank and has just completed a study of Marino Rug's cash collection system. Overall, the Bank of Montreal estimates that it can reduce Marino Rug's total float by three days with the installation of a lock-box arrangement in each of the firm's 12 regions. The lock-box arrangement would cost each region $325 per month. Any funds freed up would be added to the firm's marketable securities portfolio and would yield 9.75 percent on an annual basis. Annual sales average $6,232,375 for each regional office. The firm and the bank use a 365-day year in their analyses. Should Marino Rug's management approve the use of the proposed system?

18-2A. *(Marketable Securities Portfolio)* Allan's Fine Foods Inc. currently pays its employees on a weekly basis. The weekly wage bill is $675,000. This means that on the average the firm has accrued wages payable of ($675,000 + $0)/2 = $337,500.

Tom Smith, Jr., works as the firm's senior financial analyst and reports directly to his father, who owns all of the firm's common shares. Tom Smith, Jr., wants to move to a monthly wage payment system. Employees would be paid on Wednesday of every fourth week. The younger Smith is fully aware that the labour union representing the company's workers will not permit the monthly payments system to take effect unless the workers are given some type of fringe benefit compensation.

A plan has been worked out whereby the firm will make a contribution to the cost of life insurance coverage for each employee. This will cost the firm $50,775 annually. Tom Smith, Jr., expects the firm to earn 8.5 percent annually on its marketable securities portfolio.

- **a.** Based on the projected information, should Allan's Fine Foods Inc. move to the monthly wage payment system?
- **b.** What annual rate of return on the marketable securities portfolio would enable the firm to just break even on this proposal?

18-3A. *(Buying and Selling Marketable Securities)* Miami Dice & Card Company has generated $800,000 in excess cash that it could invest in marketable securities. In order to buy and sell the securities, the firm will pay total transactions fees of $20,000.

- **a.** Would you recommend purchasing the securities if they yield 10.5 percent annually and are held for
  1. One month?
  2. Two months?
  3. Three months?
  4. Six months?
  5. One year?
- **b.** What minimum required yield would the securities have to return for the firm to hold them for two months (what is the breakeven yield for a two-month holding period)?

18-4A. *(Cash Receipts Acceleration System)* James Waller Nail Corp. is a buyer and distributor of nails used in the home-building industry. The firm has grown very quickly since it was established eight years ago. Waller Nail has managed to increase sales and profits at a rate of about 18 percent annually, despite moderate economic growth at the national level. Until recently, the company paid little attention to cash management procedures. James Waller, the firm's president, said: "With our growth—who cares?" Bending to the suggestions of several analysts in the firm's finance group, Waller did agree to have a proposal prepared by the National Bank in Moncton. The objective of the proposal is to accelerate the firm's cash collections.

At present, Waller Nail uses a centralized billing procedure. All cheques are mailed to the Moncton office headquarters for processing and eventual deposit. Under this arrangement, all customers' remittance cheques take an average of five business days to reach the Moncton office. Once in Moncton, another two days are needed to process the cheques for deposit with the National Bank.

Daily cash remittances at Waller Nail average $750,000. The average cheque size is $3,750. The firm currently earns 9.2 percent annually on its marketable securities portfolio.

The cash-acceleration plan presented by the managers of the National Bank involves both a lock-box system and concentration banking. National would be the firm's only concentration bank. Lock boxes would be established in (1) Victoria, (2) Toronto, (3) Montreal, and (4) Moncton. This would reduce funds tied up in mail float to 3.5 days. Processing float would be totally eliminated. Funds would then be transferred twice each business day by means of automated depository transfer cheques from branch banks of the National Bank in Victoria, Toronto, and Montreal. Each depository transfer cheque (ADTC) costs $27. These transfers will occur all 270 business days of the year. Each cheque processed through the lock box will cost Waller Nail $.35.

a. What amount of cash balances will be freed if Waller Nail adopts the system proposed by the National Bank?

b. What is the opportunity cost of maintaining the current banking arrangement?

c. What is the projected annual cost of operating the proposed system?

d. Should Waller Nail Corp. adopt the system? Compute the net annual gain or loss associated with adopting the system.

18-5A. *(Lock-Box System)* Advanced Electronics is located in Montreal. The firm manufactures components used in a variety of electrical devices. All the firm's finished goods are shipped to five regional warehouses across Canada.

The Bank of Montreal is Advanced Electronics' lead bank and has recently completed a study of Advanced Electronics' cash collection system. The Bank of Montreal estimates that it can reduce Advanced Electronics' total float by 2.5 days with the installation of a lock-box arrangement in each of the firm's five regions.

The lock-box arrangement would cost each region $500 per month. Any funds freed up would be added to the firm's marketable securities portfolio and would yield 11.75 percent on an annual basis. Annual sales average $10,950,000 for each regional office. The firm and the bank use a 365-day year in their analyses. Should Advanced Electronics' management approve the use of the proposed system?

18-6A. *(Costs of Services)* The Mountain Furniture Company of Moose Jaw, Saskatchewan, may install a lock-box system in order to speed up its cash receipts. On an annual basis, Mountain Furniture receives $40 million in remittances by cheque. The firm will record and process 15,000 cheques over the year. The T-D Bank will administer the system at a cost of $.35 per cheque. Cash that is freed up by use of the system can be invested to yield 9 percent on an annual before-tax basis. A 365-day year is used for analysis purposes. What reduction in cheque collection time is necessary for Mountain Furniture to be neither better nor worse off for having adopted the proposed system?

18-7A. *(Cash Receipts Acceleration System)* Ronda Ball Kitchens Corp. is a medium-sized manufacturer, buyer, and installer of quality kitchen equipment and apparatus. The firm, which was established five years ago, has grown very quickly. Despite some rough economic times that included two recessions in three years, Ball Kitchens has managed to increase sales and profits at an approximate rate of 20 percent annually. Until recently, the firm paid no attention to its cash management procedures. Ronda Ball, the company's president, just said, "We are too busy growing to worry about those trivial things." Bending to the pleas of several analysts in the firm's finance group, Ball did agree to have a proposal prepared by the Bank of Nova Scotia. The objective of the proposal is to speed up the firm's cash collections.

At present, Ball Kitchens uses a centralized billing procedure. All cheques are mailed to the Halifax headquarters office for processing and eventual deposit. Under this arrangement all of the customers' remittance cheques take an average of four business days to reach the Halifax office. Once in Halifax, another two days are needed to process the cheques for deposit with the Bank of Nova Scotia.

The daily cash remittances of Ball Kitchens average $500,000. The average cheque size is $2,000. The firm currently earns 8.8 percent annually on its marketable securities portfolio.

The cash acceleration plan presented by the managers of the Bank of Nova Scotia involves both a lock-box system and concentration banking. The Bank of Nova Scotia would be the firm's only concentration bank. Lock boxes would be established in (1) Victoria, (2) Ottawa, (3) Montreal, and (4) Halifax. This would reduce funds tied up by mail float to 2.5 days. Processing float would be eliminated. Funds would then be transferred twice each business day by means of automated depository transfer cheques from branch banks of the Bank of Nova Scotia in Victoria, Ottawa, and Montreal. Each depository transfer cheque (ADTC) costs $22. These transfers will occur all 270 business days of the year. Each cheque processed through the lock-box system will cost Ball Kitchens $.20.

a. What amount of cash balances will be freed if Ball Kitchens adopts the system proposed by the Bank of Nova Scotia?

b. What is the opportunity cost of maintaining the current banking arrangement?

c. What is the projected annual cost of operating the proposed system?

d. Should Ball Kitchens adopt the new system? Compute the net annual gain or loss associated with adopting the system.

18-8A. *(Buying and Selling Marketable Securities)* Saturday Knights Live Products Inc. has $1 million in excess cash that it might invest in marketable securities. In order to buy and sell the securities, however, the firm must pay a transactions fee of $30,000.

a. Would you recommend purchasing the securities if they yield 11 percent annually and are held for

1. One month?
2. Two months?
3. Three months?
4. Six months?
5. One year?

b. What minimum required yield would the securities have to return for the firm to hold them for three months (what is the breakeven yield for a three-month holding period)?

18-9A. *(Lock-Box System)* Hamilton Foundry specializes in the manufacture of aluminum and steel castings, which are used for numerous automobile parts. All of the firm's output is shipped to 10 warehouses near major auto assembly plants across Canada. The Bank of Montreal is Hamilton Foundry's lead bank and has just completed a study of Hamilton Foundry's cash collection system. Overall, the Bank of Montreal estimates that it can reduce Hamilton Foundry's total float by 2.5 days with the installation of a lock-box arrangement in each of the firm's 10 regions. The lock-box arrangement would cost each region $300 per month. Any funds freed up would be added to the firm's marketable securities portfolio and would yield 10.5 percent on an annual basis. Annual sales average $5,475,000 for each regional office. The firm and the bank use a 365-day year in their analyses. Should Hamilton Foundry's management approve the use of the proposed system?

18-10A. *(Valuing Float Reduction)* Griffey Manufacturing Company is forecasting that next year's gross revenues from sales will be $890 million. The senior treasury analyst for the firm expects the marketable securities portfolio to earn 9.60 percent over this same time period. A 365-day year is used in all the firm's financial procedures. What is the value to the company of one day's float reduction?

18-11A. *(Costs of Services)* The Leblanc Lumber Company may install a lock-box system in order to speed up its cash receipts. On an annual basis, Leblanc Lumber receives $60 million of remittances by cheque. The firm will record and process 18,000 cheques over the year. The National Bank will administer the system for a unit cost of $.30 per cheque. Cash that is freed by the use of this system can be invested to yield 8 percent on an annual before-tax basis. A 365-day year is used for analysis purposes. What reduction in cheque collection time is necessary for Leblanc Lumber to be neither better nor worse off for having adopted the proposed system?

18-12A. *(Costs of Services)* Mustang Ski-Wear Inc. is investigating the possibility of adopting a lock-box system as a cash receipts acceleration device. In a typical year this firm receives remittances totalling $12 million by cheque. The firm will record and process 6,000 cheques over this same time period. The National Bank has informed the management of Mustang that it will expedite cheques and associated documents through the lock-box system for a unit cost of $.20 per cheque. Mustang's financial manager has projected that cash freed by adoption of the system can be invested in a portfolio of near-cash assets that will yield an annual before-tax return of 7 percent. Mustang financial analysts use a 365-day year in their procedures.

a. What reduction in cheque collection time is necessary for Mustang to be neither better nor worse off for having adopted the proposed system?

b. How would your solution to (a) be affected if Mustang could invest the freed balances only at an expected annual return of 4.5 percent?

c. What is the logical explanation for the difference in your answers to (a) and (b)?

18-13A. *(Valuing Float Reduction)* The Columbus Tool and Die Works will generate $18 million in credit sales next year. Collections occur at an even rate, and employees work a 270-day year. At the moment, the firm's general accounting department ties up five days' worth of remittance cheques. An analysis undertaken by the firm's treasurer indicates that new internal procedures can reduce processing float by two days. If Columbus Tool invests the released funds to earn 8 percent, what will be the annual savings?

18-14A. *(Lock-Box System)* Penn Steelworks is a distributor of cold-rolled steel products to the automobile industry. All its sales are on a credit basis, net 30 days. Sales are evenly distributed over its 10 sales regions throughout Canada. Delinquent accounts are no problem. The company has recently undertaken an analysis aimed at improving its cash management procedures. Penn determined that it takes an average of 3.2 days for customers' payments to reach the head office in Hamilton from the time they are mailed. It takes another full day in processing time prior to depositing the cheques with a local bank. Annual sales average $4,800,000 for each regional office. Reasonable investment opportunities can be found yielding 7 percent per year. To alleviate the float problem confronting the firm, the use of a lock-box system in each of the 10 regions is being considered. This would reduce mail float by 1.2 days. One day in processing float would also be eliminated, plus a full day in transit float. The lock-box arrangement would cost each region $250 per month.

- **a.** What is the opportunity cost to Penn Steelworks of the funds tied up in mailing and processing? Use a 365-day year.
- **b.** What would the net cost or savings be from use of the proposed cash- acceleration technique? Should Penn adopt the system?

18-15A. *(Valuing Float Reduction)* Next year P. F. Anderson Motors expects its gross revenues from sales to be $80 million. The firm's treasurer has projected that its marketable securities portfolio will earn 6.50 percent over the coming budget year. What is the value of one day's float reduction to the company? Anderson Motors uses a 365-day year in all of its financial analysis procedures.

18-16A. *(Cash Receipts Acceleration System)* Peggy Pierce Designs Inc. is a vertically integrated, national manufacturer and retailer of women's clothing. Currently, the firm has no coordinated cash management system. However, a proposal from the Royal Bank aimed at speeding up cash collections is being examined by several of Pierce's corporate executives.

The firm currently uses a centralized billing procedure, which requires that all cheques be mailed to the Edmonton head office for processing and eventual deposit. Under this arrangement all the customers' remittance cheques take an average of five business days to reach the head office. Once in Edmonton, another two days are required to process the cheques for ultimate deposit with the Royal Bank.

The firm's daily remittances average $1 million. The average cheque size is $2,000. Pierce Designs currently earns 6 percent annually on its marketable securities portfolio.

The cash acceleration plan proposed by managers of the Royal Bank involves both a lock-box system and concentration banking. The Royal Bank would be the firm's only concentration bank. Lock boxes would be established in (1) Vancouver, (2) Toronto, (3) Montreal, and (4) Edmonton. This would reduce funds tied up by mail float to three days, and processing float will be eliminated. Funds would then be transferred twice each business day by means of automated depository transfer cheques from branch banks of the Royal Bank in Vancouver, Toronto, and Montreal. Each depository transfer cheque (ADTC) costs $15. These transfers will occur all 270 business days of the year. Each cheque processed through the lock-box system will cost $.18.

- **a.** What amount of cash balances will be freed if Pierce Designs Inc. adopts the system suggested by the Royal Bank?
- **b.** What is the opportunity cost of maintaining the current banking setup?
- **c.** What is the projected annual cost of operating the proposed system?
- **d.** Should Pierce adopt the new system? Compute the net annual gain or loss associated with adopting the system.

18-17A. *(Marketable Securities Portfolio)* The Alex Daniel Shoe Manufacturing Company currently pays its employees on a weekly basis. The weekly wage bill is $500,000. This means that on the average the firm has accrued wages payable of ($500,000 + $0)/2 = $250,000.

Alex Daniel, Jr., works as the firm's senior financial analyst and reports directly to his father, who owns all of the firm's common shares. Alex Daniel, Jr., wants to move to a monthly wage payment system. Employees would be paid on the Wednesday of every fourth week. The younger Daniel is fully aware that the labour union representing the company's workers will not permit the monthly payments system to take effect unless the workers are given some type of fringe benefit compensation.

A plan has been worked out whereby the firm will make a contribution to the cost of life insurance coverage for each employee. This will cost the firm $35,000 annually. Alex Daniel, Jr., expects the firm to earn 7 percent annually on its marketable securities portfolio.

a. Based on the projected information, should Daniel Shoe Manufacturing move to the monthly wage payment system?

b. What annual rate of return on the marketable securities portfolio would enable the firm to just break even on this proposal?

18-18A. *(Valuing Float Reduction)* The Cowboy Bottling Company will generate $12 million in credit sales next year. Collections of these credit sales will occur evenly over this period. The firm's employees work 270 days a year. Currently, the firm's processing system ties up four days' worth of remittance cheques. A recent report from a financial consultant indicated procedures that will enable Cowboy Bottling to reduce processing float by two full days. If Cowboy invests the released funds to earn 6 percent, what will be the annual savings?

18-19A. *(Valuing Float Reduction)* Montgomery Woodcraft is a large distributor of woodworking tools and accessories to hardware stores, lumber yards, and tradesmen. All its sales are on a credit basis, net 30 days. Sales are evenly distributed over its 12 sales regions throughout Canada. There is no problem with delinquent accounts. The firm is attempting to improve its cash-management procedures. Montgomery recently determined that it took an average of 3.0 days for customers' payments to reach their office from the time they were mailed and another day for processing before payments could be deposited. Annual sales average $5,200,000 for each region, and investment opportunities can be found to return 9 percent per year. What is the opportunity cost to the firm of the funds tied up in mailing and processing? In your calculations use a 365-day year.

18-20A. *(Lock-Box System)* To mitigate the float problem discussed above, Montgomery Woodcraft is considering the use of a lock-box system in each of its regions. By so doing, it can reduce the mail float by 1.5 days and receive the other benefits of a lock-box system. It also estimates that transit float could be reduced to half its present duration of two days. Use of the lock-box arrangement in each of its regions will cost $300 per month. Should Montgomery Woodcraft adopt the system? What would the net cost or savings be?

18-21A. *(Accounts Payable Policy and Cash Management)* Bradford Construction Supply Company is suffering from a prolonged decline in new construction in its sales area. In an attempt to improve its cash position, the firm is considering changes in its accounts payable policy. After careful study it has determined that the only alternative available is to slow disbursements. Purchases for the coming year are expected to be $37.5 million. Sales will be $65 million, which represents about a 20 percent drop from the current year. Currently, Bradford discounts approximately 25 percent of its payments at 3/10, net 30, and the balance of accounts are paid in 30 days. If Bradford adopts a policy of payment in 45 days or 60 days, how much can the firm gain if the annual opportunity cost of investment is 12 percent? What will be the result if this action causes Bradford Construction suppliers to increase their prices to the company by 0.5 percent to compensate for the 60-day extended term of payment? In your calculations, use a 365-day year and ignore any compounding effects related to expected returns.

18-22A. *(Interest Rate Risk)* Two years ago your corporate treasurer purchased for the firm a 20-year bond at its par value of $1,000. The coupon rate on this security is 8 percent. Interest payments are made to bondholders once a year. Currently, bonds of this particular risk class are yielding investors 9 percent. A cash shortage has forced you to instruct your treasurer to liquidate this bond.

a. At what price will your bond be sold? Assume annual compounding.

b. What will be the amount of your gain or loss over the original purchase price?

c. What would be the amount of your gain or loss had the treasurer originally purchased a bond with a four-year rather than a 20-year maturity? (Assume all characteristics of the bonds are identical except their maturity periods.)

d. What do we call this type of risk assumed by your corporate treasurer?

18-23A. *(Marketable Securities Portfolio)* Red Raider Feedlots has $4 million in excess cash to invest in a marketable securities portfolio. Its broker will charge $10,000 to invest the entire $4 million. The president of Red Raider wants at least half of the $4 million invested at a maturity period of three months or less; the remainder can be invested in securities with maturities not to exceed six months. The relevant term structure of short-term yields follows:

Maturity	Available Yield (Annual)
One month	6.2%
Two months	6.4%
Three months	6.5%
Four months	6.7%
Five months	6.9%
Six months	7.0%

a. What should be the maturity periods of the securities purchased with the excess $4 million in order to maximize the before-tax income from the added investment? What will be the amount of the income from such an investment?

b. Suppose that the president of Red Raider relaxes his constraint on the maturity structure of the added investment. What would be your profit-maximizing investment recommendation?

c. If one-sixth of the excess cash is invested in each of the maturity categories shown above, what would be the before-tax income generated from such an action?

18-24A. *(Comparison of Yields)* A large proportion of the marketable securities portfolio of Edwards Manufacturing Inc. is invested in Government of Canada Treasury bills yielding 6.52 percent before consideration of income taxes. Cell Utilities is bringing a new issue of preferred shares to the marketplace. The preferred shares issue provides a dividend yield of 5.30 percent before taxes. The corporate treasurer of Edwards Manufacturing Inc. wants to evaluate the possibility of shifting a portion of the funds tied up in Treasury bills to the preferred shares issue.

a. Calculate the after-tax yields to Edwards from investing in each type of security. Edwards combined tax rate is 50 percent.

b. What factors apart from the available yields should be analyzed in this situation?

18-25A. *(Forecasting Excess Cash)* The C. K. S. Stove Company manufactures wood-burning stoves in the Pacific Northwest. Despite the recent popularity of this product, the firm has experienced a very erratic sales pattern. Owing to volatile weather conditions and abrupt changes in new housing starts, it has been extremely difficult for the firm to forecast its cash balances. Still, the company president is disturbed by the fact that the firm has never invested in any marketable securities. Instead, the liquid asset portfolio has consisted entirely of cash. As a start toward reducing the firm's investment in cash and releasing some of it to near-cash assets, a historical record and projection of corporate cash holdings is needed. Over the past five years sales have been $10 million, $12 million, $11 million, $14 million, and $19 million, respectively. Sales forecasts for the next two years are $23 and $21 million. Total assets for the firm are 60 percent of sales. Fixed assets are the higher of 50 percent of total assets or $4 million. Inventory and receivables amount to 70 percent of current assets and are held in equal proportions.

a. Prepare a worksheet that details the firm's balance sheets for each of the past five years and for the forecast periods.

b. What amount of cash will the firm have on hand during each year for short-term investment purposes?

INTEGRATIVE PROBLEM

New Wave Surfing Stuff Inc. is a manufacturer of surfboards and related gear that sells to exclusive surf shops along the Atlantic coast towns. The company's headquarters are located in Lawrencetown, a small Nova Scotian coastal town. True to form, the company's officers, all veteran surfers, have been somewhat laid back about various critical areas of financial management. With a economic downturn in Nova Scotia adversely affecting their business, however, the officers of the company have decided to focus intently on ways to improve New Wave's cash flows. The CFO, Wily Bonik, has been requested to forgo any more daytime surfing jaunts until he has wrapped up a plan to accelerate New Wave's cash flows.

In an effort to ensure his quick return to the surf, Wily has decided to focus on what he believes is one of the easiest methods of improving New Wave's cash collections, namely the adoption of a cash receipts acceleration system that includes a lock-box system and concentration banking. Wily is well aware that New Wave's current system leaves much room for improvement. The company's accounts receivable system currently requires that remittances from customers be mailed to the headquarters office for processing, then deposit in the local branch of the Royal Bank. Such an arrangement takes a considerable amount of time. The cheques take an average of six days to reach Lawrencetown headquarters. Then, depending on the surf conditions, processing within the company takes anywhere from three to five days, with the average from the day of receipt by the company to the day of deposit at the bank being four days.

Wily feels pretty certain that such delays are costly. After all, New Wave's average daily collections are $100,000. The average remittance size is $1,000. If Wily could get these funds into his marketable securities account more quickly, he could earn 6 percent at an annual rate on such funds. In addition, if he could arrange for someone else to do the processing, Wily could save $50,000 per year in costs related to clerical staffing.

New Wave's banker was pleased to provide Wily with a proposal for a combination of a lock-box system and a concentration banking system. The Royal Bank would be New Wave's concentration bank. Lock boxes would be established in Old Orchard Beach, Newport Beach, and Yarmouth. Each check processed through the lock-box system would cost New Wave $0.25. This arrangement, however, would reduce the mail float by an average of 3.5 days. The funds so collected would be transferred twice each day, 270 days a year, using automated depository transfer cheques from each of the local lock-box locations to the Royal Bank. Each ADTC would cost $25. The combination of the lock-box system and the concentration banking would eliminate the time it takes the company to process cash collections, thereby making the funds available for short-term investment.

- **a.** What would be the average amount of cash made available if New Wave were to adopt the system proposed by the Royal Bank?
- **b.** What is the annual opportunity cost of maintaining the current cash collection and deposit system?
- **c.** What is the expected annual cost of the complete system proposed by the Royal Bank?
- **d.** What is the net gain or loss that is expected to result from the proposed new system? Should New Wave adopt the new system?

STUDY PROBLEMS (SET B)

18-1B. *(Buying and Selling Marketable Securities)* Universal Concrete Company has generated $700,000 in excess cash that it could invest in marketable securities. In order to buy and sell the securities, the firm will pay total transactions fees of $25,000.

- **a.** Would you recommend purchasing the securities if they yield 11.5 percent annually and are held for
  1. One month?
  2. Two months?
  3. Three months?
  4. Six months?
  5. One year?
- **b.** What minimum required yield would the securities have to return for the firm to hold them for two months (i.e., what is the breakeven yield for a two-month holding period)?

18-2B. *(Cash Receipts Acceleration System)* Kobrin Door & Glass Inc. is a buyer and distributor of doors used in the home building industry. The firm has grown very quickly since it was established eight years ago. Kobrin Door has managed to increase sales and profits at a rate of about 18 percent annually, despite moderate economic growth at the national level. Until recently, the company paid little attention to cash management procedures. Charles Kobrin, the firm's president, said: "With our growth—who cares?" Bending to the suggestions of several analysts in the firm's finance group, Kobrin did agree to have a proposal prepared by the National Bank in Moncton. The objective of the proposal is to accelerate the firm's cash collections.

At present, Kobrin Door uses a centralized billing procedure. All cheques are mailed to the Moncton office headquarters for processing and eventual deposit. Under this arrangement, all customers' remittance cheques take an average of five business days to reach the Moncton office. Once in Moncton, another two days are needed to process the cheques for deposit with the National Bank.

Daily cash remittances at Kobrin Door average $800,000. The average cheque size is $4,000. The firm currently earns 9.5 percent annually on its marketable securities portfolio.

The cash acceleration plan presented by the managers of the National Bank involves both a lock-box system and concentration banking. The National Bank would be the firm's only concentration bank. Lock boxes would be established in (1) Victoria, (2) Toronto, (3) Montreal, and (4) Moncton. This would reduce the mail float by 1.5 days. Processing float would be totally eliminated. Funds would then be transferred twice each business day by means of automated depository transfer cheques from local banks in Victoria, Toronto, and Montreal to the National Bank. Each depository transfer cheque (ADTC) costs $30. These transfers will occur all 270 business days of the year. Each cheque processed through the lock box will cost Kobrin Door $.40.

- **a.** What amount of cash balances will be freed if Kobrin Door adopts the system proposed by the National Bank?
- **b.** What is the opportunity cost of maintaining the current banking arrangement?
- **c.** What is the projected annual cost of operating the proposed system?
- **d.** Should Kobrin Door & Glass adopt the system? Compute the net annual gain or loss associated with adopting the system.

18-3B. *(Lock-Box System)* Regency Components is located in Montreal. The firm manufactures components used in a variety of electrical devices. All the firm's finished goods are shipped to five regional warehouses across Canada.

The Bank of Montreal is Regency Components' lead bank and has recently completed a study of Regency's cash collection system. The Bank of Montreal estimates that it can reduce Regency's total float by 3.0 days with the installation of a lock-box arrangement in each of the firm's five regions.

The lock-box arrangement would cost each region $600 per month. Any funds freed up would be added to the firm's marketable securities portfolio and would yield 11.0 percent on an annual basis. Annual sales average $10,000,000 for each regional office. The firm and the bank use a 365-day year in their analyses. Should Regency Components' management approve the use of the proposed system?

18-4B. *(Costs of Services)* The Hallmark Technology Company of Scarborough may install a lock-box system in order to speed up its cash receipts. On an annual basis, Hallmark receives $50 million in remittances by cheque. The firm will record and process 20,000 cheques over the year. The Royal Bank will administer the system at a cost of $.37 per cheque. Cash that is freed up by use of the system can be invested to yield 9 percent on an annual before-tax basis. A 365-day year is used for analysis purposes. What reduction in cheque collection time is necessary for Hallmark to be neither better nor worse off for having adopted the proposed system?

18-5B. *(Cash Receipts Acceleration System)* Lee Collins Woodcrafts Corp. is a medium-sized manufacturer, buyer, and installer of quality kitchen equipment and apparatus. The firm, which was established five years ago, has grown very quickly. Despite some rough economic times that included two recessions in three years, Collins Woodcrafts has managed to increase sales and profits at an approximate rate of 20 percent annually. Until recently, the firm paid no attention to its cash management procedures. Lee Collins, the company's president, just said, "We are too busy growing to worry about those trivial things." Bending to the pleas of several analysts in the firm's finance group, Collins did agree to have a proposal prepared by the Bank of Nova Scotia. The objective of the proposal is to speed up the firm's cash collections.

At present, Collins Woodcrafts uses a centralized billing procedure. All cheques are mailed to the Halifax headquarters office for processing and eventual deposit. Under this arrangement all of the customers' remittance cheques take an average of four business days to reach the Halifax office. Once in Halifax, another two days are needed to process the cheques for deposit with the Bank of Nova Scotia.

The daily cash remittances of Collins Woodcrafts average $450,000. The average cheque size is $2,000. The firm currently earns 8.5 percent annually on its marketable securities portfolio.

The cash acceleration plan presented by the managers of the Bank of Nova Scotia involves both a lock-box system and concentration banking. The Bank of Nova Scotia would be the firm's only concentration bank. Lock boxes would be established in (1) Victoria, (2) Ottawa, (3) Montreal, and (4) Halifax. This would reduce funds tied up by mail float to 2.5 days. Processing float would be eliminated. Funds would then be transferred twice each business day by means of automated depository transfer cheques from local banks in Victoria, Ottawa, and Montreal to the Bank of Nova Scotia. Each depository transfer cheque (ADTC) costs $25. These transfers will occur all 270 business days of the year. Each cheque processed through the lock-box system will cost Collins Woodcrafts $.25.

- **a.** What amount of cash balances will be freed if Collins Woodcrafts adopts the system proposed by the Bank of Nova Scotia?
- **b.** What is the opportunity cost of maintaining the current banking arrangement?
- **c.** What is the projected annual cost of operating the proposed system?
- **d.** Should Collins Woodcrafts adopt the new system? Compute the net annual gain or loss associated with adopting the system.

18-6B. *(Buying and Selling Marketable Securities)* Western Photo Corp. has $1 million in excess cash that it might invest in marketable securities. In order to buy and sell the securities, however, the firm must pay a transactions fee of $35,000.

- **a.** Would you recommend purchasing the securities if they yield 10 percent annually and are held for
  1. One month?
  2. Two months?
  3. Three months?
  4. Six months ?
  5. One year?
- **b.** What minimum required yield would the securities have to return for the firm to hold them for three months (i.e., what is the breakeven yield for a three-month holding period)?

18-7B. *(Lock-Box System)* Metrocorp is located on the outskirts of Toronto. The firm specializes in the manufacture of aluminum and steel castings, which are used for numerous automobile parts. All of the firm's output is shipped to 10 warehouses near major auto assembly plants across Canada. The Royal Bank is Metrocorp's lead bank and has just completed a study of Metrocorp's cash collection system. Overall, the Royal Bank estimates that it can reduce Metrocorp's total float by 2.5 days with the installation of a lock-box arrangement in each of the firm's 10 regions. The lock-box arrangement would cost each region $300 per month. Any funds freed up would be added to the firm's marketable securities portfolio and would yield 10.0 percent on an annual basis. Annual sales average $4,475,000 for each regional office. The firm and the bank use a 365-day year in their analyses. Should Metrocorp's management approve the use of the proposed system?

18-8B. *(Valuing Float Reduction)* Brady Consulting Services is forecasting that next year's gross revenues from sales will be $900 million. The senior treasury analyst for the firm expects the marketable securities portfolio to earn 9.5 percent over this same time period. A 365-day year is used in all the firm's financial procedures. What is the value to the company of one day's float reduction?

18-9B. *(Costs of Services)* The Discount Storage Co. may install a lock-box system in order to speed up its cash receipts. On an annual basis, Discount receives $55 million of remittances by cheque. The firm will record and process 20,000 cheques over the year. The National Bank will administer the system for a unit cost of $.30 per cheque. Cash that is freed by the use of this system can be invested to yield 8 percent on an annual before-tax basis. A 365-day year is used for analysis purposes. What reduction in cheque collection time is necessary for Discount Storage to be neither better nor worse off for having adopted the proposed system?

18-10B. *(Costs of Services)* Colour Communications is investigating the possibility of adopting a lock-box system as a cash receipts acceleration device. In a typical year this firm receives remittances totalling $10 million by cheque. The firm will record and process 7,000 cheques over this same time period. The Bank of Nova Scotia has informed the management of Colour Communications that it will expedite cheques and associated documents through the lock-box system for a unit cost of $.30 per cheque. Colour Communications' financial manager has projected that cash freed by adoption of the system can be invested in a portfolio of near-cash assets that will yield an annual before-tax return of 7 percent. Colour Communications' financial analysts use a 365-day year in their procedures.

a. What reduction in cheque collection time is necessary for Colour Communications to be neither better nor worse off for having adopted the proposed system?

b. How would your solution to (a) be affected if Colour Communications could invest the freed balances only at an expected annual return of 4.5 percent?

c. What is the logical explanation for the difference in your answers to (a) and (b)?

18-11B. *(Valuing Float Reduction)* Campus Restaurants Inc. will generate $17 million in credit sales next year. Collections occur at an even rate, and employees work a 270-day year. At the moment, the firm's general accounting department ties up four days' worth of remittance cheques. An analysis undertaken by the firm's treasurer indicates that new internal procedures can reduce processing float by two days. If Campus invests the released funds to earn 9 percent, what will be the annual savings?

18-12B. *(Lock-Box System)* Alpine Systems is a distributor of refrigerated storage units to the meat products industry. All its sales are on a credit basis, net 30 days. Sales are evenly distributed over its ten sales regions throughout Canada. Delinquent accounts are no problem. The company has recently undertaken an analysis aimed at improving its cash management procedures. Alpine determined that it takes an average of 3.2 days for customers' payments to reach the head office in Edmonton from the time they are mailed. It takes another full day in processing time prior to depositing the cheques with a local bank. Annual sales average $5,000,000 for each regional office. Reasonable investment opportunities can be found yielding 8 percent per year. To alleviate the float problem confronting the firm, the use of a lock-box system in each of the ten regions is being considered. This would reduce mail float by 1.0 days. One day in processing float would also be eliminated, plus a full day in transit float. The lock-box arrangement would cost each region $225 per month.

a. What is the opportunity cost to Alpine Systems of the funds tied up in mailing and processing? Use a 365-day year.

b. What would the net cost or savings be from use of the proposed cash acceleration technique? Should Alpine adopt the system?

18-13B. *(Valuing Float Reduction)* Next year Concept Realty expects its gross revenues from sales to be $90 million. The firm's treasurer has projected that its marketable securities portfolio will earn 6.50 percent over the coming budget year. What is the value of one day's float reduction to the company? Concept Realty uses a 360-day year in all of its financial analysis procedures.

18-14B. *(Cash Receipts Acceleration System)* Carter's Bicycles Inc. is a vertically integrated, national manufacturer and retailer of racing bicycles. Currently, the firm has no coordinated cash-management system. However, a proposal from the Royal Bank aimed at speeding up cash collections is being examined by several of Carter's corporate executives.

The firm currently uses a centralized billing procedure, which requires that all cheques be mailed to the Edmonton head office for processing and eventual deposit. Under this arrangement all the customers' remittance cheques take an average of four business days to reach the head office. Once in Edmonton, another 1.5 days are required to process the cheques for ultimate deposit at the Royal Bank.

The firm's daily remittances average $1 million. The average cheque size is $2,000. Carter's currently earns 7 percent annually on its marketable securities portfolio.

The cash acceleration plan proposed by managers of the Royal Bank involves both a lock-box system and concentration banking. The Royal Bank would be the firm's only concentration bank. Lock boxes would be established in (1) Vancouver, (2) Toronto, (3) Montreal, and (4) Edmonton. This would reduce funds tied up by mail float to three days, and processing float will be eliminated. Funds would then be transferred twice each business day by means of automated depository transfer cheques from branch banks of the Royal Bank in Vancouver, Toronto, and Montreal. Each depository transfer cheque (ADTC) costs $16. These transfers will occur all 270 business days of the year. Each cheque processed through the lock-box system will cost $0.22.

a. What amount of cash balances will be freed if Carter's Bicycles adopts the system suggested by the Royal Bank?

b. What is the opportunity cost of maintaining the current banking setup?
c. What is the projected annual cost of operating the proposed system?
d. Should Carter's adopt the new system? Compute the net annual gain or loss associated with adopting the system.

18-15B. *(Marketable Securities Portfolio)* Katz Jewelers currently pays its employees on a weekly basis. The weekly wage bill is $500,000. This means that on the average the firm has accrued wages payable of ($500,000 + $0)/2 = $250,000.

Harry Katz works as the firm's senior financial analyst and reports directly to his father, who owns all of the firm's common shares. Harry Katz wants to move to a monthly wage payment system. Employees would be paid on the Wednesday of every fourth week. The younger Katz is fully aware that the labour union representing the company's workers will not permit the monthly payments system to take effect unless the workers are given some type of fringe benefit compensation.

A plan has been worked out whereby the firm will make a contribution to the cost of life insurance coverage for each employee. This will cost the firm $40,000 annually. Harry Katz expects the firm to earn 8 percent annually on its marketable securities portfolio.

a. Based on the projected information, should Katz Jewelers move to the monthly wage payment system?
b. What annual rate of return on the marketable securities portfolio would enable the firm just to break even on this proposal?

18-16B. *(Valuing Float Reduction)* Magic Hardware Stores Inc. will generate $12 million in credit sales next year. Collections of these credit sales will occur evenly over this period. The firm's employees work 270 days a year. Currently, the firm's processing system ties up 4.5 days' worth of remittance cheques. A recent report from a financial consultant indicated procedures that will enable Magic Hardware to reduce processing float by two full days. If Magic invests the released funds to earn 7 percent, what will be the annual savings?

18-17B. *(Valuing Float Reduction)* True Locksmith is a large distributor of residential locks to hardware stores, lumber yards, and tradesmen. All its sales are on a credit basis, net 30 days. Sales are evenly distributed over its ten sales regions throughout Canada. There is no problem with delinquent accounts. The firm is attempting to improve its cash management procedures. True Locksmith recently determined that it took an average of 3.0 days for customers' payments to reach their office from the time they were mailed, and another day for processing before payments could be deposited. Annual sales average $5,000,000 for each region, and investment opportunities can be found to return 9 percent per year. What is the opportunity cost to the firm of the funds tied up in mailing and processing? In your calculations use a 365-day year.

18-18B. *(Lock-Box System)* To mitigate the float problem discussed above, True Locksmith is considering the use of a lock-box system in each of its regions. By so doing, it can reduce the mail float by 1.5 days and receive the other benefits of a lock-box system. It also estimates that transit float could be reduced to half its present duration of two days. Use of the lock-box arrangement in each of its regions will cost $350 per month. Should True Locksmith adopt the system? What would the net cost or savings be?

18-19B. *(Accounts Payable Policy and Cash Management)* Meadowbrook Paving Company is suffering from a prolonged decline in new development in its sales area. In an attempt to improve its cash position, the firm is considering changes in its accounts payable policy. After careful study it has determined that the only alternative available is to slow disbursements. Purchases for the coming year are expected to be $40 million. Sales will be $65 million, which represents about a 15 percent drop from the current year. Currently, Meadowbrook discounts approximately 25 percent of its payments at 3/10, net 30, and the balance of accounts are paid in 30 days. If Meadowbrook adopts a policy of payment in 45 days or 60 days, how much can the firm gain if the annual opportunity cost of investment is 11 percent? What will be the result if this action causes Meadowbrook Paving suppliers to increase their prices to the company by 0.5 percent to compensate for the 60-day extended term of payment? In your calculations use a 365-day year and ignore any compounding effects related to expected returns.

18-20B. *(Interest Rate Risk)* Two years ago your corporate treasurer purchased for the firm a 20-year bond at its par value of $1,000. The coupon rate on this security is 8 percent. Interest payments are made to bondholders once a year. Currently, bonds of this particular risk class are yielding investors 9 percent. A cash shortage has forced you to instruct your treasurer to liquidate this bond.

a. At what price will your bond be sold? Assume annual compounding.
b. What will be the amount of your gain or loss over the original purchase price?
c. What would be the amount of your gain or loss had the treasurer originally purchased a bond with a four-year rather than a 20-year maturity? (Assume all characteristics of the bonds are identical except their maturity periods.)
d. What do we call this type of risk assumed by your corporate treasurer?

18-21B. *(Marketable Securities Portfolio)* Spencer Pianos has $3.5 million in excess cash to invest in a marketable securities portfolio. Its broker will charge $15,000 to invest the entire $3.5 million. The president of Spencer wants at least half of the $3.5 million invested at a maturity period of three months or less; the remainder can be invested in securities with maturities not to exceed six months. The relevant term structure of short-term yields follows:

Maturity	Available Yield (Annual)
One month	6.2%
Two months	6.4%
Three months	6.5%
Four months	6.7%
Five months	6.9%
Six months	7.0%

a. What should be the maturity periods of the securities purchased with the excess $3.5 million in order to maximize the before-tax income from the added investment? What will be the amount of the income from such an investment?
b. Suppose that the president of Spencer relaxes his constraint on the maturity structure of the added investment. What would be your profit-maximizing investment recommendation?
c. If one-sixth of the excess cash is invested in each of the maturity categories shown above, what would be the before-tax income generated from such an action?

18-22B. *(Comparison of Yields)* A large proportion of the marketable securities portfolio of Ramorr Manufacturing Inc., is invested in Government of Canada Treasury bills yielding 5.62 percent before consideration of income taxes. Light-on Utilities is bringing a new issue of preferred shares to the marketplace. The preferred shares issue provides a dividend yield of 4.80 percent before taxes. The corporate treasurer of Ramorr Manufacturing Inc., wants to evaluate the possibility of shifting a portion of the funds tied up in Treasury bills to the preferred shares issue.

a. Calculate the after-tax yields to Ramorr Manufacturing from investing in each type of security. Ramorr's combined tax rate is 50%.
b. What factors apart from the available yields should be analyzed in this situation?

18-23B. *(Forecasting Excess Cash)* Fashionable Floors Inc. manufactures carpeting in Victoria. Despite the popularity of this product, the firm has experienced a very erratic sales pattern. Owing to volatile weather conditions and abrupt changes in new housing starts, it has been extremely difficult for the firm to forecast its cash balances. Still, the company president is disturbed by the fact that the firm has never invested in any marketable securities. Instead, the liquid asset portfolio has consisted entirely of cash. As a start toward reducing the firm's investment in cash and releasing some of it to near-cash assets, a historical record and projection of corporate cash holdings is needed. Over the past five years sales have been $10 million, $12 million, $11 million, $14 million, and $19 million, respectively. Sales forecasts for the next two years are $24 and $20 million. Total assets for the firm are 65 percent of sales. Fixed assets are the higher of 50 percent of total assets or $4 million. Inventory and receivables amount to 70 percent of current assets and are held in equal proportions.

a. Prepare a worksheet that details the firm's balance sheets for each of the past five years and for the forecast periods.
b. What amount of cash will the firm have on hand during each year for short-term investment purposes?

CASE PROBLEM

B.C. TRANSISTOR

Cash Management

Michael Broski is the assistant treasurer for B.C. Transistor. He has been with the firm for nine years. The first six were spent within the technology ranks of the company as an electronics engineer. Broski, in fact, took his bachelor's degree in electronics engineering from a famous California-located university noted for its excellent faculty in all phases of engineering. After two years as a senior engineer for the organization, Broski indicated an interest in the administrative management of the firm. He spent one year as an analyst in the treasury department and has just completed his second as the assistant treasurer. During the latter three years, he has steeped himself in literature that focuses on the finance function of the corporate enterprise. In addition, he attended short courses and seminars by national management and accounting associations that dealt with most phases of the financial conduct of the firm. Within B.C. Transistor, other cost accountants, cost analysts, financial analysts, and treasury analysts viewed Broski's rapid grasp of finance concepts as nothing short of phenomenal.

B.C. Transistor is located on the outskirts of Vancouver, B.C. The firm began in the middle 1950s as a small manufacturer of radio and television components. By the late 1950s, it had expanded into the actual installation and service of complete communications systems on navigable vessels. Ships from the navy and from the tuna industry use the Vancouver port facilities as a major repair yard. During the early years of the firm's activity, defence contracts typically accounted for 80 to 90 percent of annual revenues. B.C. Transistor's management felt, however, that excessive reliance on defence contracts could lead to some very lean years with regard to business receipts. The company expanded during the 1960s into several related fields. Today B.C. Transistor is active in the home entertainment market as well as the defence market. Transistors and integrated circuits are produced for a wide range of final products including radios, televisions, pocket calculators, citizens band receivers, stereo equipment, and ship communications systems. Defence-related business now accounts for 25 to 30 percent of the company's annual sales. This transition to a more diversified enterprise has tended to reduce the inherent "lumpiness" of cash receipts that plagues many small firms relying on defense contracts. In the early years of B.C. Transistor's existence, progress payments on major contracts would often be months apart as the jobs moved toward completion. Currently, the firm enjoys a reasonably stable sales pattern over its fiscal year.

Broski has been studying the behaviour of the company's daily and monthly net cash balances. Over the most recent 24 months, he observed that B.C. Transistor's ending cash balance (by month) ranged between $144,000 and $205,000. Only twice during the period that he investigated had the daily closing cash balance even been as low as $100,000. Broski also noted that the firm had no outstanding short-term borrowings throughout these two years. This led him to believe that B.C. Transistor was carrying excessive cash balances; this was probably a tendency that had its roots in the period when defence contracts were the key revenue item for the company. This "first pass" analysis gained the attention of Donald Crawford, who is Broski's boss and B.C. Transistor's treasurer. Crawford is 65 years old and due to retire on July 1 of this year. He has been B.C. Transistor's only treasurer. Broski would like to move into Crawford's job upon his retirement and feels that the treasurer's recommendation might sew it up for him. He also knows that statements relating to excess cash balances being carried by the company will have to be made very tactfully and with Crawford's agreement. That bit of financial policy making has always rested with the company treasurer. Broski decided that the best plan would be to present Crawford with a solid analysis of the problem and convince him that a reduction in balances held for transactions purposes would be in the best interests of the firm. As all key managers own considerable stock options, potential increases in corporate profitability usually are well received by management.

Earlier this year Broski attended a cash management seminar in Victoria, sponsored by the commercial bank with which B.C. Transistor holds most of its deposits. The instructor spoke of the firm's cash balance as being "just another inventory." It was offered that the same principles that applied to the determination of an optimal stock of some raw material item also might be applied to the selection of an optimal average amount of transaction cash. Broski really liked that presentation. He walked away from it feeling that he could put it to work. He decided to adapt the basic economic order quantity model to his cash balance problem.

To make the model "workable," it is necessary to build an estimate of the fixed costs associated with adding to or subtracting from the company's inventory of cash. As B.C. Transistor experienced no short-term borrowings within the past two years, Broski considered this element of the problem to be the fixed costs of liquidating a portion of the firm's portfolio of marketable securities.

In Exhibit 1, Broski has summarized the essential activities that occur whenever a liquidation of a part of the securities' portfolio takes place. In all of the firm's cost analysis procedures, it is assumed that a year consists of 264 working days. An eight-hour working day is also utilized in making wage and salary cost projections. Also, minutes of labour are converted into thousandths of an hour. The assistant treasurer felt that from the information in Exhibit 1, he could make a decent estimate of the fixed cost of a security transaction (addition to the firm's cash account).

B.C. Transistor's marketable securities portfolio is usually concentrated in three major money market instruments: (1) treasury bills, (2) bankers' acceptances, and (3) prime commercial paper. A young analyst who works directly for Michael Broski supplied him with some recent rates of return on these types of securities (Exhibit 2). Broski expressed his concern to the analyst about the possibilities of high rates, such as those experienced during 1989 and 1993, continuing. His intuition or feel for the market led him to believe that short-term interest rates would be closer to 1993 levels during this next year than any other levels identified in Exhibit 2. Thus, Broski decided to use a 9 percent annual yield in this study as a reasonable return to expect from his firm's marketable securities portfolio. Then a review of cash flow patterns over the last five years, including the detailed examination of the most recent 24 months, led to a projection of a typical monthly cash outflow (or demand for cash for transactions purposes) of $800,000.

Questions

1. Determine the optimal cash withdrawal size from the B.C. Transistor marketable securities portfolio during a typical month.
2. What is the total cost (in dollars) for the use of cash held for transactions purposes during the period of analysis?
3. What will be the firm's average cash balance during a typical month?
4. Assuming that fractional cash withdrawals or orders can be made, how often will an order be placed? The firm operates continually for 30 days each month.
5. If the company's cash balance at the start of a given month is $800,000, how much of that amount would initially be invested in securities?
6. Graph the behaviour pattern of the $800,000 balance (mentioned above) over a 30-day month. In constructing your graph, round off the frequency of orders to the nearest whole day and disregard separation of the balance between cash and securities.
7. To understand the logic of the model further, provide a graph that identifies in general (1) the total cost function of holding the cash, (2) the fixed costs associated with cash transfers, and (3) the opportunity cost of earnings forgone by holding cash balances. Use dollar amounts to label the key points on the axes as they were previously computed in Questions 1 and 2. Also, identify the major assumptions of this model in a cash management setting.

Exhibit 1 B.C. Transistor: Fixed Costs Associated with Securities Liquidation

Activity	Details
1. Long-distance phone calls.	Cost: $2.75
2. Assistant treasurer's time: 22 minutes for this position	Annual salary is $27,000
3. Typing of authorization letter, with three carbon copies, and careful proofreading: 17 minutes	Annual salary for this position is $8,000.
4. Carrying original authorization letter to treasurer, who reads and signs it: 2 minutes by same secretary as above, and 2 minutes by the treasurer.	Annual salary for the treasurer is $38,000.
5. Movement of authorization letter just signed by the treasurer to the controller's office, followed by the opening of a new account, recording of the transaction, and proofing of the transaction: 2 minutes by same secretary as above, 10 minutes for account opening by general accountant, and 8 minutes for recording and proofing by the same general accountant.	Annual salary for the general accountant is $12,000.
6. Fringe benefits incurred on above times.	Cost: $4.42
7. Brokerage fee on each transaction.	Cost: $7.74

Exhibit 2 B.C. Transistor: Selected Money Market Rates (Annual Yields)

Year	Prime Commercial Paper: 90–119 Days	Bankers' Acceptances 90 Days	Three-Month Treasury Bills
1990	7.66%	7.47%	7.07%
1991	11.20	11.08	11.04
1992	13.05	13.92	10.89
1993	9.26	9.30	8.84
Simple Average	10.29	10.19	9.21

SELF-TEST SOLUTIONS

SS-1. a. Initially, it is necessary to calculate Creative Fashions' average remittance cheque amount and the daily opportunity cost of carrying cash. The average cheque size is

$$\frac{\$7{,}000{,}000}{4{,}000} = \$1{,}750 \text{ per cheque}$$

The daily opportunity cost of carrying cash is

$$\frac{0.08}{365} = 0.0002192 \text{ per day}$$

Next, the days saved in the collection process can be evaluated according to the general format (see Equation 18-1 in the text of this chapter) of

added costs = added benefits

or

$$P = (D)(S)(i) \text{ [see Equation 18-2]}$$
$$\$0.25 = (D)\ (\$1{,}750)\ (.0002192)$$
$$0.6517 \text{ days} = D$$

Creative Fashion Designs therefore will experience a financial gain if it implements the lock-box system and by doing so will speed up its collections by more than 0.6517 days.

b. Here the daily opportunity cost of carrying cash is

$$\frac{0.055}{365} = 0.0001507 \text{ per day}$$

For Creative Fashion Designs to break even, should it choose to install the lock-box system, cash collections must be accelerated by 0.9480 days, as follows:

$$\$0.25 = (D)\ (\$1{,}750)\ (.0001507)$$
$$0.9480 \text{ days} = (D)$$

c. The breakeven cash acceleration period of 0.9480 days is greater than the 0.6517 days found in (a). This is due to the lower yield available on near-cash assets of 5.5 percent annually, versus 8.0 percent. Since the alternative rate of return on the freed-up balances is lower in the second situation, more funds must be invested to cover the costs of operating the lock-box system. The greater cash-acceleration period generates this increased level of required funds.

SS-2. a.

Reduction in mail float:	(2.0 days)(\$200,000) =	\$400,000
+ reduction in processing float:	(2.5 days)(\$200,000) =	\$500,000
total float reduction		\$900,000

b. The opportunity cost of maintaining the present banking arrangement is

$$\left(\begin{array}{c}\text{forecast yield on marketable}\\ \text{securities portfolio}\end{array}\right) \times \left(\begin{array}{c}\text{total float}\\ \text{reduction}\end{array}\right)$$

$$(.106)(\$900{,}000) = \$95{,}400$$

c. The average number of cheques to be processed each day through the lock-box arrangement is

$$\frac{\text{daily remittances}}{\text{average cheque size}} = \frac{\$200{,}000}{\$800} = 250 \text{ cheques}$$

The resulting cost of the lock-box system on an annual basis is

$$(250 \text{ cheques})(\$0.25)(270 \text{ days}) = \$16{,}875$$

Next, we must calculate the estimated cost of the automated depository transfer cheque (ADTC) system. The Royal Bank will not contribute to the cost of the ADTC arrangement because it is the lead concentration bank and thereby receives the transferred data. This means that Artie Kay's Komputer Shops will be charged for six ADTCs (three locations @ 2 cheques each) each business day. Therefore, the ADTC system costs

(6 daily transfers)($20 per transfer)(270 days) = $32,400

We now have the total cost of the proposed system:

Lock-box cost	$16,875
ADTC cost	32,400
Total cost	$49,275

d. Our analysis suggests that Artie Kay's Komputer Shops should adopt the proposed cash receipts acceleration system. The projected net annual gain is $46,125 as follows:

Projected return on freed balances	$95,400
Less: Total cost of new system	49,275
Net annual gain	$46,125

SS-3. a. Here we must calculate the dollar value of the estimated return for each holding period and compare it with the transactions fee to determine if a gain can be made by investing in the securities. Those calculations and the resultant recommendations follow:

	Recommendation
1. $2,000,000 (.12) (1/12) = $20,000 < $45,000	No
2. $2,000,000 (.12) (2/12) = $40,000 < $45,000	No
3. $2,000,000 (.12) (3/12) = $60,000 > $45,000	Yes
4. $2,000,000 (.12) (6/12) = $120,000 > $45,000	Yes
5. $2,000,000 (.12) (12/12)= $240,000 > $45,000	Yes

b. Let (%) be the required yield. With $2 million to invest for three months we have

$200,000 (%) (3/12) = $45,000
$200,000 (%) (3/12) = $180,000
$200,000 (%) (3/12) = $180,000/2,000,000 = 9%
The breakeven yield, therefore, is 9%.

APPENDIX 18A:

CASH MANAGEMENT MODELS: THE SPLIT BETWEEN CASH AND NEAR CASH

In this appendix we continue our discussion of the management of the firm's cash position. We have dwelled upon the overall objectives of company cash management, described some actual liquid asset holdings of selected industries and firms, and overviewed a wide array of cash collection and disbursement procedures.

We now consider the problem of properly dividing the firm's liquid asset holdings between cash and near cash.

LIQUID ASSETS: CASH VERSUS MARKETABLE SECURITIES

The financial executive must pinpoint time periods when funds will be in either short or excess supply. One of the tools used in forecasting shortages or excesses of cash is the cash budget which will be examined in greater detail in Chapter 21. If a shortage of funds is expected, then alternative avenues of financing must be explored. On the other hand, the cash-budget projections might indicate that large, positive net cash balances in excess of immediate transactions needs will be forthcoming. In this more pleasant situation the financial manager ought to decide on the proper split of the expected cash balances between actual cash holdings and marketable securities. To hold all of the expected cash balances as actual balances would needlessly penalize the firm's profitability.

The ensuing discussion centres on various methods by which the financial manager can develop useful cash balance level benchmarks.

Benchmark 1: When Cash Need Is Certain

A basic method for indicating the proper average amount of cash to have on hand involves use of the economic order quantity concept which will appear in discussions of inventory management (Chapter 19).[17] The objective of this analysis is to balance the lost income that the firm suffers from holding cash rather than marketable securities against the transactions costs involved in converting securities into cash. The rudiments of this decision model can easily be introduced by the use of an illustration.

Suppose that the firm knows with certainty that it will need $250,000 in cash for transactions purposes over the next two months and that this much cash is currently available. This transactions demand for cash will be represented by the variable T. Let us assume, for purposes of this illustration, that when the firm requires cash for its transactions needs it will sell marketable securities in any one of five lot sizes, ranging from $30,000 to $70,000. These cash conversion (order) sizes, C, are identified in line 1 of Table 18A-1.

Line 2 shows the number of times marketable securities will be turned into cash over the next two months if a particular order size is utilized. For example, should it be decided to liquidate securities in amounts of $40,000, then the number of cash conversions needed to meet transactions over the next two months is $250,000/$40,000 = 6.25. In general, the number of cash withdrawals from the near-cash portfolio can be represented as T/C.

[17]The roots of a quantitative treatment of the firm's cash balance as just another type of inventory are found in William J. Baumol, "The Transactions Demand for Cash: An Inventory Theoretic Approach," *Quarterly Journal of Economics* 66 (November 1952), pp. 545–556.

Table 18A-1 Determination of the Optimal Cash Order Size

1. Cash conversion size (the dollar amount of marketable securities that will be sold to replenish the cash balance)	\$30,000	\$40,000	\$50,000	\$60,000	\$70,000
2. Number of cash order per time period (the time period is two months in this example) (\$250,000 ÷ line 1)	8.33	6.25	5.00	4.17	3.57
3. Average cash balance (line 1 ÷ 2)	\$15,000	\$20,000	\$25,000	\$30,000	\$35,000
4. Interest income forgone (line 3 × .01)	\$150.00	\$200.00	\$250.00	\$300.00	\$350.00
5. Cash conversion cost (\$50 × line 2)	\$416.50	\$312.50	\$250.00	\$208.50	\$178.50
6. Total cost of ordering and holding cash (line 4 + line 5)	\$566.50	\$512.50	\$500.00	\$508.50	\$528.50

Next, we assume that the firm's cash payments are of constant amounts and are made continually over the two-month planning period. This implies that the firm's cash balance behaves in the sawtooth manner shown in Figure 18A-1. The assertion of regularity and constancy of payments allows the firm's average cash balance over the planning period to be measured as *C/2* (see Figure 18A-1). When marketable securities are sold and cash flows into the demand deposit account, the cash balance is equal to *C*. As payments are made on a regular and constant basis, the cash balance is reduced to a level of zero. The average cash balance over the period is then

$$\frac{C+0}{2}=\frac{C}{2}$$

The average cash balances corresponding to the different cash conversion sizes in our example are noted in line 3 of Table 18A-1.

Line 4 measures the opportunity cost of earnings forgone based upon holding the average cash balance recorded on line 3. If the annual yield available on marketable securities is 6 percent, then over the two-month period we are analyzing,

Figure 18A-1 Cash Balances According to the Inventory Model

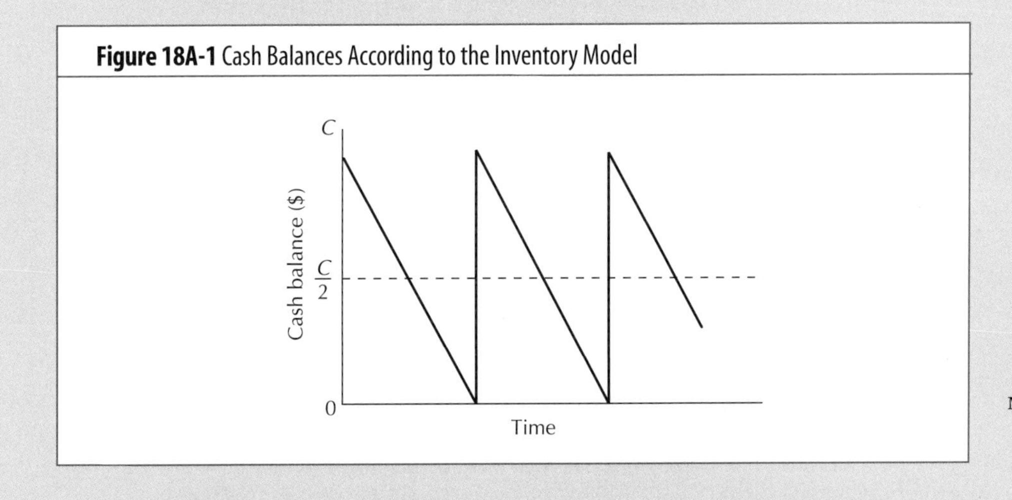

the forfeited interest rate is .06/6 = .01.[18] Multiplying each average cash balance, $C/2$, by the .01 interest rate, i, available over the two-month period produces the opportunity costs entered on line 4.

The very act of liquidating securities, unfortunately, is not costless. Transacting conversions of marketable securities into cash can involve any of the following activities, which require the time of company employees as well as direct payment by the firm for various services:

1. Assistant treasurer's time to order the trade.
2. Long-distance phone calls to effect the trade.
3. Secretarial time to type authorization letters, make copies of the letters, and forward the letters to the company treasurer.
4. Treasurer's time to read, approve, and sign the documents that authorize the trade.
5. General accountant's time to record and audit the transaction.
6. The value of fringe benefits incurred on the above times.
7. The brokerage fee on each transaction.

Suppose that the firm has properly studied the transaction costs, similar to those enumerated above, and finds they are of a fixed amount, b, per trade equal to $50. The transaction cost variable, b, is taken to be independent of the size of a particular securities order. Multiplying the transaction cost of $50 per trade by the number of cash orders that will take place during the planning period produces line 5. In general, the cash conversion cost (transaction cost) is equal to $b(T/C)$.

We are now down to the last line in Table 18A-1. This is the sum total of the income lost by holding cash rather than marketable securities and the cash-ordering costs. Line 6, then, is the total of lines 4 and 5. The inventory model seeks to minimize these total costs associated with holding cash balances. Table 18A-1 tells us, if cash is ordered on five occasions in $50,000 sizes over the two-month period, the total costs of holding an average cash balance of $25,000 will be $500. This is less than the total costs associated with any other cash conversion size.

At the beginning of the two-month planning horizon, all of the $250,000 available for transactions purposes need not be held in the firm's demand deposit account. To minimize the total costs of holding cash, only $50,000 should immediately be retained to transact business. The remaining $200,000 should be invested in income-yielding securities and then turned into cash as the firm's disbursal needs dictate.

It is useful to put our discussion of the inventory model for cash management into a more general form. Summarizing the definitions developed in the illustration, we have

C = the amount per order of marketable securities to be converted into cash
i = the interest rate per period available on investments in marketable securities
b = the fixed cost per order of converting marketable securities into cash
T = the total cash requirements over the planning period
TC = the total costs associated with maintenance of a particular average cash balance

As just pointed out, the total costs (TC) of having cash on hand can be expressed as

$$TC = \underbrace{i\left(\frac{C}{2}\right)}_{\text{total interest income forgone}} + \underbrace{b\left(\frac{T}{C}\right)}_{\text{total ordering costs}} \tag{18A-1}$$

If Equation (18A-1) is applied to the $50,000 cash conversion size column in Table 18A-1, the total costs can be computed in a direct fashion as follows:

[18]If we were studying a one-month planning period, rather than the two-month period being discussed, the annual yield would have to be stated on a monthly basis or .06/12 = .005.

$$TC = .01\left(\frac{\$50{,}000}{2}\right) + 50\left(\frac{\$250{,}000}{\$50{,}000}\right)$$
$$= \$250 + \$250 = \$500$$

You can see that the \$500 total cost is the same as was found deductively in Table 18A-1. The optimal cash conversion size, C^*, can be found by use of Equation (18A-2):

$$C^* = \sqrt{\frac{2bT}{i}} \tag{18A-2}$$

When the data in our example are applied to Equation (18A-2), the optimal cash order size is found to be

$$C^* = \sqrt{\frac{2(50)(250{,}000)}{.01}} = \$50{,}000$$

This solution to our example problem is displayed graphically in Figure 18A-2. Notice that the optimal cash order size of \$50,000 occurs at the minimum point of the total cost curve associated with keeping cash on hand.

Inventory Model Implications

The solution to Equation (18A-2) tells the financial manager that the optimal cash order size, C^*, varies directly with the square root of the order costs, bT, and inversely with the yield, i, available on marketable securities. Notice that as transactions requirements, T, increase, owing perhaps to an augmented sales demand, the optimal cash order size does not rise proportionately.

The model also indicates that as interest rates rise on near-cash investments, the optimal cash order size decreases, with the effect dampened by the square-root sign as Equation (18A-2) suggests. With higher yields to be earned on the marketable securities portfolio, the financial manager will be more reluctant to make large withdrawals because of the interest income that will be lost.

Some final perspectives on the use of the economic order quantity model in cash management can be obtained by reviewing the assumptions upon which it is derived. Among the more important of these assumptions are the following:

Figure 18A-2 Solution to the Inventory Model for Cash Management

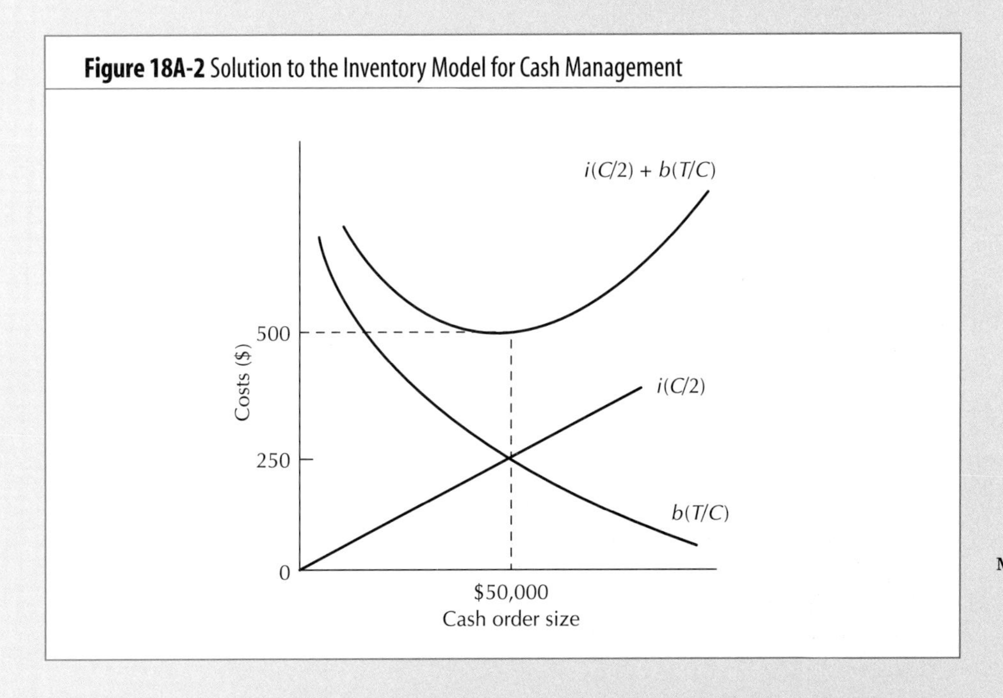

1. Cash payments over the planning period are (a) of a regular amount, (b) continuous, and (c) certain.
2. No unanticipated cash receipts will be received during the analysis period.
3. The interest rate to be earned on investments remains constant over the analysis period.
4. Transfers between cash and the securities portfolio may take place any time at a fixed cost, regardless of the amount transferred.

Clearly, the strict assumptions of the inventory model will not be completely realized in actual business practice. For instance, the amount and timing of cash payments will not be known with certainty; nor are cash receipts likely to be as discontinuous or lumpy as is implied. If relaxation of the critical assumptions is not so prohibitive as to render the model useless, then a benchmark for possible managerial action is provided. The model's output is not intended to be a precise and inviolable rule. On the other hand, if the assumptions of the model cannot be reasonably approximated, the financial manager must look elsewhere for possible guides that indicate a proper split between cash and marketable securities.

Benchmark 2: When Cash Balances Fluctuate Randomly

It is entirely possible that the firm's cash balance pattern does not at all resemble that indicated in Figure 18A-1. The assumptions of certain regularity and constancy of cash payments may be unduly restrictive when applied to some organizations.[19] Rather, the cash balance might behave more like the jagged line shown in Figure 18A-3. In this figure it is assumed that the firm's cash balance changes in an irregular fashion from day to day. The changes are unpredictable; that is, they are random. Further, let us suppose the chances that a cash balance change will be either (1) positive or (2) negative are equal at .5 each.

As cash receipts exceed expenditures, the cash balance will move upward until it hits an upper control limit, *UL*, expressed in dollars. This occurs at point A in Figure 18A-3. At such time, the financial manager will initiate an investment in marketable securities equal to *UL* – *RP* dollars, where *RP* is the cash return point.

If cash payments exceed receipts, the cash balance will move downward until it hits a lower control limit, *LL*. This situation is noted by point *B* in Figure 18A-3. When this occurs, the financial manager will sell marketable securities equal to *RP* – *LL* dollars. This will restore the cash balance to the return point, *RP*.

Figure 18A-3 Randomly Fluctuating Cash Balances

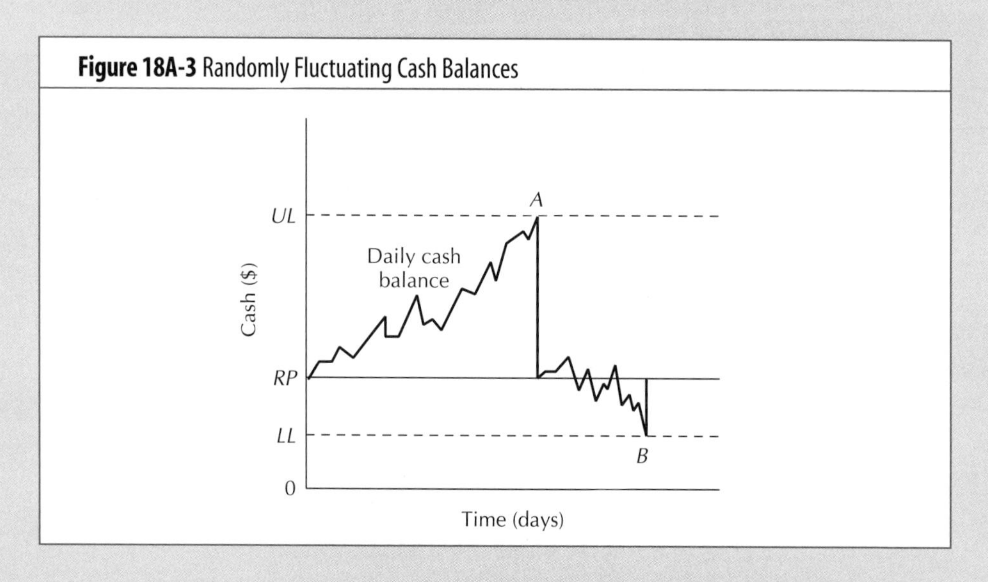

[19]This discussion is based upon Merton H. Miller and Daniel Orr, "A Model of the Demand for Money by Firms," *Quarterly Journal of Economics* 80 (August 1966), pp. 413–435.

To make this application of control theory to cash management operational, we must determine the upper control limit, UL, the lower control limit, LL, and the cash return point, RP. For the present case where a net cash increase is as likely to occur as a net cash decrease, use of the following variables will allow computation of the cash return point, RP.[20]

b = the fixed cost per order of converting marketable securities into cash
i = the daily interest rate available on investments in marketable securities
σ^2 = the variance of daily changes in the firm's expected cash balances (this is a measure of volatility of cash flow changes over time)

The optimal cash return point, RP, can be calculated as follows:

$$RP = \sqrt[3]{\frac{3b\sigma^2}{4i}} + LL \tag{18A-3}$$

The upper control limit, UL, can be computed quite simply:

$$UL = 3RP - 2LL \tag{18A-4}$$

The actual value for the lower limit, LL, is set by management. In business practice a minimum is typically established below which the cash balance is not permitted to fall. Among other things, it will be affected by (1) the firm's banking arrangements, which may require compensating balances, and (2) management's risk-bearing tendencies.

To illustrate use of the model, suppose that the annual yield available on marketable securities is 9 percent. Over a 365-day year, i becomes .09/365 = .00025 per day. Assume that the fixed cost, b, of transacting a marketable securities trade is $50. Moreover, the firm has studied its past cash balance levels and has observed that the standard deviation, σ, in daily cash balance changes is equal to $800. The firm sees no reason why this variability should change in the future. It is the firm's policy to maintain $1,000 in its demand deposit account (LL) at all times. Finally, the firm has established that each day's actual cash balance is random. The equations that comprise this control limit system can be applied to provide guidelines for cash management policy.

The optimal cash return point for transactions purposes becomes

$$RP = \sqrt[3]{\frac{3(50)(800)^2}{4(.00025)}} + 1{,}000 = (4{,}579 + 1{,}000) = \$5{,}579$$

The upper cash balance limit that will trigger a transfer of cash to marketable securities is

$$UL = 3(5{,}579) - 2(1{,}000) = \$14{,}737$$

The cash balance control limits derived from this example are graphed in Figure 18A-4. Should the cash balance bump against the upper limit of $14,737 (point A), then the financial manager is instructed to buy $9,158 of marketable securities. At the other extreme, if the cash balance should drop to the lower limit of $1,000 (point B), then the financial manager is instructed to sell $4,579 of marketable securities to restore the cash balance level to $5,579. As long as the cash balance wanders within the UL to LL range, no securities transactions take place. By acting in this manner, the financial manager will minimize the sum of interest income forgone and the costs of purchasing and selling securities.[21]

---

[20]Situations where the probabilities of cash increases and decreases are not equal are extremely difficult to evaluate within the framework of the Miller-Orr decision model. See Miller and Orr, "A Model of the Demand for Money by Firms," *Quarterly Journal of Economics* 80 (August 1966), pp. 427–429 and 433–435.

[21]As with the Baumol inventory model, the Miller-Orr control limit model seeks to minimize the total cost of managing the firm's cash balance over a finite planning horizon.

Figure 18A-4 Control Limit Mode

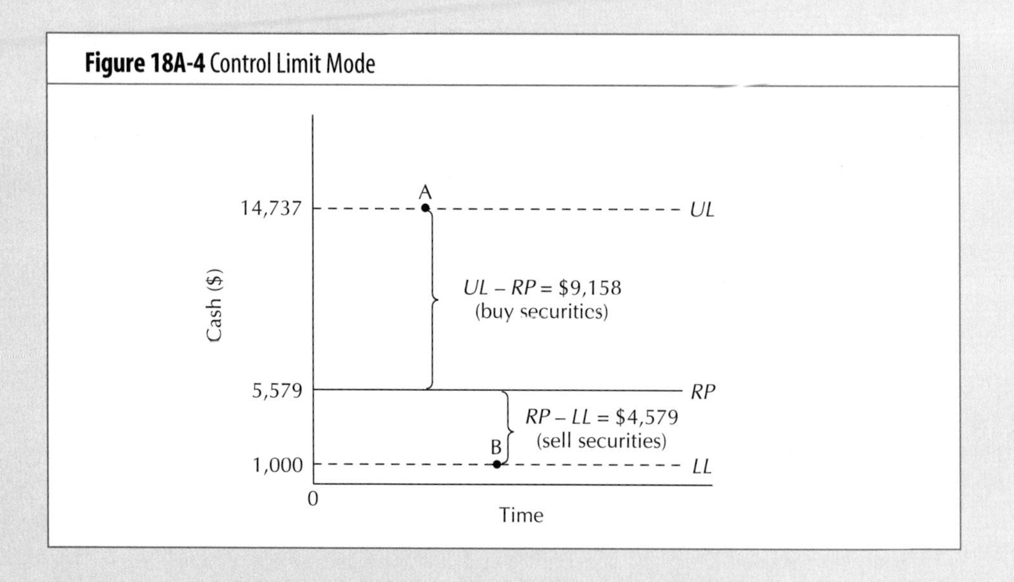

Control Limit Model Implications

Use of Equation (18A-3) results in determination of the optimal cash return point within the framework of the control limit model. Inspection of this equation indicates to the financial manager that the optimal cash return level, *RP*, will vary directly with the cube root of both the transfer cost variable, *b*, and the volatility of daily cash balance changes, σ^2. Greater transfer costs or cash balance volatility result in a greater absolute dollar spread between the upper control limit and the cash return point. This larger spread between *UL* and *RP* means that securities purchases will be effected in larger lot sizes. It is further evident that the optimal cash return point varies inversely with the cube root of the lost interest rate.

Similar to the basic inventory model reviewed in the preceding section, the control limit model implies that economies of scale are possible in cash management. In addition, the optimal cash return point always lies well below the midpoint of the range *UL* to *LL*, over which the cash balance is permitted to "walk." This means that the liquidation of marketable securities will occur (1) more frequently and (2) in smaller lot sizes than purchases of securities. This suggests that firms with highly volatile cash balances must be acutely concerned with the liquidity of their marketable securities portfolio.

Benchmark 3: Compensating Balances

An important element of the commercial banking environment that affects corporate cash management deserves special mention. This is the practice of the bank's requiring, either formally or informally, that the firm maintain deposits of a given amount in its demand deposit account. Such balances are referred to as compensating balances.

The compensating-balance requirement has become standard banking practice which allows banks to maintain favourable levels of their basic raw material—deposits. Such balances are normally required of corporate customers in three situations: (1) where the firm has an established line of credit (loan commitment) at the bank but it is not entirely used, (2) where the firm has a loan outstanding at the bank, and/or (3) in exchange for various other services provided by the bank to its customer.

As you would expect, compensating-balance policies vary among commercial banks and, furthermore, are influenced by general conditions in the financial markets. Still, some tendencies can be identified. In the case of the unused portion of a loan commitment, the bank might require that the firm's demand deposits average anywhere from 5 to 10 percent of the commitment. If a loan is currently outstanding with the bank, the requirement will probably be 10 to 20 percent of the unpaid balance. During periods when monetary policy is restrictive, so-called tight money periods, the ranges just mentioned rise by about 5 percent across the board.[22]

Instead of charging directly for certain banking services, the bank may ask the firm to "pay" for them by the compensating-balance approach. These services include cheque clearing, the availability of credit information, and any of the array of receipts-acceleration or payment-control techniques that we discussed in Chapter 18.

If the bank asks for compensation in the form of balances left on deposit rather than charging unit prices for services rendered, the requirement may be expressed as (1) an absolute amount or (2) an average amount. The latter is preferable to most firms, as it provides for some flexibility in use of the deposits. With the average balance requirement being calculated over a month, and in some instances as long as a year, the balance can be low on occasion as long as it is offset with heavy balances in other time periods.

Compensating-Balance Requirement Implications

In the analysis that the financial manager undertakes to determine the split between cash and near cash, explicit consideration must be given to compensating-balance requirements. This information can be used in conjunction with the basic inventory model (benchmark 1) or the control limit model (benchmark 2).

One approach is to use either of the models and in the calculations ignore the compensating-balance requirement. This is logical in many instances. Generally, the compensating balance required by the bank is beyond the firm's control.

In using the models, then, the focus is on the discretionary cash holdings above the required levels. Once the solution to the particular model is found, the firm's optimal average cash balance will be the higher of that suggested by the model or the balance requirement set by the bank.

A second approach is to introduce the size of the compensating balance into the format of the cash-management model and carry out the requisite calculations. We noted previously that in the control-limit model, the lower control limit (*LL*) could be the compensating-balance requirement faced by the firm as opposed to a value of zero for *LL*. In the basic inventory model the compensating balance can be treated as a safety stock. In Figure 18A-1, then, the cash balance would not touch the zero level and trigger a marketable securities purchase; rather, it would fall to the level of the compensating balance (some amount greater than zero) and initiate the securities purchase. Under ordinary circumstances, in neither model is the calculated optimal order size of marketable securities altered by consideration of compensating balances, so useful information on the proper split between near cash and discretionary cash (cash held in excess of compensating balances) is still provided by these two benchmarks.

Trends indicate that compensating balances are slowly giving way to unit pricing of bank services. Under unit pricing, the bank quotes a stated price for the service. The firm pays that rate only for the services actually used. Such a trend means that the importance of this third benchmark may diminish in the foreseeable future. At the same time, the policies suggested by application of cash management models, similar to those just presented, may well attract attention.

---

[22]See Howard D. Crosse and George H. Hempel, *Management Policies for Commercial Banks*, 2nd ed. (Englewood Cliffs, NJ: Prentice-Hall, 1973), pp. 200–202; and E. E. Reed, R. V. Cotter, E. K. Gill, and R. Smith, *Commercial Banking*, 2nd ed. (Englewood Cliffs, NJ: Prentice-Hall, 1980), pp. 246–247.

STUDY PROBLEMS

18A-1. *(The Inventory Model)* The Richard Price Metal Working Company will experience $800,000 in cash payments next month. The annual yield available on marketable securities is 6.5 percent. The company has analyzed the cost of obtaining or investing cash and found it to be equal to $85 per transaction. Since cash outlays for Price Metal Working occur at a constant rate over any given month, the company has decided to apply the principles of the inventory model for cash management to provide answers to several questions.

a. What is the optimal cash conversion size for the Price Metal Working Company?

b. What is the total cost of having cash on hand during the coming month?

c. How often (in days) will the firm have to make a cash conversion? Assume a 30-day month.

d. What will be Price Metal Working's average cash balance?

18A-2. *(The Control Limit Model)* The Edinboro Fabric Company manufactures 18 different final products, which are woven, cut, and dyed for use primarily in the clothing industry. Owing to the whimsical nature of the underlying demand for certain clothing styles, Edinboro Fabric has a most difficult time forecasting its cash balance levels. The company maintains $2,000 in its demand deposit account at all times. A detailed study of past cash balance levels has revealed that the standard deviation, σ, in daily cash balance changes has been equal to $600. The nature of the firm is not expected to undergo any structural changes in the foreseeable future, and for this reason the past volatility in cash balance levels is expected to continue in the future. Edinboro has determined that the cost of transacting a marketable securities trade is $85. Marketable securities are yielding 6 percent per annum. The firm always uses a 360-day year in its analysis procedures. Robert Cambridge, Edinboro's treasurer, has just returned from a three-day cash management seminar in Toronto. He has decided to apply the control limit model for cash management to his firm's situation.

a. What is the optimal cash return point for Edinboro?

b. What is the upper control limit?

c. In what lot sizes will marketable securities be purchased? Sold?

d. Graph your results.

CHAPTER 19

ACCOUNTS RECEIVABLE AND INVENTORY MANAGEMENT

LEARNING OBJECTIVES

After reading this chapter you should be able to

1. Discuss the determinants of the firm's investment in accounts receivable and how decisions regarding changes in credit policy are determined.
2. Discuss the reasons for carrying inventory and how inventory management decisions are made.
3. Discuss the changes that TQM and single sourcing have had on inventory purchasing.

INTRODUCTION

In the fashion industry, it is not enough for a store to have inventory on hand; it also must have what is in style. As we all know, fashion trends can change overnight. This is particularly frustrating for fashion retailers because traditionally orders must be placed at least six months in advance. As a result, most fashion retailers have quite a challenge in reacting quickly to new trends and style changes.[1]

The Limited, which has more than 3,000 retail outlets across North America, including Express and Victoria's Secret, has set up an international inventory management system that allows the fashion cycle to be cut to sixty days. It does this by examining daily reports that are taken from point-of-sale computers, with information on items sold being fed back to company headquarters. Decisions are then made on what to produce, and those orders are sent by satellite to plants located in Hong Kong. The newly produced goods are then flown back to the United States, on chartered flights that arrive four times a week. Once in the United States, the goods are sorted, priced and shipped out to stores across North America within forty-eight hours. As a result, the goods go on sale within sixty days of the order.

The Limited's inventory stocking system allows it to be successful by keeping on top of trends. This inventory system integrates the firm's global operations in such a way that it has become The Limited's most important competitive weapon. As we will see, inventory and also credit management play important roles in determining the success or failure of a company.

[1]In addition, stocking less inventory may also lead to a loss in sales and the alienation of good customers.

CHAPTER PREVIEW

In the two previous chapters, we developed a general overview of working-capital management and took an in-depth look at the management of cash and marketable securities. This chapter focuses on the management of two more working-capital items, accounts receivable and inventory. Any change in the levels of accounts receivable and inventory that are carried by a firm will affect its profitability.

In studying the management of these current assets, we first examine accounts receivable management, focusing on its importance, what determines investment in it, what the decision variables are, and how we determine them. We then turn to inventory management, examine its importance, and discuss order quantity and order point problems, which in combination determine the level of investment in inventory. We also examine the relationship between inventory and total quality management.

OBJECTIVE 1

ACCOUNTS RECEIVABLE MANAGEMENT

All firms by their very nature are involved in selling either goods or services. While some of these sales will be for cash, a large portion will involve credit. Whenever a sale is made on credit, it increases the firm's accounts receivable. Thus, the importance of how a firm manages its accounts receivable depends upon the degree to which the firm sells on credit.

Table 19-1 shows, for selected industries, the percentage of total assets made up by accounts receivable and accrued revenue. In the fourth quarter of 1995, the overall average for all industries that report accounts receivable and accrued revenue was 7.7 percent of total assets; whereas, nonfinancial industries report accounts receivable and accrued revenue as 12.6 percent of total assets. Although the data represent a snapshot at a particular point in time, it should be noted that there is considerable variation among industries. The fact that cash flows from a credit sale cannot be reinvested into the firm until the account is collected, causes control of

Table 19-1 Accounts Receivable and Accrued Revenue as a Percentage of Total Assets for Major Industries

Industries	4th Quarter 1995	4th Quarter 1994
All industries	7.7	8.2
Nonfinancial industries	12.6	13.9
Telecommunication carriers and postal and courier services	7.4	6.9
Petroleum and natural gas	8.0	9.5
Building materials and construction	18.6	18.9
Motor vehicles, parts and accessories, and tires	15.9	19.1
Other transportation equipment	18.5	18.1
Household appliances and electrical products	27.9	31.6
Printing, publishing and broadcasting	13.2	13.5
Accommodation, food and beverage, educational, health and recreational services	6.0	6.4

Source: Statistics Canada, "Quarterly Financial Statistics for Enterprises," Cat. No. 61-008, Fourth Quarter, 1995. Reproduced with the authority of the Minister of Industry, 1996.

receivables to take on added importance to the firm. Remember, an efficient collection system determines both the firm's profitability and liquidity.

Size of Investment in Accounts Receivable

The size of the investment in accounts receivable is determined by a number of factors. First, the percentage of credit sales to total sales affects the level of accounts receivable held. Although this factor certainly plays a major role in determining a firm's investment in accounts receivable, it generally is not within the control of the financial manager. The nature of the business tends to determine the blend between credit sales and cash sales. A large grocery store tends to sell exclusively on a cash basis, while most construction-lumber supply firms make their sales primarily with credit. Thus, the nature of the business, and not the decisions of the financial manager, tends to determine the proportion of credit sales.

The level of sales is also a factor in determining the size of the investment in accounts receivable. Very simply, the more sales, the greater the amount of accounts receivable. As the firm experiences seasonal and permanent growth in sales, the level of investment in accounts receivable will naturally increase. Thus, while the level of sales affects the size of the investment in accounts receivable, it is not a decision variable for the financial manager.[2]

The final determinants of the level of investment in accounts receivable are the credit and collection policies—more specifically, the terms of sale, the quality of customer, and collection efforts. The terms of sale specify both the time period during which the customer must pay and the terms, such as penalties for late payments or discount for early payments. The type of customer or credit policy also affects the level of investment in accounts receivable. For example, the acceptance of poorer credit risks and their subsequent delinquent payments may lead to an increase in accounts receivable. The strength and timing of the collection efforts can affect the period for which past-due accounts remain delinquent, which in turn affects the level of accounts receivable. Collection and credit policy decisions may further affect the level of investment in accounts receivable by causing changes in the sales level and the ratio of credit sales to total sales. However, the three credit and collection policy variables are the only true decision variables under the control of the financial manager. Figure 19-1 illustrates the determinants of investment in accounts receivable.

PERSPECTIVE IN FINANCE

As we examine the credit decision, try to keep in mind that our goal is not to minimize losses, but to maximize profits. A manager should be able to recognize those customers with the highest probability of default. This analysis is only an input into a decision based upon profit maximization. In other words, a firm with a high profit margin can tolerate a more liberal credit policy than firms with low profit margins.

Terms of Sale—Decision Variable

The **terms of sale** identify the possible discount for early payment, the discount period, and the total credit period. They are generally stated in the form *a*/*b*, net *c*, indicating that the customer can deduct *a* percent if the account is paid within *b* days; otherwise, the account must be paid within *c* days. Thus, for example, trade

Terms of Sale

The credit terms identifying the possible discount for early payment.

[2]While the major emphasis in this chapter is on the cost of credit, it should also be noted that some firms actually make money by carrying accounts receivable. For a firm that charges interest on the outstanding accounts receivable, credit sales can be more profitable than cash sales. A good example of this is General Motors Acceptance Corporation of Canada, which finances auto loans.

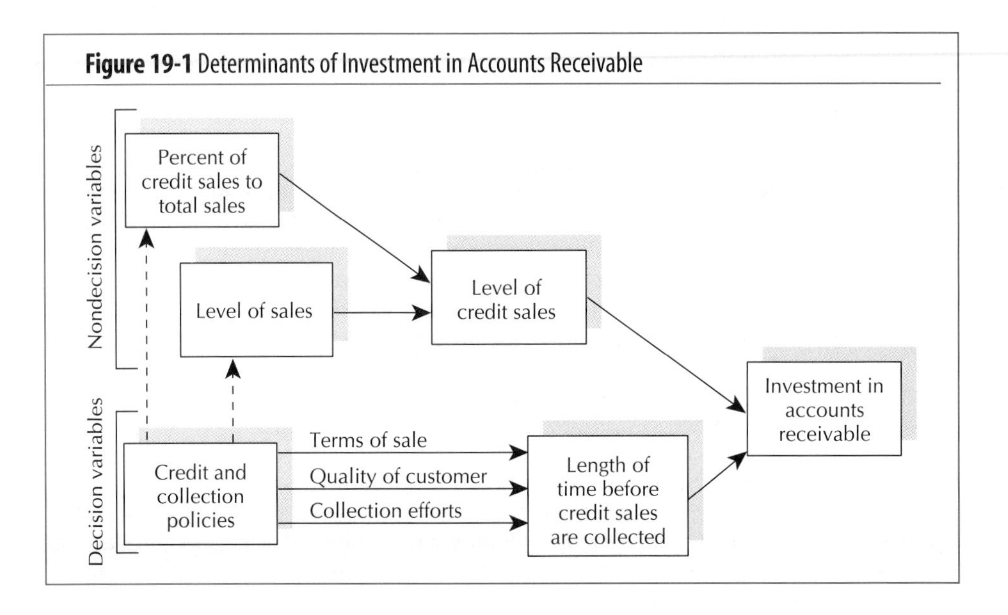

credit terms of 2/10, net 30 indicate that a 2 percent discount can be taken if the account is paid within 10 days; otherwise it must be paid within 30 days. What if the customer decides to forgo the discount and not pay until the final payment date? If such a decision is made, the customer has the use of the money for the time period between the discount date and the final payment date. However, failure to take the discount represents a cost to the customer. For instance, if the terms are 2/10, net 30, the effective annualized opportunity cost of passing up this 2 percent discount in order to withhold payment for an additional 20 days is 44.59 percent. This is determined as follows:

$$\text{Effective annual cost} = \left[1 + \left(\frac{a}{1-a}\right)\right]^{\left(\frac{365}{c-b}\right)} - 1 \qquad (19\text{-}1)$$

Substituting the values from the example, we get

$$\text{Effective annual cost} = \left[1 + \left(\frac{.02}{1-.02}\right)\right]^{\left(\frac{365}{30-10}\right)} - 1 = 44.59\%$$

In industry the typical discount ranges anywhere from one-half percent to 10 percent, while the discount period is generally 10 days and the total credit period varies from 30 to 90 days. Although the terms of credit vary radically from industry to industry, they tend to remain relatively uniform within any particular industry. Moreover, the decision to extend more favourable credit terms to one category of customers as compared to another, must be examined in terms of both costs and benefits.

PERSPECTIVE IN FINANCE

It is tempting to look at the credit decision as a single yes or no decision based upon some simple formula. However, to simply look at the immediate future in making a credit decision would be a mistake. If extending a customer credit means that the customer may become a regular customer in the future, it may be appropriate to take a risk that otherwise would not be prudent. In effect, our goal is to ensure that all cash flows affected by the decision at hand are considered, not simply the most immediate cash flows.

Type of Customer—Decision Variable

A second decision variable involves determining the *type of customer* who is to qualify for trade credit. There are several costs associated with extending credit to less credit-worthy customers. First, as the probability of default increases, it becomes more important to identify which of the possible new customers represent a high risk to the firm. Thus, more time is spent investigating the less credit-worthy customer, and the costs of credit investigation increase.

Default costs also vary directly with the quality of the customer. As the customer's credit rating declines, the chance that the account will not be paid on time increases. In the extreme case, payments are never received and the account must be written off. Thus, taking on less credit-worthy customers results in increases in default costs.

Collection costs also increase as the quality of the customer declines. More delinquent accounts force the firm to spend more time and money in collecting them. Thus an increase in the number of delinquent accounts causes an increase in the costs of credit investigation, collection, and default.

In determining whether or not to grant credit to an individual customer, we are primarily interested in the customer's short-run welfare. Thus, liquidity ratios, other obligations, and the overall profitability of the firm become the focal point in this analysis. Credit-rating services, such as Dun and Bradstreet, provide information on the financial status, operations, and payment history for most firms. Other possible sources of information would include credit bureaus, trade associations, Chambers of Commerce, competitors, bank references, public financial statements, and, of course, the firm's past relationships with the customer.

Credit scoring

The numerical credit evaluation of each candidate.

One way in which both individuals and firms are often evaluated as credit risks is through the use of credit scoring. **Credit scoring** involves the numerical evaluation of each applicant. An applicant receives a score based upon his or her answers to a simple set of questions. This score is then evaluated according to a predetermined standard, its level relative to the standard determining whether or not credit should be extended. The major advantage of credit scoring is that it is inexpensive and easy to perform. For example, once the standards are set, a computer or clerical worker without any specialized training could easily evaluate any applicant.

The techniques used for constructing credit-scoring indexes range from the simple approach of adding up default rates associated with the answers given to each question, to sophisticated evaluations using multiple discriminate analysis (MDA). MDA is a statistical technique for calculating the appropriate importance to be given to each question used in evaluating the applicant.[3] Figure 19-2 shows a credit "scorecard" used by a large automobile dealer. The weights or scores attached to each answer are based upon the auto dealer's past experience with credit sales. For example, the scorecard indicates that individuals with no telephone in their home have a much higher probability of default than those with a telephone. One caveat should be mentioned: Whenever this type of questionnaire is used to evaluate credit applicants, it should be examined carefully to be sure that it does not contain any illegal discriminatory questions.[4]

Another model that could be used for credit scoring has been provided by Edward Altman, who used multiple discriminant analysis to identify businesses that might go bankrupt. In his landmark study Altman used financial ratios to predict which firms would go bankrupt over the period 1946 to 1965. Using multiple discriminant analysis, Altman came up with the following index:

$$Z = 3.3\left(\frac{\text{EBIT}}{\text{total assets}}\right) + 1.0\left(\frac{\text{sales}}{\text{total assets}}\right) + 0.6\left(\frac{\text{market value of equity}}{\text{book value of debt}}\right) \qquad (19\text{-}2)$$

$$+ 1.4\left(\frac{\text{retained earnings}}{\text{total assets}}\right) + 1.2\left(\frac{\text{working capital}}{\text{total assets}}\right)$$

[3]Some benefits associated with the use of credit-scoring models include: consistency in credit-granting decisions; quicker decisions; ease of training individuals in the credit granting decision; quantification of risk; and identification of predictor variables.

[4]The effectiveness of implementing a credit-scoring model depends on careful collection of information to form databases which can be used for future reference.

Figure 19-2 Credit "Scorecard"

Telephone

Home	Relative	None				Score
5	1	0				

Living quarters

Own home no mortgage	Own home mortgage	Rent a house	Live with someone	Rent an apartment	Rent a room	
5	1	0	1	0	0	

Bank accounts

None	1	More than 1				
0	4	6				

Years at present address

Under 1/2	1/2-2	3-7	8 or more			
0	1	3	4			

Size of family including customer

1	2	3-6	7 or more			
2	4	3	0			

Monthly income

Under $700	$701-$900	$901-$1,200	$1,201-$1,500	More than $1,500		
0	1	2	6	8		

Length of present employment

Under 1/2 year	1/2-2 years	3-7 years	8 years or more			
0	1	2	4			

Percent of selling price on credit

Under 50	50-69	70-84	85-99			
5	3	1	0			

Interview discretionary point (+5 to –5)

Total

Credit scorecard

(Customer's name)

(Street address)

(City, Province, Postal Code)

(Home/Office telephone)

(Credit scorer)

Credit scorecard evaluation

Dollar amount: $0-$2,000

0-18	19-21	22 or more
Reject	Refer to main credit	Accept

Dollar amount: $2,001-$5,000

0-21	22-24	25 or more
Reject	Refer to main credit	Accept

Dollar amount: more than $5000

0-23	26-27	27 or more
Reject	Refer to main credit	Accept

If a previous loan customer, were payments received promptly? Yes ☐ No* ☐

Are you willing to take reponsibility for authorizing this loan? Yes ☐ No* ☐

*Refer to main credit if answer to either question is *No*.

Altman found that of the firms that went bankrupt over this time period, 94 percent had Z scores of less than 2.7 one year prior to bankruptcy and only 6 percent had scores above 2.7 percent. On the other hand, of those firms that did not go bankrupt, only 3 percent had Z scores below 2.7 and 97 percent had scores above 2.7.

Again, the advantages of credit-scoring techniques are low cost and ease of implementation. The calculations enable a manager to spot those credit risks that need more screening before credit should be extended to them.

BACK TO THE BASICS

The credit decision is another application of ***Axiom 1: The Risk-Return Tradeoff—We Won't Take on Additional Risk Unless We Expect to Be Compensated with Additional Return.*** *The risk is the chance of nonpayment while the return stems from additional sales. Although it may be tempting to look at the credit decision as a yes or no decision based on some "black box" formula, keep in mind that simply looking at the intermediate future in making a credit decision may be a mistake. If extending a customer credit means that the customer may become a regular customer in the future, it may be appropriate to take a risk that otherwise would not be prudent. In effect, our goal is to ensure that all cash flows affected by the decision at hand are considered, not simply the most immediate cash flows.*

Collection Efforts—Decision Variable

The key to maintaining control over collection of accounts receivable is the fact that the probability of default increases with the age of the account. Thus, control of accounts receivable focuses on the control and elimination of past-due receivables. One common way of evaluating the current situation is *ratio analysis.* The financial manager can determine whether or not accounts receivable are drifting out of control by examining the average collection period, the ratio of receivables to assets, the ratio of credit sales to receivables (called the *accounts receivable turnover ratio*), and the amount of bad debts relative to sales over time. In addition, the manager can use a schedule to determine the amount of aging of accounts receivable. Such a schedule provides a breakdown both in dollars and in percentage terms of the proportion of receivables that are past due. A comparison of the current aging of receivables with past data provides the manager with even more control over receivables. Table 19-2 shows an example of an aging account or schedule.

Once the delinquent accounts have been identified, the firm will make an effort to collect them. For example, the firm can send a past-due letter, called a dunning letter, if payment is not received on time, followed by an additional dunning letter

Table 19-2 Aging Account

Age of Accounts Receivable (Days)	Dollar Value (00)	Percent of Total
0–30	$2,340	39
31–60	1,500	25
61–90	1,020	17
91–120	720	12
Over 120	420	7
Total	$6,000	100

in a more serious tone if the account becomes 3 weeks past due, followed after 6 weeks by a telephone call. Finally, if the account becomes 12 weeks past due, it might be turned over to a collection agency. Again, a direct tradeoff exists between collection expenses and lost goodwill on one hand and noncollection of accounts on the other. This type of tradeoff is always part of making the decision.

Thus far, we have discussed the importance and role of accounts receivable in the firm and then examined the determinants of the size of the investment in accounts receivable. We have focused on the credit and collection policies since these are the only discretionary variables for management. In examining these decision variables, we have simply described their traits. These variables are analyzed in a decision-making process called marginal or incremental analysis.

Credit Policy Changes: The Use of Marginal or Incremental Analysis

Changes in credit policy involve direct tradeoffs between costs and benefits. When credit policies are eased, sales and profits from customers are expected to also increase. On the other hand, easing credit policies can also involve an increase in bad debts, additional funds tied up in both accounts receivable and inventory, and additional costs from customers taking a cash discount. Given these costs, when is it appropriate for the firm to change its credit policy? The answer is when the increased sales generate enough in the way of new profit to more than offset the increased costs associated with the change. Determining whether this is so is the job of **marginal or incremental analysis.** In general, there are three categories of changes in credit policy that a firm can consider: a change in the risk class of the customer, the collection process, or a change in the discount terms. The following example illustrates a change in credit policy.

Marginal or incremental analysis

A method of analysis for credit policy changes in which the incremental benefits are compared to the added costs.

BACK TO THE BASICS

Marginal or incremental analysis is a direct application of ***Axiom 4: Incremental Cash Flows—It's Only What Changes That Counts*** *into the credit analysis decision process. What we are really doing in marginal analysis is looking at all the cash flows to the company as a whole with the change in credit policy versus those cash flows without making the credit policy changes. Then, if the benefits resulting from the change outweigh the costs, the change should be made.*

EXAMPLE

Assume that Denis Electronics currently has annual sales, all credit, of \$8 million and an average collection period of 30 days. The current level of bad debts is \$240,000 and the firm's opportunity cost or required rate of return is 15 percent. Further assume that the firm produces only one product, with variable costs equalling 75 percent of the selling price. The company is considering a change in the credit terms—the current terms of net 30 to 1/30 net 60. If this change is made it is expected that half of the customers will take the discount and pay on the 30th day while the other half will not take the discount and pay on the 60th day. This will increase the average collection period from 30 days to 45 days. The major reason Denis Electronics is considering this change is that it will generate additional sales of \$1,000,000. Although the sales from these new customers will generate new profits, they will also generate more bad debts. The level of bad debts on original sales will remain constant, but the level of bad debts on the new sales will be 6 percent of those sales. In addition, the new sales will require an increase in the level of average inventory from \$1,000,000 to \$1,025,000.

General Procedure

Marginal or incremental analysis involves a comparison of the incremental profit contribution (or benefits) from new sales with the incremental costs resulting from the change in credit policy. If the benefits outweigh the costs, the change is made. If not, the credit policy remains as it is. The following procedure shows the four steps for performing marginal or incremental analysis on a change in credit policy:

Step 1: Estimate the change in profit.

Step 2: Estimate the cost of additional investment in accounts receivable and inventory.

Step 3: Estimate the cost of the discount (if a change in the cash discount is enacted).

Step 4: Compare the incremental revenues with the incremental costs.

To simplify the analysis, Table 19-3 provides a summary of the relevant information concerning Denis's proposed credit change and Table 19-4 provides the results of the incremental analysis.

In Step 1 of the analysis, the additional profits less bad debts from the new sales are $190,000. In Step 2, the additional investment in accounts receivable and inventory is determined to be $452,055. Since the pre-tax required rate of return is 15 percent, the company's required rate of return on this investment is $71,558. In Step 3, the cost of introducing a cash discount is determined to be $45,000. Finally, in Step 4 the benefits and costs are compared, and the net change in pre-tax profits is determined to be $73,442. Thus, a change to the present credit policy is warranted.

In summary, the logic behind this approach to credit policy is to examine the incremental or marginal benefits from such a change and to compare these with the incremental or marginal costs. If the change promises more benefits than costs, the change should be made. If the incremental costs are greater than the benefits, the proposed change should not be made. Figure 19-3 illustrates this process: the point where marginal costs equal marginal benefits occurs at credit policy A.

Figure 19-3 Credit Policy and Profits

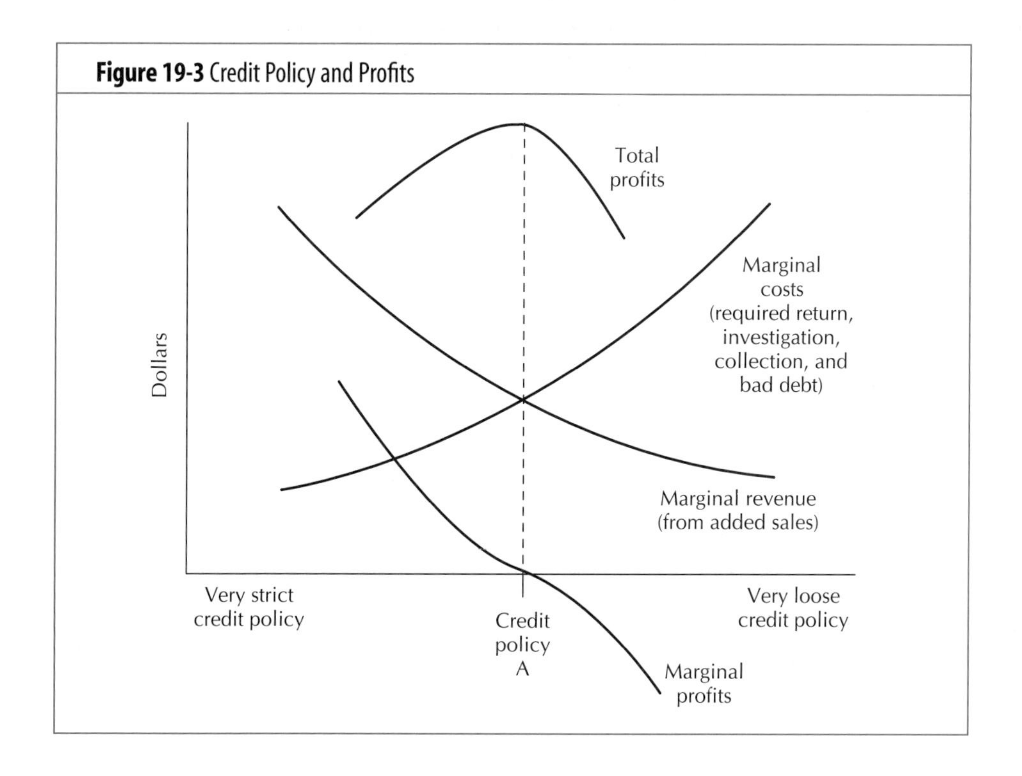

Table 19-3 Denis Electronics: Relevant Information for Incremental Analysis

New sales level (all credit):	\$9,000,000
Original sales level (all credit):	\$8,000,000
Contribution Margin:	25%
Percent bad debt losses on new sales:	6%
New average collection period:	45 days
Original average collection period:	30 days
Additional investment in inventory:	\$25,000
Pre-tax required rate of return:	15%
New percent cash discount:	1%
Percent of customers taking the cash discount:	50%

Table 19-4 Denis Electronics: Incremental Analysis of a Change in Credit Policy

Step 1: Estimate the change in profit. This is equal to the increased sales times the profit contribution on those sales less any additional bad debts incurred.

= (Incremental sales × Contribution Margin) −
(Incremental sales × Percent bad debt losses on new sales)
= (\$1,000,000 × .25) − (\$1,000,000 × .06)
= \$190,000

Step 2: Estimate the cost of any additional investment in accounts receivable and inventory. This involves first calculating the change in the investment in accounts receivable: the new and the original levels of investment in accounts receivable are calculated by multiplying the daily sales level times the average collection period. The additional investment in inventory is added to this, and the sum is then multiplied by the pre-tax required rate of return.

= (Additional accounts receivable + Additional inventory) ×
(Pre-tax required rate of return)

First, calculate the additional investment in accounts receivable:

= [(New level of daily sales) × (New average collection period)] −
[(Original level of daily sales) × (Original average collection period)]

$$= \left[\left(\frac{\$9{,}000{,}000}{365}\right) \times 45\right] - \left[\left(\frac{\$8{,}000{,}000}{365}\right) \times 30\right]$$

= \$452,055

Second, the sum of additional investments in accounts receivable and inventory is multiplied by the pre-tax required rate of return:

= (\$452,055 + \$25,000) × .15
= \$71,558

Step 3: Estimate the change in the cost of the cash discount (if a change in the cash discount is enacted). This is equal to the new level of sales times the new percent cash discount, less the original level of sales, times the original percent cash discount, times the percent of customers taking the discount.

= [(New level of sales) × (New percent cash discount) ×
(Percent of customers taking discount)] − [(Original level of sales) ×
(Original percent cash discount) × (Percent of customers taking discount)]
= [(\$9,000,000 × .01 × .50)] − [(\$8,000,000 × .00 × .00)]
= \$45,000

Step 4: Compare the incremental revenues with the incremental costs:

Net change in pre-tax profits = (Change in profits) − (Cost of new investment in accounts receivable and inventory + Cost of change in cash discount)
= (Step 1) – (Step 2 + Step 3)
= \$190,000 – (\$71,558 + \$45,000)
= \$73,442

PERSPECTIVE IN FINANCE

The calculations associated with the incremental analysis of a change in credit policy illustrate the changes that take place when the credit policy is adjusted. On the positive side, a loosening of the credit policy should increase sales. On the negative side, bad debts, investment in accounts receivable and inventory, and costs associated with the cash discount all increase. The decision then boils down to whether the incremental benefits outweigh the incremental costs.

INVENTORY MANAGEMENT

OBJECTIVE 2

Inventory management involves the control of assets that are produced to be sold in the normal course of the firm's operations. The general categories of inventory include raw materials inventory, work in process inventory, and finished goods inventory. The importance of inventory management to the firm depends upon the extent of the inventory investment.[5] For an average firm in the fourth quarter of 1995, approximately 6.0 percent of all industries' assets are in the form of inventory and 12.2 percent of nonfinancial industries' assets are in the form of inventory. Table 19-5 shows that inventory as a percentage of total assets varies widely from industry to industry. For example, it is much more important in the motor vehicles, parts and accessories, and tires industries, where inventories make up 20.3 percent of total assets, than in the accommodation, food and beverage, educational, health and recreational services industries, where the average investment in inventory is only 3.3 percent of total assets.

Inventory management

The control of the assets used in the production process or produced to be sold in the normal course of the firm's operations.

Purposes and Types of Inventory

The purpose of carrying inventories is to uncouple the operations of the firm—that is, to make each function of the business independent of each other function—so that delays or shutdowns in one area do not affect the production and sale of the final product. Since production shutdowns result in increased costs, and since

Table 19-5 Inventory as a Percentage of Total Assets for Major Industries

Industries	4th Quarter 1995	4th Quarter 1994
All industries	6.0	6.1
Nonfinancial industries	12.2	12.4
Telecommunication carriers and postal and courier services	0.9	0.8
Petroleum and natural gas	3.5	3.6
Building materials and construction	12.1	12.4
Motor vehicles, parts and accessories, and tires	20.3	21.1
Other transportation equipment	32.0	29.0
Household appliances and electrical products	27.1	26.4
Printing, publishing and broadcasting	3.8	3.7
Accommodation, food and beverage, educational, health and recreational services	3.3	3.0

Source: Statistics Canada, "Quarterly Financial Statistics for Enterprises," Cat. No. 61-008, Fourth Quarter, 1995. Reproduced with the authority of the Minister of Industry, 1996.

[5]The optimal inventory policy is specific to the firm; similar to optimal capital structure decision.

delays in delivery can lose customers, the management and control of inventory are important duties of the financial manager.

The decision-making process for an investment in inventory involves a basic tradeoff between risk and return. The risk is that if the level of inventory is too low, the various functions of business do not operate independently, and delays in production and customer delivery can result. The return results because reduced inventory investment saves money. As the size of inventory increases, storage and handling costs are expected to increase. However, the required return on capital invested in inventory is also expected to increase. Therefore, a firm will reduce the risk of running out of inventory by increasing its level of inventory, but at the cost of increasing inventory expenses. To better illustrate the uncoupling function that inventories perform, we will look at several general types of inventories.

BACK TO THE BASICS

The decision as to how much inventory to keep on hand is a direct application of ***Axiom 1: The Risk-Return Tradeoff—We Won't Take on Additional Risk Unless We Expect to Be Compensated with Additional Return****. The risk is that if the level of inventory is too low, the various functions of business do not operate independently, and delays in production and customer delivery can result. The return results because reduced inventory investment saves money. As the size of inventory increases, storage and handling costs as well as the required return on capital invested in inventory rise. Therefore, as the inventory a firm holds is increased, the risk of running out of inventory is lessened, but inventory expenses rise.*

Raw Materials Inventory

Raw materials inventory

An inventory of basic materials which is used in the firm's production operations.

Raw materials inventory consists of basic materials purchased from other firms to be used in the firm's production operations. These goods may include steel, lumber, petroleum, or manufactured items such as wire, ball bearings, or tires that the firm does not produce itself. Regardless of the type of raw materials, all manufacturing firms by definition maintain a raw materials inventory. Its purpose is to uncouple the production function from the purchasing function—that is, to make these two functions independent of each other, so that delays in shipment of raw materials do not cause production delays. In the event of a delay in shipment, the firm can satisfy its need for raw materials by liquidating its inventory.

Work in Process Inventory

Work-in-process inventory

Partially finished goods requiring additional work before they become finished goods.

Work in process inventory consists of partially finished goods requiring additional work before they become finished goods. The more complex and lengthy the production process, the larger the investment in work in process inventory. The purpose of work in process inventory is to uncouple the various operations in the production process so that machine failures and work stoppages in one operation will not affect the other operations. Assume, for example, there are ten different production operations, each one involving the piece of work produced in the previous operation. If the machine performing the first production operation breaks down, a firm with no work in process inventory will have to shut down all ten production operations. Yet if a firm has such inventory, the remaining nine operations can continue by drawing the input for the second operation from inventory rather than directly from the output of the first operation.

Finished Goods Inventory

Finished goods inventory

Goods on which the production has been completed but that are not yet sold.

The **finished goods inventory** consists of goods on which the production has been completed but that are not yet sold. The purpose of a finished goods inventory is to uncouple the production and sales functions so that it is not necessary to produce the good before a sale can occur—sales can be made directly out of inventory. For example, in the auto industry, customers would not buy from a dealer who made them wait weeks or months when another dealer could fill the order immediately.

Stock of Cash

Although we have already discussed cash management at some length in Chapter 18, it is worthwhile to mention cash again in the light of inventory management. This is because the *stock of cash* carried by a firm is simply a special type of inventory. In terms of uncoupling the various operations of the firm, the purpose of holding a stock of cash is to make the payment of bills independent of the collection of accounts due. When cash is kept on hand, bills can be paid without prior collection of accounts.

As we examine and develop inventory economic ordering quantity (EOQ) models, we will see a striking resemblance between the EOQ inventory and EOQ cash model; in fact, except for a minor redefinition of terms, they will be exactly the same.

Inventory Management Techniques

The importance of effective inventory management is directly related to the size of the investment in inventory. Since, on average, approximately 6.0 percent of a firm's assets are tied up in inventory, effective management of these assets is essential to the goal of shareholder wealth maximization. To control the investment in inventory, management must solve two problems: the order quantity problem and the order point problem.

The Order Quantity Problem

The **order quantity problem** involves determining the optimal order size for an inventory item given its expected usage, carrying costs, and ordering costs. Aside from a change in some of the variable names, it is exactly the same as the inventory model for cash management (EOQ model) presented in Chapter 18.

Order quantity problem

An inventory problem that requires determining the optimal order size for an inventory item given its usage, carrying costs, and ordering costs.

The EOQ model attempts to determine the order size that will minimize total inventory costs. It assumes that

$$\text{total inventory costs} = \text{total carrying costs} + \text{total ordering costs} \tag{19-3}$$

Assuming that inventory is allowed to fall to zero and then is immediately replenished (this assumption will be lifted when we discuss the order point problem), the average inventory becomes $Q/2$, where Q is inventory order size in units.[6] Figure 19-4 illustrates the average inventory level.

If the average inventory is Q/2 and the carrying cost per unit is C, then carrying costs become:

$$\text{total carrying costs} = \left(\text{average inventory}\right)\left(\text{carrying cost per unit}\right) \tag{19-4}$$

$$= \left(\frac{Q}{2}\right)C \tag{19-5}$$

where Q = the inventory order size in units
C = carrying costs per unit

The carrying costs on inventory include insurance fees, employee salaries associated with inventory storage, property taxes, warehouse or storage costs, wages for those who operate the warehouse, and costs associated with inventory shrinkage, obsolescence, and spoilage. Thus, carrying costs include both real cash flows and opportunity costs associated with having funds tied up in inventory.[7]

[6]The assumption of inventory being replenished immediately enables us to exclude from the model any term for losses in sales from not having enough inventory.

[7]The opportunity cost associated with having funds tied up in inventory is represented by the required rate of return on investment in inventory.

The ordering costs incurred are equal to the ordering costs per order times the number of orders.[8] If we assume total demand over the planning period is S and we order in lot sizes of Q, then S/Q represents the number of orders over the planning period. If the ordering cost per order is O, then

$$\text{total ordering costs} = \begin{pmatrix}\text{number}\\ \text{of orders}\end{pmatrix}\begin{pmatrix}\text{ordering cost}\\ \text{per order}\end{pmatrix} \tag{19-6}$$

$$= \left(\frac{S}{Q}\right)O \tag{19-7}$$

where S = total demand in units over the planning period
O = ordering cost per order

Thus, total costs in Equation (19-3) become

$$\text{total costs} = \left(\frac{Q}{2}\right)C + \left(\frac{S}{Q}\right)O \tag{19-8}$$

Figure 19-5 illustrates this equation.

What we are looking for is the ordering size, Q^*, that provides the minimum total costs. By manipulating Equation (19-8), we find that the optimal value of Q—that is, the economic ordering quantity (EOQ)—is[9]

$$Q^* = \sqrt{\frac{2SO}{C}} \tag{19-9}$$

The use of the EOQ model can best be illustrated through an example.

EXAMPLE

Suppose a firm expects total demand (S) for its product over the planning period to be 5,000 units, while the ordering cost per order (O) is \$200, and the carrying cost per unit (C) is \$2. Substituting these values into Equation (19-9) yields

$$Q^* = \sqrt{\frac{2 \cdot 5000 \cdot 200}{2}} = \sqrt{1{,}000{,}000} = 1{,}000 \text{ units}$$

Thus, if this firm orders in 1,000-unit lot sizes, it will minimize its total inventory costs.

[8]Ordering costs include: the cost of placing an order; the production set-up costs; shipping and handling costs; and quantity discount costs. An example of set-up costs in production is the cost of resetting stamping machines in the auto industry. Note that quantity discount costs are negative costs received if the purchase (order) size is at least a specified level or more.

[9]This result can be obtained through calculus

$$\text{total cost } (TC) = \left(\frac{Q}{2}\right)C + \left(\frac{S}{Q}\right)O$$

The first derivative with respect to Q defines the slope of the total cost curve. Setting this derivative equal to zero specifies the minimum point (zero slope) on the curve. Thus,

$$\frac{dTC}{dQ} = \frac{C}{2} - \frac{SO}{Q^2} = 0$$

$$Q^2 = \frac{2SO}{C}$$

$$Q^* = \sqrt{\frac{2SO}{C}}$$

To verify that a minimum point is being found, rather than a maximum point where the slope would also equal zero, we check for a positive second derivative

$$\frac{d^2TC}{dQ^2} = \frac{2SO}{Q^3} \geq 0$$

The second derivative must be positive, since SO and Q can only take on positive values; hence, this is a minimum point.

Figure 19-4 Inventory Level and the Replenishment Cycle

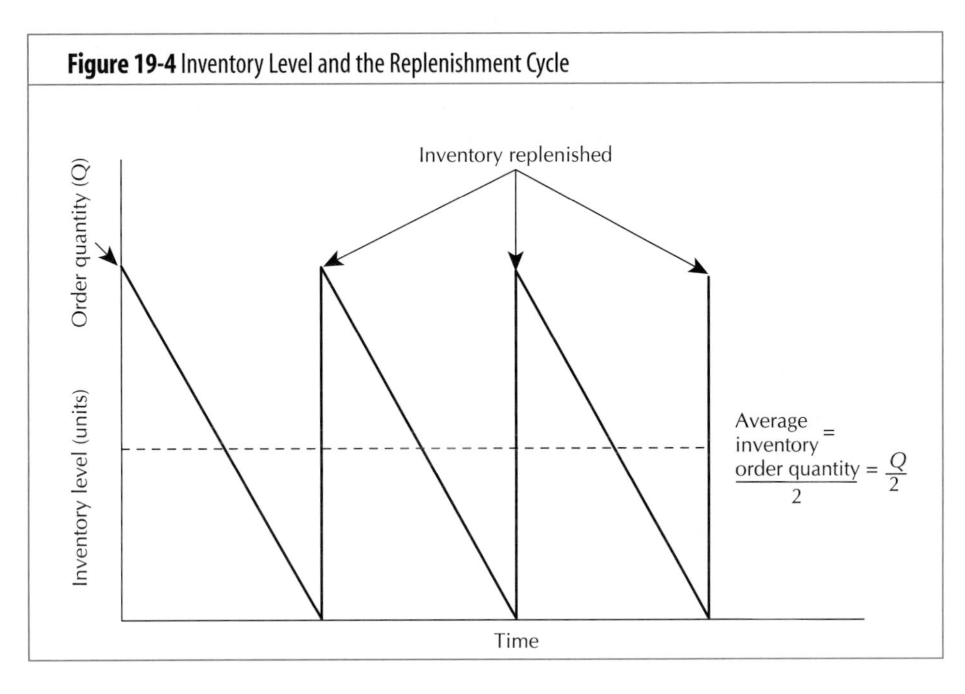

Figure 19-5 Total Cost and EOQ Determination

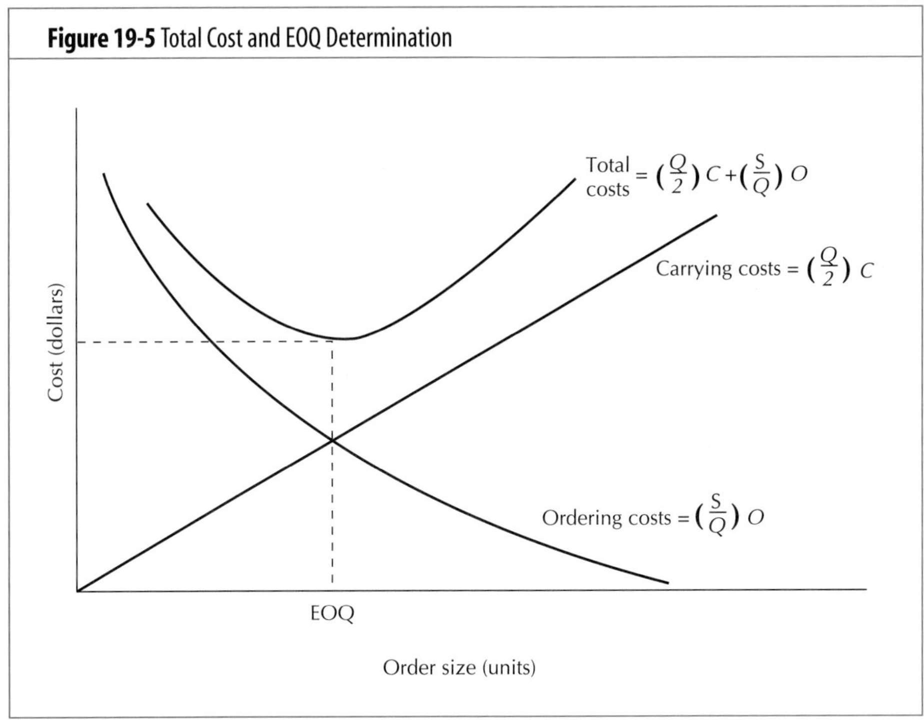

Examination of EOQ Assumptions

Despite the fact that the model tends to yield quite good results, there are major weaknesses in the EOQ model associated with several of its assumptions. If the model's assumptions are violated, the EOQ model can generally be modified to accommodate the situation. The model's assumptions are as follows:

1. **Constant or uniform demand.** Although the EOQ model assumes constant demand, demand may vary from day to day. If demand is stochastic—that is, not known in advance—the model must be modified through the inclusion of a safety stock.

2. **Constant unit price.** The inclusion of variable prices resulting from quantity discounts can be handled quite easily through a modification of the original EOQ model, redefining total costs and solving for the optimum order quantity.
3. **Constant carrying costs.** Unit carrying costs may vary substantially as the size of the inventory rises, perhaps decreasing because of economies of scale or storage efficiency or increasing as storage space runs out and new warehouses have to be rented. This situation can be handled through a modification in the original model similar to the one used for variable unit price.
4. **Constant ordering costs.** While this assumption is generally valid, its violation can be accommodated by modifying the original EOQ model in a manner similar to the one used for variable unit price.
5. **Instantaneous delivery.** If delivery is not instantaneous, which is generally the case, the original EOQ model must be modified through the inclusion of a safety stock. Safety stock represents any amount of inventory that is held to accommodate any unusually large and unexpected usage during the delivery time.
6. **Independent orders.** If multiple orders result in cost savings by reducing paperwork and transportation cost, the original EOQ model must be further modified. While this modification is somewhat complicated, special EOQ models have been developed to deal with it.[10]

These assumptions illustrate the limitations of the basic EOQ model and the ways in which it can be modified to compensate for them. An understanding of the limitations and assumptions of the EOQ model enables the financial manager to make more effective inventory decisions.[11]

The Order Point Problem

Safety stock

Inventory held to accommodate any unusually large and unexpected usage during delivery time.

Order point problem

An inventory problem that requires determining how low inventory should be depleted before it is reordered.

Delivery-time stock

The inventory needed between the order date and the receipt of the inventory ordered.

The two most limiting assumptions—those of constant or uniform demand and instantaneous delivery—are dealt with through the inclusion of **safety stock**. The decision on how much safety stock to hold is generally referred to as the **order point problem**; that is, how low should inventory be depleted before it is reordered?

Two factors go into the determination of the appropriate order point: (1) the procurement or delivery-time stock and (2) the safety stock desired. Figure 19-6 illustrates the process involved in order point determination. We observe that the order-point problem can be decomposed into its two components, the **delivery-time stock**—that is, the inventory needed between the order date and the receipt of the inventory ordered—and the safety stock. Thus, the order point is reached when inventory falls to a level equal to the delivery-time stock plus the safety stock.

$$\begin{array}{c}\text{inventory order point}\\ \text{[order new inventory when}\\ \text{the level of inventory}\\ \text{falls to this level]}\end{array} = (\text{delivery-time stock}) + (\text{safety stock}) \tag{19-10}$$

As a result of constantly carrying safety stock, the average level of inventory increases. Whereas before the inclusion of safety stock the average level of inventory was equal to EOQ/2, now it will be

$$\text{average inventory} = \frac{\text{EOQ}}{2} + \text{safety stock} \tag{19-11}$$

[10]For example, R. J. Tersire, *Material Management and Inventory Control* (New York: Elsevier–North Holland, 1976).

[11]Much of the recent criticism of the EOQ model has been based upon the inability of such a system to respond quickly to rapidly changing demand. For example, the 800-ton presses used to stamp out automobile panels take about four hours to set up. If we assume that set-up costs are equivalent to ordering costs, the EOQ model predicts that at least 10,000 identical panels must be produced before the presses be reset. It is estimated that at least two weeks of production would be required for this type of output. Unfortunately, batch sizes this large make it impossible to respond rapidly to changing demands.

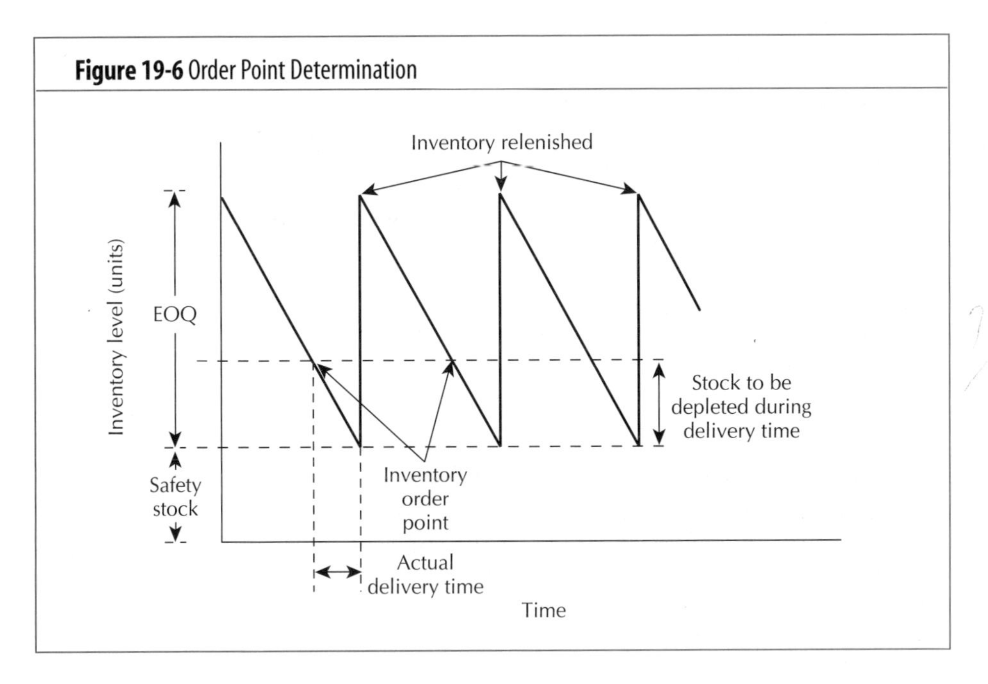

Figure 19-6 Order Point Determination

In general, several factors simultaneously determine how much delivery-time stock and safety stock should be held. First, the efficiency of the replenishment system affects how much delivery-time stock is needed. Since the delivery-time stock is the expected inventory usage between ordering and receiving inventory, efficient replenishment of inventory would reduce the need for delivery-time stock.

The uncertainty surrounding both the delivery time and the demand for the product affects the level of safety stock needed. The more certain the patterns of these inflows and outflows from the inventory, the less safety stock required. In effect, if these inflows and outflows are highly predictable, then there is little chance of any stock-out occurring. However, if they are unpredictable, it becomes necessary to carry additional safety stock to prevent unexpected stock-outs.[12]

The safety margin desired also affects the level of safety stock held. If it is a costly experience to run out of inventory, the safety stock held will be larger than it would be otherwise. If running out of inventory and the subsequent delay in supplying customers result in strong customer dissatisfaction and the possibility of lost future sales, then additional safety stock is necessary.[13] A final determinant is the cost of carrying additional inventory, in terms of both the handling and storage costs and the opportunity cost associated with the investment in additional inventory. Very simply, the greater the costs, the smaller the safety stock.

The determination of the level of safety stock involves a basic tradeoff between the risk of stock-out, resulting in possible customer dissatisfaction and lost sales, and the increased costs associated with carrying additional inventory.

Inflation and EOQ

Inflation affects the EOQ model in two major ways. First, while the EOQ model can be modified to assume constant price increases, often major price increases occur only once or twice a year and are announced ahead of time. If this is the case, the EOQ model may lose its applicability and may be replaced with **anticipatory buying**

Anticipatory buying

Buying in anticipation of a price increase to secure goods at a lower cost.

[12]Note that with a safety stock, costs will rise over what they were without a safety stock. Thus, demand uncertainty creates an additional cost to the firm.

[13]Safety stock costs include: lost sales due to stock-outs; costs associated with production stoppages owing to stock-outs; lost future sales as customers find new sources for the item during stock-out periods.

—that is, buying in anticipation of a price increase in order to secure the goods at a lower cost. Of course, as with most decisions, there are tradeoffs. The costs are the added carrying costs associated with the inventory. The benefits, of course, come from buying at a lower price. The second way inflation affects the EOQ model is through increased carrying costs. As inflation pushes interest rates up, the cost of carrying inventory increases. In our EOQ model this means that C increases, which results in a decline in Q^*, the optimal economic order quantity:

$$\downarrow Q^* = \sqrt{\frac{2SO}{C\uparrow}} \tag{19-12}$$

Reluctance to stock large inventories because of high carrying costs became particularly prevalent during the late 1970s and early 1980s when inflation and interest rates were at high levels.

Just-in-Time Inventory Control

Just-in-time inventory control system

A system that monitors inventory levels so to keep inventory to a minimum and to rely on suppliers to furnish parts "just in time."

The **just-in-time inventory control system** is more than just an inventory control system, it is a production and management system. Not only is inventory cut down to a minimum, but the time and physical distance between the various production operations are also reduced. In addition management is willing to trade off costs to develop close relationships with suppliers and promote speedy replenishment of inventory in return for the ability to hold less safety stock.

The just-in-time inventory control system was originally developed in Japan by Taiichi Okno, a vice-president of Toyota. Originally the system was called the *kanban* system after the cards placed in the parts bins which were used to call for a new supply. The idea behind the system is that the firm should keep a minimum level of inventory on hand, relying on suppliers to furnish parts "just in time" for them to be assembled. This is in direct contrast to the traditional inventory philosophy of North American firms, which is sometimes referred to as a "just in case" system, which keeps healthy levels of safety stocks to ensure that production will not be interrupted. While large inventories may not be a bad idea when interest rates are low, when interest rates are high, as they have been in recent years, they become very costly.

While the just-in-time inventory system is intuitively appealing, it has not proved easy to implement. Successful implementation of this system has been limited by the construction of plants that have too much space for storage and not enough access doors and loading docks to receive inventory. However, the just-in-time inventory system has forced the firm to change its relationship with its suppliers. Since the firm relies on suppliers to deliver high-quality parts and materials immediately, it must have a close long-term relationship with them. Despite the difficulties of implementation, many North American firms are committed to moving toward a just-in-time system. Using its own version of the Japanese inventory control system, General Motors of Canada has implemented the synchronous manufacturing system which has reduced the monthly average inventory level by 3 percent in 1990 over 1989. In this case, Woodbridge Foam corporation provides seats to GM's Sainte Therese car assembly plant. This type of inventory system has enabled the Sainte Therese plant to eliminate the stocking of car seats in their inventory.

While the just-in-time system does not at first appear to bear much of a relationship to the EOQ model, it simply alters some of the assumptions of the model with respect to delivery time and ordering costs and draws out the implications. Actually, it is just a new approach to the EOQ model that tries to produce the lowest average level of inventory possible. If we look at the average level of inventory as defined by the EOQ model, we find it to be:

$$\text{average inventory} = \frac{\sqrt{\frac{2SO\downarrow}{C}}}{2} + \text{safety stock}\downarrow \tag{19-13}$$

The just-in-time system attacks this equation in two places. First, by locating inventory supplies in convenient locations, laying out plants in such a way that it is inexpensive and easy to unload new inventory shipments, and computerizing the inventory order system, the cost of ordering new inventory, *O*, is reduced. Second, by developing a strong relationship with suppliers located in the same geographical area and setting up restocking strategies that cut time, the safety stock is also reduced. The philosophy behind the just-in-time inventory system is that the benefits associated with reducing inventory and delivery time to a bare minimum through adjustment in the EOQ model will more than offset the costs associated with the increased possibility of stock-outs.[14]

TQM AND INVENTORY-PURCHASING MANAGEMENT: THE NEW SUPPLIER RELATIONSHIPS

OBJECTIVE 3

Out of the concept of **total quality management** (TQM), which is a company-wide systems approach to quality, has come a new philosophy in inventory management of "love thy supplier." Today, the traditional antagonistic relationship between suppliers and customers, where suppliers are coldly dropped when a cheaper source can be found, is being replaced by a new order in customer-supplier relationships. In effect, what began as an effort to increase quality through closer supplier relations has turned out be have unexpected benefits. Close customer relationships have helped trim costs, in part, by allowing for the production of higher quality products. This close customer-supplier relationship has allowed the TQM philosophy to be passed across company boundaries to the supplier, enabling the firm to tap the supplier's expertise in designing higher-quality products. In addition, the interdependence between the supplier and customer has also allowed for the development and introduction of new products at a pace much faster than previously possible.

Total quality management (TQM)
A company-wide systems approach to quality.

As we have seen, inventory can make up a rather large percentage of a firm's assets. That by itself lends importance to the role of inventory management and, more specifically, purchasing. In terms of manufacturing costs, purchased materials have historically accounted for 50 percent of U.S. and about 70 percent of Japanese manufacturing costs, and most manufacturers purchase more than 50 percent of their parts. Thus it is hard to overstate the importance of purchasing to the firm.

The traditional purchasing philosophy is to purchase a part or material from a variety of different suppliers, with the suppliers contracting out to a number of different firms. In fact, many companies put an upper limit of 10 or 20 percent on the purchases of any part from a single supplier. The reasoning behind this is that the company can diversify away the effect of poor quality by any one supplier. Thus, if one supplier was unable to meet delivery schedules, delivers a poor quality batch, or even goes out of business, this affects only a small percent of the total parts or material. However, efforts to raise quality have led to a new approach to the customer-supplier relationship called **single sourcing.**

Single-sourcing
Using a single supplier as a source for a particular part or material.

Under single-sourcing, a company uses very few suppliers or, in many cases, a single supplier, as a source for a particular part or material. In this way, the company has more direct influence and control over the quality performance of a supplier because the company accounts for a larger proportion of the supplier's volume. The company and supplier can then enter into a partnership, referred to as *part-*

[14]Part of the inspiration for employing the just-in-time inventory system (besides increasing the ability to respond to changing demand) was the argument that costs other than carrying and ordering or set-up costs were affected by the batch size, namely, quality of the finished product, amount of waste, and the motivation of the workers. All these factors, say proponents of the just-in-time system, contributed to increased costs as the batch size increased.

nering, where the supplier agrees to meet the quality standards of the customer in terms of parts, material, service, and delivery. In this way the supplier can be brought into the TQM program of the customer. In return, the company enters into a long-term purchasing agreement with the supplier that includes a stable order and delivery schedule. For example, on General Motor's Quad 4 engine (its first new engine in several decades), every part except the engine block is single-sourced, resulting in only sixty-nine total suppliers, which is half the normal number for a production engine. In return for the suppliers' assurance of top quality and low cost, GM guaranteed the suppliers their jobs for the life of the engine. In the development of its new LH cars, the Chrysler Concorde, the Dodge Intrepid, and the Eagle Vision, Chrysler has trimmed its supplier base from 3,000 to a few more than 1,000, with a goal of 750 suppliers by 1995. Single-sourcing clearly creates an environment of cooperation between customers and suppliers where both share the common goal of quality.

While the partnering relationship results in higher-quality parts, it can also improve the quality of the design process by allowing for the involvement of the supplier in production design. For example, the supplier may be given the responsibility of designing a new part or component to meet the quality standards and features outlined by the company. When Guardian Industries of Northville, Michigan, developed an oversized solar glass windshield for Chrysler's new LH cars, their engineers met on almost a daily basis with the Chrysler design team to make sure the quality, features, and cost of the windshield met Chrysler standards. To produce the windshields, Guardian opened a new $35 million plant in Ligonier, Indiana, in 1991. In a similar manner, Bailey Controls and Boise Cascade entered into a pact in which Bailey is the exclusive provider of control systems for eight of ten Boise plants. The two work together, reviewing and modifying the terms of the arrangement to ensure that they reflect the ever changing business conditions and that the deal remains fair to both parties. In addition, recognizing the long-term nature of the relationship and that Boise's success would also benefit Bailey, Bailey worked with Boise—in fact, using a Boise plant on which to conduct experiments—to improve the software used to operate the Boise plants.

The concept of partnering has radically changed the way inventory is purchased. Moreover, it has turned the customer-supplier relationship from a formerly adversarial one into a cooperative one. The benefits in terms of increased quality of parts, materials, and design are so dramatic that partnering will not fade away as the fad of the 1990s, but rather continue to evolve and take on even more importance in the future.

Preventive costs

Costs resulting from design and production efforts on the part of the firm to reduce or eliminate defects.

Appraisal costs

Costs of testing, measuring, and analyzing to safeguard against possible defects going unnoticed.

Internal failure costs

Those costs associated with discovering poor quality products prior to deliver.

External failure costs

Costs resulting from a poor quality product reaching the customer's hands.

The Financial Consequences of Quality—The Traditional View

Traditionally, the cost of quality has been viewed as being made up of **preventive costs, appraisal costs,** and **failure costs.** The costs the firm incurs in running its quality management program include both preventive and appraisal costs. Preventive costs include those resulting from design and production efforts on the part of the firm to reduce or eliminate defects. Whereas preventive costs deal with the avoidance of defects the first time through, appraisal costs would include the costs of testing, measuring, and analyzing materials, parts, and products, as well as the production operations to safeguard against possible defective inventory going unnoticed. Together, preventive and appraisal costs make up much of what a typical total quality management program does.

While preventive and appraisal costs deal with the costs associated with achieving good quality, failure costs refer to the costs resulting from producing poor quality products. Failure costs can either occur within the firm, called **internal failure costs,** or once the product has left the firm, then referred to as **external failure costs.** Internal failure costs are those associated with discovering the poor-quality product prior to delivery to the final customer. Internal failure costs include the costs of reworking the product, downtime costs, the costs of having to discount poorer quality products, and the costs of scrapping the product. External failure costs, on the other hand, come as a result of a poor-quality product reaching the customer's hands. Typical external failure costs would include product return costs, warranty costs, product liability costs, customer complaint costs, and lost sales costs.

Traditionally, economists examine the tradeoffs between quality and costs looking for the point where costs are minimized. Graphically, this is done in Figure 19-7. As we can see from this diagram, preventive and appraisal costs increase as the percentage of good quality products increases. This, of course, takes place because more efforts are made in making the product right the first time. On the other hand, the cost of poor quality—failure costs—declines as the percentage of good quality products increases. The lowest total quality cost is found by adding these two cost curves together to make the total cost curve and finding the lowest point on that total cost curve. At the low point on the total cost curve, costs are minimized. Traditionally, this has been the point that economists have recommended that the firm strive for in terms of defects. As we can see by looking at Figure 19-7, the lowest point on the total cost curve allows for planned defects. In effect, the traditional point of view is that 100 percent quality is not an appropriate goal for the firm. In effect, some acceptable defect level is justified.

BACK TO THE BASICS

Axiom 5: The Curse of Competitive Markets—Why It's Hard to Find Exceptionally Profitable Projects *suggests how product differentiation can be used as a means of insulating a product from competition, thereby allowing prices to stay sufficiently high to support large profits. Producing a quality product is one of those ways to differentiate products. This strategy has been used effectively by Caterpillar Tractor, Toyota, and Honda Motors in recent years.*

The Financial Consequences of Quality—The TQM View

In the early 1980s, the notion that there is an acceptable level of defects came under attack by a number of Japanese firms in what is called the TQM view. The TQM view argues that the traditional analysis is flawed in that it ignores the fact that increased sales and market share result from better-quality products, and that this increase in sales will more than offset the higher costs associated with increased quality. In effect, the TQM view argues that because lost sales resulting from a poor quality reputation and increased sales resulting from a reputation for quality are difficult to estimate, they tend to be underestimated or ignored in the traditional approach. In addition, the TQM view argues that the cost of achieving higher quality is less than economists have traditionally estimated. In fact, the benefits from quality improvement programs seem to have spillover effects resulting in increased worker motivation, higher productivity, and improved employee relations. Moreover, large increases in quality have been achieved with very low costs when companies focus on training and educating the production workers. Thus, the TQM view concludes that traditional analysis underestimates the cost of producing a poor-quality product, while it overestimates the cost of producing a high-quality product. As a result, under this new TQM view of the quality-costs tradeoffs, the low point in the total cost curve moves to the right and for many products the optimal quality level is at 100 percent quality.

The adoption of total quality management programs by many firms has borne out this TQM view of the quality-cost relationship. For example, Xerox Corporation introduced TQM in the early 1980s, and between 1982 and 1990 experienced a decline of 90 percent in defective items. For Xerox, the end result of this was a 20 percent drop in manufacturing costs over that same period.

Figure 19-7 The Traditional Tradeoffs Between Quality and Costs

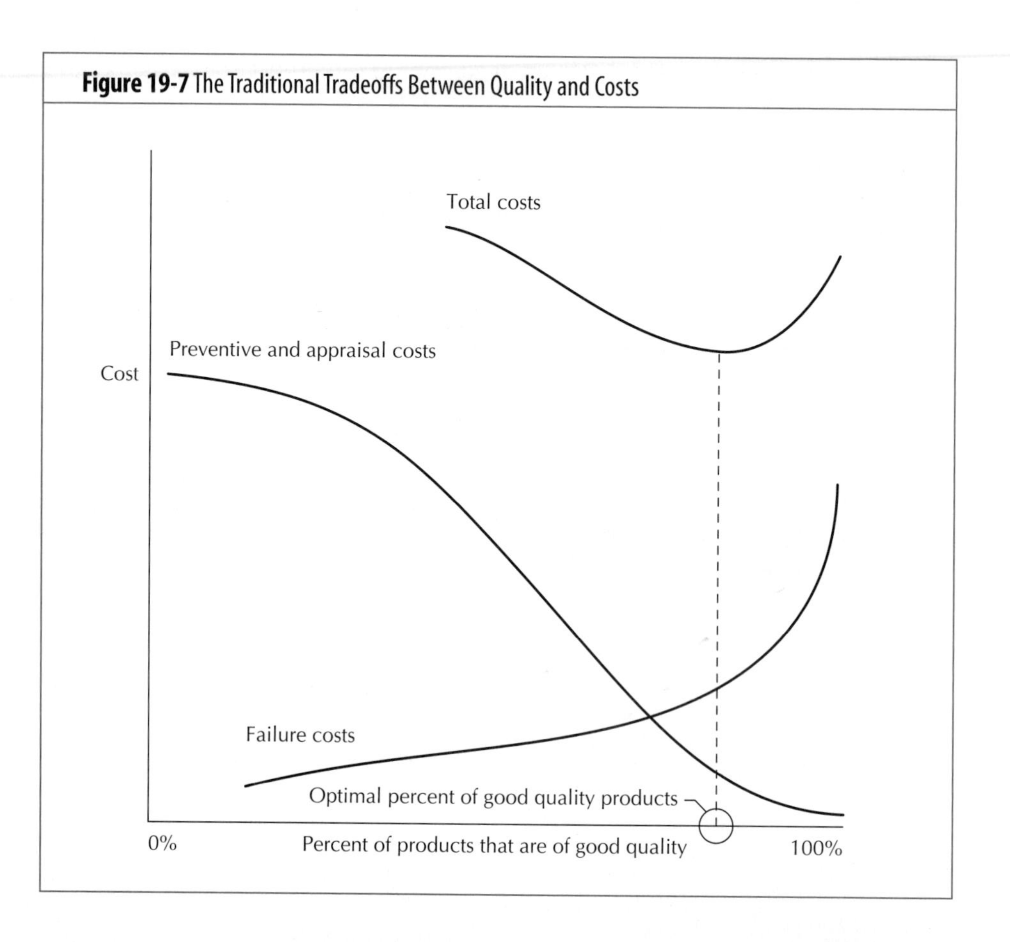

SUMMARY

Accounts Receivable Management

The size of the investment in accounts receivable depends upon three factors: the percentage of credit sales to total sales, the level of sales, and the credit and collection policies. However, only the credit and collection policies are decision variables open to the financial manager. The policies that the financial manager has control over include the terms of sale, the quality of customer, and the collection efforts.

Inventory Management

Although the typical firm has fewer assets tied up in inventory than it does in accounts receivable, inventory management and control is still an important function of the financial manager. The purpose of holding inventory is to make each function of the business independent of the other functions—that is, to uncouple the firm's operations. Inventory-management techniques primarily involve questions of how much inventory should be ordered and when the order should be placed. The answers directly determine the average level of investment in inventory. The EOQ model is employed in answering the first of these questions. This model attempts to calculate the order size that minimizes the sum of the inventory carrying and ordering costs. The order point problem attempts to determine how low inventory can drop before it is reordered. The order point is reached when the inventory falls to a level equal to the delivery-time stock plus the safety stock. In solving for the level of safety stock, the financial manager must trade off between the risk of running out of inventory and the increased costs associated with carrying additional inventory.

The New Golden Rule of Business

In that battle-scarred landscape where suppliers and their corporate customers meet and deal, lions are lying down with lambs, and swords are being beaten into plowshares. Hard-nosed businessmen use embarrassingly romantic terms to describe the new order in supplier-customer relationships. "It's like a marriage," croons a big-league purchasing executive. Companies that succeed at partnerships like this usually set them up routinely, not as exceptions, and with customers and suppliers alike. Sitting in the middle of the web of such relationships is Bailey Controls, an Ohio-headquartered, $300 million-a-year manufacturer of control systems for big factories, from steel and paper mills to chemical and pharmaceutical plants. To spin its web, Bailey, a unit of Elsag Bailey Process Automation, treats some suppliers almost like departments of its own company.

Bailey has physically plugged two of its main electronic distributors into itself. Montreal-based Future Electronics is hooked on with an electronic data interchange system. Every week, Bailey electronically sends Future its latest forecasts of what material it will need for the next six months, so that Future can stock up in time. Bailey itself stocks only enough inventory for a few days of operation, as opposed to the two to three months' worth it used to carry. Whenever a bin of parts falls below a designated level, a Bailey employee passes a laser scanner over the bin's bar code, instantly alerting Future to send the parts at once.

Arrow Electronics, headquartered on Long Island, is plugged in even more closely: it has a warehouse in Bailey's factory, stocked according to Bailey's twice monthly forecasts. Bailey provides the space, Arrow the warehouseman and the $500,000 inventory.

Source: An excerpt from Myron Magnet, "The New Golden Rule of Business," *Fortune*, February 21, 1994, pp. 60–64. Time, Inc.

The just-in-time inventory control system lowers inventory by reducing the time and distance between the various production functions. The idea behind the system is that the firm should keep a minimum level of inventory on hand and rely upon suppliers to furnish parts "just-in-time" for them to be assembled.

Total Quality Management

The adoption of a total quality management or TQM philosophy by many companies has produced dramatic changes in the way the purchasing portion of inventory management is handled. The traditional adversarial purchaser/supplier relationship has given way to close customer relationships, which have in turn helped trim costs by allowing for the production of higher-quality products. This close customer-supplier relationship has allowed the TQM philosophy to be passed across company boundaries to the supplier, and has also allowed the firm to tap the supplier's expertise in designing higher-quality products. In addition, this interdependence between the supplier and customer has allowed for the development and introduction of new products much quicker than previously possible. The use of single-sourcing (where a company uses a very few suppliers or, in many cases, a single supplier, as a source for a particular part or material) has helped align the interests of the supplier and customer.

The movement toward a policy of 100 percent quality has been fuelled by the realization that quality can be used as a means of differentiating products. The TQM view concludes that traditional quality-cost analysis underestimates the cost of producing a poor-quality product, while it overestimates the cost of producing a high-quality product. As a result, under this new TQM view of the quality-costs tradeoff, the low point in the total cost curve becomes the optimal-quality level of 100 percent.

STUDY QUESTIONS

19-1. What factors determine the size of the investment a firm makes in accounts receivable? Which of these factors are under the control of the financial manager?

19-2. What do the following trade credit terms mean?

a. 1/20, net 50

b. 2/30, net 60

c. net 30

d. 2/10, 1/30, net 60

19-3. What is the purpose of the use of an aging account in the control of accounts receivable? Can this same function be performed through ratio analysis? Why or why not?

19-4. If a credit manager experienced no bad debt losses over the past year, would this be an indication of proper credit management? Why or why not?

19-5. What is the purpose of credit scoring?

19-6. What are the risk-return tradeoffs associated with adopting a more liberal trade credit policy?

19-7. Explain the purpose of marginal analysis.

19-8. What is the purpose of holding inventory? Name several types of inventory and describe their purpose.

19-9. Can cash be considered a special type of inventory? If so, what functions does it attempt to uncouple?

19-10. In order to control investment in inventory effectively, what two questions must be answered?

19-11. What are the major assumptions made by the EOQ model?

19-12. What are the risk-return tradeoffs associated with inventory management?

19-13. How might inflation affect the EOQ model?

19-14. How do single-sourcing and closer customer-supplier relationships contribute to the firm?

19-15. What does the new TQM view of the quality-cost relationship say is misstated by the traditional economic view of tradeoffs between quality and cost, as presented in Figure 19-7?

SELF-TEST PROBLEMS

ST-1. *(EOQ Calculations)* A local gift shop is attempting to determine how many sets of wine glasses to order. The store feels it will sell approximately 800 sets in the next year at a price of $18 per set. The wholesale price that the store pays per set is $12. Costs for carrying one set of wine glasses are estimated at $1.50 per year while ordering costs are estimated at $25.00.

a. What is the economic order quantity for the sets of wine glasses?

b. What are the annual inventory costs for the firm if it orders in this quantity? (Assume constant demand and instantaneous delivery and thus no safety stock is carried.)

ST-2. *(EOQ Calculations)* Given the following inventory information and relationships for the F. Beamer Corporation:

1. Orders can be placed only in multiples of 100 units.
2. Annual unit usage is 300,000. (Assume a 50-week year in your calculations.)
3. The carrying cost is 30 percent of the purchase price of the goods.
4. The purchase price is $10 per unit.
5. The ordering cost is $50 per order.
6. The desired safety stock is 1,000 units. (This does not include delivery-time stock.)
7. Delivery time is two weeks.

Given this information,

a. What is the optimal EOQ level?
b. How many orders will be placed annually?
c. At what inventory level should a reorder be made?

STUDY PROBLEMS (SET A)

19-1A. *(Trade Credit Discounts)* If a firm buys on trade credit terms of 2/10, net 50 and decides to forgo the trade credit discount and pay on the net day, what is the effective annualized cost of forgoing the discount?

19-2A. *(Trade Credit Discounts)* If a firm buys on trade credit terms of 2/10, net 30 and decides to forgo the trade credit discount and pay on the net day, what is the effective annualized cost of forgoing the discount?

19-3A. *(Trade Credit Discounts)* Determine the effective annualized cost of forgoing the trade credit discount on the following terms:

a. 1/10, net 20
b. 2/10, net 30
c. 3/10, net 30
d. 3/10, net 60
e. 3/10, net 90
f. 5/10, net 60

19-4A. *(Altman Model)* The following ratios were supplied by six loan applicants. Given this information and the credit-scoring model developed by Altman [Equation (19-2)], which loans have a high probability of defaulting next year and thus should be avoided?

	EBIT / Total Assets	Sales / Total Assets	Market Value of Equity / Book Value of Debt	Earnings / Total Assets	Working Capital / Total Assets
Applicant 1	.2	.2	1.2	.3	.5
Applicant 2	.2	.8	1.0	.3	.8
Applicant 3	.2	.7	.6	.3	.4
Applicant 4	.1	.4	1.2	.4	.4
Applicant 5	.3	.7	.5	.4	.7
Applicant 6	.2	.5	.5	.4	.4

19-5A. (Ratio Analysis) Assuming a 365-day year, calculate what the average investment in inventory would be for a firm, given the following information in each case:

a. The firm has sales of $600,000, a gross profit margin of 10 percent, and an inventory turnover ratio of 6.
b. The firm has a cost of goods sold figure of $480,000 and an average age of inventory of 40.56 days.

c. The firm has a cost of goods sold figure of $1,150,000 and an inventory turnover ratio of 5.

d. The firm has a sales figure of $25 million, a gross profit margin of 14 percent, and an average age of inventory of 45 days.

19-6A. (*Marginal Analysis*) the Bandwagonesque Corporation is considering relaxing its current credit policy. Currently, the firm has annual sales (all credit) of $5 million and an average collection period of 60 days (assume a 365-day year). Under the proposed change the trade credit terms would be changed from net 60 days to net 90 days and credit would be extended to a more risky class of customer. It is assumed that bad debt losses on current customers will remain at their current level. Under this change, it is expected that sales will increase to $6,000,000. Given the following information, should the firm adopt the new policy?

New sales level (all credit):	$6,000,000
Original sales level (all credit):	$5,000,000
Contribution margin:	20%
Percent bad debt losses on new sales:	8%
New average collection period:	90 days
Original average collection period:	60 days
Additional investment in inventory:	$50,000
Pre-tax required rate of return:	15%

19-7A. *(Marginal Analysis)* The Foxbase Alpha Corporation is considering a major change in credit policy. The new credit policy requires extending credit to a riskier class of customer and extending their credit period from net 30 days to net 45 days. They do not expect bad debt losses on their current customers to change. Given the following information, should they go ahead with the change in credit policy?

New sales level (all credit):	$12,500,000
Original sales level (all credit):	$11,000,000
Contribution margin:	20%
Percent bad debt losses on new sales:	9%
New average collection period:	45 days
Original average collection period:	30 days
Additional investment in inventory:	$75,000
Pre-tax required rate of return:	15%

19-8A. *(Marginal Analysis)* Nirvana Inc. has annual sales of $5 million. All sales are credit and the current credit terms are 1/50, net 70. The company is studying the possibility of relaxing credit terms to 2/60, net 90 in hopes of securing new sales. Management does not expect bad debt losses on the current customers to change under the new credit policy. Given the following information, should the firm go ahead with the change in credit policy?

New sales level (all credit):	$8,000,000
Original sales level (all credit):	$7,000,000
Contribution margin:	25%
Percent bad debt losses on new sales:	8%
New average collection period:	75 days
Original average collection period:	60 days
Additional investment in inventory:	$50,000
Pre-tax required rate of return:	15%
New percent cash discount:	2%
Percent of customers taking the cash discount:	50%
Old percent cash discount:	1%
Percent of customers taking the old cash discount:	50%

19-9A. *(EOQ Calculations)* A downtown bookstore is trying to determine the optimal order quantity for a popular novel just printed in paperback. The store feels that the book will sell at four times its hardback figures. It would therefore sell approximately 3,000 copies in the next year at a price of $1.50. The store buys the book at a wholesale figure of $1. Costs for carrying the book are estimated at $.10 a copy per year, and it costs $10 to order more books.

- **a.** Determine the economic order quantity.
- **b.** What would be the total costs for ordering the books 1, 4, 5, 10, and 15 times a year?
- **c.** What questionable assumptions are being made by the EOQ model?

19-10A. *(EOQ Calculations)* The local hamburger fast-food restaurant purchases 20,000 boxes of hamburger rolls every month. Order costs are $50 an order, and it costs $.25 a box for storage.

- **a.** What is the optimal order quantity of hamburger rolls for this restaurant?
- **b.** What questionable assumptions are being made by the EOQ model?

19-11A. *(EOQ Calculations)* A local car manufacturing plant has a $75 per-unit per-year carrying cost on a certain item in inventory. This item is used at a rate of 50,000 per year. Ordering costs are $500 per order.

- **a.** What is the economic order quantity for this item?
- **b.** What are the annual inventory costs for this firm if it orders in this quantity? (Assume constant demand and instantaneous delivery.)

19-12A. *(EOQ Calculations)* Swank Products is involved in the production of camera parts and has the following inventory, carrying, and storage costs:

1. Orders must be placed in round lots of 200 units.
2. Annual unit usage is 500,000. (Assume a 50-week year in your calculations.)
3. The carrying cost is 20 percent of the purchase price.
4. The purchase price $2 per unit.
5. The ordering cost is $90 per order.
6. The desired safety stock is 15,000 units. (This does not include delivery time stock.)
7. The delivery time is one week.

Given the above information

- **a.** Determine the optimal EOQ level.
- **b.** How many orders will be placed annually?
- **c.** What is the inventory order point? (That is, at what level of inventory should a new order be placed?)
- **d.** What is the average inventory level?

19-13A. *(EOQ Calculations)* Regina Distributors has determined the following inventory information and relationships:

1. Orders can be placed only in multiples of 200 units.
2. Annual unit usage is 500,000 units. (Assume a 50-week year in your calculations.)
3. The carrying cost is 10 percent of the purchase price of the goods.
4. The purchase price is $5 per unit.
5. The ordering cost is $100 per order.
6. The desired safety stock is 5,000 units. (This does not include delivery time stock.)
7. Delivery time is four weeks.

Given this information

- **a.** What is the EOQ level?
- **b.** How many orders will be placed annually?
- **c.** At what inventory level should a reorder be made?
- **d.** Now assume the carrying costs are 50 percent of the purchase price of the goods and recalculate (a), (b), and (c). Are these the results you anticipated?

19-14A. *(Comprehensive EOQ Calculations)* Knutson Products Inc. is involved in the production of airplane parts and has the following inventory, carrying, and storage costs:

1. Orders must be placed in round lots of 100 units.
2. Annual unit usage is 250,000. (Assume a 50-week year in your calculations.)
3. The carrying cost is 10 percent of the purchase price.

4. The purchase price is $10 per unit.
5. The ordering cost is $100 per order.
6. The desired safety stock is 5,000 units. (This does not include delivery time stock.)
7. The delivery time is one week.

Given the above information

a. Determine the optimal EOQ level.

b. How many orders will be placed annually?

c. What is the inventory order point? (That is, at what level of inventory should a new order be placed?)

d. What is the average inventory level?

e. What would happen to the EOQ if annual unit sales doubled (all other unit costs and safety stocks remaining constant)? What is the elasticity of EOQ with respect to sales? (That is, what is the percent change in EOQ divided by the percent change in sales?)

f. If carrying costs double, what would happen to the EOQ level? (Assume the original sales level of 250,000 units.) What is the elasticity of EOQ with respect to carrying costs?

g. If the ordering costs double, what would happen to the level of EOQ? (Again assume original levels of sales and carrying costs.) What is the elasticity of EOQ with respect to ordering costs?

h. If the selling price doubles, what would happen to the EOQ? What is the elasticity of EOQ with respect to selling price?

INTEGRATIVE PROBLEM

Your first major assignment after your recent promotion at Ice Nine involves overseeing the management of accounts receivable and inventory. The first item that you must attend to involves a proposed change in credit policy that would relax credit terms from the existing terms of 1/50, net 70 to 2/60, net 90 in hopes of securing new sales. The management at Ice Nine does not expect bad debt losses on their current customers to change under the new credit policy. The following information should aid you in the analysis of this problem.

New sales level (all credit):	$8,000,000
Original sales level (all credit):	$7,000,000
Contribution Margin:	25%
Percent bad debt losses on new sales:	8%
New average collection period:	75 days
Original average collection period:	60 days
Additional investment in inventory:	$50,000
Pre-tax required rate of return:	15%
New percent cash discount:	2%
Percent of customers taking the cash discount:	50%
Old percent cash discount:	1%
Percent of customers taking the old cash discount:	50%

To help in your decision on relaxing credit terms, you have been asked to respond to the following questions:

a. What factors determine the size of investment Ice Nine makes in accounts receivable?

b. If a firm currently buys from Ice Nine on trade credit with the present terms of 1/50, net 70 and decides to forgo the trade credit discount and pay on the net day, what is the effective annualized cost to that firm of forgoing the discount?

c. If Ice Nine changes its trade credit terms to 2/60, net 90 what is the effective annualized cost to a firm that buys on credit from Ice Nine and decides to forgo the trade credit discount and pay on the next day?

d. What is the estimated change in profits resulting from the increased sales less any additional bad debts associated with the proposed change in credit policy?

e. Estimate the cost of additional investment in accounts receivable and inventory associated with this change in credit policy.

f. Estimate the change in the cost of the cash discount if the proposed change in credit policy is enacted.

g. Compare the incremental revenues with the incremental costs. Should the proposed change be enacted?

You have also been asked to deal with some questions involving inventory management at Ice Nine. Presently, Ice Nine is involved in the production of musical products with its German-engineered Daedlufetarg music line. Production of this product involves the following inventory, carrying, and storage costs:

1. Orders must be placed in round lots of 100 units.
2. Annual unit usage is 250,000. (Assume a 50-week year in your calculations.)
3. The carrying cost is 10 percent of the purchase price.
4. The purchase price is $10 per unit.
5. The ordering cost is $100 per order.
6. The desired safety stock is 5,000 units. (This does not include delivery-time stock.)
7. The delivery time is one week.

Given the above information

h. Determine the optimal EOQ level.

i. How many orders will be placed annually?

j. What is the inventory order point? (That is, at what level of inventory should a new order be placed?)

k. What is the average inventory level?

l. What would happen to the EOQ if annual unit sales doubled (all other unit costs and safety stocks remaining constant)? What is the elasticity of EOQ with respect to sales? (That is, what is the percent change in EOQ divided by the percent change in sales?)

m. If carrying costs double, what would happen to the EOQ level? (Assume the original sales level of 250,000 units.) What is the elasticity of EOQ with respect to carrying costs?

n. If the ordering costs double, what would happen to the level of EOQ? (Again assume original levels of sales and carrying costs.) What is the elasticity of EOQ with respect to ordering costs?

o. If the selling price doubles, what would happen to the EOQ? What is the elasticity of EOQ with respect to selling price?

p. What assumptions are being made by the EOQ model that has been used here?

q. How would the results of this model change if the carrying cost were to increase, perhaps because of increased inflation?

r. How would an improvement in the relationship that Ice Nine has with its suppliers resulting in a decrease in the average delivery time for replenishment of inventory affect your answer?

s. If Ice Nine could decrease its ordering costs, perhaps by improving its relationship with suppliers, how would this affect your answer?

STUDY PROBLEMS (SET B)

19-1B. *(Trade Credit Discounts)* If a firm buys on trade credit terms of 2/10, net 60 and decides to forgo the trade credit discount and pay on the net day, what is the effective annualized cost of forgoing the discount?

19-2B. *(Trade Credit Discounts)* If a firm buys on trade credit terms of 2/10, net 40 and decides to forgo the trade credit discount and pay on the net day, what is the effective annualized cost of forgoing the discount?

19-3B. *(Trade Credit Discounts)* Determine the effective annualized cost of forgoing the trade credit discount on the following terms:

- **a.** 1/5, net 20
- **b.** 2/20, net 90
- **c.** 1/20, net 100
- **d.** 4/10, net 50
- **e.** 5/20, net 100
- **f.** 5/30, net 50

19-4B. *(Altman Model)* The following ratios were supplied by six loan applicants. Given this information and the credit-scoring model developed by Altman [Equation (19-2)], which loans have a high probability of defaulting next year and thus should be avoided?

	EBIT / Total Assets	Sales / Total Assets	Market Value of Equity / Book Value of Debt	Earnings / Total Assets	Working Capital / Total Assets
Applicant 1	.3	.4	1.2	.3	.5
Applicant 2	.2	.6	1.3	.4	.3
Applicant 3	.2	.7	.6	.3	.2
Applicant 4	.1	.5	.8	.5	.4
Applicant 5	.5	.7	.5	.4	.6
Applicant 6	.2	.4	.2	.4	.4

19-5B. (Ratio Analysis) Assuming a 365-day year, calculate what the average investment in inventory would be for a firm, given the following information in each case.

- **a.** The firm has sales of $550,000, a gross profit margin of 10 percent, and an inventory turnover ratio of 5.
- **b.** The firm has a cost of goods sold figure of $480,000 and an average age of inventory of 35 days.
- **c.** The firm has a cost of goods sold figure of $1,250,000 and an inventory turnover ratio of 6.
- **d.** The firm has a sales figure of $25 million, a gross profit margin of 15 percent, and an average age of inventory of 50 days.

19-6B. *(Marginal Analysis)* The Hyndford Street Corporation is considering relaxing its current credit policy. Currently, the firm has annual sales (all credit) of $6 million and an average collection period of 40 days (assume a 365-day year). Under the proposed change the trade credit terms would be changed from net 40 days to net 90 days and credit would be extended to a more risky class of customer. It is assumed that bad debt losses on current customers will remain at their current level. Under this change, it is expected that sales will increase to $7,000,000. Given the following information, should the firm adopt the new policy?

New sales level (all credit):	$7,000,000
Original sales level (all credit):	$6,000,000
Contribution margin:	20%
Percent bad debt losses on new sales:	8%
New average collection period:	90 days
Original average collection period:	40 days
Additional investment in inventory:	$40,000
Pre-tax required rate of return:	15%

19-7B. *(Marginal Analysis)* The Northern Muse Corporation is considering a major change in credit policy. The new credit policy requires extending credit to a riskier class of customer and extending their credit period from net 30 days to net 50 days. They do not expect bad debt losses on their current customers to change. Given the following information, should they go ahead with the change in credit policy?

New sales level (all credit):	$18,000,000
Original sales level (all credit):	$17,000,000
Contribution margin:	20%
Percent bad debt losses on new sales:	8%
New average collection period:	50 days
Original average collection period:	30 days
Additional investment in inventory:	$60,000
Pre-tax required rate of return:	15%

19-8B. *(Marginal Analysis)* Nirvana Inc. has annual sales of $10 million. All sales are credit and the current credit terms are 1/30, net 60. The company is studying the possibility of relaxing credit terms to 2/60, net 90 in hopes of securing new sales. Management does not expect bad debt losses on the current customers to change under the new credit policy. Given the following information, should the firm go ahead with the change in credit policy?

New sales level (all credit):	$10,500,000
Original sales level (all credit):	$10,000,000
Contribution margin:	25%
Percent bad debt losses on new sales:	8%
New average collection period:	75 days
Original average collection period:	45 days
Additional investment in inventory:	$60,000
Pre-tax required rate of return:	15%
New percent cash discount:	2%
Percent of customers taking the cash discount:	50%
Old percent cash discount:	1%
Percent of customers taking the old cash discount:	50%

19-9B. *(EOQ Calculations)* A downtown bookstore is trying to determine the optimal order quantity for a popular novel just printed in paperback. The store feels that the book will sell at four times its hardback figures. It would therefore sell approximately 3,500 copies in the next year at a price of $1.50. The store buys the book at a wholesale figure of $1. Costs for carrying the book are estimated at $.20 a copy per year, and it costs $9 to order more books.

- **a.** Determine the economic order quantity.
- **b.** What would be the total costs for ordering the books 1, 4, 5, 10, and 15 times a year?
- **c.** What questionable assumptions are being made by the EOQ model?

19-10B. *(EOQ Calculations)* The local hamburger fast-food restaurant purchases 21,000 boxes of hamburger rolls every month. Order costs are $55 an order, and it costs $.20 a box for storage.

- **a.** What is the optimal order quantity of hamburger rolls for this restaurant?
- **b.** What questionable assumptions are being made by the EOQ model?

19-11B. *(EOQ Calculations)* A local car manufacturing plant has a $70 per-unit per-year carrying cost on a certain item in inventory. This item is used at a rate of 55,000 per year. Ordering costs are $500 per order.

- **a.** What is the economic order quantity for this item?
- **b.** What are the annual inventory costs for this firm if it orders in this quantity? (Assume constant demand and instantaneous delivery.)

19-12B. *(EOQ Calculations)* Swank Products is involved in the production of camera parts and has the following inventory, carrying, and storage costs:

1. Orders must be placed in round lots of 200 units.
2. Annual unit usage is 600,000. (Assume a 50-week year in your calculations.)
3. The carrying cost is 15 percent of the purchase price.
4. The purchase price is $3 per unit.
5. The ordering cost is $90 per order.
6. The desired safety stock is 15,000 units. (This does not include delivery time stock.)
7. The delivery time is one week.

Given the above information

- **a.** Determine the optimal EOQ level.
- **b.** How many orders will be placed annually?
- **c.** What is the inventory order point? (That is, at what level of inventory should a new order be placed?)
- **d.** What is the average inventory level?

19-13B. *(EOQ Calculations)* Regina Distributors has determined the following inventory information and relationships:

1. Orders can be placed only in multiples of 200 units.
2. Annual unit usage is 500,000 units. (Assume a 50-week year in your calculations.)
3. The carrying cost is 9 percent of the purchase price of the goods.
4. The purchase price is $5 per unit.
5. The ordering cost is $75 per order.
6. The desired safety stock is 5,000 units. (This does not include delivery time stock.)
7. Delivery time is four weeks.

Given this information

- **a.** What is the EOQ level?
- **b.** How many orders will be placed annually?
- **c.** At what inventory level should a reorder be made?
- **d.** Now assume the carrying costs are 50 percent of the purchase price of the goods and recalculate (a), (b), and (c). Are these the results you anticipated?

19-14B. *(Comprehensive EOQ Calculations)* Good Gravy Products Inc. is involved in the production of tractor parts and has the following inventory, carrying, and storage costs:

1. Orders must be placed in round lots of 100 units.
2. Annual unit usage is 300,000. (Assume a 50-week year in your calculations.)
3. The carrying cost is 10 percent of the purchase price.
4. The purchase price is $12 per unit.
5. The ordering cost is $100 per order.
6. The desired safety stock is 4,000 units. (This does not include delivery time stock.)
7. The delivery time is one week.

Given the above information

- **a.** Determine the optimal EOQ level.
- **b.** How many orders will be placed annually?
- **c.** What is the inventory order point? (That is, at what level of inventory should a new order be placed?)
- **d.** What is the average inventory level?
- **e.** What would happen to the EOQ if annual unit sales doubled (all other unit costs and safety stocks remaining constant)? What is the elasticity of EOQ with respect to sales? (That is, what is the percent change in EOQ divided by the percent change in sales?)
- **f.** If carrying costs double, what would happen to the EOQ level? (Assume the original sales level of 250,000 units.) What is the elasticity of EOQ with respect to carrying costs?
- **g.** If the ordering costs double, what would happen to the level of EOQ? (Again assume original levels of sales and carrying costs.) What is the elasticity of EOQ with respect to ordering costs?
- **h.** If the selling price doubles, what would happen to EOQ? What is the elasticity of EOQ with respect to selling price?

SELF-TEST SOLUTIONS

SS-1. **a.** The economic order quantity is

$$Q^* = \sqrt{\frac{2SO}{C}}$$

where, S = total demand in units over the planning period
O = ordering cost per order
C = carrying costs per unit

Substituting the values given in the self-test problem into the EOQ equation we get

$$Q^* = \sqrt{\frac{2 \cdot 800 \cdot 25}{1.50}}$$

$$= \sqrt{26,667}$$

$$= 163 \text{ units per order}$$

Thus, 163 units should be ordered each time an order is placed. Note that the EOQ calculations take place based upon several limiting assumptions such as constant demand, constant unit price, and constant carrying costs, which may influence the final decision.

b. Total costs = carrying costs + ordering costs

$$= \left(\frac{Q}{2}\right)C + \left(\frac{S}{Q}\right)O$$

$$= \left(\frac{163}{2}\right)\$1.50 + \left(\frac{800}{163}\right)\$25$$

$$= \$122.25 + \$122.70$$

$$= \$244.95$$

Note that carrying costs and ordering costs are the same (other than a slight difference caused by having to order in whole rather than fractional units). This is because the total costs curve is at its minimum when ordering costs equal carrying costs, as shown in Figure 19-5.

SS-2. a.

$$EOQ = \sqrt{\frac{2SO}{C}}$$

$$= \sqrt{\frac{2(300{,}000)(50)}{3}}$$

= 3,162 units, but since orders must be placed in 100-unit lots, the effective EOQ becomes 3,200 units

b.

$$\frac{\text{Total usage}}{EOQ} = \frac{300{,}000}{3{,}200} = 93.75 \text{ orders per year}$$

c. Inventory order point = delivery time + safety stock

$$= \frac{2}{50} \times 300{,}000 + 1{,}000$$

$$= 12{,}000 + 1{,}000$$

$$= 13{,}000 \text{ units}$$

EVALUATING FINANCIAL PERFORMANCE

CHAPTER 20

LEARNING OBJECTIVES

After reading this chapter you should be able to

1. Construct and analyze a firm's basic financial statements, including the balance sheet, the income statement, and the statement of changes in financial position.
2. Measure a firm's cash flow using information contained in a firm's reported financial statements.
3. Calculate a comprehensive set of financial ratios and use them to evaluate the financial health of a company.
4. Apply the Du Pont analysis in evaluating the firm's performance.
5. Explain the limitations of ratio analysis.

INTRODUCTION

Evaluating the performance of a firm using its financial statements can be a tricky business. The difficulty is generally not due to deliberate attempts by corporate managers and their accountants to mislead you. The problem relates to the substantial flexibility inherent in the set of rules and principles that accountants follow in preparing a firm's financial statements (Generally Accepted Accounting Principles, or GAAP). In fact, the Ontario Securities Commission (OSC) has indicated that its accounting staff is disappointed with the quality of documents being disclosed by Canadian companies.[1] In a survey of 100 annual and interim corporate reports, of which 75 were TSE 300 companies, the OSC indicated that at least 5 of the 100 companies did not adhere to GAAP. Many of these companies showed "highly aggressive" accounting practices by taking large write-downs of goodwill or provisions for restructuring costs which reduce charges against future earnings and as a consequence will result in an overstatement of future earnings. In addition, many of these companies showed significant deficiency in their level of disclosure by not providing any information above minimum requirements. These types of practices cause information, such as net earnings and operating cash flows, to be unreliable when reported in the income statement or the statement of changes in financial position. The OSC, in its function as a regulator recommends that financial information should not be taken at face value. In other words, investors should examine everything that is reported and carefully assess all

[1] Source: Adapted from: " Forecasts and Psychics," *The Financial Post*, September 2, 1995, p.14.

information so as to protect themselves from overly simplistic measures of corporate performance such as net earnings or operating cash flow.[2]

So where does this leave the analyst who attempts to evaluate the financial performance of a firm using its financial statements? The answer is that one must indeed "look through" the numbers in the statements and seek to understand subtle differences in accounting practice and the effect that they can have on reported earnings. Analyzing financial performance using accounting statements is not simply a mechanical process. It requires the analyst not only to "crunch the numbers" but to understand where the numbers come from.

CHAPTER PREVIEW

In this chapter, we review the firm's basic financial statements and discuss the use of the financial ratios to analyze them. The basic financial statements include the income statement, balance sheet, and statement of changes in financial position. We first review the basic format of the firm's financial statements with particular attention given to the statement of changes in financial position. This statement is important in financial analysis because it focuses on cash rather than income or profits. We will examine how the analysis of ratios can provide answers to questions concerning the firm's financial position. These questions focus on the following topics: liquidity, efficiency, operating profitability, financial decisions and return on equity. Finally, we apply an integrative approach to ratio analysis known as DuPont analysis.

BACK TO THE BASICS

Two axioms are especially important in this chapter: ***Axiom 3*** *tells us that* ***Cash—Not Profits—Is King****. At times, cash is more important than profits. Thus, in this chapter, considerable time will be devoted to learning how to measure cash flows.* ***Axiom 7*** *warns us there may be a conflict between management and owners, especially in large firms in which managers and owners have different incentives. That is,* ***Managers Won't Work for the Owners Unless It's in Their Best Interest.*** *In this chapter, we will learn how to use data from the firm's public financial statements to monitor management's actions.*

OBJECTIVE 1

BASIC FINANCIAL STATEMENTS

Income statement

The statement of profit or loss for the period is compromised of net revenues less expenses for the period.

An important part of a financial executive's work is to provide financial information to the firm's stakeholders. The disclosure of financial information enables management to discharge their accountability for the firm's resources to the firm's stakeholders. In other words, the financial statements contain important information about the firm's financial performance in the form of (1) **an income statement**, (2) **a balance sheet**, and (3) **a statement of changes in financial position**. We will examine each of these statements in turn.

Each of various forms of business organization which we examined in Chapter 2, report their financial performance and position through the use of three basic

[2] Source: Adapted from: "Look Beyond Corporate Reports For The True Story," *The Financial Post*, August 31, 1995, p.8.

financial statements: the balance sheet, income statement, and statement of retained earnings. However, in accordance with Canadian Generally Accepted Accounting Principles (GAAP), a business will report a fourth statement known as the statement of changes in financial position. This statement is often used by financial executives to determine the cash position of the firm.

The balance sheet and income statement for Jimco Inc., are found in Tables 20-1 and 20-2, respectively. Appendix 20A contains a review of important concepts concerning the balance sheet and income statement. We begin our discussion on financial analysis by determining the firm's cash balance through the statement of changes in financial position. The construction of this statement requires information from both the balance sheet and the income statement.

OBJECTIVE 1

Balance sheet

A statement of financial position at a particular date. The form of the statement follows the balance sheet equation: total assets = total liabilities + owner's equity

Statement of changes in financial position

The statement of changes in financial position enumerates the cash receipts and cash disbursements for a specified interval of time (usually one year).

PERSPECTIVE IN FINANCE

The double-entry system of accounting used in this country dates back to about 3600 B.C., with the first published work describing the double-entry accounting system in 1494 by Luca Pacioli in Venice. Although the details of the double-entry system can be overwhelming to the novice, the mathematical content of the basic financial statements is straightforward, as we see below:

1. Balance sheet or statement of financial position:

Assets = Liabilities + Owners' Equity

2. Income statement or statement of results from operations:

Revenues + Gains − Expenses − Losses = Income

3. Statement of changes in financial position:

Cash Inflow − Cash Outflow = Change in Cash

Statement of Changes in Financial Position

The CICA committee on accounting standards has defined the objective of the statement of changes in financial position to be that of providing information about the operating, financing, and investing activities of an enterprise and the effects of those activities on cash resources.[3] Other suitable titles for this statement include: "Cash flow statement," "Statement of operating, financing, and investment activities," or "Statement of changes in cash resources."

For the purposes of financial reporting, the statement of changes in financial position enables the financial executive to evaluate the solvency and liquidity of an

Table 20-1 Jimco Inc. Balance Sheet December 31, 1995 ($000)

Assets			
Current assets:			
Cash		$ 1,400	
Marketable securities—at cost (market value, $320)		300	
Accounts receivable		10,000	
Inventories		12,000	
Prepaid expenses		300	
Total current assets			$24,000
Fixed assets:			
Land		2,000	
Plant and equipment	$12,300		
Less: Accumulated depreciation	7,300		
Net plant and equipment		5,000	
Total fixed assets			7,000
Total assets			$31,000

[3]*CICA Handbook*, Statement of Changes in Financial Position, 1540.01.

Table 20-1 Jimco Inc. Balance Sheet December 31, 1995 ($000) (*continued*)

Liabilities and Owners' Equity		
Current liabilities:		
Accounts payable	$3,000	
Notes payable, 9%, due March 1, 1996	3,400	
Accrued salaries, wages, and other expenses	3,100	
Current portion of long-term debt	500	
Total current liabilities		$10,000
Long-term liabilities:		
Deferred income taxes	1,500	
First mortgage bonds, 7%, due January 1, 2000	6,300	
Debentures, 8%, due June 30, 2000	2,900	
Total long-term liabilities		10,700
Owners' equity:		
Common shares	100	
Contributed surplus	2,000	
Retained earnings	8,300	
Total owners' equity		10,300
Total liabilities and owners' equity		$31,000

Table 20-2 Jimco Inc. Statement of Income for the Year Ended December 31, 1995 ($000 except per-share data)

Net sales		$51,000
Cost of goods sold		(38,000)
Gross profit		$13,000
Operating expenses		
Selling expenses	$3,100	
Depreciation expense	500	
General and administrative expense	5,400	(9,000)
Net operating income (NOI)		$ 4,000
Interest expense		(1,000)
Earnings before taxes (EBT)		$ 3,000
Income taxes[a]		(1,200)
Net income (NI)		$ 1,800
Disposition of net income		
Common share dividends		$ 300
Change in retained earnings		1,500
Per-share data (dollars)		
Number of common shares		100,000 shares
Earnings per common share ($1,800,000/100,000 shares)		$ 18
Dividends per common share ($300,000/100,000 shares)		$ 3

[a]A tax rate of 40 percent on all income is assumed here for simplicity.

enterprise; the liquid resources of a firm will be measured in terms of cash and its cash equivalents.[4] To ensure measurement of these types of resources, this statement is structured to reflect the cash flows involved in the operating, financing, and investing activities of the firm. These three activities of the firm include items

[4]Cash equivalents are defined as short-term, highly liquid investments that are readily convertible into known amounts of cash and are so near to their maturity that they pose an insignificant risk of changes in value because of changes in interest rates.

such as: cash flows resulting from discontinued operations; cash flows from extraordinary items; outlays for acquisition of assets and proceeds from the disposal of assets; the issue and redemption of both debt and share capital; and the payment of dividends. As a result, this statement enables an investor to assess the firm's ability to generate cash from internal sources, to repay debt obligations, to reinvest and to make distributions to owners.

Preparing the Statement of Changes in Financial Position

OBJECTIVE 2

In Canada, most companies will construct their statements of changes of financial position by reconciling both cash and noncash items to net income. Table 20-3 shows how to prepare the statement of changes in 1995 from the comparative balance sheets of both 1994 and 1995. The purpose of comparing the various accounts of these two balance sheets is to determine how the differences in the account balances over the last fiscal period affect the firm's cash balance. For example, a

Table 20-3 Jimco Inc. Comparative Balance Sheets December 31, 1994 and 1995 ($000)

Assets	1994	1995	Changes
Current assets:			
Cash	$ 1,500	$ 1,400	$ (100)
Marketable securities	300	300	—
Accounts receivable	8,500	10,000	1,500
Inventories	11,300	12,000	700
Prepaid expenses	200	300	100
Total current assets	$21,800	$24,000	$2,200
Fixed assets:			
Land	$ 2,000	$ 2,000	$ —
Plant and equipment	11,200	12,300	1,100
Less: Accumulated depreciation	(6,800)	(7,300)	(500)
Net plant and equipment	4,400	5,000	600
Total fixed assets	6,400	7,000	600
Total assets	$28,200	$31,000	$2,800

Liabilities and Owners' Equity	1994	1995	Changes
Current liabilities:			
Accounts payable	$ 3,200	$ 3,000	$ (200)
Notes payable	900	3,400	2,500
Accrued salaries, wages, and other expenses	3,800	3,100	(700)
Current portion of long-term debt	500	500	—
Total current liabilities	$ 8,400	$10,000	$ 1,600
Long-term liabilities:			
Deferred income taxes	$ 1,400	$ 1,500	$100
First mortgage bonds	6,600	6,300	(300)
Debenture bonds	3,000	2,900	(100)
Total long-term liabilities	$11,000	$10,700	$ (300)
Owners' equity:			
Common shares	$ 100	$ 100	—
Contributed surplus	2,000	2,000	—
Retained earnings	6,700	8,200	1,500
Total owners' equity	$ 8,800	$10,300	$1,500
Total liabilities and owners' equity	$28,200	$31,000	$2,800

decrease in the level of accounts receivable over the period denotes an increase in the cash level of the firm, since in this case the firm collects more dollars from its credit accounts than it creates through new credit sales. In general, a decrease in a noncash asset balance indicates an increase in the firm's cash balance or a source of cash flow. Furthermore, an increase in a liability account signals that a net additional borrowing took place in the firm's cash balance. An increase in the levels of the common and preferred shares accounts would also increase the firm's cash balance. In contrast, a decrease in the level of the accounts payable over the period indicates a decrease in the cash level of the firm, since in this case the firm is paying its creditors immediately and is not making use of short-term credit. In general, a decrease in the liability balance would indicate a decrease in a firm's cash balance or a use of cash flow.

The column entitled "changes" in Table 20-3 provides the basis for determining Jimco's inflows (sources) and outflows (uses) of cash. Note that the balance of the cash account has decreased by $100,000 during the period. This decrease in the balance of the cash account is explained by preparing the statement of changes in financial position. Remember that the cash balance shown on 1995's balance sheet is also the cash balance at the end of the period for 1995's statement of changes in financial position; in other words, the cash balance on 1995's balance sheet can be used as a check for the accuracy of 1995's statement of changes of financial position. Table 20-3 shows no change in the balances for marketable securities. Cash and cash equivalents include cash net of short-term borrowing and temporary investments. Treasury bills, commercial paper, marketable money funds and short-term loans are examples which satisfy the criteria of liquidity and qualify as temporary investments and short-term borrowing.[5] Any buy or sell transactions in these types of temporary investments are not considered investing activities and are classified within the balance of cash and cash equivalents.

Net Cash Flows from Operating Activities

In order to reconcile net income to net cash flows from operating activities, we arbitrarily divide the items to be reconciled into two types: (1) noncash items such as depreciation, depletion, amortization, deferred income taxes, minority interest share of income or losses of subsidiaries, losses or gains from equity investments (associate companies), writedown of assets and restructuring costs, amortization of exchange losses or gains, and gains or losses from sales of businesses, investments and properties; and (2) changes in the balances of any working-capital accounts such as accounts receivable, inventories, prepaid expenses, accounts payable, accrued liabilities and income taxes payable or other taxes payable.

In this example, both the income statement and balance sheet indicate that the noncash items to be reconciled with net income include the depreciation expense of $500,000 and an increase in deferred income taxes of $100,000. Note that in this case, the increase in accumulated depreciation of $500,000 equals the depreciation expense for the period. In addition, any items of working capital which show changes in their balances must be reconciled to net income. Table 20-3 shows that the accounts receivable balance increased by $1,500,000, indicating that more credit sales were made during the period than were collected. Hence, the firm has postponed the immediate use of some of its cash inflows by granting credit to its customers in the form of accounts receivable. Likewise, inventories and prepaid expenses increased by $700,000 and $100,000, respectively. Each of these working-capital items represents an example of how the firm has postponed the immediate conversion of these assets into cash flows. Because these assets have not been

[5] Lanny G. Chasteen, Richard E. Flaherty, Melvin C. O'Connor and M. Teresa Anderson, *Intermediate Accounting* (Whitby, ON: McGraw-Hill Ryerson Limited, First Canadian Edition, 1992), pp. 933–934.

immediately converted to cash, the increase in both inventories and prepaid expenses represent examples of how the firm is using its cash. We now examine how the balances of the firm's liabilities have changed during the period. Note that the accounts payable balance has decreased by $200,000. This type of decrease indicates that the firm paid off more accounts payable than it created during the period and constitutes a use of cash flows. The accrued salaries, wages, and other expense accounts decreased by $700,000 during the period; this decrease represents an example of how the firm is using its cash flows.

Net Cash Flows from Financing Activities

Net cash flows from financing activities can arbitrarily be divided into cash inflows and outflows. Examples of cash inflows include proceeds from issuing the firm's shares, debentures, bonds, notes, and long- or short-term borrowing. Conversely, cash outflows include payments to shareholders in the form of cash dividends or to creditors in the form of principal payments to extinguish debt. In this example, notes payable increased by $2,500,000, signalling an inflow (source) of cash from short-term borrowing. However, both the first mortgage bonds and the debenture bonds decreased for the period by $300,000 and $100,000, respectively. Both decreases constitute a cash outflow (use) for the period. During the period, the common shares and contributed surplus accounts did not change which indicates that shares were neither issued nor repurchased. Jimco's retained earnings increased by $1,500,000. This change in retained earnings represents the net income for 1995 of $1,800,000 less common share dividends of $300,000. Since net income is included as a source of cash flows (as a part of cash flows provided by operations) and dividends are an outflow (use) of cash, the change in the retained earnings account is not used directly in preparing the statement of changes in financial position.

Net Cash Flows from Investing Activities

Net cash flows from investing activities can also be arbitrarily divided into cash inflows and cash outflows. Examples of cash inflows include the proceeds from the sale of property, plant, and equipment; and the proceeds from the sale of equity investments and the returns from these investments. In contrast, examples of cash outflows include any payments for either the acquisition of plant, property, and equipment or the acquisition of equity investments in associated companies. Jimco has purchased additional plant and equipment for $1,100,000, which constitutes another outflow (use) of cash.[6]

To summarize, a firm's cash flows are determined by comparing the accounts of the balance sheets between two points in time (for example, between December 31, 1994, and December 31, 1995). We classify cash flows in terms of the following categories: operating activities, investment activities, and financing activities. This procedure also enables us to determine whether the cash flows are either inflows (sources) or outflows (uses). It should be noted that the change in accumulated

[6]The use of funds attributed to the purchase of plant and equipment can also be obtained from an analysis of the change in the net plant and equipment account. For Jimco, Inc., this can be accomplished as follows:

Net plant and equipment (1995)		$5,000
Plus:	Depreciation expense for the period	500
		5,500
Less:	Net plant and equipment (1993)	(4,400)
	Net purchase (sale) of plant and equipment	$1,100

Table 20-4 Jimco Inc. Statement of Changes in Financial Position for the Year Ended December 31, 1995 ($000)

Cash flows from operating activities		
Net income (from the statement of income)	$1,800	
Adjustments for noncash items:		
Depreciation expense	500	
Increase in deferred income taxes	100	
Adjustments for noncash working capital balances:		
Increase in accounts receivable	(1,500)	
Increase in inventories	(700)	
Increase in prepaid expenses	(100)	
Decrease in accounts payable	(200)	
Decrease in accrued wages	(700)	
Net cash outflow from operating activities		($800)
Cash flows from investing activities		
Cash inflows	—	
Cash outflows		
Purchase of plant and equipment	($1,100)	
Net cash outflow from investment activities		($1,100)
Cash flows from financing activities		
Cash inflows		
Increase in notes payable	$2,500	
Cash outflows		
Decrease in mortgage bonds	(300)	
Decrease in debenture bonds	(100)	
Common dividends	(300)	
Net cash inflow from financing activities		$1,800
Effect of foreign exchange rates		—
Net increase (decrease) in cash during the period		($100)
Cash balance at the beginning of the period		$1,500
Cash balance at the end of the period		$1,400

Table 20-5 Schedule for Operating Cash Flows

Cash flows from operating activities	
Cash inflows from operating activities:	
From customers (sales + beginning A/R – ending A/R)	$ 49,500
Cash outflows from operating activities:	
Cash paid to suppliers (cost of goods sold + beginning A/P – ending A/P + beginning inventory – ending inventory)	($38,900)
Cash paid for remaining expenses (selling + gen. adm. expense + beginning accrued expenses – ending accrued expenses + ending prepaid expenses – beginning prepaid expenses)	($9,300)
Cash paid for interest expense (interest expense + beginning accrued interest expense – ending accrued interest expense)	($1,000)
Cash paid for income tax (income tax liability + beginning deferred taxes – ending deferred taxes)	($1,100)
Net cash inflow (outflow) from operating activities	($800)

depreciation in the balance sheet is disregarded. In place of this change, we use the depreciation expense for the year's income statement as a source of cash flows. Also, the change in retained earnings is not included directly in the statement.

Since a change in retained earnings equals net income less dividends paid, we prefer to list these latter items separately instead of a change in retained earnings.

The focus of the statement of changes in financial position is on cash plus cash equivalents. Table 20-4 illustrates the statement of changes in financial position for the year ended December 31, 1995. A review of Jimco's statement of changes in financial position for 1995 indicates that the firm realized a $1,800,000 positive cash flow from its financing activities, which was $100,000 less than the cash outflow experienced from the firm's operating activities plus its investments in plant, machinery, and equipment. Thus, the statement of changes in financial position indicates a net decrease in cash flow of $100,000 which reduces the cash balance to $1,400,000 at the end of the period. Furthermore, the statement of changes in financial position provides the analyst with a useful tool for determining where the firm obtained cash during a prior period and how that cash was spent.

The financial analyst may also wish to isolate the cash flows occurring from operations as shown in Table 20-5. Such a summary enables the analyst to determine the effect of the firm's primary business on cash flows.

FINANCIAL RATIO ANALYSIS

OBJECTIVE 3

Financial analysis is the assessment of a firm's past, present, and anticipated future financial condition. A major objective of financial analysis is to determine the firm's financial strengths and to identify its weaknesses. One of the principal tools of financial analysis is the use of financial ratios. Financial analysts use ratios to answer a variety of questions regarding a firm's financial well-being. This type of information is obtained by comparing financial ratios: (1) over time (say for the last five years) to identify any trends; and (2) among other firms within the industry.

The financial information extracted from a ratio analysis is dependent on the perspective of the financial analyst. For example, a financial analyst who is working for a company will perform an internal financial analysis to determine a firm's liquidity position, its debt capacity, or to examine trends in the performance of the firm's earnings. The emphasis of such an analysis is to flag any important problems so as to ensure that the firm steers away from financial difficulty. Alternatively, a financial analyst working for an investment dealer will perform a financial analysis from a position of arm's length in an effort to determine a firm's creditworthiness or investment potential. In this case, the emphasis of the analysis is to provide an opinion on the quality of the firm as an investment. Regardless of who is performing the analysis, the tools used in financial analysis are basically the same, but the information extracted will be used for different purposes depending on the perspective of the financial analyst.

The comparison for financial ratios should make use of several widely used sources of industry average ratios. Average ratios for both Canadian industrial sectors and industries are published annually by Statistics Canada. In 1996, Statistics Canada began the publication of financial performance indicators which include key ratios for various Canadian industries.[7] In the United States, Dun and Bradstreet publishes on an annual basis a set of 14 key ratios for each of the 125 lines of business. Robert Morris Associates also publishes on an annual basis a set of 16 key ratios for over 350 lines of business.

To make meaningful comparisons using ratios, we must consider the firm's size. Thus, a firm with total assets of less than $1 million should not be compared with firms having a much larger asset base. Moreover, firms which show seasonal sales should be compared on a basis that is more representative of the seasonal activity than comparisons which use an annual basis. As a result, analysts will examine ratios over the four quartiles of the year in order to determine how much variation exists within the industry in regard to each ratio.

Other tools which assist analysts in examining financial performance include the common size balance sheet and the common size income statement. The common size balance sheet simply expresses the level of each asset, liability, and owner's equity account as a percent of total assets, whereas each entry in the common size income statement is reported as a percent of total income. Thus, the levels of accounts can be compared with the levels of other accounts in the balance sheet or in the income statement.

PERSPECTIVE IN FINANCE

Mathematically, a financial ratio is nothing more than a ratio whose numerator and denominator comprise of financial data. Sound simple? Well, in theory, it is. The objective in using a ratio when analyzing financial information is simply to standardize the information being analyzed so that comparisons can be made between ratios of different firms or possibly the same firm at different points in time. Try to keep this in mind as you read through the discussion of financial ratios. All we are doing is trying to standardize financial data so that we can make comparisons with industry norms or other standards.

In learning about ratios, we could simply study the different types or categories of ratios, or we could use ratios to answer some important questions about a firm's operations. We prefer the latter approach and choose the following questions as a map in using financial ratios:

1. How liquid is the firm?
2. How efficient is the firm in using its assets?
3. How is the firm financing its assets?
4. Is the management generating adequate *operating* profits on the firm's assets?
5. Are the owners (shareholders) receiving an adequate return on their investment?

We now examine each of these questions in greater detail.

Question 1: How liquid is the firm?

Liquidity

The ability of a firm to pay its bills on time.

The **liquidity** of a business is defined as its ability to meet maturing debt obligations. In other words, does the firm have the resources to pay creditors when the debt comes due? Current liabilities represent the firm's maturing financial obligations. The firm's ability to repay these obligations when due depends largely on whether it has sufficient cash together with other assets that can be converted into cash before current liabilities mature. The firm's current assets are the primary sources of funds needed to repay current and maturing financial obligations. Thus, the current ratio is a logical measure of liquidity.

Current Ratio

The current ratio is computed as follows:

$$\textit{current ratio} = \frac{\text{current assets}}{\text{current liabilities}} \tag{20-1}$$

$$= \frac{\$24{,}000{,}000}{\$10{,}000{,}000}$$

$$= 2.40 \text{ times}$$

$$\text{industry average} = 2.01 \text{ times}$$

For 1995 Jimco's current assets were 2.40 times larger than its current liabilities. Although no firm plans to liquidate a major portion of its current assets to meet its

[7]To obtain average ratios for both Canadian industrial sectors and industries, refer to: Statistics Canada, *Financial Statistics for Enterprises*, Catalogue No. 61-008; Statistics Canada, *Financial Performance Indicators for Canadian Business*, Catalogue No. 61F0058XPE/F.

matching current liabilities, this ratio does indicate the margin of safety (the liquidity) of the firm.

Jimco's current ratio is higher than the median industry ratio of 2.01.[8] Thus, Jimco's current ratio indicates greater short-term liquidity than that of other firms in the industry.

Acid Test or Quick Ratio

Since inventories and other current liabilities such as pre-paid expenses are generally the least liquid of the firm's assets, it may be desirable to remove them from the numerator of the current ratio, thus obtaining a more refined liquidity measure. For Jimco, the *acid test ratio* is computed as follows:

$$\text{acid test ratio} = \frac{(\text{current assets} - \text{inventories} - \text{pre-paid expenses})}{\text{current liabilities}} \qquad (20\text{-}2)$$

$$= \frac{\$11,700,000}{\$10,000,000}$$

$$= 1.17 \text{ times}$$

$$\text{industry average} = 0.61 \text{ times}$$

Once again Jimco's acid test ratio is higher than the ratio for its industry of 0.61. Thus, on the basis of its current and acid test ratios, Jimco offers no visible evidence of a liquidity problem.

Question 2: How efficient is the firm in using its assets?

Efficiency ratios provide the basis for assessing how efficient the firm is in using its resources to generate sales. For example, a firm that produces $8 million in sales using $4 million in assets is certainly using its resources more efficiently than a similar firm that has $6 million invested in assets.

Efficiency ratios

Financial ratios which provide a basis for assessing how effectively the firm is using its resources to generate sales.

Efficiency ratios can be defined for each asset category in which a firm invests. Our discussion will include a limited number of key efficiency ratios, related to accounts receivable, inventories, net fixed assets, and total assets.

Average Collection Period Ratio

The *average collection period ratio* serves as the basis for determining how rapidly the firm's credit accounts are being collected. The lower this number is, other things being the same, the more efficient the firm is in managing its investment in accounts receivable. We can also think of this ratio in terms of the number of daily credit sales contained in accounts receivable. For example, the average collection period is equal to the accounts receivable balance divided by the firm's average daily credit sales. Computing the ratio for Jimco, we find

$$\text{average collection period} = \frac{\text{accounts receivable}}{\left(\dfrac{\text{annual credit sales}}{365}\right)} \qquad (20\text{-}3)$$

$$= \frac{\$10,000,000}{\left(\dfrac{\$51,000,000}{365}\right)}$$

$$= 71.57$$

$$\text{industry average} = 23.21 \text{ days}$$

[8]Note that 1994 industry ratios are used in the analysis, since more recent information was not available at the time of writing. The analyst will find that published industry averages are generally a year behind, owing to the time required to collect and publish them.

Therefore, on average, Jimco collects its credit sales every 71.57 days.

The accounts receivable turnover ratio is often interchanged with the average collection period ratio, since it contains the same information. For Jimco this ratio would be equal to

$$\text{accounts receivable turnover} = \frac{\text{credit sales}}{\text{accounts receivable}} \tag{20-4}$$

$$= \frac{\$51{,}000{,}000}{\$10{,}000{,}000} = 5.10 \text{ times}$$

Thus, Jimco is turning its accounts receivable over at a rate of 5.10 times per year. This easily translates into an average collection period of 71.57 days. For example, if Jimco's receivables turnover is 5.10 times in a 365-day year, then its average collection period must be 365/5.10 = 71.57 days.

The industry norm for the receivables turnover ratio is 15.73 times, which translates into an average collection period of 365/15.73 = 23.21 days. In terms of both standards Jimco does not compare favourably with the industry norm. This could indicate the presence of some slow-paying accounts, which calls for analysis in greater depth.[9] Before performing such an analysis, it is necessary to know whether Jimco's credit terms are longer than those of other firms in the industry. For example, if Jimco allows its customers terms calling for payment in 72 days when 23 days is the industry norm, then the longer average collection period is to be expected.

Inventory Turnover Ratio

The effectiveness or efficiency with which a firm is managing its investment in inventories is reflected in the number of times that its inventories are turned over (replaced) during the year. The *inventory turnover ratio* is defined as follows:

$$\text{inventory turnover} = \frac{\text{cost of goods sold}}{\text{inventories}} \tag{20-5}$$

$$= \frac{\$38{,}000{,}000}{\$12{,}000{,}000}$$

$$= 3.17 \text{ times}$$

$$\text{industry average} = 8.64 \text{ times}$$

Thus, Jimco turns over its inventories 3.17 times per year.[10] When quarterly or monthly information is available, an average inventory figure should be used in order to eliminate the influence of any seasonality in inventory levels from the ratio.

Jimco's inventory turnover ratio of 3.17 does not compare favourably with the industry norm of 8.64 times. In other words, Jimco invests significantly more in inventories per dollar of sales than does the average firm in its industry.

Fixed Asset Turnover Ratio

To measure the efficiency with which the firm utilizes its investment in fixed assets, we calculate the *fixed asset turnover ratio* as follows:

---

[9]Although it will not be discussed here, one tool for further assessing the liquidity of a firm's receivables is an *aging of accounts receivable schedule.* Such a schedule identifies the number and dollar value of accounts outstanding for various periods. For example, accounts that are less than ten days old, 11 to 20 days, and so forth might be examined. Still another way to construct the schedule would involve analyzing the length of time to eventual collection of accounts over a past period. For example, how many accounts were outstanding less than ten days when collected, between ten and 20 days, and so forth.

[10]Some analysts prefer the use of sales in the numerator of the inventory turnover ratio. However, cost of goods sold is used here, since inventories are stated at cost, and to use sales in the numerator would add a potential source of distortion to the ratio when comparisons are made across firms that have different "markups" on their cost of goods sold.

$$\text{fixed asset turnover} = \frac{\text{sales}}{\text{net fixed assets}} \quad (20\text{-}6)$$

$$= \frac{\$51{,}000{,}000}{\$7{,}000{,}000}$$

$$= 7.286 \text{ times}$$

$$\text{industry average} = 8.36 \text{ times}$$

Thus, Jimco has a larger investment in fixed assets relative to its sales volume than is the case for the industry norm.

Total Asset Turnover Ratio

The *total asset turnover ratio* indicates how many dollars in sales the firm squeezes out of each dollar it has invested in assets. For Jimco, we calculate this ratio as follows:

$$\text{total asset turnover} = \frac{\text{sales}}{\text{total assets}} \quad (20\text{-}7)$$

$$= \frac{\$51{,}000{,}000}{\$31{,}000{,}000}$$

$$= 1.645 \text{ times}$$

$$\text{industry average} = 2.95 \text{ times}$$

Jimco's total asset turnover ratio compares poorly with the industry norm of 2.95. This ratio indicates that compared with other firms in its industry, Jimco's management has not efficiently utilized its resources in generating sales.

Since total assets equals the sum of fixed and current assets, we can use our turnover ratios for both total and fixed assets to analyze the efficiency with which the firm manages its investment in current assets. For example, since Jimco's total asset turnover is higher than the industry norm, the lower than average fixed asset turnover indicates that Jimco's investment in current assets must be smaller, in relation to sales, than the industry norm.

Question 3: How does the firm's management finance its investments

Leverage ratios provide the basis for answering two questions: How has the firm financed its assets? and Can the firm afford the level of fixed charges associated with its use of non-owner-supplied funds such as bond interest and principal repayments? The first question is answered through the use of balance sheet leverage ratios, the second by using an income statement based on ratios, or simply coverage ratios.

Leverage ratios

Leverage ratios provide a basis to determine how a firm has financed its assets and whether the firm can afford the level of fixed charges associated with its use of non-owner-supplied funds.

At this point we will review the concept of leverage. As related to financial ratios the term will be used to mean financial leverage.[11] Financial leverage results when a firm obtains financing for its investments from sources other than the firm's owners. For a corporation, this means funds from any source other than the common shareholders. Thus, **financial leverage** is leverage that results from the firm's use of debt financing, financial leases, and preferred shares. These sources of financing share a common characteristic: They all require a fixed cash payment or return for their use. For example, debt requires contractually set interest and principal payments, leases require fixed rental payments, and preferred shares usually require a fixed cash dividend. This requirement provides the basis for the leverage in financial leverage. If the firm earns a return higher than the return which is required by the suppliers of leverage funds, then the excess goes to the common shareholders. However, should the return earned fall below the required return, then the common shareholders must make up the difference out of the returns on their invested funds. This, in a nutshell, is the concept of financial leverage.

Financial leverage

The use of securities bearing a fixed (limited) rate of return to finance a portion of the firm's assets. Financial leverage can occur from the use of either debt or preferred share financing. The use of financial leverage exposes the firm to financial risk.

[11]The concept of leverage was discussed more fully in Chapter 8.

Balance sheet leverage ratios

Financial ratios used to measure the extent of a firm's use of borrowed funds calculated using information found in the firm's balance sheet.

Balance Sheet Leverage Ratios

These ratios provide the basis for answering the question: Where did the firm obtain the financing for its investments? The label **balance sheet leverage ratios** is used to indicate that these ratios are computed using information from the balance sheet alone.

Debt Ratio The *debt ratio* measures the extent to which the total assets of the firm have been financed using borrowed funds. For Jimco, the ratio is computed as follows:

$$\text{debt ratio} = \frac{\text{total liabilities}}{\text{total assets}} \quad \text{or} \tag{20-8}$$

$$= \frac{(\text{current liabilities} + \text{noncurrent liabilities})}{\text{total assets}}$$

$$= \frac{(\$10{,}000{,}000 + \$10{,}700{,}000)}{\$31{,}000{,}000}$$

$$= .668, \textit{ or } 66.8\%$$

$$\text{industry average} = 54.9\%$$

Thus, Jimco has financed approximately 67 percent of its assets with borrowed funds. This compares with only 54.9 percent for the industry. Note that this ratio can be found by using a common size balance sheet. Simply sum the percent of total assets financed by current liabilities, long-term debt, and all other noncurrent liabilities. Jimco has relied on the use of nonowner financing to a far greater extent than the average firm in its industry. This, in turn, will mean that Jimco may have difficulty trying to borrow additional funds in the future.

Long-Term Debt to Total Capitalization The *long-term debt to total capitalization ratio* indicates the extent to which the firm has used long-term debt in its permanent financing. Total capitalization represents the sum of all the permanent sources of financing used by the firm, including long-term debt, preferred shares, and common equity. For Jimco, the ratio is computed as follows:

$$\begin{array}{c}\text{long-term debt} \\ \text{to total capitalization}\end{array} = \frac{\text{long-term (noncurrent) liabilities}}{(\text{long-term debt} + \text{preferred shares} + \text{common equity})} \tag{20-9}$$

$$= \frac{\$10{,}700{,}000}{(\$10{,}700{,}000 + 10{,}300{,}000)}$$

$$= .509, \text{ or } 50.9\%$$

$$\text{industry average} = 33.41\%$$

Therefore, Jimco has obtained a little more than half its permanent financing from debt sources.

Once again referring to a common size balance sheet for an industry norm, note that current liabilities account for 31.7 percent of total assets; thus, permanent financing is equal to $(1 - .317)$, or 68.3 percent of total assets. Furthermore, long-term debt accounts for 22.8 percent of total assets; thus it accounts for $22.8/68.3 = 33.4$ percent of the firm's total capitalization. It is evident that Jimco uses far more long-term debt in its total capitalization than is characteristic of its industry.

One final point about lease financing should be made concerning balance sheet leverage ratios. Since most firms must include the present value of long-term financial lease agreements in the assets and liabilities of the balance sheet, it is now possible to assess their impact on the firm's balance sheet leverage ratios.[12] Annual

[12]The Accounting committee of the CICA has asserted that a lease which transfers substantially all of the benefits and risks incident to the ownership of property should be accounted for as an acquisition of an asset and the occurrence of an obligation by the lessee. As a result, many firms are now forced to include the value of their lease agreements directly in the balance sheet rather than treat them as operating leases which are reported in footnotes to the balance sheet.

lease payments generally are contained in footnotes to the firm's financial statements; thus, their effect on the firm's coverage ratios, which are discussed next, can be assessed.

Coverage Ratios

These ratios are a second category of leverage ratios, and they are used to measure the firm's ability to cover the finance charges associated with its use of financial leverage. They provide the basis for answering the question of whether the firm has used too much financial leverage.

Times Interest Earned Ratio The *times interest earned ratio* indicates the firm's ability to meet its interest payments out of its annual operating earnings. The ratio measures the number of times the firm is covering its interest. Jimco's ratio is computed as follows:[13]

$$\text{times interest earned} = \frac{\text{net operating income (NOI) or earnings before interest and taxes (EBIT)}}{\text{annual interest expense}} \tag{20-10}$$

$$= \$4{,}000{,}000/\$1{,}000{,}000$$
$$= 4.00 \text{ times}$$
$$\text{industry average} = 7.62 \text{ times}$$

This ratio is much lower than the industry norm of 7.62 times, which is not surprising in light of Jimco's higher than average use of financial leverage. However, it appears that Jimco's earnings could be improved in order to afford the higher use of financial leverage.

Cash Flow Overall Coverage Ratio The purpose of this ratio is to compare the cash flow (from net operating income) available to meet fixed financial commitments against the cash flow requirements of these obligations. The financial commitments to be recognized include interest, lease payments, preferred dividends, and debt principal repayments. The cash available to pay these obligations equals net operating income plus depreciation. We also add lease payments to operating income plus depreciation, since lease payments have been deducted from revenues to calculate net operating income.[14] We adjust debt principal repayment and preferred share dividends that are not tax deductible to a before-tax basis, since we need to compute the amount of before-tax cash flows that are required to make these payments. For instance, we might assume that Jimco's current portion of long-term debt ($500,000) equals the principal repayment for the period. However, to pay $500,000 using after-tax income, we would have to earn $833,333 before taxes. For example, $833,333 income less taxes at a 40 percent rate ($333,333 = .40 × $833,333) leaves $500,000 after taxes to make the principal repayment. For preferred share dividends and debt principal repayment, we have to make the following adjustments:

$$\text{before-tax cost of preferred share dividends} = \frac{\text{preferred share dividend}}{(1 - \text{combined corporate tax rate})}$$

$$\text{before-tax debt principal repayment requirement} = \frac{\text{principal repayment}}{(1 - \text{combined corporate tax rate})}$$

[13]The interchangeable use of EBIT and NOI presumes there was no "other" income earned by the firm. If the presence of other nonoperating income is thought to be transitory, then NOI should be used; if not, then EBIT is appropriate.

[14]Lessors have frequently used the argument that lease financing is "off-balance sheet" and does not impact the firm's borrowing capacity. This argument is flawed in two major respects. First, all financial leases must be capitalized and incorporated into the firm's balance sheet. Second, lease payment commitments do impact the firm's cash flow ability to support additional borrowing.

We can now compute Jimco's cash flow coverage ratio, as follows:

$$\text{cash flow overall coverage ratio} = \frac{\text{(net operating income + lease expense + depreciation)}}{\text{[interest + lease expense + preferred dividends/(1 - combined corporate tax rate) + principal payments/(1 - combined corporate tax rate)]}} \quad (20\text{-}11)$$

$$= \frac{\$(4{,}000{,}000 + 500{,}000)}{\$1{,}000{,}000 + \dfrac{500{,}000}{(1 - 0.40)}}$$

$$= \$4{,}500{,}000/\$1{,}833{,}333 = 2.45 \text{ times}$$

Thus, Jimco's operating earnings were 4.00 times its interest expense, whereas the firm's operating cash flows were only 2.45 times its total finance charges.

One further refinement may be desirable in the cash flow overall coverage ratio. This relates to the coverage of any common dividends the firm wishes to pay. For example, should Jimco desire to pay dividends to common shareholders totalling \$300,000, then the firm must earn \$300,000/(1 − .40) = \$500,000 of income on a before-tax basis. Adding this figure for common dividends to the firm's existing finance charges reduces the coverage ratio to 1.93 times. This second version of the coverage ratio is particularly useful when the firm analyzes alternative sources of long-term financing and wishes to maintain a stable dividend payment to its common shareholders.

Summarizing the results of Jimco's leverage ratios, we have made two basic observations: first, Jimco has utilized more nonowner financing than is characteristic of its industry; second, Jimco's earnings could be improved in order to afford the higher use of financial leverage.

Question 4: Is management generating sufficient profits from the firm's assets?

Profitability ratios

Financial ratios that can be used to determine how much of each sales dollar a firm's management was able to convert into profits and how much profit the firm earned on each dollar of assets under its control.

As we have discussed, financial ratios help us answer some very important questions about the effectiveness of the firm's management of its resources to produce profits. Specifically, **profitability ratios** can be used to answer such questions as these: How much of each sales dollar was management able to convert into profits? How much profit did the firm earn on each dollar of assets under its control? For discussion purposes we will divide profitability ratios into two groups: profitability in relation to sales and profitability in relation to investment.

Profitability in Relation to Sales

These ratios can be used to assess the ability of the firm's management to control the various expenses involved in generating sales. The profit ratios discussed here are commonly referred to as profit margins and include the gross profit margin, operating profit margin, and net profit margin.

Gross Profit Margin The *gross profit margin* is calculated as follows:

$$\text{gross profit margin} = \frac{\text{gross profit}}{\text{net sales}} \quad (20\text{-}12)$$

$$= \frac{\$13{,}000{,}000}{\$51{,}000{,}000}$$

$$= .255, \text{ or } 25.5\%$$

$$\text{industry average} = 16.67\%$$

Thus Jimco's gross profit constitutes 25.5 percent of firm sales. This margin reflects the firm's markup on its cost of goods sold as well as the ability of management to minimize the firm's cost of goods sold in relation to sales (and the method for determining that cost).

Jimco's gross profit margin appears to be greater than the industry norm of 16.67 percent. Note that the gross profit margin reflects both the level of Jimco's cost of goods sold and the size of the firm's markup on those costs, or its pricing policy.

Operating Profit Margin Moving down the income statement, the next profit figure encountered is net operating income (or earnings before interest and taxes—EBIT). This profit figure serves as the basis for computing the *operating profit margin.* For Jimco this profit margin is found as follows:

$$\begin{aligned} \text{operating profit margin} &= \frac{\text{net operating income}}{\text{sales}} \qquad (20\text{-}13) \\ &= \frac{\$4{,}000{,}000}{\$51{,}000{,}000} \\ &= .0784, \text{ or } 7.84\% \\ \text{industry average} &= 5.47\% \end{aligned}$$

The operating profit margin reflects the firm's operating expenses as well as its cost of goods sold. Therefore, this ratio serves as an overall measure of operating effectiveness.

Jimco's operating profit margin is greater than the industry norm of 5.47 percent. Thus, Jimco's operating expenses per dollar of sales are below the industry norm.

Net Profit Margin The final profit margin considered involves the net after-tax profits of the firm as a percent of sales. For Jimco the *net profit margin* is computed as follows:

$$\begin{aligned} \text{net profit margin} &= \frac{\text{net income}}{\text{sales}} \qquad (20\text{-}14) \\ &= \frac{\$1{,}800{,}000}{\$51{,}000{,}000} \\ &= .035, \text{ or } 3.5\% \\ \text{industry average} &= 2.00\% \end{aligned}$$

Therefore, \$0.035 of each sales dollar is converted into profits after taxes. Note that this profit margin reflects the firm's cost of goods sold, operating expenses, finance charges (interest expense), and taxes. For the industry, profits before taxes are 3.48 percent of sales. Assuming that firms on the average pay approximately 42.6 percent of their taxable earnings in taxes, this produces a net profit margin of $.0348(1 - .426) = .020$, or 2.00 percent. Hence, Jimco's net profit margin is above the norm for its industry. In the next category of profitability ratios we investigate the firm's total investment and the return on the investment of the common shareholders.

Profitability in Relation to Investment

This category of profitability ratios attempts to measure firm profits in relation to the invested funds used to generate those profits. Thus, these ratios are very useful in assessing the overall effectiveness of the firm's management.

Operating Income Return on Investment The *operating income return on investment* reflects the rate of return on the firm's total investment before interest and taxes. For Jimco, this return measure is computed as follows:

$$\begin{aligned} \text{operating income return on investment} &= \frac{\text{net operating income}}{\text{total assets}} \qquad (20\text{-}15) \\ &= \frac{\$4{,}000{,}000}{\$31{,}000{,}000} \\ &= .129, \text{ or } 12.9\% \\ \text{industry average} &= 16.14\% \end{aligned}$$

Jimco's management produced a 12.9 percent return on its total assets before interest and taxes have been paid.[15] As a benchmark to determine whether leverage is favourable or unfavourable, the 12.9 percent rate of return can be compared with the cost of borrowed funds. If the firm is borrowing at a cost less than 12.9 percent, then leverage is favourable and will result in higher after-tax earnings to its shareholders.[16]

This rate of return is useful in assessing operating effectiveness of the firm's management. The operating return on investment does not reflect the influence of the firm's use of financial leverage. Neither the numerator (operating income) nor the denominator (total assets) is affected by the way in which the firm has financed its assets. Thus, it provides a measure of management's effectiveness in making operating decisions as opposed to financing decisions.

If the industry norm for this ratio is not readily available, we can calculate one, using the information given there. Sales divided by total assets equals 2.95; and the operating profit margin is 5.47 percent of sales for the industry. Using the following relationship, we derive an industry norm of 16.14 percent. Thus, Jimco's operating rate of return does not compare favourably with the industry norm.

$$\left(\frac{\text{operating income}}{\text{sales}}\right) \times \left(\frac{\text{sales}}{\text{total assets}}\right) = \left(\frac{\text{operating income}}{\text{total assets}}\right) \tag{20-16}$$

$$0.0547 \quad \times \quad 2.95 \quad = \quad .1614, \text{ or } 16.14\%$$

In deriving the industry norm for the operating income return on investment ratio, we have identified a very useful relationship between operating profit margin and the ratio of sales divided by total assets (which we earlier referred to as total asset turnover). For example, a firm's rate of return on investment is a function of (1) how much profit it squeezes out of each dollar of sales (as reflected in its operating profit margin), and (2) how much it has invested in assets to produce those sales (as reflected in the total asset turnover ratio). Jimco's above-average operating profit margin, combined with its lower than average turnover of total investment in assets, produces an operating income return on investment that is below its industry norm.

To understand why the firm is below average, we may separate the operating income return on investment (OIROI) into two important pieces of information: the operating profit margin and the total asset turnover. The firm's OIROI is a multiple of these two ratios and may be shown algebraically as follows:

$$\text{OIROI} = \left(\begin{array}{c}\text{operating}\\ \text{profit margin}\end{array}\right) \times \left(\begin{array}{c}\text{total asset}\\ \text{turnover}\end{array}\right) \tag{20–17}$$

or more completely,

$$OIROI = \left(\frac{\text{operating income}}{\text{sales}}\right) \times \left(\frac{\text{sales}}{\text{total assets}}\right) \tag{20-18}$$

The first component of the OIROI in Equation 20-18, the operating profit margin, is an extremely important variable in understanding a company's operating profitability. It is important that we know what drives this ratio. In coming to understand the ratio, think about the makeup of the ratio, which may be expressed as follows:

$$\frac{\begin{array}{c}\text{operating}\\ \text{income}\end{array}}{\text{sales}} = \frac{\begin{array}{c}\text{total}\\ \text{sales}\end{array} - \begin{array}{c}\text{cost of}\\ \text{goods sold}\end{array} - \begin{array}{c}\text{administrative}\\ \text{expenses}\end{array} - \begin{array}{c}\text{selling}\\ \text{expenses}\end{array}}{\text{total assets}} \tag{20-19}$$

[15]Intangible assets are often subtracted from total assets in an effort to measure the firm's return on invested capital. The lack of physical qualities of intangible assets makes evidence of their existence elusive, their value often difficult to estimate, and their useful lives indeterminable. However, since Jimco has no intangible assets, no adjustment is necessary.

[16]The concept of leverage was fully developed earlier in Chapter 8. Very simply, when a firm borrows money that requires a fixed return, then the return to the common shareholders will be enhanced only if the return the firm earns on these borrowed funds exceeds their cost. For the firm as a whole the operating income rate of return measures the before-tax-and-interest rate of return on the firm's investment. This rate must exceed the cost of borrowed funds for the firm as a whole to experience favourable financial leverage. The favourableness of financial leverage is determined by the effect of its use on earnings per share to the firm's common shareholders.

Because total sales equals the number of units sold times the sales price per unit, and the cost of goods sold equals the number of units sold times the cost of goods sold per unit, we may conclude that the driving forces of the operating profit margin are the following:

1. The number of units of product sold.[17]
2. The average selling price for each product unit.
3. The cost of manufacturing or acquiring the firm's product.
4. The ability to control general and administrative expenses.
5. The ability to control expenses in marketing and distributing the firm's product.

These influences are also apparent simply by looking at the income statement and thinking about what is involved in determining the firm's operating profits or income.[18]

Return on Total Assets or Return on Investment The *return on total assets* or *return on investment ratio* relates after-tax income to the firm's total investment in assets. For Jimco, this ratio is found as follows:

$$\text{return on total assets} = \frac{\text{net income}}{\text{total assets}} \tag{20-21}$$

$$= \frac{\$1{,}800{,}000}{\$31{,}000{,}000}$$

$$= .058, \text{ or } 5.8\%$$

$$\text{industry average} = 5.9\%$$

Again total assets are used in an attempt to measure total investment. In this analysis an industry norm was obtained in a manner similar to that used with the operating rate of return. For example,

$$\text{return on total assets} = \frac{\text{net income}}{\text{sales}} \times \frac{\text{sales}}{\text{total assets}} \tag{20-22}$$

Earlier the net income to sales ratio for the industry was estimated to be 2.0 percent. Using the industry's sales to total assets ratio of 2.95 produces an industry norm for return on total assets of $.020 \times 2.95 = .0590$, or 5.90 percent. Thus, Jimco provides a satisfactory return on its total investment when compared to other firms in its industry.[19]

---

[17]The number of units affects the operating profit margin only if some of the firm's costs and expenses are fixed. If a company's expenses are all variable in nature, then the ratio would not change as the number of units sold increases or decreases, because the numerator and the denominator would change at the same rate.

[18]We could have used the *net profit margin*, rather than the operating profit margin, which is measured as follows:

$$\text{net profit margin} = \text{net income} \div \text{sales} \tag{20-20}$$

However, because net income reflects the deduction of both operating expenses and interest expense, this ratio is influenced by operating and financing activities.

[19]The return on investment defined as the ratio of net income to total assets has long been a widely used measure with an inappropriate investment base and can produce misleading signals. The problem relates to the fact that the net income does not include interest expense, which is the return paid to the firm's creditors who financed a portion of the firm's investments. The problem can be resolved in one of two ways. First, if the analyst is interested in evaluating the performance of the firm with respect to its total asset investment, then the operating income rate of return discussed earlier is appropriate. On the other hand, if the analyst is specifically interested in how well the firm's shareholders have fared, then the return on common equity is the appropriate ratio to use.

Question 5: Are the common shareholders receiving sufficient returns on their investment?

Our one remaining question looks at the accounting return on the common shareholders' investment; that is, we want to know if the earnings available to the firm's owners or common equity investors is attractive when compared to the returns of owners of similar companies in the same industry.

Return on Equity

The *return on equity ratio* measures the rate of return earned on the shareholder's investment. Jimco earned the following rate of return for its shareholders:

$$\text{return on common equity} = \frac{\text{net income available to common}}{\text{common equity}} \quad (20\text{-}23)$$

$$= \frac{\$1{,}800{,}000}{\$10{,}300{,}000}$$

$$= .175\text{, or } 17.5\%$$

$$\text{industry average} = 13.08\%$$

The following relationship can be used to derive an industry norm from the return on total assets ratio and the debt ratio:

$$\text{return on common equity} = \frac{\text{return on total assets}}{(1 - \text{debt ratio})} \quad (20\text{-}24)$$

Recall that the debt ratio is simply total liabilities divided by total assets. The industry norm for the return on equity ratio is found as follows:

$$= \frac{0.059}{(1 - .549)}$$

$$= 0.1308 \text{ or } 13.08\%$$

Thus, Jimco's 17.5 percent return compares favourably with that of the industry. This higher-than-average return reflects both the firm's above-average profit margins (reflected in the firm's operating and net profit margins) and its above-average use of financial leverage (as reflected in its debt ratio).

The forces that drive the return on equity ratio are summarized as follows:

1. The difference between the operating income return on investment and the interest rate.
2. The amount of debt used in the capital structure relative to the size of the firm.

This return on equity ratio can also be expressed as follows:

$$\text{return on equity} = \left(\frac{\text{operating income}}{\text{return on investment}}\right) + \left[\left(\frac{\text{operating income}}{\text{return on investment}} - \frac{\text{interest}}{\text{rate}}\right) \times \frac{\text{debt}}{\text{equity}}\right]$$

In summary, Jimco has slightly below-average return on its assets; however, the firm has above-average profits in relation to its sales. This factor results from the firm's above-average operating and net profit margins and the fact that Jimco utilized more leverage than the norm for its industry.

Trend Analysis

We noted earlier that a firm's financial ratios can be compared with two types of standards and have discussed industry norms as the basis for comparison. Now, we will demonstrate the use of trend comparisons. Figure 20-1 illustrates graphs of Jimco's current ratio, acid test ratio, debt ratio, and return on total asset ratio for the past five years.

Comparing Ratios of Companies from Different Countries

Suppose that you want to analyze and compare the financial health of a group of North American companies and Dutch companies. Can you calculate a set of ratios for each of the two sets of firms and compare one set to the other? In general the answer is no. The following are examples of international accounting differences: (1) in the Netherlands assets are valued at their economic worth or replacement value rather than historical cost as is generally the case in North America and (2) intangible assets, such as goodwill from a consolidation that occurs in North America, must be amortized and expensed against reported income over a period of no more than 40 years. In Germany, the maximum amortization period is five years, while in Great Britain goodwill does not have to be expensed against reported income at all but can be charged directly against retained earnings on the firm's balance sheet. These differences in accounting practice can have a significant impact on the firm's reported earnings where all else is held constant. Finally, in Japan, income smoothing is permitted because firms are allowed to make discretionary changes in such items as depreciation and bad debts. These are just a few cases that make accounting practices across international borders difficult.

The need for uniform standards is widely recognized and the financial accounting community has long supported a move toward uniform international standards. As far back as 1973, the International Accounting Standards Committee (IASC) was formed to promote uniformity and to narrow the areas of divergence among participating nations. International accounting practices still vary widely and comparisons of firms across international borders remain difficult.

Surveying the trend in Jimco's liquidity ratios reveals that Jimco's current and acid test ratios compare very favourably with their respective industry norms from a creditor's perspective. However, the firm's management may want to question whether too much of the firm's assets are being tied up in liquid assets which yield a low return.

Jimco's debt ratio appears to have declined slightly over the past five years, with moderate interim fluctuations. However, no material change in the ratio appears to have occurred over the period. In light of Jimco's current use of leverage, however, any further increases in this ratio may be unwarranted.

Finally, the return on total assets ratio for the past five years depicts the relatively volatile nature of Jimco's business, with returns ranging from 4 percent in 1992 to 10 percent in 1994. However, based upon Jimco's return on total assets ratio for 1995 and that of the industry, it would appear that Jimco has done as well or better than the median for its industry.

Summary of Jimco's Financial Ratios

Table 20-6 summarizes Jimco's financial ratios, as well as the corresponding industry norms. Each ratio is evaluated in relation to the appropriate norm. Briefly, the results of those comparisons are as follows:

1. Jimco's liquidity position is much better than the industry.
2. Jimco has made extensive use of financial leverage. In fact, the firm has financed 67 percent of its assets with nonowner funds.

Figure 20-1 Trend Analysis Illustration

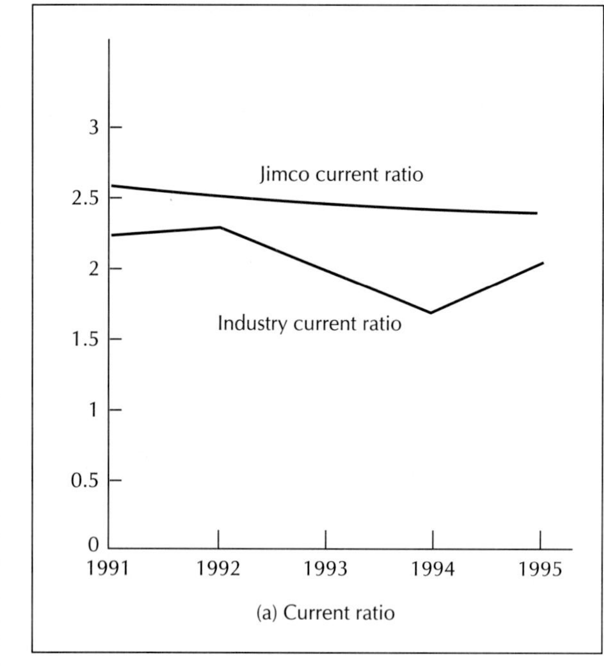

(a) Current ratio

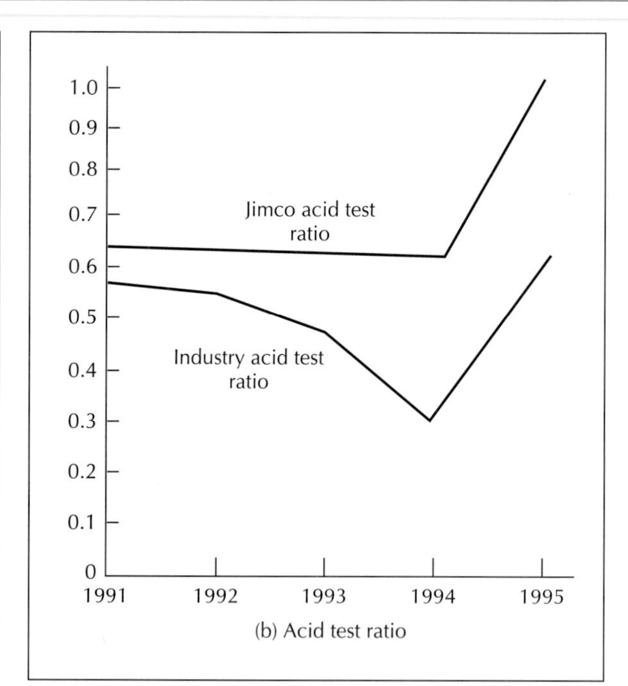

(b) Acid test ratio

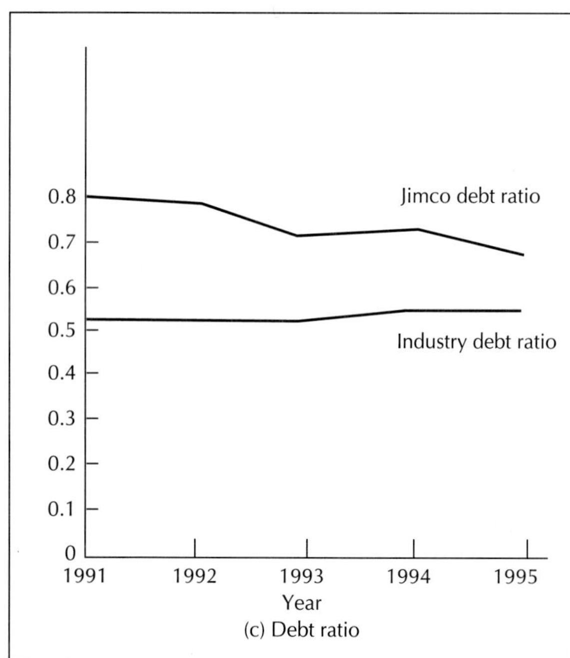

(c) Debt ratio

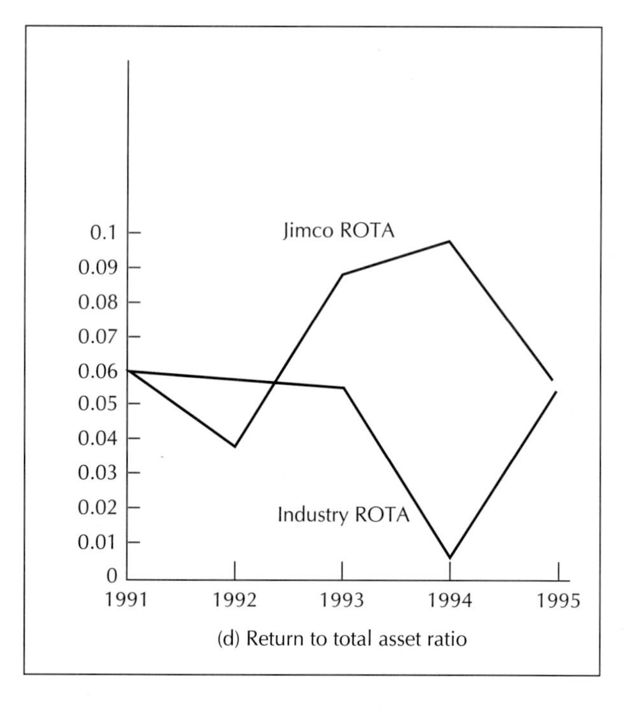

(d) Return to total asset ratio

3. The firm cannot afford any further use of financial leverage, as is indicated by the times interest earned ratio.
4. The firm must re-examine its policies for accounts receivable and inventory management, as indicated by both the average collection period and the inventory turnover ratios.
5. Jimco's profit margins are better than the respective industry norms; however, the firm has not been able to convert these profit margins into better-than-aver-

Table 20-6 Summary of Ratios for Jimco Inc.

Ratio	Formula	Calculation	Industry Average	Evaluation
Liquidity ratios				
1. Current ratio	Current assets/current liabilities	$24,000,000/10,000,000 = 2.40 times	2.01 times	Excellent
2. Acid test ratio	(Current assets − inventories − prepaid expenses)/current liabilities	$11,700,000/10,000,000 = 1.17 times	0.61 times	Good
Efficiency ratios				
3. Average collection period	Average accounts receivable/(annual credit sales/365)	$10,000,000/(51,000,000/365) = 71.57 days	23.21 days	Poor
4. Inventory turnover	Cost of goods sold/ending inventory	$38,000,000/12,000,000 = 3.17 times	8.64 times	Poor
5. Fixed asset turnover	Sales/fixed assets	$51,000,000/7,000,000 = 7.286 times	8.36 times	Satisfactory
6. Total asset turnover	Sales/total assets	$51,000,000/31,000,000 = 1.645 times	2.95 times	Poor
Leverage ratios				
7. Debt ratio	Total liabilities/total assets	$20,700,000/31,000,000 = 66.8%	54.90%	Satisfactory
8. Long-term debt to total capitalization	Long-term debt/total capitalization	$10,700,000/21,000,000 = 50.9%	33.41%	Satisfactory
9. Times interest earned	Net operating income/annual interest expense	$4,000,000/1,000,000 = 4.00 times	7.62 times	Poor
10. Cash flow overall	(NOI + lease expense + depreciation interest + lease expense + principal payments/(1 − tax rate)	$4,500,000/1,833,333 = 2.45 times	N.A.[a]	—
Profitability ratios				
11. Gross profit margin	Gross profit/sales	$13,000,000/51,000,000 = 25.5%	16.67%	Excellent
12. Operating profit margin	Net operating income/sales	$4,000,000/51,000,000 = 7.84%	5.47%	Satisfactory
13. Net profit margin	Net income/sales	$1,800,000/51,000,000 = 3.5%	2.00%	Good
14. Operating income return on investment	Net operating income/total assets	$4,000,000/31,000,000 = 12.9%	16.14%	Poor
15. Return on total assets	Net income/total assets	$1,800,000/31,000,000 = 5.8%	5.90%	Satisfactory
16. Return on common equity	Net income available to common/common equity	$1,800,000/10,300,000 = 17.5%	13.08%	Good

[a]Norm was not available.

age rates of return on investment. This resulted from the higher-than-average sales per dollar invested in assets as reflected in Jimco's efficiency ratios.

6. Finally, Jimco has benefited from the favourable use of financial leverage. The firm earned a favourable 17.5 percent return on the investment of its common shareholders, compared with 13.08 percent for the industry.

AN INTEGRATED FORM OF FINANCIAL ANALYSIS BASED ON EARNING POWER

OBJECTIVE 4

Table 20-6 presents an alternative format for analyzing financial ratios. This alternative format focuses on the firm's earning power as measured by two of the firm's profitability ratios: the operating income return on investment and the return on common equity. This method of financial analysis is particularly well suited to internal analyses carried out by the firm's management. The reason is that the analysis focuses directly on firm profitability, which in turn reflects how well the firm is being managed. In addition, the analysis of earning power is a valuable guide to analyze the firm's ability to manage effectively its financial assets from the common shareholder's perspective.

The analysis of a firm's earning power involves a two-stage procedure designed to answer two basic questions:

Stage 1: How effective has the firm's management been in generating sales using the total assets of the firm and converting those sales into operating profits?

Stage 2: How effective has the firm's management been in forming a financial structure that increases the returns to the common shareholders? Here we analyze the effect of the firm's financing decisions (that is, the mixture of debt and owner financing used by the firm) on the rate of return earned on the common shareholder's investment.

Figure 20-2 provides a template for carrying out the first stage of the analysis of Jimco's earning power. We focus our attention on the ratio of operating income return on investment. This ratio reflects the rate of return earned on the firm's investment in assets and before giving any consideration to how they have been financed. Note that the operating income return on investment can be broken down into the product of two ratios,

$$\frac{\text{Operating Income}}{\text{Return on Investment}} = \left(\frac{\text{Operating Income}}{\text{Sales}}\right) \times \left(\frac{\text{Sales}}{\text{Total Assets}}\right) = \frac{\text{Operating Income}}{\text{Total Assets}}$$

Figure 20-2 simply lays out the relationships that underlie the operating profit margin and total asset turnover ratios. The left-hand branch of the figure shows the determinants of the operating profit margin and the right-hand branch details the determinants of the total asset turnover ratio.

A total of seven ratios are calculated in the first stage of the analysis of Jimco's earning power. The first is the operating income return on investment (Step 1). Step 2 involves calculating the operating profit margin, which, along with the total asset turnover (Step 4), determines the operating income return on investment. Step 3 involves calculation of the gross profit margin, which provides the basis for assessing the impact of cost of goods sold on the operating profit margin calculated in Step 2. Steps 5, 6, and 7 involve the calculation of the fixed asset turnover, accounts receivable turnover, and inventory turnover ratios, which provide the basis for a detailed analysis of the determinants of the total asset turnover ratio (calculated in Step 4).

Note that by following the steps in Figure 20-2, the analyst is led through a detailed analysis of the determinants of the operating income return on investment. Each successive step provides the basis for understanding more about the determinants of this rate of return. For example, the total asset turnover ratio is one of the two basic determinants of the operating income return on investment (the other is the operating profit margin). By analyzing the fixed asset turnover ratio in conjunction with the total asset turnover, the analyst can determine whether fixed or current assets caused the total asset turnover ratio to deviate from the industry average. Furthermore, the accounts receivable turnover and inventory turnover ratios can be analyzed to determine the effect of the level of investment in these assets on total asset turnover, and, consequently, the observed operating income return on investment.

Figure 20-3 provides a template for use in analyzing the effect of the firm's financing decisions on the return earned on the common shareholder's investment. The analysis presented in Figure 20-3 depends on the following basic relationship:

$$\frac{\text{Return on}}{\text{Common Equity}} = \frac{\text{Net Income Available to Common Shareholders}}{\text{Common Equity}}$$

The analysis begins with the operating income return on investment ratio, which was the subject of the analysis in Figure 20-2. Next, Step 9 involves calculation of the return on total assets. This ratio is then adjusted for the influence of the firm's use of financial leverage in order to calculate the return on common equity. In Step

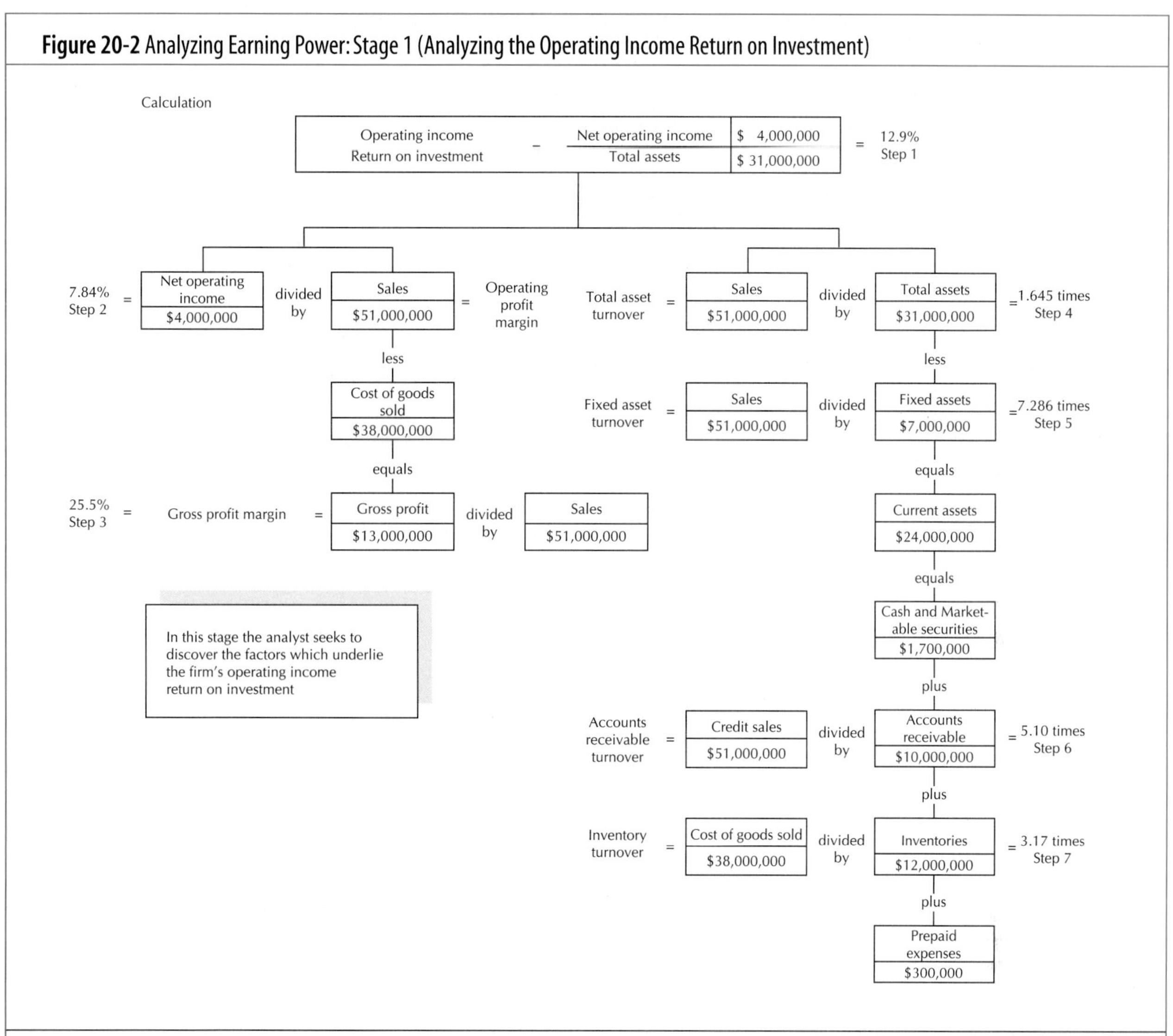

Figure 20-2 Analyzing Earning Power: Stage 1 (Analyzing the Operating Income Return on Investment)

Step	Ratio	Formula	Calculation	Industry	Evaluation
1	Operating income return on investment	$\frac{\text{Net operating income}}{\text{Total assets}}$	$\frac{\$\ 4{,}000{,}000}{\$31{,}000{,}000} = 12.9\%$	16.14%	Poor
2	Operating profit margin	$\frac{\text{Net operating income}}{\text{Sales}}$	$\frac{\$\ 4{,}000{,}000}{\$51{,}000{,}000} = 7.84\%$	5.47%	Satisfactory
3	Gross profit margin	$\frac{\text{Gross profit}}{\text{Sales}}$	$\frac{\$13{,}000{,}000}{\$51{,}000{,}000} = 25.5\%$	16.67%	Excellent
4	Total asset turnover	$\frac{\text{Sales}}{\text{Total assets}}$	$\frac{\$51{,}000{,}000}{\$31{,}000{,}000} = 1.645$ times	2.95 times	Poor
5	Fixed asset turnover	$\frac{\text{Sales}}{\text{Fixed assets}}$	$\frac{\$51{,}000{,}000}{\$\ 7{,}000{,}000} = 7.286$ times	8.36 times	Satisfactory
6	Accounts receivable turnover	$\frac{\text{Credit sales}}{\text{Accounts receivable}}$	$\frac{\$51{,}000{,}000}{\$10{,}000{,}000} = 5.10$ times	15.73 times	Poor
7	Inventory turnover	$\frac{\text{Cost of goods sold}}{\text{Inventories}}$	$\frac{\$38{,}000{,}000}{\$12{,}000{,}000} = 3.17$ times	8.64 times	Poor

Figure 20-3 Analyzing Earning Power: Stage 2 (The Return Earned on the Common Shareholder's Investment)

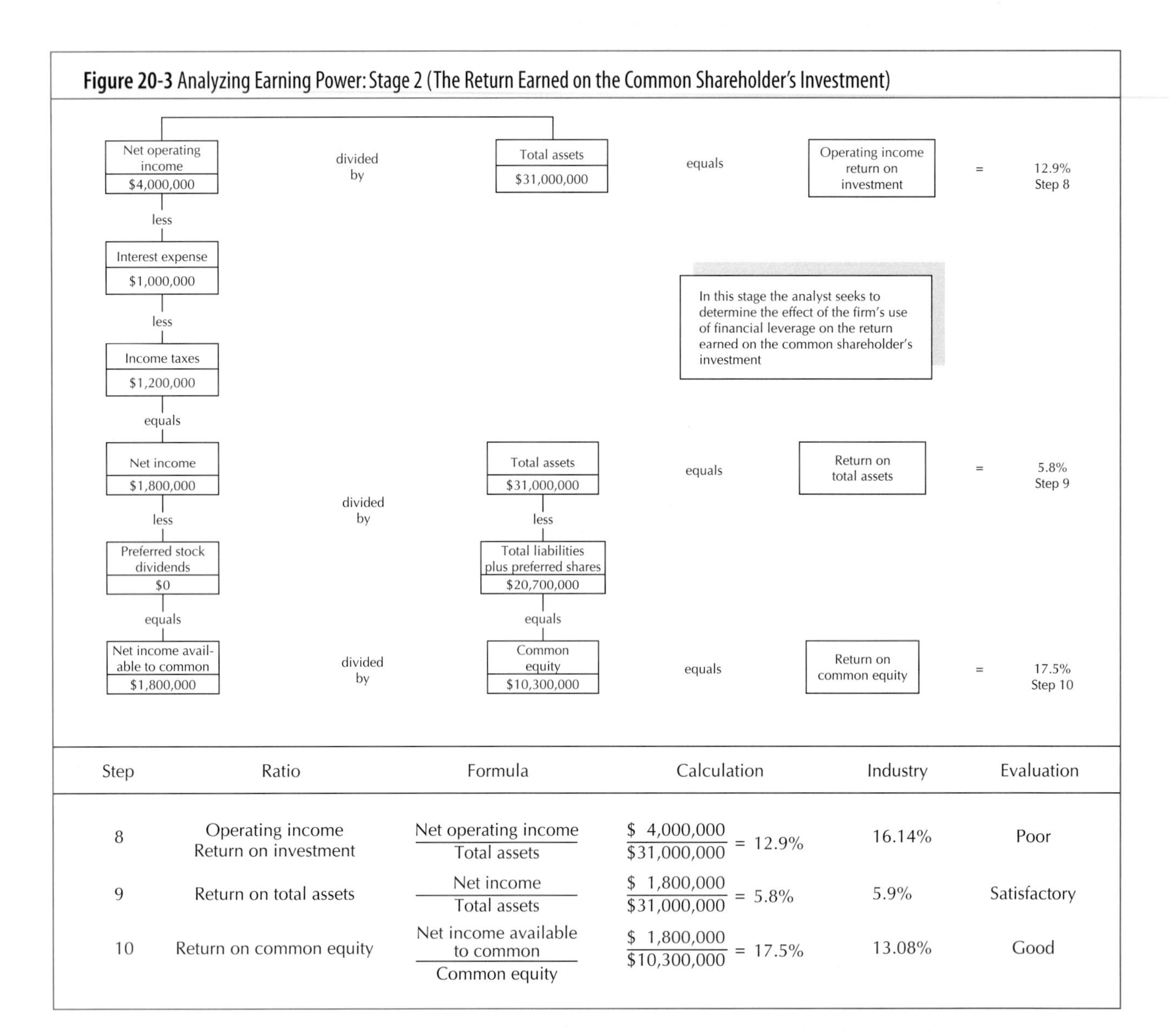

Step	Ratio	Formula	Calculation	Industry	Evaluation
8	Operating income Return on investment	$\frac{\text{Net operating income}}{\text{Total assets}}$	$\frac{\$\ 4{,}000{,}000}{\$31{,}000{,}000} = 12.9\%$	16.14%	Poor
9	Return on total assets	$\frac{\text{Net income}}{\text{Total assets}}$	$\frac{\$\ 1{,}800{,}000}{\$31{,}000{,}000} = 5.8\%$	5.9%	Satisfactory
10	Return on common equity	$\frac{\text{Net income available to common}}{\text{Common equity}}$	$\frac{\$\ 1{,}800{,}000}{\$10{,}300{,}000} = 17.5\%$	13.08%	Good

10 we measure the rate of return earned on the common shareholders' investment in the firm, which reflects both the firm's operating and financing decisions.

The ten-step procedure outlined in Figures 20-2 and 20-3 connects the return earned on common equity to the firm's use of financial leverage and the operating profitability. The operating rate of return ratio was shown to be determined by the firm's profit margins on sales (Steps 2 and 3) and the sales to asset relationship (Steps 4 through 7).

Du Pont analysis

A system of financial ratios designed to investigate the determinants of the return on equity and return on assets ratios.

Another approach frequently used to evaluate a firm's profitability and the return on equity is called the **Du Pont analysis**. Figure 20-4 shows graphically the Du Pont technique, modified somewhat from the original format developed by the management at the Du Pont Corporation. Beginning at the top of the figure, we see that the return on equity is calculated as follows:

$$\text{return on equity} = (\text{return on assets}) \div \left(1 - \frac{\text{total debt}}{\text{total assets}}\right) \tag{20-25}$$

where return on assets, or ROA, equals:

$$\text{return on assets} = \frac{\text{net income}}{\text{total assets}} \tag{20-26}$$

Figure 20-4 Du Pont Analysis

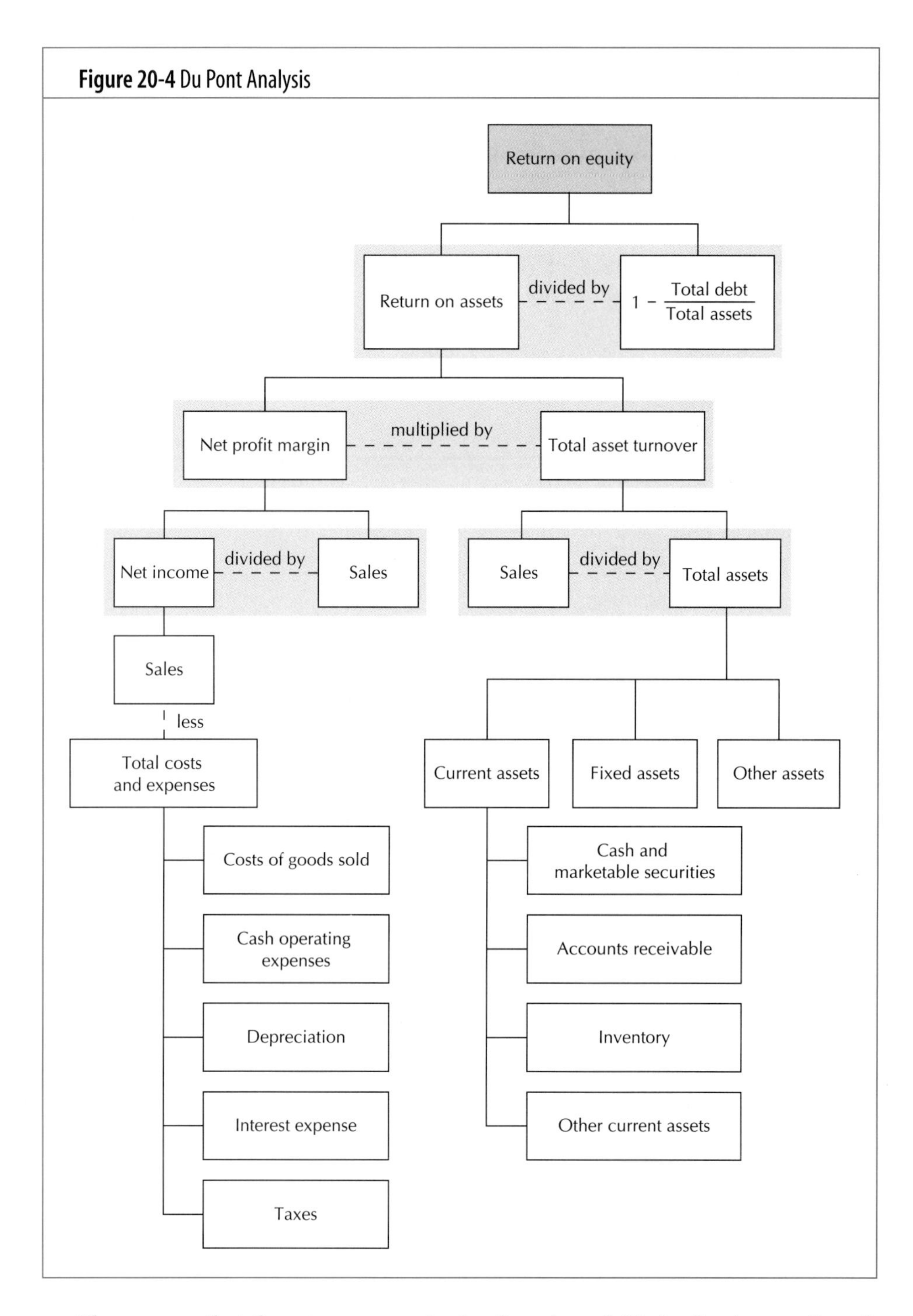

Thus we see that the return on equity is a function of: (1) the firm's overall profitability (net income relative to the amount invested in assets), and (2) the amount of debt to finance the assets. We also know that the return on assets may be represented as follows:

$$\text{return on assets} = (\text{net profit margin}) \times (\text{total asset turnover})$$

$$= \left(\frac{\text{net income}}{\text{sales}}\right) \times \left(\frac{\text{sales}}{\text{total assets}}\right) \qquad (20\text{-}27)$$

Combining Equations (20-25) and (20-27) gives us the basic Du Pont equation that shows the firm's return on equity as follows:

Figure 20-5 Du Pont Analysis: Jimco Inc.

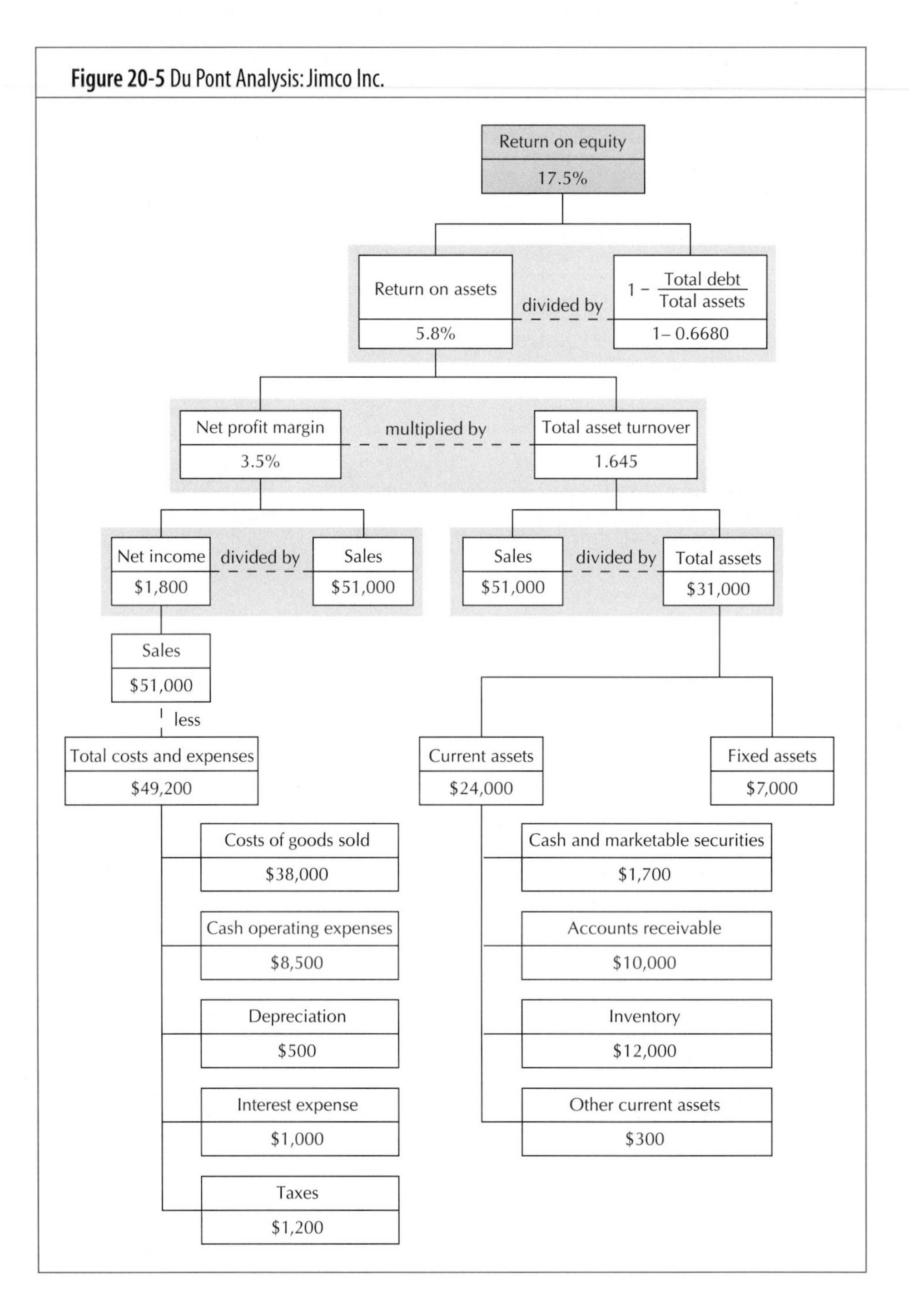

$$\text{return on equity} = \left(\begin{array}{c}\text{net profit}\\ \text{margin}\end{array}\right) \times \left(\begin{array}{c}\text{total asset}\\ \text{turnover}\end{array}\right) \div \left(1 - \frac{\text{total debt}}{\text{total assets}}\right)$$

$$= \left(\frac{\text{net income}}{\text{sales}}\right) \times \left(\frac{\text{sales}}{\text{total assets}}\right) \div \left(1 - \frac{\text{total debt}}{\text{total assets}}\right)$$

Using the Du Pont equation and the diagram in Figure 20-4 allows management to see more clearly what drives the return on equity and the interrelationships among the net profit margin, the asset turnover, and the debt ratio. Management is provided with a road map to follow in determining their effectiveness in managing the firm's resources to maximize the return earned on the owners' investment. In addition, the manager or owner can determine why that particular return was earned.

Let's return to Jimco Inc. to demonstrate the use of the Du Pont analysis. Taking the information from Jimco Inc.'s income statement (Table 20-1) and balance sheet as of December 31, 1995 (Table 20-2), we can calculate the company's return on equity as follows:

$$\text{return on equity} = \left(\frac{\text{net income}}{\text{sales}}\right) \times \left(\frac{\text{sales}}{\text{total assets}}\right) \div \left(1 - \frac{\text{total debt}}{\text{total assets}}\right)$$

$$= \left(\frac{\$1{,}800}{\$51{,}000}\right) \times \left(\frac{\$51{,}000}{\$31{,}000}\right) \div \left(1 - \frac{\$20{,}708}{\$31{,}000}\right)$$

$$= \frac{3.5 \times 1.645}{(1 - 0.668)}$$

$$= 17.5\%$$

We can also visualize the relationships graphically for Jimco Inc., as shown in 20-5.

The real value of this approach to financial analysis is its ability to demonstrate the interrelationships between the return earned on the owners' investment in the firm and a wide variety of financial attributes of the firm. The analyst is provided with a "road map" to follow in determining how successful the firm's management has been in managing its resources to maximize the return earned on the owners' investment. In addition, the analyst can determine why that particular return was earned.

Limitations of Ratio Analysis

The analyst who works with financial ratios must be aware of the limitations involved in their use. The following list includes some of the more important pitfalls that may be encountered in computing and interpreting financial ratios:

1. It is sometimes difficult to identify the industry category to which a firm belongs when the firm engages in multiple lines of business.
2. Published industry averages are only approximations and provide the user with general guidelines rather than scientifically determined averages of the ratios of all or even a representative sample of the firms within the industry. For example, Robert Morris Associates prepares a cautionary statement in conjunction with its published industry ratios (see the *International Financial Management* box).
3. Accounting practices differ widely among firms and can lead to differences in computed ratios. For example, the use of last-in, first-out (LIFO) in inventory valuation can, in a period of rising prices, lower the firm's inventory account and increase its inventory turnover ratio as compared to that of a firm that utilizes first-in, first-out (FIFO). In addition, firms may choose different methods of depreciating their fixed assets.
4. Financial ratios can be too high or too low. For example, a current ratio that exceeds the industry norm may signal the presence of excess liquidity, which results in a lowering of overall profits in relation to the firm's investment in assets. On the other hand, a current ratio that falls below the norm indicates the possibility that the firm has inadequate liquidity and may at some future date be unable to pay its bills on time.
5. An industry average may not provide a desirable target ratio or norm. At best, an industry provides a guide to the financial position of the average firm in the industry. To achieve a more meaningful comparison, a financial analyst will use industry norms according to the size of the firm.[20]

[20]A recent study (Rose and Cunningham, 1991) indicates the need to carefully consider the choice of an industry norm. In fact, your analysis may require you to construct your own norm from, say, a list of four or five firms in a particular industry that might provide the most appropriate standard of comparison for the firm being analyzed.

A Cautionary Statement on Using Industry Ratios

Industrial ratios and statement studies data should be regarded only as general guidelines and not as absolute industry norms. For example, Robert Morris Associates (RMA) provide several reasons why the data may not be fully representative of a given industry:

1. The financial statements used in the Statement Studies are not selected by any random or statistically reliable method. RMA member banks voluntarily submit the raw data they have available each year provided that the companies' total assets are less than $250 million, except for contractors' statements which have no upper size limit.
2. Many companies have varied product lines; however, the Statement Studies categorize them by their primary product Standard Industrial Classification (SIC) number only.
3. Some of our industry samples are rather small in relation to the total number of firms in a given industry. A relatively small sample can increase the chances that some of our composites do not fully represent an industry.
4. There is the chance that an extreme statement can be present in a sample, causing a disproportionate influence on the industry composite. This is particularly true in a relatively small sample.
5. Companies within the same industry may differ in their method of operations which in turn can directly influence their financial statements. Since they are included in our sample, too, these statements can significantly affect our composite calculations.
6. Other considerations that can result in variations among different companies engaged in the same general line of business are different labour markets; geographical location; different accounting methods; quality of products handled; sources and methods of financing; and terms of sale.

For these reasons, RMA does not recommend the Statement Studies figures be considered as absolute norms for a given industry. Rather the figures should be used only as general guidelines and in addition to the other methods of financial analysis. RMA makes no claim as to the representativeness of the figures printed in this book.

Source: Robert Morris Associates, One Liberty Place, Philadelphia, PA 19103.

6. Many firms experience seasonality in their operations. Thus, balance sheet entries and their corresponding ratios will vary with the time of year when the statements are prepared. To avoid this problem, an average account balance should be used (for several months or quarters during the year) rather than the year-end total. For example, an average of month-end inventory balance might be used to compute a firm's inventory turnover ratio when the firm is subject to a significant seasonality in its sales (and correspondingly in its investment in inventories).

Given their limitations, financial ratios provide the analyst with a very useful tool for assessing a firm's financial condition. The analyst should, however, be aware of these potential weaknesses when performing a ratio analysis. In many cases, the real importance of analyzing financial ratios is the framework that is provided for asking questions about the firm.

SUMMARY

Basic Financial Statements

Three basic financial statements are commonly used to describe the financial condition and performance of the firm: the balance sheet, the income statement, and the statement of changes in financial position. The balance sheet provides a picture of the firm's assets, liabilities, and owners' equity on a particular date, whereas the income statement reflects the net revenues from the firm's operations over a given period. The statement of changes in financial position combines information from both the balance sheet and income statement to describe inflows (sources) and outflows (uses) of cash for a given period in the firm's history.

Financial Ratios

Financial ratios are the principal tool of financial analysis. Sometimes referred to as benchmarks, ratios standardize financial information so that comparisons of firms of varying sizes can be made. Ratio analysis should be designed to provide information for the specific needs of an analyst.

Financial ratios are of four main kinds: (1) liquidity, (2) efficiency, (3) leverage, and (4) profitability ratios. The financial statements of Jimco, Inc., demonstrate the computation of a sample listing of ratios from each category. The set of possible ratio calculations is limited only by the analyst's imagination, and those discussed here represent only one possible listing that can be used in performing a financial analysis.

Analyzing Financial Ratios

Two methods were demonstrated for analyzing financial ratios. The first involved trend analysis for the firm over time; the second involved making ratio comparisons with industry ratios. A set of industry ratios was presented and used in the analysis of a company. In addition, an integrated form of financial analysis based upon earning power focused the user's attention on the underlying determinants of a firm's profitability.

STUDY QUESTIONS

20-1. The basic financial statements of an organization consist of the balance sheet, income statement, and statement of changes in financial position. Describe the nature of each and explain how their functions differ.

20-2. Why is it that the preferred shareholders' equity section of the balance sheet changes only when new shares are sold, whereas the common equity section changes from year to year regardless of whether new shares are bought or sold?

20-3. Discuss two reasons why net income for a particular period does not necessarily reflect a firm's cash flow during that period.

20-4. The four basic groups of financial ratios are liquidity, efficiency, leverage, and profitability ratios. Discuss the nature of each group and list two example ratios that you would use to measure that aspect of a firm's financial condition.

20-5. Discuss briefly the two sources of standards or norms that can be used in performing ratio analyses.

20-6. Where can the analyst obtain industry norms? What limitations does the use of industry average ratios suffer from? Discuss briefly.

SELF-TEST PROBLEMS

ST-1. (*Ratio Analysis and Short-Term Liquidity*) Ray's Tool and Supply Company has been expanding its level of operations for the past two years. The firm's sales have grown rapidly as a result of the expansion in the economy. However, Ray's is a privately held company, and the only source of available funds it has is a line of credit with the firm's bank. The company needs to expand its inventories to meet the needs of its growing customer base but also wishes to maintain a current ratio of at least 3 to 1. If Ray's current assets are $6,000,000 and its current ratio is now 4 to 1, how much can it expand its inventories (financing the expansion with its line of credit) before the target current ratio is violated?

ST-2. (*Ratio Analysis of Loan Request*) On February 3, 1996, Mr. Jerry Simmons, chief financial officer for M & G Industries, contacted the firm's bank regarding a loan. The loan was to be used to repay notes payable and to finance current assets. Mr. Simmons wanted to repay the loan plus interest in one year. Upon receiving the loan request, the bank asked that the firm supply it with complete financial statements for the past two years. These statements are presented below:

M & G Industries Balance Sheets at the End of Calendar Year

	1994	1995
Cash	$ 9,000	$ 500
Accounts receivable	12,500	16,000
Inventories	29,000	45,500
Total current assets	$ 50,500	$ 62,000
Land	20,000	26,000
Buildings and equipment	70,000	100,000
Less: Allowance for depreciation	− 28,000	− 38,000
Total fixed assets	$ 62,000	$ 88,000
	$112,500	$150,000
Accounts payable	$ 10,500	$ 22,000
Bank notes	17,000	47,000
Total current liabilities	$ 27,500	$ 69,000
Long-term debt	28,750	22,950
Common shares	31,500	31,500
Retained earnings	24,750	26,550
	$112,500	$150,000

M & G Industries Income Statements for Years Ending December 31

	1994	1995
Sales	$125,000	$160,000
Cost of goods sold	75,000	96,000
Gross profit	$ 50,000	$ 64,000
Operating expense:		
Fixed cash operating expense	21,000	21,000
Variable operating expense	12,500	16,000
Depreciation	4,500	10,000
Total operating expense	38,000	47,000
Earnings before interest and taxes	$ 12,000	$ 17,000
Interest	3,000	6,100
Earnings before taxes	$ 9,000	$ 10,900
Taxes	4,500	5,450
Net income	$ 4,500	$ 5,450

a. Based upon the preceding statements, complete the following table.

M & G Industries Ratio Analysis

	Industry Averages	Actual 1994	Actual 1995
Current ratio	1.80		
Acid test ratio	.70		
Average collection period[a]	37 days		
Inventory turnover[a]	2.50 times		
Debt to total assets	58%		
Long-term debt to total capitalization	33%		
Times interest earned	3.8 times		
Gross profit margin	38%		
Operating profit margin	10%		
Net profit margin	3.5%		
Total asset turnover	1.14 times		
Fixed asset turnover	1.40 times		
Operating income return on investment	11.4%		
Return on total assets	4.0%		
Return on common equity	9.5%		

[a]Based on a 365-day year and on end-of-year figures.

b. Analyze Mr. Simmons' loan request. Would you grant the loan? Explain.

ST-3. (*Statement of Changes in Financial Position*)

a. Prepare a statement of changes in financial position for M & G Industries for 1995, using information given in Self-Test Problem ST-2.

b. How does this statement supplement your ratio analysis from Self-Test Problem ST-2? Explain.

STUDY PROBLEMS (SET A)

20-1A. (*Ratio Calculation*) Given the following information for the NcNabb Construction Company, calculate the firm's total asset turnover and return on equity ratios:

Net profit margin (net income ÷ sales)	8%
Return on total assets (net income ÷ total assets)	15%
Debt ratio (total liabilities ÷ total assets)	30%

20-2A. (*Balance Sheet Analysis*) Complete the following balance sheet using the information provided below:

Cash		Accounts payable	$ 100,000
Accounts receivable		Long-term debt	
Inventory		Total liabilities	
Current assets		Common equity	
Net fixed assets	1,500,000	Total	$2,100,000
Total	$2,100,000		

Current ratio = 6.0
Inventory turnover = 8.0
Debt ratio = 20%
Total asset turnover = 1.0
Average collection period = 30 days
Gross profit margin = 15%

20-3A. (*Ratio Analysis*) The Mitchem Marble Company has a target current ratio of 2 to 1 but has experienced some difficulties financing its expanding sales in the past few months. At present the current ratio of 2.5 to 1 is based upon current assets of $2.5 million. If Mitchem expands its receivables and inventories using its short-term line of credit, how much additional funding can it borrow before its current ratio standard is reached?

20-4A. (*Ratio Analysis*) The balance sheet and income statement for the J. P. Robard Mfg. Company are as follows:

Balance Sheet ($000)	
Cash	$ 500
Accounts receivable	2,000
Inventories	1,000
Current assets	$3,500
Net fixed assets	$4,500
Total assets	$8,000
Accounts payable	$1,100
Accrued expenses	600
Short-term notes payable	300
Current liabilities	$2,000
Long-term debt	$2,000
Owners' equity	$4,000
Total liabilities and owner's equity	$8,000

Income Statement ($000)	
Net sales (all credit)	$8,000
Cost of goods sold	(3,300)
Gross profit	$4,700
Operating expenses[a]	(3,000)
Net operating income	$1,700
Interest expense	(367)
Earnings before taxes	$1,333
Income taxes (40%)	(533)
Net income	$ 800

[a]Including depreciation expense of $500 for the year.

Calculate the following ratios:

Current ratio	Gross profit margin
Debt ratio	Operating profit margin
Times interest earned	Net profit margin
Average collection period	Operating return on investment
Inventory turnover	Return on total assets
Fixed asset turnover	Return on common equity
Total asset turnover	

20-5A. (*Analyzing Profitability*) The R. M. Smithers Corporation earned a net profit margin of 5 percent based on sales of $10 million and total assets of $5 million last year.

a. What was Smithers's rate of return on total assets?

b. During the coming year the company president has set a goal of attaining a 12 percent return on total assets. How much must firm sales rise, other things being the same, for the goal to be achieved? (State your answer as an annual growth rate in sales.)

c. If Smithers finances 30 percent of its assets by borrowing, what was its return on common equity for last year? What will it be next year if the return on total asset goal is achieved?

20-6A. (*Using Financial Ratios*) The Brenmar Sales Company had a gross profit margin of 30 percent and sales of $9 million last year. Seventy-five percent of the firm's sales are on credit while the remainder are cash sales. Brenmar's current assets equal $1,500,000, its current liabilities $300,000, and it has $100,000 in cash plus marketable securities.

a. If Brenmar's accounts receivable are $562,500, what is its average collection period?

b. If Brenmar reduces its average collection period to 20 days, what will be its new level of accounts receivable?

c. Brenmar's inventory turnover ratio is nine times. What is the level of Brenmar's inventories?

20-7A. (*Ratio Analysis of Loan Request*) Pamplin Inc. has recently applied for a loan from the National Bank to be used to expand the firm's inventory of soil pipe used in construction and agriculture. This expansion is predicted on expanded sales predicted for the coming year. Pamplin's financial statements for the two most recent years are as follows:

Pamplin Inc. Balance Sheet at 12/31/93 and 12/31/94

Assets

	12/31/93	*12/31/94*
Cash	$ 200	$ 150
Accounts receivable	450	425
Inventory	550	625
Current assets	1,200	1,200
Plant and equipment	2,200	2,600
Less: Accumulated depreciation	(1,000)	(1,200)
Net plant and equipment	1,200	1,400
Total assets	$2,400	$2,600

Liabilities and Owners' Equity

	1993	*1994*
Accounts payable	$ 200	$ 150
Notes payable—current (9%)	0	150
Current liabilities	200	300
Bonds	600	600
Owners' equity		
Common shares	300	300
Contributed surplus	600	600
Retained earnings	700	800
Total owners' equity	1,600	1,700
Total liabilities and owners' equity	$2,400	$2,600

Pamplin Inc. Income Statement – Year Ended 12/31/93 and 12/31/94

Income Statement	1993	1994
Sales	$1,200	$1,450
Cost of goods sold	700	850
Gross profit	$ 500	$ 600
Operating expenses	−30	−40
Depreciation	−220	−200
Net operating income	$ 250	$ 360
Interest expense	−50	−60
Net income before taxes	$ 200	$ 300
Taxes (40%)	−80	−120
Net income	$ 120	$ 180

a. Compute the following ratios for Pamplin Inc. from the financial statements provided above:

	1993	1994	Industry Norm
Current ratio			5.0 times
Acid test (quick) ratio			3.0 times
Inventory turnover			2.2 times
Average collection period			90 days
Debt ratio			.33
Times interest earned			7.0 times
Total asset turnover			0.75 times
Fixed asset turnover			1.0 times
Operating profit margin			.20
Net profit margin			.12
Return on total assets			.09

b. Based on your answer in (a) above, what are Pamplin's financial strengths and weaknesses?

c. Would you make the loan? Why or why not?

20-8A. (*Statement of Changes in Financial Position*) Prepare a statement of changes in financial position for Pamplin, Inc., for the year ended December 31, 1994 (Problem 20-7A).

20-9A. (*Statement of Changes in Financial Position*) (a) Prepare a statement of changes in financial position for the Waterhouse Co. in the year 1994. (b) What were the firm's primary sources and uses of cash?

	1993	1994
Cash	$ 75,000	$ 82,500
Receivables	102,000	90,000
Inventory	168,000	165,000
Prepaid expenses	12,000	13,500
Fixed assets	325,500	468,000
Accumulated depreciation	94,500	129,000
Patents	61,500	52,500
	$649,500	$742,500

	1993	1994
Accounts payable	$124,500	$112,500
Taxes payable	97,500	105,000
Mortgage payable	150,000	—
Preferred shares	—	225,000
Contributed surplus—preferred	—	6,000
Common shares	225,000	225,000
Retained earnings	52,500	69,000
	$649,500	$742,500

Additional Information:

1. The only entry in the accumulated depreciation account is the depreciation expense for the period.
2. The only entries in the retained earnings account are for dividends paid in the amount of $18,000 and for the net income for the year.
3. The income statement for 1994 is as follows:

Sales	$187,500
Cost of sales[a]	141,000
Gross profit	46,500
Operating expenses	12,000
Net income	$ 34,500

[a]Includes depreciation expense of $34,500.

20-10A. (*Review of Financial Statements*) Prepare a balance sheet and income statement at December 31, 1994, for the Sharpe Mfg. Co. from the scrambled list of items below.

Accounts receivable	$120,000
Machinery and equipment	700,000
Accumulated depreciation	236,000
Notes payable—current	100,000
Net sales	800,000
Inventory	110,000
Accounts payable	90,000
Long-term debt	160,000
Cost of goods sold	500,000
Operating expenses	280,000
Common shares	320,000
Cash	96,000
Retained earnings—prior year	?
Retained earnings—current year	?

20-11A. (*Financial Ratios—Investment Analysis*) The annual sales for Salco, Inc., were $4.5 million last year. The firm's end-of-year balance sheet appeared as follows:

Current assets	$ 500,000	Liabilities	$1,000,000
Net fixed assets	$1,500,000	Owners' equity	$1,000,000
	$2,000,000		$2,000,000

The firm's income statement for the year was as follows:

Sales	$4,500,000
Less: Cost of goods sold	(3,500,000)
Gross profit	1,000,000
Less: Operating expenses	(500,000)
Net operating income	500,000
Less: Interest expense	(100,000)
Earnings before taxes	400,000
Less: Taxes	(200,000)
Net income	$ 200,000

a. Calculate Salco's total asset turnover, operating profit margin, and operating income return on investment.

b. Salco plans to renovate one of its plants, which will require an added investment in plant and equipment of $1 million. The firm will maintain its present debt ratio of .5 when financing the new investment and expects sales to remain constant, while the operating profit margin will rise to 13 percent. What will be the new operating income return on investment for Salco after the plant renovation?

c. Given that the plant renovation in part (b) occurs and Salco's interest expense rises by $50,000 per year, what will be the return earned on the common shareholders' investment? Compare this rate of return with that earned before the renovation.

20-12A. (*Statement of Changes in Financial Position*) The consolidated balance sheets of the TMU Processing Company are presented below for June 1, 1993, and May 31, 1994 (in millions of dollars). TMU earned $14 million after taxes during the year ended May 31, 1994, and paid common dividends of $10 million.

	June 1,1993	May 31,1994	Change
Cash	$ 10	$ 8	
Accounts receivable	12	22	
Inventories	8	14	
Current assets	$ 30	$ 44	
Gross fixed assets	100	110	
Less: Accumulated depreciation	(40)	(50)	
Net fixed assets	$ 60	$ 60	
Total assets	$ 90	$104	
Accounts payable	$ 12	$ 9	
Notes payable	7	7	
Long-term debt	11	24	
Common shares	20	20	
Retained earnings	40	44	
Total liabilities and owners' equity	$ 90	$104	

a. Prepare a statement of changes in financial position for TMU Processing Company.
b. Summarize your findings.

20-13A. (*Preparing the Statement of Changes in Financial Position*) Comparative balance sheets for December 31, 1994, and December 31, 1995, for Abrams Mfg. Company are found below:

	1994	1995
Cash	$ 89,000	$100,000
Accounts receivable	64,000	70,000
Inventory	112,000	100,000
Prepaid expenses	10,000	10,000
Total current assets	275,000	280,000
Plant and equipment	238,000	311,000
Accumulated depreciation	(40,000)	(66,000)
Total assets	$473,000	$525,000
Accounts payable	$ 85,000	$ 90,000
Accrued liabilities	68,000	63,000
Total current debt	153,000	153,000
Mortgage payable	70,000	0
Preferred shares	0	120,000
Common shares	205,000	205,000
Retained earnings	45,000	47,000
Total debt and equity	$473,000	$525,000

Abrams's 1995 income statement is found below:

Sales (all credit)	$184,000
Cost of sales	150,000
Gross profit	34,000
Operating, interest, and tax expenses	10,000
Net income	$ 24,000

Additional information:

a. The only entry in the accumulated depreciation account is for 1995 depreciation.

b. The firm paid $22,000 in dividends during 1995.

Prepare a 1995 statement of changes in financial position.

20-14A. (*Analyzing the Statement of Changes in Financial Position*) Identify any financial weaknesses revealed in the statement of changes in financial position for the Westlake Manufacturing Co.

Westlake Manufacturing Co. Statement of Changes in Financial Position for the Year Ended December 31, 1995 ($000)

Cash flows from operating activities		
Net income (from the statement of income)	$540,000	
Adjustments for noncash items:		
Depreciation expense	60,000	
Adjustments for noncash working capital balances:		
Decrease in accounts receivable	40,000	
Increase in inventories	(240,000)	
Increase in prepaid expenses	(10,000)	
Decrease in accrued wages	(50,000)	
Net cash inflow from operating activities		$340,000
Cash flows from investing activities		
Sale (purchase) of plant and equipment	$2,400,000	
Net cash inflow from investment activities		$2,400,000
Cash flows from financing activities		
Cash inflows		
Issuance of bonds	$1,000,000	
Cash outflows		
Repayment of short-term debt	(3,000,000)	
Payment of long-term debt	(500,000)	
Payment of common dividends	(1,000,000)	
Net cash outflow from financing activities		($3,500,000)
Net increase (decrease) in cash during the period		($760,000)

20-15A. (*Comprehensive Financial Analysis Problem*) The T. P. Jarmon Company manufactures and sells a line of exclusive sportswear. The firm's sales were $600,000 for the year just ended, and its total assets exceed $400,000. The company was started by Mr. Jarmon just ten years ago and has been profitable every year since its inception. The chief financial officer for the firm, Brent Vehlim, has decided to seek a line of credit from the firm's bank totalling $80,000. In the past the company has relied on its suppliers to finance a large part of its needs for inventory. However, in recent months tight money conditions have led the firm's suppliers to offer sizable cash discounts to speed up payments for purchases. Mr. Vehlim wants to use the line of credit to supplant a large portion of the firm's payables during the summer months, which are the firm's peak seasonal sales period.

The firm's two most recent balance sheets were presented to the bank in support of its loan request. In addition, the firm's income statement for the year just ended was provided to support the loan request. These statements are found below:

T. P. Jarmon Company Balance Sheets for 12/31/93 and 12/31/94

Assets

	1993	1994
Cash	$ 15,000	$ 14,000
Marketable securities	6,000	6,200
Accounts receivable	42,000	33,000
Inventory	51,000	84,000
Prepaid rent	1,200	1,100
Total current assets	$115,200	$138,300
Net plant and equipment	286,000	270,000
Total assets	$401,200	$408,300

Liabilities and Shareholders' Equity

	1993	1994
Accounts payable	$ 48,000	$ 57,000
Notes payable	15,000	13,000
Accruals	6,000	5,000
Total current liabilities	$ 69,000	$ 75,000
Long-term debt	$160,000	$150,000
Common shareholders' equity	$172,200	$183,300
Total liabilities and equity	$401,200	$408,300

T. P. Jarmon Company Income Statement for the Year Ended 12/31/94

Sales		$600,000
Less: Cost of goods sold		460,000
Gross profits		$140,000
Less: Expenses		
General and administrative	$30,000	
Interest	10,000	
Depreciation	30,000	
Total		70,000
Profit before taxes		$ 70,000
Less: Taxes		27,100
Profits after taxes		$ 42,900
Less: Cash dividends		31,800
To retained earnings		$ 11,100

Jan Fama, associate credit analyst for the National Bank was assigned the task of analyzing Jarmon's loan request.

a. Calculate the financial ratios for 1994 corresponding to the industry norms provided below:

Ratio	Norm	Jarmon's Ratio	Evaluation
Current ratio	1.8 times		
Acid test ratio	.9 times		
Debt ratio	.5		
Long-term debt to total capitalization	.7		
Times interest earned	10 times		
Average collection period	20 days		
Inventory turnover (based on COGS)	7 times		
Return on total assets	8.4 %		
Gross profit margin	25 %		
Net profit margin	7 %		
Operating return on investment	16.8 %		
Operating profit margin	14 %		
Total asset turnover	1.2 times		
Fixed asset turnover	1.8 times		

b. Which of the ratios reported above in the industry norms do you feel should be most crucial in determining whether the bank should extend the line of credit? What strengths and weaknesses are apparent from your analysis of Jarmon's financial ratios?

c. Based upon the ratio analysis you performed in part (b), would you recommend approval of the loan request? Discuss.

d. Prepare a statement of changes in financial position for Jarmon covering the year ended December 31, 1994. How does this statement directly support your ratio analysis of Jarmon?

e. Use the Du Pont analysis to evaluate the firm's financial position.

INTEGRATIVE PROBLEM

An integrated oil and gas company engages in petroleum exploration and production worldwide. In addition, this firm engages in natural gas gathering and processing, as well as petroleum refining and marketing. The company has three operating groups—Exploration and Production, Gas and Gas Liquids, and Downstream Operations, which encompasses Petroleum Products and Chemicals.

In the mid-eighties, a major restructuring of the company resulted in a $4.5 billion plan to exchange a package of cash and debt securities for roughly half the company's shares and to sell $2 billion worth of assets. The company's long-term debt increased from $3.4 billion in late 1986 to a peak of $8.6 billion in April 1987.

During 1994, the company was able to strengthen its financial structure by its subsidiary completing an offering of $345 million of Series A 9.32 percent Cumulative Preferred shares. As a result of these actions and the prior year's debt reductions, the company lowered its long-term debt to assets ratio dramatically. In addition, the firm refinanced over a billion dollars of its debt at reduced rates. A company spokesperson said that, "Our debt-to-capital ratio is still on the high side, and we'll keep working to bring it down. But the cost of debt is manageable, and we're beyond the point where debt overshadows everything else we do."

Highlights of the company's financial condition spanning the years 1988–95 are found below (millions of dollars). These data reflect the modern history of the company as a result of its financial structuring following the downsizing and reorganization of the company which was begun in the mid-eighties.

a. Calculate the key financial ratios for the years spanning 1988–95.

b. Evaluate any trends you observe in the company's financial condition.

Selected Accounts of Balance Sheets	1988	1989	1990	1991	1992	1993	1994	1995
Cash and equivalents	$ 1,141	$ 1,006	$ 1,079	$ 708	$ 670	$ 114	$ 131	$ 119
Accounts receivable	1,047	1,269	1,261	1,351	1,595	1,258	1,268	1,248
Inventories	507	543	562	584	704	686	664	538
Other current assets	107	67	160	233	353	401	286	288
Current assets	2,802	2,885	3,062	2,876	3,322	2,459	2,349	2,193
Fixed assets	9,601	9,601	8,906	8,380	8,808	9,014	9,119	8,675
Total assets	$15,205	$19,932	$19,197	$17,901	$18,879	$17,952	$17,703	$16,347
Current liabilities	$ 2,234	$ 2,402	$ 2,468	$ 2,706	$ 2,910	$ 2,603	$ 2,517	$ 2,271
Long-term debt	5,758	5,419	4,761	3,939	3,839	3,876	3,718	3,208
Total liabilities	10,409	10,289	9,855	9,124	9,411	8,716	8,411	7,818
Preferred shares	270	205	0	0	0	0	359	362
Common equity	1,724	1,617	2,113	2,132	2,719	2,757	2,698	2,688
Total liabilities and common equity	$15,205	$19,932	$19,197	$17,901	$18,879	$17,952	$17,703	$16,347

Income Statements	1988	1989	1990	1991	1992	1993	1994	1995
Revenues	$10,018	$10,917	$11,490	$12,492	$13,975	$13,259	$12,140	$12,545
Operating expenses	−8,659	−9,922	−9,687	−11,311	−12,168	−12,351	−11,253	−11,729
Earnings before interest and tax	1,359	995	1,803	1,181	1,807	908	887	816
Interest expense	−685	−729	−688	−645	−620	−457	−376	−278
Earnings before tax	674	266	1,115	536	1,187	451	511	538
Taxes	−446	−231	−465	−317	−646	−353	−241	−293
Net income before extraordinary items	228	35	650	219	541	98	270	245
Extraordinary items	0	0	0	0	101	213	(46)	(2)
Changes in accounting principles	0	0	0	0	137	(53)	(44)	0
Net Income	228	35	650	219	779	258	180	243
Earnings per share	0.89	0.06	2.72	0.90	2.18	0.38	1.04	0.94
Dividend per share	2.02	1.73	1.34	0	1.03	1.12	1.12	—

CASE PROBLEM

L.M. MYERS, INC.

FINANCIAL ANALYSIS

L.M. Myers Inc. is one of the three largest grain exporters in North America. The firm also engages in soybean processing and several other related activities. During fiscal 1994–95 Myers derived 72.1 percent of its sales from its exporting activities, 14.8 percent from agriproducts, and the remainder from chemical and consumer products. Myers' nonexport sales are derived almost completely from soybean processing activities, including the production and sale of a number of food ingredients. One of the most promising soybean derivatives produced by the company is a newly developed meat substitute called "Prosoy." At present Prosoy is marketed almost strictly as a ground beef substitute; however, plans are under way to market the product in a number of other forms resembling familiar cuts of meat, such as bacon and even roasts. The success that the company has enjoyed with Prosoy in its initial three years of production promises to make soybean processing an even more important segment of the firm's overall sales. Other soybean-related products produced by the firm include a number of derivatives used in animal and poultry feeds. Myers' export business primarily involves corn, wheat, and some soybeans. The principal investment made by the firm related to its exporting operations involves a chain of grain elevators at strategic locations along the seacoast. These elevators are used to store grain and load it onto ships, which deliver it all over the world.

For the fiscal year just ended Myers experienced an overall sales growth of 10 percent. This increase represented a mere 5 percent increase in export-related activities and a whopping 20 percent increase in sales related to soybean processing. This and other factors have led the company to make a commitment to expand its processing capacity by $20 million during the next three years. The firm plans to finance the expansion through an $11 million bond issue and through the retention of earnings.

Owing to the seasonal nature of its export business, Myers has had to borrow heavily during the harvest months to finance seasonal inventory buildups and then repay the loans as sales are made throughout the year. In the past the company has arranged with a group of banks for a line of credit (discussed in Chapter 18) sufficient to meet its credit needs; however, in recent years this arrangement has become increasingly more cumbersome as the firm's total needs for funds have grown. This and cost considerations have led Myers' financial vice-president, James Graham, to consider the possibility of raising all or at least a part of the firm's credit needs through a commercial paper issue (discussed in Chapter 18). Mr. Graham is somewhat concerned about his firm's creditworthiness in light of the industry norms generally used by banks and other creditors. His concern relates to the fact that Myers has never issued commercial paper and the belief that only the most creditworthy of borrowers can successfully use the commercial paper market to raise funds.

Questions

a. Using the financial statements for Myers presented below, complete the calculation of the financial ratios contained in Exhibit 1.
b. Based upon your calculated ratios and the associated industry norms, what is your financial analysis of Myers?

L. M. Myers Inc. Income Statements for Years Ended December 31 ($000)

	1994	1995
Sales (net)	$706,457	$777,104
Less: Cost of goods sold	(637,224)	(662,093)
Gross profit	$ 69,233	$115,011
Operating expenses:		
Selling and administrative expenses	(17,612)	(17,804)
Labour expense	(9,418)	(15,263)
Depreciation	(6,976)	(7,428)
Miscellaneous operating expenses	(1,887)	(2,011)
Total	($ 35,893)	($ 42,506)
Net operating income	$ 33,340	$ 72,505
Interest income	512	2,012
	33,852	74,517
Less: Interest expense	(9,127)	(8,408)
Earnings before taxes	$ 24,725	$ 66,109
Less: Taxes payable	(5,440)	(25,232)
Net income	$ 19,285	$ 40,877
Less: Preferred dividends	(58)	(56)
Net earnings available to common	$ 19,227	$ 40,821

L. M. Myers Inc. Balance Sheets for Years Ended December 31 ($000)

Assets

	1994	1995
Cash	$ 11,451	$ 12,844
Accounts receivable	64,199	52,599
Marketable securities	—	33,995
Inventories	69,814	75,366
Deferred income taxes	2,948	2,750
Prepaid expenses	1,089	1,794
Total current assets	$149,501	$179,348
Investments and advances	11,681	12,012
Other assets	14,509	3,735
Net property and equipment	118,810	153,856
Total assets	$294,501	$348,951

Liabilities and Shareholders' Equity

	1994	1995
Accounts payable	$ 34,327	$ 35,099
Notes payable	14,544	20,907
Accrued income and other taxes	28,526	40,112
Other accrued expenses	19,854	22,299
Total current liabilities	$ 97,251	$118,417
Long-term debt	75,817	67,006
Cumulative preferred shares	582	565
Common shares	26,596	26,812
Contributed surplus	2,030	2,606
Retained earnings	92,225	133,545
Total shareholders' equity	121,433	163,528
Total liabilities and shareholders' equity	$294,501	$348,951

Exhibit 1 Financial Ratios for L.M. Myers Inc.

Ratio	1994	1995	Industry Norm[a]
Acit test ratio			1.00 times
Current ratio			1.61 times
Average collection period[b]			30.00 days
Inventory turnover[b]			10.1 times
Debt ratio			49.1%
Long-term debt to total capitalization			31.0%
Times interest earned			5.87 times
Cash flow overall coverage ratio			7.42
Total asset turnover			2.1 times
Gross profit margin			11.7%
Operating profit margin			7.6%
Operating income return on investment			14.75%
Return on total assets			8.45%
Return on common equity			21.39%

[a]These industry norms pertain to Myers' grain export operations, which comprised over 70 percent of the firm's sales for 1994. Also, the industry averages are applicable to both 1994 and 1995.

[b]Compute using end-of-year figures and assuming all sales are credit sales.

STUDY PROBLEMS (SET B)

20-1B. (*Ratio Calculation*) Given the following information for the Marcus Food Distributing Company, calculate the firm's total asset turnover and return on equity ratios:

Net profit margin (net income ÷ sales)	12%
Return on total assets (net income ÷ total assets)	15%
Debt ratio (total liabilities ÷ total assets)	45%

20-2B. (*Balance Sheet Analysis*) Complete the following balance sheet using the information provided below:

Cash		Accounts payable	$ 100,000
Accounts receivable		Long-term debt	
Inventory		Total liabilities	
Current assets		Common equity	
Net fixed assets	1,000,000	Total	$1,300,000
Total	$1,300,000		

Current ratio = 3.0
Inventory turnover = 10.0
Debt ratio = 30%
Total asset turnover = 0.5
Average collection period = 45 days
Gross profit margin = 30%

20-3B. (*Ratio Analysis*) The Allendale Office Supply Company has a target current ratio of 2 to 1 but has experienced some difficulties financing its expanding sales in the past few months. At present the current ratio of 2.75 to 1 is based upon current assets of $3.0 million. If Allendale expands its receivables and inventories using its short-term line of credit, how much additional funding can it borrow before its current ratio standard is reached?

20-4B. (*Ratio Analysis*) The balance sheet and income statement for the Simsboro Paper Company are as follows:

Balance Sheet ($000)	
Cash	$1,000
Accounts receivable	1,500
Inventories	1,000
Current assets	$3,500
Net fixed assets	$4,500
Total assets	$8,000
Accounts payable	$1,000
Accrued expenses	600
Short-term notes payable	200
Current liabilities	$1,800
Long-term debt	2,100
Owners' equity	$4,100
	$8,000

Income Statement ($000)	
Net sales (all credit)	$7,500
Cost of goods sold	(3,000)
Gross profit	$4,500
Operating expenses[a]	(3,000)
Net operating income	$1,500
Interest expense	(367)
Earnings before taxes	$1,133
Income taxes (40%)	(415)
Net income	$680

[a]Including depreciation expense of $500 for the year.

Calculate the following ratios:

Current ratio	Gross profit margin
Debt ratio	Operating profit margin
Times interest earned	Net profit margin
Average collection period	Operating return on investment
Inventory turnover	Return on total assets
Fixed asset turnover	Return on common equity
Total asset turnover	

20-5B. (*Analyzing Profitability*) The R. M. Smithers Corporation earned a net profit margin of 6 percent based on sales of $11 million and total assets of $6 million last year.

a. What was Smithers' rate of return on total assets?

b. During the coming year the company president has set a goal of attaining a 13 percent return on total assets. How much must firm sales rise, other things being the same, for the goal to be achieved? (State your answer as an annual growth rate in sales.)

c. If Smithers finances 25 percent of its assets by borrowing, what was its return on common equity for last year? What will it be next year if the return on total asset goal is achieved?

20-6B. (*Using Financial Ratios*) Edgar Smith Inc. had a gross profit margin of 25 percent and sales of $9.75 million last year. Seventy-five percent of the firm's sales are on credit while the remainder are cash sales. The company's current assets equal $1,550,000, its current liabilities $300,000, and it has $150,000 in cash plus marketable securities.

a. If Smith's accounts receivable are $562,500, what is its average collection period?

b. If Smith reduces its average collection period to 20 days, what will be its new level of accounts receivable?

c. Smith's inventory turnover ratio is eight times. What is the level of the firm's inventories?

20-7B. (*Ratio Analysis of Loan Request*) The J. B. Champlin Corporation has experienced two consecutive years of improved sales and foresees the need to increase its investment in inventories and receivables to meet yet a third year of increased sales. Champlin's President, J. B. Champlin, Jr., recently approached the company's bank to discuss the possibility of extending the firm's line of credit to cover the firm's projected future funds requirements. The financial statements for the firm are found below:

J. B. Champlin Corporation Balance Sheet at 12/31/94 and 12/31/95 ($000)

Assets

	1994	1995
Cash	$ 225	$ 175
Accounts receivable	450	430
Inventory	575	625
Total current assets	$1,250	$1,230
Plant and equipment	2,200	2,500
Less: Accumulated depreciation	(1,000)	(1,200)
Net plant and equipment	1,200	1,300
Total assets	$2,450	$2,530

Liabilities and Shareholders' Equity

	1994	1995
Accounts payable	$ 250	$ 115
Notes payable—current (9%)	0	115
Current liabilities	250	230
Bonds	600	600
Owners' equity		
Common shares	300	300
Contributed surplus	600	600
Retained earnings	700	800
Total owners' equity	1,600	1,700
Total liabilities and equity	$2,450	$2,530

J. B. Champlin Corporation Income Statement
Year Ended 12/31/94 and 12/31/95

	1994	1995
Sales	$1,250	$1,450
Cost of goods sold	−700	−875
Gross profit	$ 550	$ 575
Operating expenses	−30	−45
Depreciation	−220	−200
Net operating income	$ 300	$ 330
Interest expense	−50	−60
Net income before taxes	$ 250	$ 270
Taxes (40%)	−100	−108
Net income	$ 150	$ 162

a. Compute the following ratios for Champlin from the financial statements provided above:

	1994	1995	Industry Norm
Current ratio			5.0
Acid test (quick) ratio			3.0
Inventory turnover			2.2
Average collection period			90 days
Debt ratio			.33%
Times interest earned			7.0 times
Total asset turnover			.75 times
Fixed asset turnover			1.0 times
Operating profit margin			20%
Net profit margin			12%
Return on total assets			9%

b. Based on your answer in (a) above, what are Champlin's financial strengths and weaknesses?

c. Would you make the loan? Why or why not?

20-8B. (*Statement of Changes in Financial Position*) Prepare a statement of changes in financial position for Champlin for the year ended December 31, 1995 (Problem 20-7B).

20-9B. (*Statement of Changes in Financial Position*) (a) Prepare a statement of changes in financial position for Cramer Inc. for 1995. (b) What were the firm's primary sources and uses of cash?

Cramer Inc. for 1995

	1994	1995
Cash	$ 76,000	$ 82,500
Receivables	100,000	91,000
Inventory	168,000	163,000
Prepaid expenses	11,500	13,500
Fixed assets	325,500	450,000
Accumulated depreciation	−94,500	−129,000
Patents	61,500	52,500
	$648,000	$723,500
Accounts payable	$123,000	$ 93,500
Taxes payable	97,500	105,000
Mortgage payable	150,000	—
Preferred shares	—	225,000
Contributed surplus—preferred	—	6,000
Common shares	225,000	225,000
Retained earnings	52,500	69,000
	$648,000	$723,500

Additional Information:

1. The only entry in the accumulated depreciation account is the depreciation expense for the period.
2. The only entries in the retained earnings account are for dividends paid in the amount of $20,000 and for the net income for the year.
3. The income statement for 1995 is as follows:

Sales	$190,000
Cost of sales[a]	140,000
Gross profit	50,000
Operating expenses	13,500
Net income	$ 36,500

[a]Includes depreciation of $34,500.

20-10B. (*Basic Financial Statements*) Prepare a balance sheet and income statement at December 31, 1995, for the Sabine Mfg. Co. from the scrambled list of items below.

Accounts receivable	$150,000
Machinery and equipment	700,000
Accumulated depreciation	236,000
Notes payable—current	90,000
Net sales	900,000
Inventory	110,000
Accounts payable	90,000
Long-term debt	160,000
Cost of goods sold	550,000
Operating expenses	280,000
Common shares	320,000
Cash	90,000
Retained earnings—prior year	?
Retained earnings—current year	?

20-11B. (*Financial Ratios—Investment Analysis*) The annual sales for Salco, Inc., were $5,000,000 last year. The firm's end-of-year balance sheet appeared as follows:

Current assets	$ 500,000	Liabilities	$1,000,000
Net fixed assets	$1,500,000	Owners' equity	$1,000,000
	$2,000,000		$2,000,000

The firm's income statement for the year was as follows:

Sales	$5,000,000
Less: Cost of goods sold	(3,000,000)
Gross profit	2,000,000
Less: Operating expenses	(1,500,000)
Net operating income	500,000
Less: Interest expense	(100,000)
Earnings before taxes	400,000
Less: Taxes	(250,000)
Net income	$ 150,000

a. Calculate Salco's total asset turnover, operating profit margin, and operating income return on investment.

b. Salco plans to renovate one of its plants, which will require an added investment in plant and equipment of $1 million. The firm will maintain its present debt ratio of .5 when financing the new investment and expects sales to remain constant, while the operating profit margin will rise to 13 percent. What will be the new operating income return on investment for Salco after the plant renovation?

c. Given that the plant renovation in part (b) occurs and Salco's interest expense rises by $40,000 per year, what will be the return earned on the common shareholders' investment? Compare this rate of return with that earned before the renovation.

20-12B. (*Statement of Changes in Financial Position*) The consolidated balance sheets of the SMU Processing Company are presented below for June 1, 1994 and May 31, 1995 (in millions of dollars). SMU earned $14 million after taxes during the year ended May 31, 1995, and paid common dividends of $10 million.

	June 1, 1994	May 31, 1995
Cash	$ 10	$ 8
Accounts receivable	12	22
Inventories	8	14
Current assets	$ 30	$ 44
Gross fixed assets	100	110
Less: Accumulated depreciation	(40)	(50)
Net fixed assets	$ 60	$ 60
Total assets	$ 90	$104
Accounts payable	$ 12	$ 9
Notes payable	7	7
Long-term debt	11	24
Common shares	20	20
Retained earnings	40	44
Total liabilities and owners' equity	$ 90	$104

a. Prepare a statement of changes in financial position for SMU Processing Company.

b. Summarize your findings.

20-13B. (*Preparing the Statement of Changes in Financial Position*) Comparative balance sheets for December 31, 1994 and December 31, 1995 for J. Ng Company are found below:

	1995	1994
Cash	$ 70,000	$ 89,000
Accounts receivable	70,000	64,000
Inventory	80,000	102,000
Prepaid expenses	10,000	10,000
Total current assets	230,000	265,000
Plant and equipment	301,000	238,000
Accumulated depreciation	(66,000)	(40,000)
Total assets	$465,000	$463,000
Accounts payable	$ 80,000	$ 85,000
Accrued liabilities	63,000	68,000
Total current debt	143,000	153,000
Mortgage payable	0	60,000
Preferred shares	70,000	0
Common shares	205,000	205,000
Retained earnings	47,000	45,000
Total debt and equity	$465,000	$463,000

Ng's 1995 income statement is found below:

Sales (all credit)	$204,000
Cost of sales	−160,000
Gross profit	44,000
Operating, interest, and tax expenses	− 10,000
Net income	$ 34,000

Additional information:

1. The only entry in the accumulated depreciation account is for 1995 depreciation.
2. The firm paid $32,000 in dividends during 1995.

Prepare a 1995 statement of changes in financial position.

20-14B. (*Analyzing the Statement of Changes in Financial Position*) Identify any financial weaknesses revealed in the statement of changes in financial position for the Simsboro Pulpwood Mill, Inc.

Simsboro Pulpwood Mill Inc., Statement of Changes in Financial Position for the Year Ended December 31, 1995 ($000)

Cash flows from operating activities		
Net income (from the statement of income)	$ 500,000	
Adjustments for noncash items:		
Depreciation expense	600,000	
Adjustments for noncash working capital balances:		
Decrease in accounts receivable	200,000	
Increase in inventories	(400,000)	
Increase in prepaid expenses	(100,000)	
Decrease in accrued wages	(50,000)	
Net cash inflow from operating activities		$750,000
Cash flows from investing activities		
Sale (purchase) of plant and equipment	($2,250,000)	
Net cash outflow from investment activities		($2,250,000)
Cash flows from financing activities		
Cash inflows		
Issuance of bonds	$5,000,000	
Cash outflows		
Repayment of short-term debt	(3,000,000)	
Payment of long-term debt	(500,000)	
Payment of common dividends	(1,000,000)	
Net cash inflow from financing activities		$ 500,000
Net increase (decrease) in cash during the period		($1,000,000)

20-15B. (*Comprehensive Financial Analysis Problem*) RPI Inc. is a manufacturer and retailer of high-quality sports clothing and gear. The firm was started several years ago by a group of sports enthusiasts who felt there was a need for a firm that could provide serious outdoors enthusiasts with quality products at reasonable prices. The result was RPI Inc. Since its inception the firm has been profitable with sales that last year totalled $700,000 and assets in excess of $400,000. The firm now finds its growing sales outstrip its ability to finance its inventory needs. The firm now estimates that it will need a line of credit of $100,000 during the coming year. To finance this funding requirement the management plans to seek a line of credit with its bank.

The firm's most recent financial statements were provided to its bank as support for the firm's loan request. (See the top two tables on page 771.) Joanne Peebie, a loan analyst trainee for the bank, has been assigned the task of analyzing the firm's loan request.

a. Calculate the financial ratios for 1995 corresponding to the industry norms from the table at the bottom of page 771.

b. Which of the ratios reported above in the industry norms do you feel should be most crucial in determining whether the bank should extend the line of credit? What strengths and weaknesses are apparent from your analysis of RPI's financial ratios?

c. Based upon the ratio analysis you performed in part (b), would you recommend approval of the loan request? Discuss.

d. Prepare a statement of changes in financial position for RPI covering the year ended December 31, 1995. How does this statement directly support your ratio analysis of RPI?

e. Use the Du Pont analysis to evalute the firm's financial position.

RPI Inc. Balance Sheets for 12/31/94 and 12/31/95

Assets

	1994	1995
Cash	$ 16,000	$ 17,000
Marketable securities	7,000	7,200
Accounts receivable	42,000	38,000
Inventory	50,000	93,000
Prepaid rent	1,200	1,100
Total current assets	$116,200	$156,300
Net plant and equipment	286,000	290,000
Total assets	$402,200	$446,300

Liabilities and Shareholders' Equity

	1994	1995
Accounts payable	$ 48,000	$ 55,000
Notes payable	16,000	13,000
Accruals	6,000	5,000
Total current liabilities	$ 70,000	$ 73,000
Long-term debt	$160,000	$150,000
Common shareholders' equity	$172,200	$223,300
Total liabilities and equity	$402,200	$446,300

RPI Inc. Income Statement for the Year Ended 12/31/95

Sales		$700,000
Less: Cost of goods sold		−500,000
Gross profits		$200,000
Less: Expenses		
General and administrative	$50,000	
Interest	10,000	
Depreciation	30,000	
Total		−90,000
Profit before taxes		$110,000
Less: Taxes (38%)		−27,100
Profits after taxes		$ 82,900
Less: Cash dividends		−31,800
Total retained earnings		$ 51,100

Ratio	Norm	RPI Ratio	Evaluation
Current ratio	1.8 times		
Acid test ratio	9 times		
Debt ratio	.5		
Long-term debt to total capitalization	.7		
Times interest earned	10 times		
Average collection period	20 days		
Inventory turnover (based on COGS)	7 times		
Return on total assets	8.4%		
Gross profit margin	25%		
Net profit margin	7%		
Operating return on investment	16.8%		
Operating profit margin	14%		
Total asset turnover	1.2 times		
Fixed asset turnover	1.8 times		

SELF-TEST SOLUTIONS

SS-1. Note that Ray's current ratio before the inventory expansion is as follows:

$$\text{current ratio} = \frac{\$6{,}000{,}000}{\text{current liabilities}} = 4$$

Thus, the firm's present level of current liabilities is $1,500,000. If the expansion in inventories is financed entirely with borrowed funds, then the change in inventories, ΔInv, is equal to the change in current liabilities and the firm's current ratio after the expansion can be defined as follows:

$$\text{current ratio} = \frac{(\$6{,}000{,}000 + \Delta\text{Inv})}{(\$1{,}500{,}000 + \Delta\text{Inv})} = 3$$

Note that we set the new current ratio equal to the firm's target of 3 to 1. Solving for the value of ΔInv in the above equation, we determine that the firm can expand its inventories and finance the expansion with current liabilities by $750,000 and still maintain its target current ratio.

SS-2. a.

M & G Industries Ratio Analysis

	Industry Averages	1994	1995
Current ratio	1.80	1.84	0.90
Acid test ratio	0.70	0.78	0.24
Average collection period	37 days	36.5 days	36.5 days
Inventory turnover	2.50 times	2.59 times	2.11 times
Debt ratio	58%	50%	61.3%
Long-term debt to total capitalization	33%	33.8%	28.3%
Times interest earned	3.8 times	4.0 times	2.79 times
Gross profit margin	38%	40.0%	40.0%
Operating profit margin	10%	9.6%	10.6%
Net profit margin	3.5%	3.6%	3.4%
Total asset turnover	1.14 times	1.11 times	1.07 times
Fixed asset turnover	1.40 times	2.02 times	1.82 times
Operating income return on investment	11.4%	10.6%	11.3%
Return on total assets	4.0%	4.0%	3.6%
Return on common equity	9.5%	8.0%	9.4%

b. It appears that M & G is in a very weak position to request an additional loan. An examination of the ratios computed in part (a) shows that the liquidity of M & G has decreased considerably during the past year to a point well below the industry average. In addition, its debt ratio has risen to a point above the industry average, although the difference is not particularly large. However, an additional loan would increase this difference. The times interest earned ratio has decreased significantly over the preceding year to a point well below the industry norm of 3.8 times. The firm's profit margins are very near the respective norms. The return on the owners' investment is good, but the firm's low liquidity and extensive use of financial leverage do not warrant approval of the loan.

SS-3. a.

M & G Industries Statement of Changes in Financial Position for the Year Ended December 31, 1995		
Cash flows from operating activities		
Net income (from the statement of Income)	$5,450	
Add (deduct) to reconcile net income to net cash flow		
Increase in accounts payable	$11,500	
Increase in inventories	($16,500)	
Depreciation expense	$10,000	
Increase in accounts receivable	($3,500)	
Net cash inflow from operating activities		$6,950
Cash flows from investing activities		
Cash inflows		
Cash outflows		
Purchase of land	($6,000)	
Purchase of plant and equipment	($30,000)	
Net cash outflow from investment activities		($36,000)
Cash flows from financing activities		
Cash inflows		
Increase in bank notes	$30,000	
Cash outflows		
Decrease in long-term debt	($5,800)	
Common dividends	($3,650)	
Net cash inflow from financing activities		$20,550
Effect of foreign exchange rates		—
Net increase (decrease) in cash during the period		($8,500)
Cash balance at the beginning of the period	$9,000	
Cash balance at the end of the period	$ 500	

b. The statement of changes in financial position is an important supplement to ratio analysis. This statement directs the analysts' attention to where M & G Industries obtained financing during the period and how those funds were spent. For example, 52 percent of M & G's funds came from an increase in bank notes, while 20 percent came from an increase in accounts payable. In addition, the largest uses of funds were additions to buildings and equipment and increases in inventories. Thus, M & G did little in the most recent operating period to alleviate the financial problems we noted earlier in our ratio analysis. In fact, M & G aggravated matters by purchasing fixed assets using short-term sources of financing. It would appear that another short-term loan at this time is not warranted.

APPENDIX 20A

THE LANGUAGE OF FINANCE: THE ACCOUNTING STATEMENTS

In this appendix, we will:

1. Look at the format of basic financial statements;
2. Review some of the more important accounting principles used in reporting a firm's financial activities;
3. Describe briefly some of the significant relationships in the accounting data from the perspective of a financial manager;
4. Offer a brief caveat about the appropriate use of accounting information, and introduce the Free Cash Flow concept.

BASIC FINANCIAL STATEMENTS

Balance Sheet

The **balance sheet** represents a statement of the financial position of the firm on a given date, including its asset holdings, liabilities, and owner-supplied capital. Assets represent the resources owned by the firm, whereas liabilities and owner's equity indicate how those resources were financed. Table 20A-1 gives an example balance sheet for Jimco, Inc., as of December 31, 1995. Jimco had $31 million in assets, which it financed with $10 million in current (short-term) liabilities that must be repaid within the current year, $10,700,000 in noncurrent (long-term) liabilities, and $10,300,000 in owner-supplied funds. (Each term used in the balance sheet is defined in Appendix 20B.)

Limitations of the Balance Sheet

A firm's balance sheet is typically prepared within the guidelines of "generally accepted accounting principles."[21] However, we must be aware of the following limitations of the statement:

1. The balance sheet does not reflect current value, because accountants have adopted historical cost as the basis for valuing and reporting assets and liabilities.
2. Estimates must be used to determine the level of several accounts according to different criteria. Examples include accounts receivable estimated in terms of collectibility, inventories based on saleability, and fixed (noncurrent) assets based on useful life.[22]

---

[21]The sources of accounting principles are many; however, the main contributors include: The Accounting Standards Committee, The Auditing Standards Committee, and The Accounting and Auditing Guidelines from the Canadian Institute of Chartered Accountants.

[22]In the opinion of the Accounting Standards Committee of the CICA, the preparation of financial statements requires estimating the effects of subsequent events (CICA Handbook 3820.01–.13 and 1506.21–.23). Examples of items for which estimates are necessary include doubtful accounts receivables, inventory obsolescence, the service lives and salvage values of depreciable assets, the periods expected to derive benefit from a deferred cost, liabilities under warranties, and the size and value of mineral reserves. Since subsequent events cannot be perceived with certainty, estimating requires the exercise of judgment. The implication here is that no reference guidelines can be constructed regarding these estimates; thus subjectivity enters in determining the affected accounts.

Table 20A-1 Jimco Inc. Balance Sheet December 31, 1995 ($000)

Assets			
Current assets			
Cash		$ 1,400	
Marketable securities—at cost (market value, $320)		300	
Accounts receivable		10,000	
Inventories		12,000	
Prepaid expenses		300	
Total current assets			$24,000
Fixed assets			
Land		2,000	
Plant and equipment	$12,300		
Less: Accumulated depreciation	7,300		
Net plant and equipment		5,000	
Total fixed assets			7,000
Total assets			$31,000

Liabilities and Owners' Equity		
Current liabilities		
Accounts payable	$ 3,000	
Notes payable, 9%, due March 1, 1996	3,400	
Accrued salaries, wages, and other expenses	3,100	
Current portion of long-term debt	500	
Total current liabilities		$10,000
Long-term liabilities		
Deferred income taxes	1,500	
First mortgage bonds, 7%, due January 1, 2000	6,300	
Debentures, 8%, due June 30, 2000	2,900	
Total long-term liabilities		10,700
Owners' equity		
Common shares (100,000 Shares Outstanding)	100	
Contributed Surplus	2,000	
Retained earnings	8,200	
Total owners' equity		10,300
Total liabilities and owners' equity		$31,000

3. The depreciation of long-term assets is accepted practice; however, appreciation or enhancement in asset values is generally ignored.[23] This is particularly crucial to firms holding large investments in appreciable assets such as land, timberlands, and mining properties.
4. Many items that have financial value are omitted from the balance sheet because they involve extreme problems of objective evaluation. The most obvious example consists of the human resources of the firm.[24]

[23] *CICA Handbook*, 3060.01–3060.06

[24] The subject of human resource accounting has received increased attention in recent years. For an overview of the subject, see Edwin H. Caplan and Stephen Landeckich, Human Resource Accounting: Past, Present, and Future (National Association of Accountants, 1974); and Eric Flamholtz, Human Resource Accounting (Encino, CA: Dickinson Publishing Company, 1974).

In most cases little can be done to alleviate these shortcomings; however, we should at least be aware of their existence in order to temper the analysis accordingly.

Income Statement

The **income statement** represents an attempt to measure the net results of the firm's operations over a specified interval, such as one quarter or one year. The income statement (sometimes referred to as a **profit and loss statement**) is compiled on an accrual basis rather than a cash basis. This means that an attempt is made to match the firm's revenues from the period's operations with the expenses incurred in generating those revenues. A condensed income statement for the year ended December 31, 1995, is provided in Table 20A-2 for Jimco, Inc. (The terms used in the income statement are defined in Appendix 20B.)

In looking at the income statement, it is helpful to think of it as comprising three types of activities: (1) the cost of producing or acquiring the goods or services sold; (2) the expenses incurred in marketing and distributing the product or service to the customer, along with administrative operating expenses; and (3) the financing costs of doing business, for example, interest expense and dividend payments to the preferred stockholders. Figure 20A-1 shows these three "income-statement activities."

Net Income and Cash Flow

As already suggested, the reported revenues and expenses need not represent actual cash flows for the period, so that the computed net earnings for the period do not equal the actual cash provided by the firm's operations. There are two basic reasons why the firm's net income does not equal net cash flow for the period. First, revenues and expenses are included in the income statement even though no cash flow might have occurred. For example, sales revenues consist of credit as well as cash sales. Furthermore, cash collections from prior period credit sales are not reflected in the current period's sales revenues. In addition, the expenses for the

Table 20A-2 Jimco Inc. Statement of Income for the Year Ended December 31, 1995 ($000 except per share data)

Net sales			$51,000
Cost of goods sold			(38,000)
Gross profit			$13,000
Operating expenses			
Selling expenses		$ 3,100	
Depreciation expense		500	
General and administrative expense		5,400	(9,000)
Net operating income (NOI) or Earnings before interest and taxes (EBIT)			$ 4,000
Interest expense			(1,000)
Earnings before taxes (EBT)			$ 3,000
Income tax provision[a]			(1,200)
Net income (NI)			$ 1,800
Disposition of net income			
Dividends of common shares			$ 300
Change in retained earnings			1,500
Per share data (dollars)			
Number of common shares			100,000 shares
Earnings per common share	($1,800,000 / 100,000 shares)		$ 18
Dividends per common share	($300,000 / 100,000 shares)		$ 3

[a]A tax rate of 40% on all income is assumed here for simplicity.

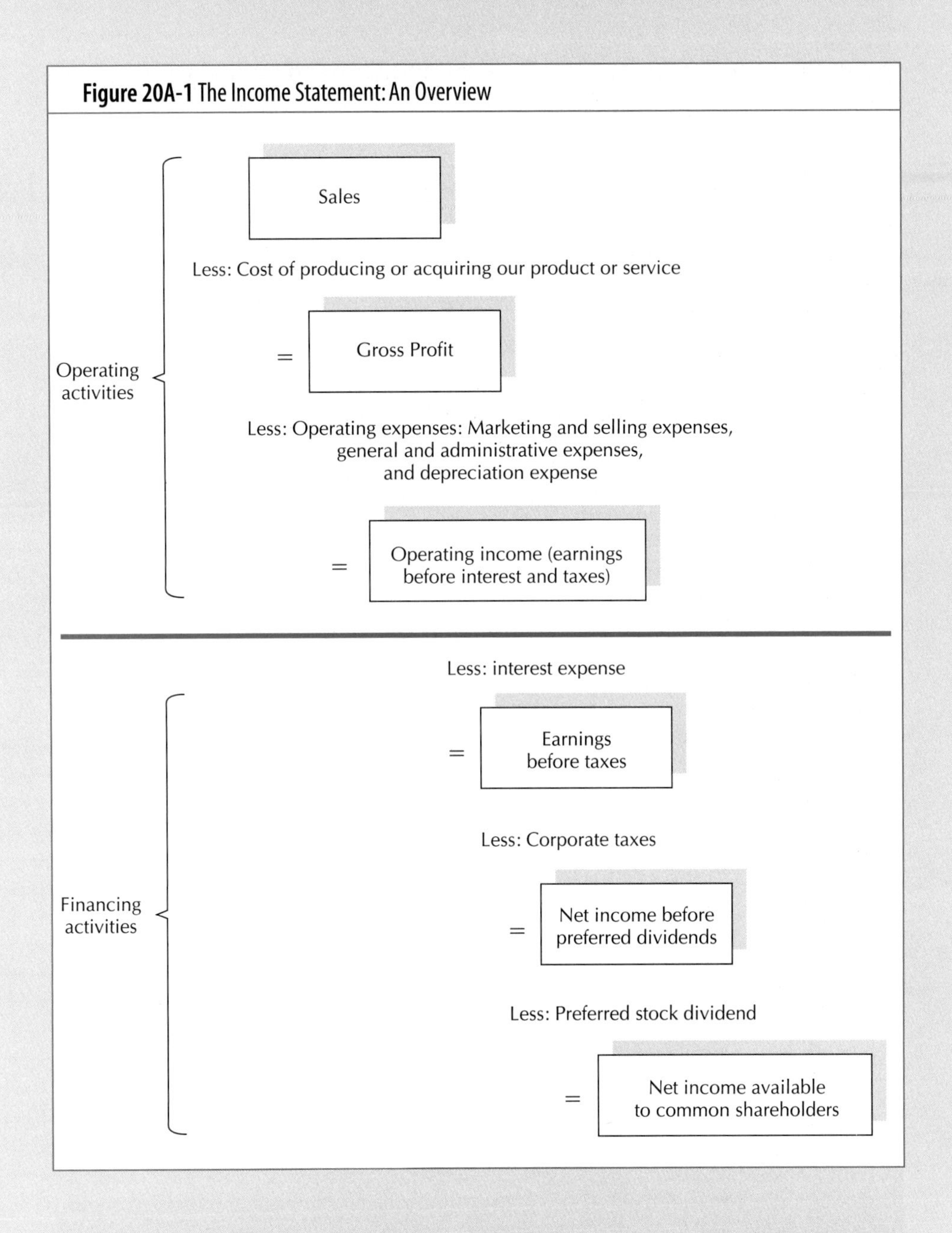

Figure 20A-1 The Income Statement: An Overview

period represent all those expenditures made in the process of generating the period's revenues. Thus, wages, salaries, utilities, and other expenses may not be paid during the period in which they are recognized in the income statement. Second, certain expenses included within the income statement are not cash expenses at all. For example, depreciation expense does not involve a cash outflow to the firm, yet it is deducted from revenues for the period in computing net income. Other examples of noncash expenses include the amortization of goodwill, patent rights, and bond discounts. Finally, some cash outflows will not be shown as "expenses" in the income statement, but will be recorded on the balance sheet as investments, such as plant and equipment. We will elaborate often on the difference between cash flow and net income, not just when we prepare the cash budget in Chapter 21 or when we prepare a statement of changes in financial position in Chapter 20, but also when we analyze investment and financing decisions.

BASIC ACCOUNTING PRINCIPLES

There are a vast number of "generally acceptable accounting principles" (GAAP) that have been developed over the years. These guidelines cover almost every imaginable situation for reporting financially related transactions of the firm. Some of the more important ones for the study of finance include: (1) the historical cost principle; (2) the accrual or matching principle; (3) inventory valuation guidelines; and (4) methods for reporting the depreciation of long-term (fixed) assets, such as plant and equipment.

Historical Cost Principle

The historical cost principle provides the basis for determining the book values of the firm's balance sheet accounts. The primary advantage of historical cost is its objectivity. Its primary disadvantage relates to the fact that the asset balances do not correspond to market values or replacement costs. Furthermore, since asset book values are not equal to market values, the book value of owners' equity does not equal its market value. In analyzing the firm's financial statements, the analyst must keep in mind that asset balances reflect historical costs of the related assets and not current market values.

Matching Principle

To compute net income for a given year's operations, accountants must identify all revenue and expense items that belong within that year. The **accrual basis of accounting** attempts to allocate revenue and expense properly among the years that an enterprise is in operation. This method utilizes the **revenue realization principle** to assure that revenue is recognized in the period in which it is earned, and it uses the **matching principle** to determine the expenses necessary during the period for the revenue to be generated. These expenses are thus matched against revenue for the period. The basic principle, therefore, involves matching expenses with the revenues, or "Let the expense follow the revenue." Thus, wage expense is not recognized when it is paid nor necessarily when work is performed, but when the work performed actually contributes to revenues.

An example of the potential difficulties posed for the analyst in implementation of the matching principle can be found in the depreciation policies followed by different firms for like assets. For example, one major food processor, referred to here as food processor A, depreciates its machines over 10 years. Another major food processor, food processor B, depreciates its machines over 16 years. Other things being equal, food processor A would report higher expenses and lower profits from its operations than would food processor B. The difficulties encountered with the different depreciation methods used for accounting purposes do not occur for tax purposes since the majority of assets are required to be depreciated by the declining balance method.

ACCOUNTING FOR INVENTORIES AND COST OF GOODS SOLD

Several methods can be used to determine a firm's cost of goods sold. Each relates to the basis used in valuing the firm's inventory, since the firm's purchases that are not passed through the income statement as cost of goods sold remain in the firm's inventory account. We will discuss two common methods for determining the cost of goods sold and consequently the value of the inventory account. The first involves assigning to the period's cost of goods sold the prices paid for the oldest items of inventory held by the firm at the beginning of the period. This is commonly referred to as the **first-in, first-out** or **FIFO** method. The second assigns the cost of the most recently purchased inventory items to the period's cost of goods sold. This is called the **last-in, first-out** or **LIFO** method. The method selected can have a material effect on the firm's computed net earnings during a period where the prices of its purchases consistently rose or fell. For example, in a period when prices have been increasing, the use of the FIFO method results in a lower cost of goods sold, a larger gross profit, a higher tax liability, a higher inventory amount, and a higher net earnings figure than LIFO; LIFO (which costs the firm's sale items using the most recent prices paid by the firm) will result in a lower inventory amount, a higher cost of goods sold figure, and, consequently, lower gross profits, lower taxes, and lower net earnings. The opposite result would follow should prices have fallen during the period.

What importance should be attached to the choice of methods for determining cost of goods sold? Under either method the cash flows that result from sales will be the same, for the actual cost of the items sold does not vary with the method chosen for computing cost of goods sold. However, a very real cash flow effect can result in terms of the amount of taxes that the firm must pay. During a period of rising prices LIFO results in lower taxes being paid than FIFO, while during a period of falling prices the opposite is true.[25] Furthermore, it should be noted that the reported inventory amounts may vary considerably with the application of one method as compared with the other—but the physical quantity and composition of the goods are not affected.

DEPRECIATION OF FIXED ASSETS

Depreciation expense represents an allocation of the cost of a fixed asset over its useful life. The objective is to match such costs with the revenues that result from their utilization in the enterprise. Furthermore, depreciation expense is used to reduce the balance sheet book value of the firm's fixed assets. Thus, the method used to determine depreciation expense also serves as the basis for determining the book value of fixed assets. The two methods that are commonly used for computing depreciation for accounting purposes are the straight line and the double-declining balance methods. However, for tax purposes, the depreciation methods for writing off the investment value of an asset are different from those methods used for accounting purposes. Depreciation for tax purposes was discussed in Chapter 2.

IMPORTANT RELATIONSHIPS IN ACCOUNTING DATA

Financial managers make extensive use of the firm's accounting information. Chapter 20 outlines in detail how the firm uses such information. However, it will be helpful to look briefly at some of the more important accounting relationships as a basis for our understanding of capital budgeting, working capital management and other key areas.

There are an almost unlimited number of relationships we could consider. However, there are at least three relationships that have an important effect on the firm and the shareholder's value.

1. **The amount of operating profits per dollar of sales.** This relationship indicates management's ability to control both cost of goods sold and operating expense (such as selling and administrative expenses) relative to sales. This relationship is important to a firm's financial well-being and bears upon the value investors are willing to pay for the company.[26]

---

[25]Revenue Canada has indicated through its Interpretation Bulletin (IT473) that the use of the LIFO assumption is not recommended for determining the cost of inventory for tax purposes. Moreover, the CICA recommends that, "the method selected for determining cost should be the one which results in the fairest matching of costs against revenues regardless of whether or not the method corresponds to the physical flow of goods" (*CICA Handbook*, 3030.09).

[26]The term "profits" is used interchangeably with "earnings" or "income."

2. **The amount of the firm's assets compared to its sales.** This relationship tells us how efficiently management is using the firm's assets to create sales. All companies must make an investment in assets in order to generate sales. You cannot have any significant amount of sales without investing in assets, such as accounts receivable, inventories, equipment, and possibly real estate. However, you would prefer to invest less rather than more.
3. **Debt to equity.** We shall see later that the use of debt increases the firm's risk and, in turn, the risk exposure for the company's investors. The increased risk would then affect the investor's required rate of return on their investments, and thereby may alter the firm's value.[27] Thus, the ratio of debt to equity is an important relationship for the financial manager.

A Word of Caution

Although accounting information is vital in our study of a company's finances, a large body of evidence suggests that investors (shareholders) look behind accounting numbers to discover the underlying economic reality of a firm's performance. Thus, the economic reality of a firm's operations is ultimately reflected in the cash flow it provides to its investors, not its reported earnings. Specifically, we will identify the firm's free cash flow to be the amount of cash that is available for distributing to its investors not only after operating expenses have been paid, but after any investments have been made that will generate future cash flows. Free cash flow is defined as follows:

$$\text{Free cash flow} = \text{Revenues} - \text{Expenses} - \text{Additional investment in assets}$$

Throughout our study, we will refer to the firm's free cash flow. It is a basic concept in financial management.

APPENDIX 20B

GLOSSARY OF ACCOUNTING TERMS

accelerated depreciation A term encompassing any method for computing depreciation expense wherein the charges decrease with time. Examples of accelerated methods include sum-of-the-years' digits and double-declining balance as contrasted with straight-line.

accounts payable A current liability representing the total amount owed by a firm from its past (unpaid) credit purchases.

accounts receivable A current asset including all monies owed to a firm from past (uncollected) credit sales.

accrual basis of accounting The method of recognizing revenues when the earning process is virtually complete and when an exchange transaction has occurred, and recognizing expenses as they are incurred in generating those revenues. Thus, revenues and expenses recognized under the accrual basis of accounting are independent of the time when cash is received or expenditures are made. This contrasts with the cash basis of accounting.

accrued salaries and wages Salary and wage expense the firm owes but has not yet paid (a current liability).

accumulated depreciation The sum of depreciation charges on an asset since its acquisition. This total is deducted from gross fixed assets to compute net fixed assets. This balance sheet entry is sometimes referred to as the reserve for depreciation, accrued depreciation, or the allowance for depreciation.

administrative expense An expense category used to report expenses incurred by the firm but not reflected in specific activities such as manufacturing or selling.

amortizing The procedure followed in allocating the cost of long-lived assets to the periods in which their benefits are derived. For capital (fixed) assets, amorti-

[27]The relationship between a firm's use of debt and the value of the firm is a complex one, which will be studied in depth in later chapters.

zation is called depreciation expense, whereas for wasting assets (natural resources), it is called depletion expense.

asset Anything owned by a firm or individual that has commercial value.

authorized share capital A term used to indicate the total number of shares of stock the firm can issue and is specified in the articles of incorporation.

bad debt expense An adjustment to income and accounts receivable reflecting the value of uncollectible accounts.

balance sheet A statement of financial position on a particular date. The balance sheet equation is as follows: total assets = total liabilities + owners' equity.

bond Long-term debt instrument carried on the balance sheet at its face amount or par value (usually $1000 per bond), which is payable at maturity. The coupon rate on the bond is the percentage of the bond's face value payable in interest each year. Bonds usually pay interest semiannually.

book value The net amount of an asset shown in the accounts of a firm. When referring to an entire firm, it relates the excess of total assets over total liabilities (also referred to as owners' equity and net worth).

capital Sometimes, the total assets of a firm; at other times, the owners' equity alone.

capital assets Identifiable assets that: (a) are held for use in the production or supply of goods and services for rental to others, for administrative purposes or for the development, construction, maintenance, or repair of other capital aasets; (b) have been acquired, constructed or developed with the intention of being used on a continuing basis; and (c) are not intended for sale in the ordinary course of business (*CICA Handbook*, 3060.04).

capital cost allowance system A system that permits a taxpayer to deduct a portion of the depreciable property's capital cost as an expense. (See Chapter 2, Appendix 2A.)

capitalization Shareholders' equity plus the par value of outstanding bonds.

cash flow The excess (deficiency) of cash receipts over cash disbursements for a given period.

common share A term used to represent the claim of residual owners of the firm. These owners have claim to earnings and asset values remaining after the claims of all creditors and preferred shareholders are satisfied.

contributed surplus A surplus paid-in by shareholders which includes premiums on shares issued, any portion of the proceeds of issue of shares without par value not allocated to share capital, gain on forfeited shares, proceeds arising from donated shares, credits resulting from redemption or conversion of shares at less than the amount set up as share capital, and any other contribution by shareholders in excess of amounts allocated to share capital. (*CICA Handbook*, 3250.05)

cost of goods sold The total cost allocated to the production of a completed product for the period.

current assets Assets that are normally converted into cash within the operating cycle of the firm (normally a period of one year or less). Such items as cash, accounts receivable, marketable securities, prepaid expenses, and inventories are frequently found among a firm's current assets.

current liabilities Liabilities or debts of the firm that must be paid within the firm's normal operating cycle (usually one year or less), such as accounts and notes payable, income taxes payable, and wages and salaries payable.

debentures Long-term debt (bonds) that are secured only by the integrity of the issuer (that is, no specific assets are pledged as collateral).

deferred income taxes Income taxes that a firm recognizes as being owed based upon its earnings but that are not payable until a later date.

depreciable life The period over which an asset is depreciated.

depreciation expense Amortization of plant, property, and equipment cost during an accounting period.

dividend A distribution of earnings to the owners of a corporation in the form of cash (cash dividend) or shares (stock dividend).

double-declining balance depreciation A method for computing declining balance depreciation expense in which the constant percentage is equal to $2/N$, where N represents the depreciable life of the asset.

earnings A synonym for net income or net profit after taxes. Owing to the ambiguity that arises in using the general term earnings, it is usually avoided in favour of more specific terms such as net operating earnings or earnings after taxes.

earnings after taxes (EAT) Other terms often used synonymously are net income and net profit after taxes.

earnings before interest and taxes (EBIT) A commonly used synonym for net operating income. Note that where other income exists, EBIT equals net operating income plus other income.

earnings before taxes (EBT) Total net earnings after the deduction of all tax-deductible expenses.

earnings per share (EPS) Net income after taxes available to the common shareholders (after preferred dividends) divided by the number of outstanding common shares.

equity financing The raising of funds through the sale of common or preferred shares.

extraordinary item A revenue or expense that is both unusual in nature and infrequent in occurrence. Such items and their tax effects are separated from ordinary income in the income statement.

FIFO A method for determining the inventory cost assigned to cost of goods sold wherein the cost of the oldest items in inventory is charged to the period's cost of goods sold (first in, first out). Ending inventories therefore will reflect the prices paid for the most recent purchases. See also the discussion in Appendix 20A.

fixed assets See **capital assets**.

general and administrative expense Expenses associated with the managerial and policy-making aspects of a business.

goodwill Included as an asset entry in the balance sheet to reflect the excess over fair market value paid for the assets of an acquired firm.

gross profit The excess of net sales over cost of goods sold.

income statement The statement of profit or loss for the period comprised of net revenues less expenses for the period. (See **accrual basis of accounting**.)

income tax An annual expense incurred by the firm based upon income and paid to a governmental entity.

intangible asset An asset that lacks physical substance, such as goodwill or a patent.

interest expense The price paid for borrowed money over some specified period of time.

inventory The balance in an asset account such as raw materials, work in process, or finished goods. (See **LIFO** and **FIFO**.)

lease A contract requiring payments by the user (lessee) to its owner (lessor) for the use of an asset. In accordance with disclosure requirements of the *CICA Handbook* (3065.21), the gross amount of assets under capital leases and related accumulated amortization should be disclosed. However for operating leases, only the future minimum lease payments, in the aggregate and for each of the five succeeding years, is required for disclosure purposes.

liability An obligation to pay a specified amount to a creditor in return for some current benefit.

LIFO A method for determining the inventory cost assigned to cost of goods sold whereby the cost of the most recent purchases of inventory is assigned to the period's cost of goods sold (last-in, first-out). Ending inventories thus contain the cost of the oldest items of inventory.

liquid assets Those assets of the firm that can easily be converted into cash with little or no loss in value. Generally included are cash, marketable securities, and sometimes accounts receivable.

long-term debt All liabilities of the firm that are not due and payable within one year. Examples include installment notes, equipment loans, and bonds payable.

marketable securities The securities (bonds and shares) of other firms and governments held by a firm.

net income See **earnings after taxes (EAT)**.

net operating income (NOI) Income earned by a firm in the course of its normal operations. Calculated as net sales less the sum of cost of goods sold and operating expenses.

net plant and equipment Gross plant and equipment less accumulated depreciation.

net sales Gross sales less returns, allowances, and cash discounts taken by customers.

notes payable A liability (normally short-term) of the firm representing a monetary indebtedness.

operating expense Any expense incurred in the normal operation of a firm.

owners' equity Total assets minus total liabilities, sometimes referred to as net worth.

par value The face value of a security.

patent The rights to the benefits of one's invention granted to the inventor by the government.

preferred share The capital share of the owners of the firm whose claim on assets and income is secondary to that of bondholders but preferred as to that of the common shareholders.

profit and loss statement Another name for the income statement.

retained earnings The sum of a firm's net income over its life less all dividends paid.

selling expense An expense incurred in the selling of a firm's product. Examples would include salespeople's commissions and advertising expenditures.

sinking fund Assets and their earnings set aside to retire long-term debt obligations of the firm. Payments into sinking funds are made after taxes and are usually described in the bond indenture (contract between the borrowing firm and the bondholders).

share capital All outstanding common and preferred shares.

stock dividend A dividend that results in a transfer of retained earnings to the capital shares and contributed surplus accounts. This contrasts with a cash dividend. (See dividend.)

CHAPTER 21

Financial Forecasting, Planning, and Budgeting

LEARNING OBJECTIVES

After reading this chapter you should be able to

1. Apply the percent of sales method to forecast the financing requirements of a firm.
2. Calculate a firm's sustainable rate of growth.
3. Explain the limitations of the percent of sales forecast of growth.
4. Prepare a cash budget so as to evaluate the amount and timing of a firm's financing needs.

INTRODUCTION

Forecasting is an integral part of the planning process, but sometimes our ability to predict the future is simply awful. An example of poor forecasting relates to projections of oil prices that were prevalent during the mid-eighties. Oil prices were roughly $30 a barrel and many firms were developing new reserves that would cost well over this amount to produce. Why? Oil prices were projected to continue to rise and many thought the price might eventually reach $50 a barrel by the end of the decade. Then in January 1986 the collapse of the oil producers' cartel in combination with the benefits of energy conservation efforts produced a dramatic drop in oil prices to below $10 a barrel. Although this case is dramatic, it is by no means unique.

The accuracy of forecasting techniques begins to break down when the environment becomes unsettled or dynamic.[1] *For example, forecasting has a dismal record in predicting nonseasonal turning points such as recessions, unusual events, discontinuities and the actions or reactions of competitors. The only major consolation for managers is that the forecasts of managers from competing firms will also tend to be as inaccurate in a dynamic environment. As a result, forecasting techniques are most accurate when the environment is static.*[2]

[1]This section is based on Essam Mahmoud, "Accuracy in Forecasting: A Survey," *Journal of Forecasting* 3(2) (1984), pp. 139–59; Reed Moyer, "The Futility of Forecasting," *Long Range Planning* (February 1984), pp. 65–72; and Ronald Bailey, "Them That Can Do, Them That Can't, Forecast," *Forbes,* December 26, 1988, pp. 94–100.

[2]Based upon an excerpt entitled "Management's Ability To Forecast Accurately," from Stephen P. Robbins, *Management,* 3rd ed. (Englewood Cliffs, NJ: Prentice-Hall Publishing Co., 1991), p. 253.

Video Case 6 

If forecasting the future is so difficult and plans are built on forecasts, why do firms engage in planning efforts? The answer, oddly enough, does not lie in the accuracy of the firm's projections, for planning offers its greatest value when the future is the most uncertain. The reason is that planning is the process of thinking about what the future might bring and devising strategies for dealing with the likely outcomes. Planning is thinking in advance so as to provide an opportunity to devise contingency plans that can be quickly and easily initiated should the need arise. A shorter response time to uncertain events means that the firm can reduce the costs of responding to adverse circumstances and quickly respond to take advantage of unexpected opportunities.

CHAPTER PREVIEW

In this chapter, we examine the role of forecasting in the firm's financial planning process. Basically, forecasts of future sales revenues and their associated expenses give the firm the information needed to project its future needs for financing. We also examine the firm's budgetary system which includes the cash budget, the pro forma (planned) income statement, and the pro forma balance sheet. The pro forma financial statements provide a useful tool for analyzing the effects of the firm's forecasts and planned activities on its financial performance, as well as the firm's needs for financing. In addition, pro forma statements can be used as a benchmark or standard to compare against actual operating results. As a result, pro forma statements are an instrument for controlling or monitoring the firm's progress throughout the planning period.

BACK TO THE BASICS

Financial decisions are made today in light of our expectations of an uncertain future. Financial forecasting involves making estimates of the future financing requirements of the firm. ***Axiom 3: Cash—Not Profit—Is King: Measuring the Timing of Costs and Benefits*** *speaks directly to this problem. Remember that effective financial management requires that consideration be given to cash flow and when it is received or disbursed.*

OBJECTIVE 1

FINANCIAL FORECASTING

A firm will use financial forecasting to estimate its future financial needs. The basic steps involved in predicting those financing needs include the following: Step 1: Project the firm's sales revenues and expenses over the planning period. Step 2: Estimate the levels of investment in current and fixed assets that are necessary to support the projected sales. Step 3: Determine the firm's financing needs throughout the planning period.

Cash flow cycle diagram

Diagram depicting a firm's operations as a large pump that pushes cash through various reservoirs, such as inventories and accounts receivable, and dispenses cash for taxes, interest, principal payments on debt, and cash dividends to the shareholders.

The Cash Flow Cycle

The entire firm's operations can be visualized through the use of a **cash flow cycle diagram**. The diagram, as shown in Figure 21-1, depicts the firm's operations as a large pump that pushes cash through various reservoirs such as inventories and accounts receivable and dispenses cash for taxes, interest and principal payments on debt, and cash dividends to the shareholders. The problem in financial forecasting is one of predicting cash inflows and outflows and the corresponding financial needs of the firm. The firm's needs are related to the size of the various reservoirs that hold cash. These reservoirs of cash represent the various assets in which the firm must make investments in order to produce the expected level of sales.

Tracing the Cash Flow Cycle

Since the firm's cash flow cycle is a continuous process, we will find it useful for illustrative purposes to consider the first cycle of a new firm. The process begins with the firm's cash reservoir, which consists initially of the owner's investment (funds raised through the issuance of shares) and funds borrowed from creditors. This cash is used to acquire plant and equipment, as well as supplies and materials. It is also used to pay wages and salaries, to pay for utilities such as heat and power, and to replace materials, plant, and equipment that are used or worn out in the process of making the firm's product. These expenditures make up the cost of goods sold for the firm's sellable product. Sales are then made either for cash or on credit (with some leakage for bad debt losses). To this cash flow stream are added cash inflows resulting from the sale of assets such as equipment, land, or securities. In the final phase of the cycle, cash is disbursed to pay obligations to suppliers for materials purchased on credit, income taxes to the government, interest and principal to the firm's creditors, and cash dividends to the firm's owners. Finally, any cash not paid in dividends is reinvested in the firm and becomes a part of the owners' investment.

Forecasting Cash Flows

The problem of financial forecasting, to summarize, consists of predicting future sales, which in turn provides the basis for predicting the level of investment in inventories, receivables, plant, and equipment required to support the firm's projected sales. Through the use of the cash budget, the information from these projections is combined to provide an estimate of the firm's future financing needs.

Figure 21-1 Cash Flow Cycle Diagram

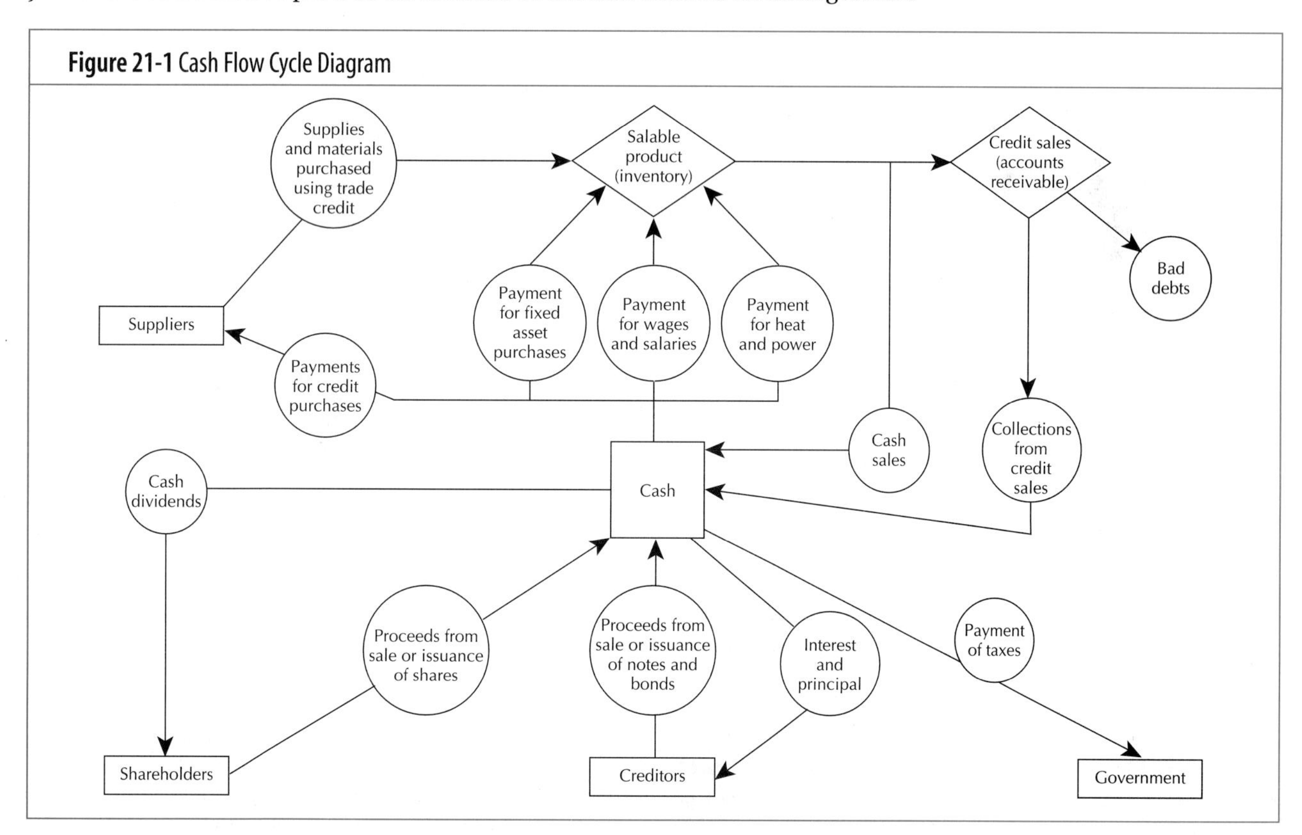

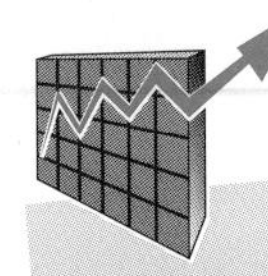

Cash Flow Rates Flawed

In these uncertain times, cash flow per share may seem like the best measure of what's really going on at a company.

But investor beware. Canada has no standard for calculating this ratio. In the U.S., the Securities and Exchange Commission won't allow cash flow per share information to be disclosed in annual reports because the ratios aren't comparable between companies.

The Ontario Securities Commission recently asked the Canadian Institute of Chartered Accountants to put cash flow per share, which is not dealt with in the *CICA Handbook*, on their agenda.

Until a standard measure is developed, the OSC wants issuers to present cash flow per share calculations in the notes, not in the face of the financial statements.

The real estate industry is one place where cash flow per share is an investment yardstick.

Maybe that's why the Canadian Institute for Public Real Estate Companies has issued a set of accounting rules which set some parameters for this calculation—take earnings from operations and add back noncash deductions such as depreciation, amortization, and deferred income taxes.

By contrast, in the oil and gas industry there are several different measures, such as discretionary cash flow per share, where the company makes adjustments for capitalized costs.

"Cash flow per share is a big headache," said one analyst. "We calculate our own cash flow from operations, which excludes extraordinary items such as cash from the sale of operations or dividends from an associated company."

Accountants may have to look at the whole idea of cash flow measurement and its presentation in the financial statements.

One would like to know the cash inflows and outflows. In other words, to be able to follow the flow of receipts from customers and payments to suppliers. Apparently, cash flow is not as straightforward as it sounds.

Source: Adapted from B. Critchley and S. Gittins, "Cash Flow Rates Flawed," *The Financial Post*, May 23, 1991, p. 44.

Sales Forecast

Sales forecast

The key ingredient in a firm's planning process which reflects (1) any past trend in sales that is expected to carry through into the new year and (2) the influence of any events that might materially affect that trend.

The key ingredient in the firm's planning process is the **sales forecast**. This projection will generally be derived using information from a number of sources. At a minimum, the sales forecast for the coming year would reflect: (1) any past trend in sales that is expected to carry through into the new year and (2) the influence of any events that might materially affect that trend.[3] An example of the latter would be the initiation of a major advertising campaign or a change in the firm's pricing policy.

[3]A complete discussion of forecast methodologies is outside the scope of this book. The interested reader will find the following references helpful: F. Gerard Adams, *The Business Forecasting Revolution* (Oxford: Oxford University Press, 1986); C. W. J. Granger, *Forecasting in Business and Economics*, 2nd ed. (Boston, Mass.: Academic Press, Inc., 1989); and Paul Newbold and Theodore Bos, *Introductory Business Forecasting* (Cincinnati, Ohio: South-Western Publishing Co., 1990).

Forecasting Financial Variables

Traditional financial forecasting takes the sales forecast as a given and estimates its impact on the firm's various expenses, assets, and liabilities. The most commonly used method for making these projections is the percent of sales method.

Percent of Sales Method of Financial Forecasting

The *percent of sales method* involves estimating the level of an expense, asset, or liability for a future period as a percent of the sales forecast. The percentage used can come from the most recent financial statement item as a percent of current sales, from an average computed over several years, from the judgment of the analyst, or from some combination of these sources.

Figure 21-2 presents a complete example of the use of the percent of sales method of financial forecasting. In that example, each item in the firm's balance sheet that varies with sales is converted to a percentage of 1995 sales. The forecast of the new balance for each item is then calculated by multiplying this percentage by the $12 million in projected sales for the 1996 planning period. This method of forecasting future financing is not as precise or detailed as the method which uses the cash budget; however, it offers a relatively low-cost and easy-to-use first approximation of the firm's financing needs for a future period. The cash budget method will be examined later on in this chapter.

Note that in the example in Figure 21-2 both current and fixed assets are assumed to vary with the level of firm sales. This means that the firm does not have sufficient productive capacity to absorb a projected increase in sales. Thus, if sales were to rise by $1, fixed assets would rise by $0.40, or 40 percent of the projected increase in sales. Note that if the fixed assets the firm presently owns had been sufficient to support the projected level of new sales, these assets should not be

Figure 21-2 Using the Percent of Sales Method to Forecast Future Financing Requirements

	Present (1995)	Percent of Sales (1995 Sales = $10 M)	Projected (Based on 1995 Sales = $12 M)	
Assets				
Current assets	$2 M	$2 M/$10 M = 20%	0.2 X $12 M =	$2.4 M
Net fixed assets	4 M	$4 M/$10 M = 40%	0.4 X $12 M =	4.8 M
Total	$6 M			$7.2 M
Liabilities and Owners' Equity				
Accounts payable	$1.0 M	$1 M/$10 M = 10%	0.10 X $12 M =	$1.2 M
Accrued expenses	1.0 M	$1 M/$10 M = 10%	0.10 X $12 M =	$1.2 M
Notes payable	0.5 M	NA[a]	no change	0.5 M
Long-term debt	$2.0 M	NA[a]	no change	2.0 M
Total liabilities	$4.5 M			$4.9 M
Common shares	$0.1 M	NA[a]		$0.1 M
Contributed surplus	0.2 M	NA[a]		$0.2 M
Retained earnings	1.2 M		1.2 M + [.05 X $12 M X (1 - .5)] =	1.5 M[b]
Common equity	$1.5 M			$1.8 M
Total	$6.0 M		Total financing provided	$6.7 M
			Discretionary financing needed	0.5 M[c]
			Total	$7.2 M

[a]Not applicable. These account balances are assumed not to vary with sales.

[b]Projected retained earnings equals the beginning level ($1.2 M) plus projected net income less any dividends paid. In this case net income is projected to equal 5 percent of sales, and dividends are projected to equal half of net income: 0.05 × $12 M × (1 – .5) = $300,000.

[c]Discretionary financing needed equals projected total assets ($7.2 M) less projected total liabilities ($4.9M) less projected common equity ($1.8 M) or $7.2 M – 4.9 M – 1.8 M = $500,000.

allowed to vary with sales. If this were the case, then fixed assets would not be converted to a percent of sales and would be projected to remain unchanged for the period being forecast.

Also, we note that accounts payable and accrued expenses are the only liabilities allowed to vary with sales. Both these accounts might reasonably be expected to rise and fall with the level of firm sales; hence the use of the percent of sales forecast. Since these two categories of current liabilities normally vary directly with the level of sales, they are often referred to as **spontaneous sources of financing**. Chapters 11 and 17 examined in greater detail working-capital management and the various sources of financing. Notes payable, long-term debt, common shares, and contributed surplus are not assumed to vary directly with the level of firm sales. These sources of financing are termed **discretionary**, in that the firm's management must make a conscious decision to seek additional financing using any one of them. Finally, we note that the level of retained earnings does vary with estimated sales. The predicted change in the level of retained earnings equals the difference in estimated after-tax profits (projected net income) of $600,000 and common share dividends of $300,000.

Spontaneous financing

Sources of financing that occur naturally during the course of business. Accounts payable is a primary example.

Discretionary financing

Sources of financing that require an explicit decision on the part of the firm's management every time funds are raised. An example is a bank note which requires that negotiations be undertaken and an agreement signed setting forth the terms and conditions for the financing.

Thus, using the example from Figure 21-2, we estimate that firm sales will increase from $10 M to $12 M, which will cause the firm's needs for total assets to rise to $7.2 M. These assets will then be financed by $4.9 M in existing liabilities plus spontaneous liabilities, $1.8 M in owner funds including $300,000 in retained earnings from next year's sales, and, finally, $500,000 in discretionary financing, which can be raised by issuing notes payable, selling bonds, offering an issue of shares, or some combination of these sources.

In summary, we can estimate the firm's needs for discretionary financing, using the percent of sales method of financial forecasting, by following a four-step procedure:

Step 1: Convert each asset and liability account that varies directly with firm sales to a percent of current year's sales.

EXAMPLE

$$\frac{\text{current assets}}{\text{sales}} = \frac{\$2\text{ M}}{\$10\text{ M}} = .2 \text{ or } 20\%$$

Step 2: Project the level of each asset and liability account in the balance sheet using its percent of sales multiplied by projected sales or by leaving the account balance unchanged where the account does not vary with the level of sales.

EXAMPLE

$$\begin{matrix}\text{projected}\\ \text{current assets}\end{matrix} = \text{projected sales} \times \frac{\text{current assets}}{\text{sales}} = \$12\text{ M} \times .2 = 2.4\text{ M}$$

Step 3: Project the level of new retained earnings available to help finance the firm's operations. This equals projected net income for the period less planned dividends from common shares.

EXAMPLE

projected addition to retained earnings =

$$\text{projected sales} \times \frac{\text{net income}}{\text{sales}} \times \left(1 - \frac{\text{cash dividends}}{\text{net income}}\right)$$

$$= \$12 \text{ M} \times .05 \times [1 - .5] = \$300{,}000$$

Step 4: Project the firm's need for discretionary financing as the projected level of total assets less projected liabilities and owners' equity.

EXAMPLE

discretionary financing needed =

projected total assets − projected total liabilities − projected owners' equity

$$= \$7.2 \text{ M} - \$4.9 \text{ M} - \$1.8 \text{ M} = \$500{,}000$$

As we noted earlier, the principal virtue of the percent of sales method of financial forecasting is its simplicity.[4] To obtain a more precise estimate of the amount and timing of the firm's future financing needs, we require a cash budget. The percent of sales method of financial forecasting does provide a very useful, low-cost forerunner to the development of the more detailed cash budget, which the firm will ultimately use to estimate its financing needs.

PERSPECTIVE IN FINANCE

Are you beginning to wonder exactly where finance comes into financial forecasting? To this point financial forecasting looks for all the world like financial statement forecasting. The reason for this is the fact that we have adopted the accountant's model of the firm, the balance sheet, as the underlying structure of the financial forecast. The key to financial forecasting is the identification of the firm's anticipated future financing requirements which is identified as the "plug figure" and balances a pro forma balance sheet.

The Discretionary Financing Needed (DFN) Model

In the preceding discussion we estimated DFN as the difference in projected total assets and the sum of projected liabilities and owner's equity. We can estimate DFN directly using the predicted change in sales (ΔS) and corresponding changes in assets, liabilities, and owner's equity as follows:

$$DFN_{t+1} = \begin{matrix}\text{projected}\\ \text{change in assets}\end{matrix} - \begin{matrix}\text{projected change}\\ \text{in liabilities}\end{matrix} - \begin{matrix}\text{projected change}\\ \text{in owner's equity}\end{matrix} \quad (21\text{-}1a)$$

or

$$DFN_{t+1} = \left[\frac{\text{assets}_t^*}{\text{sales}_t}\Delta\text{sales}_{t+1}\right] - \left[\frac{\text{liabilities}_t^*}{\text{sales}_t}\right] - [NPM_{t+1}(1 - b)(\text{sales}_{t+1}] \quad (21\text{-}1b)$$

where

DFN_{t+1} = predicted discretionary financing needed for period t+1.

[4]Although a linear type of relationship may describe a firm's inventory requirements for sales very close to the level used in estimating "%" sales, it can be a very poor predictor for large changes in sales. An improvement in this estimator might include an intercept term or in some cases a nonlinear model might be appropriate, i.e., $Y = a + b(\text{sales}) + c(\text{sales})^2$. Models of this type can be easily estimated and incorporated into financial planning using financial spreadsheet software.

$assets_t^*$ = those assets in period t that are expected to change in proportion to the level of sales. In our example we have assumed that all the firm's assets vary in proportion to sales. We will have more to say about this assumption in the next section where we consider economies of scale and lumpy fixed asset investments.

$sales_t$ = the level of sales for the period just ended.

$\Delta sales_{t+1}$ = the change in sales projected for period t+1, i.e., $sales_{t+1} - sales_t$. Note that "Δ" is the Greek symbol delta which is used here to represent "change."

$liabilities_t^*$ = those liabilities in period t that are expected to change in proportion to the level of sales. In our preceding example we assumed that accounts payable and accrued expenses varied with sales but notes payable and long-term debt did not.

NPM_{t+1} = the net profit margin (net income ÷ sales) projected for period t + 1.

b = dividends as a percent of net income or the dividend payout ratio such that (1 – b) is the proportion of the firm's projected net income that will be retained and reinvested in the firm (i.e., (1 – b) is the retention ratio).

Using the numbers from the preceding example we estimate DFN_{1996} as follows:

$$DFN_{1996} = \left[\left(\frac{\$2M + \$4M}{\$10M}\right)\$2M\right] - \left[\left(\frac{\$1M + \$1M}{\$10M}\right)\$2M\right] - [0.5(1 - 0.5)(\$12M)]$$

$$= \$0.5 \text{ million or } \$500{,}000$$

Analyzing the Effects of Profitability and Dividend Policy on DFN

Using the DFN model we can quickly and easily evaluate the sensitivity of our projected financing requirements to changes in key variables. For example, using the information from the preceding example, we evaluate the effect of net profit margins (NPM) ranging from 1 percent, 5 percent, and 10 percent in combination with dividend payout ratios of 30 percent, 50 percent, and 70 percent as follows:

Discretionary Financing Needed for Various Net Profit Margins and Dividend Payout Ratios

Net Profit Margin	Dividend Payout Ratios (Dividends ÷ Net Income)		
	30%	50%	70%
1%	$716,000	$740,000	$764,000
5%	380,000	500,000	620,000
10%	(40,000)	200,000	440,000

If these values for the net profit margin represent reasonable estimates of the possible ranges of values the firm might experience and if the firm is considering dividend payouts ranging from 30 percent to 70 percent, then we estimate that the firm's financing requirements (DFN) will range from ($40,000), which represents a surplus of $40,000, to a situation where it would need to acquire $764,000. Lower net profit margins mean higher funds requirements. Also, higher dividend payout percentages, other things remaining constant, lead to a need for more discretionary financing. This latter observation is a direct result of the fact that a high-dividend-paying firm retains less of its earnings.

Sustainable rate of growth

The maximum rate of growth in sales that the firm can sustain while maintaining its present capital structure (debt and equity mix) and without having to sell new common shares.

OBJECTIVE 2

The Sustainable Rate of Growth

The **sustainable rate of growth** (g^*) represents the rate at which a firm's sales can grow if it wants to maintain its present financial ratios and *does not* want to resort to the sale of new equity shares.[5] We can solve for the sustainable rate of growth

[5]Extensive discussion of the sustainable rate of growth concept is found in Robert C. Higgins, "Sustainable Growth with Inflation," *Financial Management* (Autumn 1981), pp. 36–40.

directly using the discretionary financing needed formula found in Equation 21-1b as we illustrate in the footnote found below.[6] Specifically, the sustainable rate of growth is the rate of sales growth for which discretionary financing needed equals zero. The resulting formula is quite simple and relies on the return on equity (ROE) ratio and dividend payout ratio (b).

$$\text{Sustainable rate of growth } (g^*) = ROE\,(1 - b) \qquad (21\text{–}2)$$

Where, *ROE* is defined as follows:

$$ROE = \frac{\text{net income}}{\text{common equity}}$$

Equation 21–2 is deceptively simple. Note that a firm's ROE is determined by a number of factors including the firm's profit margin, asset turnover, and its use of financial leverage. Specifically, recall from Chapter 20 that we developed the following relationship for ROE:

$$ROE = \left(\begin{matrix}\text{net profit}\\ \text{margin}\end{matrix}\right) \times \left(\begin{matrix}\text{total asset}\\ \text{turnover}\end{matrix}\right) \div \left(1 - \text{debt ratio}\right)$$

$$ROE = \left(\frac{\text{net income}}{\text{sales}}\right) \times \left(\frac{\text{sales}}{\text{assets}}\right) \div \left(1 - \frac{\text{total debt}}{\text{total assets}}\right)$$

Hence, the firm's sustainable rate of growth is determined by all the determinants of its return on equity (net profit margin, total asset turnover, and financial leverage) and its choice of a dividend payout ratio (b). To illustrate the calculation of the sustainable rate of growth consider the financial information for the Harris Electronics Corporation found in Table 21–1.

Harris experienced reasonably stable sustainable rates of growth ranging from a low of 5.80 percent in 1992 to a high of 6.60 percent in 1993. The reasons for the modest variation are easy to see from the data provided in Table 21–1. The firm's rate of return on common equity (ROCE) varied only slightly over the period, from a low of 9.67 percent in 1992 to a high of 11 percent in 1993, while the retention ratio $(1 - b)$ remained steady at 60 percent of earnings.

Harris' actual rate of sales growth from 1991 to 1992 was 21.74 percent which was above its sustainable rate for 1992 which was only 5.80 percent. The sustainable rate of growth applicable to 1992 is calculated using data from 1992. In this year we calculated the firm's sustainable rate of growth for 1992 to be 5.80 percent [9.67 percent (1–.40)], but its actual increase in sales for the coming year was 21.74 percent [(\$1,400 – 1,150) ÷ \$1,150]. How did Harris accommodate the financing demands during 1993? The answer can be found by examining the firm's debt-to-assets ratio and changes to the firm's common equity. We see that Harris increased its borrowing from 42.56 percent of assets to 49.48 percent without issuing any new common stock. Thus, Harris has financed its DFN using new debt issues.

---

[6] We can evaluate the impact of differing rates of sales growth on DFN by recognizing that the growth in firm sales, g, is simply the ratio of the projected change in sales ($\Delta sales_{t+1}$) divided by the most recent past level of sales ($sales_t$). Rearranging terms in Equation 21–1 and substituting *g* for

$$\frac{\Delta sales_{t+1}}{sales_t}$$

we get the following result:

$$DFN_{t+1} = (g)(\text{assets}) - (\text{liabilities})(g) - (NPM)(1 - b)(sales_{t+1})$$

Note that the sustainable rate of growth is that growth rate in firm sales (g^*) which makes $DFN_{t+1} = 0$. Thus, setting DFN_{t+1} in the above equation equal to zero and solving for *g* we get the sustainable rate of growth equation:

$$\text{Sustainable rate of growth } (g^*) = ROE(1 - b)$$

where ROE is the return on equity (net income ÷ common equity) and *b* is the fraction of firm earnings paid out in dividends or the dividend payout ratio.

Table 21-1 Harris Electronics Corporation: Sustainable Rate of Growth Calculations

	1995	1994	1993	1992	1991
Sales	$1,500	$1,450	$1,400	$1,150	$1,090
Net income	75	73	70	58	55
Assets	1,350	1,305	1,260	1,035	981
Dividends	30	29	28	23	21.8
Common equity	725	680	637	595	560
Liabilities	625	625	623	440	421
Liabilities and owner's equity	1,350	1,305	1,260	1,035	981
Sustainable rate of growth (g*)	6.21%	6.40%	6.60%	5.80%	5.84%
Actual growth rate in sales	NA*	3.45%	3.57%	21.74%	5.50%
Return on equity (ROCE)	10.34%	10.66%	11.00%	9.67%	9.73%
Retention ratio (1–b)	60.00%	60.00%	60.00%	60.00%	60.00%
Debt-to-assets ratio	46.30%	47.89%	49.48%	42.56%	42.92%
New common stock	0	0	0	0	0

*Not available or cannot be calculated without 1996 data.

OBJECTIVE 3

Percent of sales method

A method of financial forecasting that involves estimating the level of expense, asset, or liability for a future period as a percent of sales forecast.

Limitations of the Percent of Sales Forecast Method

The **percent of sales method** of financial forecasting provides reasonable estimates of a firm's financing requirements only where asset requirements and financing sources can be accurately forecast as a constant percent of sales. For example, predicting inventories using the percent of sales method involves the following predictive equation:

$$\text{inventories}_t = \frac{\text{inventories}_t}{\text{sales}_t} \times \text{sales}_t$$

Figure 21-3a depicts this predictive relationship. Note that the percent-of-sales predictive model is simply a straight line that passes through the origin (i.e., has a zero intercept). There are some fairly common instances in which this type of relationship fails to describe the relationship between an asset category and sales. Two such examples involve assets for which there are scale economies and assets that must be purchased in discrete quantities ("lumpy assets").

Economies of scale are sometimes realized from investing in certain types of assets. This means that these assets do not increase in direct proportion to sales. Figure 21-3b reflects one instance in which the firm realizes economies of scale

Figure 21-3a Percent of Sales Forecast

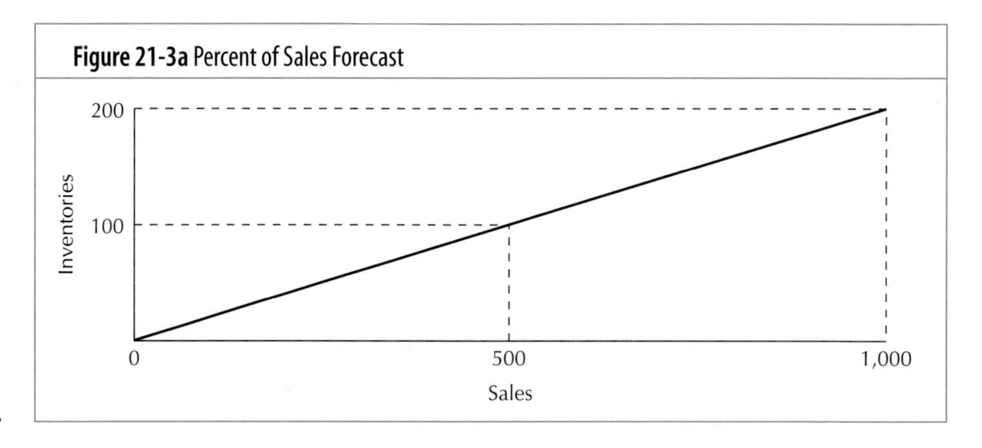

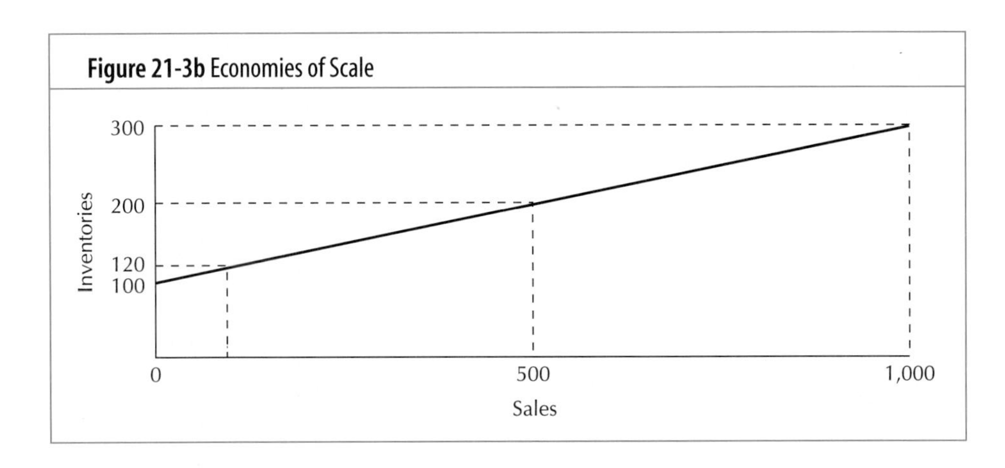

Figure 21-3b Economies of Scale

from its investment in inventory. Note that inventories as a percent of sales decline from 120 percent where sales are \$100, to 30 percent where sales equal \$1,000. This reflects the fact that there is a fixed component of inventories (in this case \$100) that the firm must have on hand regardless of the level of sales, plus a variable component (20 percent of sales). In this instance the predictive equation for inventories is as follows:

$$\text{inventories}_t = a \neq b\,\text{sales}_t$$

In this example, a is equal to 100 and b equals 0.20.[7] Figure 21-3c is an example of *lumpy assets,* that is, assets that must be purchased in large, nondivisible components. Consequently, when a block of assets is purchased it creates excess capacity until sales grow to the point where the capacity is fully used. The result is a step function like the one depicted in Figure 21-3c. Thus, if the firm does not expect sales to exceed the current capacity of its plant and equipment, there would be no projected need for added plant and equipment capacity.

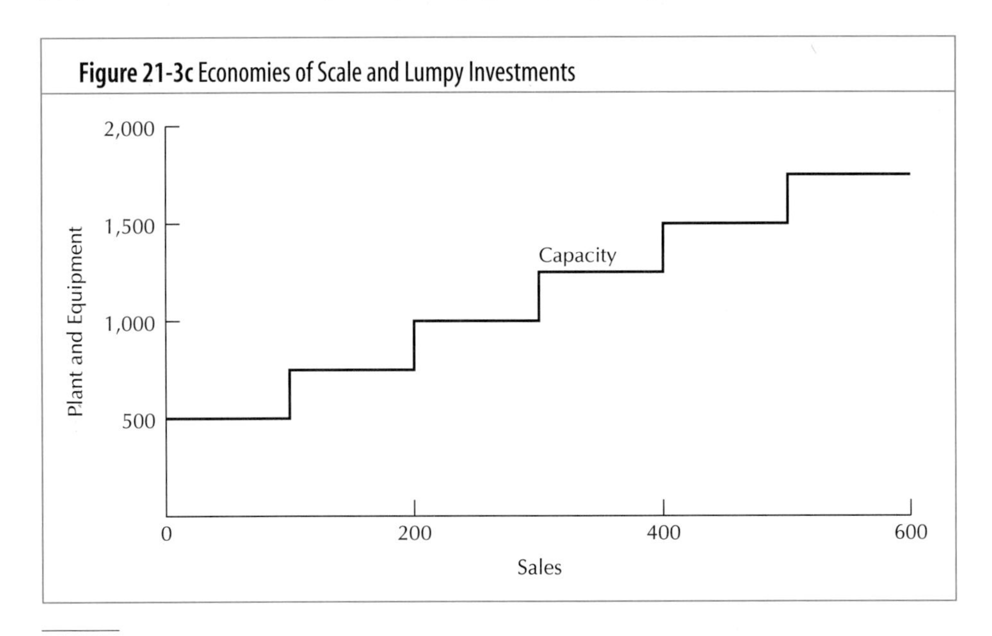

Figure 21-3c Economies of Scale and Lumpy Investments

[7]Economies of scale are evidenced here by the nonzero intercept value. However, scale economies can also result in nonlinear relationships between sales and a particular asset category. Later, when we discuss cash management, we will find that one popular cash management model predicts a nonlinear relationship between the optimal cash balance and the level of cash transactions.

OBJECTIVE 4

FINANCIAL PLANNING AND BUDGETING

Financial forecasts are put to use in constructing financial plans. These plans culminate in the preparation of a cash budget and a set of pro forma statements for a future period in the firm's operations.

BACK TO THE BASICS

Budgets have many important uses; however, their use as a tool of managerial control is critically important and often overlooked in the study of financial management. ***Axiom 7: The Agency Problem—Managers Won't Work for the Owners Unless It's in Their Best Interest*** *speaks to the root source of the problem, and budgets provide one tool for attempting to deal with it. Specifically, budgets provide management with a tool for evaluating performance and consequently maintaining a degree of control over employee actions.*

Budget Functions

Budget

A forecast of future events.

A **budget** is simply a forecast of future events. For example, students preparing for final exams make use of time budgets, which help them allocate their limited preparation time among their courses. Students also must budget their financial resources among competing uses, such as books, tuition, food, rent, clothes, and extracurricular activities.

Budgets perform three basic functions for the user. First, they indicate the amount and timing of the firm's needs for future financing. Second, they provide the basis for taking corrective action in the event that budgeted figures do not match actual or realized figures. Third, budgets provide the basis for performance evaluation. Plans are carried out by people, and budgets provide benchmarks that can be used to evaluate the performance of those responsible for carrying out those plans and, in turn, controlling their actions. Thus, budgets are valuable aids in both planning and control aspects of the firm's financial management.

In the pages that follow, we will develop an example budgetary system for a retailing firm. The primary emphasis will be on the cash budget and pro forma financial statements. These statements provide the information needed for a detailed estimate for the firm's future financing requirements.

The Budgetary System

Although our interest in financial planning focuses on the cash budget, a number of other budgets provide the basis for its preparation. This system of budgets allows planning for each source of cash flow, both inflow and outflow, that will affect the firm throughout the planning period. In general, a business will utilize four types of budgets: physical budgets, cost budgets, profit budgets, and cash budgets. Figure 21-4 presents an overview of the budgetary system.

Physical budgets include budgets for physical items as opposed to financial items. For example, physical budgets are used for unit sales, personnel, unit production, inventories, and physical facilities. Management uses these budgets as the basis for generating cost and profit budgets. *Cost budgets* are prepared for every major expense category of the firm. For example, a manufacturing firm would prepare cost budgets for manufacturing or production costs, selling costs, administrative costs, financing costs, and research and development costs. These cost budgets along with the sales budget provide the basis for preparing a profit budget. Finally, converting all the budget information to a cash basis provides the information required to prepare the cash budget.

Figure 21-4 The Budget System

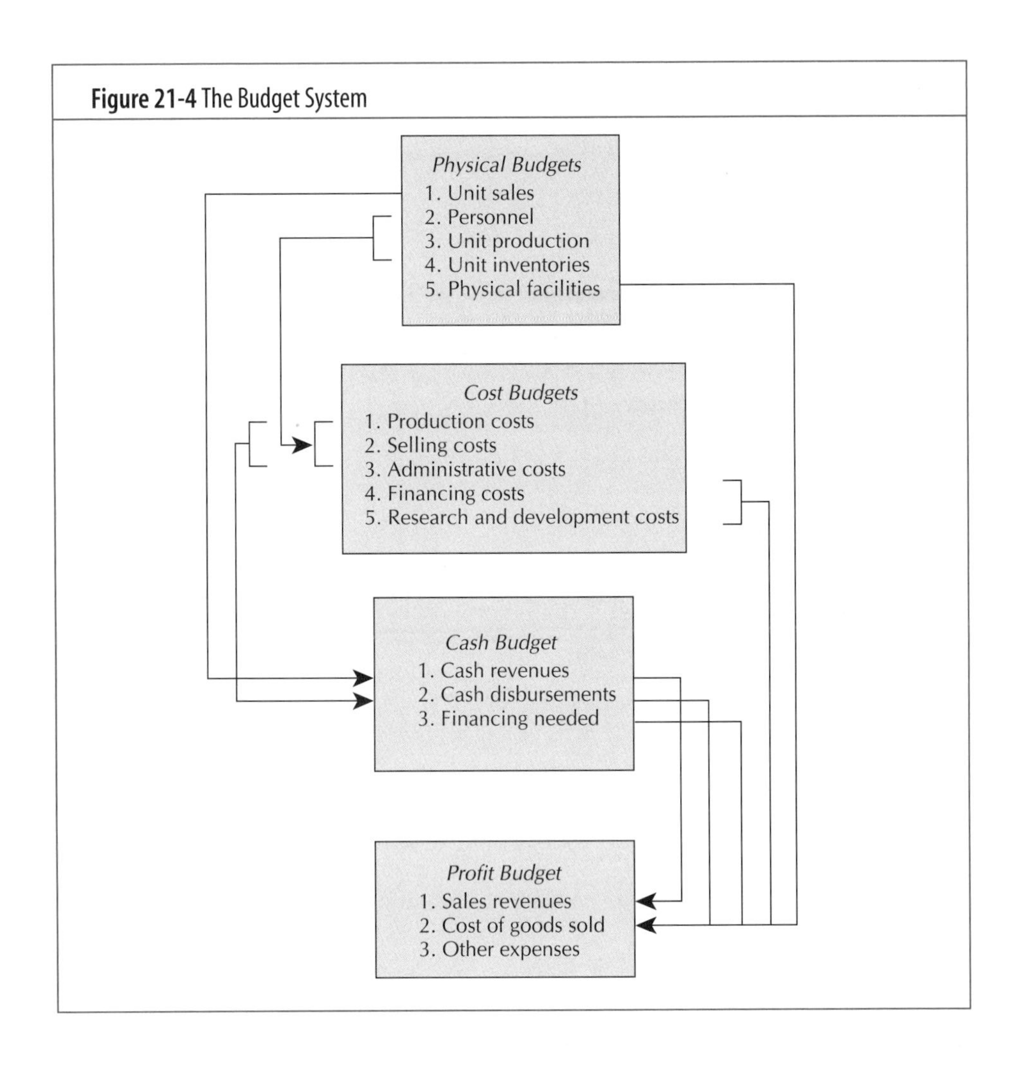

The Cash Budget

The **cash budget** represents a detailed plan of future cash flows and is composed of four elements: cash receipts, cash disbursements, net change in cash for the period, and new financing needed.

Cash budget

A detailed plan of future cash receipts and disbursements.

EXAMPLE

To demonstrate the construction and use of the cash budget, consider Salco Furniture Company Inc., a regional distributor of household furniture. Salco's sales are highly seasonal, peaking in the months of March through May. Roughly 30 percent of Salco's sales are collected one month after the sale, 50 percent two months after the sale, and the remainder during the third month following the sale.

Salco attempts to pace its purchases with its forecast of future sales. Purchases generally equal 75 percent of sales and are made two months in advance of anticipated sales. Payments are made in the month following purchases. For example, June sales are estimated at \$100,000, thus April purchases are $.75 \times \$100{,}000 = \$75{,}000$. Correspondingly, payments for purchases in May equal \$75,000. Wages, salaries, rent, and other cash expenses are recorded in Table 21-2, which gives Salco's cash budget for the six-month

Table 21-2 Salco Furniture Co. Inc. Cash Budget for the Six Months Ended June 30, 1996

Worksheet	Oct.	Nov.	Dec.
Sales	$55,000	$62,000	$50,000
Collections:			
First month (30%)			
Second month (50%)			
Third month (20%)			
Total			
Purchases			$56,250
Payments (one-month lag)			
Cash receipts:			
Collections			
Cash disbursements:			
Purchases			
Wages and salaries			
Rent			
Other expenses			
Interest expense on existing debt ($12,000 note and $150,000 in long-term debt)			
Taxes			
Purchase of equipment			
Loan repayment ($12,000 note due in May)			
Total disbursements:			
Net monthly charge			
Plus: Beginning cash balance			
Less: Interest on short-term borrowing			
Equals: Ending cash balance—no borrowing			
Financing needed[a]			
Ending cash balance			
Cumulative borrowing			

[a]The amount of financing that is required to raise the firm's ending cash balance up to its $10,000 desired cash balance.

period ended in June 1996. Additional expenditures are recorded in the cash budget related to the purchase of equipment in the amount of $14,000 during February and the repayment of a $12,000 loan in May. In June, Salco will pay $7,500 interest on its $150,000 in long-term debt for the period of January–June 1996. Interest on the $12,000 short-term note repaid in May for the period January through May equals $600 and is paid in May.

Salco presently has a cash balance of $20,000 and wants to maintain a minimum balance of $10,000. Additional borrowing necessary to maintain that minimum balance is estimated in the final section of Table 21-2. Borrowing takes place at the beginning of the month in which the funds are needed. Interest on borrowed funds equals 12 percent per annum, or 1 percent per month, and is paid in the month following the one in which funds are borrowed. Thus, interest on funds borrowed in January will be paid in February equal to 1 percent of the loan amount outstanding during January.

The financing-needed line on Salco's cash budget indicates that the firm will need to borrow $36,350 in February, $65,874 in March, $86,633 in April, and $97,599 in May. Only in June will the firm be able to reduce its borrowing to $79,875. Note that the cash budget indicates not only the amount of financing needed during the period, but also when the funds will be needed.

Jan.	Feb.	Mar.	Apr.	May	June	July	Aug.
$60,000	$75,000	$88,000	$100,000	$110,000	$100,000	$80,000	$75,000
15,000	18,000	22,500	26,400	30,000	33,000		
31,000	25,000	30,000	37,500	44,000	50,000		
11,000	12,400	10,000	12,000	15,000	17,600		
$57,000	55,400	62,500	75,900	89,000	100,600		
66,000	75,000	82,500	75,000	60,000	56,250		
56,250	66,000	75,000	82,500	75,000	60,000		
$57,000	55,400	62,500	75,900	89,000	100,600		
$56,250	66,000	75,000	82,500	75,000	60,000		
3,000	10,000	7,000	8,000	6,000	4,000		
4,000	4,000	4,000	4,000	4,000	4,000		
1,000	500	1,200	1,500	1,500	1,200		
				600	7,500		
		4,460			5,200		
	14,000						
				12,000			
$64,250	94,500	91,660	96,000	99,100	81,900		
$(7,250)	(39,100)	(29,160)	(20,100)	(10,100)	18,700		
20,000	12,750	10,000	10,000	10,000	10,000		
—	—	(364)	(659)	(866)	(976)		
12,750	(26,350)	(19,524)	(10,759)	(966)	27,724		
—	36,350	29,524	20,759	10,966	(17,724)[b]		
$12,750	10,000	10,000	10,000	10,000	10,000		
—	36,350	65,874	86,633	97,599	79,875		

[b]Negative financing needed simply means the firm has excess cash that can be used to retire a part of its short-term borrowing from prior months.

Fixed Versus Flexible Budgets

The cash budget given in Table 21-2 for Salco Inc. is an example of a *fixed budget*. Cash flow estimates are made for a single set of monthly sales estimates. Thus, the estimates of expenses and new financing needed are meaningful only for the level of sales for which they were computed. To avoid this limitation, several budgets corresponding to different sets of sales estimates can be prepared. Such a flexible budget fulfills two basic needs: first, it gives information regarding the range of the firm's possible financing needs, and second it provides a standard against which to measure the performance of subordinates who are responsible for the various cost and revenue items contained in the budget.

This second function deserves some additional comment. The obvious problem that arises relates to the fact that costs vary with the actual level of sales experienced by the firm. Thus, if the budget is to be used as a standard for performance evaluation or control, it must be constructed to match realized sales and production figures. This can involve much more than simply "adjusting cost figures up or down in proportion to the deviation of actual from planned sales." That is, costs may not vary in strict proportion to sales, just as inventory levels may not vary as a constant percent of sales. Thus, preparation of a flexible budget involves re-estimating all the cash expenses that would be incurred at each of several possible sales levels. This process might utilize a variant of the percent of sales method discussed earlier or apply a regression method to the forecasting problem.[8]

Budget Period

There are no strict rules for determining the length of the budget period. However, as a general rule it should be long enough to show the effect of management policies, yet short enough so that estimates can be made with reasonable accuracy. Applying this rule of thumb to the Salco example in Table 21-2 indicates that the six-month budget period is probably too short, in that it is not known whether the planned operations of the firm will be successful over the coming fiscal year. That is, for most of the first six-month period the firm is operating with a cash flow deficit. If this does not reverse in the latter six months of the year, then a re-evaluation of the firm's plans and policies is clearly in order.

Longer-range budgets are also prepared in the form of the capital expenditure budget. This budget details the firm's plans for acquiring plant and equipment over a five-year, ten-year, or even longer period. Furthermore, firms often develop comprehensive long-range plans extending up to ten years into the future. These plans are generally not as detailed as the annual cash budget, but they do consider such major components as sales, capital expenditures, new product development, capital funds acquisition, and employment needs.

Pro Forma Financial Statements

The final stage in the budgeting process involves construction of a set of *pro forma financial statements* depicting the end result of the planning period's operations. Salco Inc., is used to demonstrate the construction of the pro forma income statement and balance sheet. To do this, we need Salco's cash budget (found in Table 21-2) and its beginning balance sheet, which depicts the financial condition of the firm at the start of the planning period (see Table 21-3).

The Pro Forma Income Statement

The *pro forma income statement* represents a statement of planned profit or loss for a future period. For Salco a six-month pro forma income statement is constructed from the information contained in the cash budget found in Table 21-2. The final statement is presented in Table 21-4.

Net sales, found by summing the six monthly sales projections (January through June) from Table 21-2, total $533,000. Cost of goods sold is computed as 75 percent of sales, or $399,750. This figure could also have been found by summing purchases for November through April, which represent items sold from January through June. Recall that purchases are made two months in advance, so that items sold in January through June were purchased in November through April.

Depreciation expense cannot be obtained from the cash budget, since it does not constitute a cash flow. Thus, this expense must be determined from the depreciation schedules of Salco's plant and equipment. On its existing fixed assets Salco has an annual depreciation expense of $17,200. In addition, the $14,000 piece of equipment purchased at the end of February will be depreciated over a 15-year life toward a $3,650 salvage value. Using straight-line depreciation and depreciating the asset for four months of the budget period, we find this amounts to roughly $230. Thus, total depreciation expense for the period is $8,830 [or ($17,200/2) + $230].

Wages and salaries, rent, and other expenses are found by summing the relevant cash flow items from the cash budget for the months of January through June. This assumes, of course, that all these expenses are paid at the end of each month in which they are earned, rent is not paid in advance, and all other expenses are

[8]An excellent discussion of the financial forecasting problem in general is found in George G. C. Parker, "Financial Forecasting," in *Handbook of Corporate Finance*, ed. E. I. Altman. (New York: Wiley, 1986), 2.1–2.37.

Table 21-3 Salco Furniture Co. Inc. Balance Sheet December 31, 1995

Assets		
Current assets		
Cash	$ 20,000	
Accounts receivable	104,400	
Inventories	101,250	
Total current assets		$225,650
Fixed assets		
Net plant and equipment		180,000
Total assets		$405,650
Liabilities and Owners' Equity		
Current liabilities		
Accounts payable	$ 56,250	
Notes payable (due in May 1996)	12,000	
Taxes payable	4,460	
Total current liabilities		$ 72,710
Noncurrent liabilities		
Long-term debt		150,000
Shareholders' equity		
Common shares	$ 20,000	
Contributed surplus	50,000	
Retained earnings	112,940	
Total owners' equity		182,940
Total liabilities and owners' equity		$405,650

paid on a monthly basis except interest on short-term borrowing, which is paid in the month following its occurrence. Wages and salaries total $38,000, rent expense equals $24,000, and other expenses are expected to be $6,900.

Subtracting the above operating expenses from gross profit leaves a net operating income of $55,520. Interest expense of $11,764 is then deducted from net operating income to obtain earnings before taxes of $43,756.[9] Total income taxes payable are

Table 21-4 Salco Furniture Co, Inc. Pro Forma Income Statement for the Six-Month Period Ended June 30, 1996

Sales	(from cash budget—Table 21-2)		$533,000
Cost of goods sold	(75% of sales)		(399,750)
Gross profit	(calculation)		$133,250
Operating Expenses			
Depreciation	[($17,200/2) + $230]	$ 8,830	
Wages and salaries	(from cash budget—Table 21-2)	38,000	
Rent	(from cash budget—Table 21-2)	24,000	
Other expenses	(from cash budget—Table 21-2)	6,900	(77,730)
Net operating income	(calculation)		$55,520
Interest expense	(calculation—see text footnote 9)		(11,764)
Earnings before taxes	(calculation)		$43,756
Income taxes payable	(40%)		(17,502)
Net income	(calculation)		$26,254

[9]Total interest expense incurred (but not necessarily paid) during the period equals $7,500 on long-term debt plus $600 on the $12,000 note repaid in May, plus the sum of all interest incurred during the budget period on short-term borrowing. Note that we include $364 for February, $659 for March, and so forth, plus $799 for June, which was incurred but not paid until July.

found using a 40 percent corporate income tax rate. For Salco this equals a tax expense for the period of $17,502. Finally, subtracting the estimated taxes from earnings before taxes indicates net income for the period of $26,254.

Net Cash Flow Versus Net Income

The difference in the cash and accrual bases of accounting for corporate income is vividly demonstrated in the cash budget and pro forma income statements for the period. On a cash flow basis the firm has a substantial net negative cash flow, while on an accrual basis the firm earned $26,254. The difference, of course, relates to when revenues and expenses are accounted for or recognized in the two statements. In the cash budget revenues and expenses are included in the months in which cash is actually received or disbursed. In the income statement revenues and expenses are included in the month in which the corresponding sale took place, which usually is not the same month in which cash is received. The income statement is therefore prepared on an accrual basis. See Appendix 20A for added discussion of the accrual basis of accounting.

The Pro Forma Balance Sheet

We can construct the pro forma balance sheet for Salco by using information from the cash budget (Table 21-2); the December 31, 1995 balance sheet (Table 21-3); and the pro forma income statement (Table 21-4). Salco's pro forma balance sheet for June 30, 1996 is presented in Table 21-5. Estimates of the individual statement entries are provided below.[10]

Ending cash from the cash budget, $10,000, becomes the cash entry in Salco's pro forma balance sheet. The accounts receivable balance is found as follows:

Accounts receivable (12/31/95) (from Table 21-3)	$104,400
+ Credit sales (from Table 21-2)	533,000
− Collections (from Table 21-2)	(440,400)
Accounts receivable (6/30/96) (calculation)	$197,000

The beginning balance for accounts receivable is taken from the December 31, 1995 balance sheet (Table 21-3), and credit sales and collections are obtained by summing across the relevant cash budget monthly totals. Inventories are determined in a similar manner:

Inventories (12/31/95) (from Table 21-3)	$101,250
+ Purchases (from Table 21-2)	414,750
− Cost of goods sold (from Table 21-4)	(399,750)
Inventories (6/30/96) (calculation)	$116,250

Purchases were found by summing relevant monthly figures from the cash budget for all six months of the budget period; and cost of goods sold was taken from the pro forma income statement in Table 21-4. The net plant and equipment figure is found as follows:

[10]The cash flow from operating activities can be estimated from the schedule of net operating cash flows or the statement of changes in financial position; both techniques were examined in Chapter 20.

Net plant and equipment (12/31/95) (from Table 21-3)	$180,000
+ Purchases of plant and equipment (from Table 21-2)	14,000
− Depreciation expense (from Table 21-4)	(8,830)
Net plant and equipment (6/30/96) (calculation)	$185,170

Purchases of plant and equipment are reflected in the cash budget, and depreciation expense is taken from the pro forma income statement. The only changes that took place during the period involved the $14,000 purchase and depreciation expense of $8,830, leaving a net balance of $185,170. Total assets for Salco are therefore expected to be $508,420.

The liability accounts are estimated using the same basic methodology used in finding asset balances. Accounts payable is found as follows:

Accounts payable (12/31/95) (from Table 21-3)	$56,250
+ Purchases (from Table 21-2)	414,750
− Payments (from Table 21-2)	(414,750)
Accounts payable (6/30/96) (calculation)	$56,250

During June the firm had total borrowing of $77,875, on which it owes $799 in interest. However, since interest is not paid in the month in which the expense is incurred, this interest liability still exists at the end of June. We can analyze the ending balance for interest payable as follows:

Table 21-5 Salco Furniture Co. Inc. Pro Forma Balance Sheet June 30, 1996

Assets		
Current assets		
Cash	$ 10,000	
Accounts receivable	197,000	
Inventories	116,250	
Total current assets		$323,250
Fixed assets		
Net plant and equipment		185,170
Total assets		$508,420
Liabilities and Owners' Equity		
Current liabilities		
Accounts payable	$ 56,250	
Interest payable	799[a]	
Notes payable[b]	79,875	
Taxes payable	12,302	
Total current liabilities		$149,226
Noncurrent liabilities		
Long-term debt		150,000
Shareholders' equity		
Common shares	$ 20,000	
Contributed surplus	50,000	
Retained earnings	139,194	
Total owners' equity		209,194
Total liabilities and owners' equity		$508,420

[a]This contains the $799 in interest incurred on June's total borrowing of $77,875, but which will not be paid until July.
[b]Cumulative borrowing for the period was assumed to take the form of notes payable. This figure is taken from the cumulative borrowing row of cash budget contained in Table 21-2.

Interest payable (12/31/95)	$ 0
+ Interest expense (see footnote 9)	11,764
− Interest paid	10,965
Interest paid (6/30/96)	$ 799

Again, purchases and payments were taken from the cash budget for each of the six months of the budget period. Notes payable are found as follows:

Notes payable (12/31/95) (from Table 21-3)	$12,000
+ Borrowing (6/30/96) (from Table 21-2)	79,875
− Repayments (from Table 21-2)	(12,000)
Notes payable (6/30/96) (calculation)	$79,875

Here it is assumed that the total new financing needed during the period ($79,875) would be raised through notes payable. The hedging principle advocates that the firm should finance an asset with a source of financing that matches the expected life and cash-generating charactertistics of the asset as closely as possible. An accrued interest expense item of $799 is created as a result of interest expense in that amount that was incurred during June on short-term borrowing but will not be paid until July. Next, compute taxes payable as follows:

Taxes payable (12/31/95) (from Table 21-3)	$ 4,460
+ Tax liability for the period (from Table 21-2)	17,502
− Tax payments made during the period (from Table 21-2)	(9,660)
Taxes payable (6/30/96) (calculation)	$12,302

Long-term debt, common shares, and contributed surplus remain unchanged for the period, as no new shares or long-term debt was issued nor was any repurchased or retired. Finally, the retained earnings balance is found as follows:

Retained earnings (12/31/95) (from Table 21-3)	$112,940
+ Net income for the period (from Table 21-4)	26,254
− Cash dividends (from Table 21-2)	0
Retained earnings (6/30/96) (calculation)	$139,194

Since no common dividends were paid (none were considered in the cash budget—Table 21-2), the new retained earnings figure is $139,194.

Salco's management may now wish to perform a financial analysis using the newly prepared pro forma statements. Such an analysis would provide the basis for evaluating the firm's planned financial performance over the next six months. It would entail use of a set of ratios such as those discussed in Chapter 20, which could be compared with prior-year figures and industry averages. If this analysis identified any potential weakness, the firm could take steps to correct it before it became reality.

Financial Control

The pro forma statements just prepared can be used to monitor or control the firm's financial performance. One approach would involve preparing pro forma statements for each month during the planning period. Actual operating results for each month's operations could then be compared with the projected or pro forma figures. This type of analysis would provide an early warning system to detect financial problems as they develop. In particular, by comparing actual monthly (or even weekly) operating results with projected revenue and expense items (from pro forma income statements), the financial manager can maintain a very close watch on the firm's overall profitability and take an active role in determining the firm's overall performance for the planning period.

Financial Planning and Budgeting: Closing Comments

Two aspects of pro forma statements should be emphasized. First, these figures represent single-point estimates of each of the items in the entire system of budgets and the resulting pro forma statements. Although these may be the best estimates as to what the future will hold for Salco, at least two additional sets of estimates may be desired, corresponding to the very worst set of circumstances that the firm might face and the very best. These extremes provide the necessary input for formulating contingency financing plans, should a deviation from the expected figures occur. Second, notes payable was used as a plug figure for additional financing needed. The actual source of financing selected will depend on a number of factors, including: (1) the length of the period for which the financing will be needed, (2) the cost of alternative sources of funds, and (3) the preferences of the firm's management regarding the use of debt versus equity. Other factors used to select short-term financing were discussed in Chapter 11.

COMPUTERIZED FINANCIAL PLANNING

In recent years, a number of developments in both computer hardware (machines) and software (programs) have reduced the tedium of the planning and budgeting process immensely. These developments include the introduction of "user friendly" or "easy to use" computer programs that are specialized for application to financial planning.

Financial planning software packages (sometimes referred to as electronic spreadsheets) were first made popular on large mainframe computers. These packages allow even a computer novice to utilize a personal computer to construct budgets and forecasts. The real advantage of the computer occurs if there is a need for different scenarios to be evaluated quickly and at a low cost in labour-hours.

Another major development that has had a significant impact on the extent to which computers are used in the planning and budgeting process is the advent of the microcomputer. For mere hundreds of dollars the financial analyst's desk can contain the computing power it took hundreds of thousands of dollars to buy just a decade ago. The development of financial planning software has parallelled the development of microcomputer technology. The number of spreadsheet packages has mushroomed over the past five years. These packages generally sell for less than $500 and usually include graphics programs as well as elementary database management capabilities.[11] The appendix offers an example spreadsheet exercise.

[11]The number and variety of financial spreadsheet programs has expanded dramatically since the introduction of the original Visi-Calc program. These include Lotus 123 and Excel among others. In addition, there is a growing set of products referred to as "expert systems," which attempt to mimic the decisions of experts. To date these efforts have produced a limited number of financial applications software related to such things as the capital-budgeting decision, but they offer an opportunity to expand the capabilities of the financial manager of the future.

SUMMARY

This chapter has developed the role of forecasting within the context of the firm's financial planning efforts. Forecasts of the firm's sales revenues and related expenses provide the basis for projecting future financing needs. The most popular method for forecasting variables is the percent of sales method.

The percent of sales method presumes that the asset or liability being forecast is a constant percent of sales for all future levels of sales. There are cases in which this assumption is not reasonable and consequently the percent of sales method does not provide reasonable predictions. One such case occurs if there are economies of scale in the use of the asset being forecast. For example, the firm may need at least \$10 million in inventories to open its doors and operate even for sales as low as \$100 million per year. If sales double to \$200 million, inventories may only increase to \$15 million. Thus, inventories do not increase with sales in a constant proportion. A second situation where the percent of sales method fails to work properly occurs if asset sales are lumpy. In other words, the levels of plant and equipment will not remain a constant percent of sales if plant capacity must be purchased in \$50 million increments.

How serious are these possible problems and should we use the percent of sales method? Even in the face of these problems, the percent of sales method predicts reasonably well where predicted sales levels do not differ drastically from the level used to calculate the percent of sales. For example, if the current sales level used in calculating the percent of sales for inventories is \$40 million, then we can feel more comfortable forecasting the level of inventories corresponding to a new sales level of \$42 million than were the sales predicted to rise to \$60 million.

A firm's sustainable rate of growth is the maximum rate at which its sales can grow if it is to maintain its present financial ratios and not have to resort to issuing new equity. We calculate the sustainable rate of growth as follows:

$$\text{sustainable rate of growth } (g^*) = ROE(1 - b)$$

where ROE is the return earned on common equity and b is the dividend payout ratio (that is, the ratio of dividends to earnings). Consequently, a firm's sustainable rate of growth increases with ROE and decreases with the fraction of its earnings paid out in dividends.

The cash budget is the primary tool of financial forecasting and planning. This tool provides a detailed plan of future cash flow estimates and is comprised of four elements or segments: cash receipts, cash disbursements, net change in cash for the period, and new financing needed. The cash budget also serves as a tool for monitoring and controlling the firm's operations. By comparing actual cash receipts and disbursements to those in the cash budget, the financial manager can gain an appreciation for how well the firm is performing. In addition, deviation from the plan serves as an early warning system to signal the onset of financial difficulties ahead.

STUDY QUESTIONS

21-1. Discuss the shortcomings of the percent of sales method of financial forecasting.

21-2. Explain how a fixed cash budget differs from a variable or flexible cash budget.

21-3. What two basic needs does a flexible (variable) cash budget serve?

21-4. What would be the probable effect on a firm's cash position of the following events?

a. Rapidly rising sales
b. A delay in the payment of payables
c. A more liberal credit policy on sales (to the firm's customers)
d. Holding larger inventories

21-5. How long should the budget period be? Why would a firm not set a rule that all budgets be for a 12-month period?

21-6. A cash budget is usually thought of as a means of planning for future financing needs. Why would a cash budget also be important for a firm that had excess cash on hand?

21-7. Explain why a cash budget would be of particular importance to a firm that experiences seasonal fluctuations in its sales.

SELF-TEST PROBLEMS

ST-1. (*Financial Forecasting*) Use the percent of sales method to prepare a pro forma income statement for Calico Sales Co. Inc. Projected sales for next year equals $4 million. Cost of goods sold equals 70 percent of sales, administrative expense equals $500,000, and depreciation expense is $300,000. Interest expense equals $50,000 and income is taxed at a rate of 40 percent. The firm plans to spend $200,000 during the period to renovate its office facility and will retire $150,000 in notes payable. Finally, selling expense equals 5 percent of sales.

ST-2. (*Pro Forma Statements and Liquidity Analysis*) The balance sheet for Odom Manufacturing Company on December 31, 1995, is found below:

Odom Manufacturing Co. Balance Sheet December 31, 1995

Cash	$ 250,000	Accounts payable	$ 850,000
Accounts receivable	760,000	Notes payable	550,000
Inventory	860,000	Current liabilities	$1,400,000
Current assets	$1,870,000	Long-term debt	800,000
Property, plant, and equipment	1,730,000	Common shares	600,000
		Retained earnings	800,000
Total assets	$3,600,000	Total liabilities and shareholders' equity	$3,600,000

The treasurer of Odom Manufacturing wishes to borrow $500,000, the funds from which would be applied in the following manner:

1. $100,000 to reduce accounts payable
2. $75,000 to retire current notes payable
3. $175,000 to expand existing plant facilities
4. $80,000 to increase inventories
5. $70,000 to increase cash on hand

Repayment of the $500,000 will be in 20 years with interest paid annually.

a. Assuming that the loan is obtained, prepare a pro forma balance sheet for Odom Manufacturing that reflects the use of the loan proceeds.

b. Did the firm's liquidity improve after the loan was obtained and the proceeds were dispensed in the above manner? Why or why not?

ST-3. (*Cash Budget*) Stauffer Inc. has estimated sales and purchase requirements for the last half of the coming year. Past experience indicates that it will collect 20 percent of its sales in the month of the sale, 50 percent of the remainder one month after the sale, and the balance in the second month following the sale. Stauffer prefers to pay for half its purchases in the month of the purchase and the other half the following month. Labour expense for each month is expected to equal 5 percent of that month's sales, with cash payment being made in the month in which the expense is incurred. Depreciation expense is $5,000 per month; miscellaneous cash expenses are $4,000 per month and are paid in the month incurred. General and administrative expenses of $50,000 are recognized and paid monthly. A $60,000 truck is to be purchased in August and is to be depreciated on a straight-line basis over ten years with no expected salvage value. The company also plans to pay a $9,000 cash dividend to shareholders in July. The company feels that a minimum cash balance of $30,000 should be maintained. Any borrowing will cost 12 percent annually, with interest paid in the month following the month in which the funds are borrowed. Borrowing takes place at the beginning of the month in which the need for funds arises. For example, if during the month of July the firm should need to borrow $24,000 to maintain its $30,000 desired minimum balance, then $24,000 will be taken out on July 1 with interest owed for the entire month of July. Interest for the month of July would then be paid on August 1. Sales and purchase estimates are shown below. Prepare a cash budget for the months of July and August (cash on hand 6/30 was $30,000, while sales for May and June were $100,000 and purchases were $60,000 for each of these months).

Month	Sales	Purchases
July	$120,000	$50,000
August	150,000	40,000
September	110,000	30,000

STUDY PROBLEMS (SET A)

21-1A. (*Financial Forecasting*) Zapatera Enterprises is evaluating its financing requirements for the coming year. The firm has only been in business for one year, but its chief financial officer predicts that the firm's operating expenses, current assets, net fixed assets, and current liabilities will remain at their current proportion of sales.

Last year Zapatera had $12 million in sales with net income of $1.2 million. The firm anticipates that the next year's sales will reach $15 million with net income rising to $2 million. Given its present high rate of growth, the firm retains all its earnings to help defray the cost of new investments.

The firm's balance sheet for the year just ended is found below:

Zapatera Enterprises, Inc., Balance Sheet

	12/31/95	% of Sales
Current assets	$3,000,000	25%
Net fixed assets	6,000,000	50%
Total	$9,000,000	
Accounts payable	$3,000,000	25%
Long-term debt	2,000,000	NA
Total liabilities	$5,000,000	
Common shares	1,000,000	NA
Contributed surplus	1,800,000	NA
Retained earnings	1,200,000	
Common equity	4,000,000	
Total	$9,000,000	

Estimate Zapatera's total financing requirements (i.e., total assets) for 1996 and its net funding requirements (discretionary financing needed).

21-2A. (*Pro Forma Accounts Receivable Balance Calculation*) On March 31, 1995, the Syliva Gift Shop had outstanding accounts receivable of $20,000. Syliva's sales are roughly evenly split between credit and cash sales, with the credit sales collected half in the month after the sale and the remainder two months after the sale. Historical and projected sales for the gift shop are given below:

Month	Sales
January	$15,000
February	20,000
March	30,000
April (projected)	40,000

a. Under these circumstances, what should the balance in accounts receivable be at the end of April?

b. How much cash did Syliva realize during April from sales and collections?

21-3A. (*Financial Forecasting*) Sambonoza Enterprises projects its sales next year to be $4 million and expects to earn 5 percent of that amount after taxes. The firm is currently in the process of projecting its financing needs and has made the following assumptions (projections):

1. Current assets will equal 20 percent of sales while fixed assets will remain at their current level of $1 million.
2. Common equity is presently $0.8 million, and the firm pays out half its after-tax earnings in dividends.
3. The firm has short-term payables and trade credit that normally equal 10 percent of sales and has no long-term debt outstanding.

What are Sambonoza's financing needs for the coming year?

21-4A. (*Financial Forecasting—Percent of Sales*) Tulley Appliances Inc. projects next year's sales to be $20 million. Current sales are at $15 million based on current assets of $5 million and fixed assets of $5 million. The firm's net profit margin is 5 percent after taxes. Tulley forecasts that current assets will rise in direct proportion to the increase in sales but fixed assets will increase by only $100,000. At present Tulley has $1.5 million in accounts payable plus $2 million in long-term debt (due in ten years) outstanding and common equity (including $4 million in retained earnings) totalling $6.5 million. Tulley plans to pay $500,000 in dividends from common shares next year.

a. What are Tulley's total financing needs (i.e., total assets) for the coming year?

b. Given the firm's projections and dividend payment plans, what are its discretionary financing needs?

c. Based upon the projections given and assuming the $100,000 expansion in fixed assets will occur, what is the largest increase in sales the firm can support without having to resort to the use of discretionary sources of financing?

21-5A. (*Pro Forma Balance Sheet Construction*) Use the following industry average ratios to construct a pro forma balance sheet for the V. M. Willet Co.

Total asset turnover	2.0 times
Average collection period (assume a 365-day year)	9.125 days
Fixed asset turnover	5.0 times
Inventory turnover (based on cost of goods sold)	3.0 times
Current ratio	2.0 times
Sales (all on credit)	$4.0 million
Cost of goods sold	75% of sales
Debt ratio	50%

Cash		Current liabilities	
Accounts receivable		Long-term debt	
		Common shares plus	
Net fixed assets	$ ______	retained earnings	______
	$		$

21-6A. (*Cash Budget*) The Sharpe Corporation's projected sales for the first eight months of 1996 are as follows:

January	$ 90,000	May	$300,000
February	120,000	June	270,000
March	135,000	July	225,000
April	240,000	August	150,000

Of Sharpe's sales, 10 percent is for cash, another 60 percent is collected in the month following sale, and 30 percent is collected in the second month following sale. November and December sales for 1995 were $220,000 and $175,000, respectively.

Sharpe purchases its raw materials two months in advance of its sales equal to 60 percent of their final sales price. The supplier is paid one month after it makes delivery. For example, purchases for April sales are made in February and payment is made in March.

In addition, Sharpe pays $10,000 per month for rent and $20,000 each month for other expenditures. Tax prepayments of $22,500 are made each quarter, beginning in March.

The company's cash balance at December 31, 1995, was $22,000; a minimum balance of $15,000 must be maintained at all times. Assume that any short-term financing needed to maintain cash balance would be paid off in the month following the month of financing if sufficient funds are available. Interest on short-term loans (12 percent) is paid monthly. Borrowing to meet estimated monthly cash needs takes place at the beginning of the month. Thus, if in the month of April the firm expects to have a need for an additional $60,500, these funds would be borrowed at the beginning of April with interest of $605 ($0.12 \times 1/12 \times \$60{,}500$) owed for April and paid at the beginning of May.

a. Prepare a cash budget for Sharpe covering the first seven months of 1996.

b. Sharpe has $200,000 in notes payable due in July that must be repaid or renegotiated for an extension. Will the firm have ample cash to repay the notes?

21-7A. (*Financial Forecasting—Pro Forma Statements*) The Bell Retailing Company has been engaged in the process of forecasting its financing needs over the next quarter and has made the following forecasts of planned cash receipts and disbursements:

1. Historical and predicted sales:

Historical		Predicted	
April	$ 80,000	July	$130,000
May	100,000	August	130,000
June	120,000	September	120,000
		October	100,000

2. The firm incurs and pays a monthly rent expense of $3,000.
3. Wages and salaries for the coming months are estimated as follows: with payments coinciding with the month in which the expense is incurred.

July	$18,000
August	18,000
September	16,000

4. Of the firm's sales, 40 percent is collected in the month of sale, 30 percent one month after sale, and the remaining 30 percent two months after sale.
5. Merchandise is purchased one month before the sales month and is paid for in the month it is sold. Purchases equal 80 percent of sales.
6. Tax prepayments are made on the calendar quarter, with a prepayment of $1,000 in July based on earnings for the quarter ended June 30.
7. Utilities for the firm average 2 percent of sales and are paid in the month of their occurrence.
8. Depreciation expense is $12,000 annually.
9. Interest on a $40,000 bank note (due in November) is payable at an 8 percent annual rate in September for the three-month period just ended.
10. The firm follows a policy of paying no cash dividends.

Based on the above, supply the following items:

a. Prepare a monthly cash budget for the three-month period ended September 30, 1996.

b. If the firm's beginning cash balance for the budget period is $5,000 and this is its minimum desired balance, determine when and how much the firm will need to borrow during the budget period. The firm has an $80,000 line of credit with its bank with interest (12 percent annual) paid monthly (for example, for a loan taken out at the end of December, interest would be paid at the end of January and every month thereafter so long as the loan was outstanding).

c. Prepare a pro forma income statement for Bell covering the three-month period ended September 30, 1996. Use a 40 percent tax rate.

d. Given the following balance sheet dated June 30, 1996, and your pro forma income statement from (c), prepare a pro forma balance sheet for September 30, 1996.

Bell Manufacturing Co. Balance Sheet June 30, 1996

Cash	$ 5,000	Accounts payable	$104,000
Accounts receivable	102,000	Bank notes (8%)	40,000
Inventories	114,000	Accrued taxes	1,000
Current assets	221,000	Current liabilities	145,000
Net fixed assets	120,000	Common shares	100,000
		Contributed surplus	28,400
		Retained earnings	67,600
Total assets	$341,000	Total liabilities and capital	$341,000

21-8A. (*Forecasting Accounts Receivable*) Sampkin Mfg. is evaluating the effects of extending trade credit to its out-of-province customers. In the past, Sampkin has given credit to local customers but has insisted on cash upon delivery from customers in the bordering province of Manitoba. At present Sampkin's average collection period is 35 days based upon total sales; that is, including both its $900,000 in-province credit sales and its $300,000 out-of-province cash sales, the firm's accounts receivable balance is 35 days.

$$\text{average collection period} = \frac{\text{accounts receivable}}{\frac{\text{sales}}{365}}$$

$$= \frac{\$115{,}068}{\frac{\$1{,}200{,}000}{365}} = 35 \text{ days}$$

a. Calculate Sampkin's average collection period based upon credit sales alone.

b. If Sampkin extends credit to its out-of-province customers, it expects total credit sales to rise to $1,500,000. If the firm's average collection period remains the same, what will be the new level of accounts receivable?

21-9A. (*Forecasting Accounts Receivable*) The Ace Traffic Company sells its merchandise on credit terms of 2/10, net 30 (2 percent discount if payment is made within ten days or the net amount is due in 30 days). Only a part of the firm's customers take the trade discount, so that the firm's average collection period is 20 days.

a. Based upon estimated sales of $405,556, project Ace's accounts receivable balance for the coming year.

b. If Ace changes its cash discount terms to 1/10, net 30, it expects its average collection period will rise to 28 days. Estimate Ace's accounts receivable balance based on the new credit terms and expected sales of $405,556.

21-10A. (*Percent of Sales Forecasting*) Which of the following accounts would most likely vary directly with the level of firm sales? Discuss each briefly.

	Yes	No
Cash	____	____
Marketable securities	____	____
Accounts payable	____	____
Notes payable	____	____
Plant and equipment	____	____
Inventories	____	____

21-11A. (*Financial Forecasting—Percent of Sales*) The balance sheet of the Thompson Trucking Company (TTC) is found below:

Thompson Trucking Company Balance Sheet January 31, 1995 ($ millions)

Current assets	$10	Accounts payable	$ 5
Net fixed assets	15	Notes payable	0
Total	$25	Bonds payable	10
		Common equity	10
		Total	$25

TTC had sales for the year ended 1/31/95 of $50 million. The firm follows a policy of paying all net earnings out to its common shareholders in cash dividends. Thus, TTC generates no funds from its earnings that can be used to expand its operations (assume that depreciation expense is just equal to the cost of replacing worn-out assets.)

a. If TTC anticipates sales of $80 million during the coming year, develop a pro forma balance sheet for the firm for 1/31/95. Assume that current assets vary as a percent of sales, net fixed assets remain unchanged, accounts payable vary as a percent of sales, and use notes payable as a balancing entry.

b. How much "new" financing will TTC need next year?

c. What limitations does the percent of sales forecast method suffer from? Discuss briefly.

21-12A. (*Financial Forecasting—Discretionary Financing Needed*) The most recent balance sheet for the Armadillo Dog Biscuit Co. is shown in the table below. The company is about to embark on an advertising campaign, which is expected to raise sales from the present level of $5 million to $7 million by the end of next year. The firm is presently operating at full capacity and will have to increase its investment in both current and fixed assets to support the projected level of new sales. In fact, the firm estimates that both categories of assets will rise in direct proportion to the projected increase in sales.

The firm's net profits were 6 percent of the current year's sales but are expected to rise to 7 percent of next year's sales. To help support its anticipated growth in asset needs next year, the firm has suspended plans to pay cash dividends to its shareholders. In years past, a $1.50 per share dividend has been paid annually.

Armadillo Dog Biscuit Co. Inc. ($ millions)

	Present Level	Percent of Sales	Projected Level
Current assets	$2.0		
Net fixed assets	$3.0		
Total	$5.0		
Accounts payable	$0.5		
Accrued expenses	$0.5		
Notes payable	—		
Current liabilities	$1.0		
Long-term debt	$2.0		
Common shares	$0.5		
Retained earnings	$1.5		
Common equity	$2.0		
Total	$5.0		

Armadillo's payables and accrued expenses are expected to vary directly with sales. In addition, notes payable will be used to supply the added funds needed to finance next year's operations that are not forthcoming from other sources.

a. Fill in the table and project the firm's needs for discretionary financing. Use notes payable as the balancing entry for future discretionary financing needed.

b. Compare Armadillo's current ratio and debt ratio (total liabilities/total assets) before the growth in sales and after. What was the effect of the expanded sales on these two dimensions of Armadillo's financial condition?

c. What difference, if any, would have resulted if Armadillo's sales had risen to $6 million in one year and $7 million only after two years? Discuss only; no calculations required.

21-13A. (*Financial Forecasting—Changing Credit Policy*) Island Resorts Inc. has $400,000 invested in receivables throughout much of the year. It has annual credit sales of $3,650,000 and a gross profit margin of 80 percent.

a. What is Island Resorts' average collection period? [*Hint:* Recall that the average collection period = accounts receivable/(credit sales/365).]

b. By how much would Island Resorts be able to reduce its accounts receivable if it were to change its credit policies in a way that reduced its average collection period to 30 days without affecting annual credit sales?

21-14A. (*Forecasting Discretionary Financing Needs*) Fishing Charter Inc. estimates that it invests 30 cents in assets for each dollar of new sales. However, 5 cents in profits are produced by each dollar of additional sales, of which 1 cent can be reinvested in the firm. If sales rise from their present level of $5 million by $500,000 next year, and the ratio of spontaneous liabilities of sales is .15, what will be the firm's need for discretionary financing? (*Hint:* In this situation you do not know what the firm's existing level of assets is, nor do you know how they have been financed. Thus, you must estimate the change in financing needs and match this change with the expected changes in spontaneous liabilities, retained earnings, and other sources of discretionary financing.)

21-15A. (*Cash Budget*) The Cramer Enterprise projected sales for the first eight months of 1996 are as follows:

January	$100,000	May	$275,000
February	140,000	June	250,000
March	150,000	July	200,000
April	250,000	August	120,000

Of Cramer's sales, 20 percent is for cash, another 50 percent is collected in the month following sale, and 30 percent is collected in the second month following sale. November and December sales for 1995 were $220,000 and $175,000, respectively.

Cramer purchases its raw materials two months in advance of its sales equal to 65 percent of their final sales price. The supplier is paid one month after it makes delivery. For example, purchases for April sales are made in February and payment is made in March.

In addition, Cramer pays $10,000 per month for rent and $20,000 each month for other expenditures. Tax prepayments of $22,500 are made each quarter, beginning in March.

The company's cash balance at December 31, 1995 was $22,000; a minimum balance of $20,000 must be maintained at all times. Assume that any short-term financing needed to maintain cash balance would be paid off in the month following the month of financing if sufficient funds are available. Cramer has arranged with its bank the use of short-term credit at an interest rate of 12 percent per annum (1 percent per month) to be paid monthly. Borrowing to meet estimated monthly cash needs takes place at the beginning of the month but interest is not paid until the end of the following month. If in the month of April the firm expects to have a need for an additional $50,000, these funds would be borrowed at the beginning of April with interest of $500 (= .01 × $50,000) in interest during May. Finally, Cramer follows a policy of repaying any outstanding short-term loans in any month in which its cash balance exceeds the minimum desired balance of $20,000.

a. Cramer needs to know what its cash requirements will be for the next six months so that it can renegotiate the terms of its short-term credit agreement with its bank, if necessary. To evaluate this problem the firm plans to evaluate the impact of a 20 percent variation in its monthly sales effort. Prepare a six-month cash budget for Cramer and use it to evaluate the firm's cash needs.

b. Cramer has $200,000 in notes due in July. Will the firm have sufficient cash to repay the notes?

CASE PROBLEM

LINCOLN PLYWOOD INC.

FINANCIAL FORECASTING

Lincoln Plywood Inc. (LPI) owns and operates a plywood fabricating plant in northern British Columbia. LPI's primary product is 4 × 8 sheets of pine plywood used in the construction industry. The plywood is formed out of thin sheets of wood that are "peeled" from pine logs bought in the region. A number of varieties of thicknesses and grades of plywood can be made in LPI's plant, ranging from 1/4-inch sheets used as panelling in new homes to 3/4-inch sheets used in framing and heavy industrial applications. LPI's sales have grown rapidly over the past ten years at an average of 10 percent per year. This growth has resulted from the housing boom that occurred during this period and in particular the very rapid rate of regional growth in central British Columbia.

The firm's management has been pleased with the growth in sales; however, during the past six months it has begun to experience some cash flow problems that appear to have gotten out of control (LPI's current balance sheet is shown below). The firm's controller, James Christian, is aware of the decline in the firm's cash position and plans to correct the problem as soon as an appropriate course of action can be identified. One factor that he feels is surely making the problem worse is the firm's credit and collections policy. In particular, LPI recently "liberalized" its credit terms in an effort to stimulate demand in the face of stiffening competition. The full effect of the change is not known, but Christian plans to address that problem when he makes his projection of cash needs for the coming year.

Lincoln Plywood Inc. Balance Sheet for December 31, 1993

Assets		
Cash		$ 125,000
Accounts receivable	$ 517,500	
Less: Allowance for bad debts	−7,500	510,000
Inventories		725,000
Current assets		$1,360,000
Plant and equipment	$2,320,000	
Less: Accumulated depreciation	−1,280,000	
Net plant and equipment		1,040,000
Total assets		$2,400,000

Liabilities and Owners' Equity	
Accounts payable	$ 30,000
Accrued taxes	50,000
Accrued expenses	270,000
Current liabilities	$ 350,000
Common shares	$ 400,000
Retained earnings	1,650,000
Common equity	$2,050,000
Total liabilities and owners' equity	$2,400,000

LPI's cash cycle is a relatively simple one. The firm purchases timber, processes that timber by turning it into various types of plywood products, sells the plywood on credit, and collects its credit sales. The cash flow problem the firm currently faces is partly a result of its increasing the term over which credit sales can be repaid. In fact, following its recent liberalization of credit terms, the firm estimates that only 10 percent of its customers will pay within the month of the sale, 46 percent of the total will be collected in one month, and all but 2 percent of sales will be collected in the second month following the sale. Bad debt expense is estimated at 1 percent of sales in the income statement.

Christian plans to make a monthly cash budget for LPI spanning the next six months (January through June) and has already projected firm sales as follows:

Month	Projected Sales	Month	Projected Sales
January	$500,000	April	$525,000
February	300,000	May	800,000
March	250,000	June	250,000

The monthly sales estimates reflect a long-established seasonality in the firm's sales. However, the firm's purchases of timber are not closely matched to its sales. The harvest of timber is heavily seasonal due to the problems associated with getting the timber out of the woods during the rainy winter months. Since LPI contracts up to eight months in advance for its timber purchases, it has a very good idea as to what its expenditures for timber will be over the six-month planning horizon. Specifically, LPI has contracted to make the following timber purchases:

Month	Purchase Contracts	Month	Purchase Contracts
January	$120,000	April	$300,000
February	120,000	May	550,000
March	200,000	June	600,000

LPI's standard purchase contract calls for cash payment 30 days following the delivery, and its purchases during December were only $20,000. Timber purchases are added to the firm's inventories at cost plus a processing cost equal to half the cost of the timber. Thus, for the month of December LPI increased its inventories by a total of $30,000 and this is the outstanding balance of accounts payable for December. This balance will be paid in full in January.

In addition to its timber purchases, LPI plans to acquire a new chipping machine used to convert waste resulting from its processing of the timber to chips of wood that are then sold to paper mills, which convert the chips into paper stock. The new machine will cost $400,000 and will be purchased and paid for in June (payment will actually be made as late as possible in June). The firm also plans to acquire a new forklift truck in February which will cost $60,000. The purchase will be paid for in March. LPI's monthly depreciation expense for January is $15,000 but will increase by $1,000 per month in February because of the purchase of the forklift, and it will increase again in July by $8,000 per month because of the purchase of the chipper.

Christian estimates the firm's general and administrative expense to be $70,000 per month, with payment being made in the month of the expense's occurrence. The firm also has a category called miscellaneous cash expenses, which is estimated to be at $20,000 per month and paid monthly. Although LPI has no formal policy with regard to its minimum cash balance, Christian feels that a $100,000 balance is the minimum he would like to have over the planning horizon. Finally, LPI plans to make a cash dividend payment to stockholders at the end of March and June equal to $100,000 and will also make quarterly tax payments of $50,000 in the months of January and April. (For planning purposes, a 30 percent tax rate is used in estimating LPI's tax liability.)

Questions

1. Prepare a month-by-month cash budget for LPI for the next six months. You may assume that any borrowing that must be undertaken will cost 1 percent interest per month, with interest paid the month following the month in which borrowing takes place. Assume borrowing is at the beginning of the month. November and December sales for the current year were $350,000 and $400,000, respectively.
2. Prepare pro forma income statements for the three-month periods ended with March and June. You may assume that inventories consist of materials and related labour expense. Cost of goods sold equals 75 percent of sales and is comprised of materials and related labour.
3. Prepare a pro forma balance sheet for LPI for the end of March and June. You may assume that accrued expenses and common shares do not change over the six-month planning period.
4. Given LPI's financing requirements for the period, what suggestions do you have for James Christian for raising the necessary funds? Discuss your reasons for the stated plans.

INTEGRATIVE PROBLEM

The following data spanning the years 1988–94 reflect the result of a company's financial restructuring that was begun in the mid-eighties. Management is currently developing its financial plans for the next five years and wants to develop a forecast of its financial requirements. As a first approximation they have asked you to develop a model that can be used to make "ball park" estimates of the firm's financing needs under the proviso that existing relationships found in the firm's financing statements remain the same over the period. Of particular interest is whether the company will be able to further reduce its reliance on debt financing. You may assume that the company's projected sales (in millions) for 1995 through 1999 are as follows: $13,000; $13,500; $14,000; $14,500; and $15,500.

a. Project net income for 1995–99 using the percent of sales method based on an average of this ratio for 1988–94.

b. Project total assets and current liabilities for the period 1995–99 using the percent of sales method and your sales projections from part **a**.

c. Assuming that common equity increases only as a result of the retention of earnings and holding of long-term debt and preferred shares equal the 1994 balances, project the company's discretionary financing needs for 1995–99. (*Hint:* Assume that total assets and current liabilities vary as a percent of sales as per your answer to part **b** above. In addition, assume that the company plans to continue to pay its dividend of $1.12 per share in each of the next five years.)

Summary Financial Information: 1988–94: (in millions of dollars except for per share figures

Item	1988	1989	1990	1991	1992	1993	1994
Sales	10,108.00	10,917.00	11,490.00	12,492.00	13,975.00	13,259.00	12,140.00
Net Income	228.00	35.00	650.00	219.00	541.00	98.00	270.00
EPS	0.89	0.06	2.72	0.90	2.18	0.38	1.04
Current assets	2,802.00	2,855.00	3,062.00	2,876.00	3,322.00	2,459.00	2,349.00
Total assets	12,403.00	12,111.00	11,968.00	11,256.00	12,130.00	11,473.00	11,468.00
Current liabilities	2,234.00	2,402.00	2,468.00	2,706.00	2,910.00	2,603.00	2,517.00
Long-term debt	5,578.00	5,419.00	4,761.00	3,939.00	3,839.00	3,876.00	3,718.00
Total liabilities	10,409.00	10,289.00	9,855.00	9,124.00	9,411.00	8,716.00	8,411.00
Preferred shares	270.00	205.00	0.00	0.00	0.00	0.00	359.00
Common equity	1,724.00	1,617.00	2,113.00	2,132.00	2,719.00	2,757.00	2,698.00
Dividends per share[a]	2.02	1.73	1.34	0.00	1.03	1.12	1.12

[a]Number of common shares outstanding is 259,615,385

STUDY PROBLEMS (SET B)

21-1B. (*Financial Forecasting*) Hernandez Trucking Company is evaluating its financing requirements for the coming year. The firm has only been in business for three years, but its chief financial officer predicts that the firm's operating expenses, current assets, net fixed assets and current liabilities will remain at their current proportion of sales.

Last year Hernandez had $20 million in sales with net income of $1 million. The firm anticipates that the next year's sales will reach $25 million with net income rising to $2 million. Given its present high rate of growth, the firm retains all its earnings to help defray the cost of new investments.

The firm's balance sheet for the year just ended is found below:

Hernandez Trucking Company Inc. Balance Sheet

	12/31/95	% of Sales
Current assets	$ 4,000,000	20%
Net fixed assets	8,000,000	40%
Total	$12,000,000	
Accounts payable	$ 3,000,000	15%
Long-term debt	2,000,000	NA
Total liabilities	$5,000,000	
Common shares	1,000,000	NA
Contributed surplus	1,800,000	NA
Retained earnings	4,200,000	
Common equity	7,000,000	
Total	$12,000,000	

Estimate Hernandez's total financing requirements (i.e., total assets) for 1996 and its net funding requirements (discretionary financing needed).

21-2B. (*Pro Forma Accounts Receivable Balance Calculation*) On March 31, 1995 the Floydata Food Distribution Company had outstanding accounts receivable of $52,000. Sales are roughly 60 percent cash and 40 percent credit sales, with the credit sales collected half in the month after the sale and the remainder two months after the sale. Historical and projected sales for the food distribution company are given below:

Month	Sales
January	$100,000
February	100,000
March	80,000
April (projected)	60,000

a. Under these circumstances, what should the balance in accounts receivable be at the end of April?

b. How much cash did Floydata realize in April from sales and collections?

21-3B. (*Financial Forecasting*) Simpson Inc. projects its sales next year to be $5 million and expects to earn 6 percent of that amount after taxes. The firm is currently in the process of projecting its financing needs and has made the following assumptions (projections):

1. Current assets will equal 15 percent of sales while fixed assets will remain at their current level of $1 million.
2. Common equity is presently $0.7 million, and the firm pays out half its after-tax earnings in dividends.
3. The firm has short-term payables and trade credit which normally equal 11 percent of sales and has no long-term debt outstanding.

What are Simpson Inc.'s financing needs for the coming year?

21-4B. (*Financial Forecasting—Percent of Sales*) Carson Enterprises is in the midst of its annual planning exercise. Bud Carson, the owner, is a mechanical engineer by education and has only modest skills in financial planning. In fact, the firm has operated in the past on a "crisis" basis with little attention being paid to the firm's financial affairs until a problem arose. This worked reasonably well for several years until the firm's growth in sales created a serious cash flow shortage last year. Bud was able to convince the firm's bank to come up with the needed funds but an outgrowth of the agreement was that the firm would begin to make forecasts of its financing requirements annually. To support its first such effort Bud has made the following estimates for next year: Sales are currently $18 million with projected sales of $25 million for next year. The firm's current assets equal $7 million, while its fixed assets are $6 million. The best estimate Bud can make is that current assets will equal the current proportion of sales while fixed assets will rise by $100,000. At the present time the firm has accounts payable of $1.5 million, $2 million in long-term debt which is due in ten years, and common equity totalling $9.5 million (including $4 million in retained earnings). Finally, Carson's net profit margin is 5 percent after taxes and plans to continue paying its dividend of $600,000 next year.

a. What are Carson's total financing needs (i.e., total assets) for the coming year?

b. Given the firm's projections and dividend payment plans, what are its discretionary financing needs?

c. Based upon the projections given and assuming the $100,000 expansion in fixed assets will occur, what is the largest increase in sales the firm can support without having to resort to the use of discretionary sources of financing?

21-5B. (*Pro Forma Balance Sheet Construction*) Use the following industry average ratios to construct a pro forma balance sheet for the V. M. Willet Co.

Total asset turnover	2.5 times
Average collection period (assume a 365-day year)	10.139 days
Fixed asset turnover	6.0 times
Inventory turnover (based on cost of goods sold)	4.0 times
Current ratio	3.0 times
Sales (all on credit)	$5.0 million
Cost of goods sold	80% of sales
Debt ratio	60%

Cash		Current liabilities	
Accounts receivable		Long-term debt	
		Common shares plus	
Net fixed assets	$	retained earnings	
	$		$

21-6B. (*Cash Budget*) The Sharpe Corporation's projected sales for the first eight months of 1996 are as follows:

January	$100,000	May	$275,000
February	110,000	June	250,000
March	130,000	July	235,000
April	250,000	August	160,000

Of Sharpe's sales, 20 percent is for cash, another 60 percent is collected in the month following sale, and 20 percent is collected in the second month following sale. November and December sales for 1995 were $220,000 and $175,000, respectively.

Sharpe purchases its raw materials two months in advance of its sales equal to 70 percent of their final sales price. The supplier is paid one month after it makes delivery. For example, purchases for April sales are made in February and payment is made in March.

In addition, Sharpe pays $10,000 per month for rent and $20,000 each month for other expenditures. Tax prepayments of $23,000 are made each quarter, beginning in March.

The company's cash balance at December 31, 1995, was $22,000; a minimum balance of $25,000 must be maintained at all times. Assume that any short-term financing needed to maintain cash balance would be paid off in the month following the month of financing if sufficient funds are available. Interest on short-term loans (12 percent) is paid monthly. Borrowing to meet estimated monthly cash needs takes place at the beginning of the month. Thus, if in the month of April the firm expects to have a need for an additional $60,500, these funds would be borrowed at the beginning of April with interest of $605 ($0.12 \times 1/12 \times$ $60,500) owed for April and paid at the beginning of May.

a. Prepare a cash budget for Sharpe covering the first seven months of 1996.

b. Sharpe has $250,000 in notes payable due in July that must be repaid or renegotiated for an extension. Will the firm have ample cash to repay the notes?

21-7B. (*Financial Forecasting—Pro Forma Statements*) In the spring of 1995 Cherill Adison, the chief financial analyst for Jarrett Sales Company, found herself faced with a major financial forecasting exercise. She had joined the firm about three months ago and was just beginning to feel comfortable with her new duties when she was called into her supervisor's office about 3:30 p.m. on a Friday afternoon. It seemed that Margaret Simpson, the analyst who had been in charge of the firm's annual financial planning exercise, had taken maternity leave and would not be available to help prepare the firm's projected financing needs for the coming quarter. The job fell into her lap and it needed to be completed by Monday at 2:00 p.m. when the company president met with the executive committee to work out their plans for requesting a limit for the firm's line of credit with its bank for the next quarter. After reviewing Ms. Simpson's files, Cherill found that the task might not be so difficult as she had first thought. In fact, the critical forecasts had been compiled and were contained in the file. In addition, Cherill found some carefully documented assumptions that she could use to compile the necessary forecast. This information is found below:

1. Historical and predicted sales:

Historical		Predicted	
April	$ 80,000	July	$140,000
May	110,000	August	120,000
June	120,000	September	120,000
		October	100,000

2. The firm incurs and pays a monthly rent expense of $2,500.
3. Wages and salaries for the coming months are estimated as follows (with payments coinciding with the month in which the expense is incurred):

July	$20,000
August	18,000
September	13,000

4. Of the firm's sales, 40 percent is collected in the month of sale, 30 percent one month after sale, and the remaining 30 percent two months after sale.
5. Merchandise is purchased one month before the sales month and is paid for in the month it is sold. Purchases equal 80 percent of sales.
6. Tax prepayments are made on the calendar quarter, with a prepayment of $1,000 in July based on earnings for the quarter ended June 30.
7. Utilities for the firm average 2 percent of sales and are paid in the month of their occurrence.
8. Depreciation expense is $13,000 annually.
9. Interest on a $40,000 bank note (due in November) is payable at an 8 percent annual rate in September for the three-month period just ended.
10. The firm follows a policy of paying no cash dividends.

Based on the above, supply the following items:

a. Prepare a monthly cash budget for the three-month period ended September 30, 1995.

b. If the firm's beginning cash balance for the budget period is $5,000 and this is its minimum desired balance, determine when and how much the firm will need to borrow during the budget period. The firm has an $80,000 line of credit with its bank with interest (12 percent annual) paid monthly (for example, for a loan taken out at the end of December, interest would be paid at the end of January and every month thereafter so long as the loan was outstanding).

c. Prepare a pro forma income statement for Jarrett covering the three-month period ended September 30, 1995. Use a 40 percent tax rate.

d. Given the following balance sheet dated June 30, 1995, and your pro forma income statement from (c), prepare a pro forma balance sheet for September 30, 1995.

Jarrett Sales Co. Balance Sheet June 30, 1995

Cash	$ 5,000	Accounts payable	$104,000
Accounts receivable	102,000	Bank notes (8%)	40,000
Inventories	114,000	Accrued taxes	1,000
Current assets	221,000	Current liabilities	145,000
Net fixed assets	120,000	Common shares	100,000
		Contributed surplus	28,400
		Retained earnings	67,600
Total assets	$341,000	Total liabilities and capital	$341,000

21-8B. (*Forecasting Accounts Receivable*) Sam's Mfg. Co. is evaluating the effects of extending trade credit to its out-of-province customers. In the past Sam's has given credit to local customers but has insisted on cash upon delivery from customers in the bordering province of Manitoba. At present Sam's average collection period is 35 days based upon total sales; that is, including both its $700,000 in-province credit sales and its $250,000 out-of-province cash sales, the firm's accounts receivable balance is 35 days.

$$\text{average collection period} = \frac{\text{accounts receivable}}{\frac{\text{sales}}{365}}$$

$$= \frac{\$91{,}096}{\frac{\$950{,}000}{365}} = 35 \text{ days}$$

a. Calculate Sam's average collection period based upon credit sales alone.

b. If Sam's extends credit to its out-of-province customers, it expects total credit sales to rise to $1,750,000. If the firm's average collection period remains the same, what will be the new level of accounts receivable?

21-9B. (*Forecasting Accounts Receivable*) The Carter Distribution Company sells its merchandise on credit terms of 2/10, net 30 (2 percent discount if payment is made within ten days or the net amount is due in 30 days). Only a part of the firm's customers take the trade discount, so that the firm's average collection period is 20 days.

a. Based upon estimated sales of $500,000, project Carter's accounts receivable balance for the coming year.

b. If Carter changes its cash discount terms to 1/10, net 30, it expects its average collection period will rise to 28 days. Estimate Carter's accounts receivable balance based on the new credit terms and expected sales of $500,000.

21-10B. (*Percent of Sales Forecasting*) Which of the following accounts would most likely vary directly with the level of firm sales? Discuss each briefly.

	Yes	No
Cash	_____	_____
Marketable securities	_____	_____
Accounts payable	_____	_____
Notes payable	_____	_____
Plant and equipment	_____	_____
Inventories	_____	_____

21-11B. (*Financial Forecasting—Percent of Sales*) The balance sheet of the Taylor Drilling Company (TDC) is found below:

Taylor Drilling Company Balance Sheet January 31, 1995 ($ millions)

Current assets	$15	Accounts payable	$10
Net fixed assets	15	Notes payable	0
Total	$30	Bonds payable	10
		Common equity	10
		Total	$30

TDC had sales for the year ended 1/31/95 of $60 million. The firm follows a policy of paying all net earnings out to its common shareholders in cash dividends. Thus, TDC generates no funds from its earnings that can be used to expand its operations (assume that depreciation expense is just equal to the cost of replacing worn-out assets).

a. If TDC anticipates sales of $80 million during the coming year, develop a pro forma balance sheet for the firm for 1/31/95. Assume that current assets vary as a percent of sales, net fixed assets remain unchanged, accounts payable vary as a percent of sales, and use notes payable as a balancing entry.

b. How much "new" financing will TDC need next year?

c. What limitations does the percent of sales forecast method suffer from? Discuss briefly.

21-12B. *(Financial Forecasting—Discretionary Financing Needed)* Symbolic Logic Corporation (SLC) is a technological leader in the application of surface mount technology in the manufacture of printed circuit boards used in the personal computer industry. The firm has recently patented an advanced version of its original path-breaking technology and expects sales to grow from their present level of $5 million to $8 million by the end of the coming year. Since the firm is at present operating at full capacity it expects to have to increase its investment in both current and fixed assets in proportion to the predicted increase in sales.

The firm's net profits were 7 percent of current year's sales but are expected to rise to 8 percent of next year's sales. To help support its anticipated growth in asset needs next year, the firm has suspended plans to pay cash dividends to its shareholders. In years past a $1.25 per share dividend has been paid annually.

Symbolic Logic Corporation ($ millions)

	Present Level	Percent of Sales	Projected Level
Current assets	$2.5		
Net fixed assets	3.0		
Total	$5.5		
Accounts payable	$1.0		
Accrued expenses	0.5		
Notes payable	—		
Current liabilities	$1.5		
Long-term debt	2.0		
Common shares	0.5		
Retained earnings	1.5		
Common equity	$2.0		
Total	$5.5		

SLC's payables and accrued expenses are expected to vary directly with sales. In addition, notes payable will be used to supply the added funds needed to finance next year's operations that are not forthcoming from other sources.

a. Fill in the table and project the firm's needs for discretionary financing. Use notes payable as the balancing entry for future discretionary financing needed.

b. Compare SLC's current ratio and debt ratio (total liabilities/total assets) before the growth in sales and after. What was the effect of the expanded sales on these two dimensions of SLC's financial condition?

c. What difference, if any, would have resulted if SLC's sales had risen to $6 million in one year and $7 million only after two years? Discuss only; no calculations required.

21-13B. (*Financial Forecasting—Changing Credit Policy*) Carrier Pigeon Communications (CPC) has $400,000 invested in receivables throughout much of the year. It has annual credit sales of $3,500,000 and a gross profit margin of 75 percent.

a. What is CPC's average collection period? [*Hint:* Recall that average collection period = accounts receivable/(credit sales/365).]

b. By how much would CPC be able to reduce its accounts receivable if it were to change its credit policies in a way that reduced its average collection period to 30 days without affecting annual credit sales?

21-14B. (*Forecasting Discretionary Financing Needs*) Royal Charter Inc. estimates that it invests 40 cents in assets for each dollar of new sales. However, 5 cents in profits are produced by each dollar of additional sales, of which 1 cent can be re-invested in the firm. If sales rise from their present level of $5 million by $400,000 next year, and the ratio of spontaneous liabilities of sales is .15, what will be the firm's need for discretionary financing? (*Hint*: In this situation you do not know what the firm's existing level of assets is, nor do you know how they have been financed. Thus, you must estimate the change in financing needs and match this change with the expected changes in spontaneous liabilities, retained earnings, and other sources of discretionary financing.)

21-15B. (*Cash Budget*) Halsey Enterprises projected sales for the first eight months of 1996 are as follows:

January	$120,000	May	$225,000
February	160,000	June	250,000
March	140,000	July	210,000
April	190,000	August	220,000

Of Halsey's sales, 30 percent is for cash, another 30 percent is collected in the month following sale, and 40 percent is collected in the second month following sale. November and December sales for 1995 were $230,000 and $225,000, respectively.

Halsey purchases raw materials two months in advance of its sales which equal 75 percent of its final sales price. The supplier is paid one month after it makes delivery. For example, purchases for April sales are made in February and payment is made in March.

In addition, Halsey pays $12,000 per month for rent and $20,000 each month for other expenditures. Tax prepayments of $26,500 are made each quarter, beginning in March.

The company's cash balance at December 31, 1995, was $28,000; a minimum balance of $25,000 must be maintained at all times. Assume that any short-term financing needed to maintain the cash balance would be paid off in the month following the month of financing if sufficient funds are available. Halsey has arranged with its bank the use of short-term credit at an interest rate of 12 percent per annum (1 percent per month) to be paid monthly. Borrowing to meet estimated monthly cash needs takes place at the beginning of the month but interest is not paid until the end of the following month. If in the month of April the firm expects to have a need for an additional $50,000, these funds would be borrowed at the beginning of April with interest of $500 (= .01 × $50,000) in interest during May. Finally, Halsey follows a policy of repaying any outstanding short-term credit in any month in which its cash balance exceeds the minimum desired balance of $25,000.

a. Halsey needs to know what its cash requirements will be for the next six months so that it can renegotiate the terms of its short-term credit agreement with its bank, if necessary. To evaluate this problem, the firm plans to evaluate the impact of a 20% variation in its monthly sales effort. Prepare a six-month cash budget for Halsey and use it to evaluate the firm's cash needs.

b. Halsey has $200,000 in notes due in July. Will the firm have sufficient cash to repay the notes?

SELF-TEST SOLUTIONS

SS-1.

Calico Sales Co. Inc. Pro Forma Income Statement

Sales		$4,000,000
COGS (70%)		(2,800,000)
Gross profit		1,200,000
Operating expenses		
Selling expense (5%)	$200,000	
Administrative expense	500,000	
Depreciation expense	300,000	(1,000,000)
Net operating income		200,000
Interest		(50,000)
Earnings before taxes		150,000
Taxes (40%)		(60,000)
Net income		$ 90,000

Although the office renovation expenditure and debt retirement are surely cash outflows, they do not enter the income statement directly. These expenditures affect expenses for the period's income statement only through their effect on depreciation and interest expense. A cash budget would indicate the full cash impact of the renovation and debt retirement expenditures.

SS-2. a.

Odom Manufacturing Co. Pro Forma Balance Sheet

Cash	$ 320,000	Accounts payable	$ 750,000
Accounts receivable	760,000	Notes payable	475,000
Inventory	940,000	Total current liabilities	$1,225,000
Total current assets	$2,020,000	Long-term debt	1,300,000
Property, plant, and		Common shares	600,000
equipment	$1,905,000	Retained earnings	800,000
		Total liabilities and	
Total assets	$3,925,000	shareholders' equity	$3,925,000

b. In order to assess whether or not the firm's liquidity position has improved with the securing of the $500,000 loan, its current financial position must be compared with its position before the loan was granted.

	Position Before Loan	Position After Loan
Current ratio	1.34	1.65
Acid test ratio	0.72	0.88

These two ratios indicate that there has been an improvement in Odom's liquidity position and that the firm was justified in seeking the loan and in allocating the funds as previously described.

SS-3.

	May	June	July	Aug.
Sales	$100,000	$100,000	$120,000	$150,000
Purchases	60,000	60,000	50,000	40,000
Cash Receipts:				
Collections from month of sale (20%)	20,000	20,000	24,000	30,000
1 month later				
(50% of uncollected amount)		40,000	40,000	48,000
2 months later (balance)			40,000	40,000
Total receipts			$104,000	$118,000
Cash Disbursements:				
Payments for purchases—				
From 1 month earlier			$ 30,000	$ 25,000
From current month			$ 25,000	20,000
Total			$ 55,000	$ 45,000
Miscellaneous cash expenses			4,000	4,000
Labour expense (5% of sales)			6,000	7,500
General and administrative expense				
($50,000 per month)			50,000	50,000
Truck purchase			0	60,000
Cash dividends			9,000	—
Total disbursements			(124,000)	(166,500)
			(20,000)	(48,500)
Plus: Beginning cash balance			30,000	30,000
Less: Interest on short-term borrowing				
(1% of prior month's borrowing)				(200)
Equals: Ending cash balance—without borrowing			10,000	(18,700)
Financing needed to reach target cash balance			20,000	48,700
Cumulative borrowing			20,000	68,700

APPENDIX 21A:

MICROCOMPUTERS AND SPREADSHEET SOFTWARE IN FINANCIAL PLANNING

We have briefly discussed the role of personal computers and spreadsheet software in financial planning within this chapter. The purpose of this appendix is to illustrate how a financial planning model can be constructed and used. First, we examine how to design effective spreadsheets.

HOW TO DESIGN AN EFFECTIVE SPREADSHEET MODEL

Back in the dark ages of computing, when we had to program computers using languages such as Fortran, Cobol, and Pascal, we were told to "flow chart" our programs. These flow charts were like outlines that organize the programmer's work so others could more easily follow it. A current equivalent of flow charting can be of great use to spreadsheet users. The particular methodology was devised by Urschel (1987) and is summarized in the following box.[12]

Designing a Spreadsheet

Block One: *Assumptions.* These are the variables that are the object of *what if* analysis.

Block Two: *Inputs.* The data used in the model.

Block Three: *Calculations.* The mechanical section of the model containing intermediate computations but not the final output of the model.

Block Four: *Bottom Line.* The final results of the model.

The four blocks provide a useful structure for the person building the spread-sheet model. They also make it easy for others to interpret and use the model. We will illustrate this recommended structure in the development of the spreadsheet model for the cash budgeting that follows.

SPREADSHEET MODELLING EXERCISE

Cramer Enterprises has projected its sales for the first eight months of 1996 as follows:

January	$100,000	May	$275,000
February	140,000	June	250,000
March	150,000	July	200,000
April	250,000	August	120,000

Cramer collects 20 percent of its sales in the month of the sale, 50 percent is collected in the month following the sale, and the remaining 30 percent is collected two months following the sale. During November and December of 1995, Cramer's sales were $220,000 and $175,000, respectively.

Cramer purchases raw materials two months in advance of its sales equal to 65 percent of those sales. The supplier is paid one month after delivery. Thus, purchases for April sales are made in February and payment is made in March.

In addition, Cramer pays $10,000 per month for rent and $20,000 each month for other expenditures. Tax prepayments of $22,500 are made each quarter beginning in March. The company's cash balance at December 31, 1995 was $22,000 and a minimum balance of $20,000 must be maintained at all times to satisfy the firm's bank line of credit agreement. Cramer has arranged with its bank for short-term credit at an interest rate of 12 percent per annum (1 percent per month) to be paid monthly. Borrowing to meet estimated monthly cash needs takes place at the beginning of the month but interest is not paid until the end of the following month. Consequently, if the firm were to need to borrow $50,000 during the month of April then it would pay $500 (0.01 × $50,000) in interest during May. Finally, Cramer follows a policy of repaying any outstanding short-term debt in any month in which its cash balance exceeds the minimum desired balance of $20,000.

Cramer faces two related problems. First, the firm needs to know what its cash requirements will be for the next six months so that it can renegotiate the terms of its short-term credit agreement with its bank, if necessary. Second, Cramer has a $200,000 note due June 30 and needs to determine whether it will have sufficient cash to repay the loan.

[12]W. Urschel, "Worksheets by Design," *PC World* (September 1987).

Sample Spreadsheet Model

Table 21A-1 contains the cash budget for Cramer spanning the first six months of 1996. The model is composed of four blocks following the procedure discussed earlier. Note that the first block contains all those elements which the analyst might wish to use in performing a sensitivity analysis of the final outcome. The second section contains raw data inputs, the third consists of intermediate calculations necessary to support the bottom line results found in block four.

A quick analysis of the results indicate the following: Cramer will not need to borrow until March when it will need an additional $27,500 followed in April when its total borrowing needs rise to $97,025. In May the firm is able to retire $31,530 of its short-term borrowing and by June it is able to retire all of the seasonal debt it borrowed over the planning period. However, the firm's projected cash balance for June is only $33,850 which is insufficient to retire the $200,000 by June 30.

At this point you may want to explore the sensitivity of the cash requirement of the firm to changes in projected level of sales. In the Assumptions block of the model you will see a variable called "Sales Expansion %". This variable allows us to increase our January through September sales forecasts by an arbitrary percentage and then see the impact of the change on the firm's cash requirements. For example, decreasing sales by 20 percent or increasing them by 20 percent has the following impact on the firm's projected level of borrowing for the months of March, April, and May:

Sales Expansion	Cumulative Borrowing Needs		
%	March	April	May
−20%	2,900	45,229	25,681
−10%	15,200	71,127	45,588
0%	27,500	97,025	65,495
10%	39,800	122,923	85,402
20%	52,100	148,821	105,309

The effect of changes in projected sales on Cramer's cumulative borrowing needs is dramatic. A ±20% variation in sales estimates leads to an increase in the maximum level of borrowing for the planning period from $45,229 to $148,821. This is but one type of sensitivity analysis that Cramer might want to perform. The real value of having a spreadsheet model of the financial forecasting problem comes in the ease with which the user can perform such analyses.

CELL FORMULAE IN THE MODEL

Table 21A-2 contains the model's cell formulae for Blocks 3 and 4 and the month of January. Note that this spreadsheet was done using the Excel spreadsheet program, however the cell formulae would differ only slightly for Lotus 123. The principal difference that you would see in Table 21A-2 related to the fact that Lotus formulae begin with a "+" or "–" rather than "=". Of course, there are other differences in the two programs and we will point out one of them in the following discussion.

Perhaps the most interesting cell formula in the spreadsheet model is found in Block 4 of the model where the debt repayment for the month is calculated. The cell formula for cell D57 is the following:

= IF((AND(D59>0,D55>$C$7)),MIN(D59,D55 – $C$7),0)

This statement calculates the debt repayment. Note that the loan repayment is only made if two conditions are met. The first is that the cumulative borrowing be greater than zero (D59>0) and the second is that the firm has excess cash on hand which can be used to repay a portion of the debt (D55>$C$7). If both these conditions are satisfied then the debt repayment is set equal to the minimum of the amount of short-term debt owed (cell D59) or the amount of excess cash over and above the desired minimum balance (D55–$C$7). The corresponding cell formula using Lotus 123 is slightly different and appears as follows:

@IF((D59>0#AND#D55>$C$7),@MIN(D59,D55 – $C$7),0)

The result is the same in both instances, however. The cell formulae used in this example demonstrate that you can incorporate fairly sophisticated logic into your spreadsheet model.

Figure 21A-1 Cramer Enterprises Inc. Cash Budget: January–June 1996

	A	B	C	D	E	F	G	H	I
1									
2									
3									
4									
5	Block 1: Assumptions								
6									
7	Minimum Cash Balance		20,000						
8	Beginning Cash Balance		22,000						
9									
10	Sales Expansion %		0.00%		Annual Interest				
11	Purchases as a % Sales		.65 (65%)		Rate	.12 (12.00%)			
12									
13	Collections:	Current Mo.	1 Mo. Later	2 Mo. Later					
14		.2 (20%)	.5 (50%)	.3 (30%)					
15									
16	Block 2: Inputs								
17									
18	Historical Sales and Base Case Sales Predictions for Future Sales								
19	January	100,000		May	275,000				
20	February	140,000		June	250,000				
21	March	150,000		July	200,000				
22	April	250,000		August	120,000				

Table 21A-1 Continued on next page

Figure 21A-1 Cramer Enterprises Inc. Cash Budget: January–June 1996 *(Continued from previous page)*

	A	B	C	D	E	F	G	H	I
23									
24	Block 3: Calculation								
25									
26									
27		November	December	January	February	March	April	May	June
28		220,000	175,000	100,000	140,000	150,000	250,000	275,000	250,000
29	Collections								
30	Month of sales			20,000	28,000	30,000	50,000	55,000	50,000
31	First Month			87,500	50,000	70,000	75,000	125,000	137,500
32	Second Month			66,000	52,500	30,000	42,000	45,000	75,000
33	Total collections			173,500	130,500	130,000	167,000	225,000	262,500
34	Purchases	65,000	91,000	97,500	162,500	178,750	162,500	130,000	97,500
35	Payments		65,000	91,000	97,500	162,500	178,750	162,500	130,000
36									
37	*Cash Budget for January through June*			January	February	March	April	May	June
38									
39	*Cash Receipts*			173,500	130,500	130,000	167,000	225,000	262,500
40	(collections)								
41	Cash Disbursements								
42	Payments for Purchases			91,000	97,500	162,500	178,750	162,500	130,000
43	Rent			10,000	10,000	10,000	10,000	10,000	10,000
44	Other Expenditures			20,000	20,000	20,000	20,000	20,000	20,000
45	Tax Deposits					22,500			22,500
46	Interest on S-T						275	970	655
47	Borrowing								
48	*Total Disbursements*			121,000	127,500	215,000	209,025	193,470	183,155
49									
50	*Net Monthly Change*			52,500	3,000	–85,000	–42,025	31,530	79,345
51									
52	Block 4: Bottom Line								
53									
54	Beginning Cash Balance			22,000	74,500	77,500	–7,500	20,000	20,000
55	Ending Cash (No Borrowing)			74,500	77,500	–7,500	–49,525	51,530	99,345
56	Needed (No Borrowing)			0	0	27,500	69,525	0	0
57	Loan Repayment			0	0	0	0	31,530	65,495
58	Ending Cash Balance			74,500	77,500	20,000	20,000	20,000	33,850
59	Cumulative Borrowing		0	0	0	27,500	97,025	65,495	0
60									

Figure 21A-2 Cell Formulae for the Cramer Cash Budget Problem for November Through January

	A	B	C	D
23				
24	Block 3: Calculations			
25				
26				
27		November	December	January
28		220000	175000	=B19*(1+$C$10)
29	Collections			
30	Month of Sales			=$B$14*D28
31	First month			=$C$14*C28
32	Second month			=$D$14*B28
33	Total collections			=$D$26+D31+D32
34	Purchases	=$C$11*D28	=$C$11*E28	=$C$11*F28
35	Payments		=B34	=C34
36				
37	*Cash Budget*			January
38				
39	*Cash Receipts*			=D33
40	(collections)			
41	Cash Disbursements			
42	Payments for Purchases			=D35
43	Rent			10000
44	Other Expenditures			20000
45	Tax Deposits			
46	Interest on S-T			
47	Borrowing			
48	*Total Disbursements*			=SUM(D42..D47)
49				
50	*Net Monthly Change*			=D39-D48
51				
52	Block 4: Bottom Line			
53				
54	Beginning Cash Balance			=$C$8
55	Ending Cash (No Borrowing)			=D50+D54
56	Needed (Borrowing)			=IF(D55-$C$7>0,0,$C$7-D55
57	Loan Repayment			=IF((AND(D59>0,D55>$C$7)),MIN(D59,D55-$C$7),0)
58	Ending Cash Balance			=D50+D54+D56
59	Cumulative Borrowing		0	=C59+D56

CHAPTER 22

THE USE OF FUTURES AND OPTIONS TO REDUCE RISK

LEARNING OBJECTIVES

After reading this chapter you should be able to

1. Explain the difference between a commodity future and a financial future and how they might be used by a financial manager to control risk.
2. Explain what put and call options are and how they might be used by a financial manager to control risk.
3. Explain what a currency swap is and how it is used to eliminate exchange rate risk.

INTRODUCTION

As firms increasingly trade internationally, the need to protect sales against undesirable currency fluctuations becomes increasingly important. For example, Electrohome Limited might use currency options to protect sales on its projection systems, which are sold to customers in the United Kingdom. Because the projection systems are manufactured in Canada and sold abroad, its costs in labour and materials are based on the Canadian dollar. However, as the dollar fluctuates relative to the U.K. sterling pound, so must the sales price in U.K. pounds for Electrohome to receive the same amount of dollars on each sale in the United Kingdom.

Problems arise when the value of the U.K. pound decreases relative to that of the dollar. For each sale to bring the same amount of dollars back to Electrohome, the selling price in U.K. pounds would have to be increased. Unfortunately, increasing prices may lead to lost sales. To guard against this situation, Electrohome operates a hedging program consisting of call options, put options, and forward contracts to offset the foreign currency risk of U.K. pound, U.S. dollar, and Japanese yen cash flows.[1] For example, Electrohome may purchase put options on the U.K. pound to cover the anticipated United Kingdom sales. These put options give Electrohome the option to sell or convert U.K. pounds into dollars at a pre-set price. If after the put options are

[1] Electrohome Limited, *Annual Report* (1995), p. 18.

purchased, the U.K. pound decreases in value, Electrohome could keep its selling prices constant in terms of the U.K. pound and make up the loss in the currency exchange with profits from the put options. Conversely, if the value of the U.K. pound increases relative to the value of the dollar, Electrohome could lower its U.K. price, sell more projection systems, and still bring home the same dollars per sale—all that would be lost is the price paid for the put options.

CHAPTER PREVIEW

In this chapter, we examine financial instruments that are used by financial executives for the purposes of risk management.

First, we examine futures which are used to reduce the risks associated with interest, exchange rate, and commodity price fluctuations. We then examine the fundamental building blocks of options: the put and the call contracts. These options are used to create simple hedging strategies which reduce potential losses because of security price fluctuations. Finally, we examine currency swaps which are used to hedge exchange rate risk over longer periods of time.

OBJECTIVE 1

FUTURES

Both warrant and convertible debentures are financial instruments which are created by the firm. We now look at two additional financial instruments that are similar to warrants and convertibles but are not created by the firm: futures and options. Despite the fact that these instruments are not issued by the firm, it is important for us to be familiar with them for two reasons. First, these instruments can be used to reduce the risks associated with interest and exchange rate and commodity price fluctuations. Second, as you will see in future finance courses, an understanding of the pricing of options is extremely valuable because many different financial assets can be viewed as options. In fact, warrants, convertible bonds, risky bonds, common shares, and the abandonment decision can all be thought of as types of options.

PERSPECTIVE IN FINANCE

Although there are many uses for futures and options, our interest focuses on how financial managers use these financial instruments to reduce risk. Keeping this in mind, you will see how they can be used to effectively offset future movements in the price of commodities or interest rates. Options are used to eliminate the effect of unfavourable price movements, and futures are used to eliminate the effect of both favourable and unfavourable price movements.

Futures contract

A contract to buy or sell a stated commodity or financial claim at a specified price at some future, specified time.

Commodity and financial futures are perhaps the fastest-growing and in many respects most exciting new financial instrument today. Financial managers who only a few years ago would not have considered venturing into the futures market are now actively using this market to eliminate risk. As the number of participants in this market has grown, so has the number of items on which future contracts are offered, ranging from the old standbys like wheat, oats, and flaxseed to the newer ones like bankers' acceptances and the TSE 100 Index.

A **futures contract** is a contract to buy or sell a stated commodity (such as wheat and oats) or financial claim (such as Government of Canada bonds) at a specified price at some future specified time. It is important to note here that this is a contract that requires its holder to buy or sell the asset, regardless of what happens to its value during the interim. The importance of a futures contract is that it can be used by financial managers to lock in the price of a commodity or an interest rate and thereby eliminate one source of risk. For example, if a corporation is planning on issuing debt in the near future and is concerned about a possible rise in interest rates between now and when the debt would be issued, it might sell a Government of Canada bond contract with the same face value as the proposed debt offering and a delivery date the same as when the debt offering is to occur. Alternatively, Ralston-Purina or Quaker Oats can lock in the future price of corn or oats with the use of a futures contract. As a result, any costs associated with any possible rise in interest rates or commodity prices would be completely offset by the profits made by writing the futures interest rate contract. In effect, futures contracts allow the financial manager to lock in future interest and exchange rates or prices for a number of agricultural commodities like wheat, oats, and barley.

As the use of futures contracts becomes more common in the financial management of the firm, it is important for the financial manager to be familiar with the operation and terminology associated with these financial instruments.

An Introduction to the Futures Markets

The origins of the futures market go back into history to medieval times. In Canada, the origin of the first futures exchange can be traced to the Winnipeg Grain and Produce Exchange which was formed by local grain merchants and farmers in 1887 and is now known as the Winnipeg Commodity Exchange. Several futures markets were initiated in the United States during the nineteenth century, but it was not until the establishment of the Chicago Board of Trade (CBT) in 1848 that the futures markets were provided with their roots. As we will see, while these markets have been in operation for years, it was not until the early 1970s, when the futures markets expanded from agricultural commodities to financial futures, that financial managers began to regularly venture into this market.

To develop an understanding for futures markets, let us examine several features that distinguish futures contracts from simple forward contracts. To begin with, a forward contract is any contract for delivery of an asset in the future. A futures contract is a specialized form of a forward contract distinguished by (1) an organized exchange, (2) a standardized contract with limited price changes and margin requirements, (3) a clearinghouse in each futures market, and (4) daily resettlement of contracts.

The Organized Exchange

The trading of futures in Canada occurs on the following exchanges: the Winnipeg Commodity Exchange, the Toronto Futures Exchange, and the Montreal Futures Exchange. In the United States, there are over 10 different futures exchanges in operation. In North America, the Chicago Board of Trade is the oldest and largest of the futures exchanges. The importance of having organized exchanges associated with the futures market is that they provide a central trading place. If there were no central trading place, then there would be no potential to generate the depth of trading necessary to support a secondary market, and in a very circular way the existence of a secondary market encourages more traders to enter the market and in turn provides additional liquidity.

World Wide Web Site

http://www.telenium.ca

Comment: Telenium provides end-of-day stock quotes on its site from the Winnipeg Commodity Exchange, Alberta Stock Exchange, Toronto Stock Exchange, Montreal Exchange, and Vancouver Stock Exchange.

An organized exchange also encourages confidence in the futures market by allowing for the effective regulation of trading. The various exchanges set and enforce rules and collect and disseminate information on trading activity and the commodities being traded. Together, the liquidity generated by having a central trading place, effective regulation, and the flow of information through the organized exchanges has effectively fostered their development.

Standardized Contracts

A strong secondary market is developed in any security if there are a large number of identical securities—or in this case, futures contracts—outstanding. In effect, standardization of contracts leads to more frequent trades on that contract, leading to greater liquidity in the secondary market for that contract, which in turn draws more traders into the market. It is for this reason that futures contracts are highly standardized and very specific with respect to the description of the goods to be delivered and the time and place of delivery. Let's look at a Winnipeg Commodity Exchange canola contract. This contract calls for the delivery of 20 metric tonnes of No. 1 Canada grade canola to Vancouver at $309.40 per tonne. In addition, these contracts are written to come due in November, January, March, June, and September. Through this standardization of contracts, trading has built up in enough identical contracts to allow for the development of a strong and highly liquid secondary market.

To encourage investors to participate in the futures market, daily price limits are set on all futures contracts. Without these limits, there would be more price volatility on most futures contracts than many investors would be willing to accept. These daily price limits are set to protect investors, maintain order on the futures exchanges, and encourage the level of trading volume necessary to develop a strong secondary market. For example, the Winnipeg Commodity Exchange imposes a $10 per tonne ($200 per contract) price movement limit above and below the previous day's settlement price of canola contracts. This limit protects against runaway price movements. These daily price limits do not halt trading once the limit has been reached, but they do provide a boundary within which trading must occur. The price of a canola contract may rise $10 very early in the trading day—"up the limit" in futures jargon. This will not stop trading; it only means that no trade can take place above that level. As a result, any dramatic shifts in the market price of a futures contract must take place over a number of days, with the price of the contract going "up the limit" each day.

Futures Clearinghouse

The main purpose of the futures clearinghouse is to guarantee that all trades will be honoured. This is done by having the clearinghouse interpose itself as the buyer to every seller and the seller to every buyer. Because of this substitution of parties, it is not necessary for the original seller (or buyer) to find the original buyer (or seller) when he or she decides to clear his or her position. As a result, all an individual has to do is make an equal and opposite transaction that will provide a net zero position with the clearinghouse and cancel out that individual's obligation. The Winnipeg Commodity Clearing Ltd. acts as the clearinghouse for the Winnipeg Commodity Exchange.

The trading of futures contracts occur between an individual and the clearinghouse. This type of clearing mechanism ensures that buyers and sellers will realize their market gains. There are other important benefits of a clearinghouse, including providing a mechanism for the delivery of commodities and the settlement of disputed trades, but these benefits also serve to encourage trading in the futures markets and thereby create a highly liquid secondary market.

Daily Resettlement of Contracts

Another safeguard of the futures market is a margin requirement. Although margin requirements on futures resemble margin requirements on shares in that there is an initial margin and a maintenance margin that comes into play when the value of the contract declines, similarities between futures and shares margins end there.

Before we explore margin requirements on futures it would be helpful to develop an understanding of the meaning of a margin on futures. The concept of a margin on futures contracts has a meaning that is totally different from its usage in reference to common shares. The margin on common shares refers to the amount of equity the investor has invested in the shares. With a futures contract, no equity has been invested, since nothing has been bought. In fact, the only consequences of signing a futures contract are the obligation of the two parties to participate in a future transaction and the definition of the terms of that transaction. This is an important thought: There is no actual buying or selling taking place with a futures contract; it is merely an agreement to buy or sell some commodity in the future. As a result, the term **futures margin** refers to "good faith" money the purchaser puts down to ensure that the contract will be carried out.

Futures margin

Refers to money that the purchaser puts down in "good faith" to ensure that the contract will be carried out.

The initial margin required for commodities (deposited by both buyer and seller) is much lower than the margin required for common shares, generally amounting to only 3 to 10 percent of the value of the contract. For example, if December oats contracts on the Winnipeg Commodity Exchange were selling at $111.00 per metric tonne, then one contract for 5,000 metric tonnes would be selling for $111.00 × 20 = $2,220.00. The initial margin on oats is $100 per contract, which represents only about 4.5 percent of the contract price. Needless to say, the leverage associated with futures trading is tremendous—both on the up and down sides. Small changes in the price of the underlying commodity result in very large changes in the value of the futures contract, since very little has to be put down to "own" a contract. Moreover, for many futures contracts, if the financial manager can satisfy the broker that he or she is not engaged in trading as a speculator, but as a hedger, the manager can qualify for reduced initial margins. Because of the low level of the initial margin, there is also a maintenance or variation margin requirement that forces the investor or financial manager to replenish the margin account to a level specified by the exchange after any market loss.

One additional point related to margins deserves mention. The initial margin requirement can be fulfilled by supplying cash, irrevocable letters of credit, Canadian Treasury bills, and U.S. Treasury bills held in Canadian banks. The advantage of using treasury bills as margin is that the investor earns money on them, whereas brokerage firms do not pay interest on funds in commodity cash accounts. Moreover, if the financial manager is going to carry treasury bills anyway, he or she can just deposit the treasury bills with the broker and purchase the futures contracts with no additional cash outlay.

Suppose you are a manager for Ralston-Purina and are in charge of purchasing raw materials—in particular, oats. For example, a December futures contract for the delivery of oats has a price of $111.00 per metric tonne. The company requires oats for the processing of its product in December. As a manager, you feel that this is an exceptional price and that oats will probably be selling for more than that in December. Thus, you wish to lock in this price, and to do this you purchase one contract for 20 metric tonnes at $111.00 per metric tonne. Upon purchasing the December oats contract the only cash you would have to put up would be the initial margin of $100. Let's further assume that the price of oats futures then falls to a level of $110.00 per metric tonne the day after you make your purchase. In effect, you have incurred a loss of $1 per metric tonne on 20 metric tonnes, for a total loss on your investment of $20.

At this point the concept of daily resettlement comes into play. What this means is that all futures positions are brought to the market at the end of each trading day and all gains and losses, in this case a loss, are then settled. You have lost $20, which is then subtracted from your margin account, lowering it to $80 ($100 initially less the $20 loss). Since your margin account has fallen below the maintenance margin on oats, which is $100, you would have to replenish your account back to its initial level of $100. If on the following day the price of December oats contracts increased $1.50 per metric tonne, you would have made $1.50 per metric tonne on 20 metric tonnes, for a total profit of $30. This brings your margin account up from $100 to $130, which is $30 above the initial margin of $100. You can withdraw this $30 from your margin account.

Obviously, the purpose of margin requirements is to provide some measure of safety for futures traders, and despite the very small level of margin requirements imposed, they do a reasonable job. They are set in accordance with the historical price volatility of the underlying commodity in such a way that it is extremely unlikely that a trader will ever lose more than is in his or her margin account in any one day.

Commodity Futures

As mentioned previously, the futures market is well developed in North America and extends beyond agricultural commodities to commodity types, such as metals, wood products, fibres, livestock, and petroleum. A few examples of these commodities include lumber, orange juice, cotton, gold, platinum, silver, cattle, hogs, pork bellies, crude oil, heating oil, unleaded gasoline, and natural gas. For the financial manager these markets provide a means of offsetting the risks associated with future price changes. Here the financial manager is securing a future price for a good that is currently in production, or securing a future price for some commodity that must be purchased in the future. In either case, the manager is using the futures market to eliminate the effects of future price changes on the future purchase or sale of some commodity.

Financial Futures

In Canada, financial futures come in the forms of Government of Canada bonds, bankers' acceptances, the TSE 100 and the TSE 35 Indexes. However, in the United States financial futures come in a number of different forms, including Treasury bills, notes and bonds, municipal bonds, certificates of deposit, Eurodollars, foreign currencies, and stock indexes. Financial futures first appeared in North America in 1972 in the form of futures on gold and foreign currency. In Canada, the Winnipeg Commodity Exchange began trading financial futures in 1981 but withdrew their facilities in 1985. However, the Toronto Futures Exchange trades financial futures on the TSE 100 Index and the TSE 35 Index, whereas the Montreal Futures Exchange trades financial futures on Government of Canada bonds and bankers acceptances. Our discussion of financial futures will be divided into three sections: (1) interest rate futures, (2) foreign exchange futures, and (3) stock index futures.

Interest Rate Futures

Currently, Government of Canada bond futures contracts are the most popular futures contract traded in the Canadian futures market. This type of popularity is also seen in the U.S. futures market; the Treasury bond futures are the most popular of all futures contracts in terms of contracts issued. The fact that these interest rate futures contracts are risk-free, long-term bonds with a maturity of at least 15 years has been the deciding factor in making them the most popular of the interest rate futures.

For the financial manager, interest rate futures provide an excellent means for eliminating the risks associated with interest rate fluctuations. As we discovered in Chapter 5, there is an inverse relationship between bond prices in the secondary market and yields—that is, when interest rates fall bond prices rise, and when interest rates rise bond prices fall. If you think back to the chapter on valuation, you will recall that this inverse relationship between bond prices and yield is a result of the fact that when bonds are issued, their coupon rate is fixed. For the Government of Canada bond futures contract the bond's coupon interest rate is fixed at 9.00%. However, once the bond is issued it must compete in the market with other financial instruments. Since new bonds are issued to yield the current interest rates, yields on old bonds must adjust to remain competitive with the newer issues. Thus, when interest rates rise, the price of an older bond with a lower coupon interest rate must decline to increase the yield on the old bond, making it competitive with the return on newly issued bonds.

Interest rate futures offer investors a very inexpensive way of eliminating the risks associated with interest rate fluctuations. For example, banks, pension funds, and insurance companies all make considerable use of the interest rate futures market to avoid paper losses that might otherwise occur when interest rates unexpectedly increase. Corporations also use interest rate futures to lock in interest rates when they are planning to issue debt. If interest rates rise before the corporation has the opportunity to issue the new debt, the profits on the Government of

Canada bond futures contracts they have sold will offset the increased costs associated with the proposed debt offering. Several possible uses for bond futures are given in Table 22-1.

Foreign Currency Futures

Of all the financial futures, foreign currency futures have been around the longest, first appearing in 1972. In North America, the Chicago Mercantile Exchange operates the largest market for foreign currency futures. This type of future works in the same way as other futures, but in this case the commodity is Canadian dollars, German marks, British pounds, or some other foreign currency. As we will see, the similarities between these futures and the others we have examined are great. Not only do foreign currency futures work in the same way as other futures, but they also are used by financial managers for the same basic reasons—to hedge away risks, in this case to reduce the risk of fluctuations in the currency exchange rate. One of the major participants in the foreign currency futures market is the exporter who will receive foreign currency when its exported goods are finally received and who uses this market to lock in a certain exchange rate. As a result, the exporter is unaffected by any exchange rate fluctuations that might occur before receiving payment. Foreign currency futures are also used to hedge away possible fluctuations in the value of earnings of foreign subsidiaries.

In the eighties and nineties fluctuations in exchange rates became common. With foreign currency futures a financial manager could eliminate the effects—good or bad—of exchange rate fluctuation with a relatively small investment. The extremely high degree of leverage that was available coupled with the dramatic fluctuations in foreign exchange rates in the 1980s encouraged many financial managers to consider entering the currency rate futures market. One example of a dramatic price movement came as the British pound dropped to a value of just over U.S. $1.07 in early 1985. To get a feel for the degree of leverage experienced in the foreign currency futures market and the large profits and losses that can occur in this market let's look at Figure 22-1, which examines profits and losses resulting from buying and selling British pound futures.

Figure 22-1 illustrates an example of how fortunes can be made or lost by those

Table 22-1 How Features of Bond Futures Are Used

Protect Portfolio Value and Return: Pension funds, banks, corporations, insurance companies, and individual investors can use bond futures as a hedge against losses incurred on fixed-income portfolios when interest rates rise and prices decline. Careful timing in the placement of a hedge can protect the holding period return of a portfolio containing notes and other related instruments. Futures offer flexibility to money managers by allowing them to adapt to a changing market.

Protect Issuance Costs: Corporations that plan to issue intermediate-term debt can control their interest cost by selling bond futures and offsetting the position when the issue comes to market. By hedging, the issuer can take advantage of today's lower rates, thus lessening the cost of raising capital.

Transfer Risk: Underwriters of corporate issues can sell futures to transfer the risk between the time they buy the debt securities until the time they are sold to dealers and investors.

Hedge Participation in Government Auctions: Primary government securities dealers can use the new contract to hedge participation in auctions. Bond futures allow government securities dealers to remain competitive and offer a better price to their customers.

Lock in a Favorable Rate of Return: Portfolio managers can hedge the reinvestment rate of coupon income. If interest rates decline, the reinvestment rate will diminish for note holders. Buying futures today as a substitute for later investments allows the investor to hedge a drop in rates. If rates fall, the decreased return on reinvestment will be offset by a gain in the futures position. If rates rise, the loss on the futures position will be offset by a higher return on investment.

Source: *Ten Year Treasury Futures* (Chicago: Chicago Board of Trade, 1983).

investing in foreign currency futures. Some firms no doubt save themselves enormous losses by hedging away exchange rate risk, while others benefit by the dramatic swing in the exchange rate. Over just five trading days, the value of an investment in British pound futures went up almost threefold, while the return on the initial margin investment for those selling British pounds dropped almost threefold. Needless to say, the foreign currency futures market is a very risky market, characterized by extreme leverage both on the up and down side and periodic major movements in the underlying values of the foreign exchange currencies. To the financial manager, this market provides a perfect mechanism for eliminating the effect of exchange rate fluctuations.

Stock Index Futures

Stock indexes have been around for many years but it has only been recently that financial managers and investors have had the opportunity to trade futures on them directly. In North America, the Kansas City Board of Trade began trading listed stock index futures in 1982. Their index was based on the Value Line Index. It was not until 1984 that the Toronto Futures Exchange began trading the TSE 300 futures contract based on the TSE 300 Composite Index. In 1987 the Toronto Futures Exchange launched the Toronto 35 Index futures to meet the needs of professional fund managers and private investors. The creation of the Toronto 35 Index futures contract enables traders to hedge a portfolio of shares more effectively since the shares of this index belong to some of the largest companies listed on the TSE and are all actively traded. As a result, both the popularity of the Toronto 35 Index and the ease with which traders can replicate the Toronto 300 Index, caused the TSE 300 futures contract to be delisted in 1987. In the fall of 1993, the Toronto Stock Exchange launched the TSE 100 Index which contains the largest and most

Figure 22-1 Profits and Losses on British Pound Futures in March 1985

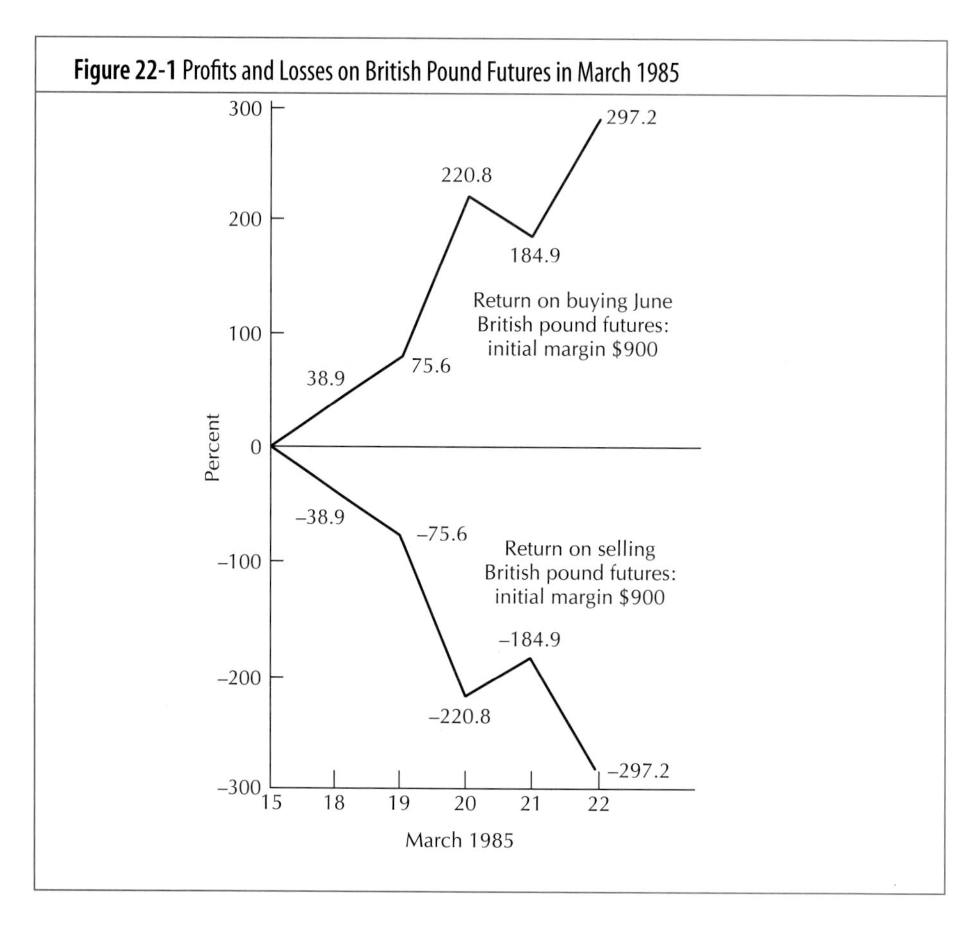

liquid stocks on the TSE 300. The TSE 100 was considered to be more focused and liquid than the TSE 300 Index. Furthermore, the TSE 100 Index would better reflect the movements of the TSE 300 Index than the more narrowly based TSE 35 Index. In an effort to assist portfolio managers in reducing the risk of their positions, the Toronto Futures Exchange began trading the TSE 100 futures and options contracts in the spring of 1994.

At this point, after looking at other futures contracts, the workings of stock index futures should be clear. They work basically the same way, with one major exception: Stock index futures contracts allow only for cash settlement. There is no delivery, since what is being traded is the future price of the index, not the underlying shares in the index. Currently there are several stock index futures available, with futures on the S&P 500 index clearly dominating in terms of volume. Table 22-2 shows the characteristics of both the TSE 35 Index and TSE 100 Index futures contracts.

Both the Toronto 35 Index and TSE 100 Index futures contracts are based on shares which have been chosen because they are some of Canada's larger publicly listed corporations and are among the most heavily traded issues on the TSE. Both these Indexes are modified market value weighted indexes. A limit of 10 percent has been placed on the weighting of any one stock, such that no one company or industry dominates the index. The contract sizes or value of each contract is $500 times the value of either the Toronto 35 Index or the Toronto 100 Index. For example, in early 1996 the values of the TSE 35 Index and the TSE 100 Index futures contracts were worth $124,245 and $143,155 respectively.

After the 1987 crash, a system of shock absorber limits and circuit breakers was introduced in most index futures markets. These serve the same purpose as do daily price limits, but are not as strict. For example, the New York Futures Exchange has 10-minute, 30-minute, one- and two-hour trading halts that result from wide swings in the stock market. The purpose of these programmed trading halts is to allow investors to rationally appraise the market during periods of large price swings.

To the financial manager, the great popularity of these financial newcomers lies in their ability to reduce or eliminate systematic risk. When we talked about the variability or risk associated with common share returns we said that there were two types of risk: systematic and unsystematic risk. Unsystematic risk accounts for a large portion of the variability of an individual security's returns and is largely eliminated in large portfolios through random diversification. Thus, portfolio man-

Table 22-2 Characteristics of TSE 35 Index and TSE 100 Index Futures Contracts

Feature	TSE 35 Index	TSE 100 Index
Location	Toronto	Toronto
Underlying market index	The Toronto 35 Index. This is a modified market value weighted index of 35 stocks from the Toronto Stock Exchange.	The Toronto 100 Index. This is a modified market value weighted index of 100 stocks from the Toronto Stock Exchange.
Contract size	Five hundred times the Toronto 35 Index	Five hundred times the Toronto 100 Index
Minimum price change	Tick size is 0.02 points. This represents a change of $10 per tick.	Tick size is 0.05 points. This represents a change of $25 per tick.
Normal daily price limits	13.50 above or below prior settlement price	16.50 above or below prior settlement price
Margins	Hedge $5,000 Speculative $9,000	Hedge $5,000 Speculative $9,000
Delivery concept	Cash settlement. Actual value of the Toronto 35 Index determines the payment. Final settlement is on the second day following the last trading day of the expiring month.	Cash settlement. Actual value of the Toronto 100 Index determines the payment. Final settlement is on the second day following the last trading day of the expiring month.

Source: Adapted from: *TSE Review*, Toronto, ON., 1994.

agers are more concerned with systematic or market risk which accounts for the variability of returns from a portfolio of securities. The variability of returns in a portfolio of securities is described by the portfolio's beta. Prior to the introduction of stock index futures, a portfolio or pension fund manager was forced to adjust the portfolio's beta if he or she anticipated a change in the direction of the market. Stock index futures allow the portfolio or pension fund manager to eliminate or mute the effects of swings in the market without the large transactions costs that would be associated with the trading needed to modify the portfolio's beta. Although stock index futures eliminate the unwanted effects of market downswings, these types of futures also eliminate the effects of market upswings. In other words, they allow the portfolio or pension fund manager to eliminate as much of the effect of the market as he or she wishes from the portfolio.

BACK TO THE BASICS

The area of risk management has grown rapidly over the last decade. In response to volatile interest rates, commodity prices, and exchange rates, financial managers are using futures, options, and swaps to hedge these risks. Once again, the financial markets demonstrated their dynamic and adaptive nature in finding new ways of reducing risk without significantly affecting return. The inspiration for such behaviour, of course, finds its roots in ***Axiom 1: The Risk-Return Tradeoff—We Won't Take on Additional Risk Unless We expect to Be Compensated with Additional Return.***

OBJECTIVE 2

OPTIONS

Option contract

An option contract gives its owner the right to buy or sell a fixed number of shares at a specified price over a limited time period.

An **option contract** gives its owner the right to buy or sell a fixed number of shares at a specified price over a limited time period. The trading of listed call options in North America was initiated by the Chicago Board Options Exchange (CBOE) in 1973. In Canada, the Montreal Exchange began trading listed call options in 1975. Several months later, the Toronto Stock Exchange also began listing options. However, in 1977 both of these exchanges agreed to form a corporation, called Trans Canada Options Inc., which would list and clear stock options on exchanges across Canada.

Obviously, there is too much going on in the options markets not to pay attention to them. Financial managers are just beginning to turn to them as an effective way of eliminating risk for a small price. As we will see, they are fascinating, but they are also confusing—with countless variations and a language of their own. Moreover, their use is not limited to speculators; options are also used by the most conservative financial managers to eliminate unwanted risk.[2] In this section we will discuss the fundamentals of options, their terminology, and how they are used by financial managers.

The Fundamentals of Options

Call option

A call option gives its owner the right to purchase a given number of shares or some other shares at a specified price over a given time period.

There are only two basic types of options: a put and a call. Everything else involves some variation. A **call option** gives its owner the right to purchase a given number

[2]Although options have been around since the time of Aristotle, the first extensive use of options occurred in Holland in the 1700s. In Holland the early options were associated with tulip bulbs, which had become very popular and were viewed by many as a speculative investment at the time. Tulip growers would buy put options from tulip dealers and thereby reduce their exposure to downward price fluctuations. Other growers would write call options and sell them to dealers, who would in turn sell bulbs for future delivery based upon the calls. Unfortunately, the market was totally unregulated and as tulipmania took hold of Holland in the mid-1700s and the price of tulip bulbs skyrocketed and then crashed in 1763, this market also crashed with many put writers either unwilling or unable to honour their commitments.

of common shares or some other asset at a specified price over a given period. This right is essentially the same as a "rain check" or a guaranteed price. As the buyer, you have the option to buy something, in this case common shares. In effect a call option gives you the right to buy, but it is not a promise to buy. Thus, if the price of the underlying common shares or asset goes up, a call buyer makes money. A **put option**, on the other hand, gives its owner the right to sell a given number of common shares or some other asset at a specified price over a given time period. A put buyer is betting that the price of the underlying common shares or asset will drop. Since these are just options to buy or sell shares or some other asset, they do not represent an ownership position in the underlying corporation, as do common shares. In fact, there is no direct relationship between the underlying corporation and the option. An option is merely a contract between two investors. As a result, we can modify the terms of an option contract for a specific set of financial conditions. For example, an **American option** can be exercised by the option holder at any time during the life of the option. A **European option**, on the other hand, can be exercised only at the time of its maturity.

Put option

A put option gives its owner the right to sell a given number of common shares or some other asset at a specified price over a given time period.

American option

An option that can be exercised by the holder at any time during the life of the option.

European option

An option that can be exercised only at the time of its maturity.

A buyer of an option can be viewed as betting against the seller or writer of the option because the option holder does not have ownership of the underlying asset. For this reason the options markets are often referred to as a zero sum game. If someone makes money, then someone must lose money; if profits and losses were added up, the total for all options would equal zero. If commissions are considered, the total becomes negative, and we have a "negative sum" game. As we will see, the options markets are quite complicated and risky. Some experts refer to them as legalized institutions for transferring wealth from the unsophisticated to the sophisticated. Just as with the call option, a put option gives the buyer the right to sell the common shares at a set price, but it is not a promise to sell.

The Terminology of Options

In order to continue with our discussion, it is necessary to define several terms that are unique to options.

The Contract An option is a contract that allows the buyer to either buy in the case of a call, or sell in the case of a put, the underlying shares or asset at a predetermined price. No asset changes hands, but the price is set for a future transaction which will take place only if and when the option buyer wants it to. In this section we will refer to the process of selling puts and calls as *writing*. The sale of options is often referred to as *shorting* or *taking a short position* in those options and buying an option is referred to as *taking a long position*.

The Exercise or Striking Price The **option striking price** is the price at which the shares or asset may be purchased from the writer in the case of a call or sold to the writer in the case of a put.

Option striking price

The price at which the shares or asset may be purchased from the writer in the case of a call, or sold to the writer in the case of a put.

Option Premium The **option premium** is merely the price of the option. It is generally stated in terms of dollars per share rather than per option contract, which covers 100 shares. Thus, if a call option premium is $2, then an option contract would cost $200 and allow the purchase of 100 shares at the exercise price.

Option premium

The price of the option.

PERSPECTIVE IN FINANCE

Remember that the option premium is what the option purchaser pays for the option, and that the option writer keeps this payment regardless of whether or not the option is ever exercised. In addition, the terms of the option do not change when the option premium is paid.

Option expiration date

The date on which the option expires.

Expiration Date The **option expiration date** is the date on which the option contract expires. An American option is one that can be exercised any time up to the expiration date. A European option can be exercised only on the expiration date.

Covered and Naked Options If a call writer owns the underlying shares or asset on which he or she writes a call, the writer is said to have written a "covered" call. However, a writer who writes a call on shares or assets that he or she does not own she is said to have written a "naked" call. The consequence of writing a naked call is that the call writer must deliver shares or assets to the buyer if the naked call is exercised.

Open Interest The term *open interest* refers to the number of option contracts in existence at a point in time. The importance of this concept comes from the fact that open interest provides the investor with some indication of the amount of liquidity associated with that particular option.

In-, Out-of, and At-the-Money A call (put) is said to be out-of-the-money if the underlying shares are selling below (above) the exercise price of the option. Alternatively, a call (put) is said to be in-the-money if the underlying shares are selling above (below) the exercise price of the option. If the option is selling at the exercise price, it is said to be selling at-the-money. For example, if Northern Telecom's common shares were selling for $45 per share, a call on Northern Telecom with an exercise price of $42 would be in-the- money, while a call on Northern Telecom with an exercise price of $48 would be out-of-the-money.

Option's intrinsic value

The minimum value of the option.

Intrinsic and Time (or Speculative) Value The term **intrinsic value** refers to the minimum value of the option—that is, the amount by which the shares are in-the-money. Thus, for a call the intrinsic value is the amount by which the share price exceeds the exercise price. If the call is out-of-the-money—that is, the exercise price is above the share price—then its intrinsic value is zero. Intrinsic values can never be negative. For a put, the intrinsic value is again the minimum value the put can sell for, which is the exercise price less the share price. For example, the intrinsic value on a Northern Telecom April 42.50 put when Northern Telecom's common shares were selling for $38.50 per share would be $4. If the put was selling for anything less than $4, investors would buy puts and sell the shares until all profits from this strategy were exhausted. Arbitrage is the process of buying and selling similar assets for different prices. This process keeps the price of options at or above their intrinsic value. If an option is selling for its intrinsic value, it is said to be selling at parity.

Option's time (or speculative) value

The amount by which the option premium exceeds the intrinsic value of the option.

The **time value** or **speculative value of an option** is the amount by which the option premium exceeds the intrinsic value of the option. The time value represents the amount above the intrinsic value of an option that an investor is willing to pay to participate in capital gains from investing in the option. At expiration, the time value of the option falls to zero and the option sells for its intrinsic value, since the chance for future capital gains has been exhausted. These relationships are as follows:

call intrinsic value = share price − exercise price
put intrinsic value = exercise price − share price
call time value = call premium − (share price − exercise price)
put time value = put premium − (exercise price − share price)

EXAMPLE

Perhaps the easiest way to gain an understanding of the pricing of options is to look at them graphically. Figure 22-2 presents a profit and loss graph for the purchase of a call for $2.30 on Northern Telecom's shares with an exercise price of $42.50. This type of option is defined as a Northern Telecom April 42.50 call. In Figure 22-2, and all other profit and loss graphs, the vertical axis represents the profits or losses realized on the option's expiration date, and the hor-

Figure 22-2 Investor's Profit/Loss Graph for Northern Telecom 42.50 Call

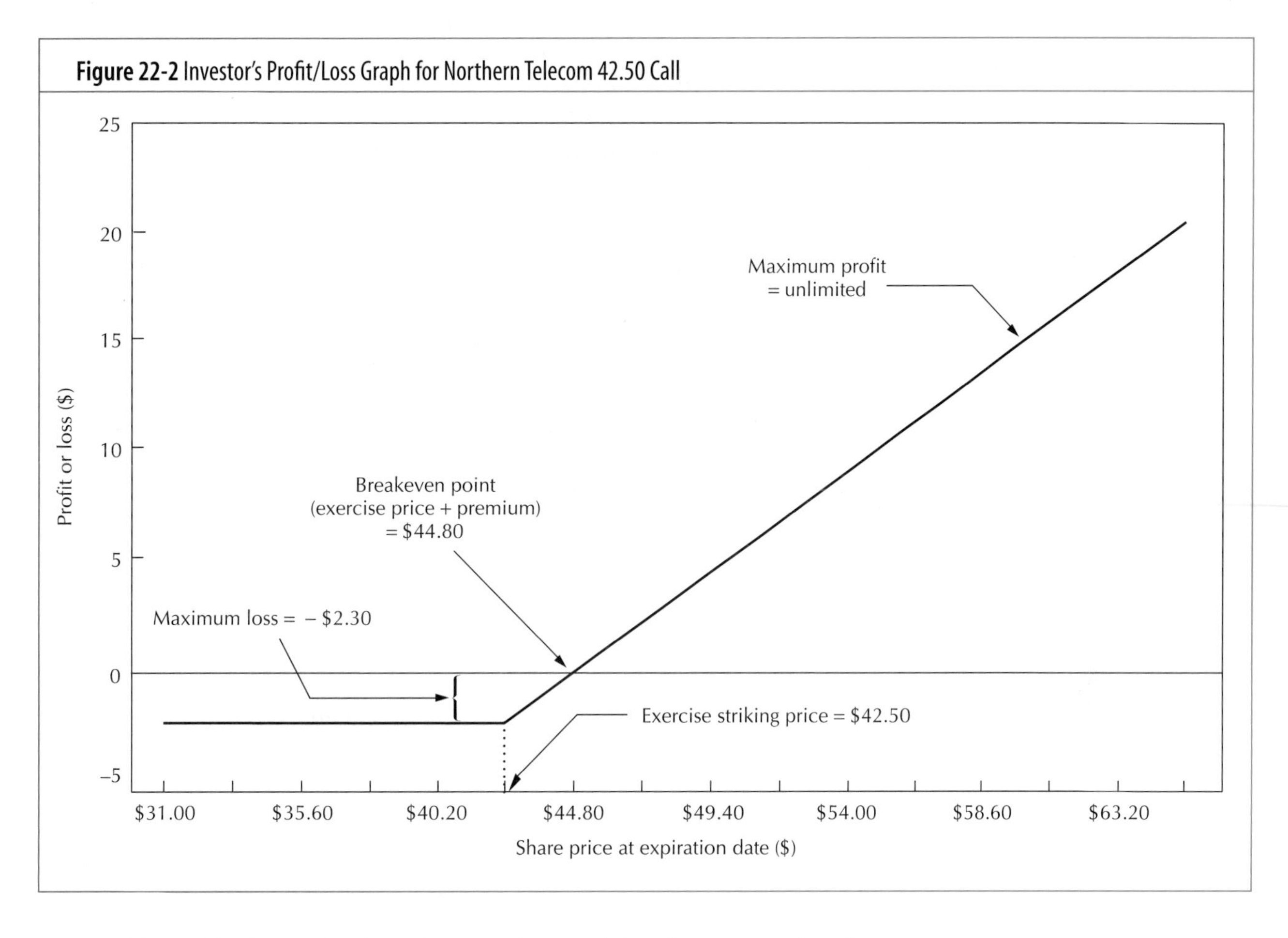

izontal axis represents the share price on the expiration date. Remember that, because we are viewing the value of the option at expiration, the option has no time value and therefore sells for exactly its intrinsic value. To keep the discussion simple, we also ignore any transaction costs.

For the Northern Telecom April 42.50 call shown in Figure 22-2, the call will be worthless at expiration if the value of the shares is less than the exercise or striking price. An individual would not exercise a call option to purchase Northern Telecom shares for $42.50 per share if the individual could buy the same Northern Telecom shares from a broker at a price less than $42.50. An option would be worthless at expiration if the share price is below the exercise price. In this case, the most an investor can lose is the option premium, that is, how much the individual paid for the option which is in this case $2.30. This amount represents all of the investment but only a fraction of the share price. Once the share price climbs above the exercise price, the call option takes on a positive value and increases in a linear one-to-one basis as the share price increases. Once the price of the Northern Telecom shares increases to $44.80 per share, the investor recovers the initial investment of $2.30, the premium paid for the option.

To the call writer, the profit and loss graph is the mirror image of the call buyer's graph. As we noted earlier, the option market is a zero sum game in which one individual gains at the expense of another. Figure 22-3 shows the profits and losses at expiration associated with writing a call option. Once again we examine the profits and losses at expiration, that point in time when an option has no time value. The maximum profit earned by the call writer is the premium paid by the buyer of the call; whereas, the call writer's maximum loss is unlimited.

Figure 22-3 Writer's Profit/Loss Graph for Northern Telecom 42.50 Call

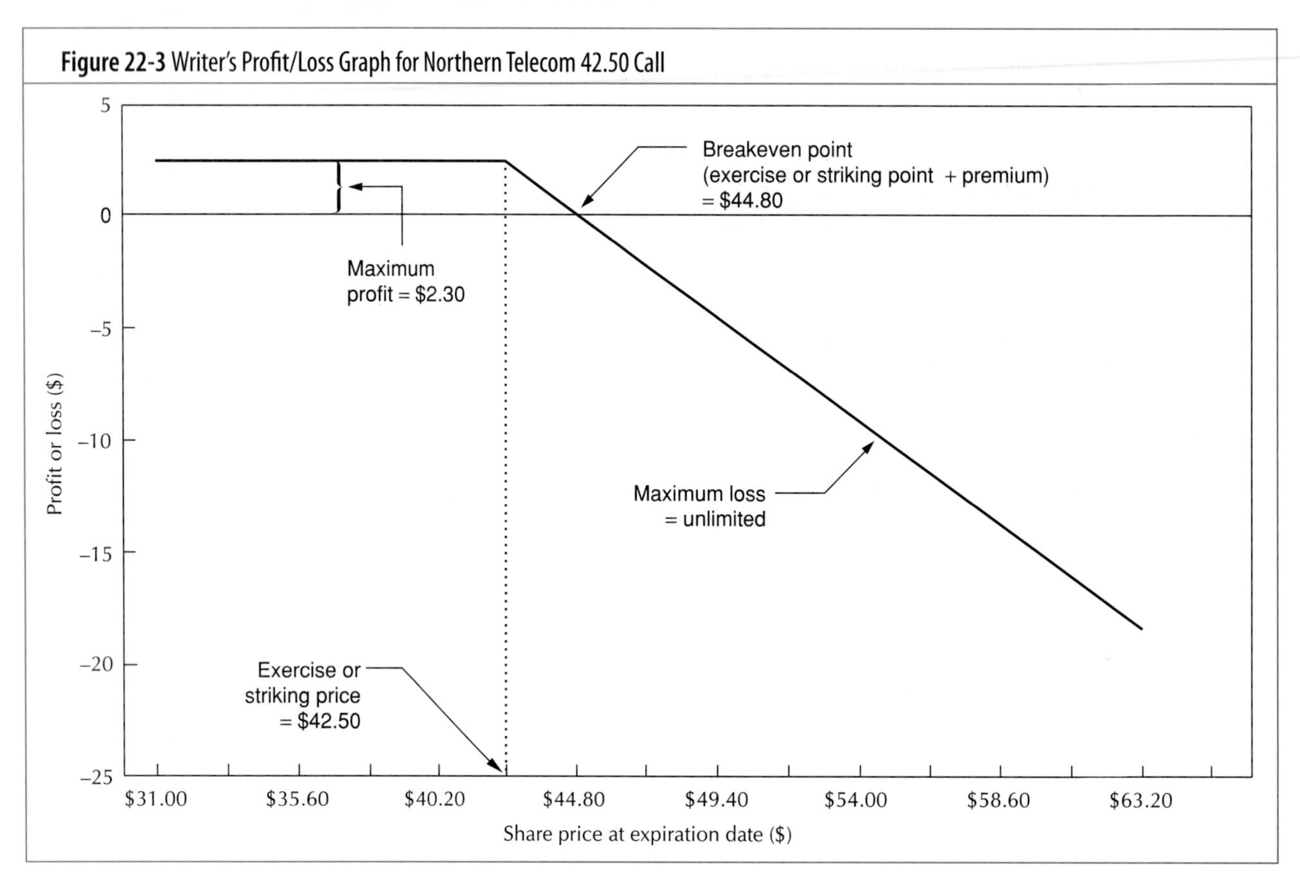

Figure 22-4 Investor's Profit/Loss Graph for Northern Telecom 42.50 Put

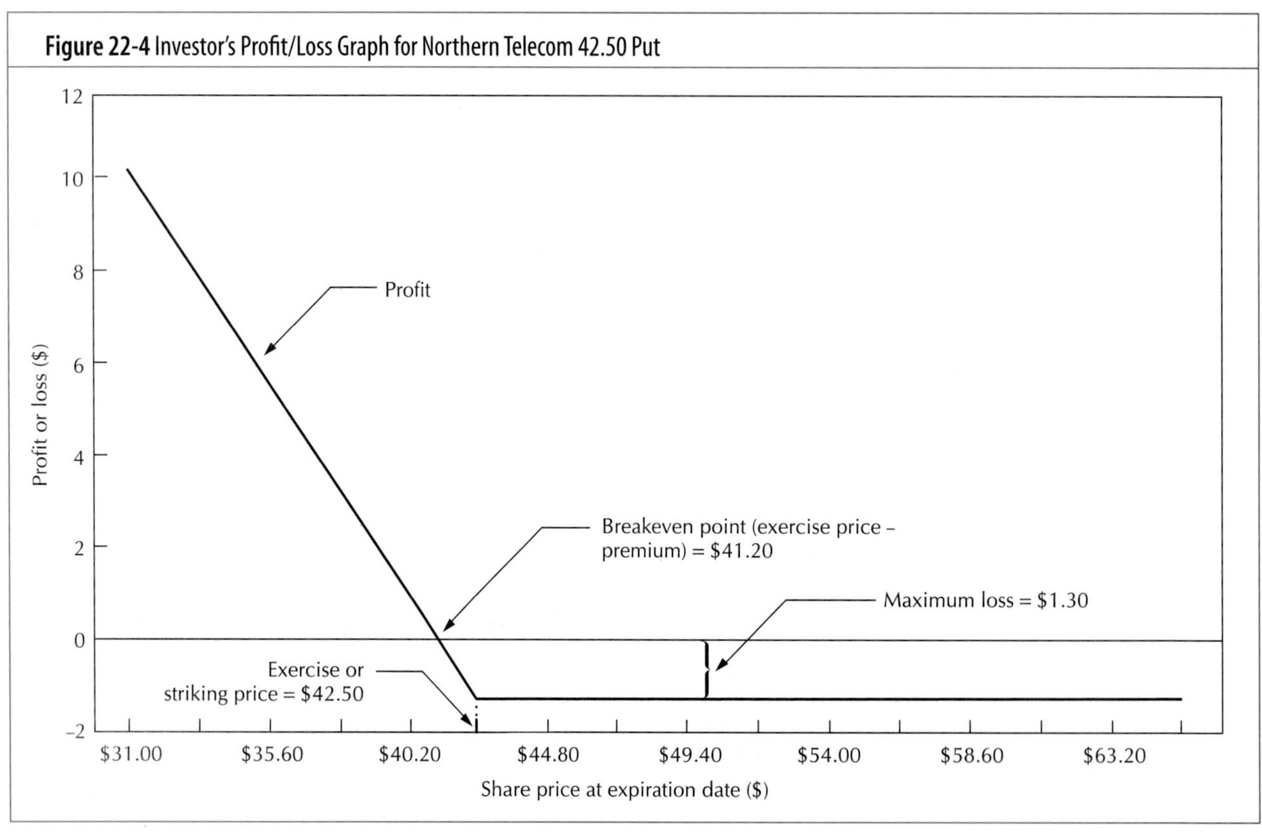

Figure 22-4 illustrates the profits and losses from the purchase of a Northern Telecom April 42.50 put that is bought for $1.30. As the price of Northern Telecom's shares decreases below $42.50, the value of the put increases. In other words, the put takes on value only if Northern Telecom's shares drop below the exercise price of $42.50 per share. The buyer of the put recovers the initial investment of $1.30, the premium of the option, if Northern Telecom's shares drop to $41.20 per share. The most this investor can lose is the premium, which is small in dollar value relative to the potential gains. The maximum gain associated with the purchase of a put is limited only by the fact that the lowest price a share can fall to is zero.

Figure 22-5 illustrates the profits and losses graph for a writer of a put. This graph is the mirror image of the graph of profits and losses of the buyer of a put. The maximum profit a put writer can earn is the premium for which the put was sold. The potential losses for the put writer are limited only by the fact that the share price cannot fall below zero.

Finally, we demonstrate one of numerous strategies which are available to an investor who wants to speculate while minimizing potential losses of an investment. In this case the investor buys one Northern Telecom April 42.50 call and one Northern Telecom April 42.50 put. Figure 22-6 illustrates the profits and losses at expiration if an investor buys the Northern Telecom April 42.50 call at a premium of $2.30 and the Northern Telecom April 42.50 put at $1.30. An investor who creates this type of position expects the share price of Northern Telecom to fluctuate dramatically in either direction from the exercise price of $42.50. Furthermore, if the investor's expectations of fluctuations in Northern Telecom's share price are not fulfilled, the investor's maximum loss is the cost

Figure 22-5 Writer's Profit/Loss Graph for Northern Telecom 42.50 Put

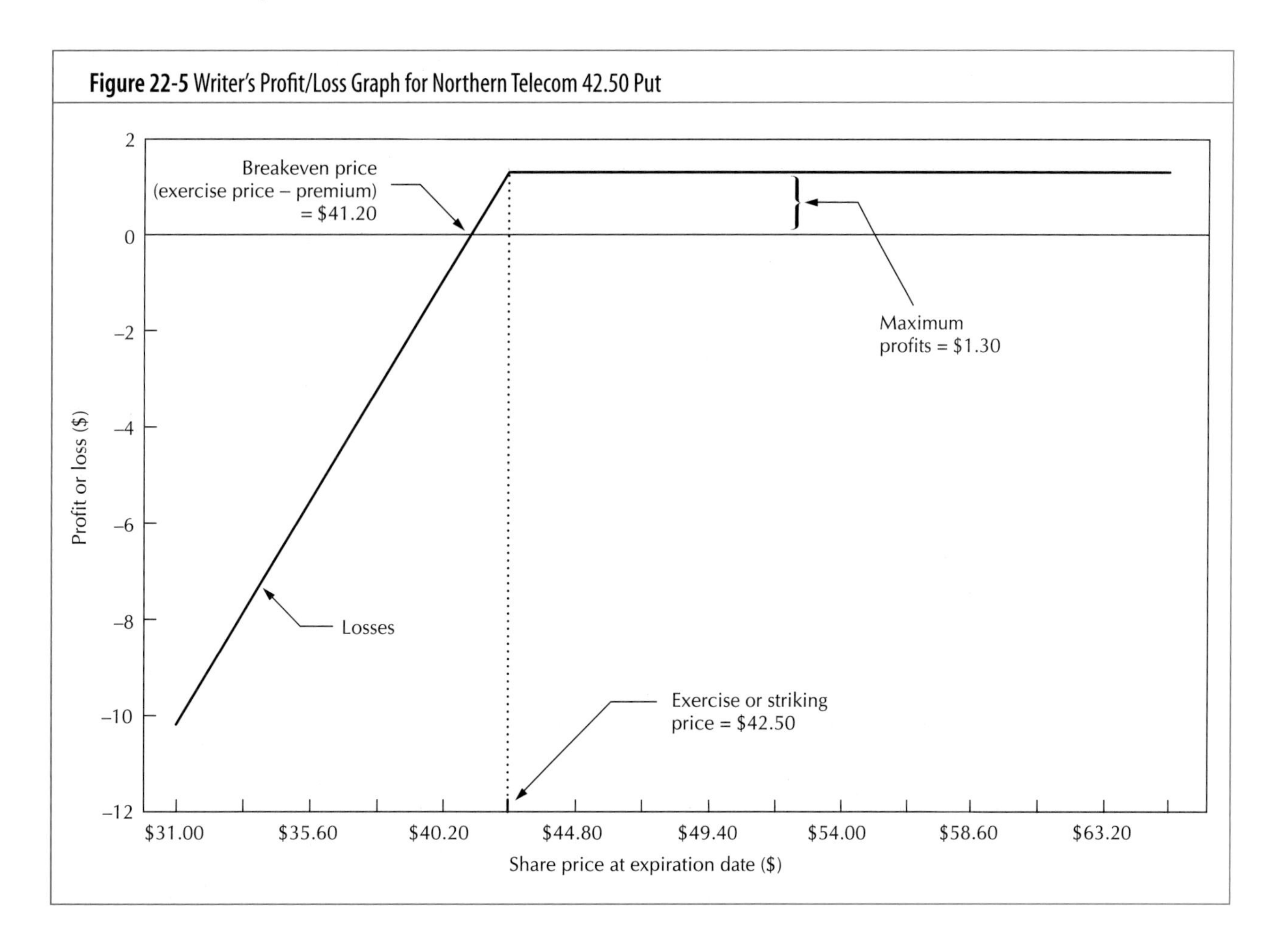

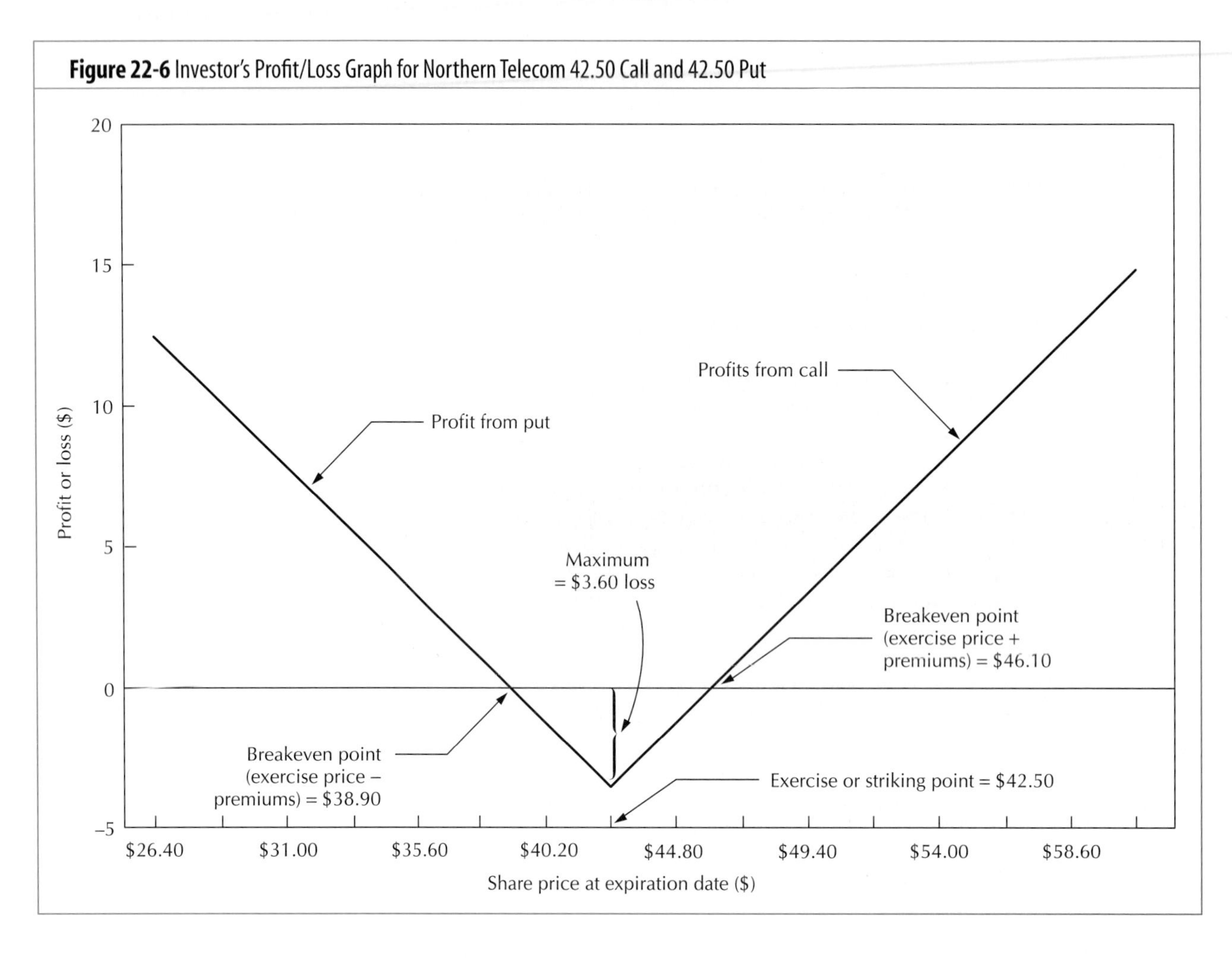

Figure 22-6 Investor's Profit/Loss Graph for Northern Telecom 42.50 Call and 42.50 Put

of buying the two options. Hence, the investor has hedged the investment from important losses if the share price fluctuation does not occur and has speculated on a dramatic fluctuation of the share price if it moves in either direction. Note that the investor does not know with certainty in which direction the share price will move. This type of strategy would be suitable in a situation in which a court decision is expected but the verdict may go in favour of or against a firm, causing the firm's share price to fluctuate in either direction.

All of our graphs have shown the intrinsic value of the option at the expiration date. When we re-examine these relationships before the expiration date, we find that options have a value greater than the intrinsic value, that is additional value that can be attributed to the life of the option. In other words, investors are willing to pay more than the intrinsic value for the option because of the uncertainty of the future share price.

PERSPECTIVE IN FINANCE

The greatest amount that an investor can lose by purchasing a put or call option is the premium or what the investor paid for it. Although this may seem rather small relative to the price of the shares, it is still 100% of his or her investment.

The Valuation of Options

An important development in the theory of finance is the Black-Scholes option pricing model. This model has determined the fair value of a call option on an underlying common share (see Appendix 22B for a description of the Black-Scholes option pricing equation). Although the derivation of this model is beyond the scope of this textbook, it should be noted that the model does indicate that the value of a call option is dependent on the following five factors:

1. The common share's current value
2. The option's exercise price
3. The option's time to expiration
4. The prevailing interest rate
5. The volatility of the underlying common share's rate of return

An important feature of this model is that the first four of the five factors on this list are directly observable; in other words, these factors can be measured from data obtained from today's newspaper. The last factor, the volatility of the underlying common share's rate of return, is unobservable which means that it must be estimated. In this case, the volatility of the underlying common share's rate of return can be measured by the standard deviation of the share's rate of return.

The option pricing model provides valuable insight as to how each of these five factors affects the value of a call option. For example, an American call option is expected to be more valuable if the price of the underlying common share price increases. In a similar manner, an American call option is more valuable when interest rates increase. In general, the value of an American call option will be greater if there is more time to expiration. We also observe that an American call option will be more valuable if there is greater variability in the returns on the underlying asset. Finally, the higher the exercise price, the lower will be the value of the American call option. The effects of these five factors on both American calls and American puts are summarized in Table 22-3.

Put-Call Parity Relationship

Another important relationship in option theory is the concept of Put-Call Parity. The concept is summarized by the following equation:

$$-PV(P) - PV\begin{pmatrix}\text{put}\\ \text{option}\end{pmatrix} = -PV\begin{pmatrix}\text{call}\\ \text{option}\end{pmatrix} - PV(EX)$$

This equation shows that the profits obtained from owning a call option can be replicated by the profits obtained from owning both the underlying asset and a put option with the same exercise price as the call option. The only difference between these two streams of profit is the time value of money to be invested in the exercise price.

Characteristics of Options

As we examine options from the point of view of the financial manager, we will see that they have some attractive features that help to explain their popularity. There are three reasons for the popularity of options:

Table 22-3 Factors Affecting the Value of American Calls and American Puts

Factors	Value of Call option	Value of of Put option
Increase in the underlying common share price	↑	↓
Increase in interest rates	↑	↓
Greater time to expiration	↑	↑
Greater variability on common share	↑	↑
Increase in exercise price	↓	↑

1. **Leverage.** Calls allow the financial manager the chance for unlimited capital gains with a very small investment. Since a call is only an option to buy, the most a financial manager can lose is what was invested, which is usually a very small percentage of what it would cost to buy the shares themselves, while the potential for gain is unlimited. As we will see, when a financial manager owns a call, he or she controls or benefits directly from any price increases in the shares. The idea of magnifying the potential return is an example of leverage. It is similar to the concept of leverage in physics, where a small amount of force can lift a heavy load. Here a small investment is doing the work of a much larger investment. Unfortunately, leverage is a double-edged sword: Small price increases can produce a large percentage profit, but small price decreases can produce large percentage losses. With an option, the maximum loss is limited to the amount invested.
2. **Financial insurance.** For the financial manager, this is the most attractive feature of options. A put can be looked on as an insurance policy, with the premium paid for the put being the cost of the policy. The transactions costs associated with exercising the put can then be looked upon as the deductible. When a put with an exercise price equal to the current share price is purchased, it insures the holder against any declines in the share price over the life of the put. Through the use of a put, a pension fund manager can reduce the risk exposure in a portfolio with little change in cost and little change to the portfolio. One dissimilarity between a put and an insurance policy is that with a put an investor does not need to own the asset, in this case the shares, before buying the insurance. A call, since it has limited potential losses associated with it, can also be viewed as an investment insurance policy. With a call option, the investor's potential losses are limited to the price of the call, which is quite a bit below the price of the shares themselves.
3. **Investment alternative expansion.** From the viewpoint of the investor, the use of puts, calls, and combinations of them can materially increase the set of possible investment alternatives available.

Financial managers who use options understand the concept of leverage and the concept of financial insurance. It is these two factors combined that allow for an expansion of the available investment alternatives. Remember, both puts and calls are merely options to buy or sell the shares at a specified price. The worst that can happen is that the options become worthless and the financial manager loses the investment.

Options Exchanges

Before a central marketplace for put and call options existed, the specifics of each option were negotiated directly between the writer and the buyer of the option. Very seldom did any two options look alike. Generally, every option written had a different expiration date and a different exercise price. As a result, there was little in the way of a secondary market for these individualized options, and the writers and buyers generally had to hold their position until expiration or until the options were exercised.

An important initiative in developing an option market in Canada was the creation of Trans Canada Options Inc. which lists all of the option trading that takes place on the Montreal, Toronto and Vancouver exchanges. The growth of the option market in Canada can be attributed to the following developments:

1. **Standardization of the option contracts.** Today, the expiration dates for all options are standardized. As a result, there is only one day per month on which a listed option on any shares can expire. The number of shares that a call allows its owner to purchase and a put allows its owner to sell has also been standardized to 100 shares per option contract. In addition, the striking prices have been stan-

dardized, generally at five-point intervals, so that there are more identical options. Through this standardization, the number of different option contracts on each stock is severely limited. The result is that more options are identical and the secondary market is made more liquid.

2. **Creation of a regulated central marketplace.** Trans Canada Options Inc. provides a listing of all option trading that occurs on the Montreal, Toronto, and Vancouver exchanges. Each exchange provides a central location for continuous trading of options listed on their exchange. In 1983, these exchanges agreed to allocate the option listings such that no option is listed more than once in Canada. In addition, these exchanges have agreed to adhere to strong surveillance and disclosure requirements.
3. **Creation of the options clearinghouse corporation.** Trans Canada Options Inc. bears full responsibility for honouring all options issued on option exchanges across Canada. In effect, all options held by individuals have been written by Trans Canada Option Inc. and alternatively, all options written by individuals are held by Trans Canada Option Inc. The purpose of creating a buffer between individual buyers and sellers of options is to provide investors with confidence in the market, in addition to facilitating the clearing and settlement of options. Because of the importance of Trans Canada Option Inc., let us look for a moment at its operation.

   When an options transaction is agreed upon, the seller writes an option contract to Trans Canada Option Inc., which in turn writes an identical option contract to the buyer. If the buyer later wants to exercise the option, he or she gives Trans Canada Option Inc. the exercise price associated with the option, which in turn provides the buyer with shares. To get the shares to cover the option, Trans Canada Option Inc. simultaneously exercises a call option it has on these shares. Because of the operation of Trans Canada Option Inc. and the strong secondary market created by the stock option exchanges in Canada, options are not exercised very frequently, but are generally sold. Rather than exercise an option, an investor or financial manager usually just sells the option to another investor, and realizes the profits in that manner. Writers of options clear their position by buying an option identical to the one they wrote. As a result the writer has two identical contracts on both sides of the market with Trans Canada Option Inc. These positions then cancel each other out.
4. **Trading was made certificateless.** Instead of issuing certificates, Trans Canada Option Inc. maintains a continuous record of traders' positions. In addition to making the clearing of positions (the cancelling out of an option writer's obligation when an identical option is purchased) easier, it has also allowed for an up-to-date record of existing options to be maintained.
5. **Creation of a liquid secondary market with dramatically decreased transactions costs.** There has also been a self-fulfilling generation of volume adding to the liquidity of the secondary market. That is, the innovations created a liquid secondary market for options, and this liquid secondary market attracted more investors into the options market, which in turn created even more liquidity in the secondary market.

Recent Innovations in the Options Market

Recently, new variations of the traditional option have appeared: the long-term equity-anticipated securities (LEAPS), the stock index option, the interest rate option, the foreign currency option, and the Canada Government bond futures option.

Long-Term Equity-Anticipated Securities

LEAPS are long-term options which can expire in 2 years or less. These instruments are advantageous to investors because they sell for less than the price of the underlying common share, which provides a greater opportunity for leverage and diversification of the investor's portfolio. For example, Trans Canada Options Inc. lists LEAPS on the common shares of the National Bank which expire in 1997 and have an exercise price of $8.00.

Stock Index Options

In Canada, the options on stock indexes were first traded on the Toronto Stock Exchange in 1984. The first stock index option was based on the TSE 300 Composite Index. However, in 1987 the Toronto Stock Exchange introduced stock index options on the Toronto 35 Index and phased out the stock index options based on the TSE 300 Composite Index. In 1994, the Toronto Stock Exchange began offering stock index options on the TSE 100 Index. In North America, there are a variety of different index options, which are based upon stock market indexes or industry indexes such as a computer industry index. However, the broader stock market indexes have carried the bulk of the popularity of index options.

The reason for this popularity is quite simple. These options allow portfolio managers and other investors holding broad portfolios cheaply and effectively to eliminate or adjust the market risk of their portfolios. When we talked about systematic and unsystematic risk, we noted that in a large and well-diversified portfolio unsystematic risk was effectively diversified away, leaving only systematic risk. Thus, the return on a large and well-diversified portfolio was a result of the portfolio's beta and the movement of the market. As a result, since the movements of the market cannot be controlled, portfolio managers periodically attempt to adjust the beta of the portfolio when they feel a change in the market's direction is at hand. Index options allow them to make this change without the massive transaction costs that would otherwise be incurred.

In general, stock index options are used in exactly the same way traditional options are used; that is, for leverage and for investment insurance. However, because of the unusual nature of the "underlying shares," these concepts take on a different meaning. In the case of leverage, the portfolio manager is speculating that the market will head either up or down and is able to cash in on any market volatility with a relatively small investment. The following example demonstrates how an investor can use this type of option and take advantage of broad market movements.

EXAMPLE

An investor decides to buy a January call on the Toronto 35 Index with an exercise price of 170. As shown in Figure 22-7 the investor pays 2.75 or $275.00 with the Toronto 35 Index priced at 169.27. Ten days later when the index closed at 178.14, the stock index option is worth 9.375, or $937.50. On the other hand, an investor who purchased a put for 3.25, or $325, would have had the value of the investment drop to $25. All this dramatic price movement was the result of only a 5.24 percent change in the underlying index.

In the case of the investment insurance motive for holding index options, the manager is really using them to eliminate the effects of a possible downward movement in the market. For example, a portfolio manager who wants to insure the portfolio against a downturn in the market might purchase a put on the Toronto 35 Index. Thus, if the market declines, the put will appreciate in value, hopefully offsetting the loss in the investor's portfolio.

In effect, index options can be used in the same way as the more traditional options. The only difference is that here the profits or losses depend upon what happens to the value of the index, rather than to one share.

Options on Canada Bond Futures

Options on Canada bond futures work the same way as any other option. The only difference between them and other bond options is that they involve the acquisi-

Figure 22-7 January 170 Option on the Toronto 35 Index

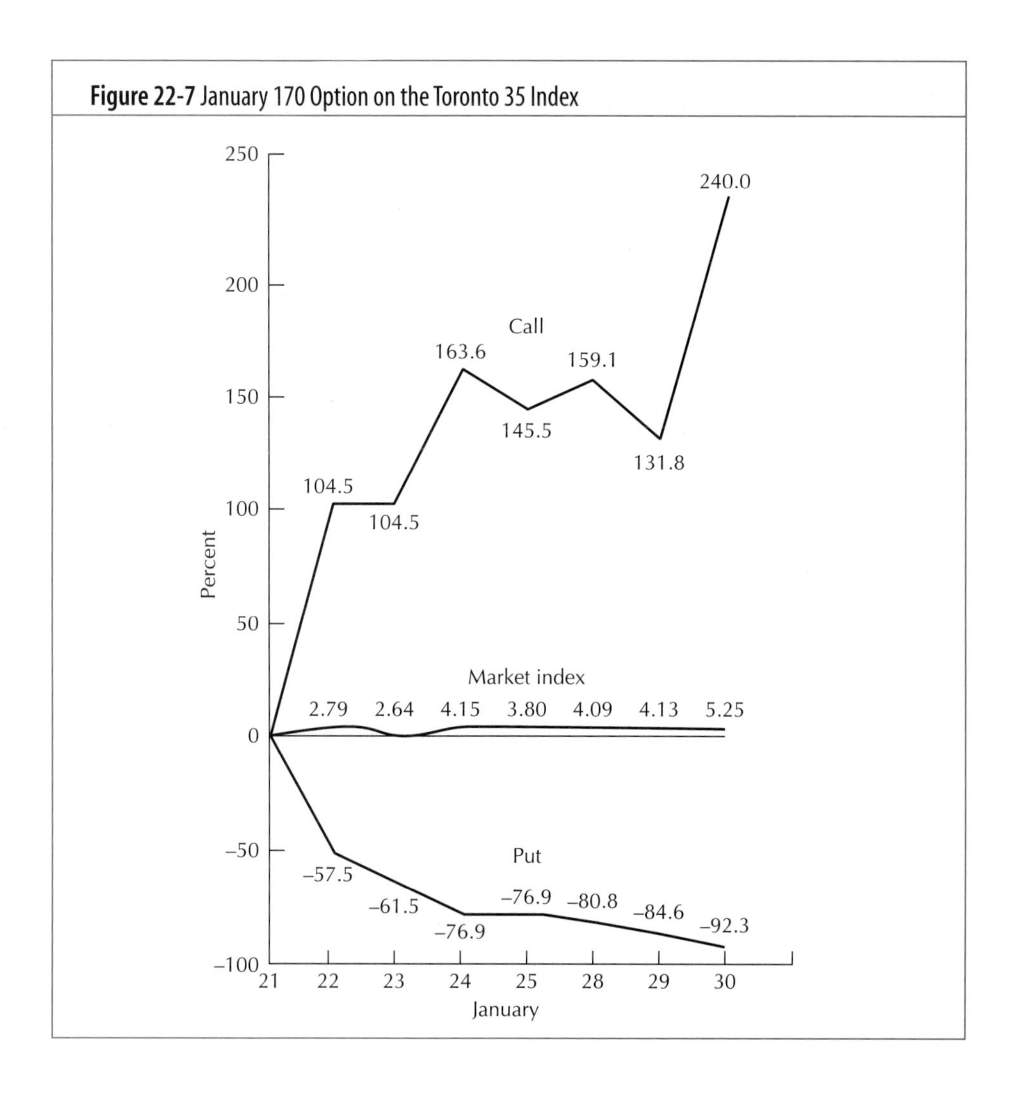

tion of a futures position rather than the delivery of actual bonds. To the creative financial manager, they provide a flexible tool to insure against adverse changes in interest rates while retaining the opportunity to benefit from any favourable interest rate movement that might occur. While a futures contract establishes an obligation for both parties to buy and sell at a specified price, an option only establishes a right. It is therefore exercised only when it is to the option holder's advantage to do so. In effect, a call option on a futures contract does not establish a price obligation, but rather a maximum purchase price. Conversely, a put option on a futures contract is used to establish a minimum selling price. The buyer of an option on a futures contract can achieve immunization against any unfavourable price movements; whereas, the buyer of a futures contract can achieve immunization against any price movements regardless of whether they are favourable or unfavourable.

In their short history, options on Canada bond futures have proved to be extremely popular. Their popularity can be traced to several important advantages they possess:

1. **Efficient price determination of the underlying instrument.** The Canada bond futures contract is well traded on the Montreal Exchange. As a result, there is a continuous stream of market-determined price information concerning these contracts. Conversely, price information on most other types of bonds is unreliable because of the substantial time intervals between trades and the wide gap between existing bid and ask prices.

2. **Unlimited deliverable supply.** Because the clearing corporation can create as many futures contracts as are needed, the process of exercising an option is made extremely simple. When an option on a futures contract is exercised, the buyer simply assumes a futures position at the exercise price of the option. Since the clearing corporation can create as many futures contracts as are needed, the market price of these contracts is not affected by the exercise of the options on them. Conversely, if an option holder on an actual bond were to exercise his or her option, he or she would have to take delivery of the underlying bond. Since the supply of any particular bond is limited, this might create serious price pressure on that bond, provided the bond does not enjoy sufficient liquidity. Thus, because of the unlimited deliverable supply of futures contracts, the exercise of options on futures does not affect the price of those futures.
3. **Greater flexibility in the event of exercise.** If the option proves to be profitable, the buyer or writer can settle the transaction in cash by offsetting the futures position acquired by exercise, or do nothing temporarily and assume the futures position and make or take delivery of the actual bonds when the futures contract comes due.
4. **Extremely liquid market.** Because of the other advantages of options on Canada bond futures, a great number of these options have been created and are traded daily. As a result of the large volume, options on Canada bond futures have developed a very liquid and active secondary market, which has encouraged other traders to enter this market.

Financial institutions seem to be major participants in the options on Canada bond futures market, although there are many potential users of financial futures. They use futures options to alter the risk-return structure of their investment portfolios and actually reduce their exposure to downside risk. A common strategy is to purchase put options and thereby eliminate the possibility of large losses while retaining the possibility of large gains. There is a cost associated with this strategy, since the option premium must be paid regardless of whether or not the option is exercised. An additional return is also generated by those who write call options against a bond portfolio. With this strategy, the premium increases the overall return if bond yields remain stable or rise; however, a maximum return is also established for the portfolio, since it is this tail of the distribution that is sold with the option.

Other examples of how options can be used on financial instruments in North America include interest rate options and foreign exchange options. Interest rate options on 30-year Treasury bonds are also traded on the Chicago Board Options Exchange. Although the trading appeal of interest rate options is somewhat limited, they do open some very interesting doors to the financial manager. In terms of the insurance and leverage traits, they allow the financial manager to insure against the effects of future changes in interest rates. We know that as interest rates rise the market value of outstanding bonds falls; thus, through the purchase of an interest rate put, the market value of a portfolio manager's bonds can be protected. Alternatively, a financial manager who is about to raise new capital through a debt offering and who is worried about a possible rise in interest rates before the offering takes place may purchase an interest rate put. This would have the effect of locking in current interest rates as the maximum level that the firm would have to pay.

Foreign currency options are the same as the other options we have examined, except that the underlying asset is the British pound, the Japanese yen, or some other foreign currency. While foreign currency options are limited to the Philadelphia Exchange, there is a considerable amount of interest in them largely due to the wide fluctuations foreign currency has had in recent years relative to the American dollar. In terms of the insurance and leverage traits, these options allow multinational firms to guard against fluctuations in foreign currencies that might adversely affect their operations. The leverage trait allows investors to speculate in possible future foreign currency fluctuations with a minimum amount of exposure to possible losses.

The Importance of Options to Financial Management

Option theory has provided several important contributions to the understanding of modern corporate finance. We now examine some of these contributions.

Shares and Bonds as Options

The shares of a leveraged firm can be viewed as a call option. In coming to this conclusion, we assume that debtholders are buying the assets of the firm from the shareholders for cash and an implied call option. The exercise price of this call option is equal to the principal and interest of the debt. Shareholders will exercise their call option if the firm is profitable and buy back the assets of the firm from the bondholder for the principal and interest of the debt. If the shareholders default because the firm is not profitable, the debtholders will exercise their call option by taking ownership of the firm. As a result, the value of the firm's shares is expressed in the following equation:

$$\begin{pmatrix}\text{value of}\\ \text{shares}\end{pmatrix} = \begin{pmatrix}\text{market value of}\\ \text{the firm's assets}\end{pmatrix} - \begin{pmatrix}\text{value of}\\ \text{riskless bonds}\end{pmatrix} + \begin{pmatrix}\text{shareholders'}\\ \text{right to default}\end{pmatrix}$$

Whereas, the value the debtholder's claim on the firm can be expressed by the following equation:

$$\begin{pmatrix}\text{value of}\\ \text{firm's debt}\end{pmatrix} = \begin{pmatrix}\text{market value of}\\ \text{the firm's assets}\end{pmatrix} - \begin{pmatrix}\text{shareholders'}\\ \text{right to default}\end{pmatrix}$$

In this equation, the shareholders' right to default is the equivalent to a put option.

Investment Decision Making as Option Pricing Problems

Capital-budgeting decisions should consider not only the value that is derived from the benefits being greater than the costs of a project, but also any value derived from options that are embedded in the project's cash flows. For example, the decision to delay processing after the initial outlay has been made so that further information can be obtained to reduce uncertainty of the operating cash flows would be considered a call option. In considering the decision to delay, management has more flexibility in making its capital-budgeting decisions. Moreover, as investments proceed, the possibility of other future investments relating to the project may appear. This possibility of future investments can be built into the capital-budgeting decision as call options. Such a relationship would expressed as follows:

$$\begin{pmatrix}\text{value of capital}\\ \text{investment}\end{pmatrix} = \begin{pmatrix}\text{NPV of}\\ \text{original project}\end{pmatrix} + \begin{pmatrix}\text{call option on}\\ \text{future investments}\end{pmatrix}$$

The Valuation of Compound Financial Instruments as Options

The Black and Scholes option pricing model can be used to determine the fair value of compound financial instruments. For example, option pricing models have been used to value convertible bonds and other claims such as loan commitments and deposit insurance. In the case of the convertible bond, this compound financial instrument is viewed as a combination of straight debt and a conversion option on the common shares of the firm. The conversion option would be valued as a call option. This relationship is shown in the following equation:

$$\begin{pmatrix}\text{value of}\\ \text{convertible debt}\end{pmatrix} = \begin{pmatrix}\text{value of}\\ \text{straight bonds}\end{pmatrix} + \begin{pmatrix}\text{value of call option}\\ \text{on firm's common shares}\end{pmatrix}$$

Here, convertible debt has been separated into both liability and equity instruments. This type of separation of compound financial instruments into basic building blocks such as assets, liabilities, and equity creates a better balance sheet because balance sheet ratios such as debt-to-asset and debt-to-equity are improved (see Appendix 22B for an example of how the Black-Scholes model can be applied to the pricing of the conversion option of convertible debt).

OBJECTIVE 3

CURRENCY SWAPS

Currency swap

An exchange of debt obligations in different currencies.

The **currency swap** is another technique for controlling exchange rate risk. Whereas options and futures contracts generally have a fairly short duration, a currency swap provides the financial executive with the ability to hedge away exchange rate risk over longer periods. As a result, currency swaps have gained in popularity. Interest rate swaps are used to provide long-term exchange rate risk hedging. Actually, a currency swap can be quite simple, with two firms agreeing to pay each other's debt obligation.

How does this serve to eliminate exchange rate risk? A Canadian firm that earns much of its income from sales in the United States can enter into a currency swap with an American firm. If the value of the American dollar depreciates from $1.38 to $1.25 in Canadian dollars, then each dollar of sales in the United States will bring fewer dollars back to the parent company located in Canada. This type of depreciation can be offset by the effects of a currency swap because it costs the Canadian firm less dollars to fulfill the American firm's interest obligations. In other words, American dollars cost less to purchase, and the interest payments owed are in American dollars. As a result, the currency swap allows the firm to engage in a long-term exchange rate risk hedging, since the debt obligation covers a relatively long period.

Needless to say, there are many variations of the currency swap. One of the more popular is the interest rate swap, whereby the principal is not included in the swap. In this case, only interest payment obligations in different currencies are swapped. The key to controlling risk is to get an accurate estimate of the net exposure level to which the firm is exposed. Then the firm must decide whether it is prudent to subject itself to the risk associated with possible exchange rate fluctuations.

SUMMARY

A futures contract is a contract to buy or sell a stated commodity (such as soybeans or corn) or financial claim (such as Government of Canada bonds) at a specified price at some future specified time. This is a contract that requires its holder to buy or sell the asset regardless of what happens to its value during the interim. The importance of a futures contract is that it can be used by financial managers to lock in the price of a commodity or an interest rate and thereby eliminate one source of risk. A futures contract is a specialized form of a forward contract distinguished by (1) an organized exchange, (2) a standardized contract with limited price changes and margin requirements, (3) a clearinghouse in each futures market, and (4) daily resettlement of contracts.

A call option gives its owner the right to purchase a given number of shares at a specified price over a given time period. Thus, if the price of the underlying common shares goes up, a call buyer makes money. A put, conversely, gives its owner the right to sell a given number of common shares at a specified price over a given time period. Thus, a put buyer is betting that the price of the underlying common shares will drop. Since options provide the right to buy or sell shares, they do not represent an ownership position in the underlying corporation, as do common shares.

A currency swap is an exchange of debt obligations in different currencies. Exchange rate variations are offset by the effects of the swap. One major advantage of a currency swap is that it allows for the hedging of the exchange rate risk over a long period of time.

STUDY QUESTIONS

22-1. What is the difference between a commodity future and a financial future? Give an example of a commodity future and a financial future.

22-2. Describe a situation in which a financial manager might use a commodity future.

- **a.** Assume that during the period following the transaction the price of that commodity went up. Describe what happened.
- **b.** Now assume that the price of that commodity went down. Now what happened?

22-3. Describe a situation in which a financial manager might use an interest rate future. Assume that during the period following the transaction the interest rates went up. Describe what happened. Now assume that interest rates went down following the transaction. Now what happened?

22-4. Define a call option.

22-5. Define a put option.

22-6. Compare the two strategies for buying a call and for writing a put. What are the differences between the two?

22-7. Draw a profit or loss graph (similar to Figure 22-2) for the buyer of a call contract with an exercise price of $65 for which a $9 premium is paid. Identify the breakeven point, maximum profits, and maximum losses. Now draw the profit or loss graph assuming an exercise price of $70 and a $6 premium.

22-8. Draw a profit or loss graph (similar to Figure 22-3) for the writer of a call contract with an exercise price of $65 for which a $9 premium is paid. Identify the breakeven point, maximum profits, and maximum losses. Now draw the profit or loss graph assuming an exercise price of $70 and a $6 premium.

22-9. Draw a profit or loss graph (similar to Figure 22-4) for the buyer of a put contract with an exercise price of $45 for which a $5 premium is paid. Identify the breakeven point, maximum profits, and maximum losses.

22-10. Draw a profit or loss graph (similar to Figure 22-5) for the writer of a put contract with an exercise price of $45 for which a $5 premium is paid. Identify the breakeven point, maximum profits, and maximum losses.

22-11. What is an option on a futures contract? Give an example of one.

22-12. What innovative developments were brought on by exchange-listed trading that led to the dramatic growth in the trading of options?

22-13. What is a currency swap and why has it gained so in popularity?

SELF-TEST PROBLEM

ST-1. As an investor, you have bought one Noranda May $28 call option contract for $1.00. Show the payoff matrix for this call option under the following price scenarios on the date of expiration: $25, $28, $29, $31, $34 (Note that an option contract is sold as a 100-share contract). Assume no transactions costs.

STUDY PROBLEMS (SET A)

For all the problems in this study set, note that an option contract is sold as a 100-share contract. Assume no transactions costs.

22-1A. (*Payoff for Buying a Call*) As an investor, you have bought one Royal Bank April $32 call option contract for $1.50. Show the payoff matrix for this call option under the following price scenarios on the date of expiration: $31, $32, $33.50, $37, $39.

22-2A. (*Payoff for Buying a Put*) As an investor, you have bought one Alcan March $42.50 put option contract for $0.80. Show the payoff matrix for this put option under the following price scenarios on the date of expiration: $35, $38, $41.70, $44, $47.

22-3A. (*Payoff for Writing a Call*) As a market maker, you have written one Royal Bank April $32.50 call option contract for $1.50. Show the payoff matrix for this call option under the following price scenarios on the date of expiration: $31, $32, $34, $37, $39.

22-4A. (*Payoff for Buying Shares and a Put*) As an investor, you have set up the following position: buy 100 shares of Alcan for $32 and buy one Alcan March $42.50 put option contract for $0.80. Show the payoff matrix if you liquidate your position under the following price scenarios on the date of expiration: $35, $38, $41.70, $44, $47.

22-5A. (*Payoff for Buying a Call and a Put*) As an investor, you have set up the following position: buy one Bombardier July $22 call option contract for $1.35 and buy one Bombardier July $22 put option contract for $1.30. Show the payoff matrix if you liquidate your position under the following price scenarios on the date of expiration: $17, $18.50, $19.35, $24.65, $27.

INTEGRATIVE PROBLEM

For your job as the business reporter for a local newspaper you are given the task of putting together a series of articles on the derivatives markets for your readers. Much recent local press coverage has been given to the dangers and the losses that some firms have experienced recently in those markets. Your editor would like you to address several specific questions in addition to demonstrating the use of futures contracts and options and applying them to several problems.

Please prepare your response to the following memorandum from your editor:

TO: Business Reporter

FROM: Perry White, Daily Planet

RE: Upcoming series on the Derivative Securities Market

In your upcoming series on the time value of money, I would like to make sure you cover several specific points. In addition, before you begin this assignment, I want to make sure we are all reading from the same script, as accuracy has always been the cornerstone of the *Daily Planet.* In this regard, I'd like a response to the following questions before we proceed:

a. What opportunities do the derivatives markets (i.e., the futures and options markets) provide to the financial markets?

b. When might a firm become interested in purchasing interest rate futures? Foreign exchange futures? Stock index futures?

c. What can a firm do to reduce exchange risk?

d. How would Government of Canada bond futures and options on Canadian Government bond futures differ?

e. What is an option on a futures contract? Give an example of one and explain why it exists?

f. Draw a profit or loss graph (similar to Figure 22-2) for the purchase of a call contract with an exercise price of $25 for which a $6 premium is paid. Identify the breakeven point, maximum profits, and maximum losses.

g. Draw a profit or loss graph (similar to Figure 22-3) for the writing of a call contract that has an exercise price of $25 for which a $6 premium is paid. Identify the breakeven point, maximum profits, and maximum losses.

h. Draw a profit or loss graph (similar to Figure 22-4) for the purchase of a put contract with an exercise price of $30 for which a $5 premium is paid. Identify the breakeven point, maximum profits, and maximum losses.

i. Draw a profit or loss graph (similar to Figure 22-5) for the writing of a put contract that has an exercise price of $30 for which a $5 premium is paid. Identify the breakeven point, maximum profits, and maximum losses.

j. What is a currency swap and who might use one?

STUDY PROBLEMS (SET B)

For all the problems in this study set, note that an option contract is sold as a 100-share contract. Assume no transactions costs.

22-1B. (*Payoff for Buying a Call*) As an investor, you have bought one Royal Bank April $33 call option contract for $1.00. Show the payoff matrix for this call option under the following price scenarios on the date of expiration: $32, $33, $34, $37, $39.

22-2B. (*Payoff for Buying a Put*) As an investor, you have bought one Alcan March $41.00 put option contract for $0.50. Show the payoff matrix for this put option under the following price scenarios on the date of expiration: $35, $38, $40.50, $43, $46.

22-3B. (*Payoff for Writing a Call*) As a market maker, you have written one Royal Bank April $31.50 call option contract for $1.00. Show the payoff matrix for this call option under the following price scenarios on the date of expiration: $28, $30, $32.50, $34, $36.

22-4B. (*Payoff for Buying Shares and a Put*) As an investor, you have set up the following position: buy 100 shares of Alcan for $34 and buy one Alcan March $42.50 put option contract for $0.90. Show the payoff matrix if you liquidate your position under the following price scenarios on the date of expiration: $35, $38, $41.60, $44, $47.

22-5B. (*Payoff for Buying a Call and a Put*) As an investor, you have set up the following position: buy one Bombardier July $22 call option contract for $1.25 and buy one Bombardier July $22 put option contract for $1.20. Show the payoff matrix if you liquidate your position under the following price scenarios on the date of expiration: $17, $18.50, $19.55, $24.45, $27.

SELF-TEST SOLUTIONS

SS-1.

	Payoff Matrix				
	$25	$28	$29	$31	$34
Cost of call option	–$100	–$100	–$ 100	–$ 100	–$ 100
Cost of purchasing shares	0	0	–$2,800	–$2,800	–$2,800
Proceeds from sale of shares	0	0	$2,900	$3,100	$3,400
Gain (Loss) from transaction	–$100	–$100	0	$ 200	$ 500

Appendix 22A:

CONVERTIBLE SECURITIES AND WARRANTS

In earlier chapters, we concerned ourselves with methods of raising long-term funds through the use of common shares, preferred shares, and short- and long-term debt. In this appendix, we will examine how convertibles and warrants can be used to increase the attractiveness of these securities. Whereas both the convertible security and the warrant are used to raise funds for a firm, futures and options are not used to raise capital, but are used for either hedging or speculative purposes.

We have grouped convertibles and warrants together in our discussion because both can be exchanged at the owner's discretion for a specified number of common shares. In fact, both the conversion feature of the convertible security, as well as the features of the warrant display characteristics that are similar to those of a long-term call option. For each of these financing alternatives, we look at its specific characteristics and purpose; then we focus on any special considerations that should be examined before the convertible security or warrant is issued.

CONVERTIBLE SECURITIES

Convertible security

Preferred shares or debentures that can be exchanged for a specified number of common shares at the will of the owner.

A **convertible security** is either a preferred share or a debt issue which can be exchanged for a specified number of common shares at the will of the owner. It provides the stable income associated with preferred shares and bonds in addition to the possibility of capital gains associated with common shares. This combination of features has led convertibles to be called *hybrid* securities.

When the convertible is initially issued, the firm receives the proceeds from the sale, less flotation costs. This is the only time the firm receives any proceeds from issuing convertibles. The firm then treats this convertible as if it were normal preferred shares or debentures, paying dividends or interest regularly. If the security owner wishes to exchange the convertible for common shares, he or she may do so at any time according to the terms specified at the time of issue.[3] The desire to convert generally follows an increase in the price of the common shares. Once the convertible owner trades the convertibles in for common shares, the owner can never trade the shares back for convertibles. From then on the owner is treated as any other common shareholder and receives only cash dividends.

Characteristics and Features of Convertibles

Conversion ratio

The number of common shares for which a convertible security can be exchanged.

Conversion price or conversion parity price

The price for which the investor in effect purchases the company's common shares when purchasing a convertible security. Mathematically it is the market price of the convertible security divided by the conversion ratio.

Conversion Ratio

The number of common shares for which the convertible security can be exchanged is set out when the convertible is initially issued. On some convertible issues this **conversion ratio** is stated directly. For example, the convertible may state that it is exchangeable for 15 common shares. Some convertibles give only a **conversion price**, stating, for example, that the security is convertible at $39 per share.[4] This tells us that for every $39 of par value of the convertible security one common share will be received.

[3]The convertibility feature is essentially a form of a call option on the firm's shares. The option has value.

[4]Note that the conversion price is similar to the exercise price of an option, and as the share price increases, the convertibility feature increases in value.

$$\text{conversion ratio} = \frac{\text{par value of convertible security}}{\text{conversion price}} \tag{22A-1}$$

In 1988 Cambridge Shopping Centres, a Canadian real estate developer, issued $110 million of convertible debentures that mature in 2003. These convertibles have a $1,000 par value, a 8½ percent coupon interest rate, and a conversion ratio of 36.9. Thus, the conversion price—the price at which the conversion to shares occurs—is $1,000/36.9 = $27.10. The security owner has the choice of holding the 8½ percent convertible debenture or trading it in for 36.9 common shares of Cambridge Shopping Centres.

Conversion Value

The **conversion value** of a convertible security is the total market value of the common shares for which the convertible security can be exchanged. This can be calculated as follows:

Conversion value

The total market value of the common shares for which a convertible security can be exchanged.

$$\begin{matrix}\text{conversion}\\ \text{value}\end{matrix} = \begin{pmatrix}\text{conversion}\\ \text{ratio}\end{pmatrix} \times \begin{pmatrix}\text{market value per}\\ \text{common share}\end{pmatrix} \tag{22A-2}$$

As an example, if Cambridge Shopping Centres' common shares were selling for $28 per share, the conversion value for the Cambridge Shopping Centres' convertible would be (36.9)($28.00) = $1,033.20. Under these conditions, the market value of the common shares for which the convertible could be exchanged would be $1,033.20. Thus, regardless of what this convertible debenture was selling for, it could be converted into $1,033.20 worth of common shares.

Security Value

The **security value** (or bond value, as it is sometimes called) of a convertible security is the price the convertible security would sell for in the absence of its conversion feature. This is calculated by determining the required rate of return on a straight (nonconvertible) issue of the same quality and then determining the present value of the interest and principal payments at this rate of return. For example, if Canadian Bond Rating Service rates Cambridge Shopping Centres' convertible with a B++ rating, an investor would expect a bond yield of 11.28 percent which is similar to yields that other B++ issues are expected to earn at this time. In determining the security value of the Cambridge Shopping Centres' debenture, we answer the following question: What is the security value required to obtain a yield of 11.28 percent, if the debenture pays semiannual coupons at a annual interest rate of 8.5 percent with 12 years left to maturity. Since this bond pays 8½ percent semiannually, it gives the investor $42.50 every six months for the next 12 years for a total of 24 payments of $42.50, and at the end of 12 years (after 24 six-month periods) it pays the investor its par value of $1,000. As shown in Equation (4-7), which examined interest with nonannual periods, we use a discount rate of 11.28/2, or 5.64 percent. Thus, the value of this bond with semiannual payments is determined as follows:

Security value

The price the convertible security would sell for in the absence of its conversion feature.

$$SV = \sum_{t=1}^{2n} \frac{\frac{IP}{2}}{\left(1 + \frac{k_b}{2}\right)^t} + \frac{P}{\left(\frac{k_b}{2}\right)^{2n}} \tag{22A-3}$$

where
SV = the security value
I = the coupon interest rate
P = the par value
n = the number of years to maturity
k_b = the required rate of return on a straight issue of the same quality

Then the value of the Cambridge Shopping Centres security as a straight debenture yields the following:

$$SV = \sum_{t=1}^{24} \frac{\$42.50}{(1 + 0.0564)^t} + \frac{\$1,000}{(1 + 0.0564)^{24}} = \$819.59$$

Thus, regardless of what happens to the value of Cambridge Shopping Centres common shares, the lowest value the convertible can drop to, assuming there is no change in interest rates, is its value as a straight bond, which is $819.59.

Conversion Period

On some issues the time period during which the convertible can be exchanged for common shares is limited. Many times conversion is not allowed until a specified number of years have passed, or is limited by a terminal conversion date. Still other convertibles may be exchanged at any time during their life. In either case the **conversion period** is specified when the convertible is originally issued.

Conversion period

The time period during which the convertible can be exchanged for common shares.

Conversion Parity Price

The *conversion parity price* is the price at which the investor, in effect, buys the company's shares:

$$\text{conversion parity price} = \frac{\text{market price of convertible security}}{\text{conversion ratio}} \tag{22A-4}$$

If the Cambridge Shopping Centres bond is selling for $1,120.00 and the investor can exchange the bond for 36.9 shares of Cambridge Shopping Centres common shares, then the conversion parity price is ($1,120.00/36.9) = $30.35. Thus, if the investor purchases this convertible and converts it to common shares, he or she is in effect buying those shares for $30.35 per share.

Conversion Premium

The **conversion premium** is the difference between the convertible's market price and the higher of its security value and its conversion value. It can be expressed as an absolute dollar value, in which case it is defined as

Conversion premium

The difference between the convertible's market price and the higher of its security value and its conversion value.

$$\begin{matrix}\text{conversion} \\ \text{premium}\end{matrix} = \begin{pmatrix}\text{market price of} \\ \text{convertible}\end{pmatrix} - \begin{pmatrix}\text{higher of the security value} \\ \text{and the conversion value}\end{pmatrix} \tag{22A-5}$$

Alternatively, the conversion premium can be expressed as a percentage, in which case it is defined as

$$\begin{matrix}\text{percentage} \\ \text{conversion} \\ \text{premium}\end{matrix} = \frac{\begin{pmatrix}\text{market price} \\ \text{of bond}\end{pmatrix} - \begin{pmatrix}\text{higher of the security value} \\ \text{and the conversion value}\end{pmatrix}}{(\text{higher of the security value and the conversion price})} \tag{22A-6}$$

The market price of the Cambridge Shopping Centres convertible was $1,120.00 while its security value was $819.59 and its conversion value was $1,033.20. Thus, its conversion premium was

$$\frac{\$1,120.00 - \$1,033.20}{\$1,033.20} = 8.4 \text{ percent}$$

In effect, an investor was willing to pay an 8.4 percent premium over the higher of its security and conversion values in order to have the possibility of capital gains from share price advances coupled with the security of the fixed interest payments associated with a debenture.

In describing convertibles, we have introduced a number of terms. To eliminate confusion, Table 22A-1 summarizes them.

PERSPECTIVE IN FINANCE

The agency problem deals with conflicts of interest between shareholders, bondholders, and managers. Convertible bonds, which allow bondholders to benefit from the price appreciation of equity, help align the interests of shareholders and bondholders. In this way convertibles help and reduce the agency problem.

Why Issue Convertibles?

The major reason for choosing to issue convertibles rather than straight debt, preferred shares, or common shares is the fact that interest rates on convertibles are indifferent to the issuing firm's risk level.

Although higher risk and uncertainty bring on higher interest costs in straight debt, this is not necessarily the case with convertibles. If we think about a convertible as a package of straight debt and a convertible feature which gives a holder the option to purchase common shares at a set price, an increase in risk and uncertainty will raise the cost of the straight-debt portion of the convertible. However, the convertibility feature benefits from this increase in risk and uncertainty and the increase in share price volatility that follows. In effect, the conversion feature will be of value at some point before the expiration date. As a result, any additional increase in risk and uncertainty will increase the value of the conversion feature of the convertible before the expiration date. Thus, the negative effect of an increase in risk and uncertainty on the straight-debt portion of a convertible is partially offset by the positive effect on the conversion feature. The result of all this is that the interest rate associated with convertible debt is, to an extent, indifferent to the risk level of the issuing firm. For example, the coupon rates for medium- and high-risk companies issuing convertibles and straight debt might be as follows:

	Company Risk	
	Medium	High
Convertible debt	8%	8.25%
Straight debt	11	13.00

In the case of a company with high risk, the conversion feature partially offsets the risk displayed by straight debt. Thus, convertible debt may allow companies with a

Table 22A-1 Summary of Convertible Terminology

Conversion ratio: the number of shares for which the convertible security can be exchanged.

$$\text{Conversion ratio} = \frac{\text{par value of convertible security}}{\text{conversion price}} \quad (22A\text{-}1)$$

Conversion value: the total market value of the common shares for which the convertible can be exchanged.

$$\text{Conversion value} = \text{conversion ratio} \times \text{market value per common share} \quad (22A\text{-}2)$$

Security value: the price the convertible security would sell for in the absence of its conversion feature.

Conversion period: the time period during which the convertible can be exchanged for common shares.

Conversion parity price: the price the investor is in effect paying for the common shares.

$$\text{Conversion parity price} = \frac{\text{market price of convertible security}}{\text{conversion ratio}} \quad (22A\text{-}4)$$

Conversion premium: the difference between the convertible's market price and the higher of its security value and its conversion value.

$$\text{Conversion premium} = \left(\begin{array}{c}\text{market price of}\\ \text{the convertible}\end{array}\right) - \left(\begin{array}{c}\text{higher of the security value}\\ \text{and the conversion value}\end{array}\right) \quad (22A\text{-}5)$$

high level of risk to raise funds at a relatively favourable rate. The implications of issuing convertible bonds to a high-risk company are twofold: if the common share price is greater than the conversion price, convertible bonds are converted to shares at bargain prices causing the company's ownership to be diluted, as well as the share value to be surrendered at a level lower than would have been achieved by the company's growth rate; if the common share price is lower than the conversion price, the company has used a cheaper source of financing since the coupon interest rate on convertible debt is lower than that of straight debt of the same quality.

Valuation of a Convertible

The valuation of a convertible depends primarily upon two factors: the value of the straight debenture or preferred shares and the value of the security if it were converted into common shares. Complicating the valuation is the fact that investors are in general willing to pay a premium for the conversion privilege, which allows them to hedge against the future. If the price of the common shares should rise, the investor would participate in capital gains; if it should decline, the convertible security will fall only to its value as a straight debenture or preferred shares.

In examining the Cambridge Shopping Centres convertible debenture we found that if it were selling as a straight debenture, its price would be $819.59. Thus, regardless of what happens to the company's common shares, the lowest the value of the convertible can drop to is $819.59. The conversion value, on the other hand is $1,033.20. Thus, this convertible is worth more as common shares than if it were straight debt. However, the real question is: Why are investors willing to pay a conversion premium of 8.4 percent over its security or conversion value for this Cambridge Shopping Centres debenture? Quite simply because investors are willing to pay for the chance for capital gains without the large risk of loss.

Figure 22A-1 illustrates the relationship between the value of the convertible and the price of its common shares. The bond value of the convertible serves as a floor for the value of the investment: When the conversion value reaches the convertible's security value (point A), the value of the convertible becomes dependent upon its conversion value. In effect the convertible security is valued as a bond when the price of the common shares is low and as a common share when the price of the common shares increases. Of course, if the firm is doing poorly and in financial distress, both the common share price and the security value will suffer. In the worst-case scenario in which a firm's total value falls to zero, the firm's common shares and debt would have no value. Although the minimum price of the convertible is determined by the higher of either the straight bond or preferred share price, or the conversion value, investors also pay a premium for the conversion option. Again, this premium results because convertible securities offer investors stable income from the debentures or the preferred shares, and thus less risk of price decline due to adverse share conditions, while retaining capital gains prospects from share price gains. In effect, downside share price variability is hedged away, and upside variability is not.

PERSPECTIVE IN FINANCE

As you look at the convertibles, warrants, options, and futures markets, keep in mind the fact that these markets are continuously evolving. For example, recently a variation of the convertible debenture called an exchangeable debenture has gained popularity. An exchangeable debenture is similar to a convertible debenture except that it can be exchanged at the option of the holder for the common shares of another company. For example, in 1992 Canadian Pacific Limited placed a $105 million issue of unsecured equity-exchangeable debentures which matured in 1995. These debentures were identical to convertibles except that they could be exchanged for common shares of United Dominion Industries. The point here is that in this chapter we are only describing several basic securities of which there are unlimited variations in the financial markets.

Figure 22A-1 Relationship Between the Market Price of the Common Shares and the Market Price of the Convertible Securities

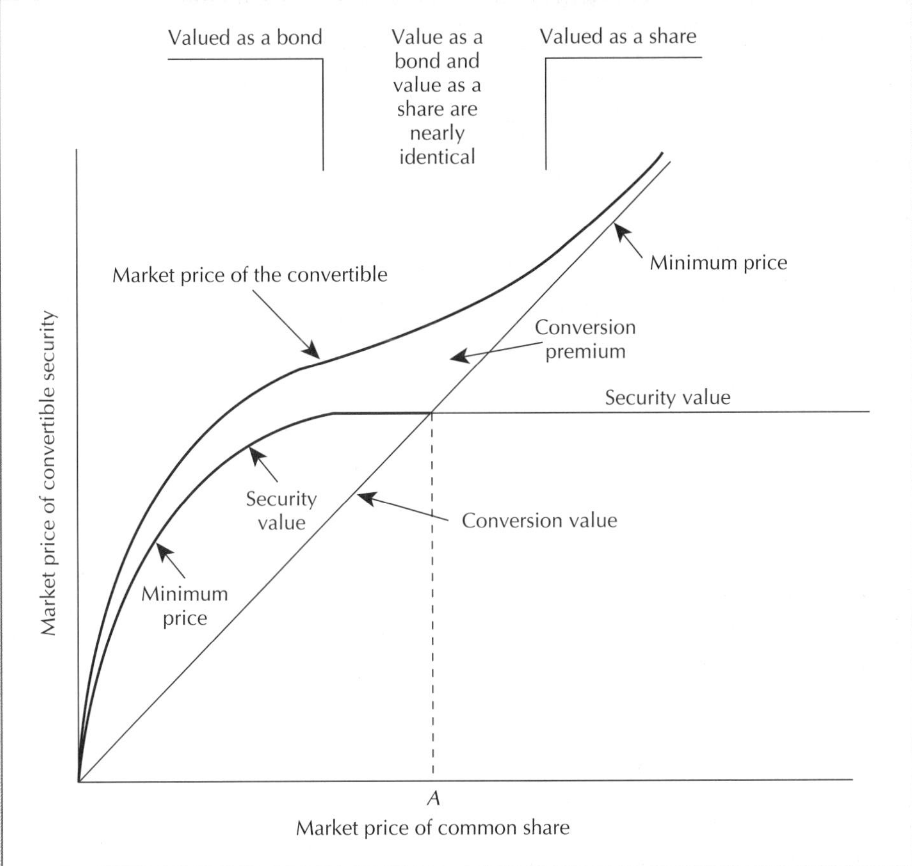

WARRANTS

A **warrant** provides the investor with an option to purchase a fixed number of common shares at a predetermined price during a specified time period. Warrants have been used in the past primarily by weaker firms as "sweetener" attachments to bonds or preferred shares to improve their marketability. In 1988, Hollinger Inc. issued warrants with adjustable rate convertible subordinated debentures. Each warrant entitles the holder to purchase one common share for $20 from October 1, 1993 to September 30, 1998. Investment dealers have also used warrants to enable companies to raise capital in the financial markets by combining common shares with warrants to form a "unit." Again, the purpose of a warrant is essentially the same as when they are issued in conjunction with debt or preferred shares; that is, to improve the reception in the market of the new offering or make a tender offer too attractive to turn down.

Warrant
An option to purchase a fixed number of common shares at a predetermined price during a specified time period.

Although warrants are similar to convertibles in that both provide investors with a chance to participate in capital gains, the mechanics of the two instruments differ greatly. From the standpoint of the issuing firm there are two major differences. First, when convertibles are exchanged for common shares, debt is eliminated and fixed finance charges are reduced; whereas, an exchange of warrants for common shares does not reduce the fixed finance charges. Second, when convertibles are exchanged there is no cash inflow into the firm—the exchange is one type

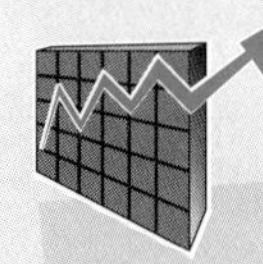

Trends in Corporate Financing

The following article describes some of the trends in corporate financing that have occurred in the early nineties.

There have been two big trends on the corporate finance scene in the early nineties. In 1991, investors wanted a little protection for the new issues they were buying, so issuers sold unit offerings, with each unit consisting of a common share plus half a share purchase warrant. A full warrant entitled the holder to purchase a common share at a premium to the issue price. In 1992, the fad was issues being done on an instalment receipt basis. On those deals, investors paid part of the purchase price when the deal closed and the rest later on. For most deals, there were two instalments; occasionally there were three. Instalments were popular on both secondary offerings (where a controlling shareholder was selling a stake) and on new offerings. British Petroleum, RTZ, LASMO PLC, Occidental Petroleum, British Gas and Interprovincial Pipe Line all used this structure for secondary issues, while Suncor used instalments for its new offering when it was effectively privatized.

New Twist on Convertible Debentures

Over 1992, raising equity using, "instalment receipts" has become the norm for large deals. That structure—where an investor pays part of the purchase price of the shares at the time the deal is done and the balance later—offers advantages to all parties.

Buyers like instalment receipts because they only put up 50% of the overall purchase price, but receive a full dividend—effectively doubling their first-year dividend return. Issuers like them because the structure makes possible a bigger deal than could otherwise be done.

Now, thanks to the creative talents of Dennis Dewan, a corporate finance professional at Midland Walwyn Capital Inc., the concept has been extended to convertible debentures. The breakthrough deal is the sale by Brascan Ltd. of $250 million worth of convertible debentures.

The instalment receipt feature is the key innovation of this issue. Brascan will receive the proceeds in two equal instalments—half when the deal closes in October and the rest in one year's time. Receiving the second instalment in a year means the overall amount Brascan will receive in today's terms has to be discounted to reflect the time value of money. Depending on the discount rate used, the gross proceeds to Brascan are probably worth about $19.50 in October terms.

Source: Adapted from: Barry Critchley, "Referendum a Block to New Stock Issues," *The Financial Post*, October 12, 1992, p. 19; and "New Twist on Convertible Debentures," *The Financial Post*, September 21, 1992, p. 26.

of security for another. But with warrants, since they are options to buy shares at a set price, a cash flow accompanies the exchange.[5]

Characteristics and Features of Warrants

Exercise price

The price at which a warrant allows its holder to purchase the firm's common shares.

Exercise Price

The **exercise price** represents the price at which the warrant allows its holder to purchase the firm's common shares. The investor trades a warrant plus the exercise price for common shares. Typically, when warrants are issued the exercise price is

[5]Warrants are essentially call options. The major differences are that warrants are issued by firms and so there is a limited supply and they have a much longer maturity (two years or more rather than nine months).

set above the current market price of the shares. Thus, if the share price does not rise above the exercise price, the warrant will never be converted. In addition there is also a step-up exercise price—that is, a warrant with which the exercise price changes over time.

Expiration Date

Although some warrants are issued with no expiration date, most warrants are set to expire after a number of years. In issuing warrants as opposed to convertibles, the firm gives up some control over when the warrants will be exercised. An issuing company can force conversion of a convertible debenture issue by calling the issue or using step-up conversion prices, whereas with warrants only the approach of the expiration date or the use of step-up exercise prices can encourage conversion.

Detachability

Most warrants are said to be detachable in that they can be sold separately from the security to which they were originally attached. Thus, if an investor purchases a primary issuance of a corporate bond with a warrant attached, he or she has the option of selling the bond alone, selling the warrant alone, or selling the combination intact. Nondetachable warrants cannot be sold separately from the security to which they were originally attached. Such a warrant can be separated from the senior security only by being exercised.

Exercise Ratio

The **exercise ratio** states the number of shares that can be obtained at the exercise price with one warrant. If the exercise ratio on a warrant were 1.5, one warrant would entitle its owner to purchase 1.5 shares of common shares at its exercise price.

Exercise ratio

The number of common shares that can be obtained at the exercise price with one warrant.

Reasons for Issuing Warrants

Sweetening Debt

Warrants attached to debt offerings provide a feature whereby investors can participate in capital gains while holding debt. The firm can thereby increase the demand for the issue, increase the proceeds, and lower the interest costs. The attachment of warrants to long-term debt provides a potential equity stake to the debt holder. As a result, a warrant increases the marketability and demand for the company's debt.

Additional Cash Inflow

If warrants are added to sweeten a debt offering, the firm will receive an eventual cash inflow when and if the warrants are exercised; a conversion feature would not provide this additional inflow.

Other Factors to Be Considered

Preservation of Debt

When warrants are exercised, the debt or security to which they were originally attached remains in existence. The process of exercising a warrant by an investor provides an additional cash inflow to the company, but this process does not eliminate the original security. In contrast, the process of exercising convertible debt causes no cash flow to occur, but the original security is eliminated. Thus, the decision between convertible and warrant financing becomes a tradeoff between the elimination of debt versus the receipt of additional cash inflows.

Dilution and Flexibility

Since present accounting standards provide that earnings per share be stated as if all the warrants outstanding had been exercised, warrants have the effect of reducing the firm's reported earnings per share. This potential dilution of earnings per share may reduce the firm's financing flexibility. Because of previously issued warrants, the issuance of additional common shares may be hindered or even prohibited. The market's reluctance to accept further equity offerings may be due not only to the potential earnings per share dilution effect of the outstanding warrants, but also to a feeling that only weaker firms have to resort to warrant financing to sweeten their senior security offerings.

Valuation of a Warrant

Since the warrant is an option to purchase a specified number of shares at a specified price for a given length of time, the market value of the warrant will be primarily a function of the common share price. To understand the valuation of warrants we must define two additional terms, the minimum price and the premium. Let us look at the Hollinger warrant that was issued in 1988 with an expiration date of September 30, 1998, an exercise ratio of 1.00, and an exercise price of \$20 through the expiration date. This means that any time until expiration on September 30, 1998, an investor with one warrant can purchase one common share of Hollinger at \$20 regardless of the market price of those shares. In 1995 these Hollinger warrants were selling at \$2.70 and the Hollinger shares were selling for \$12.75 per share.

Minimum Price The *minimum price* of a warrant is determined as follows:

$$\text{minimum price} = \left(\begin{matrix} \text{market price of} \\ \text{common shares} \end{matrix} - \begin{matrix} \text{exercise} \\ \text{price} \end{matrix} \right) \times \left(\begin{matrix} \text{exercise} \\ \text{ratio} \end{matrix} \right) \tag{22A-7}$$

In the Hollinger example the exercise price is greater than the price of the common shares (\$20 as opposed to \$12.75). In this case the minimum price of the warrant is considered to be zero, because things simply do not sell for negative prices [(\$12.75 − \$20) × 1.00 = –\$7.25]. For example, if the price of the Hollinger common shares increased to \$26 per share, the minimum price on the warrant would become (\$26 − \$20) × 1.00 = \$6. This would tell us that this warrant could not fall below a price of \$6.00, because if it did, investors could realize immediate trading profits by purchasing the warrants and converting them along with the \$20 exercise price into common shares until the price of the warrant was pushed up to the minimum price. This process of simultaneously buying and selling equivalent assets for different prices is called *arbitrage.*

Premium The *premium* is the amount above the minimum price for which the warrant sells:

$$\text{premium} = (\text{market price of warrant}) - (\text{minimum price of warrant}) \tag{22A-8}$$

In the case of the Hollinger warrant the premium is \$2.70 − \$0 = \$2.70. Investors are paying a premium of \$2.70 above the minimum price for the warrant. The potential downside of this investment to the investor is the warrant's market price which represents 21.1 percent of the common share's price. The potential return is large given that value of the warrant increases as the price of the common share increases.

The relationships among the warrant price, the minimum price, and the premium are illustrated graphically in Figure 22A-2. Point A represents the exercise price on the warrant. Once the price of the shares is above the exercise price, the warrant's minimum price takes on positive (or nonzero) values.

A hypothetical numerical example based upon Figure 22A-2 is provided in Table 22A-2. Looking at the graphic representation of the warrant and the example, we find that the warrant premium tends to drop off as the ratio of the share price to the exercise price climbs. Why? As the share price climbs, the warrant loses its

Figure 22A-2 Valuation of Warrants

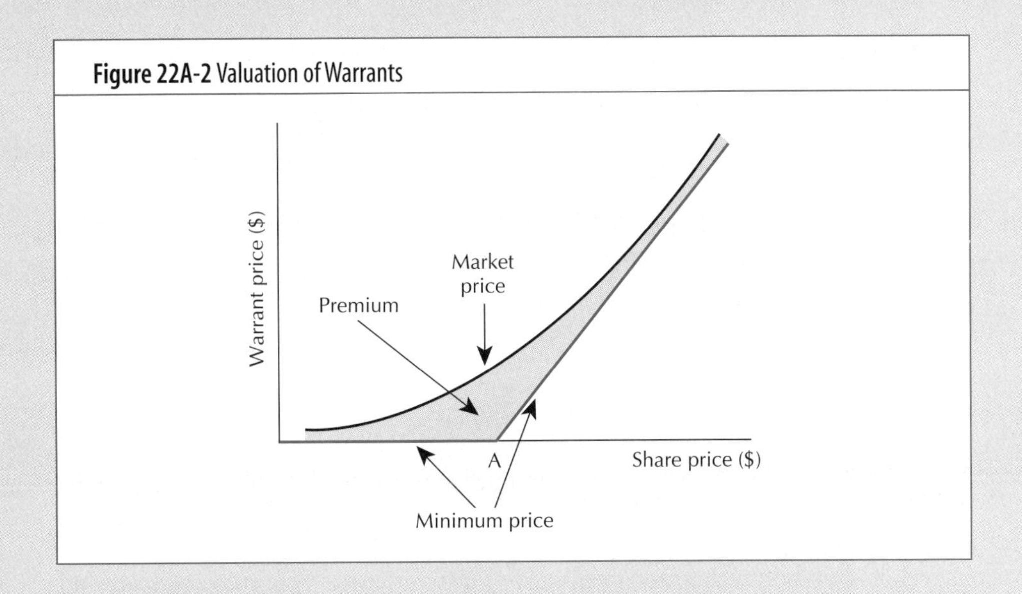

leverage ability. Looking at the numerical example, assume an investor purchased $1,000 worth of warrants and $1,000 worth of common shares when the common share price was $40 and the warrant price was $10; then the price of the shares went up 50 percent to $60 per share, which caused the price of the warrant to go up 150 percent to $25 per share. This leverage feature—the fact that the value of the warrant increases and declines by larger percentages than the value of the underlying shares and thus a small investment has large possible returns—encourages investors to purchase warrants. However, as the share price increases, the leverage ability of the warrant declines. For example, assume that an investor purchases $1,000 worth of common shares selling at $110 per share and $1,000 worth of warrants selling for $71 per share. If the share price went up to $130, the investment in shares would have returned 18 percent and the warrant price would have increased to $90 for a return of 27 percent. In the first example when the share price increased from $40 to $60 per share, the warrant resulted in profits three times greater than on the common shares. In the second example the warrant provided profits only about one and one-half times larger than the common shares. Thus, the warrant premium tends to drop off as the ratio of the share price to the exercise price climbs and the leverage ability of the warrant declines.

Table 22A-2 Hypothetical Warrant Example

Share Price, *SP*	Exercise Price, *EP*	Exercise Ratio, *ER*	Minimum Price: $(SP - EP) \times ER = MP$	Hypothetical Warrant Price, *WP*	Premium, *WP – MP*	Share Price/ Exercise Price Ratio, *SP/EP*
$ 30	$40	1	$ 0	$ 5	$ 5	75%
40	40	1	0	10	10	100
50	40	1	10	16	6	125
60	40	1	20	25	5	150
70	40	1	30	34	4	175
80	40	1	40	43	3	200
90	40	1	50	52	2	225
100	40	1	60	62	2	250
110	40	1	70	71	1	275
120	40	1	80	81	1	300
130	40	1	90	90	0	325

Although the share price/exercise price ratio is one of the most important factors in determining the size of the premium, several other factors also affect it. One such factor is the time left until the warrant expiration date. As the warrant's expiration date approaches, the size of the premium begins to shrink, approaching zero. A second factor is investors' expectations concerning the capital gains potential of the shares. If investors expect that the price of the underlying common share will increase, the premium paid for the warrant will also increase because of the relation between the price of the shares and the price of the warrant. Finally, the degree of price volatility on the underlying common shares affects the size of the warrant premium. The more volatile the common share price, the higher the warrant premium. As price volatility increases, so does the probability of and potential size of profits.

SUMMARY

Convertible securities are preferred shares or debentures that can be exchanged for a specified number of common shares. Corporations issue convertible securities to sweeten debt for marketing purposes, to delay equity financing, and to provide a source of temporarily inexpensive funds during expansions. Conversely, a corporation that issues convertible securities incurs the risk of an overhanging issue which reduces the flexibility of the firm to change its financial structure. The valuation of convertible securities is a function of both their value as straight bonds and/or preferred shares, and their value if converted into common shares. Since the convertible provides the security of debt with the capital gains potential of common shares, it generally sells for a premium above the higher of its bond or conversion value.

A warrant is an option to purchase a fixed number of common shares at a predetermined price during a specified period. Although warrants are generally issued in association with debt, most warrants are detachable in that they can be bought and sold separately from the debt to which they were originally attached. They are generally issued as a sweetener to debt in order to make it more marketable and lower the interest costs. In addition, unlike convertibles, warrants provide for an additional cash inflow when they are exercised. Conversely, the exercise of convertibles results in the elimination of debt, while the exercise of warrants does not. Thus, there is a tradeoff—additional cash inflow versus elimination of debt—involved in the warrants versus convertibles decision. Since a warrant is an option to purchase a specified number of shares during a given period, its market value is primarily a function of the price of the common shares. Warrants generally sell above their minimum price, the size of the premium is determined by the degree of leverage they provide, the time left to expiration, investors' expectations as to the future movement of the share price, and the share's price volatility.

STUDY QUESTIONS

22A-1. Define the following terms:

- **a.** *Conversion ratio*
- **b.** *Conversion value*
- **c.** *Conversion parity price*
- **d.** *Conversion premium*

22A-2. What are some reasons commonly given for issuing convertible securities?

22A-3. Why does a convertible debenture sell at a premium above its value as a bond or common shares?

22A-4. Convertible debentures are said to provide the capital gains potential of common shares and the security of bonds. Explain this statement both verbally and graphically. What happens to the graph when interest rates rise? When they fall?

22A-5. Convertible debentures generally carry lower coupon interest rates than do nonconvertible bonds. If this is so, does it mean that the cost of capital on convertible debentures is lower than on nonconvertible? Why or why not?

22A-6. Since convertible securities allow for the conversion price to be set above the current common share price, is it true that the firm is actually issuing its common shares at a price above the current market price?

22A-7. In light of our discussion of the common shareholders' preemptive right in Chapter 6, explain why convertibles are often sold on a rights basis.

22A-8. Although only the holder of a convertible has the right to convert the security, firms are often able to force conversion. Comment on this statement.

22A-9. Explain the difference between a convertible security and a warrant.

22A-10. How do firms force the exercising of warrants?

22A-11. Explain the valuation of warrants both verbally and graphically.

22A-12. What factors affect the size of the warrant premium? How?

22A-13. What is the difference between a commodity future and financial future? Give an example of a commodity future and a financial future.

22A-14. Describe a situation in which a financial manager might use a commodity future. Assume that during the period following the transaction the price of that commodity went up. Describe what happened. Now assume that the price of that commodity went down. Now what happened?

22A-15. Describe a situation in which a financial manager might use an interest rate future. Assume that during the period following the transaction the interest rates went up. Describe what happened. Now assume that interest rates went down following the transaction. Now what happened?

SELF-TEST PROBLEMS

STA-1. (*Convertible Terminology*) Winky's Cow Paste Inc. issued $10 million of $1,000 par value, 10 percent semiannual convertible debentures that mature in 20 years. The conversion price on these convertibles is $16.75 per share. The common shares were selling for $14.75 per share on a given date shortly after these convertibles were issued. These convertibles have a B+ rating, and straight B+ debentures were yielding 14 percent on that date. The market price of the convertible was $970 on that date. Determine the following:

- **a.** Conversion ratio
- **b.** Conversion value
- **c.** Security value
- **d.** Conversion parity price
- **e.** Conversion premium in absolute dollars
- **f.** Conversion premium in percentage

STA-2. (*Warrant Terminology*) Petro-Tech Inc. currently has some warrants outstanding that allow the holder to purchase, with one warrant, one common share at $18.275. If the common shares were selling at $25 per share and the warrants were selling for $9.50, what would be the

- **a.** Minimum price?
- **b.** Warrant premium?

STUDY PROBLEMS (SET A)

22A-1A. (*Convertible Terminology*) The Andy Fields Corporation issued some $1,000 par value, 6 percent convertible debentures that mature in 20 years. The conversion price on these convertibles is $40 per share. The price of the common shares is now $27.25 per share. These convertibles have a B+ rating, and straight B+ debentures are now yielding 9 percent. The market price of the convertible is now $840.25. Determine the following (assume bond interest payments are made annually):

- **a.** Conversion ratio
- **b.** Conversion value
- **c.** Security value
- **d.** Conversion parity price
- **e.** Conversion premium in absolute dollars
- **f.** Conversion premium in percentage

22A-2A. (*Convertible Terminology*) The L. Padis, Jr., Corporation has an issue of 5 percent convertible preferred shares outstanding. The conversion price on these securities is $27 per share until 9/30/99. The price of the common shares is now $13.25 per share. The preferred shares are selling for $17.75. The par value of the preferred shares is $25 per share. Similar quality preferred shares without the conversion feature are currently yielding 8 percent. Determine the following:

- **a.** Conversion ratio
- **b.** Conversion value
- **c.** Conversion premium (in both absolute dollars and percentages)

22A-3A. (*Warrant Terminology*) The T. Kitchel Corporation has a warrant that allows the purchase of one common share at $30. The warrant is currently selling at $4 and the common shares are priced at $25 per share. Determine the minimum price and the premium of the warrant.

22A-4A. (*Warrant Terminology*) Cobra Airlines has some warrants outstanding that allow the purchase of common shares at the rate of one warrant for each common share at $11.71.

- **a.** Given that the warrants were selling for $3 each and the common shares were selling for $10 per share, determine the minimum price and warrant premium as of that date.
- **b.** Given that the warrants were selling for $9.75 each, and the common shares were selling for $16.375 per share, determine the minimum price and warrant premium as of that date.

22A-5A. (*Warrant Terminology*) International Corporation has some warrants outstanding that allow the purchase of common shares at the price of $22.94 per share. These warrants are somewhat unusual in that one warrant allows for the purchase of 3.1827 common shares at the exercise price of $22.94 per share. Given that the warrants were selling for $6.25 each, and the common shares were selling for $7.25 per share, determine the minimum price and the warrant premium as of that date.

22A-6A. (*Warrants and Their Leverage Effect*) A month ago you purchased 100 Bolster Corporation warrants at $3 each. When you made your purchase, the market price of Bolster's common shares was $40 per share. The exercise price on the warrants is $40 per share while the exercise ratio is 1.0. Today, the market price of Bolster's common shares has jumped up to $45 per share, while the market price of Bolster's warrants has climbed to $7.50 each. Calculate the total dollar gain that you would have received if you had invested the same dollar amount in common shares versus warrants. What is this in terms of return on investment?

STUDY PROBLEMS (SET B)

22A-1B. (*Convertible Terminology*) The P. Mauney Corporation issued some $1,000 par value, 7 percent convertible debentures that mature in 20 years. The conversion price on these convertibles is $45 per share. The price of the common shares is now $26 per share. These convertibles have a B+ rating, and straight B+ debentures are now yielding 9 percent. The market price of the convertible is now $840.25. Determine the following (assume bond interest payments are made annually):

a. Conversion ratio
b. Conversion value
c. Security value
d. Conversion parity price
e. Conversion premium in absolute dollars
f. Conversion premium in percentage

22A-2B. (*Convertible Terminology*) The Ecotosleptics Corporation has an issue of 6 percent convertible preferred shares outstanding. The conversion price on these securities is $28 per share until 9/30/99. The price of the common shares is now $14 per share. The preferred shares are selling for $20.00. The par value of the preferred shares is $25 per share. Similar quality preferred shares without the conversion feature are currently yielding 8 percent. Determine the following:

a. Conversion ratio
b. Conversion value
c. Conversion premium (in both absolute dollars and percentages)

22A-3B. *(Warrant Terminology)* The Megacorndoodles Corporation has a warrant that allows the purchase of one common share at $32 per share. The warrant is currently selling at $5 and the common shares are priced at $24 per share. Determine the minimum price and the premium of the warrant.

22A-4B. *(Warrant Terminology)* Taco Fever has some warrants outstanding that allow the purchase of common shares at the rate of one warrant for each common share at $11.75.

a. Given that the warrants were selling for $4 each and the common shares were selling for $9 per share, determine the minimum price and warrant premium as of that date.
b. Given that the warrants were selling for $7 each, and the common shares were selling for $15.375 per share, determine the minimum price and warrant premium as of that date.

22A-5B. (*Warrant Terminology*) Fla'vo'phone Corporation has some warrants outstanding that allow the purchase of common shares at the price of $22.94 per share. These warrants are somewhat unusual in that one warrant allows for the purchase of 4.257 common shares at the exercise price of $22.94 per share. Given that the warrants were selling for $6.75 each, and the common shares were selling for $8 per share, determine the minimum price and the warrant premium as of that date.

22A-6B. (*Warrants and Their Leverage Effect*) A month ago you purchased 100 Annie Kay's Corporation warrants at $2.75 each. When you made your purchase, the market price of Annie Kay's common shares was $35 per share. The exercise price on the warrants is $35 per share while the exercise ratio is 1.0. Today, the market price of Annie Kay's common shares has jumped up to $40 per share, while the market price of Annie Kay's warrants has climbed to $6.75 each. Calculate the total dollar gain that you would have received if you had invested the same dollar amount in common shares versus warrants. What is this in terms of return on investment?

SELF-TEST SOLUTIONS

SS-1. a.

$$\text{conversion ratio} = \frac{\text{par value of convertible security}}{\text{conversion price}}$$

$$= \$1{,}000/\$16.75$$

$$= 59.70 \text{ shares}$$

b.

$$\text{conversion value} = (\text{conversion ratio}) \times \begin{pmatrix} \text{market value} \\ \text{per common share} \end{pmatrix}$$
$$= 59.70 \text{ shares} \times \$14.75/\text{share}$$
$$= \$880.58$$

c.

$$\text{security value} = \sum_{t=1}^{40} \frac{\$50}{(1 + .07)^t} + \frac{\$1{,}000}{(1 + .07)^{40}}$$
$$= \$50(13.3317) + \$1{,}000(.0668)$$
$$= \$666.59 + \$66.80$$
$$= \$733.39$$

[Note: Since this debenture pays interest semiannually, $t = 20$ years $\times\ 2 = 40$ and $k_b = 14\%/2 = 7\%$ in the calculations.]

d.

$$\text{conversion parity price} = \frac{\text{market price of convertible security}}{\text{conversion ratio}}$$
$$= \$970.00/59.70$$
$$= \$16.25$$

e.

$$\begin{matrix} \text{conversion premium} \\ \text{in absolute dollars} \end{matrix} = \begin{pmatrix} \text{market price of} \\ \text{the convertible} \end{pmatrix} - \begin{pmatrix} \text{higher of the bond value} \\ \text{and conversion value} \end{pmatrix}$$
$$= \$970.00 - \$880.58$$
$$= \$89.42$$

f.

$$\begin{matrix} \text{conversion premium} \\ \text{in absolute dollars} \end{matrix} = \frac{\begin{pmatrix} \text{market price of} \\ \text{the convertible} \end{pmatrix} - \begin{pmatrix} \text{higher of the bond value} \\ \text{and conversion value} \end{pmatrix}}{\begin{pmatrix} \text{higher of the bond value} \\ \text{and conversion value} \end{pmatrix}}$$
$$= \frac{\$970.00 - \$880.58}{\$880.58}$$
$$= 10.15\%$$

SS-2. a.

$$\text{minimum price} = \begin{pmatrix} \text{market price of} & - & \text{exercise} \\ \text{common shares} & & \text{price} \end{pmatrix} \times \begin{pmatrix} \text{exercise} \\ \text{ratio} \end{pmatrix}$$
$$= (\$25.00 - \$18.275) \times (1.0)$$
$$= \$6.73$$

b.

$$\text{warrant premium} = \begin{pmatrix} \text{market price} \\ \text{of warrant} \end{pmatrix} - \begin{pmatrix} \text{minimum price} \\ \text{of warrant} \end{pmatrix}$$
$$= \$9.50 - \$6.73$$
$$= \$2.77$$

APPENDIX 22B:

AN APPLICATION OF THE BLACK-SCHOLES OPTION PRICING MODEL

An important development in financial management is the change from classifying financial instruments on the balance sheet in terms of the instrument's legal form to one based on the economic substance of the financial instrument. The implication for compound financial instruments is that these instruments will be separated into their basic building blocks. For example, the convertible bond will be reported in terms of both its liability and equity components. As a result, option pricing models such as the Black-Scholes model are being used to determine the fair value of the conversion option of the convertible bond. We will examine how the Black-Scholes model can value the conversion option of the convertible bond as well as how the convertible bond is reported on the balance sheet.

THE BLACK-SCHOLES OPTION PRICING MODEL

The Black-Scholes option pricing model provides the following equation to value a call option:

$$PV\begin{pmatrix}\text{Call}\\\text{Option}\end{pmatrix} = P_0N(d_1) - EXe^{-R_f(t)}N(d_2)$$

where,

$$d_1 = \frac{ln\left(\frac{P_0}{EX}\right) + R_f(t)}{\sigma\sqrt{t}} + \frac{\sigma}{2}\sqrt{t}$$

$$d_2 = d_1 - \sigma\sqrt{t}$$

$N(d)$ = cumulative normal probability function
EX = exercise price of option
t = time to expiration
P_0 = current common share price
σ = standard deviation per period of (continuously compounded) rate of return of the asset
R_f = (continuously compounded) risk-free rate of interest

We now illustrate how the Black-Scholes option pricing model can be used to value the conversion option of a convertible bond.

EXAMPLE

A firm issues 2,000 convertible bonds at the beginning of year 1. These bonds have a three-year term to maturity and are issued at a par value of $1,000. Assuming taxes and transactions costs are ignored, the firm will raise $2,000,000 in capital. The annual coupon for this convertible bond yields an interest rate of 6 percent. Each bond is convertible at any time prior to maturity into 274 common shares. Straight bonds issued at the prevailing market interest rate yield 9 percent. The current market price of the common share is $2.42 and the risk-free rate is 5 percent. The standard deviation of the annual

returns on the shares is 30 percent. Determine the value of the conversion option.

$$EX = \$3.65 \text{ (\$1,000/274 common shares)}$$
$$P_0 = \$2.42$$
$$t = 3 \text{ years}$$
$$\sigma = 0.3$$
$$R_f = 0.05$$

$$d_1 = \frac{ln\left(\frac{\$2.42}{\$3.65}\right) + 0.05(3)}{0.3\sqrt{3}} + \frac{0.3}{2}\sqrt{3} = -0.2424$$

$$d_2 = -0.2424 - 0.3\sqrt{3} = -0.7620$$

$$N(d_1) = 0.4042$$

$$N(d_2) = 0.2230$$

$$PV\begin{pmatrix}\text{Call} \\ \text{Option}\end{pmatrix} = \$2.42(0.4042) - \$3.65e^{-0.05(3)}(0.2230)$$

PV(call option) = \$0.278
Value of equity = \$152,344 (\$0.278 × 274 × 2,000)
Value of debt = \$1,848,156 (\$120,000[2.5313] + \$2,000,000[0.7722])

As a result, the Black-Scholes option pricing model estimates the value of the conversion option of the convertible bond issue at $152,344. The value of the debt component, $1,848,156, was obtained by discounting the interest and principal payments at 9 percent which is the appropriate risk level for a straight bond with no conversion option attached. It should be noted that the total of the liability portion ($1,848,156) and the equity portion ($152,344) is $2,000,500. The small difference represents ($2,000,500 − $2,000,000) the error caused by estimation uncertainties. In this example, management would increase the value of debt and equity accounts by the liability portion and the conversion option of the convertible debt, respectively. This type of presentation would provide a more equitable separation of debt and equity components on the balance sheet, as well as present a better debt-to-equity ratio for the firm.

CORPORATE RESTRUCTURING: COMBINATIONS AND DIVESTITURES

CHAPTER 23

LEARNING OBJECTIVES

After reading this chapter you should be able to

1. Describe the different methods to undertake a merger.
2. Describe why mergers may create value.
3. Describe the methods used to finance a merger.
4. Explain the use of tender offers in acquiring a firm.
5. Discuss the methods used by incumbent management to resist a hostile takeover.
6. Describe the different methods used to undertake a divestiture.

INTRODUCTION

In general, the, mergers and acquisitions that occurred during the 1980s were motivated by the need for change. Specifically, this period saw some of the largest corporate giants taken private through leveraged buyouts, then dismembered and sold off piecemeal. What emerged from the process was a leaner and far more profitable corporation. Investors and the investment dealers that served them applauded these restructurings, while bondholders and employees frequently decried what they saw as unwarranted attempts by shareholders to capture a larger portion of the firm's value. Shareholders, on the other hand, simply said they were reclaiming what was theirs all along.

A breed of specialists has been created to remake staid old corporations and return them to financial health. One such individual is Albert J. Dunlap, who heads Scott Paper Company's efforts to restructure itself.[1] Mr. Dunlap developed his reputation for turning around poorly performing companies working for Sir James Goldsmith (a British takeover specialist), where he turned around Crown Zellerbach and Diamond International. So what does Mr. Dunlap (known by some as "Rambo in Stripes" and to others simply as "Chainsaw") have in mind to re-engineer Scott paper? The answer is quite simple: learn to do more with less (cut the labour force by one-third or lay off 10,500 workers) and focus on the firm's core business. Sounds simple. So why was it necessary to bring in Mr. Dunlap to accomplish the task? The answer is that the job of downsizing a firm is not a pleasant task, and few have the stomach for the trauma that it creates within an organization. Some, like Mr. Dunlap, do and they become turnaround specialists.

---

[1] Based on Glen Collins, "Tough Leader Wields Axe At Scott," *New York Times*, August 15, 1994, p. C1, C6.

CHAPTER PREVIEW

Chapter 23 presents an overview of corporate restructuring. Corporate restructuring encompasses a broad range of topics, including mergers and acquisitions, leveraged buyouts, leveraged recapitalizations, and divestitures. We begin by examining the reasons why mergers might create shareholder wealth. We then examine some of the methods used to finance a corporate takeover. Finally, we examine the different types of divestitures.

OBJECTIVE 1

MERGERS AND ACQUISITIONS[2]

Statutory amalgamation

A merger in which two firms join together to form one legal entity.

Business combination

A merger in which one firm acquires another firm, but both firms maintain their own legal identities.

In Canada, there are different methods that a manager can use to undertake a merger. For example, a **statutory amalgamation** occurs if two firms merge together to form a new legal entity with one business charter. Another example of a merger is a **business combination** in which one firm acquires another firm, but both firms maintain their own legal identities. Each of these methods of merger has different legal and tax implications. Although the terms **merger** and **acquisition** have different legal implications, we will use them interchangeably to represent a combination of two or more businesses into a single operational entity.

Company growth through the merging of two firms has been a part of the North American economic scene for many years. For example, during the 1920s a surge of combinations developed; big businesses, such as IBM and General Foods expanded into new products and/or market extensions, creating vertically integrated companies. The decade of the sixties was another period of merger activity. This period is known as the conglomerate age because firms in different industries were combined into one corporate entity, called a *conglomerate.*[3]

The creation of a conglomerate was thought to be an efficient way of monitoring individual businesses by subjecting them to regular quantitative evaluations by the central office. The conglomerate allowed funds to be reallocated from slow growing subsidiaries that generated cash, such as insurance and finance, to fast growing, high technology businesses that required investment funds.

With hindsight we now see that conglomerate acquisitions have for the most part proven unsuccessful.[4] The evidence suggests that buyers often paid too much to acquire the businesses and that mergers were frequently followed by a decline in earnings. Perhaps the most important reason may have been a disregard for the principle that specialization increases productivity. Many important business decisions were made by managers who had limited information and who divided their attention and resources between multiple businesses. Furthermore, conglomerates developed large and expensive central offices for the purpose of monitoring their

[2]Many of the comments in this section come from an article by Andrei Shleifer and Robert W. Vishny, "The Takeover Wave of the 1980s," *Journal of Applied Corporate Finance* (Fall 1991), pp. 49–56, and from J. Fred Weston, Kwang S. Chung, and Susan Hoag, Mergers, *Restructuring, and Corporate Control* (Englewood Cliffs, NJ: Prentice-Hall, Inc., 1990).

[3]The following reasons provide some insight as to why firms diversify through mergers and acquisitions: (1) the world can be seen as a portfolio, where all that matters is expected return and volatility of returns; (2) undervalued companies can be restructured to capture the "true value" of the firm; and (3) mergers and acquisitions are ways to transfer skills or share activities between two companies. The first two reasons imply that we have superior ability in finding undervalued securities. The third reason simply suggests that we can take advantage of technical skills or lower costs as a result of a merger or acquisition. This last reason has the greatest chance of success.

[4]See, for example, Amar Bhide, "The Causes and Consequences of Hostile Takeovers," *Journal of Applied Corporate Finance* (Summer 1986), pp. 6–32.

various businesses. This type of monitoring by the conglomerate's central office proved to be less effective than the market discipline to which stand-alone businesses are generally subjected. Since the subsidiaries of conglomerates are insulated from market forces, they can afford to lose money and be subsidized by other divisions. As a result, these subsidiaries are not as competitive and do not raise as much external capital as the stand-alone firms.

PERSPECTIVE IN FINANCE

Mergers and acquisitions are usually justified by management on the grounds that merging diversifies the firm, thus reducing risk. However, it may be that the shareholder can diversify personally with more ease and less expense by buying shares in the two companies. There must be other reasons for the merger.

WHY MERGERS MIGHT CREATE WEALTH

OBJECTIVE 2

During both the eighties and nineties, several large mergers and acquisitions have involved Canadian companies. For example, in the nineties, the Montreal-based Seagam Company acquired the entertainment giant MCA Inc. for US$ 5.7 billion, and Belgian brewer Interbrew paid $2.7 billion for John Labatt Ltd.[5] Examples of corporate takeovers that occurred during the eighties are the Canadian company Campeau Corporation acquiring Federated Department Stores Inc. for $6.64 billion and Allied Stores for $4.9 billion; and Conoco Canada purchasing Dome Petroleum for $4.8 billion.

Clearly, for a merger to create wealth it would have to provide shareholders with something they could not get by merely holding the individual shares of the two firms.[6] Such a requirement is the key to the creation of wealth under the capital asset pricing model. Restating the question: What benefits are there to shareholders from holding the shares of a new, single firm that has been created through a merger as opposed to holding shares in the two individual firms prior to their merger? Let's consider some of these benefits.

BACK TO THE BASICS

Once again, we see that tax policy influences business decisions. Sometimes the influence is an intended consequence of fiscal policies of government; in other cases the influence is caused by an unanticipated reaction. We find that ***Axiom 8: Taxes Bias Business Decisions*** *provides at least some partial explanation for corporate restructuring activities.*

Tax Benefits

For a business combination which maintains the legal identities of the two firms, Revenue Canada requires that each firm report their earned income and file a corporate tax return. The acquisition of control by an acquiring firm causes the acquired firm to lose its right to carry forward any net capital losses, property losses or non-capital losses. An exception to this rule occurs if the business generating the non-capital losses is continued with a reasonable expectation of profit by the

[5]Adapted from: Susan Noakes, "Mergers big news again," *The Financial Post*, December 23, 1995, p. 5.

[6]Michael C. Jensen, "The Takeover Controversy: Analysis and Evidence," *Midland Corporate Finance Journal* 4 (2) (Summer 1986), pp. 6–32.

Corporate Restructuring: The New Thinking

The fundamental assumptions about how businesses should be organized have changed significantly from what they were several years ago. Financial market behavior, economic logic and plain common sense suggest that focusing on a single business, or on a very small number of genuinely linked businesses, is the only way to build the value of a company in the long term. The tempting, even seductive, aphorisms of the past—diversify to reduce risk; diversify for higher growth; manage a balanced portfolio of businesses—are about as useful on today's corporate battlefield as is horse cavalry in a modern tank war. Sadly, these old assumptions are still cherished by some corporate managers today.

The Financial Rules of the Game

The financial markets, however, have imposed their own set of rules on corporate management. Two key points highlight the difference between the old and new thinking:

(1) Managements cannot create value by doing what shareholders can do for themselves. Unless there are real economic and business connections that lead to sustainable improvements in operating performance, managements do nothing by making acquisitions that investors cannot do for themselves by making a telephone call to a broker. Investors can diversify their risks by investing at the market price. They do not want corporate managements to do it for them by paying acquisition premiums of 50–100% above the going rate.

This sounds straightforward enough, but it actually strikes at the heart of many, if not most, of the reasons given for corporate acquisitions in the US and UK. Many managements seem to operate under the illusion that they should act as a substitute for the capital markets, making resource-allocation decisions that take them well beyond the activities in which they enjoy true operating advantages.

Few topics have been subjected to as much measurement and analysis as corporate diversification, and there is a large body of evidence that points to one conclusion: Unless an acquired business gains a sustainable operating advantage from becoming part of another company, no value for the acquirer's shareholders will be created by the acquisition. The fact that two businesses operate in adjacent or apparently similar areas of activity is not enough to justify the acquisition of one by the other. In fact, much of the restructuring we have seen over the past several years has been the unwinding of grandiose schemes devised by those self-deluding managements that have confused their own role with that of the financial markets. The markets are now exhibiting a healthy impatience with such fantasies.

(2) Shareholders will unlock values suppressed by management. If investors notice that a particular company's strategy is causing it to trade at a discount to its true value, the defenses installed by managements will not, in the long run, stop the investors from getting their way. Investors can be misled, and they frequently misprice securities; however, years of bitter experience and their increasing sophistication have made investors far less willing to accept management rationales for diversifying acquisitions. They are much more likely to seek ways to evade the various barriers erected by managements to protect their empires.

Furthermore, size is no longer a barrier. Some corporate managements acquire businesses as part of a defensive strategy, believing that bigger is, if not better, at least tougher to knock over. However, unless such acquisitions achieve sustainable operating advantages for the company, they only put off the evil day. Indeed, all such acquisitions really do is increase the size of the discount in absolute dollars and make the acquiring company even more attractive to well-financed or imaginative predators.

The New Business Environment

The changes in the ground rules mentioned above show that if companies do not restructure themselves, the financial markets will do it for them. Even without the prodding from the financial markets, however, there are compelling business reasons for companies to restructure:

(1) Competition. In a fiercely competitive environment, a broad span of businesses is a luxury that is less and less affordable. Indeed, today's managers are seeing a heightened sense of competitor performance. Absolute measures of success mean much less. Whatever the subject—whether product cost or quality, the rate of innovation, quality of service, or almost any other dimension of business—it is all relative to the competition.

(2) Flexibility. The quick and the dead. Rapidity of response is becoming crucial to corporate survival and success. Technological developments in information, production and distribution systems and an increased consumer demand for variety are causing major shifts in the pattern of business activity. As a result, diversified firms are in for a tough time. It is hard to jump in several directions at once.

(3) Excessive corporate overhead. Too much value is absorbed by unnecessary corporate expense. Such expense is often worse than unnecessary; it can be positively damaging. In order to justify their existence, corporate headquarters frequently embark upon acquisition programs that create no value for their shareholders. After all, acquisitions are exciting, dramatic and glamorous.

For example, the financial statements of one major diversified corporation reveal an item called "corporate expense" that totals slightly under $200 million. Compare this amount to the company's net income, which (after various adjustments for extraordinary items) totals about $1.3 billion. While adjusted net income has grown at slightly under 15% per annum over the past few years, corporate expense has been growing at the staggering rate of just under 50% per annum over the same period. Adjusting the corporate expense to an aftertax figure and applying the company's price/earnings ratio to the net expense produces a cost of over $1.4 billion. Even assuming that some of the expense is necessary, there is clearly a lot of value being wasted by excessive overhead.

The Role of Finance

Under these new ground rules, the CFO must play a more active role in planning the direction of the company. However, good finance is not a substitute for good strategy and operations. The CFO's role is overlapping more and more with those of the CEO and COO because of the growing interconnectedness of finance, strategy and operations. Unless this interconnectedness is acknowledged and encouraged, the best finance function in the world will not be able to make up for poor strategy or inferior performance.

Therefore, the CFO needs to be centrally involved in strategy formulation, not just to ensure that the strategy is financially sound, but to bring a financial market perspective and ensure that strategic plans translate directly into shareholder value. He also needs to be more closely involved in operations, if only to ensure that the operators are fully aware of the financial dimension of operating decisions and make certain that the real causes of poor financial performance are addressed and remedied.

Source: Simon Duffy, "Corporate Restructuring: The New Thinking," *Business International Money*, January 9, 1989, pp. 1–3.

acquiring firm. Under these circumstances, the acquired firm may carry forward its non-capital losses and investment tax credits to reduce future taxable income that is generated by the acquired business.[7]

In a statutory amalgamation the new firm can also benefit from tax losses. In general, net capital losses that could not be utilized by the predecessor firms can now take on value for the new firm. However, non-capital losses and investment tax credits can only be utilized on taxable income from the business which generated these non-capital losses and investment tax credits. Again the business generating these non-capital losses and investment tax credits must be continued and have a reasonable expectation of profit. Depreciable capital property is transferred from predecessor corporations at their original costs. Hence, the new firm will continue to benefit from tax shields which result from the depreciation of assets.[8]

[7]CCH Canadian Limited, "Acquisition of Control," in *Canadian Income Tax Act with Regulations*, 65th ed. (North York: ON: CCH Canadian Limited, 1995), section 111(5).

[8]Ibid., Section 111(4).

Reduction of Agency Costs

As we know, the agency problem can occur if management and the ownership in the firm are separated. To compensate for the agency problem, shareholders and bondholders impose a premium on funds supplied to the firm to compensate them for any inefficiency in management. A merger, particularly when it results in a holding company or conglomerate organizational form, may reduce the significance of this problem, because top management is created to monitor the management of the individual companies making up the conglomerate. As a result, management of the individual companies can be effectively monitored without any forced public announcement of proprietary information, such as new product information that might aid competitors. If investors recognize this reduction in the agency problem as material, they may provide funds to the firm at a reduced cost, no longer charging as large an "agency problem premium."

Alternatively, it can be argued that the creation of a conglomerate might result in increased agency costs. Shareholders of conglomerates may feel they have less control over the firm's managers as a result of the additional layers of management between them and the decision makers. Moreover, the resultant expenditures necessary to monitor conglomerates, because of their multi-industry nature, may give further rise to agency cost.

Free Cash Flow Problem: A Specific Case of the Agency Problem

The "free cash flow" problem was first identified by Michael Jensen in 1986. Free cash flow refers to the operating cash flow in excess of what is necessary to fund all profitable investments available to the firm; that is, to fund all projects with a positive net present value. As we know from our discussion of shareholder-wealth maximization, this free cash flow should be paid out to shareholders; otherwise it would be invested in projects returning less than the required rate of return, in effect less than shareholders could earn elsewhere.

Unfortunately, managers may not wish to pass these funds to the shareholders, because they may feel that their power would be reduced. Moreover, if they return these surplus funds they may be forced to go outside for financing if more profitable investment opportunities are identified at a later date. Certainly, what we are describing here is a form of the agency problem. However, we need to see these actions in the context of the corporate management culture rather than as an attempt by the managers to maintain their own position. In other words, as economic conditions change, managers who have successfully managed firms over the years of growing markets may have difficulty in adjusting their financial strategies to conditions in which not all cash flows can be invested at the required rate of return. Jensen argues that this was the case in the oil and gas industry in the late 1970s and resulted in much of the merger activity that took place in those markets during that period.[9] A merger can create wealth by allowing the new management to pay this free cash flow out to the shareholders, thus allowing them to earn a higher return on this excess than would have been earned by the firm.

Economies of Scale

Wealth can also be created in a merger through economies of scale. For example, administrative expenses including accounting, data processing, or simply top-management costs, may fall as a percentage of total sales as a result of sharing these resources.

The sharing of resources can also lead to an increase in the firm's productivity. For example, if two firms sharing the same distribution channels merge, distributors carrying one product may now be willing to carry the other, thereby increasing

[9]Michael C. Jensen, "The Takeover Controversy: Analysis and Evidence," *Midland Corporate Finance Journal* 4(2) (Summer 1986), pp. 6–32.

the sales outlets for the products. In effect, wealth would be created by the merger of the two firms and shareholders should benefit.

Unused Debt Potential

Some firms simply do not exhaust their debt capacity. If a firm with unused debt potential is acquired, the new management can then increase debt financing, and reap the tax benefits associated with the increased debt.

Complementarity in Financial Slack

The merger of cash-rich bidders and cash-poor targets may create wealth for the shareholders of the merged firm. For example, the merged firm would choose positive NPV projects which the cash-poor firm would have passed up. Thus, while these cash-poor firms are selling at a firm price, the discounted value of their future cash flow is below their potential price. In effect, a merger allows positive NPV projects to be accepted that would have been rejected if the merger had not taken place.

BACK TO THE BASICS

Sometimes the only hope for rejuvenating a firm to strong financial performance involves changing the firm's management. To ensure that the new management will act in the best interests of the firm's owners, these changes in management are frequently accompanied by changes in the means by which management is compensated. These changes are aimed at aligning managerial and stockholder interests such that the managers will find it in their best interest to make managerial choices that lead to the maximization of share value. Thus changes in a firm's management and the method of managerial compensation is frequently a reflection of an agency problem which forms the basis for ***Axiom 7: The Agency Problem—Managers Won't Work for the Owners Unless It's in Their Best Interest.***

Removal of Ineffective Management

Any time a merger can result in the replacement of inefficient operations, whether in production or management, shareholder wealth is created. If a firm with ineffective management can be acquired, it may be possible to replace the current management with a more efficient management team, and thereby create wealth.[10] Such is the case for firms that have grown from solely production into production and distribution companies or R&D firms that have expanded into production and distribution; the managers simply may not know enough about the new aspects of the firm to manage it effectively.

Increased Market Power

The merger of two firms can result in an increase in the market power of the two firms. While this can result in increased wealth, it may also be illegal. In Canada, the Competition Act (Bill C-91) which was enacted in 1986, makes any merger or proposed merger illegal if it prevents or lessens, or is likely to prevent or lessen competition substantially. As a result, the Competition Tribunal has been given the power to dissolve any merger that causes a reduction of competition. The Competition Tribunal is a body composed of judges from the Trial Division of the Federal

[10]Mergers may overcome the resistance to change within an organization because new management may negotiate more favourable contract concessions with unions than existing management had been able to achieve.

Court and lay experts. The judges are responsible for determining questions of law, while the lay experts provide expertise on complex economic and financial issues.

Reduction in Bankruptcy Costs

There is no question that firm diversification, when the earnings from the two firms are less than perfectly positively correlated, can reduce the chance of bankruptcy. The question is whether or not there is any wealth created by such an activity. Quite obviously, in the real world there is a cost associated with bankruptcy. First, if a firm fails, its assets in general cannot be sold for their true economic value. Moreover, the amount of money actually available for distribution to shareholders is further reduced by selling costs and legal fees that must be paid. Finally, the opportunity cost associated with the delays related to the legal process further reduces the funds available to the shareholder. Therefore, since costs are associated with bankruptcy, reduction of the chance of bankruptcy has a very real value to it.

The risk of bankruptcy also entails indirect costs associated with changes in the firm's debt capacity and the cost of debt. As the firm's cash flow patterns stabilize, the risk of default will decline, giving the firm an increased debt capacity and possibly reducing the cost of the debt. Because interest payments are tax deductible, whereas dividends are not, debt financing is less expensive than equity financing. Thus, monetary benefits are associated with an increased debt capacity. These indirect costs of bankruptcy also spread out into other areas of the firm, affecting things like production and the quality and efficiency of management. Firms with higher probabilities of bankruptcy may have a more difficult time recruiting and retaining quality managers and employees because jobs with that firm are viewed as less secure. The lack of appropriate expertise causes firms to be less productive. In addition, firms with higher probabilities of bankruptcy may have a more difficult time marketing their product because of customer concern over future availability of the product. In short, there are real costs to bankruptcy. If a merger reduces this possibility of bankruptcy, it creates some wealth.

"Chop-Shop" Approach—Buying Below Replacement Cost

The "chop-shop" approach, which will be discussed more fully later, suggests that the individual parts of a firm are worth more than the current value of the firm as a whole. In the 1980s, many corporate raiders were driven by the fact that it was less expensive to purchase assets through an acquisition than it was to obtain those assets in any other way. This was particularly true of both conglomerates and oil companies. For conglomerates, corporate raiders found that they often sold for less than the sum of the market value of their parts. Much of the merger and acquisition activity associated with oil companies was driven by the fact that it was cheaper to acquire new oil reserves by purchasing a rival oil company than it was through exploration. If assets are mispriced, as this approach seems to suggest, then identifying those assets and revealing this information about the undervalued assets to investors may result in the creation of wealth.[11]

It should be noted that the free cash flow theory could explain this creation of wealth as easily as a mispricing theory. In particular, the oil industry was characterized in the late 1970s and early 1980s by overexploration and drilling activity in the face of reduced consumption while oil price increases created large cash flows. During this time period, managers attempted to increase reserves to protect them

[11]Restructuring may be beneficial to shareholders for at least two reasons. First, restructuring may provide the means for transferring business assets to a higher-valued use or to a ore efficient user. For example, restructuring of oil and gas firms in the early 1980s helped prompt needed retrenchment in the industry which would have eventually occurred. Second, restructuring may permit improvements in management accountability and a strengthening of incentives for managers to act in the best interest of the firm's shareholders.

from possible future market fluctuations. In effect, the free cash flow problem appeared to exist in the oil industry. As mergers and restructuring raged through the oil industry, wealth was created. Again, this creation of wealth did not necessarily come about through any correction of mispricing. It may have been the elimination of wasted expenditures that created the wealth.

BASIC FINANCIAL MANAGEMENT IN PRACTICE

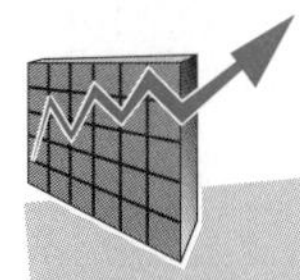

The '90s Way to Tackle the Recession

Restructuring is one of those words that doesn't ring so much as thud with the heft of purpose and deliberation. To have a structure you need a blueprint. To restructure, you need a revised blueprint. The emphasis is on design, on doing things differently, instead of just trying to do the same old things more cheaply. If there's a reason why companies choose to describe their activities as restructuring rather than cost cutting, this is probably it. In a time of increasingly testy and assertive shareholders, the word implies that management has a game plan and has thought through the contingencies. For this reason, today's corporate reorganizations more often involve a redesign of the entire company and its activities than they do paring and elimination.

Companies practise all kinds of restructuring, but most such exercises can be characterized as either calm or desperate. The calm, thoughtful approach usually involves a cash-rich company with a good balance sheet that embarks on a lengthy strategic planning process. It's a total rethink that involves all managers, analyzes markets, identifies strengths and weaknesses and decides what businesses the company should be in and the best ways to compete in them. This is textbook restructuring, with management doing precisely the job it is supposed to: looking out for the long-term health of the company. The makeovers of Alcan Aluminium Ltd. and Imperial Oil Ltd. a few years ago were good examples.

Then there is gun-to-head restructuring—the kind that companies such as Magna International Inc. and Campeau Corp. were forced to do. Atonement for past sins comes so much later in the cycle for these companies that they have no choice but to restructure. The CEOs of these unfortunate companies usually have less freedom of action than before. In many cases, there are new financial masters—such as unpaid creditors—calling the shots. These companies ignore the danger signals until the last moment and restructure as a final alternative to receivership. The two different types involve entirely different processes.

"If you're smart, you restructure yourself before someone does it for you," says Joe Wright, director with Burns Fry Ltd. "Darwin lives in business just as he lives in the jungle, and restructuring is an essential part of the process of renewal".

Canada Malting recently completed a corporate makeover begun under its own initiative in 1985. The early warning signals in this case were declining earnings on flattening sales. The company had six plants across Canada and major export markets across South America. It also had a balance sheet with underutilized assets and lots of potential. There was plenty of cash, strong cash flow and little debt, all of which could have been working much harder for the shareholders.

Using outside facilitators, Jonathan Bamberger, the company's vice president of Finance, began a three-month strategic planning exercise that had senior managers analyze the strengths and weaknesses of the company and where it stood in the markets. "The conclusion we came to," says Bamberger, "was that there was overcapacity in the world malting industry and that we were part of it. So we closed our three least productive plants, laid off 120 employees and withdrew from low-margin export markets in Central and South America where it was getting increasingly difficult to get paid."

The company then launched the proactive half of its plan—going after growth markets. Bamberger and his management team decided that their main customers, breweries, were getting bigger and fewer as the industry became more international. They also reasoned that it was increasingly important for international breweries to maintain product quality and consistency from country to country. To do this they would favour bigger suppliers such as Canada Malting, provided they could come up with the goods. The company expanded production in its more modern facilities and marketed itself as an international supplier of high-quality products.

Canada Malting's balance sheet shows the results, and few accountants would now describe its assets as underutilized. In the year of the strategic planning session, 1985, revenues for the malting operation were declining $155 million with $4 million in net earnings. In 1989, revenues topped $200 million, earnings were $16 million and both trendlines angle nicely upward. Where the company's major export market was a tottering Latin America then, it is a dynamic, beer-thirsty Japan now. Bamberger's reward was to become vice-president of the international company and general manager of the Canadian company.

This is textbook restructuring—using the luxury of time and solvency to craft a plan and get things right on the first try. The gun-to-head syndrome affords few such luxuries. Here, the company restructures in a brief window before receivership that usually requires new money for new partners and concessions from existing shareholders and creditors. Allan Crosbie, managing partner of merchant bankers Crosbie & Co. Inc., specializes in restructuring troubled companies and often takes a temporary stake in them. He says, "Companies seldom restructure and get it right the first time. It usually takes about three attempts because the first rarely goes far enough."

More often than not, the main problem is the management group that got the company group into hot water in the first place. Although the first, modest restructuring usually places limits on management's power—especially spending—the group still holds sway in the face of divided shareholders and creditors. Crosbie says the management typically downplays the severity of the company's situation, lacks the fortitude to deal with it, and assumes that the same ground rules apply for it as for sound companies. In today's financial environment, nothing could be further from the truth.

Source: Adapted from: Randall Litchfield, "The '90s Way to Tackle the Recession," *Canadian Business* (November 1990), pp. 80–84.

OBJECTIVE 3

FINANCING AND CORPORATE RESTRUCTURING[12]

Significant changes have occurred in the financing practices of North American firms, in part due to corporate restructuring. The outstanding debt of many companies increased as they borrowed heavily to finance restructuring activity. As part of this restructuring, three popular types of heavily leveraged transactions have developed: the management buyout (MBO), the leveraged buyout (LBO), and the leveraged employee stock ownership plan (leveraged ESOP). These three financing approaches are briefly described below:[13]

MBOs

An **MBO** occurs if a group of investors forms a shell corporation to acquire the target firm or buys the target firm directly. The distinguishing features of an MBO are: (1) the group of investors includes members of the management of the target company; (2) following the acquisition the newly formed company goes private (that is, common shares are delisted from the stock exchanges); and (3) the acquisition is

MBO

A management buyout or MBO involves a purchase of a firm from its shareholders by a small group of investors which includes a group from the firm's management.

[12]The ideas in this section come largely from Raymond Rupert, "The Acquirers Within: When Employees Buy the Company," *Mergers and Acquisitions in Canada* (March 1992), pp. 1–11, and Leland Crabbe, Margaret Pickering, and Stephen Prowse, "Recent Developments in Corporate Finance," *Federal Reserve Bulletin* (August 1990), pp. 595–603.

[13]Taken from John D. Martin and John W. Kensinger, *Exploring the Controversy Over Corporate Restructuring*, Morristown, NJ: Financial Executive Institute Research Foundation, 1990, p. 46.

financed primarily by debt, so the resulting firm is much more highly leveraged than most publicly traded companies.

LBOs

An **LBO** is a more general type of restructuring approach that can be distinguished from an MBO by the following characteristics: (1) the new owners do not necessarily have to include members of the original management team, and (2) the firm is not necessarily taken private following the acquisition. The majority of LBOs are MBOs, and the term LBO is generally used to describe high leverage acquisitions and restructurings.

LBO

An LBO or leveraged buyout involves a purchase of a firm from its shareholders by a small group of investors (which includes members of the firm's management in the case of an MBO). Usually the acquisition will be financed largely by debt financing, hence the term "leveraged" buyout.

Leveraged ESOPs

In a leveraged **ESOP** the rank-and-file employees participate with management in the purchase of stock in their corporation. A purchase agreement sets out the conditions for using a loan to purchase the firm's shares. The loan is repaid from the employee's retirement account. The employer corporation typically guarantees the loan. Generally, the shares are held under an escrow agreement until the participating employee retires or terminates employment. Provincial governments within Canada provide tax credits for share purchases by the employees. For example, in British Columbia there is a lifetime limit on the tax credit of $10,000 per employee. Similar tax incentives exist in other provinces such as Ontario, Quebec, Saskatchewan, and Manitoba.

ESOP

Employee stock ownership plan whereby the employees of the firm can acquire shares of their employer. In a leveraged ESOP, employees set up a trust to borrow the funds needed to acquire shares. The loan is repaid from the employees' retirement funds and the shares are credited to the employees' retirement account.

Junk Bonds

Junk bonds have been closely linked to large and at times unsuccessful mergers of the 1980s. Due in part to the conservatism of financial managers in Canada the use of these bonds has not been as prevalent in Canada as in the United States. However, junk bonds have been associated with some of the worst financial disasters in North America during the late 1980s. Examples include the failure of the Campeau retailing empire and the bankruptcy of Drexel, Burrnham, Lambert.

Junk bonds

Below investment grade bonds. These bonds are generally associated with the funding of corporate takeovers during the eighties.

Some critics blame many of the market ills on junk bonds. According to these critics, junk bonds fuelled the merger mania of the 1980s, caused the excessive use of debt, and created instability in financial markets. While these opinions are certainly understandable, the evidence does not support these extreme charges against junk bonds.[14] Junk bonds meet our expectations in terms of risk and expected returns. This type of bond is not inherently different from other securities—they have greater risk, but also greater expected return. Although the market for junk bonds in North America was significant, it was too small to account directly for the growth in corporate debt.

The junk market essentially evaporated at the conclusion of the 1980s. As a consequence of the lack of new financing in the low grade bond market, merger activity in North America declined noticeably in the early 1990s. The decline in merger activity can also be attributed to the more cautious attitude of commercial banks and the weakening in the market for asset sales. Nevertheless, well-structured acquisition proposals, especially those aimed at enhancing a firm's competitive position within its own line of business, are still well received by financial managers of the 1990s.

[14]These ideas come from Sean Beckett, "The Truth About Junk Bonds," *Federal Reserve Bank of Kansas City Economic Review* (July/August 1990), pp. 45–54.

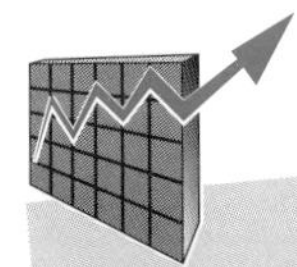

Leveraged Buyouts: How Everyone Can Win

There are many different forms of corporate acquisition. Some employ more financial leverage than others and those that use more are called "leveraged buyouts" or LBOs. There is no standardized definition of an LBO for two reasons. The degree and type of leverage can vary greatly from LBO to LBO. More importantly, the philosophy, corporate direction and strategic motivation among acquirers varies greatly.

No doubt, leveraging a company by way of an acquisition, recapitalization or otherwise, has a significant impact on its present and future business activities. It results in a financial overhead burden which can, depending on the company and its management, become either a restrictive or a positive and motivational force. Some argue that leverage is dysfunctional and does not result in any value creation. Others argue that there are material benefits and value creation opportunities in a well structured LBO acquired at the right price.

An LBO, combined with the stock ownership, can provide the catalyst to refocus management's operating efforts in a way that cannot be achieved otherwise. There is no longer room for complacency. The business must now be managed with great intensity. Opportunities for enhancing cash flow and profitability must continually be examined. And this must be done without jeopardizing the value and strategic development opportunities of the business.

The Ideal LBO Candidate

Before you can structure an LBO, you first have to identify LBO targets. The ideal LBO candidate has most of these eleven characteristics:

1. An experienced and strong management team;
2. A history of consistent and proven profitability;
3. Maintainable profit margins;
4. Strong and predictable cash flows;
5. Intrinsic financial strength;
6. Quality assets, both hard as well as those of a less tangible proprietary nature;
7. A distinctive market position with well known brand or productive names;
8. Competitive distinction through low cost producer advantage, information technology or differentiation;
9. Growth potential within the business as well as by acquisition;
10. Low technological risk to products or services;
11. Non-cyclical, non-commodity or non-regulated business.

Once acquired, the LBO must be subjected to a specific structural discipline if all LBO stakeholders are to be winners.

Management Must Have Ownership Interest

We noted earlier that acquirers all have their own philosophies and strategic directions. Some are going to be more successful than others. Onex, for example, is a diversified company that uses the leverage buyout technique as a way to enhance returns for its shareholders. It believes that the management of the operating companies must have a material financial interest in their business.

Ownership by management increases the focus on value creation. It ensures that management is properly motivated and rewarded through their performance and successful achievement of operating and financial targets. Management stands to accumulate substantial wealth through an increase in the value of their shares. Ownership establishes a direct linkage between success of the LBO and wealth creation for the management and all other shareholders.

Controlling Expenses and Managing Working Capital

Overhead costs and operating expenses are an important source of potential savings. It is amazing how these costs tend to increase over the years and often compromise non-essential and difficult-to-justify items. Elimination of these costs results in a leaner and more effective organization. Payment for performance in cost reductions is a good way to reinforce management's thrust and focus. This ensures that goals are communicated to all levels of management and is particularly effective in motivating those who do not have stock ownership in the company. This establishes a direct relationship between effective performance and financial returns.

An explicitly defined financial framework establishes financial covenants and investor return objectives. These covenants and objectives in turn focus the management team on a more efficient utilization of working capital. This means managing the business in a way that maximizes cash flow. However, great skill is required to do so without impairing the strength of the business or negating the opportunities for further growth and development. To achieve this, improved inventory and receivable turnover as well as negotiated/extended payment terms from suppliers have yielded commensurate savings in interest costs.

Better Focus on Capital Expenditures

Capital expenditures often consume great amounts of cash without producing the requisite or targeted rates of return. Subsidiary companies of large multinational corporations especially tend to spend money whenever they can order to obtain their share of group resources. But when such a subsidiary moves to a stand-alone basis, a more judicious consumption of capital is a *sine qua non*. A disciplined approach to improved cash management and debt reduction discourages undesirable or unnecessary capital expenditures.

Achieving an acceptable return on investment becomes an imperative when the alternative is, on a zero risk basis, to use the cash to pay down debt and to reduce interest costs. This serves to encourage better maintenance programs which are an effective way to avoid or at least defer replacement capital expenditures. This enables capital rationing to be focused on new equipment that can better achieve efficiencies and cost reductions, thereby leading to value creation and competitive differentiation.

Source: Adapted from Tony Melman, "Leveraged Buyouts: How Everyone Can Win," *Canadian Investment Review* (Spring 1990), pp. 67–69.

TENDER OFFER

OBJECTIVE 4

As an alternative approach in purchasing another firm, the acquiring corporation may consider using a tender offer. A **tender offer** involves a bid by an interested party for controlling interest in another corporation. The prospective purchaser approaches the shareholders of the firm, rather than the management, to encourage them to sell their shares, typically at a premium over the current market price. For instance, in late January of 1988 Campeau Corporation offered $47 for each share of Federated Department Stores Inc., a general retailer based in Cincinnati. The Federated stock sold for $35.87 on the New York Stock Exchange the preceding Friday. This bid represented a premium of 31 percent to Federated shareholders. Several days later, Federated rejected Campeau's offer indicating that it was grossly inadequate since market forces had pushed the share price to the 60 dollar range. In early April of 1988, Campeau Corporation accomplished the takeover of Federated Department Stores Inc. with a tender offer of $73.50 for each of the 90.4 million fully diluted shares outstanding, a value of $6.64 billion.

Tender offer

A bid by an interested party, usually a corporation, for a controlling interest in another corporation.

Since the tender offer is a direct appeal to the shareholders, prior approval is not required by the management of the target firm. However, the acquiring firm may choose to approach the management of the target firm. If the two managements are unsuccessful in negotiating the terms, a tender offer may then be attempted. Alternatively, a firm's management interested in acquiring control of a second corporation may try a surprise takeover without contacting the management of the latter company.

OBJECTIVE 5

RESISTING THE MERGER

In response to "unfriendly" merger attempts, especially tender offers, the management of the firm under attack will frequently strike back. A number of defense tactics are used to counter tender offers. These defensive maneuvers include white knights, shark repellents, poison pills, and golden parachutes. Let us examine these more closely.

White knight

A defensive tactic to a tender offer whereby an able company comes to the rescue of the firm targeted for takeover.

A **white knight** is a company that comes to the rescue of a corporation that is being targeted for a takeover. For example, when Campeau Corporation made a tender offer for Federated Department Stores Inc., Federated's management opposed the Campeau attempt. To prevent the takeover, Federated's management encouraged R. H. Macy and Co. Inc., a large U.S. retailer, to make a tender offer of cash and stock, in a friendly takeover bid.

Shark repellents

Any of a variety of legalistic means used to counteract a tender offer by the firm under attack.

Shark repellents are used to discourage unfriendly takeovers. As an example, a firm may revise its bylaws to stagger the terms of directors so that only a few come up for election in any one year. A firm making a tender offer would have to wait at least two years before gaining a majority of board members. Other tactics used to discourage unfriendly takeovers include the introduction of a majority voting amendment and a fair price amendment. The former amendment requires that any takeover or change to the board of directors be approved by the super majority of preferred and common share voters. The latter amendment requires the introduction of a fair price for tender offers. In certain cases a tender offer may make a bid for the shares at two different prices or tiers. The fair price amendment would disallow a second tier that is lower than the first tier.[15]

Halpern has observed that most of these tactics known as shark repellents are not as common in Canada as in the United States.[16] The following reasons were given for this observation: most takeover bids would have to be negotiated with the majority shareholder since most Canadian companies are closely held and the provincial securities regulations in certain provinces preclude two-tier tender offers.

Poison pill

A tactic used by the target of a hostile takeover which involves triggering a very unfavourable event for the takeover firm should it win the struggle for control of the firm. Examples include the sale of a large block of newly issued shares to a friendly party at a very low price.

In Canada a tactic that was used to defend a firm from hostile takeovers in the late eighties was the **poison pill.** Inco Ltd. was the first company in Canada to introduce the poison pill in 1988. The purpose of the poison pill is to discourage an unfriendly party trying to acquire the firm and to give the board of directors time to follow other alternatives to maximize shareholder value. For instance, management might devise a plan whereby all the firm's debt becomes due, if the management is removed. Another example of a management plan that can be used is the *golden parachute,* which stipulates that the acquiring company must pay the executives of the acquired firm a substantial sum of money to "let them down easy" as new management is brought into the company. Advocates of the poison pill indicate that the poison pill is in the best interest of the shareholder since it ensures a better deal for the shareholder. The following article summarizes the arguments of Professor MacIntosh, a law professor at the University of Toronto, against the use of the poison pill.

The disadvantages of the unfriendly takeover are readily apparent from the preceding examples. If the target firm's management attempts to block it, the costs of executing the offer may increase significantly. Also, the purchasing company may fail to acquire a sufficient number of shares to meet the objective of controlling the firm. On the other hand, if the offer is not strongly contested, it may possibly be less expensive than the normal route for acquiring a company, in that it permits control by purchasing a smaller portion of the firm. Also, the tender offer has proven somewhat less susceptible to judicial inquiries regarding the fairness of the purchase price, since each shareholder individually decides the fairness of the price.

[15]Paul Halpern, "Poison Pills: Whose Interests Do They Serve?," *Canadian Investment Review* (Spring 1990), pp. 57–66.

[16]Ibid.

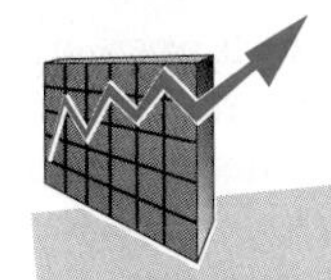

Poison Pill Always Bad News, Professor Says

Professor MacIntosh, a law professor at the University of Toronto, believes takeovers have two benefits. They bring shareholders a premium in excess of the market price of their shares. The threat of a takeover encourages management efficiency.

Poison pills discourage takeovers because they give too much power to management, which almost always has an interest in stopping a change of control. He said, "Officers of a corporation often lose their job after a takeover so they have an interest in resisting any bid. Officers, especially the CEO, have great influence over the board of directors, so good bids may be resisted."

MacIntosh doesn't attribute such resistance to bad faith by managers: "It's just human nature to fight some hostile force threatening to seize your company," he said.

In 1989 MacIntosh published an investigation of the effects of poison pills in the U.S. "I found that adopting a poison pill caused share prices to go down."

Toronto securities lawyer Philip Anisman has just published a report examining the 19 pills which have been adopted in Canada and he came to the same conclusion, MacIntosh said.

"Even if we consider a case where there is good, well-intentioned management, and the pill is in the interests of the shareholders of the target company, the availability of poison pills is not in the interests of society as a whole," MacIntosh said.

He believes poison pills remove management from the discipline of the market. "The possibility of takeovers keeps management on its toes," he said.

If poison pills give management too much control and they cause a drop in share prices, why do shareholders vote to ratify them?

Two phenomena, the "collective action problem" and co-opting of institutional investors contribute to the adoption of unfavourable measures, according to MacIntosh.

The collective action problem occurs when small investors feel it is not worth their while to study complicated procedures and try to weigh the merits of management proposals. Consequently, they tend just to go along with the proposal.

Institutional investors, on the other hand, have a strong incentive to side with management if they do business with the company. "Institutional investors were co-opted into voting approval for the Inco shareholders' rights plan, even though it was against their interest as shareholders," said MacIntosh.

Poison pills can be improved, if they are giving a short life. "If their horizon were limited so that they had to be voted in every five years, management would be more accountable."

He concedes that good management may feel more secure with pills in place, and that may allow for long-term planning. "But it is impossible for shareholders to determine when a plan will be in their interest. Even if there is good management now, that may change, and once a pill is in place there may be no getting rid of inefficient officers. It is still better to avoid the pill."

Despite his reservations, MacIntosh would not support an outright ban on poison pill provisions. "This is a tough question, but I am a noninterventionist" he said. "I still believe in a free market. Shareholder approval mechanisms can be corrupted but, in the end, we must have faith in investors' ability to protect their own interests."

Source: Adapted from James Carlisle, "Poison Pills Always Bad News, Professor Says," *The Financial Post*, February 15, 1990, p. 15.

OBJECTIVE 6

DIVESTITURES

Divestitures

The removal of a division or subsidiary from the company. Typically, the part of the firm being separated is viewed as not contributing to the firm's basic purposes.

While the mergers-and-acquisition phenomenon has been a major influence in restructuring the corporate sector, **divestitures**, or what we might call "reverse mergers," may have become an equally important factor. In fact, preliminary research to date would suggest that we may be witnessing a "new era" in the making—one where the public corporation is becoming a more efficient vehicle for increasing and maintaining shareholder wealth.[17] Donald Chew Jr. calls it the "new math," when he writes that a new kind of arithmetic has come into play. Whereas corporate management once seemed to behave as if 2 + 2 were equal to 5, especially during the conglomerate heyday of the '60s, the wave of reverse mergers seems based on the counter-proposition that 5 − 1 is 5. And the market's consistently positive response to such deals seems to be providing broad confirmation of the "new math."[18]

A successful divestiture allows the firm's assets to be used more efficiently, and therefore to be assigned a higher value by the market forces. It essentially eliminates a division or subsidiary that does not fit strategically with the rest of the company; that is, it removes an operation that does not contribute to the company's basic purposes.

The different types of divestitures may be summarized as follows:

Sell-off

The sale of a subsidiary, division, or product line by one firm to another.

Spin-off

The separation of a subsidiary from its parent, with no damage in the equity ownership. The management of the parent company gives up operating control over assets involved in the spin-off but the shareholders retain ownership, albeit through shares of the newly created spin-off company.

1. **Sell-off.** A sell-off is the sale of a subsidiary, division, or product line by one company to another. For example, Joseph E. Seagram & Sons sold seven of its product lines to American Brands Inc., for $372.5 million in 1991.
2. **Spin-off.** A spin-off involves the separation of a subsidiary from its parent, with no change in the equity ownership. The management of the parent company gives up operating control of the subsidiary, but the shareholders retain the same percentage ownership in both firms. New shares representing ownership in the diverted assets are issued to the original shareholders on a pro-rata basis. For example, in 1993 John Labatt Ltd. announced that it would spin-off its dairy operations, Ault Foods Ltd. in Canada to form a stand-alone company.[19]
3. **Abandonment.** The assets of the company are sold to another company, and the proceeds are distributed to the company's shareholders as a liquidation dividend.
4. **Going private.** Going private results when a company whose shares are traded publicly is purchased by a small group of investors, and the shares are no longer bought and sold on a public exchange. The ownership of the company is transferred from a diverse group of outside shareholders to a small group of private investors, usually including the firm's management.
5. **Leveraged buyout.** The leveraged buyout is a special case of going private. The existing shareholders sell their shares to a small group of investors. The purchasers of the shares use the firm's unused debt capacity to borrow the funds to pay for the shares. Thus, the new investors acquire the firm with little, if any, personal investment. However, the firm's debt ratio may increase by as much as tenfold.

Reorganization

A procedure, administered by the courts, that attempts to revitalize a firm by changing its operating procedures and capital structure.

A more drastic type of divestiture occurs if a company declares voluntarily or is forced involuntarily by its creditors into bankruptcy. As a result, the firm is either reorganized or liquidated. In Canada, a **reorganization** of a company can occur either through the Bankruptcy and Insolvency Act (Bill C-22) or through the Companies' Creditors Arrangement Act. Both of these Acts have as their objective to revitalize the firm by changing its operating procedures and capital structure. However, the Companies' Creditors Arrangement Act provides for the reorganization of financially troubled companies in the courts, but with fewer statutory requirements

[17]See G. Alexander, P. Benson, and J. Kampmeyer, "Investigating the Valuation Effects of Voluntary Corporate Sell-offs," *Journal of Finance* 39 (1984), pp. 503–517; and D. Hearth, "Voluntary Divestitures and Value," *Financial Management* (1984).

[18]Joel M. Stern, and Donald H. Chew, Jr. (eds.), *The Revolution in Corporate Finance* (New York: Basis Blackwell, 1986), p. 416.

[19]S. Feschuk, "Labatt Switches Dairy Plans," *The Globe and Mail,* March 13, 1993, p. B5.

KKR Learns ABCs of LBOs, 1977

Kohlberg Kravis Roberts & Co. on April 7, 1977, used mostly bank loans to buy a small truck-suspension maker, thus launching the leveraged buy-out's modern era.

Such buy-outs by a few investors using loans date back to obscure transactions in the late 1940s, but KKR gave them prominence in the late 1970s. And new twists: Buy a public company with bank loans and high-interest junk bonds; sell off parts of it to reduce the debt; and then resell the company to the public or to another firm. LBO makers gain whopping profits from the later sale, plus fees along the way.

KKR, formed by three former Bear Stearns partners in May 1976 with $120,000 in capital, set up in spartan Manhattan offices and became by far the No. 1 LBO firm of the 200 or so firms in the hot field in the mid-1980s.

In that first $25.6 million KKR buy-out of A.J. Industries in 1977, KKR partners invested only $1.7 million; they sold it in 1985 for $75 million. KKR has since done more than 30 LBOs.

Within seven years KKR—especially aided since 1981 by big state employee pension plan investors—was buying companies totalling more than $1 billion in size. And early this year it bought RJR Nabisco for a record takeover tag of $25 billion, picking up $75 million in fees.

KKR's managing partners became wealthy. Henry Kravis pledged $10 million in 1987 to the Metropolitan Museum of Art to complete the Henry R. Kravis Wing by 1990. As to criticism that LBOs only line their makers' pockets, Kravis, in a Fortune interview, said: "Greed really turns me off. To me, money means security."

Recently, LBOs have been tested by the weakness in junk bonds, the souring of some debt-plagued LBOs and the failure of banks to provide the financing for UAL's planned LBO. But LBOs have bounced back before; so experts aren't predicting their end.

Source: "KKR Learns ABCs of LBOs, 1977," *Wall Street Journal*, November 6, 1989, B1.

and regulations, allowing for greater flexibility than the Bankruptcy and Insolvency Act. Amendments to the Companies' Creditors Arrangement Act are expected in 1996 and will harmonize the two Acts further.

Video Case 7

Reorganization

A **reorganization** can be accomplished through the use of a proposal. The provisions of the proposal have been set out in the Bankruptcy and Insolvency Act. The proposal represents a compromise between the debtor and the firm's creditors. If the proposal is accepted, it is legally binding on all creditors who participated in its development. In effect, a proposal gives a financially unsound business that is viable an opportunity to save the business and provide creditors with a return greater than that expected from a liquidation.

In 1992, the Bankruptcy and Insolvency Act (Bill C-22) gave the debtor the right to be heard and more time to develop a reorganization proposal before assets can

Trustee

The third party responsible for disposal of the debtor firm's assets and distribution of the proceeds to the creditors in an assignment.

be seized by secured creditors. The debtor,[20] **trustee** in bankruptcy, liquidator, or receiver under the Bankruptcy and Insolvency Act can file a notice of intention to submit a proposal with the Official Receiver. The debtor is also permitted to apply for extensions of 45 days up to a maximum of five months after the initial 30-day stay in order to develop a definitive proposal for reorganization.[21] In addition, the law enables the debtor to revoke leases with landlords and sets the maximum rental payable at six months if the lease is broken. However during this period, the debtor must provide detailed cash flow statements to the official receiver and the trustee at the beginning of this process, and the trustee monitors cash flows throughout the reorganization process.

Reorganization Through the Companies' Creditors Arrangement Act

The Companies' Creditors Arrangement Act (CCAA) was created to facilitate the financial reorganization of larger companies which have outstanding issues of secured and unsecured debt. The CCAA provided provisions for a debtor to deal with claims of secured creditors. However, amendments to the Bankruptcy and Insolvency Act now enable debtors to deal with the claims of secured creditors. As a result of these amendments, both large and small firms are able to deal with their secured creditors under both Acts. Both Acts differ significantly in their approach to developing a framework for bankruptcy proceedings. The Bankruptcy and Insolvency Act provides detailed and specific provisions for proposals; whereas the CCAA requires that procedures be developed through a series of court orders. This lack of framework in the CCAA favours large corporations which have the legal resources to develop the required procedures in court.

Arrangement

A reorganizational plan involving a petition to the court requesting that the firm be required to make only partial or a delayed payment to its creditors.

The CCAA enables management of the debtor company to develop a plan of **arrangement** which deals with making the payments required under the new capital structure and paying any deferred obligations as they mature. Again, the court may prohibit legal suits by both secured and unsecured creditors before a definitive plan of arrangement is filed with the courts. The plan for arrangement generally includes all creditors. Since all creditors do not have the same priority in claim, the CCAA allows creditors to be classified according to the nature of their claims. However, the Act does not provide any direction as to how to classify creditors' claims. A fair and feasible compromise must be obtained between the debtor and creditors since the debtor's plan of arrangement must be approved by 50 percent of the creditors who have at least 75 percent of the dollar value of the creditors voting on the plan. The court provides final approval of the plan of arrangement which is binding on all participating creditors.

The reorganization of Olympia and York Development Ltd. (O&Y) is an example of a company that used the CCAA to file for protection from its creditors. The Canadian court action protects O&Y from claims on $8.4 billion in secured loans and about $200 million in unsecured bills. As a consequence of filing for creditor protection, the privately held O&Y had to release their financial statements which disclosed details of specific debts, cash flow levels and even staff salaries.

In March of 1992, O&Y publicly admitted to a liquidity crisis which was caused by a depressed North American real estate industry. O&Y filed for creditor protection under the CCAA on May 14, 1992. Over the next eight months, the company produced several debt-restructuring proposals, but these proposals were rejected by secured creditors. In late October of 1992, a proposal was made public which

[20]The term *debtor* in this context means insolvent persons, bankrupts, trustee in bankruptcy, and liquidator or receiver under the Bankruptcy and Insolvency Act.

[21]For a more detailed discussion on the 1992 amendments to the Bankruptcy and Insolvency Act, see Stephen H. Barnes, "What Do You Propose?" *CGA Magazine* (November 1992), pp. 34–37.

outlined a plan to restructure the real estate empire's massive debt. The major points of this restructuring plan were:[22]

- The Reichmanns would give up 90 percent of the privately owned O&Y if unsecured creditors were not fully repaid within five years. In the interim, O&Y would pay no interest on unsecured creditors' claims and these claims would be exchanged for company bonds.
- Three of O&Y's major properties in Toronto would be placed under the control of a joint venture owned 20 percent by O&Y Properties and 80 percent by the Bank of Nova Scotia, the Canadian Imperial Bank of Commerce, the National Bank, and the Royal Bank. In exchange for this concession, the proposal required lower interest rates on outstanding bank loans, conversion of $131 million in loans to capital and a new credit line to attract tenants.
- The spinning off of the company into three companies: a Canadian real estate holding company which holds the main Canadian buildings; a U.S. real estate and investment company holding U.S. real estate properties; and a residual Canadian company holding investments and some real estate.

The reorganization plan was dependent on O&Y retaining its investment in Gulf Canada Resources Ltd. and Abitibi-Price Inc., and on office rental rates rising about 33 percent over the period 1992 to 1998. This reorganization plan was the basis for developing a final reorganization plan on which both secured creditors and unsecured creditors voted in late January, 1993. Eight classes of secured creditors voted to turn down the reorganization plan and won their right to foreclose on their collateral which included a $475 million mortgage on O&Y's flagship First Canadian Place in Toronto. However, the plan was structured such that only the unsecured creditors class could veto the reorganization. Rejection of the proposal by this class of unsecured creditors would have meant liquidation of the firm's assets through the Bankruptcy and Insolvency Act. On January 26, 1993, the unsecured creditors voted to accept the reorganization plan. The Reichmanns ended up owning 100 percent of a new company called, O&Y Properties, which would manage the former real estate assets under the control of secured creditors.[23]

Liquidation

If the likelihood is small that reorganization of an insolvent firm will result in a profitable business, the firm should be liquidated. In this situation the liquidation value exceeds the going-concern value, and it is to the investor's advantage to terminate the business by liquidating its assets. Continuing the operation at this point generally results only in further losses. A firm's management can voluntarily declare bankruptcy by filing for an assignment in bankruptcy or the firm can be forced involuntarily into bankruptcy by creditors filing a petition. The objective of either procedure is to distribute any proceeds from the liquidation to the creditors.

Bankruptcy by Assignment

If an **assignment** is used to liquidate the business, the final settlement is done privately with a minimum amount of court involvement. The debtor transfers the title of the firm's assets to a third party, the trustee, who has been appointed by the Official Receiver. The trustee administers the liquidation process and sees that an appropriate distribution is made of the proceeds. Under an assignment, the debtor is not

Assignment

A method of liquidating a business that has been declared insolvent. To minimize court involvement (and therefore costs), a third party receives title for the firm's assets and is responsible for the disposal of the assets and distribution of proceeds. Creditors have the right to refuse such a settlement, and if they so choose, to force the debtor into a court-administered liquidation.

[22] *The Gazette*, "Reichmanns Would Give Up 90 Per Cent of Olympia & York," October 29, 1992, pp. E1–E2; and Dan Westell, "Reichmanns Release Details of O&Y Restructuring Plan," *The Globe and Mail*, November 11, 1992, p. B2.

[23] Canadian Press, "O&Y Merger to Proceed," *The Globe and Mail*, January 27, 1993, p. B11; and Margaret Philip, "O&Y Plan Comes to Crunch," *The Globe and Mail*, January 26, 1993, pp. B1, B2.

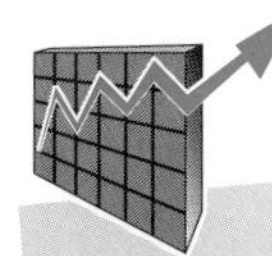

Bankruptcy Pays — for Trustees, Lawyers

The fees of trustees and lawyers eat up 44 cents of every $1 in assets left on the books of firms that declare bankruptcy, says a new report.

While the professionals who wrap up the affairs of failed companies are well-paid, the study of 417 businesses that filed for bankruptcy from 1977 to 1987 found that creditors are not so fortunate.

Unsecured creditors got nothing in 77% of bankruptcies, and even preferred creditors came out empty-handed in 53% of the cases studied.

The average payoff on unsecured claims was 2.5% of the money owed, while the average collected by preferred creditors was 23.2% of the value of the claim.

"Trustees and legal experts appear to be the biggest beneficiaries of bankruptcy," concluded the author of the study, Jocelyn Martel, a University of Montreal economist.

Despite the substantial losses sustained by creditors, Martel raised doubts about the 1992 reform of the Bankruptcy Act aimed at promoting early steps to save floundering firms.

Based on the heavy debt burdens typically carried by the companies he reviewed, Martel concluded that bankruptcy "was most likely to be the best decision for the vast majority of them. Encouraging their reorganization may be inefficient and costly for the Canadian economy."

Source: Adapted from: John Geddes, "Bankruptcy Pays—for Trustees, Lawyers," *The Financial Post*, October 11th, 1995, p.6.

automatically discharged of the remaining balance due a creditor. When the liquidation funds are distributed to the creditors, a notation on the cheques should be made indicating that endorsement of the cheque represents the creditor's acceptance of the money as full settlement of the claim. However, even then a creditor can refuse to accept the payment, considering that his or her interests would be better served by a court-administered liquidation. Owing to the possible nonacceptance of the assignment by one or more creditors, the debtor may attempt to reach a prior agreement with the creditors that the assignment will represent a complete settlement of claims.

An assignment has several advantages over a formal bankruptcy procedure. The assignment is usually quicker, is less expensive, and requires less legal formality. The trustee can take quicker action to avoid further depletion in the value of inventories or equipment and typically is more familiar with the debtor's business, which should improve the final results. However, the assignment does not legally discharge the debtor from the responsibility of unpaid balances, and the creditor is not protected against fraud; consequently, formal bankruptcy proceedings may be preferred by either the debtor or the creditors.

Bankruptcy by Petition

A petition for a receiving order that adjudges a company bankrupt is initiated by filing a petition with the Bankruptcy Court. A petition may be filed by the firm's management (voluntary) or by its creditors (involuntary). To initiate an involuntary reorganization, the creditors must show that the debtor has debts outstanding of at least $1,000 and has committed an act of bankruptcy within the six months preceding the date of the petition. The Bankruptcy and Insolvency Act contains several provisions defining what constitutes an act of bankruptcy. Examples of these provisions include the debtor's (1) ceasing to meet liabilities as the come due, (2) giving notice to suspend payment of debts, (3) failing to meet maturing obligations, or (4) committing an act of fraud toward a creditor.

After the filing and approval of the bankruptcy petition by the Official Receiver, the Bankruptcy Court will issue a receiving order which formally declares the firm bankrupt and authorizes liquidation of the company's assets. The court will also appoint an interim receiver, who serves as temporary caretaker of the company's assets, until a trustee can be appointed by the creditors. The interim receiver then requests from the debtor a list of assets and liabilities.

A meeting of the creditors is called by the interim receiver. At the meeting, the creditors will be required to appoint a trustee and a board of inspectors. This board of inspectors will represent the creditors throughout the bankruptcy proceedings by providing advice to the trustee on behalf of the creditors. In addition, both the financial position of the debtor company and a list of its assets and liabilities will be disclosed at the meeting. The debtor may be questioned for additional information relevant to the proceedings.

The trustee initiates plans to liquidate the company's assets. As a part of this conversion process, individuals owing money to the debtor are contacted in an effort to collect these outstanding balances. Appraisers are selected to determine a value for the property, and these values are used as guidelines in liquidating assets. After all assets have been converted into cash through private sales or public auctions, all expenses incurred in the bankruptcy process are paid. The remaining cash is distributed on a pro rata basis to the creditors according to the provisions of the Bankruptcy and Insolvency Act.

Priority of Claims

Following the provisions of the Bankruptcy and Insolvency Act, claims are honoured in the following order:

1. Secured creditors, with the proceeds from the sale of the specific property going first to these creditors. If any portion of the claim remains unpaid, the balance is treated as an unsecured loan.
2. Expenses incurred in administering the bankrupt estate.
3. Expenses incurred after the bankruptcy petition has been filed but prior to a trustee being appointed.
4. Salaries and commissions not exceeding $2,000 per employee that were earned within the three months preceding the bankruptcy petition.
5. Unsecured creditors.
6. Preferred shares.
7. Common shares.

When the distribution of the proceeds is completed the trustee provides a final accounting to the creditors and files a final report to the Superintendent of Bankruptcy. If all claims are satisfied in full and the company is not to be dissolved, the trustee may apply to the court on behalf of the debtor's company for a discharge from bankruptcy. Once the discharge is granted, the directors and officers of the corporate entity regain the right to sign legal documents and their powers to perform their business functions. However, if the claims of the creditors are not satisfied in full, the company will remain in the state of bankruptcy until it ceases to exist under the Canada Business Corporation Act.

An example will illustrate the order of payments from a liquidation.

EXAMPLE

The Poverty Stricken Corporation has been judged bankrupt; its balance sheet is shown in Table 23-1. Although the historical cost of the assets was $100 million, the trustee in bankruptcy was able to realize only $44 million from their sale. In addition, costs of $8 million were incurred in administering the bankruptcy.

The order of settling the claims is listed in Table 23-2. As the table shows, the first mortgage bondholders are entitled to receive the net proceeds resulting from the sale of specific property identified in the lien (an office building). We are assuming that the sale price of this property is $10 million, which leaves an unpaid balance of $4 million to be treated as an unsecured claim. Next, the administrative costs associated with filing and executing the bankruptcy petition, the employees' salaries, and the tax liabilities are to be paid.

After all payments have been made for prior claims, $22,400,000 is available for general unsecured creditors, whose claims total $56 million; dividing 22.4 million by 56 million, we find that the claimants are entitled to 40 percent of the original loan. This percentage is computed from the amount available for general creditors, $22,400,000, relative to the total claims of $56 million. However, an adjustment must be made to recognize the subordination of the debentures to the notes payable. In essence, the owners of debentures must relinquish their right to any money until the notes payable have been settled in full. In this situation, $12 million of the note balance remains unpaid. As a consequence, the $4 million originally shown to be received by the owners of debentures must be paid to the investors owning the notes payable. The actual distributions to the unsecured creditors are shown in the last column of Table 23-3.

Table 23-1 Poverty Stricken Corporation Balance Sheet ($000)

Assets			
Current assets			$ 20,000
Net plant and equipment			80,000
Total assets			$100,000
Liabilities			
Current liabilities			
Accounts payable	$22,000		
Accrued wages[a]	3,600		
Notes payable	20,000		
Total current liabilities		$45,600	
Long-term liabilities			
First mortgage bonds[b]	$14,000		
Subordinated debentures[c]	10,000		
Total long-term liabilities		$24,000	
Total liabilities			$ 69,600
Equity			
Preferred shares	$ 4,000		
Common shares	26,400		
Total equity			$ 30,400
Total liabilities and equity			$100,000

[a]No single claim exceeds $2,000.
[b]These bonds have a first lien on an office building.
[c]Subordinated to the notes payable.

Table 23-2 Poverty Stricken Corporation Priority of Claims ($000)

Distribution of Proceeds		
Liquidation value of assets		$44,000
Priority of claims:		
1. First mortgage receipts from the sale of an office building	$10,000	
2. Administrative expenses	8,000	
3. Salaries due employees	3,600	
Total prior claims		–$21,600
Amount available for settling claims of general creditors		$22,400

Table 23-3 Poverty Stricken Corporation Claims of General Creditors

General Claims	Claim	40% of Claims	Adjustment for Subordination
Accounts payable	$22,000,000	$ 8,800,000	$ 8,800,000
Notes payable	20,000,000	8,000,000	12,000,000
Remaining portion of first mortgage bond	4,000,000	1,600,000	1,600,000
Subordinated debenture	10,000,000	4,000,000	0
Totals	$56,000,000	$22,400,000	$22,400,000

SUMMARY

Business combinations historically have represented a major influence in the growth of firms within North America. This avenue for growth has been particularly important during selected periods, such as the latter part of the 1980s.

The assertion that merger activity creates wealth for the shareholder cannot be maintained with certainty. Only if the merger provides something that the investor cannot do on his or her own can a merger or acquisition be of financial benefit. There may in fact be certain benefits that accrue to the shareholders, but we must be careful to take a position that can be justified.

Financing the Merger

Financial innovation frequently has come in conjunction with corporate restructuring; the most notable examples include MBOs, LBOs, and leveraged ESOPs. These concepts are not new, but the extent of their application has increased significantly. In more recent years, we have seen a marked increase in the use of debt. Before the takeovers of the 1980s, managers endeavoured to maintain a targeted capital structure. In the 1980s, the use of an LBO sometimes meant that an acquired firm's debt might be increased tenfold; the intent was not to maintain that high level of debt, but rather to bring it down to a more acceptable level as soon as possible. In the 1990s, the failure of some large firms to meet their large debt obligations brought on by mergers has caused the use of equity to be the favoured way to finance a merger.

Tender Offers

Normally, the invested purchaser approaches the management of the firm to be acquired. Alternatively, the purchasing firm can directly approach the shareholders of the firm under consideration. This bid for ownership, called a tender offer, has been used with increasing frequency. This approach may be cheaper, but often managers of the target firm attempt to block it.

Divestitures

A divestiture represents a variety of ways to eliminate a portion of the firm's assets. It has become an important vehicle in restructuring the corporation into a more efficient operation.

A more drastic type of divestiture occurs if a company declares voluntarily or is forced involuntarily by its creditors into bankruptcy. As a result, a firm has to be reorganized or liquidated. In the case of a reorganization, an insolvent company may choose to follow the procedures that are specified in either the Bankruptcy and Insolvency Act or the Companies' Creditors Arrangement Act. If the insolvent firm cannot regain profitability, the firm should be liquidated. A firm's management can voluntarily declare bankruptcy by filing an assignent in bankruptcy or the firm can be forced involuntarily into bankruptcy by creditors filing a petition.

STUDY QUESTIONS

23-1. What is the difference between a statutory amalgamation and a business combination?

23-2. Why might merger activities create wealth?

23-3. Describe an MBO, an LBO, and a leveraged ESOP.

23-4. What are the prerequisites for an earn-out plan to be effective?

23-5. Under the base-period earn-out, the amount of future payments depends on what factors?

23-6. What are the disadvantages of the tender offer?

23-7. Explain the different types of divestitures.

23-8. How does the Companies' Creditors Arrangement Act differ from the Bankruptcy and Insolvency Act?

23-9. Explain the provisions of the Bankruptcy and Insolvency Act for the priority of claims.

23-10. Explain the process of liquidation by assignment.

SELF-TEST PROBLEM

ST-1. (*Distribution of Proceeds in Bankruptcy*) Pioneer Enterprises is bankrupt and being liquidated. The liabilities and equity portion of its balance sheet are given below. The book value of the assets was $60 million, but the realized value when liquidated was only $31.58 million, of which $7.55 million was from the sale of the firm's office building. Administrative expenses associated with the liquidation were $4.8 million. Determine the distribution of the proceeds.

Pioneer Enterprises Liabilities and Equities

Current debt		
Accounts payable	$6,500,000	
Accrued wages[a]	400,000	
Notes payable	19,250,000	
Total current debt		$26,150,000
Long-term debt		
Bonds[b]	$14,000,000	
Subordinated debentures[c]	8,500,000	
Total long-term debt		22,500,000
Equity		
Preferred shares	$2,350,000	
Common shares	9,000,000	
Total equity		11,350,000
Total debt and equity		$60,000,000

[a]No single claim exceeds $2,000.
[b]Have first lien on the office building.
[c]Subordinated to notes payable.

STUDY PROBLEMS (SET A)

23-1A. (*Distribution of Proceeds in Bankruptcy*) Loggins Industries has filed a bankruptcy petition and is in the process of being liquidated. Its liabilities and equity are shown below. The book value of the assets is $80 million, but the liquidation value is $42 million. Administrative expenses of the liquidation procedures were $2 million. What is the distribution of proceeds? (See table below).

23-2A. (*Distribution of Proceeds in Bankruptcy*) Heinburner Inc., a local manufacturing concern, is bankrupt and is being liquidated. The book value of the assets is $1.5 million. When the firm was liquidated, $1.25 million was realized. Administrative costs for the liquidation were $150,000. Determine the distribution of proceeds. (See table at top of page 896.)

Loggins Industries Liabilities and Equities

Current debt		
Accounts payable	$10,000,000	
Accrued wages[a]	1,000,000	
Notes payable	21,000,000	
Total current debt		$32,000,000
Long-term debt		
Bonds	$25,000,000	
Subordinated debentures[b]	18,000,000	
Total long-term debt		43,000,000
Equity		
Preferred shares	$ 1,000,000	
Common shares	4,000,000	
Total equity		5,000,000
Total debt and equity		$80,000,000

[a]No single claim exceeds $2,000.
[b]Subordinated to notes payable.

Heinburner Inc. Liabilities and Equities		
Current debt		
Accounts payable	$200,000	
Accrued wages[a]	10,000	
Notes payable	490,000	
Total current debt		$700,000
Long-term debt		$300,000
Equity		
Preferred shares	$100,000	
Common shares	400,000	
Total equity		500,000
Total debt and equity		$1,500,000

[a]No single claim exceeds $2,000.

STUDY PROBLEMS (SET B)

23-1B. (*Distribution of Proceeds in Bankruptcy*) Grigsby Industries has filed a bankruptcy petition and is in the process of being liquidated. Its liabilities and equity are shown below. The book value of the assets is $84 million, but the liquidation value is $43 million. Administrative expenses of the liquidation procedures were $2 million. What is the distribution of proceeds?

Grigsby Industries Liabilities and Equities		
Current debt		
Accounts payable	$12,000,000	
Accrued wages[a]	1,000,000	
Notes payable	21,000,000	
Total current debt		$34,000,000
Long-term debt		
Bonds	$25,000,000	
Subordinated debentures[b]	20,000,000	
Total long-term debt		45,000,000
Equity		
Preferred shares	$1,000,000	
Common shares	4,000,000	
Total equity		5,000,000
Total debt and equity		$84,000,000

[a]No single claim exceeds $2,000.
[b]Subordinated to notes payable.

23-2B. (*Distribution of Proceeds in Bankruptcy*) Scoggins Inc., a local manufacturing concern, is bankrupt and is being liquidated. The book value of the assets is $1.5 million. When the firm was liquidated, $1.25 million was realized. Administrative costs for the liquidation were $150,000. Determine the distribution of proceeds.

Scoggins Inc. Liabilities and Equities

Current debt		
Accounts payable	$190,000	
Accrued wages[a]	12,000	
Notes payable	490,000	
Total current debt		$692,000
Long-term debt		250,000
Equity		
Preferred shares	$100,000	
Common shares	500,000	
Total equity		600,000
Total debt and equity		$1,542,000

[a]No single claim exceeds $2,000.

SELF-TEST SOLUTION

SS-1.

Pioneer Enterprises Distribution of Proceeds

Liquidation value of assets		$31,580,000
Priority of claims		
1. Administrative expenses	$4,800,000	
2. Wages payable	400,000	
3. First mortgage receipts from building sale	7,550,000	
Total prior claims		−12,750,000
Amount available to general creditors		$18,830,000

Claims of General Creditors

Creditors	Amount	46.26% of Claim	Claim After Subordination
Accounts payable	$ 6,500,000	$ 3,007,225	$ 3,007,225
Notes payable	19,250,000	8,906,015	12,838,715
Remainder of first mortgage bonds	6,450,000	2,984,060	2,984,060
Subordinated debentures	8,500,000	3,932,700	–0–
	$40,700,000	$18,830,000	$ 18,830,000

Percentage of claims = amount available to general creditors/amount of claims
= $18,830,000/$40,700,000
= 46.26%

CHAPTER 24

INTERNATIONAL BUSINESS FINANCE

LEARNING OBJECTIVES

After reading this chapter you should be able to

1. Explain how the globalization of product and financial markets has affected business.
2. Explain the concept of exchange rate and how its risk affects investments.
3. Explain the concept of interest rate parity.
4. Explain the purchasing power parity theory and the law of one price.
5. Explain what exchange rate risk is and how it can be controlled.
6. Identify how working-capital management techniques can be used by international businesses to reduce exchange risk and potentially increase profits.
7. Explain how the financing sources available to MNCs differ from those available to domestic firms.
8. Explain the risks involved in direct foreign investments.

INTRODUCTION

Today, international economic boundaries no longer exist. As a result, we now seek answers to the following set of questions in financial management: Where can funds be raised least expensively? Where can this new product be produced least expensively? Where will our major competition come from? The answers will probably involve a foreign country. If we are going to be successful in business, we must have an international perspective in our decision making.

The case of the biotechnology industry typifies how internationalization of business, coupled with recent technological breakthroughs, has created both competition and opportunities for Canadian firms. In this industry, financial executives must be able to look at new projects and examine their risks and returns from an international perspective.

One business battle presently being fought in the international arena is for dominance in biotechnology products. Not only are the profits from success going to come from sales around the world, but the competition is also international in perspective. Although North American companies have been at the forefront of biotechnology, Japanese companies are quickly gaining ground. Drawing on a strategy that has served them well in other industries, Japanese companies are relying heavily on "piggybacking." The key to this strategy is to use their considerable financial assets to supply needed cash to biotechnology laboratories around the world, which then grant the Japanese access to technological breakthroughs. By riding "piggyback" on basic research conducted in other countries, the Japanese companies, in effect, augment their own research budgets, freeing resources for perfecting products that have the best chance of dominating markets around the world. For example, Takeda has signed an agreement for a joint venture with Abbott Labs and forged alliances in Germany, France, and Italy, as well as funded research at Harvard University. These types of strategies demonstrate that businesses are no longer bound by international boundaries and that we must be able to apply our trade around the world.

CHAPTER PREVIEW

This chapter highlights the complications that an international business faces when it operates in environments that have different currencies. We examine strategies for the reduction of foreign exchange risk. We also examine how working-capital management and capital structure decisions are made in an international context. Finally, the decision of a firm to make a direct foreign investment is treated as a capital-budgeting decision in an international business environment.

PERSPECTIVE IN FINANCE

This chapter will highlight multiple currencies as the major dimension of international finance that firms must consider. Effective strategies for the reduction of foreign exchange risk will be discussed. Working-capital management and capital-structure decisions in the international context will also be covered. For the international firm, direct foreign investment is a capital-budgeting decision—with some additional complexities.

OBJECTIVE 1

GLOBALIZATION OF PRODUCT AND FINANCIAL MARKETS

World trade has grown much faster over the last few decades than world aggregate output. Global exports and imports were about one-fifth of global aggregate output in 1962. This figure had increased to about one-fourth by 1972, to more than one-third by 1982, and it continues to increase. However, Figure 24-1 illustrates that the proportion of both exports and imports has averaged about one-fourth of Canadian GDP during the period 1973 to 1995. The value of world exports has grown from US$ 129 billion in 1962 to US$ 3.3 trillion in 1990. This remarkable increase in international trade is reflected in the increased openness of almost all national economies to international influences. In fact, the value of exports of goods and services in Canada has increased from $7,939 million in 1962 to $253,536 million in 1995.

Some industries are even more dependent on the international economy. As an example, the electronic consumer products and automobile industries are now widely considered global industries. There has also been a rise in the global level of international portfolio and direct investment. As an example, both direct and portfolio investment in Canada have been increasing faster than Canadian investment overseas. **Direct investment** occurs when the **multinational corporation (MNC)** has control over the investment, such as when it builds an off-shore manufacturing facility. **Portfolio investment** involves financial assets with maturities greater than one year, such as the purchase of foreign shares and bonds. Both portfolio and direct foreign investments in Canada now exceed such Canadian investments overseas.

Direct investment

Physical assets, such as plant and equipment, acquired outside a corporation's home country but operated and controlled by that corporation.

Multinational corporation (MNC)

A corporation with holdings and/or operations in one or more countries.

Portfolio investment

An investment whereby financial assets are added to a corporation's portfolio of assets (for example, when a multinational company invests in foreign securities).

A major reason for long-run overseas investments of Canadian companies is the high rates of return obtainable from investments in other countries. The amount of

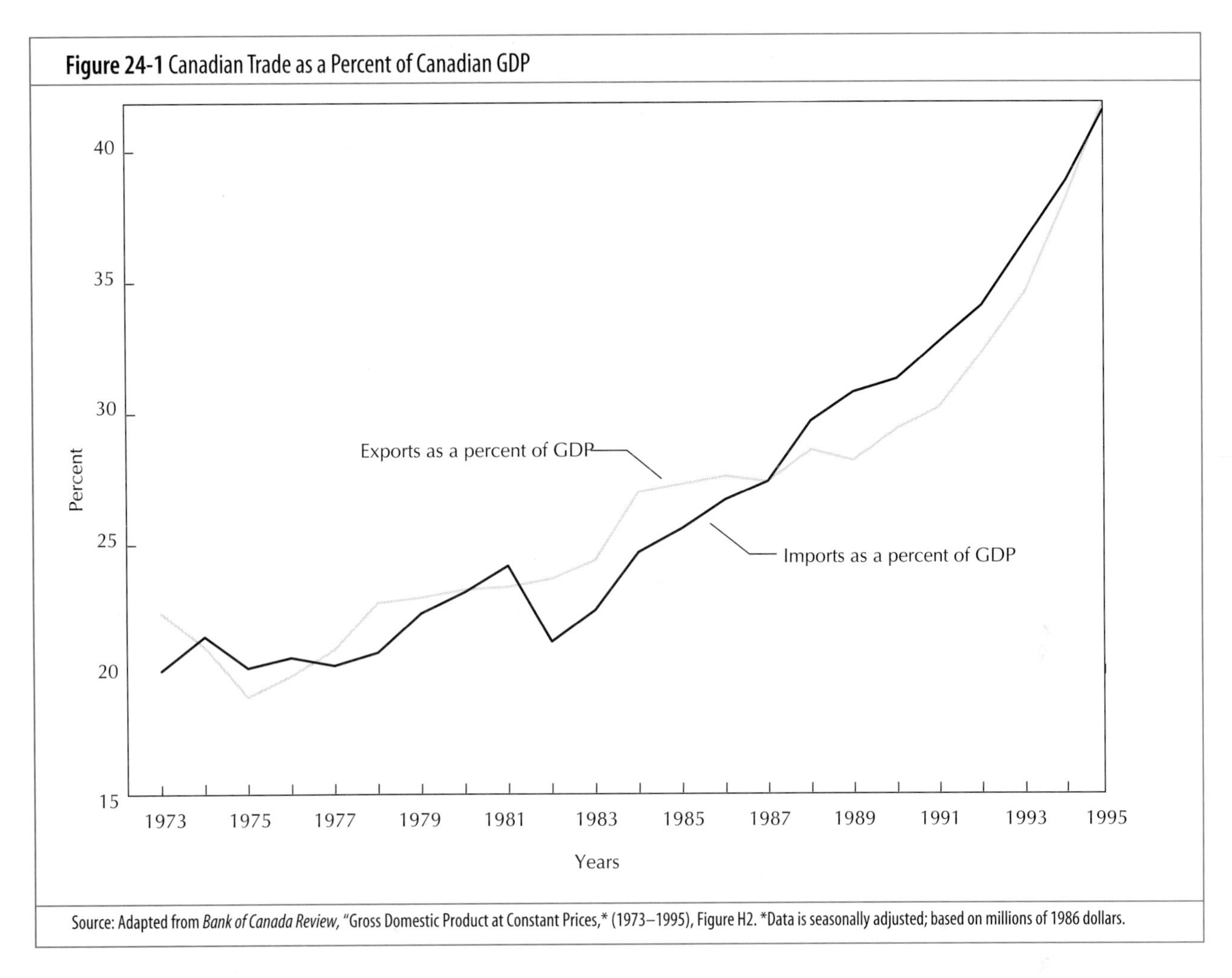

Figure 24-1 Canadian Trade as a Percent of Canadian GDP

Source: Adapted from *Bank of Canada Review*, "Gross Domestic Product at Constant Prices,* (1973–1995), Figure H2. *Data is seasonally adjusted; based on millions of 1986 dollars.

Canadian direct foreign investments (DFI) abroad is large and growing. A large number of Canadian corporations are now highly dependent upon their overseas investments. Significant amounts of the total assets, sales, and profits of Canadian MNCs are attributable to foreign investments and their foreign operations. Direct foreign investments are not limited to Canadian firms. Many European and Japanese firms have operations abroad, too. During the last decade these firms have been increasing their sales and setting up production facilities in Canada and abroad.

The basic problems facing international companies differ from those facing domestic companies. Examples of additional complexities of conducting international business would include the following:

1. **Multiple currencies.** Revenues may be denominated in one currency, costs in another, assets in a third, liabilities in a fourth, and share price in a fifth. Thus, in maximizing shareholder wealth, a firm will use several different currency values.
2. **Differing legal and political environments.** International variations exist in tax laws, depreciation allowances, and other accounting practices, as well as in government regulation and control of business activity. Repatriation of profits may be a problem in certain countries.[1]
3. **Differing economic and capital markets.** The extent of government regulation and control of the economy and capital markets may differ greatly across nations. For example, the ability of a foreign company to raise different types and amounts of capital may be restricted.

[1]Repatriation of profits refers to the withdrawal of profits of foreign operations to the home country of the MNC.

4. **Internal control challenge.** It may be difficult to organize, evaluate, and control different divisions of a company when they are separated geographically and when they operate in different environments.

Capital flows between countries for international financial investment purposes have also been increasing. Many firms, investment companies, and individuals invest in the capital markets of foreign countries. The motivation is twofold: to obtain returns higher than those obtainable in the domestic capital markets and to reduce portfolio risk through international diversification. The increase in world trade and investment activity is reflected in the recent globalization of financial markets. The Eurodollar market is larger than any domestic financial market. Companies are increasingly turning to this market for funds. Even companies and public entities that have no overseas presence are beginning to rely on this market for financing.

In addition, most national financial markets are becoming more integrated with global markets because of the rapid increase in the volume of interest rate and currency swaps. Because of the widespread availability of these types of swaps, the currency denomination and the source country of financing for many globally integrated companies are dictated by accessibility and relative cost considerations regardless of the amount of currency needed by the firm.

The foreign exchange markets have also grown rapidly, and the weekly trading volume in these globally integrated markets, between $3 and $5 trillion, exceeds the annual trading volume on the world's securities markets. Even a purely domestic firm that buys all its input and sells all its output in its home country is not immune to foreign competition, nor can it totally ignore the workings of the international financial markets.

OBJECTIVE 2

PRINCIPLES OF INTERNATIONAL TRADE

In order to fully understand the financial dimensions of international trade we must develop an understanding of the exchange or conversion of one currency for another. In so doing we will consider exchange rate risk, which arises where a firm in one country sells goods or services and in exchange receives payment in the currency of the purchasing nation.

Recent History of Exchange Rates

Between 1949 and 1970 the exchange rates between the major currencies were fixed. However, the Government of Canada pursued both a floating rate policy and a fixed rate policy during the period 1949 to 1972. All countries were required to set a specific parity rate for their currency vis-à-vis the U.S. dollar.[2] For example, consider the German currency, the Deutschemark (DM). In 1949 the parity rate was set at 4.0 DM to $1. The actual exchange rate prevailing on any day was allowed to lie within a narrow band around the parity rate. The DM was allowed to fluctuate between 4.04 and 3.96 per U.S. dollar. A country could effect a major adjustment in the exchange rate by changing its parity rate with respect to the U.S. dollar. When the currency was made cheaper with respect to the U.S. dollar, this adjustment was called a *devaluation*. A *revaluation* resulted when a currency became more expensive with respect to the dollar. In 1969 the DM parity rate was adjusted to 3.66 per U.S. dollar. This adjustment was a revaluation of the DM parity by 9.3 percent. The new bands around the parity were 3.7010 and 3.6188 DM per U.S. dollar.

[2]From 1870 until World War I, the world economic system operated under a gold standard. All settlements between countries had to be met in gold. Thus, a country that ran a trade deficit would lose gold reserves. The system of fixed rates was established by the Bretton Woods Conference in July 1944. Under this system all currencies were pegged to the dollar and the dollar was pegged to gold.

Since 1973 a floating-rate international currency system has been operating. For most currencies, there are no parity rates and no bands within which the currencies fluctuate.[3] Most major currencies, including the Canadian dollar, fluctuate freely, depending upon their values as perceived by the traders in foreign exchange markets.[4] The country's relative economic strengths, its level of exports and imports, the level of monetary activity, and the deficits or surpluses in its balance of payments (BOP) are all important factors in the determination of exchange rates.[5] Short-term, day-to-day fluctuations in exchange rates are caused by changing supply and demand conditions in the foreign exchange market.

The Foreign Exchange Market

The foreign exchange market provides a mechanism for the transfer of purchasing power from one currency to another. This market is not a physical entity like the Toronto Stock Exchange; it is a network of telephone and cable connections among banks, foreign exchange dealers, and brokers.[6] The market operates simultaneously at three levels. At the first level, customers buy and sell foreign exchange through their banks. At the second level, banks buy and sell foreign exchange from other banks in the same commercial centre. At the last level, banks buy and sell foreign exchange from banks in commercial centres in other countries. Some important commercial centres for foreign exchange trading are New York, London, Zurich, Frankfurt, Hong Kong, Singapore, Tokyo, Toronto, and Montreal.

An example will illustrate this multilevel trading. A manufacturer in Ontario may buy foreign exchange (yen) from the Royal Bank for payment to a Japanese supplier against some purchase made. The Royal Bank may buy the yen from another bank in New York or from a bank in Tokyo.

Since this market provides transactions in a continuous manner for a very large volume of sales and purchases, the exchange market is efficient. In other words, it is difficult to make a profit by shopping around from one bank to another. Minute differences in the quotes from different banks are quickly eliminated by arbitrageurs. In fact, the arbitrage mechanism ensures that quotes to different buyers in Tokyo and New York are likely to be the same.

Two major types of transactions are carried out in the foreign exchange markets: *spot* and *forward transactions.*

Spot Exchange Rates

A typical spot transaction involves a Canadian firm buying foreign currency from its bank and paying for it in Canadian dollars. The price of foreign currency in terms of the domestic currency is the exchange rate. Another type of spot transaction occurs if a Canadian firm receives foreign currency from abroad. The firm typically would sell the foreign currency to its bank for Canadian dollars. These are both **spot transactions**, because one currency is exchanged for another currency today. The actual exchange rate quotes are expressed in several different ways, as discussed below.

Spot transaction

A transaction made immediately in the market place at the market price.

Direct and Indirect Quotes

On the spot exchange market the quoted exchange rate is typically called a **direct quote**. A direct quote indicates the number of units of the home currency required

Direct quote

The exchange rate that indicates the number of units of the home currency required to buy one unit of foreign currency.

[3]The system of floating rates is referred to as the "floating-rate regime."

[4]In practice, the floating rate policy may be misleading since the Government of Canada actively supports the Canadian dollar by selling and buying its foreign money reserves.

[5]The balance of payments reflects the difference between the imports and the exports of goods (the trade balance) and services (capital inflows and outflows are tabulated under services).

[6]The foreign exchange market is an over-the-counter market.

Table 24-1 Foreign Exchange Rates (Toronto Selling Rate) December 29, 1995

(1) Country	(2) Contract	(3) Direct Quote (Canadian $ Equivalent)	(4) Indirect Quote (Currency per Canadian $)
U.S. (dollar)	Spot	1.3652	0.7325
	1-month forward	1.3653	0.7324
	6-month forward	1.3663	0.7319
	12-month forward	1.3700	0.7299
Britain (pound)	Spot	2.1202	0.4716
	1-month forward	2.1187	0.4720
	6-month forward	2.1129	0.4733
	12-month forward	2.1086	0.4743
Japan (yen)	Spot	0.013220	75.6430
	1-month forward	0.013281	75.2955
	6-month forward	0.013566	73.7137
	12-month forward	0.013909	71.8959

Adapted from *The Globe and Mail*, December 30, 1995.

to buy one unit of the foreign currency. That is, in Toronto the typical exchange-rate quote indicates the number of Canadian dollars needed to buy one unit of a foreign currency: dollars per pound, dollars per mark, and so on. The spot rates in Table 24-1 are the direct exchange quotes taken from *The Globe and Mail* on December 30, 1995. In order to buy one U.S. dollar on December 29, 1995, 1.3652 Canadian dollars were needed. In order to buy one pound and one yen, 2.1202 dollars and 0.013220 dollars were needed, respectively. The quotes in the spot market in Tokyo are given in terms of yen and those in London in terms of pounds.

Indirect quote

The exchange rate that expresses the required number of units of foreign currency to buy one unit of home currency.

An **indirect quote** indicates the number of units of foreign currency that can be bought for one unit of the home currency. This reads as U.S. dollars per dollar, pounds per dollar, and so on. Some indirect quotes are given in Table 24-1.

In summary, a direct quote is the dollar/foreign currency rate, and an indirect quote is the foreign currency/dollar rate. Therefore, an indirect quote is the reciprocal of a direct quote and vice versa. The following example illustrates the computation of an indirect quote from a given direct quote.

EXAMPLE

Compute the indirect quotes from the direct quotes of spot rates for U.S. dollars, pounds, and yen given in Table 24-1. The direct quotes are: U.S. dollar, 1.3652; pound, 2.1202; and yen, 0.013220. The related indirect quotes are computed as follows:

$$\text{indirect quote} = \frac{1}{\text{direct quote}}$$

Thus

U.S. dollars:

$$\frac{1}{1.3652(\$/\text{U.S. dollar})} = 0.7325(\text{U.S. dollar}/\$)$$

pounds:

$$\frac{1}{2.1202\ (\$/£)} = 0.4716\ (£/\$)$$

Yen:

$$\frac{1}{0.013220} (\$/\text{Yen}) = 75.6430 (\text{Yen}/\$)$$

Notice that the above direct quotes and indirect quotes are identical to those shown in Table 24-1.

Direct and indirect quotes are useful in conducting international transactions as the following examples show.

EXAMPLE

A Canadian business must pay 100,000 yen to a Japanese firm on December 29, 1995. How many dollars would have been required for this transaction?

0.013220($/yen) × 100,000 (yen) = $1,322

EXAMPLE

A Canadian business must pay $2,000 to a British resident on December 29, 1995. How many pounds does the British resident receive?

0.4716 (£/$) × 2,000 ($) = £943.2

Exchange Rates and Arbitrage

The foreign exchange quotes in two different countries must be in line with each other. For example, the direct quote for Canadian dollars in London is given in dollars/pound. Since the foreign exchange markets are efficient, the direct quotes for the Canadian dollar in London, on December 29, 1995, must be very close to the indirect rate of 0.4716 pound/dollar prevailing in Toronto on that date.

If the exchange-rate quotations between the London and Toronto spot exchange markets were out of line, then an enterprising trader could make a profit by buying in the market where the currency was cheaper and selling it in a market where the currency is more expensive. Such a buy-and-sell strategy would involve a zero net investment of funds for a very short time, and no risk bearing would provide a sure profit. Such a person is called an **arbitrageur**, and the process of buying and selling in more than one market to make a riskless profit is called **arbitrage**. Spot exchange markets are efficient in the sense that arbitrage opportunities do not persist for any length of time. That is, the exchange rates between two different markets are quickly brought in line, aided by the arbitrage process. Simple arbitrage eliminates exchange rate differentials across the market for a single currency, as in the above example for the Toronto and London quotes. Triangular arbitrage does the same across the markets for all currencies. Covered interest arbitrage eliminates differentials across currency and interest rate markets.

Arbitrageur

A person involved in the process of buying and selling in more than one market to make riskless profits.

Arbitrage

The process of buying and selling in more than one market in order to make a riskless profit.

Suppose that London quotes 0.4900 £/$ instead of 0.4716 £/$. If you simultaneously bought a pound in Toronto for 0.4716 £/$ and sold a pound in London for 0.4900 £/$ you would have done the following: taken a zero net investment position since you bought one pound and sold one pound, locked in a sure profit of .0184 £/$ no matter which way the pound subsequently moves, and set in motion the forces that will eliminate the different quotes in Toronto and London. As others in the marketplace learn of your transaction, they will attempt to make the same transaction. The increased demand to buy pounds in Toronto will lead to a higher quote there and the increased supply of pounds will lead to a lower quote in London. The workings of the market will produce a new spot rate that lies between 0.4716 £/$ and 0.4900 £/$ and is the same in Toronto and in London.

Ask and Bid Rates

In the spot exchange market two types of rates are quoted: the ask and the bid rates. The *ask rate* is the rate the bank or the foreign exchange trader "asks" the customer to pay in home currency for foreign currency when the bank is selling and the customer is buying. The ask rate is also known as the *selling rate* or the *offer rate.* The *bid rate* is the rate at which the bank buys the foreign currency from the customer by paying in home currency. The bid rate is also known as the buying rate. Note that Table 24-1 contains only the selling, offer, or ask rates, and not the buying rate.

The bank sells a unit of foreign currency for more than it pays for it. Therefore, the direct ask quote ($/foreign currency) is greater than the direct bid quote. The difference is known as the **bid-ask spread.** Thus, the indirect ask quote must be less than the indirect bid quote. For example, assume the direct ask and bid quotes are 1.2 and 1.15, respectively. The indirect ask quote will be $1/1.2 = .8333$. The indirect bid quote will be $1/1.15 = .8696$. The indirect bid is greater than the indirect asked quote.

Bid-ask spread

The difference between the ask quote and the bid quote.

When there is a large volume of transactions and the trading is continuous, the spread between the ask and the bid rates can be as small as 0.5 percent for major currencies. The spread is much higher for infrequently traded currencies than for frequently traded currencies. The spread exists to compensate the banks for holding the risky foreign currency and for providing the service of converting currencies.[7]

Cross Rates

A **cross rate** is the exchange rate for a currency derived from the exchange rates of two other currencies. The following example illustrates how this works.

Cross rate

The indirect computation of the exchange rate of one currency from the exchange rates of two other currencies.

EXAMPLE

The dollar/pound and the yen/dollar rates are given in Table 24-1. From this information, we could determine the yen/pound and pound/yen exchange rates. We see that

$$\left(\frac{\$}{£}\right) \times \left(\frac{\text{Yen}}{\$}\right) = \left(\frac{\text{Yen}}{£}\right)$$

or

$$2.1202 \times 75.6430 = 160.3783\left(\frac{\text{Yen}}{£}\right)$$

Thus, the pound/yen exchange rate is

$$\frac{1}{160.3783} = 0.006235\left(\frac{£}{\text{Yen}}\right)$$

Cross-rate computations make it possible to use quotations in Toronto to compute the exchange rate between pounds, marks, and francs. Arbitrage conditions hold in cross rates, too. For example, the pound exchange rate in Tokyo yen (the direct quote or the Yen/£) must be 160.3783. The yen exchange rate in London must be 0.006235 £/yen. If the rates prevailing in Tokyo and London were different from the computed cross rates, using quotes from Toronto, a trader could use three different markets and lock in arbitrage profits through this process. The arbitrage condition for the cross rates is called triangular arbitrage.

Forward Exchange Rates

A forward exchange contract requires delivery, at a specified future date, of one currency for a specified amount of another currency. The exchange rate for the

[7]The spread can also be seen as the bank's compensation for bearing exchange rate risk as well as holding both long and short positions simultaneously.

future transaction is agreed upon today; the actual payment of one currency and the receipt of another currency take place at the future date. For example, a 1-month contract on the 1st of March is good for delivery on the 1st of April. Note that the forward rate is not the same as the spot rate that will prevail in the future. The actual spot rate that will prevail is not known today; only the forward rate is known. The actual spot rate will depend upon the market conditions at that time; it may be more or less than today's forward rate. **Exchange rate risk** is the risk that tomorrow's exchange rate will differ from today's rate.

Exchange rate risk

The risk that tomorrow's exchange rate will differ from today's rate.

As indicated earlier, it is extremely unlikely that the future spot rate will be exactly the same as the forward rate quoted today. Assume that you are going to receive a payment denominated in pounds from a British customer in 1 month. If you wait for 1 month and exchange the pounds at the spot rate, you will receive a dollar amount reflecting the exchange rate 1 month hence (that is, the future spot rate). As of today, you have no way of knowing the exact dollar value or your exact future dollar receipts. Consequently, you cannot make precise plans about the use of these dollars. If, on the other hand, you buy a forward contract, then you know the exact dollar value of your future receipts, and you can make precise plans concerning their use. Therefore, the forward contract can reduce your uncertainty about the future. This reduction of uncertainty or risk reduction is the major advantage of using the forward market.

Forward contracts are usually quoted for periods of 1, 3, 6, and 12 months. A contract for any intermediate date can be obtained, usually with the payment of a small premium. Contracts for periods greater than one year can be costly.

Forward rates, like spot rates, are quoted in both direct and indirect form. The direct quotes for the 1-month, 6-month, and 12-month forward contracts on U.S. dollars, pounds, and yen are given in Table 24-1. The indirect quotes for forward contracts are reciprocals of the direct quotes. The indirect quotes are indicated in Table 24-1. The direct quotes are the dollar/foreign currency rate, and the indirect quotes are the foreign currency/dollar rate similar to the spot exchange quotes.

The 1-month forward quote for pounds is \$2.1187 per pound. This means that if one purchases the contract for forward pounds on December 29, 1995, the bank will deliver a pound against the payment of \$2.1187 on January 29, 1996. The bank is contractually bound to deliver the pound at this price, and the buyer of the contract is legally obligated to buy it at this price on January 29, 1996. Therefore, this is the price the customer must pay regardless of the actual spot rate prevailing on January 29, 1996. If the spot price of the pound is less than \$2.1187, then the customer pays more than the spot price. If the spot price is greater than \$2.1187, then the customer pays less than the spot price.

The forward rate is often quoted at a premium or discount from the existing spot rate. For example, the 1-month pound forward rate may be quoted at a .0015 discount (2.1202 spot rate − 2.1187 forward rate). If the forward contract is selling for fewer dollars than the spot—that is, a smaller direct quote—the forward contract is said to be selling at a discount. This example implies that the dollar is expected to be strengthened against the pound. In other words, the market expects that in the future fewer dollars will be required to buy a pound. Conversely, the pound is expected to be worth fewer dollars in the future. When the forward contract sells for more dollars than the spot—a larger direct quote—the forward rate is said to be at a premium from the spot rate. Notice in Table 24-1 that the forward contracts are selling at a premium over the spot for U.S. dollars and Japanese yen. This premium or discount is also called the *forward-spot differential.*

The relationship may be written as

$$F - S = P\,(or\,D) \tag{24-1}$$

where
F = the forward rate, direct quote
S = the spot rate, direct quote
P = the premium, when F is greater than S
D = the discount, when S is greater than F

The premium or discount can also be expressed as an annual percent rate, computed as follows:

$$\frac{(F-S)}{S} \times \frac{12}{n} \times 100 = P \text{ (or } D) \qquad (24\text{-}2)$$

where F, S = the forward and spot rates defined earlier
n = the number of months of the forward contract
P = the annualized percentage premium on forward, if $F > S$
D = the annualized percentage discount on forward, if $S > F$

EXAMPLE

Compute the percent-per-annum premium on the 1-month forward pound.

***Step 1:* Identify F, S, and n.**

$F = 2.1187$, $S = 2.1202$, $n = 1$ month

***Step 2:* Since S is greater than F, we compute D via Equation (24-2).**

$$D = \frac{(2.1202 - 2.1187)}{2.1202} \times \frac{12 \text{ months}}{1 \text{ month}} \times 100 = 0.849\%$$

The percent-per-annum discount on the 1-month pound is 0.849 percent.

The percent-per-annum premiums or discounts on the 1-month and 6-month U.S. dollar, pound, and yen contracts are computed similarly. The results are given in Table 24-2.

Examples of Exchange Rate Risk

The concept of exchange rate risk applies to all types of international businesses. The measurement of these risks, and the type of risk, may differ among businesses. Let us see how exchange rate risk affects international trade contracts, international portfolio investments, and direct foreign investments.

Exchange Rate Risk in International Trade Contracts

The idea of exchange rate risk in trade contracts is illustrated in the following situations.

Case I A Canadian automobile distributor agrees to buy a car from the manufacturer in Oshawa. The distributor agrees to pay $9,500 upon delivery of the car, which is expected to be one month from today. The car is delivered one month from today and the distributor pays $9,500. Notice that, from the day this contract was written until the day the car was delivered, the buyer knew the exact dollar amount of the liability. There was, in other words, no uncertainty about the value of the contract.

Case II A Canadian automobile distributor enters into a contract with a British supplier to buy a car from the United Kingdom for 5,000 pounds. The amount is payable on the delivery of the car, six months from today. From Figure 24-2, we see the range of spot rates that we believe can occur on the date the contract is consummated. Six months from today, the Canadian importer will pay some amount in the range of $9,937 (5,000£ × 1.9874$/£) to $11,192 (5,000£ × 2.2384$/£)

Table 24-2 Percent-per-Annum Premium (Discount)

	1-month	6-month
Pound	(0.849)%	(0.689)%
U.S. Dollar	0.088	0.161
Yen	5.537	5.234

Figure 24-2 Subjective Probability Distribution of the Pound Exchange Rate Six Months into the Future

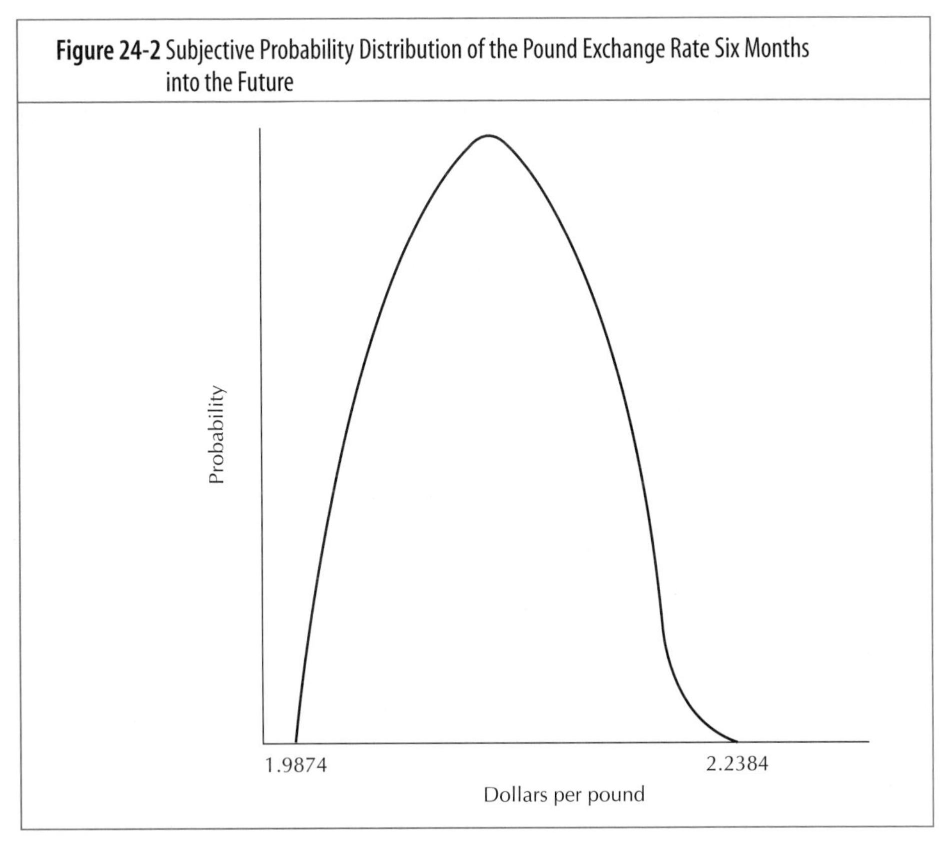

for the car. Today, the Canadian firm is not certain what its future dollar outflow will be six months hence. That is, the dollar value of the contract is uncertain.

These two examples help illustrate the idea of foreign exchange risk in international trade contracts. In the domestic trade contract the exact dollar amount of the future dollar payment is known today with certainty. In the case of the international trade contract, it is written in the foreign currency—the exact dollar amount of the contract is not known. The variability of the exchange rate induces variability in the future cash flow.

Exchange rate risk exists when the contract is written in terms of the foreign currency or denominated in foreign currency. There is no direct exchange risk if the international trade contract is written in terms of the domestic currency. If the contract in Case II was written in dollars, the Canadian importer would face no direct exchange risk. With the contract written in dollars, the British exporter would bear all the exchange risk because the British exporter's future pound receipts would be uncertain. That is, the British exporter would receive payment in dollars, which would have to be converted into pounds at an unknown (as of today) pound-dollar exchange rate. In international trade contracts of the type discussed here, at least one of the two parties to the contract always bears the exchange risk.

Certain types of international trade contracts are denominated in a third currency, different from either the importer's or the exporter's domestic currency. In Case II the contract might have been denominated in, say, the Deutsche mark. With a mark contract, both importer and exporter would be subject to exchange rate risk.

Exchange risk is not limited to two-party trade contracts, this type of risk also exists in foreign portfolio investments and direct foreign investments.

Exchange Risk in Foreign Portfolio Investments

Let us look at an example of exchange risk in the context of portfolio investments. A Canadian investor buys a German security. The exact return on the investment in the security is unknown. Thus, the security is a risky investment. The investment return in the holding period of, say, three months stated in marks could be anything from –2 to +8 percent. In addition, the mark-dollar exchange rate may depreciate by 4 percent or appreciate by 6 percent in the three-month period during which the investment is held. The return to the Canadian investor, in dollars, will therefore be in the range of –6 to +14 percent.[8] Notice that the return to a German investor, in marks, is in the range of –2 to +8 percent. Clearly, for the Canadian investor, the exchange factor induces a greater variability in the dollar rate of return. Hence, the exchange rate fluctuations may increase the riskiness of the investments.

Exchange Risk in Direct Foreign Investment

The exchange risk of a direct foreign investment (DFI) is more complicated. In a DFI the parent company invests in assets denominated in a foreign currency. That is, the balance sheet and the income statement of the subsidiary are written in terms of the foreign currency. The parent company receives the repatriated profit stream in dollars. Thus, the exchange risk concept applies to fluctuations in the dollar value of the assets located abroad as well as to the fluctuations in the home-currency-denominated profit stream. Exchange risk not only affects immediate profits, but also affects the future profit stream as well.

BACK TO THE BASICS

In international transactions, just as in domestic transactions, the key to value is the timing and amounts of cash flow spent and received. However, economic transactions across international borders add an element of risk in that cash flows are denominated in the currency of the country in which business is being transacted. Consequently, the dollar value of the cash flows will depend on the exchange rate that exists at the time cash changes hands. The fact remains, however, that it's cash spent and received that matters. This is the point of ***Axiom 3: Cash—Not Profits—Is King: Measuring the Timing of Costs and Benefits.***

Relevant Measures of Exchange Risk

The relevant measures of exchange risk differ for different types of businesses. Exchange risk represents the variability in expected returns caused by exchange rate fluctuations. Again, total exchange risk can be separated into diversifiable exchange risk and non-diversifiable risk. For example, a Canadian trader who does not trade on international markets frequently faces total exchange risk. Because this trader is not well diversified in trading on international markets, total exchange risk is equal to the total variability in expected returns due to exchange rate fluctuations. For an export-import firm that does continual business in a number of countries, the relevant risk measure is non-diversifiable risk of exchange rate fluctuations. The diversification argument also applies to a multinational corporation that has plants and operations in numerous countries. For corporations with very large investments in a single foreign country, the total variability of the exchange rate is the appropriate measure of exchange risk.

---

[8]Example: Assume the spot exchange rate is .50 dollars per mark. In three months the exchange rate would be .50 × (1 – .04) = .48 to .50 × (1 + .06) = .53. A $50 investment today is equivalent to a 100-mark investment. The 100-mark investment would return 98 to 108 marks in three months. The return, in the worst case, is 98 marks × .48 = $47.04. The return, in the best case, is 108 marks × .53 = $57.24. The holding-period return, on the $50 investment, will be between –6 percent ($47.04 – $50)/$50) and +14 percent ($57.24 – $50)/$50).

THE INTEREST PARITY THEORY

OBJECTIVE 3

All the forward rates given in Table 24-1 entail a premium or a discount relative to current spot rates. However, these forward premiums and discounts differ between currencies and maturities (see Table 24-2). These differences depend solely on the difference in the level of interest rates between the two countries, called the *interest rate differential.* The value of the premium or discount can be theoretically computed from the **interest rate parity theory** (IRPT). This theory states that (except for the effects of small transactions costs) the forward premium or discount should be equal and opposite in sign to the difference in the interest rates of securities that have the same maturity.

Interest rate parity theory (IRPT)

States that (except for the effect of small transaction costs) the forward premium or discount should be equal and opposite in size to the differences in the national interest rates for securities of the same maturity.

Specifically, the premium or discount on the percent-per-annum basis should be equal to the following:[9]

$$P\,(or\;D) = \left(-\frac{I^f - I^d}{1 + I^f}\right) = \left(\frac{I^d - I^f}{1 + I^f}\right) \qquad (24\text{-}3a)$$

where P (or D) = the percent-per-annum premium (or discount) on the forward rate

I^f = the annualized interest rate on a foreign instrument having the same maturity as the forward contract

I^d = the annualized interest rate on a domestic instrument having the same maturity as the forward contract

Note that in order to compute the forward premium on, say, a six-month forward U.S. dollar contract, we need to have the six-month Treasury bill rate in Canada and its counterpart in the United States, both expressed as annual rates.

The IRPT states that the P (or D) calculated by the use of Equation (24-3a) should be the same as the P (or D) calculated by using Equation (24-2), except for small variations due to political and commercial risks and transactions costs. When the interest rates are relatively low, Equation (24-3a) can be approximated by:

$$P\,(\text{or } D) \cong -(I^f - I^d) = (I^d - I^f) \qquad (24\text{-}3b)$$

Notice in Table 24-2 that the 6-month forward U.S. dollar is selling at a premium of approximately 0.161 percent. The IRPT says that the six-month interest rate in the United States must be approximately 0.161 percent (annualized) less than the six-month Treasury bill rate in Canada, as indicated by Equation (24-3a).

Covered Interest Arbitrage

The rationale for the interest rate parity theory is provided by the covered interest arbitrage argument. This argument states that if the premiums (or discounts) in forward rates are not exactly equal to the interest rate differential, then arbitrage or riskless profits can be made.[10] The arbitrage mechanics involved here are substantially more complicated than the "simple" and "triangular" arbitrage discussed earlier. The covered interest arbitrage argument is best explained by the following examples.

EXAMPLE

The U.S. dollar spot and six-month forward rates are $1.3652 and $1.3663, respectively, on December 29, 1995. The 6-month Canada Treasury bill rate on December 29, 1995, expressed as an annual rate was 5.61 percent. Assume now that the rate on six-month American instruments on December 29, 1995, was

[9]Note that the equation implies an equilibrium condition that eliminates any arbitrage opportunities and says nothing about expected future exchange rates. In reality, the equilibrium condition may not hold true at every particular point in time.

[10]Interest rate differential is approximately equal to the right-hand side of Equation (24-3b).

5.04 percent, annualized. Given these data, does a riskless profit (or arbitrage) opportunity exist?

***Step 1*: Compute the forward premium using Equation (24-2).**

$P = 0.161$ percent

***Step 2*: Compute the interest rate differential by Equation (24-3a).**

$$P = \frac{(0.0561 - 0.0504)}{(1 + 0.0504)} = 0.00543 = 0.543\%$$

The premiums computed by the use of the two equations are not identical. The IRPT is violated. Profit opportunities exist.[11]

Since profit opportunities exist, what can the arbitrageur do to make a profit? A Canadian arbitrageur can take the following steps.

On December 29, 1995:

***Step 1:* Borrow 1 million U.S. dollars for six months from the United States money markets at a 5.04 percent annual rate of interest. For six months the interest is 2.52 percent.[12] On June 29, 1996, the arbitrageur will need to repay 1 million U.S. dollars × 1.0252 = 1,025,200 U.S. dollars.**

***Step 2:* Exchange the borrowed U.S. dollars into dollars at the spot exchange rate of $1.3652 per U.S. dollar on December 29, 1995. The arbitrageur receives 1 million U.S. dollars × $1.3652 = $1,365,200.**

***Step 3:* Invest the dollars in the Canadian money market for six months at a 5.61 percent annual rate. For six months the rate is 2.805 percent. Receive, on June 29, 1996, $1,365,200 × 1.02805 = $1,403,494.**

***Step 4:* Enter into a six-month forward contract to buy 1,025,200 U.S. dollars on June 29, 1996. On June 29, 1996, the arbitrageur will need to pay 1,025,200 × 1.3663 = $1,400,731 to receive the 1,025,200 U.S. dollars needed to repay the loan principal and interest on the 1 million U.S. dollar loan taken in Step 1.**

On June 29, 1996 the arbitrage position is closed out as follows:

***Step 1:* Receive $1,403,494 from the Canadian money market investment ($1,365,200 invested plus $38,294 interest at a monthly rate of 2.805 percent).**

***Step 2:* Pay $1,400,731 to the bank to obtain 1,025,200 U.S. dollars at a 1.3663 exchange rate.**

***Step 3:* Pay 1,025,200 U.S. dollars toward the principal and interest on the 1 million U.S. dollar loan.**

On June 29, 1996, the Canadian arbitrageur's net position is no liability and no assets. However, a net profit has been made of $2,763 ($1,403,494 – $1,400,731). None of the arbitrageur's own funds were invested, there was no risk (since the arbitrageur knew payments and receipts exactly on December 29, 1995), yet the arbitrageur made a net profit of $2,763. This arbitrage was accomplished by simultaneously borrowing in one market, investing in another money market, and covering the exchange position in the forward exchange market. The entire process is known as covered interest arbitrage.

Note that the arbitrage profit was possible because the forward premium was not equal to the interest rate differential. Under such circumstances, arbitrageurs enter the market, increase the demand for the forward foreign currency, and drive up the price of the forward contract. The equilibrium price of the forward contract again obeys the interest parity theory.

[11]Note that the interest rate differential was greater than the forward premium and it was better to hold Canadian dollars. If the interest rate differential was less, then one would hold the foreign currency.

[12]The six-month rate, as an approximation, is equal to the annual rate × 6/12.

EXAMPLE

Assume that the interest rates indicated in the previous example are the actual rates prevailing in Canada and the United States. What should be the "correct" price of the forward contract, consistent with the interest rates in the two countries?

***Step 1:* Compute the premium using Equation (24-3a).**

$$P = \frac{(0.0561 - 0.0504)}{(1 + 0.0504)} = 0.00543 = 0.543\%$$

***Step 2:* Using this premium in Equation (24-2), compute the forward rate, *F*.**

$$\frac{(F - 1.3652)}{1.3652} \times \frac{12 \text{ months}}{1 \text{ month}} \times 100 = 0.543\%$$

$$F = \$1.3658\%$$

The example illustrates the technique for computing the correct price of the forward contract. If the quote is less than the computed price, the forward contract is *undervalued.* If the quote is greater than the computed price, the forward contract is *overvalued.* In our example, the forward contract is undervalued.

The forward markets are *efficient* in the sense that the quotes in the market represent the "correct" price of the contract. The markets' efficiency also implies that no profit can be made by computing the prices at every instant and buying the forward contract when they appear undervalued. Some minor deviations from the computed correct price may exist for short periods. These deviations, however, are such that after the transactions costs involved in the four separate arbitrage steps have been recognized, no net profit can be made. Numerous empirical studies attest to the efficiency of the forward markets.

In the previous example the correct price of the forward U.S. dollar was computed under the assumption that the six-month interest rate in the United States was 5.04 percent. All the other data for the examples were obtained from the actual market quotes in Toronto. We now ask, given the data on spot rate, forward rate, and the Canadian treasury bill rate, can we compute the interest rate for the six-month United States instrument? The answer is given in the next example.

EXAMPLE

The spot rate and forward rate for the U.S. dollar are 1.3652 and 1.3663, respectively. The six-month Canadian Treasury bill rate is 5.61 percent annualized. What is the six-month American rate?

***Step 1:* Compute the premium using Equation (24-2). From previous computations,**

$$P = 0.161\%$$

***Step 2:* Using Equation (24-3a), solve for the unknown foreign interest rate.**

$$0.00161 = \frac{(0.0561 - I^f)}{(1 + I^f)}$$

$$I^f = 0.0544, \text{ or } 5.44\%$$

Thus, the actual six-month rate in the United States on December 29, 1995, must have been 5.44% annualized.

OBJECTIVE 4

PURCHASING POWER PARITY

Purchase power parity (PPP) theory

In the long run, exchange rates adjust so that the purchasing power of each currency tends to remain the same. Thus, changes in exchange rate tend to reflect international differences in inflation rates. Countries with high rates of inflation tend to experience declines in the value of their currency.

Long-run changes in exchange rates are influenced by international differences in inflation rates and the purchasing power of each nation's currency. Exchange rates of countries with high rates of inflation will tend to decline. According to the **purchasing power parity (PPP) theory**, exchange rates will tend to adjust in such a way that each currency will have the same purchasing power (especially in terms of internationally traded goods). Thus, if the United Kingdom experiences a 10 percent rate of inflation in a year that Germany experiences only a 6 percent rate, the U.K. currency (the pound) will be expected to decline in value approximately by 4 percent (10 percent − 6 percent) against the German currency (the mark). More accurately, according to the PPP

$$S_{t+1} = \frac{S_t(1 + P_d)}{(1 + P_f)^n} \tag{24-4}$$

$$\cong S_t(1 + P_d - P_f)^n$$

where S_t = the direct exchange rate (units of domestic currency per unit of the foreign currency) at time t
P_f = the foreign inflation rate
P_d = the domestic inflation rate
n = number of time periods

Thus, if the beginning value of the mark was 0.40, with a 6 percent inflation rate in Germany and a 10 percent inflation rate in the U.K., according to the PPP, the expected value of the mark at the end of that year will be 0.40 × [1.10/1.06], or 0.4151.

The Law of One Price

Underlying the PPP relationship is the Law of One Price. This law is actually a proposition that in competitive markets where there are no transportation costs or barriers to trade, the same good sold in different countries sells for the same price if all the different prices are expressed in terms of the same currency. The idea is that the worth, in terms of marginal utility, of a good does not depend on where it is bought or sold. Since inflation will erode the purchasing power of any currency, its exchange rate must adhere to the PPP relationship if the Law of One Price is to hold over time. The following example illustrates the workings of this proposition.

EXAMPLE

Given the information above, assume that today a widget costs DM 100.00 in Germany and £40.00 in the U.K. At the current exchange rate the Law of One Price appears to be holding since Germans can buy widgets for DM 100.00 (× £0.40/DM = £40) and the British can buy German widgets for £40 (× DM/£.40 = DM 100). It is possible to determine how the exchange rate must change if the Law of One Price holds in relative form—for example, over the next year. We might expect that the exchange rate will change to reflect the decline in each currency's domestic purchasing power caused by next year's inflation. Today we know:

DM 100 = 1 widget = £40.00

At this point, Germans would be indifferent between buying German widgets or U.K. widgets. Given that inflation in the U.K. is expected to be 10 percent, one U.K. widget can be expected to cost £44.00 = (£40.00)(1 + .10). At the current exchange rate, in one year U.K. widgets would cost Germans DM 110.00:

DM 110 = 1 widget = £44.00

$$= £44.00 \times \left(\frac{\text{DM}}{.40}\right)$$

We will ignore the effects of German inflation for the moment. Germans would not buy U.K. widgets for DM 110.00 in one year; they would buy German widgets for DM 100.00. Thus at the current exchange rate, PPP would not hold. Germans might, however, be willing to continue spending DM 100.00 one year from now for U.K. widgets. This could happen only if the DM were to strengthen against the pound. The new exchange rate implied by U.K. inflation is:

DM 100 = 1 widget = £44.00

$$\left(\frac{£}{DM}\right) = \frac{£44.00}{100.00\ DM} = \frac{£.44}{DM}$$

or:

$$\left(\frac{£}{DM}\right) = £0.40 \times (1 + P_{U.K.}) = £0.40 \times 1.10$$

Note that the terms of trade are important—Germans want the same amount of goods for the same number of DM before and after the U.K. inflation. After the U.K. inflation, the DM buys more pounds at the new exchange rate. In addition, the British will also be paying £40.00 per widget in today's pounds—all U.K. goods will cost 10 percent more next year because of inflation. In other words, widgets will cost more in nominal terms next year in the U.K., but the real cost of widgets will be unchanged.

Now we will consider German inflation. If since the DM's purchasing power is also expected to decline over the next year, Germans would be willing to pay DM 106.00 (= DM 100.00 × (1 + P_{DM})) for a U.K. widget in one year since this is what they would have to pay for a German widget at that time. The new exchange rate becomes the same as the PPP estimate calculated above:

DM 106 = 1 widget = £44.00

$$\left(\frac{£}{DM}\right) = \frac{£44.00}{DM\ 106.00} = \frac{£.4151}{DM}$$

or:

$$\left(\frac{£}{DM}\right) = (£0.40)\frac{(1 + P_{U.K.})}{(1 + P_{DM})} = \frac{£0.4151}{DM}$$

International Fisher Effect

According to the Fisher effect, interest (I) rates reflect the expected inflation rate (P) and a real rate of return (I_r). In other words

$$1 + I = (1 + P)(1 + I_r)$$

and

$$I = P + I_r + PI_r$$

While there is mixed empirical support for the Fisher effect internationally, it is widely commented that for the major industrial countries, I_r is about 3 percent, when a long-term period is considered. In such a case, with the previous assumption regarding inflation rates, interest rates in the U.K. and Germany would be (1+ 0.10) (1 + 0.03) − 1 or 13.3 percent and (1 + 0.06) (1 + 0.03) − 1 or 9.18 percent, respectively.

In addition, according to the IRPT, the expected premium for the mark forward rate should be (0.133 − 0.0918)/1.0918 or 3.774 percent (rounded off). Starting with a mark value of 0.40 gives us a one-year forward rate of £0.40 × (1.03774) = £0.4151. As you may notice, this one-year forward rate is exactly the same as the PPP expected spot rate one year from today. In other words, if the real rate (I_r) is the same in both Germany and the U.K. and expectations regarding inflation rates hold true, today's one-year forward rate is likely to be the same as the future spot rate one year from now. Thus, it is contended that in efficient markets, with rational

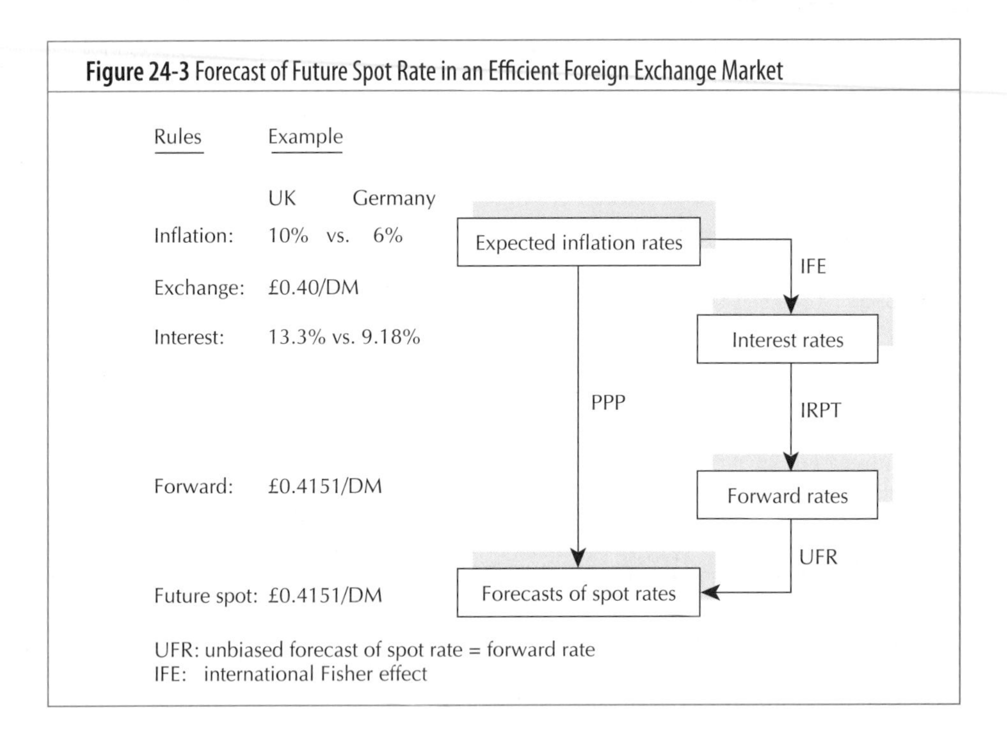

Figure 24-3 Forecast of Future Spot Rate in an Efficient Foreign Exchange Market

expectations, the forward rate is an unbiased (not necessarily accurate) forecast of the future spot rate. These relationships between inflation and interest rates, and spot and forward rates are depicted in Figure 24-3.

The actual future spot rate, as of today, is a random variable. In Figure 24-2 we indicated a possible hypothetical distribution of the pound spot rate on June 29, 1996. This distribution is what we, on December 29, 1995, subjectively believed the future spot rate to be. Notice in Figure 24-2 that the exchange rate may be as low as 1.9874 dollars or as high as 2.2384 dollars per pound with different probabilities.

Yet given the available information, we have done our best to accurately forecast the future spot exchange rate. Such forecasts are useful in eliminating exchange risk for near-term international transactions, called *hedging*. However, there is no easy way to hedge long-term cash flows for international operations. The problem is that the terms of trade can also change; exchange rate changes can go beyond PPP-induced changes, leading to real exchange rate changes. Currency gains and losses from nominal exchange rate changes will generally be offset over time by differences in relative rates of inflation between two countries if PPP holds and there are no real exchange rate changes.

BACK TO THE BASICS

Although exchange-risk can be a serious complication in international business activity, remember ***Axiom 1: The Risk-Return Tradeoff: We Won't Take on Additional Risk Unless We Expect to Be Compensated with Additional Return.*** *In the international market, traders and corporations find numerous reasons that the returns from international transactions outweigh the risks.*

OBJECTIVE 5

EXPOSURE TO EXCHANGE RATE RISK

An asset which is denominated in or valued in terms of foreign currency cash flows will lose value if that foreign currency declines in value. It can be said that such an asset is exposed to exchange rate risk. However, this possible decline in asset value

Hedging Foreign Exchange Risk

Every form of business activity contains risk; it can never be completely eliminated. In finance, we classify the risk that a firm is exposed to as either commercial (business) or financial risk. We include interest rate risk and foreign exchange risk as components of financial risk. In this excerpt from the article entitled, "Hedging: A User's Manual," we examine how hedging techniques are used to reduce foreign exchange risk.

Frequently, a commercial or financial transaction must be settled in foreign currency. An increase in the exchange rate means that the cost of borrowing or of buying on credit increases. Conversely, a decrease in the exchange rate reduces the return on credit extended and increases the cost of credit used.

A Strategy to Manage Financial Risk

Finding ways to protect an entity against financial risks demands strong technical knowledge, serious planning and an ability to anticipate events. Clairvoyants excepted, those determined to establish an effective hedging strategy must take three steps: define the objectives, measure exposure to financial risk and apply the strategy.

Defining the objectives—The objectives of a hedging strategy should be defined in terms of amount, duration and nature of coverage. What's the proportion of the overall value of the financial instrument that will be hedged and what level of risk is acceptable? Is complete or partial coverage needed or can the company opt instead for a form of self insurance? How long are the financial markets expected to be unstable?

Following that, the type of coverage must be settled. When will it take effect? Will the value of the financial instrument be guaranteed at a given date or will the coverage be continuous, ensuring that the value of the instrument will always exceed a predetermined minimum value? Will it be used to freeze a value or a profit, or to curb losses while making it possible to take advantage of a favourable market swing?

Measuring the exposure—Start by calculating the difference between the market value of the assets whose value is influenced by the risk factor identified (the exchange rate for example) and the market value of the liabilities whose value is influenced by this same risk factor. Possible variations of the risk factor are then applied to the difference measured in order to estimate the gains or losses that could result from the exposure.

Choosing the Instrument

Hedging Against Foreign Exchange Risk

There are many instruments available to freeze or guarantee exchange risk.

Freezing an exchange rate—With a freeze, when an agreement is made to conclude a foreign currency transaction, the exchange rate for the future transaction is predetermined. Forward contracts and cross-currency swaps are among the instruments that can be used for this purpose.

Foreign exchange contracts result from an agreement under which a bank guarantees its client that, at a specified date or within a specified time period, it will exchange a certain quantity of a given currency at the specified rate. The buyer has to proceed with the exchange and cannot benefit under any circumstances from favourable fluctuations in currency rates. In effect, a forward contract defers the payment of the foreign currency.

Suppose that at the beginning of February a company agreed to purchase $1 million (US) in goods from a supplier in the United States. Payment will be made in US dollars at the beginning of June. The company wants to freeze the exchange rate that will apply when its debt matures and enters into a forward contract to buy, on June 1, $1 million (US) for $1.21 million (Cdn). The US dollar is currently exchanged for $1.19 (Cdn).

Supposing that, on the settlement date, the US dollar is worth $1.24 (Cdn), the forward contract will have generated the following saving:

Loss avoided	
(($1.24 - $1.21) x $1 million)	$30,000
Cost of hedge	
(($1.21 - $1.19) x $1 million)	$20,000
Amount saved:	$10,000

Currency futures are more like a loan than insurance. Using a straightforward hedge, the buyer could have bought $1 million

(US) at the beginning of February and invested it in a term deposit until the debt matured. The bank merely performed this transaction on the client's behalf and charged $20,000 for the service.

A cross-currency swap is a transaction executed in three stages, and is intended to eliminate the foreign exchange risk inherent to a foreign currency loan or investment. The swap begins by a translation of the capital value of the financial instrument involved. The prevailing exchange rate as well as the performance of the instrument to be translated are taken into account. This value will be used as a reference to determine the interest expense or income for the client. At the same time, the client and the financial institution agree that, when the financial instrument matures, the client's position will be settled at the rate used for the initial translation, without taking into account the actual exchange rate at maturity.

For the duration of this agreement, the bank will pay or receive the interest on the instrument in foreign currency, while the client will pay or receive the interest applied to the instrument in Canadian dollars.

Here's an example of how it works. Suppose a company purchases a capital asset that results in a 5-million French franc (FF) debt for the year at 10%, at a time when 5 (FF) are worth $1 (Cdn). The company wants to protect itself against the effects of a possible devaluation of the Canadian dollar by entering into a cross-currency swap and informs the bank of its intention. The bank translates the debt in French francs into Canadian dollars, applying the current exchange rate. It then enters into a forward contract on French francs with the company and increases the capital value of the loan by an amount equal to the premium charged for a 12 month futures contract. The rate is 4.75 (FF), which means that the loan's capital value will be set at $1,052,000 (Cdn). The company's interest expense in Canadian dollars will be calculated on the basis of this value.

Once the translation of the principal is completed, the next step is the interest swap. The company will pay the interest charged by the bank in Canadian dollars, and the bank will give it enough French francs to pay the interest related to the debt.

The third stage concerns the principal repayment. On the basis of the initial forward contract, the bank will hand over, in exchange for $1,052,000 (Cdn), 5 million (FF) which will then be used to pay off the loan.

With a cross-currency swap, the value of the foreign financial instrument and its performance are frozen.

Guaranteeing an exchange rate—Currency options are used to minimize foreign exchange risk while reserving the possibility of taking advantage of a favourable evolution of exchange rates. An option allows a buyer to purchase or sell the underlying currency at a predetermined price in consideration for a premium established on the basis of the risks taken by the seller (usually a financial institution).

A Canadian company lends $2 million (US) to an American company that will reimburse the amount three months from now in US dollars. When the loan is made, the US dollar is worth $1.25 (Cdn), and the call options on Canadian dollars at $1.25 (Cdn) for $1 (US) are worth $0.01 (Cdn) per $1 (US). The company wants to protect itself against a potential decrease in the price of the US dollar without freezing its exchange rate. It therefore buys a Canadian-dollar call option at a total cost of $20,000 ($2 million x 0.01).

Three months later, the US dollar is worth $1.28 (Cdn). The company relinquishes the option and cashes the holding gain. In this case, the option would be worth exercising only if the US dollar was worth less than $1.25 (Cdn).

Source: Adapted from: Sylvie Léger and Jacques Fortin, "Hedging: A User's Manual," *CAmagazine* (April 1994), pp. 20–24.

may be offset by the decline in value of any liability that is also denominated or valued in terms of that foreign currency. Thus, a firm would normally be interested in its net exposed position (exposed assets – exposed liabilities) for each period in each currency.

Although expected changes in exchange rates can often be included in the cost-benefit analysis relating to such transactions, in most cases there is an unexpected component in exchange rate changes and often the cost-benefit analysis for such transactions does not fully capture even the expected change in the exchange rate. For example, price increases for the foreign operations of many MNCs often have to be less than those necessary to fully offset exchange rate changes due to the competitive pressures generated by local businesses.

A company can follow one of three policies with regard to the management of its foreign currency exposure. In the two simplest policies, the firm covers all exposed positions or covers no exposed positions. A firm may cover all exposed foreign currency positions if it considers the generally small but positive difference between the average cost of cover and the average expected loss due to exchange rate changes to be less important than the need to maintain stability in the dollar value of its foreign currency cash flows. Such a policy is usually suitable for small firms with limited expertise and/or little business in foreign currencies. If the firm covers no foreign currency exposure, it generally expects foreign exchange losses and gains to offset each other, with the net balance representing one of the costs of doing business. Such a policy may be particularly suitable for large, well-diversified companies. Other companies would have to consider adopting the third policy of using various hedging or covering alternatives for most but not all of their currency exposure.

The three measures of foreign exposure are: translation exposure, transaction exposure, and economic exposure. Translation exposure arises because the foreign operations of MNCs have accounting statements denominated in the local currency of the country in which the operation is located. For Canadian MNCs, the reporting currency for its consolidated financial statements is the Canadian dollar, so assets, liabilities, revenues, and expenses of the foreign operations must be translated into dollars. International transactions often require a payment to be made or received in a foreign currency in the future, so these transactions are exposed to exchange rate risk. Economic exposure exists over the long term since the value of future cash flows in the reporting currency from foreign operations is exposed to exchange rate risk. Indeed, the whole stream of future cash flows is exposed. The three measures of exposure are now examined more closely.

Translation Exposure

Foreign currency assets and liabilities are considered exposed if their foreign currency value for accounting purposes is to be translated into the parent company currency. The currency rate used for translation of monetary assets is the current exchange rate—the exchange rate in effect on the balance sheet date. Other assets and liabilities and equity amounts that are generally translated at the historic exchange rate—the rate in effect when these items were first recognized in the company's accounts—are not considered to be exposed. The rate (current or historic) used to translate various accounts depends on the translation procedure used.

Although transaction exposure can result in exchange rate change-related losses and gains that are realized and have an impact on both reported and taxable income, translation exposure results in exchange rate losses and gains that are reflected in the company's accounting books, but that are unrealized and have little or no impact on taxable income. Thus, if financial markets are efficient and managerial goals are consistent with owner wealth maximization (agency and signalling costs are negligible), a firm should not have to waste real resources hedging against possible paper losses caused by translation exposure. However, if there are significant agency or information costs or if markets are not efficient (that is, if translation losses and gains raise information costs for investors, or if they endanger the firm's ability to satisfy debt or other covenants, or if the evaluation of the firm's managers depends on translated accounting data), a firm may indeed find it economical to hedge against translation losses or gains.

Transactions Exposure

Accounts receivable, accounts payable, and fixed priced sales or purchase contracts are examples of foreign currency transactions whose monetary value was fixed at a time different from the time when these transactions are actually completed.

Transactions exposure is a term that describes net contracted foreign currency transactions for which the settlement amounts are subject to changing exchange rates over a period too short to allow fully for compensating changes in prices. A company normally must set up an additional reporting system to track transactions exposure, since a number of these amounts are not recognized in the accounting books of a firm.

Exchange risk may be neutralized or hedged by a change in the asset and liability position in the foreign currency. An exposed asset position, such as an account receivable, can be hedged or covered by creating a liability of the same amount and maturity denominated in the foreign currency. The liability used for hedging could be a forward contract to sell the foreign currency. An exposed liability position, such as an account payable, can be covered by acquiring assets of the same amount and maturity in the foreign currency. The asset could be a forward contract to buy the foreign currency. The objective is to have a zero net asset position in the foreign currency.[13] This eliminates exchange risk, since the loss (gain) in the liability (asset) is exactly offset by the gain (loss) in the value of the asset (liability) when the foreign currency appreciates (depreciates). Two popular forms of hedge are the money-market hedge and the exchange-market or forward-market hedge. In both types of hedge the amount and the duration of the asset (liability) positions are matched. The next two sections demonstrate how IRP assures that each hedge provides the same cover.

Money-Market Hedge

In a money-market hedge, the exposed position in a foreign currency is offset by borrowing or lending in the money market. Consider the case of the Canadian firm with a net liability position of 3,000 pounds. The firm knows the exact amount of its pound liability in one month, but it does not know the liability in dollars. Assume that the one-month money-market rates in Canada and United Kingdom are one percent for lending and 1.5 percent for borrowing. The Canadian business can take the following steps:

Step 1: Calculate the present value of the foreign currency liability (3,000 pounds) that is due in one month. Use the money-market rate applicable for the foreign country (1 percent in the United Kingdom). The present value of 3,000 pounds is 2,970.30 pounds, computed as follows: $3000/(1 + .01)$.

Step 2: Exchange dollars on today's spot market to obtain the 2,970.30 pounds. The dollar amount needed today is \$6,297.63 (2,970.30 × \$2.1202).

Step 3: Invest 2,970.30 pounds in a United Kingdom one-month money-market instrument. This investment will compound to exactly 3,000 pounds in one month. The future liability of 3,000 pounds is covered by the 2,970.30 pounds investment.[14]

Note: If the Canadian business does not own this amount today, it can borrow it from the Canadian money market at the going rate of 1.5 percent. In one month the Canadian business will need to repay \$6,392.09 [\$6,297.63 × (1 + .015)].

Assuming that the Canadian business borrows the money, its management may base its calculations on the knowledge that the British goods, upon delivery in one month, will cost it \$6,392.09. The British business will receive 3,000 pounds. The Canadian business need not wait for the future spot exchange rate to be revealed. On today's date, the future dollar payment of the contract is known with certainty. This certainty helps the Canadian business in making its pricing and financing decisions.

The mechanics of the money-market hedge indicate that the Canadian firm, on today's date, knows the exact dollar amount of its future liability. Thus, the exchange risk is entirely eliminated. Many businesses hedge in the money market.

[13]Note that the hedge will work well only to the extent that both the assets and liabilities move together.

[14]Observe that 2,970.30 pounds × (1 + .01) = 3,000 pounds.

The firm creates a liability by borrowing in one market, lending or investing in the other money market, and using the spot exchange market on today's date. The mechanics of covering a net asset position in the foreign currency are the exact reverse of the mechanics of covering the liability position. With a net asset position in pounds: Borrow in the United Kingdom money market in pounds, convert to dollars on the spot exchange market, invest in the Canadian money market. The cost of hedging in the money market is the cost of doing business in three different markets. Information about the three markets is needed, and analytical calculations of the type indicated here must be made.

Many small and infrequent traders find the cost of the money-market hedge prohibitive, owing especially to the need for information about the market. These traders use the exchange or the forward-market hedge, which has very similar hedging benefits.

The Forward-Market Hedge

The forward market provides a second possible hedging mechanism. It works as follows: A net asset (liability) position is covered by a liability (asset) in the forward market. Consider again the case of the Canadian firm with a liability of 3,000 pounds that must be paid in one month. The firm may take the following steps to cover its liability position.

Step 1: Buy a forward contract today to purchase 3,000 pounds in one month. The one-month forward rate is, say, $2.1187 per pound.

Step 2: On the thirtieth day pay the banker $6,356.10 (3,000 × $2.1187) and collect 3,000 pounds. Pay these pounds to the British supplier.

By the use of the forward contract the Canadian business knows the exact worth of the future payment in dollars ($6,356.10). The exchange risk in pounds is totally eliminated by the net asset position in the forward pounds. In the case of a net asset exposure, the steps open to the Canadian firm are the exact opposite: Sell the pounds forward, and on the future day receive and deliver the pounds to collect the agreed-upon dollar amount.[15]

The use of the forward market as a hedge against exchange risk is simple and direct. This type of position requires that the liability or asset position be matched against an offsetting position in the forward market. For example, a firm directs its banker to buy or sell a foreign currency on a future date, and the banker gives a forward quote.

The forward hedge and the money-market hedge give an identical future dollar payment (or receipt) if the forward contracts are priced according to the interest rate parity theory. The alert student may have noticed that the dollar payments in the money-market hedge and the forward-market hedge examples were, respectively, $6,856.98 and $6,356.10. Recall from our previous discussions that in efficient markets, the forward contracts do indeed conform to the IRPT. However, the numbers in our example are not identical because the forward rate used in the forward hedge is not exactly equal to the interest rates in the money-market hedge.

Currency Future and Option Contracts

The forward-market hedge is not adequate for some types of exposure. If the foreign currency asset or liability position occurs on a date for which forward quotes are not available, the forward hedge cannot be accomplished. In certain cases the forward hedge may cost more than the money-market hedge. In these cases a corporation with a large amount of exposure may prefer the money-market hedge. In addition to forward and money-market hedges, a company can also hedge its exposure by buying (or selling) some relatively new instruments—foreign currency

[15]The use of forward contracts provides the firm with flexibility since this type of contract is tailor made to the needs of the firm. However, the firm will incur higher transaction costs.

futures contracts and foreign currency options. While futures contracts are similar to forward contracts in that they provide fixed prices for the required delivery of foreign currency at maturity, options permit fixed (strike) price foreign currency transactions anytime prior to maturity. Futures contracts and options differ from forward contracts in that, unlike forward contracts, which are customized with regard to amount and maturity date, futures and options are traded in standard amounts with standard maturity dates. In addition, while forward contracts are written by banks, futures and options are traded on organized exchanges, and individual traders deal with the exchange-based clearing organization rather than with each other. The purchase of futures requires the fulfillment of margin requirements (about 5 to 10 percent of the face amount), while the purchase of forward contracts requires only good credit standing with a bank. The purchase of options requires an immediate outlay that reflects a premium above the strike price and an outlay equal to the strike price when and if the option is exercised.

Economic Exposure

The economic value of a company can vary in response to exchange rate changes. This type of change in value may be caused by a rate-change-induced decline in the level of expected cash flows and/or by an increase in the riskiness of these cash flows. Thus, economic exposure refers to the overall impact of exchange rate changes on the value of the firm and therefore includes not only the strategic impact of changes in competitive relationships between alternative foreign locations that arise from exchange rate changes, but also the economic impact of transactions exposure and the economic impact, if any, of translation exposure.

Economic exposure to exchange rate changes depends on the competitive structure of the markets for a firm's inputs and its outputs and how these markets are influenced by changes in exchange rates. This influence, in turn, would depend on a number of economic factors, including price elasticities of the products, the degree of competition from foreign markets and direct (through prices) and indirect (through incomes) impact of exchange rate changes on these markets. Assessing the economic exposure faced by a particular firm thus depends on the ability to understand and model the structure of the markets for its major inputs (purchases) and outputs (sales).

A company need not engage in any cross-border business activity in order to be exposed to losses due to exchange rate changes, because product and financial markets in most countries are related and influenced to a large extent by the same global forces. The output of a company engaged in business activity only within one country may be competing with imported products, or it may be competing for its inputs with other domestic and foreign purchasers. The degree of competition in these markets may change with changes in exchange rates. For example, a Canadian chemical company that did no cross-border business nevertheless found that its profit margins depended directly on the Canadian dollar–Japanese yen exchange rate. The company used coal as an input in its production process, and the Canadian price of coal was heavily influenced by the extent to which the Japanese bought Canadian coal, which in turn depended on the yen-dollar exchange rate.

Although translation exposure need not be managed, it might be useful for a firm to manage its transaction and economic exposures since they affect firm value directly. In most companies, transaction exposure is generally tracked and managed by the office of the corporate treasurer. Economic exposure is difficult to define in operating terms, and very few companies manage it actively. In most companies, economic exposure is generally considered part of the strategic planning process, rather than as a treasurer's or finance function.

Access to International Financial Markets

A corporate treasurer must have access to international financial markets to manage the firm's transaction and economic exposure. Examples of financial institutions which provide access to international markets include the International Banking Centres, the International Financial Offices and the International Financial Centres. The following article discusses the objectives of these financial institutions.

Around the World in 80 Ways

International Financial Centres (IFC), International Financial Offices (IFO), and International Banking Centres (IBC) are all businesses that specialize in international financial transactions, and get tax breaks for doing so. The differences are in eligible transactions and in the tax laws that govern them. IFCs, the creation of the Quebec government, operate only in that province. IFOs are the BC equivalent. IBCs, under the federal Income Tax Act, can operate in Vancouver and Montreal and get federal as well as provincial tax breaks.

In addition to the political and economic factors coming into play in Montreal and Vancouver (the only two Canadian cities in which international financial institutions are eligible for tax relief), the governments involved had five objectives in creating these institutions:

1. To increase the number and size of international financial transactions carried out in Canada, as opposed to New York, London or elsewhere.
2. To attract specialists in international financial markets to Canada and consequently to develop Canadian expertise in those areas.
3. To make foreign capital more accessible for the different governments involved, as well as for Canadian companies and individuals.
4. To give Canada a higher profile on international financial markets.
5. To diversify Canadian financial activity, by developing poles in the west (Vancouver) and the east (Montreal).

Why were Montreal and Vancouver chosen to host these enterprises? Because they are major centres with established stock exchanges. It may also have been an effort by the three governments to decentralize stock trading and financial activities. Others suggest that Montreal is the Canadian doorway to European markets, and Vancouver the gateway to Asia.

These institutions can help companies wishing to obtain financing for projects abroad. For example, an IFC can help a Canadian company that has just obtained a contract to build a turnkey plant in Australia to finance the project from down under, in Australian dollars. They can also help clients reduce their vulnerability to exchange fluctuations, diversify their portfolios, learn to play the exchange market, or obtain capital more easily and at more affordable rates. Even small investors can think of acquiring foreign debt securities or shares in deutsche marks, yen or other foreign currencies.

There are many other commercial services that IFCs, IBCs, and IFOs can now offer, including the issue of letters of credit for foreign companies, purchases and sales of foreign currencies, forward exchange contracts, and loans in other currencies. They are generally specialized in those fields, and can even offer competitive rates and conditions.

Any company financing a major project may want to investigate the advantages of using foreign capital. Having access to experts in Canada who can evaluate such projects, and are familiar with the other country's economy and political and financial situation, can make locating these new sources of capital much easier.

All that having been said, IFCs, IBCs and IFOs must nevertheless meet various criteria before their transactions are eligible for tax breaks. For a transaction to be "international," and hence eligible for these institutions, it must either be conducted with a nonresident or involve a security issued on foreign markets, regardless of the other party to the transaction. The security may have been issued by a Canadian body (government, company, or individual), provided that it can be traded on a market outside Canada.

For example, a Canadian resident purchases shares of Société Générale, a French bank whose shares are listed on the Paris stock market; a Japanese resident purchases Bell Canada shares; a Canadian company building a plant in Australia wishes to have its projects financed in Australian dollars; a Canadian manufacturer sells goods to Italians and enters into a forward exchange contract to sell Italian lira that it will receive two months later.

The federal government has considerably limited the eligible transactions of IBCs, however—as compared with provincially governed IFCs and IFOs—because of lobbying from Toronto. For an IBC to be exempt from income tax, for example, its total eligible deposits must represent at least 96 percent of its eligible loans, for each working day. If this condition is not met, the amount eligible for exemption will be determined on a pro rata basis. Because the daily matching of daily deposits and loans is a fairly laborious exercise, the condition complicates the centre's operations and can cut substantially into its profitability.

In 1989, IFCs, and IFOs had a tax rate of only 29 percent, the lowest in North America, because of their exemption from provincial income tax. It is surprising, then, that so few companies have requested certification as international centres. From a taxation point of view, Montreal and Vancouver are the most profitable North American cities in which to conduct certain international transactions—such as loans to or deposits by nonresidents when the IFC or IFO is also an IBC. In that case, the net revenue from the transaction is tax-free in Canada. In fact, Montreal and Vancouver can be considered true tax-free zones for international financial transactions.

Source: Excerpts from: Michel Drouin and Evelyn Paquin, "Around the World in 80 Ways," *CA Magazine* (October, 1991), pp. 28–34.

OBJECTIVE 6

MULTINATIONAL WORKING-CAPITAL MANAGEMENT

The basic principles of working-capital management for a multinational corporation are similar to those for a domestic firm. However, tax and exchange rate factors are additional considerations for the multinational corporation (MNC). For an MNC with subsidiaries in many countries, the optimal decisions in the management of working capital are made by considering the company as a whole. The global or centralized financial decision for an MNC is superior to the set of independent optimal decisions for the subsidiaries. This is the control problem of the MNC. If the individual subsidiaries make decisions that are best for them individually, the consolidation of such decisions may not be best for the MNC as a whole. In order to effect global management, sophisticated computerized models—incorporating a large number of variables for each subsidiary—are solved to provide the best overall decision for the MNC.

Before considering the components of working-capital management, we examine two techniques that are useful in the management of a wide variety of working-capital components.

Lead and Lag Strategies

Techniques used to reduce exchange risk where the firm maximizes its asset position in the stronger currency and its liability position in the weaker currency.

Leading and Lagging

Two important risk-reduction techniques for many working-capital problems are called leading and lagging. Often, forward- and money-market hedges are not available to eliminate exchange risk. Under such circumstances, leading and lagging may be used to reduce exchange risk.

Recall that a net asset (long) position is not desirable in a weak or potentially depreciating currency. If a firm has a net asset position in such a currency, it should expedite the disposal of the asset. The firm should get rid of the asset earlier than it otherwise would have, or **lead**, and convert the funds into assets in a relatively stronger currency. By the same reasoning, the firm should **lag** or delay the collection against a net asset position in a strong currency. If the firm has a net liability (short) position in the weak currency, then it should delay the payment against the liability, or lag, until the currency depreciates. In the case of an appreciating or strong foreign currency and a net liability position, the firm should lead the payments—that is, reduce the liabilities earlier than it would otherwise have.

These principles are useful in the management of the working capital of an MNC. However, management cannot eliminate the foreign exchange risk. If exchange rates change continuously, managers find it impossible to determine by what amount or at what time the currency will depreciate or appreciate. As a result, the risk of exchange rate changes cannot be eliminated. Nevertheless, the reduction of risk, or the increase in the amount of gain from exchange rate changes, via the lead and lag is useful for cash management, accounts receivable management, and short-term liability management.

Cash Management and Positioning of Funds

Positioning of funds takes on an added importance in the international context. Funds may be transferred from a subsidiary of the MNC in country A to another subsidiary in country B such that the foreign exchange exposure and the tax liability of the MNC as a whole are minimized. It bears repeating that, owing to the global strategy of the MNC, the tax liability of the subsidiary in country A may be greater than it would otherwise have been, but the overall tax payment for all units of the MNC is minimized.

The transfer of funds among subsidiaries and the parent company is done by royalties, fees, and transfer pricing. A transfer price is the price a subsidiary or a parent company charges other companies that are part of the MNC for its goods or services. A parent that wishes to transfer funds from a subsidiary in a depreciating-currency country may charge a higher price on the goods and services sold to this subsidiary by the parent or by subsidiaries from strong-currency countries.

Centralized cash management of all the affiliates at the global level, achieved with the help of computer models, reduces both the overall cost of holding cash and the foreign exchange exposure of the MNC as a whole with respect to cash. The excess cash balance of one subsidiary is transferred to a cash-deficit subsidiary in the guise of a loan. The optimal holdings of cash in different currencies are calculated in a manner similar to the optimal portfolio problem discussed in Appendix 18A.

INTERNATIONAL FINANCING AND CAPITAL STRUCTURE DECISIONS

OBJECTIVE 7

A multinational corporation has access to many more financing sources than a domestic firm. It can tap not only the financing sources in its home country that are available to its domestic counterparts, but also sources in the foreign countries in which it operates. Host countries often provide access to low-cost subsidized financing in order to attract foreign investment. In addition, the MNC may enjoy preferential credit standards because of its size and investor preference for its home currency. In addition, an MNC may be able to access capital markets in countries which it does not operate but which may have large, well-functioning capital markets. Finally, an MNC can also access external currency markets: either the Eurodollar, Eurocurrency, or Asian dollar markets. These external markets are unregulated, and because of their lower spread, can offer very attractive rates for financing and for investments. With the increasing availability of interest rate and currency swaps, a firm can raise funds in the lowest-cost maturities and currencies and swap them into funds with the maturity and currency denomination it requires. MNCs have a more continuous access to external finance and a lower cost of capital as compared to a domestic company for the following reasons: the MNCs have the ability to tap a larger number of financial markets and the MNCs are able to avoid the problems or limitations of any one financial market.

Access to national financial markets is regulated by governments. For example, in Canada, access to capital markets is governed by both federal and provincial regulations. Access to Japanese capital markets is governed by regulations issued by the Ministry of Finance. Some countries have extensive regulations; other countries have relatively open markets. These regulations may differ depending on the legal residency terms of the company raising funds. A company that cannot use its local subsidiary to raise funds in a given market will be treated as foreign. In order to increase their visibility in a foreign capital market, a number of MNCs are now listing their equities on the stock exchanges of many of these countries.

The external currency markets are predominantly centred in Europe. In fact, most external currency markets can be characterized as Eurodollar markets. We define the Eurodollar (Euro-Canadian dollar) as a U.S. dollar deposit (Canadian dollar deposit) made with a European bank. The Eurodollar market is an international market which consists of an active short-term money market and an intermediate-term capital market with maturities ranging up to 15 years and averaging about 7 to 9 years. The intermediate-term market consists of the Eurobond and the Syndicated Eurocredit markets. Eurobonds are usually issued as unregistered bearer bonds and generally tend to have higher flotation costs but lower coupon rates compared to similar bonds issued in Canada. A Syndicated Eurocredit loan is simply a large term loan that involves contributions by a number of lending banks. Most large Canadian banks are active in the external currency markets.

In arriving at its capital-structure decisions, an MNC has to consider a number of factors. First, the capital structure of its local affiliates is influenced by local norms regarding capital structure in that industry and in that country. Local norms for companies in the same industry can differ considerably from country to country. Second, the local affiliate capital structure must also reflect corporate attitudes toward exchange rate and political risk in that country, which would normally lead to higher levels of local debt and other local capital. Third, local affiliate capital structure must reflect home country requirements with regard to the company's consolidated capital structure. Finally, the optimal MNC capital structure should reflect its wider access to financial markets, its ability to diversify economic and political risks, and its other advantages over domestic companies.

OBJECTIVE 8

DIRECT FOREIGN INVESTMENT

The MNC makes direct foreign investments abroad in the form of plants and equipment. The decision process for this type of investment is very similar to the capital-budgeting decision in the domestic context—with some additional twists. Most real-world capital-budgeting decisions are made with uncertain future outcomes. Recall that a capital-budgeting decision has three major components: the estimation of the future cash flows (including the initial cost of the proposed investment), the estimation of the risk in these cash flows, and the choice of the proper discount rate. We will assume that the net-present-value criterion is appropriate as we examine (1) the risks associated with direct foreign investment and (2) factors to be considered in making the investment decision that may be unique to the international scene.

BACK TO THE BASICS

Investments across international boundaries give rise to special risks not encountered when investing domestically. Specifically, political risks and exchange rate risk are unique to international investing. Once again, ***Axiom 1: The Risk-Return Tradeoff—We Won't Take on Additional Risk Unless We Expect to Be Compensated with Additional Return*** *provides a rationale for evaluating these considerations. When added risks are present, added rewards are necessary to induce investment.*

Risks in Direct Foreign Investments

Risks in domestic capital budgeting arise from two sources: business risk and financial risk. The international capital-budgeting problem has these risks as well as political risk and exchange risk.

Business Risk and Financial Risk

International business risk is due to the response of business to economic conditions in the foreign country. Thus, the Canadian MNC needs to be aware of the business climate in both Canada and the foreign country. Additional business risk is due to competition from other MNCs, local businesses, and imported goods. Financial risk refers to the risks introduced in the profit stream by the firm's financial structure. The financial risks of foreign operations are not very different from those of domestic operations.

Political Risk

Political or sovereignty risk arises because the foreign subsidiary conducts its business in a political system different from that of the home country. Many foreign governments, especially those in the Third World, are less stable than the Canadian government. A change in a country's political setup frequently brings a change in policies with respect to businesses, and especially with respect to foreign businesses. An extreme change in policy might involve nationalization or even outright expropriation of certain businesses. These are the political risks of conducting business abroad. A business with no investment in plants and equipment is less susceptible to these risks. Some examples of political risk are listed below:

1. Expropriation of plants and equipment without compensation
2. Expropriation with minimal compensation that is below actual market value
3. Nonconvertibility of the subsidiary's foreign earnings into the parent's currency—the problem of blocked funds
4. Substantial changes in the laws governing taxation
5. Governmental controls in the foreign country regarding the sale price of the products, wages, and compensation to personnel, hiring of personnel, making of transfer payments to the parent, and local borrowing
6. Some governments require certain amounts of local equity participation in the business. Some require that the majority of the equity participation belong to their country.

All these controls and governmental actions introduce risks in the cash flows of the investment to the parent company. These risks must be considered before the foreign investment decision. An MNC may decide against investing in countries with risks of types 1 and 2. Other risks can be borne—provided that the returns from the foreign investments are high enough to compensate for them. Insurance against some types of political risks may be purchased from private insurance companies. It should be noted that while an MNC cannot protect itself against all foreign political risks, political risks are also present in domestic business.

Exchange Risk

The exposure of the fixed assets is best measured by the effects of the exchange rate changes on the firm's future earnings stream: that being economic exposure rather than translation exposure. For instance, changes in the exchange rate may adversely affect sales by making competing imported goods cheaper. Changes in the cost of goods sold may result if some components are imported and their price in the foreign currency changes because of exchange rate fluctuations. The thrust of these examples is that the effect of exchange rate changes on income statement items should be properly measured to evaluate exchange risk. Finally, exchange risk affects the dollar-denominated profit stream of the parent company, whether or not it affects the foreign-currency profits.

International Capital Budgeting

Capital budgeting for international operations involves more extensive analysis than does capital budgeting for domestic operations. As in a domestic company, an MNC must develop estimates of the investment needed and the net cash flow generated by a proposed foreign capital investment. It must also assess the riskiness of these cash flows. Finally, it must summarize the cash flow and risk analysis information into a measure of desirability for the project, such as the net present value. The company can then compare the proposed investment with other possibilities and make a decision.

While this basic outline of steps necessary to evaluate a specific long-term foreign investment is not much different from that followed for evaluating a domestic capital investment, the cross-border nature of the foreign investment may require the consideration of a number of additional factors, such as the impact of currency and political risks on project value. Here we discuss some of the issues that may have to be considered in international capital budgeting.

For example, should the proposed capital investment be evaluated from the viewpoint of the local firm or from the viewpoint of the parent? The objective of maximizing shareholder wealth would indicate that the parent company perspective be used. However, the parent company may have supplied only part or none of the initial investment for the project. When using the parent company perspective, the project should be evaluated in terms of the parent's currency with the parent company's net initial investment and its net receipt of cash flows from the project discounted at a rate appropriate for the risks of the project cash flows with the inflation rate in the parent's currency. However, cash flows to the parent depend not just on the performance of the project, but also on how the local cash flows are managed. For example, because of tax considerations, it may not be optimal to remit all project cash flows to the parent only to have to send some or all of them back to be reinvested in the original foreign affiliate. In addition, value added by a project to a foreign affiliate will normally be reflected in the value of the parent company.

Thus, it may be better, even for shareholder wealth maximization, to evaluate a foreign capital proposal from the perspective of its local affiliate, using the total investment for the project and all of its cash flows discounted at a rate that reflects the local inflation rate and the riskiness of the project. In practice, most MNCs use at least the local perspective to evaluate a foreign project, while some supplement the local perspective with present value analysis from the parent's perspective. In such a case, the foreign affiliate perspective is the primary basis for a decision, while the parent perspective is used to ensure that the structure of the investment and the disposition of the cash flows are consistent with corporate objectives.

It is important that cash flow projections be realistic and reflect local conditions regarding competition, inflation, price controls, and other government regulations. When taking the local currency perspective, the discount rate should also reflect a realistic assessment of the expected local inflation rate. This may be especially difficult, given that capital markets in many foreign countries are often not free or efficient and interest rates do not reflect market expectations. Project cash flows should also reflect the advantages of using subsidized financing that may have been used to finance the project. When using the parent currency perspective, the discount rate should reflect the expected inflation rate in the parent currency, and foreign currency cash flows should be converted to parent currency cash flows using projected exchange rates. These projections are often based on the assumption that purchasing power parity holds during the life of the project. In such a case, it may be possible to take initial-year cash flows and assume that increases because of local inflation will be offset exactly by purchasing power parity-related declines in the exchange rate. However, if significant deviations from purchasing power parity are expected, such a simplifying assumption would be inappropriate. In addition, high inflation rates may be accompanied by government price controls. A company may face a lag between allowed price increases and inflation, or may face other economic and political instabilities. In such cases, simply assuming that purchasing power parity will hold may be insufficient, and additional analysis to reflect these factors should be undertaken.

In the domestic case, an NPV less than zero leads to rejection of a project. In the international context, an NPV less than zero leads to rejection of the direct foreign investment; however, the MNC then has other options open to it.

If the foreign sales volume is expected to be low, the MNC may consider setting up a sales office in the foreign country. The product may be exported to the foreign country from production facilities in the home country or from some other foreign subsidiary. An NPV calculation may now be employed. The acceptance of this scheme is ensured, since no direct capital investment is needed. If the estimated sales levels are high enough that the establishment of a plant in the foreign country appears profitable, owing to the potential savings in the transportation costs, yet the NPV of the direct foreign investment (DFI) is negative, the MNC may consider licensing or an affiliate arrangement with a local company. The MNC provides the technology, and the interested domestic firm finances and sets up the plant. The MNC does not bear the risks of a DFI, but receives a royalty payment from the sales of the affiliate company instead.

SUMMARY

The growth of the global economy, the increase in the number of multinational companies and the increase in foreign trade underscore the importance of the study of international finance.

Exchange rate mechanics were discussed in the context of the prevailing floating rates. Under this system, exchange rates between currencies vary in an apparently random fashion in accordance with the supply and demand conditions in the exchange market. Important economic factors affecting the level of exchange rates include the relative economic strengths of the countries involved, the balance-of-payments mechanism, and the countries' monetary policies. The asked and the bid rates are terms which describe the selling and buying rates of currencies. The direct quote is the units of home currency per unit of foreign currency, while the indirect quote is the reciprocal of the direct quote. Cross-rate computations reflect the exchange rate between two foreign currencies. Finally, simple arbitrage for indirect quotes and triangular arbitrage for cross rates were shown to hold. The efficiency of spot exchange markets implies that no arbitrage (riskless) profits can be made by buying and selling currencies in different markets.

The forward-exchange market provides a valuable service by quoting rates for the delivery of foreign currencies in the future. The foreign currency is said to sell at a premium (discount) forward from the spot rate when the forward rate is greater (less) than the spot rate, in direct quotation. The computation of the percent-per-annum deviation of the forward from the spot rate was used to demonstrate the interest rate parity theory (IRPT), which states that the forward contract sells at a discount or premium from the spot rates, owing solely to interest rate differential between the two countries. The IRPT was shown to hold by means of the covered interest arbitrage argument. In addition, the influences of purchasing power parity (PPP) and the international Fisher effect (IFE) in determining the exchange rate were discussed. In rational and efficient markets, forward rates are unbiased forecasts of future spot rates that are consistent with the PPP.

Exchange risk exists because the exact spot rate that prevails on a future date is not known with certainty today. The concept of exchange risk is applicable to a wide variety of businesses, including export-import firms and firms involved in making direct foreign investments or international investments in securities. Exchange exposure is a measure of exchange risk. There are different ways of measuring the foreign exposure, including the net asset (net liability) measurement. Different strategies are open to businesses to counter the exposure to this risk, including the money-market hedge, the forward-market hedge, futures contracts, and options. Each involves different costs.

In discussing working-capital management in an international environment we found leading and lagging techniques useful in minimizing exchange risks and increasing profitability. In addition, funds positioning is a useful tool for reducing exchange risk exposure. The MNC may have a lower cost of capital because it has access to a larger set of financial markets than a domestic company. In addition to the home, host, and third country financial markets, the MNC can tap the rapidly growing external currency markets. In making capital-structure decisions, the MNC must consider political and exchange risks and host and home country capital-structure norms.

The complexities encountered in the direct foreign investment decision include the usual sources of risk—business and financial—and additional risks associated with fluctuating exchange rates and political factors. Political risk is due to differences in political climates, institutions, and processes between the home country and abroad. Under these conditions the estimation of future cash flows and the choice of the proper discount rates are more complicated than for the domestic investment situation. Rejection of a DFI proposal may lead either to the setting up of a sales office abroad or to an affiliate arrangement with a foreign company.

STUDY QUESTIONS

24-1. What additional factors are encountered in international as compared with domestic financial management? Discuss each briefly.

24-2. What different types of businesses operate in the international environment? Why are the techniques and strategies available to these firms different?

24-3. What is meant by *arbitrage profits*?

24-4. What are the markets and mechanics involved in generating (a) simple arbitrage profits, (b) triangular arbitrage profits, (c) covered interest arbitrage profits?

24-5. How do purchasing power parity, interest rate parity, and the Fisher effect explain the relationships between the current spot rate, the future spot rate, and the forward rate?

24-6. What is meant by (a) exchange risk, (b) political risk?

24-7. How can exchange risk be measured?

24-8. What are the differences between transaction, translation, and economic exposures? Should all of them ideally be reduced to zero?

24-9. What steps can a firm take to reduce exchange risk? Indicate at least two different techniques.

24-10. How are the forward-market and the money-market hedges affected? What are the major differences between these two types of hedges?

24-11. Compare and contrast the use of forward contracts, futures contracts, and options to reduce foreign exchange exposure. When is each instrument most appropriate?

24-12. In the Toronto exchange market, the forward rate for the Indian currency, the rupee, is not quoted. If you were exposed to exchange risk in rupees, how could you cover your position?

24-13. Indicate two working-capital management techniques that are useful for international businesses to reduce exchange risk and potentially increase profits.

24-14. How do the financing sources available to an MNC differ from those available to a domestic firm? What do these differences mean for the company's cost of capital?

24-15. What risks are associated with direct foreign investment? How do these risks differ from those encountered in domestic investment?

24-16. How is the direct foreign investment decision made? What are the inputs to this decision process? Are the inputs more complicated than those to the domestic investment problem? If so, why?

24-17. A corporation desires to enter a particular foreign market. The DFI analysis indicates that a direct investment in the plant in the foreign country is not profitable. What other course of action can the company take to enter the foreign market? What are the important considerations?

24-18. What are the reasons for the acceptance of a sales office or licensing arrangement when the DFI itself is not profitable?

SELF-TEST PROBLEMS

The data for self-test Problems ST-1 and ST-2 are given in the following table:

Selling Quotes for the German Mark in Montreal

Country	Contract	$/DM
Germany—mark	Spot	0.9520
	1-month forward	0.9537
	6-month forward	0.9614

ST-1. You own $10,000. The dollar rate on the German mark is 1.0604. The German mark rate is given in the table above. Are arbitrage profits possible? Set up an arbitrage scheme with your capital. What is the gain (loss) in dollars?

ST-2. If the interest rates on the one-month instruments in Canada and Germany are 6 and 8 percent (annualized), respectively, what is the correct price of the one-month forward mark?

STUDY PROBLEMS (SET A)

The data for Study Problems 24-1A through 24-10A are given in the following table:

Selling Quotes for Foreign Currencies in Montreal

Country	Contract	$/Foreign Currency
U.S.—dollar	Spot	1.1963
	1-month	1.1977
	6-month	1.2038
Japan—yen	Spot	0.009706
	1-month	0.009711
	6-month	0.009760
Switzerland—franc	Spot	0.9498
	1-month	0.9475
	6-month	0.9359

24-1A. A Canadian business needs to pay (a) 10,000 American dollars, (b) 2 million yen, and (c) 50,000 Swiss francs to businesses abroad. What are the dollar payments to the respective countries?

24-2A. A Canadian business pays $10,000, $15,000, and $20,000 to suppliers in, respectively, Japan, Switzerland, and the U.S. How much, in local currencies, do the suppliers receive?

24-3A. Compute the indirect quote for the spot and forward American dollar, yen, and Swiss franc contracts.

24-4A. The spreads on the contracts as a percent of the ask rates are 2 percent for yen, 3 percent for American dollars, and 5 percent for Swiss francs. Show, in a table similar to the one above, the bid rates for the different spot and forward rates.

24-5A. You own $10,000. The dollar rate in Tokyo is 106.0454. The yen rate in Montreal is given in the table above. Are arbitrage profits possible? Set up an arbitrage scheme with your capital. What is the gain (loss) in dollars?

24-6A. Compute the American dollar/yen and the yen/Swiss franc spot rate from the data in the table above.

24-7A. Compute the simple premium (discount) on the one-month and six-month yen, Swiss franc, and the American dollar quotes. Tabulate the percent-per-annum deviations as in Table 24-2.

24-8A. Assume that the interest rate on the Canadian one-month Treasury bill is 5 percent (annualized). The corresponding American rate is 3 percent. The spot and the forward rates are shown in the table above. Can a Canadian trader make arbitrage profits? If the trader had $100,000 to invest, indicate the steps he or she would take. What would be the net profit? (Ignore transactions and other costs.)

24-9A. If the interest rates on the one-month instruments in Canada and Japan are 5 and 2 percent (annualized), respectively, what is the correct price of the one-month forward yen? Use the spot rate from the table.

24-10A. The one-month Treasury bill rate in Canada is 5 percent annualized. Using the one-month forward quotes, compute the 30-day interest rates in the U.S., Switzerland, and Japan.

INTEGRATIVE PROBLEM

For your job as the business reporter for a local newspaper you are given the task of putting together a series of articles on the derivative markets for your readers. Much recent local press coverage has been given to the losses in the foreign exchange markets by JGAR, a local firm that is a subsidiary of Daedlufetarg, a large German manufacturing firm. Your editor would like you to address several specific questions dealing with multinational finance. Please prepare your response to the following memorandum from your editor:

TO: Business Reporter

FROM: Perry White, Daily Planet

RE: Upcoming Series on the Derivative Securities Market

In your upcoming series on the time value of money, I would like to make sure you cover several specific points. In addition, before you begin this assignment, I want to make sure we are all reading from the same script, as accuracy has always been the cornerstone of the *Daily Planet.* In this regard, I'd like a response to the following questions before we proceed:

- **a.** What new problems and factors are encountered in international as opposed to domestic financial management?
- **b.** What does the term *arbitrage profits* mean?
- **c.** What can a firm do to reduce exchange risk?
- **d.** What are the differences between a forward contract, futures contract and options?

Use the following data in your response to the remaining questions:

Selling Quotes for Foreign Currencies in Toronto

Country	Contract	$/Foreign Currency
U.S.—dollar	Spot	1.3368
	1-month	1.3429
	6-month	1.3520
Japan—yen	Spot	0.012820
	1-month	0.012960
	6-month	0.013660
Switzerland—franc	Spot	1.1840
	1-month	1.1880
	6-month	1.2068

e. A Canadian business needs to pay (1) 10,000 American dollars, (2) 2 million yen, and (3) 50,000 Swiss francs to businesses abroad. What are the dollar payments to the respective countries?

f. A Canadian business pays $10,000, $15,000, and $20,000 to suppliers in, respectively, Japan, Switzerland, and the U.S. How much, in local currencies, do the suppliers receive?

g. Compute the indirect quote for the spot and forward American dollar, yen, and Swiss franc contracts.

h. You own $10,000. The dollar rate in Tokyo is 106.0454. The yen rate in Montreal is given in the table above. Are arbitrage profits possible? Set up an arbitrage scheme with your capital. What is the gain (loss) in dollars?

i. Compute the American dollar/yen and the yen/Swiss franc spot rates from the data in the table above.

STUDY PROBLEMS (SET B)

The data for Study Problems 24-1B through 24-10B are given in the following table:

Selling Quotes for Foreign Currencies in Montreal

Country	Contract	$/Foreign Currency
U.S.—dollar	Spot	1.2024
	1-month	1.2046
	6-month	1.2130
Japan—yen	Spot	0.009642
	1-month	0.009652
	6-month	0.009706
Switzerland—franc	Spot	0.8861
	1-month	0.8836
	6-month	0.8710

24-1B. A Canadian business needs to pay (a) 15,000 American dollars, (b) 1.5 million yen, and (c) 55,000 Swiss francs to businesses abroad. What are the dollar payments to the respective countries?

24-2B. A Canadian business pays $20,000, $5,000, and $15,000 to suppliers in, respectively, Japan, Switzerland, and the U.S. How much, in local currencies, do the suppliers receive?

24-3B. Compute the indirect quote for the spot and forward American dollar, yen, and Swiss franc contracts.

24-4B. The spreads on the contracts as a percent of the ask rates are 4 percent for yen, 3 percent for American dollars, and 6 percent for Swiss francs. Show, in a table similar to the one above, the bid rates for the different spot and forward rates.

24-5B. You own $10,000. The dollar rate in Tokyo is 106.3649. The yen rate in Montreal is given in the table above. Are arbitrage profits possible? Set up an arbitrage scheme with your capital. What is the gain (loss) in dollars?

24-6B. Compute the American dollar/yen and the yen/Swiss franc spot rate from the data in the table above.

24-7B. Compute the simple premium (discount) on the one-month and six-month yen, Swiss franc, and American dollar quotes. Tabulate the percent-per-annum deviations as in Table 24-2.

24-8B. Assume that the interest rate on the Canadian one-month Treasury bill is 7 percent (annualized). The corresponding American rate is 4 percent. The spot and the forward rates are shown in the table above. Can a Canadian trader make arbitrage profits? If the trader had $100,000 to invest, indicate the steps he or she would take. What would be the net profit? (Ignore transactions and other costs.)

24-9B. If the interest rates on the one-month instruments in Canada and Japan are 6 and 3 percent (annualized), respectively, what is the correct price of the one-month forward yen? Use the spot rate from the table.

24-10B. The one-month Treasury bill rate in Canada is 6 percent annualized. Using the one-month forward quotes, compute the one-month interest rates in the U.S., Switzerland, and Japan.

SELF-TEST SOLUTIONS

SS-1. The German rate is 1.0604 DM/$1, while the (indirect) Toronto rate is

$$\frac{1}{0.9520} = 1.0504 \frac{\text{DM}}{\$}$$

Assuming no transaction costs, the rates between Germany and Toronto are out of line. Thus, arbitrage profits are possible.

Step 1: Since the mark is cheaper in Germany, buy $10,000 worth of marks in Germany. The number of marks purchased would be

$$\$10{,}000 \times 1.0604 = 10{,}604 \text{ DM}$$

Step 2: Simultaneously sell the marks in Toronto at the prevailing rate. The amount received upon the sale of the marks would be:

$$10{,}604 \text{ DM} \left(\frac{\$0.9520}{\text{DM}}\right) = \$10{,}095.01$$

net gain is $10,095.01 − $10,000 = $95.01

SS-2. *Step 1:* Compute the percent-per-annum premium on the forward rate using Equation (24-3a):

$$D = \frac{I^f - I^d}{1 + I^f} \quad (24\text{-}3a)$$

where I^d = the annualized interest rate on a domestic instrument having the same maturity as the forward contract

I^f = the annualized interest rate on a foreign instrument having the same maturity as the forward contract

$$\text{Thus,} \quad D = \frac{0.08 - 0.06}{1.08} = \frac{0.02}{1.08} = 0.01852 = 1.852\%$$

Step 2: Using this premium in Equation (24-2), compute the forward rate F:

$$\frac{(S - F)}{S} \times \frac{12 \text{ months}}{1 \text{ month}} \times 100 = D \quad (24\text{-}2)$$

where F = the forward rate, direct quote

S = the spot rate, direct quote

n = the number of months of the forward contract

D = the annualized percent discount

$$\frac{(0.9520 - F)}{0.9520} \times \frac{12 \text{ months}}{1 \text{ month}} \times 100 = 1.852\%$$

$$\frac{(1{,}142.40 - 1{,}200F)}{0.9520} = 1.852\%$$

$$1{,}142.40 - 1200F = 1.7631$$

$$F = 0.9505$$

Thus the correct price of the forward rate is $0.9505.

FINANCIAL MANAGEMENT: SMALL AND MID-SIZED BUSINESS PERSPECTIVE

CHAPTER 25

LEARNING OBJECTIVES

After reading this chapter you should be able to

1. Distinguish between small and mid-sized business firms.
2. Describe the differences between small and large business firms.
3. Explain the importance of cash management in a small business firm.
4. Explain how small business firms can make capital investment decisions.
5. Describe the various sources of financing for small business firms.
6. Explain the prerequisites for a business to go public.

INTRODUCTION

The importance of the small business sector to Canada's economy is self-evident: small and mid-sized enterprises contribute to some 57 percent of the total private-sector gross domestic product; 99 percent of businesses in Canada have fewer than 100 employees; and approximately 15 percent of the total work force is self-employed.[1] Despite this impressive performance, small business firms indicate that their growth is being impeded by the rapid increase in the Government of Canada's debt. Large government deficits leads to high taxes rates and high real interest rates. Both these rates cause small business firms to reduce the scale of their long-term expansion plans. The Small Business Working Committee, a private sector advisory group, also shares the same view. For example, this committee has indicated that the percentage of Canadian firms carrying out research and development is lower than that in other countries. This type of expenditure is considered critical to small and mid-sized firms in technology-driven industries. In fact, a recent study published by Statistics Canada reports that the following corporate strategies differentiate between whether small and mid-sized enterprises are more successful or less successful in capturing market share: research and development capability; access to markets; technological ability, government assistance, marketing, access to capital, and the cost of capital.[2] Clearly, there is a link between long-term corporate strategies and the success of small and mid-sized firms. Although many of these corporate strategies are beyond the scope of this textbook, we will examine how financing and the cost of capital affect small and mid-sized enterprises.

---

[1]Entrepreneurship and Small Business Office, Industry Canada, *Small Business in Canada: A Statistical Overview* (Ottawa: Minister of Supply and Services of Canada, 1995), p.1.

[2]John R. Baldwin, *Strategies for Success: A Profile of Growing Small and Medium-sized Enterprises (GSMEs) in Canada* (Ottawa: Statistics Canada Catalogue No. 61-532R, 1994).

CHAPTER PREVIEW

In this chapter, we examine how the principles of financial management are applied to small and mid-sized businesses. We begin by distinguishing the characteristics of a small business firm. We then examine how small business firms manage their cash, make decisions on capital investments and raise funds for making investments. Finally, we describe the prerequisites that a business firm must have in order to go public.

OBJECTIVE 1

WHAT IS A SMALL BUSINESS?

Small business firm

Any firm with fewer than 50 paid employees and annual revenues of less than $5 million. Note that a firm in the manufacturing sector is a small business if the firm has fewer than 100 paid employees and annual revenues of less than $5 million.

Mid-sized business firm

Any firm with more than 100 paid employees and fewer than 500 paid employees.

Many definitions of "small business firm" have been offered; however, the definition and size restrictions used by Statistics Canada are probably the most widely used.[3] Statistics Canada defines a **small business firm** as any firm with fewer than 50 paid employees and annual revenues less than $5 million. The exception to this definition is the manufacturing sector where a small business is any firm with fewer than 100 paid employees. A **mid-sized business** is defined as a firm with more than 100 paid employees and fewer than 500 paid employees.

In this chapter it is not necessary to define a small business in the precise manner indicated by Statistics Canada. In general, we will define a small business as one that is growth-oriented and does not have easy access to the capital and money markets.

OBJECTIVE 2

SMALL VERSUS LARGE: IS THERE A DIFFERENCE?

World Wide Web Site

http://info.ic.gc.ca/opengov/cbsc/english/
Comment: This site provides information about federal and provincial government programs, services, and regulations. This site also provides the contact addresses for the Canadian Business Service Centres in each province.

Unquestionably there are differences between the financial management practices of small and large business firms.[4] For example, the decision-making process of small business firms is influenced by a different set of factors than those affecting large business firms. The following factors influence the financial management practices of small business firms:

1. Small firms are less liquid than their larger counterparts; hence, short-term cash flow patterns become critical to the success of the small business.
2. Small companies have more volatile profits and cash flows over time, owing to the greater business and financial risk experienced by smaller entities.
3. Small firms use relatively large amounts of debt in the financing of their business activities, which may be the consequence of entrepreneurs' greater propensity to assume risk. However, it is more likely that the issue relates to the small company's lack of access to the public capital markets, especially the equity markets.
4. The lack of access of the small corporation to the capital markets means the value of the firm is more difficult to determine. Also, the absence of market data for the firm's securities makes determining the cost of capital a difficult, if not impossible, job.

There are some underlying reasons for these differences. It has been said that "a small company is not just a little big business," the issue is more than a matter of

[3]Industry Canada, "Small Business: A Progress Report" (Minster of Supply & Services Canada, 1995), p. 1.

[4]Walker and Petty examine the quantitative differences between large and small firms. For further information, see: Ernest W. Walker and J. William Petty II, "Financial Differences Between Large and Small Firms," *Financial Management* (Winter 1978), pp. 61–68.

the number of zeros.[5] In other words, there are reasons why traditional financial analysis may not tell the story at times. The following reasons can be cited:

1. The owner may not always be a value maximizer. There may be other personal goals that are of equal, if not greater importance. Personal life styles which are realized through the company may distort the economic content of the financial statements, such as a firm owning a house on a lake. If the small firm is an extension of the owner, it becomes difficult to separate the firm from the owner in a financial context without causing distortions. For example, in making an investment decision, the need to create autonomy outside the firm may be as important as the investment's net present value.
2. For the small company, the goal of survival may supersede ideal financial practices, due largely to the small firm's limited access to the public capital markets, and bankers' insistence on looking to the owner's personal guarantees in addition to the firm's financial position.
3. Traditional definitions of debt and equity, as we have studied them, may not apply. For instance, loans to the owners may in reality be a form of equity, if the firm is unable to repay the loan. In fact, the owner's personal preference for financial risk may be more important than minimizing the firm's cost of capital when determining the desired level of debt.

The implications of these differences are truly significant to the small business owner. In making financial decisions, the owner should seek to maximize the total value of both corporate and individual wealth, subject to personal life-style preferences. Second, small business owners should make every effort to maintain flexibility when dealing with bankers and other providers of capital. Finally, the owner should always prepare a cash budget along with the income statement, and contingency planning should be a constant.

CASH MANAGEMENT IN THE SMALL FIRM

OBJECTIVE 3

The nature and importance of cash management were described in Chapter 18. Although it is important that a "mature" firm be cognizant of its cash requirements, it is even more important for a firm encountering change and for a firm that is undercapitalized, both of which are frequently characteristic of small businesses. For instance, an undercapitalized enterprise results in a working-capital shortage such that the turnover of working capital from the point of investing in raw materials to the final collection of receivables continues to plague the management of the company. In such an atmosphere, the executive is compelled to keep close tabs on the company's cash flow.

The second factor associated with many small businesses that should encourage the implementation of careful cash management is the growth element. The ownership of a small business facing an expansion has to be more cognizant of the cash inflows and the necessary outflows. Contrary to the belief of many small business owners, growth is not synonymous with financial prosperity. A growth firm is especially susceptible to financing difficulties, and thus requires a closer watch on funds flows. The belief that cash flows in a business equal profits plus depreciation gives the impression that a profitable business will not experience cash flow difficulties, which is wrong and misleading when applied to a growth firm.

A third instance in which the small firm may find cash planning essential is in approaching parties who represent potential sources of financing. The financing for the small business usually comes from the banker. In requesting a bank loan that entails a substantial amount, the cash plan becomes an important tool in two ways. First, the small business owner-manager is more familiar with the realities of

[5]See J. A. Welsh and J. F. White, "A Small Business Is Not a Little Big Business," *Harvard Business Review* (July–August 1981), pp. 18–32 and Richard Levin and Virginia Travis, "Small Company Finance. What the Books Don't Say," *Harvard Business Review* (November–December, 1987), pp. 30–35.

the loan request, especially in having a better understanding of how the loan is to be repaid. This issue is generally a key concern to the banker. Second, and of equal importance, is the opportunity for the small business owner to establish credibility with the banker if, when requesting the loan, he or she can explicitly set forth detailed projected cash flows.[6]

Despite the importance of careful cash management for the small firm, large firms typically make more extensive use of cash plans. Most managers of small firms believe they do not have the time to spend on forecasting. Besides, they view themselves as being close enough to the events that they are able to plan adequately in an informal fashion, without having to prepare a written budget. Such an argument may be true of a few business owners; however, for most of them time has a way of making financial details hazy. As a result, the benefit of past experience is not considered as critical. An understanding of the cash flow cycle, as well as the process for implementing a forecast, is particularly relevant for the small firm.

OBJECTIVE 4

CAPITAL BUDGETING AND SMALL BUSINESS FIRMS

We have examined decision-making techniques in Chapters 13, 14, and 15, which applied discounted cash flow (DCF) techniques to the valuation of capital investment projects. However, the application of these decision-making techniques by small business firms to capital investments is influenced by factors which are unique to this type of business organization form. We briefly examine some of the factors which may affect the application of these decision techniques by small business firms.

1. The firm's activities and owner's personal goals are inseparable. As a consequence, nonfinancial variables may play a significant part in their decisions. For instance, the desire to be viewed as a respected part of the community may be more important to the owner than the present value of a decision.
2. The frequent undercapitalization and liquidity problems of the small firm impact directly on the decision-making process within the small firm, where survival becomes the top priority.
3. The greater uncertainty of cash flows within the small firm makes long-term forecasts and planning unappealing, and even viewed as a waste of time. The owner simply has no confidence in his or her ability to reasonably predict cash flows beyond two or three years. Thus, calculating the cash flows for the entire life of a project is viewed as an effort in futility.
4. Since the value of a closely held firm is not as observable as that of a publicly held firm, where the market value of the firm's securities is actively traded in the marketplace, the owner of the small firm may consider the market value rule of maximizing net present values as irrelevant. In this environment, estimating the firm's cost of capital is also difficult. If computing the large firm's cost of capital is difficult at best, the measurement for the small firm becomes virtually impossible.
5. The smaller size of projects of a small firm may make net present value computations not feasible in practice. Much of the time and costs required to analyze a capital investment are fixed; thus, the small firm incurs a diseconomy of scale in evaluation costs.
6. Management talent within a small firm is a scarce resource. Also, the training of the owner-managers is frequently of a technical nature, as opposed to a business or finance orientation. The perspective of these owners is influenced greatly by their backgrounds.

[6]With increased risk and the limited access to the financial markets, the liquidity position of a small firm contributes further to the importance of an open relationship between a small firm and its bank.

The foregoing characteristics of a small business, and equally important, the owners, have a significant impact on the decision-making process within the small firm, whether or not we agree with the logic. The result is a short-term mind set, somewhat by necessity and partly by choice. Nevertheless, given the nature of the environment, what could we recommend to the owner-managers of the small business firm? What process can owner-managers use to make investment decisions for the firm?

A Better Way of Capital Budgeting for the Small Business Firm

Management of a large firm can sometimes become more concerned with its own priorities than serving the owners' best interests. This type of agency problem must be considered if we are to understand the financial nature of the firm. Owners cannot ignore potential conflicts if they develop within a large firm. In the case of the small firm, we have an owner-manager who is one and the same. The decision to be wealth satisfiers rather than wealth maximizers cannot be criticized. If the owners are maximizing their utility, which includes more than financial considerations, who is to say they are wrong? What finance can do, however, is address the potential need of the small firm to consider liquidity on an equal footing with value maximization, as well as the problem of estimating cash flows in the long-term future. Furthermore, we may at least partially resolve the difficulty in measuring the firm's cost of capital.

BACK TO THE BASICS

If we are to make intelligent capital-budgeting decisions we must accurately measure the timing of the benefits and costs; that is, when we receive money and when it leaves our hands. ***Axiom 3: Cash— Not Profits—Is King: Measuring the Timing of Costs and Benefits*** *speaks directly to this. Remember, it is cash inflows that can be reinvested and cash outflows that involve paying out money. Furthermore, we must incorporate* ***Axiom 2: The Time Value of Money—A Dollar Received Today Is Worth More Than a Dollar Received in the Future*** *into our capital-budgeting decision. If we are at all to make rational business decisions we must recognize that money has a time value. Finally, we want our capital-budgeting decision to measure the benefits and costs of projects on an incremental basis, which relates directly to* ***Axiom 4: Incremental Cash Flows—It's Only What Changes that Counts.*** *In effect, we will ask ourselves what the cash flows will be if the project is taken on versus what they will be if the project is not taken on.*

The Need for Liquidity

While not an ideal answer, a case may be developed for the small firm to use a discounted payback period in evaluating a proposed investment.[7] The payback method was shown in Chapter 13 to have limitations when applied to investment proposals. However, this method does give us an indication of how long our funds are tied up in an investment. This type of indication gives us some measure of liquidity, which may be vitally important for the small firm. Also, while we had faulted the payback approach for ignoring cash flows beyond the payback period, such a limitation may have less significance for the small firm, since the cash flows are more uncertain over the long run.

EXAMPLE

To illustrate the use of the discounted payback period, we have estimated the cash flows for a project for the first five years of its life. The project is expected to cost $50,000, and the owners have a required rate of return of 15 percent. The expected cash flows for the first five years are as follows:

[7]See Richard Wacht, "Capital Investment Analysis for the Small Firm," Working Paper (February 1988)

Years	Expected Cash Flows	Present Value of Expected Cash Flows
1	$12,000	$10,435
2	14,000	10,586
3	17,000	11,178
4	20,000	11,435
5	20,000	9,944
	Total present value	$53,578

The payback period for the investment would be 3.35 years ($43,000 received in three years and the remaining $7,000 recouped in 0.35 years, i.e., $7,000/$20,000). However, when we recognize the owner's required rate of return of 15 percent, we find the present value of the first five years' expected cash flows to be $53,578. Using these present values, we see that the owners recoup their investment in 4.64 years ($43,634 of the investment received in four years and the remaining $6,366 in 0.64 year, i.e., $6,366/$9,944). By comparing projects in this manner, we give consideration both to the present value criterion and the liquidity of the project.

Measuring the Cost of Capital: A Direct Approach

The methodology used in Chapter 12 for measuring the cost of capital cannot be applied to a small firm. We simply do not have access to market data for the small firm as we do for the large corporation whose shares are traded on the Toronto Stock Exchange. Thus, we must resort to some alternative method.

As noted in Chapter 12, the cost of capital is an opportunity-cost concept. Shareholders should receive from their investment in the firm an amount at least equal to the rate of return available in the capital markets, given the level of risk. We use the market data of the firm's securities to estimate these opportunity costs. However, no such information is available for most small companies. We therefore need to modify our approach.

BACK TO THE BASICS

All the methods used to compensate for risk in capital budgeting find their roots in ***Axiom 1: The Risk-Return Tradeoff—We Won't Take on Additional Risk Unless We Expect to Be Compensated with Additional Return.*** *In this case, we compensate for risk in the capital-budgeting decision by using a competitive rate of return from the marketplace and substituting it for the owner's required rate of return.*

We will assume that the small business is either family-owned or closely held. Since the shares of this firm are not publicly traded, we cannot use this type of data to estimate the owner's opportunity cost of investing in the next best project. However, the marketplace will give a well-informed owner a competitive rate of return which can be used as a proxy for the owner's required rate of return. We would then suggest using the "residual NPV approach," which compares the cash flows going to the shareholders to their required rate of return, rather than using a weighted cost of capital for all investors.

The Residual Net Present Value

The residual net present value is a slight modification of the conventional weighted cost of capital approach. Rather than computing the present value of the expected cash flows going to all investors discounted at the weighted cost of capital, we estimate the cash that will flow to the owners after all debt and preferred shareholders have been paid their returns. We then discount these flows at the owners' required rate of return. As a financial advisor to the small firm, we recommend the following method:

1. Determine a target position for the firm's debt-equity ratio.
2. Compute the residual after-tax cash flows available to the owners of the company, net of interest expense, debt principal repayment, and preferred dividend payments. In Chapters 13, 14, and 15 we used after-tax operating cash flows (cash flows available to all investors of the firm, that is, debt, preferred shares, and common shares) in our net present value calculations. However, with the weighted cost of capital, our discount rate recognized the required rates of return for all these investors as well.
3. Have the owners of the company decide what is a fair rate of return for the project, given its level of business risk. This rate may be somewhat subjective; however, an appropriate rate can be determined by recognizing alternative uses of the funds and by allowing for any personal factors that affect the owners' total utility from operating the business.
4. Calculate the present value of the residual cash flows going to the owners and determine the project's discounted payback period. Use these results to evaluate the proposed investment against other investments under consideration or investments that have been recently accepted.[8]

EXAMPLE

Assume the Arganes Corporation is contemplating an investment that costs \$55,000. The project would have an expected life of about 15 years; however, the uncertainty of these cash flows makes management uncomfortable at projecting the flows beyond six years. The firm has a 40 percent target debt ratio; the interest rate on the debt is expected to be 10 percent; and the principal on the debt is to be repaid by reducing the balance by 10 percent at the end of each year. The asset will be depreciated on a straight-line basis over 10 years for accounting purposes. However, for tax purposes the asset will be depreciated using the declining balance method at a CCA rate of 30 percent. The asset class is expected to remain open and the asset is not expected to have a salvage value (SV = \$0) after its useful life of 15 years. The firm's tax rate is 22 percent. The residual cash flows which accrue to the owners in each year are computed in Table 25-1.

In the case of the small business, we are interested in cash flows accruing to the owners of the firm. These cash flows accrue to the owners only after interest payments and debt principal are paid. This condition requires that we adjust for the tax benefits from interest payments by adding the interest tax shields to net operating income. This type of adjustment is similar to that of adding the depreciation tax shields to net operating income. Finally, the payments which decrease debt principal are subtracted from the operating cash flows after tax. The residual cash flows accrue to the owners of the firm and should be discounted at the owner's required rate of return (15%) to obtain their present value. The last row of Table 25-1 shows the present value of these residual cash flows.

From the present values of the annual cash flows in Table 25-1, we can determine the discounted payback period as follows:

$$\text{discounted payback period} = \text{(4 years of cash flows with present value of \$52,227)} + .276 \text{ years to receive remaining cash flow of \$2,773 or \$2,773/10,061} = 4.276 \text{ years}$$

Thus, in 4.276 years, the owners may expect to recoup their original investment, while earning their required rate of return of 15 percent. The decision to accept or reject the project would come only after comparing it with other alternative uses of the funds, and the impact of the project on key nonfinancial variables that only the owners can know, based upon their preferences.

[8] If the project payback period is shorter than the expected life of the investment, we may be assured that the project will have a positive net present value.

Table 25-1 Arganes Corporation Project Analysis

	Years 1	2	3	4	5	6
Operating revenue	$40,000	$48,000	$52,000	$52,000	$55,000	$55,000
Operating expenses (including accounting expenses)	–25,500	–30,500	–32,500	–32,500	–33,500	–33,500
Accounting depreciation	5,500	5,500	5,500	5,500	5,500	5,500
Operating income (EBIT)	20,000	23,000	25,000	25,000	27,000	27,000
Taxes (@22%)	–4,400	–5,060	–5,500	–5,500	–5,940	–5,940
Net operating income	15,600	17,940	19,500	19,500	21,060	21,060
Interest tax shields[a]	484	436	392	353	318	286
Depreciation tax shields[a]	1,815	3,086	2,160	1,512	1,058	741
Cash before debt payment	17,899	21,462	22,052	21,365	22,436	22,087
Debt payment	–2,200	–2,200	–2,200	–2,200	–2,200	–2,200
Cash flow to owners	$15,699	$19,262	$19,852	$19,165	$20,236	$19,887
Present value	$13,651	$14,565	$13,053	$10,958	$10,061	$ 8,598
Cumulative present value	$13,651	$28,216	$41,269	$52,227	$62,288	

[a]Schedule for Depreciation and Interest Tax Shields

Year	UCC (Beg)	CCA	CCA (T)	UCC (End)	Loan Balance (Beg)	I	I (T)	Loan Balance (End)
1	$27,500	$ 8,250	$1,815	$46,750	$22,000	$2,200	$484	$19,800
2	46,750	14,025	3,086	32,725	19,800	1,980	436	17,820
3	32,725	9,818	2,160	22,907	17,820	1,782	392	16,038
4	22,907	6,872	1,512	16,035	16,038	1,604	353	14,434
5	16,035	4,811	1,058	11,224	14,434	1,443	318	12,991
6	11,224	3,367	741	7,857	12,991	1,299	286	11,692

BASIC FINANCIAL MANAGEMENT IN PRACTICE

Factors for Small Business Survival

Money helps: 84 percent survival rate for businesses with initial investment of $50,000 or more; 74 percent survival rate for businesses with initial investment of $20,000 or less.

Bigger is better: 82 percent survival rate for companies starting with six or more employees; 71 percent survival rate for companies starting with fewer than two employees.

Work hard, but don't overwork: 80 percent success rate for owners who work between 60 and 69 hours per week; 75 percent success rate for owners who work more than 69 hours per week.

Source: "Factors for Small Business Survival: Profiles of Success," an American Express Study of New Business as reported in *The Dallas Morning News* (October 11, 1989), p. D1.

We will now direct our attention to the valuation of the small firm, which has been a matter of great interest to all involved in the management of the small firm.

FINANCING THE SMALL BUSINESS FIRM

OBJECTIVE 5

In studying how a small firm is financed, we will first describe the financing stages a firm experiences during its business life. We will then look at a special group of investors called "venture capitalists" who may be particularly significant for a small growth business, but not for a large business. Next, we will examine the different sources of financing in an effort to identify the more important ones for the small firm. Naturally, the conventional sources of short-term and long-term debt and common shares apply to all firms, small and large. However, the desirability and the availability of these various funds for the small firm may be different from those for the large firm. Finally, the decision to "go public" may be a serious consideration for a successful growth firm that has expanded beyond the financing capabilities of its owners. The advantages and disadvantages of issuing shares to the public are discussed in the concluding section of the chapter.

Stages of Financing

As shown in Figure 25-1, the business firm goes through three primary financing stages in the earlier segment of its business life cycle. Phase one, the initial investment, consists of the owner's personal capital and the credit provided by the chartered banker, and a host of miscellaneous sources. Examples of miscellaneous sources in this phase would include (1) friends and relatives, (2) leasing companies, and (3) trusts, credit unions, caisses populaires, and commercial finance companies.

Phase two of a young firm's financial existence may be referred to as *the gap*. During this period the firm has grown beyond the level where the owners have the capability to finance all investments; however, the firm is not large enough to justify a public offering. At this point it must sell securities to private individuals or groups. As shown in Figure 25-2 a number of sources for private placements exist and may be directly placed to come through an investment dealer. For instance, financial institutions (banks and trusts or credit unions), investor groups, and venture capitalists may be approached either directly or through an investment dealer acting as a liaison.[9] Large corporations may provide financing for a small firm but are generally approached directly. Furthermore, the kind of financing available from these respective sources varies, with financial institutions providing primarily debt financing and large corporations providing equity financing. Between the two extremes, a mixture of debt and equity capital would be available.

The final stage of development for the small firm in terms of financing sources is the raising of funds in the public markets. As explained in Chapter 7, a firm's shares are first traded publicly over the counter, then move on to one of the major Canadian exchanges such as the Vancouver Stock Exchange or the Toronto Stock Exchange.

Before looking at specific sources of financing, we will discuss the nature of venture capital.

Venture capitalists

Investors interested in supplying capital to particularly high-risk situations, such as startups or firms denied conventional financing.

Venture Capital and the Small Firm

The **venture capitalist** has increasingly become a source of financing for small businesses. A definition of venture capital investing is difficult, in that venture cap-

[9]The term *venture capitalist* is used to describe professional investors interested in investing in high-risk firms with potentially large returns. A more detailed explanation is provided in the next section.

Figure 25-1 Business Financing Stages

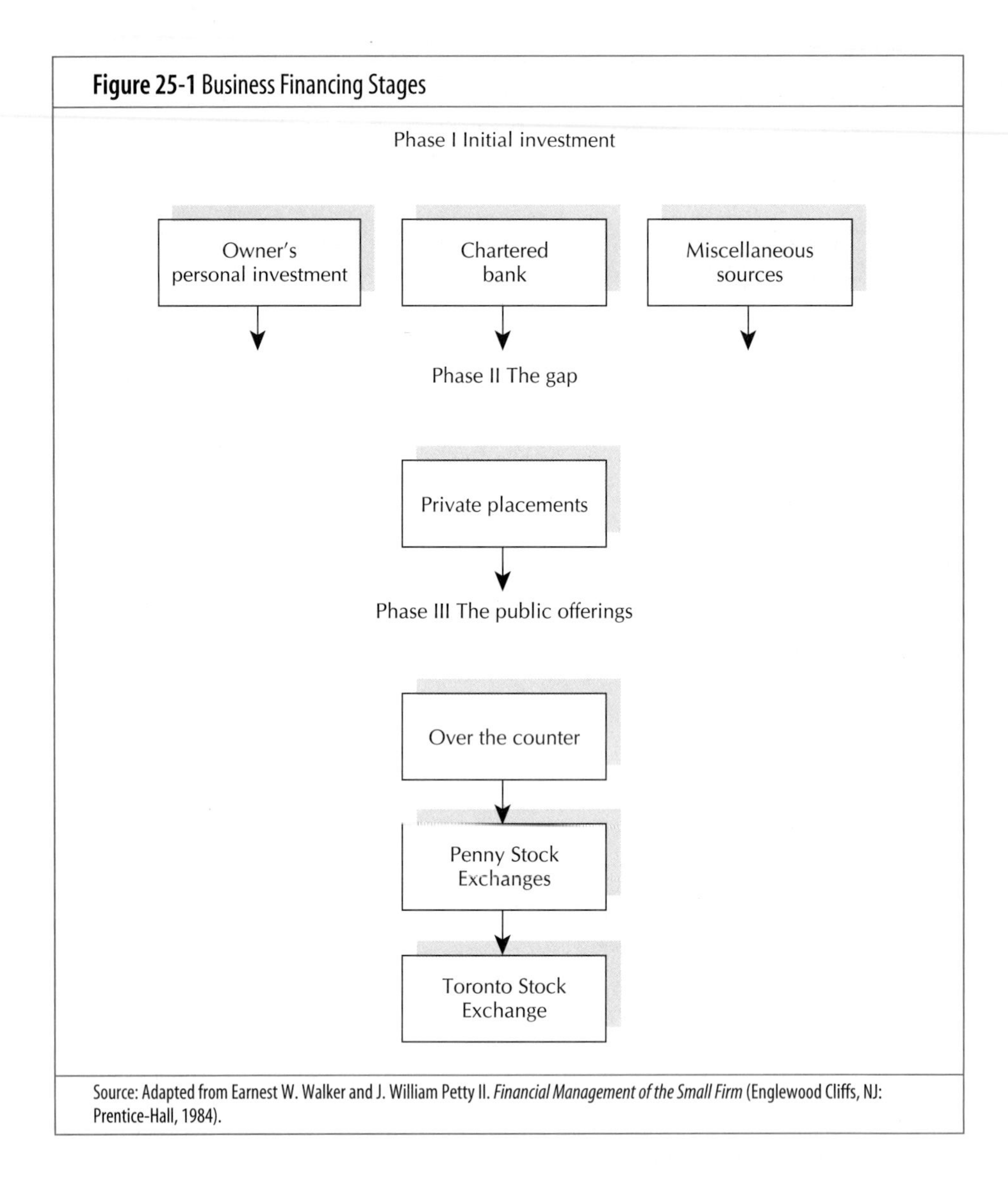

Source: Adapted from Earnest W. Walker and J. William Petty II. *Financial Management of the Small Firm* (Englewood Cliffs, NJ: Prentice-Hall, 1984).

italists have a broad range of interests and activities. Most venture capitalists, however, would fall somewhere along the following spectrum of investment interests:

1. Providing capital for any high-risk financial venture;
2. Providing seed capital for a start-up situation;
3. Investing in a firm that is unable to raise capital from conventional sources;
4. Investing in large, publicly traded corporations where risk is significant.[10]

Which definition is most descriptive depends in part on the eye of the beholder. For instance, some venture capitalists believe that providing seed money is too risky. Others, however, believe that such start-up capital is the primary intent of their investments. Thus, generalities become difficult, but three underlying attributes normally prevail with venture capital investing. First, the investor is usually "locked in" to the firm for some duration of time, without an opportunity to sell shares quickly in response to a change in the firm's position, either favourable or unfavourable. Second, the venture capitalist normally represents the first equity

[10]See Patrick R. Liles, "Venture Capital: What It Is and How to Raise It," *New Business Ventures and the Entrepreneur* (Homewood, IL: Richard D. Irwin, 1974), p. 461.

Figure 25-2 Sources of Private Placements

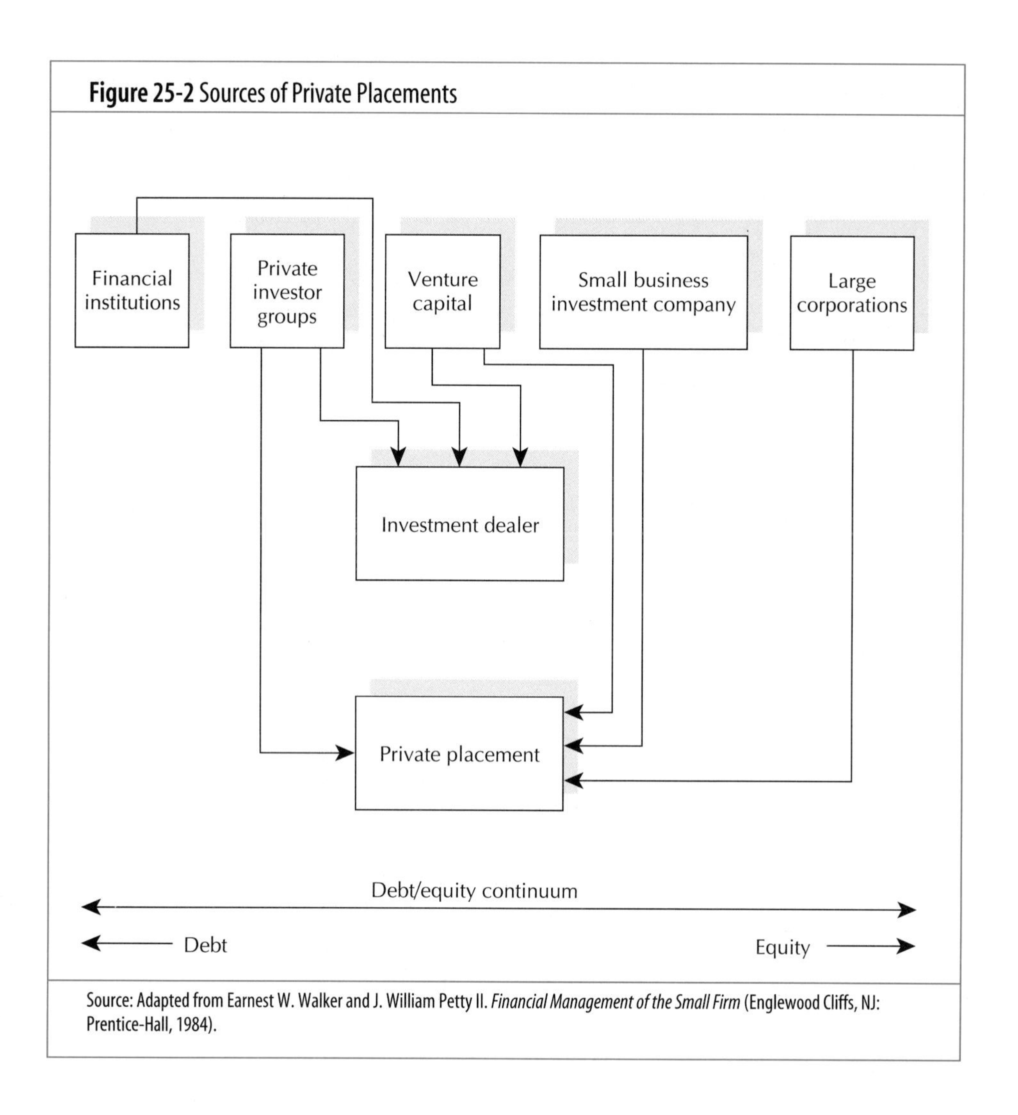

Source: Adapted from Earnest W. Walker and J. William Petty II. *Financial Management of the Small Firm* (Englewood Cliffs, NJ: Prentice-Hall, 1984).

financing from an "independent" party, as opposed to a friend or relative. However, additional financing from the venture capitalist may be required at a later date. Finally, venture capital investors have divergent interests. An investment having no appeal to one venture capital group might be of extreme interest to another group. For this reason, the management of a firm seeking funds should be aware of the compatibility of interests of the firm and the venture capitalist being approached.

If a financing package comprised of debt financing is presented to a venture capitalist, two major issues are involved. First, the venture capitalist may insist upon protective covenants if the financial condition of the firm deteriorates or does not materialize as anticipated. These covenants, similar to the protective restrictions of a bank loan, are particularly common in start-up situations, with effective control of the organization being more easily lost through default on these long-term covenants than through the loss of majority ownership of the equity shares.

Second, the venture capitalist considers the tradeoff between the interest rate and the conversion privilege. The use of debt by venture capital investors generally involves a convertibility feature or bonds with warrants attached. By attaching either warrants or a conversion privilege, a significant reduction in interest charges usually results. For a small growth business a reduction in interest payments is not only desirable but essential, since most small businesses may have difficulty meeting the interest requirements associated with straight debt. The excerpt on page 946 profiles the venture capital industry.

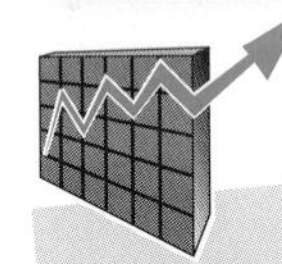

Venture Capital

Financial institutions now offer many competitive services, and the traditional boundaries among them are becoming blurred. Three elements differentiate venture capital from other sources of financing: the investment involves equity or equity-like participation; the investment is long-term; and investors are actively involved in the companies they finance.

Venture capital firms provide long-term financing by supporting businesses through a broad range of transactions. These transactions range from seed investment for the development of new products to the acquisition of established firms. The venture capitalist generally assumes greater risks than other agents in the financial system, and venture capital firms fill critical roles in the economy by taking equity positions in companies where there are insufficient assets to use as collateral for loans. For new or rapidly growing firms with relatively few assets, venture capital firms may be the only source of financing available. Venture capital firms make a commitment for a longer time frame than some other financial institutions and are usually prepared to provide subsequent financing to companies already in their portfolios.

Frequently, in contrast to other sources of financing, venture capital firms offer strategic management support for the companies in which they invest. These efforts to ensure the realization of the company's potential growth are often referred to as the value-added brought to the transaction by the venture capitalist and may include advising the firm through representation on the board of directors or assisting the personnel of the firm in order to strengthen the management team. If the portfolio company prospers, the venture capitalist can "exit" positively from the investment in a number of ways: generally the equity stake is sold as part of a public offering of shares, purchased by another firm, or repurchased by the portfolio firm.

When viewed in an international context, Canadian venture capital firms are notably smaller than foreign counterparts. The average capital base of the Canadian firms is equal to approximately two-thirds of the average for American firms and one-third that of the British firms. Canadian venture capital firms average four professionals each, although some companies have a staff of 20.

Four subgroups make up the Canadian professional venture capital industry: independent firms, venture affiliates or subsidiaries of Canadian corporations, Crown-related venture groups, and "hybrids" having a mixture of government and private sector support.

The past number of years has seen a significant change in the sources of funds for the professional venture capital industry. As a result of federal initiatives to stimulate institutional investment activity and the higher profile that venture investing thus received, pension funds tremendously increased their funding of independent venture firms, and they became the dominant financiers of these firms. Corporations, insurance companies, and individuals still invested in venture capital firms, but their overall significance was lessened.

This shift in the source of funds for venture firms brought about a refocusing of types of investments. Venture investing was new to pension fund managers during the mid-1980s. To allay their concerns about risk, venture portfolios were often structured to focus more on established firms and away from investments in newer, high-risk company start-ups. This move was reflected in the surge of buy-outs of established firms. Such deals involve relatively low risk and are often larger and more profitable than the average venture placement.

Independent venture capital funds usually have a fixed lifetime. Those that have matured recently or are maturing in the early 1990s have so far performed below investor expectations in general, as a result of abnormally high investment losses and high transaction costs in the mid 1980s.[11] The lower returns have resulted in a notable slowdown in the flow of money from major institutional investors to new

[11]A. G. Fells, "Venture Capital: A Five-Year Update," (Toronto: SB Capital Corporation Ltd. 1988), p. 8.

Canadian venture capital funds, as evidenced by the absence of growth in the industry's capital during 1990. However, the general impact of this slowdown over the medium term may not be great. The flow of resources to venture capital funds tends to be cyclical. A troubling aspect of the situation is that a large proportion of the industry's funds available for investment are concentrated in firms that focus on Quebec and Alberta, thus creating liquidity concerns in other parts of the country.

Source: Adapted from *Industry Profile, 1990–1991, Venture Capital* (Communications Branch, Industry, Science and Technology Canada, 1991).

Debt as Source of Financing

Assuming that the business owners have depleted all available equity funds or want to take advantage of what they hope will be favourable financial leverage, management will no doubt search for debt financing. Potential sources of such financing for a small firm include (1) commercial banks, (2) governmental assistance programs, and (3) insurance companies.

Chartered Banks

The chartered bank is the leading source of borrowed funds available to the small business. The loans available through the bank are primarily of two types, short-term and mortgage loans. However, the short-term instruments, although formally defined as short-term, provide considerable long-term financing for the small business entity. And although statistics indicate that the bank is principally a provider of short-term capital, such figures can be misleading.[12] For instance, a loan with a maturity of 90 or 180 days, although classified as short-term, provides long-term financing through regular renewals as the indebtedness approaches the due date. Turning the debt over frequently is a technique by which the owners of a small firm convert short-term financing to an equivalent long-term status. This approach carries one distinct disadvantage in that the firm's financing is at the discretion of the banker each time the debt matures, which may cause some anxiety for the small business owner. The second and more formal source of long-term financing is the mortgage loan. These obligations relate to a specific purchase of real estate that is used as collateral in securing the financing arrangement.

Bank relations with small business have been improved by the introduction of a business code of conduct that sets out minimum service standards. In addition, the banks have introduced an Alternative Dispute Resolution (ADR) system that mediates through a neutral party any credit-related complaints that cannot be resolved through internal procedures. Banks are expected to notify customers of reasons for credit refusals, the steps required to meet approval and where appropriate, the availability of alternative financing sources. Banks are also expected to provide a minimum of 15 days calendar notice before a credit line is reduced or terminated.

In relying on the bank as a source of funds, the small business owner should develop a relationship with the banker prior to the actual need for funds. Ideally, the interaction should be on a professional and a personal basis. As part of the relationship, the business owner should be completely open with the banker. Only in this way can the banker be of maximum effectiveness in providing services. Also, in maintaining an association with the banker, the owner should continue to examine

[12]In general, banks provide short-term financing by extending open lines of credit; however, banks do provide longer-term financing from three to five years via revolving lines of credit.

the benefits being received. He or she should be aware of the services being provided by competitive banks, as well as the costs related to these services.

The affiliation with a banker is not purely a function of the interest rate being charged, but involves a number of factors. The mere fact that a bank is charging a slightly higher interest rate does not necessarily justify changing banks. A long-term relationship is important, especially if monetary conditions are tight or if the firm temporarily encounters adversity. Finally, an inquiry should be made as to the banker's expertise and the bank's capabilities in terms of loan size. The firm's owner should seek a bank having particular expertise in the firm's type of operation. That is, not only is the accessibility of capital important, but a source of counselling may be of substantial benefit.

Governmental Sources of Assistance[13]

The federal government provides assistance in the startup and the improvement of small and mid-size businesses through governmental bodies such as Industry Canada, the Business Development Bank of Canada (BDC), and the Canadian International Development Agency (CIDA). Industry Canada has been created with a mandate to improve the competitiveness of Canadian industry. With respect to the small business area, Industry Canada has the responsibility to "develop and implement national policies to foster entrepreneurship and the startup, growth and expansion of small business." Industry Canada has implemented several programs to provide business intelligence, technology, and financial assistance. In passing the Small Business Loans Act in 1961, all chartered banks and other financial institutions are authorized to make business improvement loans to assist in the start up and the financing of fixed assets needs. For example, floating rate loans will cost interest equal to prime plus 3 percent, while a fixed rate loan will cost interest equal to the residential mortgage rate plus 3 percent. The Business Development Bank of Canada, formerly the Federal Business Development Bank, is a Crown corporation which has a mandate to promote the creation and development of business in Canada. The Business Development Bank of Canada provides assistance to owners of small businesses through management and financial services. For example, the corporation's financial services include loans, working-capital loans and loan guarantees for expansion, acquisition of fixed assets and the purchase of existing businesses. The Canadian International Developmental Agency has a mandate to administer Canada's development assistance to developing countries which includes the Canadian private sector's involvement in these countries. CIDA's assistance includes financial assistance that supports long-term business relationships with developing countries through joint ventures and licensing arrangements. The agency will also finance and provide professional advice on conducting preliminary and detailed studies of capital projects to potential clients in developing countries.

World Wide Web Site

http://info.ic.ga.ca/ic-data/icon.html
Comment: This bulletin board service provides access to Industry Canada documents and publications. This service also provides information on Industry Canada's programs and services.

Insurance Companies

The executive of the small firm may find that an insurance company is a source of financing. As a financial intermediary, an insurance company has large sums of money available for investments. Only a small portion of such funds is available for the small firm, but to the extent that such funds may be used, the extension of credit generally comes in the form of a mortgage loan. Hence, a life insurance company may be a potential long-term source of financing for a small firm planning the construction of a real estate investment, such as a plant or an office. In examining the request for a loan, the insurance company looks closely at the type of building being constructed, with a preference for a facility having potential multiple uses. As to the size of loans, most insurance companies are interested in making loans in large amounts; however, loans for as little as $10,000 have been granted. In general, mort-

[13]Government of Canada, *Helping Small Business: A Guide to Federal Support* (Ottawa: Minister of Supply and Services Canada, 1995), pp. 5-7.

gage loans extend over a 10- to 20-year period, with the amount of the mortgage being limited to approximately 70 percent of the appraised value of the property.

In addition to the mortgage loan, the insurance company may issue credit on an unsecured long-term basis to the borrower who has an excellent credit rating. Such loans, if granted, generally extend for periods of 10 years with an interest rate exceeding that of the bank by 1 to 2 percent.

Common Shares as a Source of Financing

In examining common shares as a source of financing for the small firm, three areas are of primary concern: (1) the problems a small firm may encounter in issuing common shares, (2) the forms of common shares used by small firms, and (3) a review of specific sources of common equity that may be available to small firms.

Factors Affecting the Use of Equity Capital by Small Business Firms

Chapter 6 discussed the basic characteristics of common shares, along with their advantages and disadvantages as a source of financing. The following factors affect the use of common shares as a means to finance the small business firm:

1. Because of the lack of marketability of most securities of small firms, particularly common shares, it may be necessary to price the common share so that it will yield a significantly higher return than that of a similar firm having a ready market for its shares. Naturally, the result is an increased cost of common capital. It should be noted, however, that shares of qualified small business corporations are eligible for a $500,000 lifetime capital gain exemption.[14]
2. The flotation cost of selling a small issue, particularly in the public markets, becomes almost prohibitive for the small firm. For instance, such expenses may approach 25 percent of the issue size.
3. Issuing common shares to external investors permits a potential dilution of control for existing shareholders. Although such a loss may only be psychological, most small owners are averse to placing themselves in a position that even slightly compromises their control over the firm.

Forms of Common Shares

Common shares can be classified in terms of the shareholders' claim on income, claim on assets, and voting rights. Typically, Class A shares have a higher claim on dividend income and assets but less extensive voting power, even to the point of voting being nonexistent. In contrast, Class B common shares have a lesser claim on income and assets but greater voting power. This classification mechanism permits the owners to maintain greater control or be able to attract outside investors by providing greater income and protection upon liquidation. In the same context, "founders' shares" have been used by small business. These shares are similar to Class B shares except that the owners of the shares maintain the sole voting rights but generally have no right to dividends for a specified number of years. The objective of the founders' shares is to permit the organizers of a company to maintain complete control of the company during the early years of the operation, often with a minimal investment.

Beyond relatives and friends, the owner of the small business must look to venture capitalists for common share financing. Specific sources are described on page 951.

[14]For the purposes of the Income Tax Act a qualified small business corporation must be a Canadian-controlled private corporation of which all or substantially all of the fair market value of its assets are used in active business carried on primarily in Canada. In addition, shares must be owned by the taxpayer or related person for at least 24 months preceding the disposition and during this 24 month period, more than 50 percent of the fair market value of the corporation's assets must be used in active business carried on primarily in Canada.

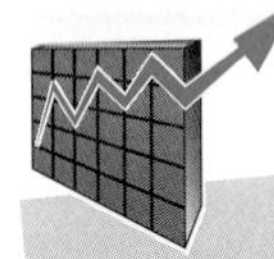

Investing in Mid-Sized Firms: Unique Problems, Unique Opportunities

Mid-sized firms are critically important in market oriented economies with an increasing orientation towards knowledge and service industries, and where manufacturing industries are experiencing extensive restructuring. This is because it is in the mid-sized firm sector of such economies that entrepreneurship, innovation, human and financial resources and market base reach a joint critical mass. In a Canadian context, it is in the mid-sized firm sector especially where we must hold our own under the North American Free Trade Arrangement and in a liberalized world trading environment generally.

A striking feature of our mid-sized firms is the extensive use of debt in their capital structures. The use of debt as described by the debt/equity ratio can have a range over 150 percent for Canadian-controlled mid-sized firms, double that of mid-sized foreign-controlled firms and large Canadian-controlled firms. The Canadian Chamber of Commerce investment study FOCUS 2000 comments: "While Canada has one of the most sophisticated capital markets in the world for debt, its ability to provide equity is not well developed." Indeed, Canada's bankers and mezzanine financiers do an excellent job getting debt money to mid-sized firms.

The Shape of Things to Come: A Growing Private Placements Market

Besides the volatility of equity markets and LBO-related factors, there are other factors pointing to a growing private placement market at the expense of the public markets. From an institutional perspective, private placements are simple, easily negotiated because only a small a group of professional investors is involved. More and more frequently, direct placements will be done without the use of an investment dealer, eliminating the intermediary fee. Also, with private placements the company does not have to deal with the securities commissions, does not have to worry about public statements, and is only answerable to a board which consists of friendly institutions or their representatives. These considerations have a great deal to commend to investors and to large and small investees alike.

The disadvantages, of course, are an absence of liquidity and the fact that there is no measurement of success in the public market. Also, a quick exit through the public markets if the company gets into trouble is no longer available. In the new world, exits are going to be outright sales of companies through private transactions, recapitalizations where companies acquire an enhancing debt raising capacity over a period and raise money by way of debt to pay out the original equity holders, or redemptions or buyouts by formula which give institutional investors a negotiated yield over a defined investment period. These kinds of solutions will likely flourish and grow.

Private Placements and Canadian Mid-Sized Companies

Today most investments in mid-sized companies are done through intermediaries such as venture capitalists, leveraged buy-out funds, or mezzanine finance pools. The reason is that many mid-sized companies which raise money need nursing. Institutional investors such as pension funds have thus far been ill-equipped to do this. But there are so many pools of funds out there that entrepreneurs are increasingly using intermediaries to make the right match-up with the right capital pool.

Between the fees charged by capital pool managers and by the matchmakers, the cost of putting out the money to the entrepreneur has been rising, reducing investor returns. In the U.S. this has caused large institutional investors such as Prudential to eliminate all intermediaries and take the requisite skills in-house. It is only a matter of time before large Canadian institutions begin to do the same thing on a large scale.

Source: Adapted from: Gordon R. Sharwood, "Investing In Mid-sized Firms: Unique Problems, Unique Opportunities," *Canadian Investment Review* (Spring 1990), pp. 27–30.

Sources of Common Equity for the Small Business Firm

Publicly Owned Venture Capital Investment Companies These organizations are publicly owned by a broad base of investors, including companies listed on the Toronto Stock Exchange.

Private Venture Capital Investors The private venture capital investor is primarily concerned with the development of small enterprises needing long-term capital for growth, companies for sale due to the death or retirement of the owner, or firms requiring professional management. These types of private venture investors operate under a high degree of flexibility and therefore may be interested in long-term capital appreciation, in which funds are frequently invested in participation with others to enable the small private venture company to maintain diversity. Thus, private venture capital investors may possibly offer not only another source of venture capital, but also an avenue for increased flexibility relative to the publicly owned companies.

Canada Community Investment Plan (CCIP) The Canada Community Investment Plan has been initiated to help communities which have firms that are looking for small amounts of equity to network with local sources of venture capital. The federal government has launched this program to provide investment-development training and tools to interested communities.

Atlantic Investment Fund (AIF) The Atlantic Canada Opportunities Agency (ACOA) in partnership with the four provinces, major banks and other private sector investors established an investment fund which provides equity or quasi-equity investments to growth-oriented firms in the Atlantic region. This investment fund is operated by the private sector on a full commercial basis and at arm's length from governments.

Business Development Bank of Canada The Business Development Bank of Canada (BDC) provides to companies with high growth potential the opportunity of obtaining access to capital markets through its lending and venture capital programs. The BDC provides very small businesses micro-business loans of up to $25,000 as well as loans which top up a company's existing credit line so that it can take advantage of growth opportunities. This bank can also purchase shares in a business venture or facilitate other financial institutions helping clients obtain equity financing. This corporation has also initiated a program which enables owners of small businesses to acquire equity capital for important development projects. Such a program enables the owner of a small business to obtain venture financing without relinquishing control or diluting ownership.

Wealthy Individuals Wealthy individuals within the community should be recognized as a viable source of venture capital. If relatively small amounts are required (less than $100,000) and the management of the business desiring the funds is well known within the business community, wealthy individuals may be quite receptive to incurring significant risk for the purpose of achieving high expected returns. The attraction becomes even greater if the returns are tax-sheltered. The advantage in developing such arrangements relates to the informality and the often negligible formal reporting requirements. Typically, the investors are relying on their "faith" in the chief executive and not on detailed accounting reports. Yet management should proceed with caution in the selection of local investors, because such individuals occasionally develop a strong propensity for participating in the management of the firm.

Large Corporations Major corporations have developed formal venture capital operations.[15] Large companies have entered the venture capital field for two reasons: the hopes of striking a capital gains bonanza and the search for technology, with a potential acquisition in mind. A prime example of corporate involvement is Westinghouse Canada.

Retention of Earnings as a Source of Financing

The retention of earnings is a major source of equity financing for any organization, regardless of size. Even large corporations rely heavily on the retention of profits as a means of building an equity base for growth. Yet, in comparison, the small business firm draws even more heavily on profits in financing its growth.[16] An important reason for this type of financing is government incentives such as the small business deduction (SBD). This tax incentive reduces the basic federal corporate tax by 16 percentage points on the first $200,000 of active business income earned by Canadian-controlled private corporations (CCPC).[17] Any income earned above $200,000 is taxed at the normal marginal tax rate of 28 percent. The Canadian government estimates that the SBD provides $2 billion of assistance to CCPCs on an annual basis. It should be noted that the accumulated retained earnings do not represent a large proportion of the small companies' capital structure; this lower percentage has to be due, at least in part, to the fewer years of existence of most smaller businesses. In summary, small business firms rely more extensively on retained earnings as a source of equity financing.

The relative retention rates hold implications regarding the dividend policies of the different-sized companies. In Chapter 10 we saw that the company's dividend policy is believed to be a function of two factors. First, the company's investment opportunities should be a key input into the dividend decision. If the expected returns from prospective investments exceed the company's cost of capital, an incentive exists to retain greater amounts and therefore pay lower dividends. On the other hand, poor investments suggest higher dividend payments. Second, the common shareholders' preference for dividends versus capital gains is an important consideration in ascertaining the appropriate dividend strategy. Inducements frequently noted for the distribution of profits include the favourable informational content associated with an investor's receiving a dividend, the resolution of uncertainty, and the preference for current income.

The incentives for paying dividends for a large company having a broad base of shareholders are not as significant for the small company. For instance, the informational content from dividends may be important to a "distant" investor in a

---

[15]Some of the venture capital firms are subsidiaries or affiliates of Canadian corporations and financial institutions. These firms are funded, at least initially, by parent organizations, which include such well-known Canadian businesses as the Royal Bank of Canada, the Toronto Dominion Bank, BCE, and Noranda. A major reason for the existence of the bank affiliates is the regulation under the Bank Act that banks engaging in venture capital activities must do so through subsidiary operations.

[16]Although small business firms with less than $5 million in annual revenues draw more heavily on internally generated funds than external funds for financing, the proportion will vary in any given year. For example, the percentage of funds obtained from operations in 1991 was 22.7 percent as compared to 36.5 percent in 1989; whereas, the percentage of funds obtained from the sale of assets and fixed assets was 25.2 percent in 1991 as compared to 18.3 percent in 1989. In another study, the three major sources that small and mid-sized firms, having less than $25 million in annual revenues in 1992, used for financing were identified as retained earnings (29.9 percent), financial institutions (27.7 percent) and suppliers (23.7 percent). For further information, see: Entrepreneurship and Small Business Office, Industry Canada, *Reference Handbook on Small Business Statistics* (Ottawa: Minister of Supply and Services of Canada, 1993), p.15; and John R. Baldwin, *Strategies for Success: A Profile of Growing Small and Medium-sized Enterprises (GSMEs) in Canada* (Ottawa: Statistics Canada Catalogue No. 61-532R, 1994).

[17]For a more precise definition as to how to calculate the small business deduction for CCPCs, as well as some restrictions on its use, see Chapter 2.

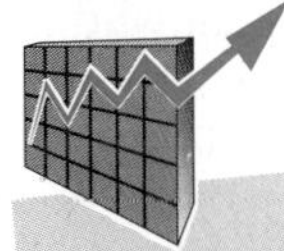

Beyond the Bounds of Traditional Lending

That small business is the engine for employment growth has become accepted as reality. And it is a fact that this is the fastest growing segment of the Canadian economy. The small business workforce is still considerably smaller than those of the public and corporate sectors. But the trend is clear. At a time when the corporate and public sectors have been downsizing, the small business sector has been growing—with a net gain during the recession of about 10 per cent or 73,000 new businesses between 1989 and 1992—bringing the total to 850,000 small businesses.

During the 1993 federal election campaign in Canada, politicians of all stripes called for greater financial support for small business in Canada—and targeted the banks as the major funder. There is no question that the banks, as low-risk lenders, have criteria that may exclude many worthwhile enterprises—and there have no doubt been cases where overanxious bankers have made mistakes in their credit decisions. However, there is no real evidence that bank credit, or the lack thereof, is the single key to small business growth. Nor is there any evidence of a surge in demand for credit from the bank's small business client base.

Availability of small business financing is a broader problem. Canada suffers from a shortage of venture capital and equity financing sources in the area of business start-ups.

Given the level of government debt and deficits, the public sector is not in a position to make major pools of capital available to provide high-risk financing for small business start-ups and risk ventures. However, the federal government took a major step forward in 1993 with improvements—recommended by the banks—to the Small Business Loan Act, which provides for government guarantees for term loans for fixed assets. These improvements have made this financing source more accessible to a broader range of businesses.

The banking industry has been working together with government at all levels to come up with new approaches to meet the financing gap. A recent success story is the $70 million Ontario Lead Investment Fund (OLIF), which was launched in November 1993. A co-operative program between the private sector and the provincial government. OLIF provides long-term capital to innovative growth companies in Ontario. The financial services sector, represented primarily by the banks, are contributing $42 million—or 60 percent—of the funding for OLIF.

As low-risk lenders in a low-margin business, the banks target loan losses of only 0.5 percent to 1.0 percent of their total loan portfolio. They need to stay in that range in order to pay interest depositors, cover expenses and provide a return to shareholders. In 1991 and 1992, small business loan losses represented 1.3 percent for Canada's six largest banks—up from the more acceptable level of 0.7 percent in 1990. The issue is not just the level of loan losses, but the cost of servicing small business borrowers. It may be as labour intensive for banks to review, approve, secure and service a $50,000 loan as a $5 million loan—for one-tenth the proceeds.

Ironically, banks are contending with a low-yield business at the same time as they are perceived to be ignoring the financing needs of the small business sector.

In fact, banks are by far the major source of debt financing for small business—providing over 80 percent of financing needs to this sector. And during the height of the recession—between 1991 and 1992—the banks' total small business portfolio actually grew by 3.1 percent—reaching close to $26 billion.

The largest challenge to successful lending is the turnover rate of small business. The reality is that fewer than half of all small businesses survive beyond the three-year mark. Between 1989 and 1991, 73,000 new businesses were created—but that was *net* new businesses. Such a high growth rate meant that as many as 200,000 businesses opened and folded within that time.

Due to the level of risk, the banks tend to avoid start-up financing where there is no proven record, collateral or equity in place. This is where other financing sources are essential.

Small business closures have as much to do with competitive forces in the marketplace, poor management, or a career change of the principal, as they do with financing issues. This is one of the reasons why banks must look beyond their traditional financing role to find other ways to support small business success and growth.

Source: Adapted from: Joanne De Laurentiis, "Beyond the Bounds of Traditional Lending," *Canadian Business Review* (Spring 1994), pp. 19–21.

large business but meaningless to shareholders in a small business firm who are also actively involved in its management. Second, the small business owner would not be as concerned about the resolution of uncertainty as the individual who is strictly an investor seeking a given return. The owner-manager continues to have the opportunity to act as a decision maker with respect to the risk-return relationship without having to receive the monies in the form of dividends. Thus, on a purely intuitive basis, the potential lack of a preference for dividends, complemented by a strong incentive to minimize taxes, would apparently explain the retention of earnings by small business firms.

When a firm has successfully passed through the early stages of financing, the owners may want to consider the option of going public.

OBJECTIVE 6

GOING PUBLIC

In analyzing the process of going public, we should first define the term. What is *going public*? This terminology, in a legal context, is difficult to define because of the numerous peripheral technical questions. In general, however, going public simply relates to the procedures involved in selling the securities of a privately owned company. In the article on page 955, we examine factors which affect the initial public offering subscription price.

The securities sold may be shares belonging to the owners of the company, a secondary distribution, or the company itself may sell additional shares to raise new capital. The latter acquisition of public owners is of prime concern to the financial manager as a source of funds to finance growth. In this section we will highlight thc prerequisites for the company seeking public funds.

Prerequisites for Going Public

A viable candidate for going public can be identified by the following prerequisites: size, marketability, earnings record, management, growth potential, and the products or services. With these six variables providing the structure, the prerequisites for a public offering can be explained as follows.

Size Constraint

As would be expected, the management of the small enterprise has to be concerned with the question: "Is the firm large enough to go public?" The size criterion generally involves two aspects: (1) the level of earnings and (2) the size of the issue. With regard to the size of the issue (earnings capacity is addressed in a later section), not only may a small issue in terms of dollars be infeasible because of high costs, but a related factor, the number of shares, also has to be considered. Formulating a general guideline, however, is difficult, with the minimal size depending on the philosophy of the investment dealer. Regardless of the number of shares being quoted by an individual dealer, an important standard is the potential for developing an active market.[18] Specifically, without a reasonable number of shares outstanding for active trading, the investment dealer has difficulty justifying involvement in the financing.

Marketability

A second item of concern to investment dealers is the creation of a substantial market. All else being constant, they prefer to develop a market sufficiently large to pro-

[18]Small company shares which trade publicly at prices less than a dollar per share are called "penny stocks." The centre of trading activity for penny stocks in Canada is the Vancouver Stock Exchange.

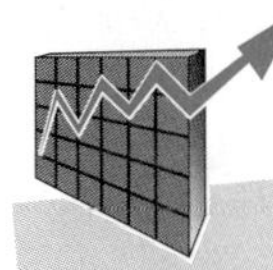

Factors Affecting the Initial Public Offering Subscription Price

Information would seem a critical Initial Public Offering (IPO) pricing determinant. We review the theoretical and empirical relevance of three categories of information. These are the direct accounting disclosures contained in the offering prospectus, the quality and fee structure of the auditors that attest to the validity of disclosed financial data and the underwriters that manage the disclosure of information to investors, and the "signals" about expected future performance that are conveyed to outsiders in the offering prospectus.

Direct Disclosures About Past Performance and Operations

For seasoned firms, the value of historical accounting information is believed to lie primarily in the areas of monitoring and in the confirmation of prior expectations. It contains only limited information which is useful for predictive purposes. In contrast, potential investors in IPOs had no need to monitor or to form expectations regarding management performance prior to the offering. Anyway, very little information is publicly available, since the investment community does not normally gather information about firms which are privately held.

Hence, investors may rely on accounting information relating to past performance, since the new equity issue involves acquiring claims on firm assets and establishing a relationship with the firm's management team. Historical financial data may provide outside investors with measures of the firm's success in initiating and managing operations. A high profit margin on existing operations may also alleviate concerns over the historical level of managerial shirking and/or perquisite consumption.

In U.S. studies of IPOs, sales and earnings from existing operations were found to be significant in explaining initial firm value. In a recent Canadian study, however, no significant relationship between initial firm value and either historical sales revenue or the profit margin of existing operations was detected. The magnitude of assets in place proved to be the only direct accounting disclosure that was relevant to the valuation process. That is, after controlling for other information available to investors, firms with more assets in place prior to the offering are valued more highly. This finding may indicate that the lack of information problem is more severe for small firms undertaking initial offerings, implying a size-related bias in the capital acquisition process.

The Quality and Fee Structures of Auditors and Underwriters

Direct disclosures in the offering prospectus are intermediated by the issuing firm's auditors and underwriters. The quality of the information intermediaries can be described in terms of the amount and accuracy of the information supplied to investors. The higher the quality of the information provider, the more accurately can capital suppliers estimate a firm's future cash flows. Outsiders place greater confidence in disclosures verified by prestigious intermediaries, since they are expected to disclose favourable information only for high quality investment projects. An owner with more favourable projects is believed to select higher-quality investment dealers and auditors, and the higher the quality, the higher the price at which the new issue can be sold. Entrepreneurs with less favourable information are not expected to select high quality intermediation services, since they would incur the added cost and risk but cannot expect to receive as large a marginal benefit by doing so.

Consistent with this view is the observation that many firms switch from local to national auditors when going public. This raises two empirical questions. First, whether the reputation of information provided enters into the fee structure for their service and second, whether initial firm value is related to the selected quality of information intermediaries after controlling for other variables. In fact, higher quality auditors and underwriters have been shown to charge higher fees, after controlling for measures of marginal cost of performing information intermediation services. That is, they are able to charge a fee for their reputation capital. Further, as expected, initial firm value has been shown to be higher for those issuing firms that employ the services of prestigious intermediaries.

Signals of Expected Future Performance by the Owners

There is a general assumption that entrepreneurs have better information than outsiders about the expected value of their projects, but there are moral hazard problems associated with the direct disclosure of this information. Indirect disclosure may be achieved through the observable and promised financial actions of inside owners. Three such disclosures are the percentage of ownership insiders retain in the firm, the proposed use of issue proceeds and the promised future dividend policy.

Initial firm value should be a positive function of the fraction of equity in the firm retained by the entrepreneurs. Since entrepreneurs are risk averse and must hold imperfectly diversified personal investment portfolios, it is only in their interest to invest a large fraction of their wealth in ventures they view very favourably. Thus, insiders' willingness to invest serves as a signal for the investment market's assessment of the offer price. If the offer price is below the value at which the shares trade immediately after issuance, then the initial returns to IPOs investors are positive and the issue is said to be "underpriced."

Source: An excerpt from: Wendy Rotenberg, "Pricing Initial Public Equity Offerings: Who Wins, Who Loses And Why?" *Canadian Investment Review* (Spring 1990), pp. 17–24.

vide the investor additional liquidity in terms of the ability to buy and sell securities as needed. To investment dealers, the depth and liquidity of the market are crucial. Without the possibility of developing such a market, investment dealers are usually not interested.

Established Earnings Record

Management of a firm entertaining the thought of going public must be aware that the size, quality, and trend of earnings are of utmost importance in public financing. The size and quality of earnings must also be considered, with the quality being a function of the stability and the source of earnings. Finally, the trend of earnings is related to the likelihood that a long-term upward trend from the current source of income will continue to exist. The earnings trend must be "significant," must have grown at a steady rate during the most recent years, and must be expected to continue into the future.

The Management Team

The investigation into a company preparing for public financing includes an analysis of the basic quality and philosophy of top management. In examining the quality of management, the factors to consider include experience, integrity, and depth of the corporate leadership. While no one of these three factors can be emphasized at the expense of the others, depth of management is often a difficult problem for the small business. A family-owned company typically relies on the capabilities of the second generation. Thus, the second generation has to be able and willing to continue the operation, or management personnel external to the family must be available within the organization.

Growth Potential

Although "growth" and public markets are not necessarily mutually exclusive, the owners of an enterprise interested in making a public offering should be aware of the uphill battle to be fought if growth is not evident. When competing for funds, an organization having negligible growth potential can anticipate serious difficulties in attracting the public investor. Hence, steady growth patterns expected to continue into the future are a strong factor in the company's favour when entering the public market. In addition, the continued growth of the firm's operation requires that careful attention be given to the public disclosure requirements. For instance, accounting information being reported to the public is costly, requiring increased supervision and monitoring by accountants and managers.

Products and Services

The reputation and image of an entity's products or services should also be considered. Concisely stated, the sound reputation of an organization's products or services is an essential ingredient for a successful public offering.

SUMMARY

In this chapter, we examined the differences between the financing of large corporations and the financing of small and mid-sized businesses. We observed that small and mid-sized businesses will (1) tend to rely more heavily on the retention of earnings as a way to build equity, (2) have less liquidity, (3) use greater amounts of debt, and (4) experience more business risk. We then considered the implications of these differences with respect to cash management, capital budgeting, and sources of financing.

As small and mid-sized firms mature over time, the owner-manager may want to raise funds in the public equity markets. The prerequisites for going public include: size, marketability, earnings record, management, growth potential, and the product or services. Going public remains a major business decision for most owners of small firms.

STUDY QUESTIONS

25-1. Describe the financial difference between a small and a large firm.

25-2. Why is cash management important for the small firm?

25-3. What factors unique to the small firm must be considered in analyzing capital investments?

25-4. Explain the stages of financing that a firm may experience during its growth.

25-5. What are the prerequisites for a firm to go public?

SOLUTIONS FOR SELECTED END-OF-CHAPTER PROBLEMS

Chapter 2

2-1A. Nominal rate = 11.28%
Inflation rate = 7.28%

2-3A Nominal rate = 8.15%
Inflation rate = 3.15%

2-6A. Capital cost allowance = \$2,000
Undepreciated capital cost = \$8,000

2-8A. a. Capital cost allowance = \$138,000
Undepreciated capital cost = \$382,000
b. Capital cost allowance = \$114,600
Undepreciated capital cost = \$267,400

2-11A. a. Taxable income = \$17,500.00
b. Taxable income = \$24,167.50
c. Capital gains = \$28,890.00

Appendix 2A

2A-1A. a. Net Income = \$292,050
b. Taxable Income = \$518,750

2A-3A. a Net Income = \$220,000
b. Taxable Income = \$350,000
c. Total taxes payable = \$157,920

2A-5A. a. Net Income = –\$5,500
b. Taxable Income = –\$25,000

2A-7A. a. Net Income = \$89,100
b. Taxable Income = \$160,000

2A-9A. a. Total Tax Payable = \$163,244
b. Combined corporate tax rate = 44.12%

Chapter 3

3-1A. $\bar{k} = 9.1\%$
$\sigma = 3.06\%$
No, Pritchard should not invest.

3-3A. Security A $\bar{k} = 16.7\%$
$\sigma = 10.13\%$
$\gamma = 0.61$
Security B $\bar{k} = 9.2\%$
$\sigma = 3.57\%$
$\gamma = 0.39$
Choose security A

3-5A. Aproximately 0.5

3-7A. Required rate of return = 10.56%

3-9A. Holding period return

Time	Asman	Salinas
2	20.00%	–6.67%
3	–8.33	14.29
4	18.18	9.38

3-11A. a. $\bar{k}_p = 15.8\%$
$\beta_p = 0.945$
c. Winners; 1, 3, and 5
Losers; 2 and 4

Appendix 3B

3B-1 *Average return—Arka* = 12.25%
Average return—TSE 300 = 15.25%
Standard deviation—Arka = 5.32%
Standard deviation—TSE 300 = 7.37%
$\alpha = 1.31\%$
$\beta = 0.72$

Appendix 3C

3C-1 $k_A = 15.60\%$
$k_B = 20.84\%$
$k_C = 15.34\%$

Chapter 4

4-1A. a. FV_{10}= \$12,969
c. FV_{12}= \$3,019.40

4-3A. a. i = 12%
c. i = 9%

4-5A. a. FV_{10} = \$6,289
c. FV_{10} = \$302.89

4-7A. a. FV_1 = \$10,600
FV_5 = \$13,382
FV_{15} = \$23,966
b. at 8%
FV_1 = \$10,800
FV_5 = \$14,693
FV_{15} = \$31,722
at 10%
FV_1 = \$11,000
FV_5 = \$16,105
FV_{15} = \$41,773

4-9A. a. FV_5 = \$6,691
c. FV_5 = \$8,812
FV_5 = \$8,955
FV_5 = \$9,057

4-11A. FV_1 = 18,000
FV_2 = 21,600
FV_3 = 25,920

4-13A. PMT = \$262.79

4-15A. i = 8%

4-17A. PMT = \$658,202.19

4-19A.

Investment	A	B	C
PV_0	\$29,906	\$16,039.89	\$26,693

4-21A. a. PV_0 = \$3,750
c. PV_0 = \$1,111.11

4-23A. Approximately 9 years

4-25A. PV = \$31,868

4-27A. i = 20%

4-29A. $PMT = \$6{,}935.20$

4-31A. $PMT = \$3{,}148.17$

4-33A. $PMT = \$7{,}924.22$

4-35A. at 10%; $FV = \$17{,}531.14$
at 15%; $FV = \$23{,}349.26$

4-37A. $PV = \$3{,}153.41$

4-39A. a. $PMT = \$1{,}989.73$
c. $PMT = \$1{,}206.89$

4-41A. a. $PV = \$316{,}900$

Chapter 5

5-1A. $V_b = \$752.25$

5-3A. $k_b = 9.61\%$

5-5A. $k_b = 5.29\%$

5-7A. a. $V_b = \$863.78$
b. $V_b = \$707.63$
$V_b = \$1{,}171.19$
d. $V_b = \$927.90$
$V_b = \$832.39$
$V_b = \$1{,}079.85$

5-9A. a. \$1,182.57
b. $V_b = \$925.59$
$V_b = \$1{,}573.50$

Chapter 6

6-1A. $V_p = \$50$

6-3A. $V_p = \$116.67$

6-5A. a. $k_p = 8.5\%$

6-7A. a. $k_c = 18.9\%$
b. $P_0 = \$28.57$

6-9A. $g = 7.2\%$

6-11A. $P_0 = \$39.96$

6-13A. a. $k_c = 10.91\%$
b. $P_0 = \$36$

Chapter 8

8-1A. Breakeven quantity = 40,000 bottles

8-3A. a.

	Jake's	Sarasota	Jefferson
EBIT	\$154,067.40	\$480,000	\$28,970

c.

	Jake's	Sarasota	Jefferson
DOL_s	1.78 times	2.77 times	4.09 times

8-5A. a. Breakeven quantity = 10,000 pairs of shoes

c.

	12,000 pairs of shoes	15,000 pairs of shoes	20,000 pairs of shoes
EBIT	\$108,000	\$270,000	\$540,000

8-7A. a.

	Blacksburg	Lexington	Williamsburg
EBIT	\$163,750	\$620,000	\$261,500

c.

	Blacksburg	Lexington	Williamsburg
DOL_s	1.21 times	1.16 times	1.27 times

8-9A. a. $F = \$780{,}000$
b. Breakeven point = \$1,560,000

8-11A. Breakeven point = 7,500,000 units

8-13A. a.

	Oviedo	Gainesville	Athens
EBIT	\$255,000	\$600,000	\$325,000

c.

	Oviedo	Gainesville	Athens
DOL_s	1.098 times	1.167 times	1.108 times

8-15A. a. $F = \$454{,}545$
b. Breakeven volume = 18,939 units
Breakeven dollars = \$1,515,150

8-17A. a. $P = \$6.875$

8-19A. a. Breakeven points = \$1,200 units
c. $DOL_s = 1.316$ times

8-21A. a. $F = \$168{,}000$

8-23A. a. $DOL_s - 2.5$ times
c. $DCL_s = 4$ times

8-25A. $F = \$400{,}000$

8-27A. a. $DOL_s = 2$ times
c. $DCL_s = 3.2$ times

8-29A. Breakeven point = \$93,631

Chapter 9

9-1A. Minimum level on cash collections from sales = \$1,280,000

9-3A. a. EBIT indifference level = \$2,000,000
b. EPS will be \$1.00 for each plan
d. Plan B

9-5A. a. $EBIT = \$240{,}000$
b. EPS will be \$2.34 for each plan

9-6A. a. $EBIT = \$220{,}000$
b. Plan B

9-8A. a. \$640,000
b. $EPS = \$3.20$

9-10A. a. \$300,000
b. $EPS = \$3.00$

9-12A. a. Plan A = \$7.68
b. 10.378

9-16A. a. \$20,000,000
b. $k_c = 25\%$; $K_0 = 25\%$

9-18A. a. Plan A vs. Plan B = \$9,000
Plan A vs. Plan C = \$18,000

Chapter 10

10-1A. Number of shares = 83,214 shares
Dollar issue size = \$7,073,171

10-3A. Dividends available = \$16,000

10-5A. Value of stock for both plans = \$31.76

10-7A. a. Net Gain = \$0
b. Net Gain = \$24,500

10-9A. a. Repurchase price = \$52
b. Number of shares = 9,615 shares

10-11A. Available dividends = \$90,000

Chapter 11

11-1A. a. Average monthly payables balance = \$19,726.05
b. Average monthly payables balance = \$59,178.15
c. *EAR* = 13.01%

11-3A. *EAR* = 13.83%

11-4A. a. *EAR* = 44.59%
b. *EAR* = 109.84%
c. *EAR* = 44.86%
d. *EAR* = 17.81%

11-5A. a. *Rate* = 37.25%
b. *Rate* = 75.26%
c. *Rate* = 37.63%
d. *Rate* = 16.55%

11-6A. a. *EAR* = 16.3%
b. *EAR* = 19.7%

11-8A. a. *EAR* = 17.01%

11-10A. a. *EAR* = 22.9%

Chapter 12

12-1A. a. k_b = 9.9%
K_b = 5.54%
b. K_{nc} = 15.57%
c. K_c = 15.14%
d. K_p = 8.77%
e. K_b = 6.72%

12-3A. a. K_b = 4.48%
b. K_{nc} = 9.85%
c. K_b = 6.74%
d. K_p = 8.24%
e. K_c = 9.40%

12-5A. K_p = 7.69%

12-7A. K_b = 7.20%

12-9A. K_c = 18.74%

12-10A. a. V_b = \$1,320.52
b. NP_0 = \$1,181.87
c. Number of bonds = 423 bonds
d. K_b = 6.12%

12-12A. a. K_c = 17.59%
b. K_{nc} = 18.25%

12-14A. K_0 = 15.07%

12-16A.

Investment	Required rate of return	Decision
A	16.50%	Accept
B	15.36	Reject
C	17.41	Reject
D	14.21	Reject

12-18A.

Financing amounts	Weighted marginal cost of capital
\$0–\$388,889	10.16%
\$388,890–\$666,667	10.48
\$666,668–\$800,000	10.70
\$800,001–\$1,000,000	12.20
\$1,000,001–\$1,500,000	12.30
over \$1,500,000	14.35

12-20A.

Financing amounts	Weighted marginal cost of capital
\$0–\$450,000	10.94%
\$450,001–\$627,907	11.03
\$627,908–\$1,756,757	11.33
\$1,756,758–\$2,432,432	11.47

Investments to be accepted A, B, E.

Chapter 13

13-1A. a. $AROR_A$ = 63.89%
$AROR_B$ = 62.5%
b. Select project A

13-4A. a. *IRR* = 7%
b. *IRR* = 17%
c. *IRR* = 13%
d. *IRR* = 11%

13-5A. a. *IRR* = 19%
b. *IRR* = 30%
c. *IRR* = 11%

13-7A. a. *NPV* = –\$10,070
b. *PI* = 0.748
c. *IRR* = 9.315%

13-9A. a. NPV_A = –\$668
NPV_B = –\$16,557
b. PI_A = 0.9866
PI_B = 0.7635
c. IRR_A = 11.53%
IRR_B = 3.18%
Reject projects

13-11A. a. IRR approximately 23%
b. *IRR* = 8%
c. *IRR* = 16%

13-13A. a. *MIRR* = 16.9372%
b. *MIRR* = 18.0694%
c. *MIRR* = 19.2205%

Chapter 14

14-1A. a. Total taxes payable = $9,564.75
b. Total taxes payable = $5,627.25
c. Total taxes payable = $1,127.25
d. Total taxes payable = $222.75

14-3A. a. *NPV* = $17,422
b. *PI* = 1.09
c. *IRR* = 18.58%

14-5A. a. $165,121
b. *PI* = 1.35
c. *IRR* = 21.5%

14-7A. a. *IO* = –$40,000
c. Terminal cash flow = $15,653
e. *NPV* = $6,884
g. *IRR* = 26.8%

14-8A. *IO* = $50,000
Year 1; $15,475
Year 2; $17,838
Year 3; $16,116
Terminal cash flow = $6,248

14-9A. a. *NPV* = –$8,350
b. *PI* = 0.833
c. *IRR* = 5.24%

14-11A. a. *Payback(A)* = 1.589 years
Payback(B) = 3.019 years
c. IRR_A = 40%
IRR_B = 30%
e. NPV_A = $18,279
NPV_B = $11,615

14-13A. a. NPV_A = $136.30
NPV_B = $454
b. PI_A = 1.2726
PI_B = 1.0908
c. IRR_A = 40%
IRR_B = 20%

14-15A.

	Steel Mills	Honeybilt Jet
NPV	$20,627	$4,406
EAA	$ 4,237	$ 905

Chapter 15

15-1A. a. Expected value of cash flows (project A) = $5,000
Expected value of cash flows (project B) = $6,000
b. NPV_A = $8,025
NPV_B = $10,112

15-3A. NPV_A = $726,380
NPV_B = $501,420

15-5A. NPV_A = –$20,950
NPV_B = $22,145

15-6A. NPV_A = $ 9,813.25
NPV_B = $20,506.50

Chapter 16

16-1A. *NPV* = $23,051,836

16-3A. a. *NPV(Purchase)* = $32,725.48
b. *NPV(Leasing)* = –$3,340

16-5A. Average theoretical value = $6,083

16-7A. *NPV* = –$57.78

Chapter 17

17-1A. a. Return on equity = 19.48%
b. Return on equity = 14.05%

17-2A a. Current ratio = 1.0
Net working capital = 0
Return on total assets = 10%
Return on common equity = 25%
b. Current ratio = 5.0
Net working capital = 400,000
Return on total assets = 11.02%
Return on common equity = 13.78%
c. Current ratio = 9.0
Net working capital = 800,000
Return on total assets = 6.59%
Return on common equity = 11.525%

Chapter 18

18-1A. Net annual savings = $13,133

18-3A. a.

Holding period	Justification	Recommendation
1	$7,000 < $20,000	No
2	$14,000 < $20,000	No
3	$21,000 > $20,000	Yes
6	$42,000 > $20,000	Yes
12	$84,000 > $20,000	Yes

b. Breakeven yield = 15%

18-5A. Net annual savings = $14,063

18-7A. a. Total float reduction = $1,750,000
b. Opportunity cost = $154,000
c. Total cost = $49,140
d. Net annual gain = $104,860

18-9A. Net annual savings = $3,375

18-11A. Reduction in cheque collection time = 0.4106 days

18-13A. Annual savings = $10,667

18-15A. Value of one day's flaot reduction = $14,247

18-17A. a. Net annual savings = $17,500
b. Break-even rate of return = 4.67%

18-19A. Opportunity cost of mail and processing float = $61,545

18-21A. Loss to Bradford by adopting 45 days payment policy = $37,911
Gain to Bradford by adopting 60 days payment policy = $145,633
Price increase = $187,500;
Loss to Bradford = $41,867

18-24A. $K_b = 3.26\%$
$K_p = 5.30\%$

Appendix 18

18A-1. a. Optimal cash conversion size = \$158,698
b. Total cost = \$856.97
c. Frequency of cash conversion = 5.95 days
d. Average cash balance = \$79,349

Chapter 19

19-1A. *EAR* = 20.24%

19-3A. a. *EAR* = 44.32%
b. *EAR* = 44.59%
c. *EAR* = 74.35%
d. *EAR* = 24.9%
e. *EAR* = 14.91%
f. *EAR* = 45.42%

19-5A. a. Average investment in inventory = \$90,000
b. Average investment in inventory = \$53,333
c. Average investment in inventory = \$230,000
d. Average investment in inventory = \$2,654,321

19-7A. Incremental net income = \$13,870

19-9A. a. Economic order quantity = 775 units
b.

	Total costs
Order one time	\$160
Order four times	\$77.50
Order five times	\$80
Order ten times	\$115
Order fifteen times	\$160

19-11A. a. *EOQ* = 816 units
b. Total costs = \$61,237

19-13A. a. *EOQ* = 14,142 units
b. Orders per year = 35.2
c. Inventory order point = 45,000 units
d-a.
EOQ = 6,325 units
d-b.
Orders per point = 78.1
d-c.
Inventory order point = 45,000 units

Chapter 20

20-1A. Net Profit margin = 8.0%

20-5A. a. Total asset turnover = 2x
c. Operating return on investment = 35%

20-7A. a.

Ratio	1993	1994
Current ratio	6.0	4.0
Acid-test ratio	3.25x	1.92x
Inventory turnover	1.27x	1.36x
ACP	136.9 days	107.0 days
Debt ratio	33%	34.6%
Times interest earned	5.0x	6.0x
Total asset turnover	0.5x	0.56x
Fixed asset turnover	1.0x	1.04x
Operating profit margin	20.8%	24.80%
Net profit margin	10.0%	12.40%
Return on total assets	5.0%	6.90%

20-9A. Net cash inflow from operating activities \$ 87,000
Net cash outflow from investing activities –\$147,500
Net cash inflow from financing activities \$ 63,000
Net cash flow \$ 7,500

20-11A. a. total asset turnover = 25%
operating profit margin = 2.25x
operating income return on investment = 11.11%
b. Operating income return on investment = 19.5%
c. return on common equity = 14.5%

Chapter 21

21-1A. Discretionary financing needed = \$1.5 million

21-3A. Total assets = \$1.8 million
Capital to be raised = \$0.5 million

21-8A. ACP = 40.6 days
Accounts receivable = \$166,849

21-9A. Accounts receivable = \$22,222
Accounts receivable = \$31,111

21-12A. a.

	Projected level
Current assets	\$2.8m
Net fixed assets	\$4.2m
Accounts payable	\$0.7m
Accrued expenses	\$0.7m
Notes Payable	\$1.11m
Current liabilities	\$2.51m
Long-term debt	\$2.00m
Common shares	\$0.50m
Retained earnings	\$1.99m

b.

	Before	After
Current ratio	2 times	1.12 times
Debt ratio	60%	64.4%

21-14A. Projected change in total assets $150,000
Projected change in spontaneous liabilities $ 75,000
Projected new retained earnings $ 55,000
Discretionary financing $ 20,000

Chapter 22

22-1A.

Price	$ 31	$ 32	$33.50	$ 37	$ 39
Gain (Loss)	–$150	–$150	$ 0	$350	$550

22-3A.

Price	$ 31	$ 32	$34	$ 37	$ 39
Gain (Loss)	$150	$150	$ 0	–$300	–$500

22-5A

Price	$ 17	$18.50	$19.35	$24.65	$ 27
Gain (Loss)	$235	$ 85	$ 0	$ 0	$235

Appendix 22A

22A-1A a. Conversion ratio = 25 shares
b. Conversion value = $681.25
c. Security value = $725.74
d. Conversion parity price = $33.61
e. Conversion premium in absolute dollars = $114.51
f. Conversion premium in percentage = 15.78%

22-3A. Minimum price = $0
Warrant premium = $4

22-5A. Minimum price = $0
Warrant premium = $6.25

Chapter 23

23-1A. Administrative expenses $ 2,000,000
Accrued wages 5,000,000
Amount available to general creditors $35,000,000

Chapter 24

24-1A. US; $11,963
Japan; $19,412
Switzerland; $47,490

24-3A.

	US$	Yen	Franc
Spot	0.8359	103.0291	1.0529
30-day	0.8493	102.9760	1.0554
60-day	0.8307	102.4590	1.0685

24-5A. Net gain = $292.77

24-7A.

	Simple Premium (Discount) 1 month	6 month
US	0.0014	0.0075
Japan	0.000005	0.000054
Switzerland	(0.0023)	(0.0139)

	Percent-Per-Annum Deviation 1 month	6 month
US	1.404%	1.2539%
Japan	0.6181	1.1127
Switzerland	(2.9058)	(2.9269)

24-9A. Annual Premium = 2.941%
Forward Price = 0.0097298

SUBJECT INDEX

A

B

*Key terms are given in boldface

C

D

E

G

H

I

Corporate Name Index

APPENDIX *A*

Compound Sum of \$1: $\$(1 + i)^n$

Period	1%	2%	3%	4%	5%	6%	7%	8%	9%	10%	12%	14%	15%	16%	18%	20%	24%	28%	32%	36%
1	1.0100	1.0200	1.0300	1.0400	1.0500	1.0600	1.0700	1.0800	1.0900	1.1000	1.1200	1.1400	1.1500	1.1600	1.1800	1.2000	1.2400	1.2800	1.3200	1.3600
2	1.0201	1.0404	1.0609	1.0816	1.1025	1.1236	1.1449	1.1664	1.1881	1.2100	1.2544	1.2996	1.3225	1.3456	1.3924	1.4400	1.5376	1.6384	1.7424	1.8496
3	1.0303	1.0612	1.0927	1.1249	1.1576	1.1910	1.2250	1.2597	1.2950	1.3310	1.4049	1.4815	1.5209	1.5609	1.6430	1.7280	1.9066	2.0972	2.3000	2.5155
4	1.0406	1.0824	1.1255	1.1699	1.2155	1.2625	1.3108	1.3605	1.4116	1.4641	1.5735	1.6890	1.7490	1.8106	1.9388	2.0736	2.3642	2.6844	3.0360	3.4210
5	1.0510	1.1041	1.1593	1.2167	1.2763	1.3382	1.4026	1.4693	1.5386	1.6105	1.7623	1.9254	2.0114	2.1003	2.2878	2.4883	2.9316	3.4360	4.0075	4.6526
6	1.0615	1.1262	1.1941	1.2653	1.3401	1.4185	1.5007	1.5869	1.6771	1.7716	1.9738	2.1950	2.3131	2.4364	2.6996	2.9860	3.6352	4.3980	5.2899	6.3275
7	1.0721	1.1487	1.2299	1.3159	1.4071	1.5036	1.6058	1.7138	1.8280	1.9487	2.2107	2.5023	2.6600	2.8262	3.1855	3.5832	4.5077	5.6295	6.9826	8.6054
8	1.0829	1.1717	1.2668	1.3686	1.4775	1.5938	1.7182	1.8509	1.9926	2.1436	2.4760	2.8526	3.0590	3.2784	3.7589	4.2998	5.5895	7.2058	9.2170	11.703
9	1.0937	1.1951	1.3048	1.4233	1.5513	1.6895	1.8385	1.9990	2.1719	2.3579	2.7731	3.2519	3.5179	3.8030	4.4355	5.1598	6.9310	9.2234	12.166	15.916
10	1.1046	1.2190	1.3439	1.4802	1.6289	1.7908	1.9672	2.1589	2.3674	2.5937	3.1058	3.7072	4.0456	4.4114	5.2338	6.1917	8.5944	11.805	16.059	21.646
11	1.1157	1.2434	1.3842	1.5395	1.7103	1.8983	2.1049	2.3316	2.5804	2.8531	3.4785	4.2262	4.6524	5.1173	6.1759	7.4301	10.657	15.111	21.198	29.439
12	1.1268	1.2682	1.4258	1.6010	1.7959	2.0122	2.2522	2.5182	2.8127	3.1384	3.8960	4.8179	5.3502	5.9360	7.2876	8.9161	13.214	19.342	27.982	40.037
13	1.1381	1.2936	1.4685	1.6651	1.8856	2.1329	2.4098	2.7196	3.0658	3.4523	4.3635	5.4924	6.1528	6.8858	8.5994	10.699	16.386	24.758	36.937	54.451
14	1.1495	1.3195	1.5126	1.7317	1.9799	2.2609	2.5785	2.9372	3.3417	3.7975	4.8871	6.2613	7.0757	7.9875	10.147	12.839	20.319	31.691	48.756	74.053
15	1.1610	1.3459	1.5580	1.8009	2.0789	2.3966	2.7590	3.1722	3.6425	4.1772	5.4736	7.1379	8.1371	9.2655	11.973	15.407	25.195	40.564	64.358	100.71
16	1.1726	1.3728	1.6047	1.8730	2.1829	2.5404	2.9522	3.4259	3.9703	4.5950	6.1304	8.1372	9.3576	10.748	14.129	18.488	31.242	51.923	84.953	136.96
17	1.1843	1.4002	1.6528	1.9479	2.2920	2.6928	3.1588	3.7000	4.3276	5.0545	6.8660	9.2765	10.761	12.467	16.672	22.186	38.740	66.461	112.13	186.27
18	1.1961	1.4282	1.7024	2.0258	2.4066	2.8543	3.3799	3.9960	4.7171	5.5599	7.6900	10.575	12.375	14.462	19.673	26.623	48.038	85.070	148.02	253.33
19	1.2081	1.4568	1.7535	2.1068	2.5270	3.0256	3.6165	4.3157	5.1417	6.1159	8.6128	12.055	14.231	16.776	23.214	31.948	59.567	108.89	195.39	344.53
20	1.2202	1.4859	1.8061	2.1911	2.6533	3.2071	3.8697	4.6610	5.6044	6.7275	9.6463	13.743	16.366	19.460	27.393	38.337	73.864	139.37	257.91	468.57
21	1.2324	1.5157	1.8603	2.2788	2.7860	3.3996	4.1406	5.0338	6.1088	7.4002	10.803	15.667	18.821	22.574	32.323	46.005	91.591	178.40	340.44	637.26
22	1.2447	1.5460	1.9161	2.3699	2.9253	3.6035	4.4304	5.4365	6.6586	8.1403	12.100	17.861	21.644	26.186	38.142	55.206	113.57	228.35	449.39	866.67
23	1.2572	1.5769	1.9736	2.4647	3.0715	3.8197	4.7405	5.8715	7.2579	8.9543	13.552	20.361	24.891	30.376	45.007	66.247	140.83	292.30	593.19	1178.6
24	1.2697	1.6084	2.0328	2.5633	3.2251	4.0489	5.0724	6.3412	7.9111	9.8497	15.178	23.212	28.625	35.236	53.108	79.496	174.63	374.14	783.02	1602.9
25	1.2824	1.6406	2.0938	2.6658	3.3864	4.2919	5.4274	6.8485	8.6231	10.834	17.000	26.461	32.918	40.874	62.668	95.396	216.54	478.90	1033.5	2180.0
26	1.2953	1.6734	2.1566	2.7725	3.5557	4.5494	5.8074	7.3964	9.3992	11.918	19.040	30.166	37.856	47.414	73.948	114.47	268.51	612.99	1364.3	2964.9
27	1.3082	1.7069	2.2213	2.8834	3.7335	4.8223	6.2139	7.9881	10.245	13.110	21.324	34.389	43.535	55.000	87.259	137.37	332.95	784.63	1800.9	4032.2
28	1.3213	1.7410	2.2879	2.9987	3.9201	5.1117	6.6488	8.6271	11.167	14.421	23.883	39.204	50.065	63.800	102.96	164.84	412.86	1004.3	2377.2	5483.8
29	1.3345	1.7758	2.3566	3.1187	4.1161	5.4184	7.1143	9.3173	12.172	15.863	26.749	44.693	57.575	74.008	121.50	197.81	511.95	1285.5	3137.9	7458.0
30	1.3478	1.8114	2.4273	3.2434	4.3219	5.7435	7.6123	10.062	13.267	17.449	29.959	50.950	66.211	85.849	143.37	237.37	634.81	1645.5	4142.0	10143.
40	1.4889	2.2080	3.2620	4.8010	7.0400	10.285	14.974	21.724	31.409	45.259	93.050	188.88	267.86	378.72	750.37	1469.7	5455.9	19426.	66520.	*
50	1.6446	2.6916	4.3839	7.1067	11.467	18.420	29.457	46.901	74.357	117.39	289.00	700.23	1083.6	1670.7	3927.3	9100.4	46890.	*	*	*
60	1.8167	3.2810	5.8916	10.519	18.679	32.987	57.946	101.25	176.03	304.48	897.59	2595.9	4383.9	7370.1	20555.	56347.	*	*	*	*

*FVIF > 99,999.

APPENDIX *B*

Present Value of \$1: $\frac{\$1}{(1 + i)^n}$

Period	1%	3%	5%	6%	7%	8%	9%	10%	11%	12%	13%	14%	15%	16%	17%	18%	19%	20%	24%	28%
1	.9901	.9709	.9524	.9434	.9346	.9259	.9174	.9091	.9009	.8929	.8850	.8772	.8696	.8621	.8547	.8475	.8403	.8333	.8065	.7813
2	.9803	.9426	.9070	.8900	.8734	.8573	.8417	.8264	.8116	.7972	.7831	.7695	.7561	.7432	.7305	.7182	.7062	.6944	.6504	.6104
3	.9706	.9151	.8638	.8396	.8163	.7938	.7722	.7513	.7312	.7118	.6930	.6750	.6575	.6407	.6244	.6086	.5934	.5787	.5245	.4768
4	.9610	.8885	.8227	.7921	.7629	.7350	.7084	.6830	.6587	.6355	.6133	.5921	.5718	.5523	.5336	.5158	.4987	.4823	.4230	.3725
5	.9515	.8626	.7835	.7473	.7130	.6806	.6499	.6209	.5934	.5674	.5428	.5194	.4972	.4761	.4561	.4371	.4190	.4019	.3411	.2910
6	.9420	.8375	.7462	.7050	.6663	.6302	.5963	.5645	.5346	.5066	.4803	.4556	.4323	.4104	.3898	.3704	.3521	.3349	.2751	.2274
7	.9327	.8131	.7107	.6651	.6627	.5835	.5470	.5132	.4187	.4523	.4251	.3996	.3759	.3538	.3332	.3139	.2959	.2791	.2218	.1776
8	.9235	.7894	.6768	.6274	.5820	.5403	.5019	.4665	.4339	.4039	.3762	.3506	.3269	.3050	.2848	.2660	.2487	.2326	.1789	.1388
9	.9143	.7664	.6446	.5919	.5439	.5002	.4604	.4241	.3909	.3606	.3329	.3075	.2843	.2630	.2434	.2255	.2090	.1938	.1443	.1084
10	.9053	.7441	.6139	.5584	.5083	.4632	.4224	.3855	.3522	.3220	.2946	.2697	.2472	.2267	.2080	.1911	.1756	.1615	.1164	.0847
11	.8963	.7224	.5847	.5268	.4751	.4289	.3875	.3505	.3173	.2875	.2607	.2366	.2149	.1954	.1778	.1619	.1476	.1346	.0938	.0662
12	.8874	.7014	.5568	.4070	.4440	.3971	.3555	.3186	.2858	.2567	.2307	.2076	.1869	.1685	.1520	.1372	.1240	.1122	.0757	.0517
13	.8787	.6810	.5303	.4688	.4150	.3677	.3262	.2897	.2575	.2292	.2042	.1821	.1625	.1452	.1299	.1163	.1042	.0935	.0610	.0404
14	.8700	.6611	.5051	.4423	.3878	.3405	.2992	.2633	.2320	.2046	.1807	.1597	.1413	.1252	.1110	.0985	.0876	.0779	.0492	.0316
15	.8613	.6419	.4810	.4173	.3624	.3152	.2745	.2394	.2090	.1827	.1599	.1401	.1229	.1079	.0949	.0835	.0736	.0649	.0397	.0247
16	.8528	.6232	.4581	.3936	.3387	.2919	.2519	.2176	.1883	.1631	.1415	.1229	.1069	.0930	.0811	.0708	.0618	.0541	.0320	.0193
17	.8444	.6050	.4363	.3714	.3166	.2703	.2311	.1978	.1696	.1456	.1252	.1078	.0929	.0802	.0693	.0600	.0520	.0451	.0258	.0150
18	.8360	.5874	.4155	.3503	.2959	.2502	.2120	.1799	.1528	.1300	.1108	.0946	.0808	.0691	.0592	.0508	.0437	.0376	.0208	.0118
19	.8277	.5703	.3957	.3305	.2765	.2317	.1945	.1635	.1377	.1161	.0981	.0829	.0703	.0596	.0506	.0431	.0367	.0313	.0168	.0092
20	.8195	.5537	.3769	.3118	.2584	.2145	.1784	.1486	.1240	.1037	.0868	.0728	.0611	.0514	.0433	.0365	.0308	.0261	.0135	.0072
25	.7798	.4776	.2953	.2330	.1842	.1460	.1160	.0923	.0736	.0588	.0471	.0378	.0304	.0245	.0197	.0160	.0129	.0105	.0046	.0021
30	.7419	.4120	.2314	.1741	.1314	.0994	.0754	.0573	.0437	.0334	.0256	.0196	.0151	.0116	.0090	.0070	.0054	.0042	.0016	.0006
40	.6717	.3066	.1420	.0972	.0668	.0460	.0318	.0221	.0154	.0107	.0075	.0053	.0037	.0026	.0019	.0013	.0010	.0007	.0002	.0001
50	.6080	.2281	.0872	.0543	.0339	.0213	.0134	.0085	.0054	.0035	.0022	.0014	.0009	.0006	.0004	.0003	.0002	.0001	*	*
60	.5504	.1697	.0535	.0303	.0173	.0099	.0057	.0033	.0019	.0011	.0007	.0004	.0002	.0001	.0001	*	.0000	*	*	*

*The factor is zero to four decimal places.